Advanced Topics in Mechanics of Materials, Structures and Construction
AToMech1-2023

AToMech1-2023
International Conference on Advanced Topics in Mechanics of Materials, Structures and Construction
March 12-14, 2023, College of Engineering, Prince Mohammad Bin Fahd University, Al Khobar, Saudi Arabia

Editors
Erasmo Carrera
Faramarz Djavanroodi
Muhammad Asad

Peer review statement

All papers published in this volume of "Materials Research Proceedings" have been peer reviewed. The process of peer review was initiated and overseen by the above proceedings editors. All reviews were conducted by expert referees in accordance to Materials Research Forum LLC high standards.

Published under License by **Materials Research Forum LLC**
Millersville, PA 17551, USA

Published as part of the proceedings series
Materials Research Proceedings
Volume 31 (2023)

ISSN 2474-3941 (Print)
ISSN 2474-395X (Online)

ISBN 978-1-64490-258-5 (Print)
ISBN 978-1-64490-259-2 (eBook)

Distributed worldwide by

Materials Research Forum LLC
105 Springdale Lane
Millersville, PA 17551
USA
https://www.mrforum.com

Manufactured in the United State of America
10 9 8 7 6 5 4 3 2 1

Table of Contents

Preface
Committees

Response of GFRP bars at different temperatures
Farid Abed, Zeinah Elnassar, Wael Abuzaid, Yazan Alhoubi 1

Characteristics of electrorheological fluids under single and mixed modes
Ali K. El Wahed 7

Experimental characterization and environmental impact of different grades of CRM-modified asphalt binders
Mohammad Ali Khasawneh, Samar D. Dernayka, Saidur R. Chowdhury 16

Progressive failure analysis of postbuckled plates via mixed formulation
Riccardo Vescovini, Mohammad Reza Najafian Zadeh Najafabadi 29

High dimensional model representation for the probabilistic assessment of seismic pounding
R. Sinha, B.N. Rao 38

Static and free vibration analysis of laminated sandwich shell with double curvature considering the effect of transverse normal strain
Ajim S. Shaikh, Atteshamuddin S. Sayyad 46

New approach for modeling randomly distributed CNT reinforced polymer nanocomposite with van der Waals interactions
Umut Caliskan, Hilal Gulsen, Mustafa Kemal Apalak 55

Effect of shearing conditions and initial aggregates' state on the mechanical behavior of cellular glass foam
Layal Jradi, Bassel Seif El Dine, Tahar Ayadat, Saidur Schowdhury, Andi Asiz 66

Static response of functionally graded porous spherical shells using trigonometric shear deformation theory
Rupali B. Tamnar, Atteshamuddin S. Sayyad 76

Assessment of recent methods for determination of soil final settlement using field data
Mohamed Aymen Dali, Ahcène Benamghar 86

The flexoelectric effect for interface cracks between two dissimilar materials
Tomas Profant, Jan Sladek, Vladimir Sladek 99

Comprehensive review of soil stabilization agents
Batool Al-Shorman, Mousa Bani Baker, Tahar Ayadat 109

Static response of FGM shell using refined higher-order shear and normal deformation theory
Sumit S. Kolapkar, Atteshamuddin S. Sayyad 117

Constitutive modeling of dynamic strain aging in niobium
Modified VA Model, Dynamic Strain Aging, Thermal Activation Energy, Niobium 128

Best theory diagram using genetic algorithms for composite plates
M.A. Hinostroza, J.L. Mantari ... 135

AI-based prediction of strength and tensile properties of expansive soil stabilized with recycled ash and natural fibers
Abolfazl Baghbani, Firas Daghistani, Katayoon Kiany, Mohamad Mahdi Shalchiyan ... 146

Performance of self-compacting concrete based on fine recycled concrete aggregate incorporating polyethylene terephtalate fibers
Bayah Meriem, Debieb Farid ... 155

Preparation and characterization of nanoparticles for encapsulation and delivery vehicles
Megdi Eltayeb ... 165

Isogeometric topology optimization of auxetic materials based on moving morphable components method
Xiaoya Zhai, Yundong Gai, Liuchao Jin, Wei-Hsin Liao, Falai Chen, Ping Hu ... 172

Biomechanical model for musculoskeletal simulation
Megdi Eltayeb ... 187

Recycled smelter slag as an engineering material - opportunity and sustainability
Saidur R. Chowdhury ... 194

Structure, electronic properties of AlAs using first-principles calculations
Sansabilla Bouchareb, Rachid Tigrine ... 204

A comparative study on the use of laser beam and abrasive water jet in hole making process of woven laminated GFRP
Hussein Mohammad Ali ... 212

Influence of canard geometrical configuration on the radar cross section
Gauri Patki, Attada Phanendra Kumar, Dineshkumar Harursampath ... 227

Investigating the fluid-structure interaction of L-shaped pipe bends using machine learning
Pratik Punj, Md Adil ... 240

Fatigue life evaluation of asphalt mixtures containing natural river sands and designed by bailey method
Mohammad Ahmad Alsheyab, Mohammad Ali Khasawneh ... 248

Stone dust as replacement for fine aggregate in cellular lightweight concrete (CLC): Volume weight and compressive strength
Parea R. Rangan1, M. Tumpu, Y. Sunarno, Mansyur ... 262

Contemporary conceptual models for predicting carrying capacity of multi helices screw piles
Tahar Ayadat, Hocine Lefkir, Zahar Said Haddad, Danish Ahmed, Andi Asiz ... 269

Size-effects due to Burgers tensor in classical deformation of isotropic thermoplastic materials
A.S. Borokinni, A.P. Akinola, O.P. Layeni, O.O. Fadodun, B.A. Olokuntoye ... 281

Aeroponic tower garden solar powered vertical farm
Omar Yahya, Gharam Yahya, Ahmad Al-Omair, Emad Yahya, Esam Jassim, Faramarz Djavanroodi ... 287

Using powder metal gears in industrial applications- A review
Omar D. Mohammed, QatrAlnada Mahmood ... 299

Preliminary investigation and SWOT analysis of sustainable mortar utilizing Arabic Gum
Muhammad Nasir, Walid A. Al-Kutti, A.B.M. Saiful Islam, Khalid Saqer Alotaibi, Ammar Al Eid, Yousef Khalid Algatam, Ali Alqahtani ... 306

A numerical study on the effect of basement excavation and fibre reinforced concrete as tunnel lining material
Farzana Iqbal, Malavika Varma ... 313

Estimation of failure load of adhesively bonded composite joints with embedded crack in adherends: bond EM
James Polagangu, Byji Varughese ... 324

Finite electroelastic deformation of dielectric semilinear hyperelastic tubes
Odunayo O. Fadodun, Bolanle A. Olokuntoye, Patrick O. Layeni and Adegbola P. Akinola ... 336

Design of a solar PV power plant at KFU premises
Mahdi Alnasser, Hassan Alghazal, Amjad, Almeihan, Mounir Bouzguenda, Mohammed Abdulaziz Alaqil ... 347

Experimental investigations on the thermo-mechanical properties of carbon-basalt-aramid/epoxy and glass-basalt-aramid/epoxy hybrid interply composites under different aging environments
Zekiye Erdoğan, Munise Didem Demirbaş ... 356

Critical review on ethanol producing feedstock and methods
Jaanvi Garg, Akshay Jain, Avadhoot Abaso Mohite, Bhaskor Jyoti Bora, Prabhakar Sharma, Bhaskar Jyoti Medhi, Debrata Barik ... 366

Investigation of blast resistance of cladding with square dome-shaped Kirigami folded structures as core
Vishal M Birari, B N Rao ... 375

Von mises stress distribution along the cruciform specimen under biaxial loading: geometric variation effect
Abdul Mateen Mohammed, Jaya KandaSamy, Bhavani Venkata Yashwanth Dasari, Lingala Syam Sundar ... 385

Effect of steel fibres on the mechanical strengths of fly ash/GGBS based geopolymer concrete under ambient curing condition
Musa Adamu, Tejas D Doshi, Veerendrakumar C. Khed, Yasser E. Ibrahim ... 398

Design of an automated cleaning system for solar panels
Mounir Bouzguenda, Bader Bashmail, Mohammed Ben Youssef Al-Khaldi, Nasser Al-Jafar, Saad Al-Jasim, Mohammed A. Al Aqil ... 408

Strength properties of self-compacting concrete incorporating iron ore tailings
AJAGBE Wasiu Olabamiji, OKPANACHI Robert Ojonugwa, GANIYU Abideen Adekunle, ABDULKAREEM Muyideen Oladimeji ... 415

Tensor as a tool in engineering analysis
A.P. Akinola, A.S. Borokinni, O.O. Fadodun 422

Bayesian regularization optimization algorithm for the experimental thermophysical property for 80:20% water and ethylene glycol based ZrO2 nanofluids
M. Manzoor Hussain, L. Syam Sundar, Feroz Shaik 437

Modelling the effect of tool material on material removal rate in electric discharge machining
Umair Arif, Imtiaz Ali Khan, Faisal Hasan 448

Energy harvesting and energy conversion in an electromechanical coupling acoustic black hole beam
ZHANG Linli, SHENG Xiaozhen, LI Meng 456

Bi-axial behavior of glass/epoxy pipe subjected to internal pressure
Abdul Mateen Mohammed, Tajuddin Mohammed, Ravi Shankar D.V., Manzoor Hussain M., Puneetha 466

Manufacturing strategy for additive manufacturing of a piping component for oil and gas application
Meshal Alsaiari, Mushtaq Khan, Faramarz Djavanroodi 475

A homogenized finite element analysis of the deformation of axially-loaded thin-walled epoxy/coir fibre-reinforced aluminum 6063 composite tubes
B.O. Malomo 485

Stabilization of a swelling soil using three different chemical additives
Tahar Ayadat, Ammar Noui, Layal Jradi, Danish Ahmed, Muhammad Ajmal, Andi Asiz 495

One dimensional oedometer laboratory testing for expansive clay submerged with hydrocarbon fluids
Mousa Bani Baker, Adel Hanna, Maria Elektorowicz, Tahar Ayadat, Batool Al-Shorman, Raed Abendeh 504

Design and simulation of a large-scale telescopic paraboloidal solar concentrator: Dual receiver
Barena Bekalo Betela, Venkata Ramayya Ancha, Lingala Syam Sundar 514

The Innovation in wire arc additive manufacturing (WAAM): A review
Saleh Fahad Alaboudi, Muhammad Azhar Ali Khan, Muhammad Asad, Mushtaq Khan, Faramarz Djavanroodi 522

Technical and economic feasibility of solar PV systems supported by energy storage in hospitals in KSA
Mounir Bouzguenda, Abdulaziz Almulhim, Abdulrahman Shawki, Ahmed Al-Baa'dani, Ibrahim Ahmed, Mohammed Alkulaib, Mohammed Al-Aqil 531

Design of brake pads from economical materials
NOUARI Abdessamed, TAFRAOUI Ahmed, TAFRAOUI Salima, MALLEM Mohamed 538

Recycling of machining waste fibers in the formulation of new concrete
Salima TAFRAOUI, Ahmed TAFRAOUI, Naima SEGHOUANI 550

Numerical simulation of high-pressure gas-jet impingement on the adjacent equipment
Esam I. Jassim, Muhammad Asad, Bashar Jasem 564

Ultrasonic pulse velocity (UPV) and initial rate of water absorption (IRA) of foam concrete containing blended cement
Y. Sunarno, M.W. Tjaronge, R. Irmawaty, A.B. Muhiddin, M.A. Caronge, M. Tumpu............571

Comparison study of performance and heat leak factor of three types of heat exchangers operated with nanofluid
Esam I. Jassim, Bashar I. Jasem............581

Dynamic deformation analysis of a spherical cavity explosion
Olawanle Patrick LAYENI, Adegbola Peter AKINOLA............589

Linear response of thin axysimmetric cross-ply structure under a static load: Numerical and analythical comparisons
Salvatore Saputo, Erasmo Carrera, Volodymyr V. Zozulya............596

Interaction of a rectangular jet with a slotted plate in presence of a control mechanism: Experimental study of the aeroacoustic field
Marwan Alkheir, Hassan Assoum, Nour Eldin Afyouni, Bilal El Zohbi, Kamel Abed Meraim, Anas Sakout, Mouhammad El Hassan............614

Investigation of the influence of external edge restraint on reinforced concrete walls
Muhammad Kashif Shehzad, John P. Forth, Nikolaos Nikitas, Imogen Ridley, Robert Vollum, Abo Bakr Elwakeel, Karim El Khoury, Bassam Izzuddin............622

Passive vibration damping of a plate interacting with a flowing fluid using shunted piezoelectric element
Sergey Lekomtsev, Valerii Matveenko, Alexander Senin............634

Predicting the strength of recycled glass powder-based geopolymers for improving mechanical behavior of clay soils using artificial intelligence
Abolfazl Baghbani, Firas Daghistani, Hasan Baghbani, Katayoon Kiany............646

Utilizing machine learning techniques for predictive modelling of absorptivity in I-shaped metamaterials
Md Adil, Pratik Punj............656

Modeling and simulation of modular agricultural robot flexible production systems
Mohammed Aldossary, Yousuf Alsuliman, Abdul Aziz Afzal, Stephen Limbos, Megdi Eltayeb............666

Numerical study on the effect of real gas model on the flow structure and shock location of Laval nozzle
Bashar I. Jasem, Esam I. Jassim............674

Investigating the influence of alternative fuels on the properties of sand-bentonite liners
Mousa Bani Baker, Adel Hanna, Maria Elektorowicz, Tahar Ayadat, Batool Al-Shorman............683

On the deformation of layered composite arches using exponential shear and normal deformation theory
Valmik M. Mahajan, Amit Sharma............693

Mechanics and failure analysis of stitched sandwich structures damaged by the impact of various energies
Amina Assouli, Bounoua Tab, Fodil Hammadi, Izzeddine Bekkar............704

Magnetoelastic deformation of conductive semilinear hyperelastic solids
Odunayo Olawuyi Fadodun ... 715

Tensile behavior of functionally graded sandwich PLA-ABS produced via fused filament fabrication process
Umut Caliskan, Caglar Sevim, Munise Didem Demirbas ... 726

Artificial intelligence-based prediction of geotechnical impacts of polyethylene bottles and polypropylene on clayey soil
Abolfazl Baghbani, Firas Daghistani, Hasan Baghbani, Katayoon Kiany, Jafar Bolouri Bazaz 734

Static analysis of cross-ply laminated spherical shells using a new hyperbolic shear deformation theory
Anuja S. Jape, Atteshamuddin S. Sayyad ... 745

A comprehensive analysis of the erosion in a carbon steel boiler tube elbow through the use of 3D mapping of the corroded surface and CFD modelling
Nikolay Bukharin, M.A. Farrokhzad, Mouhammad El Hassan ... 755

Structural and environmental safety study of the holy mosque area using CFD
Mohamed Farouk ... 765

Physical-chemical characterization and exploitation of an ultra fine local in mortars
ZAOUI Miloud, TAFRAOUI Ahmed, BEKKAR Izzeddine ... 787

The state of 3D printing technologies: A review
Omar Yahya, Gharam Yahya, Faramarz Djavanroodi, Mushtaq Khan ... 799

Analyzing the electromagnetic forming process of SS304 sheet using AA6061-T6 driver through a fully coupled numerical model and experimentally validation
Zarak Khan, Mushtaq Khan, Faramarz Djavanroodi, A Samad Danish ... 806

Keyword index

Preface

On behalf of the conference committee, I would like to thank all the participants in the International Conference on Advanced Topics in Mechanics of Materials, Structures and Construction (AToMech1-2023) which had been held at Prince Mohammad Bin Fahd University (PMU) Khobar, Kingdom of Saudi Arabia on March, 2023. I was honored to serve as Chairman of this important event. We are all aware that research plays an important role in its contribution to the knowledge, development and shaping of the future directions.

While globalization appeared to have significant impacts for the worldwide society, AToMech1 provided a platform to promote sustainable development, international networking, for researchers, practitioners and educators all over the world. The conference offered a truly comprehensive view while inspiring the attendees to come up with solid recommendations to tackle hot-topic challenges. Finally, I thank the keynote speakers; presenters and authors for contribution.

Dr. Faisal Yousif Al Anezi
Conference Chair

Committees

Honory Chair

Issa H. Alansari, President, Prince Mohammed Bin Fahd University

Conference Chair

Dr. Faisal Yousif Al Anezi, Vice President, Prince Mohammed Bin Fahd University

Conference Co – Chair

Professor Erasmo Carrera, Prince Mohammed Bin Fahd University

Technical Program Committee

Dr. J. N. Reddy, Texas A&M University, USA
Dr. Jamal Nayfeh, Prince Mohammad Bin Fahd University, KSA
Dr. Erasmo Carrera, Politecnico di Torino, Italy
Dr. Adel M. Hanna, University of Concordia, Canada
Dr. Chandra Sekhar Matli, National Institute of Technology, India
Dr. Siti Noor Linda bt Taib, Universiti Malaysia Sarawak, Malaysia
Dr. Faramarz Djavanroodi, Prince Mohammad Bin Fahd University, KSA
Dr. Weiqiu Chen, Zhejiang University, PR China
Dr. Muhammad Asad, Prince Mohammad Bin Fahd University, KSA
Dr. Volodymyr Zozulya, Kharkiv State Technical University, Ukraine
Dr. Esam Jassim, Prince Mohammad Bin Fahd University, KSA
Dr. Muhammad Azhar Khan, Prince Mohammad Bin Fahd University, KSA
Dr. Mabrouk Touahmia, University of Hail, KSA
Dr. Muhammad Kalimur Rahman, King Fahd University of Petroleum and Minerals, KSA
Dr. Mustafa Aytekin, University of Bahrain, Bahrain
Dr. Xin Ren, Nanjing Tech University, PR China
Dr. Jose Luis Mantari Laureano, Universidad De Ingeniera Y Tecnologia, Peru
Dr. Mushtaq Khan, Prince Mohammad Bin Fahd University, KSA
Dr. Adegbola Akinola, Obafemi Awolowo University, Nigeria
Dr. Bouchaib Zazoum, Prince Mohammad Bin Fahd University, KSA
Dr. O. P. Layeni, Obafemi Awolowo University, Nigeria
Dr. Andi Asiz, Prince Mohammad Bin Fahd University, KSA
Dr. Xiangyang Xu, Soochow University, PR China
Dr. Li xiangyu, Southwest Jiaotong University, PR China
Dr. Alfonso Pagani, Politecnico di Torino, Italy
Dr. Mário Rui Arruda, Instituto Tecnico Lisboa, Portugal
Dr. Tahar Ayadat, Prince Mohammad Bin Fahd University, KSA
Dr. Xin Fang, National University of Defence Technology, PR China
Dr. Maiaru Marianna, University of Massachusetts, USA
Dr. Feroz Shaik, Prince Mohammad Bin Fahd University, KSA
Dr. Mouhammad El Hassan, Prince Mohammad Bin Fahd University, KSA
Dr. Yousef Alshammari, Prince Mohammad Bin Fahd University, KSA
Dr. Malek Brahimi, New York City College of Technology, USA,
Dr. Syed Sohail Akhtar, King Fahd University of Petroleum and Minerals, KSA

Technical Program Chairs

1. **Session Title: Novel Metamaterial Structures**

Organizer: Dr. Xin Ren, xin.ren@njtech.edu.cn Nanjing Tech University, PR China

2. **Session Title: Structural analysis with machine learning.**

Organizer:Dr. Jose L. Mantari Laureano, josemantari@gmail.com Universidad De Ingenieriay Technologia, Perù

3. **Session Title: Tensor Application and Constitutive Equation Modelling in Continuum Mechanics and Engineering Design**

Organizers: Dr. Adegbola Akinola, aakinola@oauife.edu.ng . Dr. O. P. Layeni, olayeni@oauife.edu.ng Obafemi Awolowo University, Nigeria

4. **Session Title: Intelligent monitoring of engineering structure**

Organizer: Dr. Xiangyang Xu, xuxiangyang_ca@163.com Soochow University, PR, China

5. **Session title: Mechanics of functional materials**

Organizer: Dr. Li xiangyu, lixiangyu@swjtu.edu.cn Southwest Jiaotong University, PR China

6. **Session Title: Nonlinear structure mechanics and failure analysis**

Organizer: Dr. Alfonso Pagani, alfonso.pagani@polito.it Politecnico di Torino, Italy Dr. Mário Rui Arruda, mario.rui.arruda@etecnico.ulisboa.pt Instituto Tecnico Lisboa, Portugal

7. **Session Title: Programmable gear-based mechanical metamaterials**

Organizer: Dr. Xin Fang, xinfangdr@sina.com National University of Defence Technology, PR China

8. Session Title: Multiscale Analysis of Structures and Materials

Organizer: Dr. Maiaru Marianna, Marianna_Maiaru@uml.edu University of Massachusetts, USA

9. Session Title: Mechanics of materials and structures produced by additive manufacturing technologies

Organizer: Dr. Mohammad Asad, masad@pmu.edu.sa Prince Mohammad Bin Fahad University, KSA

10. Session Title: Numerical modelling and experimental testing of materials and structures

Organizer: Dr. Muhammad Azhar Khan, mkhan6@pmu.edu.sa Prince Mohammad Bin Fahad University, KSA

11. Session Title: Advanced Construction Materials in Geotechnical Applications, Structures and Foundation Engineering

Organizer: Dr. Tahar Ayadat, tayadat@pmu.edu.sa Prince Mohammad Bin Fahad University, KSA

12. Session Title: Future Sustainable Infrastructure and Construction Materials

Organizer: Dr. Andi Asiz, aasiz@pmu.edu.sa Prince Mohammad Bin Fahad University, KSA

13. Session Title: Fluid-Structure Interaction

Organizer: Dr. Esam Jassim, ejassim@pmu.edu.sa Prince Mohammad Bin Fahad University, KSA

14. Session Title: Modelling and Optimization of Materials Manufacturing Processes

Organizer: Dr. Kuldeep Saxena, saxena0081@gmail.com Research and Development, Lovely Professional University, Phagwara, India

Response of GFRP bars at different temperatures

Farid Abed[1,a,*], Zeinah Elnassar[2,b], Wael Abuzaid[3,c], and Yazan Alhoubi[1,d]

[1]Civil Engineering Department, American University of Sharjah, Sharjah, United Arab Emirates

[2]Materials Science and Engineering Department, American University of Sharjah, Sharjah, United Arab Emirates

[3]Mechanical Engineering Department, American University of Sharjah, Sharjah, United Arab Emirates

[a]fabed@aus.edu, [b]g00062755@aus.edu, [c]wabuzaid@aus.edu, [d]b00052990@aus.edu

Keywords: GFRP, DIC, Compression, Elevated Temperature

Abstract. Owing to their numerous advantages, industries have been adopting fiber-reinforced polymer (FRP) bars as reinforcement in concrete members due to their superior mechanical properties and high durability. More specifically, glass FRP (GFRP) bars have been increasingly adopted for use in harsh environmental conditions due to their excellent corrosion resistance. However, the compressive response of the GFRP bars at high temperatures has yet to be studied. This paper presents the results of an experimental study that aims to investigate the mechanical behavior of glass fiber reinforced polymers (GFRP) bars subjected to compressive loads at various elevated temperatures and deformation rates. Four different temperatures, ranging between ambient and 150°C, and two deformation rates were considered. The compressive strength and the modulus of elasticity for each bar sample are measured and the failure modes are demonstrated. It was observed that the compressive strength of GFRP bars is similar under both loading rates for all temperature conditions. However, the elastic modulus of the bars showed a slight discrepancy between the different load rates, especially at an elevated temperature of 150°C. The test results indicate a linear degradation in the mechanical properties as the temperature is raised. In addition, the observed failure mode for the tested samples is splitting of the bars.

Introduction

Over recent years, polymers have been replacing metals and ceramics in engineering applications. This is primarily attributed to some of the polymers' impressive properties which include being lightweight, easily fabricated, and their ability to be processed at low temperatures [1]. An important characteristic of polymers which makes them attractive in the construction industry, is their resistance against corrosion. The deterioration of the infrastructure owing to the corrosion of steel reinforcement has been one of the major concerns in the construction industry [2]. To avoid infrastructure deterioration and loss of durability caused by corrosion of steel rebars in concrete structures, FRP bars have been used as an alternative to steel reinforcement in reinforced concrete structures due to their many advantages which include being corrosion-resistant and non-conductive. The noncorrosive FRP bars have shown promise as a way to further improve and protect concrete structures like bridges, that are directly affected by the damaging outcomes of corrosion. In addition, GFRP possesses the advantage of being more economical than other types of FRP, including aramid and carbon, thus, making this specific FRP bar type more attractive to be used for infrastructure applications [2].

Although several research studies have been conducted to characterize GFRP as longitudinal reinforcement, little information have been provided about their compressive strength, in particular, as a function of temperature and deformation rate. Extensive research studies have been conducted on the behavior of FRP bars under tension [3,4] and the performance of FRP-reinforced beams under shear [5-7]. As a general design practice, the compressive strength of GFRP bars is neglected during the design stage, according to ACI 440.11-22 [8]. This is due to the gap in literature regarding the response of such bars under compressive loading and due to the GFRP's elastic modulus, that is as low as that of concrete.

Alnajmi and Abed [9] investigated the performance of FRP bars under compression and examined the behavior of the bars as main reinforcement in columns. The experimental study consisted of a series of compression tests conducted on GFRP and BFRP bars. The results showed an increase of up to 35% of the failure loading in columns reinforced with FRP bars.

Khan et al. [10] investigated the behavior of GFRP bars under compression and tension with bar specimens having 12 mm diameter and 48 mm length. All the specimens failed prematurely due to splitting at the bar ends. The results showed that the compressive average maximum load of the specimens was 60% less than that of the average maximum load recorded under tensile loading. Abed et al. [11] conducted a series of quasi-static and dynamic load tests on GFRP and basalt FRP (BFRP) bars having different diameters (12, 17, 21, and 27 mm) with different loading rates (0.1, 1.0, and 10.0 mm/min) in order to investigate the compressive strength of such bars. The results showed that the compressive strength of GFRP and BFRP increased as the loading rate was increased. Also, the compressive strength increases as the bar diameter increases, for all FRP specimens. Similar to the results obtained from [10], the failure mode of the specimens was a premature failure at the bar ends. Thiyagarajan et al. [12] conducted an experimental study on the characteristics of basalt FRPs (BFRP) bars with different diameters of 8, 10, and 12 mm. The study revealed that the compressive strength of the BFRP bars was about 50% of the tensile strength. In addition, the ultimate compressive strength of the BFRP bars showed a slight increase as the diameter of the bar increases.

Despite the scarcity of available experimental studies investigating the compression response of FRP bars, many researchers have already accepted the use of FRP bars as a reinforcement in concrete structural members.

This paper aims to investigate the compressive properties of GFRP bars at high temperatures ranging from 25°C to 150°C. The recent ACI 440.11-22 code [8] disregards the compressive strength of GFRP reinforcement at high temperatures due to the loss of stiffness of the polymer which occurs at elevated temperature and results in buckling of the fibers. The motivation behind this work is to fill the gap in the literature regarding the compressive strength of GFRP bars at up to 150°C temperatures and provide experimental results that can be used to study the thermal response of FRP-reinforced concrete structures.

Test Set-Up and Procedure

In this research, GFRP bars of 12 mm diameter were tested under compression at four distinct temperatures of room-temperature (RT), 50, 100, and 150°C. The length of the specimens was taken to be twice the diameter of the GFRP bars (i.e., around 24 mm) in order to prevent buckling.

All tests were conducted using the universal screw-driven INSTRON test machine equipped with an environmental chamber. A high-temperature black paint was used to generate a suitable speckle pattern for full-field deformation measurements using Digital Image Correlation (DIC). Compression displacement-controlled loading was applied at two constant rates (i.e., crosshead speeds) of 0.5 and 1000 mm/min until failure. Prior to loading, the environmental chamber was used to heat the specimen to the desired deformation temperature followed by a 30-minute soaking time to ensure a homogenous temperature distribution for the entire GFRP specimens. During the tests, images of the specimens' surface were captured, and a commercial DIC software (Vic-2D 6, Correlated solutions, USA) was used the measure the displacement fields and calculate the resulting strains.

The test setup is shown in Fig. 1 and a representative specimen installed between the compression grips is shown in Fig. 2.

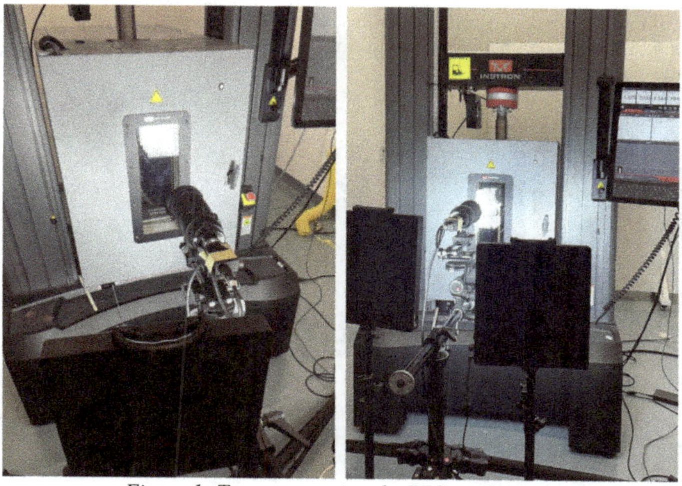

Figure 1: Test set-up using the DIC machine.

Figure 2: GFRP bar subjected to compressive loading.

Results and Discussion

As mentioned in the previous sections, this paper focuses on evaluating the temperature-dependent mechanical properties (compressive strength and elastic modulus) of GFRP bars subjected to compressive loading at various temperatures. Fig. 3 and Fig. 4 depict the compressive strength and elastic modulus for the 12 mm GFRP bars under various temperatures, ranging from room temperature (RT) to 150°C, and for the slow and fast rates, respectively. It is important to note that the values of the compressive strength and elastic modulus represent the average values obtained from about 4 different specimens for each testing condition.

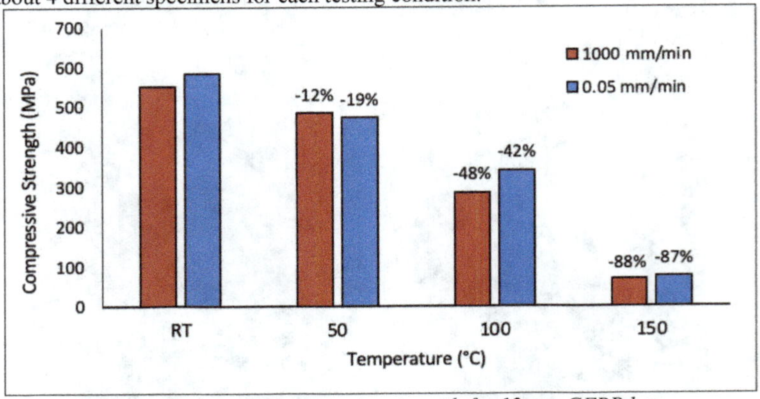

Figure 3: Average compressive strength for 12 mm GFRP bar.

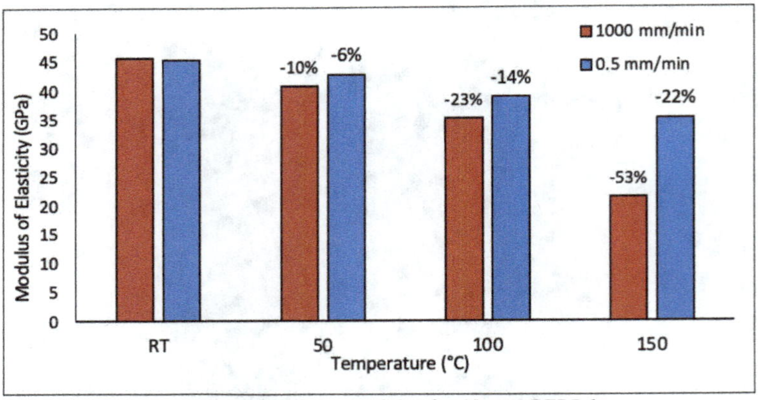

Figure 4: Average elastic modulus for 12 mm GFRP bar.

At room temperature, the compressive strength of GFRP at a loading rate of 0.5 mm/min is around 550 MPa, and it degrades by 19% and 42% at elevated temperatures of 50°C and 100°C, respectively. At 150°C temperature, the strength significantly decreases by up to 87%, as this temperature exceeds the glass transition temperature (T_g) of GFRP. The modulus of elasticity of the GFRP bars at ambient temperature for the 0.5 mm/min loading rate is recorded as 45 GPa and it decreased by 6%, 14%, and 22% as the temperature reached 50°C, 100°C, and 150°C, respectively. By comparing the rates shown in Figure 3, one can observe that the loading rate does not play a significant role in the performance of the bars, considering the strength perspective. In other words, the compressive strength of GFRP bars under a loading rate of 1000 mm/min depicted a great agreement to that of the slow rate (0.5 mm/min), in which the strength was reduced by 12%, 48%, and 88% when subjected to 50°C, 100°C, and 150°C, respectively. However, the modulus of elasticity of GFRP (shown in Figure 4) subjected to a loading rate of 1000 mm/min displayed a greater reduction of 53% at a temperature of 150°C, when compared to the 22%

reduction for the 0.5 mm/min rate. The observed failure modes of some tested specimens are shown in Fig. 5. It is evident from the figure that the governing failure mode of the specimens is splitting of the bar in half. Splitting occurred after the crushing load exceeds the strength of the bar. In addition, buckling was not observed as the length-to-diameter ratio was kept as 2.0.

Figure 5: Failure modes of the tested specimens subjected to different temperatures.

Summary and Conclusions

GFRP bars were tested at different temperatures ranging from ambient (25°C) to 150°C in order to investigate the degradation in the mechanical properties (compressive strength and modulus) of the bars at elevated temperature, while simultaneously being subjected to compressive loads. The different parameters examined in this paper are the temperature and rate of loading.

The following observations and conclusions were drawn after the test results:

1. Generally, the modulus of elasticity and the compressive strength of GFRP bars degrade linearly as the temperature is raised from 25°C to 150°C.
2. For the compressive strength, the experimental results reported similar average stress values as the rate of loading increased from 0.5 mm/min to 1000 mm/min.
3. The average compressive strength at 150°C for the 0.5 mm/min and 1000 mm/min rates are 75.4 MPa and 68.4 MPa, respectively. However, the elastic modulus at 150°C with the 0.5 mm/min rate is 35.2 GPa and 21.53 GPa with the 1000 mm/min for the fast rate. This

further proves a discrepancy between the elastic modulus at different loading rates, as opposed to the compressive strength.

4. Splitting of the bars without premature crushing or buckling was the governing observed failure mode for the tested specimens.

References

[1] T.E. Glodek, S.E. Boyd, I.M. Mcaninch, J.J. Lascala, Properties and Performance of Fire Resistant Eco-Composites Using Polyhedral Oligomeric Silsesquioxane (POSS) Fire Retardants, J. Compos. Sci. Technol. (2008) 2994-3001. https://doi.org/10.1016/j.compscitech.2008.06.019

[2] B. Benmokrane, E. El-Salakawy, A. El-Ragaby, and T. Lackey, Designing and testing of concrete bridge decks reinforced with Glass Frp Bars, J. of Bridge Eng. (2006) 217-229. https://doi.org/10.1061/(ASCE)1084-0702(2006)11:2(217)

[3] A. Serbescu, M. Guadagnini, K. Pilakoutas, Mechanical Characterization of basalt FRP rebars and Long-Term strength Predictive model, J. Compos. Constr. 12 (2014) 1-13. https://doi.org/10.1061/(ASCE)CC.1943-5614.0000497

[4] M. Al Rifai, H. El-Hassan, T. El-Maaddawy, F. Abed, Durability of basalt FRP reinforcing bars in alkaline solution and moist concrete environments. J. Constr. and Build. Materials. (2020). https://doi.org/10.1016/j.conbuildmat.2020.118258

[5] F. Abed, A. El Refai, S. Abdalla, Experimental and finite element investigation of the shear performance of BFRP-RC short beams, J. Struct. 20 (2019) 689-701. https://doi.org/10.1016/j.istruc.2019.06.019

[6] F. Abed, H. El-Chabib, M. AlHamaydeh, Shear characteristics of GFRP-reinforced concrete deep beams without web reinforcement, J. Reinf. Plast. Compos. 31 (16) (2012) 1063-1073. https://doi.org/10.1177/0731684412450350

[7] A. Al-Tamimi, F. Abed, A. Al-Rahmani, Effects of harsh environmental exposures on the bond capacity between concrete and GFRP reinforcing bars. Advances in concrete construction (2014). https://doi.org/10.12989/acc2014.2.1.001

[8] American Concrete Institute, "Building Code Requirements for Structural Concrete Reinforced with Glass Fiber-Reinforced Polymer (GFRP) Bars," ACI 440.11-22, ACI Committee 440, Farmington Hills, MI, USA, 2022.

[9] L. AlNajmi, F. Abed, Evaluation of FRP bars under compression and their performance in RC columns, J. Materials (2020). https://doi.org/10.3390/ma13204541

[10] Q.S. Khan, M.N. Sheikh, M.N. Hadi, Tension and compression testing of fibre reinforced polymer (FRP) bars, 12th Int. Symp. Fiber Reinf. Polym. Reinf. Concr. Struct. FRPRCS-12 5th Asia-Pac. Conf. Fiber Reinf. Polym. Struct. APFIS-2015 Jt. Conf., Nanjing, China: Z. Wu, G. Wu & X. Wang; (2015).

[11] F. Abed, Z. Mehaini, C. Oucif, A. Abdul-Latif, R. Baleh, Quasi-static and dynamic response of GFRP and BFRP bars under compression, J. Compos. Part C Open Access 2 (2020) 100034. https://doi.org/10.1016/j.jcomc.2020.100034

[12] P. Thiyagarajan, V. Pavalan, R. Sivagamasundari, Mechanical characterization of basalt fiber reinforced polymer bars for reinforced concrete structures, Int. J. Appl. Eng. Res. 13 (8) (2018) 5858-5862.

Characteristics of electrorheological fluids under single and mixed modes

Ali K. El Wahed

Department of Mechanical Engineering, College of Engineering, Prince Mohammad Bin Fahd University, Alkhobar, Saudi Arabia

Keywords: Electrorheological Fluids, Mode of Operation

Abstract. This paper is concerned with an experimental assessment of the rheological performance of Electrorheological (ER) fluids when utilized under single and mixed modes of operation. The experimental facility, which incorporated a dedicated ER cell, was developed to enable the instantaneous measurement of the mechanical and electrical responses of the ER fluid. The ER cell comprises a cylinder, which provides the reservoir for the ER fluid, and a piston. The cylinder is subjected to an oscillatory sinusoidal motion while the piston is fixed. The current ER cell was designed to permit the ER fluid, which is sandwiched between its cylindrical and circular gaps, to be energized either separately or simultaneously to simulate the fluid operation under either a single shear or a single squeeze or a mixed shear and squeeze modes, respectively. The transient rheological characteristics of the fluid were determined for various mechanical and electrical input conditions, which were done using a combination of displacement, force, velocity and acceleration transducers. The results have shown that the force transmitted across the fluid in squeeze is greater than that transmitted when the fluid is in shear. However, the transmitted force level was further enhanced when the fluid was utilised under mixed shear and squeeze mode of operation. In addition, the implications of the results to vibration control, where the ER fluid is employed in an engine mount, are discussed.

Introduction

Electrorheological (ER) fluids involve suspensions of semiconducting solid particulates, typically micrometres in size, in dielectric liquids. Their response to an applied electric field is the familiar chaining of the particles in the direction of the field and the resulting 'solidification' or increase in their apparent viscosity. The fast and reversible field-induced rheological changes of ER fluids have presented them as a solution to advance the performance of many electromechanical devices that are potentially useful in the automotive, aerospace, structural, medical, and other industries. Since their discovery by Winslow [1] in the forties of last century, numerous investigations have been carried out in order to improve the mechanical, chemical and electrical characteristics of ER fluids and increase their potential for industrial applications [2-6]. Vibration control has been recognised as one of the most promising areas of industrial applications of ER fluids, which is due mainly to the fact that damping levels required by many vibration suppression applications fall within the capability of commercially available ER fluids [7]. There are, however, many other areas where the static yield stresses exhibited by these fluids have failed to meet industrial requirements [8].

The majority of ER fluids have been applied in either a simple shear or flow mode of operation [9-12]. Alternatively, a squeeze mode of operation, in which the fluid is subjected to oscillatory compression and subsequent tensile stresses (resulting in a variable fluid gap) was identified [13] and found to deliver a yield strength with an order of magnitude higher than that in shear mode [14]. As a result, systematic experimental, theoretical and numerical studies have been carried out by the author and others to evaluate the mechanical and electrical properties of ER fluids in squeeze, which have helped to provide design information on the application of ER fluids in the area of short-stroke damping (see for example [15-19]).

This paper is concerned with an experimental assessment of the comparative performance of an ER fluid in dynamic shear, squeeze and, mixed shear and squeeze modes, in which the fluid was energised by a constant electric field module. Also, the relevance of this work in the application to vibration control using a short-stroke damper is discussed.

Experimental Arrangement

The experimental rig (Figure 1) consists of a Ling Dynamic Systems electromagnetic shaker (Model No. V450) which is capable of providing vertical oscillatory motion with a maximum amplitude of 19 mm peak-peak (P-P) over a frequency in the range D.C to 7.5 kHz.

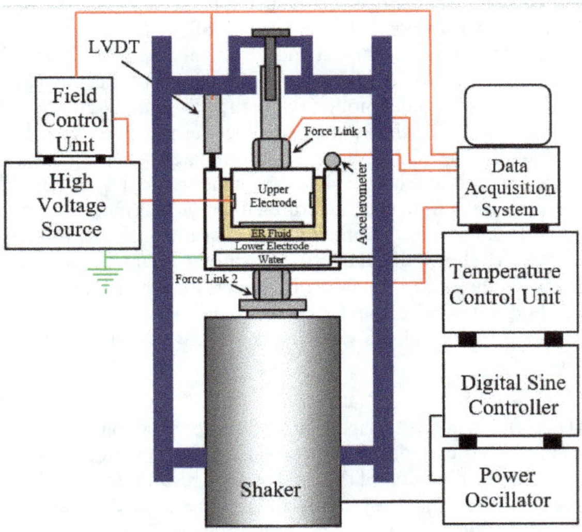

Figure 1. Experimental arrangement.

The shaker head is attached rigidly to a Kistler (Model No. 9311A) piezoelectric force link and an earthed brass electrode having a recessed cylindrical cavity of diameter 74 mm, which provides the reservoir for the ER fluid. The high voltage upper electrode consists of a circular brass disc and a circular brass cylinder of diameters 56 and 70 mm, respectively. The height of the latter electrode was chosen to be about 11 mm so that its exposed area is equivalent to that of the former electrode. The circumferential edge and rear face of both electrodes are surrounded and supported by a PTFE collar, Figure 2, which is rigidly attached to a second identical force link and positioning assembly to the supporting frame.

The instantaneous displacement and velocity of the lower electrode are measured using an RDP (Type GTX 2500) LVDT and an RDP (Type 240A0500) self-induced velocity sensor, respectively whilst its acceleration is measured using an Endevco (Type 7254-100) accelerometer. These three sensors are attached to the upper surface of the lower electrode. Electrical excitation of the ER fluid is achieved by means of a Trek (Model 664) high voltage amplifier, driven by a Thander (Model TG102) function generator.

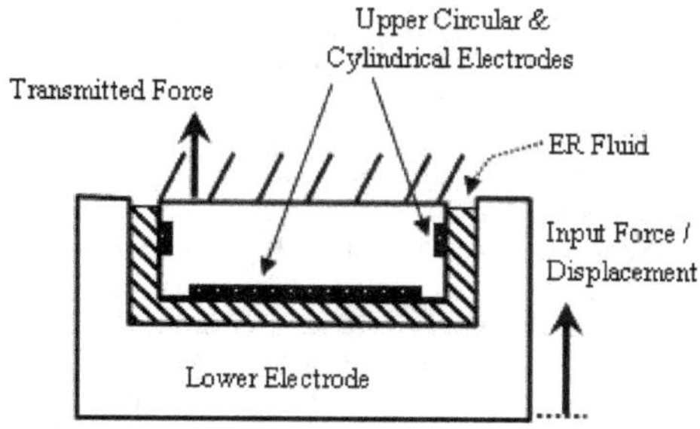

Figure 2. ER fluid cell.

In addition, a National Instruments analogue-to-digital converter, model NI9205 in conjunction with a cDAQ-9172 CompactDAQ USB chassis capable of simultaneous sampling, which was controlled by a National Instruments data acquisition software, type LabVIEW running on an IBM compatible personal computer, was used to collect and record the data from the sensors. Feedback control was not imposed in these tests although it could be achieved, if required, using a Ling Dynamic Systems (Model DSC4) digital sine controller employing as input the signal from the accelerometer. However, an electric field strength controller is used to control the electric field across the gap between the circular electrodes, which varies under the oscillatory motion input. This field controller uses the LVDT signal to determine the instantaneous gap and then control the Trek high-voltage output to maintain the required electric field strength across the fluid gap [20]. In order that meaningful comparisons could be made between the results of the various tests, the ER fluid temperature was controlled by re-circulating water through a closed cavity in the lower electrode using a Grant Instruments (Model LTD6) temperature controller.

Both force links were calibrated statically by sequential loading using small weights, while the LVDT was calibrated using a micrometre calibration unit that was specifically designed for the accurate calibration of LVDTs. The accelerometer, pre-calibrated by the manufacturer, was found to function as specified to within $\pm$ 0.6% when its maximum transverse sensitivity was checked using the digital sine controller. As an additional check, the displacement signal from the LVDT was differentiated twice using central differences and the resulting signal was found to compare well with that from the accelerometer. The data acquisition system was checked against a direct current signal supplied by a millivolt calibration unit (Time Electronics Ltd, Model 404S) and was found to be accurate to within $\pm$ 0.5%.

The ER fluid used in this investigation is a suspension of agglomerated calcium alumina silicate in silicone oil. The solid phase was supplied with an average diameter of 90 μm and was subsequently ground and sieved, using an Endecots Ltd laboratory test sieve system in conjunction with a Fritsch analysette (Type 03502) mechanical vibrator, to produce particulates in the range 10 to 28 μm. The electrical conductivity of the solid phase was measured in a dedicated cell and found to be 4.0×10^{-10} S/m. The kinematic viscosity of the oil at 20 °C was 50 cSt and the weight fraction of the solid phase was chosen to be 57%. A 5% by weight stabiliser was added to the fluid to reduce the solid phase sedimentation.

Results and Discussion

The performance of ER fluids in dynamic shear, squeeze and mixed shear and squeeze modes was systematically assessed for a range of electrical and mechanical input conditions using a dedicated cell that simulates a short-stroke damper.

In this experimental investigation, simultaneous measurements of the input force delivered by the shaker, the transmitted force across the fluid, the applied voltage, and the current passing through the fluid, in addition to the displacement, velocity and acceleration of the cylinder were carried out. Using the employed data acquisition system, these measurements were collected at a sampling frequency of 5 kHz when the shaker was set to deliver sinusoidal input vibrations with mechanical frequencies in the range 2 to 18 Hz. The input displacement amplitude of the cylinder, for the electrically unstressed fluid, was chosen as 0.6 mm (P-P) at the resonant frequency of the system, which was found to be 7 Hz. The mean gap between the plane circular electrodes was set at 2.0 mm, which was also maintained in the gap between the cylindrical electrodes of the ER cell. In the adopted experimental set up of this investigation, the fluid was energised by a constant electric field in the range zero to 2.0 kV/mm whilst the fluid temperature was maintained at 30 °C throughout the tests.

For a mechanical frequency of 7 Hz, which is the resonant frequency of the system, the variation of the input displacement of the lower assembly of the ER cell is shown in Figure 3 for the three cases when the ER fluid was utilised under a single shear, single squeeze, and mixed shear and squeeze operation modes.

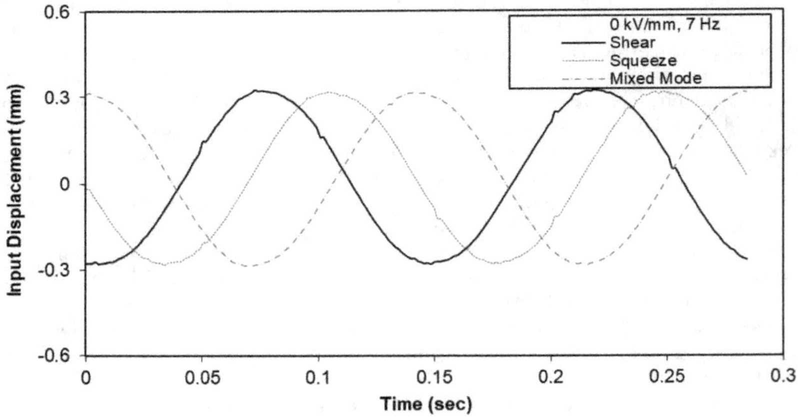

Figure 3. Variation of input displacement with time (mech Freq = 7 Hz; applied field = 0 kV/mm)

When a constant electric field of about 0.75 kV/mm was applied across the fluid (Figure 4), the zero-field displacement of the lower assembly of the ER cell was reduced by 25%, 42% and 79% when the fluid was utilized under shear, squeeze and mixed modes of operation, respectively.

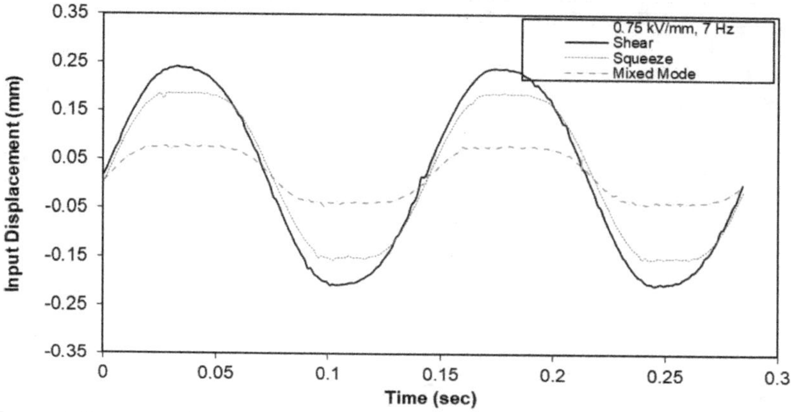

Figure 4. Variation of input displacement with time (mech freq = 7 Hz; applied field = 0.75 kV/mm)

Figure 5 shows the reduction in the input velocity of the lower assembly of the ER cell when a constant field of about 0.75 kV/mm was applied across the ER fluid. It can be seen that the largest reduction in the input velocity was achieved when the fluid was utilised in a mixed shear and squeeze mode, indicating that the ER effect is larger with this mode of operation in comparison with the cases when the fluid was employed under single modes.

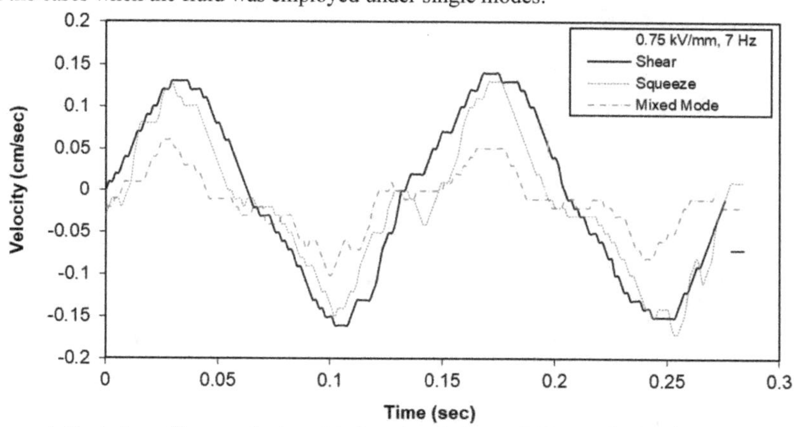

Figure 5. Variation of input velocity with time (mech freq = 7 Hz; applied field = 0.75 kV/mm)

The variations of the transmitted force across the ER fluid under a mechanical frequency of 7 Hz and electrical excitation of 0.75 kV/mm is shown in Figure 6. It can be seen that the fluid utilised under the mixed mode produced a P-P transmitted force in excess of 15 N, which was about 47% and 64% greater than the forces that were caused under the squeeze and shear modes, respectively.

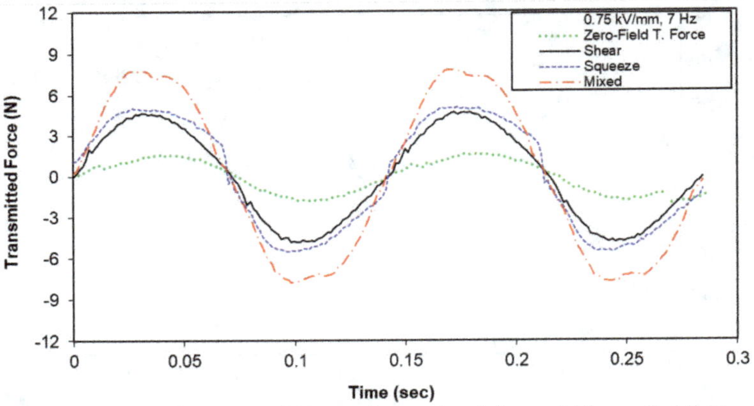

Figure 6. Variation of transmitted force with time (mech freq = 7 Hz; applied field = 0.75 kV/mm)

The zero-field transmitted force is also shown in this figure, which is caused mainly by the viscous effect of the fluid.

A comparative assessment of the performance of the ER fluid under the shear mode of operation is shown in Figure 7 where the variation of the transmitted force with input displacement is presented for a range of electrical field strengths, namely between 0 and 1.5 kV/mm.

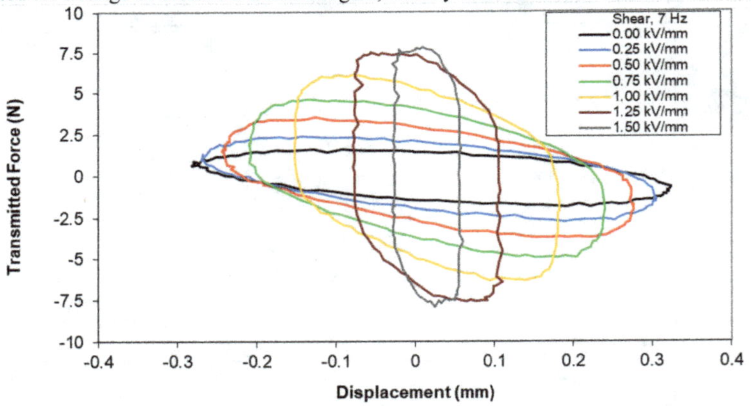

Figure 7. Transmitted force versus displacement for the ER fluid under shear mode

In this figure, the shape of the individual curves, which represent one cycle of the vibration delivered by the shaker, are likely to be influenced by the inherent nonlinearity of ER fluids, which were also shown by other smart fluids when utilised under similar conditions as part of their short-stroke damping applications [21].

For the fluid utilised in the shear mode and under the same above applied conditions, the variation of the transmitted force with the input velocity is shown in Figure 8. It should be noted that for clarity, data for some of the applied electric fields were removed from this figure, but the retained curves should clearly show the effect of increasing the electric field strength on the behaviour of the ER cell. In particular, the results confirm that as the electric field is increased, the fluid transmits higher forces, which is attributed to the fact that the fluid develops higher yield stresses with increasing field strength.

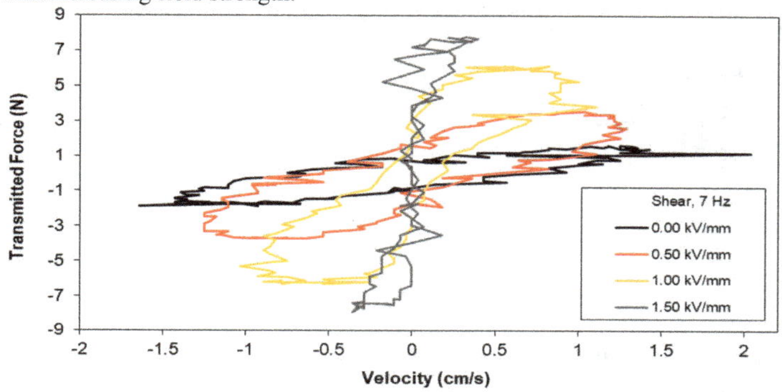

Figure 8. Transmitted force versus velocity for the ER fluid under shear mode

It is worth mentioning that the same trend in the results was found when the ER fluid was employed under the squeeze mode. However, the results have shown that the best fluid performance was again obtained under the mixed shear and squeeze mode operation, which can be seen in Figures 9 and 10 where the variations of the transmitted force was plotted against the input displacement and velocity, respectively for a range of applied electric field strengths between zero and 0.75 kV/mm.

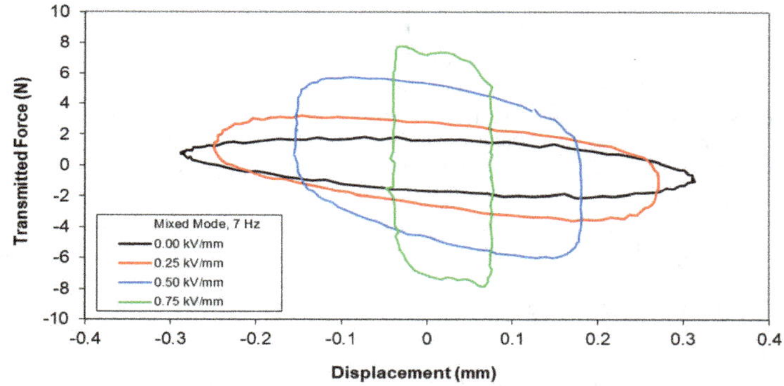

Figure 9. Transmitted force versus displacement for the ER fluid under mixed mode

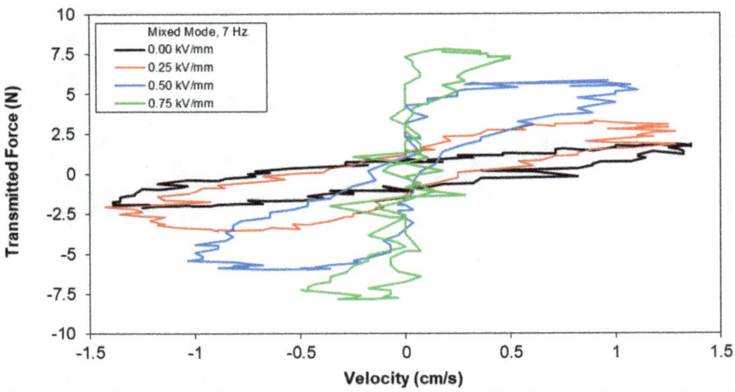

Figure 10. Transmitted force versus velocity for the ER fluid under mixed mode

It is clear that under the mixed mode of operation, the fluid exhibits solid-body characteristics when the fluid is subjected to high electric field excitations, which caused the shaker motion to be almost arrested. The electric field magnitude that produced such performance was found to be lower than those caused similar rheological properties under the single squeeze or single shear modes.

Conclusions

This paper is concerned with an investigation of the comparative performance of electrorheological (ER) fluids, applied in a specially designed short-stroke vibration isolation cell, under dynamic shear, squeeze and mixed shear and squeeze modes. The results have confirmed that the best fluid performance was achieved under the mixed shear and squeeze mode, since the transmitted force was higher than that developed by the fluids in a single shear or squeeze modes. This was also evident, when the fluid employed under the mixed mode operation caused the largest reduction in the zero-field displacement of the lower assembly of the ER cell, which was about 79% in comparison with 25% and 42% that were achieved with the shear and squeeze fluid operation, respectively.

The mixed mode of operation of ER fluids with its enhanced damping characteristics is to be investigated further with the aim to develop a smart device that is capable of the reduction of transient disturbances in short-stroke vibration isolation over a range of operating conditions within which an automotive engine mount is likely to operate.

References

[1] Winslow, W. M., 1949, J. of Applied Physics, Vol. 20, pp. 1137-1140.https://doi.org/10.1063/1.1698285

[2] Block H. and Kelly J. P., 1988, J. Phys. D: Appl. Phys., Vol. 21, pp. 1661-77.https://doi.org/10.1088/0022-3727/21/12/001

[3] Gamota, D. R., Wineman A. S. and Filisko F. E., 1993, J. Rheol. Vol. 37, pp. 919-33.https://doi.org/10.1122/1.550403

[4] Wen, W.J., Huang, X.X., Yang, S.H., Lu, K. and Sheng, P., 2003, Nature Materials, Vol. 2, pp. 727.https://doi.org/10.1038/nmat993

[5] Dhar, J., Bandopadhyay, A. and Chakraborty, S., 2015, J. Non-Newtonian Fluid Mechanics, Vol. 223, pp. 165-175.https://doi.org/10.1016/j.jnnfm.2015.07.001

[6] Sun, W., Zheng, H., Ma, J., Xi, Z., Wang, B. and Hao, C., 2020, Colloids and Surfaces A: Physicochemical and Engineering Aspects, Vol. 601, art. no. 125055.https://doi.org/10.1016/j.colsurfa.2020.125055

[7] Stanway, R., Sproston, J.L. and El Wahed, A.K., 1996, Smart Materials and Structures, Vol. 5, pp. 464-482.https://doi.org/10.1088/0964-1726/5/4/011

[8] Choi, S.B. and Gołdasz, J., 2018, In Proceedings of the ACTUATOR 2018 - 16th International Conference on New Actuators, Bremen, Germany, 25-27 June 2018; pp. 1-9.

[9] Helal, A., Qian, B., McKinley, G.H. and Hosoi, A.E., 2016, Physical Review Applied, Vol. 5, art. no. 064011.https://doi.org/10.1103/PhysRevApplied.5.064011

[10] Ivanov, K.V., Ivanova, O.S., Agafonov, A.V. and Kozyukhin, S.A., 2017, Colloid Journal, Vol. 79, pp. 204-211.https://doi.org/10.1134/S1061933X17020041

[11] Youngwook, P.S., Sangsok, H., Jihun, K., Hyoung, J.C. and Yongsok, S., 2020, Rheologica Acta, Vol. 59, pp. 415-423.https://doi.org/10.1007/s00397-020-01205-9

[12] Li, C., Chen, B. and Huang, Y., 2022, Frontiers in Materials, Vol. 9, art. no. 854872.

[13] Stanway, R., Sproston, J. L., Prendergast, M. J., Case, J. R. and Wilne, C. E., 1992, J. of Electrostatics, Vol. 28, pp. 89-94.https://doi.org/10.1016/0304-3886(92)90029-S

[14] Monkman G. J., 1995, J. Phys. D: Appl. Phys, Vol. 28, pp. 588-93.https://doi.org/10.1088/0022-3727/28/3/022

[15] El Wahed A. K., Sproston J. L. and Stanway R., 1996, Proc. ASME, Vol. 217, pp. 277-86.

[16] El Wahed, A.K., Sproston, J.L. and Williams, E.W., 2000, J. Physics D: Applied. Physics, Vol. 33, pp. 2995-3003.https://doi.org/10.1088/0022-3727/33/22/319

[17] El Wahed, A. K., Sproston, J.L., Stanway, R. and Williams, E.W., 2003, J. S & V, Vol. 268, no. 3, pp. 581.https://doi.org/10.1016/S0022-460X(03)00374-2

[18] Zhang, X. and Yu, T., 2021, Encyclopedia of Smart Materials, pp. 152-170.https://doi.org/10.1016/B978-0-12-803581-8.11721-7

[19] Huang, Y., Sun, Y., Ding, J., Yuan, S., Zhao, J., Wang, M., Luo, J. and Pu, H., 2022, J. Sound and Vibration, Vol. 527, art. no. 116864.https://doi.org/10.1016/j.jsv.2022.116864

[20] El Wahed, A.K., Sproston, J.L. and Stanway, R., 1998, J. Physics D: Applied Physics, Vol. 31, pp. 2964-2974.https://doi.org/10.1088/0022-3727/31/20/033

[21] El Wahed, A.K. and Balkhoyor, L., 2015, Smart Structures and Systems, Vol. 16, pp. 961-980.https://doi.org/10.12989/sss.2015.16.5.961

Experimental characterization and environmental impact of different grades of CRM-modified asphalt binders

Mohammad Ali Khasawneh[1,a *], Samar D. Dernayka[1,b] and Saidur R. Chowdhury[1,c]

[1]Department of Civil Engineering, Prince Mohammad Bin Fahd University, Al Khobar, Saudi Arabia

[a]mkhasawneh@pmu.edu.sa, [b]sdernayka@pmu.edu.sa, [c]schowdhury1@pmu.edu.sa

Keywords: Crumb Rubber Modifier (CRM), Asphalt Binder, Environmental Impact Analysis (EIA), Experimental, Statistical Analysis

Abstract. The use of Crum Rubber Modifier (CRM) with asphalt binders improves their properties by reducing their temperature susceptibility; this improvement is dependent on the way CRM and binder react with each other. The CRM particles tend to absorb some oils from asphalt binder and swell, resulting in a higher viscosity and thus stiffer binder. This paper discusses the effect of adding CRM to asphalt binder by applying three different percentages of CRM on two asphalt binders, viz., 60-70 and 85-100. A total of 120 specimens were prepared to test their basic physical properties such as penetration, softening point, ductility, flash and fire points, and rotational viscosity. It was found that the physical properties of asphalt binder are influenced by the percentage of crumb rubber used. The modified asphalt binders with higher crumb rubber content have lower ductility and penetration number, but higher softening point and viscosity values. Thus, it could be expected that they will have improved rutting resistance and extended service life. Furthermore, different analytical methods were utilized to validate the experimental results. It was reported that only penetration displayed change in behavior between 60-70 and 85-100 asphalt binders. CRM Content was found to be influential on all performance-related physical tests. Finally, it was obviously shown that the combined effect of Asphalt-Types and CRM Content is statistically significant on softening point, ductility and rotational viscosity test results. The incorporation of CRM in asphalt mix has multiple environmental benefits as well. The energy saving coupled with other benefits including less cracking and maintenance makes rubberized asphalt very attractive and beneficial in highway construction. Rubberized asphalt technique reduced the environmental impact of highway construction by 5% to 10% compared to traditional practice.

Introduction

Crumb rubber is a term that refers to recycled rubber from vehicular tires; it has been used widely as a modifier in the asphalt mixes since 1930, due to its great effect in improving their performance. Recently, it is recognised as being one of the most innovative sustainable binders used in the pavement industries. Actually, the incorporation of Crum Rubber Modifier (CRM) in various asphalt mixtures offers different manifold environmental benefits [1, 2] for the pavement as well for the surrounding area. These effects are influenced by many factors such as the method of production of crumb rubber, the approach of asphalt mix, the CRM content, and the availability of other additives within the mix.

The CRM usage in asphalt mixtures contributes to the recycling of millions of scrap tires that are discarded annually without proper treatment [3]. Solid tire wastes are materials that cannot be decomposed naturally. Every year, about one billion tires complete the end of their useful lives, and it is estimated to reach at least 1.2 billion end-of-life tires' worldwide by the year 2030 [4,5]. Further, tires can leak chemicals into the air, ground, and water that can disrupt the ecosystem when they accumulate in landfills or junkyards. A used tire emits methane gas into the atmosphere just by being in the sun. Our carbon footprint grows as a result of this greenhouse gas, which may also contribute to climate change. Moreover, in the event that the tire catches fire, poisonous black smoke may be released into the atmosphere. Burning tire wastes also worsen air, water, and soil pollution [6]. Therefore, it is necessary to find alternative options to handle this enormous quantity of tire debris in order to reduce the ecological effect and the depletion of available disposal sites. Over the past few decades, there has been a lot of interest in the use of recycled tire rubber in asphalt pavements [7]. Crumb Rubber Modifier (CRM) has a great role in improving the asphalt binder properties such as rutting resistance, fatigue/reflective cracking resistance, and durability due to the noticeable increase in the viscosity and softening point of the modified asphalt binders [8-12]. In addition, many other advantages were achieved such as reducing temperature susceptibility, traffic noise, pavement maintenance costs and pollution. [13, 14-18].

Previous investigation found that recycled tire rubber could be able to decrease the permanent deformation of flexible pavements [19, 20], improve the pavement's resistance to rutting, lower the cost of constructing and maintaining the pavement, and increase the pavement's resilience to fatigue damage [21,5,4].) In the present study, it is clear that using CRM can improve asphalt binding properties, binder stiffness, asphalt pavement durability and it represents a cheaper substitute for polymers used to modify bitumen. Moreover, reclaimed asphalt surface material with CRM (obtained from waste tires and waste rubber powder) has potential to contribute to environmentally sustainable construction by way of saving energy consumption, less greenhouse gas (GHG) emission, solid waste reduction as well as lowering the cost of hot mix asphalt (HMA) pavement rehabilitation [22,5]. Besides, the study conducted in [23] have referred to the great enhancement of the environmental noise performance of flexible pavements when crumb rubber is included in HMA mixes [23].

On the other hand, the environmental impact of CRM asphalt mixes is influenced by the method of production [24], whether it is ambient mechanical grinding or cryogenic grindingwhich is more expensive and produces smaller and smoother crumbs. The ambient mechanical grinding is a multi-step process in which the scrap tires are broken up at or above the room temperature, it uses the tires in the form of shred or chips by passing them through a shredder. While, in cryogenic grinding process, the grinding of used tires is done at extremely low temperatures near -80 oC so that the rubber becomes brittle and easily crushed. This method requires less energy and produces much finer particles. When quantifying the environmental negative impacts of CRM asphalt mixes in [25], it was found that the production of CRM using the ambient grinding method results in the least score (3 – 20%) while the asphalt mixture production and base course construction present the higher rate that reaches 93%.

The interaction between CRM and other asphalt binders is another factor that would affect the improved properties of produced CRM, and this is related to the method of CRM incorporation in road paving materials. Previous studies proved that the wet technology is much favourable in terms of environmental drawbacks. The CRM asphalt mixture generated with wet process stands behind the lower environmental burdens as per [26] and [24]; although it presents higher annual energy use (by wet process-high viscosity). While dry mixtures induce notable rise in scores of all negative impact categories. Since the rubber absorbs the lighter parts of bitumen, the amount of bitumen augments with the increase of CRM content.

Other aspects of CRM such as CRM content, size particles have been examined in other studies. As per [27 – 29], the improvement of the modified binder performance depends on many factors such as size of crumb rubber particles, blending conditions, surface characteristics of crumb rubber particles, physical and chemical properties and source of original asphalt binder. In terms of blending, [24] found that CRM incorporation requires less energy than virgin asphalt binder in the two ways of blending 1) the terminal blending of rubberized asphalt (TBRA) and 2) field blending of rubberized asphalt (FBRA)by using 18 % and 20% CRM content. However, the mixing process of rubberized asphalt under high temperature, emissions increase, more specifically the cumulative energy consumption is 7.26% lower for TBRA, while it is 1.25% higher than conventional HMA. Therefore, the TBRA blending seems to be more environmentally effective in terms of energy use and GHG emissions.

Finally, the effect of CRM incorporation on the rheological properties of asphalt mixes has been quite a popular topic within the literature. Roberts et al. (1991) examined the effects of incorporating crumb rubber into asphalt mixes and notes that the crumb rubber increases viscosity at high service temperature to improve rutting and shoving resistance. In addition, it increases the relaxation properties at low temperatures, which helps decrease thermal cracking and increases adhesion between asphalt binder and aggregate particles in the presence of moisture to prevent stripping [30]. Navarro et al. (2004) studied the thermo-rheological behavior and the storage stability of the ground tire rubber-modified binder. He recommended to use small CRM particle sizes (less than 0.35 mm) with high shear rates in manufacturing operations. As storage temperature and particle size increase, storage stability of CRM decreases [31]. On the other hand, Navarro et al. (2005) indicated that the rheology, storage stability, and microstructure of the modified binders depend largely on the rubber content concentration. The thermal susceptibility of the modified binder is reduced clearly when adding more rubber and decreases the viscous and elastic moduli at low temperatures and increases binder flexibility, while at high temperatures both moduli were increased significantly. The maximum amount of CRM should not exceed 9% to have a binder adequate for paving applications [32]. Ghavibazoo et al. (2013) showed that CRM dissolution in asphalt with an intermediate interaction temperature of 190 °C and high mixing speed of 50 Hz produces more homogenous CRM asphalt binder with better high service temperature properties. However, more investigation is still needed to butter comprehend other performance related properties such as storage stability, low temperature performance of the asphalt and aging susceptibility [33]. According to Xiang et al. (2009) results, the CRM asphalt binders have better performance compared with matrix asphalt binders, the performance of CRM asphalt binders are improved with increasing ash content and acetone extract [34]. Shen et al. (2009) found that the surface area of the ambient CRM was two times larger than the cryogenic CRM, with a much higher phase angle and G* of the CRM binders. Also the G* and phase angle were both affected by average size and the surface area of CRM particles. Results showed that the ambient CRM binders were 3 °C higher in temperature grade than cryogenic CRM binders [35]. Simultaneously, Thodesen et al. (2009) showed that base binder viscosity and the grinding procedure have the greatest effect on CRM binder viscosity. However, arising changes are due to physical interaction between binder and CRM particles rather than chemical interactions [36]. According to S. Liu et al. (2009), the variance analysis shows that the main influencing factor is the crumb rubber content, then comes the crumb rubber type and last in order the particle size. As the crumb rubber content increases, softening point and penetration index increase, and the low temperature ductility decreases. Results of Bending Beam Rheometer (BBR) and Dynamic Shear Rheometer (DSR) tests reveal that the optimum CRM content is 20% for tread rubber (TR-CRM) asphalt, 15% for heavy truck (HT-CRM) asphalt and 10 - 15% for small truck (ST-CRM) asphalt [37].

The main objective of this paper is to discuss the effect of adding CRM to the asphalt binder and to check if it always improves basic physical properties of asphalt or if it actually has some negative impacts on asphalt. In addition, the motivation of this study was also underlined due to the contradiction present in existing studies regarding the effect of this commonly and widely used modifier on paving materials behaviour. Besides, previous studies did not explicitly clarify the optimum rubber content that best improves asphalt performance, which was evidently assessed in this study. Temperature susceptibility evaluation of the two asphalt grades after rubber modification was another main area of focus in this study. Furthermore, there is an environmental performance evaluation of the suggested samples.

Methodology, Materials and Preparation of CRM Binders

Two binders were used in this study graded as 60-70 and 85-100. The physical properties of the asphalt samples are given in Table 1.

Table 1: Physical properties of asphalt binder

Properties of asphalt binder	Grade 60-70	Grade 85-100	Specification Standard	References
Penetration (25°C,100g,5s)	69.67	85	ASTM D 5	[38]
Softening point(°C)	48.8	38.2	ASTM D 36	[39]
Ductility (25°C, 5 cm/min)	76.3	76	ASTM D 113	[40]
Flash point	263.67	256.67	ASTM D 92	[41]
Fire point	289	286.33	ASTM D 92	[41]
Specific gravity	1.015	1.01	ASTM D 70	[42]

The CRM particles used in this study were prepared using the ambient mechanical grinding process. The physical properties of CRM particles are as follow: size ranges from 0.177 mm to 0.25 mm and specific gravity of 1.056.

The wet process approach was used to mix CRM particles with asphalt binder. In this method, CRM particles dispersed into the asphalt binder and mixed at an elevated temperature of 170 °C, a speed of 1200 rpm, and a mixing time of 60 minutes. In total, 120 specimens were prepared, 30 of them were control specimens. Then these specimens were tested for penetration, softening point, ductility, rotational viscosity, flash and fire points test. Additionally, temperature susceptibility was evaluated.

Three different percentages of CRM were tried (10%, 12.5%, and 15%) on 60-70 and 85-100 asphalt binder grades (the most commonly used grades worldwide and are typically used in the United States). The selected percentages were a result of intensive search in the available literature [28, 29]. The grades 60-70 and 85-100 are characterized with moderate relative consistency that is suitable for intermediate climatic conditions (neither very cold nor very hot). A hundred and twenty (120) specimens were prepared to test their basic physical properties such as penetration, softening point, ductility, flash point, and rotational viscosity. Fig. 1 shows a flow chart of the testing procedure.

Preliminary Results and Discussion

Physical Tests Results - Penetration

It can be seen that as the percentage of CRM used increases, the penetration number for both types of binders decrease due to the increase in the stiffness of the modified binder (Fig. 2). Thus, the consistency of the original asphalt changes and becomes harder. The reduction percentage in penetration number for modified binders is almost 24% of the original 60-70 and 22% of the original 85-100 with the addition of 15% CRM.

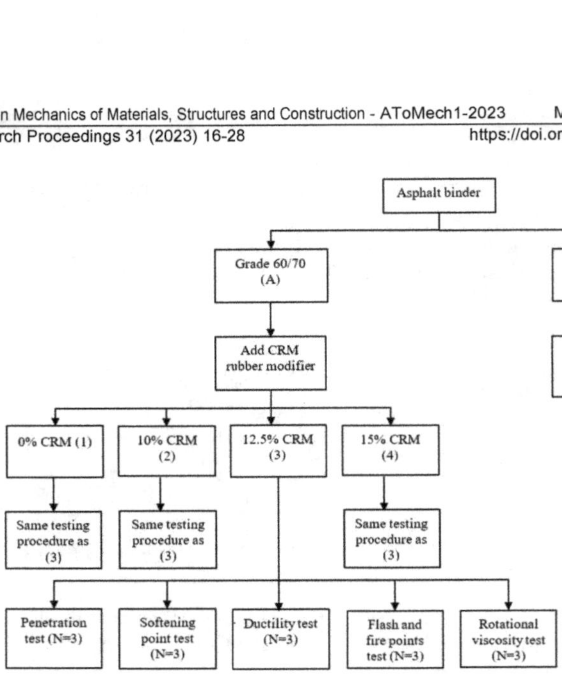

Figure 1: Flow Chart of Testing Procedure

Physical Tests Results - Softening Point
It is obvious that there is an increase in softening point temperature of the CRM modified asphalt as the percentage of CRM added to the base asphalt increases (Fig. 3). The increase in softening point values of modified asphalt with addition of 15% CRM is almost 40% of original asphalt 60-70 and 72% of the original asphalt 85-100, consequently higher softening point is required for achieving better asphalt pavement performance.

Physical Tests Results - Ductility Test
It can be seen that original asphalt ductility appears to decrease as the crumb rubber percentage added increases (Fig. 4). The ductility value of base asphalt (60-70) is 76.3 cm on average, which reduced to 55.5 cm by the addition of 15% CRM; this reduction is almost 27% of base asphalt value. The reduction percentage in ductility for 85-100 original binder is 23 %. This reduction in ductility specifies that there is a significant loss in the flexibility of the original asphalt occurred by the incorporation of such small quantity of crumb rubber.

Physical Tests Results - Flash and Fire Points Test
Flash and fire points are also affected by the addition of CRM as shown in Fig. 5 and Fig. 6. he increment in both flash and fire points with addition of 12.5% CRM to base asphalt 60-70 is nearly 18% and 23%, respectively, and 19% and 22.2% for the addition of CRM to the 85-100 asphalt.

Physical Tests Results - Rotational Viscosity
In order to ensure that the binder is sufficiently workable, a cylindrical spindle is submerged in asphalt binder at a constant temperature of 135°C which simulates the workability at mixing and laydown, then the torque that is required to maintain a constant rotational speed of 20 rpm of the spindle is measured and converted to viscosity [30]. The addition of CRM increases the kinematic viscosity (see Fig. 7). As can be seen the rotational viscosity for original asphalt 60-70 increases averagely from 494.8 mPa.s to 1890 mPa.s with the addition of 15% CRM and increased from 312

mPa.s to 930 mPa.s for original binder 85-100. This result indicates that using CRM is effective in improving the viscosity property of asphalt binder. Rutting is more of a concern early in the asphalt pavement life to mid-life especially at high service temperatures, whereas fatigue cracking is of great concern older in the pavement life at intermediate service temperatures.

Physical Tests Results - Temperature Susceptibility

The temperature susceptibility of asphalt binder means the change in its consistency due to changes in temperature. It is considered one of the control parameters during mixing, compaction, placing and an important tool for evaluating asphalt binder performance. There are several methods to calculate temperature susceptibility, two of which are explained herein:

Penetration Index (PI) method: by using the resulted softening point temperatures and the penetration value at 25°C the PI can be obtained from a nomograph and then the temperature susceptibility can be calculated using the Eq. 1 [31]:

$$\mathrm{PI} = \frac{20(1-25A)}{(1+50A)}. \quad (1)$$

Where,
PI = Penetration Index
A = Temperature susceptibility

Alternatively, for any conventional paving grade bitumen, the softening point temperature is the same as that which gives a penetration of 800 d-mm. This, along with the penetration at 25°C, can be used to compute temperature susceptibility (A) according to Eq. 2:

$$A = \frac{\log(\text{pen at} 25°\text{C}) - \log(800)}{25 - \text{ASTM softening point}} \quad (2)$$

A. Penetration-Viscosity Number (PVN) Method: by using penetration at 25 °C and viscosity at (135 °C or 60 °C), which are usually specification requirement for viscosity graded asphalt cement. Eq. 3 is used to calculate PVN [31]:

$$\mathrm{PVN} = \frac{L-X}{L-M}(-1.5) \quad (3)$$

Where,
X = the logarithm of viscosity in centistoke measured at 135 °C
L= the logarithm of viscosity at 135 °C for a PVN value of 0.0 and,
M= the logarithm of viscosity at 135 °C for a PVN value of -1.5.

It can be seen (refer to Fig. 8) that the addition of CRM modifier to asphalt binders reduces the temperature susceptibility, but the values still in the range since they vary from 0.0015 for highly blown low temperature binders (high PI) to 0.06 for high temperature susceptible binders (low PI).

Temperature susceptibility has an inverse relationship with PVN values. Usually, PVN varies between +0.5 to -2 for most paving asphalt cements. For the CRM-modified asphalt materials in this study one may refer to Fig. 9 for PVN results. Although the PI changes with aging, during mixing and subsequently in service, PVN remains substantially the same.

Materials Research Proceedings 31 (2023) 16-28 https://doi.org/10.21741/9781644902592-3

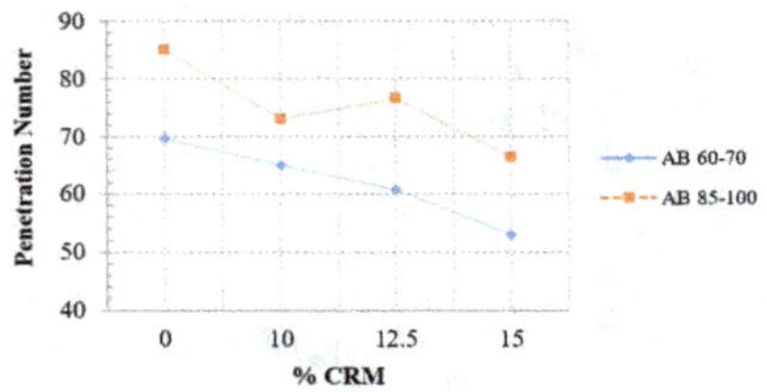

Figure 2: Penetration Number vs. % of CRM

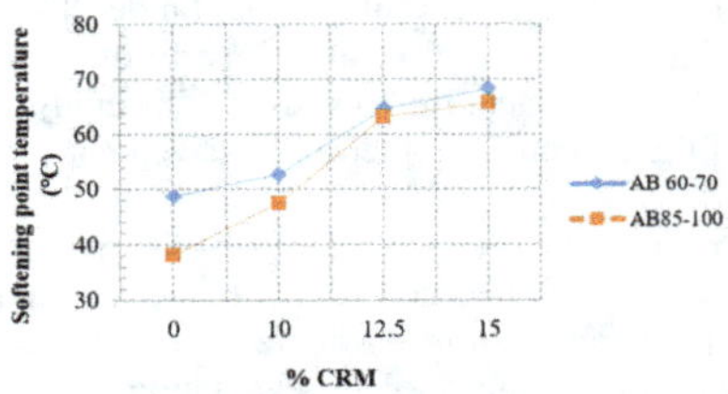

Figure 3: Softening Point Temperature vs. % of CRM

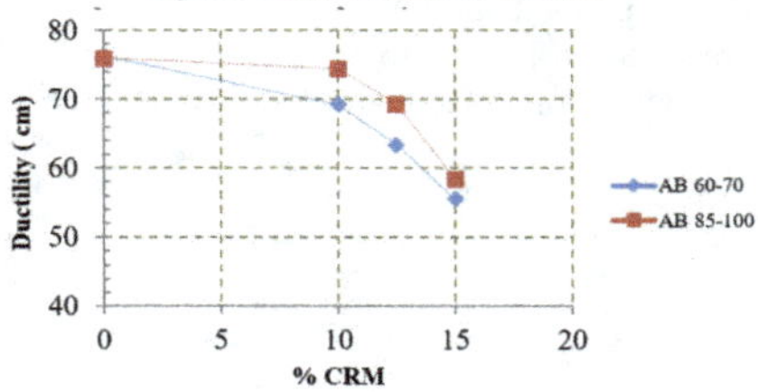

Figure 4: Ductility vs. % of CRM

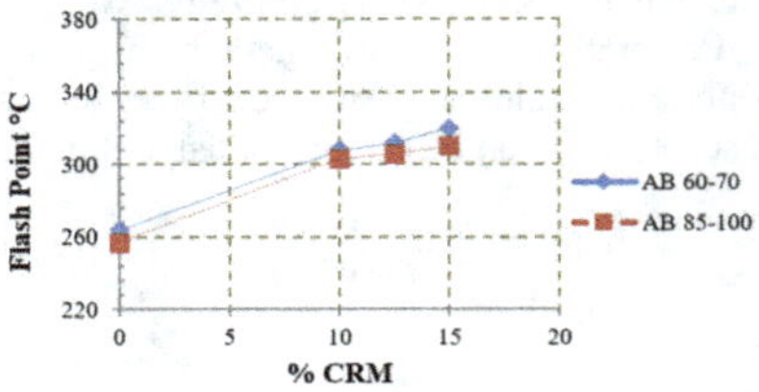

Figure 5: Flash Point vs. % of CRM

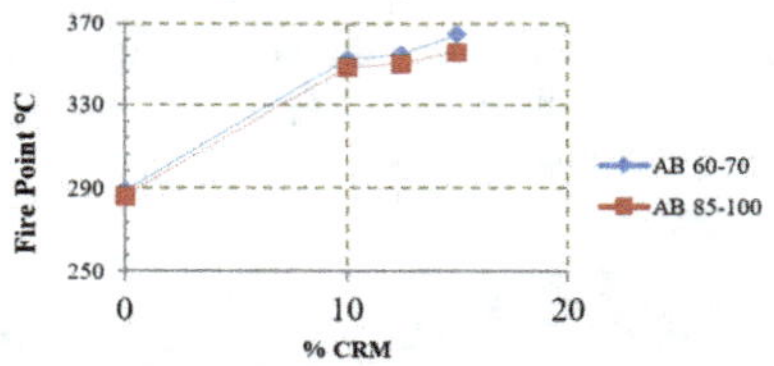

Figure 6: Fire Point vs. % of CRM

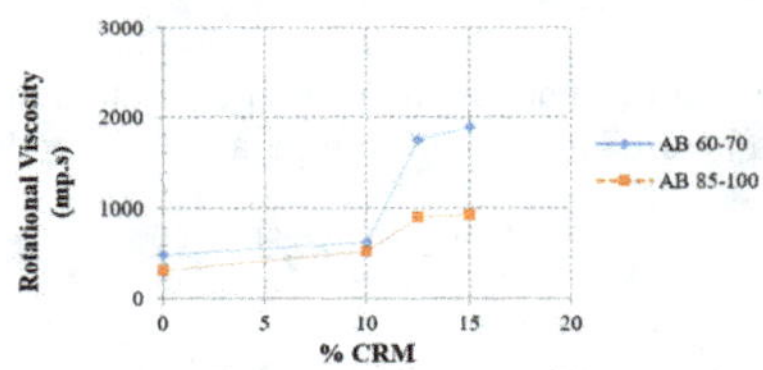

Figure 7: Rotational Viscosity vs. % of CRM

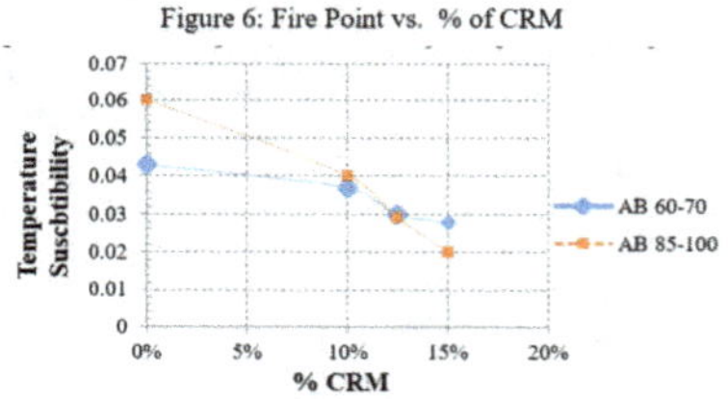

Figure 8: Temperature Susceptibility vs. % of CRM (PI Method)

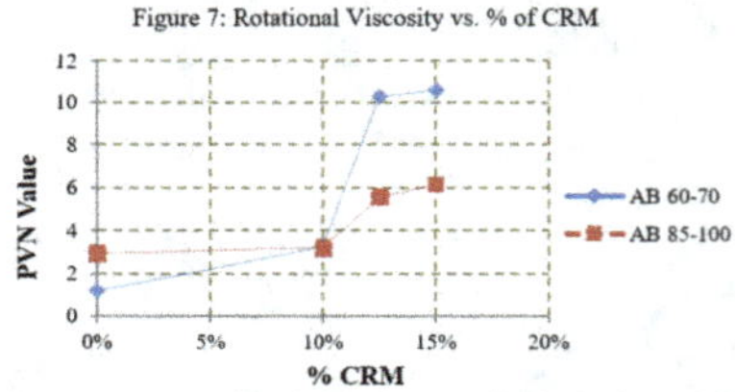

Figure 9: Temperature Susceptibility vs. % of CRM (PVN Method)

Statistical Analysis

The experimental data was statistically analyzed using the Statistical Package for Social Sciences (SPSS Inc. Version 23). Descriptive statistics, independent samples t-test, One-Way Analysis of Variance (ANOVA) or the so called Completely Randomized Design (CRD) and Two-Way ANOVA (AKA the Two Factor Factorial Design; TFFD) were used. Table 2 provides a summary of the descriptive statistics pertinent to all laboratory work conducted as part of this study.

In this study the basis for comparison was the asphalt types (called factor) in which there are two types (called groups), 60-70 and 85-110. It was found that the homogeneity of variance assumption is met (p-value greater than 0.05) for all asphalt physical tests except the softening point and RV tests where the p-value is less than 0.05. From the t-test for Equality of Means, only penetration showed statistically significant difference in behavior for 60-70 asphalt binder as compared to 85-100.

Table 2: Overall Descriptive Statistics

	N	Min.	Max.	Mean	Std.Deviation	Variance	Skewness	Kurtosis
Penetration	24	44.0	91.0	68.7	10.8	117.3	0.17	0.40
Softening Point	24	37.0	69.2	56.2	10.5	110.3	-0.41	-1.19
Ductility	24	53.2	80.2	67.8	7.9	62.6	-0.33	-1.01
Flash Point	24	255.0	322.0	297.4	22.6	512.1	-1.08	-0.53
Fire Point	24	285.0	368.0	337.5	29.9	894.4	-1.10	-0.62
RV	24	310.0	1920.0	930.2	560.5	314171.0	0.82	-0.84

One-Way ANOVA hypothesis tests the difference in population means for more than two groups based on one characteristic or factor. Four groups are used in this single-factor analysis called CRM Content (%). Groups are 0%, 10%, 12.5% and 15%. It was found that the homoscedasticity assumption check is satisfied for all asphalt binder physical tests except softening point and RV tests. Not meeting the test of homogeneity of variances assumption does not matter much as ANOVA test is robust and can overcome this violation.

The one-way ANOVA test with CRM Content in percent having four levels (0%, 10%, 12.5% and 15%) considered as the factor whose effect on penetration, softening point, ductility, flash point, fire point and RV values was evaluated. Results revealed that changing CRM Content has a statistically significant effect on all physical performance-related values on the 0.05 significance level.

Two-Way ANOVA tests perform comparisons between populations based on multiple characteristics or factors. In this study, the two factors considered are Asphalt-Types (60-70 and 85-100) and CRM Content in percent (0%, 10%, 12.5% and 15%).

Table 3 summarizes the two-way ANOVA results for all tests; viz. penetration, softening point, ductility, flash point, fire point and RV tests. It is obviously shown that the combined effect of Asphalt-Types and CRM Content is statistically significant, on the 0.05 significance level, only for softening point, ductility and RV tests. Consequently, the combined effect is NOT statistically significant when considering penetration, flash and fire point tests. Besides, Table 3 presents the individual effects of Asphalt-Types and CRM Content on asphalt physical properties.

Table 3: Tests of Between-Subjects Effects for the Two-Way ANOVA

Dependent Variable: Penetration					
Source	**Sum of Squares**	**Degrees of freedom**	**Mean Square**	**F**	**Sig.**
Asphalt-Types	1040.1	1	1040.1	25.1	.000
CRM Content (%)	937.3	3	312.4	7.5	.002
Asphalt-Types * CRM Content (%)	59.1	3	19.7	0.4	.702
Dependent Variable: Softening Point					
Source	**Sum of Squares**	**Degrees of freedom**	**Mean Square**	**F**	**Sig.**
Asphalt-Types	153.0	1	153.0	46.8	.000
CRM Content (%)	2259.8	3	753.2	230.8	.000
Asphalt-Types * CRM Content (%)	72.3	3	24.1	7.3	.003
Dependent Variable: Ductility					
Source	**Sum of Squares**	**Degrees of freedom**	**Mean Square**	**F**	**Sig.**
Asphalt-Types	23.8	1	23.8	3.9	.065
CRM Content (%)	1235.3	3	411.7	68.1	.000

Asphalt-Types * CRM Content (%)	84.4	3	28.1	4.6	.016
Dependent Variable: Flash Point					
Source	**Sum of Squares**	**Degrees of freedom**	**Mean Square**	**F**	**Sig.**
Asphalt-Types	308.1	1	308.1	87.0	.000
CRM Content (%)	11389.8	3	3796.6	1071.9	.000
Asphalt-Types * CRM Content (%)	23.1	3	7.7	2.1	.130
Dependent Variable: Fire Point					
Source	**Sum of Squares**	**Degrees of freedom**	**Mean Square**	**F**	**Sig.**
Asphalt-Types	145.0	1	145.0	10.0	.006
CRM Content (%)	20165.7	3	6721.9	466.2	.000
Asphalt-Types * CRM Content (%)	30.4	3	10.1	0.7	.563
Dependent Variable: RV					
Source	**Sum of Squares**	**Degrees of freedom**	**Mean Square**	**F**	**Sig.**
Asphalt-Types	1109185.0	1	1109185.0	1366.0	.000
CRM Content (%)	4716445.8	3	1572148.6	1936.1	.000
Asphalt-Types * CRM Content (%)	1387310.3	3	462436.7	569.5	< .001

Environmental Impact Analysis

When compared to energy recovery and landfill disposal, the use of crumb rubber in asphalt has significant environmental and energy savings advantages. Due to the decrease in raw materials and increased pavement service life, it solves the issue of disposing of used tires and produces material savings and environmental advantages. During the rehabilitation stage, using rubberized asphalt hot mixes could use less energy than using conventional asphalt mixtures, saving landfill space and reducing transportation costs.

Wang et al. (2018) claimed that the longer service life and lower raw material requirements of rubberized asphalt applications resulted in energy savings. Energy consumption could also be reduced due to less engineering material uses such as aggregate savings. Due to overall savings, energy usage might be decreased as well. Using rubberized asphalt technique instead of traditional HMA solution resulted in energy savings of 28.651 kWh/m^3 [23]. Application of rubberized asphalt resulted in energy savings of 47% of the total energy used during the repair or rehabilitation stage [43]. In addition, less landfill space is needed and recovered rubberized asphalt does not need to be carried away to disposal site, thus, consume less energy than conventional asphalt mixtures during the rehabilitation stage [23].

Rubberized asphalt is particularly appealing and useful in highway construction because of the energy savings and additional advantages like fewer maintenance and cracking. Rubberized asphalt has better mechanical performance, which allows for lower maintenance and conservation costs over the course of the service cycle, resulting in significant energy and natural resource savings. Rubberized asphalt technique reduced the environmental impact of highway construction by 5% to 10% compared to traditional technology. The benefit came about as a result of the rubberized asphalt road's reduced thickness and the materials, energy, and fuel that were saved during construction [44]. Better structural and functional performance from rubberized asphalt could significantly lessen the environmental impact.

Additionally, the traditional method of producing rubberized asphalt requires a lot of energy because the material's high viscosity necessitates a higher mixing temperature and longer mixing time. However, the Wet Mix Asphalt (WMA) technology has the potential to reduce rubberized asphalt's viscosity and further lower the mixing temperature and duration, resulting in a 20% to 25% fuel savings [23]. The inclusion of rubber powder may reduce the frequency of noise at the tire-to-pavement interaction [45]. One of the advantages of rubberized asphalt pavement for the environment is the noise reduction it provides. When compared to conventional pavement, asphalt rubber can lower tire-pavement noise by 40% to 88% in the frequency range of 500 to 4000 Hz [23].

Conclusions

To examine the physical properties of modified asphalt binders, three percentages of CRM rubber were used to modify asphalt binders 60-70 and 85-100. Based on the work carried out in this study, the following conclusions can be made:

1. The addition of CRM rubber modifier to asphalt binder at increased proportions does not always improve the behaviour of bitumen, it may reflect negatively on its properties. The reduction percentage in penetration number and ductility for modified binders is significant and as it is known that the bitumen with insufficient penetration or ductility gets cracked when subjected to repeated traffic loading. This result can be explained by the fact that the CRM rubber particles tend to absorb the oils from asphalt binder and swell.
2. Adding CRM to the base asphalt has shown an improvement in viscosity, softening point values and safety characteristics. Consequently, improved rutting resistance and extended service life are expected.
3. According to SuperPave specifications, the asphalt binder rotational viscosity must not exceed 3 Pa.s. The addition of CRM rubber modifier increases the binder viscosity, but it can be noticed that the viscosity still in the acceptable range.
4. Different statistical methods have been utilized and provided evidence of the following:
    a. According to the results of the independent sample t-tests, penetration is the only physical test that showed statistically significant difference in behaviour between 60-70 and 85-100 asphalt binders.
    b. One-way ANOVA results revealed that using CRM Content has a statistically significant influence on all performance-related physical tests.
    c. Two-way ANOVA results exposed that the combined effect of Asphalt-Types and CRM Content is statistically significant only for softening point, ductility and RV tests.
5. Rubberized asphalt technique reduced the environmental impact of highway construction by 5% to 10% compared to traditional practice.

References

[1] L. M. Khiong, M. Safiuddin, M. Abdul Mannan and Resdiansyah, Material Properties and Environmental Benefits of Hot-Mix Asphalt Mixes Including Local Crumb Rubber Obtained from Scrap Tires, Environments (2021). https://doi.org/10.3390/environments8060047

[2] I. M. Khan, S. Kabir, M. A. Alhussain, F. F. Almansoor, Asphalt Design using Recycled Plastic and Crumb-rubber Waste for Sustainable Pavement Construction, Procedia Engineering 145 (2016) 1557 – 1564. https://doi.org/10.1016/j.proeng.2016.04.196

[3] B.S. Thomas, R.C. Gupta, V.J. Panicker, Recycling of waste tire rubber as aggregate in concrete: Durability-related performance. J. Clean. Prod., 112 (2016) 504–513. https://doi.org/10.1016/j.jclepro.2015.08.046

[4] P.R. Shakya, P. Shrestha, C.S. Tamrakar, P.K. Bhattarai, Studies on potential emission of hazardous gases due to uncontrolled open-air burning of waste vehicle tyres and their possible https://doi.org/10.1016/J.ATMOSENV.2008.04.013

[5] S.A. Alfayez, A.R. Suleiman, and M.L. Nehdi, Recycling Tire Rubber in Asphalt Pavements: State of the Art. Sustainability 12, (2020) 9076. DOI:10.3390/su12219076.

[6] F. Cotana, S. Vittori, G. Marseglia, C.M. Medaglia, V. Coccia,; A. Petrozzi, A. Nicolini, G. Cavalaglio, Pollutant emissions of a biomass gasifier inside a multifuel energy plant. Atmos. Pollut. Res. 10, (2019) 2000–2009. https://doi.org/10.1016/j.apr.2019.09.007

[7] L.P.G. Santos, S.D. Capitão, J.M.c. Neves, Crumb rubber asphalt mixtures: A literature review. Construction and Building Materials 247(2020) 118577. https://doi.org/10.1016/j.conbuildmat.2020.118577

[8] F. J. Navarro, P. Partal, F. Martınez-Boza, C. Valencia, & C. Gallegos, Rheological characteristics of ground tire rubber-modified bitumens, Chemical Engineering Journal, 89(1) (2002) 53-61. https://doi.org/10.1016/S1385-8947(02)00023-2

[9] H. Y. Katman, M. R. Ibrahim, M. R. Karim, & A. Mahrez, Resistance to Disintegration of the Rubberized Porous Asphalt, Proceedings of the Eastern Asia Society for Transportation Studies Vol. 7 (The 8th International Conference of Eastern Asia Society for Transportation Studies, (2009), 286-286, Eastern Asia Society for Transportation Studies.

[10] A. Mahrez, & M. Rehan, Rheological evaluation of aging properties of crumb rubber-modified bitumen, Journal of the Eastern Asia Society for Transportation Studied, 5 (2003) 820-833.

[11] L. P. Fontes, G. Triches, J. C. Pais, & P. A. Pereira, Evaluating permanent deformation in asphalt rubber mixtures, Construction and Building Materials, 24(7) (2010) 1193-1200. https://doi.org/10.1016/j.conbuildmat.2009.12.021

[12] C. Wong, C, & W. G. Wong, Effect of crumb rubber modifiers on high temperature susceptibility of wearing course mixtures, Construction and Building Materials, 21(8), (2007) 1741-1745. https://doi.org/10.1016/j.conbuildmat.2006.05.020

[13] J. Emery, Evaluation of rubber modified asphalt demonstration projects, Transportation Research Record, 1515(1995) 37-46.

[14] C. Thodesen, K. Shatanawi, and S. Amirkhanian, Effect of crumb rubber characteristics on crumb rubber modified (CRM) binder viscosity, Construction and Building Materials, 23(1) (2009) 295-303. https://doi.org/10.1016/j.conbuildmat.2007.12.007

[15] S. Liu, W. Cao, J., Fang, and S. Shang, Variance analysis and performance evaluation of different crumb rubber modified (CRM) asphalt, Construction and Building Materials, 23(7), (2009) 2701-2708. https://doi.org/10.1016/j.conbuildmat.2008.12.009

[16] S. J. Lee, S. Amirkhanian, and K. Shatanawi, Effects of crumb rubber on aging of asphalt binders, Proceedings of Asphalt Rubber, (2006)779-795.

[17] R. Liang, and S. Lee, Short-term and long-term aging behavior of rubber modified asphalt paving mixture, Transportation Research Record: Journal of the Transportation Research Board, (1530), (1996) 11-17. https://doi.org/10.1177/0361198196153000102

[18] B. E. Ruth and R. Roque, Crumb rubber modifier (CRM) in asphalt pavements. In Transportation Congress, Volumes 1 and 2: Civil Engineers—Key to the World's Infrastructure (1995) 768-785, ASCE.

[19] F.G. Praticò, M. Giunta, M. Mistretta, Maria Gulotta, T.M, Energy and environmental life cycle assessment of sustainable pavement materials and technologies for urban roads, Sustainability 12 (2020) 704. https://doi.org/10.3390/su12020704

[20] A. Riekstins, V. Haritonovs, V. Straupe, Economic and Environmental Analysis of crumb rubber modified asphalt Construction, Building Materials 335 (2022). https://doi.org/10.1016/j.conbuildmat.2022.127468

[21] National Institute for Occupational Safety and Health (NIOSH), Crumb-Rubber Modified Asphalt Paving: Occupational Exposures and Acute Health Effects, HETA #2001-0536-2864. (2001) Cincinnati, Ohio, USA.

[22] M. Russell, J.S. Uhlmeyer, J. DeVol, C. Johnson, J. Weston, Evaluation of Hot In-Place Recycle, WA 09-01(2010) Washington State Dept. of Transportation. USA.

[23] T. Wang, F. Xiao, X. Zhu, B. Huang, J. Wang, S. Amirkhanian, Energy consumption and environmental impact of rubberized asphalt pavement, J. Clean. Prod. 180, (2018) 139–158. https://doi.org/10.1016/j.jclepro.2018.01.086

[24] M. R. Pouranian and M. Shishehbor, Sustainability Assessment of Green Asphalt Mixtures: A Review, Environments (2019). https://doi.org/10.3390/environments6060073

[25] S. Bressi, J. Santos, M. Oreskovic and M. Losa, A comparative environmental impact analysis of asphalt mixtures containing crumb rubber and reclaimed asphalt pavement using life cycle assessment, International Journal of Pavement Engineering (2019). https://doi.org/10.1080/10298436.2019.1623404

[26] M. Puccini, P. Leandri, A. L. Tasca, L. Pistonesi and M. Losa, Improving the Environmental Sustainability of Low Noise Pavements: Comparative Life Cycle Assessment of Reclaimed Asphalt and Crumb Rubber Based Warm Mix Technologies, Coatings (2019). https://doi.org/10.3390/coatings9050343

[27] Y. Yildirim, Polymer modified asphalt binders. Construction and Building Materials, 21(1) (2007) 66-72. https://doi.org/10.1016/j.conbuildmat.2005.07.007

[28] C. R. C. Rongsheng, Laboratory study on process parameters of asphalt rubber and their effects on performance, Journal of Southeast University (Natural Science Edition*)*, *2* (2008) 016.

[29] H. U. Bahia, and R. Davies, Effect of crumb rubber modifiers (CRM) on performance related properties of asphalt binders, Asphalt paving technology*, 63* (1994) 414-414.

[30] F. L. Roberts, P. S. Kandhal, E. R. Brown, D. Y. Lee, and T. W. Kennedy, Hot mix asphalt materials, mixture design and construction, (1991).

[31] F. J. Navarro, P. Partal, F. Martınez-Boza, & C. Gallegos, Thermo-rheological behaviour and storage stability of ground tire rubber-modified bitumens, *Fuel*, *83*(14) (2004) 2041-2049. https://doi.org/10.1016/j.fuel.2004.04.003

[32] F. J. Navarro, P. Partal, F. Martinez-Boza, & C. Gallegos, Influence of crumb rubber concentration on the rheological behavior of a crumb rubber modified bitumen, Energy & fuels, 19(5) (2005) 1984-1990. https://doi.org/10.1021/ef049699a

[33] A. Ghavibazoo, M. Abdelrahman, & M. Ragab, Mechanism of crumb rubber modifier dissolution into asphalt matrix and its effect on final physical properties of crumb rubber-modified binder, Transportation Research Record: Journal of the Transportation Research Board, (2370), (2013) 92-101. https://doi.org/10.3141/2370-12

[34] L. Xiang, J. Cheng, and G. Que, Microstructure and performance of crumb rubber modified asphalt, Construction and Building Materials, 23 (12), (2009) 3586-3590. https://doi.org/10.1016/j.conbuildmat.2009.06.038

[35] J. Shen, S. Amirkhanian, F. Xiao, and B. Tang, Influence of surface area and size of crumb rubber on high temperature properties of crumb rubber modified binders, Construction and Building Materials, *23*(1), (2009) 304-310. https://doi.org/10.1016/j.conbuildmat.2007.12.005

[36] C. Thodesen, K. Shatanawi, and S. Amirkhanian, Effect of crumb rubber characteristics on crumb rubber modified (CRM) binder viscosity, Construction and Building Materials, *23*(1), (2009) 295-303. https://doi.org/10.1016/j.conbuildmat.2007.12.007

[37] S. Liu, W. Cao, J. Fang, and S. Shang, Variance analysis and performance evaluation of different crumb rubber modified (CRM) asphalt. Construction and Building Materials, *23*(7), (2009) 2701-2708. https://doi.org/10.1016/j.conbuildmat.2008.12.009

[38] ASTM D5 / D5M-13, Standard Test Method for Penetration of Bituminous Materials, ASTM International, West Conshohocken, PA, 2013

[39] ASTM D36 / D36M-14e1, Standard Test Method for Softening Point of Bitumen (Ring-and-Ball Apparatus), ASTM International, West Conshohocken, PA, 2014

[40] ASTM D113-07, Standard Test Method for Ductility of Bituminous Materials (Withdrawn 2016), ASTM International, West Conshohocken, PA, 2007

[41] ASTM D92-16b, Standard Test Method for Flash and Fire Points by Cleveland Open Cup Tester, ASTM International, West Conshohocken, PA, 2016

[42] ASTM D70-09e1, Standard Test Method for Density of Semi-Solid Bituminous Materials (Pycnometer Method), ASTM International, West Conshohocken, PA, 2009

[43] I. Antunes, and A. Murachelli, Analysis of environmental sustainability in the rehabilitation of existing pavements using Asphalt Rubber hot mixes, Proceedings of the Asphalt Rubber 2009 Conference, Nanjing, China (2009) 2-4.

[44] I. Bartolozzi, I. Antunes, F. Rizzi, The environmental impact assessment of asphalt rubber: life cycle assessment, Proceedings, 5th Asphalt Rubber" Roads of the Future" International Conference, (2012)799-819.

[45] F.M. Tehrani, Noise abatement of rubberized hot mix asphalt: a brief review, Inter. J. Pavement Res. Technol. 8 (1),(2015) 58-61. Doi:10.6135/ijprt.org.tw/2015.8(1).58

Advanced Topics in Mechanics of Materials, Structures and Construction - AToMech1-2023 Materials Research Forum LLC
Materials Research Proceedings 31 (2023) 29-37 https://doi.org/10.21741/9781644902592-4

Progressive failure analysis of postbuckled plates via mixed formulation

Riccardo Vescovini[1,a *], Mohammad Reza Najafian Zadeh Najafabadi[1,b]

[1]Politecnico di Milano Department of Aerospace Science and Technology, Via La Masa 34 20156, Milano, Italy

[a]riccardo.vescovini@polimi.it, [b]mohammadreza.najafian@polimi.it

Keywords: Progressive Failure Analysis, Mixed Formulation, Semi-Analytical Methods, Ritz Method, Elastic Couplings

Abstract. This paper presents a strategy for Progressive Failure Analysis (PFA) of laminated plates, including Variable Stiffness ones, based on the Ritz method. The formulation is developed using a mixed variational principle, where the unknowns are the stress function and the out-of-plane deflections. A linear degradation model is implemented to account for damage evolution. To ensure accurate predictions, emphasis is placed on the method's ability to correctly handle the laminate's elastic couplings. The iterative nature of this type of analysis is effectively managed due to the reduced number of degrees of freedom required. Exemplary results are reported, and comparisons with Abaqus results are provided to demonstrate the quality of the predictions.

Introduction

Thin plates are elementary structural components of many engineering structures, including those employed in aerospace constructions. In this field, lightweight designs are of crucial importance and, for this purpose, it is beneficial to exploit the plate load-carrying capabilities in the postbuckling field. Indeed, after the panel buckles, internal stress redistribution allows further loads to be sustained. The failure mechanisms of composite panels can be relatively complex and may involve geometrically and materially nonlinear phenomena such as mode-jumping, intra- and inter-laminar failures – independent on each other or in combination. The behavior can be even more complex due to internal load redistribution mechanisms arising from the interaction with the surrounding parts of the structure, such as in the case of the local buckling of stringers [1].

Simplified approaches, such as the ones based on the effective width concept, are a useful mean for predicting the failure load in the preliminary design phases. These strategies offer the potential to obtain an estimate of the loads, but, for many purposes, tend to be oversimplified. The designers would benefit from the availability of more advanced tools, where insight can be gathered into the underlying mechanisms leading the structure to the failure. These considerations are even more true for the composite structures of the next generation based on the Variable Stiffness (VS) concept [2,3]. The availability of more design variables, as well as the possibility of tailoring the internal load paths, make the adoption of fast yet accurate tools even more appealing.

Continuum Damage Mechanics (CDM) is a well-consolidated framework for modeling damage response at macro-scale level and predicting failure in composites. Matzenmiller et al. [4] proposed a degradation procedure for fiber-reinforced composites exhibiting elastic-brittle behavior. Based on his work, Lapczyk and Hurtado [5] presented a linear damage evolution law with focus on finite element implementation. A few studies in the past have shown that CDM can be successfully coupled also with semi-analytical methods to simulate progressive damage modeling of postbuckled composite plates [6,7]. All previous works in this context are restricted to isotropic plates and classical straight-fiber composite laminates. To the best of the authors' knowledge, a recent work by Campagna et al. [8] is the only attempt to account for VS configurations too. The Ritz model of [8] refers to a displacement-based approach and allows for improved computational efficiency. In the present effort, a Ritz method is developed along with continuum damage capabilities. As opposed to [8], the formulation proposed here refers to a unitary variational principle based on a mixed formulation [9,10,11], the main advantage being the reduction of the unknown fields from five to two. This framework is extended here, for the first time, to account for a CDM model aiming at achieving improved failure load prediction capabilities. The proposed extension not only involves the implementation of the damage model, but includes a novel strategy to appropriately account for elastic couplings that can affect the accuracy of the predicted failure load and the identification of the critical spots.

Mixed Formulation

Goal of this investigation is the failure analysis of Variable Stiffness plates operating in the postbuckling field. These plates are characterized by nonuniform elastic properties, hence appropriate methods capable of capturing any relevant elastic coupling effect are of concern. The plate is assumed to be thin such that Kirchhoff assumptions can be applied. The dimensions are defined with l_x and l_y, while the thickness is h. The layup is symmetric, so membrane and bending anisotropy effects can be considered. These effects play an important role in the plate failure mechanisms, as discussed later.

The plate model aims to represent the skin of a stiffened panel, such as those employed in aeronautical structures. Any set of flexural boundary conditions can be considered, i.e. clamped, free or simply-supported, while in-plane ones are based on the assumption of unloaded edges free to translate but forced to remain straight. Loading conditions of uniaxial compression are considered, and the load is introduced via prescribed displacement Δu. A sketch of the structure is reported in Figure 1.

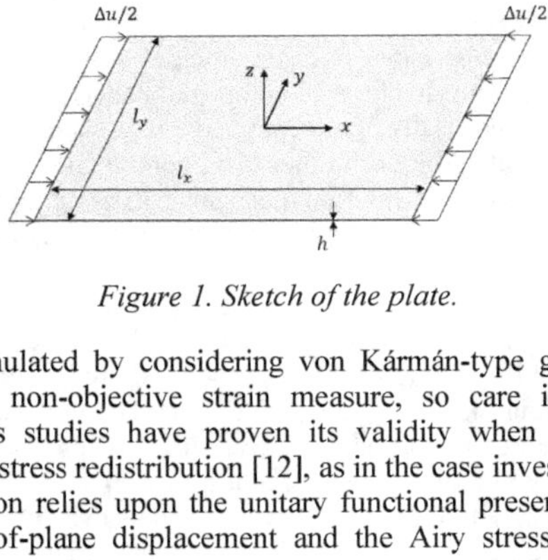

Figure 1. Sketch of the plate.

The problem is formulated by considering von Kármán-type geometric nonlinearity. This assumption leads to a non-objective strain measure, so care is needed when using this approximation. Previous studies have proven its validity when boundary conditions allow significant postbuckling stress redistribution [12], as in the case investigated here.

The mixed formulation relies upon the unitary functional presented in [9,10,11], where the unknowns are the out-of-plane displacement and the Airy stress function. Specifically, the functional reads:

$$\Pi^* = \Pi_m + \Pi_b + \Pi_{nl} + \Pi_{imp} + \Pi_{load} \tag{1}$$

with:

$$\begin{aligned}\Pi_m = -\frac{1}{2}\int_{\bar{S}} [& r^4 a_{11}(\xi,\eta) F_{,\eta\eta}^2 + 2r^2 a_{12}(\xi,\eta) F_{,\xi\xi} F_{,\eta\eta} + a_{22}(\xi,\eta) F_{,\xi\xi}^2 + r^2 a_{66}(\xi,\eta) F_{,\xi\eta}^2 \\ & - 2r^3 a_{16}(\xi,\eta) F_{,\eta\eta} F_{,\xi\eta} - 2r a_{26}(\xi,\eta) F_{,\xi\xi} F_{,\xi\eta}] d\bar{S}\end{aligned} \tag{2}$$

$$\Pi_b = \frac{1}{2}\int_{\bar{S}}[D_{11}(\xi,\eta)w_{,\xi\xi}^2 + 2r^2 D_{12}(\xi,\eta)w_{,\xi\xi}w_{,\eta\eta} + r^4 D_{22}(\xi,\eta)w_{,\eta\eta}^2 + 4r^2 D_{66}(\xi,\eta)w_{,\xi\eta}^2 + 4rD_{16}(\xi,\eta)w_{,\xi\xi}w_{,\xi\eta} + 4r^3 D_{26}(\xi,\eta)w_{,\eta\eta}w_{,\xi\eta}]d\bar{S} \tag{3}$$

$$\Pi_{nl} = \frac{1}{2}r^2\int_{\bar{S}}\left[F_{,\eta\eta}w_{,\xi}^2 + F_{,\xi\xi}w_{,\eta}^2 - 2F_{,\xi\eta}w_{,\xi}w_{,\eta}\right]d\bar{S} \tag{4}$$

$$\Pi_{imp} = -r^2\int_{\bar{S}}[F_{,\eta\eta}w_0 w_{\xi\xi} - 2F_{,\xi\eta}w_0 w_{\xi\eta} + F_{,\xi\xi}w_0 w_{\eta\eta}] \tag{5}$$

$$\Pi_{load} = -\frac{l_x r^2}{4}\left(\int_{-1}^{1}F_{,\eta\eta}\ \ |_{\xi=-1}d\eta + \int_{-1}^{1}F_{,\eta\eta}\ \ |_{\xi=1}d\eta\right)\Delta u \tag{6}$$

where $r=l_x/l_y$ is the plate aspect ratio, w_0 represents the initial imperfection, while D_{ik}, a_{ik} are the bending stiffness and the membrane compliance according to the standard notation used for laminate analysis. Integration is carried out in the nondimensional domain $(\xi,\eta) \in [-1\ 1] \times [-1\ 1]$.

The variational principle based on Eq. (1) requires the first variation of Π^* to be zero. This leads to the governing equations for the problem, which are the out-of-plane equilibrium and the in-plane compatibility.

Solution by Accounting for Anisotropy Effects

Previous studies suggested the use of the Ritz method as an effective mean for obtaining an approximate solution with improved computational efficiency. In particular, Wu et al. [10] proposed an approach based on Legendre polynomials. Despite its effectiveness, this approach cannot account for membrane anisotropy. Strategies for overcoming this restriction have been discussed in [11] based on Lagrange multipliers. Here, we propose a relatively straightforward approach that can be used to account for the above mentioned elastic couplings with no need to resort to Lagrange multipliers.

The out-of-plane deflection is approximated as presented in [10,11]. A new strategy is proposed for the description of the Airy stress function. In particular, the following requirements need to be fulfilled:

1. the edge force N_{x0} must be allowed to have different distributions at $\xi = \pm 1$. Similarly, N_{y0} must be allowed to have different distributions at $\eta = \pm 1$.
2. shearing force at all the edges, N_{xy0}, must be equal to zero.
3. the lateral edges are free to move but constrained to remain straight, thus: $\int_{-1}^{1} N_{y0}d\xi = 0$

An expansion respectful of these three conditions is carried out as:

$$F(\xi,\eta) = F_0(\xi,\eta) + F_1(\xi,\eta) + F_2(\xi,\eta) \tag{7}$$

where:

$$F_{0,\xi}(\xi,\eta) = -\frac{1}{2}X_1(\eta)(1-\xi^2)\sum_{j=0}^{J} d_{1j}M_j(\xi) - \frac{1}{2}X_2(\eta)(1-\xi^2)\sum_{j=0}^{J} d_{2j}M_j(\xi) \tag{8}$$

$$F_{1,\eta}(\xi,\eta) = \frac{l_y^2}{4}\bar{N}_x\eta - \frac{1}{2}X_1(\xi)(1-\eta^2)\sum_{k=0}^{K} c_{1k}L_k(\eta) - \frac{1}{2}X_2(\xi)(1-\eta^2)\sum_{k=0}^{K} c_{2k}L_k(\eta) \tag{9}$$

and:

$$X_1(\xi) = \begin{cases} 1 - \dfrac{(\xi+1)^2}{2}, & -1 \le \xi < 0 \\ \dfrac{(\xi-1)^2}{2}, & 0 \le \xi \le 1 \end{cases} \tag{10}$$

$$X_2(\xi) = \begin{cases} \dfrac{(\xi+1)^2}{2}, & -1 \le \xi < 0 \\ 1 - \dfrac{(\xi-1)^2}{2}, & 0 \le \xi \le 1 \end{cases} \tag{11}$$

Materials Research Proceedings 31 (2023) 29-37 https://doi.org/10.21741/9781644902592-4

The functions X_1 and X_2 can be either 1 or 0 at $\xi = \pm 1$, and their derivatives at $\xi = \pm 1$ are zero. These properties are exploited to model the skewness in the internal forces and to satisfy boundary conditions, too.

The coefficients $d_{1j}, d_{2j}, c_{1k}, c_{2k}, \bar{N}_x$ are the new unknowns to be calculated by minimizing the unitary functional.

Damage Model

The Ritz formulation presented above allows postbuckling simulations to be carried out. Aiming at gathering insights into the failure mechanisms of postbuckled panels, a damage model is implemented. In this regard, the failure load of the structure can be estimated with an approach that is more physically sound than simpler strategies, such as the first ply failure criterion.

The constitutive law of each orthotropic ply composing the stacking is expressed as:

$$\boldsymbol{Q}_d = \frac{1}{D}\begin{bmatrix} (1-d_f)E_1 & (1-d_f)(1-d_m)\nu_{21}E_1 & 0 \\ (1-d_f)(1-d_m)\nu_{12}E_2 & (1-d_m)E_2 & 0 \\ 0 & 0 & D(1-d_s)G_{12} \end{bmatrix} \tag{12}$$

where three damage variables, d_f, d_m, d_s are introduced to represent fiber, matrix, and shear failure modes, and:

$$D = 1 - (1-d_f)(1-d_m)\nu_{12}\nu_{21} \tag{13}$$

It is assumed that the damage variable for the shear, d_s, is not independent and is expressed as a function of two other damage variables:

$$d_s = 1 - (1-d_f)(1-d_m) \tag{14}$$

Hashin and Rotem criteria are employed to predict damage initiation. This criterion has been suggested in past studies, e.g. [21,23,24]. Specifically, the Hashin and Rotem criterion distinguishes between failure modes based on the following set of subcriteria:

$$\begin{aligned} &\sigma_{11} \geq 0 && F_{ft} = \left(\frac{\sigma_{11}}{X_t}\right)^2 = 1 \\ &\sigma_{11} < 0 && F_{fc} = \left(\frac{\sigma_{11}}{X_c}\right)^2 = 1 \\ &\sigma_{22} \geq 0 && F_{mt} = \left(\frac{\sigma_{22}}{Y_t}\right)^2 + \left(\frac{\tau_{12}}{S_L}\right)^2 = 1 \\ &\sigma_{22} < 0 && F_{mc} = \left(\frac{\sigma_{22}}{Y_c}\right)^2 + \left(\frac{\tau_{12}}{S_L}\right)^2 = 1 \end{aligned} \tag{15}$$

where X_t, X_c, Y_t, Y_c, S_L are the strength of material in the directions fiber extension, fiber compression, matrix tension, matrix compression and shear, respectively.

As per the approach proposed in [5], the material properties are decreased linearly once damage initiation occurs. The evolution law is expressed in terms of equivalent stress, σ_{eq}, and equivalent strain, ϵ_{eq}. For both the fiber and matrix in tension and compression, they are defined as:

$$\begin{aligned} &\sigma_{11} \geq 0 && \epsilon_{eq.ft} = \langle \epsilon_{11} \rangle && \sigma_{eq.ft} = \frac{\langle \sigma_{11} \rangle \langle \epsilon_{11} \rangle}{\epsilon_{eq.ft}} \\ &\sigma_{11} < 0 && \epsilon_{eq.fc} = \langle -\epsilon_{11} \rangle && \sigma_{eq.fc} = \frac{\langle -\sigma_{11} \rangle \langle -\epsilon_{11} \rangle}{\epsilon_{eq.ft}} = \langle -\sigma_{11} \rangle \\ &\sigma_{22} \geq 0 && \epsilon_{eq.mt} = \sqrt{\langle \epsilon_{22} \rangle^2 + \gamma_{12}^2} && \sigma_{eq.mt} = \frac{\langle \sigma_{22} \rangle \langle \epsilon_{22} \rangle + \tau_{12}\gamma_{12}}{\epsilon_{eq.mt}} \\ &\sigma_{22} < 0 && \epsilon_{eq.mc} = \sqrt{\langle -\epsilon_{22} \rangle^2 + \gamma_{12}^2} && \sigma_{eq.mc} = \frac{\langle -\sigma_{22} \rangle \langle -\epsilon_{22} \rangle + \tau_{12}\gamma_{12}}{\epsilon_{eq.mc}} \end{aligned} \tag{16}$$

The Macaulay bracket operator is denoted by ⟨⟩.The damage variables d_i evolve based on the stress-strain relation shown in Figure 2, where ϵ_{eq}^{0} is the equivalent strain at which the damage criteria for the corresponding mode is satisfied, and ϵ_{eq}^{f} is the one at which the material is completely damaged for that mode. The area below the graph is the fracture energy, G^c.

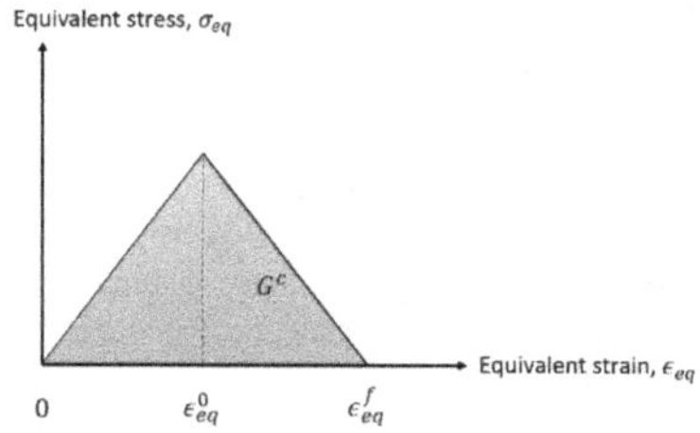

Figure 2. Equivalent strain/stress for linear damage evolution law.

The strain ϵ_{eq}^{f} can be related to ϵ_{eq}^{0} either by specifying the energy dissipation during damage, G^c, or simply with a factor α:

$$\epsilon_{eq}^{f} = \alpha\epsilon_{eq}^{0} \tag{17}$$

In this work, a value of $\alpha = 2$ has been used for each of failure mode [6]. When damage is detected in the fiber or matrix direction, depending on whether the stress state is tensile or compressive, the corresponding damage variable, $d_i$is calculated with the following formulation:

$$d_i = \frac{\epsilon_{eq}^{f}(\epsilon_{eq} - \epsilon_{eq}^{0})}{\epsilon_{eq}(\epsilon_{eq}^{f} - \epsilon_{eq}^{0})} \tag{18}$$

Results

In this section, exemplary results are presented to illustrate the capabilities of the method. In the first part, postbuckling results are presented to show the ability of the proposed method to capture skewness in postbuckling membrane resultants. This effect is crucial to guarantee accurate failure predictions. In the second part, a parametric study is conducted on different VS configurations and failure loads are reported based on the application of PFA.

In all the tests, the material properties are those summarized in Table 1, while the ply strengths are given in Table 2 [1,13].

Table 1. Carbon/epoxy engineering properties.

E_{11} (GPa)	E_{22}(MPa)	G_{12}(MPa)	ν_{12}
150	9080	5290	0.32

Table 2. Ply strenghts.

X_T(MPa)	X_C(MPa)	Y_T(MPa)	Y_C(MPa)	S_L(MPa)
2323	1200	160.2	199.8	130.2

The plates are square with dimension equal to 1000 mm and total thickness fixed to 9.6 mm. Simply-supported boundary conditions are considered. Imperfections are introduced with a shape equal to the first buckling mode and maximum amplitude equal to 0.096 mm, corresponding to 1% of the plate thickness.

Materials Research Proceedings 31 (2023) 29-37 https://doi.org/10.21741/9781644902592-4

Example 1

In this test case, the plate is characterized by the stacking $[\pm\langle 0|45\rangle]_{3S}$, where use is made of the notation proposed in [3]. This configuration is characterized by a non-null degree of bending/twisting coupling, that is quantified by referring to the nondimensional parameters proposed by Nemeth [14]:

$$\gamma = \frac{D_{16}}{\sqrt[4]{D_{11}^3 D_{22}}}, \qquad \delta = \frac{D_{26}}{\sqrt[4]{D_{11} D_{22}^3}} \tag{19}$$

For the laminate at hand, these nondimensional parameters are function of the position due to fiber steering. At the plate center they are null, while they are equal to 0.1895 at the plate loaded edges. These values suggest a significant degree of flexural anisotropy, hence leading to skew buckled shapes. Due to nonlinear coupling between in-plane and out-of-plane response, skew waves promote nonsymmetric membrane resultant redistribution. This effect is illustrated in Figure 3, where the membrane resultant is reported in the deep postbuckling range, at a load level equal to $\Delta u = 10\Delta u_{cr}$. The results are presented by considering two Ritz simulations, both performed with J=K=17: the first, denoted as "Ritz–full", is the one proposed in this work; the second, "Ritz–simplified", is the formulation obtained by neglecting the corrections proposed in Eqs. (8) and (9) and corresponds to the approach proposed in **[Error! Bookmark not defined.]**. The comparison of Figure 3 reports also the results obtained via Abaqus simulations.

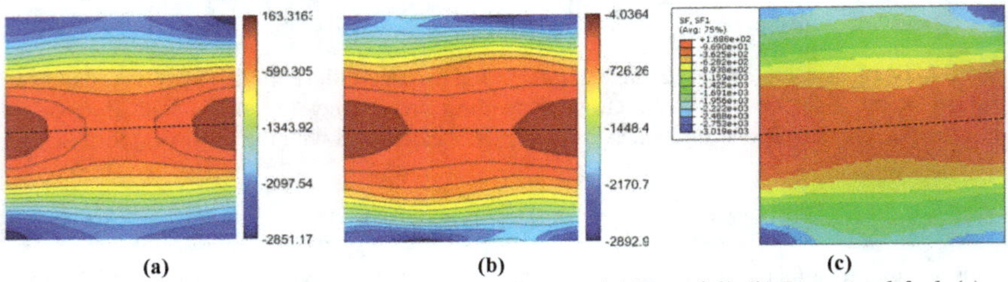

Figure 3. Membrane resultant N_{xx} at $\boldsymbol{\Delta u = 10\Delta u_{cr}}$: (a) Ritz– full, (b) Ritz–simplified, (c) Abaqus.

As seen, the pattern is characterized by a certain degree of skewness that is not captured by the Ritz–simplified approach. On the contrary, the proposed formulation allows for an accurate prediction of this effect. The contours illustrate that different membrane resultant patterns are achieved. This, in turn, determines different spots to be the most heavily loaded ones. An appropriate prediction of damage onset is then affected by the ability to account for this elastic coupling effect.

In addition to the mentioned local effects, a proper description of the postbuckling stress redistribution has an effect on the panel global response in the deep postbuckling range. The plot of Figure 4 illustrates the force-shortening curve obtained with different strategies.

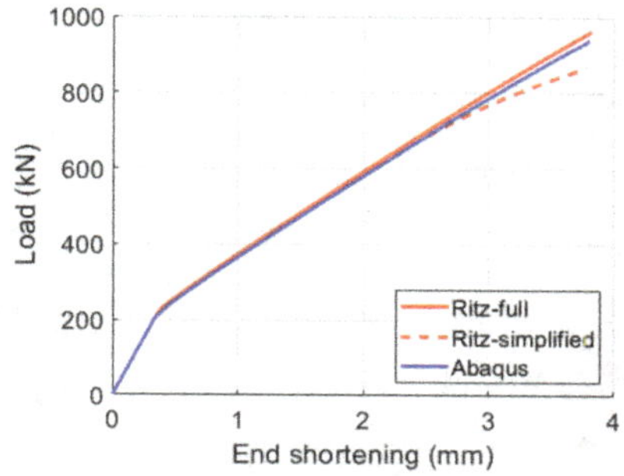

Figure 4. Load-shortening curve.

The curves of Figure 4 illustrate the close agreement between the Ritz– full formulation and the Abaqus results. On the contrary, noticeable discrepancies are observed in the deep postbuckling field when the Ritz-simplified approach is employed.

Example 2
The formulation is applied here to predict the failure loads of different VS configurations using Progressive Failure Analysis. The geometry and the material properties of the plate are the same of the previous example.

The layup is now given by the stacking of 24 plies oriented at $[\pm[\langle 45|T\rangle]_{6S}$, where T is the orientation at the plate edge ranging from 0 to 90 degrees with steps of 10 degrees. Each configuration is associated with different stiffness distributions. In particular, different bending stiffnesses have influence on the buckling load; a combined effect of bending and membrane stiffnessess determines different postbuckling responses. The results are summarized in Figure 5, where Ritz–full computations are compared against costly Abaqus PFA simulations. For the former, results are obtained by using 19 trial functions along both directions.

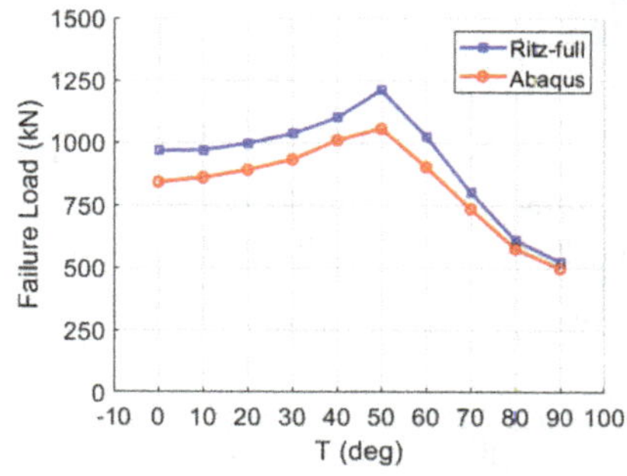

Figure 5. Failure loads for layups $[\pm[\langle \mathbf{45}|\boldsymbol{T}\rangle]_{\mathbf{6S}}$.

The results demonstrate a good matching between Ritz and FEM simulations. Maximum discrepancies of approximately 10% can be noted, which is believed satisfactory owing to the complex combined effects of material and geometric nonlinearities. More important, the Ritz approach is able to predict the trend correctly. This feature is of interest from a design perspective. The proposed tool can be used for preliminary studies aimed at understanding how the response of the plate is affected by changing one or more design parameters. For the problem at hand, the maximum failure load is achieved for T=50°. Clearly, this configuration does not represent an absolute optimum. Many other requirements – e.g. linear stiffness, postbuckling stiffness, buckling load – would be part of a more realistic design scenario.

For the laminate corresponding to T=0°, the plot of the membrane resultant is reported in Figure 6 at the load level corresponding to the laminate failure.

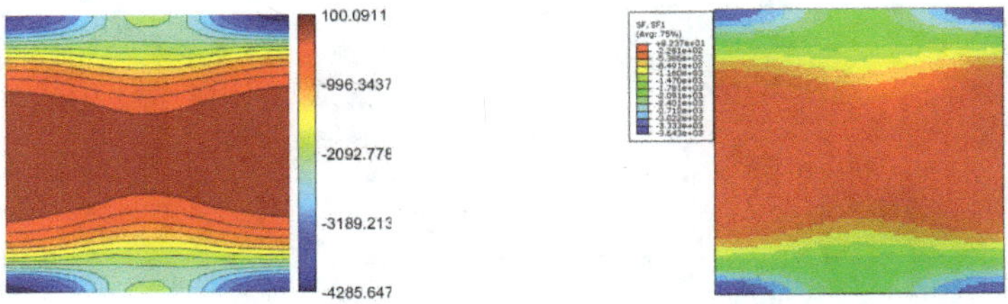

(a) **(b)**

Figure 6. Membrane resultant N_{xx} at failure: (a) Ritz–full, (b) Abaqus.

In this case, the laminate undergoes a damage mechanism that is driven by matrix failure in tension. Specifically, large tensile forces develop at the middle of the transverse edges as a response to the straightness condition. The plot of the corresponding damage variables, d_m, is reported in Figure 7 for the first ply of the stack.

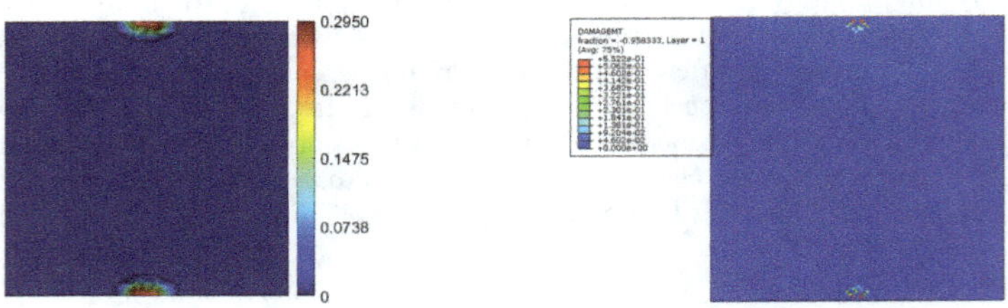

(a) **(b)**

Figure 7. Damage variable d_m: (a) Ritz–full, (b) Abaqus.

The contours of Figure 7 demonstrate the ability of the method to identify the critical spots involved in the failure mechanism. Indeed, both Ritz and Abaqus predictions display similar damaged patterns. In this regards, the proposed Ritz method is a useful mean not only to predict the failure load, but also to gather insight into the whole failure process.

Conclusions

This work presented a novel approach to perform fast preliminary evaluations to estimate the failure load of laminated plates. Advanced configurations with fibers running along curvilinear paths can be considered within the proposed framework. The approach is developed on the basis of a mixed variational approach, and the Ritz method is combined with a linear degradation model.

This approach can accurately predict the laminate response, not only in the initial postbuckling field, but up to the deep postbuckling range, where material failure usually takes place. To achieve this capability, a refined approximation of the stress function has been proposed to accurately capture stress redistribution effects induced by bending/twisting coupling. This capability is of crucial importance to guarantee an accurate prediction of the failure load, as well as identifying the regions involved in the failure process. The comparisons against Abaqus simulations demonstrate the excellent quality of the predicted internal stresses and failure loads. The potential of the approach to investigate the effects of fiber steering on the laminate failure load has been shown with a parametric study. More realistic design scenarios including multiple design requirements will be investigated as part of future investigations.

References

[1] C. Bisagni, R. Vescovini and C.G. Dávila, Single-stringer compression specimen for the assessment of damage tolerance of postbuckled structures, Journal of Aircraft 48 (2011) 495-502. https://doi.org/10.2514/1.C031106
[2] M.W. Hyer and R.F. Charette, Use of curvilinear fiber format in composite structure design, AIAA Journal 29 (1991) 1011-1015. https://doi.org/10.2514/3.10697
[3] Z. Gürdal, B.F. Tatting and C.K. Wu, Variable stiffness composite panels: effects of stiffness variation on the in-plane and buckling response. Composites Part A: Applied Science and Manufacturing 39 (2008) 911-922. https://doi.org/10.1016/j.compositesa.2007.11.015
[4] A. Matzenmiller, J. Lubliner and R.L. Taylor, A constitutive model for anisotropic damage in fiber-composites, Mechanics of Materials 20 (1995) 125-152. https://doi.org/10.1016/0167-6636(94)00053-0
[5] I. Lapczyk and J.A. Hurtado, Progressive damage modeling in fiber-reinforced materials, Composites Part A: Applied Science and Manufacturing 38 (2007) 2333-2341. https://doi.org/10.1016/j.compositesa.2007.01.017
[6] Q.J. Yang and B. Hayman, Simplified ultimate strength analysis of compressed composite plates with linear material degradation, Composites Part B: Engineering 69 (2015) 13-21. https://doi.org/10.1016/j.compositesb.2014.09.016
[7] S.A.M. Ghannadpour and M. Shakeri, Application of a new energy-based collocation method for nonlinear progressive damage analysis of imperfect composite plates, Thin-Walled Structures 147 (2020) 106369. https://doi.org/10.1016/j.tws.2019.106369
[8] D. Campagna, A. Milazzo, I. Benedetti and V. Oliveri, A non-linear Ritz method for progressive failure analysis of variable angle tow composite laminates, Mechanics of Advanced Materials and Structures (2022) 1-14. https://doi.org/10.1080/15376494.2022.2134951
[9] C. Bisagni and R. Vescovini, Analytical formulation for local buckling and post-buckling analysis of stiffened laminated panels, Thin-Walled Structures 47 (2009) 318-334. https://doi.org/10.1016/j.tws.2008.07.006
[10] Z. Wu, G. Raju and P.M. Weaver, Postbuckling analysis of variable angle tow composite plates, International Journal of Solids and Structures 50 (2013) 1770-1783. https://doi.org/10.1016/j.ijsolstr.2013.02.001
[11] Z. Wu, G. Raju and P.M. Weaver, Buckling of VAT plates using energy methods, 53rd AIAA/ASME/ASCE/AHS/ASC Structures, Structural Dynamics and Materials Conference 20th AIAA/ASME/AHS Adaptive Structures Conference 14th AIAA, 2012. https://doi.org/10.2514/6.2012-1463
[12] G. Garcea, A. Madeo and R. Casciaro. Nonlinear fem analysis for beams and plate assemblages based on the implicit corotational method, Journal of Mechanics of Materials and Structures 7 (2012) 539-574. https://doi.org/10.2140/jomms.2012.7.539
[13] P.P. Camanho, P Maimí and C.G. Dávila, Prediction of size effects in notched laminates using continuum damage mechanics, Composites Science and Technology 67 (2007) 2715-2727. https://doi.org/10.1016/j.compscitech.2007.02.005
[14] M.P. Nemeth, Importance of anisotropy on buckling of compression-loaded symmetriccomposite plates, AIAA Journal 24 (2012), 1831-1835. https://doi.org/10.2514/3.9531

High dimensional model representation for the probabilistic assessment of seismic pounding

R. Sinha[1,a*], B.N. Rao[1,b]

[1] Structural Engineering Division, Department of Civil Engineering, Indian Institute of Technology Madras, Chennai, India – 600036

[a] rohit.sinha2096@gmail.com, [b] bnrao@iitm.ac.in

Keywords: Fragility Curves, HDMR, MCS, Structural Pounding, Response Surface Method, Meta-Model

Abstract: The study presented herein aims to analyse the seismic performance of a two-dimensional eight-storey non-ductile reinforced concrete frame against structural pounding with an adjacent three-storey stiff frame having different storey heights. The examined case of pounding refers to the extremely critical floor-to-column structural pounding for three different initial separation gaps between the said structures. Seismic vulnerability analysis is usually performed by way of developing fragility curves for a set of damage and intensity measures using a suitable fragility curve generation technique. For this study, damage measures are characterized by the percentage maximum inter-storey drifts of the taller, flexible frame while the peak ground accelerations of the ground motion data are used as the corresponding intensity measures. Displacement-based fragility curves were generated for 9 sampling points using the High Dimensional Model Representation (HDMR) technique and the results were compared with actual probabilistic data obtained using Monte-Carlo Simulations (MCS). The results of this study imply that the proposed use of HDMR provides excellent fragility curves for the estimation of pounding risks with a significant reduction in the number of simulations required, thereby reducing the computational cost by huge margins. Results also indicate that fragility curves for target separation distances can also be obtained using HDMR without performing additional simulations. This can further be used for the mitigation of pounding risks and for the reliability-based design of buildings for target separation distances and damage measures.

Introduction

Structural pounding between adjacent buildings with insufficient separation distance is an undesirable event and has often been a cause of severe structural damages [1]. This issue particularly prevails in metropolitan cities where land resources available for construction practices are limited [2].

Seismic Pounding has been proven to be detrimental to structural systems rather than benefit them. This is especially true for buildings present on corners of a series configuration [1]. Its main effects include an increase in the acceleration and drift demand at different storey levels [3, 4]. The past three decades have seen extensive research being conducted to develop ways of mitigating pounding risks and over the years, numerous ways of minimizing these effects have been suggested [5-7]. However, modern seismic codes have opted to adopt the simplest approach of minimizing the risks associated with pounding by prescribing a minimum separation gap between adjacent buildings. This approach even though efficient, lacks elegance since the prescribed clearance may not always be available. Such a procedure is also characterized by uncertain levels of safety and does not allow explicit control over the pounding risk [8].

This highlights the need for advanced probabilistic methods to accurately estimate damage levels. Conventionally, seismic vulnerability is represented by either Damage Probability Matrices (DPMs) [9] or fragility curves. DPMs describe discrete relationships between damage and intensity measures whereas, fragility curves are continuous representations of the cumulative probability distributions of performance limits and prove to be useful tools for the estimation of the probability of structural damage.

The simplest and most straightforward method to obtain failure probabilities is the Monte-Carlo Simulation (MCS) technique. It is the most accurate methodology but is known to be computationally burdensome due to the substantial number of simulations it requires for probability estimation. Other efficient approaches to generate fragility curves within modern performance-based frameworks [10] such as the Probabilistic Seismic Demand Model (PSDM) [8] have also been developed and widely used [11-13]. A problem with the PSDM framework, however, is the homoscedasticity assumption associated with it. Such an assumption may lead to major disparities where the variance of error terms is not constant.

Hence, there appears to be a need to develop modern methods that maintain a higher level of coherence with the actual data. A simple meta-model based approach to do so is the use of response surface methodology (RSM) [14] which is particularly an efficient technique for representing multivariate responses. RSMs possess the distinct advantage of representing complex and implicit phenomena as simple polynomial expansions that are easier to work with. This study attempts to introduce the concept of response surface based fragility curve generation for the pounding risk of adjacent structures using High Dimensional Model Representation (HDMR). As a response surface meta-model, HDMR represents a large set of data in the form of simple closed-form multivariate polynomial equations. HDMR has been adopted in previous studies to develop seismic fragility curves [15, 16] with a prominent level of accuracy and minimal computational cost.

The present study adopts the nonlinear time history analyses (NLTHA) technique to simulate pounding effects on an eight-storey Reinforced Concrete Ordinary Moment Resisting Frame (RC-OMRF) against a shorter and stiffer three-storey RC frame. A series of 831 nonlinear dynamic analyses have been performed on this configuration using a suite of 20 real accelerograms for three initial separation distances, d_g = 0.0 cm, 5.5 cm, and 11.0 cm. The peak ground accelerations (PGAs) of the accelerograms have been scaled in the range between 0.005g and 0.7g. Displacement-based fragility curves have been generated using HDMR for the percentage maximum inter-storey drifts (IDR_{max}) at the level of Immediate Occupancy (IO) for PGA as the intensity measure (IM). These fragility curves have been compared with those obtained using MCS. Additionally, fragility curves have also been approximated for randomly chosen initial gaps of d_g = 2.6 cm and 6.1 cm to verify the validity of HDMR-based fragility curves in the estimation of failure probabilities for target separation distances.

The primary aim of this study is to introduce HDMR as a suitable, accurate and computationally efficient methodology to generate fragility curves for the mitigation of risks associated with the seismic pounding.

Overview of HDMR

HDMR is a response surface methodology describing a family of multivariate representations to capture the input-output relationships of complex high-dimensional systems with many input variables. This is an efficient technique that systematically reveals the hierarchical correlations amongst input random variables. The general foundations of HDMR were laid by Rabitz [17] and it has since been actively applied in various disciplines [18-20]. The meta-models obtained by using HDMR are not only simpler than the original complex and nonlinear systems but are also accurate and computationally efficient in the uncertainty analysis of the computationally burdensome models.

HDMR is a general set of quantitative model assessment and analysis tools for capturing the high dimensional relationships between sets of input-output variables [21]. Since the effects of input random variables may or may not be independent, for an N-dimensional vector of input

Materials Research Proceedings 31 (2023) 38-45 https://doi.org/10.21741/9781644902592-5

variables $X = \{x_1, x_2, x_3, \dots\dots, x_N\}$, HDMR inherently expresses the output $f(x)$ as a hierarchical correlated expansion to account for the cooperative effects of all inputs. Generally, an HDMR expansion up to the second order (Eq. 1) is sufficient to describe output responses.

$$f(x) = f_0 + \sum_{i=1}^{N} f_i(x_i) + \sum_{i=1}^{N}\sum_{j=1}^{N} f_{ij}(x_i, x_j) \tag{1}$$

here, the constant term f_0 represents the mean response to $f(x)$, and $f_i(x_i)$ and $f_{ij}(x_i, x_j)$ represent the first and second-order terms of the HDMR expansion, respectively. The first-order term considers only the individual contribution of each input variable, while the cooperative effects of a pair of input variables are accounted for by the second-order term.

HDMR can be broadly classified into (1) ANOVA-HDMR; (2) Cut-HDMR [16]. Among these, the Cut-HDMR methodology has been adopted for this study. In Cut-HDMR, the convergence limit is invariant to the choice of reference point and thus, it returns exact results along the lines, planes, volumes etc. through and around the reference point $c = \{c_1, c_2, c_3, \dots\dots, c_N\}$ defined in the variable space. The expansion terms are determined using the following equations:

$$f_0 = f(c) \tag{2}$$

$$f_i(x_i) = f(x_i, c^i) - f_0 \tag{3}$$

$$f_{ij}(x_i, x_j) = f(x_i, x_j, c^i) - f_i(x_i) - f_j(x_j) - f_0 \tag{4}$$

In every higher-order term, the previous lower-order terms are subtracted. This is done to remove their dependence and provide a unique contribution from the new expansion function. Since the increment of the order of HDMR expansion makes it more computationally expensive than the previous, this study has been limited to the use of second-order HDMR expansion only.

Estimating HDMR-based fragility curves

The first task in estimating seismic fragility curves is to define the input and response variables. This is followed by choosing appropriate limit states corresponding to the chosen damage measure. For this study, the damage measure, or Engineering Demand Parameter (EDP) of IDR_{max} has been chosen as the output response while d_g, and the IM - PGA have been chosen as the input random variables. Aleatory uncertainties from earthquakes are implicitly accounted for by using a sufficiently large suite of real accelerograms.

Various combinations of input variables representing different scenarios of earthquake-structure interaction are generated. These serve as the sampling points for HDMR expansion. The mean and standard deviation for the ground motion records (20 records for this study) are calculated and meta-models are formulated by applying the HDMR technique. The polynomials so obtained can then be used for identifying reliability indices using FORM/SORM which will yield the failure probabilities.

Application: Pounding Risk of Adjacent Buildings

Structural Layout of the Building Frame: Fig. 1 (a) shows the details of the eight-storey RC-OMRF used for this study. Each storey is 3 m high and spans three 6 m wide bays. The supports are assumed to be fixed. The studied frame has been designed using the commercially available finite element software package CSI SAP2000 with two-dimensional beam and column elements. In addition to the self-weight of structural members, a live load of 18 kN/m has also been considered in the design of the frame. Nonlinear hinges have been provided at the possible locations of yielding. Nonlinear constitutive relations for concrete and steel have been defined, with concrete following the modified model of Mander *et al.* [22], and steel following the constitutive relation given in IS 456:2000 [23].

Damage and Intensity Measures: Fig. 1 (b) shows the examined case of the non-eccentric pounding of the flexible eight-storey RC-OMRF against a short and stiff three-storey RC frame with different storey heights. Uncertainties in system response are taken into account by using Fault Normal (FN) and Fault Parallel (FP) components of a suite of ten pairs of three-component ground motions extracted from the PEER NGA-West2 Database [24] for the target elastic response spectrum given in the Indian Standard Code IS 1893-1:2002 [25]. NLTHA using this suite of ground motion data provides the storey displacement records which have been used to evaluate the IDR_{max} for PGA values ranging between 0.005g and 0.7g. Thus, values of IDR_{max} and PGA serve as the EDP-IM pairs for the current study.

Performance Limit States: To indicate the level of structural distress, it is imperative to define performance limit states. Due to the non-ductile nature of the flexible frame studied, limit states for maximum storey drifts (θ_{LS}) defined by Ghobarah [26] have been found to be the most consistent. These limit states have been summarized in Table 1.

Table 1: *Limit States for various performance levels of IDR_{max}*

Limit State	ND	LD	IO	LS	CP
Damage Description	No Damage	Light Damage	Immediate Occupancy	Life Safety	Collapse Prevention
$\boldsymbol{\theta_{LS}}$(%)	0.1	0.2	0.5	0.8	1.0

This paper focuses on the fragility curves obtained for a threshold level of 0.5% storey drift corresponding to the performance level of IO only.

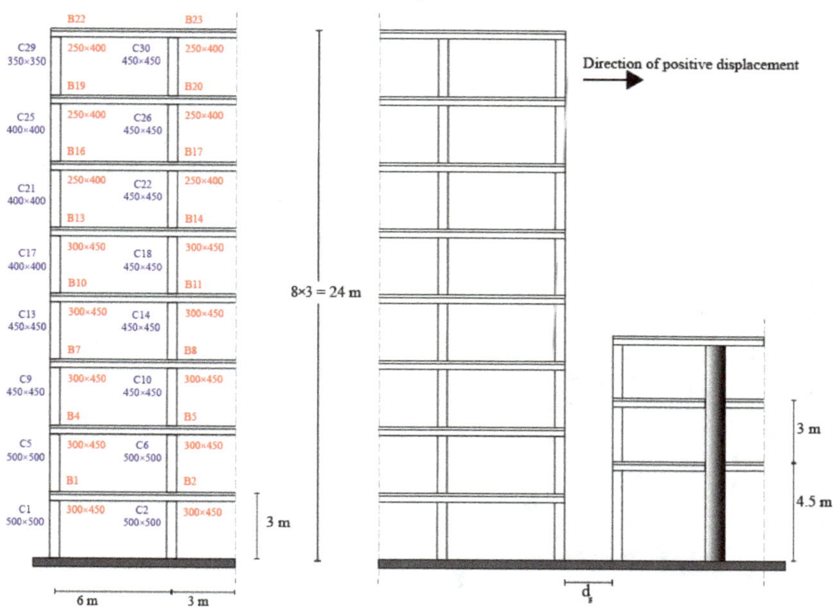

Figure 1: *(a) Skeletal Framework of the designed eight-storey RC-OMRF (b) Floor-to-Column Pounding Case under consideration*

Results

This section illustrates the seismic risk evaluation for the inter-storey pounding case demonstrated by Fig. 1(b). Risk estimates have been found for three initial separation gaps of (1) $d_g = 0.0$ cm (signifying direct contact of buildings built in tight spaces); (2) $d_g = 5.5$ cm (mean value of the minimum and maximum separation gaps considered); and (3) $d_g = 11.0$ cm (minimum recommended separation gap as per IS 1893-1: 2002). HDMR is employed to evaluate failure probabilities for IDR_{max} values exceeding the limit state of IO. These failure probabilities are used to develop failure fragility curves against the actual fragility curves obtained by applying MCS to the acquired dataset.

Fig. 2 shows the comparison of fragility curves obtained using HDMR with those obtained by using MCS. It is to be noted that the fragility curve obtained by using MCS gives the exact representation of failure probabilities, but is used only as a reference and not for fragility curve generation in general due to the high computational cost associated with it.

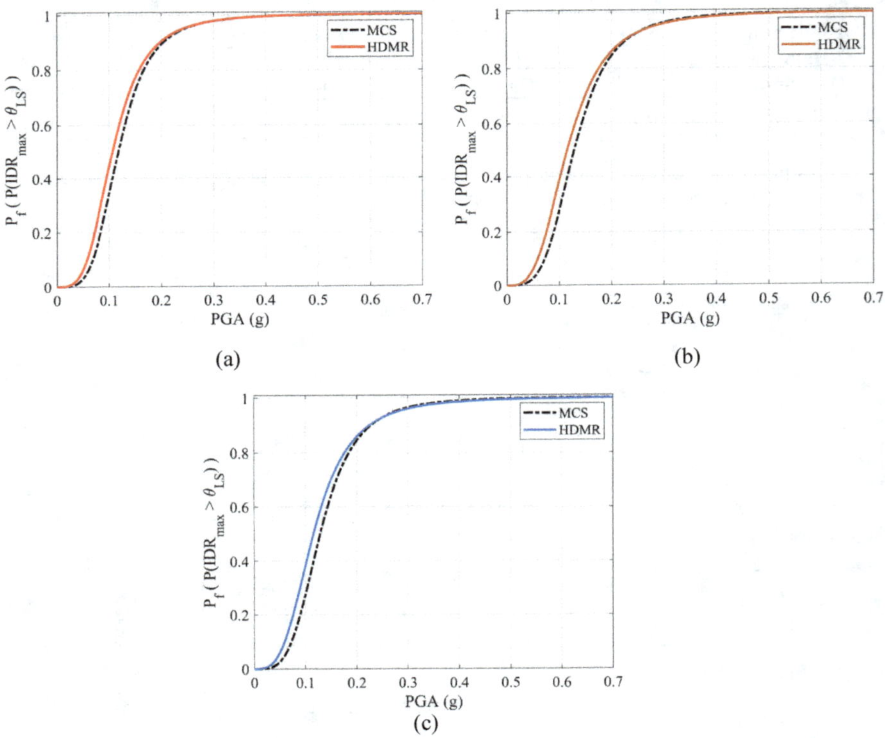

Figure 2: *Fragility Curves for IDR_{max} using HDMR vs MCS at separation distances (a) $d_g = 0.0$ cm, (b) $d_g = 5.5$ cm and (c) $d_g = 11.0$ cm*

From Fig. 2, it is clear that HDMR shows a remarkably close relationship with the actual probability data. For small values of PGA, the HDMR-based fragility curves not only lie close to the actual MCS-based failure probabilities displaying only slight overestimates but also continue with the same slope as their MCS counterpart until convergence. As an illustration of the estimation accuracy of HDMR, an error of only 5.46% was observed in the estimated failure probabilities at a PGA level of 0.1685g for d_g = 11 cm. For higher PGAs, HDMR-based fragility curves usually end up merging with the MCS data demonstrating a supreme level of accuracy. It thus proves to be an efficient tool to develop pounding-based seismic fragility curves.

Fig. 3 shows the comparison of HDMR-based fragility curves with their MCS counterparts for randomly chosen d_gs. This test in particular focuses on the fact that the recommended separation gap between adjacent buildings may not always be available, and that the field personnel will have to operate with whatever amount of space is available on site. Target separation gaps randomly chosen to be d_g = 2.6 cm and 6.1 cm were used to perform 554 simulations in addition to the 831 nonlinear dynamic analyses already performed. Structural responses obtained from the extra NLTHAs were used to develop actual fragility curves using MCS as shown in Fig. 3. However, instead of the data so obtained, the HDMR meta-model already formulated (from the previous 831 NLTHAs for separation gaps d_g = 0.0 cm, 5.5 cm, and 11.0 cm) was used to derive failure probabilities for the randomly selected target separation distances.

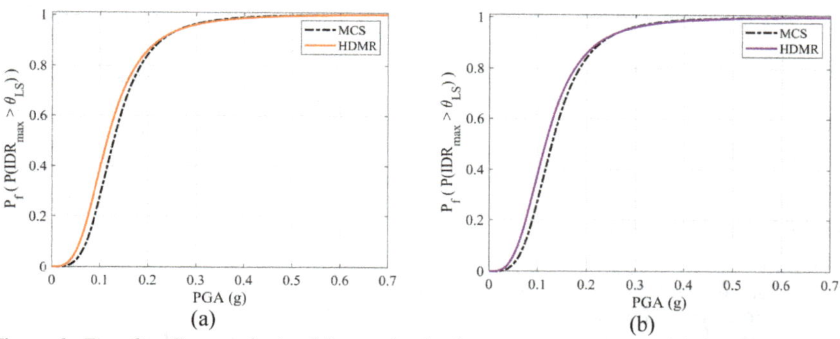

Figure 3: *Fragility Curves obtained for randomly chosen separation gaps (a) d_g = 2.6 cm, (b) d_g = 6.1 cm using HDMR (without performing simulations) vs MCS (after performing simulations)*

As expected, results for the randomly chosen d_gs showed similar trends as those for the separation distances for which NLTHAs were initially performed. For lower PGAs, HDMR-based fragility curves were again close to the actual failure probabilities and showed the same slope for most of the regime, beyond which, HDMR and MCS-based fragility curves began to converge. The use of HDMR thus saved 60% of computational effort. With a similar level of accuracy, the widely used PSDM model would have required either additional simulations or regression analysis to identify failure probabilities for target separation distances. With HDMR, it also becomes easier to include numerous variable characteristics affecting seismic responses in the same equation. This can be used to mitigate seismic risks related to pounding by estimating accurately the probabilities of failure for target separation distances in the pre-construction phases which can be utilized in reliability-based structural design for target failure probabilities.

Summary and Conclusions

This paper presents HDMR as an accurate and computationally efficient technique to assess the seismic vulnerability of adjacent structures subject to seismic pounding. This was the first time that this response surface methodology was used for the development of fragility curves for seismic pounding risks. The case study was based on a structural system consisting of an eight-storey RC-OMRF subject to floor-to-column structural pounding against a shorter and stiff three-storey RC frame. Displacement-based fragility curves (HDMR-based fragility curves were made against MCS-based fragility curves) for IDR_{max} were developed for the IM ranging between PGA values of 0.005g and 0.7g.

It was observed that fragility curves developed by using HDMR show a consistent level of efficiency and accuracy when compared with the actual probabilistic data obtained from MCS estimation. It was also observed that at low PGAs, the HDMR-based fragility curves showed errors as small as 5.46% and maintained nearly similar slopes as their MCS-based equivalents. Results established also implied that at higher levels of PGA, HDMR-based fragility curves tend to converge well with their MCS-based counterpart. Similar levels of accuracy and efficiency were obtained for randomly chosen separation distances within the range of the study. It was inferred that the use of HDMR eliminates the need to perform additional NLTHAs for target separation distances or to use regression techniques on the existing fragility curves to estimate failure probabilities for a target d_g. A 60% reduction in the computational effort was obtained for two randomly chosen target separation distances. It is proposed that due to such a level of accuracy, efficiency, and computational economy, HDMR can be used to obtain excellent fragility curves for the estimation of pounding risks. It is proposed that this method could be used for the determination of critical separation distances between adjacent building structures, as well as for performing the reliability-based design of buildings for actual on-site available separation spaces thereby mitigating pounding risks even in the pre-construction phases. The present study used only the separation distance between the adjacent structures and the peak ground acceleration of seismic excitations as the input random variables. Future studies could include a greater number of random parameters including real characteristics of the structural systems to extract a much higher accuracy out of the studied methodology. Future studies could also ponder upon the usage of HDMR for curvature-based fragility curve generation for local structural responses to establish an overall supremacy or limitation of the studied method.

References

[1] Anagnostopoulos, S. A. 1988. “Pounding of buildings in series during earthquakes.” Earthq. Eng. Struct. Dyn., 16 (3): 443–456. https://doi.org/10.1002/eqe.4290160311.

[2] Maison, B. F., and K. Kasai. 1990. “Analysis for a Type of Structural Pounding.” J. Struct. Eng., 116 (4): 957–977. https://doi.org/10.1061/(asce)0733-9445(1990)116:4(957).

[3] Polycarpou, P. C., and P. Komodromos. 2010. “Earthquake-induced poundings of a seismically isolated building with adjacent structures.” Eng. Struct., 32 (7): 1937–1951. Elsevier Ltd. https://doi.org/10.1016/j.engstruct.2010.03.011.

[4] Skrekas, P., A. Sextos, and A. Giaralis. 2014. “Influence of bi-directional seismic pounding on the inelastic demand distribution of three adjacent multi-storey R/C buildings.” Earthq. Struct., 6 (1): 71–87. https://doi.org/10.12989/eas.2014.6.1.071.

[5] Wolf, J. P., and P. E. Skrikerud. 1980. “Mutual pounding of adjacent structures during earthquakes.” Nucl. Eng. Des., 57 (2): 253–275. https://doi.org/10.1016/0029-5493(80)90106-5.

[6] Hideo Takabatake and Masaaki Yasui and Yoshihisa Nakagawa and Akiko, K. 2007. “Relaxation method for pounding action between adjacent buildings at expansion joint.” Pacific Conf. Earthq. Eng., (056): 1–6. https://doi.org/10.1002/eqe.2402.

[7] Tubaldi, E. 2011. “Dynamic behavior of adjacent buildings connected by linear viscous/viscoelastic dampers.” Struct. Control Heal. Monit., (May 2011). https://doi.org/10.1002/stc.1734.

[8] Tubaldi, E. and Freddi, F. and Barbato, M. 2007. “Probabilistic seismic demand model for pounding risk assessment.” Pacific Conf. Earthq. Eng., (056): 1–6. https://doi.org/10.1002/eqe.2725.

[9] Whitman, R. V, J. M. Biggs, J. E. Brennan, C. A. Cornell, R. L. de Neufville, and E. H.

Vanmarcke. 1975. "Seismic Design Decision Analysis." J. Struct. Div., 101 (5): 1067–1084. https://doi.org/10.1061/JSDEAG.0004049.

[10] Porter, K. A. 2003. "An Overview of PEER's Performance-Based Earthquake Engineering Methodology." 9th Int. Conf. Appl. Stat. Probab. Civ. Eng., 273 (1995): 973–980.

[11] Gardoni, P., K. M. Mosalam, and A. Der Kiureghian. 2003. "Probabilistic seismic demand models and fragility estimates for RC bridges." J. Earthq. Eng., 7: 79–106. https://doi.org/10.1080/13632460309350474.

[12] Ramamoorthy, S. K., P. Gardoni, and J. M. Bracci. 2006. "Probabilistic Demand Models and Fragility Curves for Reinforced Concrete Frames." J. Struct. Eng., 132 (10): 1563–1572. https://doi.org/10.1061/(asce)0733-9445(2006)132:10(1563).

[13] Flenga, M. G., and M. J. Favvata. 2021. "Probabilistic seismic assessment of the pounding risk based on the local demands of a multistory RC frame structure." Eng. Struct., 245 (July). https://doi.org/10.1016/j.engstruct.2021.112789.

[14] Box, G. E. P., and K. B. Wilson. 1951. "On the Experimental Attainment of Optimum Conditions." J. R. Stat. Soc. Ser. B, 13 (1): 1–38. https://doi.org/10.1111/j.2517-6161.1951.tb00067.x.

[15] Bhasker, R., and A. Menon. 2022. "A seismic fragility model accounting for torsional irregularity in low-rise non-ductile RC moment-resisting frames." Earthq. Eng. Struct. Dyn., 51 (4): 912–934. https://doi.org/10.1002/eqe.3597.

[16] Chowdhury, R., B. N. Rao, and A. M. Prasad. 2009. "High-dimensional model representation for structural reliability analysis." (April 2008): 301–337. https://doi.org/10.1002/cnm.1118.

[17] Rabitz, H., and Ö. F. Aliş. 1999. "General foundations of high-dimensional model representations." J. Math. Chem., 25 (2–3): 197–233. https://doi.org/10.1023/a:1019188517934.

[18] Rabitz, H., Ö. F. Aliş, J. Shorter, and K. Shim. 1999. "Efficient input-output model representations." Comput. Phys. Commun., 117 (1): 11–20. https://doi.org/10.1016/S0010-4655(98)00152-0.

[19] Li, G., C. Rosenthal, and H. Rabitz. 2001a. "High dimensional model representations." J. Phys. Chem. A, 105 (33): 1–26. https://doi.org/10.1021/jp010450t.

[20] Li, G., S. W. Wang, C. Rosenthal, and H. Rabitz. 2001b. "High dimensional model representations generated from low dimensional data samples. I. mp-Cut-HDMR." J. Math. Chem., 30 (1): 1–30. https://doi.org/10.1023/A:1013172329778.

[21] V. U. Unnikrishnan, A. M. Prasad, and B. N. Rao. 2007. "Development of fragility curves using high-dimensional model representation." Pacific Conf. Earthq. Eng., (056): 1–6. https://doi.org/10.1002/eqe.2214.

[22] Mander, J. B., M. J. N. Priestley, and R. Park. 1988. "Theoretical Strss-Strain Model for Confined Concrete." J. Struct. Eng, 114 (8): 1804–1826.

[23] Bureau of Indian Standards New Delhi. 2000. "IS 456 - Plain and Reinforced Concrete - Code of Practice." 1–114.

[24] PEER Ground Motion Database. 2011.

[25] Bureau of Indian Standards New Delhi. 2002. "IS 1893 Part 1 - Criteria for Earthquake Resistant Design of Structures - General Provisions and Buildings Part-1." Bur. Indian Stand. New Delhi, Part 1 (1): 1–39.

[26] Ghobarah, A. 2004. "On drift limits associated with different damage levels." Int. Work. performance-based Seism. Des. concepts Implement., (February): 321–332.

Static and free vibration analysis of laminated sandwich shell with double curvature considering the effect of transverse normal strain

AJIM S. SHAIKH[1,a*], ATTESHAMUDDIN S. SAYYAD[2,b]

[1*]Research Scholar, Department of Civil Engineering, Sanjivani College of Engineering, Savitribai Phule Pune University, Kopargaon-423603, Maharashtra, India

[2]Professor, Department of Structural Engineering, Sanjivani College of Engineering Kopargaon, Savitribai Phule Pune University, Kopargaon-423603, Maharashtra, India

[a]shaikhajims38@gmail.com, [b]attu_sayyad@yahoo.co.in

Keywords: Static Analysis, Sandwich Shells, Transverse Normal Strain, Hyperbolic Shell Theory

Abstract. For the static bending analysis of sandwich spherical shells, higher-order closed-form solutions are provided in the current study based on a new hyperbolic shell theory considering the effects of transverse normal strain. A shell consists of three layers wherein the top and the bottom layers (face sheets) are made up of hard material and the middle layer (core) is made up of soft material. The governing equations and associated boundary conditions of the theory are produced by employing the principle of virtual work. Semi-analytical closed-form solutions for the static problem are produced by the Navier technique for simply supported boundary conditions of the shell. The present results are compared with results that have already been published in order to confirm the accuracy and efficacy of the current higher-order hyperbolic shell theory.

Introduction

Fiber-reinforced polymer composites are the most important kind of composite material. The most significant characteristics of fiber-reinforced polymer composite materials are their high strength-to-weight and stiffness-to-weight properties. Consequently, these are being employed more frequently in a variety of technical applications. Lightweight composite panels called laminated sandwich shells have a soft inner core between two thin, stronger skins. Its function is to transfer loads from the face sheets to the core structure, and if it fails, the structure will no longer function. Inflicted stresses on sandwich structures are distributed between the face sheets and the core structure according to their material properties and thicknesses. Sandwich panels are frequently utilized in a variety of engineering fields, including civil, mechanical, aerospace, marine, and offshore. Mourtiz et al. [1] described composite shell structures are widely used in different engineering sectors for many years, including the naval, aerospace, automotive, and construction sectors, as well as sporting goods, medical devices, and many other areas. Mallikarjuna and Kant [2] provided a critical review and some results of recently developed refined theories of fiber-reinforced laminated composites and sandwiches and this review is limited to linear free vibration and transient dynamic analyses, and geometric nonlinear transient response of multilayer sandwich/fiber-reinforced composite plates. Ferreira et al. [3] presented Non-linear analysis of sandwich shells and the effect of core plasticity using first order shear deformation theory based on the finite element method and uses the Ahmad shell element with five degrees of freedom per node. Kant and Swaminathan [4] presented analytical solutions for the static analysis of laminated sandwich plates using higher order refined theory based on Navier's solution technique. The theoretical model presented by the author incorporates laminate deformations which account for the effects of transverse shear deformation. Hohe and Librescu [5] presented a nonlinear theory for doubly curved anisotropic sandwich shells with transversely compressible core using an

Materials Research Proceedings 31 (2023) 46-54 https://doi.org/10.21741/9781644902592-6

advanced geometrically nonlinear shell theory of doubly curved structural sandwich panels with transversely compressible core is presented based on Kirchhoff theory. This theory accounts for dynamic effects as well as for initial geometric imperfections. Zhong and Reimerdes [6] presented stability behavior of cylindrical and conical sandwich shells with flexible core using a higher-order theory based on three-layer model and solved by numerical integration. Khare et al. [7] presented solutions for thick laminated sandwich shells using higher order theory based on closed form solutions. Closed-form formulations of 2D higher-order shear deformation theory are presented for the thermo-mechanical and free vibration analysis of simply supported, cross-ply, laminated sandwich, doubly thick curved shells. Results on static and dynamic problems of double core sandwich shell are not presented in the paper. Garg et al. [8] presented Solutions for free vibration of laminated composite and sandwich shells using higher-order closed-form. It described free vibration characteristics of simply supported, laminated cross-ply, composite, and sandwich shell panels using the various higher-order theories, which account for the effects of transverse shear strains/stresses and the transverse normal strain/stress. Results on a multilayered sandwich shell analysis are not presented by the authors. Turkin [9] presented a technique for calculating rational design parameters of a sandwich shell with account of thermal loading using nonlinear theory of thin elastic shells.

Objective of the Present Study

Based on the aforementioned literature review, it is found that the literature on the mechanics of sandwich shells considering the effects of transverse normal strain is limited. Hence, the objective of the present study is to carry out static bending analysis of laminated sandwich shells using a new hyperbolic shell theory. A semi-analytical solution for the static problem is obtained using the Navier method.

Kinematic formulation

Fig. 1 shows a differential shell element considered in the (x, y, z) coordinate system. The x and y curves depicted here are lines of substantial curvature on the mid-plane of the laminate. The downward z-direction is seen to be positive. R_1 and R_2, respectively, stand for the primary radii of curvature of the mid-plane along the x and y axes. Layers of orthotropic composite material that are thought to be suitably linked together make up a laminate. On the top surface of a laminate, that is, $z = -h/2$, a transverse load of q (x, y) is applied.

Following is the displacement field assumed for the current hyperbolic shell theory.

$$u(x,y,z) = \left(1+\frac{z}{R_1}\right)u_0(x,y) - z\frac{\partial w_0}{\partial x} + f(z)\theta_x(x,y)$$
$$v(x,y,z) = \left(1+\frac{z}{R_2}\right)v_0(x,y) - z\frac{\partial w_0}{\partial y} + f(z)\theta_y(x,y) \quad (1)$$
$$w(x,y,z) = w_0(x,y) + f'(z)\theta_z(x,y)$$

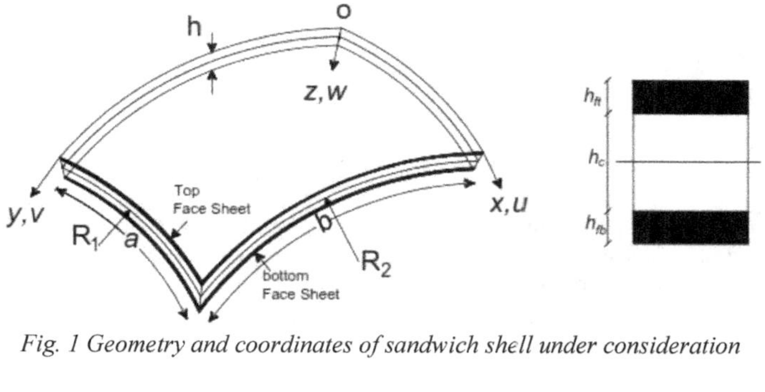

Fig. 1 Geometry and coordinates of sandwich shell under consideration

where, *u, v, w* are the displacements in *x, y, z* directions respectively; $\theta_x, \theta_y, \theta_z$ are the shear slopes in *x, y* and *z* direction respectively; u_0, v_0, w_0 are the mid-plane displacements in *x, y, z* direction respectively. Using the linear theory of elasticity, the normal and shear strains associated with the displacement field can be calculated as follows:

$$\begin{aligned}
\varepsilon_x &= \left(\frac{\partial u_0}{\partial x} + \frac{w_0}{R_1}\right) - z\frac{\partial^2 w_0}{\partial x^2} + f(z)\frac{\partial \theta_x}{\partial x} + \frac{f'(z)}{R_1}\theta_z \\
\varepsilon_y &= \left(\frac{\partial v_0}{\partial y} + \frac{w_0}{R_2}\right) - z\frac{\partial^2 w_0}{\partial y^2} + f(z)\frac{\partial \theta_y}{\partial y} + \frac{f'(z)}{R_2}\theta_z \\
\varepsilon_z &= f''(z)\theta_z \\
\gamma_{xy} &= \left(\frac{\partial u_0}{\partial y} + \frac{\partial v_0}{\partial x}\right) - 2z\frac{\partial^2 w_0}{\partial x \partial y} + f(z)\left(\frac{\partial \theta_x}{\partial y} + \frac{\partial \theta_y}{\partial x}\right) \\
\gamma_{xz} &= f'(z)\theta_x + f'(z)\frac{\partial \theta_x}{\partial x} \\
\gamma_{yz} &= f'(z)\theta_y + f'(z)\frac{\partial \theta_z}{\partial y}
\end{aligned} \tag{2}$$

where

$$f(z) = \left[z\cosh\left(\frac{\xi}{2}\right)\right] - \left[\left(\frac{h}{\xi}\right)\sinh\left(\frac{\xi z}{h}\right)\right],$$
$$f'(z) = \left[\cosh\left(\frac{\xi}{2}\right)\right] - \left[\cosh\left(\frac{\xi z}{h}\right)\right] \text{ where } \xi = 2.634. \tag{3}$$

Stresses are calculated using the Hooke's law from the 3D elasticity problem for cross-ply laminated shells.

$$\begin{Bmatrix} \sigma_x \\ \sigma_y \\ \sigma_z \\ \tau_{xy} \\ \tau_{xz} \\ \tau_{yz} \end{Bmatrix}^k = \begin{bmatrix} Q_{11} & Q_{12} & Q_{13} & 0 & 0 & 0 \\ Q_{12} & Q_{22} & Q_{23} & 0 & 0 & 0 \\ Q_{13} & Q_{23} & Q_{33} & 0 & 0 & 0 \\ 0 & 0 & 0 & Q_{66} & 0 & 0 \\ 0 & 0 & 0 & 0 & Q_{55} & 0 \\ 0 & 0 & 0 & 0 & 0 & Q_{44} \end{bmatrix}^k \begin{Bmatrix} \varepsilon_x \\ \varepsilon_y \\ \varepsilon_z \\ \gamma_{xy} \\ \gamma_{xz} \\ \gamma_{yz} \end{Bmatrix}^k \tag{4}$$

where $(\sigma_x, \sigma_y, \sigma_z, \tau_{xy}, \tau_{xz}, \tau_{yz})$ are the normal and shear stresses, and $(\varepsilon_x, \varepsilon_y, \varepsilon_z, \gamma_{xy}, \gamma_{yz}, \gamma_{xz})$ are the normal and shear strains. $(Q_{11}, Q_{12}, Q_{13}, Q_{22}, Q_{23}, Q_{33}, Q_{44}, Q_{55}, Q_{66})^k$ are the reduced stiffness coefficients.

$$\begin{aligned}
Q_{11} &= \frac{E_1(1-\mu_{23}\mu_{32})}{\Delta};\ Q_{12} = \frac{E_1(\mu_{21}+\mu_{31}\mu_{23})}{\Delta};\ Q_{13} = \frac{E_1(\mu_{31}+\mu_{21}\mu_{32})}{\Delta}; \\
Q_{22} &= \frac{E_2(1-\mu_{13}\mu_{31})}{\Delta};\ Q_{23} = \frac{E_2(\mu_{32}+\mu_{12}\mu_{31})}{\Delta};\ Q_{33} = \frac{E_3(1-\mu_{12}\mu_{21})}{\Delta}; \\
Q_{44} &= G_{23};\ Q_{55} = G_{13};\ Q_{66} = G_{12}; \\
\Delta &= 1 - \mu_{12}\mu_{21} - \mu_{23}\mu_{32} - \mu_{13}\mu_{31} - 2\mu_{21}\mu_{32}\mu_{13}
\end{aligned} \tag{5}$$

Principle of Virtual Work

$$\int_0^a \int_0^b \int_{-h/2}^{h/2} (\sigma_x \delta\sigma_x + \sigma_y \delta\sigma_y + \sigma_z \delta\sigma_z + \tau_{xy}\delta\tau_{xy} + \tau_{xz}\delta\tau_{xz} + \tau_{yz}\delta\tau_{yz})\, dzdydx = \int_0^a \int_0^b \int_{-h/2}^{h/2} q(x,y)\, dzdydx \tag{6}$$

The virtual work principle can be used to generate the six variationally-consistent governing equations and boundary conditions. The resulting governing equations can be visualized as the following in terms of stress resultants.

$$\delta u_0: \frac{\partial N_{xx}}{\partial x} + \frac{\partial N_{xy}}{\partial y} = 0$$
$$\delta v_0: \frac{\partial N_{xy}}{\partial x} + \frac{\partial N_{yy}}{\partial y} = 0$$
$$\delta w_0: \frac{\partial^2 M_{xx}^b}{\partial x^2} + \frac{\partial^2 M_{yy}^b}{\partial y^2} + 2\frac{\partial^2 M_{xy}^b}{\partial x\, \partial y} - \frac{N_{xx}}{R_1} - \frac{N_{yy}}{R_2} + q = 0$$
$$\delta \theta_x: \frac{\partial M_{yy}^S}{\partial x} + \frac{\partial M_{xy}^S}{\partial y} - Q_{xz}^S = 0 \tag{7}$$
$$\delta \theta_{y:}: \frac{\partial M_{yy}^S}{\partial y} + \frac{\partial M_{xy}^S}{\partial x} - Q_{yz}^S = 0$$
$$\delta \theta_z: \frac{\partial Q_{xz}^S}{\partial x} + \frac{\partial Q_{yz}^S}{\partial y} - \frac{V_{xx}^S}{R_1} - \frac{V_{yy}^S}{R_2} - V_{zz}^S = 0$$

where

$$\left(N_{xx}, N_{yy}, Ns_{xy}, M_{xx}^b,\ M_{yy}^b, M_{xy}^b\right) = \int_{-h/2}^{h/2} \left[\sigma_x, \sigma_y, \tau_{xy}, z\sigma_x,\ z\sigma_y, z\tau_{xy}\right] dz$$
$$\left(M_{xx}^S, M_{yy}^S, M_{xy}^S\right) = \int_{-h/2}^{h/2} \{[f(z)(\sigma_x, \sigma_y, \tau_{xy})]\}\ dz$$
$$\left(Q_{xz}^S, Q_{yz}^S\right) = \int_{-h/2}^{h/2} \{\ [f'(z)(\tau_{xz}, \tau_{yz})]\}\ dz \tag{8}$$
$$\left(V_{xx}^S, V_{yy}^S\right) = \int_{-h/2}^{h/2} \{\ [f'(z)(\sigma_x, \sigma_y)]\}\ dz$$
$$\left(V_{zz}^S\right) = \int_{-h/2}^{h/2} \{[f''(z)]\sigma_z\ \}\ dz$$

Closed-Form Solution

The following solution form is assumed for the unknown variables in accordance with Navier's solution process, and it precisely satisfies the boundary conditions that are easily supported.

$$(u_0,\ \theta_x) = (u_{mn}, \theta_{xmn})\cos\alpha x\ \sin\beta y$$
$$(v_0,\ \theta_y) = (v_{mn}, \theta_{ymn})\sin\alpha x\ \cos\beta y \tag{9}$$
$$(w_0,\ \theta_z) = (w_{mn}, \theta_{zmn})\sin\alpha x\ \sin\beta y$$

where, $u_{mn}, \theta_{xmn}, v_{mn}, \theta_{ymn}, w_{mn}, \theta_{zmn}$ are the unknown coefficients. The expression for the transverse sinusoidal load is expressed as,

$$q(x, y) = q_0 \sin\alpha x \sin\beta y \tag{10}$$

where $\alpha = \pi/a,\ \beta = \pi/b$, and q_0 is the maximum intensity of the sinusoidal load. Substituting Eqs. (9) – (10) into Eq. (7) and the resultant equations are expressed in matrix form as,

$$[K]\{\Delta\} = \{f\} \tag{11}$$

where $[K]$, $\{f\}$ and $\{\Delta\}$ represent the stiffness matrix, force vector and the vector of unknowns. The elements of these matrices are as follows.

$$K_{11} = -A_{11}\alpha^2 - A_{66}\beta^2, \quad K_{12} = K_{21} = -A_{12}\alpha\beta - A_{66}\alpha\beta,$$
$$K_{13} = K_{31} = \frac{A_{11}}{R_1}\alpha + \frac{A_{12}}{R_2}\alpha + B_{11}\alpha^3 + B_{12}\alpha\beta^2 + 2B_{66}\alpha\beta^2,$$
$$K_{14} = K_{41} = -C_{11}\alpha^2 - C_{66}\beta^2, \; K_{15} = K_{51} = -C_{12}\alpha\beta - C_{66}\alpha\beta,$$
$$K_{16} = K_{61} = \left[\frac{F_{11}}{R_1}\alpha + \frac{F_{12}}{R_2}\alpha + D_{13}\alpha\right], \; K_{22} = -A_{22}\beta^2 - A_{66}\alpha^2,$$
$$K_{23} = K_{32} = B_{22}\beta^3 + [B_{12} + 2B_{66}]\alpha^2\beta + \left[\frac{A_{12}}{R_1} + \frac{A_{22}}{R_2}\right]\beta$$
$$K_{24} = K_{42} = -C_{21}\alpha\beta - C_{66}\alpha\beta, K_{25} = K_{52} = -C_{22}\beta^2 - C_{66}\alpha^2,$$
$$K_{26} = K_{62} = \left(D_{23} + \frac{F_{21}}{R_1} + \frac{F_{22}}{R_2}\right)\beta,$$
$$K_{33} = -(H_{11}\alpha^4 + H_{22}\beta^4) - 2\alpha^2\beta^2(H_{12} + 2H_{66}) - 2\alpha^2\left(\frac{B_{11}}{R_1} + \frac{B_{12}}{R_2}\right) - 2\beta^2\left(\frac{B_{12}}{R_1} + \frac{B_{22}}{R_2}\right)$$
$$-\left(\frac{2A_{12}}{R_1R_2} + \frac{A_{11}}{R_1^2} + \frac{A_{22}}{R_2^2}\right),$$
$$K_{34} = K_{43} = I_{11}\alpha^3 + I_{21}\alpha\beta^2 + 2I_{66}\alpha\beta^2 + \frac{C_{11}}{R_1}\alpha + \frac{C_{21}}{R_2}\alpha,$$
$$K_{35} = K_{53} = I_{12}\alpha^2\beta + I_{22}\beta^3 + 2I_{66}\alpha^2\beta + \frac{C_{12}}{R_1}\beta + \frac{C_{22}}{R_2}\beta,$$
$$K_{36} = K_{63} = \left\{\begin{matrix} -K_{13}\alpha^2 - K_{23}\beta^2 - \frac{D_{13}}{R_1} - \frac{D_{23}}{R_2} - \left(\frac{J_{11}}{R_1} + \frac{J_{12}}{R_2}\right)\alpha^2 - \left(\frac{J_{21}}{R_1} + \frac{J_{22}}{R_2}\right)\beta^2 \\ -\left(2\frac{F_{12}}{R_1R_2} + \frac{F_{22}}{R_2^2} + \frac{F_{11}}{R_1^2}\right) \end{matrix}\right\},$$
$$K_{44} = -L_{11}\alpha^2 - L_{66}\beta^2 - O_{55}, K_{45} = K_{54} = -(L_{12} + L_{66})\alpha\beta, \tag{11}$$
$$K_{46} = K_{64} = \left(N_{13} - O_{55} + \frac{M_{11}}{R_1} + \frac{M_{12}}{R_2}\right)\alpha,$$
$$K_{55} = -L_{66}\alpha^2 - L_{22}\beta^2 - O_{44},$$
$$K_{56} = K_{65} = \left(-O_{44} + N_{23} + \frac{M_{21}}{R_1} + \frac{M_{22}}{R_2}\right)\beta,$$
$$K_{66} = \left(-O_{55}\alpha^2 - O_{44}\beta^2 - S_{33} + 2\frac{P_{23}}{R_2} - 2\frac{P_{13}}{R_1} - \frac{O_{11}}{R_1^2} - 2\frac{O_{12}}{R_1R_2} - \frac{O_{22}}{R_2^2}\right),$$

where

$$(A_{ij}, B_{ij}, H_{ij}, C_{ij}, F_{ij}\, I_{ij}) = Q_{ij}\int_{-h/2}^{h/2}[1.0,\, z, z^2, f(z), f'(z), zf(z)]\, dz;$$
$$L_{ij} = Q_{ij}\int_{-h/2}^{h/2}\{[f(z)]^2\}\, dz; O_{ij} = Q_{ij}\int_{-h/2}^{h/2}\{[f'(z)]^2\}\, dz;$$
$$(D_{ij}, S_{ij}, P_{ij}) = Q_{ij}\int_{-h/2}^{h/2}\{[f''(z)], [f''(z)]^2, [f'(z)]\}\, dz; \tag{12}$$
$$(K_{ij}, N_{ij}) = Q_{ij}\int_{-h/2}^{h/2} f''(z)[z, f(z)]\, dz;$$
$$(J_{ij}, M_{ij}) = Q_{ij}\int_{-h/2}^{h/2} f'(z)[z, f(z)]\, dz;$$

Numerical Result and Discussion

To carry out the static analysis of the sandwich spherical shells, the following material characteristics stated in Eq. (13)-(14) and non-dimensional parameters stated in Eq. (15) are taken into account.

Face Sheet Properties

$$\frac{E_1}{E_2} = 25, \frac{E_3}{E_2} = 1, \frac{G_{12}}{E_2} = \frac{G_{13}}{E_2} = 0.5, \frac{G_{23}}{E_2} = 0.2, \mu_{12} = \mu_{13} = \mu_{23} = 0.2 \tag{13}$$

Core Properties

$$\boldsymbol{E_1 = E_2 = 0.04,\ E_3 = 0.5,\ G_{13} = G_{23} = 0.06, G_{12} = 0.016,} \\ \boldsymbol{\mu_{12} = \mu_{32} = \mu_{31} = 0.25} \tag{14}$$

and

$$\begin{gathered} \bar{u}\left(0,\frac{b}{2},\frac{z}{h}\right) = \frac{h^2E_3}{q_0a^3}u, \quad \bar{w}\left(\frac{a}{2},\frac{b}{2},\frac{z}{h}\right) = \frac{100h^3E_3}{q_0a^4}w, \\ (\bar{\sigma}_x,\bar{\sigma}_y)\left(\frac{a}{2},\frac{b}{2},\frac{z}{h}\right) = \frac{h^2}{q_0a^2}(\sigma_x,\sigma_y),\ \bar{\tau}_{xy}\left(0,0,\frac{z}{h}\right) = \frac{h^2}{q_0a^2}\tau_{xy} \\ \bar{\tau}_{zx}\left(0,\frac{b}{2},\frac{z}{h}\right) = \frac{h}{q_0a}\tau_{zx},\ \bar{\tau}_{yz}\left(\frac{a}{2},0,\frac{z}{h}\right) = \frac{h}{q_0a}\tau_{yz}. \end{gathered} \tag{15}$$

where, E_3 is modulus of elasticity of the middle layer i.e. the core of the sandwich shell.

Sandwich plate consists of face sheets with thickness of $0.1h$ and the core with thickness $0.8h$ where h is the total thickness of the sandwich shells. The material properties used for the sandwich shell are mentioned in Eq. (13)-(14) whereas non-dimensional parameters are stated in Eq. (15). Table 1 shows the comparison of non-dimensional displacements and stresses obtained using the current theory with those presented by third-order theory of Reddy [10], first-order theory of Mindlin [13] and Exact elasticity solution by Pagano [12] wherever applicable. Through-the-thickness distributions of displacements and stresses are plotted in Fig. 2. The numerical results are presented for R/a = 5, 10, 20, 50, 100 and ∞. Table 1 reveals that the current theory predicts the numerical results in good agreement with existing literature. Transverse shear stresses are discontinuous at the layer interface because those are calculated using constitutive relations. To get the stress continuity at the layer interface, those must be calculated using the equilibrium equations of the theory of elasticity.

Table 1 *Non-dimensional displacements and stresses of three-layer (0^0/core/0^0) sandwich spherical shells under sinusoidal load ($a = 10h$, $R_1 = R_2 = R$).*

R/a	Theory	$\bar{u}$	$\bar{w}$	$\bar{\sigma}_x$	$\bar{\sigma}_y$	$\bar{\tau}_{xy}$	$\bar{\tau}_{xz}$	$\bar{\tau}_{yz}$
5	Present ($\varepsilon_z \neq 0$)	0.0124	0.9965	1.0655	0.0772	0.0926	0.3108	0.0558
	Reddy [10] ($\varepsilon_z = 0$)	0.0131	1.0063	1.0733	0.0745	0.0932	0.2956	0.0486
	Mindlin [13] ($\varepsilon_z = 0$)	0.0109	0.7122	1.0147	0.0607	0.0715	0.3096	0.0384
10	Present ($\varepsilon_z \neq 0$)	0.0099	1.0152	1.1002	0.0921	0.0805	0.3166	0.0569
	Reddy [10] ($\varepsilon_z = 0$)	0.0102	1.0250	1.1081	0.0891	0.0812	0.3011	0.0495
	Mindlin [13] ($\varepsilon_z = 0$)	0.0088	0.7215	1.0385	0.0708	0.0628	0.3137	0.0389
20	Present ($\varepsilon_z \neq 0$)	0.0085	1.0199	1.1128	0.0993	0.0739	0.3181	0.0571
	Reddy [10] ($\varepsilon_z = 0$)	0.0086	1.0298	1.1207	0.0962	0.0747	0.3025	0.0497
	Mindlin [13] ($\varepsilon_z = 0$)	0.0077	0.7238	1.0471	0.0757	0.0582	0.3147	0.0390
50	Present ($\varepsilon_z \neq 0$)	0.0076	1.0213	1.1247	0.1035	0.0698	0.3185	0.0572
	Reddy [10] ($\varepsilon_z = 0$)	0.0077	1.0312	1.1267	0.1003	0.0707	0.3029	0.0498
	Mindlin [13] ($\varepsilon_z = 0$)	0.0070	0.7245	1.0512	0.0786	0.0553	0.3150	0.0390
100	Present ($\varepsilon_z \neq 0$)	0.0073	1.0215	1.1205	0.1048	0.0685	0.3186	0.0572
	Reddy [10] ($\varepsilon_z = 0$)	0.0073	1.0314	1.1284	0.1017	0.0693	0.3029	0.0498
	Mindlin [13] ($\varepsilon_z = 0$)	0.0068	0.7246	1.0524	0.0795	0.0544	0.3150	0.0390
Plate	Present ($\varepsilon_z \neq 0$)	0.0069	1.0215	1.1220	0.1062	0.0671	0.3186	0.0572
	Reddy [10] ($\varepsilon_z = 0$)	0.0070	1.0315	1.1300	0.1030	0.0679	0.3029	0.0498
	Mindlin [13] ($\varepsilon_z = 0$)	0.0066	0.7246	1.0535	0.0805	0.0534	0.3151	0.0390
	Exact [12]	0.0071	1.1002	1.1518	0.1098	0.0706	0.2997	0.0526

Conclusions

In this work, a new hyperbolic higher-order shell theory is used to obtained closed-form solutions for the static analysis of a sandwich spherical shell under sinusoidal loading. To take into account the effect of cross-sectional deformation, the theory is formulated by adding hyperbolic kinematic function in terms of thickness coordinates. Using the principle of virtual work, the governing equation and the associated boundary condition are obtained. Navier's solution method is used as solution technique. Based on the numerical results and discussion it is concluded that the present theory accurately predict the static response of sandwich spherical shells under transverse loading.

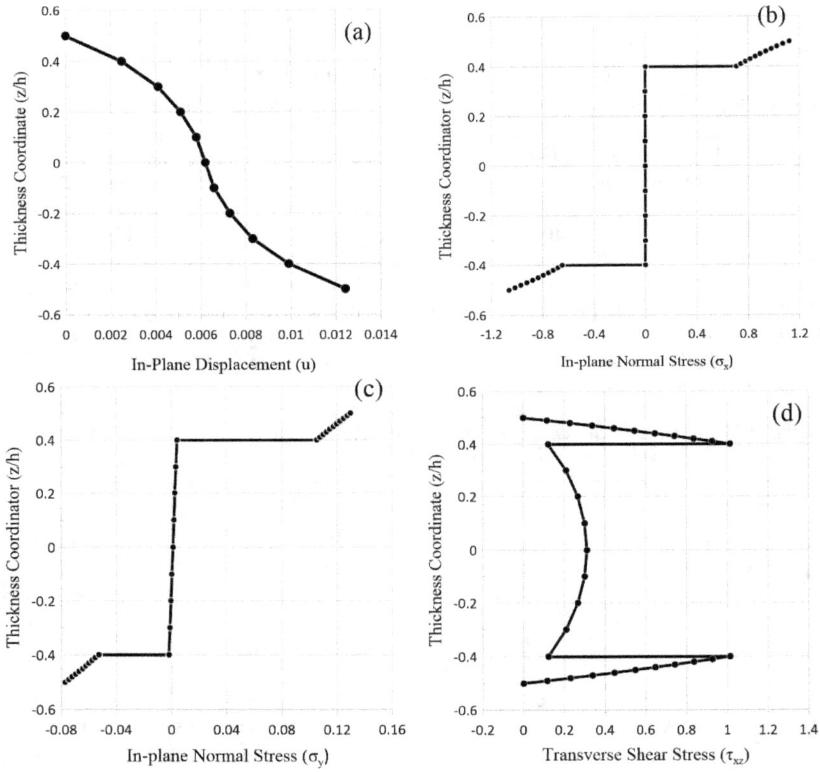

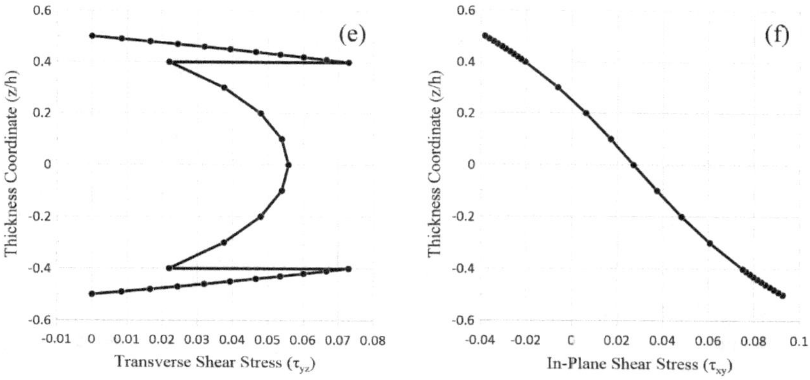

Fig. 2 Through-the-thickness plots of displacements and stresses (R/a = 5, a/h = 10)

References

[1] A.P. Mouritz, E. Gellert, P. Burchill, K. Challis, Review of advanced composite structures for naval ships and submarines, Compos. Struct. 53 (2001) 21-41. https://doi.org/10.1016/S0263-8223(00)00175-6

[2] Mallikarjuna, T. Kant, A critical review and some results of recently developed refined theories of fiber-reinforced laminated composites and sandwiches, Compo. and Struct. 23 (1993) 293-312. https://doi.org/10.1016/0263-8223(93)90230-N

[3] A.J.M. Ferreira, J.T. Barbosa, A. T. Marques, J. C. De Sâ, Non-linear analysis of sandwich shells: the effect of core plasticity, Compos. Struct. 76 (2000) 337-346. https://doi.org/10.1016/S0045-7949(99)00156-X

[4] T. Kant, K. Swaminathan, Analytical solutions for the static analysis of laminated composite and sandwich plates based on a higher order refined theory, Compos. Struct. 56 (2002) 329- 344. https://doi.org/10.1016/S0263-8223(02)00017-X

[5] J. Hohe, L. Librescu, A nonlinear theory for doubly curved an isotropic sandwich shells with transversely compressible core, I. J. Sol. and Struct. (40) 2003 1059-1088. https://doi.org/10.1016/S0020-7683(02)00656-X

[6] C. Zhong, H.G. Reimerdes, Stability behaviour of cylindrical and conical sandwich shells with flexible core, Sandwich Structures 7: Adv. Aircr. Spacecr. Sci. An. Int. J. (2005) 241-250. https://doi.org/10.1007/1-4020-3848-8_24

[7] R.K. Khare, V. Rode, A.K. Garg, S. P. John, Higher-order closed-form solutions for thick laminated sandwich shells, J. Sandw. Struct. Mater. (2005) 335-358. https://doi.org/10.1177/1099636205050260

[8] A.K. Garg, R.K. Khare, T. Kant, Higher-order closed-form solutions for free vibration of laminated composite and sandwich shells, J. Sandw. Struct. Mater. 8(2006) 205-235. https://doi.org/10.1177/1099636206062569

[9] I.K. Turkin, A technique for calculating rational design parameters of a sandwich shell with account of thermo force loading, Russ. Aero. 54(2) (2011) 209-212. https://doi.org/10.3103/S1068799811020164

[10] B.M. Shinde, A.S. Sayyad, Analysis of laminated and sandwich spherical shells using a new higher-order theory, Adv. Aircr. Spacecr. Sci. An. Int. J. 7(1) (2020) 19-40.

[11] N.J. Pagano, Exact solutions for bidirectional composites and sandwich plates, J Compos Mater 4 (1970) 20-34. https://doi.org/10.1177/002199837000400102

[12] R.D. Mindlin, Influence of rotatory inertia and shear on flexural motions of isotropic elastic plates, ASME J. Appl. Mech. 18(1951) 31-8. https://doi.org/10.1115/1.4010217

[13] A. Bhimaraddi, A higher order theory for free vibration analysis of circular cylindrical shells, Int. J. Solids Struct. 20(7) (1984) 23-30. https://doi.org/10.1016/0020-7683(84)90019-2

[14] A. Bhimaraddi, Three-dimensional elasticity solution for static response of orthotropic doubly curved shallow shells on rectangular platform, Compos. Struct. 24(1) (1993) 67-77. https://doi.org/10.1016/0263-8223(93)90056-V

[15] K.P. Soldatos, T.A. Timarci, unified formulation of laminated composite shear deformable, five-degrees-of-freedom cylindrical shell theories, Compos. Struct. 25 (1993) 165-71. https://doi.org/10.1016/0263-8223(93)90162-J

[16] J.N. Reddy, A simple higher order theory for laminated composite plates, J. Appl. Mech. 51 (1984) 745-52. https://doi.org/10.1115/1.3167719

[17] J.N. Reddy, Exact solutions of moderately thick laminated shells, ASCE J. Eng. Mech. 110(5) (1984) 794-809. https://doi.org/10.1061/(ASCE)0733-9399(1984)110:5(794)

[18] Y.M. Ghugal, A.S. Sayyad, Stress analysis of thick laminated plates using trigonometric shear deformation theory, Int. J. Appl. Mech. 5(1) (2013) 1-23. https://doi.org/10.1590/S1679-78252013000500008

[19] A.S. Sayyad, Y.M. Ghugal, Effect of stress concentration on laminated plates, Cambridge J. Mech. 29(2) (2013) 241-52. https://doi.org/10.1017/jmech.2012.131

[20] A.S. Sayyad, Y.M. Ghugal, On the buckling of isotropic, transversely isotropic and laminated composite rectangular plates, Int. J. Struct. Stab. Dyn. 14(6) (2014) 1-32. https://doi.org/10.1142/S0219455414500205

[21] A.S. Sayyad, Y.M. Ghugal, Flexure of cross-ply laminated plates using equivalent single layer trigonometric shear deformation theory, Struct. Eng. Mech. 51(5) (2014) 867-91. https://doi.org/10.12989/sem.2014.51.5.867

[22] A.S. Sayyad, Y.M. Ghugal, A new shear and normal deformation theory for isotropic, transversely isotropic, laminated composite and sandwich plates, Int. J. Mech. Mater. Des. 10(3) (2014) 247-67. https://doi.org/10.1007/s10999-014-9244-3

[23] A.S. Sayyad, Y.M. Ghugal, Cylindrical bending of multi-layered composite laminates and sandwiches, Adv. Aircr. Spacecr. Sci. An. Int. J. 3(2) (2016) 113-48. https://doi.org/10.12989/aas.2016.3.2.113

[24] A.S. Sayyad, Y.M. Ghugal, On the free vibration of angle-ply laminated composite and soft core sandwich plates, J. Sandw. Struct. Mater. 19(6) (2017) 679-711. https://doi.org/10.1177/1099636216639000

Advanced Topics in Mechanics of Materials, Structures and Construction - AToMech1-2023 Materials Research Forum LLC
Materials Research Proceedings 31 (2023) 55-65 https://doi.org/10.21741/9781644902592-7

New approach for modeling randomly distributed CNT reinforced polymer nanocomposite with van der Waals interactions

Umut Caliskan[1,a *], Hilal Gulsen[2,b] and Mustafa Kemal Apalak[3,c]

[1]Department of Mechanical Engineering, Erciyes University, Kayseri 38039, Turkey

[2]Graduate School of Natural and Applied Sciences, Erciyes University, Kayseri 38039, Turkey

[3]Department of Mechanical Engineering, Erciyes University, Kayseri 38039, Turkey

[a]ucaliskan@erciyes.edu.tr, [b]hilalgulsen97@gmail.com, [c]apalakmk@erciyes.edu.tr

Keywords: Carbon Nanotube, Nanocomposite, Multi-Scale Model, Randomly Distribution, van der Waals Interactions

Abstract. In this paper, using molecular and micromechanics methods, a new approach for the prediction of the stiffness of randomly distributed CNT/polymer nanocomposites with the van der walls interactions is presented. A multi-scale modeling technique was designed for CNT nanoparticles randomly embedded in the polymer using AMBER force field. This multi-scale model constitutes a representative volume element. The representative volume element consists of polymer, CNT nanoparticle, CNT-polymer interfacial region and van der waals bonds. A programming code was developed that randomly distributes nanoparticles according to the desired volume fraction. Python scripting language was used for the modeling technique performed in a finite element environment. By modeling the interfacial regions around randomly distributed CNTs, van der Waals bonds are modeled stochastically. In this study, the subject of interest is the number of CNTs positioned in the RVE according to the volume ratio. These numbers were determined at the level allowed by finite element equations and computational solvers and their effects were investigated by calculated stiffness behavior.

Introduction

With the discovery of nanomaterials, polymer nanocomposites are a new type of material whose advanced properties can be used in the polymer industry and new applications. Among nanomaterials, carbon nanotubes (CNTs) and graphene sheets are very important due to their superior mechanical, thermal and electrical superior and unique properties. Carbon nanotubes (CNTs) were discovered by Lijima in 1991 [2]. Carbon nanotubes have attracted the attention of scientists as reinforcement materials in nanocomposites with their low densities, high strength and high stiffness[3]. For example Qian et al.[4] when added only 1% by weight of carbon nanotubes to the polyester resin, they obtained a 35-42% increase in modulus of elasticity and a 25% increase in strength. In carbon nanotube-reinforced nanocomposites, the diameter and chirality of the carbon nanotube, its distribution, orientation and the general behavior of the polymer affect the properties of the nanocomposite. Therefore, modeling and simulations are very important in investigating the effect of adding carbon nanotubes to polymers. Epoxy polymers are well known for their excellent adhesion, significant mechanical and high electrical insulating properties, and relatively good chemical resistance. Epoxy is now widely used in various applications.

Tserpes et al. [5] investigated a multi-scale representative volume element for modeling the tensile behavior of carbon nanotube-reinforced composites. The RVE integrates nanomechanics and continuum mechanics, thus bridging the length scales from the nano- through the mesoscale. A progressive fracture model based on the modified Morse interatomic potential was used for simulating the behavior of the isolated carbon nanotubes and the FE method for modeling the matrix and building the RVE. The effect of interfacial shear strength on the tensile behavior of the nanocomposite was also studied. Stiffness was unaffected while tensile strength significantly decreased with decreasing the interfacial shear strength.

Khalili and Haghbin [6] to investigate the role of nanotube specifications on the nonlinear tensile behavior of nanocomposites, multi-scale material modeling was used. The effect of diameter, chirality and volume fraction of nonlinearly modeled. SWCNTs was studied on their nanocomposites. Multi-scale modeling was applied to assemble various RVEs composed of different SWCNTs embedded in a polymer. Continuum mechanics based on their atomic structures in the case of space frame structures were used to model nanotubes. Attained stress-strain curves of modeled nanocomposites revealed that using Armchair SWCNTs rather than Zigzags made nanocomposites tougher in tensile loading. Also, the diameter of CNT had an inverse effect on the stress-strain curve level. Using CNTs with smaller diameters, regardless of the chirality and type, made nanocomposites stronger in tension. Furthermore, the effect of diameter was more obvious in higher volume fractions of CNTs. Ayatollahi et al.[7] presented multiscale modeling for the nonlinear properties of a polymer/single wall carbon nanotube (SWNT) nanocomposite under tensile, bending and torsional loading condition. They used a finite element (FE) model based on the theory of molecular mechanics to predict the mechanical properties of both armchair and zigzag SWNTs. The equivalent beam element was then used to build a cylindrical representative volume element (RVE) where the effects of the interphase between SWNT and the polymer on mechanical response could be studied. The results showed that while the interphase had a small effect on the nanocomposite stiffness, the ratio of (SWNT length)/(RVE length) dramatically affected nanocomposite stiffness. For nano-structured materials a multi-scale model which is based on the equvalent continuum modeling technique was introduced by Odegard et al [8]. In their work the elastic parameters of the model were obtained through the same loading conditions in both the continuum model in finite element and the discrete model in molecular mechanics.

In this paper present a new multi-scale modeling approach including randomly-distributed carbon nanotube/epoxy nanocomposites with interfacial interactions to predic elastic properties and mechanical behavior of nanocomposites with single-walled carbon nanotube reinforcement. All modeling procedure and analyses were performed in ABAQUS finite element software [9].

Description of Multi-Scale Model

Geometry of Carbon Nanotubes

The CNT is a lattice of hexagonal bonds between carbon atoms, where each atom is bonded to the three nearest neighbors through very strong covalent bonds, 1.421 Å length. The mechanical and electrical properties of CNT depend on the direction and length of the vector, which describes the atomic arrangement of the bonds. This vector, in a word chirality vector or chirality index C_h, is produced from two vectors a_1 and a_2 of the unit cell in the CNT. The chiral vector consists of n units of $\vec{a}_1$ and m units of a_2 as seen in Figure 1.

$$C_h = n\vec{a}_1 + m\vec{a}_2 \text{ , bn} \geq m \quad (1)$$

Chirality index specified by (n,m), represents length and direction of chiral vector on a CNT. Three types of CNT are defined regarding the chiral index. Armchair geometry are defined by (n,n) ve Zigzag geometry by (n,0). Chiral CNT have different chirality indexes.

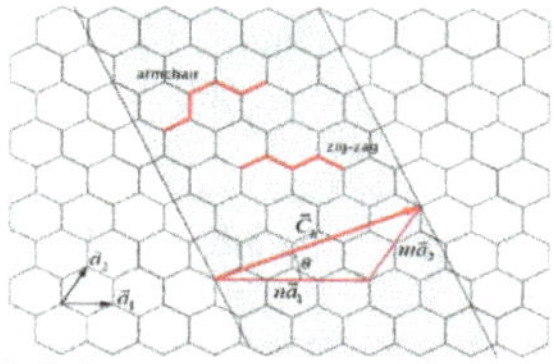

Figure 1. Schematic diagram of CNT geometry

The chiral vector with the length of L is rolled up and forms the nanotube circumference; the nanotube diameter can be obtained by the following equation [10]

$$D = \frac{L}{\pi} = \frac{a(3(n^2+m^2+nm))^{\frac{1}{2}}}{\pi} \tag{2}$$

where a is the carbon–carbon bond length and n, m are chiral indices.

Molecular Structural Mechanics Modelling

In computational nanomechanics, the energy of a system that includes hundreds of billions of atoms is evaluated by internal potential. The total potential energy function of an n-body structure is always clearly understood to address the configurational potential energy that can be the state of the coordinate of the constituent atom. The potential energy function explains as follows [10].

$$H_I = \frac{1}{2!}\sum_i \sum_{j\neq i} V_2(r_i, r_j) + \frac{1}{3!}\sum_i \sum_{j\neq i} \sum_{k\neq i,j} V_3(r_i, r_j, r_k) + \ldots \tag{3}$$

Where V_n, is n-body interatomic potential functions. The potentials are functions of atomic coordinates however these are stated in terms of interatomic distances in practice. Therefore, the potential energy depends on interatomic separation and the angle between bonds connecting individual atoms. The atomistic field of force consists of two basic potentials. These are potentials that occur between bonded atoms (V_{bonded}) and potentials that occur between atoms that are not bonded to each other ($V_{non\text{-}bonded}$). Figure 2 shows the atomic force field due to this work potential [10].

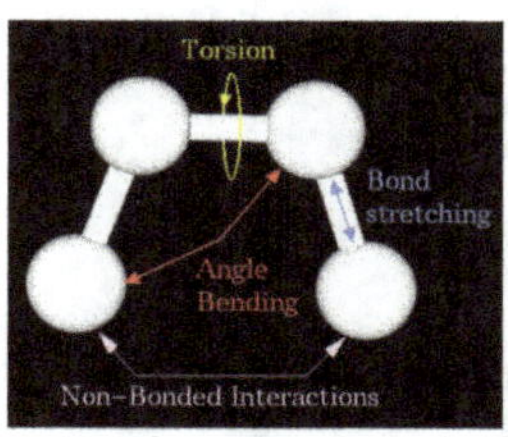

Figure 2. Atomic force field

$$U = U_{bonded} + U_{non\text{-}bonded} \tag{4}$$

In molecular mechanics, CNTs are acceptable as great molecules including carbon atoms. Thus, the potential energy function is expressed as steric potential energy, which depends only on the position of the nuclei of the carbon atoms. The general form of force field of a nanostructure explains such as [10]

$$U = \sum U_r + \sum U_\theta + \sum U_\emptyset + \sum U_W + \sum U_{vdw} \tag{5}$$

where represents the bonded interatomic potential, $\mathbf{U_r}$,is bond stretching, $\mathbf{U_\theta}$, is bond bending angle, $\mathbf{U_\emptyset}$ is dihedral torsion bending, $\mathbf{U_W}$, out of plane torsion and $\mathbf{U_{vdw}}$ is Van der Waals interactions potential between non-bonding atoms.

The molecular mechanical modeling method is intended to simulate the atomic properties of the CNT as a space framework in classical structural mechanics. Therefore, as seen in Figure 3 carbon atoms on CNT are acceptable as load-carrying beam elements. Simple harmonic functions and an Amber force field are applied to explain terms of potential energy in molecular mechanics [10].

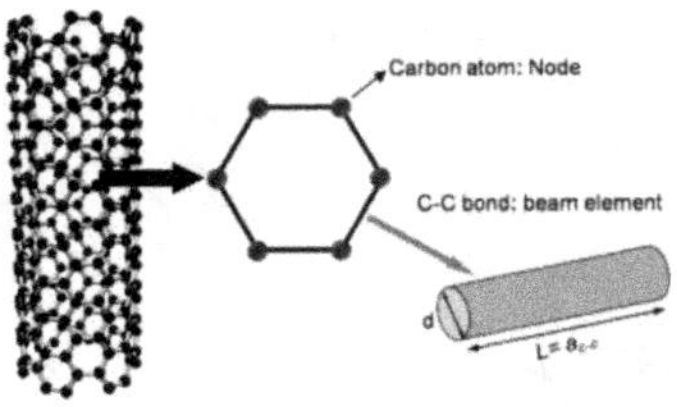

Figure 3. Model of CNT as a space-frame structure

$$\text{Bond Stretching Energy}: U_r = \frac{1}{2}k_r(r - r_0)^2 = \frac{1}{2}k_r(\Delta r)^2 \tag{6}$$

$$\text{Bond Angle Bending Energy}: U_\theta = \frac{1}{2}k_\theta(\theta - \theta_0)^2 = \frac{1}{2}k_\theta(\Delta\theta)^2 \tag{7}$$

$$\text{Bond Torsion Energy}: U_\tau = \frac{1}{2}k_\tau(\tau - \tau_0)^2 = \frac{1}{2}k_\tau(\Delta\tau)^2 \tag{8}$$

where, $\mathbf{k_r}$, $\mathbf{k_\theta}$, $\mathbf{k_\tau}$ are bond stretching force, bond angle bending force and torsional preventive constant, respectively, and the increase in these forces is expressed by the following symbols, respectively, $\Delta \boldsymbol{r}$, $\Delta\boldsymbol{\theta}$ and $\Delta\boldsymbol{\tau}$. In structural mechanics, the tensile energy of a regular beam element representing a carbon bond of length L due to deformations is expressed as [10]:

$$U_A = \frac{1}{2}\int_0^L \frac{N^2}{EA}dL = \frac{1}{2}\frac{N^2L}{EA} = \frac{1}{2}\frac{EA}{L}(\Delta L)^2 \tag{9}$$

$$U_T = \frac{1}{2}\int_0^L \frac{T^2}{GJ}dL = \frac{1}{2}\frac{T^2L}{GJ} = \frac{1}{2}\frac{GJ}{L}(\Delta\beta)^2 \tag{10}$$

$$U_M = \frac{1}{2}\int_0^L \frac{M^2}{EI}dL = \frac{2EI}{L}\alpha^2 = \frac{1}{2}\frac{EI}{L}(2\alpha)^2 \tag{11}$$

where $\mathbf{U_A}$, $\mathbf{U_T}$, $\mathbf{U_M}$ are the strain energies of a beam element due to axial force N, causing bending moment M, causing torsion T respectively. In relation to them, ΔL, α, $\Delta\beta$ are axial stretching deformation, rotating angle at the end of the beam and the relative rotation between the ends of the beam, respectively.

$$\frac{EA}{L} = k_r, \frac{EI}{L} = k_\theta, \frac{GJ}{L} = k_\tau, \tag{12}$$

Beam elements are expressed from a circular cross-section. Taking into account the molecular mechanical parameters from the amber force field, the beam elements E, G and diameter (d) are made available for application in structural mechanics modeling with the following relations:

$$d = 4\sqrt{\frac{k_\theta}{k_r}}, E = \frac{k_r^2 L}{4\pi k_\theta}, G = \frac{k_r^2 k_\tau L}{8\pi k_\theta^2} \tag{13}$$

Table 1. Amber Force Field Constants [10]

k_r,	938 kcal mol^{-1} A^{-2}	6.52×10^{-7} N nm^{-1}rad^{-2}
k_θ,	126 kcal mol^{-1}rad^{-2}	8.76×10^{-10} N nm rad^{-2}
k_τ,	40 kcal mol^{-1} rad^{-2}	2.78×10^{-10} N nm rad^{-2}

The mechanical properties of the beam elements are as follows:

Table 2. Mechanical properties of beam elements

d	0.147 nm
E	5.49 TPa
G	0.87 TPa

This formulation was applied in the finite element method by solving the system of matrices that are derived for circular two-node beam elements. Each node has six degrees of freedom (DOF), three translational (x, y, z) and three rotational degrees of freedom (DoF).

Modelling of Carbon Nanotubes with Representative Volume Element (RVE)

In this study linear Euler-Bernoulli beam elements with six degrees of freedom are used for molecular mechanic modeling. This element, is based on known elastic parameters such as the Young Modulus (E) and shear modulus (G) and also on the diameter obtained by the molecular structural modeling method. Beam elements are applied to a space-frame structure which is modeled geometrically with coordinates of carbon atoms in its molecular structure. Coordinates are obtained by using a nanomaterial modeling program. After that, beam elements are placed between two carbon atoms respecting the chirality of CNT. ABAQUS finite element software was used to model CNT nanocomposites as shown in Figure 4.

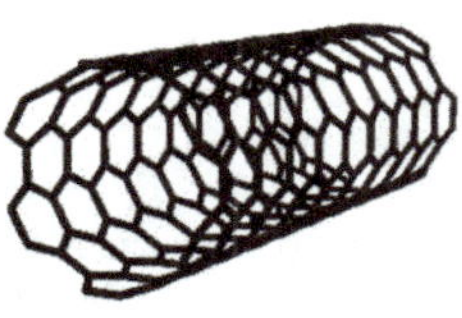
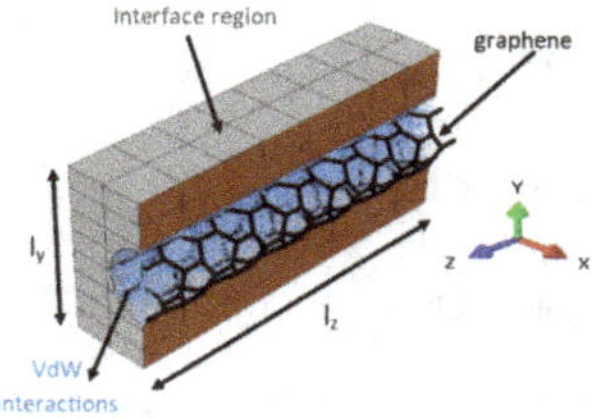

Figure 4. Carbon nanotube and RVE modeled in ABAQUS

Randomly Distributed Algorithm

The realistic behavior of CNT reinforced nanocomposite was modeled using a new random distribution algorithm proposed by this study. This algorithm was developed with the open-source programming language Python, in which the finite element code is also written. The random distribution algorithm of nanoparticles in a polymer matrix without intersecting each other was created with the following steps:

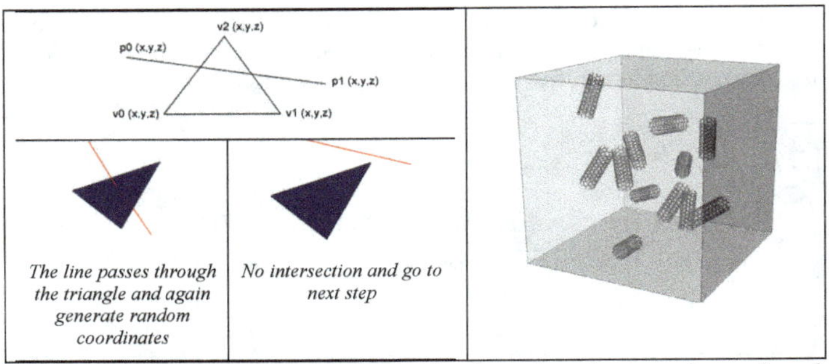

Figure 5. Checking intersection of a surface with a line, plotted core and transferred to finite element code

- Input step: Input information is needed for algorithm logic to work.
- Drawing step: According to the control volume dimensions received as input from the user, a center in 3-dimensional space is determined.
- Generating random coordinate step: In this stage, random coordinates need for the positioning of the drawn geometries within the control volume.
- Moving step: CNT particles in the center of the control volume are positioned according to randomly generated coordinates considering reference point in the center of the CNT.
- Checking intersection step: The steps of the developed algorithm up to this step include classical methods. After the CNT is positioned randomly within the control volume, it is necessary to check whether any of CNT cross-sections intersect with the control volume.
- Output step: The coordinates according to the written algorithm are imported into a text file.

Interactions Between CNT and Epoxy

Interfacial interactions between matrix and reinforcement have a great effect on the general properties of nanocomposites. Electrostatic interactions are negligible compared to Van der Waals interactions as the Van der Waals interaction contributes three orders of magnitude higher than electrostatic energy. Non-linear Van Der Waals force, with respect to interatomic distance, explained by the 6/12- Lennard-Jones equation [11];

$$F_{udW} = \frac{48e}{r}\left[-\left(\frac{\sigma}{r}\right)^{12} + 0.5\left(\frac{\sigma}{r}\right)^{6}\right] \tag{14}$$

Lennard-jones parameters, e and σ, are 0.4396 $kJmol^{-1}$ and 0.3851 nm, respectively [12]. The interactions between the carbon atoms of the CNT and the nodes of the inner surface of the surrounding polymers, Van der Waals interaction will be modeled using the axial translation connector type (CONN3D2). These linear elements, which have different names in the finite element method, are named 'connector elements' in the ABAQUS software. In this way, only the Van der Waals interaction between the CNT node (atom) and the adjacent nodes of the polymer will be created. Van der Waals force is a non-linear force and can be neglected when the interatomic distance is greater than or equal to 0.85 nm as shown in Figure 5[12].

Between these nodes, a code would be written to form an element with distances less than 0.85 nm, and van Der Waals bonds would be formed only between the carbon atoms of the carbon nanotube or graphene and the inner surface of the polymer.

Materials Research Proceedings 31 (2023) 55-65 https://doi.org/10.21741/9781644902592-7

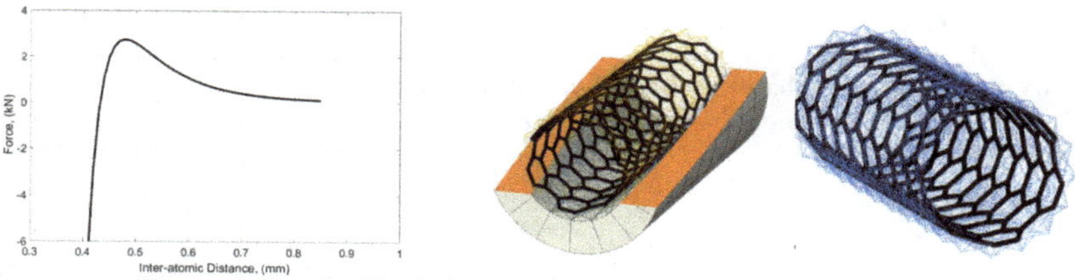

Figure 6. Van der Waals forces-interatomic distance variation

Stochastic RVE configurations

In this study, the polymer was considered as a linear elastic material with a Young's modulus of 2 GPa and a Poisson ratio of 0.35. Linear cubic eight-node elements with six degrees of freedom for each node in space were used to connect the polymer part in the finite element. CNTs were located in the middle of the representative volume element of the same length as the polymer interface. The CNT volume ratio was calculated by ABAQUS and the thickness of the CNT was taken to be equal to the interlayer space in the graphite sheets, i.e. 0.34 nm. Stochastic is an adjective that means variable, random, unpredictable. Stochastic processes are especially important in engineering. Transferring the stochastic process to numerical work is important for more accurate results. The stochastic process to be represented in this study is based on the unpredictable dispersal of CNT and the unpredictable Van der Waals interactions. At this point, there is another issue that makes the study more original and difficult. A bond was established between CNT atoms and polymer atoms, but these bonds form an interfacial layer on polymer atoms and the mechanical behavior of this interface region is different from the[13]. Interface material is also included in this unpredictable distribution. The interface material was modeled as embedded in the polymer material and included in the stochastic model. In the study, the transfer of mechanical properties from the interface material to the polymer was considered linear. This may result in transferring the high mechanical properties of nanomaterials to the polymer more than necessary. A description of the multi-scale model is given in Figure 7.

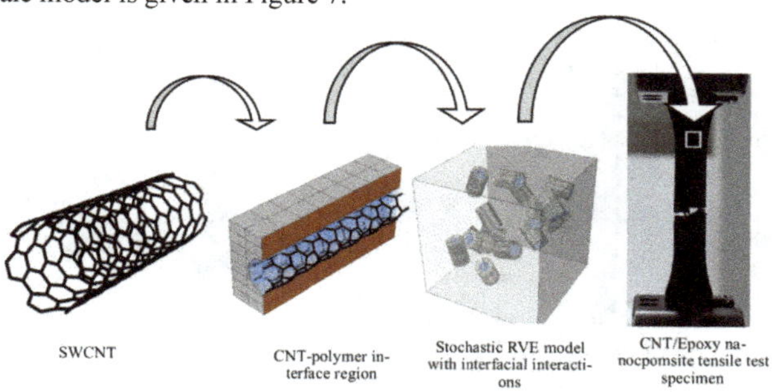

Figure 7. description of the multi-scale model

In this study, after the above-described tube processes have been carried out, there is only one point left to be decided. This situation constitutes the aim and originality of the study. The number of CNT to be used in the analysis according to the determined volume ratio is the first parameter that will affect the results. As this number increases, the interfaces and vdw interactions will also increase. Its solution is related to its solution capabilities. In this study, the case of randomly dispersed 5, 10 and 15 CNT in a representative volume element at 1% volume was investigated for all direction. Under these conditions, dimensions were formed according to the determined volume ratio, CNT moved to randomly determined positions in the cubic volume element, interfacial phases were formed around them first, and van der walls interactions were formed for each CNT with its own phase under the specified conditions. This process is explained in detail in Figure 7 and the information about all RVE models were given in the table 3.

Table 3. Information about the RVE models

	Volume Fraction	CNT (Å)	CNT Volume (Å^3)	Epoxy (Å)	Number of VdW İnteractions
CNT-5		$C_h(6x6)$	5x835.21	75x75x75	22924
CNT-10	0.5, 1, 2 %	L_{CNT}=24	10x835.21	95x95x95	45844
CNT-15		$d_{(C-C)}$=0.735	15x835.14	108x108108	68768

Figure 8 shows the proposed RVE models for multi-scale stochastic modeling of CNT nanocomposites. Along with investigating the effect of CNT number in RVE, the effect of CNT volume ratio on mechanical properties and modulus of elasticity was also investigated. For the tensile behavior of the CNT nanocomposite, the total force was calculated by applying 10% displacement to the RVE. The finite element model is shown in Figure 9.

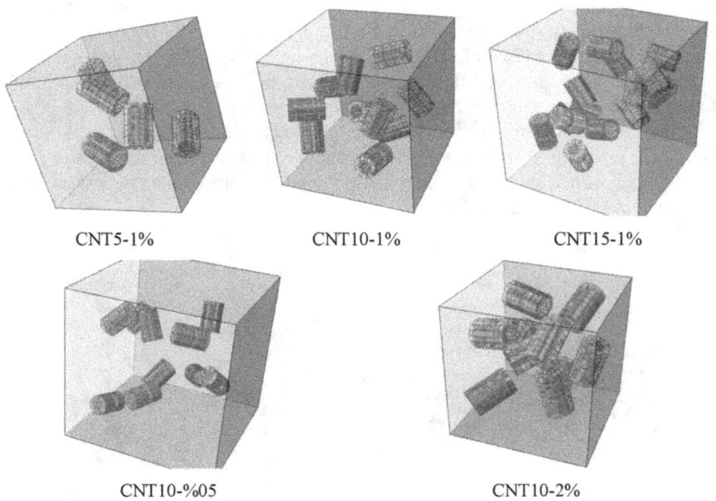

Figure 8. Proposed RVE models for multi-scale stochastic modeling of CNT nanocomposites

Advanced Topics in Mechanics of Materials, Structures and Construction - AToMech1-2023 Materials Research Forum LLC
Materials Research Proceedings 31 (2023) 55-65 https://doi.org/10.21741/9781644902592-7

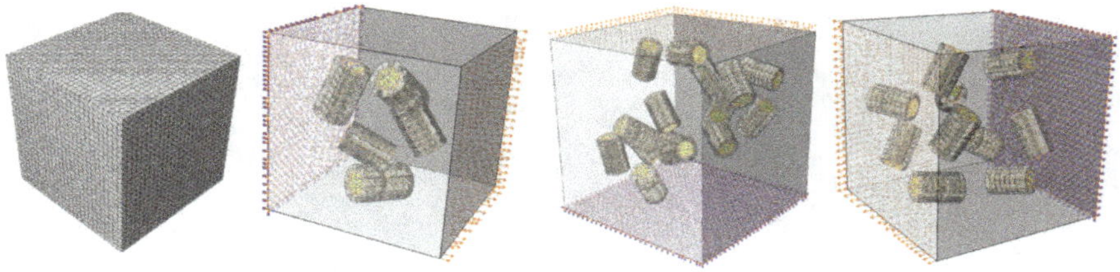

Figure 9. FE model of RVE

Results &Conclusion

Multi-scale material modeling with a stochastic approach with van der walls interaction was applied to investigate the effect of CNT quantity in an epoxy matrix, volume fraction and random distribution by direction of CNT. First, the effect of the number of CNTs in epoxy at the same volume fractions was investigated on RVEs under tensile load. Stress-strain curves and modulus of elasticity were calculated. Models named pure epoxy, CNT5, CNT10 and CNT15 were compared in the graphics. The general result in nanocomposites modeled at 1% volume ratio; the strength increases as the number of CNTs in the epoxy increases (Figure 10). The model with a CNT number of 5 showed approximately two times more resistance than pure epoxy. However, this increased rate decreased as the number of CNTs increased. With the increase in the number of CNTs in the models created to the extent of computational ability, the models will be able to give stable results. In order to examine the results of the random distribution effect, tensile loads were applied separately in 3 different axes, and the results were compared. Stress-Strain behaviors showed similar results. However, differences emerged in the Elasticity module. While the X and Y directions show similar results, the Z direction is calculated differently.

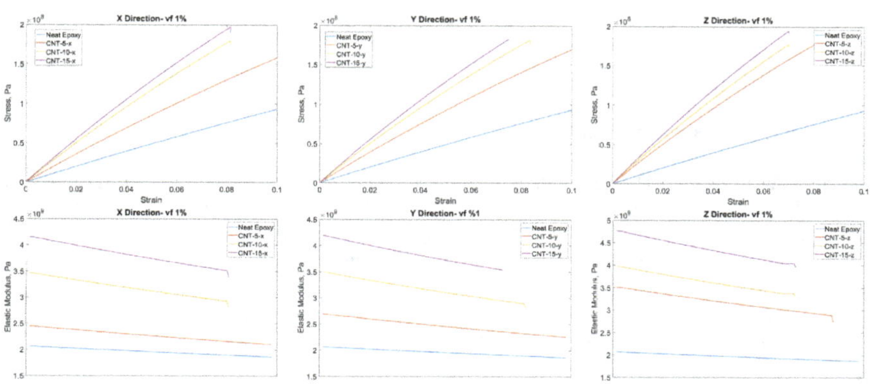

Figure 10. The effect of CNT quantity inside epoxy on the stress-strain behavior and elastic modulus for x, y and z direction.

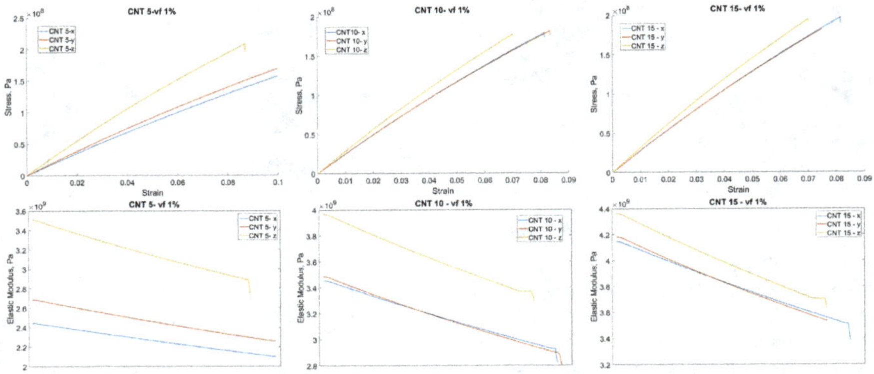

Figure 11. The effect of tensile direction on the stress-strain behavior and elastic modulus for different CNT quantity inside epoxy.

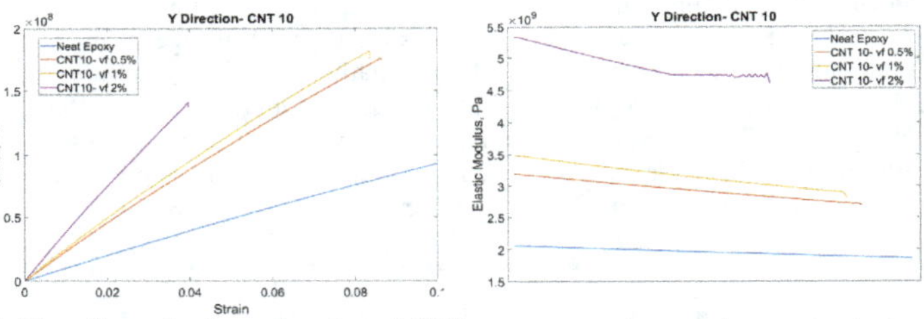

Figure 12. The effect of volume fraction of CNT nanocomposites on the stress-strain curve and elastic modulus variation.

In Figure 11, comparisons of the mechanical behavior of models with the same CNT number under loading in the x, y and z directions are given. Especially the distribution results in x and y directions were similar. The distribution results in the Z direction revealed higher stress levels and modulus of elasticity. In Figure 12, the results of the CNT volume ratio effect in the nanocomposite are given. The mechanical behavior of pure epoxy was improved with the use of CNT. 0.5% and 1 gave approximate results, while 2% gave very high results.

Acknowledgment
The authors would like to thank the Scientific Research Project Division of Erciyes University for the financial support under contract: FBA-2022-11927.

References

[1] A. Hussein, B. Kim, Micromechanics based FEM study on the mechanical properties and damage of epoxy reinforced with graphene based nanoplatelets, Composite Structures, 215 (2019), 266-277. https://doi.org/10.1016/j.compstruct.2019.02.059

[2] S. Iijima, Helical microtubules of graphitic carbon. Nature, 354 (1991), 56-8. https://doi.org/10.1038/354056a0

[3] D. Qian, G. J. Wagner, W. K. Liu, M. F. Yu, R. S. Ruoff, Mechanics of carbon nanotubes, Applied Mechanics Reviews, 55(2022), 495-533. https://doi.org/10.1115/1.1490129

[4] D. Qian, E. Dickey, R. Andrews, T. Rantell, Load transfer and deformation mechanisms in carbon nanotube-polystyrene composites, Appl Phys Lett, 76(2000), 2868-70. https://doi.org/10.1063/1.126500

[5] K.I. Tserpes, P. Papanikos, G. Labeas, and Sp.G. Pantelakis, Multi-scale modeling of tensile behavior of carbon nanotube-reinforced composites, Theoretical and Applied Fracture Mechanics, 49(2008), 51-60. https://doi.org/10.1016/j.tafmec.2007.10.004

[6] S.M.R. Khalili and A. Haghbin, Multi-scale modeling of nonlinear tensile behavior in single-walled carbon nanotube reinforced nanocomposites, International Journal of Modeling and Optimization,1(2011), 199-204. https://doi.org/10.7763/IJMO.2011.V1.35

[7] M.R. Ayatollahi, S. Shadlou and M.M. Shokrieh, Multiscale modeling for mechanical properties of carbon nanotube reinforced nanocomposites subjected to different types of loading, Composite Structures, 93(2011), 2250-2259. https://doi.org/10.1016/j.compstruct.2011.03.013

[8] G.M. Odegard, T.S. Gates, K.E. Wise, C. Park, E.J. Siochi, Constitutive Modeling of Nanotube-Reinforced Polymer Composites, Composites Science and Technology, 63 (2003), 671-1687. https://doi.org/10.1016/S0266-3538(03)00063-0

[9] ABAQUS/Standart, User's manual, Finite Element Software. www.simulia.com

[10] S.M.R. Khalili and A. Haghbin, Investigation on design parameters of single-walled carbon nanotube reinforced nanocomposites under impact loads, Composite Structures, 98(2013), 253-260. https://doi.org/10.1016/j.compstruct.2012.09.049

[11] L. Battezzati, C. Pisani, F. Ricca, Equilibrium conformation and surface motion of hydrocarbon molecules physisorbed on graphite J. Chem. Soc., Faraday Trans. 2: Mol. Chem. Phys., 71 (1975), 1629-1639. https://doi.org/10.1039/f29757101629

[12] Z. Shokrieh, M.M. Shokrieh, A new model to simulate the creep behavior of graphene/epoxy nanocomposites, Polymer Testing, 75 (2019), 321-326. https://doi.org/10.1016/j.polymertesting.2019.02.032

[13] Shokrieh, M., M., Rafiee, R. Investigation of nanotube length effect on the reinforcement efficiency in carbon nanotube based composites, Composite Structures, 92 (2010) (10), 2415-2420. https://doi.org/10.1016/j.compstruct.2010.02.018

Effect of shearing conditions and initial aggregates' state on the mechanical behavior of cellular glass foam

Layal Jradi[1], Bassel Seif El Dine[2], Tahar Ayadat[1], Saidur Schowdhury[1], Andi Asiz[1]

[1]Prince Mohammad Bin Fahd University, Saudi Arabia

[2]Lebanese University, Lebanon

Keywords: Cellular Glass Aggregates, Triaxial Testing, Monotonic Shear, Dry Density, Mechanical Properties

Abstract. Cellular glass aggregates made from recycled glass are increasingly being used in civil engineering and infrastructure applications. This contemporary material is relatively new in civil engineering applications. This paper investigates the effect of shearing conditions and initial aggregates' state on the mechanical properties of cellular glass foam. A series of monotonic large-scale triaxial tests (300 mm diameter and 600 mm height) was performed in order to investigate the influence of initial dry density, initial aggregates' state (i.e. flawless, weathered, and compacted), and consolidation stresses on the mechanical behavior of cellular glass foam. The material's behaviour observed under monotonic shear is of contracting type, with significant evolution of the material during shear (grain crushing) and a strongly strain-hardening character. The failure criterion of the studied material is similar to the failure criterion used for granular soils for which a shearing resistance angle and apparent cohesion is perceived.

Introduction

Cellular glass aggregates has many interests, in particular, its lightweight (ten times lighter than the gravel), its self-stability and drainage. Therefore, they are increasingly being used in civil engineering and infrastructure applications such as, lightweight cellular cemented clays, expanded polystyrene, and lightweight concrete [1,2,3,4,5,6].

Foam glass aggregate, is an artificial mixture produced by transferring finely ground glass powder from different glass sources (domestic wastes, industrial wastes, etc.) into glass foam. The condition for the usability of glass is that it should not contain heavy metals or other substances posing a health risk. The glass is grinded to the size of dust where it is spread on a belt conveyor running through high temperature ovens. In the passage through the long furnaces with temperatures of 900°C–1000°C, the glass powder expands about four times upon leaving the furnace, the expanded compound cools down from about 900°C to ambient temperatures. This change in temperature leads to the crack of the product into smaller pieces. The temperature, the thickness of the powder layer and the time spent in the furnaces define the final properties of the material.

In the literature, there are several studies on the properties and mechanical behavior of cellular glass. The NPRA (Norwegian Public road administration) has initiated a program to investigate the possible use of cellular glass material (Hasopor) for road construction applications where this material has been applied on about 25 road projects in Norway. Deformations of about 1-2 % were recorded for a relatively short time. Observations over time (3 years) indicate that further crushing and deformations tend to be negligible [7].

Mechanical characterization of cellular foam has been studied by various authors under different laboratory testing. Mear [8] presented the results of mechanical testing of the material against flexural and compressive strengths. According to the results obtained on the deformation variation of the material, three distinguished deformational zones were observed (zone I: an elastic deformation for low compression values, zone II: a discontinuity in deformation indicating the rupture of the weakest element of the material, and zone III: a local damage which led to total fracture). Arulrajah [9] performed shear box testing on foamed recycled glass. It was observed that the shear resistance at large shear displacement is high leading to an apparent cohesion $c = 23.36$ kPa and an angle of shearing resistance $\varphi = 54.7^{\circ}$. Furthermore, several authors studied the influence of the type of the experimental apparatus on the mechanical characterization of cellular glass foam [10,11]. Seif El Dine [12] studied the effect of the testing set-up on the mechanical characterization of matrix coarse-grained soils. The results obtained indicated that the direct shear box gives higher values of the mechanical properties of the tested material than those obtained by triaxial apparatus (i.e. the direct shear box overestimates the characteristics of the studied material).

As mentioned previously, this material is relatively still considered as new construction material. It is believed that, its mechanical characterization under different state and testing conditions is crucial. The objective of this experimental work is to investigate the effect of shearing conditions and initial aggregates' state on the mechanical properties of cellular glass foam. A series of monotonic large-scale triaxial tests (300 mm diameter and 600 mm height) was performed in order to investigate the influence of initial dry density, initial aggregates' state (i.e. flawless, weathered, and compacted), and consolidation stresses on the mechanical behavior of cellular glass foam.

Experimental Investigation

A large scale triaxial apparatus, which allow the shearing of large soil samples, is adopted in this study to determine the mechanical properties of the cellular glass aggregates. This device is accommodated to perform monotonic tests at different confining pressures, rates of deformation and loading, and saturation conditions which simulate field conditions. The dimensions of the samples used were 300 mm in diameter and 600 mm in height. The setup is composed of a large size triaxial cell incorporated in four columns loading frame, equipped with a 500 kN hydraulic actuator for the application of axial forces. The confining pressure applied to the triaxial cell is also controlled through a second hydraulic actuator. The sample is first prepared outside the frame, and then the cell is positioned under the hydraulic actuator which applies the axial force. A general view of the experimental setup is shown in Figure 1*a*.

The reconstitution process of the samples consisted of weighing the required quantity of the material according to a preselected dry density value. This quantity is then splitted into six buckets of equal parts for the purpose to prepare the samples in six compacted layers. A neoprene membrane is then inserted into the mold covered by a geotextile film that is used to protect the membrane from any damage that may result during the compaction process. The mold is then filled up with the aggregates which were compacted in layers of 10 cm height each. Finally a porous disk and top end plate are placed on the top of the specimen. Upon preparing the specimen, a 50 kPa vacuum is applied to the specimen in order to remove the mold. This value should always be less than the confining pressure value in order to prevent any over consolidation of the specimen. The triaxial cell is fixed on a movable support that allows its displacement in both translation and rotation. Once the cell is adjusted under the actuator, the test can be launched (Figure 1*b*).

After the preparation of the specimens and the assembly of the apparatus, some preliminary tests were undertaken in order to gain some familiarity with the procedure followed in setting up and carrying out the tests and to ensure the repeatability of the tests. For this purpose, three tests were carefully repeated for consolidation stress values of 25 kPa, 50 kPa and 100 kPa at a dry density of $\rho_d = 212$ kg/m^3. The results obtained showed good consistency between the different tests. %). Consequently, it may be concluded that the test apparatus gives repeatable and reproducible results under the same conditions.

(a) (b)

Figure 1. *General view of the experimental setup*

Test Results and Analysis

In order to evaluate the behavior of cellular glass aggregates under monotonic shear, a series of 14 triaxial tests (M1 to M14) were performed. The main test characteristics are summarized in Table 1. All tests are displacement controlled tests. They were run on dry material at an axial deformation rate of 0.2 %/mn, up to a maximum value of axial deformation of approximately 15%. Most tests were run at a dry specific density of 212 kg/m^3 which was the value originally chosen. In order to evaluate the influence of the dry density (ρ_d) on the behavior of the aggregates, two tests (M12 and M13) were carried out for ρ_d of 230 kg/m^3. Five values of isotropic consolidation stress (25, 50, 100 and 200 kPa) were used in order to evaluate the failure characteristics as well as Mohr-Coulomb failure parameters of the aggregates.

Table 1 – Experimental program carried out for monotonic shear tests.

Test	ρ_d (kg/m³)	σ'_c (kPa)	$\varepsilon_{a,max}$ (%)	q_{max} (kPa)
Intact material				
M1	142	50	15	214
M2	212	100	15	373
M3	212	100	15	396
M4	212	50	15	259
M5	212	50	15	291
M6	212	100	15	404
M7	212	25	15	243
Aged material				
M8	212	50	15	257
M9	212	25	15	166
M10	212	25	18	194
M11	212	200	16	452
M12	230	50	18	290
M13	230	50	18	296
M14	212	200	16	493

Influence of Consolidation Stress

The influence of the consolidation stress on the mechanical properties of the cellular glass aggregates were considered in the present experimental program. Figure 2 shows the results obtained for a dry density of 212 kg/m^3 and four different consolidation stresses, in this case 25 kPa, 50 kPa, 100 kPa and 200 kPa. This figure indicates that, for all the tests, the deviator stress still keeps increasing after an axial deformation of 15%. However, the higher the consolidation stress and the higher the axial deformation needed to reach a deviator stress plateau. For the highest consolidation stress of 200 kPa, the deviator stress is still strongly increasing after the axial deformation of 15%. As expected, the shear strength of the material increases almost proportionally with the consolidation stress, which is fairly similar to the behavior observed in sands and gravels and should yield a failure criterion of the Mohr-Coulomb type.

Two sieve analyses were performed in order to evaluate the degradation of the material after the shear test for both the lowest and the highest consolidation stresses (i.e. 25 and 200 kPa). The results were compared to the grading curve of the material before shearing and compacted at the same dry density ρ_d = 212 kg/m^3 (Figure 3). The grading curves of the material before and after shear under a consolidation stress of 25 kPa are almost identical, indicating that the shear test has had a very low effect on the degradation of the material. On the other hand, a clear evolution of the grading curve is observed for the 200 kPa consolidation shear test. In this case, all the elements sizes are affected by the degradation, especially the elements of intermediate size (i.e. from 5 mm to 31.5 mm), where the proportion was increased by 30 %.

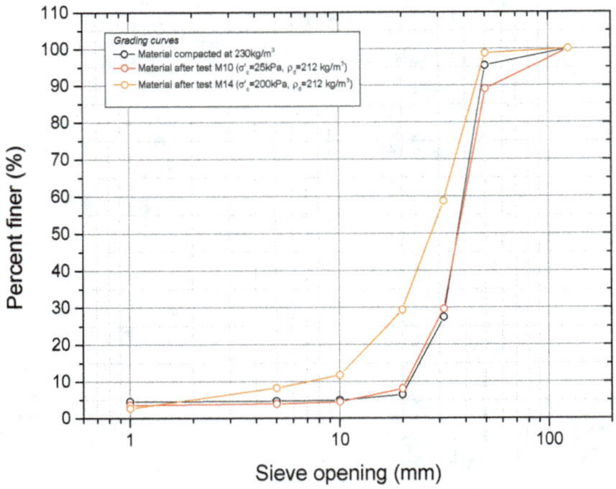

Figure 2. Shear curves obtained on cellular glass for four consolidation stresses: 25kPa, 50kPa, 100 kPa, and 200kPa (for ρ_d =212 kg/m^3))

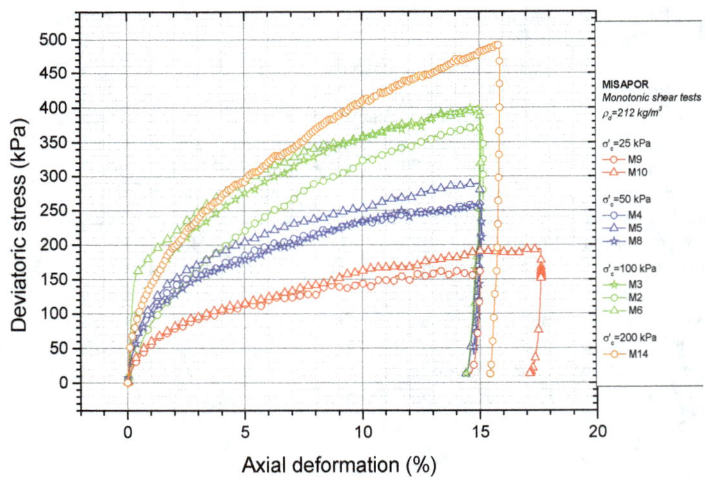

Figure 3. Evolution of the grading curve of cellular glass after shear tests M10 and M14

The Mohr circles at failure were plotted for the different tests as shown in Figure 4. Failure is defined here as the maximum deviator stress reached during shear and therefore corresponds to the maximum axial deformation reached. As indicated on this figure, two different failure regimes were observed. For low values of consolidation stress, it can be noted a relatively high value of angle of shearing resistance (φ') and a rather low apparent cohesion (c'). For higher values of the consolidation stress, the angle of shearing resistance could appear lower with a higher value of apparent cohesion. A nonlinear failure criterion, based on the nonlinear envelope curve of the failure Mohr circles, can be defined. This observation can be attributed to an increase in level of grain crushing for higher values of consolidation stress.

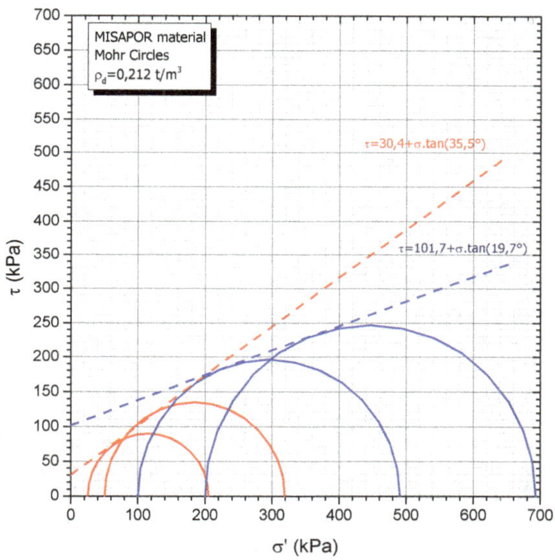

Figure 4. Evaluation of Mohr-Coulomb failure parameters

Influence of dry density

In order to investigate the influence of the initial dry density on the mechanical properties of the cellular glass aggregates; two different values of dry density were considered (ρ_d = 212 kg/m^3 and ρ_d = 230 kg/m^3). These two densities correspond respectively to a compression ratio, with respect to the loose state, of 1.18 and 1.28. Figure 5 shows the results obtained at a consolidation stress of 50 kPa. The differences observed between the results corresponding to the two densities are not very significant, where a discrepancy of about 7% was noted.

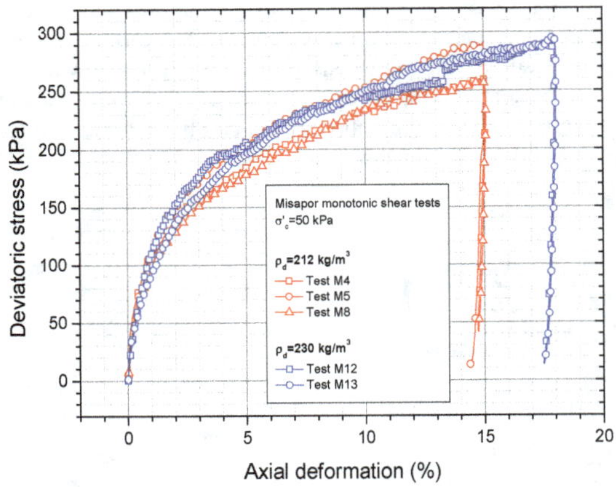

Figure 5. Shear curves obtained at σ'_c = 50 kPa for two different dry densities, in this case ρ_d = 212 kg/m³ (M4, M5 and M8) and ρ_d = 230 kg/m³ (M12 and M14)

Influence Initial Aggregates' State

It is believed that the initial aggregates' state plays an important role in the shearing characteristics of cellular glass foam. For this reason, it was decided to investigate the influence of the aggregates' state on the initial dry density and the particle size distributions of this material. Three different aggregates' states were considered and studied, in this respect flawless or intact, weathered, and compacted state. For testing, the material was divided into six packs; where each pack was poured into the mold from a constant drop height. This procedure is repeated three times. The average dry density ρ_d was found to be 178 kg/m^3, which corresponds to a loose state. Knowing that, the lowest dry density that can be reached for a standard Fontainebleau sand (e_{max} = 0.87, ρ_s = 2650 kg/m^3) is ρ_d = 1417 kg/m^3, it can be confirms that this material is lightweight (i.e. eight times lighter than Fontainebleau sand).

It is well known that using cellular glass in the field involves several phases of handling and transportation that may damage the material. In order to take this aspect into account, a weathering process has been used, based on mixing or shaking the material in a cement mixer.

The cement mixer was filled at 45 % of its maximum capacity with the intact material. The mixer was inclined at 45° with respect to the horizontal and the material was mixed or shacked for 2 minutes. The weathered material is then collected and sieved in order to determine the impact of the weathering protocol on the material. Figure 6 shows a comparison of grading curves obtained for the intact and the weathered materials, showing a moderate decrease of the large elements proportion, decreasing from 21 % to 15 % and an increase by 4 points of the proportion of the particles smaller than 31.5 mm. The grading parameters needed to classify the weathered material are reported in Table 2. They differ slightly from those of the intact material, showing a low impact of the weathering process on the material, which is still classified as a poorly graded gravel (GP).

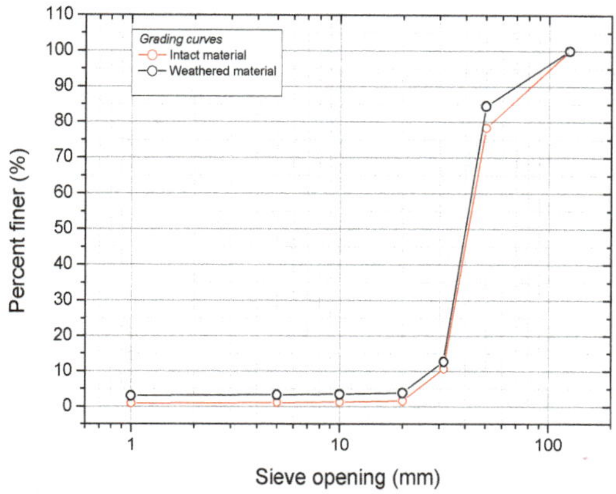

Figure 6. Comparison between the grading curves obtained for the intact and weathered material

Table 2. Values of d_{10}, d_{50}, d_{60} and C_u for cellular glass after the weathering process

Type of material	Dry density ρ_d (kg/m^3)	d_{10} (mm)	d_{50} (mm)	d_{60} (mm)	$c_u = d_{60}/d_{10}$
Loose material	179	30	41	45	1.50
Aged Material	179	27	40	43	1.59

For industrial applications of cellular glass, compaction is usually necessary for the implementation of the material. It is therefore important to characterize and quantify the compaction process. Different compaction tests have been performed on the material in order to set up a well-defined protocol and evaluate the impact of this protocol on the evolution (damaging) of the material. The tests have been carried out for two specific dry densities, $\rho_d = 212$ kg/m^3 and $\rho_d = 230$ kg/m^3, which correspond, respectively, to a compaction ratio (with respect to natural state) of 1.19 and 1.29. In order to ensure a good homogeneity of the material density within the mold, the specimens were fabricated by successively compacting six layers of 10 cm thickness each. A PVC disc was used to level each layer and homogenize the compaction energy within the layer. The compaction tool was dropped on the plastic plate from a constant drop height and the thickness of the layer was measured after each drop. This process was repeated until a layer thickness of 10 cm was reached. Figure 7 shows the state of the material after compaction. Clearer zones are visible, corresponding to contact surfaces with the compacting plate, where the material was locally crushed. Also, some big size elements were broken down into smaller elements. In order to better quantify the material damage induced by compaction, the grading curves of the material after compaction, for both densities, are shown in Figure 6 and compared with the grading curve of the intact material. The evolution of the grading curve appears mainly on the elements bigger than 20 mm and it may be seen that the compacting process mainly breaks the biggest elements down to intermediate ones, while creating a small amount of fines. A synthesis of the effects of the compaction on the material is given in the Table 3.

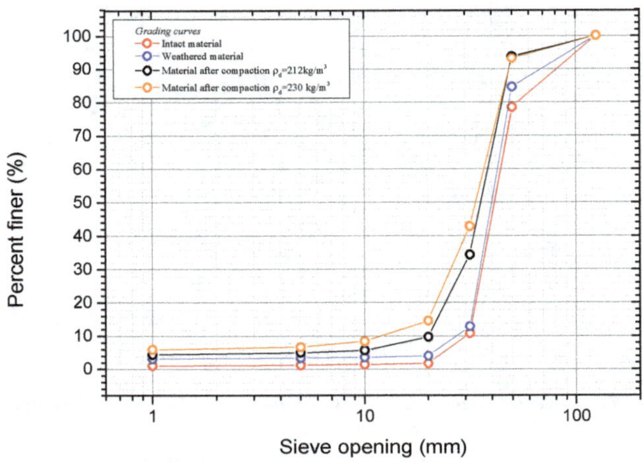

Figure 7. Comparison of grading curves obtained for cellular glass: intact, weathered and compacted material

Table 3. Values of d_{10}, d_{50}, d_{60} and c_u for cellular glass in a loose and compacted state

Type of material	Dry density ρ_d (kg/m^3)	d_{10} (mm)	d_{50} (mm)	d_{60} (mm)	$c_u = d_{60}/d_{10}$
Bulk material	179	30	41	45	1.50
Compacted material	212	21	36	39	1.87
Compacted material	230	20	35	38	1.90

Conclusion

Cellular glass material is considered as relatively new and innovative material. It is therefore important and essential to obtain mechanical characterization of this material under different testing conditions. In this experimental program, the influence of initial dry density, initial aggregates' state (i.e. flawless, weathered, and compacted), and consolidation stresses on the mechanical behavior of cellular glass foam were investigated. It can be conclude that the material's behaviour observed under monotonic shear is of contracting type, with significant evolution of the material during shear (grain crushing) and a strongly strain-hardening character. The failure criterion of the studied material is similar to the failure criterion used for granular soils for which a shearing resistance angle and apparent cohesion is perceived.

It is important to mention that; cellular glass waste recycling can also reduce costs associated with obtaining natural aggregates and reduce the environmental impact associated with solid waste disposal. Recycling of wastes glasses could reduce the demand for virgin natural resources and help to dispose of them effectively. The innovation in the circular economy is crucial to the transition to more sustainable infrastructure development.

References

[1] Disfani, M.M., Arulrajah, A., Haghighi, H., Mohammadinia, A., Horpibulsuk, S., 2014. Flexural beam fatigue strength evaluation of crushed brick as a supplementary material in cement stabilized recycled concrete aggregates. Constr. Build. Mater. 68, 667-676. https://doi.org/10.1016/j.conbuildmat.2014.07.007

[2] Athanasopoulos-Zekkos, A., Lamote, K., Athanasopoulos, G.A., 2012. Use of EPS geofoam compressible inclusions for reducing the earthquake effects on yielding earth retaining structures. Soil Dyn. Earthq. Eng. 41, 5-71 https://doi.org/10.1016/j.soildyn.2012.05.004

[3] Deng, A., Xiao, Y., 2010. Measuring and modeling proportion-dependent stress- strain behavior of EPS-sand mixture. Int. J. Geomech. 10, 214-222. https://doi.org/10.1061/(ASCE)GM.1943-5622.0000062

[4] Chindaprasirt, P., Rattanasak, U., 2011. Shrinkage behavior of structural foam lightweight concrete containing glycol compounds and fly ash. Mater. Des. 32, 723-727. https://doi.org/10.1016/j.matdes.2010.07.036

[5] Wang, G., Liu, Y., Cui, Y., 2012. Performance Studies of Lightweight Concrete Mixtures Made with Rigid Polyurethane Foam Wastes, pp. 4007-4010. https://doi.org/10.4028/www.scientific.net/AMM.204-208.4007

[6] Zach, J.; Sedlmajer, M; Bubenik, J.; and Drdlova, M. (2019). Development of lightweight composites based on foam glass aggregate. IOP Conf. Series: Materials Science and Engineering 583 (2019) 012016, doi:10.1088/1757-899X/583/1/012016. https://doi.org/10.1088/1757-899X/583/1/012016

[7] Aabøe Roald, 1980, "Lette fyllmasser I vegbygging" (lightweight filling materials in road construction - text in Norwegian), Intern rapport 954, Norwegian Road Research Laboratory, Oslo

[8] Yot, P., Cambon, M., Ribes, M., & Me, F. (2005). Elaboration and characterisation of foam glass from cathode ray tubes, 104(3), 123-131.http://doi.org/10.1179/174367605X20171 https://doi.org/10.1179/174367605X20171

[9] Arulrajah, A., Disfani, M. M., & Maghoolpilehrood, F. (2015). Engineering and environmental properties of foamed recycled glass as a lightweight engineering material. Journal of Cleaner Production, 94, 369-375. https://doi.org/10.1016/j.jclepro.2015.01.080

[10] Afriani, L. 2003. Essais de cisaillement direct des sols grossiers : Incidences des procédures d'essai et effets d'échelle. Ph.D. the- sis, University of Caen, France.

[11] Vallé, N. 2001. Propriétés mécaniques d'un sol grossier d'une terrasse alluvionnaire de la Seine. Ph.D. thesis, University of Caen, Caen, France.

[12] Seif El Dine, B. 2007. Etude du comportement mécanique de sols grossiers à matrice. Ph.D. thesis, Ecole Nationale des Ponts et Chaussées, Paris.

Static response of functionally graded porous spherical shells using trigonometric shear deformation theory

Rupali B. Tamnar[1,a *], Atteshamuddin S. Sayyad[2,b]

[1]Research Scholar, Department of Civil Engineering, Sanjivani College of Engineering, Savitribai Phule Pune University, Kopargaon-423603, Maharashtra, India

[2]Professor, Department of Structural Engineering, Sanjivani College of Engineering, Savitribai Phule Pune University, Kopargaon-423603, Maharashtra, India

[a]tamnar1991@gmail.com, [b]attu_sayyad@yahoo.co.in

Keywords: Trigonometric Shear Deformation Theory, Functionally Graded Shells, Porosity, Static Response

Abstract. In this study, a static response of FGM shells containing even distribution of porosity is investigated using a trigonometric shear deformation theory accounts for effects of transverse shear and normal strains. The principal of virtual work is used for obtaining governing equations and boundary conditions of the current theory. The theory satisfies zero transverse shear stress conditions at the top and the bottom surfaces of the shell. The simply-supported FGM shell is analyzed in the present study using the Navier method. The present results of displacements and stresses in FGM shells are obtained and compared with other higher order theories available in the literature to verify the current theory.

Introduction

Functionally graded materials (FGMs) is a new class of material which has many applications in the field of aerospace, civil, mechanical, biomedical, chemical, nuclear, mining, and power plant industries. Functionally graded materials has many advantages such as smooth stress distribution, less stress concentration and high joint strength of different materials. The porosities inside FGM can occur during the fabrication and lead to occurrence of micro-voids in the material and therefore reduction the density of the material and ultimately strength of the material. Therefore, many researchers have worked on static problems of FGM shells considering the effects of porosity. Kirchhoff [1] and Mindlin [2] have developed the classical shell theory (CST) and the first-order shear deformation theory (FSDT) respectively for the analysis of shear deformable beams, plates, and shells. But these theories are assumption based theories, and not appropriate for the accurate analysis of thick plates/shells made up of advanced composite materials such as FGM. Therefore, researchers have developed refined higher-order shell theories which capture the bending behaviour accurately and predicts safe design of the composite shells. Sayyad and Ghugal [3] have presented static and free vibration analysis of FGM shells with double curvature using various types of higher-order shell theories uisng a generalized shear deformation theory. Shinde and Sayyad [4, 5] have developed a new fifth-order shear and normal deformation theory for the static and vibration analysis of FGM sandwich plates and shells. Yan and Zu [6] focused on the large amplitude vibration of sigmoid functionally graded materials thin plates with porosities. Wattanasakulpong and Ungbhakorn [7] studied the vibration properties of FGM porous beams using the differential transformation approach with various types of elastic supports. Yan and Zu [8] have studied the vibration behaviors of rectangular FGM plates with porosities and moving in thermal environment.Wang et al. [9] performed the vibration analysis of longitudinal traveling FGM porous plates considering even and uneven porosity distributions. Zhu et al. [10] presented the static and dynamic responses of functionally graded material pipes with porosities and geometric imperfections. Zenkour [11] studied the generalized shear deformation theory for bending analysis of functionally graded plates. Dharan et al. [12] studied a higher order shear deformation model for functionally graded plates. Hadji et al. [13] performed bending and free vibration analysis for FGM plates containing various distribution shapes of porosity.

Geometric configuration and material properties

For the sake of theoretical formulation, a spherical shell element on a rectangular platform in the Cartesian coordinate system (*x, y, z*) is taken into consideration. A shell has thickness h in the z-direction and curved dimensions a and b in the x and y directions, respectively. The principle radii of curvature along the x and y axes of the mid-plane are represented by R_1 and R_2, respectively. The transverse load $q(x, y)$ is applied to the upper surface of the shell i.e. $z = -h/2$.

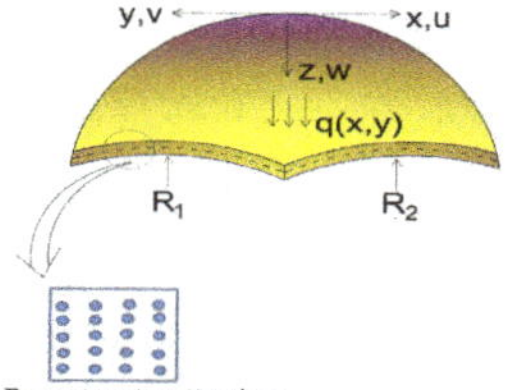

Fig 1. Geometric configuration and porosity distribution of FGM shells

In this study, the authors have considered an FGM shell with the homogenous porosity distribution (($\alpha << 1$). The modified mixture rule proposed by Wattanasakulpong and Ungbhakornb [7] is as follows.

$$P(z) = (P_c - P_m)\left(\frac{z}{h}+\frac{1}{2}\right)^p + P_m - (P_c + P_m)\frac{\alpha}{2}. \quad (1)$$

where, $P(z)$ represents the material properties, p is the power-law index that takes values greater than or equals to zero; α is the porosity distribution factor. The FGM shell becomes a fully ceramic when $p = 0$ and fully metal when $p = \infty$. The FGM shell is called perfect when $\alpha = 0$ and imperfect when $\alpha \neq 0$. The Eq. (1) for the modulus of elasticity of the material is modified as.

$$E(z) = (E_c - E_m)\left(\frac{z}{h}+\frac{1}{2}\right)^p + E_m - (E_c + E_m)\frac{\alpha}{2}. \quad (2)$$

Development of theory

The displacement field of the current trigonometric shear deformation theory for the FGM shell is as follows.

$$\begin{aligned} u(x,y,z) &= \left(1+\frac{z}{R_1}\right)u_0(x,y) - z\frac{\partial w_0}{\partial x} + \frac{h}{\pi}\sin\left(\frac{\pi z}{h}\right)\theta_x \\ v(x,y,z) &= \left(1+\frac{z}{R_2}\right)v_0(x,y) - z\frac{\partial w_0}{\partial y} + \frac{h}{\pi}\sin\left(\frac{\pi z}{h}\right)\theta_y \quad (3) \\ w(x,y,z) &= w_0(x,y) + c_1\cos\left(\frac{\pi z}{h}\right)\theta_z. \end{aligned}$$

where *u, v, w* are the displacements of any point of the shell in *x, y, z* directions respectively; (u_0, v_0, w_0) are the mid-plane displacements in *x, y, z* directions; $(\theta_x, \theta_y, \theta_z)$are the shear rotations; $\frac{h}{\pi}sin\frac{\pi z}{h}$ represent the shape function associated with the realistic distribution of transverse shear strain across the thickness of the shell; c_1 is the arbitrary constant included to consider or not to consider the effects of transverse normal strain. Using the linear theory of elasticity, the normal and shear strains associated with the current theory are obtained

$$\begin{gathered}
\varepsilon_x = \left(\frac{\partial u_0}{\partial x} + \frac{w_0}{R_1}\right) - z\frac{\partial^2 w_0}{\partial x^2} + f(z)\frac{\partial \theta_x}{\partial x} + C_1\frac{f'(z)}{R_1}\theta_z \\
\varepsilon_y = \left(\frac{\partial v_0}{\partial y} + \frac{w_0}{R_2}\right) - z\frac{\partial^2 w_0}{\partial y^2} + f(z)\frac{\partial \theta_y}{\partial y} + C_1\frac{f'(z)}{R_2}\theta_z \\
\varepsilon_z = c_1 f''(z)\theta_z \\
\gamma_{xy} = \frac{\partial u_0}{\partial y} + \frac{\partial v_0}{\partial x} - 2z\frac{\partial^2 w_0}{\partial x \partial y} + f(z)\left(\frac{\partial \theta_x}{\partial y} + \frac{\partial \theta_y}{\partial x}\right) \\
\gamma_{xz} = f'(z)\theta_x + f'(z)\frac{\partial \theta_x}{\partial x} \\
\gamma_{yz} = f'(z)\theta_y + f'(z)\frac{\partial \theta_z}{\partial y}.
\end{gathered} \tag{4}$$

where

$$f(z) = \frac{h}{\pi}\sin\left(\frac{\pi z}{h}\right),\ f'(z) = \cos\left(\frac{\pi z}{h}\right),\ f''(z) = -\sin\left(\frac{\pi z}{h}\right)\left(\frac{\pi}{h}\right). \tag{5}$$

The following Hooke's law is used to calculate the normal and transverse stresses of the FGM shell.

$$\begin{Bmatrix} \sigma_x \\ \sigma_y \\ \sigma_z \\ \tau_{xy} \\ \tau_{xz} \\ \tau_{yz} \end{Bmatrix} = \begin{bmatrix} Q_{11} & Q_{12} & Q_{13} & 0 & 0 & 0 \\ Q_{12} & Q_{22} & Q_{23} & 0 & 0 & 0 \\ Q_{13} & Q_{23} & Q_{33} & 0 & 0 & 0 \\ 0 & 0 & 0 & Q_{66} & 0 & 0 \\ 0 & 0 & 0 & 0 & Q_{44} & 0 \\ 0 & 0 & 0 & 0 & 0 & Q_{55} \end{bmatrix} \begin{Bmatrix} \varepsilon_x \\ \varepsilon_y \\ \varepsilon_z \\ \gamma_{xy} \\ \gamma_{xz} \\ \gamma_{yz} \end{Bmatrix} \tag{6}$$

where $(\sigma_x, \sigma_y, \sigma_z, \tau_{xy}, \tau_{xz}, \tau_{yz})$ represent the normal and shear stresses, $(\varepsilon_x, \varepsilon_y, \varepsilon_z, \gamma_{xy}, \gamma_{yz}, \gamma_{xz})$ represent the normal and shear strains, and Q_{ij} represents the reduced stiffness coefficients.
where

$$\begin{gathered}
Q_{11} = Q_{22} = Q_{33} = \left(\frac{E(1-\mu)}{(1+\mu)(1-2\mu)}\right) \\
Q_{12} = Q_{13} = Q_{23} = Q_{21} = Q_{31} = Q_{32} = \frac{\mu}{(1-\mu)}\left(\frac{E(1-\mu)}{(1+\mu)(1-2\mu)}\right) \\
Q_{44} = Q_{55} = Q_{66} = \frac{1-2\mu}{2(1-\mu)}\left(\frac{E(1-\mu)}{(1+\mu)(1-2\mu)}\right).
\end{gathered} \tag{7}$$

The governing equations of the current trigonometric theory are derived by using the principle of virtual work.

$$\int_0^a \int_0^b \int_{-h/2}^{h/2} \left(\sigma_x \delta\varepsilon_x + \sigma_y \delta\varepsilon_y + \sigma_z \delta\varepsilon_z + \tau_{xy}\delta\gamma_{xy} + \tau_{xz}\delta\gamma_{xz} + \tau_{yz}\delta\gamma_{yz}\right) dx\, dy\, dz - \int_0^a \int_0^b q(x,y)\,\delta w\, dx\, dy = 0 \tag{8}$$

Substituting the values of strains and stresses from Eq. (4) and Eq. (6) into Eq. (8), integrating Eq. (8) by parts and collecting the coefficients of unknown variables, the following six governing equations are derived.

$$\begin{gathered}
\delta u_0 : A_{12}\left(\frac{\partial^2 u_0}{\partial x^2} + \frac{1}{R_1}\frac{\partial w_0}{\partial x}\right) - B_{11}\frac{\partial^3 w_0}{\partial x^3} + C_{11}\frac{\partial^2 \theta_x}{\partial x^2} + A_{12}\left(\frac{\partial^2 v_0}{\partial x \partial y} + \frac{1}{R_1}\frac{\partial w_0}{\partial x}\right) - B_{12}\frac{\partial^2 w_0}{\partial x \partial y^2} \\
+ \left(\frac{f_{11}}{R_1} + \frac{f_{12}}{R_2}\right) c_1 \frac{\partial \theta_z}{\partial x} + c_{12}\frac{\partial^2 \theta_y}{\partial x \partial y} + D_{13}\, c_1 \frac{\partial \theta_z}{\partial x} + A_{66}\left(\frac{\partial^2 u_0}{\partial y^2} + \frac{\partial^2 v_0}{\partial x^2}\right) - 2\, B_{66}\frac{\partial^3 w_0}{\partial x \partial y^2} + C_{66}\left(\frac{\partial^2 \theta_x}{\partial y^2} + \frac{\partial^2 \theta_y}{\partial x \partial y}\right)
\end{gathered} \tag{9}$$

$$\begin{gathered}
\delta v_0 : A_{21}\left(\frac{\partial^2 u_0}{\partial x \partial y} + \frac{1}{R_1}\frac{\partial w_0}{\partial y}\right) - B_{21}\frac{\partial^3 w_0}{\partial x^2 \partial y} + C_{21}\frac{\partial^2 \theta_x}{\partial x \partial y} + A_{22}\left(\frac{\partial^2 v_0}{\partial y^2} + \frac{1}{R_2}\frac{\partial w_0}{\partial y}\right) - B_{22}\frac{\partial^3 w_0}{\partial y^3} + \\
\left(\frac{F_{21}}{R_1} + \frac{F_{22}}{R_2}\right) c_1 \frac{\partial \theta_z}{\partial y} + c_{22}\frac{\partial^2 \theta_y}{\partial y^2} + D_{23}\, c_1 \frac{\partial \theta_z}{\partial y} + A_{66}\left(\frac{\partial^2 u_0}{\partial x \partial y} + \frac{\partial^2 v_0}{\partial x^2}\right) - 2\, B_{66}\frac{\partial^3 w_0}{\partial x^2 \partial y} + C_{66}\left(\frac{\partial^2 \theta_x}{\partial x \partial y} + \frac{\partial^2 \theta_y}{\partial x^2}\right)
\end{gathered} \tag{10}$$

$$\begin{aligned}
&\partial w_0: B_{11}\left(\frac{\partial^3 u_0}{\partial x^3}+\frac{1}{R_1}\frac{\partial^2 w_0}{\partial x^2}\right)-H_{11}\frac{\partial^4 w_0}{\partial x^4}+I_{11}\frac{\partial^3\theta_x}{\partial x^3}+\left(\frac{J_{11}}{R_1}+\frac{J_{12}}{R_2}\right)c_1\frac{\partial^2\theta_z}{\partial x^2}\\
&+B_{12}\left(\frac{\partial^3 v_0}{\partial x^2\partial y}+\frac{1}{R_2}\frac{\partial^2 w_0}{\partial x^2}\right)-H_{12}\frac{\partial^4 w_0}{\partial x^2\partial y^2}+I_{12}\frac{\partial^3\theta_y}{\partial x^2\partial y}+K_{13}c_1\frac{\partial^2\theta_z}{\partial x^2}+B_{21}\frac{\partial^3 u_0}{\partial x\,\partial y^2}-H_{21}\frac{\partial^4 w_0}{\partial x^2\,\partial y^2}\\
&+I_{21}\frac{\partial^3\theta_x}{\partial x\,\partial y^2}+B_{21}\frac{1}{R_1}\frac{\partial^2 w_0}{\partial y^2}+J_{21}\frac{c_1}{R_1}\frac{\partial^2\theta_z}{\partial y^2}+B_{22}\frac{\partial^3 v_0}{\partial y^3}-H_{22}\frac{\partial^4 w_0}{\partial y^4}+I_{22}\frac{\partial^3\theta_y}{\partial y^3}+B_{22}\frac{1}{R_2}\frac{\partial^2 w_0}{\partial y^2}\\
&+J_{22}\frac{c_1}{R_2}\frac{\partial^2\theta_z}{\partial y^2}+K_{23}c_1\frac{\partial^2\theta_z}{\partial y^2}+2B_{66}\frac{\partial^3 u_0}{\partial x\,\partial y^2}+2B_{66}\frac{\partial^3 v_0}{\partial x^2\,\partial y}-4H_{66}2B_{66}\frac{\partial^4 w_0}{\partial x^2\,\partial y^2}+2C_{66}\frac{\partial^3\theta_x}{\partial x\partial y^2}\\
&+2C_{66}\frac{\partial^3\theta_y}{\partial x^2\partial y}+\left(-\frac{A_{11}}{R_1}\left(\frac{\partial u_0}{\partial x}+\frac{w_0}{R_1}\right)\right)+\frac{B_{11}}{R_1}\frac{\partial^2 w_0}{\partial x^2}-\frac{C_{11}}{R_1}\frac{\partial\theta_x}{\partial x}-\frac{1}{R_1}\left(\frac{F_{11}}{R_1}+\frac{F_{12}}{R_2}\right)c_1\theta_z-\\
&\frac{A_{12}}{R_1}\left(\frac{\partial v_0}{\partial y}+\frac{w_0}{R_2}\right)+\frac{B_{12}}{R_1}\frac{\partial^2 w_0}{\partial y^2}-\frac{C_{12}}{R_1}\frac{\partial\theta_y}{\partial y}-\frac{D_{13}}{R_1}c_1\theta_z+\left(-\frac{A_{21}}{R_1}\left(\frac{\partial u_0}{\partial x}+\frac{w_0}{R_1}\right)\right)+\frac{B_{21}}{R_2}\frac{\partial^2 w_0}{\partial x^2}\\
&-\frac{C_{21}}{R_2}\frac{\partial\theta_x}{\partial x}-\frac{1}{R_2}\left(\frac{F_{21}}{R_1}+\frac{F_{22}}{R_2}\right)c_1\theta_z-\frac{A_{22}}{R_2}\left(\frac{\partial v_0}{\partial y}+\frac{w_0}{R_2}\right)+\frac{B_{22}}{R_2}\frac{\partial^2 w_0}{\partial y^2}-\frac{C_{22}}{R_2}\frac{\partial\theta_y}{\partial y}-\frac{D_{23}}{R_2}c_1\theta_z=-q
\end{aligned} \tag{11}$$

$$\begin{aligned}
&\delta\theta_x: C_{11}\left(\frac{\partial^2 u_0}{\partial x^2}+\frac{1}{R_1}\frac{\partial w_0}{\partial x}\right)-I_{11}\frac{\partial^2 w_0}{\partial x^3}+L_{11}\frac{\partial^2\theta_x}{\partial x^2}+C_{12}\left(\frac{\partial^2 v_0}{\partial x\partial y}+\frac{1}{R_2}\frac{\partial w_0}{\partial x}\right)-I_{12}\frac{\partial^3 w_0}{\partial x\partial y^2}\\
&+L_{12}\frac{\partial^2\theta_y}{\partial x\,\partial y}+N_{13}\frac{\partial\theta_z}{\partial x}+C_{66}\left(\frac{\partial^2 u_0}{\partial y^2}+\frac{\partial^2 v_0}{\partial x\partial y}\right)-2I_{66}\frac{\partial^3 w_0}{\partial x\,\partial y^2}+C_{66}\left(\frac{\partial^2\theta_x}{\partial y^2}+\frac{\partial^2\theta_y}{\partial x\partial y}\right)-\\
&O_{55}\theta_x-O_{55}\frac{\partial\theta_z}{\partial x}+\frac{\partial\theta_z}{\partial x}\left(\frac{M_{11}}{R_1}+\frac{M_{12}}{R_2}\right)-\frac{F_{55}}{R_1}u_0=0
\end{aligned} \tag{12}$$

$$\begin{aligned}
&\delta\theta_y: C_{21}\left(\frac{\partial^2 u_0}{\partial x\,\partial y}+\frac{1}{R_1}\frac{\partial w_0}{\partial y}\right)-I_{21}\frac{\partial^3 w_0}{\partial x^2\partial y}+L_{21}\frac{\partial^2\theta_x}{\partial x\partial y}+C_{22}\left(\frac{\partial^2 v_0}{\partial y^2}+\frac{1}{R_2}\frac{\partial w_0}{\partial y}\right)-I_{22}\frac{\partial^3 w_0}{\partial y^3}\\
&+\frac{\partial\theta_z}{\partial y}\left(\frac{M_{21}}{R_1}+\frac{M_{22}}{R_2}\right)c_1+L_{12}\frac{\partial^2\theta_y}{\partial y^2}+N_{23}c_1\frac{\partial\theta_z}{\partial y}+C_{66}\left(\frac{\partial^2 u_0}{\partial x\partial y}+\frac{\partial^2 v_0}{\partial x^2}\right)-2I_{66}\frac{\partial^3 w_0}{\partial x^2\,\partial y}\\
&+L_{66}\left(\frac{\partial^2\theta_x}{\partial x\,\partial y}+\frac{\partial^2\theta_y}{\partial x^2}\right)-O_{44}\theta_y-O_{44}c_1\frac{\partial\theta_z}{\partial y}=0
\end{aligned} \tag{13}$$

$$\begin{aligned}
&\delta\theta_z: O_{55}\left(\frac{\partial\theta_x}{\partial x}c_1+c_1^2\frac{\partial\theta_z}{\partial x^2}\right)+O_{44}\left(\frac{\partial\theta_y}{\partial y}c_1+c_1^2\frac{\partial\theta_z}{\partial y^2}\right)+\left(-\frac{F_{11}}{R_1}c_1\left(\frac{\partial u_0}{\partial x}+\frac{w_0}{R_1}\right)\right)+\frac{J_{11}}{R_1}c_1\frac{\partial^2 w_0}{\partial x^2}\\
&-\frac{M_{11}}{R_1}c_1\frac{\partial\theta_x}{\partial x}-\frac{F_{12}}{R_1}c_1\left(\frac{\partial v_0}{\partial y}+\frac{w_0}{R_2}\right)+\frac{J_{12}}{R_1}c_1\frac{\partial^2 w_0}{\partial y^2}-\frac{1}{R_1}\left(\frac{O_{11}}{R_1}+\frac{O_{12}}{R_2}\right)c_1^2\theta_z-\frac{M_{12}}{R_1}c_1\frac{\partial\theta_y}{\partial y}\\
&-\frac{P_{13}}{R_1}c_1^2\theta_z-\frac{M_{22}}{R_2}c_1\frac{\partial\theta_y}{\partial y}-\frac{P_{23}}{R_2}c_1^2\theta_z+\left(-D_{13}c_1\left(\frac{\partial u_0}{\partial x}+\frac{w_0}{R_1}\right)\right)+K_{31}c_1\frac{\partial^2 w_0}{\partial x^2}-N_{31}c_1\frac{\partial\theta_x}{\partial x}\\
&-D_{32}c_1\left(\frac{\partial v_0}{\partial y}+\frac{w_0}{R_2}\right)+K_{32}c_1\frac{\partial^2 w_0}{\partial y^2}-\left(\frac{P_{31}}{R_1}+\frac{P_{32}}{R_2}\right)c_1^2\theta_z-N_{32}c_1\frac{\partial\theta_y}{\partial y}-S_{33}c_1^2\theta_z=0
\end{aligned} \tag{14}$$

where

$$\begin{aligned}
&\left(A_{ij}, B_{ij}, HH_{ij}, C_{ij}, F_{ij}\, I_{ij}\right)=Q_{ij}\int_{-h/2}^{h/2}[1,\ z, z^2, f(z), f'(z), zf(z)]\,dz,\\
&\left(L_{ij}\right)=Q_{ij}\int_{-h/2}^{h/2}\{[f(z)]^2\}\ dz,\ \left(O_{ij}\right)=Q_{ij}\int_{-h/2}^{h/2}[f'(z)]^2\,dz,\\
&\left(D_{ij}, S_{ij}, P_{ij}\right)=Q_{ij}\int_{-h/2}^{h/2}\{f''(z), [f''(z)]^2, [f''(z)f'(z)]\}\,dz,\\
&\left(KK_{ij}, N_{ij}\right)=Q_{ij}\int_{-h/2}^{h/2}f''(z)[z, f(z)]\,dz,\ \text{and}\ \left(J_{ij}, M_{ij}\right)=Q_{ij}\int_{-h/2}^{h/2}f'(z)[z, f(z)]\ dz
\end{aligned} \tag{15}$$

Static analysis using the Navier method

According to the literature, the Navier method is the simplest and most popularly used semi-analytical method for the analysis of simply-supported boundary conditions of FGM shell. As per the assumption in the Navier method, the following solution form for the unknown variables is assumed which satisfying the simply-supported boundary conditions exactly.

$(u_0,\ \theta_x)=(u_{mn}, \theta_{xmn})\cos\lambda x\ \sin\beta y$

$$(v_0, \theta_y) = (v_{mn}, \theta_{ymn})\sin\lambda x \cos\beta y \tag{16}$$
$$(w_0, \theta_z) = (w_{mn}, \theta_{zmn})\sin\lambda x \sin\beta y.$$

where $u_{mn}, v_{mn}, w_{mn}, \theta_{xmn}, \theta_{ymn}, \theta_{zmn},$ are the unknown coefficients to be determine; $\lambda = \frac{m\pi}{a}, \beta = \frac{n\pi}{b}$. The expression for the transverse sinusoidal load is expressed as.

$$q(x, y) = q_0 \sin\lambda x \sin\beta y. \tag{17}$$

where q_0 is Fourier coefficient of load. Substitution of Eqs. (16) and (17) into Eqs. (9) – (14) leads to the six simultaneous equations which are written in the following matrix form.

$$[K]\{\Delta\} = \{f\} \tag{18}$$

where $[K]$ is the stiffness matrix, $\{f\}$ is the force vector and $\{\Delta\}$ is the vector of unknowns.

Numerical Result and Discussion

In this study, static analysis of simply-supported FGM porous shell is presented using the trigonometric shear deformation theory. The effects of homogenous porosity distribution on the static response of FGM shell are investigated. In order to verify the current theory, comparisons of the present results are done with those available in the literature. A FGM shell is made up of the following material.

Metal (Aluminium, Al): $E_m = 70\ GPa, \mu=0.3 E_m = 70\ GPa$

Ceramic (Alumina, Al_2O_3): $E_c = 380\ GPa$, $\mu=0.3$

The numerical results are presented in the following dimensionless form for the comparison purpose.

$$\bar{w} = 10\frac{E_c h^3}{q_0 a^4} w\left(\frac{a}{2}, \frac{b}{2}\right), \bar{u} = 100\frac{E_c h^3}{q_0 a^4} u\left(0, \frac{b}{2}, \frac{-h}{4}\right), \bar{v} = 100\frac{E_c h^3}{q_0 a^4} w\left(\frac{a}{2}, 0, \frac{-h}{6}\right)$$
$$\bar{\sigma}_x = \frac{h}{a q_0}\sigma_x\left(\frac{a}{2}, \frac{b}{2}, \frac{h}{2}\right), \bar{\sigma}_y = \frac{h}{a q_0}\sigma_y\left(\frac{a}{2}, \frac{b}{2}, \frac{h}{3}\right), \tag{19}$$
$$\bar{\tau}_{xz} = \frac{h}{a q_0}\tau_{xz}\left(0, \frac{b}{2}, 0\right), \bar{\tau}_{yz} = \frac{h}{a q_0}\tau_{yz}\left(\frac{a}{2}, 0, \frac{h}{6}\right).$$

Table 1 shows the comparison of in-plane and transverse displacements of FGM plates and porous shells subjected to sinusoidal load. Table 1 examine the effects of the power-law index and the porosity distribution factor on the dimensionless displacements. Table 1 reveals that the increase in the power-law index and the porosity distribution factor increases the values of dimensionless displacements. The present theory is in close agreement with the existing literature. Fig. 2 plots through-the-thickness variations of in-plane displacements for different values of the power-law index and the porosity distribution factor. Table 2 shows comparison of in-plane normal stresses of FGM plates and porous shells subjected to sinusoidal loading. Examination of Table 2 shows that increase in the values of the power-law index and the porosity factor also increases the values of dimensionless in-plane stresses. Fig. 3 shows through-the-thickness distributions of in-plane stresses in FGM porous shells for different power-law index whereas Fig. 4 shows through-the-thickness distributions of in-plane stresses in FGM porous shells for different porosity distribution factors.

Materials Research Proceedings 31 (2023) 76-85 https://doi.org/10.21741/9781644902592-9

Table 1 Effects of the power-law index and the porosity distribution factor on the dimensionless displacements of a FGM plates and shells subjected to sinusoidal loading (a/h = 10, R/a = 1)

p	Theory	α	Plate (R_1=R_2 =∞)		Spherical Shells (R_1=R_2 =R)	
			$\overline{u}$	$\overline{w}$	$\overline{u}$	$\overline{w}$
0	Present	$\alpha = 0$	0.21810	0.29603	0.20586	0.77894
		$\alpha = 0.1$	0.23150	0.31710	0.23152	0.20951
		$\alpha = 0.2$	0.24651	0.34265	0.23323	0.90182
	Hadji et al. [13]	$\alpha = 0$	0.21816	0.29604	-	-
		$\alpha = 0.1$	0.23189	0.31468	-	-
		$\alpha = 0.2$	0.24746	0.33581	-	-
	Dharan et al. [12]	$\alpha = 0$	0.21805	0.29423	-	-
	Zenkour [11]	$\alpha = 0$	0.23090	0.29600	-	-
1	Present	$\alpha = 0$	0.64097	0.58810	0.40649	0.34539
		$\alpha = 0.1$	0.77030	0.68758	0.46073	0.38848
		$\alpha = 0.2$	0.96276	0.85487	0.53045	0.44553
	Hadji et al. [13]	$\alpha = 0$	0.64112	0.58893	-	-
		$\alpha = 0.1$	0.77156	0.68318	-	-
		$\alpha = 0.2$	0.96748	0.81924	-	-
	Dharan et al. [12]	$\alpha = 0$	0.64258	0.59059	-	-
	Zenkour [11]	$\alpha = 0$	0.66260	0.58890	-	-
2	Present	$\alpha = 0$	0.89746	0.75714	0.54258	1.99248
		$\alpha = 0.1$	1.18013	0.96435	0.64490	0.52566
		$\alpha = 0.2$	1.69307	1.51949	0.77297	0.65214
	Hadji et al. [13]	$\alpha = 0$	0.89793	0.75733	-	-
		$\alpha = 0.1$	1.18383	0.94196	-	-
		$\alpha = 0.2$	1.73521	1.28005	-	-
	Dharan et al. [12]	$\alpha = 0$	0.90220	0.76697	-	-
	Zenkour [11]	$\alpha = 0$	0.92810	0.75730	-	-
5	Present	$\alpha = 0$	1.06620	0.91216	0.73865	2.40040
		$\alpha = 0.1$	1.50131	1.34216	0.93113	3.53200
		$\alpha = 0.2$	2.77676	1.65803	1.40101	4.36323
	Hadji et al. [13]	$\alpha = 0$	1.06620	0.91171	-	-
		$\alpha = 0.1$	1.52547	1.19970	-	-
		$\alpha = 0.2$	2.70313	1.87542	-	-
	Dharan et al. [12]	$\alpha = 0$	1.06786	0.94325	-	-
	Zenkour [11]	$\alpha = 0$	1.11580	0.91180	-	-

Table 2 Effects of the power-law index and the porosity distribution factor on the dimensionless in-plane normal stresses of a FGM plates and shells subjected to sinusoidal loading (a/h = 10, R/a = 1)

			Plate (R_1=R_2=∞)		Spherical Shells (R_1=R_2=R)	
p	Theory	α	$\bar{\sigma}_x$	$\bar{\sigma}_y$	$\bar{\sigma}_x$	$\bar{\sigma}_y$
0	Present	$\alpha = 0$	1.99370	1.31090	0.87764	0.04216
		$\alpha = 0.1$	1.99700	1.31040	0.87618	0.04506
		$\alpha = 0.2$	2.00150	1.30920	0.87391	0.04956
	Hadji et al. [13]	$\alpha = 0$	1.99515	1.31219	-	-
		$\alpha = 0.1$	1.99515	1.31219	-	-
		$\alpha = 0.2$	1.99515	1.31219	-	-
	Dharan et al. [12]	$\alpha = 0$	1.98915	1.31035	-	-
	Zenkour [11]	$\alpha = 0$	1.99550	1.31210	-	-
1	Present	$\alpha = 0$	3.27700	1.31090	1.34432	0.04290
		$\alpha = 0.1$	3.27130	1.51480	1.39291	0.04330
		$\alpha = 0.2$	3.55570	1.55050	1.45108	0.03950
	Hadji et al. [13]	$\alpha = 0$	3.08640	1.48950	-	-
		$\alpha = 0.1$	3.26288	1.51850	-	-
		$\alpha = 0.2$	3.51847	1.56040	-	-
	Dharan et al. [12]	$\alpha = 0$	3.07011	1.48935	-	-
	Zenkour [11]	$\alpha = 0$	3.08700	1.48940	-	-
2	Present	$\alpha = 0$	3.60550	1.38750	1.66029	0.05110
		$\alpha = 0.1$	3.99110	1.41052	1.76702	0.04740
		$\alpha = 0.2$	4.86129	1.44234	1.88560	0.05180
	Hadji et al. [13]	$\alpha = 0$	3.60856	1.39575	-	-
		$\alpha = 0.1$	3.96831	1.41036	-	-
		$\alpha = 0.2$	4.61670	1.43564	-	-
	Dharan et al. [12]	$\alpha = 0$	3.58089	1.39680	-	-
	Zenkour [11]	$\alpha = 0$	3.60940	1.39540	-	-
5	Present	$\alpha = 0$	4.24726	1.06810	2.24739	0.03880
		$\alpha = 0.1$	4.88748	1.01340	2.48838	0.03880
		$\alpha = 0.2$	5.60382	0.94690	2.97257	0.03970
	Hadji et al. [13]	$\alpha = 0$	4.24758	1.10329	-	-
		$\alpha = 0.1$	4.74916	1.03851	-	-
		$\alpha = 0.2$	5.78994	0.88676	-	-
	Dharan et al. [12]	$\alpha = 0$	4.19547	1.10870	-	-
	Zenkour [11]	$\alpha = 0$	4.24880	1.10290	-	-

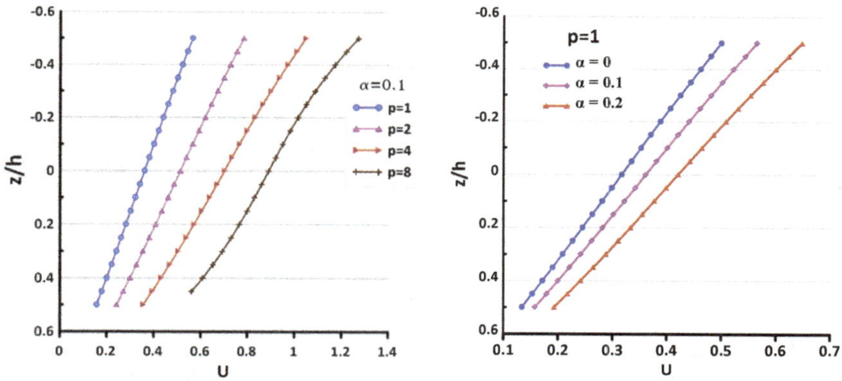

Fig. 2. Effects of the power-law index and the porosity distribution factor on the in-plane displacement of FGM spherical shell (*a/h*=10, *R/a*=1)

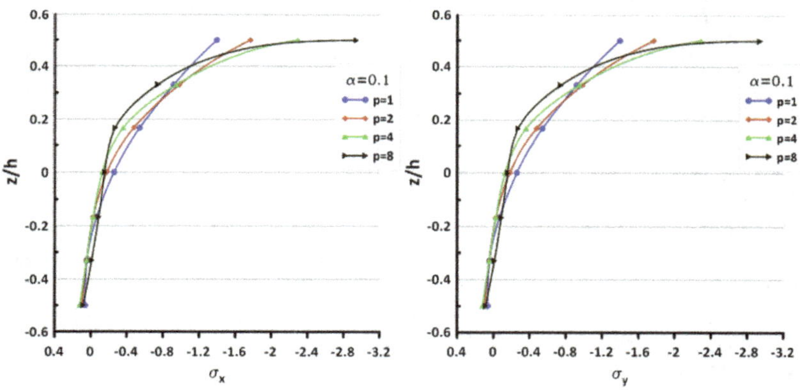

Fig. 3. Effects of the power-law index on the through-the-thickness distributions of in-plane normal stresses for FGM spherical shell (a/h=10, R/a=1).

Conclusions

This paper focused on investigating a static response of FGM porous shells using a trigonometric shear deformation theory accounts for effects of transverse shear and normal strains. The simply-supported FGM shell is analyzed in the present study using the Navier method. Based on the numerical results and the discussion it is concluded that the present theory is accurate enough to capture the static bending response of FGM shells with homogenous porosity effects. It is also concluded that the values of dimensionless displacements and in-plane stresses are directly proportional to the power-law index as well as the porosity distribution factor.

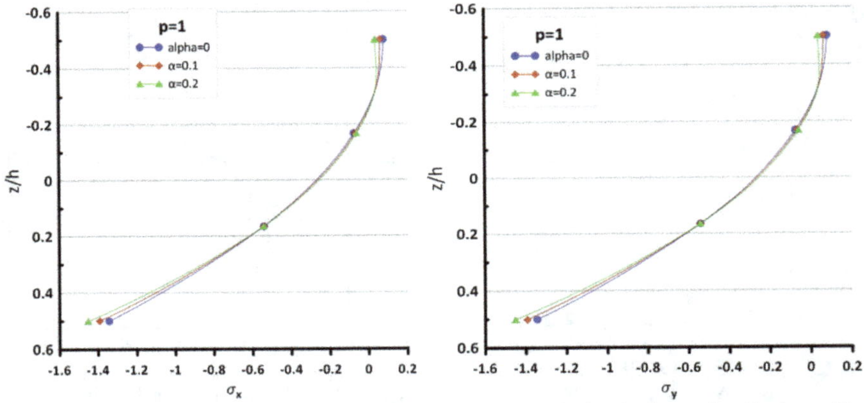

Fig. 4. Effects of the porosity distribution factor on the through-the-thickness distributions of in-plane normal stresses for FGM spherical shell (a/h=10, R/a=1)

References

[1] G. R. Kirchhoff, Uber das Gleichgewicht und die Bewegung einer Elastischen Scheibe, J. fur Reine Angew. Math. 40(1850) 51-88. https://doi.org/10.1515/crll.1850.40.51

[2] R. D. Mindlin, Influence of rotatory inertia and shear on flexural motions of isotropic elastic plates, J. Appl. Mech. 18 (1951) 31-38. https://doi.org/10.1115/1.4010217

[3] A. S. Sayyad and Y. M. Ghugal, Static and free vibration analysis of doubly-curved functionally graded material shells, Compos. Struct. 269 (2021) 1-17. https://doi.org/10.1016/j.compstruct.2021.114045

[4] B. M. Shinde and A. S. Sayyad, A new higher order theory for the static and dynamic responses of sandwich FG plates, Comput. Mech 52(1) (2021) 102-125.

[5] B. M. Shinde and A. S. Sayyad, A new higher order shear and normal deformation theory for FGM sandwich shells, Compos. Struct. 280 (2021) https://doi.org/10.1016/j.compstruct.2021. 114865

[6] Y. Q. Wang and J. W. Zu, Large amplitude vibration of sigmoid functionally graded thin plates with porosities, THIN WALL STRUCT 119 (2017) 911-924.https://dx.doi.org/10.1016 /j.tws.2017.08.012

[7] N. Wattanasakulponga and V. Ungbhakornb, Linear and nonlinear vibration analysis of elastically restrained ends FGM beams with porosities, Aerosp Sci Technol 32(1) (2014) 111-120. https://doi.org/10.1016/j.ast.2013.12.002

[8] Y. Q. Wang, J.W. Zu, Vibration behavior of functionally graded rectangular plates with porosities and moving in thermal environment, Aerosp Sci Technol 69 (2017) 550-562. https://doi.org/10.1016/j.ast.2017.07.023

[9] Y. Q. Wang, H. Wan, Y.F. Zhang, Vibration of longitudinally traveling functionally graded material plates with porosities, Eur. J. Mech. 66 (2017) 55-58. https://doi.org/10.1016/j.euromechsol.2017.06.006

[10] B. Zhu, X. C. Chen, Y. Guo, Y. Li, Static and dynamic characteristics of the post-buckling of fluid conveying porous functionally graded pipes with geometric imperfections, IJMS 189(2020) 1-57 https://doi.org/10.1016/j.ijmecsci.2020.

[11] A. M. Zenkour, Generalised shear deformation theory for bending analysis of functionally graded plates, Appl. Mathe. Modell.30 (2006) 67-84. https://doi.org/10.1016/j.apm.2005.03.009

[12] S. Dharan, S. Prakash, V. S. Savithri, A higher order shear deformation model for functionally graded plates, ICTT (2010).

[13] L. Hadji, F. Bernard, A. Safa and A. Tounsi, Bending and free vibration analysis for FGM plates containing various distribution shape of porosity Adv Mat Res.10 (2) (2021) 115-135. https://doi.org/10.12989/amr.2021.10.2.115

Assessment of recent methods for determination of soil final settlement using field data

Mohamed Aymen Dali[1,a *], Ahcène Benamghar[1,b]

[1] Ecole Nationale Supérieure des Travaux Publics, Laboratoire LTPiTE, Algiers, Algeria

[a]ma.dali@enstp.edu.dz, [b]a.benamghar@gmail.com

Keywords: Soil Settlement, Coefficient of Consolidation, Hyperbolic Method, Asaoka's Method, Guo, Prediction

Abstract. The problem of soil consolidation was first studied by Terzaghi who formulated the well-known one-dimensional consolidation theory. However, this widely used method was found to provide poorly accurate estimations, since Terzaghi considered the assumption of constant permeability and consolidation coefficient. The poor accuracy is also due to the low resemblance between laboratory results and field actual measurements. In order to overcome this issue, many researchers proposed new approaches to estimate and predict soil settlement more accurately. Among these approaches, field-based methods are particularly promising. For instance, the Asaoka method (1978), Sridharan (1987) method, Tan (1995) method, Chunlin's method (2014), and recent methods of Guo et al. (2017, 2018b) and Guo et al. 2018(a,). This paper aims to assess and discuss the results of some recent methods using a field consolidation monitoring data set.

Introduction

The prediction of soil settlement has always been an issue in soil mechanics. Terzaghi (1948) formulated the famous one-dimensional consolidation theory, which revealed to be not that accurate in terms of estimating consolidation settlement, mainly because the coefficient of consolidation obtained in laboratory is usually different from the field value. Hence, many researchers attempted to propose new approaches in order to improve the accuracy of the prediction of soil settlement. In the beginning, many works focused on a way to determine a value of coefficient of consolidation that is close to the field value. For instance, the standard methods of Casagrande logarithmic of time fitting method and Taylor square root of time fitting method can be mentioned. More recently, promising approaches based on field monitoring data were proposed. The idea of using real field settlement data has been explored by Asaoka (1978), Sridharan (1987), Tan (1995), Chunlin (2014), Guo & Chu (2017) and Guo et al. (2018(a, b)). Asaoka's method which is based on Mikasa's consolidation theory uses the concept of linear regression. Tan's hyperbolic and Sridharan rectangular hyperbola methods are curve fitting methods that assume the curve $\boldsymbol{U_v}$ vs $\boldsymbol{T_v}$ ($\boldsymbol{U_v}$: degree of vertical consolidation, $\boldsymbol{T_v}$: vertical factor of time) as an hyperbolic curve. Chunlin's method is based on Terzaghi's 1-D consolidation theory which allows a better estimation of the settlement without using the initial consolidation. Guo & Chu (2017) and Guo et al. (2018(b)) proposed a method for predicting the soil final settlement using linear regression, by fitting the settlement curve to the Chapman–Richards equation. And more recently, Guo et al. (2018(a)) published a modification of the procedure of the hyperbolic method with a new chart.

Among all these new methods and approaches one can legitimately ask which one should be used, and in which situation.

This paper aims to show and share a comparison between Guo & Chu (2017), Guo et al. (2018(b)), and Guo et al.(2018(a)) methods, using a set of field soil settlement monitoring data. As a reference, Asaoka's method was also processed for the same data set.

Asaoka's method:
The method proposed by Asaoka (1978) is an observational procedure, in which the future settlement is predicted using previous settlement observations.
Asaoka adopted Mikasa's equation:

$$\dot{\varepsilon} = c_v \ \varepsilon_{zz} \tag{1}$$

$\varepsilon(t, z)$ is the vertical strain, where $\dot{\varepsilon} = \partial\varepsilon/\partial t$ and $\varepsilon_{zz} = \partial^2\varepsilon/\partial z^2$

By introducing two function of time, the solution of the previous equation is written as:

$$\varepsilon(t, z) = T + \frac{1}{2!}\left(\frac{z^2}{c_v}\dot{T}\right) + \frac{1}{4!}\left(\frac{z^4}{c_v^2}\ddot{T}\right) + \cdots + zF + \frac{1}{3!}\left(\frac{z^3}{c_v}\dot{F}\right) + \frac{1}{5!}\left(\frac{z^5}{c_v^2}\ddot{F}\right) + \cdots \tag{2}$$

Where c_v is the coefficient of vertical consolidation,$T = \varepsilon(t, z = 0)$ and $F = \frac{\partial}{\partial z}\ \varepsilon(t, z = 0)$.

By considering the two boundary conditions; drainage from both top and bottom boundaries, and upward drainage the following equations were derived:

For double drainage:

$$\delta + \frac{1}{3!}\left(\frac{H^2}{c_v}\dot{\delta}\right) + \frac{1}{5!}\left(\frac{H^4}{c_v^2}\ddot{\delta}\right) + \cdots = \frac{H}{2}(\bar{\varepsilon} + \underline{\varepsilon}) \tag{3}$$

Where :δ is the settlement, $\varepsilon(t, z = 0) = \bar{\varepsilon}$: constant and $\varepsilon(t, z = H) = \underline{\varepsilon}$: constant.

After neglecting the higher order differential terms, Asaoka (1978) adopted the following first order approximation:

$$\delta + c_1\ \dot{\delta} = C \tag{4}$$

Where $c_1, c_2, \dots, c_n$ and C are unknown constants

By introducing discrete time $t_j = \Delta t . j$ with Δt : constant

Eq.4 can be expressed as:

$$\delta_j = \beta_0 + \beta_1\delta_{j-1} \tag{5}$$

Where δ_j and δ_{j-1} are the settlement at time j and j-1, β_0 and β_1 are unknown parameters.

When the settlement approaches its final value δ_{ult}, a stable state is observed which can be expressed as:

$$\delta_j = \delta_{j-1} = \delta_{ult} \tag{6}$$

Form Eq.6, when the values of δ_j are plotted against the values of δ_{j-1}, the final settlement δ_{ult} can be determined by identifying graphicly the intersection of the plot δ_j vs δ_{j-1} with the straight 45° line presented by the equation (y = x).

From Eq.5 and Eq.6 The final settlement can be predicted also by finding the values of β_1 and β_0 from the extrapolated plot, β_1 corresponding to its slope, and β_0 corresponding to the intersection with the ordinate axis. And then the final settlement can be calculated by substituting the values of β_1 et β_0 in Eq.5. this leads to the equation below:

$$\delta_{ult} = \frac{\beta_0}{1-\beta_1} \tag{7}$$

Hence, the settlement at time $t = \Delta t \cdot j$ can be expressed by the following equation

$$\delta_j = \frac{\beta_0}{1-\beta_1} - \left(\frac{\beta_0}{1-\beta_1} - \delta_0\right)(\beta_1)^j \tag{8}$$

The observational method of Guo et al.(2017,2018(b)) [5], [6]:
For vertical drainage, the relationship of U_v vs T_v from Terzaghi one dimensional consolidation theory, is given as (Terzaghi et al. 1996):

$$U_v = 1 - \frac{4}{\pi^2}\sum_{n=0}^{n=\infty}\frac{2}{(2n+1)^2}\exp(-M^2T_v) \quad (9)$$

Where $M = (2n+1) \times \pi/2$, (n: integer)

T_v: The Factor of time ($T_v = c_v t/H^2$)

t : Consolidation time

H: the thickness of the layer

The equation for calculating the average degree of consolidation for pure horizontal drainage case was given by Hansbo (1981) as

$$U_h = 1 - \exp\left(-\frac{8T_h}{\mu}\right) \quad (10a)$$

Where the time factor is given by:

$$T_h = \frac{c_h t}{D^2} \quad (10b)$$

And μ:

$$\mu = \frac{n^2}{n^2-1}\log n - \frac{3n^2-1}{4n^2} \quad (10c)$$

n: drainage spacing ration D/d; D: diameter of an equivalent cylinder of soil influenced by each drain, $D = 1.13s$ for a square pattern and $= 1.05s$ for a triangular pattern (s: vertical drain spacing), d: diameter of a sand drain $d = 2(b + t')/\pi$ ($b = \text{width}, t' =$ thickness of drain cross section).

Carrillo (1942) proposed an expression of the average degree of consolidation for combined vertical and horizontal drainage U_{vh} as:

$$U = 1 - (1 - U_v)(1 - U_h) \quad (11)$$

The expression of U_{vh} is then obtain by substituting Eq.9 and Eq.10a into Eq.11 which gives:

$$U_{vh} = 1 - \sum_{n=0}^{n=\infty}\frac{2}{M^2}\exp[-M^2T_v - 2\nu_{hv}T_v] \quad (12)$$

Where: ν_{hv} is the ratio of time factors in horizontal and vertical directions, according to Guo et al. 2018(a,b), It is calculated as:

$$\nu_{hv} = \frac{4}{\mu}\frac{c_h}{c_v}\frac{H^2}{D^2} \quad (13)$$

Solving Eq.12 for different values of ν_{hv} gives the U_{vh} vs T_v curves as shown in Fig.1(a). The case of Terzaghi's one dimensional equation is represented by the curve where $\nu_{hv} = 0$.

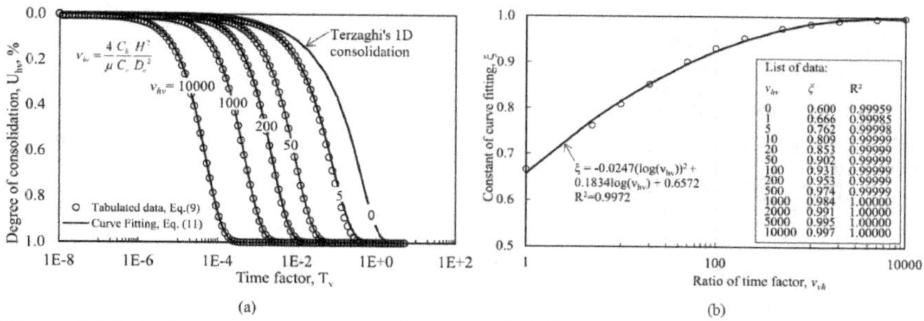

*Fig.1: **a)** Effect of ν_{hv} on the U_{vh} versus T_v curves. **b)** Effect of ν_{hv} on the constants of curve fitting ξ { Guo et al. 2018(b)}*

In this method, Guo et al. (2017 and 2018(b)) adopted the model Chapman-Richards to fit the $U_{vh} - T_v$ curve. The mathematical expression of Chapman-Richards' model can be written as:

$$y = \eta[1 - \kappa \exp(-\mu t)]^{\lambda} + \varepsilon \tag{14}$$

Where: η is the amplitude of the curve. ε is the offset from zero. κ, μ, and λ are rate constants.

Guo et al. (2017 and 2018(b)) found out that the values of κ, η, and ε are 1.0, 1.0 and 0 respectively. The value of μ is 2. Therefore, when Eq.14 is used to fit Terzaghi's consolidation curve, Eq.14 becomes of the following form:

$$U_{vh} = [1 - \exp(-2(\nu_{hv} + 1)T_v)]^{\xi} \tag{15}$$

Where ξ is a curve fitting constant.

The relationship between the constant ξ and ν_{vh} is illustrated in Fig.1(b) . It should be noted that the case where $\nu_{vh} = 0$ and $\xi = 1$, represent the case of Terzaghi's one dimensional consolidation curve as pointed out by Guo & Chu (2017).

By combining Eq.13 and Eq.15, the relationship between soil settlement and time can be written as:

$$\delta = \delta_{ult}[1 - \exp(-2(\nu_{hv} + 1)T_v)]^{\xi} \tag{16}$$

According to Guo et al. (2018b), Eq.15 was derived using excess pore pression distribution. Thus, the settlement estimations by Eq.16 are more accurate where the degree of consolidation is above 40%.

Eq.16 is then used as an observational model to predict δ_{ult} and c_v using the observed settlement data. Guo & Chu (2017) adopted the same procedure as Asaoka, selecting settlement data $\delta_1, \delta_2, \ldots, \delta_n$ at constant time intervals $\Delta t = t_{n+1} - t_n$, and then expressing the relationship between δ_{n+1} and δ_n.This yields to:

$$\delta_i = \delta_{ult}\left[1 - \exp\left(-2(\nu_{hv} + 1)\frac{c_v}{H^2}t_i\right)\right]^{\xi} \tag{17}$$

$$\delta_{i+1} = \delta_{ult}\left[1 - \exp\left(-2(\nu_{hv} + 1)\frac{c_v}{H^2}t_{i+1}\right)\right]^{\xi} \tag{18}$$

δ_i: settlement at time $t_i = \Delta t.i$ (i is an integer)

After combining Eq.17 and Eq.18 the relationship between δ_{n+1} and δ_n can be expressed as:

$$\delta_{i+1}^{1/\xi} = \alpha + \beta\delta_i^{1/\xi} \tag{19}$$

Where:

$$\alpha = (1-\beta)\rho_{ult}^{1/\xi} \tag{20}$$

$$\beta = \exp\left(-2(\nu_{hv}+1)\frac{c_v}{H^2}\Delta t\right) \tag{21}$$

From Eq.19, when the curve of $\delta_{n+1}^{1/\xi}$ vs $\delta_n^{1/\xi}$ is plotted, a straight line is obtained where β is the slope and α is the intercept. Therefore, the final settlement δ_{ult} can be estimated by the following expression:

$$\delta_{ult} = \left(\frac{\alpha}{1-\beta}\right)^{\xi} \tag{22}$$

Furthermore, the coefficient of consolidation in the vertical direction can be estimated by:

$$c_v = -\frac{H^2 \ln\beta}{2\Delta t}\frac{1}{1+\nu_{hv}} \tag{23}$$

And the coefficient of consolidation by:

$$c_h = -\frac{\mu D_e^2 \ln\beta}{8\Delta t}\frac{\nu_{hv}}{1+\nu_{hv}} \tag{24}$$

Eq.19 to Eq.24 can be used as observational model to predict the values of δ_{ult}, c_v, and c_h based on monitored settlement data. The procedure is similar to that used in Asaoka's method, it consists of the following steps:

- Calculate the ratio of time factors in horizontal and vertical directions ν_{hv}, and Identify the value of ξ from Fig.1(b)
- Select settlement data set ($\delta_1, \delta_2, ... \delta_n$) where δ_n is the settlement at time t_n, in a way that $\Delta t = t_n - t_{n-1}$ is constant
- Plot $\delta_{n+1}^{\frac{1}{\xi}}$ vs $\delta_n^{\frac{1}{\xi}}$ curve and determine the slope β and the intercept α
- Calculate the final settlement δ_{ult}, c_v, and c_h using Eq.22, Eq.23 and Eq.24

Guo & Chu(2017) also adopted the same aspect as Edil et al. (1991), where the number of samples needed to achieve 95% of consolidation j_{95} is used. And proposed N_{90} , which is defined as the number of samples needed to achieve a 90% degree of consolidation, where:

$$\left[1 - exp\left(-2(\nu_{hv}+1)\frac{c_v}{H^2}N_{90}\Delta t\right)\right]^{\xi} = 90\% \tag{25}$$

From Eq.11 and Eq.15 the number of sampling points N_{90} is $N_{90} = \frac{ln\,(1-0.9^{\frac{1}{\xi}})}{ln\,\beta}$

Although Guo & Chu (2017) found out that the sampling period doesn't have much of an effect on the prediction, yet the number of sampling points N_{90} should have a value greater than 20, in order to achieve a high value of the coefficient of regression for the least-squares linear regression.

The modified hyperbolic method of Guo et al. (2018(a)) [7]:

This method is based on the hyperbolic method developed by (Sridharan et al., 1987; Tan et al.,1991). In this method a new parameter ν_{hv} was introduced, which is defined as the ratio of time factors in the horizontal and vertical directions. This parameter is used to redefine the relationship between U_{vh} and T_v.

Let's remind that the hyperbolic approach uses the linear settlement segment produced when settlement data between U_{60} and U_{90} are plotted as t/δ vs t, this segment represent the relationship:

$$\frac{t}{\delta} = \alpha t + \beta \tag{26}$$

Where α is the slope, and β is the intercept of the linear segment between U_{60} and U_{90}, t is the consolidation time, and δ is the monitored settlement.
From Eq.26 the final consolidation settlement can be obtained as:

$$\delta_{ult} = \frac{1}{\alpha} \tag{27}$$

Furthermore, when radiating lines are drawn from the origin to U_{60} and U_{90} points, the slopes of these lines are 1/0.6 and 1/0.9 respectively. By plotting these lines, direct identifications of U_{60} and U_{90} are performed. Once δ_{60} and δ_{90} are identified from t/δ vs t plot, the estimation of the final settlement is possible by using the equations:

$$\delta_{ult} = \frac{1}{0.6} * \delta_{60} \tag{28}$$

And

$$\delta_{ult} = \frac{1}{0.9} * \delta_{90} \tag{29}$$

The advantage of the hyperbolic method is that it can be extended to clay deposits treated with vertical drains. According to Tan (1996) in reference to Hansbo (1981), for ideal drains the average degree of consolidation can be described as in Eq12a , Eq12b, and Eq12c.

Tan (1993) used the equation for the average degree of consolidation obtained from Carrillo (1942), which is expressed by Eq.11 above.

Using Terzaghi's solution for U_v , Eq.10a, Eq.10b, Eq.10c, and Eq.11, Tan (1993,1994,1995) used different documented history cases in order to produce theoretical hyperbolic settlement-time curves for any practical vertical drain problem. From these case histories. And thus, Fig.2 was obtained.

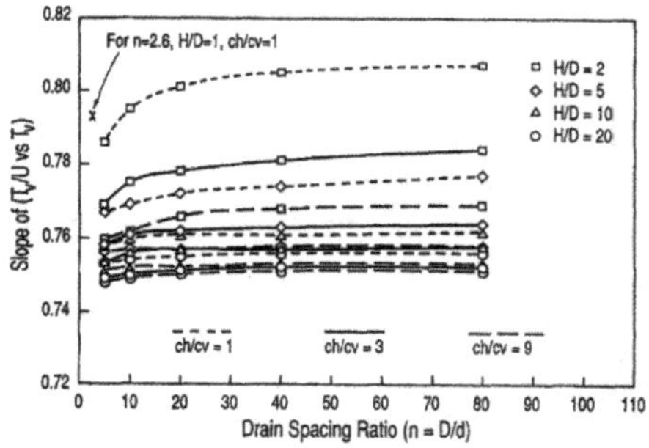

Fig. 2. *Relationship of slopes (α) of initial linear (between U_{60} and U_{90}) segments of theoretical hyperbolic plots with parameters n, H/D and c_h/c_v. {From Tan.1995}*

The slopes α for the first theoretical first linear segment between U_{60} and U_{90} are dependent on soil and drain parameters, the slopes can be obtained from Fig.2 (Tan1995). Therefore, if the initial slopes of segments of the field hyperbolic plots between U_{60} and U_{90} can be determined through observations, then the slopes of radiating lines are given by:

$$S_{60} = \frac{1}{0.6}\frac{S_i}{\lambda} \tag{30}$$

And

$$S_{90} = \frac{1}{0.9}\frac{S_i}{\lambda} \tag{31}$$

λ: the slope of the theoretical segment between U_{60} and U_{90}

S_i: the slope of the experimental segment between δ_{60} and δ_{90}

S_{60} and S_{90} are the slopes of the radiating lines used to identify δ_{60} and δ_{90}

By constructing radiating lines with the slopes described above, δ_{60} and δ_{90} can be located. Hence, the ultimate settlement can be predicted using the equations Eq.28, Eq.29 and the following equation:

$$\delta_{ult} = \frac{\lambda}{S_i} \quad (32)$$

For vertical drainage, the relationship of U_v vs T_v from Terzaghi one dimensional consolidation theory, is given as:

$$U_v = 1 - \sum_{n=0}^{n=\infty} \frac{2}{M^2} exp(-M^2 T_v) \quad (33)$$

The expression of U_{vh} is then obtained as in Eq.12, where:

$$U_{vh} = 1 - \sum_{n=0}^{n=\infty} \frac{2}{M^2} exp[-M^2 T_v - 2\nu_{hv} T_v] \quad (34)$$

Fig.2 shows plots of U_{vh} vs T_v calculated from Eq.34 with $n = 0$ *to* 50 for different values of ν_{hv}.

Just like in the hyperbolic method, the relationship of $U_{vh} - T_v$ is plotted as T_v/U_{vh} vs T_v, so λ can be identified between U_{60} and U_{90} , and thus λ can be used to predict the final settlement in the hyperbolic method. Values of λ were calculated for different cases of ν_{hv} values, Guo et al. (2018(a)) reported a curve representing the relationship λ vs ν_{hv} illustrated in Fig.3

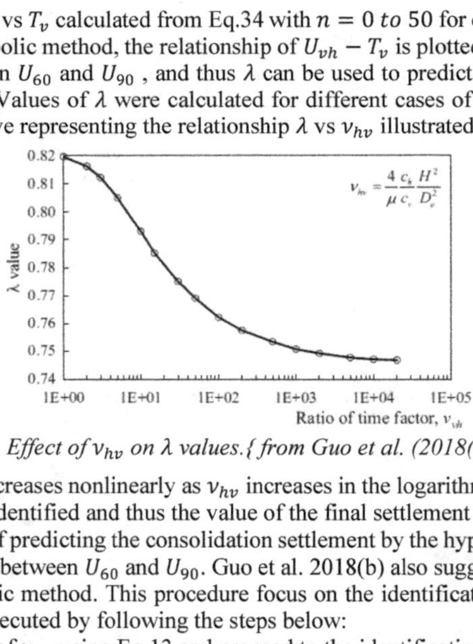

Fig. 3. *Effect of ν_{hv} on λ values.{ from Guo et al. (2018(a)) }*

Fig.3 shows that λ decreases nonlinearly as ν_{hv} increases in the logarithmic scale. By using the curve in Fig.3 λ can be identified and thus the value of the final settlement can be predicted.

Since the procedure of predicting the consolidation settlement by the hyperbolic method is only valid for settlement data between U_{60} and U_{90}. Guo et al. 2018(b) also suggested a new procedure to improve the hyperbolic method. This procedure focus on the identification of the slope of the theoretical curve, it is executed by following the steps below:

1. Calculate the value of ν_{hv} using Eq.13 and proceed to the identification of λ from Fig.3 .
2. Plot the t/δ vs t curve, select a linear segment in the curve, and by a linear regression determine the slope S_i of the linear segment.
3. Substitute the values of λ and S_i in the equations: $S_{60} = \frac{1}{0.6} \frac{S_i}{\lambda}$ and $S_{90} = \frac{1}{0.9} \frac{S_i}{\lambda}$.
4. Draw two lines from the origin with the slopes S_{60} and S_{90} and locate the intersection of these lines with the t/δ vs t curve to obtain the values of t_{60} and t_{90} corresponding to U_{60} and U_{90} respectively.

Materials Research Proceedings 31 (2023) 86-98 https://doi.org/10.21741/9781644902592-10

5. Check if the linear segment selected in step 2 falls in the range between t_{60} and t_{90} from step 4. If there is a disagreement, reselect a linear segment to get a new S_i and repeat steps 2 to 4, until a good agreement between what was selected and the determined intervals is obtained.
6. The slope achieved from the last step is the targeted slope. Calculate ρ_f by substituting the obtained λ and S_i in the Eq.31.

Case study

The three methods discussed above have been processed on a real observational soil settlement data set. The studied soil is under an embankment on the motorway penetrating Bejaia, Algeria.

The studied segment is part of the section 1 (Pk0 to Pk21+400) between the port of Bejaia and the famous Algerian east-west highway . In this particular zone, the Soumam River is limited from the east and the west, respectively, by the national road 75 (RN75), the RN12 and the railway. The route passes close to the industrial facilities of Sonatrach Petroleum Company, Bejaia airport and El Kseur city industrial area. The layout is a 2x3 lane, which runs entirely on embankments.

The section of the motorway follows the Soummam River and passes through swampy agricultural areas requiring, sometimes the use of massive backfill on soft ground. The main problem posed by these embankments is the settlement of the foundation soils. This vertical movement must be known with precision for backfill preparation and implementation. This is why a test embankment was carried out, and instrumented over a period of 360 days; on a marshy area of this section in order to study in situ the real settlement.

The Bejaia motorway test embankment was realized and instrumented for 360 days. The obtained data enabled us to carry out this comparative study in order to better understand the phenomenon of consolidation settlement, in particular, which remains the most difficult to estimate accurately.

The geotechnical in situ survey has been carried out at the Pk13+000. The foundation is mainly composed, top down, of:

- A thin layer of topsoil with a thickness of 0 to 1m.
- A layer of wet and slightly plastic clay with thickness variable from 1 to 3m
- A layer of saturated and poorly plastic clay exceeding 15m of thickness.

It should be noted that the water table is, on average, at 2.16m depth in the study area.

The settlement observations carried on the embankment started on 22/04/2014 and stopped on 17/04/2015 at this point the settlements were already stabilized at 40.64 cm. the settlement observations curve and Soil parameters are illustrated in Fig.5.

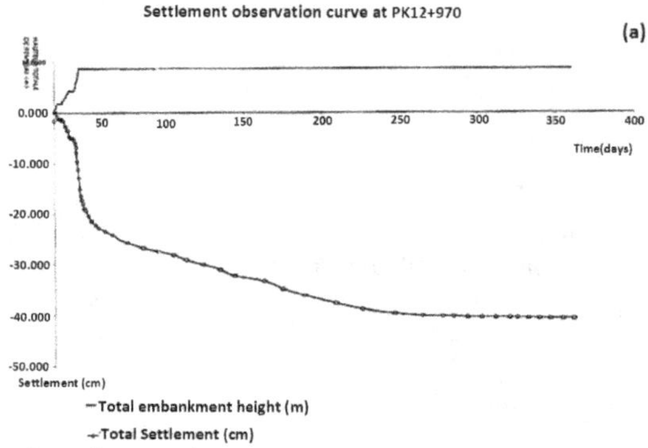

Fig. 4. *settlement observation curve at PK 12+970 (LCTP, 2013).*

Computation of stress and settlement due to embankment by Terzaghi's method:
According to the theory of Terzaghi, the total or final settlement δ_{ult} of a soil consists of primary settlement δ_p, secondary settlement δ_s and settlement induced by lateral displacements of the considered soil δ_{lat}. Also, the primary settlement has two components: an immediate settlement δ_i and a deferred settlement associated with consolidation δ_c. Hence the overall formula (Costet and Sanglerat, 1981):

$$\delta_{ult} = \delta_p + \delta_s + \delta_{lat} = \delta_i + \delta_c + \delta_s + \delta_{lat} \quad (35)$$

Where: δ_p: primary settlement, δ_s : secondary settlement, δ_{lat}: lateral displacement induced settlement, δ_i : immediate settlement, δ_c: consolidation settlement.

The consolidation settlement is calculated using the expressions from (Costet and Sanglerat, 1981), and the soil stresses are calculated using the expressions from Holtz, et al. (1981).
Taking into account the correction of Skempton and Bjerrum (1957), the calculated values of the components of settlement are: $\delta_c = 40.15cm$, $\delta_i = 12.4cm$, $\delta_s = 7.1cm$, $\delta_{lat} = 3.3cm$
Hence, the final settlement is $\delta_{Terzaghi} = 62.95\ cm$
The degree of consolidation at 360 days can be calculated using the coefficient of consolidation, as (Terzaghi, 1925):$T_v = \frac{3.28*10^{-7}}{\left(\frac{27.8}{2}\right)^2} * 360 * 24 * 3600 = 0.052$
In this case $T_v < 0.5$, hence, according to the approximation of Casagrande: $U = 26\ \%$
The settlement at 360 days, is given by:$\delta_{360} = U \times (S_c) + S_i = 0.26 \times (40.15) + 12.4 = 22.84$ cm

Predicting final settlement by the procedure of Guo et al. (2017&2018(b))
For this procedure, consecutive settlement should be considered. For this case, successive settlement measurements were taken every day, so $\Delta t = 1\ day$, therefore, Fig.5 is obtained, and then used to predict the final settlement.

For this case, no drains were installed in the field, Therefore, this is a Terzaghi one dimensional consolidation case. So, the values of v_{vh} and ξ are ($v_{vh} = 0$, and $= 0.6$). Hence, for the observation procedure, Eq.17 is used to obtain Fig.5.

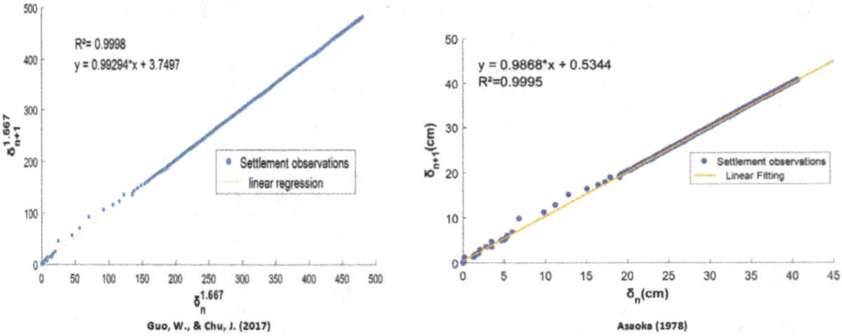

***Fig. 5.** Regression curve (using MATLAB) obtained by the application of the method of Guo & Chu (2017) and Asaoka(1978) {data up to 100% were used}*

From Fig. 5, and by a graphical identification, the line formed by settlement data for the method Guo & Chu (2017), has the following equation:

$$\delta_{n+1}^{1.667} = 0.99294\ \delta_n^{1.667} + 3.7497 \quad (36)$$

Where the slope β=0.99294, and the intercept $\alpha = 3.7497$.
the line formed by settlement data for Asaoka's method the line formed by settlement data for:

$$\delta_n = 0.9868\,\delta_{n-1} + 0.5344 \tag{37}$$

Where: $\beta_1 = 0.9868\ and\ \beta_0 = 0.5344$

From Eq.36 the predicted final settlement is 43.16 cm. comparing it to the measured settlement at 360 days indicates an over estimation of 6%.

The error was calculated as follows: $Error\% = \left(\frac{\delta_e - \delta_{real}}{\delta_{real}}\right) \times 100;\ \delta_e\text{: estimated settlement.}$

The number of sampling point N_{90} is greater than 20 which means that the condition for this procedure to achieve a higher accuracy is respected.

On the other hand, the prediction using Asaoka's method yielded a value of 40.48 cm. when compared with the measured settlement value at 360 days, the predicted value has an error of 0.4%.

Predicting settlement by the procedure of Guo et al. 2018(a)

In this method, Settlement data are plotted as the t/δ vs t. The procedure concentrates on the linear part of the obtained curve, which has the form of Eq.26.

The obtained curve and the selected linear segment are illustrated in Fig.6 **(a)**. The segment has the following expression: $Y = 0.02489\,x + 0.8448$.

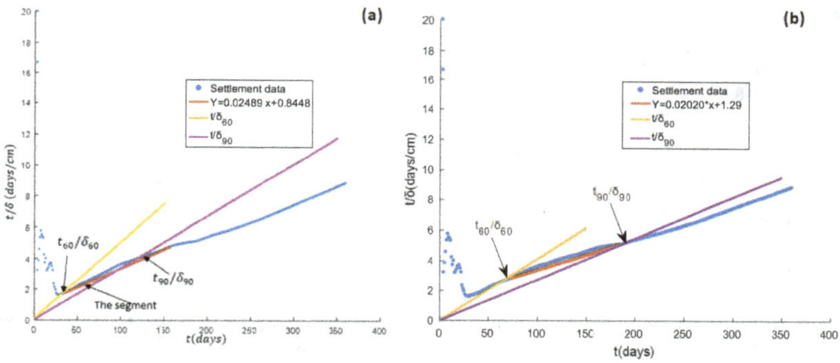

***Fig. 6.** curve of the suggested procedure for the hyperbolic method (Guo et al. 2018) used to predict the final settlement **(a)** first attempt **(b)** final phase*

In this case of embankments, no vertical drains were installed. So, the slope λ of the theoretical curve would be equal to that of the case of the one-dimensional consolidation of Terzaghi which is according to Guo et al. 2018(a) equal to 0.82. having the value of λ will allow us to construct the lines needed to locate δ_{60} and δ_{90}. And thus Fig.6 (a) is obtained, where it's clear that the segment doesn't fall in the range of δ_{60} and δ_{90} which requires a reselection of the segment as suggested in the procedure of the method.

After few attempts of the reselection of the segment a successful attempt curve is obtained (Shown in Fig.6**(b)**), in which the selected segment has a slope equal to 0.02020, it is located between U_{60} and U_{90}, where the lines constructed to locate δ_{60} and δ_{90} have the slopes $S_{60} = 0.0410$ and $S_{90} = 0.0273$. Hence, the final settlement can be predicted using either Eq.30, Eq.31, or Eq.32. So:

$$\delta_{ult} = \frac{\lambda}{S_i} = \frac{0.82}{0.0202} = 40.59cm \tag{38}$$

Results and discussions

Table 1 shows the comparison results between the methods processed above. It shows the predicted final settlement with the associated error compared to field value. Also, the intermediate settlement at (360 days) estimates for each method are given.

***Table 1:** results of the prediction of the final settlement and estimation of settlement after 360 days by the methods of (Terzaghi (1948), Asaoka (1978), Guo, & Chu (2017,2018b), Guo et al. (2018a)*

Methods	Terzaghi (1948)	Asaoka (1978)	Guo et al. [5]&[6]	Guo et al. [7]	Field final settlement δ_{ult}
Predicted δ_{ult} (cm)	62.95	40.48	43.16	40.59	40.64
Error % $\left\|\frac{\delta_{estimated}-\delta_{real}}{\delta_{real}}\right\| \times 100$	54.89%	0.4%	6.2%	0.1%	-
Settlement at 360 days (estimated) cm	22.84	40.12	41.10	34.47	-
Measured settlement at 360 days	40.64 cm				

For the estimation of the settlement at 360 days Eq. 37 was used for Asaoka's method by considering the settlement at 10 days as a starting point, Eq.16 for Guo et al (2017,2018b), and for Guo et al. 2018 (a) the following expression was used:

$$\delta(t) = \frac{\lambda \times t}{\alpha \times t + \beta} \tag{39}$$

From Table 1, it can be noted that the observational methods (Asaoka (1978), Guo et al. (2017,2018b), Guo et al. (2018a) show better accuracy in predicting the final settlement than Terzaghi's method which has an error of 54.89%.

As for the observational methods, it can be noted that their errors associated with these methods are comparable, although Guo et al. 2018a is better.

Now, concerning the observational methods' results, it can be concluded that the methods of Asaoka (1978) and Guo et al. (2017,2018b) seems to be better than the method of Guo et al. 2018 (a), due to the fact that the first two methods are less demanding in terms of observational data. Indeed, these approaches are functional even if fewer measurements of settlement are available, meanwhile Guo et al. 2018a requires at least settlement observations beyond 60% of consolidation.

On the other hand, for the intermediate settlement at 360 days, the estimation by Terzaghi's method yielded value that is equivalent to 56% of the actual measured value. For the method of Guo et al. 2018 (a) it yielded a value of 34.5 cm which represents 85% of the actual settlement although the predicted final settlement is the most accurate. As for Asaoka (1978), and Guo et al. (2017,2018b) the estimated settlements at 360 days are reasonably accurate, which confirms the previous conclusion.

Conclusion

In this paper, two recent and promising methods for final soil settlement prediction were tested on a real settlement data set, and compared with the prediction by Terzaghi's, and Asaoka's method. At first, it was clear that the methods based on field data have better accuracy than those based on laboratory test in term of predicting the final soil settlement. It is also noticed that the prediction error for the final settlement by Guo et al. (2018a) method was far smaller than Guo et al. (2017, 2018b). Furthermore, for the intermediate settlement at 360 days the estimations showed that Guo et al. (2018a) yielded an underestimation, unlike the method of Guo et al. (2017, 2018b) which is found to be more accurate.

It can be concluded that the method of Guo et al. (2017, 2018b) is preferred for the estimation of intermediate settlements, whereas, Guo et al. (2018a) is preferable for the prediction of final settlements, despite its complicated procedure.

References

[1] Akira Asaoka. "Observational procedure of settlement prediction". In: Soils and foundations 18.4 (1978), pp. 87-101. https://doi.org/10.3208/sandf1972.18.4_87

[2] Nabor Carrillo. "Simple two and three dimensional case in the theory of consolidation of soils". In: Journal of Mathematics and Physics 21.1-4 (1942), pp. 1-5. https://doi.org/10.1002/sapm19422111

[3] Jean Costet et al. Cours pratique de mécanique des sols. Dunod Paris, 1969.

[4] J Michael Duncan. "Limitations of conventional analysis of consolidation settlement". In: Journal of geotechnical engineering 119.9 (1993), pp. 1333-1359. https://doi.org/10.1061/(ASCE)0733-9410(1993)119:9(1333)

[5] W Guo and J Chu. "New observational method for prediction of one-dimensional consolidation settlement". In: Géotechnique 67.6 (2017), pp. 516-522. https://doi.org/10.1680/jgeot.16.P.089

[6] Wei Guo, Jian Chu, and Wen Nie. "An observational method for consolidation analysis of the PVD-improved subsoil". In: Geotextiles and Geomembranes 46.5 (2018), pp. 625-633. https://doi.org/10.1016/j.geotexmem.2018.04.014

[7] Wei Guo, Jian Chu, and Wen Nie. "Design chart for the modified hyperbolic method". In: Soils and Foundations 58.2 (2018), pp. 511-517. https://doi.org/10.1016/j.sandf.2018.02.014

[8] S Hansbo,MJamiolkowski, and L Kok. "Consolidation by vertical drains". In: Geotechnique 31.1 (1981), pp. 45-66. https://doi.org/10.1680/geot.1981.31.1.45

[9] Robert D Holtz,William D Kovacs, and Thomas C Sheahan. An introduction to geotechnical engineering. Vol. 733. Prentice-Hall Englewood Cliffs, 1981.

[10] Chunlin Li. "A simplified method for prediction of embankment settlement in clays". In: Journal of Rock Mechanics and Geotechnical Engineering 6.1 (2014), pp. 61-66. https://doi.org/10.1016/j.jrmge.2013.12.002

[11] A Sridharan, NS Murthy, and K Prakash. "Rectangular hyperbola method of consolidation analysis". In: Geotechnique 37.3 (1987), pp. 355-368. https://doi.org/10.1680/geot.1987.37.3.355

[12] Siew-Ann Tan. "Hyperbolic method for settlements in clays with vertical drains". In: Canadian Geotechnical Journal 31.1 (1994), pp. 125-131. https://doi.org/10.1139/t94-014

[13] Siew-Ann Tan. "Ultimate settlement by hyperbolic plot for clays with vertical drains". In: Journal of geotechnical engineering 119.5 (1993), pp. 950-956. https://doi.org/10.1061/(ASCE)0733-9410(1993)119:5(950)

[14] SIEWANN TAN. "Validation of hyperbolic method for settlement in clays with vertical drains". In: Soils and Foundations 35.1 (1995), pp. 101-113. https://doi.org/10.3208/sandf1972.35.101

[15] SIEWANN TAN and SOONHOE CHEW. "Comparison of the hyperbolic and Asaoka observational method of monitoring consolidation with vertical drains". In: Soils and Foundations 36.3 (1996), pp. 31-42. https://doi.org/10.3208/sandf.36.3_31

[16] Thiam-Soon Tan, Toshiyuki Inoue, and Seng-Lip Lee. "Hyperbolicmethod for consolidation analysis". In: Journal of geotechnical engineering 117.11 (1991), pp. 1723-1737. https://doi.org/10.1061/(ASCE)0733-9410(1991)117:11(1723)

[17] Karl Terzaghi and RALPH B Peck. "Soil Mechanics". In: Engineering Practice. John Wiley and Sons, Inc., New York (1948).

[18] Karl Terzaghi, Ralph B Peck, andGholamrezaMesri. Soilmechanics in engineering practice. John Wiley & Sons, 1996.

[19] Alfredo Urzua, Charles 358 C Ladd, and John T Christian. "New approach to analysis of consolidation data at early times". In: Journal of Geotechnical and Geoenvironmental Engineering 142.10 (2016), p. 06016009. https://doi.org/10.1061/(ASCE)GT.1943-5606.0001523

[20] Yang-Ping Yao et al. "Postconstruction settlement prediction of high embankment of silty clay at Chengde airport based on one-dimensional creep analytical method: case study". In: International Journal of Geomechanics 18.7 (2018), p. 05018004. https://doi.org/10.1061/(ASCE)GM.1943-5622.0001191

The flexoelectric effect for interface cracks between two dissimilar materials

Tomas Profant[1,2,a], Jan Sladek[1,b], and Vladimir Sladek[1,c*]

[1]Institute of Construction and Architecture, Slovak Academy of Sciences, 84503 Bratislava, Slovakia

[2]Institute of Solid Mechanics, Brno University of Technology, Technicka 2, 61669 Brno, Czech Republic

[a]profant@fme.vutbr.cz, [b]jan.sladek@savba.sk, [c]vladimir.sladek@savba.sk

Keywords: Dielectric Material, Gradient Theory, Flexoelectricity, Asymptotic Solution, Mixed FEM

Abstract. It is developed the procedure of the assessment of the amplitude factors in the asymptotic solution for the interface crack between two flexoelectric materials. The stress exponents with the appropriate eigenvectors of the regular and auxiliary solutions are evaluated from the eigenvalue problem assembled from the boundary conditions prevailing at the tip of the crack. The amplitude factors of the asymptotic solution are computed from the two-state integrals in which the regular, auxiliary, and finite element solution represent the independent equilibrium states. The obtained results show the capability of the two-state integrals to extract the dominant terms of the asymptotic solution from the weak solution of the fracture problem represented by the finite element method. The amplitude factors representing the most singular terms of the asymptotic solution at the crack tip are quantities playing an important role in the problems of crack stability criterions.

Introduction

The influence of the flexoelectricity on behaviour of the interface crack between two dissimilar dielectric materials under a mechanical load is investigated in the present paper. The direct flexoelectricity is an electro-mechanical coupling phenomenon, where the electric polarization is induced by a strain gradient in dielectrics [1-3]. For a long time, the polarization has been observed in nature only in non-centrosymmetric arrangement of atoms in crystals and this effect is called as piezoelectricity. After experimental discovery of flexoelectricity, we can see that strain gradients may break the inversion symmetry in centrosymmetric crystals and electric polarization might be observed in all dielectric materials. A significant flexoelectricity is expected at the crack tip vicinity, where the strain gradients are large [2,3]. Earlier research results showed a strong influence of the strain gradients on fields at the crack tip vicinity and the finite value of the Cauchy stresses at the crack tip contrary to the classical elasticity [4], if strain gradients are considered in constitutive equations. Gourgiotis and Georgiadis [4] derived asymptotic expressions for stress and displacement fields around the crack tip within a strain gradient continuum model. Recently, authors [5] have derived asymptotic fields at the crack tip vicinity in a homogeneous material with a direct flexoelectric effect.

Layered composite structures are frequently employed in engineering to utilize excellent properties of individual constituents. Due to a poor adhesion of layers, the interface crack can often be observed there. Therefore, behaviour of interface cracks should be studied to increase the safety of layered structures. Gradient theories better describe fields at the crack tip vicinity since they consider large strain gradients counterpart to conventional continuum models. The flexoelectric phenomenon affects the mechanical fields and appeared also to be helpful in the self-repairing of the human bones [5]. In the present paper, asymptotic expressions for displacements, electric potentials, stresses at the crack tip vicinity are derived for an interface crack between two dissimilar dielectric materials under the crack mode I. The asymptotic expansion method is applied to get these expressions. The results for asymptotic field distributions are compared with numerical results obtained by the mixed finite element method (FEM) [6].

Governing equations

It is considered an isotropic dielectric solid with direct flexoelectricity effect. Because of the size dependency and electro-mechanical coupling [7], the constitutive equations must combine the infinitesimal strain tensor ε_{ij}, its gradient η_{ijk} and the electric field E_i. These are defined as $\varepsilon_{ij} = (u_{i,j} + u_{j,i})/2$, $\eta_{ijk} = \varepsilon_{ij,k}$ and $E_i = -\varphi_{,i}$, where u_i is the displacement vector and φ is the electric potential. The isotropic flexoelectric solid is characterized by the Lamé coefficients λ and μ, the length scale parameter l, two flexoelectric constants f_1 and f_2, and the dielectric permittivity κ. The following form of the constitutive equations describe all these circumstances [8],

$$\sigma_{ij} = \lambda\varepsilon_{ll}\delta_{ij} + 2\mu\varepsilon_{ij}, \tag{1}$$
$$\tau_{ijk} = l^2\left(\lambda\varepsilon_{ll,k}\delta_{ij} + 2\mu\varepsilon_{ij,k}\right) - f_1E_k\delta_{ij} - f_2E_i\delta_{jk} - f_2E_j\delta_{ik}, \tag{2}$$
$$D_i = \kappa E_i + f_1\varepsilon_{ll,i} + 2f_2\varepsilon_{li,l}, \tag{3}$$

where the Einstein notation is used, σ_{ij} is the Cauchy stress tensor, τ_{ijk} is the higher order stress tensor, D_i is the electric displacement vector and δ_{ij} is the Kronecker delta. The governing equations inside a domain Ω are [7-9]

$$\sigma_{il,l} - \tau_{ilm,lm} = 0, \tag{4}$$
$$D_{l,l} = 0 \tag{5}$$

with the associated essential boundary conditions

$$u_i = \overline{u}_i \text{ on } \partial\Omega_u, \tag{6}$$
$$\varphi = \overline{\varphi} \text{ on } \partial\Omega_\varphi \tag{7}$$
$$v_i = u_{i,l}n_l = \overline{v}_i \text{ on } \partial\Omega_v, \tag{8}$$

or the natural boundary equations

$$t_i = n_l\left(\sigma_{il} - \tau_{ilm,m}\right) - \pi_l\rho_{i,l} + \textstyle\sum_c |\rho_i(\boldsymbol{x}^c)|\delta(\boldsymbol{x} - \boldsymbol{x}^c) = \overline{t}_i \text{ on } \partial\Omega_t, \tag{9}$$
$$R_i = n_ln_m\tau_{ilm} = \overline{R}_i \text{ on } \partial\Omega_R, \tag{10}$$
$$Q = n_lD_l = \overline{Q} \text{ on } \partial\Omega_Q. \tag{11}$$

Besides modification of the classical elasticity expression for traction vector $\boldsymbol{t}_i$ in considered model used in classical elasticity, there are also additional boundary densities in our mathematical model, such as the normal derivative of displacements $\boldsymbol{v}_i$, traction vector of the higher order of stresses $\boldsymbol{R}_i$ and electric charge $\boldsymbol{Q}$. The vectors $\boldsymbol{n}_i$ and $\boldsymbol{\pi}_i$ are the unit normal and tangent, respectively, on the boundary $\boldsymbol{\partial\Omega}$. The sum in Eq. 9 means the jumps at possible corners $\boldsymbol{c}$ on the boundary $\boldsymbol{\partial\Omega}$ with magnitude

$$|\rho_i(\boldsymbol{x}^c)| = \rho_i(\boldsymbol{x}^c - 0) - \rho_i(\boldsymbol{x}^c + 0), \tag{12}$$

where $\boldsymbol{\rho_i} = \boldsymbol{n_l\pi_m\tau_{ilm}}$. It is worth to note that $\boldsymbol{\partial\Omega_u} \cup \boldsymbol{\partial\Omega_t} = \boldsymbol{\partial\Omega_v} \cup \boldsymbol{\partial\Omega_R} = \boldsymbol{\partial\Omega_\varphi} \cup \boldsymbol{\partial\Omega_Q} = \boldsymbol{\partial\Omega}$ and $\boldsymbol{\partial\Omega_u} \cap \boldsymbol{\partial\Omega_t} = \boldsymbol{\partial\Omega_v} \cap \boldsymbol{\partial\Omega_R} = \boldsymbol{\partial\Omega_\varphi} \cap \boldsymbol{\partial\Omega_Q} = \emptyset$ must be satisfied. The substituting of the constitutive equations Eq. 1-Eq. 3 to the governing equations Eq. 4 and Eq. 5 allows one to derive the following Navier-type equations, written in the vector form [8]

$$\Delta[\kappa\varphi + f\mathrm{div}(\boldsymbol{u})] = 0, \tag{13}$$
$$(\lambda + \mu)(1 - l_1^2\Delta)\mathrm{grad}\big(\mathrm{div}(\boldsymbol{u})\big) + \mu(1 - l^2\Delta)\Delta\boldsymbol{u} = 0, \tag{14}$$

where Δ is the Laplacian operator, $f = f_1 + 2f_2$ and

$$l_1^2 = l^2 + f^2\kappa^{-1}(\lambda + \mu)^{-1}. \tag{15}$$

Recall that f and l_1 are used just for abbreviation and they are not extra material coefficients.

General asymptotic solution

A standard Knein-Williams asymptotic technique [4,10,11] is employed to establish the general solution of the stress, displacement and electric field at the crack tip vicinity with consideration of large strain gradients. It is necessary to implement the polar coordinate system (r, θ) at the crack tip and consider the distance from it so small $(r \to 0)$ that only the dominant part of Eq. 14 can be retained

$$(\lambda + \mu)l_1^2\Delta\big[\mathrm{grad}\big(\mathrm{div}(\boldsymbol{u})\big)\big] + \mu l^2\Delta(\Delta\boldsymbol{u}) = 0. \tag{16}$$

The solution of Eq. 13 and Eq. 16 for displacement components and electric potential, u_r, u_θ and φ, respectively, are found as an asymptotic series of separated variable terms

$$u_r = r^p\{A_1\cos(p-1)\theta + B_1\sin(p-1)\theta + A_2\cos(p+1)\theta + B_2\sin(p+1)\theta + A_3\cos(p-3)\theta + B_3\sin(p-3)\theta\}, \tag{17}$$
$$u_\theta = r^p\{B_4\cos(p-1)\theta + A_4\sin(p-1)\theta + B_2\cos(p+1)\theta - A_2\sin(p+1)\theta + \gamma[B_3\cos(p-3)\theta - A_3\sin(p-3)\theta]\}, \tag{18}$$
$$\varphi = r^{p-1}\kappa^{-1}\{E_1\cos(p-1)\theta + F_1\sin(p-1)\theta\} + \kappa^{-1}f\mathrm{div}(\boldsymbol{u}), \tag{19}$$

where

$$\gamma = \frac{l_1^2(p+1)(\lambda+\mu)+4l^2\mu}{l_1^2(p-3)(\lambda+\mu)-4l^2\mu}. \tag{20}$$

The exponent p and the coefficients A_i and B_i are generally complex. Consequently, the solution with their complex conjugate values also satisfies Eq. 13 and Eq. 16 and the real valued solution is the superposition of both ones. It can be simply verified that the solution with exponent $p^* = 2 - p$ and corresponding coefficients A_i^* and B_i^* also satisfies Eq. 13 and Eq. 16. This solution is so-called auxiliary solution and it plays an important role in the evaluation procedure of the regular coefficients A_i and B_i.

Asymptotic solution for an interface crack

Let the interface semi-infinite crack lies along the negative x axis and the polar coordinate system (r, θ) is introduced to its tip. The material in the upper half plane $y > 0$ is denoted by the superscript I and the material in the lower half plane $y < 0$ is denoted by the superscript II. The asymptotic solutions Eq. 17 and Eq. 18 for both materials can be rewritten to the matrix form

$$\boldsymbol{u}^{I,II} = r^p\{\boldsymbol{U}_a^{I,II}(\theta;p)\boldsymbol{a}^{I,II} + \boldsymbol{U}_b^{I,II}(\theta;p)\boldsymbol{b}^{I,II}\}, \tag{21}$$

where

$$\boldsymbol{a}^{I,II} = \left[A_1^{I,II}, A_2^{I,II}, A_3^{I,II}, A_4^{I,II}, E_1^{I,II}\right]^T, \boldsymbol{b}^{I,II} = \left[B_1^{I,II}, B_2^{I,II}, B_3^{I,II}, B_4^{I,II}, F_1^{I,II}\right]^T \tag{22}$$

and

$$\boldsymbol{U}_a^{I,II}(\theta;p) = \begin{bmatrix} \cos(p-1)\theta & \cos(p+1)\theta & \cos(p-3)\theta & 0 & 0 \\ 0 & -\sin(p+1)\theta & -\gamma^{I,II}\sin(p-3)\theta & \sin(p-1)\theta & 0 \end{bmatrix}, \tag{23}$$

$$\boldsymbol{U}_b^{I,II}(\theta;p) = \begin{bmatrix} \sin(p-1)\theta & \sin(p+1)\theta & \sin(p-3)\theta & 0 & 0 \\ 0 & \cos(p+1)\theta & \gamma^{I,II}\cos(p-3)\theta & \cos(p-1)\theta & 0 \end{bmatrix}. \tag{24}$$

Similarly the electric potential φ^J from Eq. 19 and remaining densities $\boldsymbol{v}^J$, $\boldsymbol{t}^J$, $\boldsymbol{R}^J$ and Q^J in Eq. 8-Eq. 11 can be written as

$$\varphi^{I,II} = r^{p-1}\{\boldsymbol{\phi}_a^{I,II}(\theta;p)\boldsymbol{a}^{I,II} + \boldsymbol{\phi}_b^{I,II}(\theta;p)\boldsymbol{b}^{I,II}\}, \tag{25}$$
$$\boldsymbol{v}^{I,II} = r^{p-1}\{\boldsymbol{V}_a^{I,II}(\theta;p)\boldsymbol{a}^{I,II} + \boldsymbol{V}_b^{I,II}(\theta;p)\boldsymbol{b}^{I,II}\}, \tag{26}$$
$$\boldsymbol{t}^{I,II} = r^{p-3}\{\boldsymbol{T}_a^{I,II}(\theta;p)\boldsymbol{a}^{I,II} + \boldsymbol{T}_b^{I,II}(\theta;p)\boldsymbol{b}^{I,II}\}, \tag{27}$$
$$\boldsymbol{R}^{I,II} = r^{p-2}\{\boldsymbol{R}_a^{I,II}(\theta;p)\boldsymbol{a}^{I,II} + \boldsymbol{R}_b^{I,II}(\theta;p)\boldsymbol{b}^{I,II}\}, \tag{28}$$
$$Q^{I,II} = r^{p-2}\{\boldsymbol{q}_a^{I,II}(\theta;p)\boldsymbol{a}^{I,II} + \boldsymbol{q}_b^{I,II}(\theta;p)\boldsymbol{b}^{I,II}\}. \tag{29}$$

The vectors $\boldsymbol{\phi}_{a,b}^{I,II}(\theta;p)$, $\boldsymbol{q}_{a,b}^{I,II}(\theta;r)$ and matrices $\boldsymbol{V}_{a,b}^{I,II}(\theta;p)$, $\boldsymbol{T}_{a,b}^{I,II}(\theta;p)$, $\boldsymbol{R}_{a,b}^{I,II}(\theta;p)$ are not shown here due to their complexity. The boundary conditions for the traction-free and isolated crack faces at $\theta = \pm\pi$ of material *I* and *II* are

$$\boldsymbol{t}^I(r,\pi) = \boldsymbol{R}^I(r,\pi) = 0 \text{ and } Q^I(r,\pi) = 0, \tag{30}$$
$$\boldsymbol{t}^{II}(r,-\pi) = \boldsymbol{R}^{II}(r,-\pi) = 0 \text{ and } Q^{II}(r,-\pi) = 0 \tag{31}$$

with compatibility conditions along the interface for $\theta = 0$

$$\boldsymbol{u}^I(r,0) = \boldsymbol{u}^{II}(0;p), \varphi^I(r,0) = \varphi^{II}(r,0), \boldsymbol{v}^I(r,0) = -\boldsymbol{v}^{II}(r,0), \tag{32}$$
$$\boldsymbol{t}^I(r,0) = -\boldsymbol{t}^{II}(r,0), \boldsymbol{R}^I(r,0) = \boldsymbol{R}^{II}(r,0), Q^I(r,0) = -Q^{II}(r,0). \tag{33}$$

The substitution of Eq. 21 and Eq. 25-Eq. 29 into Eq. 30-Eq. 33 can be written formally as follows

$$\boldsymbol{B}^{I,II}(p)\boldsymbol{g}^{I,II} = 0, \tag{34}$$
$$\boldsymbol{C}^I(p)\boldsymbol{g}^I = \boldsymbol{C}^{II}(p)\boldsymbol{g}^{II}, \tag{35}$$

where $\boldsymbol{g}^{I,II} = [\boldsymbol{a}^{I,II}, \boldsymbol{b}^{I,II}]^T$ and

$$\boldsymbol{B}^{I,II}(p) = \begin{bmatrix} \boldsymbol{T}_a^{I,II}(\pm\pi;p) & \boldsymbol{T}_b^{I,II}(\pm\pi;p) \\ \boldsymbol{R}_a^{I,II}(\pm\pi;p) & \boldsymbol{R}_b^{I,II}(\pm\pi;p) \\ q_a^{I,II}(\pm\pi;p) & q_b^{I,II}(\pm\pi;p) \end{bmatrix}, \boldsymbol{C}^{I,II}(p) = \begin{bmatrix} \boldsymbol{U}_a^{I,II}(0;p) & \boldsymbol{U}_b^{I,II}(0;p) \\ \boldsymbol{\phi}_b^{I,II}(0;p) & \boldsymbol{\phi}_b^{I,II}(0;p) \\ \boldsymbol{V}_b^{I,II}(0;p) & \boldsymbol{V}_b^{I,II}(0;p) \\ \boldsymbol{T}_a^{I,II}(0;p) & \boldsymbol{T}_b^{I,II}(0;p) \\ \boldsymbol{R}_a^{I,II}(0;p) & \boldsymbol{R}_b^{I,II}(0;p) \\ q_a^{I,II}(0;p) & q_b^{I,II}(0;p) \end{bmatrix}. \tag{36}$$

The relations Eq. 35 allow one to establish the following eigenvalue problem

$$\boldsymbol{A}(p)\boldsymbol{g}^{I} = 0, \tag{37}$$

in which

$$\boldsymbol{A}(p) = \begin{bmatrix} \boldsymbol{B}^{I}(p) \\ \boldsymbol{B}^{II}(p)[\boldsymbol{C}^{II}(p)]^{-1}\boldsymbol{C}^{I}(p) \end{bmatrix}. \tag{38}$$

The condition for the existence of the nontrivial solution of Eq. 37 is

$$\det[A(p)] = 0, \tag{39}$$

which is the characteristic equation for the complex exponent p. It is worth to note that the most singular admissible real value of the exponent p is $\Re\{p\} = 3/2$ or $1 < p < 2$ in the case of real p. This is due to the requirement of the finite value of the potential energy of the cracked bimaterial. The eigenvalue $p = 1$ also satisfies Eq. 39, but the corresponding lower-order terms Eq. 17-Eq. 18 and their counterparts Eq. 1-Eq. 3 do not contribute to the J-integral and to the crack opening displacement [4]. Taking into account the lower-order and the most singular terms, the regular solution for displacement and electric field for material I and II and for $r \to 0$ can be written as

$$\begin{aligned}\boldsymbol{u}^{I,II} &= r\{\Gamma_1^{I,II} + \Gamma_2^{I,II}\cos 2\theta + \Gamma_3^{I,II}\sin 2\theta\} + H_0 r^{p_0}\{\boldsymbol{U}_a^{I,II}(\theta;p_0)\boldsymbol{a}_0^{I,II} + \boldsymbol{U}_b^{I,II}(\theta;p_0)\boldsymbol{b}_0^{I,II}\} + \\ &\textstyle\sum_{i=1}^{4} H_i r^{p_i}\{\boldsymbol{U}_a^{I,II}(\theta;p_i)\boldsymbol{a}_i^{I,II} + \boldsymbol{U}_b^{I,II}(\theta;p_i)\boldsymbol{b}_i^{I,II}\} = r\{\Gamma_1^{I,II} + \Gamma_2^{I,II}\cos 2\theta + \Gamma_3^{I,II}\sin 2\theta\} + \\ &\textstyle\sum_0^4 H_i \boldsymbol{G}_i^{I,II}(r,\theta;p_i),\end{aligned} \tag{40}$$

$$\begin{aligned}\varphi^{I,II} &= \kappa^{-1}\Gamma_0^{I,II} + H_0 r^{p_0-1}\{\boldsymbol{\phi}_a^{I,II}(\theta;p_0)\boldsymbol{a}_0^{I,II} + \boldsymbol{\phi}_b^{I,II}(\theta;p_0)\boldsymbol{b}_0^{I,II}\} + \\ &\textstyle\sum_{i=1}^{4} H_i r^{p_i-1}\{\boldsymbol{\phi}_a^{I,II}(\theta;p_i)\boldsymbol{a}_i^{I,II} + \boldsymbol{\phi}_a^{I,II}(\theta;p_i)\boldsymbol{b}_i^{I,II}\} + (\kappa^{I,II})^{-1} f^{I,II}\mathrm{div}(\boldsymbol{u}^{I,II}) = \kappa^{-1}\Gamma_0^{I,II} + \\ &\textstyle\sum_{i=0}^{4} H_i F_i^{I,II}(r,\theta;p_i),\end{aligned} \tag{41}$$

where H_0 and H_i are unknow amplitude factors. Except the exponent $p_0 = 3/2$, the remaining ones p_i can be real or complex with their conjugates. The eigenvectors $\boldsymbol{a}_i^{I,II}$ and $\boldsymbol{b}_i^{I,II}$ for $i = 0,1,\dots,4$ are obtained from the eigenvalue problem Eq. 37 and consequently from the relation Eq. 35 after the back substitution of the associated exponents p_i. The factors $\Gamma_i^{I,II}$ for $i = 0,\dots,3$ are evaluated from the finite values of the electric potential and components of the Cauchy stresses given by Eq. 1 at the tip of the crack as follows [12]

$$\Gamma_0^{I,II} = (\kappa^{I,II})^{-1}\varphi^{I,II}|_{r=0} - 2f^{I,II}\Gamma_1^{I,II}, \tag{42}$$

$$\Gamma_1^{I,II} = \frac{(\sigma_{xx}^{I,II} + \sigma_{yy}^{I,II})|_{r=0}}{4(\lambda^{I,II} + \mu^{I,II})},\ \Gamma_2^{I,II} = \frac{(\sigma_{xx}^{I,II} - \sigma_{yy}^{I,II})|_{r=0}}{4\mu^{I,II}},\ \Gamma_3 = \frac{\sigma_{xy}^{I,II}|_{r=0}}{2\mu^{I,II}}. \tag{43}$$

Although the case $p < 1$ is excluded from the regular solution, the so-called auxiliary solutions including the exponents $p_i^* = 2 - p_i$ for $i = 0,1,\dots,4$ are also taken into account in the following and can written in the form

$$\boldsymbol{u}_i^{*I,II} = r^{p_i^*}\{\boldsymbol{U}_a^{I,II}(\theta;p_i^*)\boldsymbol{a}_i^{*I,II} + \boldsymbol{U}_b^{I,II}(\theta;p_i^*)\boldsymbol{b}_i^{*I,II}\}, \tag{44}$$

$$\varphi_i^{*I,II} = r^{p_i^*-1}\{\boldsymbol{\phi}_a^{I,II}(\theta;p_i^*)\boldsymbol{a}_i^{*I,II} + \boldsymbol{\phi}_a^{I,II}(\theta;p_i^*)\boldsymbol{b}_i^{*I,II}\} + (\kappa^{I,II})^{-1} f^{I,II}\mathrm{div}(\boldsymbol{u}^{*I,II}). \tag{45}$$

The auxiliary solutions Eq. 44 and Eq. 45 are the mathematical tool that plays a key role in the evaluation of the amplitude factors H_i that appear in the asymptotic solution Eq. 40 and Eq. 41. They can extract the magnitude of the amplitude factors H_i from the exact complete solution of the interface crack problem through the Betti contour integral. The extension of the Betti's reciprocal theorem to the direct flexoelectricity problem is derived in [13].

Two-state integrals

Assuming that the body forces are absent, the two-state integral as the consequence of the Betti reciprocal theorem [13] can be defined as

$$\mathcal{H}\left(\boldsymbol{u}^{(1)},\varphi^{(1)};\boldsymbol{u}^{(2)},\varphi^{(2)}\right)=\int_\Gamma[\boldsymbol{t}^{(1)}\cdot\boldsymbol{u}^{(2)}+\boldsymbol{R}^{(1)}\cdot\boldsymbol{v}^{(2)}-\varphi^{(1)}Q^{(2)}-\boldsymbol{t}^{(2)}\cdot\boldsymbol{u}^{(1)}-\boldsymbol{R}^{(2)}\cdot\boldsymbol{v}^{(1)}+\varphi^{(2)}Q^{(1)}]\mathrm{d}\Gamma, \tag{46}$$

where the dot symbol means the dot product of the vectors. Two sets of forces $\left(\boldsymbol{t}^{(1)},\boldsymbol{R}^{(1)},Q^{(1)}\right)$ and $\left(\boldsymbol{t}^{(2)},\boldsymbol{R}^{(2)},Q^{(2)}\right)$ act through kinematical fields $\left(\boldsymbol{u}^{(1)},\boldsymbol{v}^{(1)},\varphi^{(1)}\right)$ and $\left(\boldsymbol{u}^{(2)},\boldsymbol{v}^{(2)},\varphi^{(2)}\right)$, respectively, along the closed smooth contour Γ in a linear elastic planar body. The two-state integral Eq. 46 is path independent [13] in the case, when the path Γ encircles the crack tip. Consider the first state $\mathcal{H}\left(\boldsymbol{u}^{I,II},\varphi^{I,II};\boldsymbol{u}_i^{*I,II},\varphi_i^{*I,II}\right)$ of integral Eq. 46 is to be evaluated on the integration path shrunk to the tip of the crack, in which the exact solutions $\boldsymbol{u}^{(1)}$ and $\varphi^{(1)}$ are approximated by the asymptotic solutions $\boldsymbol{u}^{I,II}$ and $\varphi^{I,II}$ from Eq. 40 and Eq. 39, respectively. Next, the second state $\mathcal{H}\left(\boldsymbol{u}^{h},\varphi^{h};\boldsymbol{u}_i^{*I,II},\varphi_i^{*I,II}\right)$ of integral Eq. 46 is evaluated along the remote integration path Γ and the exact solution $\boldsymbol{u}^{(1)}$ and $\varphi^{(1)}$ is approximated by a finite element solution $\boldsymbol{u}^h$ and φ^h from [7]. The auxiliary solution $\boldsymbol{u}_i^{*I,II}$ and $\varphi_i^{*I,II}$ from Eq. 44 and Eq. 45 are substituted for $\boldsymbol{u}^{(2)}$ and $\varphi^{(2)}$ in both states, respectively. Then, due to the integral path independency, both states of integral Eq. 46 must give the same value [13], i.e.

$$\mathcal{H}\left(\boldsymbol{u}^{I,II},\varphi^{I,II};\boldsymbol{u}_i^{*I,II},\varphi_i^{*I,II}\right)=\mathcal{H}\left(\boldsymbol{u}^{h},\varphi^{h};\boldsymbol{u}_i^{*I,II},\varphi_i^{*I,II}\right)\text{ for } i=0,1,\dots,4. \tag{47}$$

The auxiliary solution Eq. 44 and Eq.45 appropriate to the exponent $p_i^*=2-p_i$ in the integral on the left-hand side of Eq. 47 causes that all terms of the asymptotic solutions Eq. 40 and Eq. 41 vanish except those with argument p_i. Consequently, the amplitude factor H_i can be extracted from the integral on the left-hand side of Eq. 47 and evaluated as

$$H_i=\frac{\mathcal{H}\left(\boldsymbol{u}^h,\varphi^h;\boldsymbol{u}_i^{*I,II},\varphi_i^{*I,II}\right)}{\mathcal{H}\left(\boldsymbol{G}_i^{I,II}(r,\theta;p_i),\boldsymbol{F}_i^{I,II}(r,\theta;p_i);\boldsymbol{u}_i^{*I,II},\varphi_i^{*I,II}\right)}. \tag{48}$$

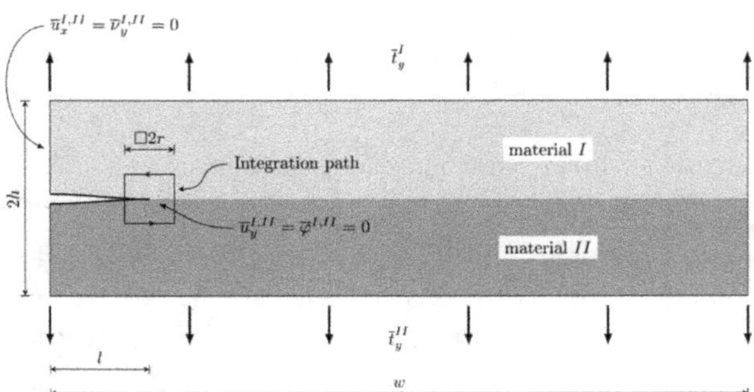

Figure 1: A symmetric part of the cracked strip under a uniform axial tension.

Table 1: The characteristics of the bimaterial

	material *I*	**material *II***
λ [Pa]	7.40×10^{10}	8.88×10^{10}
μ [Pa]	2.56×10^{10}	3.07×10^{10}
l [m]	1.00×10^{-8}	0.80×10^{-8}
κ [$C^2N^{-1}m^{-2}$]	1.50×10^{-8}	1.80×10^{-8}
f_1 [Cm^{-1}]	1.00×10^{-9}	0.80×10^{-9}
f_2 [Cm^{-1}]	0	0

Numerical results

It is considered a plane fracture problem of the of the bimaterial strip $2h \times 2w$ under the $\bar{t}_y^{I,II}$ external loadings with the interfacial crack whose length is $2l$. Due to the symmetry, the right half of the strip is taken into account only, as it is shown in Fig. 1. The dimensions of the crack and the strip are $l = h = 1 \times 10^{-7}$ m and $w = 1.25 \times 10^{-6}$ m. The material properties of the bimaterial are given in Tab. 1 and the external loading is $\bar{t}_y^{I} = \bar{t}_y^{II} = 1 \times 10^7$ Pa. The shape of the integration path in the amplitude factor evaluation procedure influences the quality of the final numerical results due to the higher-order derivatives with respect to normal $\boldsymbol{n}$ appearing in the expressions of Eq. 9. Hence, the FEM approximations of modified traction vector $\boldsymbol{t}^{I,II}$ and traction vector of the higher order of stresses $\boldsymbol{R}^{I,II}$ along the rectangular integration path, as it is shown in Fig. 1, significantly increase the accuracy of further numerical calculations.

The real and imaginary part of the characteristic function $f(p) \equiv \det[A(p)]$ for $\Re\{p\} = 3/2$ is depicted in Fig. 2, where the condition Eq. 39 is fulfilled at the intersections of both curves on the axis representing $\Im\{p\}$. These intersections give one real and two pairs of complex conjugate exponents p_i for $i = 0,1,\dots,4$, which are given in Tab. 2. The insertion of p_i into Eq. 35 and Eq. 37 leads to the evaluation of the associated eigenvectors $\boldsymbol{g}_i^{I}$ and $\boldsymbol{g}_i^{II}$ for $i = 0,1,\dots,4$ for both materials of the strip. It is worth noting that the auxiliary eigenvectors $\boldsymbol{g}_i^{*I}$ and $\boldsymbol{g}_i^{*II}$ are established in the same way as the regular ones.

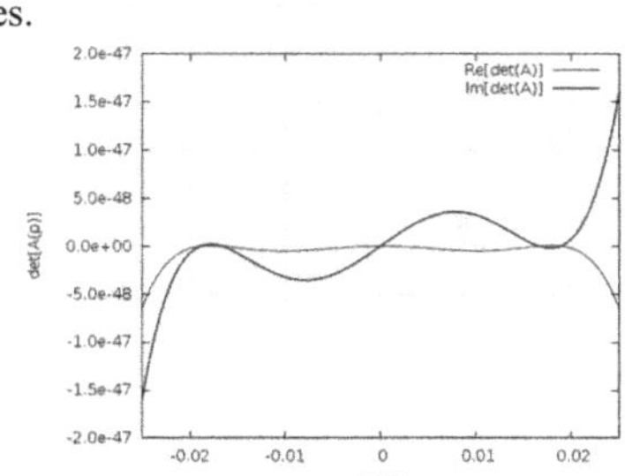

Figure 2: The real and imaginary part of the characteristic function $f(p) \equiv det[A(p)]$ for $\Re\{p\} = 3/2$.

The knowledge of exponents p_i and regular eigenvectors $\boldsymbol{g}_i^{I}$, $\boldsymbol{g}_i^{II}$ as well as the auxiliary eigenvectors $\boldsymbol{g}_i^{*I}$ and $\boldsymbol{g}_i^{*II}$ for $i = 0,1,\dots,4$ and the FEM approximations of $\boldsymbol{u}^h$ and φ^h, allows one to evaluate the amplitude factors H_i applying the formulae Eq. 48. For the chosen external loading $\bar{t}_y^{I} = \bar{t}_y^{II} = 1 \times 10^7$ Pa the appropriate values of H_i for $i = 0,1,\dots,4$ are calculated on the rectangular path Γ enclosing the crack in the distance $r = 4 \times 10^{-9}$ m from its tip, as it shows Fig. 1.

Table 2: Exponents of the regular asymptotic solution.

p_0	1.500
p_1, p_2	$1.500 \pm i1.891 \times 10^{-2}$
p_3, p_4	$1.500 \pm i1.663 \times 10^{-2}$

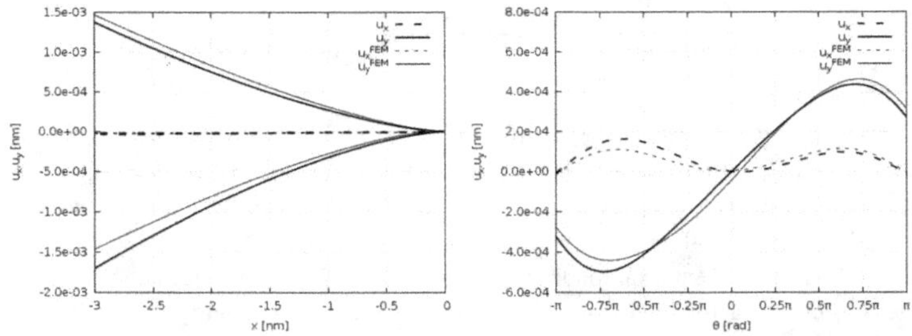

Figure 3: The crack opening displacements of the asymptotic solution and FEM solution at the right crack tip.

The comparison of the asymptotic and FEM solution of the components of the displacement vector u_x and u_y, components of Cauchy stress tensor σ_{xx}, σ_{xy} and σ_{yy} and electric potential φ in front of the crack and in the distance $r = 1 \times 10^{-9}$ m from its tip is shown in Fig. 3-Fig. 5. There is very good agreement between the asymptotic and FEM solution. On the other hand, the mismatch appears between the asymptotic and the FEM solution for increasing distance $r = x$ from the crack tip, especially for electric potential φ. This is because of the construction of the asymptotic solution Eq. 40 and Eq. 41. They are derived from the Eq. 13 and Eq. 16 at the point in whose enclosing domain the strain gradients prevail. Moreover, only the most singular terms of the asymptotic solution are taken into account. In points far from the crack tip vicinity there are significant also non-singular terms which are not considered in the asymptotic solution. The graphs in Fig. 3-Fig. 5. show that this assumption is fulfilled very close to the crack tip. The higher order terms should be added to the asymptotic solution Eq. 13 and Eq. 16 to approach the FEM solution for the larger distances r from the crack tip.

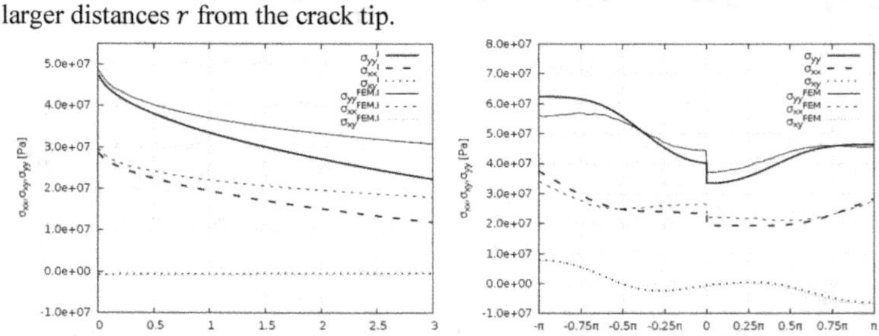

Figure 4: The component of Cauchy stress tensor of the asymptotic solution and FEM solution at the right crack tip.

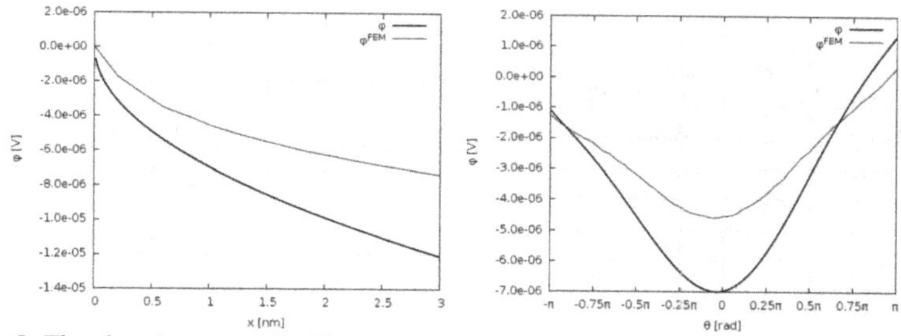

Figure 5: The electric potential of the asymptotic solution and FEM solution at the right crack tip.

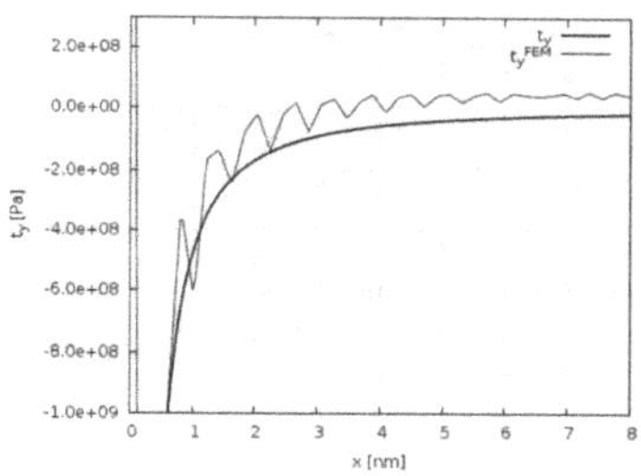

Figure 6: The component $\mathbf{t}_y$ of the modified traction vector t of the asymptotic solution and FEM solution ahead of the right crack tip.

The component $t_y = t_y^I = t_y^{II}$ of the modified traction vector $\boldsymbol{t}$ ahead of the crack tip is shown in Fig. 6. The graph depicts the strong singularity $r^{-3/2}$ of $\boldsymbol{t}$ and consequently the failing of the FEM approximation for $r = x \to 0$. Nevertheless, it is seen that the FEM solution becomes to deviate from the analytical solution at $x \approx 1$ nm because of the missing of the higher order terms in the asymptotic solution. Hence the asymptotic solutions Eq. 13 and Eq. 16 describe precisely the singular behavior of the higher order stresses very close to the crack tip and the two-state integral seems to be a very strong tool allowing the extraction of these singular terms from the comprehensive weak solution represented by the FEM.

Summary

The procedure for the assessment of the amplitude factors appearing in the asymptotic solution of the interface crack in flexoelectric bimaterial was discussed. This work follows up and expands the application of the two-state integrals in the fracture mechanics of the material governed by strain gradient elasticity and flexoelectric effect. The obtained results show the ability of the two-state integrals to extract the dominant terms of the asymptotic solution from the weak solution of the fracture problem represented by the FEM. The evaluated amplitude factors of the asymptotic solution are important parameters, which can be used as the inputs to the fracture criterions to predict the interface crack behavior in the flexoelectric materials.

Acknouwledgement

The authors acknowledge the support by the Slovak Science and Technology Assistance Agency registered under number APVV-18-0004, and the support by the Ministry of Education, Science, Research and Sport of the Slovak Republic under grant number VEGA-2/0061/20.

References

[1] P. Yudin, A. Tagantsev, Fundamentals of flexoelectricity in solids, Nanotechnology 24 (2013) 432001. https://doi.org/10.1088/0957-4484/24/43/432001

[2] H. Wang, X. Jiang, Y. Wang, R.W. Stark, van Aken, J. Mannhart, Direct observation of huge flexoelectric polarization around crack tips, Nano Letter 20 (2020) 88-94. https://doi.org/10.1021/acs.nanolett.9b03176

[3] X. Tian, M. Xu, H. Zhou, Q. Li, J. Sladek, V. Sladek, Analytical studies on mode III fracture in flexoelectric solids, Journal of Applied Mechanics 89 (2022) 041006. https://doi.org/10.1115/1.4053268

[4] P.A. Gougiotis, H.G. Georgiadis, Plane-strain crack problems in microstructured solids governed by dipolar gradient elasticity, J. Mech. Phys. Solids 57 (2009) 1898-1920. https://doi.org/10.1016/j.jmps.2009.07.005

[5] F. Vasquez-Sancho, A. Abdollahi, D. Damjanovic, G. Catalan, Flexoelectricity in bones, Advanced Materials 30 (2018) 1705316.

[6] X. Tian, J. Sladek, V. Sladek, Q. Deng, Q. Li, A collocation mixed finite element method for the analysis of flexoelectric solids, Int. J. Solids Struct. 217 (2021) 27-39. https://doi.org/10.1016/j.ijsolstr.2021.01.031

[7] J. Sladek, V. Sladek, M. Hrytsyna, T. Profant, Application of the gradient theory to interface crack between two dissimilar dielectric materials, Engineering Fracture Mechanics 276(B) (2022), 108895. https://doi.org/10.1016/j.engfracmech.2022.108895

[8] F. Deng, Q. Deng, W. Yu, S. Shen, Mixed Finite Elements for Flexoelectric Solids, Journal of Applied Mechanics 84(8) (2017) 081004. https://doi.org/10.1115/1.4036939

[9] J. Sladek, V. Sladek, P. Stanak, C. Zhang, C.-L. Tan, Fracture mechanics analysis of size-dependent piezoelectric solids, Int J Solids Struct 113-114 (2017) 1-9. https://doi.org/10.1016/j.ijsolstr.2016.08.011

[10] M.L. Williams, Stress singularities resulting from various boundary conditions in angular corners of plates in extension, Journal of Applied Mechanics 74 (1952) 526–528.

[11] M. Kotoul, T. Profant, Asymptotic solution for interface crack between two materials governed by dipolar gradient elasticity, Engineering Fracture Mechanics 201 (2018) 80-106. https://doi.org/10.1016/j.engfracmech.2018.05.002

[12] T. Profant, J. Sládek, V. Sládek. M. Kotoul, Asymptotic solution for interface crack between two materials governed by dipolar gradient elasticity: Amplitude factor evaluation, Theoretical and Applied Fracture Mechanics 120 (2022) 103378. https://doi.org/10.1016/j.tafmec.2022.103378

[13] S. Mao, P. K. Purohit, Insights Into Flexoelectric Solids From Strain-Gradient Elasticity, Journal of Applied Mechanics 81(8) (2014) 081004(10). https://doi.org/10.1115/1.4027451

Comprehensive review of soil stabilization agents

Batool Al-Shorman[1,a], Mousa Bani Baker[2,b,*], Tahar Ayadat[3,c]

[1] Department of Civil Engineering, Jordan University of Science and Technology, Irbid, Jordan

[2] Department of Civil and Infrastructure Engineering, Al-Zaytoonah University of Jordan, Amman, Jordan

[3] Department of Civil Engineering, Prince Mohammad Bin Fahd University, Al-Khobar, Saudi Arabia

[a]byalsharman17@eng.just.edu.jo, [b]m.banibaker@zuj.edu.jo, [c]tayadat@pmu.edu.sa

Keywords: Soil, Stabilization, Additives, Traditional Additives, Non-Traditional Additives, Clay, Chemical Additives, Nanomaterials

Abstract. Soil stabilization is a technique that is used in most construction projects to enhance the geotechnical and engineering properties of soil. There is a wide range of research studies related to soil stabilization techniques and agents, these studies discussed the effects of the different types of soil stabilization on soil, the most suitable agent type regarding soil classification, and the challenges that were founded during the application of these processes. Stabilization agents include traditional and non-traditional additives with their different categories were reviewed and discussed in this paper by presenting the results of the recent studies concerned with various types of soil stabilization agents in different laboratories and project tests with highlights on the enhancement of soil properties. In addition to increasing the compressive and shear strength parameters, Maximum Dry Density (MDD), and California Bearing Ratio (CBR) of the soil, the soil stabilization agents play a great role in decreasing the soil plasticity index, swelling, compressibility, porosity, permeability, and Optimum Moisture Content (OMC).

Introduction

Soil is one of the most essential and plentiful construction materials, its role cannot be ignored and should be employed acceptably. Some of the soil's engineering and geotechnical properties have to be improved to participate in the position in which it is involved such as; sub-grade and sub-base constructions, embankments and foundation constructions, road and rail constructions, backfill for retaining walls and bridges abutments, etc. Soil properties enhancement is signified by improving soil engineering properties and overcoming some of their flaws to avoid replacing them with others that satisfy the required engineering standards for the purpose it is used for. Particularly, the main engineering properties desired from the engineering material are dependent on their functions, type of load which they will withstand, and the environment which they will face. The most central soil engineering properties is soil mechanical strength (shear strength and compressive strength) which is the most important property for soil stabilization, on the other hand, soil density, permeability, durability, plasticity, and compressibility are extremely significant properties [1].

The additives which were studied in civil engineering research related to soil stabilization are categorized as traditional and non-traditional additives. Traditional additives include lime, cement, bituminous, waste/ recycled materials (e.g., shredded tire, plastic bottle strips, crushed glass, etc.), fly ash, slag, etc [2-4], While non-traditional additives consist of various combinations such as nano additives, acids, ions, liquid polymers, fibres, enzymes, resins, petroleum emulsions, silicates, and lignin derivatives. Some of the additives are more common than others, simultaneously, the type of additive added to the soil depend on many factors, such as; the soil characteristics (eg. Plasticity behaviour, organic matter content, grain size distributions, mineralogy and chemical compositions, etc.), stabilization agent characters and the quantity required, and the quality and method of construction. Furthermore, the proper use of additives including the curing time and effective compaction of soil affected the efficiency of the soil stabilization.

The main objective of this paper is to introduce a comprehensive review of the following subjects: (a) present the most common types of soil additives and their compositions, (b) provide how these stabilizer agents assist to enhance the soil geotechnical and engineering properties depending on the findings of recent researches and studies, and (c) introduce the difference between the traditional and nontraditional soil stabilization agents. This paper is provided to display the importance, processes, and types of soil additives in one paper by consolidating the recent studies on this field to be a reference for professional researchers who are concerned with this type of research to improve the knowledge of soil additives and to help in selecting the accurate type of additives for different types of soils.

General types of soil stabilization agents

Soil stabilization can be defined as the process in which natural soil and cementing materials are added to advance one or more of its engineering properties. physically mixing the desired soil with stabilizing agents together to reach a homogeneous mixture or by mixing stabilizing material to in-situ soil deposits and making interaction by letting it saturate through soil voids [5].

Stabilizing agents lay under many categories which differ from each other by their compositions, cost, availability as a natural material or artificial material, and the ease of usage due to required experts in the construction site. Table 1 shows the most common types of soil stabilizers, their composition, and if they are natural or artificial.

Table 1. *Types of soil stabilization agents.*

Stabilizer	Composition	Natural or Artificial
Cement	Combination of calcium, silicon, aluminium, iron and other ingredients	Artificial
Lime	Combination of oxides, and hydroxide, usually calcium oxide and/ or calcium hydroxide.	Natural
Glass Fibers	Consist of numerous extremely fine fibres of glass.	Artificial
Chemical Additives	Calcium, sodium hydroxide, zycobond.	Artificial
Polymers	Poly (vinyl alcohol) , Butanetetracarboxylic acid.	Artificial
Nano additives	Nano silica, nano clay.	Artificial
Organic Additives	Coal humate,peat-gel, biochar.	Natural
Recycled Materials	Plastic bottles strips, crushed glass, shredded tires.	Artificial

Selecting the accurate soil stabilization method and stabilizer may be challenging. It is essential to keep in mind that there isn't a gold set in regards to soil stabilization, this indicates that each construction or project and each soil type will have dissimilar requirements for soil stabilization. It is valuable to be acquainted with the different stabilizing agents that are on hand, then the type of soil, the project category longevity of the project, budget, and environmental concerns may help to choose the proper type of soil stabilizing agent. On the other hand, the method of providing the soil stabilizing agent to the soil may differ from one case to another. Ultrasonic dispersion, direct mixing, and solution mixing are the most frequently used methods for this purpose.
Following are explanations of the effect of traditional and non-traditional stabilizer agents on enhancing the soil properties in detail.

Traditional soil stabilization agents

Chemical soil stabilization agents

Cement

Using cement as a soil stabilizer agent is the most common product used among many projects due to its availability, cost efficiency, long-term performance record, and its effectiveness in bonding the soil particles by the hydration process, the cement grows as crystals then these crystals interlock to give a high compressive strength. To achieve a strong bond the cement particles should coat most of the soil particles and assure good contact between cement and soil particles; proficient soil cement stabilization, mixing soil and cement properly is extremely necessary [6].

The chemical reaction between the Portland cement which is considered a calcium-based stabilizer and soil (mainly clayey soil) begins once the Portland cement is mixed with soil and may take a specific time to finish, the reaction starts with hydration which may be obtained by one month after mixing, and once the concentration of (OH^-) increase the alkalinity of the soil increase and led to develop the pozzolanic reaction. The results of the pozzolanic reaction are seen by increasing the concentration of (Ca^{+2}) (which is supplied from the Portland cement stabilizer) and then reacting with (Si and Al) to create a cementitious material as calcium aluminate hydrates (CAH), calcium alumino-silicate hydrates (CASH), and calcium silicate hydrates (CSH) [7]. [8] provided a study on improving the engineering properties of soils taken from Modinagr UP, ABU PUR, in India. The soil samples were tested before and after adding cement as a soil stabilizer for the projects related to railway track base courses, sub-grade, and sub-base courses. Cement was mixed with the soil to reach the preferred properties. The experimental tests which were obtained to study the effect of soil stabilizer (cement) are; grain size distribution, Atterberg'slimits (Liquid, Plastic, and Shrinkage) limits, Proctor Compaction Test, California Bearing Ratio Test, and Direct Simple Shear test. The soil understudy was Silty Sand/ Clayey Sand (SC/SM) soil, the tests showed an improvement of the engineering soil properties after using cement agents with percents (2%, 4%, and 6%) as shown in Table 2.

Lime

Lime material is used for both soil modification and soil stabilization depending on two main reactions between the soil and lime, the first one is short-term or quick reactions and the second one is a pozzolanic or long-term reaction which needs time to present the results of improving the engineering soil properties [9]. The reaction between the soil particles and lime require water to occur accurately and the mechanism is completed within four parts; 1. Cations/ions exchange; 2. Agglomeration and flocculation; 3. Pozzolanic reaction; and 4. Carbonation. It is worth mentioning that the carbonation process is not mostly needed because it does not have a significant effect on improving the soil strength. The exothermic reaction between the soil-water system and quick lime (CaO) forms hydrated lime ($Ca(OH)_2$) and increase the workability of the soil, on the other hand, the cations and anions (Ca^{+2}, OH^-) appear after this reaction, then the cations and anions exchange with clay lattice take place to enhance the clay properties by increasing the pH due to increasing the OH^- ions. The quick reactions help to reduce the Diffuse Double Layers (DDL) which make extremely valuable effects in reducing the swelling behaviour of the clayey soil [10].

To present the effect of lime as a soil stabilization agent [11], treated an expansive soil from the Belgaum district of Karnataka state in India utilizing the lime material, the clayey soil contained quartz, montmorillonite, and aluminium oxide. Hydrated lime ($Ca(OH)_2$); which is available as a commercial soil stabilizer was employed, results on treated soil samples showed significant improvement in the engineering properties of the clayey soil, the plasticity behaviour of the clayey soil was enhanced due to reduction in the thickness of DDL because of the raise of electrolyte concentration of pore water and the substitute of monovalent cation (Na^+) with the divalent cation

(Ca^{+2}), this led to a considerable improvement in the swelling behaviour of the clayey soil. Moreover, testing the strength of the treated soil by lime presents an increase in the Unified Compression Strength (UCS) and raises the alkalinity after curing the lime-treated soil for different periods. Table 3 shows the improvement of the lime-treated soil UCS with different curing periods.

Table 2. *Cement-treated soil tests results [8].*

Test	Notes
Proctor Compaction Test	Maximum dry density increased while the optimum moisture content decreased.
Direct shear test	The cohesion and internal friction angles were as; c=1.6, Φ=38. With cement percentage (2%, 4%, 6%) the cohesion and internal friction angle were equal to, (c=1.1, Φ=40), (c=0.6, Φ=41), (c=0.6, Φ=40), representivly.
California Bearing Ratio test	Max subgrade CBR values are5.07%, 6.62%, 8.23% and 10.15% for untreated soil sample, 2%, 4%,and 6% cement contents respectively
The cement increased the bearing capacity ofSC/SM soil effectively, so it is recommended to be used for subgrade of railway track with 6% cement content or more.	

Table 3. *Strength and pH of lime-treated soil with curing periods [11].*

Mixes	Curing periods/day	UCS, kPa	pH value
Soil	-	312.04	6.04
Soil+ 6% lime	0	267.68	12.04
	7	634.38	11.78
	14	971.80	11.77
	28	1350.00	11.65
	90	1520.00	10.80
	180	1700.00	10.75
	360	1656.67	10.90

Sodium hydroxide additives

Sodium hydroxide can be described as an odourless, non-volatile, white solution. Which is a highly reactive material, its reactions with water produce heat which is sufficient to set fire to the nearby flammable materials. The useful property due to sodium hydroxide is reacting with soil containing aluminium in presence of water to increase the density of soil in small efforts [12], [13]studied the effect of sodium hydroxide as a chemical soil stabilization agent on the clayey soil rich with kaolinite clay minerals which were taken from the Odooru area in Ogbomoso, Oyo State. The tests were carried through three parts, first as dried samples, second as soaked samples, and at the end, they carried the tests on the cycling samples. Adding sodium hydroxide to the soil enhance each of the following engineering properties as shown in Table 4; compressive strength, the density of the soil, and soil porosity (soil's water absorption). The results showed that the dry specimen provide better strength than wet and cycled specimens. The water absorption less or more raises with increases in the NaOH percentage. even though the increase is not standardized and is relatively small in contrast with other variations. On the other hand, densities of all specimens increased after moulding with increased with all NaOH percentages, this results from the bonding generated by subsequent isomorphous substitution ofAluminium (Al) and alkaline solution.

Sodium silicates

[14] used multistep techniques to rehabilitate and stabilize liners (bentonite clay and sand mixtures) contaminated with hydrocarbon fluids using silicate grout solution and pretreatment under the effect of electrokinetic phenomena. Silica grout formulations were developed and adequate curing periods were established for the electro–silicatization process. Results showed that hydraulic conductivity was reduced fourfold in the case of using three-step electro

rehabilitation for alternative fuels under the pressure of 40 kPa, and reduced threefold in the case of 100 kPa pressure on a liner.

Table 4. *Tests results [13].*

NaOH (%)	Compressive strength (MPa)			Soil porosity (%)		Density (g/cm3)	
	dry	cycling	wet	Water Absorption for Wet Sample	Open Porosity for Wet Sample	Density after Moulding	Density after curing at 80°c
7	29.26	5.25	4.48	8.360	17.54	2,183	2,032
10	36.43	10.08	11.35	9.093	17.92	2,184	2,016
13	40.38	11.97	19.06	10.151	20.68	2,196	1,994
16	47.03	18.3	21.72	10.068	20.01	2,195	1,998

Recycled material (waste plastic bottle strips)

One of the beneficial usages of recycled material is to utilize the harmful material to the environment as an agent provide an important role in another place in the same environment as a reverse reaction, it is obvious that plastic materials are the most common type of materials should be recycled due to our daily massive production and usage of it as an essential need. [15] investigate the effect of adding the waste plastic bottle strips on the improvement of the soil engineering properties. The natural soil utilized in the study is from the plain of Bihar (Patna), they mixed it with plastic bottle strips (15mm*25mm strips size) which mainly consist of Polyethylene Terephthalatewhich which is produced from petroleum hydrocarbons. A series of CBR tests, Direct Simple Shear tests, and compaction tests were made on the natural soil and the reinforced soil with plastic strips, the results showed a considerable improvement of the engineering soil properties as shown in Table 5. the development in shear strength parameters is most favourable at 0.4% plastic content. The shear stress increases due to the allocation of plastic pieces in various directions along the surface of shear linking two halves of direct shear boxes.

Non-Traditional soil stabilization agents

Polymers and glass fibres

Polymers can be defined as every artificial or natural substance collected from extremely big molecules which are called macromolecules which are consisting of multiples of unpretentious chemical units (monomers). The usage of polymers increased recently due to their high-performance related to structural materials, one of the disadvantages of polymers is their sensitivity to environmental factors such as; exposure to gases, liquids, radiation, electrical field, and temperature. Polymers are one of the stabilization agents added to the soil to develop its engineering properties, their role can be summarized by the adhesion of the clayey soil particles together then the strength of the soil will be improved, the plasticity and swelling behaviour will be reduced, and the workability will be increased due to forming the flocculated particles [16], [17] studied the effect of polymers (Poly(vinyl alcohol)([-CH2CHOH-]n) which is called (PVA) and 1,2,3,4-Butanetetracarboxylic acid ($C_8H_{10}O_8$) which is called (BTCA)) on the soil samples classified as highly plastic clayey soil and the soil activity ratio is 0.5 with specific gravity equal to 2.71 from Township in Central Queensland of Australia. Adding PVA and BTCA to the clayey soil with different percentages provides a significant effect on improving the UCS and ductility of the soil; treating the clayey soil with 1% of PVA led to a reasonable raise of the UCS of the clayey soil sample with properties (dry unit weigth= 16.2 kN/m3, water content=16.8% and initial void ratio=0.64), on the other hand, by increasing the percentage of PVA up to 1.5% the UCS raise from 10 kPa to 116 kPa, this show that adding PVA as a soil stabilization agent is very effective due to its availability as an organic material and its nontoxicity. It is worth to be mentioned that the BTCA does not have any reflect on improving the soil strength but was used to make PVA insoluble in water.

Table 5. *Test results of reinforced soil with plastic contents for strip size of (15 mm 9 25 mm) [15].*

Percent of plastic content for strip size (15 * 25) mm	Compaction parameters		Shear strength Parameters		CBR Ratio
	OMC(%)	MDD(kN/m3)	c (kN/m2)	Φ(degree)	(%)
0%	16.8	16.75	19	23.1	3.3
0.2%	16.1	17.5	28	28.7	7.1
0.4%	14.2	18.4	34	32.8	16.5
0.6%	15	18.1	18	27	13.7
0.8%	16.4	17.3	13	25	12.5

The influences of fibre glass with nanocomposite are can be obvious on the clay distribution, the mixing of clay with different percentages of fibre glass gave a significant effect on the thermal and mechanical properties of the clayey soil [18]. On the other hand [19] investigated the effect of fibre glass on the durability of clay by treating it with nanophase fibre glass in different conditions; cold and wet, dry with high temperature, wet with high temperature, and cold and dry, they observed an increase of flexural strength and fibre glass modulus

Nano materials

One of the most recent common used types of soil stabilization additives are nano materials, which may be mixed directly in different percentages with the soil or used as agents with other traditional soil stabilization agents like cement, lime, bentonite clay for improvement the different engineering soil properties such as shear strength of sand [20]. Nano-Clay and Nano-Silica (pozzolanic material), are the most important nano additives, but Carbon-Nanotube, Nano-alumina, Nano-MgO, and Nano-CuO are also used as nano material soil additives and there are research studies concerned with them.

The definition of nano clay can be summarized as the layered silicates material with the double-layer thickness equal to 1nm and single-layer thickness equal to 0.7nm, the modification of the interlayer of the nano clay made for swelling and plasticity behaviour improvement. Nature of surface atoms, interlayer exchangeable cations, and nanosheet charge are the factors that have influenced the properties of the nano clay particles, the negative charge of the nano clay is due to the isomorphous substitution between the silicon and cations ($Mg^{+2}$and Al^{+3}) [21-23]. [24] treated soft soil with three types of nano materials which are; nano clay, nano MgO, and nano CuO, they studied the effect of the nano materials on the clayey soil behaviour and properties such as; Atterberg's limits, shear strength, and compaction parameters (MDD, OMC). The soil engineering properties are enhanced by adding the nano materials (there was a reduction in plasticity index and OMC, and increase in the MDD), this improvement was obvious up to exceeding the optimum contents then the clay particles agglomeration which affects negatively. furthermore, adding nano materials with a percent not more than 1% has a significant effect on improving the compressive strength of the tested clayey soil. Clay materials such as bentonite clay are also used to improve and stabilize various engineerings properties of other engineering materials such as asphalt concrete mixtures concrete, and celleular concrete by the addition of clay in optimum amounts which improved stability of asphalt concrete, its resistance to rapid freeze and thaw, and improved the compressive and flexural strengths of concrete and celleular concrete mixtures [24,26].

Conclusions

This paper presents a comprehensive review of a range of soil stabilization agents. The following conclusions can be drawn:

1. The two major types of soil stabilization agents can be classified into traditional and non-traditional soil stabilization agents which have a large effect by mixing them with the soil in different amounts and efforts, every additive enhances a specific or more than one specific property of the soil some was used to increase the strength of the soil and others were concerned with other soil properties.
2. Chemical soil stabilization agents (cement, lime, and sodium hydroxide, silicates) provide a considerable enhancement on the soil properties (mainly on the clayey soil), but the Portland cement stabilizer offers the best strength properties in a short curing time in compassion with the chemical soil stabilization agents, because of its rapid hydration reaction.
3. Recycled materials additives enhance the strength of the soil by reinforcing the soil particles. Other recycled materials may be used such as; cardboard, cereal boxes, aluminium cans, steel cans, and used gallons.
4. Polymers and Nano-materials such as; Nano-Clay and Nano-Silica, provide a significant effect on the soil strength properties, permeability, and compressibility of the soil by enhancing the microstructure of the soil, because of their high cation exchange capacity and their high specific surface area which help them to interact with soil particles.

References

[1] Behnood, A. (2018). Soil and clay stabilization with calcium- and non-calcium-based additives: A state-of-the-art review of challenges, approaches and techniques. Transportation Geotechnics. doi:10.1016/j.trgeo.2018.08.002. https://doi.org/10.1016/j.trgeo.2018.08.002

[2] B.V., Rao, V., Reddy, M., Muttharam, 2001. The impact of cyclic wetting and drying on the swelling behaviour of stabilized expansive soils.Eng. Geolo. 60(14), 223-233. https://doi.org/10.1016/S0013-7952(00)00103-4

[3] Z., Nalbantoglu, 2004. Effectiveness of Class C fly ash as an expansive soil stabilizer. Construction and Building Materials 18(6), 377-381. https://doi.org/10.1016/j.conbuildmat.2004.03.011

[4] R.N., Yong, V.R., Ouhadi, 2007. Experimental study on instability of bases on natural and lime/cement-stabilized clayey soils. Applied Clay. https://doi.org/10.1016/j.clay.2006.08.009

[5] Abood, Tamadher T, Anuar Bin Kasa and Zamri Bin Chik, 2007. "Stabilization of Silty ClaySoil Using Chloride Compounds" Int J Eng 2: 102-103.

[6] Habiba, Afrin, 2017. "A Review on Different Types of Soil Stabilization Techniques" Int JTranspEng Tech 3: 19-24. https://doi.org/10.11648/j.ijtet.20170302.12

[7] Phoo-ngernkham T, Hanjitsuwan S, Li L, Damrongwiriyanupap N, Chindaprasirt P, 2018. Adhesion characterisation of Portland cement concrete and alkali-activated binders. Forthcoming in AdvCem Res:1-11.

[8] Saksham, S, Jagdish Y and Rajat P, 2018. "Analysis of Stabilization of Soil Cement for Base of Railway Track and Subgrade" Int J EngRes: 2321-9939.

[9] S.K., Dash, M. Hussain,2011. Lime stabilization of soils: reappraisal. J Mater CivEng 24(6):707-714. https://doi.org/10.1061/(ASCE)MT.1943-5533.0000431

[10] N.C., Consoli, LDS. Jr., Lopes, PDM.,Prietto, L .Festugato, R.C., Cruz, 2010. Variables controlling stiffness and strength of lime-stabilized soils. J GeotechGeoenvironEng 137(6):628-632. https://doi.org/10.1061/(ASCE)GT.1943-5606.0000470

[11] Jha, A. K., &Sivapullaiah, P. V. (2019). Lime Stabilization of Soil: A PhysicoChemical and Micro-Mechanistic Perspective. Indian Geotechnical Journal. doi:10.1007/s40098-019-00371-9. https://doi.org/10.1007/s40098-019-00371-9

[12] Olaniyan, O. S., 2008. Stabilization of kaolinitc soil from Nigeria for construction purposes.Unpublished Soil Stabilization Practical Exercise, Vrije Universiteit Brussels.

[13] Olaniyan, O.S., Olaoye, R.A, Okeyinka, O.M, and Olaniyan, D.B, 2011. Soil Stabilization Techniques Using Sodium Hydroxide Additives. International Journal of Civil & Environmental Engineering IJCEE-IJENS Vol: 11 No: 06.

[14] Bani Baker, M., Elektorowicz, M., Hanna, A. "Electrokinetic nondestructive in-situ technique for rehabilitation of liners damaged by fuels". Volume 359 (2018), Pages 510-515. https://doi.org/10.1016/j.jhazmat.2018.07.113. https://doi.org/10.1016/j.jhazmat.2018.07.113

[15] Peddaiah, S., Burman, A., &Sreedeep, S. (2018). Experimental Study on Effect of Waste Plastic Bottle Strips in Soil Improvement. Geotechnical and Geological Engineering, 36(5), 2907-2920. doi:10.1007/s10706-018-0512-0. https://doi.org/10.1007/s10706-018-0512-0

[16] A., Langroudi, S., Yasrobi, 2009. A micro-mechanical approach to swelling behavior of unsaturated expansive clays under controlled drainage conditions. Applied Clay Science 45(12), 8-19. https://doi.org/10.1016/j.clay.2008.09.004

[17] Mirzababaei, M., Arulrajah, A., and Ouston, M. (2017). Polymers for Stabilization of Soft Clay Soils. Procedia Engineering, 189, 25-32. doi:10.1016/j.proeng.2017.05.005. https://doi.org/10.1016/j.proeng.2017.05.005

[18] Lin L.Y., Lee J.H., Hong C.E., Yoo G.-H., Advani S.G.,2006. Compos. Sci. Technol. 66: 2116. https://doi.org/10.1016/j.compscitech.2005.12.025

[19] Zainuddin S., Hosur M.V., Zhou Y., Kumar A., Jeelani S.,2010. Mater. Sci. Eng. A; 527: 3091. https://doi.org/10.1016/j.msea.2010.02.022

[20] Bani Baker, M., Abendeh, R., Sharo, A., Hanna, A. (2022) Stabilization of Sandy Soils by Bentonite Clay Slurry at Laboratory Bench and Pilot Scales. Coatings 12 (12). https://doi.org/10.3390/coatings12121922

[21] Srithammaraj, K., Magaraphan, R., &Manuspiya, H. (2011). Modified Porous Clay Heterostructures by Organic-Inorganic Hybrids for Nanocomposite Ethylene Scavenging/Sensor Packaging Film. Packaging Technology and Science, 25(2), 63-72. doi:10.1002/pts.958. https://doi.org/10.1002/pts.958

[22] Sharo, A.A., Khasawneh, M.A., Bani Baker, M., Al Tarawneh, D.M.Sonicated waves procedure effect on stabilizing expansive soil by nano-clay: Treat with cause. Frontiers in Built Environment, 2022, 8, 975993 https://doi.org/10.3389/fbuil.2022.975993

[23] Zaid, H. M., Mohd, R. T., Ibtehaj. T. J., 2014. "Stabilization of Soft Soil Using Nanomaterials". Research Journal of Applied Sciences, Engineering and Technology.

[24] MI Bani Baker, RM Abendeh, TAS Obaidat. Employing natural bentonite clay as partial replacement of mineral filler in asphalt mixtures. Journal of Materials in Civil Engineering 30 (8), 04018167. https://doi.org/10.1061/(ASCE)MT.1943-5533.0002375

[25] MI Bani Baker, RM Abendeh, MA Khasawneh. Freeze and Thaw Effect on Asphalt Concrete Mixtures Modified with Natural Bentonite Clay, Coatings 12 (11), 1664. https://doi.org/10.3390/coatings12111664

[26] Mousa Bani Baker, Raed Abendeh, Batool Al-Shorman. Density and Strength Performance of Cellular Concrete Blended with Bentonite Clay and Polypropylene Fiber. Key Engineering Materials, 2023, Volume 944. Pages 219-230. https://doi.org/10.4028/p-v85in5

Advanced Topics in Mechanics of Materials, Structures and Construction - AToMech1-2023 Materials Research Forum LLC
Materials Research Proceedings 31 (2023) 117-127 https://doi.org/10.21741/9781644902592-13

Static response of FGM shell using refined higher-order shear and normal deformation theory

Sumit S. Kolapkar[1,2,a *], Atteshamuddin S. Sayyad[2,b]

[1]Research Scholar, Department of Civil Engineering, Sanjivani College of Engineering, Savitribai Phule Pune University, Kopargaon-423603, Maharashtra, India

[2]Department of Structural Engineering, Sanjivani College of Engineering, Savitribai Phule Pune University, Kopargaon-423603, Maharashtra, India

[a]sumitkolapkar@gmail.com*, [b]attu_sayyad@yahoo.co.in

Keywords: Static Response, FGM Shell, Higher-Order Shear And Normal Deformation Theory

Abstract. In the present study, a static response of functionally graded plate and spherical shells is investigated using higher-order trigonometric shear and normal deformation theory. A need of the shear correction factor is obviated and the effect of actual cross-sectional warping has been considered to get the realistic behaviour of transverse shear stresses across the thickness of the shell. The Navier solution technique has been used to analyse the simply-supported boundary conditions of the shell. To verify the theory, the numerical results obtained using the present theory are compared with other higher-order shear deformation theories available in the literature. The numerical results are obtained with and without considering the effects of transverse normal strain (ε_z).

Introduction

The development of functionally graded materials (FGM) has broad applications in the fields of off-shore structures, aerospace, rocket casing, nuclear mining, power plants, etc. It consists of two distinct materials that behave separately to obtain specific desired features depending on the application for which FGM is used. These materials have quite varied engineering properties. It is a substance whose physical characteristics, such as its density and coefficient of thermal expansion, gradually change along a single (usually along its thickness) or multiple directions. A material's gradation in a specific direction improves properties such as thermal conductivity, corrosion resistance, hardness, stiffness, weldability, etc. The most often used FGM composites are made of ceramic and metal, where the former offers strong thermal insulation and corrosion resistance, and the latter offers good fracture and toughness coupled with weldability. Classical shell theory [1] neglects the effects of transverse shear stresses therefore provides inaccurate results of displacements and stresses for thick plates and shells. Mindlin [2] has developed a first order shear deformation theory which considers the effect of transverse shear stresses for thin and moderately thick plates and shells, but it does not satisfy realistic shear stress conditions at the top and the bottom surfaces of the shell. This necessitated the development of higher-order refined shear deformation theories which consider the effects of both transverse shear and normal deformations. A new sinusoidal shear deformation theory is developed by Thai and Vo [3] for the bending, buckling, and vibration analysis of FGM plates. Shyang and Yen [4] investigated the elastic behaviour of moderately thick, rectangular, simply-supported FGM plates. Mantari et al. [5] used the Carrera unified formulation for the static analysis of FGM single-layer and multi-layered sandwich plates including the trigonometric, exponential, and hyperbolic type functions in the displacement fields. Response of FGM plate under thermo-mechanical loading is investigated by Bhandari and Purohit [6, 7] for varied boundary conditions and aspect ratios. The impact of the shear correction factor on the static behaviour of porous FGM plates is evaluated by Mota et al. [8]. With the use of two refined higher-order models, Punera and Kant [9] examined the elastostatics behaviour of laminated and FGM sandwich cylindrical shells. For the thermal analysis of FGM plates, Swaminathan and Sangeetha [10] explored several modelling methodologies and its solutions, using the four variable refined plate theory, Zidi et al. [11] investigated the bending response of a FGM plate supported by an elastic foundation and exposed to hygrothermal and mechanical loadings. A literature on the stress, vibration, and buckling analysis of FGM plates is reviewed by Swaminathan et al. [12]. The effectiveness of a novel fifth-order shear and normal deformation theories for the static and dynamic responses of sandwich FGM plates and shells is examined by Shinde and Sayyad [13, 14]. A bending response is studied for FG beam using semi-analytical and by shear deformation theory under transverse loading conditions by Yadav et al. [15]. A static and free vibration analysis of doubly-curved FGM shells using different types of higher-order shell theories via unified formulation is presented by Sayyad and Ghugal [16]. Analysis of FG sandwich plates using a fifth order shear deformation theory is carried out by Thai et al. [17]. For the bending analysis of FGM plates based on the four variable plate theory, Demirhan and Taskin [18] employed the Levy solution. The observations that need further study are based on the literature review that is done.

1. Laminated composite and FGM plates' mechanical analyses have been extensively studied in the literature. However, there is very little study on mechanical analysis of FGM shells in the entire body of literature.
2. Researchers are paying more attention to cylindrical shell analyses. There is, however, a dearth of information on mechanical analysis of doubly curved FGM shells.
3. Many theories available in the literature ignores the effects of transverse normal strain on the analysis of FGM shells which plays an important role for the accurate structural analysis of FGM shells under mechanical/environmental loading conditions.

Methodology

The analysis in the current study takes into account a single layer FGM shell that is simply-supported at edges as shown in Fig. 1. It has curved dimensions of "*a*" and "*b*" along *x-y* directions and "*h*" is the thickness of the shell along *z*-direction. The principle radii of curvature of the mid-plane along the *x-y* directions are R_1 and R_2, respectively. The distance between the shell's top and bottom surfaces from the mid-plane surface is ($z = -h/2$) and ($z = +h/2$), respectively. A transverse load of intensity $q(x,y)$ is applied to the top surface of the shell.

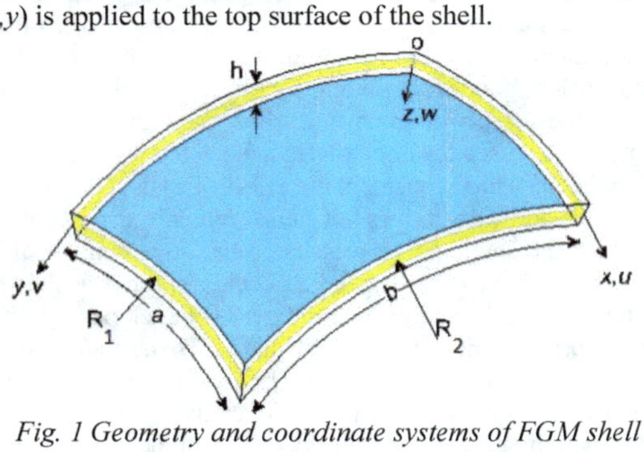

Fig. 1 Geometry and coordinate systems of FGM shell

The FGM used to create the shell has gradations of material properties in the direction of thickness i.e. z-direction according to the power-law. Following is the simple form of power law.

$$E(z) = E_m + (E_c - E_m)V_f. \tag{1}$$

where, $E(z)$ is the value of modulus of elasticity at any point of the shell in the z-direction; E_m and E_c are the modulus of elasticity of metal and ceramic, respectively; V_f is the volume fraction which is given as

$$V_f = \left(0.5 + \frac{z}{h}\right)^p. \tag{2}$$

where p is the power law factor. When $p = 0$ shell is fully ceramic whereas for $p = \infty$ it is fully metallic.

Formulation

The displacement field of the present higher-order trigonometric shear and normal deformation theory is written as

$$\begin{aligned} U(x,y,z) &= \left(1+\frac{z}{R_1}\right)u_0(x,y) - z\frac{\partial w_0}{\partial x} + \frac{h}{\pi}\sin\left(\frac{\pi z}{h}\right)\theta_x, \\ V(x,y,z) &= \left(1+\frac{z}{R_2}\right)v_0(x,y) - z\frac{\partial w_0}{\partial y} + \frac{h}{\pi}\sin\left(\frac{\pi z}{h}\right)\theta_y, \\ W(x,y,z) &= w_0(x,y) + C_1\cos\left(\frac{\pi z}{h}\right)\theta_z. \end{aligned} \tag{3}$$

where, U, V, W are the displacements of any point in the x-, y-, and z- directions respectively; $u_0(x,y), v_0(x,y)$and$w_0(x,y)$are the displacement of a point on the mid-plane in x-, y-, and z-directions respectively; $\theta_x(x,y), \theta_y(x,y)and\theta_z(x,y)$are the shear slopes. $\left(\frac{h}{\pi}sin\frac{\pi z}{h}\right)$ is the trigonometric function associated with the transverse shear strain. $(1 + z/R_1)$ and $(1 + z/R_2)$ are the Lame's constant; and C_1 is the arbitrary constant account for the consideration of effects of transverse normal strain. When C_1=0, $\varepsilon_z = 0$ and when $C_1 = 1$, $\varepsilon_z \neq 0$. Using linear theory of elasticity, the normal strain and shear strain components are obtained.

$$\begin{aligned} \varepsilon_x &= \frac{\partial u_0}{\partial x} - z\frac{\partial^2 w_0}{\partial x^2} + f(z)\frac{\partial \theta_x}{\partial x} + \frac{w_0}{R_1} + C_1\frac{f'(z)\theta_z}{R_1}, \\ \varepsilon_y &= \frac{\partial v_0}{\partial y} - z\frac{\partial^2 w_0}{\partial y^2} + f(z)\frac{\partial \theta_y}{\partial y} + \frac{w_0}{R_2} + C_1\frac{f'(z)\theta_z}{R_2}, \\ \varepsilon_z &= C_1 f''(z)\theta_z, \\ \gamma_{xy} &= \frac{\partial u_0}{\partial y} + \frac{\partial v_0}{\partial x} - 2z\frac{\partial^2 w_0}{\partial x \partial y} + f(z)\frac{\partial \theta_x}{\partial y} + f(z)\frac{\partial \theta_y}{\partial x}, \\ \gamma_{xz} &= f'(z)\theta_x + C_1 f'(z)\frac{\partial \theta_z}{\partial x}, \\ \gamma_{yz} &= f'(z)\theta_y + C_1 f'(z)\frac{\partial \theta_z}{\partial y}. \end{aligned} \tag{4}$$

where

$$f(z) = \frac{h}{\pi}\left(\sin\frac{\pi z}{h}\right), f'(z) = \cos\left(\frac{\pi z}{h}\right), f''(z) = -\sin\left(\frac{\pi z}{h}\right)\frac{\pi}{h}. \tag{5}$$

Stress components associated with the strain components mentioned in Eq. (4) are obtained using the following stress-strain relationship which is also called as generalized Hooke's law.

$$\begin{Bmatrix} \sigma_x \\ \sigma_y \\ \sigma_z \\ \tau_{xy} \\ \tau_{xz} \\ \tau_{yz} \end{Bmatrix} = \begin{bmatrix} Q_{11} & Q_{12} & Q_{13} & 0 & 0 & 0 \\ Q_{12} & Q_{22} & Q_{23} & 0 & 0 & 0 \\ Q_{13} & Q_{23} & Q_{33} & 0 & 0 & 0 \\ 0 & 0 & 0 & Q_{66} & 0 & 0 \\ 0 & 0 & 0 & 0 & Q_{44} & 0 \\ 0 & 0 & 0 & 0 & 0 & Q_{55} \end{bmatrix} \begin{Bmatrix} \varepsilon_x \\ \varepsilon_y \\ \varepsilon_z \\ \gamma_{xy} \\ \gamma_{xz} \\ \gamma_{yz} \end{Bmatrix}. \tag{6}$$

where

$$\begin{gathered} Q_{11} = Q_{22} = Q_{33} = \frac{E(z)(1-\mu)}{(1+\mu)(1-2\mu)}, \\ Q_{12} = Q_{13} = Q_{21} = Q_{23} = Q_{31} = Q_{32} = \frac{E(z)\mu}{(1+\mu)(1-2\mu)}, \\ Q_{44} = Q_{55} = Q_{66} = \frac{E(z)}{2(1+\mu)}. \end{gathered} \tag{7}$$

where, $(\sigma_x, \sigma_y, \sigma_z, \tau_{xy}, \tau_{xz}, \tau_{yz})$are the normal and shear stress components, $(\varepsilon_x, \varepsilon_y, \varepsilon_z, \gamma_{xy}, \gamma_{yz}, \gamma_{xz})$are the normal and shear strain components, $(Q_{11}, Q_{12}, Q_{13}, Q_{22}, Q_{23}, Q_{33}, Q_{44}, Q_{55}, Q_{66})$are the reduced stiffness matrix components, $E(z)$ is the modulus of elasticity and μ is the Poisson's ratio. The principle of virtual work is used to derive the governing differential equations and boundary conditions associated with the present theory.

$$\int_0^a \int_0^b \int_{-h/2}^{+h/2} \left(\sigma_x \delta\varepsilon_x + \sigma_y \delta\varepsilon_y + \sigma_z \delta\varepsilon_z + \tau_{xy}\delta\gamma_{xy} + \tau_{xz}\delta\gamma_{xz} + \tau_{yz}\delta\gamma_{yz}\right) dxdydz = \int_0^a \int_0^b q\delta w dxdy. \tag{8}$$

Substituting Eqs. (4) - (6) in Eq. (8), performing triple integration by parts and collecting the coefficients of unknown variables, the following six governing differential equations are derived.

$$\begin{gathered} \delta u_0: \frac{\partial N_x}{\partial x} + \frac{\partial N_{xy}}{\partial y} = 0, \\ \delta v_0: \frac{\partial N_y}{\partial y} + \frac{\partial N_{xy}}{\partial x} = 0, \\ \delta w_0: \frac{\partial^2 M_x^b}{\partial x^2} + \frac{\partial^2 M_y^b}{\partial y^2} + 2\frac{\partial^2 M_{xy}^b}{\partial x \partial y} - \frac{N_x}{R_1} - \frac{N_y}{R_2} + q = 0, \\ \delta\theta_x: \frac{\partial M_x^s}{\partial x} + \frac{\partial M_{xy}^s}{\partial y} - Q_x^S = 0, \\ \delta\theta_y: \frac{\partial M_y^s}{\partial y} + \frac{\partial M_{xy}^s}{\partial x} - Q_y^S = 0, \\ \delta\theta_z: \frac{\partial Q_x^S}{\partial x} + \frac{\partial Q_y^S}{\partial y} - \frac{S_x}{R_1} - \frac{S_y}{R_2} - S^S = 0. \end{gathered} \tag{9}$$

where

Materials Research Proceedings 31 (2023) 117-127 https://doi.org/10.21741/9781644902592-13

$$\begin{gathered}
\left(N_x, N_y, N_{xy}, M_x^b, M_y^b, M_{xy}^b\right)=\int_{-h/2}^{+h/2}\left(\sigma_x, \sigma_y, \tau_{xy}, z\sigma_x, z\sigma_y, z\tau_{xy}\right) dz, \\
\left(M_x^s, M_y^s, M_{xy}^s\right)=\int_{-h/2}^{+h/2} f(z)\left(\sigma_x, \sigma_y, \tau_{xy}\right) dz, \\
\left(S_x, S_y, Q_x^S, Q_y^S\right)=\int_{-h/2}^{+h/2} f'(z)\left(\sigma_x, \sigma_y, \tau_{xz}, \tau_{yz}\right) dz, \\
S^S=\int_{-h/2}^{+h/2} f''(z)\sigma_z\, dz.
\end{gathered} \tag{10}$$

where the in-plane force resultants are (N_x, N_y, N_{xy}); the shear force resultants are (S_x, S_y, Q_x^s, Q_y^s, S^s); the resultant bending moments are (M_x^b, M_y^b, M_{xy}^b); and the resultant shear moments are (M_x^s, M_y^s, M_{xy}^s). The superscript "b" stands for traditional bending effects, while the superscript "s" stands for shear effects. After substitution of stress resultant expressions from Eq. (10) into six governing equation stated in Eq. (9), one can derive governing equations in terms of unknown variables.

$$\begin{gathered}
\delta u_0: A_{11}\left(\frac{\partial^2 u_0}{\partial x^2}+\frac{1}{R_1}\frac{\partial w_0}{\partial x}\right)-B_{11}\frac{\partial^3 w_0}{\partial x^3}+C_{11}\frac{\partial^2\theta_x}{\partial x^2}+A_{12}\left(\frac{\partial^2 v_0}{\partial x\partial y}+\frac{1}{R_2}\frac{\partial w_0}{\partial x}\right)-B_{12}\frac{\partial^3 w_0}{\partial x\partial y^2}+C_{12}\frac{\partial^2\theta_y}{\partial x\partial y} \\
+\left(\frac{F_{11}}{R_1}+\frac{F_{12}}{R_2}\right)\frac{\partial\theta_z}{\partial x}C_1+D_{13}\frac{\partial\theta_z}{\partial x}C_1+A_{66}\left(\frac{\partial^2 u_0}{\partial y^2}+\frac{\partial^2 v_0}{\partial x\partial y}\right)-2B_{66}\frac{\partial^3 w_0}{\partial x\partial y^2}+C_{66}\left(\frac{\partial^2\theta_x}{\partial y^2}+\frac{\partial^2\theta_y}{\partial x\partial y}\right)=0.
\end{gathered} \tag{11}$$

$$\begin{gathered}
\delta v_0: A_{12}\left(\frac{\partial^2 u_0}{\partial x\partial y}+\frac{1}{R_1}\frac{\partial w_0}{\partial y}\right)-B_{12}\frac{\partial^3 w_0}{\partial x^2\partial y}+C_{12}\frac{\partial^2\theta_x}{\partial x\partial y}+A_{22}\left(\frac{\partial^2 v_0}{\partial y^2}+\frac{1}{R_2}\frac{\partial w_0}{\partial y}\right)+\left(\frac{F_{12}}{R_1}+\frac{F_{22}}{R_2}\right)\frac{\partial\theta_z}{\partial y}C_1 \\
-B_{22}\frac{\partial^3 w_0}{\partial y^3}+C_{22}\frac{\partial^2\theta_y}{\partial y^2}+D_{23}\frac{\partial\theta_z}{\partial y}C_1+A_{66}\left(\frac{\partial^2 u_0}{\partial x\partial y}+\frac{\partial^2 v_0}{\partial x^2}\right)-2B_{66}\frac{\partial^3 w_0}{\partial x^2\partial y}+C_{66}\left(\frac{\partial^2\theta_x}{\partial x\partial y}+\frac{\partial^2\theta_y}{\partial x^2}\right)=0.
\end{gathered} \tag{12}$$

$$\begin{gathered}
\delta w_0: B_{11}\left(\frac{\partial^3 u_0}{\partial x^3}+\frac{1}{R_1}\frac{\partial^2 w_0}{\partial x^2}\right)-H_{11}\frac{\partial^4 w_0}{\partial x^4}+I_{11}\frac{\partial^3\theta_x}{\partial x^3}+\left(\frac{J_{11}}{R_1}+\frac{J_{12}}{R_2}\right)C_1\frac{\partial^2\theta_z}{\partial x^2}+B_{12}\left(\frac{\partial^3 v_0}{\partial x^2\partial y}+\frac{1}{R_2}\frac{\partial^2 w_0}{\partial x^2}\right) \\
-H_{12}\frac{\partial^4 w_0}{\partial x^2\partial y^2}+I_{12}\frac{\partial^3\theta_y}{\partial x^2\partial y}+K_{13}\frac{\partial^2\theta_z}{\partial x^2}C_1+B_{12}\left(\frac{\partial^3 u_0}{\partial x\partial y^2}+\frac{1}{R_1}\frac{\partial^2 w_0}{\partial y^2}\right)-H_{12}\frac{\partial^4 w_0}{\partial x^2\partial y^2}+I_{12}\frac{\partial^3\theta_x}{\partial x\partial y^2}+ \\
\left(\frac{J_{12}}{R_1}+\frac{J_{22}}{R_2}\right)C_1\frac{\partial^2\theta_z}{\partial y^2}+B_{22}\left(\frac{\partial^3 v_0}{\partial y^3}+\frac{1}{R_2}\frac{\partial^2 w_0}{\partial y^2}\right)-H_{22}\frac{\partial^4 w_0}{\partial y^4}+I_{22}\frac{\partial^3\theta_y}{\partial y^3}+K_{23}\frac{\partial^2\theta_z}{\partial y^2}C_1 \\
+2B_{66}\left(\frac{\partial^3 u_0}{\partial x\partial y^2}+\frac{\partial^3 v_0}{\partial x^2\partial y}\right)-4H_{66}\frac{\partial^4 w_0}{\partial x^2\partial y^2}+2I_{66}\left(\frac{\partial^3\theta_x}{\partial x\partial y^2}+\frac{\partial^3\theta_y}{\partial x^2\partial y}\right)-\frac{A_{11}}{R_1}\left(\frac{\partial u_0}{\partial x}+\frac{w_0}{R_1}\right)+\frac{B_{11}}{R_1}\frac{\partial^2 w_0}{\partial x^2} \\
-\frac{C_{11}}{R_1}\frac{\partial\theta_x}{\partial x}-\frac{1}{R_1}\left(\frac{F_{11}}{R_1}+\frac{F_{12}}{R_2}\right)C_1\theta_z-\frac{A_{12}}{R_1}\left(\frac{\partial v_0}{\partial y}+\frac{w_0}{R_2}\right)+\frac{B_{12}}{R_1}\frac{\partial^2 w_0}{\partial y^2}-\frac{C_{12}}{R_1}\frac{\partial\theta_y}{\partial y}-\frac{D_{13}}{R_1}C_1\theta_z-\frac{A_{12}}{R_2}\left(\frac{\partial u_0}{\partial x}+\frac{w_0}{R_1}\right) \\
+\frac{B_{12}}{R_2}\frac{\partial^2 w_0}{\partial x^2}-\frac{C_{12}}{R_2}\frac{\partial\theta_x}{\partial x}-\frac{1}{R_2}\left(\frac{F_{12}}{R_1}+\frac{F_{22}}{R_2}\right)C_1\theta_z-\frac{A_{22}}{R_2}\left(\frac{\partial v_0}{\partial y}+\frac{w_0}{R_2}\right)+\frac{B_{22}}{R_2}\frac{\partial^2 w_0}{\partial y^2}-\frac{C_{22}}{R_2}\frac{\partial\theta_y}{\partial y}-\frac{D_{23}}{R_2}C_1\theta_z=-q.
\end{gathered} \tag{13}$$

$$\begin{gathered}
\delta\theta_x: C_{11}\left(\frac{\partial^2 u_0}{\partial x^2}+\frac{1}{R_1}\frac{\partial w_0}{\partial x}\right)-I_{11}\frac{\partial^3 w_0}{\partial x^3}+L_{11}\frac{\partial^2\theta_x}{\partial x^2}+C_{12}\left(\frac{\partial^2 v_0}{\partial x\partial y}+\frac{1}{R_2}\frac{\partial w_0}{\partial x}\right)-I_{12}\frac{\partial^3 w_0}{\partial x\partial y^2} \\
+\left(\frac{M_{11}}{R_1}+\frac{M_{12}}{R_2}\right)\frac{\partial\theta_z}{\partial x}C_1+L_{12}\frac{\partial^2\theta_y}{\partial x\partial y}+N_{13}\frac{\partial\theta_z}{\partial x}C_1+C_{66}\left(\frac{\partial^2 u_0}{\partial y^2}+\frac{\partial^2 v_0}{\partial x\partial y}\right)-2I_{66}\frac{\partial^3 w_0}{\partial x\partial y^2}+ \\
L_{66}\left(\frac{\partial^2\theta_x}{\partial y^2}+\frac{\partial^2\theta_y}{\partial x\partial y}\right)-O_{55}\theta_x-O_{55}\frac{\partial\theta_z}{\partial x}C_1=0.
\end{gathered} \tag{14}$$

$$\begin{gathered}
\delta\theta_y: C_{21}\left(\frac{\partial^2 u_0}{\partial x\partial y}+\frac{1}{R_1}\frac{\partial w_0}{\partial y}\right)-I_{21}\frac{\partial^3 w_0}{\partial x^2\partial y}+L_{21}\frac{\partial^2\theta_x}{\partial x\partial y}+C_{22}\left(\frac{\partial^2 v_0}{\partial y^2}+\frac{1}{R_2}\frac{\partial w_0}{\partial y}\right)-I_{22}\frac{\partial^3 w_0}{\partial y^3} \\
+\left(\frac{M_{21}}{R_1}+\frac{M_{22}}{R_2}\right)\frac{\partial\theta_z}{\partial y}C_1+L_{22}\frac{\partial^2\theta_y}{\partial y^2}+N_{23}\frac{\partial\theta_z}{\partial y}C_1+C_{66}\left(\frac{\partial^2 u_0}{\partial x\partial y}+\frac{\partial^2 v_0}{\partial x^2}\right) \\
-2I_{66}\frac{\partial^3 w_0}{\partial x^2\partial y}+L_{66}\left(\frac{\partial^2\theta_x}{\partial x\partial y}+\frac{\partial^2\theta_y}{\partial x^2}\right)-O_{44}\theta_y-O_{44}\frac{\partial\theta_z}{\partial y}C_1=0.
\end{gathered} \tag{15}$$

$$\begin{aligned}\delta\theta_z: & O_{55}\left(\frac{\partial\theta_x}{\partial x}C_1+\frac{\partial^2\theta_z}{\partial x^2}C_1^2\right)+O_{44}\left(\frac{\partial\theta_y}{\partial y}C_1+\frac{\partial^2\theta_z}{\partial y^2}C_1^2\right)-\frac{F_{11}}{R_1}C_1\left(\frac{\partial u_0}{\partial x}+\frac{w_0}{R_1}\right)+\frac{J_{11}}{R_1}C_1\frac{\partial^2 w_0}{\partial x^2}\\ & -\frac{M_{11}}{R_1}\frac{\partial\theta_x}{\partial x}C_1-\frac{F_{12}}{R_1}C_1\left(\frac{\partial v_0}{\partial y}+\frac{w_0}{R_2}\right)+\frac{J_{12}}{R_1}C_1\frac{\partial^2 w_0}{\partial y^2}-\frac{1}{R_1}\left(\frac{O_{11}}{R_1}+\frac{O_{12}}{R_2}\right)C_1^2\theta_z-\frac{M_{12}}{R_1}\frac{\partial\theta_y}{\partial y}C_1\\ & -\frac{P_{13}}{R_1}C_1^2\theta_z-\frac{F_{12}}{R_2}C_1\left(\frac{\partial u_0}{\partial x}+\frac{w_0}{R_1}\right)+\frac{J_{12}}{R_2}C_1\frac{\partial^2 w_0}{\partial x^2}-\frac{M_{12}}{R_2}\frac{\partial\theta_x}{\partial x}C_1-\frac{F_{22}}{R_2}C_1\left(\frac{\partial v_0}{\partial y}+\frac{w_0}{R_2}\right)+\frac{J_{22}}{R_2}C_1\frac{\partial^2 w_0}{\partial y^2}\\ & -\frac{1}{R_2}\left(\frac{O_{12}}{R_1}+\frac{O_{22}}{R_2}\right)C_1^2\theta_z-\frac{M_{22}}{R_2}\frac{\partial\theta_y}{\partial y}C_1-\frac{P_{23}}{R_2}C_1^2\theta_z-D_{13}C_1\left(\frac{\partial u_0}{\partial x}+\frac{w_0}{R_1}\right)+K_{13}C_1\frac{\partial^2 w_0}{\partial x^2}-N_{13}\frac{\partial\theta_x}{\partial x}C_1\\ & -D_{23}C_1\left(\frac{\partial v_0}{\partial y}+\frac{w_0}{R_2}\right)+K_{23}C_1\frac{\partial^2 w_0}{\partial y^2}-\left(\frac{P_{13}}{R_1}+\frac{P_{23}}{R_2}\right)C_1^2\theta_z-N_{23}\frac{\partial\theta_y}{\partial y}C_1-S_{33}C_1^2\theta_z=0.\end{aligned} \tag{16}$$

where

$$\begin{aligned}\left(A_{ij},B_{ij},C_{ij},D_{ij},F_{ij}\right)&=Q_{ij}\int_{-h/2}^{+h/2}[1,z,f(z),f''(z),f'(z)]dz\,,\\ \left(H_{ij},I_{ij},J_{ij},K_{ij},L_{ij}\right)&=Q_{ij}\int_{-h/2}^{+h/2}[z^2,f(z)z,f'(z)z,f''(z)z,f(z)^2]dz\,,\\ \left(M_{ij},N_{ij},O_{ij},P_{ij},S_{ij}\right)&=Q_{ij}\int_{-h/2}^{+h/2}[f'(z)f(z),f''(z)f(z),f'(z)^2,f''(z)f'(z),f''(z)^2]dz.\end{aligned} \tag{17}$$

The Navier Solution

The closed-form semi-analytical solution for the static analysis of FGM shell for a simply-supported boundary condition can be obtained by using the Navier solution technique. Following are the simply supported boundary conditions associated with the present theory.
At the edge $x = 0$ and $x = a$

$$v_0 = w_0 = M_x^b = M_x^s = N_x = 0 \tag{18}$$

At the edge $y = 0$ and $y = b$

$$u_0 = w_0 = M_y^b = M_y^s = N_y = 0 \tag{19}$$

Therefore, the unknown variable according to the Navier technique in the form of double trigonometric series satisfying boundary conditions stated in Eqs. (18) and (19) can be expressed as

$$\begin{aligned}u_0 = u_{mn}\cos\alpha x\sin\beta y, v_0 = v_{mn}\sin\alpha x\cos\beta y, w_0 = w_{mn}\sin\alpha x\sin\beta y,\\ \theta_x = \theta_{x_{mn}}\cos\alpha x\sin\beta y, \theta_y = \theta_{y_{mn}}\sin\alpha x\cos\beta y, \theta_z = \theta_{z_{mn}}\sin\alpha x\sin\beta y.\end{aligned} \tag{20}$$

where $\alpha=m\pi/a$, $\beta=n\pi/b$, and $\left(u_{mn}, v_{mn}, w_{mn}, \theta_{xmn}, \theta_{ymn}, \theta_{zmn}\right)$ are the unknown coefficients. The uniform mechanical load acting on the top surface of the shell is also expressed in double trigonometric series as follows.

$$q(x,y) = q_{mn}\sin\alpha x\sin\beta y. \tag{21}$$

where, q_{mn} is Fourier coefficient of load; for sinusoidal load $q_{mn} = q_0$ where q_0 is the maximum intensity of the load with m =1 and n = 1. Substituting Eqs. (20) and (21) in to Eqs. (11) – (16) leads to the six simultaneous equations which can we written in the following matrix form.

$$[K]\{\Delta\} = \{f\}. \tag{22}$$

where matrices $[K]$, $\{f\}$ and $\{\Delta\}$ are the stiffness matrix, the force vector and the vector of unknowns, respectively.

Numerical Results and Discussion

In this section, the accuracy of the present higher-order trigonometric shear and normal deformation theory is verified by applying it for the static analysis to FGM plates and shells. The FGM shell is made up from the following materials.

$$E_c = 380\,GPa, E_m = 70\,GPa, \text{and } \mu = 0.3$$

where E_c and E_m are the modulus of elasticity for ceramic (alumina) and metal (aluminium) respectively.
Following non-dimensional forms are used to present the numerical values of displacements and stresses.

$$\overline{w}\left(\frac{a}{2},\frac{b}{2},0\right) = \frac{10E_c h^3}{q_0 a^4}, w^*\left(\frac{a}{2},\frac{b}{2},0\right) = \frac{100E_0 h^3}{q_0 a^4}, \overline{\sigma}_x\left(\frac{a}{2},\frac{b}{2},-\frac{h}{2}\right) = \frac{h}{q_0 a}\sigma_x, \overline{\tau}_{xz}\left(0,\frac{b}{2},\frac{z}{h}\right) = \frac{h}{q_0 a}\overline{\tau}_{xz}. \quad (23)$$

For a single layer FGM plate and FGM spherical shell, the transverse displacement, in-plane, and transverse shear stresses are calculated with and without considering the effects of transverse normal strain. The FGM plate and FGM spherical shell results are shown here for a range of R/a ratios and for an a/h value of 10. The results obtained are contrasted with those of Shinde and Sayyad [14], Thai et al. [17], and Demirhan and Taskin [18]. Table 1 shows the comparison of displacements and stresses for FGM plate under sinusoidal loading with those presented by Shinde and Sayyad [14], Thai et al. [17], and Demirhan and Taskin [18]. Table 1 show that the transverse displacements increase whereas stresses are decreases with increase in the power-law index which is the effect of decrease in stiffness of material with increase in the power-law index. Fig. 2 shows through-the-thickness distributions of stresses in FGM plate. Table 2 shows the effects power law factor and radius of curvature on the dimensionless transverse displacement of FGM spherical shell under sinusoidal load. Table 2 reveals that the increase in radius of curvature and the power-law index increase the non-dimensional displacements and stresses in FGM spherical shell under the action of transverse sinusoidal load. Fig. 3 shows distribution of stresses through-the-thickness of the spherical shell.

***Table 1** Comparison of non-dimensional transverse displacements and stresses in FGM plate*

p	Theory	$\bar{w}$	$\bar{\sigma}_x$ (h/3)	$\bar{\tau}_{xz}$ (h/6)
1	Present ($\varepsilon_z \neq 0$)	0.5680	1.4158	0.2613
	Present ($\varepsilon_z = 0$)	0.5889	1.4893	0.2621
	Shinde and Sayyad [14] ($\varepsilon_z \neq 0$)	0.5695	1.4588	0.2607
	Thai et al. [17] ($\varepsilon_z = 0$)	0.5875	1.5062	0.2510
	Demirhan and Taskin [18] ($\varepsilon_z = 0$)	0.5889	1.4894	0.2622
	Thai et al. [17] ($\varepsilon_z = 0$)	0.5890	1.4898	0.2599
2	Present ($\varepsilon_z \neq 0$)	0.7198	1.3041	0.2754
	Present ($\varepsilon_z = 0$)	0.7573	1.3953	0.2763
	Shinde and Sayyad [14] ($\varepsilon_z \neq 0$)	0.7225	1.4588	0.2763
	Thai et al. [17] ($\varepsilon_z = 0$)	0.7570	1.5062	0.2510
	Demirhan and Taskin [18] ($\varepsilon_z = 0$)	0.7573	1.4894	0.2622
	Thai et al. [17] ($\varepsilon_z = 0$)	0.7573	1.3960	0.2721
4	Present ($\varepsilon_z \neq 0$)	0.8402	1.0801	0.2572
	Present ($\varepsilon_z = 0$)	0.8818	1.1782	0.2580
	Shinde and Sayyad [14] ($\varepsilon_z \neq 0$)	0.8429	1.1456	0.2630
	Thai et al. [17] ($\varepsilon_z = 0$)	0.8823	1.1985	0.2362
	Demirhan and Taskin [18] ($\varepsilon_z = 0$)	0.8819	1.1783	0.2580
	Thai et al. [17] ($\varepsilon_z = 0$)	0.8815	1.1794	0.2519
8	Present ($\varepsilon_z \neq 0$)	0.9431	0.8634	0.2114
	Present ($\varepsilon_z = 0$)	0.9750	0.9463	0.2120
	Shinde and Sayyad [14] ($\varepsilon_z \neq 0$)	0.9466	0.9088	0.2145
	Thai et al. [17] ($\varepsilon_z = 0$)	0.9738	0.9687	0.2262
	Demirhan and Taskin [18] ($\varepsilon_z = 0$)	0.9750	0.9466	0.2121
	Thai et al. [17] ($\varepsilon_z = 0$)	0.9747	0/9477	0.2087

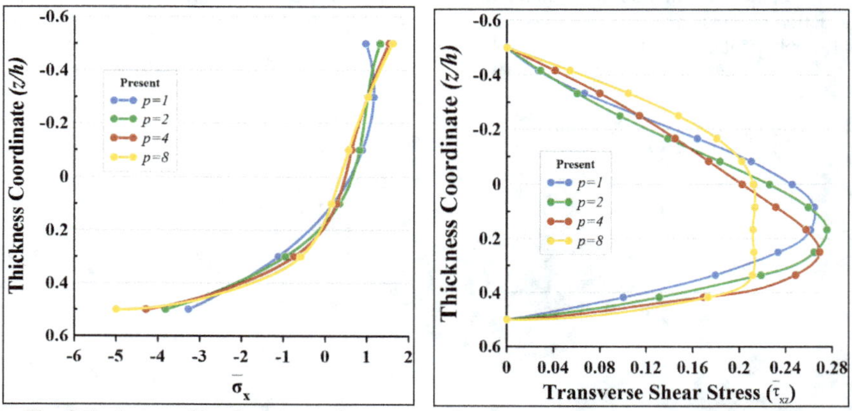

***Fig. 2** Variation of in-plane normal stresses and transverse shear stresses with respect to thickness coordinate for FGM plate at various power law index.*

Table 2 *Effects of radii of curvature and various power law factors on a non-dimensional transverse displacement and stresses in FGM spherical shell.*

		C1=1 ($\varepsilon_z \neq 0$)			C1=0($\varepsilon_z = 0$)		
p	*R/a*	w^*	$\bar{\sigma}_x$	$\bar{\tau}_{xz}$	w^*	$\bar{\sigma}_x$	$\bar{\tau}_{xz}$
0	1	0.182285	0.8867	0.059227	0.196709	0.877621	0.062162
	2	0.426848	1.616985	0.136448	0.447698	1.5721	0.141477
	5	0.683527	2.161396	0.217494	0.696549	2.048913	0.220116
	10	0.747757	2.210737	0.237775	0.756631	2.081884	0.239102
	20	0.765746	2.185615	0.243455	0.773307	2.054303	0.244372
	50	0.770939	2.153269	0.245095	0.778108	2.022706	0.245889
	100	0.770939	2.139642	0.245331	0.778799	2.009705	0.246107
1	1	0.333162	1.340047	0.054467	0.345392	1.344237	0.059081
	2	0.818613	2.534497	0.138501	0.827979	2.435844	0.141629
	5	1.345933	3.434034	0.233552	1.360064	3.225955	0.232645
	10	1.467366	3.479844	0.256964	1.497545	3.267515	0.256162
	20	1.494501	3.410171	0.26295	1.536371	3.206274	0.262803
	50	1.497965	3.33761	0.264315	1.547605	3.141506	0.264725
	100	1.496925	3.308512	0.264385	1.549224	3.115356	0.265001
2	1	0.436394	1.661697	0.057442	0.447408	1.660349	0.062031
	2	1.062913	3.088568	0.146551	1.069408	2.952679	0.148269
	5	1.722104	4.084018	0.244961	1.751018	3.836549	0.242771
	10	1.867198	4.097459	0.268509	1.926424	3.854851	0.26709
	20	1.897397	3.996052	0.274366	1.975908	3.766159	0.273951
	50	1.899644	3.900158	0.275612	1.990222	3.68	0.275935
	100	1.897695	3.862666	0.275638	1.992284	3.645959	0.276221
4	1	0.558303	2.093204	0.06188	0.571133	2.08663	0.066412
	2	1.307231	3.711913	0.151393	1.314225	3.552994	0.152819
	5	2.033347	4.674429	0.242421	2.067377	4.410728	0.240396
	10	2.184533	4.629893	0.263082	2.251721	4.376198	0.261832
	20	2.215168	4.496298	0.268138	2.303061	4.257186	0.267802
	50	2.21703	4.380945	0.269192	2.317858	4.152421	0.269522
	100	2.214852	4.336984	0.269206	2.319988	4.112156	0.26977
8	1	0.677125	2.620321	0.054154	0.697486	2.62067	0.058081
	2	1.524411	4.474527	0.125934	1.536758	4.314154	0.127968
	5	2.293287	5.459686	0.193726	2.317603	5.18519	0.19299
	10	2.450173	5.382528	0.208615	2.498998	5.116218	0.208095
	20	2.483345	5.227416	0.21229	2.548872	4.976183	0.212248
	50	2.486733	5.097692	0.213098	2.563196	4.858044	0.213441
	100	2.485059	5.048693	0.213129	2.565255	4.813208	0.213613

Conclusions

In the current study, the static response of FGM plate and shell is studied using refined higher-order trigonometric shear and normal deformation theory. The currently used theory yields realistic boundary condition at the top and the bottom surfaces of the shell. The Navier technique has been used to calculate transverse displacements and stresses for FGM plates and spherical shell under the action of sinusoidal load. The obtained results are compared with previously published results. It is concluded that, the present results of static analysis of FGM plate and shell shows a good agreement with the previously published results when the effects of transverse normal strain is considered. The other theories available in the literature overestimate the results due to neglecting the effects of transverse normal strain.

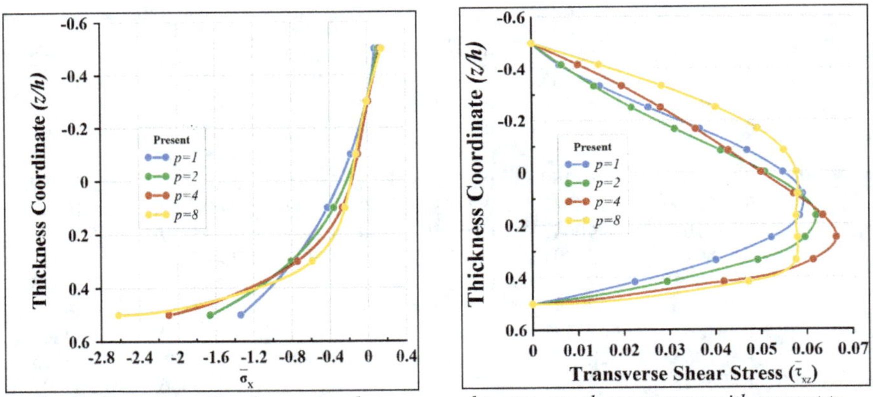

Fig. 3. *Variation of in-plane normal stresses and transverse shear stresses with respect to thickness coordinate for FGM spherical shell at various power law index (R/a = 1,* $\boldsymbol{\varepsilon_z} \neq \mathbf{0}$*).*

References

[1] G. Kirchhoff, Über das Gleichgewicht und die Bewegung, J Reine Angew Math. 40 (1850) 51-88. https://doi.org/10.1515/crll.1850.40.51

[2] R. D. Mindlin, Influence of Rotatory Inertia and Shear on Flexural Motions of Isotropic, Elastic Plates, J. Appl. Mech. vol. 18, no. 1(Mar. 1951) 31-38. doi: 10.1115/1.4010217. https://doi.org/10.1115/1.4010217

[3] H. T. Thai and T. P. Vo, A new sinusoidal shear deformation theory for bending, buckling, and vibration of functionally graded plates, Appl. Math. Model. vol. 37, no. 5 (2013) 3269-3281. doi: 10.1016/j.apm.2012.08.008. https://doi.org/10.1016/j.apm.2012.08.008

[4] S. H. Chi and Y. L. Chung, Mechanical behavior of functionally graded material plates under transverse load-Part I: Analysis, Int. J. Solids Struct. vol. 43, no. 13 (2006) 3657-3674. doi: 10.1016/j.ijsolstr.2005.04.011. https://doi.org/10.1016/j.ijsolstr.2005.04.011

[5] J. L. Mantari, I. A. Ramos, E. Carrera, and M. Petrolo, Static analysis of functionally graded plates using new non-polynomial displacement fields via Carrera Unified Formulation, Compos. Part B Eng. 89 (2016) 127-142. doi: 10.1016/j.compositesb.2015.11.025. https://doi.org/10.1016/j.compositesb.2015.11.025

[6] M. Bhandari and K. Purohit, Response of Functionally Graded Material Plate under Thermomechanical Load Subjected to Various Boundary Conditions, Int. J. Met. 2015 (2015) 1-16. doi: 10.1155/2015/416824. https://doi.org/10.1155/2015/416824

[7] M. Bhandari and K. Purohit, Static Response of Functionally Graded Material Plate under Transverse Load for Varying Aspect Ratio, Int. J. Met. 2014 (2014) 1-11. doi: 10.1155/2014/980563. https://doi.org/10.1155/2014/980563

[8] A. F. Mota, M. A. R. Loja, J. I. Barbosa, and J. A. Rodrigues, Porous Functionally Graded Plates: An Assessment of the Influence of Shear Correction Factor on Static Behavior, Math.

Comput. Appl. vol. 25, no. 2 (2020) 1-26. doi: 10.3390/mca25020025. https://doi.org/10.3390/mca25020025

[9] D. Punera and T. Kant, Elastostatics of laminated and functionally graded sandwich cylindrical shells with two refined higher order models, Compos. Struct. vol. 182, no. January (2017) 505-523. doi: 10.1016/j.compstruct.2017.09.051. https://doi.org/10.1016/j.compstruct.2017.09.051

[10] K. Swaminathan and D. M. Sangeetha, Thermal analysis of FGM plates - A critical review of various modeling techniques and solution methods, Compos. Struct. 160 (2017) 43-60. doi: 10.1016/j.compstruct.2016.10.047. https://doi.org/10.1016/j.compstruct.2016.10.047

[11] M. Zidi, A. Tounsi, M. S. A. Houari, E. A. Adda Bedia, and O. Anwar Bég, Bending analysis of FGM plates under hygro-thermo-mechanical loading using a four variable refined plate theory, Aerosp. Sci. Technol. vol. 34, no. 1 (2014) 24-34. doi: 10.1016/j.ast.2014.02.001. https://doi.org/10.1016/j.ast.2014.02.001

[12] K. Swaminathan, D. T. Naveenkumar, A. M. Zenkour, and E. Carrera, Stress, vibration and buckling analyses of FGM plates-A state-of-the-art review, Compos. Struct. 120 (2015) 10-31. doi: 10.1016/j.compstruct.2014.09.070. https://doi.org/10.1016/j.compstruct.2014.09.070

[13] B. M. Shinde and A. S. Sayyad, A new higher order shear and normal deformation theory for FGM sandwich shells, Compos. Struct. 280 (2022) 1-25. doi: 10.1016/j.compstruct.2021.114865. https://doi.org/10.1016/j.compstruct.2021.114865

[14] B. M. Shinde and A. S. Sayyad, A new higher-order theory for the static and dynamic responses of sandwich FG plates, J. Comput. Appl. Mech. vol. 52, no. 1 (2021) 102-125. doi: 10.22059/JCAMECH.2020.313152.569.

[15] S. Yadav, S. Damse, S. Pendhari, K. Sangle, and A. S. Sayyad, Comparative studies between Semi-analytical and shear deformation theories for functionally graded beam under bending, Forces Mech. 8 (2022). 01-12. doi: 10.1016/j.finmec.2022.100111. https://doi.org/10.1016/j.finmec.2022.100111

[16] A. S. Sayyad and Y. M. Ghugal, Static and free vibration analysis of doubly-curved functionally graded material shells, Compos.Struct. 269 (2021) 1-17. doi:10.1016/j.compstruct.2021.114045. https://doi.org/10.1016/j.compstruct.2021.114045

[17] H.-T. Thai, T.-K. Nguyen, T. P. Vo, and J. Lee, Analysis of functionally graded sandwich plates using a new first-order shear deformation theory, Eur. J. Mech. - A/Solids. 45 (2014) 211-225. doi: 10.1016/j.euromechsol.2013.12.008. https://doi.org/10.1016/j.euromechsol.2013.12.008

[18] P. A. Demirhan and V. Taşkin, Static analysis of simply supported functionally graded sandwich plates by using four variable plate theory, Tek. Dergi/Technical J. Turkish Chamb. Civ. Eng. vol. 30, no. 2 (2019) 8987-9007. doi: 10.18400/tekderg.396672. https://doi.org/10.18400/tekderg.396672

Advanced Topics in Mechanics of Materials, Structures and Construction - AToMech1-2023 Materials Research Forum LLC
Materials Research Proceedings 31 (2023) 128-134 https://doi.org/10.21741/9781644902592-14

Constitutive modeling of dynamic strain aging in niobium

Arhum Hassan[1, a], and Farid Abed[1, 2, b*]

[1] Materials Science and Engineering Program, College of Arts and Sciences, American University of Sharjah, Sharjah 26666, United Arab Emirates

[2] Department of Civil Engineering, American University of Sharjah, Sharjah 26666, United Arab Emirates

[a] b00090812@aus.edu, [b] fabed@aus.edu

Keywords: Modified VA Model, Dynamic Strain Aging, Thermal Activation Energy, Niobium

Abstract. As the temperature rises, metals should lose strength. However, under some combinations of strain rate and temperature, they show a dramatic increase in strength due to the interaction of impurity/solute atoms with the dislocations, a phenomenon known as dynamic strain aging (DSA). Thermomechanical stress-strain curves have been modeled using a variety of numerical approaches, but accurately modeling DSA activation remains a challenge. The activation free energy for dislocation movement rises during DSA, as there are more solute atoms concentrated at the local barriers. As a result, we see an increase in strength. This work modifies the physical parameters-based VA model for commercially pure bcc Niobium, which is known to display DSA at low strain rates and elevated temperatures, in order to create a constitutive model that can accurately capture DSA.

Introduction

Constitutive model development aims for wide applicability and complicated material response. Successful models capture material static and dynamic behavior and obtain model constants from limited experimental data. Furthermore, metals with different crystal structures have been extensively researched for thermomechanical response over a wide strain rate and temperature range. JC [1], ZA [2], Nemat Nasser [3], and VA model [4] are some of the models devised by researchers to capture the thermomechanical response of materials. These models very well evidence the regular thermomechanical response. But when it comes to simulating dynamic strain aging (DSA), which is a sudden increase in the strength of a material at certain configurations of temperature and strain rates, these models have not yielded satisfactory results.

Applied stress causes shear in the material along with a movement of dislocations that is discrete in nature as the dislocations stop for a certain amount of waiting time (t_w) at every obstacle. During this time, a dislocation core atmosphere is formed, in which solute atoms diffuse into the mobile dislocations and create pinning effect. Breaking past this pinned dislocation barrier requires higher stress, thus increasing material strength known as DSA [5]. This is usually seen as a unique response, and it has been observed in many studies [6, 7]

The model must depend on microscopic constituents because solute atoms interact with dislocations to induce DSA. As discussed before, thermomechanical response is well-described by many models but they cannot capture DSA [1-4, 8-10], and have been modified in numerous works to model DSA. One such example is the recent work reported by Devotta et al. [11]. They modified the JC model for a ferritic-pearlitic C45E steel using a second-degree polynomial fit to account for the DSA. However, this approach was not adopted due to the empirical nature of the JC model.

Cheng and Nemat [12] presented a dislocation trough model for titanium and a constitutive model for predicting DSA by developing a relationship between the overall activation free energy, and the solute atom concentration. However, their model is only applicable at higher strain rates and is inapplicable at lower strain rates, whereas DSA is generally reported at lower strain rates [13, 14]. In contrast, the VA model is applicable over a wide range of temperatures and strain rates [4, 15-17]. It is a constitutive model based on physical parameters. However, it has been reported that the model could not be tested for DSA [4, 16, 17] as it does not consider the dependence of activation free energy (G_o) on the concentration of solute atoms at local obstacles. The solute concentration has a direct effect on the activation energy at local barriers and is responsible for the activation of dynamic strain aging in metals under thermomechanical loading conditions, and therefore, the VA model is modified in the current work to simulate DSA in commercially pure Niobium at low strain rates, as detailed later in this work.

In the Numerical Model Section, the VA model and its dependency on physical parameters are discussed. Later, the modification in the model is detailed. The application of the modified VA model to capture DSA in Nb is described in the Application, Comparisons and Discussion Section. In the same section, a brief description of how to obtain the various parameters for the modified VA model is provided, followed by the results and discussions. Finally, the study is concluded in the last section.

Constitutive Model

Voyiadjis-Abed (VA) Model. During plastic deformations, the movement of dislocations is constrained by either long- or short-range obstacles, resulting in an increase in the material's strength. Depending on the nature of obstacles, these can be overcome with or without thermal assistance. For bcc metals, if the barrier is long-ranged, thermal assistance is not necessary, but if it is short-ranged, it cannot be overcome without thermal influence. Plasticity in a metal is the flow of dislocations through both these long-range and short-range obstacles. VA model for bcc metals is developed considering these phenomena of flow stress by decomposing it into two components i.e. the athermal stress (σ_{ath}) component (based on long range obstacles) and the thermal stress (σ_{th}) component (based on short range obstacles) [4, 16] as shown in Eq. (1):

$$\sigma = \underbrace{Y_a + B\varepsilon_p^n}_{athermal\ stress} + \underbrace{\hat{Y}\left(1 - \left(\beta_1 T - \beta_2 T \ln\dot{\varepsilon}_p\right)^{\frac{1}{q}}\right)^{\frac{1}{p}}}_{thermal\ stress} \tag{1}$$

In athermal stress component (σ_{ath}), (Y_a) is the temperature-insensitive yield strength of the material, and B and n are temperature-insensitive thermal hardening constants. On the other hand, $\hat{Y}$ in the thermal stress component (σ_{th}) represents the minimum stress necessary to breach the Peierls barrier at 0 K. T is the operating temperature, whereas β_i is essentially the thermal activation parameter that depends on Boltzmann's constant (k), reference strain rate ($\dot{\varepsilon}_{po}^{i}$), and activation free energy (G_o). $\dot{\varepsilon}_p$ represents the desired strain rate. (1)The physical dependence of all the material parameters given in Eq. (1) on different micro and nano structural quantities (including grain size, dislocation densities and burgers vector) is detailed in the works by Abed et al. [4, 16, 18].

Dynamic Strain Aging. As discussed so far, the VA model falls short in detecting DSA activation because it does not account for the effect of dislocation's pinning due to variations in solute atom concentration on the activation free energy.

As noted previously, dynamic strain aging is caused by the interaction between impurity/solute atoms and dislocations. The diffusion of solute atoms impedes the mobility of these dislocations, resulting in an increase in the needed waiting time. To overcome these obstacles, a higher activation energy is required, resulting in an increase in the strength of the metal. In a metallic structure, the impurity elements such as C, N, and O are scattered in minute amounts throughout the available interstitial spaces of the structure. Due to a scarcity of available space, these atoms are under pressure, and they will always seek a larger site to minimize their energy whenever possible.

When stressed, metals shear. The deformation creates new sites that impurity atoms quickly occupy and pins down the moving dislocations. Consequently, the dislocation waiting time and the activation energy increase. Thus, under specific combinations of strain rate and temperature, a higher strength is observed.

In light of the preceding, the VA model has been modified to accommodate the DSA. It has been found that the thermal activation energy remains unaltered in the absence of DSA [4], whereas DSA enhances the thermal activation energy [12]. Cheng and Nasser [12] linked the binding free energy (G_B) per unit length to the activation free energy (G_o) as follows:

$$G_o \propto \sqrt{G_B G_D} w \tag{2}$$

The model states that the activation free energy is directly affected by the trough's effective width (w) and dislocation free energy per unit length (G_D)., the trough's effective width (w) and the free energy per unit length of dislocation (G_D) directly effects the activation free energy. However, we are interested in how binding free energy (G_B) depends on the concentration of solute atom (C). Cheng and Nasser represented this dependence as a product of the strength of interaction for each solute atoms (G_{BO}) with C as shown in Eq. (3).

$$G_B \propto G_{BO} C \tag{3}$$

Keeping all other parameters constant in Eqs. (2) and (3), it is evident that increasing solute concentration increases the activation free energy. The higher activation free energy, the greater the stress needed to displace dislocations. So, due to impurity/solute concentration, DSA is observed. A modification in Eq. (1) is utilized in the current work to capture DSA using this concept. The VA model modification captured DSA that was not captured when the total activation energy was constant. Next section explains this.

Applications, Comparisons and Discussions

The constitutive model from the previous section is compared to Nemat-Nasser and co-workers' experimental data for commercially pure Niobium [14]. The evaluation of model parameters is discussed first in this section. Later in this section, application of the model for estimating the flow stress and comparisons to experimental results is presented for Niobium.

Model parameters evaluation. Thermomechanical experimental data from Nemat Nasser and coworkers [14] determines all the parameters in Eq. (1). Generally, an increase in temperature decreases material strength until a critical temperature is reached, beyond which the strength becomes insensitive to temperature and the athermal stress component is defined at this temperature. Afterwards, the thermal hardening component is determined at a lower temperature where DSA is not observed. The details to find these parameters is given in the works by Voyiadjis and coworkers [4, 13, 15, 16, 18]. The dependency of the change in activation energy with the temperature and strain rates is beyond the scope of current work due to limited data available. The values for different parameters of VA model is given in Table 1.

Application and Comparisons. The VA model modification was verified by the thermomechanical modelling of Niobium (Nb), a silvery bcc metal, which is utilized to strengthen alloys in carbon and alloy steels. Nemat Nasser [14] and Voyiadjis et al. [4] successfully modeled Nb at higher strain rates in the absence of DSA but their model was incompatible at lower strain rates where DSA was observed.

Table 1 tabulates the parameter values for Niobium. As mentioned, to capture DSA, the total activation energy changed with temperature at the strain rate (0.001/s) as shown in Table 2. The activation free energy increased in the DSA range, rising till the peak position, and decreasing afterwards due to high temperature's assistance in dispersing solute atoms from dislocation core. Figure 1 shows the impact of this activation energy change. VA model could not reflect DSA, but the modification made it successful in doing so.

Table 1 Model parameters for Niobium.

Parameter	Y_a [MPa]	B [MPa]	n	$\hat{Y}$ [MPa]	p	q	$\dot{\varepsilon}^i_{po}$ [s^{-1}]	β_2 [K^{-1}]
Value	25	300	0.23	1450	0.36	1.84	7.07 x 10^6	5.49 x 10^{-5}

Table 2 Thermal Activation Energy (G_o) for the VA model to capture DSA at $\dot{\varepsilon}_p$ = 0.001/s

Temperature [K]	293, 400 and 800	500	600	700
G_o **Value [eV]**	1.57	2.2	3.4	2.8

Figure 2 (a-f) displays the results of a comparison between the experimental data and the numerical models with and without the influence of DSA. It is evident that the VA model does not account for the activation of DSA between 500 and 700 K. But when the proposed modification was accounted for, the model successfully predicted DSA activation (Figure 2 (c-e)). The proposed modification captured DSA activation quite well for Nb, as evidenced by the comparison given.

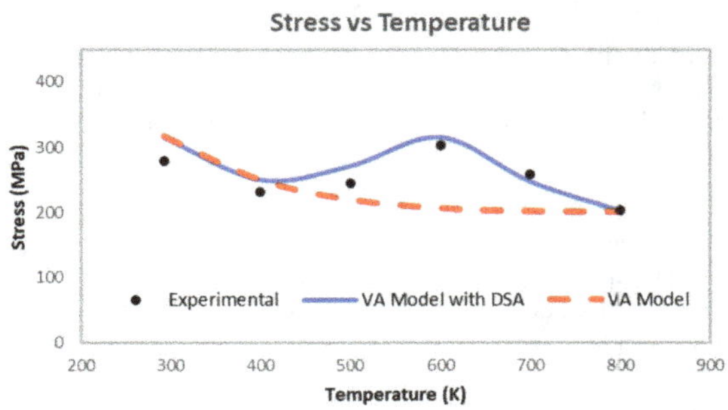

Figure 1. Stress-temperature experimental results compared to VA and modified VA model.

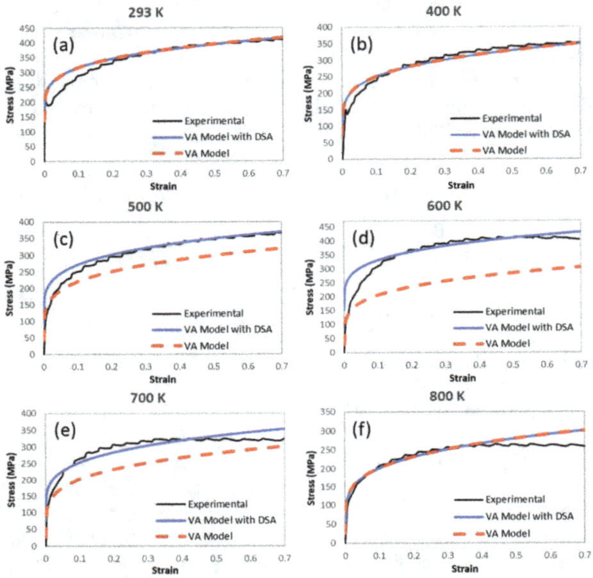

Figure 2. VA model, modified VA model, and experimental stress-strain curves at $\dot{\varepsilon}_p$ = 0.001/s and T = (a) 293 K, (b) 400 K, (c) 500 K, (d) 600 K, (e) 700 K, and (f) 800 K.

Figure 2 shows that VA model results fitted experimental data when DSA was not activated. On the other hand, the modified VA model with DSA inclusion captures DSA inactivation and activation for commercially pure Niobium at low strain rates very efficiently.

Conclusions

This article proposes a VA model modification to capture DSA in commercially pure Niobium using experimental data from Dr. Nemat Nasser. The highlights include:

- An accumulation of solute atoms at local barriers slows down dislocation motion. To get around this, more activation energy is needed, leading to DSA.
- Dislocations pin at thermal stress controlling short-range obstacles in bcc metals and activate DSA.
- Niobium showed DSA at 500–700 K and $\dot{\varepsilon}_p$ = 0.001/s.
- During DSA, activation energy rises as temperature rises, reaching a maximum at the temperature at which the effect is most pronounced. Then, it begins to decline as the higher temperatures facilitate the dispersion of solute atoms from the obstacles.

The modified VA model accurately predicted the stress increase necessary to break through the pinned dislocation-solute atmosphere barrier during DSA. With the model's eventual incorporation into commercial finite element software, it will be possible to forecast a wide variety of scenarios using DSA, greatly increasing the field's potential applications.

References

[1] Johnson, G.R. and W.H. Cook, Fracture characteristics of three metals subjected to various strains, strain rates, temperatures and pressures. Engineering fracture mechanics, 1985. 21(1): p. 31-48. https://doi.org/10.1016/0013-7944(85)90052-9

[2] Zerilli, F.J. and R.W. Armstrong, Dislocation-mechanics-based constitutive relations for material dynamics calculations. Journal of applied physics, 1987. 61(5): p. 1816-1825. https://doi.org/10.1063/1.338024

[3] Nemat-Nasser, S. and J. Isaacs, Direct measurement of isothermal flow stress of metals at elevated temperatures and high strain rates with application to Ta and TaW alloys. Acta Materialia, 1997. 45(3): p. 907-919. https://doi.org/10.1016/S1359-6454(96)00243-1

[4] Voyiadjis, G.Z. and F.H. Abed, Effect of dislocation density evolution on the thermomechanical response of metals with different crystal structures at low and high strain rates and temperatures. Archives of Mechanics, 2005. 57(4): p. 299-343.

[5] Van den Beukel, A., Theory of the effect of dynamic strain aging on mechanical properties. Physica status solidi (a), 1975. 30(1): p. 197-206. https://doi.org/10.1002/pssa.2210300120

[6] Ghazisaeidi, M., L. Hector Jr, and W. Curtin, Solute strengthening of twinning dislocations in Mg alloys. Acta materialia, 2014. 80: p. 278-287. https://doi.org/10.1016/j.actamat.2014.07.045

[7] Mulford, R. and U. Kocks, New observations on the mechanisms of dynamic strain aging and of jerky flow. Acta Metallurgica, 1979. 27(7): p. 1125-1134. https://doi.org/10.1016/0001-6160(79)90130-5

[8] Abed, F., A. Abdul-Latif, and G.Z. Voyiadjis, Performance of MMFX steel rebar at elevated temperatures. Journal of Engineering Mechanics, 2020. 146(11): p. 04020126. https://doi.org/10.1061/(ASCE)EM.1943-7889.0001862

[9] Abed, F.H. and G. Voyiadjis, Adiabatic shear band localizations in BCC metals at high strain rates and various initial temperatures. International Journal for Multiscale Computational Engineering, 2007. 5(3-4). https://doi.org/10.1615/IntJMultCompEng.v5.i3-4.120

[10] Voyiadjis, G. and F. Abed, Transient localizations in metals using microstructure-based yield surfaces. Modelling and Simulation in Materials Science and Engineering, 2006. 15(1): p. S83. https://doi.org/10.1088/0965-0393/15/1/S08

[11] Moris Devotta, A., et al., A modified Johnson-Cook model for ferritic-pearlitic steel in dynamic strain aging regime. Metals, 2019. 9(5): p. 528. https://doi.org/10.3390/met9050528

[12] Cheng, J. and S. Nemat-Nasser, A model for experimentally-observed high-strain-rate dynamic strain aging in titanium. Acta materialia, 2000. 48(12): p. 3131-3144. https://doi.org/10.1016/S1359-6454(00)00124-5

[13] Abed, F.H., et al., Flow stress and damage behavior of C45 steel over a range of temperatures and loading rates. Journal of Engineering Materials and Technology, 2017. 139(2). https://doi.org/10.1115/1.4035488

[14] Nemat-Nasser, S. and W. Guo, Flow stress of commercially pure niobium over a broad range of temperatures and strain rates. Materials Science and Engineering: A, 2000. 284(1-2): p. 202-210. https://doi.org/10.1016/S0921-5093(00)00740-1

[15] Abed, F.H. and G.Z. Voyiadjis, Plastic deformation modeling of AL-6XN stainless steel at low and high strain rates and temperatures using a combination of bcc and fcc mechanisms of

metals. International Journal of Plasticity, 2005. 21(8): p. 1618-1639. https://doi.org/10.1016/j.ijplas.2004.11.003

[16] Voyiadjis, G.Z. and F.H. Abed, Microstructural based models for bcc and fcc metals with temperature and strain rate dependency. Mechanics of Materials, 2005. 37(2-3): p. 355-378. https://doi.org/10.1016/j.mechmat.2004.02.003

[17] Abdul-Latif, A., et al., Effect of various temperatures and strain-rates combinations on the thermomechanical behavior of 42CrMo steel. Archives of Mechanics, 2022. 74(5).

[18] Abed, F.H., S.I. Ranganathan, and M.A. Serry, Constitutive modeling of nitrogen-alloyed austenitic stainless steel at low and high strain rates and temperatures. Mechanics of Materials, 2014. 77: p. 142-157. https://doi.org/10.1016/j.mechmat.2014.07.007

Advanced Topics in Mechanics of Materials, Structures and Construction - AToMech1-2023 Materials Research Forum LLC
Materials Research Proceedings 31 (2023) 135-145 https://doi.org/10.21741/9781644902592-15

Best theory diagram using genetic algorithms for composite plates

M.A. Hinostroza[1*], J.L. Mantari[1,2]

[1]Faculty of Mechanical Engineering, National University of Engineering, Av. Túpac Amaru 210, Rimac, Lima, Perú

[2]Department of Science, University of Engineering and Technology (UTEC), Medrano Silva 165, Barranco, Lima, Peru

Keywords: Composite Plates, Genetic Algorithm, Best Theory Diagram, Carrera Unified Formulation (CUF), Machine Learning

Abstract. Composite structures offer a practical approach for many engineering applications, but their design is complex and can result in excessive sizing due to limitations in current modeling techniques. BTDs minimize the number of unknown variables in a kinematic theory for desired accuracy or for a fixed error in the Carrera Unified Formulation. This paper presents a method for computing Best Theory Diagrams (BTDs) for laminated composite plates using Genetic Algorithms (GA). As reported in previous papers by the authors, a multi-objective optimization technique using a GA is applied to build BTDs for a given structural problem. The plate models stresses and displacements are compared to those of a reference solution, and a plate model performance is quantified in terms of the number of unknown variables, the mean error and standard deviation of the stresses and displacements. Also, with the objective of reducing the computational time, a Neural-Networks (NN) was trained to reproduce the mean error and standard deviation of the stresses and displacements for any plate model refined from a reference plate model is addressed. Numerical simulations were computed for laminated composite plates with previously uninvestigated boundary conditions and compare computational time for BTD calculation. The preliminary results show that the use of multi-objective GA plus NN method reduces considerably the computation time to build BTDs.

Introduction

To accurately predict stress distributions in modern composite structures, appropriate modeling formulations are necessary to account for their complex mechanical behavior, including both normal and transverse stress components. It is essential to have analysis tools that balance accuracy with computational efficiency. Consequently, the literature contains numerous articles on high-order models for composite structures [1]. The present work is embedded in the framework of the Carrera Unified Formulation (CUF). According to CUF, the displacement field for plate analysis is defined as an expansion of the thickness coordinate [2]. High-order theories are beneficial for response analysis, but they come at the cost of high computational expense. To address this issue, Carrera and Petrolo [3] developed the AXiomatic/Asymptotic method, which identifies unnecessary terms in a plate model for a specific output, such as displacement or stress. By eliminating these terms, a refined plate model with fewer unknown variables can be obtained without sacrificing accuracy. However, these refined plate models are problem-specific. Carrera and Miglioretti [4] extended the AXiomatic/Asymptotic method to create the Best Theory Diagram (BTD), which evaluates all combinations of terms in a full plate model. The BTD represents refined plate models in a plot that shows the number of terms versus the error. The "best" refined plate models with the least error for a fixed number of terms form the BTD.

Machine learning techniques such as neural networks and genetic algorithms have become increasingly popular in recent years for solving problems that involve a large number of variables, high levels of uncertainty, and rapidly changing behavior. These techniques have been applied in numerous fields, including computational mechanics, where they have been used to develop multiscale elements and data-driven solvers. In Ref. [4], a genetic algorithm was employed to construct Best Theory Diagrams (BTDs) with lower computational cost. Yarasca et al. [6] introduced a Multi-objective Optimization Method to create BTDs for sandwich plates, and later Mantari et al. [7] investigated the use of neural networks to reduce computation time in BTD construction.

In this study, a method for computing Best Theory Diagrams (BTDs) for composite plates using Genetic Algorithms (GA) is presented. This study uses mathematical formulation and benchmarks from prior research [7], with a new GA technique [8] and cost function for optimization. The study will perform numerical simulations for laminated composite plates with previously uninvestigated boundary conditions and compare computational time for BTD calculation. Also, the use of Neural Network models combined with Genetic algorithms is investigated.

Preliminaries

The Best Theory Diagrams (BTDs):

The construction of reduced models through axiomatic/asymptotic methods, can lead to a diagram in which, for a given problem, each reduced model is associated with the number of active terms and its error computed on a reference solution. This diagram, Fig.1, allows editing an arbitrary given theory to get a lower number of terms for a given error, or to increase the accuracy while keeping the computational cost constant. Considering all the reduced models, it is possible to recognize that some of them provide the lowest error for a given number of terms. These models represent a Pareto front for this specific problem. As in Ref. [4], the Pareto front is defined as the best theory diagram (BTD). This curve is case-dependent since it changes for several problems, i.e., different materials, geometries, boundary conditions, and output parameters. If a single output parameter is selected (only one displacement or one component of the stress tensor), the BTD considers the number of active terms and the error of the selected output parameter computed on a reference solution. Investigations of this sort have been reported in Ref. [9]. It is worth noticing that, although the output parameter may be freely selected, only one at a time can be investigated. This latter limitation has been removed by the multi-objective BTDs proposed by Mantari et al. [10], where multiple output parameters can be investigated in a single analysis.

Methodology and Proposed Algorithm

Mathematical Model: Carrera Unified Formulation for plates:

Figure 2 illustrates the geometry and coordinate system of a multilayered plate comprising L layers, where the in-plane coordinates are denoted by x and y, and the thickness coordinate is denoted by z. The layer number, denoted by the integer k, represents the layer's position in the multilayered plate, counting from the bottom to the top surface.

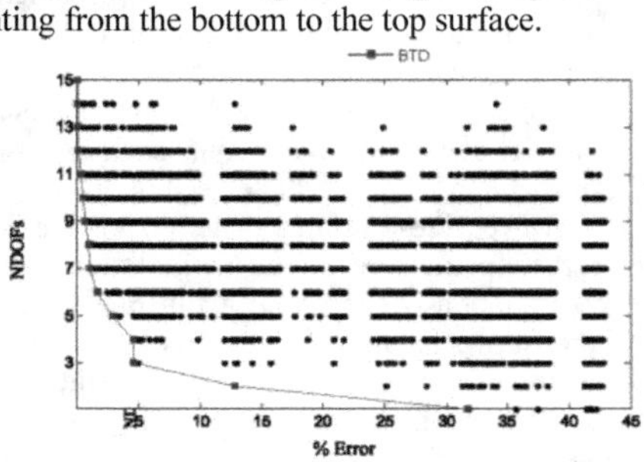

Figure 1: Body Theory Diagram on ED4, aluminum path, a/h=2.5, Mantari et al. [10].

According to CUF, the displacement field of a plate structure can be written as follows:

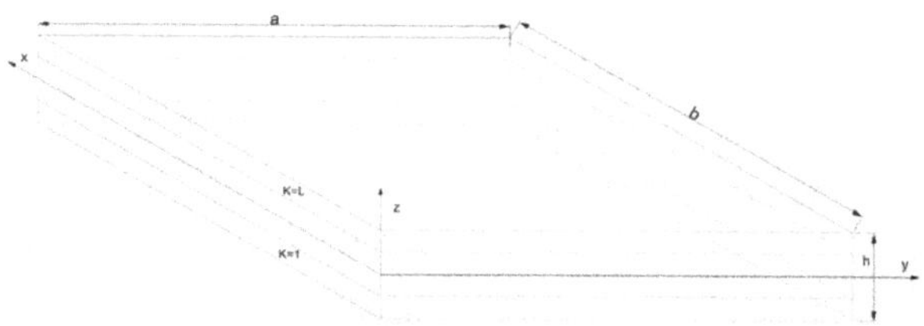

Fig. 2. Plate geometry and reference system.

$$\begin{cases} u_x(x,y,z) = F_1(z)u_{x_1}(x,y) + F_2(z)u_{x_2}(x,y) + \cdots + F_{N_{\exp}}(z)u_{x_{N_{eqp}}}(x,y), \\ u_y(x,y,z) = F_1(z)u_{y_1}(x,y) + F_2(z)u_{y_2}(x,y) + \cdots + F_{N_{exp}}(z)u_{y_{N_{eq}}}(x,y) \\ u_z(x,y,z) = F_1(z)u_{z_1}(x,y) + F_2(z)u_{z_2}(x,y) + \cdots + F_{N_{\exp}}(z)u_{z_{\exp}}(x,y) \end{cases} \quad (1)$$

In compact form:

$$\boldsymbol{u}(x,y,z) = F_\tau(z) \cdot \boldsymbol{u}_\tau(x,y)\tau = 1, z, \dots, N_{\exp} \quad (2)$$

where $\boldsymbol{u}$ is the displacement vector whose components (u_x; u_y; u_z) are the displacements along the x, y and z reference axes. F_τ are the expansion functions and $\boldsymbol{u}_\tau$ (u_{x_τ} ; u_{y_τ} ;u_{z_τ}) are the displacements variables. $N_{\exp}$ is the number of terms of the expansion. According to the equivalent single layer scheme, a multilayered heterogeneous plate is analyzed as a single equivalent lamina. In this case, F_τ functions can be considered as functions of z defined as $F_\tau = z^{\tau-1}$. The number of unknown variables is independent of the number of plate layers. The equivalent single layer models are indicated as EDN, where N is the expansion order. In this paper, the 'zig-zag' function proposed is employed. The equivalent single layer models considering Murakami 'zig-zag' function are indicated as EDZN. An example of an EDZ4 displacement field is reported as:

$$\begin{aligned} u_x &= u_{x_1} + zu_{x_2} + z^2u_{x_3} + z^3u_{x_4} + (-1)^k\zeta_k u_{x_5} \\ u_y &= u_{y_1} + zu_{y_2} + z^2u_{y_3} + z^3u_{y_4} + (-1)^k\zeta_k u_{y_5} \\ u_z &= u_{z_1} + zu_{z_2} + z^2u_{z_3} + z^3u_{z_4} + (-1)^k\zeta_k u_{z_5} \end{aligned} \quad (3)$$

where $\zeta_k = 2z_k/h_k$ is a non-dimensional layer coordinate and h_k the thickness of the k-layer. On the other hand, layer-wise models can be conveniently built using Legendre's polynomials expansions in each layer. Detailed description of equation derivation can be found in [7].
In this paper, layer-wise models are denoted by the acronym as LDN, where N is the expansion order. An example of LD4 layer displacement field:

$$\begin{aligned} u_x^k &= F_t u_{x_t}^k + F_2 u_{x_2}^k + F_3 u_{x_3}^k + F_4 u_{x_4}^k + F_b u_{x_b}^k \\ u_y^k &= F_t u_{y_t}^k + F_2 u_{y_2}^k + F_3 u_{y_3}^k + F_4 u_{y_4}^k + F_b u_{y_b}^k \\ u_z^k &= F_t u_{z_t}^k + F_2 u_{z_2}^k + F_3 u_{z_3}^k + F_4 u_{z_4}^k + F_b u_{z_b}^k \end{aligned} \quad (4)$$

Finite element approximation:
A classical Finite Element technique is used to easily deal with arbitrary shaped cross-sections. The generalized displacement vector is given by:

$$\boldsymbol{u}_\tau(y) = N_i(y)\boldsymbol{q}_{\tau i} \quad (5)$$

where N_i are the shape functions and $q_{\tau i}$ is the nodal displacement vector:

$$\boldsymbol{q}_{\tau i} = \left\{q_{u_{x_{\tau i}}} q_{u_{y_{\tau i}}} q_{u_{z_{\tau i}}}\right\}^T \tag{6}$$

For the sake of brevity, the functions are not listed here, they can be found in Carrera Formulation. The functions are defined in the natural coordinates and transpose in the real coordinate in according with the isoperimetric formulation. The stiffness matrix of the elements and the external loadings are obtained via the Principle of Virtual Displacements (PVD):

$$\delta L_{int} = \int_V (\delta \boldsymbol{\varepsilon}_p^T \boldsymbol{\sigma}_p + \delta \boldsymbol{\varepsilon}_n^T \boldsymbol{\sigma}_n) dV = \delta L_{ext} \tag{7}$$

Where L_{int} stands for the strain energy, L_{ext} is the work of external loadings and δ stands for virtual variation. The PVD for a multilayered plate structure reads, a detailed derivation of equation can be found n [7]:

$$\delta \boldsymbol{q}^{ksj}: \qquad \boldsymbol{K}^{k\tau sij} \boldsymbol{q}^{k\tau i} = \boldsymbol{P}^{ksj} \tag{8}$$

where $\boldsymbol{P}^{ksj}$ is a 3 x 1 matrix, called fundamental nucleus of the external load. $\boldsymbol{q}^{k\tau i}$ and $\delta \boldsymbol{q}^{ksj}$ are the nodal displacements and its variation respectively.

Proposed Genetic Algorithm Optimization Method:

Refined plate theories provide improved accuracy and the ability to detect non-classical effects, but the higher number of displacement variables required leads to higher computational costs. To minimize the computational cost required to construct a Best Theory Diagram (BTD), a genetic algorithm (GA) is employed. The GA evaluates a set of random refined models, referred to as the population, over multiple generations until the final generation's BTD converges. The Axiomatic/Asymptotic method, developed by Carrera and Petrolo [3], addresses this issue by enabling the identification of the role of each variable in computing a specific displacement stress output variable. This method involves evaluating every potential plate model combination resulting from deactivating each term. Hence, the number of evaluations needed is potential, 2^M, where M the number of deactivated terms. A graphical notation is introduced to represent the results. This consists of a table with three rows, and some columns equal to the number of the displacement variable used in the expansion. As an example, an LD4 model for a two-layer plate is shown in Table 1 (full model). Table 1 also shows a refined model in which the term in the first, where, squared symbols (■) means Non-deactivable term (due to material continuity), empty-triangle (△) is Inactive term, and filled-triangle (▲) is active term.

Table 1: Example model representation.

Full model representation									Reduced model representation								
■	▲	▲	▲	■	▲	▲	▲	■	■	▲	▲	▲	■	△	▲	▲	■
■	▲	▲	▲	■	▲	▲	▲	■	■	▲	▲	▲	■	▲	▲	▲	■
■	▲	▲	▲	■	▲	▲	▲	■	■	△	▲	▲	■	▲	▲	▲	■

For multiple displacement/stress outputs, each output parameter has a given error which is computed simultaneously. The optimization method objectives functions are the number of terms in the refined models, the mean error and the standard deviation of the stresses and displacements. After the Axiomatic/Asymptotic method is employed, the mean error and the standard deviation is replaced by a new objective function denoted by $\%Error_{ul}$. $\%Error_{ul}$ is the sum of the mean error and the standard deviation. In this study the error is defined as:

Materials Research Proceedings 31 (2023) 135-145 https://doi.org/10.21741/9781644902592-15

$$\% \, Error = 100 W_r \frac{\sum_{i=1}^{N_p} |Q^i - Q_{ref}^i|}{max|Q_{ref}| \cdot N_p} \tag{9}$$

where W_r is the vector of weights for the new optimization function, as has the value of:

$$W_r = \begin{cases} 2, & if\ r = 1,2,3 \\ 1, & otherwise \end{cases}$$

This weight gives priority to the u_x, u_y and u_z error: The used of weighs to improve estimations was proposed in [11].

In this way, a 2-dimensional Pareto front, the so-called BTD, is built selecting the best plate theories considering the objective functions: number of terms and $\%Error_{ul}$. It is important to remark that $\%Error_{ul}$ is not exactly the upper limit error of the output parameters, but an indicator employed for comparative purposes.

In this study, the Kerry-Lothrop GA from [8] is utilized due to its demonstrated high performance across various research areas. Typically, the classical approach involves converting all variable types to binary design variables, either explicitly or through a user interface. This conversion process was implemented using the Matlab Genetic Algorithm Optimization Toolbox. The Multi-objective optimization technique flowchart is presented in Fig. 3. In this procedure, the most time-consuming step is the FEA of the refined plate models. If the number of generations and/or the population size is too large, the procedure can become unviable. For this reason, a Neural-Network (NN) is implemented to replace the FEA, as was addressed in Yarasca et al. [7]. The NN consists of simple processing units, the neurons, and directed, weighted connections between those neurons. The neurons are connected to different layers. Once the number of layers and neurons per layer are set, the NN starts the training procedure. The training is an iterative process. To initiate the training of a neural network (NN), a set of initial weights is randomly selected in the first iteration. A fixed number of random cases with predetermined inputs and outputs are used to compute the weights. Next, the NN output for the given inputs is compared to the desired outputs, and the errors are then propagated back to the NN to adjust the weights. This process continues for each iteration until the NN reaches an acceptable global error and converges. In the current study, a population of 1000 and 100 generations were used for the genetic algorithm (GA), and the NN utilized 2000 training samples. The NN architecture consisted of 15 neurons with three layers for EDZ4 models and four layers for LD4 models.

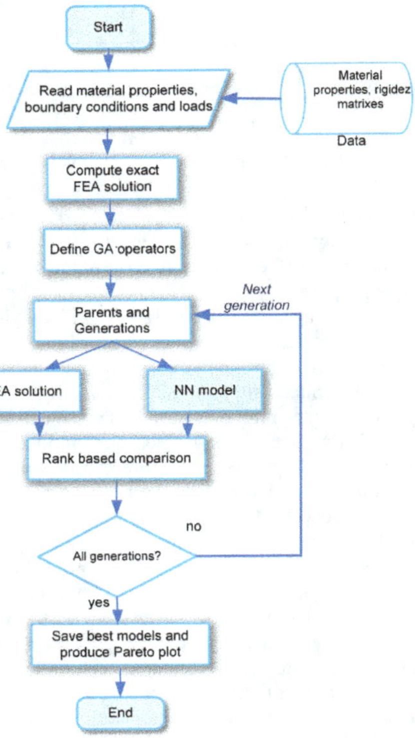

Figure 3: Multi-objective optimization technique flowchart, using FEA and Neural Networks

Numerical Results and Discussion:

This section presents the results obtained from the new multi-objective optimization technique. A transverse uniformly distributed pressure was applied at the top surface of a square composite plate with equal side lengths *(a = b)* and thickness (*h*), as shown in Figure 2. The study focused on a laminated composite plate subjected to three different sets of boundary conditions: simple support on all four sides, clamped on all four sides, and clamped-free with opposing sides having the same boundary condition (CFCF). Due to space limitations, this paper only presents the results for CFCF, while the complete set of results will be presented in a forthcoming publication.The reduced models are developed for the displacements u_x, u_y, u_z, and the stresses $\sigma_{xx}, \sigma_{yy}, \sigma_{zz}, \tau_{xy}, \tau_{xz}$ and τ_{yz}. The following normalized quantities are defined for the displacements and stresses:

$$\bar{u}_x = \frac{u_x \cdot E_2^{k=1} \cdot h^2}{\bar{p}_z \cdot a^3}, \bar{u}_y = \frac{u_y \cdot E_2^{k=1} \cdot h^2}{\bar{p}_z \cdot a^3}, \bar{u}_z = \frac{u_z \cdot 100 \cdot E_2^{k=1} \cdot h^3}{\bar{p}_z \cdot a^4} \tag{10}$$

$$\bar{\sigma}_{xx,yy} = \frac{\sigma_{xx,yy}}{\bar{p}_z \cdot (a/h)^2}, \bar{\sigma}_{zz} = \frac{\sigma_{zz}}{\bar{p}_z}, \bar{\tau}_{xy} = \frac{\tau_{xy}}{\bar{p}_z \cdot (a/h)^2}, \bar{\tau}_{xz,yz} = \frac{\tau_{xz,yz}}{\bar{p}_z \cdot (a/h)},$$

where k =*1* identifies the bottom layer; $\bar{u}_x$ and $\bar{\tau}_{xz}$ are calculated at *x=0, y=b/2*; $\bar{u}_y$ and $\bar{\tau}_{yz}$ are calculated at *x=a/2, y=0*; $\bar{u}_z, \bar{\sigma}_{xx}, \bar{\sigma}_{yy}$ and $\bar{\sigma}_{zz}$ are calculated at *x= a/2, y=b/2* and $\bar{\tau}_{xy}$ is calculated at *x=y=0*. The stresses $\bar{\tau}_{xz}$, $\bar{\tau}_{yz}$ and $\bar{\sigma}_{zz}$ obtained from the EDZ4 and LD4 plate model were computed via the indefinite equilibrium equations of 3D elasticity. In this paper, an LD4 model was employed as the reference solution. Also, comparisons with the work presented at Mantari et al. [7] are addressed. The results reported in Refs. [12] show that the LD4 plate model is in good agreement with the three-dimensional exact elasticity result. The GA used here was designed by Lothrop (2003) and suffered only minor changes in order to be applied to this problem. The floating-point representation was chosen for all variables. Those parameters that were problem specific are provided in Table 2.

Table 2: Considered genetic algorithms parameters.

Population	1000
Generations	30
Xover	70% for 30 pairs
Mutate	5%

Laminated Plates:
Laminated composited plates with different length-to-thickness ratios, boundary conditions, and lamination sequences were investigated using an EDZ4 plate model. In the examples considered, the individual laminae were considered of equal thickness and the following set of material properties was used for each lamina: $E_{L/E_T} = 25; G_{LT}/E_T = 0.5;\ G_{TT}/E_T = 0.2; \nu_{LT} = \nu_{TT} = 0.25$. The length-to-thickness ratio *a/h=5* and lamination sequences 0/90, and the combinations of boundary conditions *CFCF*, were studied. The BTDs for the case studies are presented in Fig. 4. From this plot as was expected, GA+NN is faster but has the drawback of being a little more errors in the models optimized. The refined plate model's accuracies are reported in Table 3. Selected displacements and stress through the thickness distributions are shown in Figs. 5. The notation used is the following: the refined models built are indicated as N-*hybrid* refined model (N-HRM); where *N* is the number of variables in the HRM. The reference solution (in this case, LD4) is included for comparison purposes. To verify the correct convergence of the GA and NN to the Axiomatic/Asymptotic method's BTD, a comparison between the different BTDs is shown in Fig. 4. The BTD denoted by AAM (Axiomatic/ Asymptotic method) is built evaluating every refined model from the full EDZ4 plate model. As can be observed, the BTDs obtained are in complete agreement. Fig. 4 shows that the NN precision improves for high length-to-thickness and simply supported boundary conditions.

For the sake of reproducibility, the selected refined plate models from the BDTs are reported in Tables 3. The number of active terms is indicated by M_E. This is expected since for high length-to-thickness ratio and simply supported boundary conditions, the number of refined plate models with adequate results in the population increases. Therefore, the NN prediction improves because of the reduction in population performance variability. For that reason, hereafter only low length-to-thickness ratios are investigated.

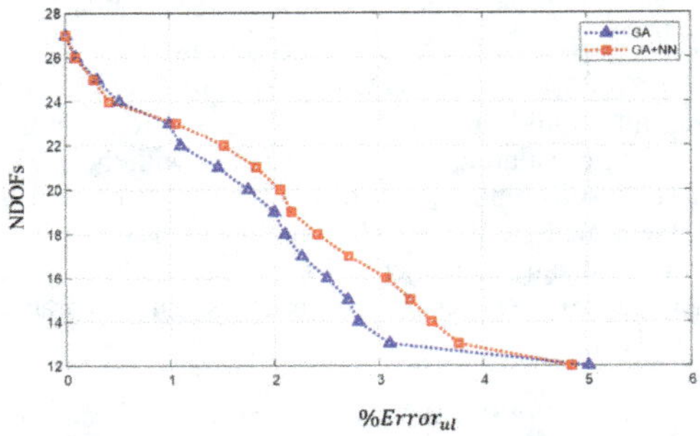

Figure 4. BTDs for laminated composite plate, 0/90, a/ h=5, CFCF.

Table 3: Refined EDZ4 models, laminated composite plate, 0/90, a/h=5, CFCF.

M_E=15/27									M_E=21/27								
■	▲	▲	△	■	△	▲	△	■	■	▲	▲	▲	■	△	▲	▲	■
■	▲	△	△	■	▲	△	△	■	■	▲	△	▲	■	▲	△	△	■
■	▲	△	△	■	△	△	△	■	■	▲	▲	△	■	△	▲	▲	■

M_E=24/27									M_E=26/27								
■	▲	▲	▲	■	▲	△	▲	■	■	▲	▲	▲	■	▲	▲	▲	■
■	▲	▲	△	■	▲	▲	△	■	■	▲	▲	▲	■	▲	▲	△	■
■	▲	▲	▲	■	▲	▲	▲	■	■	▲	▲	▲	■	▲	▲	▲	■

The NN can predict the results in terms of mean error and standard deviation with acceptable accuracy. Concerning the EDZ4 plate model, Fig. 5 and Table 4 show that EDZ4 refined plate models are insufficient to simulate laminated plates with CFCF boundary conditions and asymmetric laminations such as 0/90. The NN implementation makes the multi-objective optimization method practical in terms of computational cost. In this study, the computation cost for the GA population presented in Table 2 and FEA solution are presented in Table 5. In this table for $M_E = 5$ is 16.1 minutes and $M_E = 15$ is 23.3 minutes. The computational time is reduced when NN, The NN can predict the results in terms of mean error and standard deviation with acceptable accuracy, the Neural Network herein used in a NN trained to reproduce the mean error and standard deviation of the stresses and displacements, the mode is composed by set to 15 neurons with 3 layer. The new computation time are for $M_E = 5$ is 8.2 minutes and $M_E = 15$ is 9.6 minutes. Table 5 also shows a comparison of the computational cost between the present work and the method presented in Mantari et al. [7]. From this table, it is possible to observe a better performance of the presented work. However, it is important to notice that this method computes one theory for a determined number of active terms, while the Mantari et al. [7] method performs all the theories at once.

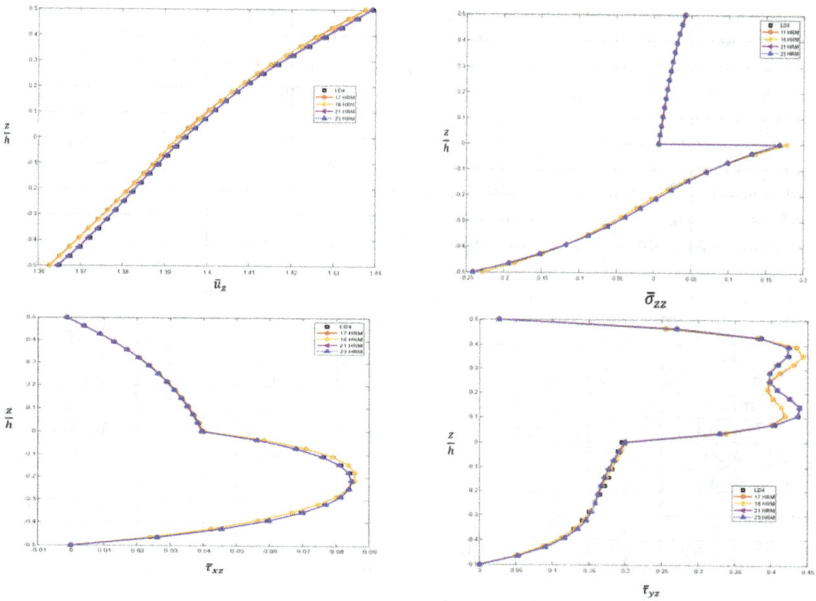

Figure 5: Selected displacement and stresses distribution for laminated composite plate, 0/90, a/h=5,CFCF.

Table 4: Error percentages of the refined EDZ4 models, laminated plate, 0/90, a/ h=5, CFCF.

a/h=5									
CFCF									
M_E	$\bar{u}_x$	$\bar{u}_y$	$\bar{u}_z$	$\bar{\sigma}_{xx}$	$\bar{\sigma}_{yy}$	$\bar{\tau}_{xy}$	$\bar{\tau}_{xz}$	$\bar{\tau}_{yz}$	$\bar{\sigma}_{zz}$
5	0.853	1.299	0.717	1.031	1.619	1.752	1.164	3.785	3.007
7	0.741	0.062	0.128	1.064	0.138	1.688	1.221	1.641	2.401
10	0.360	0.013	0.033	0.157	0.024	1.009	0.217	0.489	1.628
12	0.176	0.009	0.021	0.101	0.011	0.348	0.158	0.488	0.499
15	0.135	0.000	0.000	0.002	0.000	0.226	0.002	0.001	0.003

Table 5. Computation time for laminated composite plate, 0/90, a/ h=5, CFCF

Number of active terms	GA+FEA	GA+NN
M_E=5	16.1 min	8.2 min
M_E=15	23.3 min	9.6 min
Mantari et al. [7], raw	90 min	6.71 min
Mantari et al. [7], optimized	90 min	4.42 min

Conclusions and Future work

In this article, we describe the development of equivalent single layer and layer-wise plate models for laminated composite plates, which offer 3D-like accuracy and optimized computational cost. The Best Theory was built using the axiomatic/asymptotic method, a genetic algorithm, and the neural network, with output parameters including all displacements and stresses. By replacing finite element analysis with a neural network, computational time is drastically reduced. Our results present the BTDs, displacements and stresses for a 0/90 Laminated composite plate CFCF boundary condition. The main conclusion drawn from our findings is that the implementation of the neural network significantly reduces the computational time required to build BTDs. Also, this study proposes a different GA algorithm [8] with new optimization function, using weights, than in Mantari [7], and better performance in computational cost was found. Future work will be concentrated is carry out numerical simulations for more benchmarks, different boundary conditions and different materials.

Acknowledgments

This paper was written in the context of the project: "Desarrollo de un algoritmo autónomo y óptimo de mecánica computacional para un análisis de estructuras complejas impresa con tecnología 3D, utilizando inteligencia artificial y algoritmos genéticos" founded by CONCYTEC under the contract number N°060-2021-PROCIENCIA. The authors of this manuscript appreciate the financial support from the Peruvian Government.

References

[1] Noor AK, Burton WS. Assessment of shear deformation theories for multilayered composite plates. Appl Mech Rev 1989;42:1-13. https://doi.org/10.1115/1.3152418

http[2] Carrera E, Brischetto S, Nali P. Plates and shells for smart structures classical and advanced theories for modeling and analysis. Wiley; 2011. https://doi.org/10.1002/9781119950004

http[3] Carrera E, Petrolo M. Guidelines and recommendation to construct theories for metallic and composite plates. AIAA J 2012;48(12):2852-66. https://doi.org/10.2514/1.J050316

http[4] Carrera E, Miglioretti F. Selection of appropriate multilayered plate theories by using a genetic like algorithm. Compos Struct 2012;94(3):1175-86. https://doi.org/10.1016/j.compstruct.2011.10.013

http[5] Koutsourelakis Phaedon-Stelios. Stochastic upscaling in solid mechanics: an exercise in machine learning. J Comput Phys 2007;226(1):301-25. https://doi.org/10.1016/j.jcp.2007.04.012

http[6] Yarasca J, Mantari JL. N-objective genetic algorithm to obtain accurate equivalent single layer models with layerwise capabilities for challenging sandwich plates. Aero Sci Technol 2017;70:170-88. https://doi.org/10.1016/j.ast.2017.07.035

http[7] Mantari, J. L., J. Yarasca, F. G. Canales, and R. A. Arciniega. "New methodology for the construction of best theory diagrams using neural networks and multi-objective genetic algorithm." Composites Part B: Engineering 176 (2019): 107126. https://doi.org/10.1016/j.compositesb.2019.107126

http[8] Lothrop, K.W. (2003). Conceptual design optimization of a Cis-Lunar transportation architecture using genetic algorithms. Thesis.

http[9] M. Petrolo, A. Lamberti, and F. Miglioretti, "Best theory diagram for metallic and laminated composite plates," Mech. Adv. Mater. Struct., vol. 23, no. 9, pp. 1114-1130, 2016. https://doi.org/10.1080/15376494.2015.1121525

http[10] Mantari, Jose Luis, Jorge Yarasca, Marco Petrolo, and Erasmo Carrera. "On the effects of trigonometric and exponential terms on the best theory diagrams for metallic, multilayered, and functionally graded plates." Mechanics of Advanced Materials and Structures 27, no. 5 (2020): 426-440. https://doi.org/10.1080/15376494.2018.1478048

http[11] Hinostroza, M. A., and C. Guedes Soares. "Parametric estimation of the directional wave spectrum from ship motions." International Journal of Maritime Engineering 158, no. A2 (2016). https://doi.org/10.5750/ijme.v158iA2.981

http[12] Carrera E, Brischetto S. A survey with numerical assessment of classical and refined theories for the analysis of sandwich plates. Appl Mech Rev 2009;62:1-17. https://doi.org/10.1115/1.3013824

Advanced Topics in Mechanics of Materials, Structures and Construction - AToMech1-2023 Materials Research Forum LLC
Materials Research Proceedings 31 (2023) 146-154 https://doi.org/10.21741/9781644902592-16

AI-based prediction of strength and tensile properties of expansive soil stabilized with recycled ash and natural fibers

Abolfazl Baghbani[1,a *], Firas Daghistani[2,b], Katayoon Kiany[3,c] and Mohamad Mahdi Shalchiyan[4,d]

[1]School of engineering, Deakin University, Geelong, VIC, Australia

[2]Department of Civil Engineering, La Trobe University, Melbourne, VIC, Australia; Also with Civil Engineering Department, University of Business and Technology, Jeddah, Saudi Arabia

[3]School of Design, University of Melbourne, Melbourne, VIC, Australia

[4]School of Engineering, Guilan University, Rasht, Iran

[a]abaghbani@deakin.edu.au, [b]f.daghistani@latrobe.edu.au, [c]kianyk@student.unimelb.edu.au, [d]mohamad.shalchiyan@gmail.com

Keywords: Artificial Intelligence; Soil Stabilizer; Recycled Ash, Recycled Fiber, PLS, CRRF

Abstract. This study investigated the uniaxial compressive strength (UCS) and split tensile strength of a mixture of soil and recycled ash and natural fibers using two different methods, partial least squares (PLS) and classification and regression random forest (CRRF). The study analyzed a dataset of 20 sets with five inputs and two outputs, and the importance of the input parameters was evaluated. The performance of the PLS and CRRF models was assessed, and it was found that the CRRF model outperformed the PLS model. The study also revealed the most and least important parameters in predicting the split tensile strength and UCS in both models. The findings of this study have implications for the use of soil and recycled ash mixtures with natural fibers in construction applications.

Introduction

Expansive soils are a type of soil that undergoes significant volume changes due to fluctuations in moisture content [1-3]. These volume changes can cause significant damage to buildings and infrastructure, making the stabilization of expansive soils an essential area of research. In recent years, researchers have been investigating the use of recycled ash and natural fibers as stabilizers for expansive soils to improve their strength and durability [4].

Expansive soils are a type of soil that undergoes significant volume changes due to fluctuations in moisture content. These soils are characterized by high clay content and can cause significant damage to buildings and infrastructure due to swelling and shrinking [1]. The swelling and shrinking of expansive soils are due to the clay particles' ability to absorb and release water, causing significant stresses and strains [5-8].

Recycled ash is a waste material generated from the combustion of coal in thermal power plants [8]. It contains various elements that can be beneficial for soil stabilization, including silica, alumina, and iron oxide. The use of recycled ash in soil stabilization has been studied extensively, and its potential as a soil stabilizer has been recognized [9]. Natural fibers, such as coir, sisal, jute, and kenaf, are obtained from plant sources and have been used in various applications, including soil stabilization [10]. Natural fibers are biodegradable and eco-friendly, making them a sustainable alternative to synthetic materials.

Several recent studies have investigated the use of recycled ash and natural fibers as stabilizers for expansive soil [11-13]. Tiwari et al. [12] investigated the use of recycled ash and natural fibers to control the strength and durability of expansive soil. The soil was stabilized with different percentages of bottom ash and reinforced with coir fibers. The results showed that the use of bottom ash and coir fibers improved the soil's strength and durability, and the approach is a sustainable and economical solution for stabilizing pavement subgrades. Another study by Punthutaecha et al. [11] investigated the use of recycled ash and natural fibers to control the strength and durability of expansive soil. The soil was stabilized with different percentages of bottom ash and reinforced with coir fibers. The study assessed the swelling behavior, mechanical and chemical properties, and durability of the stabilized soil through freeze-thaw cycles. The results showed that the use of bottom ash and coir fibers improved the soil's strength and durability, and the approach is a sustainable and economical solution for stabilizing pavement subgrades.

A variety of factors, including soil density, fly content can affect the strength of soil with recycled ash and natural fibers. By using traditional methods, it is impossible to produce a comprehensive model with sufficient input parameters due to the multiplicity of parameters and their non-linearity. Artificial intelligence is one of the modern solutions to this problem. Using artificial intelligence methods, relationships between parameters can be found with a high degree of accuracy without prior knowledge [14]. In various civil engineering fields, such as slope stability [15-18], road construction and tunnelling [19-20], soil cracking [21-23] and recycled materials [24-27], artificial intelligence methods have been used successfully. The strength of mixture of soil and recycled ash and natural fibers has not yet been predicted using AI. For the first time, this study examined the uniaxial compressive strength (UCS) and split tensile strength of mixture of soil and recycled ash and natural fibers using a statistical method, i.e., partial least squares (PLS), and an artificial intelligence method, i.e., classification and regression random forest (CRRF). In this study, a 20-sets database with five inputs, including bottom ash, coir fibers (CF) content, electrical conductivity (EC), PH and Calcium content, and two outputs, namely uniaxial compressive strength (UCS) and split tensile strength, were considered. The importance of input parameters has also been evaluated after modelling.

Database Collection and Processing

Experiment and data collection

In order to investigate the effect of recycled ash and natural fibers on soil, a database consisting of 20 tests to determine the soil uniaxial compressive strength (UCS) and split tensile strength of the mixture was collected from the study conducted by Tiwari et al. [12]. Table 1 presents the statistical information from the collected database. The study conducted by Tiwari et al. [12] investigated the effects of combining recycled ash with natural fibers on soil strength and durability. The bottom ash (BA) was used for chemical treatment of the soil and coal fibers (CF) were used as reinforcement against tensile cracks.

Table 1. Statistical information of database

Variable	Observations	Minimum	Maximum	Mean	Std. deviation
Split tensile strength (kPa)	20	37.500	180.000	101.450	44.609
UCS (kPa)	20	0	20	140.000	470.000
Bottom Ash (%)	20	0.000	20.000	10.000	7.255
Coir Fibers (CF) (%)	20	0.000	1.000	0.438	0.379
Electrical Conductivity, EC (mS/cm)	20	1.750	5.500	4.100	1.447
PH	20	10.400	13.230	12.186	1.037
Calcium Content (%)	20	0.000	7.500	4.420	2.876

Preparation of the data for AI modelling

A linear normalization of the database was performed before it was used for modelling in AI methods. The linear normalization is represented by Eq. 1. A parameter's maximum value is considered equal to 1, its minimum value is considered equal to 0, and the rest of the values are distributed between 0 and 1. It is important to normalize data in order to improve model accuracy since different parameters have different units.

$$X_{norm} = \frac{X - X_{min}}{X_{max} - X_{min}} \quad (1)$$

where X_{max}, X_{min}, X and X_{norm} are maximum, minimum, actual, and normalized values, respectively.

A database is divided into two parts: training (80%) and testing (20%). A statistical analysis of two databases is presented in Tables 2 and 3. Statistics in two databases are generally similar. AI Model accuracy can be improved as a result of this similarity.

Table 2. Statistical information of training database

Variable	Observations	Minimum	Maximum	Mean	Std. deviation
Split tensile strength (kPa)	16	37.500	180.000	97.063	41.751
UCS (kPa)	16	140.000	470.000	265.938	103.073
Bottom Ash (%)	16	0.000	20.000	9.688	7.181
Coir Fibers (CF) (%)	16	0.000	1.000	0.375	0.365
Electrical Conductivity, EC (mS/cm)	16	1.750	5.500	4.047	1.427
PH	16	10.400	13.230	12.162	1.016
Calcum Content (%)	16	0.000	7.500	4.297	2.846

Table 3. Statistical information of testing database

Variable	Observations	Minimum	Maximum	Mean	Std. deviation
Split tensile strength (kPa)	4	46.000	175.000	119.000	58.086
UCS (kPa)	4	155.000	430.000	306.250	117.995
Bottom Ash (%)	4	0.000	20.000	11.250	8.539
Coir Fibers (CF) (%)	4	0.250	1.000	0.688	0.375
Electrical Conductivity, EC (mS/cm)	4	1.750	5.500	4.313	1.737
PH	4	10.400	13.230	12.283	1.281
Calcum Content (%)	4	0.000	7.500	4.913	3.388

Data-driven modeling

Partial least squares (PLS)

PLS (partial least squares) is a statistical approach commonly used for multivariate data analysis. This is a type of regression analysis used to model the relationship between independent variables (X) and dependent variables (Y). PLS is particularly useful when there is a large number of predictor variables, which can cause problems with traditional regression methods, such as multicollinearity. It is possible to use PLS for both linear and nonlinear relationships between X and Y. In PLS, latent variables are extracted from the predictor variables (X) and the response variables (Y) that explain the maximum covariance between them. PLS components are linear combinations of the original variables, which are calculated by maximizing the covariance between X and Y. A PLS can be applied to a wide variety of fields, including engineering, chemistry, biology, finance, and marketing. In fields with large amounts of data, this method is used to identify the most significant variables that are driving the relationship between the predictor variables and the response variables.

Classification and regression random forest (CRRF)

The random forest algorithm is a powerful machine learning algorithm that can be used for both classification and regression tasks. Using this method, multiple decision trees are combined to create a more accurate and stable prediction.

In a random forest classification model, multiple decision trees are generated, each trained on a randomly selected subset of training data and a randomly selected subset of predictor variables. To determine the final predicted class, the decision trees are combined by taking a majority vote based on their predictions. The method is effective at reducing overfitting as well as improving the model's accuracy and generalizability.

Additionally, in a random forest regression model, the algorithm creates multiple decision trees, each trained on a randomly selected subset of the training data and a randomly selected subset of the predictor variables. By taking the average of the predictions made by each decision tree, the final predicted value is determined.

- There are several advantages to using random forest models, including:
- In contrast to other machine learning algorithms, they are robust against noise and overfitting.
- Large datasets with a large number of predictor variables can be handled by these programs.
- This allows them to identify important predictor variables that are driving the prediction.

In various fields such as engineering, finance, healthcare, marketing, and natural language processing, random forest models are commonly used. Additionally, they are used in feature engineering, where they can identify the most important variables to include in a model.

Results

Partial least squares (PLS)

After implementing a large number of PLS models and changing various parameters, the best and most optimal PLS model was identified. According to Figs 1 and 2, the actual values are compared to the values predicted by the best PLS model for split tensile strength and UCS, respectively. The PLS model was able to predict both outputs with an almost perfect accuracy based on the obtained results.

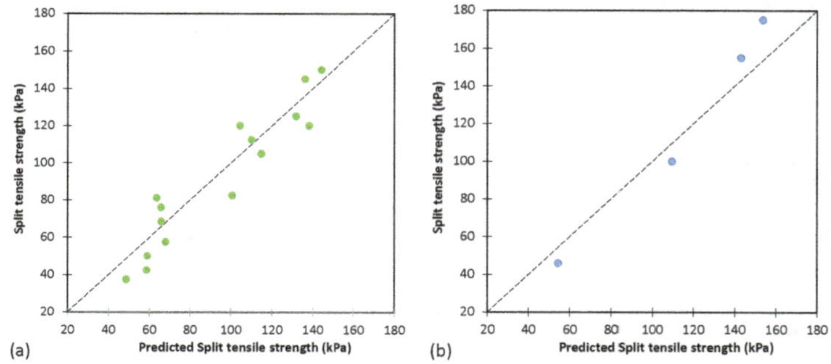

Fig. 1. The results of PLS for predicting split tensile strength

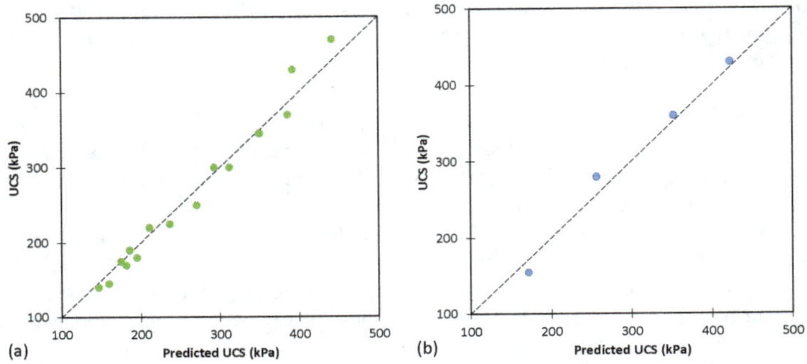

Fig. 2. The results of PLS for predicting UCS

In Table 4, the model's accuracy is demonstrated by the values of performance metrics for predicting split tensile strength and UCS, respectively. The results provide performance metrics of a predictive model on both training and testing data. The PLS model performed well with high accuracy and good generalization to unseen data. The metrics include Mean Absolute Error (MAE) and R^2. According to these metrics, the model performs well both on training and testing databases, with testing data showing slightly worse performance. In general, the PLS model is an effective tool for predicting outputs.

Table 4. The performance of PLS model

Performance metrics	Split tensile strength		UCS	
	Training	Testing	Training	Testing
MAE	10.789	12.808	13.932	12.748
R^2	0.914	0.925	0.977	0.975

Classification and regression random forest (CRRF)

The performance of the CRRF model can be affected by a variety of parameters. The effective parameters of the CRRF model were repeatedly changed to find the best model. Table 5 presents the values of the best CRRF model. Figs 3 and 4 illustrate the actual results of the test compared to the values predicted by the CRRF model. According to the results, the CRRF model is relatively accurate and is capable of predicting both split tensile strength and UCS.

Table 5. The specifications of the best CRRF.

Trees parameters					Forest parameters		
Min. node size	Min. son size	Max depth	Mtry	CP	Sampling	Sample size	Number of trees
2	1	12	2	0.00001	Random with replacement	16	1000

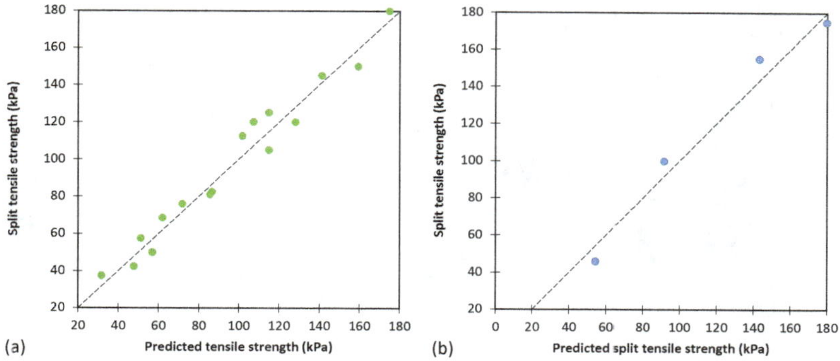

Fig. 3. The results of CRRF for predicting split tensile strength

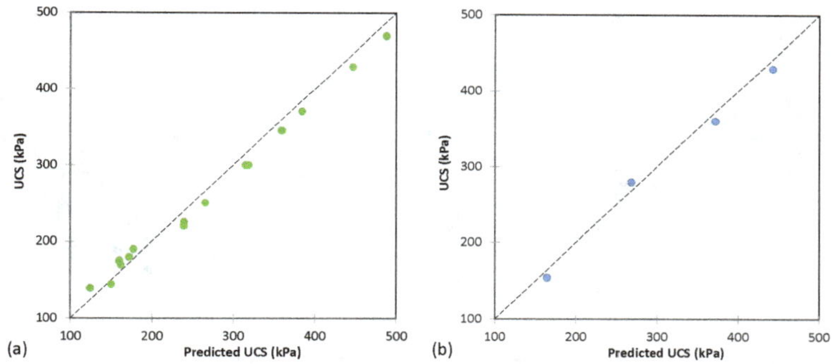

Fig. 4. The results of CRRF for predicting UCS

Table 6 illustrates the accuracy results of the CRRF model based on some parameters. The CRRF model performed well in both training and testing databases with high R^2 values. While the CRRF model slightly overfits on the training dataset, the overall results indicate that the model is accurate and effective in predicting outputs.

Table 6. The performance of CRRF model

Performance metrics	Split tensile strength		UCS	
	Training	Testing	Training	Testing
MAE	11.352	13.880	7.107	9.305
R^2	0.988	0.979	0.965	0.965

The variable importance of input parameters

Modeling artificial intelligence methods requires checking the importance of input parameters. Figs 5 and 6 illustrate the importance of input parameters in determining two outputs in both models. According to Figure 5, based on both models, the most important parameter for predicting split tensile strength was bottom ash content, and the least important was coir fiber content based on the PLS model and EC based on the CRRF model. For both models, for predicting UCS,

Calcium content was the most important parameter to predict the UCS and coir fiber was the least important parameter to predict the UCS.

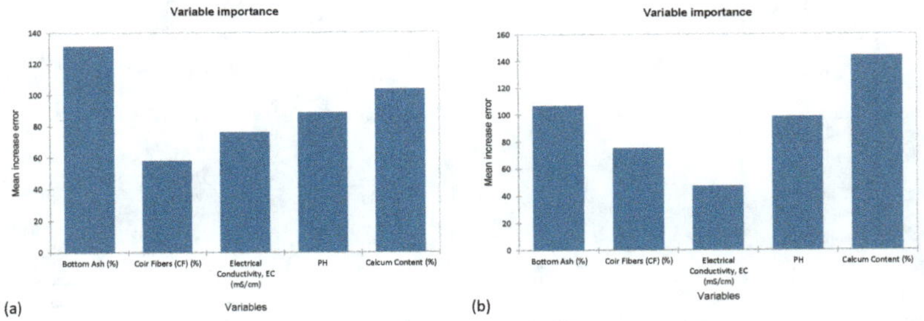

Fig. 5. The importance of parameters to predict split tensile strength based on (a) PLS and (b) CRRF

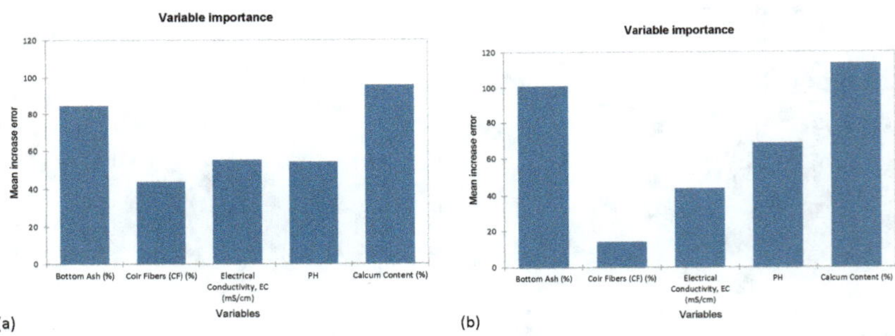

Fig. 6. The importance of parameters to predict UCS based on (a) PLS and (b) CRRF

Conclusion

This study has examined the uniaxial compressive strength (UCS) and split tensile strength of a soil mixture containing recycled ash and natural fibers using two different methods, i.e., partial least squares (PLS) and classification and regression random forest (CRRF). The study has utilized a 20-sets database consisting of five inputs and two outputs, and the importance of input parameters has been evaluated. The results have shown that the best CRRF model has superior performance metrics compared to the PLS model. According to both models, bottom ash content is the most important parameter for predicting split tensile strength, whereas Calcium content is the most significant parameter in predicting UCS. In contrast, coir fiber content is found to be the least important parameter in both models. The results suggest that bottom ash and Calcium content are critical factors that should be considered when designing a soil mixture for achieving higher strength.

References

[1] Baghbani A, Baumgartl T, Filipovic V. Effects of Wetting and Drying Cycles on Strength of Latrobe Valley Brown Coal. Copernicus Meetings; 2023 Feb 22. https://doi.org/10.5194/egusphere-egu23-4804

[2] Sahebzadeh S, Heidari A, Kamelnia H, Baghbani A. Sustainability features of Iran's vernacular architecture: A comparative study between the architecture of hot-arid and hot-arid-windy regions. Sustainability. 2017 May 4;9(5):749. https://doi.org/10.3390/su9050749

[3] Baghbani A, Costa S, O'Kelly BC, Soltani A, Barzegar M. Experimental study on cyclic simple shear behaviour of predominantly dilative silica sand. International Journal of Geotechnical Engineering. 2022 Oct 23:1-5. https://doi.org/10.1080/19386362.2022.2135226

[4] Puppala AJ, Congress SS, Banerjee A. Research advancements in expansive soil characterization, stabilization and geoinfrastructure monitoring. InFrontiers in geotechnical engineering 2019 Feb 12 (pp. 15-29). Singapore: Springer Singapore. https://doi.org/10.1007/978-981-13-5871-5_2

[5] Nguyen MD, Baghbani A, Alnedawi A, Ullah S, Kafle B, Thomas M, Moon EM, Milne NA. Investigation on the suitability of aluminium-based water treatment sludge as a sustainable soil replacement for road construction. Transportation Engineering. 2023 Mar 28:100175. https://doi.org/10.1016/j.treng.2023.100175

[6] Estabragh AR, Pereshkafti MR, Parsaei B, Javadi AA. Stabilised expansive soil behaviour during wetting and drying. International Journal of Pavement Engineering. 2013 Apr 1;14(4):418-27. https://doi.org/10.1080/10298436.2012.746688

[7] Baghbani A, Choudhury T, Samui P, Costa S. Prediction of secant shear modulus and damping ratio for an extremely dilative silica sand based on machine learning techniques. Soil Dynamics and Earthquake Engineering. 2023 Feb 1;165:107708. https://doi.org/10.1016/j.soildyn.2022.107708

[8] Nguyen MD, Baghbani A, Alnedawi A, Ullah S, Kafle B, Thomas M, Moon EM, Milne NA. Experimental Study on the Suitability of Aluminium-Based Water Treatment Sludge as a Next Generation Sustainable Soil Replacement for Road Construction. Available at SSRN 4331275.

[9] Asokan P, Saxena M, Asolekar SR. Coal combustion residues-environmental implications and recycling potentials. Resources, Conservation and recycling. 2005 Feb 1;43(3):239-62. https://doi.org/10.1016/j.resconrec.2004.06.003

[10] Leao, A.L., Rowell, R. and Tavares, N., 1998. Applications of natural fibers in automotive industry in Brazil-thermoforming process. Science and Technology of Polymers and Advanced Materials: Emerging Technologies and Business Opportunities, pp.755-761. https://doi.org/10.1007/978-1-4899-0112-5_66

[11] Punthutaecha K, Puppala AJ, Vanapalli SK, Inyang H. Volume change behaviors of expansive soils stabilized with recycled ashes and fibers. Journal of materials in Civil Engineering. 2006 Apr;18(2):295-306. https://doi.org/10.1061/(ASCE)0899-1561(2006)18:2(295)

[12] Tiwari N, Satyam N, Puppala AJ. Strength and durability assessment of expansive soil stabilized with recycled ash and natural fibers. Transportation Geotechnics. 2021 Jul 1;29:100556. https://doi.org/10.1016/j.trgeo.2021.100556

[13] Puppala A, Hoyos L, Viyanant C, Musenda C. Fiber and fly ash stabilization methods to treat soft expansive soils. InSoft ground technology 2001 (pp. 136-145). https://doi.org/10.1061/40552(301)11

[14] Baghbani A, Choudhury T, Costa S, Reiner J. Application of artificial intelligence in geotechnical engineering: A state-of-the-art review. Earth-Science Reviews. 2022 May 1;228:103991. https://doi.org/10.1016/j.earscirev.2022.103991

[15] Baghbani A, Costa S, Faradonbeh RS, Soltani A, Baghbani H. Experimental-AI Investigation of the Effect of Particle Shape on the Damping Ratio of Dry Sand under Simple Shear Test Loading. 2023, preprint. https://doi.org/10.20944/preprints202303.0021.v1

[16] Baghbani, A., Daghistani, F., Naga, H.A. and Costa, S., 2022. Development of a Support Vector Machine (SVM) and a Classification and Regression Tree (CART) to Predict the Shear Strength of Sand Rubber Mixtures. In Proceedings of the 8th International Symposium on Geotechnical Safety and Risk (ISGSR), Newcastle, Australia. https://doi.org/10.3850/978-981-18-5182-7_00-08-004.xml

[17] Baghbani A, Costa S, Choudhury T. Developing Mathematical Models for Predicting Cracks and Shrinkage Intensity Factor During Clay Soil Desiccation. Available at SSRN 4408164.

[18] Baghbani A, Costa S, Faradonbeh RS, Soltani A, Baghbani H. Modeling the Effects of Particle Shape on Damping Ratio of Dry Sand by Simple Shear Testing and Artificial Intelligence. Applied Sciences. 2023 Mar 29;13(7):4363. https://doi.org/10.3390/app13074363

[19] Lin SS, Shen SL, Zhang N, Zhou A. Modelling the performance of EPB shield tunnelling using machine and deep learning algorithms. Geoscience Frontiers. 2021 Sep 1;12(5):101177. https://doi.org/10.1016/j.gsf.2021.101177

[20] Baghbani A, Costa S, Choundhury T, Faradonbeh RS. Prediction of Parallel Desiccation Cracks of Clays Using a Classification and Regression Tree (CART) Technique. In Proceedings of the 8th International Symposium on Geotechnical Safety and Risk (ISGSR), Newcastle, Australia 2022. https://doi.org/10.3850/978-981-18-5182-7_00-08-003.xml

[21] Xu JJ, Zhang H, Tang CS, Cheng Q, Tian BG, Liu B, Shi B. Automatic soil crack recognition under uneven illumination condition with the application of artificial intelligence. Engineering Geology. 2022 Jan 1;296:106495. https://doi.org/10.1016/j.enggeo.2021.106495

[22] Sahebzadeh S, Rezaeiha A, Montazeri H. Impact of relative spacing of two adjacent vertical axis wind turbines on their aerodynamics. InJournal of Physics: Conference Series 2020 Sep 1 (Vol. 1618, No. 4, p. 042002). IOP Publishing. https://doi.org/10.1088/1742-6596/1618/4/042002

[23] Baghbani A, Costa S, Lu Y, Soltani A, Abuel-Naga H and Samui P. Effects of Particle Shape on Shear Modulus of Sand Using Dynamic Simple Shear Testing. Arabian journal of geosciences. 2023.

[24] Baghbani A, Daghistani F, Baghbani H, Kiany K, Bazaz JB. Artificial Intelligence-Based Prediction of Geotechnical Impacts of Polyethylene Bottles and Polypropylene on Clayey Soil. EasyChair; 2023 Feb 19.

[25] Baghbani A, Daghistani F, Baghbani H, Kiany K. Predicting the Strength of Recycled Glass Powder-Based Geopolymers for Improving Mechanical Behavior of Clay Soils Using Artificial Intelligence. EasyChair; 2023 Feb 19.

[26] Baghbani, A.; Nguyen, M.D.; Alnedawi, A.; Milne, N.; Baumgartl, T.; Abuel-Naga, H. Improving Soil Stability with Alum Sludge: An AI-Enabled Approach for Accurate Prediction of California Bearing Ratio. Preprints 2023, 2023030197. https://doi.org/10.20944/preprints202303.0197.v1

[27] Baghbani A.; Abuel-Naga H.; Shirani Faradonbeh R.; Costa S.; Almasoudi R. Ultrasonic Characterization of Compacted Salty Kaolin–Sand Mixtures Under Nearly Zero Vertical Stress Using Experimental Study and Machine Learning, Geotechnical and geological engineering, 2023. https://doi.org/10.1007/s10706-023-02441-5

Advanced Topics in Mechanics of Materials, Structures and Construction - AToMech1-2023 Materials Research Forum LLC
Materials Research Proceedings 31 (2023) 155-164 https://doi.org/10.21741/9781644902592-17

Performance of self-compacting concrete based on fine recycled concrete aggregate incorporating polyethylene terephtalate fibers

Bayah Meriem[1,2,a*], Debieb Farid[1,b]

[1]LME Laboratory, Faculty of Technology, UniversityYahia Fares of Medea, Algeria

[2]Civil Engineering departement, Faculty of Technology, UniversitySaad Dahleb of Blida1, Algeria

[a]meriembayah@gmail.com, [b]debieb.farid@univ-medea.dz

Keywords: Self-Compacting Concrete, Fiber-Reinforced Concrete, Recycled Concrete, Polyethylene Terephthalate Fiber, Mechanical Resistance, Fresh Properties

Abstract. This experimental research aims to investigate the effect of adding polyethylene terephthalate plastic fibers (PETF) on the behavior of recycled self-compacting concrete (RSCC) based on recycled fine concrete aggregates (RFCA). Twenty RSCC mixes were made for this study. RFCA was obtained from the laboratory demolition of a moderate concrete slab and substituted by natural fine aggregates (NFA) at various mass fractions (0%, 25%, 50%, 75%, and 100%). Furthermore, four volumetric fractions (V_f) of plastic fibers (0.3%, 0.5%, 1%, and 1.2%) were added and sorted from plastic bottle recycling. The properties of the fresh and hardened new composite (RSCC made with PETF and RFCA contents) are analyzed and compared. The results showed that the mechanical performances of RSCC in terms of flexural strength and elasticity modulus were improved, where the compressive strength decreased with an increase in the V_f content of PETF and RFCA. The incorporation of 100% RFCA combined with 1.2% of PETF can enhance both flexural strength and modulus of elasticity of concrete up to 9% and 24%. This type of concrete can be recommended for structural repair applications.

Introduction

Concrete has become the most material of construction used in the world, which is made with a high number of natural resources (aggregates). Therefore the consumption of the natural resources in concrete creates environmental and economic problems and also effect the reduction of these natural resources protection[1-2]. But currently, the using of waste in construction has become a necessary solution to countries' environmental and economic problems, particularly in third-world countries[3].In this context, reflection on the search for new concretes capable of solving the economic and technical problems encountered in the construction industry begins. These materials include self-compacting concrete (SCC), it can be flowing freely in narrow gaps between reinforcement bars and consolidate under its weight without any external vibration with superior deformability and high resistance to segregation[4-5]. Also, there are many studies show that we can use different waste (construction and/or demolition waste (CDW), glass, plastic) to replace those aggregates with the coarse and/or fine natural aggregates (NA) in concrete design[6-7].
The CDW is considered the waste most widely used in the concrete in form of aggregates. There are many studies on the effect of recycled concrete aggregates (RCA) on the properties of concrete [8]. In this among, Sasanipour et al [9]. Assessed the effect contents 25%, 50%, 75%, and 100% of fine and coarse RCA as a partial substitute of natural coarse and fine aggregates (NCA, NFA) on the mechanical properties (compressive, flexural strengths and ultrasonic pulse velocity) of SCC. They found that the partial use of RCA and RFCA as a substitute for NCA decreased the mechanical behavior of specimens by 43% and 52% respectively for those mixes by replacing 75%. Silva et al [10] also showed that the increase in the replacement of RCA led to a decrease in

the mechanical properties of concrete this can be attributed to the porosity of the mortar adhering to RCA, their high water absorption, and the formation of the interface transition zones (ITZ)[11]. The researchers were significantly investigating; they can produce the concrete with RCA by controlling the water-cement ratio, proper use of chemical and mineral admixtures as well as appropriate preparation methods before using RCA[12]. The influence of fine RCA on the fresh properties and the microstructural characteristics of mortar were studied by Carro-López [13]. The obtained results of this study showed that the porosity increased with the percentage of fine RCA. And also Nieto et al [14] determined the durability (water absorption and carbonatation) and mechanical characteristics of SCC produced by coarse RCA as a substitute for natural coarse aggregates (NCA). For this aim, NCA was replaced by coarse RCA at 0%, 20%, 40%, 60%, 80%, and 100% and water-cement ratios of 0.45, 0.50, and 0.55. They found that using up to 20% coarse RCA has little influence on the characteristics of the RSCC.

According to the literature, the using of RCA decreased the mechanical properties of concrete. Therefore, the researchers are working to improve the behavior of RCA concrete by using additional materials. A research study found that using fiber in concrete, enhanced the mechanical characteristic of concrete[15]. Aslani et al.[16].Assessed the impact of polypropylene fiber (PPF) and steel fiber (SF) on the fresh property and hardened mechanical of recycled aggregates self-compacting concrete (RASCC). A total of RCA replacement contents of NCA were considered 100% in terms of mass. In addition, PPF was added at 0%, 0.1%, 0.15%, 0.2%, and 0.25% in terms of volume, and SF at 0.25, 0.5, 0.75, and 1% in terms of volume. They found that the addition of fibers leads to enhancing the tensile strength of concrete.

The using of RCA as a partial replacement of both NCA and NFA in previous studies has indicated a negative effect on the workability and mechanical behavior of SCC. On the other hand, adding fibers in terms of volume in the concrete can improve the mechanical properties of concrete. The main objective of this study is to investigate the mechanical properties of SCC incorporated with up to 25%, 50%, 75%, and 100% of RFCA and containing different volume fractions of PETF.

Experimental program

Materials

An industrial Portland cement type CEM II/B 42.5N product by Lafarge-M'sila Company in Algeria, with a density of 3.08 kg/m^3 and a specific surface of 3900 cm^2/g, was used for all concrete mixtures. It is one of the most frequent cement for concrete structures in Algeria. **Table1** shows the chemical composition of this cement. Two types of NFA and RFCA (0/4 mm) obtained from the demolition of moderate concrete slab made in the laboratory are used. The NFA is partially replaced by various fractions (25%, 50%, 75%, and 100%) of RFCA. Two types of NCA (3/8 and 8/15 mm) are used. The chemical composition and characteristics of coarse and fine aggregate are shown in Table1, 2.

Table1*.Chemical composition of fine natural, recycled fine concrete aggregates and cementitious.*

Composition(%)	SiO_2	CaO	CO_2	Fe_2O_3	Al_2O_3	MgO	Na_2O	K_2O	SO_3
Cement	15.96	62.17	11.46	2.75	3.75	1.33	0.12	0.52	2.30
NFA	92.86	0.37	0.64	0.25	2.52	0.13	0.04	1.01	0.30
RFCA	9.58	49.40	38.76	0.91	2.43	1.54	-	0.20	2.75

***Table2**. Physical properties of used aggregates.*

Type of Aggregate	Maximum Size (mm)	Apparent density (g/ cm^3)	Bulk density (g/cm^3)	Water absorption (%)	Fineness Modulus	Los Angeles Abrasion (%)
CNA	15	1.52	2.61	0.59	-	26.5
CNA	8	1.50	2.63	0.79	-	27.5
NFA	4	1.59	2.65	0.65	2.24	-
RFCA	4	1.24	2.40	8.5	3.01	-

The used PETF in this research was produced by recycling waste plastic bottles. The physical and mechanical properties of this fiber are presented in Table 3. Therefore, we used PETF at four volumetric percentages of 0.3%, 0.5%, 1%, and 1.2%. Superplasticizer is based on Polycarboxylates which significantly improves the properties of concrete. It is made by Granitex-NP Company in Algeria.

***Table 3.** Main properties of used PET fiber.*

Length (mm)	Diameter (mm)	Elongation to break (%)	Density (20 °C) (g/cm^3)	Tensile strength (MPa)	Young's modulus (GPa)
30	0.44	53	0.690	383	7.17

Concrete mixes design

We created twenty concrete mixtures for this study, which are shown in Table 4. Three steps of mixing are involved in the creation of SCC mixes. The cement and aggregates were first briefly mixed for 30 seconds. Following the addition of 70% of the mixing water, the mixture was mixed for 1 minute. It was then mixed for a minute with the remaining 30% of the superplasticizer-containing water. This procedure continued for another 5 minutes before coming to an abrupt halt for 2 minutes. Before discharge, the SCC was mixed once more for 30 seconds to ensure uniformity of mixtures.

***Table 4.** Concrete mixes design (kg/m^3).*

Specimens	Cement	NFA	RFCA	NCA	PETF	SP	W/B
ontrol mix	483.69	922.43	-	750.04	-	4.60	0.4
5RFCA-0PETF	483.69	691.82	230.60	750.04	-	4.60	0.4
5RFCA-0.3PETF	483.69	691.82	230.60	750.04	2.07	4.69	0.4
5RFCA-0.5PETF	483.69	691.82	230.60	750.04	3.45	4.69	0.4
5RFCA-1PETF	483.69	691.82	230.60	750.04	6.9	4.84	0.4
5RFCA-1.2PETF	483.69	691.82	230.60	750.04	8.28	4.84	0.4
0RFCA-0PETF	483.69	461.21	461.21	750.04	-	4.60	0.4
0RFCA-0.3PETF	483.69	461.21	461.21	750.04	2.07	4.69	0.4
0RFCA-0.5PETF	483.69	461.21	461.21	750.04	3.45	4.69	0.4
0RFCA-1PETF	483.69	461.21	461.21	750.04	6.9	4.84	0.4
0RFCA-1.2PETF	483.69	461.21	461.21	750.04	8.28	4.84	0.4
5RFCA-0PETF	483.69	230.60	691.82	750.04	-	4.60	0.4
5RFCA-0.3PETF	483.69	230.60	691.82	750.04	2.07	4.69	0.4
5RFCA-0.5PETF	483.69	230.60	691.82	750.04	3.45	4.69	0.4
5RFCA-1PETF	483.69	230.60	691.82	750.04	6.9	4.84	0.4
5RFCA-1.2PETF	483.69	230.60	691.82	750.04	8.28	4.84	0.4

100RFCA-0PETF	483.69	-	922.43	750.04	-	4.60	0.4
100RFCA-0.3PETF	483.69	-	922.43	750.04	2.07	4.69	0.4
100RFCA-0.5PETF	483.69	-	922.43	750.04	3.45	4.69	0.4
100RFCA-1PETF	483.69	-	922.43	750.04	6.9	4.84	0.4
100RFCA-1.2PETF	483.69	-	922.43	750.04	8.28	4.84	0.4

Testing procedures

The fresh concrete state is the main property of SCC in this study, it is descript the filling ability, passing ability, and stability of self-compacting concrete is measured at different tests:

- The slump flow diameter (EN 12350-8)[17];
- The V-funnel flow time (EN 12350-9) [17];
- The L-box height ration (EN 12350-10)[18];
- The sieve stability (EN 12350-11) [19].

The compressive strength and flexural strength were measured at 28 and 90 days. For each mixture, three cubic specimens 100×100×100 mm were used to measure compressive strength, according to EN 12390-3[20], the flexural strength was tested under three-point flexural load by using three prismatic samples 70×70×280 mm according to EN 12390-5[21]. The static modulus of elasticity was determined by using cylindrical specimens of 100×200 mm at the curing age of 90 days according ASTM C469-02[22].

Results and discussion

Fresh properties

The results from the slump flow test are shown in Fig1.(a), and all mixtures indicated the best deformability in the slump flow diameter between 767.5 and 650mm and compliance with the EFNARC recommendation[23].The replacement of RFCA presents a remarkable reduction in the diameter of diffusion of SCC. The reduction in the slump flow diameter for RFCA75%, and RFCA100% compared to NFA was 8.15% and 10.09% respectively. This reduction in the workability is attributed to its specific surface and also due to its high water absorption of RFCA compared to that of natural aggregates[13]. The analysis of the results also shows the use of the PETF which presented a less decrease in the diameter of all mixtures of RSCC which contained 0.3% and 0.5% of PETF compared to those with the volume 1% and 1.2% of fibers, this was due to the difference in the dosage of SP, and also can be attributed to the reason for the large specific surface area of these fibers[24].

Fig1.(b) shows the time results of the (V-Funnel) test. We observe from these resultsV-Funnel time that all SCC mixtures values are acceptable by the EFNARC recommendations[23]. It can be concluded that the replacement of NFA by RFCA increases the flow time of RSCC compared to the reference mixture, which is in the range of 2 and 6 seconds, The mixture contained 100% RFCA shows a high plastic viscosity that could result in a loss of flow capacity property and this is due to the powder content and the increase in the specific surface area of RFCA[10]. According to Fig1.(b), we observed that the addition in PETF increases the flow time (V-Funnel) of RSCC mixes compared to the SCC reference, but still the recorded flow times that respond to the properties of the SCC by the increase in SP dosage, which showed the best filling ability without any sign of blocking. The reason for this increment is the decrease in the thickness of the surrounding aggregate layer (cement paste) and more friction between the particles and fibers which leads to an increase the viscosity of the concrete mixture[16].

The Fig1.(c) showed the ratio of H2/H1 in the L-box test can be taken as an indicator of the ability to pass SCC. According to the recommendation of the EFNARC[23],this ratio should not be less than 0.8 to ensure an adequate capacity of passage of the SCC. The substitution of NFA by RFCA resulted in a reduction of H2/H1 measured in this study. The reduction rate ranged from about 2% with 25% RFCA to 13% with 100% RFCA substitution. this reduction can due to the high values of water absorption of RFCA compared with natural aggregates. It was also noted that the present of PETF in the mixtures of RSCC, which gives a small reduction of H2/H1 between the two volume increments of fibres (about 0.1% to 0.2%), and the mix with 100%RFCA+1.2%PETF showed a severe reduction in the passing ability.

Fig1.(D) shows the influence of PETF and the replacement of RFCA on the sieve stability of SCC. These results showed that all the mixtures of SCC containing only RFCA have a segregation rate between 6-13% which indicates good stability,while these percentages decreased for the mixes incorporating RFCA and PETF together . In the end, they noticed that there was a decrease in the segregation ratio of RSCC by the PETF addition. This result could be attributed to the large amount of fins in the particulate of RFCA, and due to the higher water absorption rate of these aggregates [11] ,and we can due to the high amount of fiber in the mix.

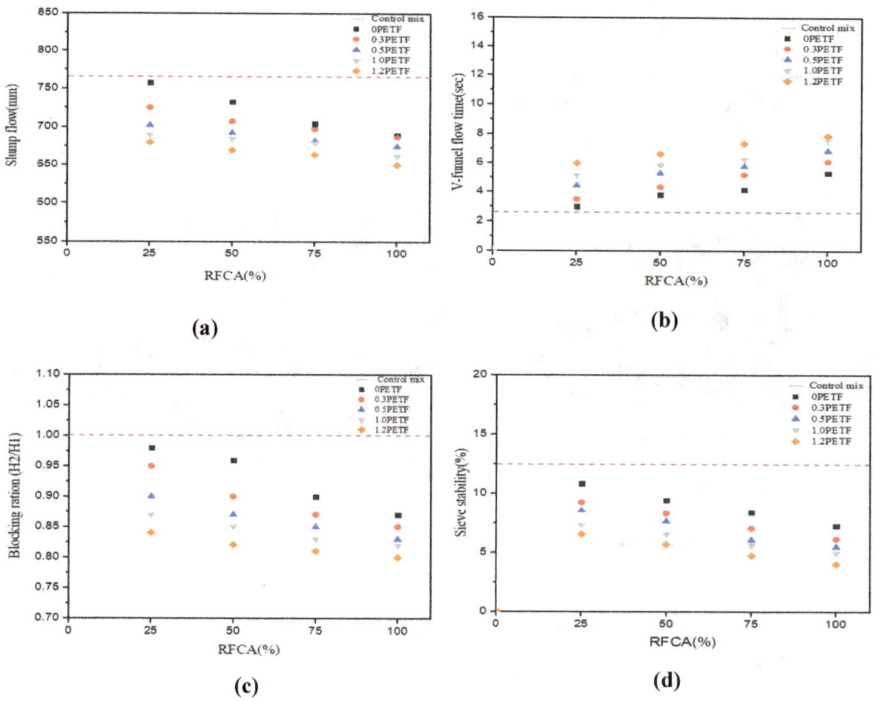

Fig. 1. *Influence of RFCA and PETF on: (a) slump flow;(b) V-funnel flow time; (c) L-box ratio;(D) sieve stability.*

Mechanical properties

Compressive strength

Fig.2 shows the effect of PETF and RFCA in the SCC mixture on the compressive strength at 28, and 90 days after curing. As shown in this figure, the compressive strength results of SCC increase with the age of curring and also showed there was a reduction in the compressive strength with the use of RFCA at all ages of hardening. At 28 days, the compressive strength of the control mix is 48.11 Mpa. On the replacement of NFA with the RFCA, the compressive strength decreased about 1.59%, 3.93 %,6.80%,and 9.33% for the mixtures made with 25%,50%,75%, and 100% of RFCA respectively. This reduction in the compressive strength with the presence of RFCA on the SCC is due to the intrinsic properties of RFCA and attributed to the higher absorption capacity of the adhered mortar in the RFCA, which provides worse adhesión between the cement paste and the RFCA compared with NFA[1]. The addition of PETF in RSCC decreases compressive strength. The 90days compressive strength of RSCC containing PETF had a rate decreasing from 2%, 4%1% ,and 3% for mixtures 0.3 PETF+ 100 RFCA, 0.5 PETF+ 100 RFCA, 1 PETF+ 100 RFCA, and 1.2 PETF+ 100 RFCA. This reduction is explained by the bundling of fibers during mixing and casting. This assembly of fiber leads to grow a weak point in concrete between the area and the fiber surface and this area make microcracks and macrocracks caused by compression loading [25].

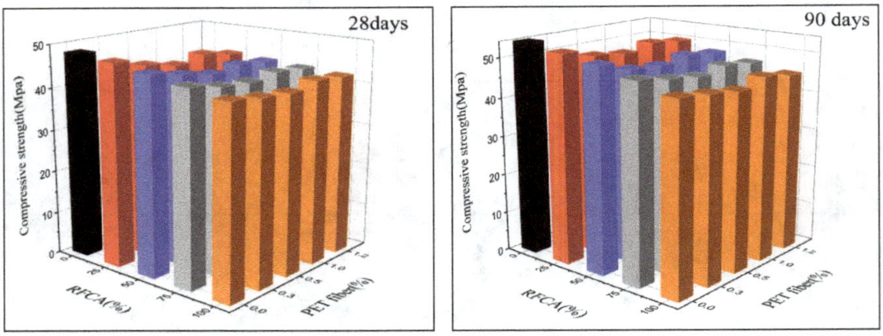

***Fig.2.** Influence of RFCA and PETF on the compressive strength.*

Flexural strength

The flexural strength results of RFCA and PETF contents at 28 and 90 days after curing are illustrated in Fig.3. Similarly to compressive strengths, the flexural resistance was significantly decreased with using RFCA. Therefore, the flexural strength was decreased by 4%, 11%, 16% , and 20% when 25%, 50%, 75% and100% RFCA was used as a replacement for NFA, The same results we observed with Mahakavi et al.Previous studies showed that reduction can be due to the low strength of the mortar adhered to RCA [8] . However, Fig.3 showed also that adding PETF in the RSCC increase the flexural strength, the addition of 0.3, 0.5,1 and1.2% PETF with 100% RFCA improved this property respectively by 2%, 5%, 8% ,and 9% because the presence of fiber in the concrete enhances to fill the cracks during the occurrence of tensile stresses.

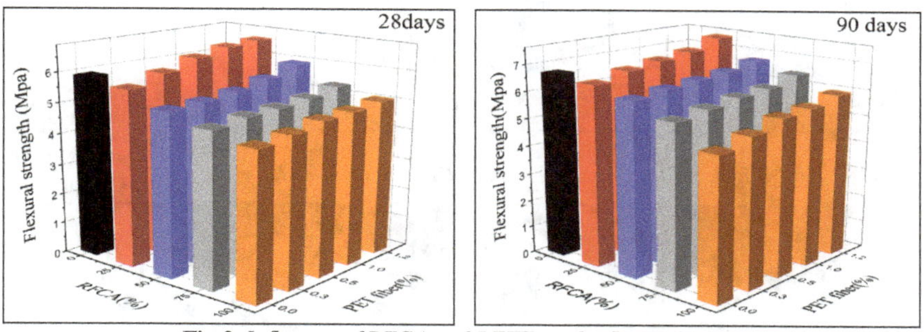

Fig.3. *Influence of RFCA and PETF on the flexural strength.*

Modulusof elasticity

Fig.4 presents the effect of RFCA and PETF on the elastic modulus of all the SCC mixtures. Regarding this figure, the elastic modulus was decreased with an increase in RFCA contents. Therefore, in specimens without PET fiber, the elastic modulus was decreased by 5.01%, 7.41%, 14.10% ,and 19.30% when 25%, 50%, 75% ,and 100% RFCA were used as a substitute for NFA, respectively. The same reduction in the elastic modulus of concrete was shown by previous investigations[26].Conversely, this test was improved with the addition of PETF and the elastic modulus was further enhanced with an increase in PETF content. Additionally, the highest modulus elastic was obtained for all specimens with 1.2% PETF contents. Therefore, the addition of 0.3%, 0.5%, 1% ,and 1.2% PETF with 100% RFCA enhanced the modulus elasticity respectively by 11.40%, 13.29%, 17.21%, and 24.87%, in comparison with the control sample with 100% RFCA.

Microstructure analysis

Fig.5 (a), (b), (c), (d) presents the list of scanning electron microscopy (SEM) images for the different fractured surfaces of SCC with RFCA and PETF. The SEM images in fig.10(c) showed that the surface of 100% RFCA has a large number of pores areas and cracks in the interfacial transition zone (ITZ), which analyzes of porosity higher in this mixture. These areas (ITZ) consider the most easily damaged in the matrix due to the subjected internal stress, which is the main reason for the low strength of the sample (100% RFCA). Fig.5 (b) shows that the through cracks and the number of connected pores are not present in the 1.2 PETF matrix significantly. This may explain why the sample had a uniform and denser structure between the paste and fiber, which make a small decrease in the compressive strength. On the other hand, Fig 5. (d) presents weak points PETF in the sample with 100% of RFCA, so we observed some pores In the area between fiber and cement paste, this explains the weak adherent of fiber to cement paste in concrete. It can be engendered by the rippling surface of PETF. Presence of the fibers in these samples trigger an increase in the strength of RFCA so that it doesn't break easily into two pieces when it is subjected to flexural or splitting tensile strength. It can be from the friction strength between the surfaces of fiber and cement paste around the fibers.

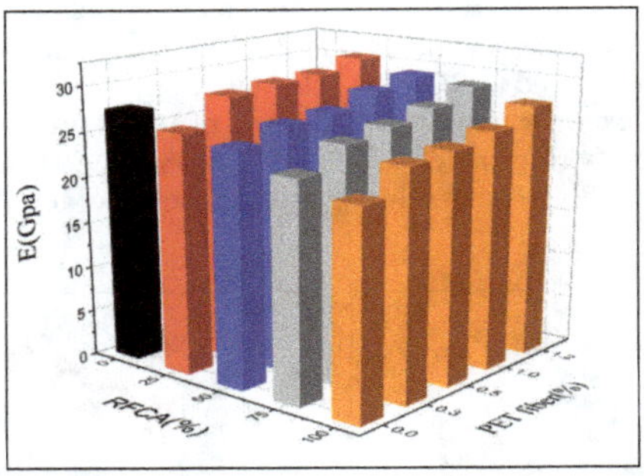

Fig.4. *Influence of RFCA and PETF on the modulus elastic.*

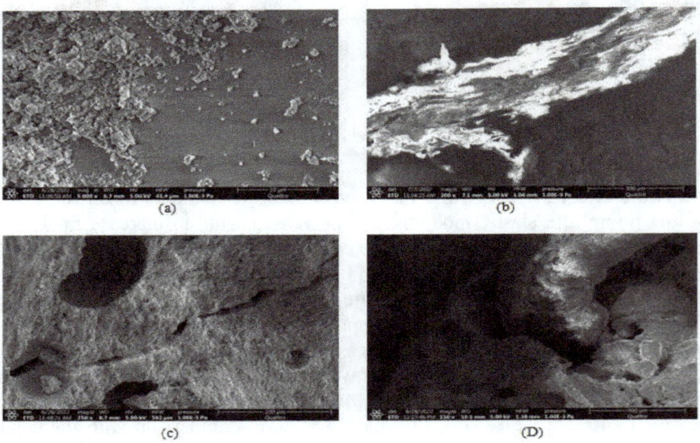

Fig. 5. *SEM images of typical specimens a) 0RFCA-0PETF, b) 0RFCA-1.2PETF, c) 100RFCA-0PETF and D) 100RFCA-1.2PETF.*

Conclusions

In this study, were investigated the effect of PETF and RFCA on the fresh property and mechanical properties in self-Compacting Concrete (SCC). This part of the study examined the use of RFCA as a replacement for NFA: 25%, 50%, 75%, and 100% in terms of mass, and were also the PETF added in the SCC at four volumetric percentages: 0.3%, 0.5%, 1% , and 1.2%. Additionally, were proposed to study the mechanical characteristics of RFCA concrete containing various PETF contents. According to the experimental results, the following conclusion can be drawn:

- The replacement of RFCA by the NFA and the adding PETF decrease the flowability of SCC, but this decreasing could be improved by using of low dosage of SP. The high-water absorption of RFCA and the specific surface of PETF are the main reason for this decrease in the flowability. Eventually, the RFCA and PETF have a negative effect on the fresh properties of SCC;

- The compressive strength of concrete was decreased with the use of RFCA as different partial replacement contents for NFA, and also the use of PETF decreased these test results of concrete, that the replacement of 25% and 50% of RFCA with NFA present a slight decrease. Adding 0.3%, 0.5%, 1%, and 1.2% PETF with the substituting 100% RFCA by NFA in the mixtures made a little drop in the compressive strength of the concrete. Therefore, these mixtures decreased by 2%, 3%,1% and 2% respectively;
- The flexural strengths and modulus elasticity of concrete containing different RFCA can be significantly enhanced with the use of PETF. Therefore, the replacement of RFCA by the NFA without PETF decreased the flexural strength by about 3.39%, 11.03%, 16.46%, and 20.54% for the mixtures 25%, 50%, 75%, and 100%, respectively. In addition, the simultaneous use of 100% RFCA with 0.3%, 0.5%, 1% and 1.2% PETF increased the flexural strength by 2.35%, 5.76%, 6.88%, and 8.76%, correspondingly.

References

[1] A. Karimipour and M. Edalati, "Influence of untreated coal and recycled aggregates on the mechanical properties of green concrete," J. Clean. Prod., vol. 276, p. 124291, 2020, https://doi.org/10.1016/j.jclepro.2020.124291

[2] M. Ghalehnovi, A. Karimipour, A. Anvari, and J. de Brito, "Flexural strength enhancement of recycled aggregate concrete beams with steel fibre-reinforced concrete jacket," Eng. Struct., vol. 240, no. February, p. 112325, 2021, https://doi.org/10.1016/j.engstruct.2021.112325

[3] A. Karimipour, M. Edalati, and J. de Brito, "Influence of magnetized water and water/cement ratio on the properties of untreated coal fine aggregates concrete," Cem. Concr. Compos., vol. 122, no. February, p. 104121, 2021, https://doi.org/10.1016/j.cemconcomp.2021.104121

[4] S. Mahesh, "Self compacting concrete and its properties," Int. J. Eng. Res. Appl., vol. 4, no. 8, pp. 72–80, 2014.

[5] M. Okamura, H; Ouchi, "Self Compacting Concrete - research paper," Journal of Advanced Concrete Technology, vol. 1, no. 1. pp. 5–15, 2003.

[6] V. Revilla-cuesta, V. Ortega-lópez, M. Skaf, and J. Manuel, "Effect of fine recycled concrete aggregate on the mechanical behavior of self-compacting concrete," vol. 263, 2020, https://doi.org/10.1016/j.conbuildmat.2020.120671

[7] A. A. Mohammed, I. I. Mohammed, and S. A. Mohammed, "Some properties of concrete with plastic aggregate derived from shredded PVC sheets," Constr. Build. Mater., vol. 201, no. July, pp. 232–245, 2019, https://doi.org/10.1016/j.conbuildmat.2018.12.145

[8] F. Debieb, L. Courard, S. Kenai, and R. Degeimbre, "Mechanical and durability properties of concrete using contaminated recycled aggregates," Cem. Concr. Compos., vol. 32, no. 6, pp. 421–426, 2010, https://doi.org/10.1016/j.cemconcomp.2010.03.004

[9] H. Sasanipour and F. Aslani, "Durability properties evaluation of self-compacting concrete prepared with waste fine and coarse recycled concrete aggregates," Constr. Build. Mater., vol. 236, p. 117540, 2020, https://doi.org/10.1016/j.conbuildmat.2019.117540

[10] Y. F. Silva, R. A. Robayo, P. E. Mattey, and S. Delvasto, "Properties of self-compacting concrete on fresh and hardened with residue of masonry and recycled concrete," Constr. Build. Mater., vol. 124, pp. 639–644, 2016, https://doi.org/10.1016/j.conbuildmat.2016.07.057

[11] O. K. Djelloul, B. Menadi, G. Wardeh, and S. Kenai, "Performance of self-compacting concrete made with coarse and fine recycled concrete aggregates and ground granulated blast-furnace slag," Adv. Concr. Constr., vol. 6, no. 2, pp. 103–121, 2018,

https://doi.org/10.12989/acc.2018.6.2.103

[12] I. González-Taboada, B. González-Fonteboa, F. Martínez-Abella, and S. Seara-Paz, "Analysis of rheological behaviour of self-compacting concrete made with recycled aggregates," Constr. Build. Mater., vol. 157, pp. 18–25, 2017, https://doi.org/10.1016/j.conbuildmat.2017.09.076

[13] D. Carro-López, B. González-Fonteboa, J. De Brito, F. Martínez-Abella, I. González-Taboada, and P. Silva, "Study of the rheology of self-compacting concrete with fine recycled concrete aggregates," Constr. Build. Mater., vol. 96, pp. 491–501, 2015, https://doi.org/10.1016/j.conbuildmat.2015.08.091

[14] D. Nieto, E. Dapena, P. Alaejos, J. Olmedo, and D. Pérez, "Properties of Self-Compacting Concrete Prepared with Coarse Recycled Concrete Aggregates and Different Water:Cement Ratios," J. Mater. Civ. Eng., vol. 31, no. 2, p. 04018376, 2019, https://doi.org/10.1061/(asce)mt.1943-5533.0002566

[15] I. Irki, F. Debieb, E. H. Kadri, O. Boukendakdji, M. Bentchikou, and H. Soualhi, "Effect of the length and the volume fraction of wavy steel fibers on the behavior of self-compacting concrete," J. Adhes. Sci. Technol., vol. 31, no. 7, pp. 735–748, 2017, https://doi.org/10.1080/01694243.2016.1231394

[16] F. Aslani, L. Hou, S. Nejadi, J. Sun, and S. Abbasi, "Experimental analysis of fiber-reinforced recycled aggregate self-compacting concrete using waste recycled concrete aggregates, polypropylene, and steel fibers," Struct. Concr., vol. 20, no. 5, pp. 1670–1683, 2019, https://doi.org/10.1002/suco.201800336

[17] BS EN 12350-8:2010, "BSI Standards Publication Testing fresh concrete," Br. Stand., no. April, p. 18, 2010.

[18] N. F. EN, "12350–10, Novembre 2010," Partie Bét. autoplaçant–essai à la boite en L.

[19] 2010 BS EN12350-11:, "BSI Standards Publication Testing fresh concrete Part 11: Self-compacting concrete -- sieve Segregation test," BSI Stand. Publ., 2010.

[20] B. S. I. BSI, "12390-3 Testing hardened concrete Compressive strength of test specimens," Aberdeen's Concr. Constr., vol. 38, no. 10, 1993.

[21] British Standards Institute, "BS EN 12390-5:2009 Testing hardened concrete — Part 5: Flexural strength of test specimens," BSI Stand. Publ., no. August, pp. 1–22, 2009.

[22] ASTM C469/C469M, "Standard Test Method for Static Modulus of Elasticity and Poisson's Ratio of Concrete in Compression," ASTM Stand. B., pp. 1–5, 2014, https://doi.org/10.1520/C0469

[23] P. and U. EFNARC The European Guidelines for Self-Compacting Concrete: Specification, "The European Guidelines for Self-Compacting Concrete: Specification, Production and Use," Eur. Guidel. Self Compact. Concr., no. May, p. 68, 2005.

[24] S. Benimam, M. Bentchikou, F. Debieb, S. Kenai, and M. Guendouz, "Physical and mechanical properties of cement mortar with LLDPE powder and PET fiber wastes," Adv. Concr. Constr., vol. 12, no. 6, pp. 461–467, 2021, https://doi.org/10.12989/acc.2021.12.6.461

[25] M. A. Samsudin et al., "Investigation on Polyethylene Terephthalate (PET) Waste Fiber Performances in Concrete Material," vol. 2, no. 1, pp. 682–690, 2021.

[26] Y. Wang, H. Zhang, Y. Geng, Q. Wang, and S. Zhang, "Prediction of the elastic modulus and the splitting tensile strength of concrete incorporating both fine and coarse recycled aggregate," Constr. Build. Mater., vol. 215, pp. 332–346, 2019, https://doi.org/10.1016/j.conbuildmat.2019.04.212

Advanced Topics in Mechanics of Materials, Structures and Construction - AToMech1-2023 Materials Research Forum LLC
Materials Research Proceedings 31 (2023) 165-171 https://doi.org/10.21741/9781644902592-18

Preparation and characterization of nanoparticles for encapsulation and delivery vehicles

Megdi Eltayeb[1,a *]

[1]Department of Mechanical Engineering, Prince Mohammad Bin Fahad University

[a]meltayeb@pmu.edu.sa

Keywords: Electrohydrodynamic, Polymer, Nanoparticles, Controlled Release, Encapsulation

Abstract. This research creates nanoparticles that are tunable, have a high active ingredient content, and are highly encapsulated and controlled. We use model materials to determine the ethylcellulose-vanillin ratios as well as the effective processing conditions required for nanoparticle formation using ethylcellulose and vanillin as active ingredients. Nanoparticles with a mean size of 45 to 64 nm were prepared at a rate of approximately 1.3 million nanoparticles per minute using a different polymer:vanillin ratio. The encapsulation efficiency and loading of vanillin at different concentrations in nanoparticles from solutions ranged from 75 to 94 percent and 66.90 to 86.54 percent, respectively, indicating that the loading and encapsulation efficiency of the nanoparticles decreased as the amount of vanillin increased.

Introduction

Recent years have seen nanotechnology used extensively in the pharmaceutical, cosmetic, and food industries for encapsulating and releasing active ingredients under controlled conditions [1–7]. The nanoparticles used for these applications must be nontoxic, stable, have high encapsulation efficiency, and have high loading capacity, and must also be capable of being precisely tailored to meet the release kinetics of the flavour component.

There are several different types of polymer-based colloidal particles, e.g. solid lipid nanoparticles, liposomes, and micelles. A variety of different purposes can be accomplished with these capsules, including encapsulating flavor, preservatives, vitamins, and other "nutraceuticals" [8–11] Lipophilic and hydrophilic favor components can be loaded into them, and studies have demonstrated that encapsulating materials increases bioavailability as compared to unencapsulated material. Food ingredients containing liposomes have already been approved or have entered clinical trials. Liposomes are limited for use in food encapsulation, however, due to their low loading capacity, rapid release of favor components, and instability during storage [12]. Solid lipid nanoparticles are made up of solid lipids as opposed to lamellar lipid structures. There has been evidence of greater stability and more sustained release problems in the materials encasing a liquid core. In literature, three methods for flavour encapsulation have been reported: (i) homogeneous matrix; (ii) enriched shell with flavor components; and (iii) enriched core with flavour components. Matrix particles homogeneous in composition have a uniform distribution of the favour component. In this way, a long-term favor release (over weeks) may be achieved along with protection against oxidation and high encapsulation capacity. Alternatively, particle shells possessing a high proportion of favored components exhibit a rapid release rate (in minutes). Fick's law of dissolution governs the release rates of particles with enriched favour components and is affected by physicochemical properties (the partition coefficient) of the molecules and the thickness and density of the outer lipid shell. The preparation of solid lipid nanoparticles has been carried out using a variety of techniques. Hot homogenization and cold homogenisation are the two main classes . Using the latter method, the solid lipid nanoparticles will exhibit poor control over their chemical and physical properties, resulting in the release of the active ingredient to the surrounding medium or to recrystallization. In contrast, cold homogenisation promotes the formation of particles with enriched shells containing favor components. In order to overcome these drawbacks, alternative methods of fabricating solid lipid nanoparticles are desirable. In this study, electrohydrodynamic techniques are examined for use in this application. Electrohydrodynamic atomisation or electrospraying involves an electrically induced focus and breakup of a liquid jet in order to produce droplets of micrometers to nanometres in diameter. As far as particle size and uniformity are concerned, physical properties of the materials being processed, their low rate(s), and the applied electrical potential difference play the largest roles. A wide range of nano-, micro-, and electrohydrodynamic structures have been produced using electrohydrodynamic techniques. It is easier to control particle size, shape, and uniformity, and it does not require multiple processing steps or high temperatures or pressure. The production of multilayer particles has also been shown to be feasible by single and coaxial spraying devices. An electrohydrodynamic process was used in this study to produce nanoparticles of ethylcellulose encapsulating vanillin. In the food industry, both materials have a wide range of applications and are typical hydrophobic coatings and hydrophilic favor components respectively [13]. The purpose of this study is to analyze the rate of release of active ingredients so that particle characteristics can be tailored to a particular application. In addition to demonstrating that electrohydrodynamic techniques [14] can be applied to many combinations of materials, our kinetic analysis has been significantly improved.

Materials and methods

Materials

All the consumables were purchased from Sigma-Aldrich (Poole, Dorset, UK), including ethylcellulose, vanillin (3-ethoxy-4-hydroxybenzaldehyde), and 95% (v) ethanol. All experiments used double distilled (DD) water as the release medium.

Preparation and characterisation of spraying solutions

Various concentrations of ethylcellulose and vanillin were dissolved in ethanol (Table 1) (from 1 to 4 weight%).

Table 1: Characteristics of ethylcellulose only (mean ± S.D., n = 5).

Solutions	Polymer concentration (wt %)	Density (kg/m^{-3})	Surface tension (mN m^{-1})	Viscosity (mPa s)	Electrical conductivity (mS m^{-1})
S1	0.25	759	22.2	4.4	2.97
S2	0.50	761	22.1	3.7	3.5
S3	0.75	763	22.1	3.2	5.16
S4	1.00	765	21.9	2.8	6.3
S5	1.25	767	21.8	2.7	6.91
S6	1.50	769	21.7	2.5	7.57
S7	1.75	771	21.5	2.3	8.33
S8	2	773	21.3	2	8.56
S9	2.25	777	20.1	1.6	8.78
S10	2.5	779	20	1.5	9.3

Table 2: Characteristics of ethylcellulose and vanillin (mean ± S.D., n = 5).

Solutions	Polymer concentration (wt %)	Density (kg/m^{-3})	Surface tension (mN m^{-1})	Viscosity (mPa s)	Electrical conductivity (mS m^{-1})
S1	0.25	764	22.1	4.2	3.1
S2	0.50	768	22	3.6	4.35
S3	0.75	770	21.9	3.1	5.24
S4	1.00	772	21.7	2.8	7.01
S5	1.25	774	21.6	2.6	7.21
S6	1.50	776	21.5	2.3	7.93
S7	1.75	779	21.4	2.1	8.39
S8	2	781	21.2	1.6	9.17
S9	2.25	784	20	1.5	9.49
S10	2.5	786	19.9	1.3	10.13

A magnetic stirrer was used to mix the solutions until they became optically transparent at ambient temperature (25C): this took around 10 minutes. From three repeat measurements of each solution, surface tension and viscosity were determined by a Kruss tension meter (Model-K9, Kruss GmbH, Germany) and a U-tube viscometer (75 mlnCannon-Fenske Routine Viscometer, Cannon Instruments, USA).

Electrohydrodynamic processing

Silicon tubing was used to feed the solution into a stainless steel needle with internal diameter 450 m from a 10 ml plastic syringe driven by Harvard PHD 4400 syringe pump (Edenbridge, UK) at a rate of 15 l min. In the experiment, a high voltage power supply was used to apply a potential difference between a needle electrode and a ground electrode between 13 and 15 kV (Glassman Europe Ltd., Tadley, UK). After an optimal cone jet voltage of 14:5 kV was achieved, nanoparticles were collected on glass microscope slides or in vials containing DD water for the release studies. Both vials were maintained at a distance of 100mm from the needle tip. A video camera (Leica S6D JVC-color) was used to observe the jet created at the tip of the needle when nanoparticles were produced. Before the current study, a study was carried out to determine how operating parameters and formulation could affect nanoparticle physical properties and vanillin release rate [50, 56]. In the present study, operating parameters (i.e. polymer concentration, low solution rate, distance between needle tip and collector) were selected to produce an effective cone jet and control the nanoparticle formation process [50, 56].

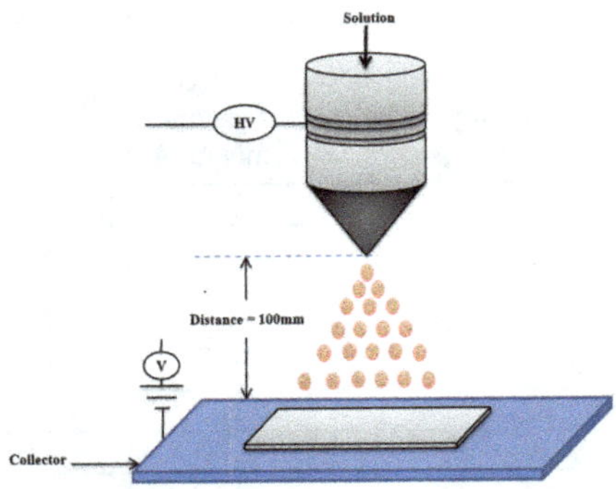

Figure 1: An illustration of an EHD spraying setup used for nanoparticle preparation.

Scanning electron microscopy (SEM)
By using SEM (Model JEOL JSM 3600, UK), the polymer nanoparticles were characterised in terms of size and morphology. After vacuum sputter coating dried polymer nanoparticles with gold for 2 min, with 40 mA, and mounting them on aluminium stubs with double-sided carbon tape, they were subjected to high voltage SEM examination (5 kV). The standard deviation and mean of the nanoparticle diameter were calculated using ImageJ (National Institutes of Health, NIH, Maryland, USA). For each set of processing conditions, approximately 300 nanoparticles were measured.

Table 3: Nanoparticle mean sizes, standard deviations (nm) and polydispersivity index.

Solutions	nanoparticle mean size (nm)	polydispersivity index	SD	Min	Max	PDI
S1	64.39	20.74	13.36	39.98	97.15	0.21
S2	62.51	17.31	10.82	35.85	81.98	0.17
S3	61.11	30.64	12.11	35.94	82.81	0.22
S4	57.66	16.50	17.67	38.89	92.05	0.31
S5	56.57	21.02	11.89	40.54	78.43	0.21
S6	55.00	22.91	12.60	34.48	83.05	0.23
S7	53.54	31.45	15.08	30.76	89.75	0.28
S8	51.86	23.89	16.31	29.73	91.89	0.31
S9	46.96	31.12	14.61	24.36	73.12	0.31
S10	45.31	33.82	10.82	24.47	70.27	0.24

Determination of encapsulation efficiency, loading capacity and release profile
Determination of vanillin encapsulation efficiency, loading capacity and release profile. A total of 3 ml of the collected suspensions were divided equally among three tubes for evaluation of vanillin encapsulation efficiency, loading capacity, and release rate of vanillin. The tubes were filtered using 200nm syringe filters in 10 ml of DD water at ambient temperature (25C). Each tube of polymer nanoparticles was mixed with equal amounts of DD water to disperse the polymer nanoparticles. A UV spectrophotometer (Perkin Elmer, Lambda 35, UV/Vis spectrophotometer, Waltham, USA) was used to determine the free vanillin content in each tube. Vanillin's absorbance at 278 nm was measured using a UV spectrophotometer. To calculate vanillin encapsulated in the

polymer nanoparticles, we subtracted the amount in the DD water from the total amount in the polymer nanoparticle solution. Equations 1 and 2 were used to determine encapsulation efficiency and loading capacity:

$$\%\ \text{Encapsulation efficiency} = \frac{\text{Amount of dried nanoparticles}}{\text{Total active ingredient used}} \quad (1)$$

$$\%\ \text{loading capacity} = \frac{\text{weight of dry nanoparticles}}{\text{weight of dried ethylcellulose and vanillin}} \times 100 \quad (2)$$

This equation expresses the weight of vanillin in nanoparticles, the weight of vanillin in supernatants, the total weight of polymer nanoparticles as NP, and the total weight of electrosprayed nanoparticles as DNP. Each set of polymer nanoparticles was released three times over 240 minutes, with the cumulative vanillin release rate plotted as a function of time.

Table 4: Nanoparticle loading capacity and entrapment efficiency obtained by electrohydrodynamic flow at 10, 25 and 50μl/min.

Solutions	Polymer concentration (w%)	Active component loading (%)	Flow rate (μl/min)	Loading capacity (%)	Entrapment efficiency (%)	Burst (%)
S1	10.0	10	25	86.54	0.94	5.57
S2	10.0	10	20	88.57	0.95	4.33
S3	10.0	10	15	90.51	0.97	2.80
S4	10.0	25	15	86.98	0.94	5.83
S5	10.0	50	15	64.61	0.70	27.39
S6	7.5	10	15	79.21	0.88	10.59
S7	7.5	25	15	74.54	0.83	15.56
S8	5.0	10	15	73.71	0.81	16.89
S9	7.5	50	15	70.96	0.80	18.24
S10	2.5	10	15	66.90	0.75	22.00

This equation expresses the weight of vanillin in nanoparticles, the weight of vanillin in supernatants, the total weight of polymer nanoparticles as NP, and the total weight of electrosprayed nanoparticles as DNP. Each set of polymer nanoparticles was released three times over 240 minutes, with the cumulative vanillin release rate plotted as a function of time.

Results and Discussion

This study was conducted at a variety of solution flow rates ranging from 10 to 25 1 min1. This was discovered to be necessary for the formation of uniform nanoparticles. The voltage required to generate nanoparticles increased as polymer concentration increased, and the jet eventually became unstable [15]. The polymer concentration is known to affect the surface tension, viscosity, and electrical conductivity of the liquid phase, all of which affect the EHD process [15]. Initial research was conducted to optimize the polymer concentration, and it was discovered that the viscosity and surface tension of the solution increased with polymer concentration Table 3 and

Table 2), resulting in an increase in mean nanoparticle size (Table 3). Furthermore, as the polymer concentration increased, the electrical conductivity of the solutions decreased noticeably (table 1), increasing nanoparticle size. Depending on the operating conditions, the diameter of the polymer nanoparticles ranged between 45.31 and 64.39 nm. With a polydispersity index ranging between 16.50 and 33.82 percent, the corresponding size distributions were relatively narrow (see Table 3). Table 3 shows how the size of the nanoparticles varied with polymer concentration and operating conditions. When the flow rate was increased from 10 to 25, the mean size of the nanoparticles increased by 19%, as did the polydispersity index. This finding is consistent with previous research [16]. The size of nanoparticles increased with polymer concentration (Table 3), which is most likely due to an increase in viscosity. Increasing the vanillin concentration reduced the mean size of the nanoparticles. This could be explained by a decrease in the liquid's electrical conductivity. However, the effect was less pronounced than for the corresponding polymer concentration, which is consistent with previous work by the authors [17]. Table 2 shows that the encapsulation efficiency and loading of vanillin at different concentrations in nanoparticles from solutions P1 - P10 ranged from 75 percent and 94 percent and 66.90 percent to 86.54 percent respectively, indicating that the nanoparticles' loading and encapsulation efficiency decreased as the amount of vanillin increased. 1wt% vanillin was encapsulated in nanoparticles with high encapsulation efficiency. The effect of vanillin loading capacity is consistent with previous studies [18] and can be explained by the fact that the decreased active component loading increases the relative amount of ethylcellulose acting as a diffusional barrier. By changing the polymer concentration of the polymer solution, the vanillin loading had a similar but smaller influence on nanoparticle mean size, with a large amount of active component loading causing an increase in mean particle size. As expected, flow rate has an effect on nanoparticle size. This is not surprising given that ethylcellulose is nearly insoluble in water.

Conclusions

Electrospraying has been shown to produce polymer nanoparticles containing vanillin. The diameter and size distribution of the nanoparticles were controlled by varying the polymer:vanillin concentration ratio in the sprayed solution. The thickness of the outer polymer layer of the nanoparticles was influenced by the concentration of polymer in the processed solution. Similarly, encapsulation efficiency and loading capacity were both affected by the polymer-vanillin ratio, though there was a limit beyond which both began to decline. A polymer-based nanoparticle that can be separated electrohydrodynamically in one step may be of considerable use for mass production of polymer based nanoparticles with different characteristics for food or medical applications.

References

[1] Attaalmanan, E.M., Mohammed-Ahmed, H.K., Hussein, S.A., Farag, M., Eltayeb, M., Fragoon, A.: The Biostability of Gold Nanoparticles for Cancer Treatment. J. Clin. Eng. 45, 66–71 (2020).

[2] Woering, M.H., Depreitere, B., Vander Sloten, J.: Estimated and underreported parameters in report based vehicle-bicycle accident reconstructions have a significant influence. Accid. Anal. Prev. 150, 105903 (2021). https://doi.org/10.1016/j.aap.2020.105903

[3] Yahya, I., Eltayeb, M.: Computational Modeling of Gold Nanoparticle in Lung Cancer Photothermal Therapy. J. Clin. Eng. 44, 157–164 (2019).

[4] Atif, R., Yahya, I., Eltayeb, M.: Simulation of drug release from biodegradable and erodible polymers. In: Proceedings of the International Conference on Computer, Control, Electrical, and Electronics Engineering 2019, ICCCEEE 2019. pp. 1–6 (2019). https://doi.org/10.1109/ICCCEEE46830.2019.9071247

[5] Ahmed, L., Yahya, I., Eltayeb, M.: Computational modeling of drug release profiles from swellable polymeric nanoparticles. In: Proceedings of the International Conference on Computer, Control, Electrical, and Electronics Engineering 2019, ICCCEEE 2019. pp. 1–5 (2019). https://doi.org/10.1109/ICCCEEE46830.2019.9070894

[6] Ahmed, L., Atif, R., Eldeen, T.S., Yahya, I., Omara, A., Eltayeb, M.: Study the Using of Nanoparticles as Drug Delivery System Based on Mathematical Models for Controlled Release. Int. J. Latest Technol. Eng. Manag. Appl. Sci. 8, 52–56 (2019).

[7] Yahya, I., Atif, R., Ahmed, L., Eldeen, T.S., Omara, A., Eltayeb, M.: Utilization of solid lipid nanoparticles loaded anticancer agents as drug delivery systems for controlled release. Int. J. Eng. Appl. Sci. Technol. 3, 7–16 (2019).

[8] Eltayeb, M.: Nanoencapsulation of ethylvanillin using Electrohydrodynamic technology. IOP Conf. Ser. Mater. Sci. Eng. 1150, 012017 (2021). https://doi.org/10.1088/1757-899x/1150/1/012017

[9] Yahya, I., Atif, R., Ahmed, L., Eldeen, T.S., Omara, A., Eltayeb, M.: Polymeric Nanoparticles as Drug Delivery Systems for Controlled Release. Adv. Sci. Eng. Med. 12, 263–270 (2020).

[10] Eltayeb, M., Yahya, I.: Computational approaches for investigating different shapes of nanoparticles-based drug delivery. bioRxiv. (2020).

[11] Yahya, I., Eltayeb, M.: Modeling of nano-carriers for vascular-targeted delivery for blood clots treatment. BioRxiv. (2020).

[12] Yahya, I., Atif, R., Ahmed, L., Eldeen, T.S., Omara, A., Eltayeb, M.: Mathematical Modeling of Diffusion Controlled Drug Release Profiles from Nanoparticles. Int. J. Res. Sci. Innov. (IJRSI),. 6, 287–291 (2019).

[13] Eltaye, M., Saleh, M.E.: Edible hybrid coating for food applications. In: AIP Conference Proceedings. p. 20009. AIP Publishing LLC (2022).

[14] Eltayeb, M.: Preparation and Characterization of Electrosprayed Nanoparticles containing Ethyl Maltol Flavor. In: Materials Science Forum. pp. 131–136. Trans Tech Publ (2021).

[15] Eltayeb, M., Stride, E., Edirisinghe, M., Harker, A.: Electrosprayed nanoparticle delivery system for controlled release. Mater. Sci. Eng. C. 66, 138–146 (2016). https://doi.org/10.1016/j.msec.2016.04.001

[16] Eltayeb, M., Stride, E., Edirisinghe, M.: Electrosprayed core-shell polymer-lipid nanoparticles for active component delivery. Nanotechnology. 24, (2013). https://doi.org/10.1088/0957-4484/24/46/465604

[17] Eltayeb, M., Bakhshi, P.K., Stride, E., Edirisinghe, M.: Preparation of solid lipid nanoparticles containing active compound by electrohydrodynamic spraying. Food Res. Int. 53, 88–95 (2013). https://doi.org/10.1016/j.foodres.2013.03.047

[18] Eltayeb, M., Stride, E., Edirisinghe, M.: Preparation, characterization and release kinetics of ethylcellulose nanoparticles encapsulating ethylvanillin as a model functional component. J. Funct. Foods. 14, 726–735 (2015). https://doi.org/10.1016/j.jff.2015.02.036

Isogeometric topology optimization of auxetic materials based on moving morphable components method

Xiaoya Zhai[1,2,a], Yundong Gai[3,4,b,*], Liuchao Jin[2], Wei-Hsin Liao[2,*], Falai Chen[1], Ping Hu[4]

[1] School of Mathematical Sciences, University of Science and Technology of China, Hefei, Anhui, China

[2] Department of Mechanical and Automation Engineering, The Chinese University of Hong Kong, Hong Kong, China

[3] Hithink Royal Flush Information Network Co., Ltd., Hangzhou, Zhejiang, China

[4] State Key Laboratory of Structural Analysis for Industrial Equipment, School of Automotive Engineering, Dalian University of Technology, Dalian, Liaoning, China

[a]xiaoya_zhai@outlook.com, [b]yd_gai@foxmail.com

Keywords: Auxetic Materials Design, Isogeometric Topology Optimization, Microstructures Design, Moving Morphable Method, Homogenization Theory

Abstract. Auxetic materials are a class of materials that exhibit a negative Poisson's ratio. They have held a major interest in academics and engineering focusing on finding the material distribution and examining the mechanisms, properties, and applications. Inverse homogenization theory is taken as an effective material design tool and has been applied to optimize various metamaterials. In this paper, we derive and implement the energy-based isogeometric homogenization to generate auxetic materials. Numerical examples show that the homogenized elasticity matrix obtained by the energy-based isogeometric homogenization method is almost the same as that obtained by the finite element homogenization method within a tolerated error. On this basis, we applied the isogeometric Moving Morphable Components (MMC) method to the optimization design of auxetic materials which is named the TOP-IGA-MMC method. We further make a comparison of the Solid Isotropic Material with the Penalization (SIMP) method and the TOP-IGA-MMC method in the geometries and properties of the final optimal auxetic materials. Parameter tests and physical tests are also introduced to verify the robustness and effectiveness of the proposed method.

Introduction

Metamaterials are engineered materials with properties usually not seen in nature such as negative Poisson's ratio (also named auxetic material) [1]. In recent years, metamaterials such as left-handed materials [2] and invisibility cloak [3] have been increasingly prominent in the fields of optics, communications, national defense, and other applications.

Generally, metamaterials are artificially optimized to achieve extraordinary properties based on topology optimization. The grounding breaking work by Sigmund [4] is assumed that microstructures are arranged periodically throughout the space, while their length scale is much smaller than those of macroscopic space. After that, topology optimization has been rapidly developed, and researchers have proposed a variety of different topology optimization methods such as **S**olid **I**sotropic **M**aterial with **P**enalization (SIMP) method [5], **E**volutionary **S**tructural **O**ptimization (ESO) method [6], **L**evel **S**et **M**ethod (LSM) [7], etc. However, for these existing methods, the topology optimization methods are implemented in an implicit way. The topologies only have implicit descriptions which means that the final optimized structure is detected either from a higher dimensional level set function (LSM) or a black-white binary field (SIMP and ESO methods). In addition, for the density-based method, sensitivity analyses are carried out on a fixed finite element grid. The accuracy of analysis results highly depends on the resolution of the finite element meshes.

In order to obtain explicit boundary representations, Guo et al. [8] established an explicit topology optimization method based on the moving morphable components (MMC) method. The method takes the geometric parameters related to components as design variables. By optimizing these geometric parameters, the components can be moved, stretched, and overlapped to change the topologies. Further, an open-source 188- line MATLAB code for the MMC framework was given [9]. Later, Zhang et al. [10, 11] proposed another explicit topology optimization method in that they adopted closed B-splines and Boolean operations to represent the boundaries of holes. The topologies changed by evolving and merging their boundaries. This method has been gradually improved by [12–14]. For these explicit topology optimization methods, design variables of topology optimization are replaced by controlled parameters, and the number of design variables is reduced.

In the previously mentioned topology optimization methods, the traditional Finite Element Method [15] was adopted in numerical analysis. Gao et al. [16] summarized three deficiencies of FEM which are inexact representation, low continuity between neighboring finite elements, and low calculation efficiency to gain a high quality of the finite element mesh. To address these issues, IsoGeometric Analysis (IGA) is proposed by Hughes et al. [17] to perform the numerical analysis that is represented by non-uniform rational B-splines (NURBS). IGA overcomes these limitations by using the same basis functions to define both the geometry and the finite element model of the structure, resulting in a seamless connection between geometry and analysis. Additionally, the use of NURBS functions allows for the use of high-order splines, which can provide improved accuracy and convergence in the optimization process. In addition, many geometric standards in the industry regard NURBS as a powerful tool for geometric design. NURBS also plays an important role in CAD/CAE because of its good mathematical properties and efficient algorithms.

Inspiring above merits, some researchers combined IGA with topology optimization named Isogeometric Topology Optimization (ITO) which is a computational method used to optimize the design of mechanical structures by combining traditional numerical optimization techniques with the use of NURBS geometry representation such as isogeometric-based SIMP method [18], isogeometric-based moving morphable components [19], ITO using trimmed spline surfaces [20] and so on. It has been applied to various fields such as structural vibration [21], shell optimization [22] and meta-materials design [23].

In this work, we propose an explicit isogeometric topology optimization approach named TOP-IGAMMC, and applied this approach to the field of material design to generate auxetic materials. We first derive the isogeometric homogenization theory and compare the numerical equivalent stiffness matrix with the results of traditional homogenization theory. The matrixes are almost the same. Then we apply the isogeometric homogenization theory to do the inverse design of auxetic materials based on MMC methods. Finally, we compare our optimized results with those generated from the isogeometric SIMP method and then do parameter tests and physical tests. The results show the effectiveness and robustness of our method.

The outline of the paper is as follows. In Section 2, the theories of the MMC method and IGA-based homogenization are presented. Thereafter, the formulations of topology optimization of auxetic material are proposed in Section 3. In Section 4, numerical examples are illustrated and compared with SIMP methods. Parameter selection is also discussed. More experiments are conducted to demonstrate the performance in TOP-IGA-MMC for auxetic materials design. Section 5 closes the paper with some concluding remarks.

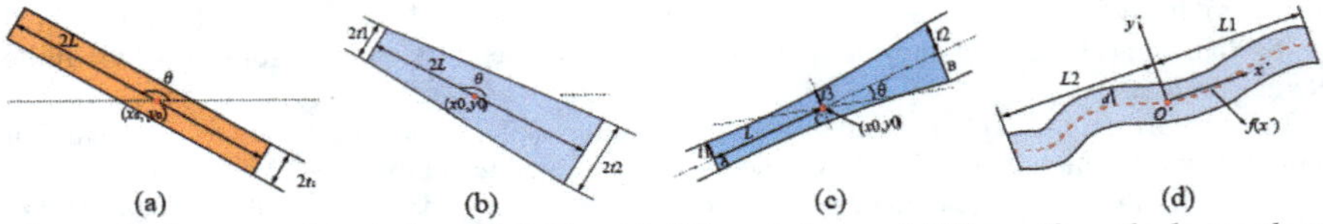

Figure 1: Geometry description of MMCs with different shapes. (a) The uniform thickness along a straight skeleton. (b) Linearly varying thickness. (c) Quadratically varying thickness. (d) The uniform thickness along the curved skeleton.

Preliminaries

Geometry description of moving morphable components. Let $\mathcal{D}$ represent a prescribed design domain and $\Omega^s = \{\Omega_1, \cdots, \Omega_n\}$ be a subset of $\mathcal{D}$. Ω^s denotes a collection of n components. For the MMC method, there are four types of components illustrated in Paper [9] and shown in Figure 1. The components with quadratically varying thicknesses (Figure 1(c)) are chosen in this paper. For one component $\Omega_i \in \Omega^s$, the solid and void region is defined as follows

$$\begin{cases} \phi_i(\boldsymbol{x}) = 0; & \text{if } \boldsymbol{x} \in \partial\Omega_i; \\ \phi_i(\boldsymbol{x}) > 0; & \text{if } \boldsymbol{x} \in \Omega_i; \\ \phi_i(\boldsymbol{x}) < 0; & \text{otherwise}; \end{cases} \tag{1}$$

where $\phi_i(\boldsymbol{x})$ ($\forall \boldsymbol{x} = (x, y) \in \mathcal{D} \subset \mathbb{R}^2$) is an explicit topology description function, which can be defined as

$$\begin{aligned} &\phi_i(\boldsymbol{x}) = \phi_i\,(x, y) = \left(\frac{x'}{L_i}\right)^p + \left(\frac{y'}{f_i(x')}\right)^p - 1, \\ &\begin{bmatrix} x' \\ y' \end{bmatrix} = \begin{bmatrix} \cos\theta_i & \sin\theta_i \\ -\sin\theta_i & \cos\theta_i \end{bmatrix} \begin{bmatrix} x - x_{0i} \\ y - y_{0i} \end{bmatrix}, \end{aligned} \tag{2}$$

where (x', y') is the coordinate transformation of (x, y). p is set as 6. (x_{0i}, y_{0i}), L_i and θ_i represent the coordinates of a center point O_i, half of the total length, and the inclination angle in the i-th component Ω_i, respectively. $f_i(\boldsymbol{x}')$ describes the shape of the i-th component. For components with quadratically varying thickness (Figure 1(c)), $f_i(\boldsymbol{x}')$ is defined as

$$f_i(\boldsymbol{x}') = \frac{t_{1i}+t_{2i}-2t_{3i}}{2L_i^2}(x')^2 + \frac{t_{2i}-t_{1i}}{2L_i}x' + t_{3i}. \tag{3}$$

where $t_i = \{t_{1i}, t_{2i}, t_{3i}\}$ is the thickness vector of the i-th component

These components $\Omega^s = \{\Omega_1, \cdots, \Omega_n\}$ is controlled by the parameters $\boldsymbol{L} = \{L_1, \cdots, L_n\}$, $\boldsymbol{t} = \{t_{11}, t_{12}, t_{13}, \cdots, t_{1n}, t_{2n}, t_{3n}\}$ and $\boldsymbol{\theta} = \{\theta_1, \cdots, \theta_n\}$. n is the number of components. A 2D component is represented by $\boldsymbol{X} = \{x_0, y_0, L, t_1, t_2, t_3, \theta\}$. Thus, the final structure is determined by $\boldsymbol{X} = \{\boldsymbol{X}_1^T, \cdots, \boldsymbol{X}_n^T\}$. When two or more components overlap, the maximum value of the topological description function $\phi(\boldsymbol{x})$ ($\forall \boldsymbol{x} \in \mathcal{D}$) of these overlapping components can be used to describe the occupied area. The final topological description function can be described as $\phi^s = \max_i(\phi_i(\boldsymbol{x}))$, $i = 1, \cdots, n$ which is represented as

$$\begin{cases} \phi^s(\boldsymbol{x}) = 0; & \text{if } \boldsymbol{x} \in \partial\Omega^s; \\ \phi^s(\boldsymbol{x}) > 0; & \text{if } \boldsymbol{x} \in \Omega^s; \\ \phi^s(\boldsymbol{x}) < 0; & \text{otherwise}; \end{cases} \tag{4}$$

Isogeometric homogenization theory. The homogenization method [24, 25] is a numerical method to calculate the equivalent macroscopic properties of composites. The square domain is discretized with 4-node brick elements represented by the biquadratic NURBS elements. Given a set of points $\mathbf{P}_{ij} \in \mathbb{R}^2$ and the corresponding weights w_{ij}, $(i = 0, 1, \ldots, m,\ j = 0, 1, \ldots, n)$, a tensor product NURBS surface (the square domain) of bi-degree (p, q) is defined by

$$\boldsymbol{S}(\zeta, \xi) = \sum_{i=0}^{m} \sum_{j=0}^{n} R_{ij}^{pq}(\zeta, \xi)\mathbf{P}_{ij},$$

$$R_{ij}^{pq}(\zeta, \xi) = \frac{w_{ij}\, N_i^p(\zeta)\, N_j^q(\xi)}{\sum_{k=0}^{m} \sum_{l=0}^{n} w_{kl}\, N_k^p(\zeta) N_l^q(\xi)} \tag{5}$$

where $\zeta \in [\zeta_p, \zeta_{m+1}], \xi \in [\xi_q, \xi_{n+1}], i = 0, 1, \ldots, m, j = 0, 1, \ldots, n$ are bivariate NURBS basis functions defined over the knot vectors $\zeta = \{\zeta_0, \zeta_1, \cdots, \zeta_{m+p+1}\}$ and $\xi = \{\xi_0, \xi_1, \cdots, \xi_{n+q+1}\}$ with weights $\{w_{ij}\}_{i=0,j=0}^{m,n}$. $N_i^p(\zeta)$ and $N_j^q(\xi)$ are the B-spline basis function defined by the Cox-de Boor algorithm [26]. In addition, the macroscopic equivalent properties of the unit cell are calculated based on NURBS basis. Thus, the isogeometric homogenization theory is derived.

The homogenized equivalent elastic matrix $\mathbf{D}_{ijkl}^H$ is obtained by averaging the integral over Y in the Einstein index, there is

$$\mathbf{D}_{ijkl}^H = \frac{1}{|Y|} \int_Y \mathbf{D}_{pqrs} \left(\epsilon_{pq}^{0(ij)} - \epsilon_{pq}^{*(ij)}\right)\left(\epsilon_{rs}^{0(kl)} - \epsilon_{rs}^{*(kl)}\right) \mathrm{dY}, \tag{6}$$

where $\mathbf{D}_{pqrs}$ is a locally varying elasticity tensor. $|Y|$ is the volume of the unit cell. $\epsilon_{rs}^{0(kl)}$ is the predefined strain field, $\epsilon_{rs}^{*(kl)}$ is the Y-periodic solution of the following equation

$$\int_Y \mathbf{D}_{ijrs} \epsilon_{ij}(v) \epsilon_{rs}^{*(kl)} \mathrm{dY} = \int_Y \mathbf{D}_{ijrs} \epsilon_{ij}(v) \epsilon_{rs}^{0(kl)} \mathrm{dY}, \tag{7}$$

where v is a virtual displacement field. $\epsilon_{rs}^{(kl)} = \epsilon_{rs}^{0(kl)} = \epsilon_{rs}^{*(kl)}$ is superimposed strain field. In the isogeometric homogenization, one unit cell is modeled by NURBS representation with open node vectors and can be rewritten as the discrete form

$$\mathbf{D}_{ijkl}^H = \frac{1}{|Y|} \sum_{e=1}^{N_e} d_e^{(ij)} \mathbf{K}_e d_e^{(kl)} \tag{8}$$

where $\mathbf{K}_e = \rho_e^p \mathbf{K}_e^0$, ρ_e is the density of the NURBS element. p is the penalty parameters, $\mathbf{K}_e^0$ is the solid stiffness matrix. N_e is the number of elements. $d_e^{(ij)}$ is the displacement field defined on the NURBS elements. We modified the Eq. (8) by replacing engineering notation instead of Einstein notation using $11 \rightarrow 1$, $22 \rightarrow 2$, $12 \rightarrow 3$. So Eq. (8) is rewritten as

$$\mathbf{D}_{ij}^H = \frac{1}{|Y|} \sum_{e=1}^{N_e} d_e^{(i)} \mathbf{K}_e d_e^{(j)} \tag{9}$$

Under the periodic assumption, the displacement field d_i of one unit cell can be expressed as the sum of the macroscopic displacement field $\epsilon_{ij}^0 y_j$ and the microscopic periodic perturbation displacement d^*, that is

$$d_i = \epsilon_{ij}^0 y_j + d^* \tag{10}$$

where d_1 and d_2 represent the displacement along x and y directions. For a 2D unit cell, the displacement of the pairs of opposite sides is respectively

$$\begin{cases} d_i^{k+} = \epsilon_{ij}^0 y_j^{k+} + d^* \\ d_i^{k-} = \epsilon_{ij}^0 y_j^{k-} + d^* \end{cases} \tag{11}$$

where the superscripts $k+$ and $k-$ represent the top (right) and bottom (left) lines perpendicular to the k-th direction ($k = 1$ (x direction) or 2 (y direction)), respectively. Since the periodic perturbation displacement field d^* is unknown, d^* can be eliminated by calculating the displacement difference between the opposite side $k+$ and $k-$, then

$$w_i^k = d_i^{k+} - d_i^{k-} = \epsilon_{ij}^0 (y_j^{k+} - y_j^{k-}) = \epsilon_{ij}^0 \Delta y_j^k \tag{12}$$

For the NURBS patch, the first and last control points in the k-th direction are interpolated on the cell boundary, thus $\Delta y_1^1 = y_1^0$, $\Delta y_1^2 = 0$, $\Delta y_2^1 = 0$, and $\Delta y_2^2 = y_2^0$. In the horizontal direction, the displacement difference between the left and right sides of a unit cell is $w_i^1 = \epsilon_{ij}^0 y_1^0$; Vertically, the difference between the top and bottom of the cell is $w_i^2 = \epsilon_{i2}^0 y_2^0$.

Methods

The TOP-IGA-MMC method is applied to the topology optimization for auxetic materials design. The proposed model is given in subsection 3.1. Then the detailed sensitivity analysis is introduced. Finally, the optimization process is summarized in algorithm 1.

Algorithm 1: The TOP-IGA-MMC method

input : Numbers of elements along x and y direction are $nelx$ and $nely$; the upper limit of the volume fraction is set as V^* and the penalty factor p;
output: Optimized components.

```
Initialization: The parameters of initial components X_k = {x_0k, y_0k, L_k, t_1k, t_2k, t_3k, θ_k}
  ∀k; the iteration index i = 0; design change Δ = 1.0; threshold ϵ = 10^-3;
while Δ > ϵ & i < I_max or i ≤ I_min do
  | i = i+1;
  | Calculate the equilibrium equation via. Eq. (16).
  | Compute objective function and volume constraints via Model (13);
  | Evaluate ∂J/∂X via Eq.(18);
  | Update X^[i] via MMA solver;
  | Δ = max_∀k(|X_k^[i] − X_k^[i−1]|);
end
```

Model formulation. In TOP-IGA-MMC, MMCs are used as the basic units to describe the structural topologies. The component with quadratically varying thickness is adopted. The geometric description parameters $\boldsymbol{X} = \{\boldsymbol{X}_1^T, \cdots, \boldsymbol{X}_n^T\}$ are taken as the design variables. The model is formulated by optimizing the negative Poisson's ratio under the predefined volume V^*. The objective function follows the relaxation form in Paper [27]. Therefore, the optimization model of topology optimization for auxetic materials can be written as

$$\begin{aligned} &\text{find:} && \boldsymbol{X} = \{\boldsymbol{X}_1^T, \cdots, \boldsymbol{X}_n^T\} \\ &\min_{\boldsymbol{X}}: && J(\mathbf{D}^H(\boldsymbol{\rho})) = D_{12}^H - \gamma^l (D_{11}^H + D_{22}^H) \\ &\text{s.t.} && \mathbf{K}\boldsymbol{d} = \mathbf{F} \\ & && \sum_e \frac{\rho_e v_e}{|Y|} \leq V^* \\ & && 0 \leq \rho_e \leq 1 \end{aligned} \tag{13}$$

where $X_i = \{x_{0i}, y_{0i}, L_i, t_{1i}, t_{2i}, t_{3i}, \theta_i\}$ $(i = 1, \cdots, n)$ is the geometric parameters. In the objective function J, l is the number of iteration steps and $\gamma = 0.8$. $\mathbf{D}^H(\boldsymbol{\rho})$ is the homogenized equivalent elastic matrix based on engineering notation calculated by Eq. (9). The element density ρ_e is represented as

$$\rho_e = \sum_{j=1}^{N_g} \kappa_j \left(\sum_{i=1}^{N_{cp}} R_i(\zeta_j, \xi_j) H(\phi_{i,e}) \right) \tag{14}$$

where N_g is the number of Gauss points in one element. κ_j is the coefficients of the Gaussian integral. $\phi_{i,e}$ $(i = 1, \cdots, N_{cp})$ is value of topological description function $\phi_s(x)$ at the i-th control point of the e-th element. Ncp is the number of control points. $R_i(\zeta_j, \xi_j)$ is non-uniform rational B-splines basis in Eq. (5). The Heaviside function $H(x)$ is

$$H_\epsilon(x) = \begin{cases} 1, & \text{if} \quad x > \epsilon \\ \frac{1(1-\alpha)}{4}\left(\frac{x}{\epsilon} - \frac{x^3}{3\epsilon^3}\right) + \frac{1+\alpha}{2}, & \text{if} -\epsilon < x \leq \epsilon \\ \alpha, & \text{otherwise} \end{cases} \tag{15}$$

where α is a constant value that is very small to avoid the singularity of the stiffness matrix. ϵ is used to describe the function regularization. v_e is volume of the e-th element. $|Y|$ is the volume of the single cell. V^* is the upper limit of the volume fraction. Young's modulus is represented as $E_e(\rho_e) = E_{min} + \rho_e^p(E_0 - E_{min})$. E_0 and $E_{min} = 10^{-9}$ are Young's modulus of solid and void materials. p is the penalty factor. $\mathbf{K}\boldsymbol{d} = \mathbf{F}$ is equilibrium equation. The displacement $\boldsymbol{d}$ of all control points is divided into four parts. Among them, $\overline{\boldsymbol{d}}_1$ represents the known displacement of the four corner control points; $\boldsymbol{d}_3$ and $\boldsymbol{d}_4$ represent the displacement of other control points on the boundary of one unit cell, which should satisfy $\boldsymbol{d}_4 = \boldsymbol{d}_3 + \overline{\boldsymbol{w}}$; $\overline{\boldsymbol{w}}$ can be obtained from Eq. (12); $\boldsymbol{d}_2$ represents the unknown displacement of all internal control points. Then the equilibrium equation can be written as

$$\begin{bmatrix} \mathbf{K}_{11} & \mathbf{K}_{12} & \mathbf{K}_{13} & \mathbf{K}_{14} \\ \mathbf{K}_{21} & \mathbf{K}_{22} & \mathbf{K}_{23} & \mathbf{K}_{24} \\ \mathbf{K}_{31} & \mathbf{K}_{32} & \mathbf{K}_{33} & \mathbf{K}_{34} \\ \mathbf{K}_{41} & \mathbf{K}_{42} & \mathbf{K}_{43} & \mathbf{K}_{44} \end{bmatrix} \begin{bmatrix} \overline{\boldsymbol{d}}_1 \\ \boldsymbol{d}_2 \\ \boldsymbol{d}_3 \\ \boldsymbol{d}_4 \end{bmatrix} = \begin{bmatrix} \boldsymbol{F}_1 \\ \boldsymbol{F}_2 \\ \boldsymbol{F}_3 \\ \boldsymbol{F}_4 \end{bmatrix} \tag{16}$$

where $\mathbf{K}$ is the global stiffness matrix, and $\boldsymbol{F}_1$ is the reaction force of angular control point under a given displacement. $\boldsymbol{F}_2 = 0$, and $\boldsymbol{F}_3 = -\boldsymbol{F}_4$ under the periodic assumption. Then we have

$$\begin{bmatrix} \mathbf{K}_{22} & \mathbf{K}_{23} + \mathbf{K}_{24} \\ \mathbf{K}_{32} + \mathbf{K}_{42} & \mathbf{K}_{33} + \mathbf{K}_{34} + \mathbf{K}_{43} + \mathbf{K}_{44} \end{bmatrix} \begin{bmatrix} \boldsymbol{d}_2 \\ \boldsymbol{d}_3 \end{bmatrix} = -\begin{bmatrix} \mathbf{K}_{21} \\ \mathbf{K}_{31} + \mathbf{K}_{41} \end{bmatrix} \boldsymbol{d}_1 - \begin{bmatrix} \mathbf{K}_{24} \\ \mathbf{K}_{34} + \mathbf{K}_{44} \end{bmatrix} \overline{\boldsymbol{w}} \tag{17}$$

Solve the above linear equations, the displacement $\boldsymbol{d}$ is obtained. It is worth noting that all variables are defined at the control points of NURBS elements.

Sensitivity analysis. We take b as the design variables in $\boldsymbol{X} = \{x_0, y_0, L, t_1, t_2, t_3, \theta\}$, then the sensitivity of the objective function $J(D_{ij}^H)$ to the design variable b is

$$\frac{\partial J}{\partial b} = \sum_{i=1}^{3} \sum_{j=1}^{3} \frac{\partial J}{\partial D_{ij}^H} \frac{\partial D_{ij}^H}{\partial b}$$
$$\frac{\partial D_{ij}^H}{\partial b} = \frac{p}{|Y|} \sum_{e=1}^{N_e} \rho_e^{p-1} \left[\sum_{j=1}^{N_g} \frac{1}{N_g} \left(\sum_{i=1}^{N_{cp}} R_i(\zeta_j, \xi_j) \frac{\partial H(\phi_{i,e})}{\partial b} \right) \right] d_e^{(i)} \mathbf{K}_e d_e^{(j)} \tag{18}$$

The sensitivity of the volume constraint function V to the design variable b is

$$\frac{\partial V}{\partial b} = \frac{v_e}{|Y|} \sum_{e=1}^{N_e} \left[\sum_{i+g=1}^{N_g} \frac{1}{N_g} \left(\sum_{i=1}^{N_{cp}} \frac{\partial H(\phi_{i,e})}{\partial b} \right) \right] \tag{19}$$

Algorithm. The size of one unit cell is set as $L_x \times L_y = 1 \times 1$, the parameter $\gamma = 0.8$ in the objective function $J(D^H(\boldsymbol{\rho}))$ in Model (13). The convergence criterion is the maximum change $\triangle$ of the design variables between two iterations is less than ϵ, that is, $\triangle \leq \epsilon$. The Greville points are used as the control points of the elements. Young's modulus and Poisson's ratio of the solid material are set as $E_0 = 1$ and $\mu = 0.3$. In order to avoid premature convergence conditions, the minimum iteration steps are set as $I_{min} = 50$ and the maximum iteration times as $I_{max} = 500$. To solve the optimization problem numerically, the Method of Moving Asymptotes (MMA) [28] is used as the optimizer.

Discussion

To verify the effectiveness of the proposed TOP-IGAMMC method for auxetic material design, we make a comparison between the TOP-IGA-MMC method and another density-based method in the numerical performances, material properties, and geometries. All the examples are run on a desktop computer: the CPU is the 11th Gen Intel(R) Core(TM) i7-1165G7 2.80GHz, the RAM is 64GB, the OS is Windows 10, and the software environment is MATLAB 2022b. The code for this paper is at https://github.com/xiaoyazhai/TOPIGA- MMC-Micro.

Comparisons. In order to compare the results of the proposed algorithm with those of the SIMP method, we first briefly introduce the model based on the isogeometric SIMP method named TOP-IGA-SIMP for material design. In the TOP-IGA-SIMP method, design variables are the element densities ρ_e and the objective function and constraints are the same as those in Model (13). To avoid the checkerboard patterns and obtain better structural topology, filtered sensitivity or density is often used to participate in the optimization process. For implementation details, please refer to Supplementary 5.

The optimal design of auxetic materials obtained by the TOP-IGA-MMC method and TOP-IGA-SIMP method is presented in Figure 2. The number of elements N of these two methods is set as 50 × 50. The penalty factor is 5, the volume fraction is 0.3. The number of components is 40. For all elements, $t_1 = t_2 = t_3 = 0.05$ and $\theta = 0$ are the initial inputs. Thus, the initial parameter combination is $(N, p, V^*, L, t, \theta) = (50 \times 50, 5, 0.3, 0.25, 0.05, 0)$. These two methods are compared in three aspects: numerical performances, material properties, and geometries. For the numerical performances, there are some oscillations during the first 100 iterations. Large structure changes (shown in intermediate structures of Iter=2, 8, 25, 50, 80) cause this phenomenon. The Poisson's ratio of these two optimized structures is −0.6124 and −0.3166. Our method can get better results under the same number of elements. In Figure 2(b), the final result has a large number of gray densities, which will cause the instability of the results. We further increase the number of elements of the TOP-IGA-SIMP method to 100 × 100 and 200 × 200, and the Poisson's ratio values of the final results are −0.4988 and −0.6026, respectively. When the number of elements is 200×200, the Poisson's ratio of the optimized structure is very similar to that of our method. But the time consumption is about 2.5 times that of our algorithm. In order to show more comparative results, we optimize auxetic materials under different volume fraction constraints (shown in Figure 3).

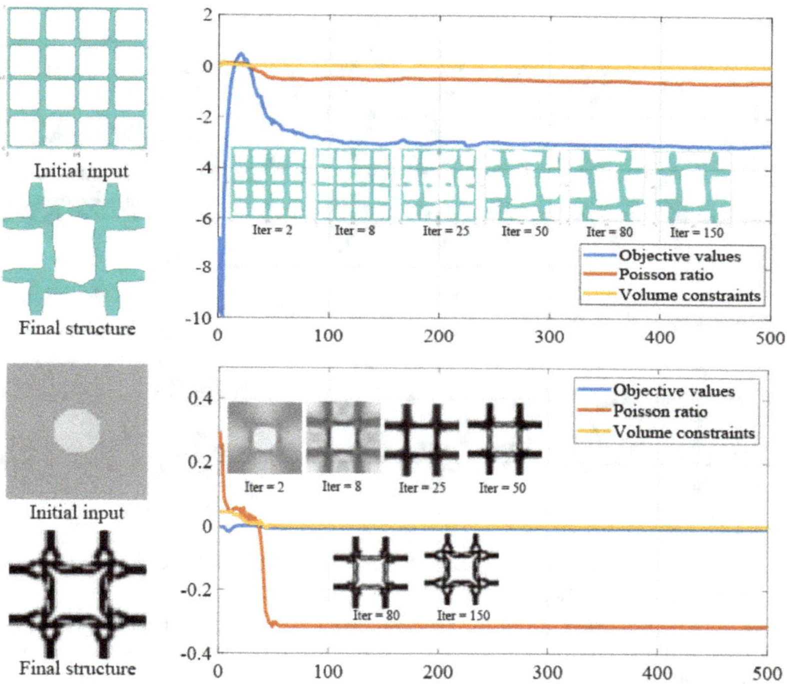

Figure 2: Plots of iteration curves of TOP-IGA-MMC method and TOP-IGA-SIMP method under the volume fraction 30%. The initial input and final optimized auxetic material are given on the left. The objective functions, Poisson's ratio, and volume constraints are plotted in the red, blue, and yellow curves. The initial parameter combination is $(N, p, V^*, L, t, \theta) = (50 \times 50, 5, 0.3, 0.25, 0.05, 0)$.

In Figure 3, the volume fraction constraints increase from 30% to 70%. Figure 3 (a) represents the results obtained based on the TOPIGA-MMC method. The initial parameters for the component are the same as in Figure 2. Figure 3(b)-(c) show the results under different element numbers (50×50 vs. 100×100 vs. 200×200) of the TOP-IGA-SIMP method. With the increase of elements numbers, the gray densities gradually decrease and there are more and more details in the final optimized structures. Whereas very small structures are not conducive to manufacturing, the result of our method is more suitable for manufacturing due to the final structures of components consolidation and clear boundaries. A discussion of the element numbers of the TOP-IGA-MMC method is covered in Section 4.2.2.

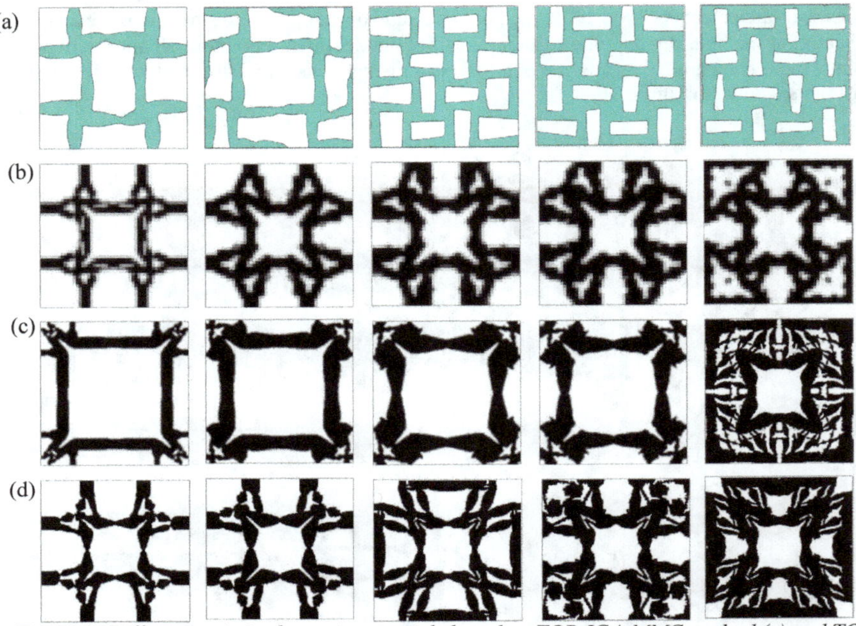

Figure 3: Different optimized auxetic materials based on TOP-IGA-MMC method (a) and TOP-IGA-SIMP method (b)-(d). The volume fraction increases gradually from left (30%) to right (70%). The number of elements is 50×50*(a)(b),* 100×100*(c),* 200×200 *(d).*

Numerical tests. In this section, we first verify the calculation of the elastic matrix based on isogeometric homogenization theory by comparing with calculation methods in Paper [27] and Paper [29]. In addition, there are many local minima in the topological optimization materials design due to optimization parameters such as penalty factor p, and initialization inputs (L, t and θ). Different optimization parameters and the initial components are set to analyze the impact on the generation of auxetic materials.

Numerical validation. A square cell with a central hole is taken as a calculation example. The size of the cell is $L_x \times L_y = 1 \times 1$, and the radius of the central hole is $\frac{\min(\text{nelx},\text{nely})}{3}$ and $\frac{\min(\text{nelx},\text{nely})}{6}$. nelx and nely are the numbers of elements along x and y directions. The initial parameter combination for TOP-IGA-MMC method is $(N, p, V^*, L, t, \theta) = (50 \times 50, 5, 0.3, 0.25, 0.05, 0)$. Homogenize function in [29], topX function in [27], and the proposed isogeometric homogenization method are used to calculate an equivalent elastic matrix, respectively. The comparison results are given in Table 1. It can be seen that the equivalent elastic matrices obtained by different methods are almost the same, and the maximum error is less than 10^{-4}.

Table 1: The homogenized effective elasticity matrices obtained from different methods (TopX [27] vs. Homogenization [29] vs. Ours).

Initial inputs	TopX [27]	Homogeniztion [29]	Ours
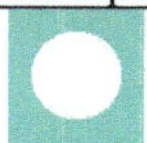	$\begin{bmatrix} 0.0732 & 0.0186 & 0.0000 \\ 0.0186 & 0.0732 & 0.0000 \\ 0.0000 & 0.0000 & 0.0199 \end{bmatrix}$	$\begin{bmatrix} 0.0732 & 0.0185 & 0.0000 \\ 0.0185 & 0.0732 & 0.0000 \\ 0.0000 & 0.0000 & 0.0199 \end{bmatrix}$	$\begin{bmatrix} 0.0732 & 0.0186 & -0.0000 \\ 0.0186 & 0.0732 & -0.0000 \\ -0.0000 & -0.0000 & 0.0199 \end{bmatrix}$
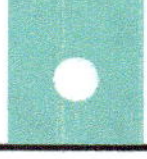	$\begin{bmatrix} 0.1157 & 0.0344 & 0.0000 \\ 0.0344 & 0.1157 & 0.0000 \\ 0.0000 & 0.0000 & 0.0394 \end{bmatrix}$	$\begin{bmatrix} 0.1157 & 0.0343 & 0.0000 \\ 0.0343 & 0.1157 & -0.0000 \\ 0.0000 & -0.0000 & 0.0394 \end{bmatrix}$	$\begin{bmatrix} 0.1158 & 0.0344 & -0.0000 \\ 0.0344 & 0.1158 & -0.0000 \\ -0.0000 & -0.0000 & 0.0394 \end{bmatrix}$

Initial inputs. Number of components. The initial components and final structural topologies are shown in Figure 4. The initial parameter combination of these two cases is $(N, p, V^*, L, t, \theta) = (40 \times 40, 3, 0.5, 0.5, 0.08, 0)$ and $(40 \times 40, 3, 0.5, 0.25, 0.08, 0)$. The number of components is related to the components' thickness L. In Figure 4(a), the number of components is 12, and since each component can be described with 7 parameters $\boldsymbol{X} = \{x_0, y_0, L, t_1, t_2, t_3, \theta\}$, the number of design variables is 84. In Figure 4(b), the number of components is 40 and the number of design variables is 280. The final topologies are very similar but the value of negative Poisson's ratios are -0.4996 and -0.3307, respectively. Poisson's ratio is sensitive to the small difference between structures, but the different number of components has little effect on the topologies of the structures.

(a)

(b)

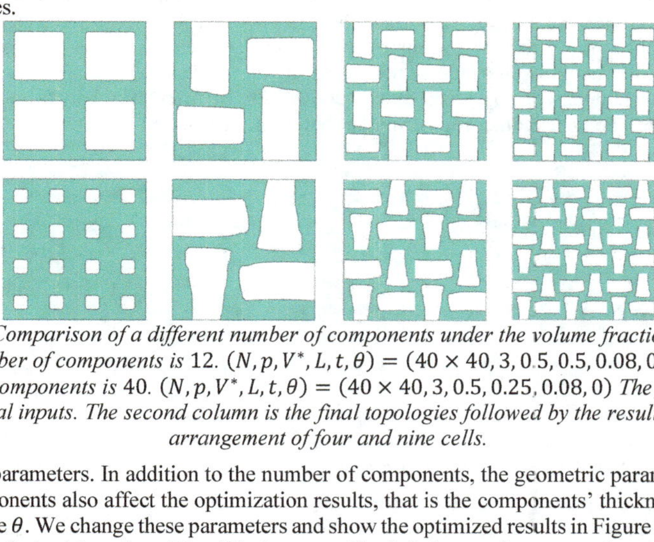

Figure 4: Comparison of a different number of components under the volume fraction 50%. (a) The number of components is 12. $(N, p, V^*, L, t, \theta) = (40 \times 40, 3, 0.5, 0.5, 0.08, 0)$ *(b) The number of components is* 40. $(N, p, V^*, L, t, \theta) = (40 \times 40, 3, 0.5, 0.25, 0.08, 0)$ *The first column is initial inputs. The second column is the final topologies followed by the results of an arrangement of four and nine cells.*

Geometric parameters. In addition to the number of components, the geometric parameters of the initial components also affect the optimization results, that is the components' thickness t_1, t_2, t_3 and the angle θ. We change these parameters and show the optimized results in Figure 5. Parameter testing is optimized based on 50×50 elements. The left three columns are the test results for different thicknesses t, where the $\theta = 0$ and the predefined volume fraction is 35%. The Poisson's ratio values are -0.5413, -0.3302 and -0.6641. The right three columns are results from different angles θ, where the component has a thickness of 0.05 $(t_1 = t_2 = t_3)$ and a predefined volume fraction is 40%. The penalty factor of both tests is 5. The Poisson's ratio values are -0.3146, -0.3157, and -0.5115. The first row in Figure 5 shows the initial inputs. The structures shown in the second row of Figure 5 are the optimization results periodically arranged by 3×3.

Different parameter selections will affect the geometry and Poisson's ratio of the final result. After testing, the value range of thickness is $t \in [0, 0.1]$, and the value range of angle is $\theta \in [0, 0.1]$.

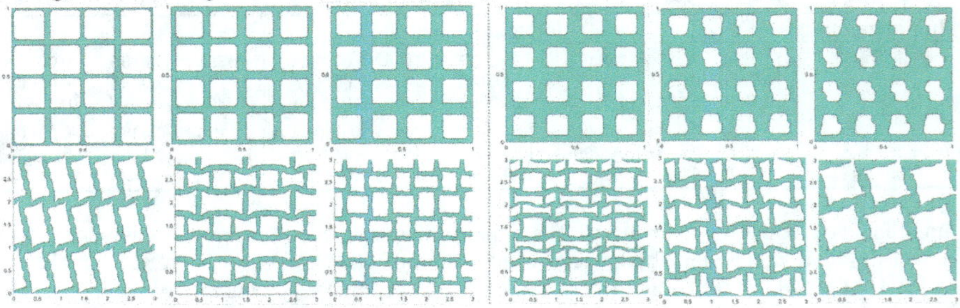

(a) Different component thickness t. (b) Different component angle θ.

Figure 5: Illustration of different geometric parameters of components.

Other parameters. The number of elements and the penalty factors are further tested and shown in Figure 6 under the same volume fraction 50%, the same initial inputs component length $L = 0.25$, thickness $t_1 = t_2 = t_3 = 0.05$, and the same initial input component angle $\theta = 0$. Since geometries of the TOP-IGAMMC method are controlled by the optimized components not defined on the elements, there are no grey densities in the final results. The increase in the element number helps to improve the calculation accuracy. The number of elements has little effect on the geometric results, but Poisson's ratio values are sensitive to the small difference between different structures. In addition, we also give the optimization results of different penalty factors ($p = 3$ and 5) in Figure 6. The final geometries are similar, but the performance is improved.

(a)

(b)

Figure 6: Different optimized auxetic materials under the volume fraction 50% by testing different element numbers and penalty factors (from left to right: 40×40, 50×50, 60×60, 70×70, *and* 80×80*). (a) penalty factor* $p = 3$*; (b) penalty factor* $p = 5$.

Physical tests. In order to compare experimental measurements with numerical results, we print the two optimized auxetic structures by thermoplastic polyurethane (TPU). CMT5105 universal Electromechanical Testing Machine from MTS Systems Corporation in Canada is adopted to test the Poisson's ratio. The tensile tests were conducted at predefined speeds 50 mm/min. The patterns in Figure 7(a) are the optimized structures based on the TOP-IGA-MMC method. The parameters are the same as Figure 2 and Figure 4 (b). Figure 7(b) is the experimental operating status. In addition, we also simulate the deformation and equivalent stresses of these two auxetic structures, as shown in Figure 7(c) and (d). After experimental tests and numerical simulation, the designed structures do have the property of negative Poisson's ratio.

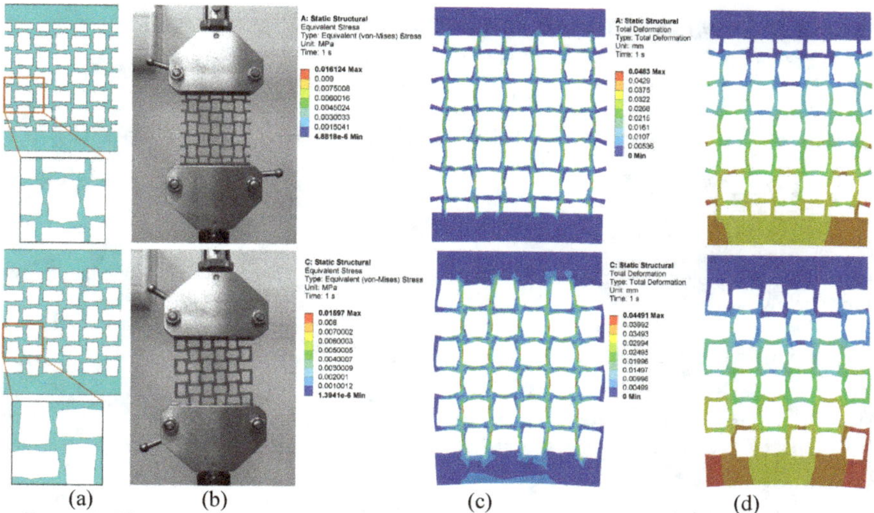

Figure 7: Physical tests and numerical simulations of two auxetic structures generated by the proposed TOP-IGA-MMC method. (a) The optimized auxetic structures and print models. (b) Physical tests. (c) Equivalent stress distribution of maximum deformation status. (d) Colormaps of total deformation distribution.

Conclusion and Future Work

In this work, an explicit isogeometric topology optimization method, TOP-IGA-MMC, is developed. Compared with implicit topology optimization methods (such as the SIMP method), explicit topology optimization methods can avoid checkerboard patterns, gray densities, boundary diffusion, and other numerical instability phenomena. The proposed method inherits the feature that the MMC topology optimization method can express the structure topology explicitly by fewer design variables. At the same time, it has the advantages of high continuity and high calculation accuracy of isometric analysis based on higher-order NURBS basis function.

2D Auxetic materials are generated by the TOP-IGAMMC method. For an equivalent elastic matrix calculation, the isogeometric homogenization theory is illustrated. In order to verify the accuracy of the calculation, we compare it with the algorithms proposed [27, 29]. The maximum error is less than 10−4. Then, the TOPIGA-SIMP method is also compared with ours in numerical performances, material properties, and geometric topologies. Furthermore, the computational accuracy is improved by IGA to obtain a more accurate objective function and constraints. Hence, the proposed TOP-IGA-MMC is efficient and accurate and has great potential to solve 2D topology optimization problems for material design.

Due to the heavier computational burden of solving 3D topology optimization problems, our algorithm will be extended to 3D in future work based on some acceleration algorithms [30–32]. In addition, the extension of multi-material topology optimization based on the proposed TOP-IGA-MMC method will also be considered in the future.

Supplementary

For the TOP-IGA-SIMP method, the problem of material design can be solved using the optimality criteria (OC) [25]. The density iteration formula is

$$p_e^{new} = \begin{cases} \max(0, \rho_e - m), & \text{if } \rho_e B_e^\eta \leq \max(0, \rho_e - m) \\ \min(0, \rho_e + m), & \text{if } \rho_e B_e^\eta \leq \min(0, \rho_e + m) \\ \rho_e B_e^\eta, & \text{otherwise} \end{cases} \tag{20}$$

where m $(= 0.1)$ is the step limit, η $(= 1)$ is the damping coefficient, and B_e^η is obtained by

$$B_e^\eta = -\left(\frac{\frac{\partial J}{\partial \rho_e}}{\lambda \frac{\partial V}{\partial \rho_e}}\right)^\eta. \tag{21}$$

where the Lagrange multiplier λ can be obtained by dichotomy. The sensitivity calculations are shown below:

$$\frac{\partial J}{\partial \rho_e} = \sum_{i=1}^{3} \sum_{i=1}^{3} \frac{\partial J}{\partial D_{ij}^H} \frac{1}{|Y|} p \rho_e^{p-1} (E_0 - E_{min}) \left(d_e^{(i)}\right)^T \mathbf{K}_e d_e^{(j)}. \tag{22}$$

where $\mathbf{K}_e^0$ is the stiffness matrix of the NURBS element with Young's modulus E_0. For more details on the OC criteria approach, see the literature [25, 33].

Acknowledgements
This work is supported by the Provincial Natural Science Foundation of Anhui (2208085QA01), the Fundamental Research Funds for the Central Universities (WK0010000075), the National Natural Science Foundation of China (No. 61972368), and Research Grants Council (C4074-22GF), Innovation and Technology Commission (MRP/030/21, under PiH/273/22), Hong Kong Special Administrative Region, China.

References

[1] R. Jafari Nedoushan, Y. An, W.-R. Yu, New auxetic materials with stretch-dominant architecture using simple trusses, Mechanics of Advanced Materials and Structures (2021) 1-17.

[2] D. R. Smith, N. Kroll, Negative refractive index in left-handed materials, Physical Review Letters 85 (14) (2000) 2933. https://doi.org/10.1103/PhysRevLett.85.2933

[3] T. Ergin, N. Stenger, P. Brenner, J. B. Pendry, M. Wegener, Three-dimensional invisibility cloak at optical wavelengths, Science 328 (5976) (2010) 337-339. https://doi.org/10.1126/science.1186351

[4] M. P. Bendsøe, N. Kikuchi, Generating optimal topologies in structural design using a homogenization method, Computer Methods in Applied Mechanics and Engineering 71 (2) (1988) 197-224. https://doi.org/10.1016/0045-7825(88)90086-2

[5] M. P. Bendsøe, Optimal shape design as a material distribution problem, Structural Optimization 1 (1989) 193-202. https://doi.org/10.1007/BF01650949

[6] X. Huang, M. Xie, Evolutionary topology optimization of continuum structures: methods and applications, John Wiley & Sons, 2010. https://doi.org/10.1002/9780470689486

[7] S. Osher, J. A. Sethian, Fronts propagating with curvaturedependent speed: Algorithms based on hamilton-jacobi formulations, Journal of Computational Physics 79 (1) (1988) 12-49. https://doi.org/10.1016/0021-9991(88)90002-2

[8] X. Guo, W. Zhang, W. Zhong, Doing topology optimization explicitly and geometricallya new moving morphable components based framework, Journal of Applied Mechanics 81 (8) (2014). https://doi.org/10.1115/1.4027609

[9] W. Zhang, J. Yuan, J. Zhang, X. Guo, A new topology optimization approach based on moving morphable components (mmc) and the ersatz material model, Structural and Multidisciplinary Optimization 53 (2016) 1243-1260. https://doi.org/10.1007/s00158-015-1372-3

[10] W. Zhang, W. Yang, J. Zhou, D. Li, X. Guo, Structural topology optimization through explicit boundary evolution, Journal of Applied Mechanics 84 (1) (2017). https://doi.org/10.1115/1.4034972

[11] W. Zhang, L. Zhao, T. Gao, S. Cai, Topology optimization with closed b-splines and boolean operations, Computer Methods in Applied Mechanics and Engineering 315 (2017) 652-670. https://doi.org/10.1016/j.cma.2016.11.015

[12] W. Zhang, Y. Liu, P. Wei, Y. Zhu, X. Guo, Explicit control of structural complexity in topology optimization, Computer Methods in Applied Mechanics and Engineering 324 (2017) 149-169. https://doi.org/10.1016/j.cma.2017.05.026

[13] S. Zhang, A. L. Gain, J. A. Norato, A geometry projection method for the topology optimization of curved plate structures with placement bounds, International Journal for Numerical Methods in Engineering 114 (2) (2018) 128-146. https://doi.org/10.1002/nme.5737

[14] W. Zhang, D. Li, J. Zhou, Z. Du, B. Li, X. Guo, A moving morphable void (mmv)-based explicit approach for topology optimization considering stress constraints, Computer Methods in Applied Mechanics and Engineering 334 (2018) 381-413. https://doi.org/10.1016/j.cma.2018.01.050

[15] T. J. Hughes, The finite element method: linear static and dynamic finite element analysis, Courier Corporation, 2012.

[16] J. Gao, M. Xiao, Y. Zhang, L. Gao, A comprehensive review of isogeometric topology optimization: methods, applications and prospects, Chinese Journal of Mechanical Engineering 33 (1) (2020) 1-14. https://doi.org/10.1186/s10033-020-00503-w

[17] T. J. Hughes, J. A. Cottrell, Y. Bazilevs, Isogeometric analysis: Cad, finite elements, nurbs, exact geometry and mesh refinement, Computer Methods in Applied Mechanics and Engineering 194 (39-41) (2005) 4135-4195. https://doi.org/10.1016/j.cma.2004.10.008

[18] J. Gao, L. Gao, Z. Luo, P. Li, Isogeometric topology optimization for continuum structures using density distribution function, International Journal for Numerical Methods in Engineering 119 (10) (2019) 991-1017. https://doi.org/10.1002/nme.6081

[19] W. Hou, Y. Gai, X. Zhu, X. Wang, C. Zhao, L. Xu, K. Jiang, P. Hu, Explicit isogeometric topology optimization using moving morphable components, Computer Methods in Applied Mechanics and Engineering 326 (2017) 694-712. https://doi.org/10.1016/j.cma.2017.08.021

[20] Y.-D. Seo, H.-J. Kim, S.-K. Youn, Isogeometric topology optimization using trimmed spline surfaces, Computer Methods in Applied Mechanics and Engineering 199 (49-52) (2010) 3270- 3296. https://doi.org/10.1016/j.cma.2010.06.033

[21] J. A. Cottrell, A. Reali, Y. Bazilevs, T. J. Hughes, Isogeometric analysis of structural vibrations, Computer Methods in Applied Mechanics and Engineering 195 (41-43) (2006) 5257-5296. https://doi.org/10.1016/j.cma.2005.09.027

[22] P. Kang, S.-K. Youn, Isogeometric topology optimization of shell structures using trimmed nurbs surfaces, Finite Elements in Analysis and Design 120 (2016) 18-40. https://doi.org/10.1016/j.finel.2016.06.003

[23] J. Gao, L. Wang, M. Xiao, L. Gao, P. Li, An isogeometric approach to topological optimization design of auxetic composites with tri-material micro-architectures, Composite Structures 271 (2021) 114163. https://doi.org/10.1016/j.compstruct.2021.114163

[24] P. Theocaris, G. Stavroulakis, P. Panagiotopoulos, Negative poisson's ratios in composites with star-shaped inclusions: a numerical homogenization approach, Archive of Applied Mechanics 67 (1997) 274-286. https://doi.org/10.1007/s004190050117

[25] B. Hassani, E. Hinton, A review of homogenization and topology optimization ihomogenization theory for media with periodic structure, Computers & Structures 69 (6) (1998) 707-717. https://doi.org/10.1016/S0045-7949(98)00131-X

[26] W. B¨ohm, Generating the b'ezier points of b-spline curves and surfaces, Computer-Aided Design 13 (6) (1981) 365-366. https://doi.org/10.1016/0010-4485(81)90213-X

[27] L. Xia, P. Breitkopf, Design of materials using topology optimization and energy-based homogenization approach in matlab, Structural and Multidisciplinary Optimization 52 (6) (2015) 1229-1241. https://doi.org/10.1007/s00158-015-1294-0

[28] K. Svanberg, The method of moving asymptotesa new method for structural optimization, International Journal for Numerical Methods in Engineering 24 (2) (1987) 359-373. https://doi.org/10.1002/nme.1620240207

[29] E. Andreassen, C. S. Andreasen, How to determine composite material properties using numerical homogenization, Computational Materials Science 83 (2014) 488-495. https://doi.org/10.1016/j.commatsci.2013.09.006

[30] J. Mart'ınez-Frutos, D. Herrero-P'erez, Large-scale robust topology optimization using multi-gpu systems, Computer Methods in Applied Mechanics and Engineering 311 (2016) 393-414. https://doi.org/10.1016/j.cma.2016.08.016

[31] V. J. Challis, A. P. Roberts, J. F. Grotowski, High resolution topology optimization using graphics processing units (gpus), Structural and Multidisciplinary Optimization 49 (2) (2014) 315-325. https://doi.org/10.1007/s00158-013-0980-z

[32] D. Zhang, X. Zhai, X.-M. Fu, H. Wang, L. Liu, Large-Scale Worst-Case Topology Optimization, Computer Graphics Forum (2022). doi:10.1111/cgf.14698. https://doi.org/10.1111/cgf.14698

[33] S. S. Rao, Engineering optimization: theory and practice, John Wiley & Sons, 2019. https://doi.org/10.1002/9781119454816

Biomechanical model for musculoskeletal simulation

Megdi Eltayeb[1,a] *

[1]Department of Mechanical Engineering, Prince Mohammad Bin Fahad University

[a]meltayeb@pmu.edu.sa

Keywords: Biomechanical, Musculoskeletal, Modeling, OpenSim

Abstract. Musculoskeletal modeling is a technique for studying joint contact forces and moments during a movement. Subject-specific models can achieve high accuracy in estimating joint contact forces. Construction of subject-specific models, on the other hand, remains costly and time-consuming. The objective of this study was to determine what changes could be made to generic musculoskeletal models to improve the estimation of joint contact forces. The effect of these changes on the accuracy of the estimated joint contact forces was evaluated. A variety of change strategies were discovered, including muscle models (e.g., muscle length), joint angle models (e.g., angle, number of degrees of freedom), moments and optimization problems (e.g. objective function, constraints, design variables). All of these changes had an effect on joint contact force accuracy, demonstrating the potential for improving model predictions without requiring time-consuming and expensive medical techniques. However, due to inconsistencies in the literature evidence about this effect, and despite the high quality of the reviewed studies, no trend defining which change had the greatest effect could be identified.

Introduction

Movement is essential to human and animal life and is produced by the interaction of complex neural, muscular, and skeletal systems. Movement research draws on and contributes to diverse disciplines, including neuroscience, biology, robotics and biology. OpenSim combines methods from these fields to generate accurate and fast movement simulations, enabling two essential tasks. First, building a model can predict new trends. Second, compute variables that are hard to calculate during movement. OpenSim is a software package that is extensible and user-friendly, based on decades of experience in computational simulation and modelling of biosystems. It allows computational researchers to create new innovative tools and use these tools in research and applications. OpenSim helps a large and growing community of scientists exchange models for reproducing and growing research. It is possible to discover strategies to improve performance and prevent injury by studying the biomechanical structures that control underlying movement. The benefits are improved rehabilitation for patients after a stroke [1]; musculoskeletal analysis and for optimizing assistive devices [2, 3]; musculoskeletal injuries [4, 5]; as well as biomechanical interpretations [6, 7]. In order for scientists to make progress in the field of movement science, they have to develop tools that integrate computational modeling and simulation tools from a variety of disciplines such as robotics, mechanics, and computer science. There are a number of open-source software packages available for collecting and analyzing experimental movement data (e.g., OpenMA [4] and BTK [5]) but this kind of software is less suited for simulations and optimizations.

In addition to providing research with biomechanical models and simulation tools, OpenSim facilitates the advancement of movement science. There are several main areas of functionality within OpenSim. It is possible for users to create and manipulate biomechanical models. As an example, musculoskeletal models can be constructed using this software [6]. It is also necessary to simulate the dynamics of the musculoskeletal system. It is possible for researchers to conduct experiments that are impossible to conduct experimentally through simulations. Researchers are investigating how humans and animals exploit tendon elasticity to make running more efficient [7-9] and optimizing the design of implantable mechanisms and assistive devices [10-13]. Additionally, it is capable of predicting novel movements and adaptations to novel conditions without the need for any experiments. Through the study of loaded and inclined walking, this capability has allowed a deeper understanding of muscle coordination [14]. Furthermore, it provides insight into the limitations of reflexes when it comes to the prevention of ankle injuries when landing [20]. In addition, it provides suggestions for enhancing jumping performance with optimum device design [15]. In addition, improved computational models, numerical methods, and simulation tools will be developed and shared to extend the capabilities of the software.

Methods

OpenSim software is licensed under the Apache License 2.0, which permits its reuse for a variety of purposes, including non-profit and commercial applications. Community members are encouraged to contribute to the source code, which is available for free and anonymously on GitHub. The musculoskeletal model, freely available in OpenSim (http://opensim.stanford.edu/) was evaluated in this study. This model included a generic gait model. The generic model was gait2392 [16]. This model is based on gait2392 with the addition of the obturator and rectus abdominus muscles. Adjustments were also made to the muscle geometry so that the moment arms of the model matched experimental measurements. Model generalized coordinates that best reproduced the experimental marker coordinate data for each trial were computed by solving an inverse kinematics problem. This global optimization algorithm is formulated as a least-squares problem that minimizes the differences between the measured marker locations and the model's virtual marker locations, subject to joint constraints. Inverse dynamics was performed using the measured ground reaction forces and inverse kinematics results to calculate the intersegmental moments. Muscle forces were estimated using static optimization to minimize the sum of squares of muscle activation at each instant in time. The peak and root-mean-square of each reserve actuator were verified to be less than 5% of the net moment calculated via inverse dynamics. Estimated muscle activations were then verified via qualitative comparisons between the model-based predicted activations and experimental electromyographic data provided in the HIP98 dataset as per current recommendations. Due to concern that the magnetic field of the coil used for powering the prosthesis could affect the electromyographic signal in the HIP98 dataset, we also compared our model-based activations to experimental electromyographic data available in the literature [16-19]. This process of scaling, inverse kinematics, inverse dynamics, static optimization, and joint reaction analysis was repeated for the model.

Result and Discussion

Musculoskeletal models allow us to investigate neuromuscular coordination, assess athletic performance, and calculate musculoskeletal loads. OpenSim is free and open-source software that allows users to create, visualize and analyze models of the musculoskeletal system as well as create dynamic movement simulations [1]. Researchers can also look into joint kinematics, musculoskeletal geometry, and muscle-tendon properties to see how they affect the joint moments and forces that muscles can generate. The objective of OpenSim is to provide a framework for researchers to create models and dynamic simulation tools for studying and quantifying human movement. The goal of this research is to study the relationship between muscle-tendon lengths and moment arms on limb models; find restrictions of musculoskeletal models; and study differences between bi-articular and uni-articular muscles. The goal of this research is to load a lower extremity model [2] into OpenSim and make it walk. The model depicts an adult subject standing 1.8 m tall and weighing 75 kg. The model is composed of 13 rigid body segments and 92 muscle action lines (43 per leg and 6 at the torso) Figure 1a. In OpenSim, muscle-tendon paths are represented by a series of points connected by line segments. Multiple lines of action are represented by the gluteus medius in this model. (e.g., glut_med1_r, glut_med2_r, glut_med3_r).

Examine how muscle-tendon lengths and moment arms vary according to limb models. Musculoskeletal geometry is critical to muscle function and the development of quantitative musculoskeletal models. Muscle-tendon forces are proportional to muscle-tendon length, and joint moments are proportional to both muscle-tendon forces and moment arms. As a result, accurate musculoskeletal geometry specification is critical in improving an accurate model for predicting joint moments and muscle-tendon forces. The Opensim Plotter displays muscle-tendon properties like length, force, joint moment, and moment arm.

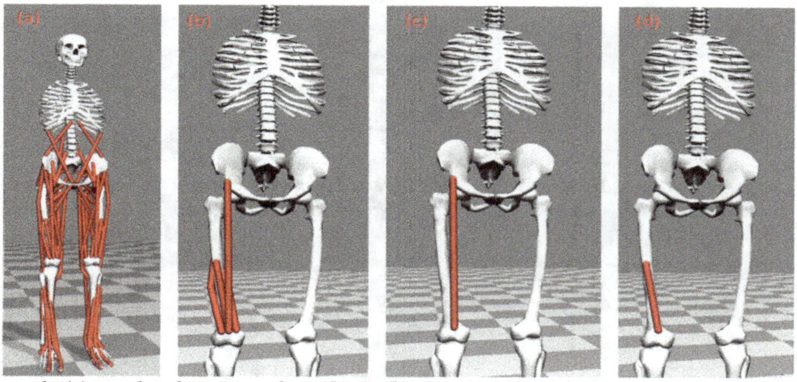

Figure 1: (a) graphical user interface (GUI), (b), (c) rectus femoris, and (d) vastus intermedius muscles

OpenSim software has been used to assess muscle fibre length changes. Fibre length was reported to be shortened during knee extension (Figure 2). This is the study to investigate effects of right hip extension angle on fiber lengths of rectus femoris muscle. It shows that the changes in hip extension angle, did not affect the fiber length of vastus intermedius muscle fibers (Figure 3).

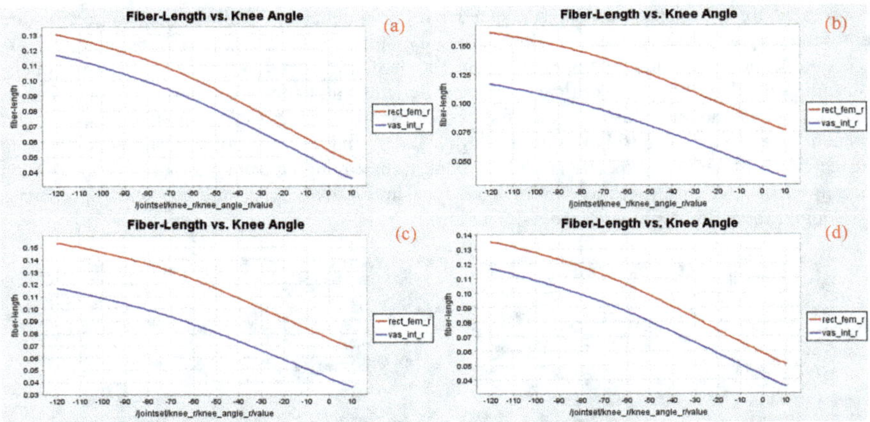

Figure 2: Fiber-Length vs. Knee Angle (right rectus femoris (rect_fem_r) and right vastus intermedius muscles (vas_int_r)) for right hip flexion to (a) 10, (b) 20, (c) 30, and (d) 40 degrees

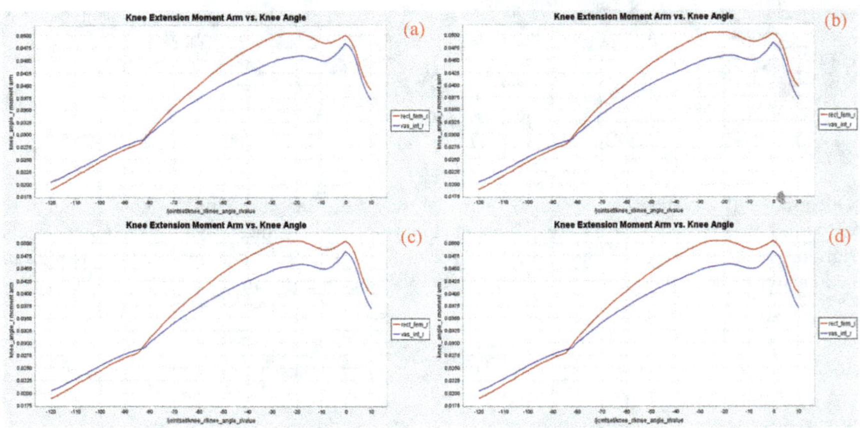

Figure 3: The knee extension moment arm vs. knee angle for the right rectus femoris and vastus intermedius muscles for right hip flexion to (a) 10, (b) 20, (c) 30, and (d) 40 degrees

Crouch gait is characterized by extreme knee flexion during the terminal swing and stance phases, making it one of the most common movement abnormalities in people with cerebral palsy. In many cases, abnormal hamstrings caused by spasticity [5-7] or static contracture [7, 8] are thought to be the cause of excessive knee flexion. Thus, surgical hamstring lengthening is commonly used to treat crouch gait, usually in conjunction with other orthopaedic procedures. Unfortunately, predicting which patients will benefit from hamstring surgery is difficult. OpenSim can predict which patients will benefit, change the speed, and rotate the models.

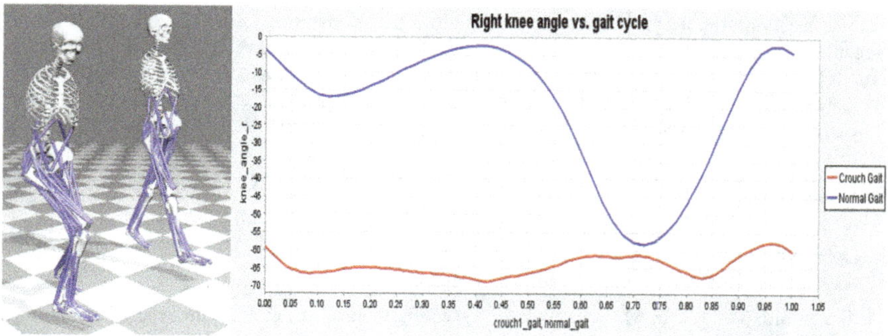

Figure 4: (a) Normal gait and crouch gait, and (b) Right knee angle vs gait cycle for normal gait and crouch gait.

Use OpenSim to study a possible cause of crouch gait, one of the most common walking abnormalities among people with cerebral palsy. It is distinguished by excessive knee flexion during the stance phase, which is frequently accompanied by exaggerated flexion and internal rotation of the hip. Short hamstrings are one of the hypothesized causes of crouch gait, and in order to improve a patient's gait, surgeons may lengthen the hamstrings. There could be other causes of excessive knee flexion. (e.g., weak ankle plantarflexors), and lengthening the hamstrings can impair these muscles' ability to generate force [3]. Develop a musculoskeletal model and compare the length of the hamstrings during the patient's crouch gait cycle to the length of the hamstrings during a normal gait cycle to determine whether a patient's hamstrings are shorter than normal Figure 4.

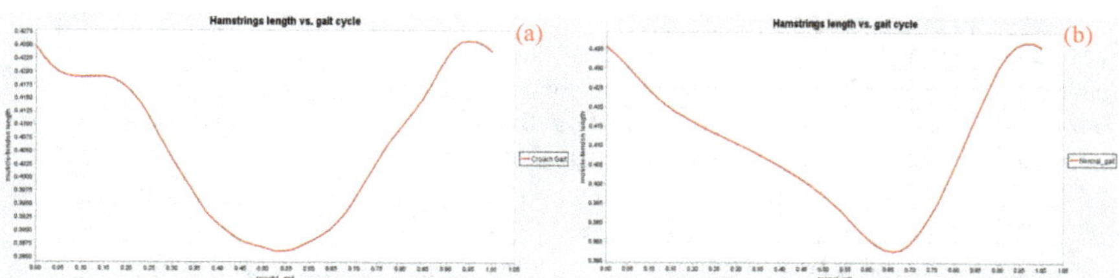

Figure 5: The length of the semitendinosus muscle-tendon during the (a) crouch and (b) gait cycles.

Figure 5 shows the length of the semitendinosus muscle-tendon over the crouch and gait cycles, compares the two motions, and describes the general differences in kinematics between the normal and crouch gait motions qualitatively. In addition, the length of the hamstrings during a patient's crouch gait cycle should be compared to the length of the hamstrings during a normal gait cycle.

Conclusion

In conclusion, the study developed a musculoskeletal full-body model. Despite some limitations, validation studies show that this model is suitable for predicting hamstrings length, muscle-tendon lengths, and muscle fiber length. When aiming at investigating hamstrings length, muscle-tendon lengths, and muscle fiber length, the model can serve as a basis for the normal and crouch gait models as well as for comparative purposes. In this study, the parameters and implementation choices of a generic musculoskeletal model have been modified. These include the number of degrees of freedom, the length and moment of the muscles, constraints, and the angle of the joints. As a result of these changes, the joint contact forces were more accurate than before, demonstrating that model predictions may be improved without the use of expensive and time-consuming medical tools. It was difficult to identify a trend defining which change had the greatest impact despite the high quality of the reviewed studies. In addition, there were discrepancies in the reported evidence about this impact.

References

[1] J. W. Krakauer, "Motor learning: its relevance to stroke recovery and neurorehabilitation," Curr. Opin. Neurol., vol. 19, no. 1, pp. 84-90, 2006. https://doi.org/10.1097/01.wco.0000200544.29915.cc

[2] A. S. Arnold, M. Q. Liu, M. H. Schwartz, S. Ounpuu, and S. L. Delp, "The role of estimating muscle-tendon lengths and velocities of the hamstrings in the evaluation and treatment of crouch gait," Gait Posture, vol. 23, no. 3, pp. 273-281, 2006. https://doi.org/10.1016/j.gaitpost.2005.03.003

[3] D. J. J. Bregman, M. M. Van der Krogt, V. De Groot, J. Harlaar, M. Wisse, and S. H. Collins, "The effect of ankle foot orthosis stiffness on the energy cost of walking: a simulation study," Clin. Biomech., vol. 26, no. 9, pp. 955-961, 2011. https://doi.org/10.1016/j.clinbiomech.2011.05.007

[4] A. Barre and S. Armand, "Biomechanical ToolKit: Open-source framework to visualize and process biomechanical data," Comput. Methods Programs Biomed., vol. 114, no. 1, pp. 80-87, 2014. https://doi.org/10.1016/j.cmpb.2014.01.012

[5] A. Barre, F. Y. Leboeuf, and S. Armand, "OpenMA: a new open source library for working and altering the conventional gait model," Gait Posture, vol. 49, p. 260, 2016. https://doi.org/10.1016/j.gaitpost.2016.07.314

[6] M. Domalain, A. Bertin, and G. Daver, "Was Australopithecus afarensis able to make the Lomekwian stone tools? Towards a realistic biomechanical simulation of hand force capability in fossil hominins and new insights on the role of the fifth digit," Comptes Rendus Palevol, vol. 16, no. 5-6, pp. 572-584, 2017. https://doi.org/10.1016/j.crpv.2016.09.003

[7] E. M. Arnold, S. R. Hamner, A. Seth, M. Millard, and S. L. Delp, "How muscle fiber lengths and velocities affect muscle force generation as humans walk and run at different speeds," J. Exp. Biol., vol. 216, no. 11, pp. 2150-2160, 2013. https://doi.org/10.1242/jeb.075697

[8] J. W. Rankin, J. Rubenson, and J. R. Hutchinson, "Inferring muscle functional roles of the ostrich pelvic limb during walking and running using computer optimization," J. R. Soc. Interface, vol. 13, no. 118, p. 20160035, 2016. https://doi.org/10.1098/rsif.2016.0035

[9] T. K. Uchida, J. L. Hicks, C. L. Dembia, and S. L. Delp, "Stretching your energetic budget: how tendon compliance affects the metabolic cost of running," PLoS One, vol. 11, no. 3, p. e0150378, 2016. https://doi.org/10.1371/journal.pone.0150378

[10] T. Homayouni, K. N. Underwood, K. C. Beyer, E. R. Martin, C. H. Allan, and R. Balasubramanian, "Modeling implantable passive mechanisms for modifying the transmission of forces and movements between muscle and tendons," IEEE Trans. Biomed. Eng., vol. 62, no. 9, pp. 2208-2214, 2015. https://doi.org/10.1109/TBME.2015.2419223

[11] D. J. Farris, J. L. Hicks, S. L. Delp, and G. S. Sawicki, "Musculoskeletal modelling deconstructs the paradoxical effects of elastic ankle exoskeletons on plantar-flexor mechanics

and energetics during hopping," J. Exp. Biol., vol. 217, no. 22, pp. 4018-4028, 2014. https://doi.org/10.1242/jeb.107656

[12] T. K. Uchida, A. Seth, S. Pouya, C. L. Dembia, J. L. Hicks, and S. L. Delp, "Simulating ideal assistive devices to reduce the metabolic cost of running," PLoS One, vol. 11, no. 9, p. e0163417, 2016. https://doi.org/10.1371/journal.pone.0163417

[13] R. W. Jackson, C. L. Dembia, S. L. Delp, and S. H. Collins, "Muscle-tendon mechanics explain unexpected effects of exoskeleton assistance on metabolic rate during walking," J. Exp. Biol., vol. 220, no. 11, pp. 2082-2095, 2017. https://doi.org/10.1242/jeb.150011

[14] T. W. Dorn, J. M. Wang, J. L. Hicks, and S. L. Delp, "Predictive simulation generates human adaptations during loaded and inclined walking," PLoS One, vol. 10, no. 4, p. e0121407, 2015. https://doi.org/10.1371/journal.pone.0121407

[15] C. F. Ong, J. L. Hicks, and S. L. Delp, "Simulation-based design for wearable robotic systems: an optimization framework for enhancing a standing long jump," IEEE Trans. Biomed. Eng., vol. 63, no. 5, pp. 894-903, 2015. https://doi.org/10.1109/TBME.2015.2463077

[16] S. L. Delp, J. P. Loan, M. G. Hoy, F. E. Zajac, E. L. Topp, and J. M. Rosen, "An Interactive Graphics-Based Model of the Lower Extremity to Study Orthopaedic Surgical Procedures," IEEE Trans. Biomed. Eng., vol. 37, no. 8, pp. 757-767, 1990, doi: 10.1109/10.102791. https://doi.org/10.1109/10.102791

[17] A. Seth, et al., "OpenSim: Simulating musculoskeletal dynamics and neuromuscular control to study human and animal movement," 2018, doi: 10.1371/journal.pcbi.1006223. https://doi.org/10.1371/journal.pcbi.1006223

[18] S. L. Delp et al., "OpenSim: Open-Source Software to Create and Analyze Dynamic Simulations of Movement," IEEE Trans. Biomed. Eng., vol. 54, no. 11, pp. 1940-1950, 2007, doi: 10.1109/TBME.2007.901024. https://doi.org/10.1109/TBME.2007.901024

[19] A. S. Arnold, M. Liu, S. Ounpuu, M. Swartz, and S. L. Delp, "The role of estimating hamstrings lengths and velocities in planning treatments for crouch gait," Gait Posture, vol. 23, no. 3, pp. 273-281, 2006. https://doi.org/10.1016/j.gaitpost.2005.03.003

Materials Research Proceedings 31 (2023) 194-203 https://doi.org/10.21741/9781644902592-21

Recycled smelter slag as an engineering material - opportunity and sustainability

SAIDUR R. Chowdhury[1*]

[1]Saidur R. Chowdhury, Civil Engineering Department, Prince Mohammad Fahd University, PO BOX 1664, Al Khobar 31952; Kingdom of Saudi Arabia; Phone: 966138494357;

schowdhury1@pmu.edu.sa

Keywords: Slag, Adsorbent, Construction, Recycling, Nickel (Ni)

Abstract. Slags obtained from the Vale Copper Cliff smelter in Sudbury, Ontario, Canada, were investigated as sustainable engineering materials in this study. The recycled smelter waste products can remove toxic contaminants from the aqueous environment as well as be used in the construction industry (as aggregates, cement admixtures, filling materials), soil improvement for agricultural purposes, and other value-added applications and products. The removal mechanisms of the heavy metals (such as Zn, Pb, and Cu, etc.) from aqueous solutions could be physical or chemical adsorption, ion exchange, oxidation-reduction, etc. At the same time, using recycled smelter slags in various engineering applications can help with waste reduction, disposal cost reduction, resource recovery, and increased reused activities. The present study helps explore the scope of using recycled materials in the treatment or construction industry. Using industrial smelter slag as a recycling or renewable resource rather than a waste product has environmental and economic benefits. The study also specifically discusses Ni smelter slag's composition, application, treatment efficiency, opportunity, economic benefits, and circularity for sustainable management.

Introduction

Slags, waste products generated from metal mining activities or the metallurgical industry, can be treated and recycled for other applications [1-2]. Water and wastewater treatment using low-cost by-products from the agricultural, domestic, and industrial sectors has been acknowledged as a viable option [3]. They make it possible to remove pollutants from aqueous solutions while also contributing to waste minimization, recovery, and reuse. The current study investigated the removal efficiency of various contaminants using low-cost waste products as well as other engineering applications from Ni smelter slags. It considered recycled smelter slags consisting of iron oxides, iron silicates, and different oxide minerals. Slags are a potential adsorbent or reactive media in water and wastewater treatment [1–3]. Other applications for recycled smelter slags include aggregates, cement admixtures, filling materials, fertilizers, and other value-added applications and products [4]. Vale, a major nickel producer in the world, discards 1.2 million tonnes of slag annually, whereas Inco's entire slag inventory in the Sudbury region is 115 million tonnes after about 90 years of business operations [5]. Huge amounts of abandoned slag result in leaching of metals and the waste of valuable metals, as well as environmental contamination from the oxidation of entrained sulfur to sulphuric acid [5].

To provide justification, a lab-based experimental investigation using recycled Ni smelter slags was conducted to explore the physical and chemical properties as well as the adsorption capacity and kinetics of toxic Zn, Cu, and Pb from water solutions. Furthermore, the primary motivation was to evaluate the applicability options of recycled Ni smelter slags in different engineering applications in order to assess the circularity of the materials in the sustainable development process. The study explored the benefits of using discarded waste products in the development of sustainable green technology.

Materials Research Proceedings 31 (2023) 194-203 https://doi.org/10.21741/9781644902592-21

Physical and chemical characteristics of Ni smelter slags

In the present study, the Ni smelter slags obtained from the Vale Copper Cliff smelter in Sudbury, Ontario, Canada were stored in the Environmental Engineering Lab at Western University in Canada. These slag particles were crushed into smaller fragments and later more finely ground into finer particles with a pestle and mortar. It was kept in airtight containers for experimental purposes. The composition of the bulk smelter slag was also measured using PAN-analytical PW-2400 Wavelength and Dispersive XRF (X-Ray Fluorescence). Moreover, the Surpass Anton Analyzer was used to determine surface charge and charge changes at different pH values. Figure 1 shows the change of surface charges with different pHs. Slag particles larger than 25 μm were used to conduct the experiment. The operational conditions for the Anton analyzer were 115 to 230 V, 50 to 60 Hz, and 200 VA. From XRF tests, the results showed that the major compositions of the slag were iron oxides and silicon dioxides, at 59% and 39-41%, respectively.

In the study, the major components of the slag particles were found to be Fe_2O_3, SiO_2, Al_2O_3, CaO, and MgO. Previous researchers have used SiO_2, Al_2O_3, Fe_2O_3, MgO, CaO, and TiO_2 for metal removal [2, 6-9]. According to Liu et al. [9] and Thakur et al. [2] steel slag containing 10 to 40% of Fe_2O_3, SiO_2 and CaO was capable of reducing 99% of Pb from aqueous solution. Perederiy [5] reported that the composition of iron and silicon are the main elements (e.g., 57.5% wt) of the Ni smelter slag. In addition to iron and silicon, Ni and Cu are found to be around 1% of the slag mass, together with some cobalt (0.24%wt). Chowdhury et al. [10] found the presence of adsorbed Cu (1.6% wt) and Ni (0.2% wt) in the recycled slag materials, indicating the affinity for Cu adsorption. The Surpass Anton Analyzer was used to determine the variations in surface charge on Ni smelting slag as a function of pH. The findings revealed that before a pH value of 3.1, recycled Ni smelting slag had a positive charge. However, it had a negative surface charge after pH 3.1 (as shown in Figure 1), indicating the positive cation affinity. From three sets of experiments, the standard deviation of each point for Figure 1 was found to be ±1 to ± 3.

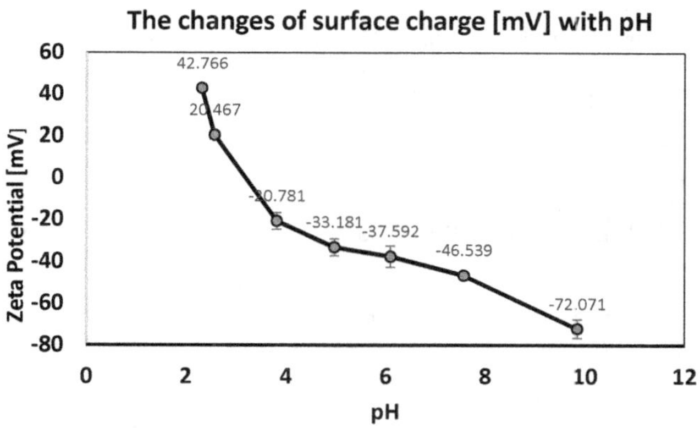

Figure 1: The change of surface charges of Ni smelter slag with different pHs

The slag was further analyzed by using X-ray based techniques, such as X-ray photoelectron and Raman spectroscopies. As reported by Chowdhury et al. [10] the XPS Fe mutiplet analysis showed that the Ni smelter slag surface would consist of a mixture of iron oxides such as magnetite, maghemite, and goethite, etc. More specifically, it contains 50% magnetite, 28% maghemite, and 22% goethite. Besides, 21% of SiO_4 was detected, indicating the presence of fayalite ($FeSiO_4$) compound. Thus, in conclusion, the major composition would be mixed iron oxides and silicon dioxide. Table 1 shows the characteristics and possible applications of Ni smelter slag in different engineering fields. The preliminary characterization reveals that it can be used as aggregates for concrete, an admixture for cement, fertilizer and soil improvement, raw material for cement, possible CO_2 capture, flue gas desulfurization, and an adsorbent for waste water treatment.

Table 1: *Characteristics and possible applications of Ni smelter slag*

Slags Properties	Possible Engineering Applications
Porous, alkaline	Waste water treatment
Al_2O_3, FeO, SiO_2	Adsorbents
Hard, wear-resistant,	Aggregates for concrete
Adhesive, rough, Cementitious components (e.g. dicalcium silicate, tricalcium silicate)	Aggregates, admixture for cement
FeOx, $FeSiO_2$ and Fe components	Adsorbent in Water Treatment, Iron reclamation
CaO, SiO_2, MgO, FeO	Fertilizer and soil improvement, raw material for cement
CaO, MgO components	Possible CO_2 capture and flue gas desulfurization

In this study, Laser Tests were used to determine the concentrations of crystalline, trace, and bulk substances. After crushing, a Malvern Laser 2000 was used to assess the fresh slag's specific surface area and particle size distribution. The average size and specific surface area of the slag materials were found to be 55 μm and 0.6 m^2/g, respectively. It was found that the slag grain size distribution ranged from 0.3 to 155 μm.

Utilization of Ni smelter Slag materials

Water treatment

Slag has a porous structure, large surface area, and is quite dense, making it simple to separate from water, alkaline and consisting of FeO, Al_2O_3, FeO, SiO_2 etc. As a result, the use of steel slag in the treatment of industrial waste water has garnered a lot of interest recently.

Different Metals Removal by Ni smelter slags

Batch experimental methods

A batch experiment was conducted by using known amounts of nickel slag with 50 mL metal solutions in the 125 to 300 mL Erlenmeyer flask at 23°C ± 2°C. This test measured the adsorption efficiency of smelter slag. All batch experiments were conducted three times on the rotary shaker in certain environmental conditions. A speed of 170 rpm was used during batch experiments since good contact between the adsorbent and adsorbate was achieved. In all experiments, control blanks were also used for quality control and assurance purposes. In this study, the batch tests were conducted in different experimental conditions. Solution samples were collected after the completion of adsorption and then filtered with 0.2 μm filters. Metals' concentrations were measured using the Varian Vista Pro inductively coupled plasma optical emission spectroscopy (ICP-OES). The filtrates were acidified with 2 % HNO_3 for the determination of metal concentration. In this study, the batch adsorption experiment was performed at 0,1 to 10 mg/L of Pb As, Cu solution with 10 g/L of slag particles for 10 hours at pH 2–11 in order to find out the adsorption capacity.

Experimental results

The extent of adsorption and the effect of pH on Cu and Pb removal by Ni smelter slag were determined in this study. Figure 2 shows the results of the possible removal of Pb and Cu from the series of experiments in the different pH ranges. The experimental results were found to be 98% of Pb removal from water in the pH range of 4 to 9 and more than 90% of Cu removal was achieved in the same pH range. In Fig. 2, the range of standard deviations for the removal efficiency (%) of Pb and Cu was found to be small, that is, ±0.08 to ± 2.5.

The effect of pH on cations' removal could depend on the point of zero charge [11]. The point of zero charge of the nickel smelting slag was around 3.1 in this study (as shown in Figure 1). The surface of the adsorbent is negative below the point of zero charge. Thus, cations (Me^{2+}, $Me(OH)^{+}$) would be attached to the surface through ion exchange with H^{+} [12]. This explains why metal removal rates at pH 5 and higher are so high in the case of Cu or Pb removal by Ni smelter slag.

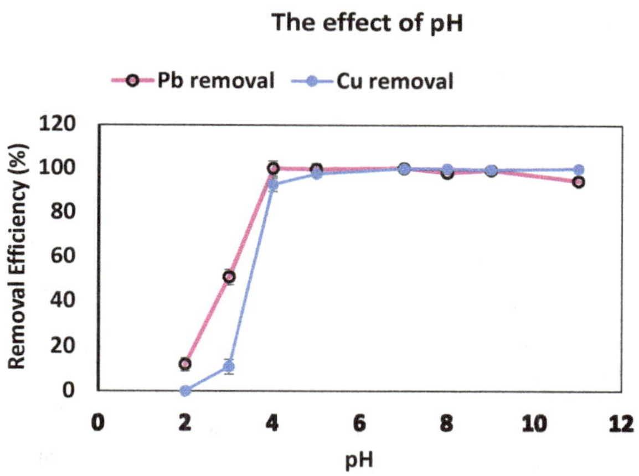

Figure 2: *The removal of Pb and Cu by Ni smelter slag with different pH*

The pseudo first order and second order kinetic models were used to investigate the sorption kinetics of Pb and Cu on Ni smelting slag. Figure 3 shows the fitted lines obtained from the second order kinetic rate models for Pb and Cu removal from aqueous solution. From the results, it can be concluded that the pseudo second order model was better fitted to the experimental data than the pseudo first order model. An R^2 value of 1 for both Pb and Cu was found.

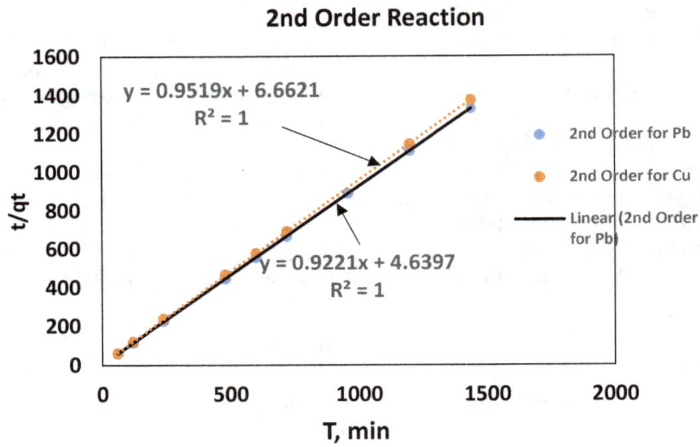

Figure 3: *Pseudo second order plot for Cu and Pb removal by Smelter Slag*

The pseudo-second order describes chemisorption, which includes both ion exchange and covalent interaction between an adsorbate and an adsorbent [7,12-13]. The results from this study showed that the adsorption of Pb and Cu by the Ni smelter slag was due to chemical interaction between the adsorbate and the adsorbent (e.g., reactive Ni smelter slag). Thus, it is clear that the removal mechanism of Pb and Cu could be physio-chemisorption.

The optimum contact time and the kinetics of the removal of Pb and Cu with the recycled Ni smelter slag were also assessed using batch experiments. The results showed that the removal was very fast in the initial stages of the experiments and then gradually decreased at the end of the experiments. 55.85 % and 91.58 % of Pb were adsorbed in the first 5 and 20 minutes while 34.55 % and 88.65 % of Cu was removed in the experimental condition. The study was conducted at pH 5, 10 mg/L of initial concentration and 10 g/L of slag doses.

The effect of slag size on the removal of Pb or Cu by smelter slag was investigated. The batch experiment was conducted at an initial concentration of 10 mg/L for both metals, 10 hours of contact time, 170 rpm of shaking speed, and 250C. Different particle sizes of slag at a dose of 10 g/L were used to measure the effect of particle size on metal removal. The removal of toxic metals was shown to increase with a decrease in particle size (as shown in Figure 4). Removal between 10% and 100% was achieved for Cu and 16 % to 100% for Pb over the size range studied. A significant increase in adsorption was achieved when the particle size was reduced from 200 μm to 128 μm for both Cu and Pb removal. More than 99% of both toxic metals' removal was achieved when particle size was kept below 90 μm.

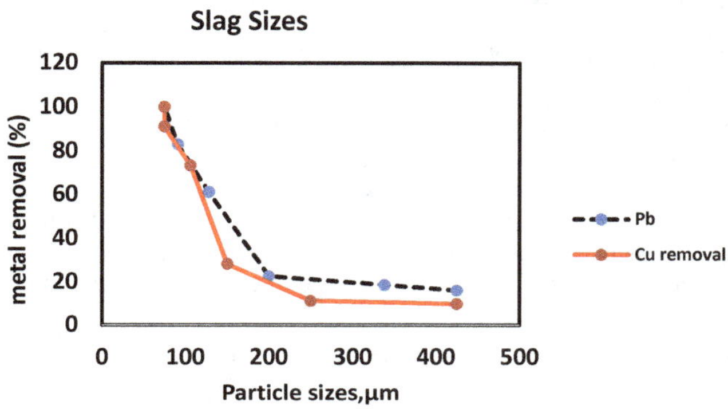

Figure 4: *Effect of slag sizes on removal of Pb or Cu by smelter slag (Initial concentration: 10 mg/L, contact time: 10 hours, shaking speed: 170 rpm, temperature: 25^0C)*

Utilization of Slags in Construction

The cement industry consumes a lot of energy and raw materials. Therefore, it has a major impact on the environment because it releases a lot of greenhouse gases, like carbon dioxide, during the manufacturing of cement clinker and energy use. The need to limit the emission of industrial pollutants through the utilization of industrial byproducts like ferrous and nonferrous slag materials. With an annual production of more than 3 billion tonnes, cement is the most commonly used building material [4, 19, 20]. The production of cement is an energy-intensive process and produces large amounts of CO_2 emission, which causes serious environmental problems such as global warming and climate change [20]. Therefore, it is necessary to make cement-based products more environmentally friendly [19–20]. Due to their comparable composition and morphology to cement or sand, some metallurgical slags, such as steel, copper, slags, such as steel, copper, nickel, lead-zinc, and electric furnace ferro-nickel slag, have the potential to be employed in cement-based materials [4, 9–21]. A partial substitution of metallurgical slags for cement or sand can significantly reduce CO_2 emissions, save natural resources, and minimize the discharge and pollution of waste slags due to the enormous use of building materials.

The chemical properties of smelter slags (e.g., Ni smelter slags) show some advantages to use as an additive, admixture or cementing substances for the cement and concrete construction. The presence of calcium, alumina and silicate endorse the Ni smelter slags cementitious properties [21]. It is generally agreed that the cementitious properties of ferrous and nonferrous slag increase with its basicity. Thus, slag materials crushed into fine materials could be used as cement additives and concrete admixtures. The potential cementitious property of slag particles considerably increases with their fineness. The fineness of the slag is another major element determining the activity [21-22].

Slags from nickel smelters can also be utilized as an aggregate in concrete for refractory and high strength applications. Because slags are hard and have a poor grind-ability index, it is important to build highly effective and energy-efficient pulverizing mill equipment for crushing [21]. According to Qasrawi, et al. [23], steel slag used as fine aggregates increased the compressive strength of concrete by 1.1 to 1.3 times. Papayianni et al. [24] created a high-strength (>70MPa) concrete from Iron slags. Zhao et al. [19] used Cu smelter slags for concrete construction and achieved maximum compressive strength to be 40 MPa (after 28 days). Therefore, it is expected that similar application in the construction sector can be achieved by Ni smelter slags. In addition, in many situations, the slag material may represent the most effective and affordable long-term solution in the waste management economy, filling the market gap in light of the depleting anthropogenic resources.

Fertilizer from Ni smelter Slags

CaO, SiO_2, and MgO are fertilizer constituents found in Ni smelter slag. Thus it can be used for a variety of agricultural uses because it comprises these three components as well as others including FeO, MnO, and P_2O_5. Its alkaline characteristics treat acidic soil [19]. Moreover, converter slags are used to create siliceous fertilizer, phosphorus fertilizer, and micronutrient fertilizer in developed nations including Germany, the USA, France, and Japan [25].

Possible application in CO_2 capture

One of the main greenhouse gases, CO_2, has a significant impact on climate change. Therefore, research on CO_2 reduction technologies has been concentrated on carbon capture and storage (CCS). At present, mineral CO_2 sequestration is one of the most widely used methods of CO_2 sequestration due to its advantages, such as being ecologically friendly, permanently trapping CO_2 in the form of carbonate, and obviating the need for post-storage surveillance for CO_2 leakage [26]. Magnesium or calcium oxides in silicate minerals are promoted to react with carbon dioxide and create carbonates in order to store CO_2 gas in minerals carbonation [27]. Ni smelter slag contains CaO, MgO as well as SiO_2. Thus, it is possible to store CO_2 in carbonates forms using Ni slag slag slurry with room temperature and CO_2 pressure [28]. With regard to reaction time, liquid-to solid ratio (L/S), temperature and initial pH, Chang, et al. [29] studied the technological conditions of CO_2 sequestration with steel slag slurry. Sun, et al. [30] in his experimental study showed that the maximum CO_2 capture capacity could reach to 211kg CO_2/ton steel slag. The results found that the consideration of the contribution $Mg(HCO_3)_2$ in capturing CO_2. The precipitate produced under ideal carbonation conditions contained a high amount of $CaCO_3$, up to 96 2 wt% in composition [30].

Technical benefits and sustainable management

Recycling is one of the options for waste reduction. This can be done by recovering products from waste outputs. It is clear that smelter slag is a promising reactive material for metal adsorption. Slags can be used as a recycling or renewable resource rather than as a waste product, which has environmental and economic benefits. The results of the present study indicate the potential use of recycled smelter slags as adsorbents. Although the maximum adsorption capacity of slag was less than many conventional adsorbents, the results indicated that the slag adsorbed As (V), Pb, and Cu from aqueous solutions and thus could be used in the treatment of metal contaminated water. They can be considered circular products by applying recycling at the source, recovering the precise compounds before final disposal, and using the products in different engineering applications. The amount of industrial smelter waste for disposal would be reduced if it is used in different engineering applications. The results from the different scientific studies suggested that the application of recycled waste materials for site treatment as well as contaminated water treatment would be sustainable and promising.

The Ni smelter slag can be used for in-situ treatment of the mine effluent or contaminated groundwater [31] by using permeable reactive barrier treatment since the present study showed the adsorption efficiencies for As, Pb, and Cu. For every successful engineering solution to be implemented, cost is a critical consideration. Proper installation, operation, and maintenance costs should be determined for any proposed technology. Smelter slags are crushed into desired sizes prior to use as a reactive medium in any treatment process. The major cost factor associated with the use of slag particles as an adsorbent would be the crushing of the slag particle sizes [31]. Using crushed slag in water and wastewater treatment is beneficial both in terms of recycling and economics. Figure 5 shows Ni smelter slags production and the possible applications. The possible application of recycled Ni smelter slag materials suggest as potential circular engineering materials which could be used in contaminated water treatment, and site remediation engineering. It can be used in construction work as well as soil improvement purposes [4, 19].

Conclusions

In conclusion, the present study examined an overview of the adsorption process between the crushed waste product and the contaminant. The application of recycled smelter products in the treatment process can contribute to a circular economy, sustainable technology and a green environment. In addition, recycled slags can also reduce costs associated with obtaining natural aggregates and reduce the environmental impact associated with solid waste disposal. It can be used as admixture of cement and can be constructed concrete blocks or aggregate for construction projects. Recycling of waste slags could reduce the demand for virgin natural resources. By lowering the amount of new materials and products being produced, recycling and upcycling of used or discarded products have a significant potential to ease the burden on our planet. The innovation in the circular economy are crucial to the transition to more sustainable infrastructure development.

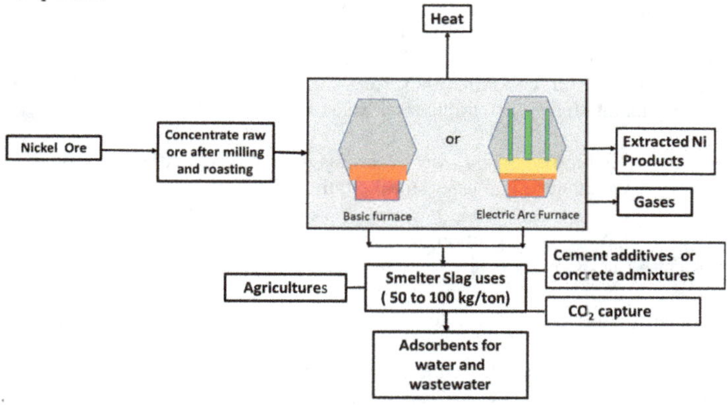

Figure 5: Ni smelter slags production and the possible applications [32]

Acknowledgement

The author is very thankful for the continuous support received from the Department of Civil Engineering at Prince Mohammad Bin Fahd University, and Civil and Environmental Engineering Lab at the University of Western (Ontario in Canada) during the collection and preparation of all necessary data and information for the research.

References

[1] T.C. Nguyen, P. Loganathan, T.V. Nguyen, J. Kandasamy, R. Naidu, R., S.Vigneswaran, Adsorptive removal of five heavy metals from water using blast furnace slag and fly ash'. Environ Sci. Pollut. Res. 25 (2008), 20430–20438. https://doi.org/10.1007/s11356-017-9610-4

[2] A.K. Thakur, T.R. Singh, R.T. Pullela, V. Pundir, Green adsorbents for the removal of heavy metals from Wastewater: A review. Materials Today: Proceedings, (2021), https://doi.org/10.1016/j.matpr.2021.11.373

[3] S.D. Gisi, G. Lofrano, M. Grassi, M. Notarnicola, Characteristics and adsorption capacities of low-cost sorbents for wastewater treatment: A review. Sustainable Materials and Technologies 9 (2016) 10–40.
[4] M. Oge, D. Ozkan, M.B. Celik, M.S. Gok, A.C. Karaoglanli, An Overview of Utilization of Blast Furnace and Steelmaking Slag in Various Applications. Materials Today: Proceedings 11 (2019) 516–525.
[5] I. Perederiy, Dissolution of Valuable Metals from Nickel Smelter Slags by Means of High Pressure Oxidative Acid Leaching. PhD Thesis. Department of Chemical Engineering and Applied Chemistry University of Toronto (2011).
[6] R. Chakraborty, A. Asthana, A.K. Singh, B. Jain, A.B.H. Susan, Adsorption of heavy metal ions by various low-cost adsorbents: a review. International Journal of Environmental Analytical Chemistry 102,(2), (2022) https://doi-org.library.pmu.edu.sa/10.1080/03067319.2020.1722811.
[7] H.K. Boparai, M. Joseph, and D.M. Carroll, Kinetics and thermodynamics of cadmium ion removal by adsorption onto nano zerovalent. J. Hazmat. 186(1), (2011) 458-65 https://doi.org/10.1016/j.jhazmat.2010.11.029
[8] D. Feng, C. Aldrich, H. Tan, H, Treatment of acid mine water by use of heavy metal precipitation and ion exchange. Minerals Engineering, 13(6), (2000)623-642.
[9] S. Liu, J. Gao, Y. Yang, Y. Yang, and Z. Ye, Adsorption intrinsic kinetics and isotherms of lead ions on steel slag. Journal of Hazardous Materials, 173(1–3), (2010)558-562. https://doi.org/http://dx.doi.org/10.1016/j.jhazmat.2009.08.122
[10] S.R. Chowdhury, E.K. Yanful, and A.R. Pratt, Recycling of nickel smelter slag for arsenic remediation - an experimental study'. Environmental Science and Pollution Research. 21, (17), (2014)10096- 10107.
[11] S.R. Chowdhury, Application of mixed iron oxides in subsurface remediation technologies. London, Ont.: School of Graduate and Postdoctoral Studies, University of Western Ontario. Canada (2013).
[12] F.Y. Wang, H.Wang, J.W. Ma, Adsorption of cadmium (II) ions from aqueous solution by a new low-cost adsorbent—Bamboo charcoal'. Journal of Hazardous Materials 177, (2010) 300–306.
[13] Ahmaruzzaman, M. Industrial wastes as low-cost potential adsorbents for the treatment of wastewater laden with heavy metals. Advances in Colloid and Interface Science, 166(1–2), (2011) 36-59. https://doi.org/http://dx.doi.org/10.1016/j.cis.2011.04.005.
[14] G. McKay, M.S. Otterburn, and A.G. Sweeney, Surface mass transfer processes during colour removal from effluent using silica. Water Research, 15 (1980), 327-331.
[15] Y. Chen, W. Ye, M. Yang, H. Yong, Effect of contact time, pH, and ionic strength on Cd (II) adsorption from aqueous solution onto bentonite from Gaomiaozi, China. Environmental Earth Sciences 64(2) (2011): 329-336.
[16] D.L. Sparks, Environmental soil chemistry. Amsterdam; Boston: Academic Press. 2003.
[17] S.K. Srivastava, V.K. Gupta, D. Mohan, Removal of lead and chromium by activated slag-a blast-furnace waste. J Environ. Eng. 123(5) (1997):461–46
[18] Y. Zhou, and R.J. Haynes, Sorption of heavy metals by inorganic and organic components of solid wastes: Significance to use of wastes as low-cost adsorbents and immobilizing agents'. Critical Reviews in Environmental Science and Technology, 40(11), (2010) 909- 977. https://doi.org/10.1080/10643380802586857
[19] Q. Zhao, L. Pang, D. Wang, Adverse Effects of Using Metallurgical Slags as Supplementary Cementitious Materials and Aggregate: A Review. Materials 15 (2022) 3803. https://doi.org/10.3390/ ma15113803

[20] M. Czop, and B.L. Piekarczyk, Use of Slag from the Combustion of Solid Municipal Waste as A Partial Replacement of Cement in Mortar and Concrete. Materials, 13 (2020), 1593; https://doi.org/10.3390/ma13071593
[21] H. Yi, G. Xu, H. Cheng, J. Wang, Y. Wan, H. Chen, An overview of utilization of steel slag. Procedia Environmental Sciences 16 (2012) 791 – 801.
[22] Q. Wanga, P.Y. Yana, J.W. Feng, A discussion on improving hydration activity of steel slag by altering its mineral compositions. J. Hazard Mater.186(2011)1070-1075.
[23] H. Qasrawi, F. Shalabi, I. Asi, Use of low CaO unprocessed steel slag in concrete as aggregate. Constr Build Mater 23 (2009)1118-1125.
[24] I. Papayianni, E. Anastasiou, Production of high-strength concrete using high volume of industrial by-products. Constr Build Mater, 24 (2010),1412-7.
[25] Z.H. Wu, Z.S. Zou, C.Z. Wang, Application of converter slags in agriculture. Multipurpose Util Min Resour (in Chinese), 6 (2005), 25-8.
[26] Y. Sun, M.S. Yao, J.P. Zhang, G. Yang, Indirect CO_2 mineral sequestration by steelmaking slag with NH4Cl as leaching solution. Chem Eng J.173 (2011), 437-445.
[27] S. Eloneva, S. Tei, J. Salminen, C.J. Fogelholma, R. Zevenhoven. Fixation of CO_2 by carbonating calcium derived from blast furnace slag. Energ. 33 (2008)1461-7.
[28] C. Kunzler, N. Alves, E. Pereira, J. Nienzewksi, R. Ligabue, S. Einloft, et al., CO_2 storage with indirect carbonation using industrial waste. Energ Procedia, 4 (2011)1010-7.
[29] E.E. Changa, S.Y. Panb, Y.H. Chen, H.W. Chu, C.F. Wang, P.C. Chiang, CO_2 sequestration by carbonation of steelmaking slags in an autoclave reactor. J Hazard Mater. 195 (2011) 107-114.
[30] Y. Sun, M.S. Yao, J.P. Zhang, G. Yang. Indirect CO_2 mineral sequestration by steelmaking slag with NH4Cl as leaching solution. Chem Eng J. 173 (2011) 437-445.
[31] S.R. Chowdhury, E.K. Yanful, Application of recycling waste products for exsitu and in situ water treatment methods'. Environmental Technology, 41 (2020) (7), 878-889. https://doi.org/10.1080/09593330.2018.1513078
[32] S.R. Chowdhury, Recycled Smelter Slags for In Situ and Ex Situ Water and Wastewater Treatment— Current Knowledge and Opportunities. Processes 11 (2023), 783. https://doi.org/10.3390/pr11030783

Structure, electronic properties of AlAs using first-principles calculations

Sansabilla Bouchareb[1,a *], Rachid Tigrine [2,b]

[1]Laboratoire, d'énergie , environnement et systèmes d'informations ,université Ahmed Draya, Adrar, Algérie

[2]Laboratoire, De Physique et Chimie Quantiques ,Université Mouloud Mammeri, Tizi Ouzou, Algérie

[a] sansabilla123@gmail.com, [b] tigriner@univ-adrar.edu.dz

Keywords: Electronic Properties, Local Density, Generalized Gradient Approximations, Generalized Gradient Approximation Given by Perdew-Burke-Ernzerhof

Abstract. Density functional theory has been used to study the structural and electronic properties of AlAs using generalized gradient approximation implemented through the WIEN2k code by the local density (LDA), generalized gradient approximations (GGAPBE), and generalized gradient approximation is given by Perdew-Burke-Ernzerhof (GGAPBE Sol) methods .The AlAs's cubic structure is mechanically stable in the Fd-3m space group and exhibits semi-conducteur behavior with a 2.259 eV indirect energy band gap (E_G) along the *Γ-X.* The optimized lattice is constant (a_0=5,6605Å). The properties suggest that the material is also suitable for use as a photoconductor such as diodes, laser diode, optoelectronic devices, and such as light emitting diodes.

Introduction

Knowing the structural properties in the physics of materials is important especially if it is related to simulation because it gives us information about the microstructure of the material. In order to study the physical properties of the material (electronic, elastic, compressive...) it is necessary to go through the structural properties. In this work, we will study the physical properties of AlAs using DFT and its approximations (LDA, GGA PBE, GGA PBE SOL) and using the FP-LAPW method included in the Wien2K program that we used in our study.

We mention that the aim of this study is not to study the matter as much as a comparative study of the validity of the derivatives of the density function theory in the study of the properties of the material, so we chose a binary alloy with a cubic structure.

Semiconductors are materials that are classified between conductors and insulators in terms of electrical conductivity.

The bands, which are the transport band and the valence band between them are a forbidden band. Semiconductors are affected by temperature, light, and magnetic field [1]. They are binary materials based on aluminum, AlX, which are in group III-V. They are specific to semiconductors, i.e. they are the family found in columns III and V of the periodic table of elements. They have been studied It has become the subject of many experimental and theoretical studies, both under normal conditions and at high pressure, because it is widely used in electronic devices and optoelectronic devices.

The III-V elements have less electronic properties compared to the I-VII and II-VI elements. Most compounds similar in composition to semiconductors take the role of insulators, and from it can be said that the forbidden band is relatively small. As for the III-V compounds, they are among the best examples of the elements that It is characterized by partially ionic or partially covalent bonds [2].

These alloys are in normal (normal) conditions (of the zinc blende structure) case B3; symmetry group F43m ((N°=216) knowing that the zinc blende network has a face-centered cubic structure (cfc) consisting of two atoms Al and X that are centered in the positions [3]:

Al : (0,0, 0) ;(1/2,1/2,0) ; (1/2, 0,1/2) ; (0,1/2,1/2).

X : (1/4,1/4,1/4) ; (3/4,3/4,1/4) ; (3/4,1/4,3/4) ; (1/4,3/4,3/4).

Computational Details

In this paper, we present the structural, electronic, magnetic properties of a AlAS. The calculations were carried out on the basis of density functional theory. The semi-conducteur properties have been calculated by the procedure followed in the WIEN2k [4] code based on the full potential linearized augmented plane wave (FPLAPW) method [5].

The generalized gradient approximation given by Perdew-Burke-Ernzerhof (PBE-GGA) has been treated within the exchange and correlation operations [6,7]. R MT × K Max was set to be 7 in the plane wave augmentation. G Max, which is the optimal magnitude in Fourier expanded density of charge, was taken to be 12 a.u^{-1}. The order of energy was set to 0.00001 Ry for convergence. 500 k mesh points were used for integration in the Brillouin zone.

Results and Discussion

Structural Properties

The aim of studying the structural properties of these compounds is to find the crystal lattice constant, and in order to study the structural properties, we decided to calculate the lattice constant and compressibility coefficient. By using the data recorded in the table

Table 1. Structure, special group and experimental crystal lattice constant of AlAs.

/	Structure, special group	experimental crystal lattice constant a(Å)
AlAs	Fd-3m 216 Al (0, 0, 0) As (1/4,1/4,1/4)	5,6605[1]

In general, this compound crystallizes into a coherent and stable cubic structure, so the microstructure of the alloy is according to the Fig. 1

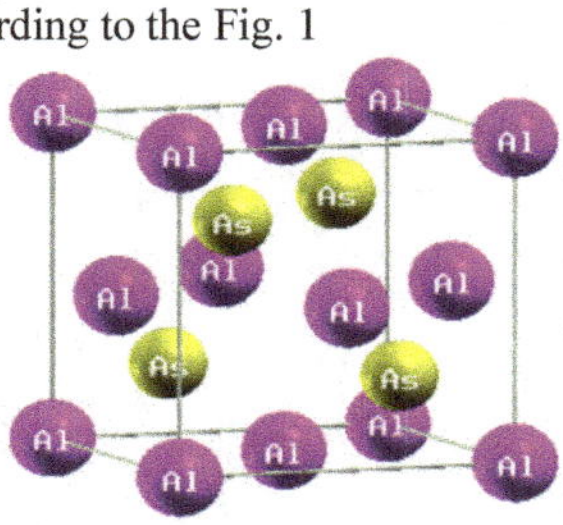

Fig. 1: Lattice view of AlAS.

The crystal structure of the AlAS, which is a semi-conducteur, is shown in Fig. 1, where the results of our calculations are given. In this structure

Al : (0,0, 0) ;(1/2,1/2,0) ; (1/2, 0,1/2) ; (0,1/2,1/2).

X : (1/4,1/4,1/4) ; (3/4,3/4,1/4) ; (3/4,1/4,3/4) ; (1/4,3/4,3/4).

In order to calculate the structural properties in the equilibrium state, experimental values of the crystal lattice constant should be established. By calculating the total energy for different values of the crystal lattice constants a (Å) next to the value of a0, then drawing the curve of the energy values in terms of volume a3 then adapting them to the different state equations.

The results are presented in the form of four equations of state which are mentioned below and are listed in the table.

state equations

Define The **Murnaghan** equation given below has been used for volume optimization.

$$E(V) = E_0 + \frac{B_0 V}{\acute{B}}\left[(V/V_0)^{-\acute{B}}\frac{1}{\acute{B}-1} + 1\right] - \frac{B_0 V_0}{\acute{B}-1} \quad (1)$$

B_0 and B' compressibility modulus and the first derivative of the compressibility modulus at equilibrium

V_0 is the volume at equilibrium

The coefficient that expresses B is defined by the following expression:

$$B = V\,\frac{\delta^2 E}{\delta V^2}$$

The **Birch-Murnaghan** equation given below has been used for volume optimization

$$E(V) = E_0 + \frac{9\beta_0 V_0}{16}\left\{\left[\left(\frac{V}{V_0}\right)^{-2/3} - 1\right]^3 \acute{\beta} + \left[\left(\frac{V}{V_0}\right)^{-2/3} - 1\right]^2\left[6 - 4\left(\frac{V}{V_0}\right)^{-2/3}\right]\right\} \quad (2)$$

The **Vinet-Rose** equation given below has been used for volume optimization

$$E(V) = E_0 + \frac{4B_0V_0}{(\acute{B}-1)^2}\left\{1 - \left[\frac{3}{2}\left\{1-\left(\frac{V}{V_0}\right)^{1/3}\right\}(\acute{B}-1)\right].exp\left[1 - \left[\frac{3}{2}\left\{1-\left(\frac{V}{V_0}\right)^{1/3}\right\}(\acute{B}-1)\right]\right]\right\} \quad (3)$$

The **Poirier-Tarantola** equation given below has been used for volume optimization

E=E0+(B/14703.6)*(V0/2)*(ln(eta))**2+(B/14703.6)*(V0/6)*(Bp-2)*(ln(eta))**3 (4)

We calculated the crystal lattice constant a(Å), the compressibility modulus β and the first derivative of the compressibility modulus β'

So we get the results according to the table:

Table 1. Structure, special group and experimental crystal lattice constant of AlAs.

	LDA			GGAPBE			GGAPBESol		
	a(Å)	β	β'	a(Å)	β	β'	a(Å)	β	β'
AlAs	5.6338[(1)] 5.6339[(2)] 5.6339[(3)] 5.6340[(4)]	75.5430 75.8273 75.9565 76.1201	4.5553 4.5159 4.4983 4.4561	5.7332[(1)] 5.7330[(2)] 5.7329[(3)] 5.7328[(4)]	71.8550 72.0910 72.2046 72.3747	4.3825 4.4511 4.4841 4.5113	5.6790[(1)] 5.6789[(2)] 5.6788[(3)] 5.6788[(4)]	71.8558 72.0910 72.2046 72.3747	4.3825 4.4511 4.4841 4.5113

(1) The Murnaghan equation (3) The Vinet-Rose equation
(2) The Birch-Murnaghan equation (4) The Poirier-Tarantola equation

The selection of the Murnaghan, Birch-Murnaghan, Vinet-Rose, and Poirier-Tarantola equations was not important because of the convergence of the results, as most of the results obtained with the four equations were convergent on the order of 10^{-3}.

Therefore, we compared the values obtained by the Murnaghan state equation for various approximations of the state density function with the experimental values, and we summarized the results in Table 2, and drew the curve that represents the total energy changes in terms of volume for the element AlAs

We completed the calculation with three approximations (LDA), (GGA-PBE), (GGA-PBEsol) (and represented the energy curve in terms of volume as shown in documents No. IV.1, where we noticed that the energy changes curve decreases to a lower value and then increases) i.e. there is a limit value Lower Emin corresponds to a specific volume (this value corresponds to the value of the particle's ground state density since all physical properties are related to this state

$$E[\rho_0(r)] = minE[\rho(r)]$$

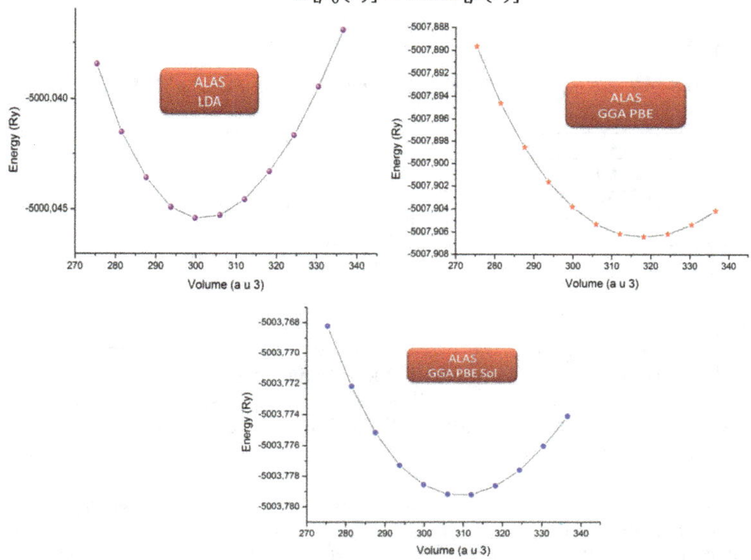

Fig. 2: Total energy changes as a function of volume for AlAs by the LDA, GGAPBE, and GGAPBE Sol methods.

We compared the results of the crystal lattice constant computationally obtained by the methods, (LDA) (GGA-PBE) and (GGA-PBEsol) (AlAs). With the experimental results, we chose to compare the corresponding results of the Murnaghan equation (Table IV.5) and recorded the following observations:

-Most of the values of the crystal lattice constant are close to the experimental results.

- The calculated values of the compressibility coefficient and the derivative of the compressibility coefficient by different methods were good and close to the theoretical values available in the references.

Table 3. Results of the crystal lattice constant by methods (LDA), (GGA-PBE), (GGA-PBEsol).

the theoretical values		GGA PBE Sol			GGA PBE			LDA			
β β'	a(Å)	β'	β	a(Å)	β'	B	a(Å)	β'	B	a(Å)	
74±4 [8] 5.01± 1 [8]	5.6605 [1]	4.3825	71.8558	5.6790	4.3825	71.8550	5.7332	4.5553	75.5430	5.6338	AlAs

Table 4. The error ratio of the crystal lattice constant by (LDA), (GGA-PBE), (GGA-PBEsol) methods.

AlAs			/
5.6605			**a_0(Å)**
GGA-PBEsol	**GGA-PBE**	**LDA**	**method**
5.6790	5.7332	5.6338	**a(Å)**
0.33	1.28	0.47	**The error ratio %**

Electronic Properties

The importance of the electronic properties of solid materials is reflected in determining the quality of the studied material, knowing the nature of the different bonds between its atoms, these properties are explained by the structure of the bands and the state density. We used the method of linear plane waves and the full potential (FP-LAPW) in order to calculate the energy bands of the material (AlAs).) using some DFT derivatives represented in (LDA) (GGA-PBE), and (GGA-PBEsol).

In solid physics, the theory of energy bands is the formation of energy by taking the internal electrons, from the diffusion equation E(K) and from this equation important values can be found, the most important of which are the energy (Eg, the energy gap or the energy gap), the effective mass, the valence band width, electron transmission, etc. ....

Using the value of the energy interval, Eg, it is possible to know the quality of the material, whether it is an insulator, a metal, a semi-conductor, or a semi-metal.... and justify it by studying the density of the state.

It is possible to know the quality of the materials, whether they are insulators, metals, semi-conductors, or semi-metals, and justify them by studying the density of the state.

The energy bands of AlAs are represented in Fig. 3 from the energy field boundaries [-12,14] eV.

In the case of LDA, the bands are grouped into two separate groups that share a point, the first group being in the [-12, 0] energy field.

The second group is confined from the Pharmi level up to 12 eV What we notice well is the presence of an overlap between the energy band and the valence band.

As for (GGA-PBE), (GGA-PBEsol), we notice that the bands were collected in two separate groups separated by an energy barrier:

The first group in the energy field [-12, 0]

The second group is in the energy range from 1.459 eV to 12 eV and they are separated by a relatively small indirect energy gap (less than 4 eV).

The presence of a small energy separation means that the element (AlAs) has a semi-conducteur property.

The results obtained for the energy interval (energy gap) Eg for the basic energy band X, Γ for the LDA approximation) were non-existent. As for the two approximations (GGA-PBE), (GGA-PBEsol), its value was= 1.609 eV, Eg = 1.459 eV respectively. We note that these values were less than the experimental value Eg(exp) = 2.22 eV Therefore, we decided to recalculate the value of the energy interval using (TB-mBJ) in order to approximate(GGA-PBEsol), its value was Eg = 2.259eV, and thus we approached the experimental value with an error of 1.76%.

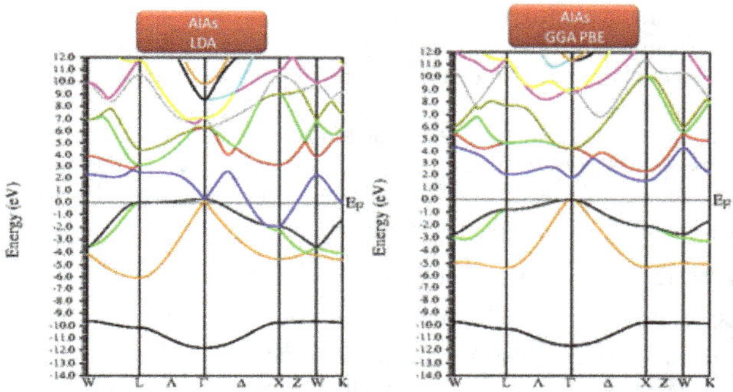

Fig. 3a: Variation of band gap energy for AlAs by the LDA, GGAPBE methods.

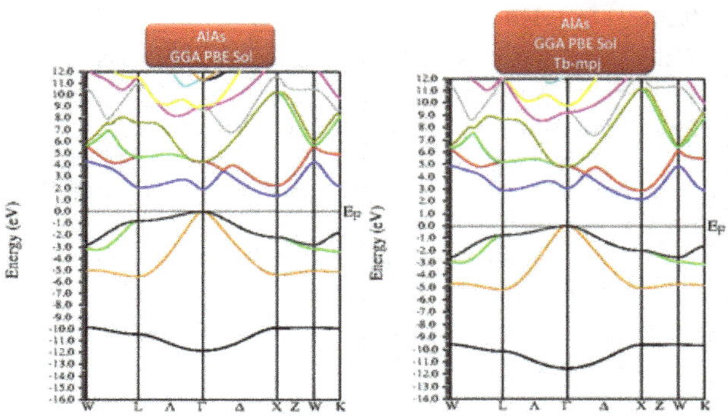

Fig. 3b: Variation of band gap energy for both methods, GGA-PBEsol, TB-mBJ.

We summarize all results in the following table:

able 5. Results of gap energy by methods (LDA), (GGA-PBE), (GGA-PBEsol) (Tb-mbj).

	E_g (LDA) eV	E_g (GGA PPE) eV	E_g (GGA PBE Sol) eV	E_g (exp) eV	E_g (Tb-mbj) eV
AlAs	0	1.609	1.459	2.22 [9]	2.259

Conclusion

In this work, we have computed structural and electronic properties of AlAs semiconductor material. We employed first-principles calculations based on density functional theory and a systematically study. It was found that, the Lattice parameter constant and Band gap energy are consistent with previous studies.

The choice of the Murnaghan, Birch-Murnaghan, Vinet-Rose, equation of state was not Poirier-Tarantola, it is important because of the convergence of the results, as most of the results obtained with the four equations were convergent on the order of 10^{-3}.

Calculate the structural properties from the crystal lattice constant a(Å), the compressive modulus β and the first derivative of the compressive modulus β', is not affected by the method used from the derivatives of the density function (LDA) (GGA-PBE), and (GGA-PBEsol), this means that the calculation of the structural properties is not affected by the choice of the density function of the state.

For the electronic study, Eg should be calculated for AlAs. The (LDA) method did not give any results compared to (GGA-PBE) derivatives. It gave close results that called us to use the TB-mBJ correction, which showed very close values to the experimental values.

References

[1] Annane F, Meradji H, Ghemid S and Hassan F E H 2010 Comput. Mater. Sci. 50 274 Phys. B 22 026201. https://doi.org/10.1088/1674-1056/22/2/026201

[2] Rehamn G, Shafiq M, Saifullah M, Ahmad M, Asadadabadi S J, Maqbool M et al 2016 J. Electron. Mater. 45 3314; DOI: https://doi.org/10.1007/s12034-018-1702-3

[3] Ziane M I, Bensaad Z, Labdelli B and Bennacer H 2014 Sens. Transducers J. 27 374. https://doi.org/10.1063/1.5006170

[4] Ferreira L, Wei S H and Zunger A 1998 Phys. Rev. B 40 3197 Phys. Rev. B 40, 3197 – Published 15 August 1989. https://doi.org/10.1103/PhysRevB.40.3197

[5] D. Singh, Planes Waves, Pseudo-Potentials and the LAPW Method, Kluwer Academic Publishers, Boston, Dortrecht, London, 1994. Hieder Al-Shami, Ahmed M. Salah, Mohamed Fathy Adel Ali Open Journal of Modern Neurosurgery Vol.10 No.3, May 8, 2020. https://doi.org/10.4236/ojmn.2020.103034

[6] J.P. Perdew, K. Burke, Y. Wang, Phys. Rev. B 54 (1996) 16533. Received 14 February 1996. https://doi.org/10.1103/PhysRevB.54.16533

[7] J.P. Perdew, S. Burke, M. Ernzerhof, Phys. Rev. Lett. 77 (1996) 3865. Received 21 May 1996. https://doi.org/10.1103/PhysRevLett.77.3865

[8] R.G. Greene, H. Luo, T. Li, A. Ruoff, Phys. Rev. Lett. 72, 2045 (1994) High pressure structural study of AlSb to 50 Gpa. https://doi.org/10.1016/0022-3697(94)00231-2

[9] B. Monemar, Phys. Rev. B 8, 5711 (1973) . Received 7 May 1973. https://doi.org/10.1103/PhysRevB.8.5711

[10] Joyce J H, Docherty C J, Gao Q, Tan H H, Jagadish C, Hughes, J L et al 2013 Nanotechnology 24 214006. https://doi.org/10.1007/s12034-018-1702-3

[11] Garriga M, Lautenschlager P, Cardona M and Ploog K 1987 Solid State Commun. 63 157 Journal of Applied Physics 42, 3499 (1971). https://doi.org/10.1063/1.1660760

[12] Adachi S 1987 Phys. Rev. B 35 7454 Phys. Rev. B 35, 7454 – Published 15 May 1987 Received 10 November 1986. https://doi.org/10.1103/PhysRevB.35.7454

[13] Joyce J H, Docherty C J, Gao Q, Tan H H, Jagadish C, Hughes J L et al 2013 Nanotechnology 24 214006 Semiconductor Science and Technology, Volume 31, Number 10Citation Hannah J Joyce et al 2016 Semicond. Sci. Technol. 31 103003. https://doi.org/10.1088/0268-1242/31/10/103003

[14] Garriga M, Lautenschlager P, Cardona M and Ploog K 1987 Solid State Commun. 63 157. https://doi.org/10.1007/BF00615025

[15] Djurisic A B, Rakic A D, Kwok P C K, Li E H and Majewski M L 1999 J. Appl. Phys. 85 3638 Journal of Applied Physics 88, 2172 (2000). https://doi.org/10.1063/1.1305544

[16] Ahmed R, Hashemifar S J, Akbarzadeh H, Ahmad M and Aleem F 2007 Comput. Mater. Sci. 39 580. https://doi.org/10.1063/1.4757439

A comparative study on the use of laser beam and abrasive water jet in hole making process of woven laminated GFRP

Hussein Mohammad Ali

Ph.D in Mechanical Engineering; Department of Power Mechanic Techniques Engineering, Technical Engineering College of Mosul, North Technical University, Mosul, Iraq

alabadi.hussein@ntu.edu.iq. hrlha@yahoo.com

Keywords: AWJ, CO_2 Laser, Dimensional Accuracy, Strength

Abstract. In the manufacturing process, the designed part will be presented in a drawing with all dimensions normally given within a certain range of tolerances. The tolerance defines the limits of induced deviation for which allowance should be made in the design, and within which actual size is acceptable. In laser and abrasive water jet cutting, dimensional accuracy is one of the important parameters to define the quality of produced part. The aim of the present work is to compare experimentally the influence of cutting parameters on dimensional accuracy and strength of hole making in GFRP by using (LBM) and (AWJM) cutting technologies. Full factorial design was used as a statistical method to study the effects of control parameters on the response variables. The results show that abrasive water jet cutting gives a less out of roundness in cutting hole diameter, less reduction in strength and large difference between upper and lower diameter compared to the laser cutting technology of hole making in the type of the GFRP composite material used in the present work.

Nomenclature

LBM	Laser beam machining
AWJM	Abrasive water jet machining
D	nominal hole diameter
t	material thickness
Vc	cutting feed
LP	Laser power
P	water jet pressure
Sod	Stand of distance
O.O.R	out of roundness
Du-DL	Difference between upper and lower diameter
T.S	Tensile strength

Introduction

Glass fiber reinforced polymer (GFRP) composites are used in a large number of industrial applications because of the advantages they have compared to other materials. These advantages are high strength to weight ratio, high modulus, high fracture toughness, and corrosion and thermal resistance. As well as the relative ease of manufacture of components using GFRPs. [1]. As structural materials, joining composite laminates to other metal materials structures could not be avoided [2], and bolt joining efficiency and quality depend critically on the quality of machined holes. Various cutting processes are extensively used for producing riveted and bolted joints during assembly operation of composite laminates with other components. For rivets and bolted joints, damaged-free and precise holes must be made in the components to ensure high joint strength and precision. [3,4]. Conventional machining of hole making in fiber-reinforced composites is difficult due to diverse fiber and matrix properties, fiber orientation, inhomogeneous nature of the material, and the presence of high-volume fraction (volume of fiber over total volume) of hard abrasive fibers in the matrix. Abrasive water jet machining (AWJM) & Laser beam machining (LBM) processes have been used for processing composite materials because of the advantages offered by these technologies as compared to traditional techniques of processing. Laser beam machining (LBM) process has a wide range of applications in different manufacturing processes in industry due to its advantages of high cut quality and cost effectiveness through mass- production rate [5]. LBM is particularly suitable for making accurately placed holes. The material to be cut is locally melted by the focused laser beam. The melt is then blown away with the aid of assist gas, which flow coaxially with the laser beam, in the cutting procedures, different types of assist gases are used such as oxygen and nitrogen. It is suitable for fine cutting of sheet metal at high speed [6]. Abrasive water jet machining (AWJM) has been used also for processing composite materials because of the advantages offered by this technology as compared to traditional techniques of processing. Many researchers carry out the studies on AWJM & LBM of composite materials. Ho-Cheng [7] discussed an analytical approach to study the delamination during drilling by water jet piercing. Their model predicted an optimal water jet pressure for no delamination as a function of hole depth and material parameter. Ramulu et al. [8] reviewed and investigated the AWJ drilling for various materials (steel, aluminum, glass, titanium and polycarbonate). He was found that water pressure, abrasive flow rate and drilling time significantly affected the dimensions and accuracy of the AWJ drilled holes. Hocheng and Sao [9] studied various non-traditional drilling techniques and observed that WJ drilling can be effectively used to make fine holes of medium to large diameter, by contour cutting very speedily. They found that delamination could be eliminated by reducing the jet speed while the piercing capability deteriorates.

This research presents approach to select optimal cutting parameters for high dimensional accuracy and strength, of hole making in laminate GFRP composite by using AWJM and LBM processes. A numerical optimization has been performed using Derringer-Suich multi-criteria decision modeling approach. ANOVA is a basic statistical technique was used for determining the proportion of influence of an input parameter on total variation of response parameters. A set of experiments regarding the two machining technique were conducted, with cutting parameters prefixed on glass fiber reinforced plastic (GFRP) laminate.

Experimental Work

Cutting Mechanism by LBM and AWJM

Glass fiber reinforced plastic (GFRP) composite materials are the combination of two materials, glass fiber and polymer matrix, that have significant different characteristics. Since each of these materials oxidizes at a different temperature, the laser beam process used to cut the glass fibers would cause the epoxy resin to decompose and melt resulting in a flow of the fibers within the resin and charring and tearing of the resin layer [12]. While abrasive water jet cutting technology uses a jet of high pressure, velocity water and abrasive slurry to cut the target material by means of erosion. It was shown from the scanning electron microscopy (SEM) analysis for the cut surfaces of polymer matrix composites that the erosive process for the matrix material (resin) involves shearing and ploughing as well as intergranular cracking. Shearing or cutting was found to be the dominant process for cutting the fibers in the upper cutting region, but the fibers are mostly pulled out in the lower region of the cutting surface [7].

Material

For the experimental study, a sheet of woven laminated glass fiber reinforced plastic (GFRP), Type 3240 produced by Jinhao Material Co. / China was used as shown in Figs. (1&2). This material is mainly used in aerospace, transportation tools and electrical appliances as insulation materials. The major properties of the laminated GFRP material used are listed in Table 1.

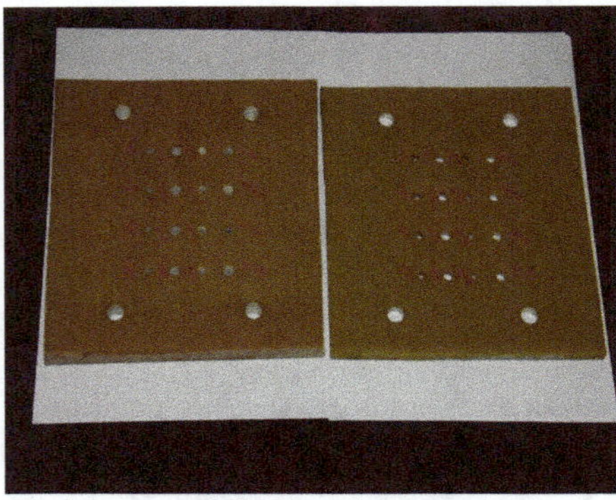

***Fig.1** Laminated GFRP with the two thicknesses.*

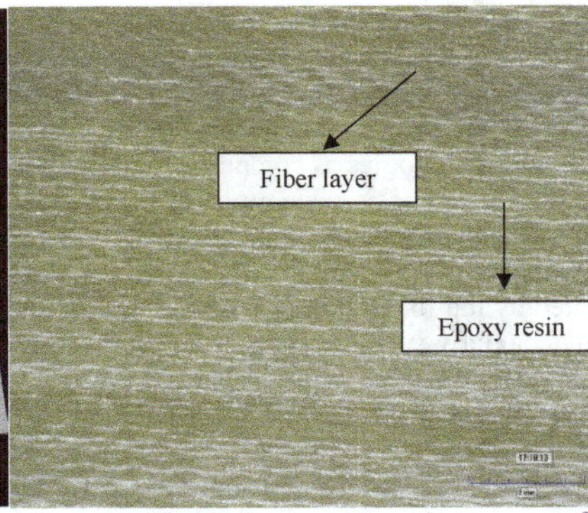

***Fig. 2** Cross-sectional view.* 3mm

Table 1 *Major properties of Laminated GFRP Type 3240*

	Property	Value/unit
1.	Fiber density	0.82 gm/cm^3
2.	Fiber volume fraction	45%
3.	Max.working temperature	200 °C
4.	Tensile strength	295.45 MPa
5.	Layer thickness	0.5 mm

Design of Experiments

The control parameters are selected based on the available literature, availability of speed and feed rate on the machines, the control parameters ranges are carefully provided between the levels for comparison purpose. A five factors, two-level, full-factorial design of experiments (2^5= 32 tests) was developed for LBM and AWJM cutting process. High and low level of control parameters for AWJM and LBM is shown in tables 2 and 3.

The following is description of response variables (performance measures) to be measured in the tests:

1. Dimensional accuracy in term of out of roundness (O.O.R) as in Eq.1 and the difference between the upper & lower diameter (Du-DL) Fig.3. High out of roundness and high difference between the upper & lower diameter represent low dimensional accuracy.

   $$O.O.R = L_1+L_2+L_3 / 3 \quad (1)$$

   Where:
   L_1, L_2 and L_3 is the deviation distance at three different points measured from the optical microscope picture for each hole in the two types of cutting technologies as shown in fig.3.
2. Tensile strength, measured in MPa. 32 tensile test of hole specimens (16 holes cut each by AWJM and LBM) according to ASTM D5766 [13] was carried using Universal Tensile Testing Machine, Type WDW-300, made by Changchun Kexin Com. / China. Fig.4.shows this setup.

Materials Research Proceedings 31 (2023) 212-226 https://doi.org/10.21741/9781644902592-23

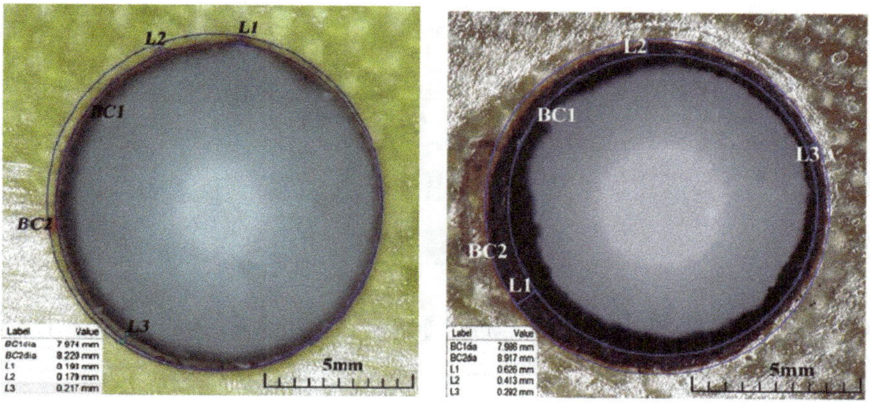

***Fig.3** Optical microscope picture for cutting hole by a: AWJM, b: LBM showing L_1, L_2 and L_3.*

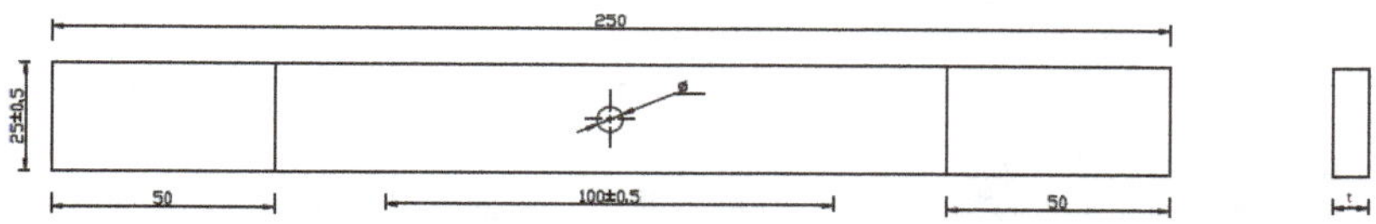

(a)

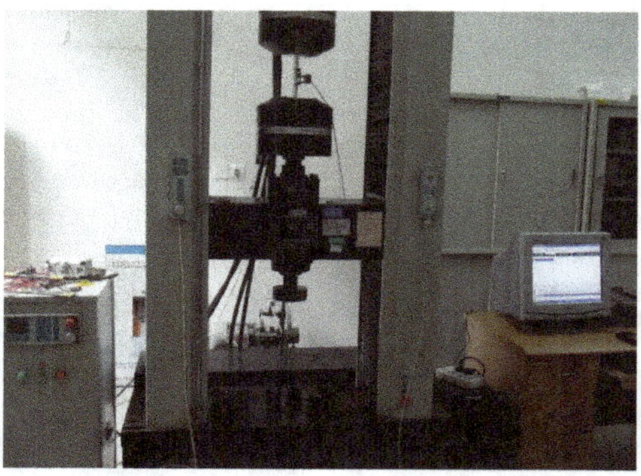

(b)

***Fig.4** (a) Standard hole specimen for tensile test. (b) Universal Tensile Testing Machine*

***Table 2.** High and Low setting of control parameters in (AWJM).*

code	Input factor	Unit	Level 1	Level 2
A	Nominal hole diameter (D)	mm	6	8
B	Material thickness (t)	mm	8	16
C	Cutting feed (Vc)	m/min	0.2	0.3
D	Jet pressure (P)	Mpa	150	200
E	Standoff distance (Sod.)	mm	2	3

***Table 3.** High and Low setting of control parameters in (LBM).*

code	Input factor	Unit	Level 1	Level 2
A	Nominal hole diameter (D)	mm	6	8
B	Material thickness (t)	mm	8	16
C	Cutting feed (Vc)	m/min	0.2	0.3
D	Laser power (LP)	Kw	1.5	2
E	Standoff distance (Sod.)	mm	1	2

Experimental Setup

The AWJM experiments was conducted on Ultra – high pressure water cutting machine produced by Nanjing Hezhan Microtechnic.Co. Ltd./China with a maximum jet pressure of 220-230 Mpa, abrasive flow rate 3.7 lit/min,water flow rate 3.5-3.7 lit/hr and type of abrasive is Garnet . In all the tests, the nozzle diameter used was 1-mm. LBM experiments were conducted on Rw – 6015 X cantilevered flight optical path laser cutter produced by Nanjing Nanchuan Laser Equipment Co. Ltd. With laser power 2-4 kW, max.speed 50 m/min, table size 2500/1250mm. In all the laser experiments the nozzle (orifice) diameter used was 1.5 mm. The dimensions of the work piece material to be cut in the two types of cutting process were (200 × 200 × 8) mm and (200 × 200 ×16) mm. Optical Microscope type Leica DVM500, having accuracy 0.001 mm was used to measure the cut profile and hole diameter. The experimental setup is presented in Fig.3

(a) AWJM setup

(b) LBM setup

(c) Optical microscope setup

***Fig. 3** Experimental setup.*

Experimental Results

The experimental layout and results for the two types of cutting processes are presented in Tables 4 and 5 respectively.

***Table 4.** Experimental results for AWJM*

	Control Variables					Response Variables		
S.N	D (mm) **A**	t (mm) **B**	Vc (m/min) **C**	P MPa **D**	Sod (mm) **E**	O.O.R (mm)	Du-DL (mm)	T.S (Mpa)
1	6	8	0.2	150	2	0.111	-0.133	283.95
2	8	8	0.2	150	2	0.141	-0.265	283.53
3	6	16	0.2	150	2	0.145	-0.021	285.23
4	8	16	0.2	150	2	0.114	+0.097	246.76
5	6	8	0.3	150	2	0.161	+0.164	294.74
6	8	8	0.3	150	2	0.135	+0.174	269.12
7	6	16	0.3	150	2	0.062	+0.133	249.67
8	8	16	0.3	150	2	0.094	+0.13	291.32
9	6	8	0.2	200	2	0.16	+0.277	115.20
10	8	8	0.2	200	2	0.131	+0.305	115.51
11	6	16	0.2	200	2	0.071	+0.142	371.12
12	6	8	0.3	200	2	0.154	+0.069	274.61
13	8	16	0.2	200	2	0.105	+0.143	274.60
14	8	8	0.3	200	2	0.133	+0.316	98.01
15	6	16	0.3	200	2	0.084	+0.086	291.78
16	8	16	0.3	200	2	0.174	+0.474	224.71
17	6	8	0.2	150	3	0.222	+0.225	280.39
18	8	8	0.2	150	3	0.229	+0.309	263.97
19	6	16	0.2	150	3	0.088	+0.086	235.36
20	6	8	0.3	150	3	0.102	+0.234	98.95
21	6	8	0.2	200	3	0.275	+0.286	281.71
22	8	16	0.2	150	3	0.069	+0.015	284.67
23	8	8	0.3	150	3	0.122	+0.269	114.78
24	8	8	0.2	200	3	0.167	+0.378	119.34
25	6	16	0.3	150	3	0.122	+0.385	301.45
26	6	16	0.2	200	3	0.138	+0.288	291.45
27	6	8	0.3	200	3	0.153	+0.343	71.97

28	8	8	0.3	200	3	0.198	+0.389	119.56
29	8	16	0.3	150	3	0.124	+0.074	279.34
30	8	16	0.2	200	3	0.125	+0.216	287.21
31	6	16	0.3	200	3	0.128	+0.479	288.22
32	8	16	0.3	200	3	0.196	+0.237	281.73

Table 5. *Experimental results for LBM.*

	Control Variables					Response Variables		
S.N	D (mm) **A**	t (mm) **B**	Vc (m/min) **C**	LP (kW) **D**	Sod (mm) **E**	O.O.R (mm)	Du-DL (mm)	T.S. (Mpa)
1	6	8	0.1	1.5	1	0.251	-0.158	159.08
2	8	8	0.1	1.5	1	0.163	+0.045	246.99
3	6	16	0.1	1.5	1	0.251	-0.039	160.53
4	8	16	0.1	1.5	1	0.260	-0.206	169.93
5	6	8	0.2	1.5	1	0.166	+0.448	108.42
6	8	8	0.2	1.5	1	0.134	+0.061	104.41
7	6	16	0.2	1.5	1	0.393	+0.001	159.54
8	8	16	0.2	1.5	1	0.303	-0.109	138.27
9	6	8	0.1	2	1	0.188	-0.022	71.51
10	8	8	0.1	2	1	0.092	-0.053	70.37
11	6	16	0.1	2	1	0.081	-0.102	65.89
12	6	8	0.2	2	1	0.19	-0.09	77.11
13	8	16	0.1	2	1	0.290	-0.402	65.73
14\	8	8	0.2	2	1	0.136	+0.06	79.26
15	6	16	0.2	2	1	0.110	-0.207	70.07
16	8	16	0.2	2	1	0.393	-0.106	69.69
17	6	8	0.1	1.5	2	0.219	-0.316	88.49
18	8	8	0.1	1.5	2	0.292	-0.053	84.19
19	6	16	0.1	1.5	2	0.21	-0.178	136.51
20	6	8	0.2	1.5	2	0.195	-0.009	262.24
21	6	8	0.1	2	2	0.201	-0.022	68.49
22	8	16	0.1	1.5	2	0.220	-0.156	130.18
23	8	8	0.2	1.5	2	0.25	+0.003	94.56
24	8	8	0.1	2	2	0.093	-0.058	73.75
25	6	16	0.2	1.5	2	0.277	-0.256	172.01
26	6	16	0.1	2	2	0.192	-0.256	62.56
27	6	8	0.2	2	2	0.28	+0.09	75.99
28	8	8	0.2	2	2	0.147	-0.015	78.09
29	8	16	0.2	1.5	2	0.459	-0.022	131.32
30	8	16	0.1	2	2	0.325	-0.363	65.84
31	6	16	0.2	2	2	0.185	-0.032	70.18
32	8	16	0.2	2	2	0.333	-0.048	69.88

Results and discussion

Experimental data have been analyzed using ANOVA (Analysis of Variance) and numerical optimization has been performed using Derringer-Suich multi-criteria decision modeling approach. ANOVA is a basic statistical technique for determining the influence of an input parameter on response parameter(s). In Derringer-Suich, multi-criteria optimization technique different desirability functions are assigned to maximization/minimization the response parameters (variables). Further details can be read from reference [10]. All the statistical analyses, including ANOVA and numerical optimization, were performed using commercial statistical software called Design-Expert®. The detail is presented in upcoming sub-sections.

Analysis of Variance:

Tables 6, 7 present ANOVA performed on the data related to the response variables in hole making by AWJM and LBM. The effects of all the individual input variables have been shown. The effects of all the possible interactions among the input variables were analyzed and only the significant interactions have been shown in the plots. This is to be mentioned, with respect to ANOVA table, that effect of any parameter is considered to be significant if $p\text{-}value \leq 0.05$. F and P values only were included in ANOVA tables. F-value is the ratio between mean square of the input parameter to the mean square of error while, P-value is the probability of a test statistics. The bold numbers of p- values represent the significant parameters and insignificant if otherwise.

Analysis of hole making by AWJM and LBM processes:

Tables 7&8 presents ANOVA performed on data related to response variables for the AWJM & LBM.

Table 6. *ANOVA details for Ra, O.O.R, DU-DL and T.S. and identification of significant input parameters in AWJM process.*

Source	O.O.R		DU-DL		T.S.	
	F-value	P- value	F-value	P- value	F-value	P- value
Model	2.03	0.0863	4.12	0.0038	2.75	0.0263
A- (D)	0.13	0.7208	0.51	0.4871	1.02	0.3282
B- (t)	11.5	**0.0037**	11.07	**0.0043**	17.73	**0.0007**
C- (Vc)	0.45	0.5129	3.07	0.0990	2.24	0.1543
D- (P)	2.48	0.1345	12.41	**0.0028**	3.59	0.0763
E- (Sod)	4.71	**0.0455**	3.89	0.0663	1.33	0.2657
A×B	1.21	0.2875	2.30	0.1492	0.24	0.6295
A×C	2.32	0.1474	0.010	0.9210	0.079	0.7825
A×D	0.052	0.8217	1.40	0.2543	1.29	0.2720
A×E	0.12	0.7340	9.88	**0.0063**	0.65	0.4320
B×C	3.34	0.0863	7.59	**0.0141**	1.61	0.2231
B×D	0.061	0.8080	5.21	**0.0365**	5.69	**0.0297**
B×E	0.81	0.3801	0.21	0.6496	1.63	0.2197
C×D	1.21	0.2875	2.71	0.1195	0.073	0.7906
C×E	0.71	0.4134	0.77	0.3944	3.24	0.0907
D×E	1.29	0.2726	0.86	0.3675	0.88	0.3610

Table 7. *ANOVA details for Ra, O.O.R, DU-DL and T.S. and identification of significant input parameters in LBM process.*

Source	O.O.R		DU-DL		T.S.	
	F-value	P- value	F-value	P- value	F-value	P- value
Model	1.76	0.1371	1.22	0.3498	5.14	0.0012
A- (D)	0.46	0.5088	0.028	0.8689	0.46	0.5081
B- (t)	6.18	**0.0244**	1.72	0.2079	0.052	0.8228
C- (Vc)	1.45	0.2464	3.56	0.0776	0.22	0.6462
D- (LBP)	4.76	**0.0445**	0.40	0.5346	63.54	**0.0001**
E- (Sod)	1.82	0.1962	0.29	0.5991	1.00	0.3333
A×B	9.81	0.0064	2.45	0.1373	0.020	0.8904
A×C	0.19	0.6707	0.52	0.4829	2.02	0.1742
A×D	0.22	0.6457	0.22	0.6490	0.81	0.3814
A×E	5.666×10^{-4}	0.9409	1.72	0.2081	1.32	0.2680
B×C	0.82	0.3777	0.91	0.3551	3.859×10^{-4}	0.9951
B×D	0.028	0.8699	1.42	0.2500	1.85	0.1927
B×E	0.12	0.7345	2.04	0.1720	0.11	0.7428
C×D	0.25	0.6215	1.10	0.3105	0.24	0.6314
C×E	0.060	0.8093	1.79	0.1992	4.62	**0.0472**
D×E	0.19	0.6654	0.091	0.7667	0.80	0.3855

The columns F-value and p-value in table 6, which is show the identification of significant input parameters in AWJM process, suggest that effect of material thickness to be cut and stand of distance are significant upon out of roundness. Whereas the significant factors upon the difference in hole diameter are the thickness of the material, pressure of water jet, interaction between hole diameter and stand of distance, interaction between material thickness and cutting feed and finally the interaction between material thickness and jet pressure. The analysis shows also that the significant factors upon the tensile strength are the material thickness to be cut and the interaction between material thickness and jet pressure. While the columns F-value and p-value in table 7, which is show the identification of significant input parameters in LBM process, suggest that material thickness and laser power are significant factors upon out of roundness. The analysis shows also that the significant factors upon the tensile strength are laser beam power and the interaction between laser beam power and the stand of distance (Sod). The analysis shows that there are no significant factors upon the difference between upper and lower diameter.

Figs. 4, 5 and 6 shows, in graphical form, the effects of influential parameters upon out of roundness, difference between upper and lower diameter and tensile strength respectively in AWJM.

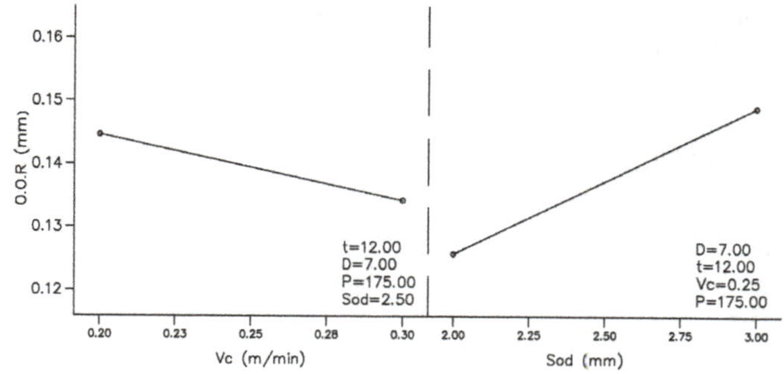

(a) (b)

***Fig. 4** Factorial plots showing the effects of (a) cutting feed (b) standoff distance upon out of roundness in AWJM*

It is clear from graph (a) that as the cutting feed of abrasive water jet increased, the quality characteristic (reducing out of roundness) of the cutting surface will improve. This phenomenon is depending on kinetic energy absorption by work piece due to hydrodynamic friction of abrasive water jet [14]. While graph (b) shows that by increasing the standoff distance the material surface is exposed to the downstream of the jet. At downstream, the jet starts to diverge losing its coherence thereby reducing the effective cutting area that directly affects the kerfs taper angle [13].

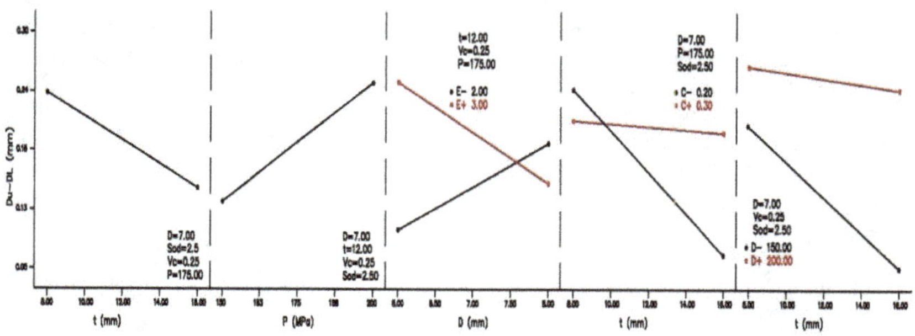

(a) (b) (c) (d) (e)

***Fig. 5** Factorial plots showing effects of (a) material thickness (b) water jet pressure (c) interaction between nominal hole diameter and standoff distance (d) interaction between material thickness and cutting feed (e) interaction between material thickness and water jet pressure upon difference between upper and lower diameter in AWJM.*

It is clear from graphs (a, b, c, d & e) that the difference between upper and lower diameter is increased as material thickness to be cut decrease and water jet pressure increase this is because, the taper geometry directly depends on the shape of the jet, which is not similar to the shape of a fixed geometry tool. In fact, due to hydrodynamic characteristics of the jet, it is geometry significantly influenced by pressure, cutting feed, standoff distance. Through cutting factors, created tool (water jet) hits the work piece at the upper erosion base, where erosion process begins[14]. When the water jet pressure is increased, the jet kinetic energy increase that leads to a high momentum transfer of the abrasive particles, generating a wider-bottom kerf. leading to a decrease in kerf taper angle[13].

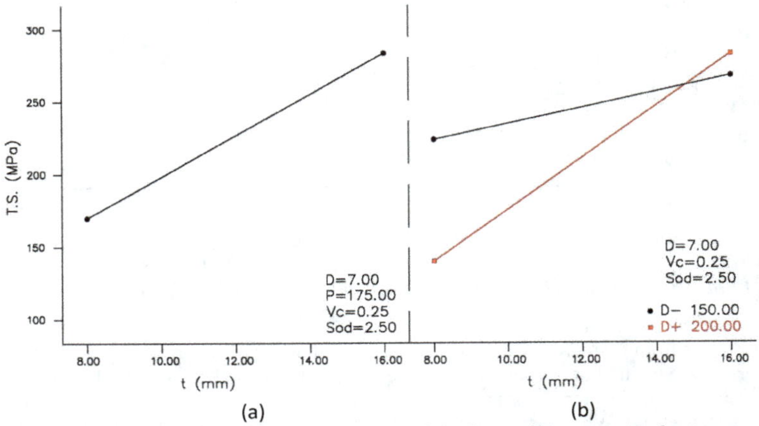

Fig. 6 *Factorial plots showing effects of (a) material thickness (b) interaction between material thickness and water jet pressure on tensile strength in AWJM.*

It is clear from graph (a) that the thickness of material to be cut and the interaction between material thickness and water jet pressure are affected factors on the strength of the composite. material. Reduction in the strength of the composite is decreased as the thickness of the composite increase. This is related to the formula of calculating the strength of hole specimen which is define as:

Ultimate strength of hole specimen = max.force carried by the test specimen before failure/ gross cross-sectional area (mm^2).

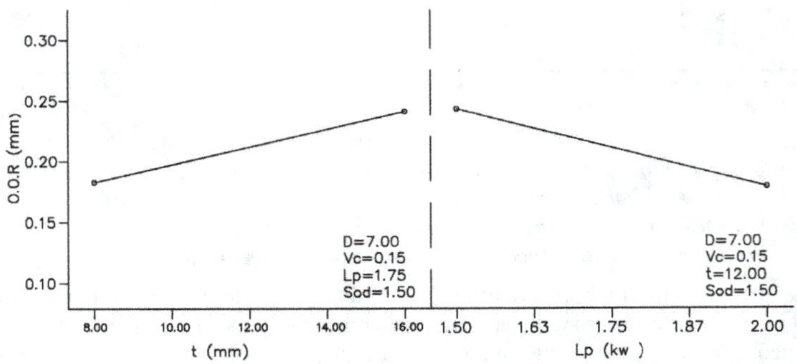

(a) (b)

Fig. 7 *Factorial plots showing effects of (a) material thickness (b) laser power upon out of roundness in LBM.*

It is clear from graph (a), that an increase in material thickness for the same laser power, cutting feed and standoff distance results in higher out of roundness (cut path deviation) at the cut region around the hole. This is due to higher input energy required for a larger volume of material removal. While graph (b) shows, that an increase in the laser power with constant cutting feed and a given thickness of material results in lower out of roundness. This is due to the reduction in the cutting duration and the entrance angle with respect to the surface becomes higher. [11].

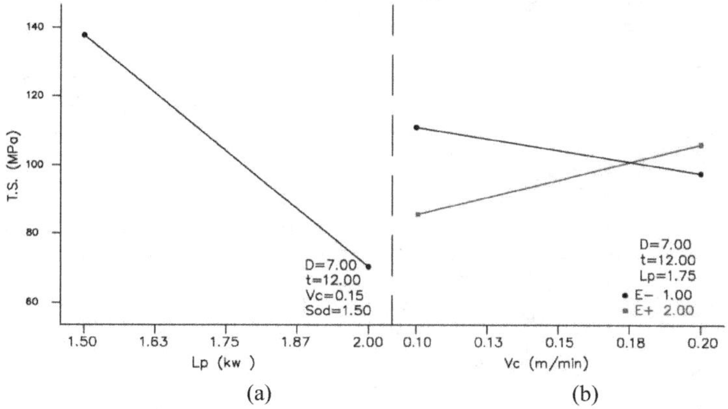

***Fig. 8** Factorial plots showing effects of (a) laser power (b) interaction between cutting feed and standoff distance on tensile strength in LBM.*

It is shown from graph (a) that, the strength of the composite decreases with the increase of laser power. This is because with increasing the laser power, the heat-affected zone (HAZ) is increased and a large volume of fibers in the composite is vaporized, this causes reduction in the strength of the composite. While graph (b) shows less reduction in the strength with increasing the cutting feed. This is because with increasing the cutting feed, the heat-affected zone (HAZ) is decreased [12].

Numerical Optimization

The AWJM and LBM processes have been widely used in industry. The two technologies have procured many overlapping applications and it is thus important for the industry to understand both processes, in order to select the optimum method in different situations. The comprehensive knowledge on dimensional accuracy and strength of hole making in GFRP, would help the users to judge which method is more appropriate for each type of application. The target of numerical optimization in the comparison related to the present study could be any of the following three objectives:

1. Minimize the difference between upper & lower diameter.
2. Minimize out of roundness.
3. Maximize tensile strength (i.e. reducing the reduction in tensile strength)

***Table 9**. Recommendations and predictions of multi-objective optimization against each set of objectives and comparison with experimental results in AWJM process.*

	Fixed Parameters		Optimized Parameters				
Objectives	t (mm)	D (mm)	Vc (m/min)	P (MPa)	Sod (mm)	Predicted values	Experimental values
Minimize (O.O.R)						0.113mm	0.110mm
Minimize(D_U-D_L)	12	7	0.3	150	2	0.147mm	0.151mm
Maximize (T.S)						275.539MPa	271.614 MPa

***Table 10**. Recommendations and predictions of multi-objective optimization against each set of objectives and comparison with experimental results in LBM process*

	Fixed Parameters		Optimized Parameters				
Objectives	t (mm)	D (mm)	Vc (m/min)	LP (Kw)	Sod (mm)	Predicted values	Experimental values
Minimize (O.O.R)			0.1	2	1	0.146mm	0.150mm
Minimize(D_U-D_L)	12	7	0.2	1.5	2	0.019mm	0.014mm
Maximize (T.S)			0.1	1.5	1	165.376MPa	168.265 MPa

Tables 9 & 10 presents optimized values (within tested range) of the predictor variables for different objectives in the two cutting technologies. Last column of the table shows the actual results of confirmation experiments performed against each optimized values.

Table 9 shows that minimum out of roundness, minimum difference between upper and lower diameter and maximum tensile strength in AWJM can be achieved by cutting at high settings of cutting feed, low settings of jet pressure and low settings of stand of distance. Table 10 shows that minimum value of out of roundness can be achieved by cutting at low settings of cutting feed, low settings of stand of distance and high settings of laser power. Minimum value of the difference between upper and lower diameter of the cutting hole can be achieved by high setting of cutting feed, high stand of distance and low setting of laser power. Finally, reducing the reduction in strength will be achieved if low setting of cutting feed, laser power and stand of distance is applied.

Conclusions

This work is intended to provide initial technical information relating to the dimensional accuracy and strength of hole making in GFRP by AWJM and LBM. The work presented comprehensive statistical analysis of effects of major AWJM & LBM cutting parameters on out of roundness, difference between upper & lower diameter of the cutting hole, difference between. Thirty-two tests following full factorial design of experiments were performed on the laminated GFRP. The following conclusions can be drawn with regard to the AWJM & LBM of GFRP:

1. In AWJM process, improving dimensional accuracy (reducing out of roundness, difference between upper and lower diameter of the cutting hole) can be done by increasing the cutting feed and reducing the jet pressure and stand of distance.
2. In AWJM process, reducing the reduction in the strength of the cutting material can be achieved by reducing the increasing the thickness of material to be cut and reducing the jet pressure.
3. In LBM process, improving dimensional accuracy (reducing out of roundness) can be done by reducing cutting feed, stand of distance and increasing laser power whereas reducing the difference in the upper & lower diameter of the cutting hole can be done by increasing cutting feed, stand of distance and decreasing laser power.
4. In LBM process, reducing the reduction in the strength of the cutting material can be achieved by reducing the laser power, cutting feed and standoff distance.

Acknowledgements
The authors would like to acknowledge the help and support provided by the staff of College of Mechanical and Electrical Engineering / Nanjing University of Aeronautic and Astronautic in performing experimental work there.

References

[1] E. Lemma, L. Chen and E. Siores, Study of Cutting Fiber-Reinforced Composites by Using Abrasive Water-Jet with Cutting Head Oscillation. Composite Structures. 2002, 57:1- 4, pp. 297-303. https://doi.org/10.1016/S0263-8223(02)00097-1

[2] W. Hufenbach, D. ski LA, M.Gude, J.Konieczny, A.Czulak. Optimization of rivet joints of CFRP composite material and aluminum alloy. J Achieve MaterManuf Eng 2007, 20:119-22.

[3] R.Teti . Machining of composite materials. CIRP Annals - Manu Techno 2002, 51:611-34 https://doi.org/10.1016/S0007-8506(07)61703-X

[4] A. Abrao , P. Faria, R.Campos, P. Reis , J. Davim. Drilling of fiber reinforced plastics: a review. J Mater Process Techno 2007, 186:1-7. https://doi.org/10.1016/j.jmatprotec.2006.11.146

[5] H. Eltawahni an, M.Hagino b, K.Y.Benyounis d, T.Inoue b, A.G.Olabic. Effect of CO2 laser cutting process parameters on edge quality and operating cost of AISI316L. Optics & Laser Technology 2012, 44 :1068-1082 https://doi.org/10.1016/j.optlastec.2011.10.008

[6] A. Kumar, Y. Vinod. Laser beam machining-A review International Journal of Machine Tools & Manufacture. 2008, 48: 609-62 https://doi.org/10.1016/j.ijmachtools.2007.10.017

[7] H. Hocheng, A failure analysis of water jet drilling in composite laminates, International Journal of Machine Tools and Manufacture, Vol. 30, No. 3, 1990, pp.423-429. https://doi.org/10.1016/0890-6955(90)90186-M

[8] M. Ramulu, M, P. Young, H. Kao, Drilling of Graphite/Bismaleimide Composite Material, Journal of Materials Engineering and Performance, Vol.8, No.3, 1999, pp.330-338. https://doi.org/10.1361/105994999770346882

[9] H. Hocheng, C.C. Tsao, The path towards delamination-free drilling of composite materials, Journal of Materials Processing Technology, 2005,167, pp.251-264. https://doi.org/10.1016/j.jmatprotec.2005.06.039

[10] G. Derringer, R. Suich, Simultaneous optimization of several response variables, Journal of Quality Technology1980, 12: 214-219 https://doi.org/10.1080/00224065.1980.11980968

[11] I.A.Choudhury, S.Shirley, Laser cutting of polymeric materials: An experimental investigation, Journal of Optics & Laser Technology, Optics & Laser Technology 42 (2010) 503-508 https://doi.org/10.1016/j.optlastec.2009.09.006

[12] A.Cenna, P.Mathew, E Evaluation of Cut Quality of Fiber reinforced plastic, A review, Int.J.Mach.Tools Manufact.Vol.37, No.6, pp.723-736, 1997 https://doi.org/10.1016/S0890-6955(96)00085-5

[13] D. Shanmugam, S.H. Masood. An investigation on kerf characteristics in abrasive waterjet cutting of layered composites, journal of materials processing technology 2 0 9 (200 9) 3887-3893 https://doi.org/10.1016/j.jmatprotec.2008.09.001

[14] S. Hloch, Stanislav FABIAN, Miroslav RIMÁR, Design of Experiments Applied on abrasive water factors sensitivity identification. Nonconventional Technologies Review - no. 2/2008

Advanced Topics in Mechanics of Materials, Structures and Construction - AToMech1-2023 Materials Research Forum LLC
Materials Research Proceedings 31 (2023) 227-239 https://doi.org/10.21741/9781644902592-24

Influence of canard geometrical configuration on the radar cross section

Gauri Patki[1], Attada Phanendra Kumar[2], Dineshkumar Harursampath[2]

[1]Department of Aerospace Engineering, IIT Kharagpur, Kharagpur-721302, India

[2]Department of Aerospace Engineering, IISc Bengaluru, Karnataka, India

Keywords: Radar Cross Section, Canard Configuration, Physical Optics, Monostatic RCS, Far Field Approximation

Abstract. The Radar Cross Section (RCS) of fighter jet hold a lot of importance in determining its air superiority. Electromagnetic simulations can be performed by employing numerical methods such as MoM, ray tracing methods or Physical Optics. Configuration of the wing is an important factor in determining the overall RCS of the aircraft. The current study explores the RCS of canard as placed in Chengdu J-20. The parameters are varied, the models are created and the simulations are conducted for monostatic RCS using far field approximation over a continuous frequency range using Physical Optics in FEKO simulator. The trends in results are analyzed and promising configurations are identified.

Introduction

Stealth technology, or Low Observable technology as it is alternatively called, is a combination of military tactics and active and passive countermeasures that aim at reducing a fighter aircraft's visibility on radar, infrared, sonar or any other methods of detection [1]. Complete invisibility to the radar through stealth technology is an ideal situation. But the development and progress in the field has rendered 5th and higher generation aircrafts much less visible to the radar as compared to the older generations. Stealth technology mainly involves reduction in electromagnetic reflections and infrared footprints by reducing thermal emission from thrusts, reduction in aircraft's radar cross section and visual camouflage by blending the aircraft with the background sky. Development of specific shapes such as large surfaces, avoiding sharp curves and right angles helps in improving the stealth abilities. A combination series of curved surfaces having complex radius or several small, carefully oriented flat planes may combine to generate an exterior shape. Vertical surfaces are avoided as they significantly enhance the radar cross section. Radiation absorbent material (RAM) absorbs the EM radiation sent towards the target by the source radar or sonar. Metals and conductors have a higher reflectivity than non-conductors, so the latter is generally used as an RAM. Some examples are iron ball paint absorber in which microscopic spheres coated in carbonyl iron are suspended in the paint. They convert radiation into heat and dissipate it. Carbon nanotube is an excellent RAM but coating the entire surface with it is not feasible. The use of RAM in the early aircrafts made them heavy and not strong even for structural use at high altitudes. Hence, the focus remains mainly on reducing the RCS of the aircraft.

Research comprising of simulation and analysis of different generation of aircrafts such as Valkyrie and F-16 for RCS calculation have been conducted. It was seen in these studies that the Valkyrie aircraft had a large RCS despite being ahead of performance in time. On the other hand, the F-16 aircraft had a very low RCS and has survived a long time being harder to spot making it a prolific fighter. Further in the study, tailless configuration was considered as a candidate for stealth analysis. This configuration gave almost zero RCS except for a few angles in roll, pitch and yaw planes [2].

While there has been a certain progress in designing the shape of the aircraft, the methods for RCS calculation have also been improved. In order to devise better methods for RCS estimation, calculation abilities based of numerical schemes such as MoM, MLFMM and HOBF have also been explored and researched [3]. Studies comparing models for dynamic RCS calculation have shown that Gaussian Mixture Density Model (GMDM) gives better approximation compared to Chi-Square and Log-normal distribution [4]. The sparse-matrix method (SMM) algorithm generates matrices that take shorter running time than those generated by Method of Moments (MOM) [5]. RCS modelling validation has also been done by experimenting on a full- scale aircraft, the data being obtained from the experimentation, processed, analyzed and compared with the results from simulations for validation [6].

Different aircraft configurations arose in the past decades as a result of efforts towards obtaining least radar visibility. While most fighter jets adopt the conventional layout, the Chinese 5th generation fighter jet, Chengdu J-20 deviates with its duck layout, i.e., the presence of canards on the frontal part of the aircraft. It has been observed that addition of canards gives the aircraft an aerodynamic advantage [7]. However, it is also equivalent to adding a scattering component in the front as the movable gap between the fuselage and the canard is directly exposed in the front. The deflection of canards thus increases the RCS of the nose of the aircraft significantly.

The previous studies conducted in the field of RCS explore the measurement and calculation of RCS via different method and application of various algorithm. There have been comparisons between conventional and duck layout to understand how differing from conventionality affects the stealth and if the effect is drastic or not [7]. This work attempts to understand how change in certain parameter at a time affect the RCS of the aircraft considering every other aspect to remain the same. Forming the duck layout, a canard poses as a means to bring significant change in the RCS of the aircraft on the whole. Hence, it is required to find geometrical parameters such that this effect is not drastic.

Objective

The current study explores the RCS of canard having original parameters as that of J-20 and then proceeds to calculate the RCS of canards modelled by varying parameters. The CAD models are created in SOLIDWORKS and simulated using Altair FEKO simulation software. The graphs are then obtained. The focus of the study is upon the structural aspects of the canard as the only target under observation without considering the attachment to the fuselage. How the RCS is affected by variation in the geometrical parameters of the canard is observed.

Radar Cross Section (RCS) For Different Cross-Section Profiles

Radar cross section is basically the cross section of a perfectly reflecting sphere that will reflect the same strength of signal as the aircraft in question. RCS describes the amount of scattered power from the target towards the Radar. It is the measure of the ratio of backscatter power per steradian (unit solid angle) in the direction of the radar to power density that is intercepted by the target [8]. Higher the RCS, more detectable is the aircraft.

The basic equation for RCS calculation is given as:

$$\sigma = \lim_{R \to \infty} 4\pi R^2 \frac{E_S^2}{E_0^2}$$

where,

σ = radar cross section

R = range of the target

E_S = amount of reflected power from the target

E_0 = power density of wave incident on the target

RCS is measured in dBsm (decibel square meter). The unit conversion from dBsm to m2 is given as:

$$\text{dBsm} = 10 \times \log_{10}\left(\frac{RCS}{1\ m^2}\right)$$

Materials Research Proceedings 31 (2023) 227-239 https://doi.org/10.21741/9781644902592-24

Mathematical formulations for the RCS of some basic shapes such as sphere, cylinder, ellipsoid and triangular plate have been reported [4].

1. **Sphere**

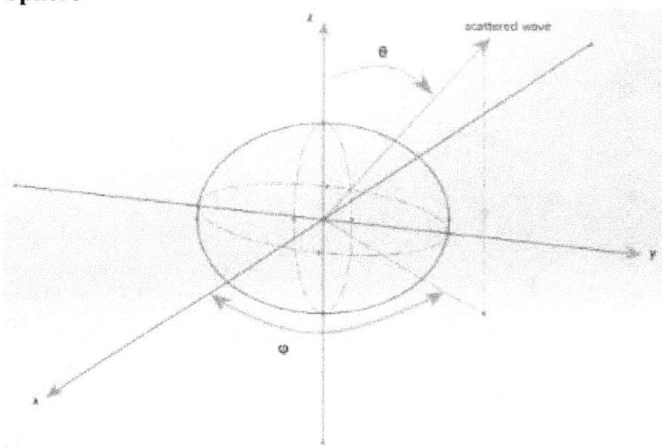

$$\frac{\sigma}{\pi r^2} = \frac{1}{kr}\sum_{n=1}^{\infty}(-1)^n\,(2n+1)\left[\left(\frac{kr\, j_n(kr) - n\, j_n(kr)}{krH_n - 1(kr) - nH_n^1(kr)}\right) - \left(\frac{j_n(kr)}{H_n^1(kr)}\right)\right]$$

Where,
$k = 2\pi/\lambda$, λ is the wavelength
j_n= spherical Bessel of the first kind of order n
H_n^1= Hankel function of order n and is given by
$H_n^1(kr) = j_n(kr) + jy_n(kr)$
y_n= Spherical Bessel function of the second kind of order n

2. **Cylinder**

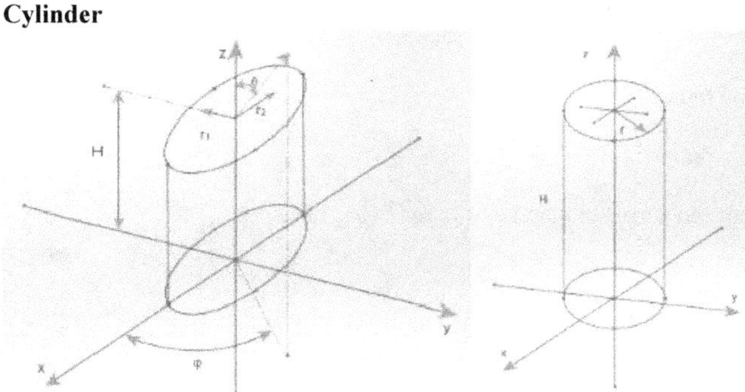

$$\sigma_{\theta n} = \frac{2\pi H^2 r_1^2 r_2^2}{\lambda(r_1^2(cos\varphi)^2 + r_2^2(sin\varphi)^2)^{1.5}}$$

$$\sigma = \frac{\lambda r_1^2 r_2^2}{8\pi(cos\theta)^2(r_1^2(cos\varphi)^2 + r_2^2(sin\varphi)^2)}$$

gives RCS for an incident wave other than normal.

Materials Research Proceedings 31 (2023) 227-239 https://doi.org/10.21741/9781644902592-24

The equations can be further reduced to obtain mathematical formulation for circular cylinder as

$$\sigma_{\theta n} = \frac{2\pi L^2 r}{\lambda}$$

$$\sigma = \frac{\lambda r sin\theta}{8\pi (cos\theta)^2}$$

where,
H = height of the cylinder
R = radius of the cylinder

3. Ellipsoid

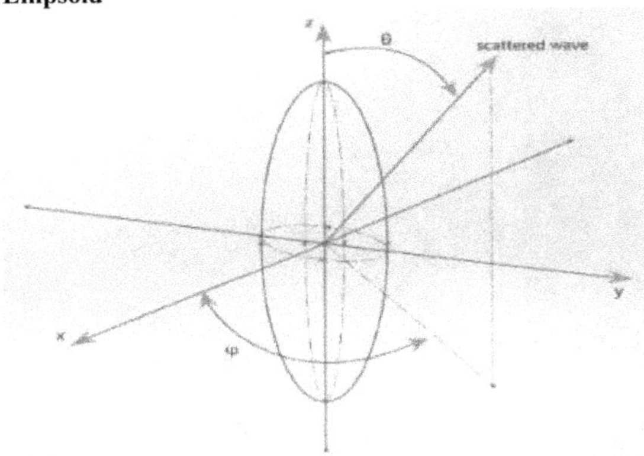

$$\sigma = \frac{\pi a^2 b^2 c^2}{(a^2 (sin\theta)^2 (cos\varphi)^2 + b^2 (sin\theta)^2 (sin\varphi)^2 + c^2 (cos\theta)^2)^2}$$

$\sigma = \frac{\pi b^4 c^2}{(a^2 (sin\theta)^2 + c^2 (cos\theta)^2)^2}$, when a = b as ellipsoid becomes roll symmetric and RCS is independent of φ.

When a = b = c, σ = πc2 which is backscattered RCS of sphere.
where,
a = ellipsoid a radius
b = ellipsoid b radius
c = ellipsoid c radius
φ = roll angle
θ = angle between z axis and direction to receiving radar

Materials Research Proceedings 31 (2023) 227-239 https://doi.org/10.21741/9781644902592-24

4. **Triangular plate**

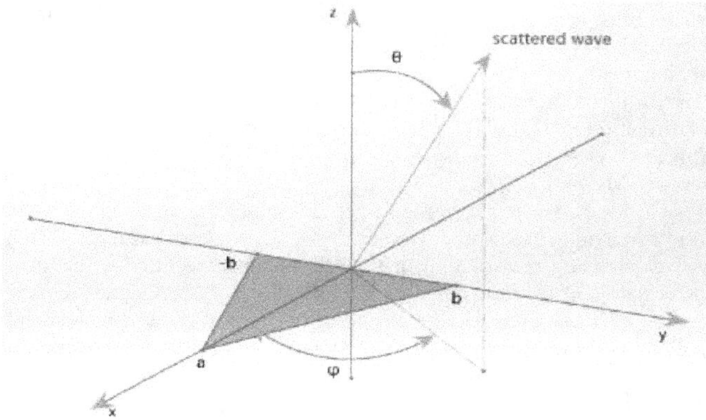

$$\sigma = \frac{4\pi\lambda^2}{\lambda^2} cos\theta\sigma_0$$

$$\sigma_0 = \frac{\sigma_{01}+[(sin\alpha)^2-(sin^{\beta}/_2)^2]^2}{(\alpha^2-(^{\beta}/_2)^2)^2}$$

$$\sigma_{01} = 0.25(sin\varphi)^2(\frac{2a}{b}cos\varphi sin\beta - sin\varphi sin2\alpha)^2$$

α = k a sinθ cosφ, β = k a sinθ sinφ

for φ = 0,

$$\sigma = \frac{4\pi A^2}{\lambda^2}(cos\theta)^2[\frac{(sin\alpha)^4}{\alpha^4} + \frac{(sin2\alpha-2\alpha)^4}{4\alpha^4}]$$

$$A = \frac{ab}{2}$$

For $\varphi = \frac{\pi}{2}$,

$$\sigma = \frac{4\pi A^2}{\lambda^2}(cos\theta)^2[\frac{(sin\beta/2)^4}{(\beta/2)^4}]$$

Methods of Computation and RCS Calculation

Computer simulations use numerical methods for the RCS prediction for arbitrary dimensional target such as modelled aircrafts. Some of the popular methods are:

1. **Method of Moments (MoM)**

   The most common yet rigorous, full-wave numerical technique to solve open boundary electromagnetic problems is Method of Moments. It solves the integral form of Maxwell's equations to predict the RCS, so, it must satisfy the said equations and relevant boundary conditions. MoM takes the currents and the fields on surface of the structure as unknowns. The structure is immersed in free space, the background medium of which is modelled using free-space Green's function [5].

   The technique is highly accurate but produces large matrices by reducing the operator equations into a system of linear equations.

2. **Ray Tracing Methods**

   Ray tracing methods consist of a collection of techniques that can be used individually or in conjunction with each other to analyse electrically large and arbitrarily shaped targets. Two most commonly used methods are Geometrical Optics (GO) and Geometrical Theory

of Diffraction (GTD). Geometrical Optics assumes that the photons or rays are reflected at the complement of the angle between the incident ray and the surface normal [6]. Thus the monostatic RCS arises only from the rays normal to the specular components such as edges, vertices and dihedrals.
Other ray tracing methods include Uniform Theory of Diffraction (UTD) and Physical Theory of Diffraction (PTD). These too analyse electrically large targets taking into account diffracted rays.

3. **Finite Difference Method**
Finite Difference Methods discretize the target and apply the approximated differential operators in either time or frequency domain to the said discretized target [7]. The time domain involves stepping the solution in time throughout the grid on the surface. The Fourier transformation of the time domain gives the data in the frequency domain.
Finite difference methods provide rigorous solution. However, gird refinement for the required level of accuracy makes it computationally expensive as the execution time increases.
4. **Physical Optics (PO)**
Physical Optics is a method of estimation of electromagnetic scattering using induced currents assuming that the surface is perfectly conducting. An incident source illuminates the scatterer surface and induced electric currents from tangential magnetic fields are considered. These currents reradiate to produce the scattered field [13]. Physical Optics approximation assumes that the radiation from the illuminated part is directly proportional to the incident magnetic field intensity while that from the rest is zero [14].
The simulations here solve the model using Physical Optics- Always illuminated solver in Altair FEKO software. This assumes that all the triangles on which PO approximation is made are illuminated [15].

Results and Discussion

The canard models were first created using SOLIDWORKS and then imported in CADFEKO. The incident plane wave was set to loop over multiple directions with $\theta = 90^\circ$ and φ between 0° and 180° with an increment of 30°. The frequency was set to be a continuous range between 1.7 GHz and 5.6 GHz [7]. Fields were calculated in the direction of plane wave incidence. Monostatic RCS is calculated using far-field approximation.

The parameters considered while modelling the canard were wing span, wing chord at root, wing chord at tip, airfoil shape and angle of incidence. Additionally, canard lift coefficient and angle of attack affect the utility of canards on an aircraft. We performed simulations over 24 models varying wing span and wing chords at the root and tip.

The tip of the root chord was placed at the origin with the root airfoil in XZ plane. The span is along the Y- axis, positive z direction id downwards positive x axis from leading edge to the trailing edge of the root airfoil.

Different geometrical specifications of canard

1. Wing chord at root and tip of the canard are approximated by scaling the dimensions of the main wing from the available information.
2. Angle of incidence of the canard in J-20 can be varied. It is taken to be 0 in all cases here.
3. Information on the type of wing airfoil of J-20 is unknown/ classified. NACA airfoils of required chord lengths, 100% thickness, 5° pitch are used to create the SOLIDWORKS models.
4. The aspect ratio (AR) varies with the models as the parameters keep changing.

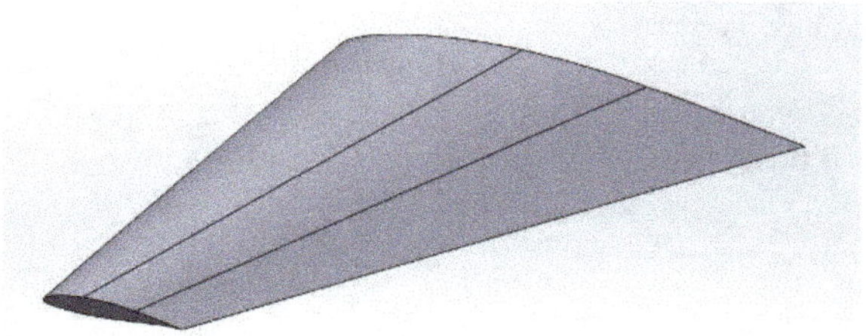

Fig 1. Canard model in SOLIDWORKS Original Canard Parameters from J20:

Wing span = 7.62m (Single canard measures 3.81m)
Wing chord (root) = 2.286 m
Wing chord (tip) = 0.6858 m
Wing aspect ratio = 2.3
Wing airfoil – unknown
Angle of Incidence – variable
(The dimensions specified are approximately scaled using the available information.)

First order GO approximations were used to obtain theoretical values of the RCS. At $\varphi = 0^{\circ}$ for 1.7 GHz frequency, considering the effective edge length to be 4.1324 m, the RCS comes out to be

$\sigma = \frac{L_{eff}^2}{\pi}$ [11]

$\sigma = 7.35255$ dBsm

The result from simulation for the same specified condition is $\sigma = 7.95864$ dBsm. While the result from the theoretical calculation was not accurate, it was still close enough. The simulation used Physical Optics coupled with other numerical techniques to increase the accuracy of the results. This might be the reason for discrepancy.

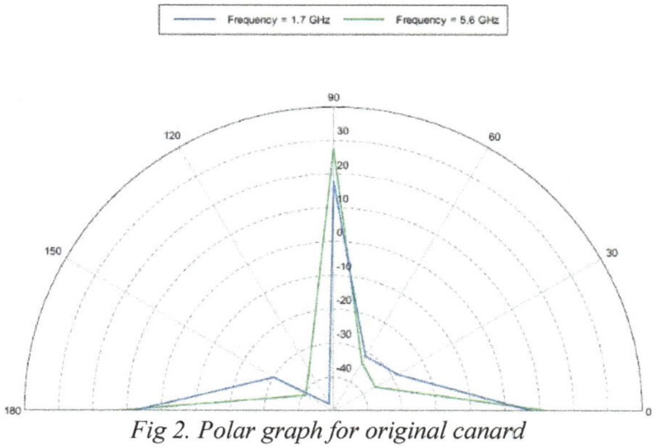

Fig 2. Polar graph for original canard

Variable Chord at Tip

Keeping all other parameters constant, 8 models were created with varying values of chord at the tip. They were simulated and the RCS was compared with that of the original canard.

The parameters used are given below:

Table 1. Parameters of canards with variable chord at tip

	Wing Span (one wing)	**Wing Chord (root)**	**Wing Chord (tip)**	**Aspect Ratio**
C1	3.81	2.286	0.65	**2.59**
C2	3.81	2.286	0.6	**2.64**
C3	3.81	2.286	0.55	**2.68**
C4	3.81	2.286	0.5	**2.73**
C5	3.81	2.286	0.7	**2.55**
C6	3.81	2.286	0.75	**2.50**
C7	3.81	2.286	0.8	**2.46**
C8	**3.81**	**2.286**	**0.85**	**2.42**

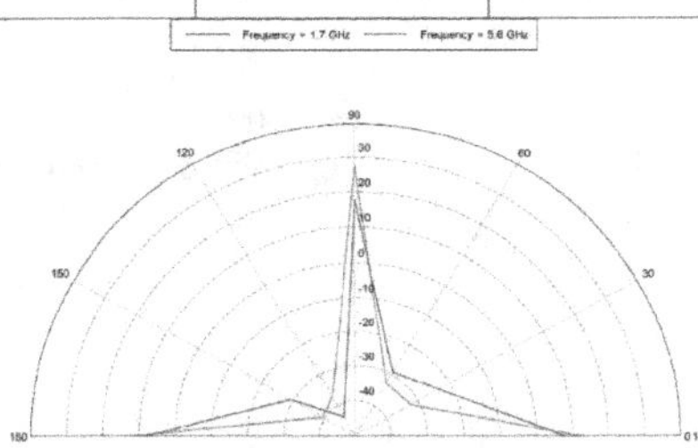

Fig 3. Polar graph for C1

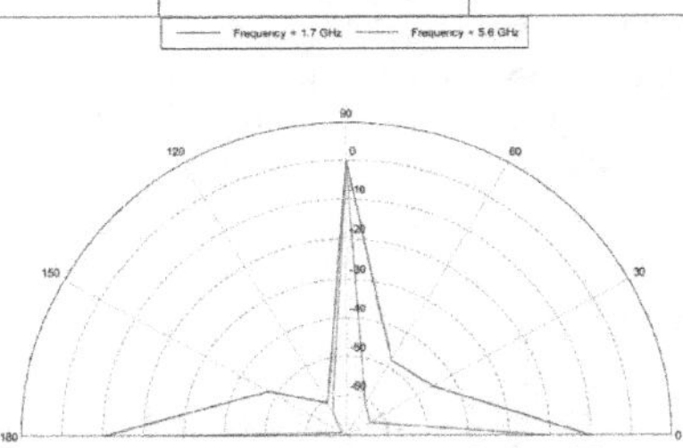

Fig 4. Polar graph for C2

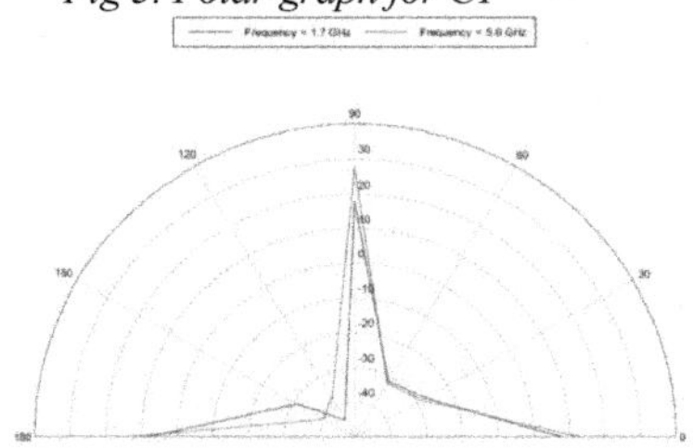

Fig 5. Polar graph for C3

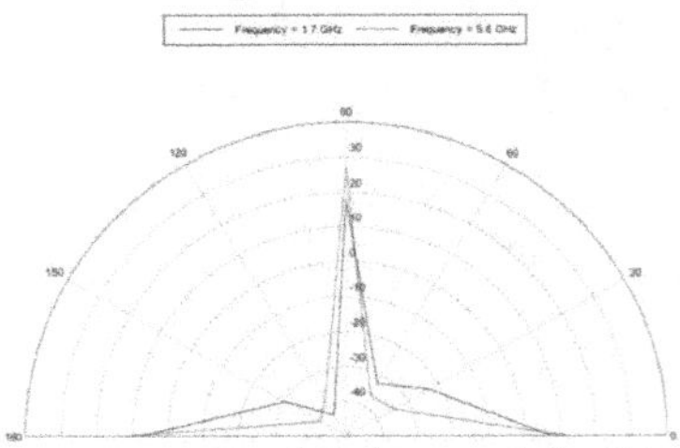

Fig 6. Polar graph for C4

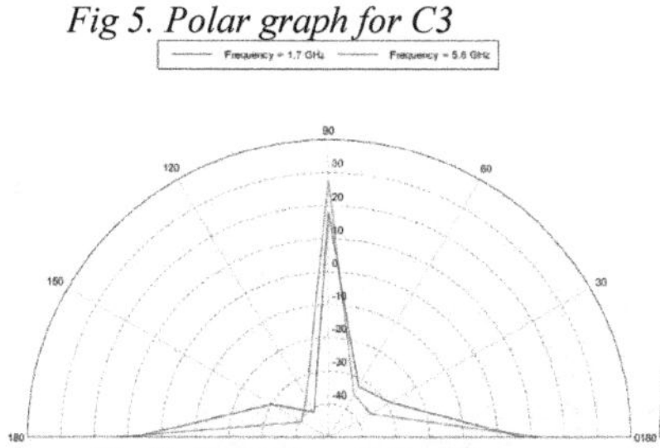

Fig 7. Polar graph for C5

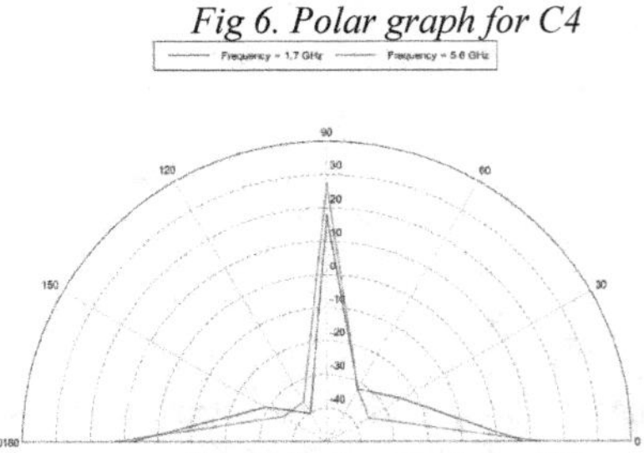

Fig 8. Polar graph for C6

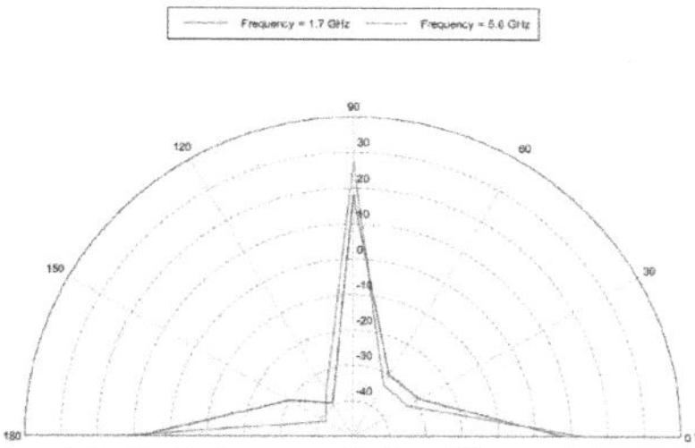
Fig 9. Polar graph for C7

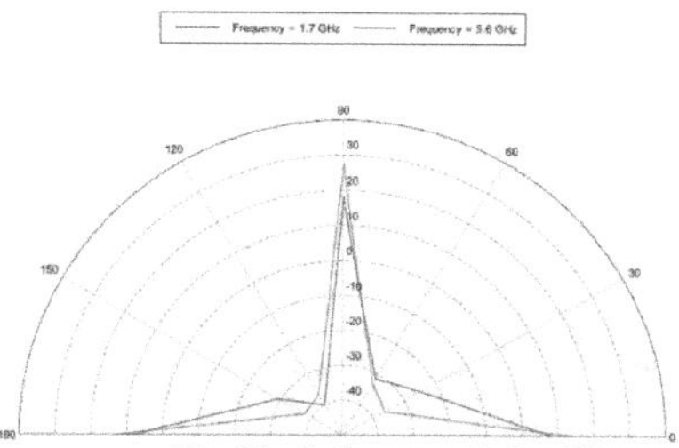
Fig 10. Polar graph for C8

As can be observed from the curves, at φ = 90˚, there is a sharp spike after the RCS being considerably low for φ = 30˚ and 60˚ and then declines sharply for 120˚ before increasing again at 150˚ and spiking at 180˚. This is the general trend for both high and low frequencies.

The notable configuration here is that of C2. While all other configurations result in RCS being between -50 dBsm and 30 dBsm, the RCS of C2 lies between 0 and -50 dBsm. The significance of RCS being below 0 dBsm is that RCS remains less than 1m2 in the entire frequency range. The RCS goes as low as -60 dBsm for φ = 150˚ at 5.6 Hz frequency. This is lower than all others where the minimum value is -40 dBsm.

Variable Chord at Root

Next, all other parameters constant were again kept constant and 8 models were created with varying values of chord length at the root. They were simulated and the RCS was compared with that of the original canard.

The parameters used are given below:

Table 2. Parameters of canards with variable chord at root

	Wing Span (one wing)	**Wing Chord (root)**	**Wing Chord (tip)**	**Aspect Ratio**
C9	3.81	2.0	0.6858	2.83
C10	3.81	1.8	0.6858	3.06
C11	3.81	1.5	0.6858	3.48
C12	3.81	1.2	0.6858	4.04
C13	3.81	2.4	0.6858	2.46
C14	3.81	2.7	0.6858	2.25
C15	3.81	3.0	0.6858	2.06
C16	3.81	3.3	0.6858	1.91

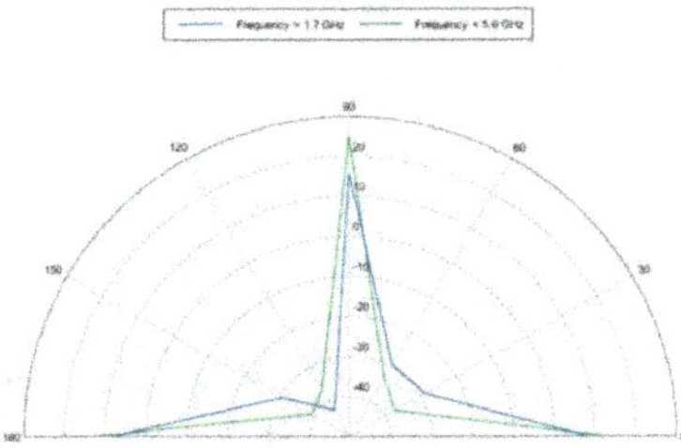

Fig 11. Polar graph for C9

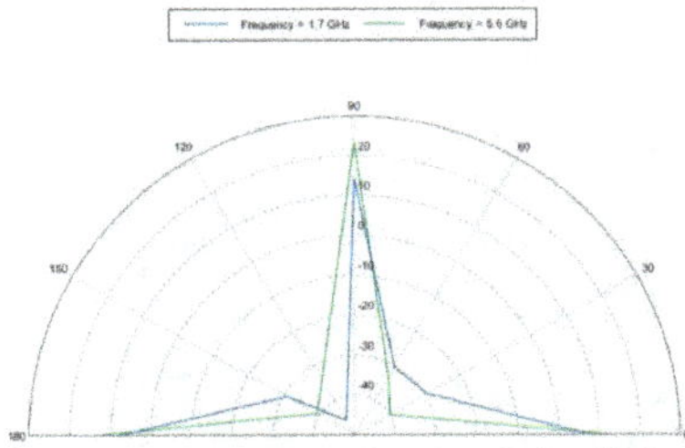

Fig 12. Polar graph for C10

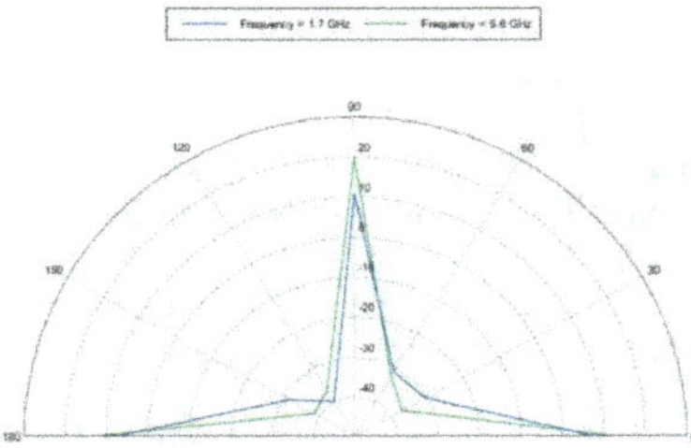

Fig 13. Polar graph for C11

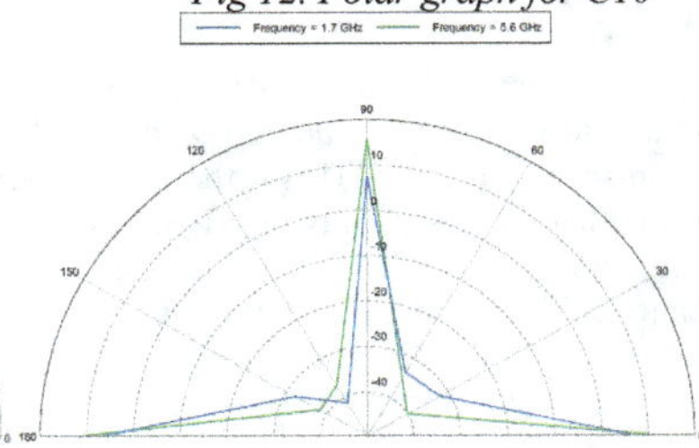

Fig 14. Polar graph for C12

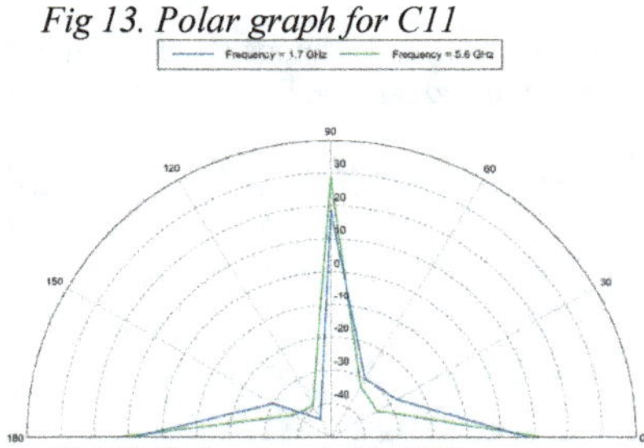

Fig 15. Polar graph for C13

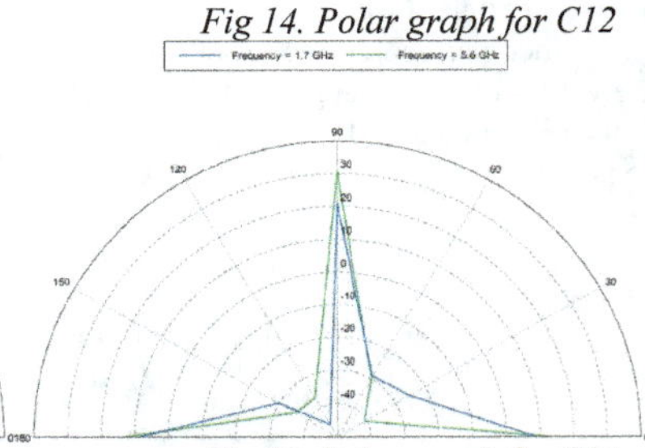

Fig 16. Polar graph for C14

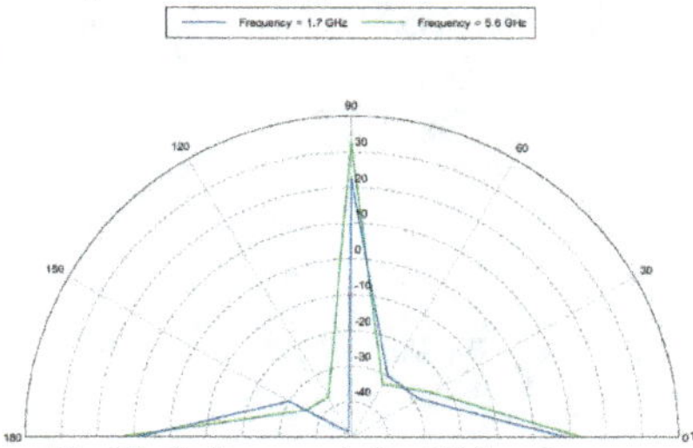

Fig 17. Polar graph for C15

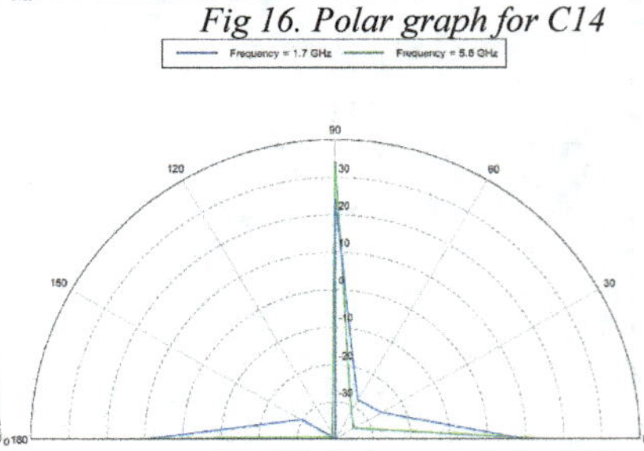

Fig 18. Polar graph for C16

Varying the chord at root sees a general trend between φ = 0˚ and 90˚. For 1.7 GHz frequency, the RCS increases and a spike occurs between 60˚ and 90˚. As the value of root chord increases, a sharper dip in RCS can be seen between φ = 90˚ and 120˚. It increases for 150˚ and spikes again for 180˚.

The high frequency curve follows irregular pattern for varying parameters. C16 configuration stands out for 5.6 GHz frequency RCS where a spike can be seen between 30° and 90° and an equally sharp dip from 90° to 150°, spiking again at 180°. The minimum value is -39.21 dBsm at 5.6 GHz and below -40 dBsm at 1.7 GHz.

Variable Wing Span

For the third case, the span was varied keeping all other parameters constant. The simulations results were compared with that obtained from the original parameters.

The span length used is specified below:

Table 3. Parameters of canard with variable span

	Wing Span (one wing)	Wing Chord (root)	Wing Chord (tip)	Aspect Ratio
C17	3.7	2.286	0.6858	2.49
C18	3.6	2.286	0.6858	2.42
C19	3.5	2.286	0.6858	2.35
C20	3.4	2.286	0.6858	2.28
C21	3.9	2.286	0.6858	2.62
C22	4.0	2.286	0.6858	2.69
C23	4.1	2.286	0.6858	2.75
C24	4.2	2.286	0.6858	2.82

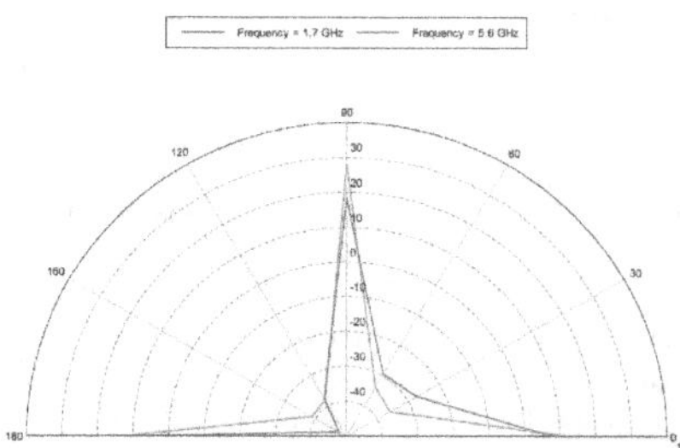

Fig 19. Polar graph for C17

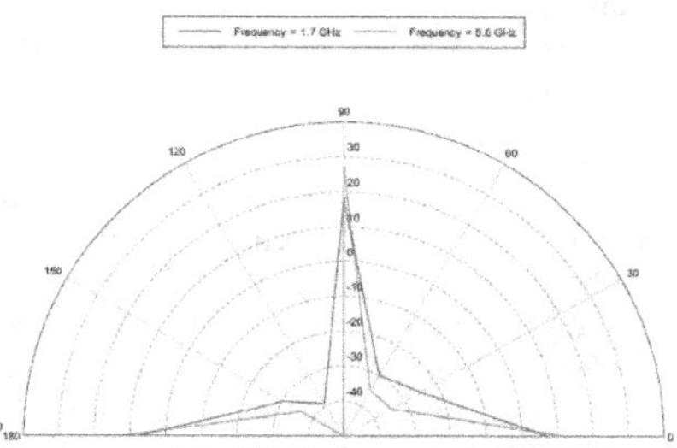

Fig 20. Polar graph for C18

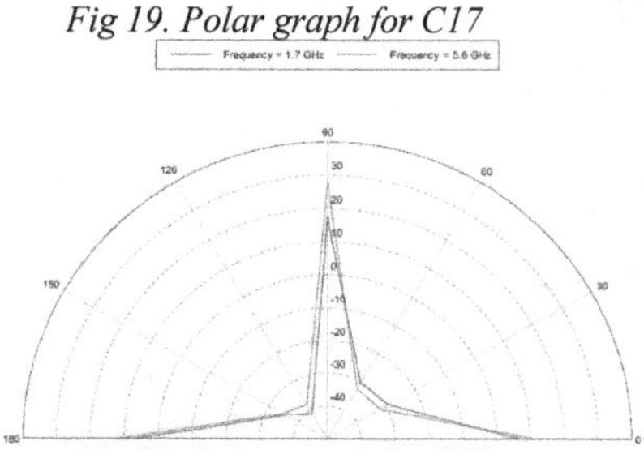

Fig 21. Polar graph for C19

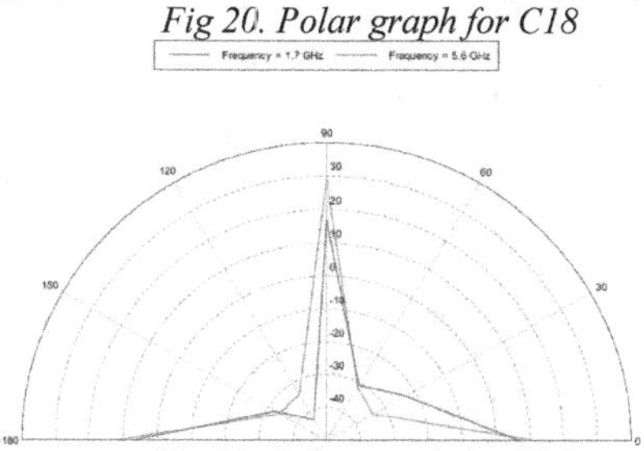

Fig 22. Polar graph for C20

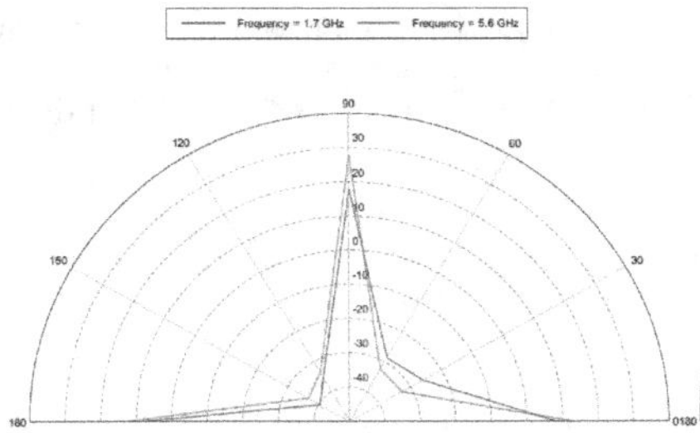

Fig 23. Polar graph for C21

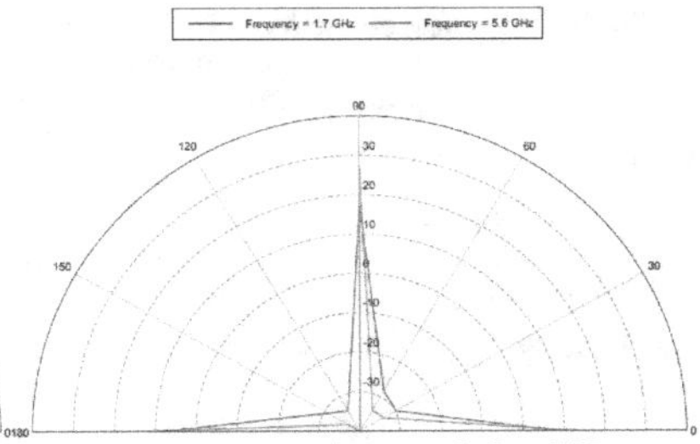

Fig 24. Polar graph for C22

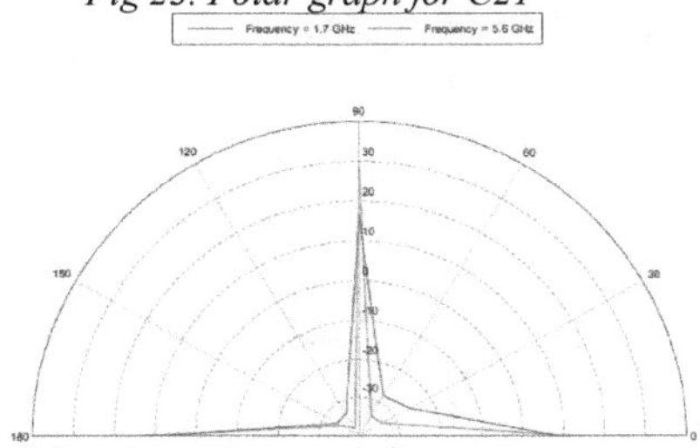

Fig 25. Polar graph for C23

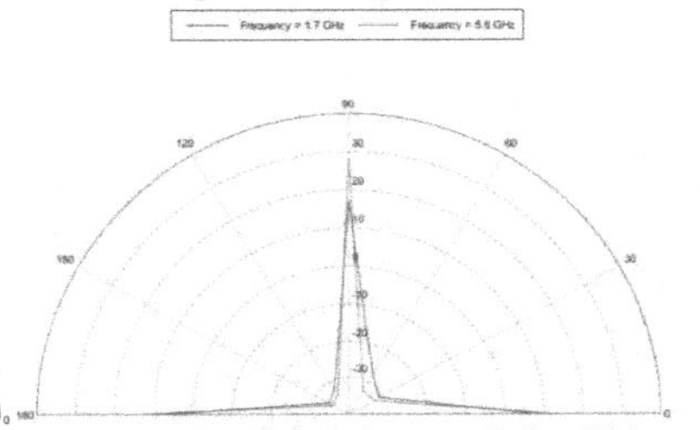

Fig 26. Polar graph for C24

For frequency 1.7 GHz, the longer canards see a steeper rise in RCS between 30˚ - 90˚ and similar sharp dip between 90˚ to 150˚. The dip reaches even lower values for longer canards in 5.6 GHz frequency. The canards having span lower than original one do not follow any trend as such. However, the RCS reaches as low as -50 dBsm for low frequency in C17 at 150˚ and high frequency in C18 at 120˚.

Conclusion

A general trend can be spotted irrespective of the constant and variable parameters for φ in the range 0˚-90˚ at 1.7GHz frequency (exceptions being C1, C3, C6, C8, C18 and C20). The RCS increases between 0˚- 30˚, then there is steeper rise between 30˚- 60˚ and then it spikes up to 90˚. Varying wing span does not affect the trend in general but longer wings see a direct and much steep rise between 30˚- 90˚. Variation in root chord of the wing gives different RCS trends, not following a specific pattern as such. Tip chord variation follows the same trend, the RCS increases between 0˚- 30˚, then there is steeper rise between 30˚- 60˚ and then it spikes up to 90˚ with the rise between 30˚- 90˚ being steeper and sharper than the others.

The C2 configuration stands out in all of the 24 models. The RCS for these parameters remains less than 1 m2 in the entire frequency range. The RCS is 1.37 × 10-7 m2 at φ = 150˚ for 5.6 GHz frequency and 8.71 × 10-7 m2 at φ = 120˚ for 1.7 GHz frequency, compared to the original parameters, where RCS reaches the minimum of 1.71 × 10- 5 m2 for high frequency.

Nomenclature

AR	Aspect ratio	--
θ	Angle between z axis and xy plane	(°)
φ	Roll angle	(°)

References

[1] "Stealth Technology". Wikipedia. 2022.

[2] S. M., Vaitheeswaran & Gowthami, T. "RCS of the Tail Swept Flying Wing and Comparison with Milestone Aircrafts". 2016.

[3] Michishita, Naobumi & Nguyen, Quoc Dinh & Yamada, Yoshihide. "Simulation and Measurement Methods for RCS Estimations of a Scale Model Airplane". REV Journal on Electronics and Communications. 5. 10.21553/rev-jec.93. 2016. https://doi.org/10.21553/rev-jec.93

[4] Ya-Qiang Zhuang, Chen-Xin Zhang & Xiao-Kuan Zhang. "Dynamic RCS Statistical Characterization of Aircraft Target Using Gaussian Mixture Density Model (GMDM)".

[5] Zhenghong, Gao, and Wang Mingliang. "An efficient algorithm for calculating aircraft RCS based on the geometrical characteristics." Chinese Journal of Aeronautics 21.4, 2008, pp. 296-303. https://doi.org/10.1016/S1000-9361(08)60039-4

[6] Van Schalkwyk, Richard F., and Johan C. Smit. "Dynamic radar cross section measurements of a full- scale aircraft for RCS modelling validation." 2017. https://doi.org/10.1049/cp.2017.0429

[7] Gou Zhanzhi, Chen Yingwen, MA Lianfeng. Radar cross-section effect of canard[J]. ACTA AERONAUTICAET ASTRONAUTICA SINICA, 2020, 41(6): 523485-523485.

[8] "Radar Cross Section (RCS)". Electronic Warfare and Radar Systems Engineering Handbook. https://www.rfcafe.com/references/electrical/ew- radar-handbook/radar-cross-section.htm

[9] Mahafza, B.R., "Radar Systems Analysis and Design using Matlab," CRC Press, 2000 Chapman &hall /crc, Boca Raton, London, New York, Washington, D.C. https://doi.org/10.1201/9781584888543

[10] "Basic Principles of The Method of Moments." Emagtech Wiki. 2018. http://www.emagtech.com/wiki/index.php/Basic_Principles_of_The_Method_of_Moments

[11] David L. Lynch Jr., "Introduction to stealth systems". An Introduction to RF Stealth (Radar, Sonar and Navigation), Institution of Engineering and Technology, 2021, pp. 42. https://doi.org/10.1049/SBRA540E

[12] Chibuisi, I. "Simulation of Radar Cross Section (RCS) of Spherical Objects". International Journal of Trend in Research and Development. Vol 2(5), 2015.

[13] "Basic Principles of Physical optics". Emagtech Wiki. 2018. http://www.emagtech.com/wiki/index.php/Basic_Principles_of_Physical_Optics

[14] Richardson, M. "Physical Optics Based Methods for Scattering Analysis". Thesis. Stellenbosch University. March 2018.

[15] Altair Engineering Inc, Altair FEKO 2021.1.2 User Guide. 2021.

Investigating the fluid-structure interaction of L-shaped pipe bends using machine learning

Pratik Punj[1,a *], Md Adil[2,b]

[1*]Department of Mining Machinery Engineering, IIT (ISM) Dhanbad, Dhanbad, India

[2]Department of Mechanical Engineering, Jadavpur University, Kolkata, India

[a*]pratik10singh@gmail.com, [b]adil.alam45@gmail.com

Keywords: FSI, L-Shaped Bent, Machine Learning, CFD, Structural Simulation

Abstract. The fluid Structure interface is an important area of research for its challenges in fluid structure dynamics in understanding the effect of fluid on motion and deformation of structures. In the current study, we used the L-Shaped pipe bent and did a CFD simulation at the velocity inlet condition of the range 1-3 m/s with keeping adiabatic wall condition and environmental pressure at the outlet. The reason for choosing L-Shaped bent is that it creates a sharp change in the flow direction, which leads to complex vortices, turbulence and pressure distribution. It also puts a significant mechanical load on the structure due to this change in flow, resulting in a large structural deformation. The result of CFD simulation is used to do the structural simulations at different material types, lengths of both arms, keeping the diameter, angle and fillet radius of the bent at a constant value. The database created is then used as an input to the machine learning (ML) model to predict for an arbitrary material and at any length of the bent without doing all the simulations. The simulation results also help to co-relate the impact of variation in length with the bent's stress, strain and displacement.

Introduction

Fluid-structure dynamics refers to studying the interactions between fluid flow and structures. This is an important area of research in various fields, including engineering, physics, and biology. One of the key challenges in fluid structure dynamics is understanding how fluid flow can affect the motion and deformation of structures. This is particularly relevant in the design of aircraft, ships, and other vehicles that must withstand the forces of fluid flow.

One important tool for studying fluid-structure dynamics is computational fluid dynamics (CFD), which uses computer simulations to model the flow of fluids and their interactions with structures. CFD can provide valuable insights into the behavior of fluid-structure systems, allowing engineers to optimize designs and predict performance. Many factors can influence fluid-structure dynamics, including the shape and size of the structure, the properties of the fluid, and the flow conditions. Research in this area often involves developing mathematical models and computational methods to accurately predict the behavior of these systems.

The IB method, introduced by Charles Peskin in 1972, revolutionized the study of the interaction between flexible structures and viscous, incompressible fluids. The method, described in [1, 2], uses a fully coupled computational analysis that represents the fluid domain using an Eulerian mesh and the immersed structure using a Lagrangian grid. It employs the Dirac-delta function to transfer forces from the boundary to the fluid and velocity from the fluid to the boundary. Despite its limitations in only being applicable to flexible boundaries, researchers have made advancements in the method. Goldstein et al. [3] proposed a feedback forcing scheme to determine the fluid's external force, while Saiki & Biringen [4] used a discrete hat function for transfer of force and velocity information. Mohd-Yusof [5] took it further by creating a direct forcing formulation, eliminating the need for discrete functions or feedback forcing, using the pseudo-spectral method. The IB method has also been applied to other areas, such as the analysis of blood clotting by Fogelson & Guy [6], where they modeled a fluid containing suspended platelets and included chemical reaction equations to study their response to stimuli.

Significant progress has been made in the field of fluid-structure interaction (FSI) by combining the Navier-Stokes (N-S) and Euler-Bernoulli (E-B) equations. These equations are utilized to model an incompressible fluid flow and a flexible immersed boundary respectively, as described in [7-10]. The N-S equations control the behavior of the fluid by considering mass and momentum conservation, while the E-B theory predicts small deflections of structures subjected to pointwise or distributed lateral loads. This theory is founded on a fourth-order differential equation that links the transverse displacement of a structure to the force applied to it. The major challenge in FSI lies in managing the N-S and E-B equations together. Energy transport between a beam-like structure and a Newtonian fluid is explored in [11]. Pontaza & Menon [12] introduced an FSI problem of a flexible pipe in a viscous fluid, modeled as an E-B beam, to determine its response to vortex-induced vibrations in the time domain.

Recently, the use of physics-guided machine learning (ML) approaches has become prevalent due to their ability to combine data-driven methods with physical knowledge to build descriptive models, carry out efficient simulations, and identify input-output relationships. For instance, deep neural networks (NNs) have been utilized to approximate partial differential equations by training on extensive datasets [13,14]. ML techniques have been explored in different areas such as structural dynamics, fluid mechanics, and FSI problems. In [15], recurrent NNs and multi-layer perceptrons were merged with domain knowledge to enhance structural dynamics simulations. For fluid domains with complex boundary conditions, a hybrid network, V2P-Net, was designed to predict pressure from observed velocity fields [15]. Furthermore, a novel hydro-elastic reduced order FSI model using ML was proposed to overcome instability issues associated with traditional Galerkin Projection method [15]. These works motivated the researchers to use ML to significantly decrease the computational cost in traditional FSI analysis.

In this analysis we use Deep learning method to predict the structural behaviors of a L-Shaped bend based on FSI analysis at different inlet flow rate, for the different upper as well as lower length and construction material.

Materials and methods

For the current study we use a L-shaped bend with the dimension of a 10 mm diameter, 25mm upper length and 50mm lower length with a fillet radius of 5 mm at the bent given by figure 1.

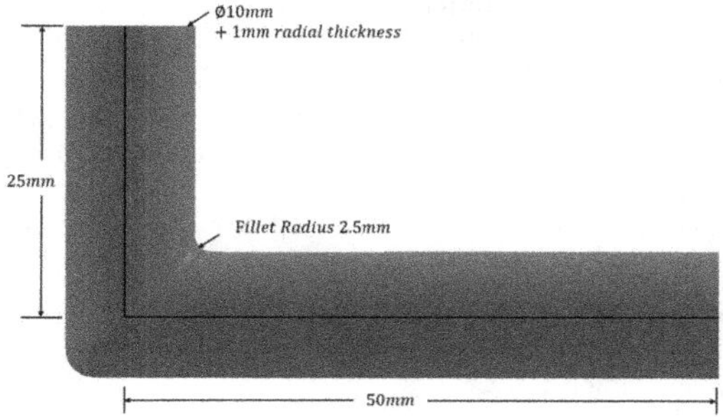

Figure 1: Dimension of L-Shaped Bent for FSI analysis

The FSI is achieved performing CFD analysis at first and then static structural analysis with pressure, temperature distribution as an external load. This FSI analysis is carried out for 15 different structural material (namely, SS 304, Alloy Steel, Plain Carbon Steel, ductile Iron, Grey Cast Iron, 1060 Aluminum Alloy, Aluminum Bronze, Brass, Copper, Manganese Bronze, Magnesium Alloy, Monel(R) 400, ABS, Nickel and PET) and a database is created by varying different upper, lower arm lengths and for different materials.

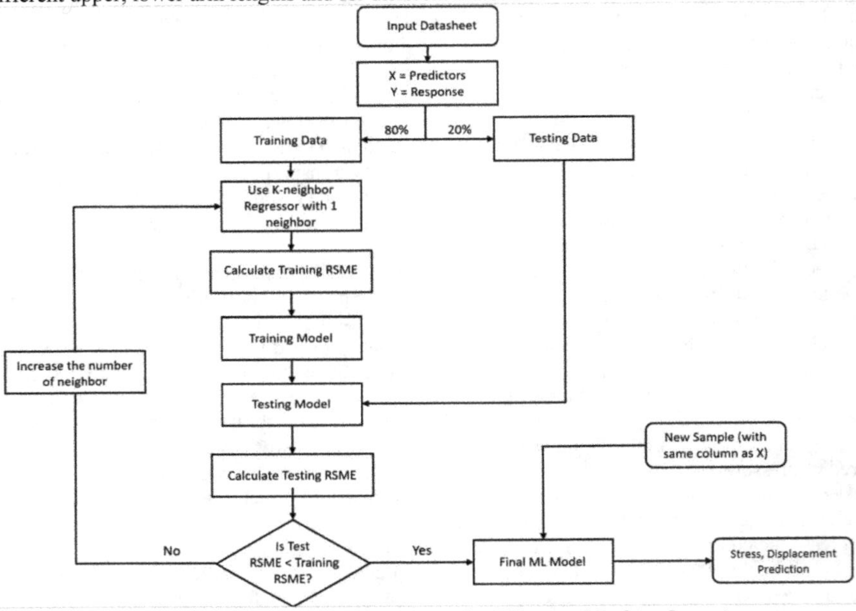

Figure 2: Algorithm for Investigating the FSI using Machine Learning

This database is use as input to the machine learning algorithm (KNN-Regressor) in python and with that the ML model will be able to predict the Stress and Displacement of bend for unknown material (alloy) by providing Young's Modulus and Poisson's ratio. Figure 2 explains the ML algorithm to predict the FSI results for the L-shaped bend with the of a flow chart. The predictor and response is selected and split in training and testing datasets using test_train_split python function. Training data is fed to KNN Regressor with 1 neighbour and then training and testing RMSE (Root Mean Squared Error) is calculated. If the $RMSE_{Test} < RMSE_{Training}$ then the model can be used for making prediction on new sample and if $RMSE_{Test} > RMSE_{Training}$ then the number of neighbors is increased until $RMSE_{Test} < RMSE_{Training}$. Now, the Stress and displacement of the bend is obtained with the new sample in the same sequence as that of predictor.

Results and discussion

To begin the FSI analysis the CFD simulation is done with the following conditions using SOLIDWORKS Flow simulation software:

Table 1: CFD simulation parameters

SL NO.	PARAMETER	VALUE
1.	Fluid Property	Water (20.5 °C, 1 atm)
2.	Inlet velocity	1m/s, 2m/s
3.	Outlet Condition	1 atm (abs)
4.	Surface Roughness	0 micrometre
5.	Wall Condition	Adiabatic
6.	Mesh	15008

This Pressure distribution obtained for the pipe at the particular condition is imported to SolidWorks Simulation software, and constraining the upper length of the bend while keeping the lower length free, the static, steady-state simulation is executed with figure 2 showing fluid pressure distribution and Von-Misse Stress for 1m/s of fluid inlet condition and AISI SS 304 as pipe material. It is evident from the figure that the maximum fluid pressure is at the bent but for the static stress it is maximum at the fixed geometry and minimum at the free end with accounting the same fluid pressure distribution as external load in static stress analysis.

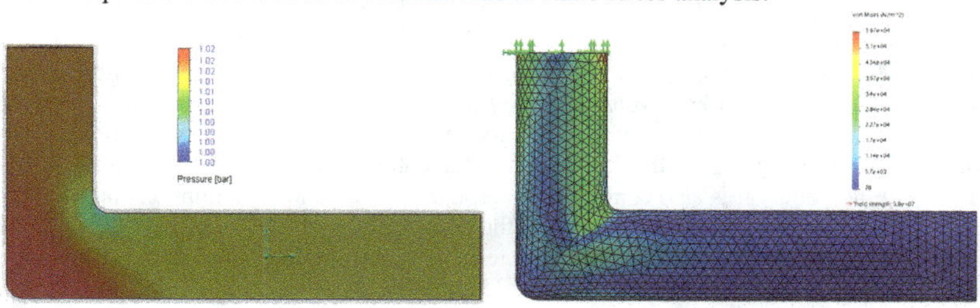

(a) (b)

Figure 2: Schematic diagram of CFD simulation (a. pressure distribution at 1m/s velocity input) and the respective static structural analysis (b. Von-Mises Stress distribution with the external load as pressure distribution from CFD simulation)

Table 2 shows the meshing result for static stress analysis for the initial geometry (i.e. upper length 25mm and lower length 50mm).

Table 2: Meshing result for static structural analysis

Static Structural Meshing Result	
Mesh Type	Solid Mesh
Mesher Used	Blended curvature-based mesh
Max Element size	1.14408mm
Min Element size	0.381356mm
Total Nodes	26056
Total Elements	12946
Max Aspect Ratio	6.0982

Figure 3 shows the variation of displacement for different material at 1m/s inlet velocity (figure 3 (a)) and the displacement for different inlet velocities with SS 304 and it can be seen that as the length increases the displacement also increases not linearly but quadratically (different quadratic function for different material).

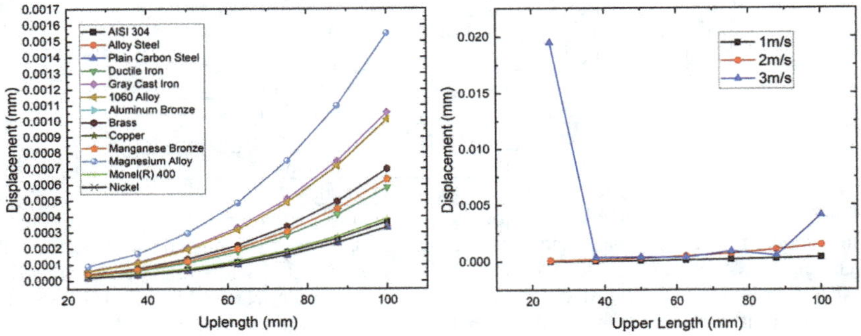

(a) **(b)**

Figure 3: Variation of displacement with upper arm length for different material (a), and for different inlet velocities with SS 304 material (b).

Figure 4 shows the variation in displacement with change in lower arm length for different materials as well as for different inlet velocities also. And it is evident from figure 3 and 4 that at inlet velocity of 3 m/s the pipe with SS304 goes to the plastic deformation (which is shown in figure 5) hence it is eliminated from inclusion in the database for Machine Learning predictions.

From figure 3 and 4 it is also concluded that the maximum and minimum displacement for Magnesium Alloy and Carbon Steel for both the cases i.e., upper arm length as well as lower arm length. The database is created by varying materials of construction (15 materials, namely Young's Modulus and Poisson's ratio), upper length (25-100mm) and lower length (50-150mm) and the respective Stress, Stain, Displacement. **Figure 6** shows a snapshot of the database with first 5 data points.

As for KNN Regression Machine learning method the input i.e. Predictors should be a numerical value so the we need to exclude the material column. Hence the predictors are "uplength, lowerlength, YM, PR and velocity" while the response for the model are "Displacement1 and Stress2".

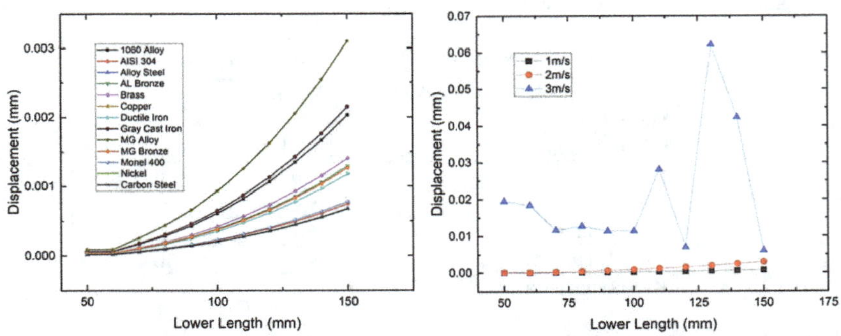

(a) (b)

Figure 4: Variation of displacement with lower arm length for different material (a), and for different inlet velocities with SS 304 material (b).

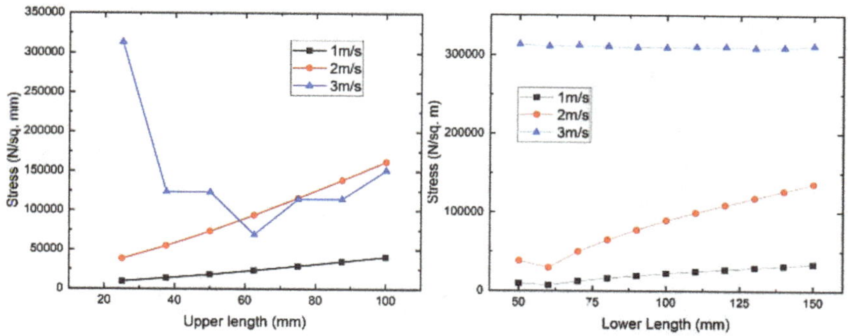

Figure 5: Variation in stress with (a) change in upper arm length, (b) change in lower arm length for SS304

	Material	uplength	lowerlength	YM	PR	Velocity	Displacement1	Stress2
0	1060 Alloy	25.0	50	66178000000	0.27	1	0.000059	9549
1	1060 Alloy	25.0	60	66178000000	0.27	1	0.000060	7294
2	1060 Alloy	25.0	70	66178000000	0.27	1	0.000169	12320
3	1060 Alloy	25.0	80	66178000000	0.27	1	0.000287	16110
4	1060 Alloy	25.0	90	66178000000	0.27	1	0.000432	19250

Figure 6: Snapshot of the database with first five data points for Jupiter notebook

$$X = \{uplenth, lowerlength, YM, PR, Velocity\}$$
$$Y = \{Displacement1, Stress2\}$$

In order to determine the best value of nearest neighbours using *for* loop the prediction with neighbours in range of 1 to 10 is executed and the neighbour with least Mean Absolute Percentage Error (MAPE) is chosen. Figure 7 shows the values of MAPE with different neighbours and according to this the nearest neighbour of 4 is chosen for our model. Taking KNN Regressor with 4 Nearest-Neighbours the testing and the training MAPE and R2 Value is shown in table 3 and from that it is evident that our ML model is a very good match for the current scenario and will show a very good prediction.

Table 3: Training and Testing MAPE and R^2 Value with 4 neighbour KNN Regressor

Training		Testing	
MAPE	0.11	MAPE	0.155
R^2 Value	0.9848	R^2 Value	0.9769

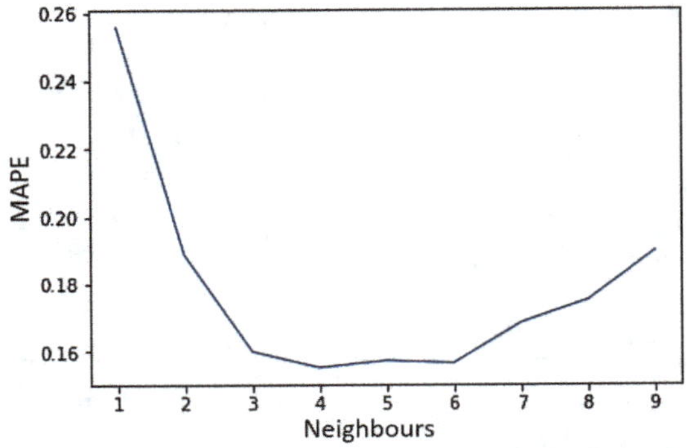

Figure 7: Mean Absolute Percentage Error (MAPE) for different neighbours.

Based on the accuracy of the KNN- Regression model a plot is created showing the comparison between original and the predicted values of stress and displacement for ABS material and keeping upper length as constant at 50mm as shown in figure 8 and the results are quite impressive.

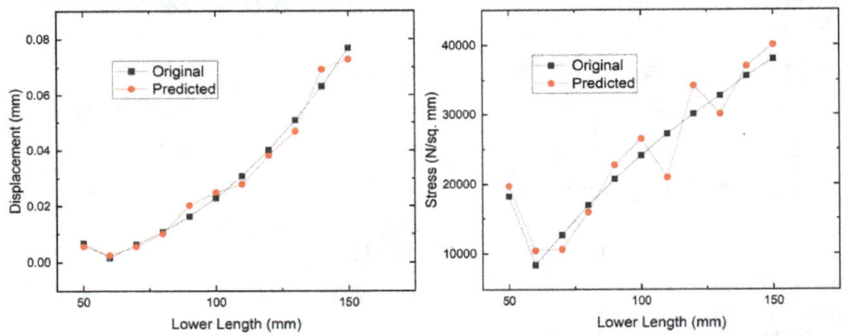

Figure 8: Displacement and Stress comparison between Original and Predicted values for ABS material keeping upper arm length constant at 50mm.

Conclusion

As the L-shaped bend suffer high mechanical stress and deformation due to high fluid momentum and in order to determine the FSI using ML model a database is required for the different model to learn and do the prediction without the further use of simulation. For achieving this goal, the KNN Regression model is selected because of its ability to predict multiple responses, in current case displacement and stress, while keeping the understanding and implementation of code as simple as possible. The FSI process and the algorithm for ML prediction is discussed earlier and a snapshot of the database (figure 6) is also provided. The KNN Regression model with 4 neighbours shows a very good fitness with the current database with training and testing R^2 Value of 0.9848 and 0.9769 respectively (table 3). And for the final evidence for the fitness of the model, figure 8 shows the comparison between original and predicted data for a section of ABS material with the fluid pressure distribution of inlet condition at 1m/s and it is observed that the ML model describes the original data nature quiet accurately.

References

[1]. Peskin CS. 1972 Flow patterns around heart valves: a numerical method. J. Comput. Phys. 10, 252 271. (doi:10.1016/0021- 9991(72)90065-4) https://doi.org/10.1016/0021-9991(72)90065-4

[2]. Peskin CS. 2002 The immersed boundary method. Acta Numer. 11, 479-517. https://doi.org/10.1017/S0962492902000077

[3]. Goldstein D, Handler R, Sirovich L. 1993 Modeling a no-slip flow boundary with an external force field. J. Comput. Phys. 105, 354-366. https://doi.org/10.1006/jcph.1993.1081

[4]. Saiki EM, Biringen S. 1996 Numerical simulation of a cylinder in uniform flow: application of a virtual boundary method. J. Comput. Phys. 123, 450-465. https://doi.org/10.1006/jcph.1996.0036

[5]. Mohd-Yusof J. 1997 Combined immersedboundary/B-spline methods for simulations of flow in complex geometries. Center Turbul. Res. Annu. Res. Briefs 161, 317-327.

[6]. Fogelson AL, Guy RD. 2008 Immersedboundary-type models of intravascular platelet aggregation. Comput. Methods Appl. Mech. Eng. 197, 2087-2104. https://doi.org/10.1016/j.cma.2007.06.030

[7]. Raymond JP. 2010 Feedback stabilization of a fluid-structure model. SIAM J. Control Optim. 48, 5398-5443. https://doi.org/10.1137/080744761

[8]. Faria CT, Inman DJ. 2014 Modeling energy transport in a cantilevered Euler-Bernoulli beam actively vibrating in Newtonian fluid. Mech. Syst. Signal Process. 45, 317-329. https://doi.org/10.1016/j.ymssp.2013.12.003

[9]. Pontaza JP, Menon RG. 2009 Prediction of VIV response of a flexible pipe by coupling a viscous flow solver and a beam finite element solver. In Int. Conf. on Offshore Mechanics and Arctic Engineering, ASME, Jan 1, vol. 43451. pp. 303- 309. https://doi.org/10.1115/OMAE2009-79150

[10]. Diop M, Mbaye I. 2017 Numerical method for unsteady fluid structure interaction problem. Appl. Math. Sci. 11, 1835-1844. https://doi.org/10.12988/ams.2017.75185

[11]. Long Z, Lu Y, Ma X, Dong B 2018 Pde-net. Learning pdes from data. In Int. Conf. on Machine Learning, PMLR, July 3, pp. 3208- 3216.

[12]. Long Z, Lu Y, Dong B. 2019 PDE-Net 2.0: learning PDEs from data with a numericsymbolic hybrid deep network. J. Comput. Phys. 399, 108925. https://doi.org/10.1016/j.jcp.2019.108925

[13]. Yu Y, Yao H, Liu Y. 2020 Structural dynamics simulation using a novel physics-guided machine learning method. Eng. Appl. Artif. Intell. 96, 103947. https://doi.org/10.1016/j.engappai.2020.103947

[14]. Shit S, Das D, Ezhov I, Paetzold JC, Sanches AF, Thuerey N, Menze BH. 2021 Velocity-to-pressure (V2P)-net: inferring relative pressures from time-varying 3D fluid flow velocities. In Int. Conf. on Information Processing in Medical Imaging, pp. 545-558. Cham, Switzerland: Springer. https://doi.org/10.1007/978-3-030-78191-0_42

[15]. Whisenant MJ, Ekici K. 2020 Galerkin-free technique for the reduced-order modeling of fluid-structure interaction via machine learning. In AIAA Scitech 2020 Forum Jan 5,1637. https://doi.org/10.2514/6.2020-1637

Advanced Topics in Mechanics of Materials, Structures and Construction - AToMech1-2023 Materials Research Forum LLC
Materials Research Proceedings 31 (2023) 248-261 https://doi.org/10.21741/9781644902592-26

Fatigue life evaluation of asphalt mixtures containing natural river sands and designed by bailey method

Mohammad Ahmad Alsheyab[1,a] and Mohammad Ali Khasawneh[2,b *]

[1] Civil, Construction and Environmental Engineering, Iowa State University, 813 Bissell Road, Ames, IA, United States

[2] Civil Engineering, Prince Mohammad Bin Fahd University, Al Azeziya, Eastern Province, Kingdom of Saudi Arabia.

alsheyab@iastate.edu, [b]mkhasawneh@pmu.edu.sa

Keywords: Bailey Gradation, Asphalt mix, Natural Sand, Fatigue Life

Abstract. This study aims to investigate the efficiency of Bailey method of optimizing the fatigue life of asphalt mixtures when natural sand is included in the mix at two coarseness levels of aggregate gradations: Fine-Graded (FG) and Coarse-Graded (CG), with three mixes which varied with the percentage of the natural river sand, were prepared at each coarseness level, namely: Corse-Graded with Quarry Sand only (CG-QS), Corse-Graded with Natural Sands only (CG-NS), Corse-Graded with Quarry and Natural Sands (CG-QNS), Fine-Graded with Quarry Sand only (FG-QS), Fine-Graded with Natural Sands only (FG-NS), and Fine-Graded with Quarry and Natural Sands (CG-QNS). The portions of the natural sand either in CG-QNS and FG-QNS mixes were minimized as possible without violating the Bailey ratios. The Beam Fatigue (BF) test was used to evaluate the performance of each mixture at a strain level of 1000 micro strain. The sensitivities of the volumetric measures with N_f were evaluated. The study's findings indicate that the Number of Cycles to Failure (N_f) was generally decreasing with the increase of the natural sand in the mix at any strain levels. The Dust Proportion (DP) was the most significant volumetric. The Bailey gradation method successfully provided a similar gradation coarseness for CG-QNS compared to CG-QS, which resulted in comparable N_f and indicates a similar aggregate interlock.

Introduction and Background

Understanding the material characteristics that define the performance of the asphalt over the lifespan of the pavement is necessary for the design of asphalt pavements. The type of aggregate and its gradation have a considerable impact on the workability and performance of asphalt concrete pavements in use. Mineral aggregates make up 90–95% of the mixed weight [1]. In essence, the type of binder used as well as the shape, toughness, durability, and gradation of the aggregate is mainly accountable for the permanent deformation and cracking of asphalt mixtures. Additionally, excessive amounts of fine aggregate in rounded shapes are primarily diagnosed for early failures in asphalt concrete mixtures [2,3]. Pavement agencies prefer natural fine aggregates over crushed sand because they are more readily available, cost-effective, and workable when producing Hot Mix Asphalt (HMA). Contrarily, crushed fine aggregate must be quarried, processed, cleaned, and separated into distinct size fractions, all of which raise the price of making asphalt concrete. As a result, several experiments have been done to assess how well asphalt concrete performs when natural sands are added to the mix.

Most of these studies had inconsistent results, where the evaluation of the inclusion of natural sand was arbitrarily or without following a systematic procedure, and percentages of natural sands included in the mix were relatively high. This can be seen in a number of studies including [4, 5, 6, 7, 8, 9, 10]. The results of the aforementioned studies lack consistency since not much consideration was given to aggregates particles shapes, packing, and gradations. On the other hand,

Ramli [11] and Park and Lee [12], Stuart and Mogawer [13], and Freeman and Kuo [14] suggested identifying poor quality sand from good quality to be used in the asphalt mix based on the combined effect of sand or fine aggregate texture, shape, gradation, and quantity which eventually affect the packing densities of sand. This signifies the necessity of following a systemic procedure to assess and design the aggregate structure for asphalt mixes. Bailey method is a systematic procedure that can evaluate and design blends of aggregates and provides a practical tool assisting the designer to ensure that the pavement obtains the required durability with the available aggregates. The method evaluates how the particles pack together to provide a good interlock between the particles. Bailey proposed three ratios were to evaluate the gradation of asphalt mixture. Two principles were considered in Bailey method to evaluate a mixture: the first one is aggregate packing which considers that aggregates cannot completely fill a certain volume. The second principle is defining coarse and fine aggregate. Coarse particles are essential to provide the aggregate interlock and deformation resistance and are defined as the particles retained on a pre-specified sieve size (primary control sieve) and fine aggregates are defined as particles that fill the voids between coarse aggregates. Bailey method has proven its superiority over any other aggregate gradation design and resulted in a better performance mix. This can be shown in studies conducted by Ghuzlan et al.[15]; Sivasubramaniam et al.[16]; Zhu et al.[17]; and Vavrik et al.[18]. There is a need for a more inclusive approach to evaluate the utilization of natural sands at different levels of gradation coarseness as well as studying the most influencing volumetric measure on the performance of the compacted asphalt mixtures.

To this end, all of the aforementioned studies indicate the necessity of evaluating the aggregate surface textures, packing, and gradation especially when natural sand is utilized in the mix. The Bailey method is the most reliable method to achieve an optimized gradation so far. Henceforth, the study aims to investigate the fatigue performance of asphalt mixtures designed by following Bailey method when natural sand is included in the mixture at different coarseness levels of aggregate gradations. Two levels of gradations were evaluated: Fine and coarse, and three mixes which varied with the percentage of the natural river sand were prepared at each coarseness level. The Beam Fatigue (BF) was used to evaluate the fatigue performance of the mixtures.

Materials and Methods

This section describes the material properties used in producing the HMA specimens; including asphalt, aggregate, and the HMA mixture with all the tests and specifications used in characterizing these paving materials in accordance with the Superpave mix design procedure described in the Asphalt Institute (AI) MS-2 manual [19]. Six types of HMA mixes were utilized in this study, namely: Corse-Graded with Quarry Sand only (CG-QS), Corse-Graded with Natural Sands only (CG-NS), Corse-Graded with Quarry and Natural Sands (CG-QNS), Fine-Graded with Quarry Sand only (FG-QS), Fine-Graded with Natural Sands only (FG-NS), and Fine-Graded with Quarry and Natural Sands (CG-QNS). A traffic level of 0.3 to 3 million Equivalent Single Axle Load (ESAL) was chosen in this study to mimic the low level of traffic. The Nominal Maximum Aggregate Size (NMAS) for aggregate blends was chosen to be 12.5 mm. The Beam Fatigue (BF) test was conducted to evaluate the inclusion of the Natural Sand (NS) on the performance of the asphalt mixes.

Asphalt Binder. The asphalt binder used in this study had a penetration grade of 60/70. The asphalt binder was further tested to determine its performance grade (PG) and to ensure that it meets the Superpave performance requirements. Accordingly, by conducting several asphalt binder tests, it was found that the proper Superpave asphalt grade is PG 70–10. The results are summarized in Table 1.

Table 1. Results of asphalt binder test

Test	Test Method	Test Temp. (°C)	Parameter	Criterion	Results
Rotational Viscometer (RV)	[20]	135	Viscosity (Pa.s)	≤ 3	0.204
Dynamic Shear Rheometer (DSR; unaged)	[21]	64	G*/sin δ (kPa)	≥ 1	1.610
		70			0.772
Rolling Thin Film Oven (RTFO)	[22]	163	% Mass Loss	≤ 1	0.13
Direct Shear Rheometer (DSR; RTFO residue)	[21]	64	G*/sin δ (kPa)	≥ 2.2	2.452
		70			1.112
Pressure Aging Vessel (PAV)	[23]	100	-	-	-
Direct Shear Rheometer (DSR; PAV residue)	[21]	31	G*×sin δ (kPa)	≤ 5000	261.54
		34			172.79
Bending Beam Rheometer (BBR)	[24]	0	S(t) at t = 60s (Mpa)	≤ 300	7.62
			m value at t = 60s	≥ 0.3	0.502

Aggregate. Several aggregates stockpiles of crushed limestone were used in this study. The aggregate stockpiles were 12.5 mm Aggregate, 9.5 mm Aggregate, Quarry-made Sand (QS), and River Natural Sand (NS). Fig. 1 shows the gradations of each stockpile utilized to prepare the proposed mixtures. Table 1 illustrates the test procedures used to characterize the coarse and fine aggregates, respectively along with the experimental results of each test. Accordingly, the tests included flat and elongated particles for coarse aggregate, coarse aggregate angularity, fine aggregate angularity, and bulk-specific gravity. These properties are significant to provide a reliable aggregate interlock and proper compaction without braking and endure the traffic loading over the life span of the pavement. It is important to note that significant portions of each stockpile were taken, combined, and then sieved (i.e., fine and coarse) to measure the consensus properties of the aggregate. Table 2 shows the results of the aggregate properties.

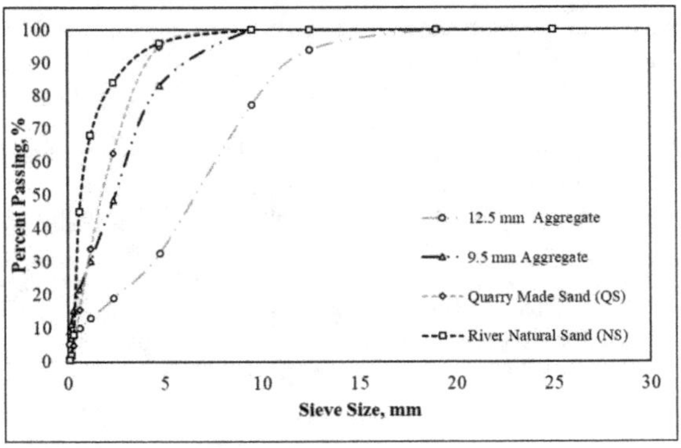

Figure 1. Aggregate stockpiles gradations

Table 2. Aggregate consensus properties

Property	Coarse aggregate		Fine aggregate				Superpave acceptable range (0.3 – 3 million ESALs)	Test method
	12.5 mm	9.5 mm	QS	NS	12.5 mm	9.5 mm		
Flat and elongated,%	3	2	-	-	-	-	-	[25]
Coarse aggregate angularity, %	92/97	94/98	-	-	-	-	≥50/-	[26]
Fine aggregate angularity, %	-	-	43	40	42		≥40	[27]
Bulk Specific Gravity	2.684	2.681	2.679	2.652	2.789	2.686	-	[28]

As shown in Vavrik et al. [18], there are three control sieves in Bailey gradation used to determine the optimized blend, namely: Primary Control Sieve (PCS) that separates the coarse aggregate portion from the fine portion, Secondary Control Sieve (SCS) that splits the fine part from the previous equation (passing the PCS) into two further fractions, and then Tertiary Control Sieve (TCS) where the materials passing SCS are further spliced with the Tertiary Control Sieve (TCS). The calculations of PCS, SCS, and TCS are shown below in Eq. 1, 2, and 3, respectively.

$$PCS = NMAS \times 0.22 \quad (1)$$
$$TCS = PCS \times 0.22 \quad (2)$$
$$SCS = TCS \times 0.22 \quad (3)$$

This method also suggests three ratios to evaluate the aggregate blend (Vavrik et al., 2002). They are used to understand and analyze the structure of the aggregate gradation. Accordingly, aggregates in the blend are classified into three portions, the coarse portion which retained on the PCS, the coarser part of fine aggregates which passes the PCS and retained on the SCS, and the finer part of fine aggregates which passes through the SCS and retained on the TCS. Based on these definitions, Bailey suggested three ratios; Coarse Aggregate Ratio (CA), Fine Aggregate Coarse Ratio (FA_c), and Fine Aggregate Fine Ratio (FA_f). In essence, the CA Ratio obtains the interlock between coarse particles by introducing a Half Sieve (HS), which is simply half of the NMAS. The particles with sizes between HS and PCS are called interceptors.

Changing the percentages of interceptors affects the VMA of the mixture because they cannot fill the voids between the larger particles. Voids will also be found between the particles of the coarser part of the fine proportion, and the finer parts of the fine proportion will fill these voids. The ratio of these fractions is described by FA_c. Finally, FA_f ratio is significant for understanding some mixture properties such as binder content and volume of voids, because it is concerned with the very finest material content (including filler) in the mixture.

Vavrik et al. [18] suggest different ranges of CA, FA_c, and FA_f based on the coarseness of gradation (fine or coarse) and NMAS. Coarse-graded the follows Bailey guidelines generally provide higher resistance to deformation when compared to fine-graded mixes [16]. In fine mixes, the fine aggregate has to be noticeably strong in order to compensate for the absence of the larger aggregate portions to resist different types of deformation. In dense-graded mixtures, the gradation coarseness is defined by the position of the gradation line with reference to the maximum density line. According to AASHTO M323 [29] and Khasawneh and Alsheyab [30], when the gradation line mostly lies above the maximum density line and above PCS, the gradation is defined as fine-graded. In contrast, when the gradation line mostly lies below the density line and below PCS, the gradation is defined as coarse-graded. The calculations of CA, FA_c, and FA_f are shown below in

Eq.4, 5, and 6, respectively. Table 3 shows the sieves and formulas used for the calculations of the Bailey ratio for NMAS 12.5 mm as reported in Vavrik et al. [18]. Note for fine-graded mixtures the calculations and the sieves included in Bailey ratios calculations are slightly modified as reported in Vavrik et al. [18] as if PCS is considered the NMAS for the fine-graded mixtures.

$$CA = \frac{\%\text{Passing HS - }\%\text{Passing PCS}}{100\text{-}\%\text{Passing HS}} \tag{4}$$

$$FA_c = \frac{\%\text{Passing SCS}}{\%\text{Passing PCS}} \tag{5}$$

$$FA_f = \frac{\%\text{Passing TCS}}{\%\text{Passing SCS}} \tag{6}$$

Table 4. Sieves and formulas used in Baily ratio calculation for NMAS 12.5 mm

Gradation Coarseness	Ratio		
	CA	FA_c	FA_f
Coarse-graded	$\frac{(6.25\text{ mm - }2.36\text{ mm})}{(100\%\text{ - }6.25\text{ mm})}$	$\frac{0.60\text{ mm}}{2.36\text{ mm}}$	$\frac{0.150\text{ mm}}{0.60\text{ mm}}$
Fine-graded	$\frac{(1.18\text{ mm - }0.60\text{ mm})}{(100\%\text{ - }1.18\text{ mm})}$	$\frac{0.150\text{ mm}}{0.60\text{ mm}}$	N/A

A comprehensive procedure for designing and identifying the optimized blend from different stockpiles can be found elsewhere in Vavrik et al. (2002). Accordingly, the loose and rodded unit weights of the aggregate stockpiles need to be determined in order to obtain the proportions from each stockpile to finally determine the mix design blend for Bailey gradations. The results of unit weights are summarized in Table 5.

Table 5. Summary of stockpiles unit weights

Unit weight	Stockpiles				Test Method
	12 mm	9.5 mm	Quarry Sands (QS)	Natural Sand (NS)	
Loose	1.683	1.635	1.533	1.495	[31]
Rodded	1.811	1.706	1.692	1.594	

By using the unit weights summarized in Table 4 and following the procedure in Vavrik et al. [18]. The proportions that contributed to obtaining the final blend of Baily gradations are shown in Table 6. The results of calculated Bailey ratios for each gradation are shown in Table 7.

All of the gradations were designed and evaluated to meet the Bailey ratios constraints and requirements as shown in Vavrik et al [18]. Accordingly, six gradations were evaluated in this study, namely Corse-Graded with Quarry Sand only (CG-QS), Corse-Graded with Natural Sands only (CG-NS), Corse-Graded with Quarry and Natural Sands (CG-QNS), Fine-Graded with Quarry Sand only (FG-QS), Fine-Graded with Natural Sands only (FG-NS), and Fine-Graded with Quarry and Natural Sands (CG-QNS). The portions of NS either in CG-QNS and FG-QNS mixes were minimized as possible without violating the Bailey ratios.

Fig. 2 shows the gradations lines of the six mixes used in the study. Concerning course-graded mixes, it can be seen that CG-QS and CG-QNS gradation lines are not significantly different. In CG-QNS, the inclusion of NS decreased the need for QS significantly, slightly changed the proportions taken from each stockpile, and finally decreased the portion of the total sands in the mix. Also, Bailey ratios are somewhat similar for both CG-QS and CG-QNS. Moreover, the CG-NS was the finest compared to either CG-QS or CG-QNS. On the other hand, for fine-graded mixes, it can be seen that the differences between FG-QS, FG-QNS, and FG-NS gradation lines are more evident. Accordingly, FG-NS was the finest followed by FG-QNS, and FG-QS, respectively. It can be seen that the inclusion of NS slightly increased the portion of the total sands for either FG-QNS or FG-NS. Also, the difference in Bailey ratios is higher in fine-graded gradations when compared to coarse-graded ones. It is important to mention that the proportions taken from each stockpile are highly dependent on the gradation of each stockpile and constrained by Bailey ratios, and it may be challenging to design a gradation curve unless the constraints are met or meet a certain or specified Baily ratio.

Table 6. Summary of the proportions taken from each stockpile

Gradations	Stockpiles			
	12 mm	9.5 mm	Quarry Sands (QS)	Natural Sand (NS)
CG	60.3	25.1	14.6	0.0
CG-QNS	64.4	26.6	6.7	2.3
CG-NS	56.3	33.2	0	10.5
FG	7.7	68.1	24.2	0.0
FG-QNS	9.5	65.4	13.5	11.6
CG-NS	5.4	72.2	0	22.4

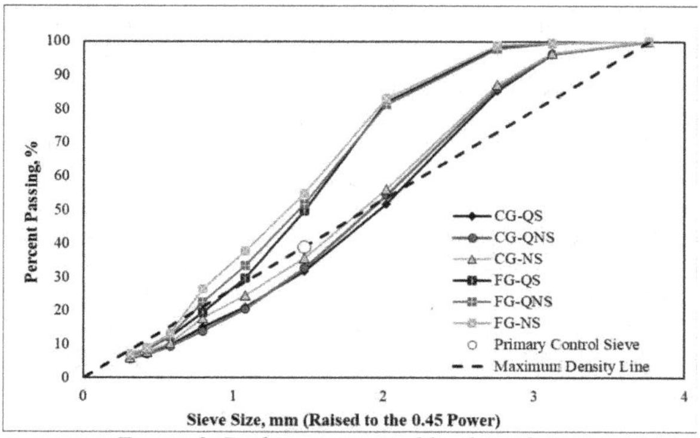

Figure 2. Bailey aggregate blend gradation

Table 7. Summary of calculated Bailey ratios

Gradations		Ratios		
		CA	FA_c	FA_f
Coase-gradaed	CG	0.54	0.47	0.50
	CG-QNS	0.54	0.42	0.50
	CG-NS	0.57	0.50	0.43
Constraint		0.5-0.65	0.35-0.5	0.35-0.5
Fine-graded	FG	0.60	0.44	-
	FG-QNS	0.62	0.37	-
	FG-NS	0.63	0.35	-
Constraint		0.6-1	0.35-0.5	-

Determination of the Volumetric Properties. The determination of the Optimum Asphalt Content (OAC) was carried out following the procedure described in MS-2 (Asphalt Institute, 2014). Accordingly, the mixing and the compaction temperatures were 161 °C and 152 °C, respectively. Loose test specimens were conditioned for 2 h after mixing at the specified compaction temperature (152 °C) to ensure mixture aging, then eventually the loose mixtures were compacted at a gyration number (Ndesign) of 75 gyrations to produce the compacted test specimens having a total mixture air voids percent (VTM) of 4% and having the desired volumetric properties of each mixture as recommended by Superpave specifications.

The 75 gyrations are intended to simulate the low traffic level which is equivalent to 0.3 to 3 million ESALs. Table 7 shows the Superpave average volumetric measures of the compacted test specimens for each mix. As can be concluded from both Table 8, the compacted test specimens' volumetric measures have met the Superpave volumetric criteria. The calculations of the volumetric measures are shown in Eq. 7 to 10. More details regarding the calculations of the volumetric measures can be found elsewhere [19, 32]. On average, coarse-graded mixes had higher OAC when compared to fine-graded. That is, more asphalt is needed to be absorbed by the larger aggregate particles to reach the same VTM as fine mixes (4%). However, larger air volumes (V_a) were formed, and lower asphalt portions covering the surface of the aggregate (V_{be}) were needed to reach a VTM of 4% for coarse-graded mixes due to lower surface area of the aggregate.

The DP is inversely proportional to the V_{be}, and therefore, the coarse-graded mixes had lower results of DP. The values of VMA and VFA are dependent on both V_a and V_{be}, and the trends of VMA and VFA of all mixes are challenging to identify. Concerning the coarse-graded mixes, the CG-NS had the highest OAC, highest V_{be}, and lowest V_a. The V_a values were very similar for both CG-QS and CG-QNS mixes, but OAC for CG-QS is slightly higher than CG-QNC which may be attributed to the fact that both gradations lines of CG-QS and CG-QNS were very similar. Concerning the fine-graded level, the FG-NS mix had the highest values OAC and V_{be} as well as the lowest V_a followed by FG-QNS, and FG-QS, respectively.

It is challenging to identify a specific trend for HMA volumetrics. However, the combination of NS and QS at the coarse-graded level reduced the V_{be} when compared to mixes that had completely either NS or QS in the mix. However, this was not noticed at the fine-graded level where it is clearly shown that the replacement of QS by NS always results in a higher aggregate surface area, lower OAC, lower V_a and higher V_{be}.

$$\text{VTM} = \frac{V_a}{\text{Total mix volume}} \times 100 \quad (7)$$

$$VMA = \frac{V_a+V_{be}}{\text{Total mix volume}} x\ 100 \quad (8)$$

$$VFA = \frac{VMA-VTM}{VMA} x\ 100 \quad (9)$$

$$DP = \frac{\text{Percent passing sieve No. 200 x Total mix volume}}{V_{be}} \quad (10)$$

Table 8. Summary of the volumetric properties

Volumetric Property	Mix Type						Constraint (0.3-3 million ESALs)
	Coarse-graded			Fine-graded			
	CG-QS	CG-QNS	CG-NS	FG-QS	FG-QNS	FG-NS	
OAC,%	4.36	4.30	4.44	4.18	4.21	4.31	-
VTM, %	4	4	4	4	4	4	4
VMA, %	15.3	15.1	15.0	15.0	15.2	14.6	14
VFA, %	73.9	73.5	73.3	73.3	73.7	72.6	70-80
DP	0.61	0.64	0.69	0.77	0.81	0.85	0.6-1.2
V_a, cm³	83.91	83.89	82.96	82.82	80.61	79.82	-
V_{be}, cm³	214.7	215.2	217.5	216.22	220.31	225.72	-

Beam Fatigue Test. Three-point flexural fatigue bending test was conducted to evaluate the fatigue performance of asphalt mixtures. By using the pneumatic repeated load system, the test was performed in strain-controlled mode with a flexural strain level of 1000 micro strain. The type of loading was haversine where the ultimate indirect tensile strength was applied at the frequency 10 Hz with 0.1 s loading and 0.9 s unloading times. All tests were conducted at 25°C on beam specimens of 76 mm (3 in) x 76 mm (3 in) x 381 mm (15 in) prepared according to the method described in AASHTO T 321 [33]. In the fatigue test, the initial Elastic Modulus (E_i) has been determined at the 200th repetition by using Eq. 11 shown below, and the E_i was plotted versus the number of repetitions to failure. The collapse of the beam was defined as a failure when E_i reaches the half indicating the Numbers of Cycles to Failure (N_f), and the plot can be approximated by a straight line and has the form shown below in Eq. 12.

$$\varepsilon = \frac{\sigma}{E} = \frac{12h\Delta}{3L^2-b^2} \quad (11)$$

$$N_f = k_1(E)^{-k_2} \quad (12)$$

Where ε is the tensile strain; σ is the flexural stress; E is the stiffness modulus based on center deflection; h is the height of the beam; Δ is the deflection at the center of the beam; L is the length of span between fixed supports; b is the distance from support to the load point (L/3); N_f is the numbers of cycles to failure; k_1 and k_2 are regression coefficients. The N_f is generally explained by the values of k_1 and k_2. k_1 is a constant directly related to N_f. On the other hand, k_2 represents the deterioration rate of the beam. That is, when k_1 value increases or k_2 value decreases, the N_f value increases indicating a longer life (more fatigue resistant) of the asphalt concrete.

Results and Discussion

Tables 9 and 10 show the results of the BF results for CG mixtures and FG mixtures, respectively. It can be seen from both tables that the increase of NS content in mixtures always results in lower k_1 and E_i as well as a higher k_2 which resulted in lower N_f overall. Moreover, the CG mixtures resulted in higher N_f when compared to FG mixtures with an average increase of 104%. This is due to the improved aggregate interlock in CG mixtures caused by the presence of the coarse aggregate.

The CG-QNS had a comparable Nf with CG-QS with a percentage difference of 0.1% which indicates the proficiency of Bailey method of creating similar gradations and similar aggregate interlock at the coarse gradation level. However, the FG-QNS didn't have a comparable Nf with FG-QS where the percentage difference was 18%. This could be attributed to the absence of the course aggregate which creates a considerable interlock in the aggregate structure and improves the stability of the asphalt concrete. That is, Bailey method is effective if used only in a coarse mix design.

Table 9. BF Results of CG mixtures

Property	Mixture Type					
	CG-QS		CG-QNS		CG-NS	
	Sample 1	Sample 2	Sample 1	Sample 2	Sample 1	Sample 2
k_1	2319.15 x 10^{20}	2310.14 x 10^{20}	2301.13 x 10^{20}	2305.64 x 10^{20}	2111.85 x 10^{20}	1922.57 x 10^{20}
k_2	6.616	6.619	6.622	6.621	6.656	6.689
E_i, MPa	1409.2	1404.3	1399.4	1401.85	1380.55	1361.7
N_f	33386	33362	33338	33350	27485	21632
$N_{f\,average}$	33374		33344		24559	

Table 10. BF Results of FG mixtures

Property	Mixture Type					
	FG-QS		FG-QNS		FG-NS	
	Sample 1	Sample 2	Sample 1	Sample 2	Sample 1	Sample 2
k_1	1769.82 x 10^{20}	1846.2 x 10^{20}	1675.48 x 10^{20}	1722.65 x 10^{20}	1218.50 x 10^{20}	1446.99 x 10^{20}
k_2	6.771	6.73	6.793	6.782	6.810	6.8015
E_i, MPa	1287.5	1324.6	1276.7	1282.1	1259.8	1268.25
N_f	16922	19277	14824	15873	10181	12503
$N_{f\,average}$	18100		15349		11342	

Statistical Assessment. The volumetric properties of a compacted mixture are important criteria by which the quality of an asphalt mixture is evaluated. Volumetrics historically provided a reliable indication of the mixture's probable performance during its service life. It is important to note that all of the asphalt mix volumetrics are mathematically interrelated. However, they may differ in their significance and corresponding sensitivities to the performance tests.

An attempt was made in the current study to identify the significance and the sensitivity (importance) of different volumetrics including DP, VFA, VMA, Va, and Vbe to each performance test output. Fig. 3 shows the trend lines of a number of volumetric measures with Nf including Va, Vbe, VMA, VFA, and DP. It can be seen that Nf was inversely related to both Vbe and DP. In contrast, Nf had a direct relationship Va, VMA nad VFA. Moreover, the data points dispersion is lower Va, Vbe, and DP indicating an improved relationship with Nf.

Fig.4 show the sensitivities and the significance of each volumetric explained by the tornado chart. The importance of each volumetric measure is based on the correlation coefficient (r), and the significance of each volumetric measure is based on the p-value at a risk level of 0.05. In other words, the higher the correlation, the higher the sensitivity and the higher the significance. As can be seen from Figure 4, all volumetric measures were significant at a risk level of 0.05 (p-value ≤ 0.05), but varied in importance or sensitivities. That is, although the correlation values were close due to the mathematical interrelationships between them, these mathematical relations have minor differences that lead to different correlations with the output of any performance test. Eq. 7 to 10 again show the calculations of the volumetric measures. In essence, DP was the most sensitive (important) to N_f followed by V_a, V_{be}, VMA, and VFA, respectively. This is because DP results in more distinctive values when compared to other volumetrics and the parameters used to calculate are mainly shared and included in the calculations of all volumetric measures.

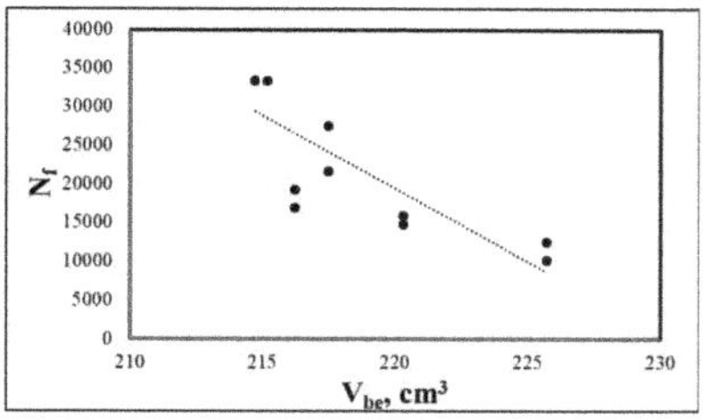

(a) V_{be}

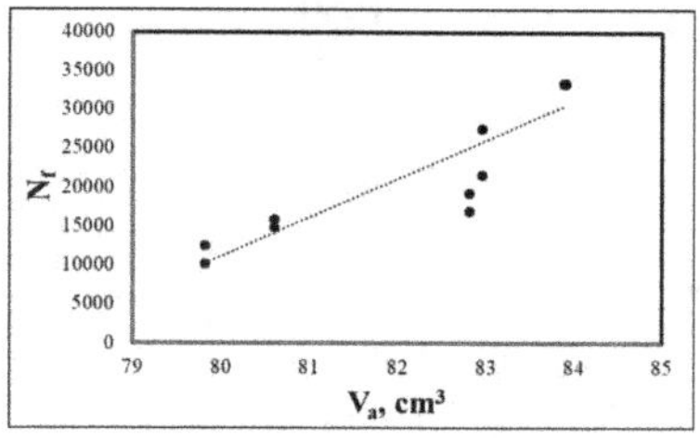

(b) V_a

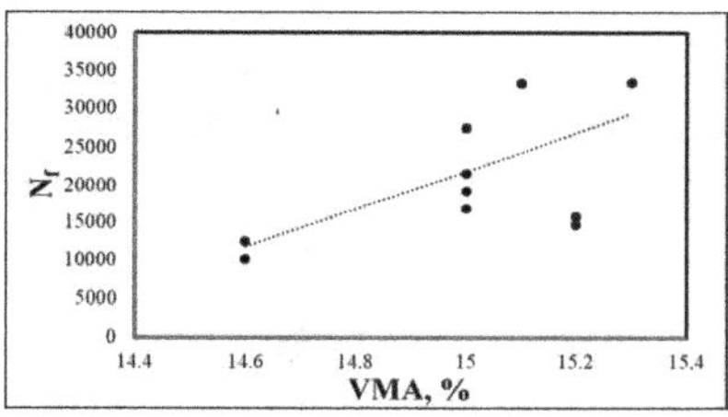

(c) VMA

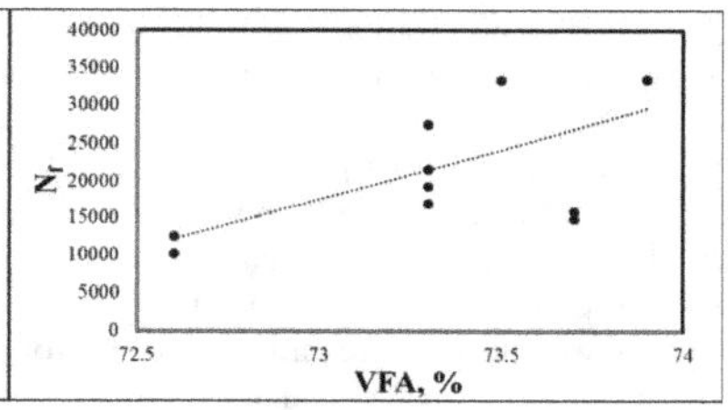

(b) VFA

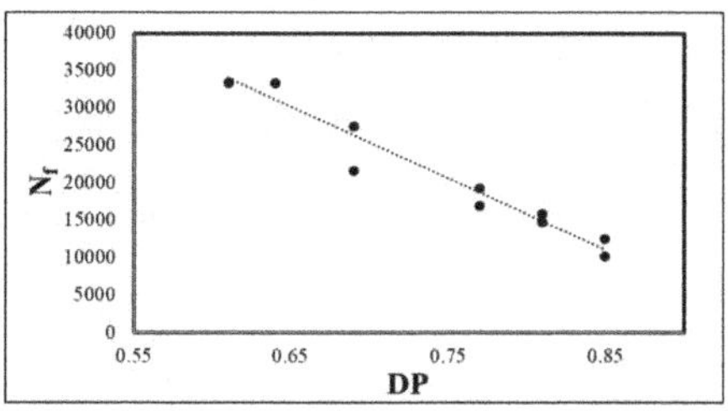

(e) DP

Figure 3. Volumetric measures relationship with N_f: (a) V_{be}, (b) V_a, (c) VMA, (d) VFA, and (e) DP

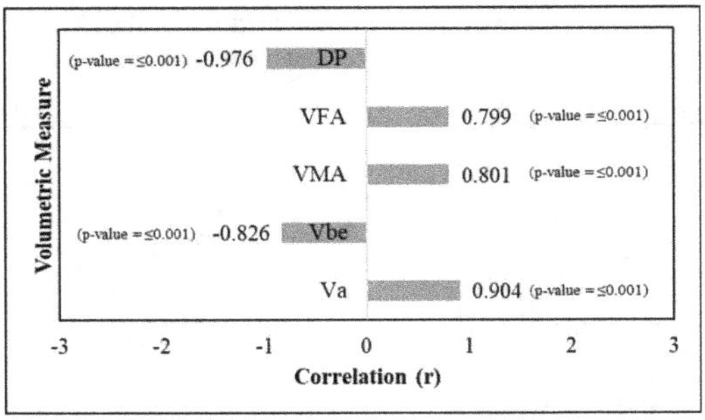

Figure 4. Volumetric measures sensitivities and significance

To this end, it is shown that N_f is highly dependent on aggregate structure that is correspondent to the volumetric measures which, in turn, had a significant influence on N_f. Therefore, it would be advantageous to generate statistical relationships between the volumetric measures and N_f. Khasawneh et al. [34] showed that curve-fitting regression can improve prediction and provide more reliable results than linear regression. The non-linear curve fitting regression analysis was carried out. To run the nonlinear analysis, curve fitting software is needed to select the best fit for each predictor from a pool of existing functions. MATLAB R2022a and CurveExpert Professional 2.7 are powerful software and easy to use for this purpose. Multiple functions are fitted with data and the best-fit function can be chosen based on the Coefficient of Determination (R^2) value. Accordingly, Table 10 shows the relationships between V_{be} and each performance test output. The R^2 values are also provided in the table. It can be noticed R^2 values are consistent with the correlation values shown in Figure 4. That is, the DP model had the highest R^2 followed by V_a, V_{be}, VMA, and VFA, respectively. That is, the DP was the best predictor for N_f since DP results in more distinctive values when compared to other volumetrics and the parameters used to calculate is mainly shared and included in the calculations of all volumetric measures.

Table 10. Statistical relationships of different volumetric measures with N_f

Volumetric Measure	Model	R^2
DP	N_f=219.7- 6923.4 ln(DP)	0.956
V_{be}	$N_f = (-2.194 \times 10^{-7} + 1.0292 \times 10^{-9} V_{be})^{-\frac{1}{1.93}}$	0.817
V_a	$N_f = (8.36 \times 10^{-7} - 1.0187 \times 10^{-8} V_a)^{-\frac{1}{1.782}}$	0.930
VMA	$N_f = 1.12135 \times 10^{12}\, e^{-\frac{266.7}{VMA}}$	0.425
VFA	N_f=- 4206864 + 984603 ln(VFA)	0.418

Conclusions

1. The fine-graded mixes have lower optimum asphalt content compared to coarse-graded ones. That is the surface area for fine-graded mixes are higher than coarse-graded, and this resulted in less need for asphalt to reach the same air voids level of 4%.
2. Increasing NS content in the mixture reduced the optimum asphalt content, since the presence of the natural sand fills the voids between angular aggregate (fine and coarse).

3. DP, V_a and V_{be} result in more distinctive values than VTM, VMA, DP and VFA and are mainly included in the calculations of all volumetric measures. However, DP had more distinctive values than V_a and V_{be}.
4. The increase of NS content in mixtures always results in lower k_1 and E_i as well as a higher k_2 which resulted in lower N_f overall.
5. The CG mixtures resulted in higher N_f when compared to FG mixtures with an average increase of 104% due to the enhanced aggregate interlock when the coarser aggregate are used in the mixture.
6. The CG-QNS had a comparable N_f with CG-QS with a percentage difference of 0.1% which indicates the proficiency of Bailey method of creating similar gradations and similar aggregate interlock.
7. The FG-QNS didn't have a comparable N_f with FG-QS where the percentage difference was 18%.
8. The volumetric measures including VMA, DP, VFA, V_a and V_{be} were significant to the performance indices. However, DP was the most significant and sensitive followed by V_a, V_{be}, VMA, and VFA, respectively.
9. Bailey method is effective if used only in a coarse mix design. That is, the absence of the course aggregate creates a considerable interlock in the aggregate structure and improves the stability of the asphalt concrete.

References

[1] Pouranian, M. Reza, and John E. Haddock. "Determination of voids in the mineral aggregate and aggregate skeleton characteristics of asphalt mixtures using a linear-mixture packing model." *Construction and Building Materials* 188 (2018): 292-304. https://doi.org/10.1016/j.conbuildmat.2018.08.101

[2] Kamaruddin, Ibrahim, Madzlan Napiah, and Yasreen Gasm. "The effect of fine aggregate properties on the rutting behavior of the conventional and polymer modified bituminous mixtures using two types of sand as fine aggregate." (2010): 89-97. https://doi.org/10.3141/2205-03

[3] Arabani, M., and A. R. Azarhoosh. "The effect of recycled concrete aggregate and steel slag on the dynamic properties of asphalt mixtures." *Construction and Building Materials* 35 (2012): 1-7. https://doi.org/10.1016/j.conbuildmat.2012.02.036

[4] Almadwi, Fathi S., and Gabriel J. Assaf. "Finding an optimal bitumen and natural sand balance for hot mix asphalt concrete in hot and arid regions." In *Solving Pavement and Construction Materials Problems with Innovative and Cutting-edge Technologies: Proceedings of the 5th GeoChina International Conference 2018–Civil Infrastructures Confronting Severe Weathers and Climate Changes: From Failure to Sustainability, held on July 23 to 25, 2018 in HangZhou, China*. Springer International Publishing (2019): 1-12. https://doi.org/10.1007/978-3-319-95792-0_1

[5] Albayati, Amjad H., and Harith Abdulsattar. "Performance evaluation of asphalt concrete mixes under varying replacement percentages of natural sand." *Results in Engineering* 7 (2020): 100131. https://doi.org/10.1016/j.rineng.2020.100131

[6] Ahlrich, Randy C. *The effects of natural sands on asphalt concrete engineering properties*. Army engineer waterways experiment station vicksburg ms geotechnical lab (1991).

[7] Abedali, Abdulhaq Hadi, and Yasir Mohammed Jebur. "Reduced Permanent Deformation of Asphalt Pavement by Enhancing Aggregate Gradation." In *IOP Conference Series: Materials*

Science and Engineering, vol. 671, no. 1, p. 012082. IOP Publishing (2020). https://doi.org/10.1088/1757-899x/671/1/012082

[8] Fernandes Jr, Jose Leomar, and L. T. D. Gouveia. "Limitation of the Fine Aggregate Angularity (FAA) Test to Predict the Behavior of Asphalt Mixtures." *Department of Transportation, University of Sao Paulo, Brazil* (2003): 1-9.

[9] Shen, Der-Hsien, Ming-Feng Kuo, and Jia-Chong Du. "Properties of gap-aggregate gradation asphalt mixture and permanent deformation." *Construction and Building Materials* 19, no. 2 (2005): 147-153. https://doi.org/10.1016/j.conbuildmat.2004.05.005

[10] Moghaddam, Taher Baghaee, Mohamed Rehan Karim, and Mahrez Abdelaziz. "A review on fatigue and rutting performance of asphalt mixes." *Scientific Research and Essays* 6, no. 4 (2011): 670-682. https://doi.org/10.14311/ee.2016.015

[11] Ramli, Izzul. "Effect of Fine Aggregate Angularity on Rutting Resistance of Asphalt Concrete AC10." PhD diss., Universiti Teknologi Malaysia (2013).

[12] Park, Dae Wook, and Hyung Seok Lee. "Test methods for fine aggregate angularity considering resistance of rutting." *KSCE Journal of Civil Engineering* 6 (2002): 421-427.

[13] Stuart, Kevin D., and Walaa S. Mogawer. *Evaluation of natural sands used in asphalt mixtures*. No. 1436 (1994).

[14] Freeman, Reed B., and Chun-Yi Kuo. "Quality control for natural sand content of asphalt concrete." *Journal of transportation engineering* 125, no. 6 (1999): 539-546. https://doi.org/10.1061/(asce)0733-947x(1999)125:6(539)

[15] Ghuzlan, Khalid A., Al-Mistarehi Bara'W, and Ahmed S. Al-Momani. "Rutting performance of asphalt mixtures with gradations designed using Bailey and conventional Superpave methods." *Construction and Building Materials* 261 (2020): 119941. https://doi.org/10.1016/j.conbuildmat.2020.119941

[16] Sivasubramaniam, Sivaranjan, Khaled A. Galal, A. Samy Noureldin, Thomas D. White, and John E. Haddock. "Laboratory, prototype, and in-service accelerated pavement testing to model permanent deformation." *Transportation Research Record* 1896, no. 1 (2004): 189-198. https://doi.org/10.3141/1896-19

[17] Zhu, Wei, Hong Zhen Li, Shi Rang Ma, and Da Bin Liu. "Application of Bailey Method for Aggregate Grading Design of Continuous Dense Gradation Asphalt Mixture." In *Advanced Materials Research*, vol. 413, pp. 154-159. Trans Tech Publications Ltd, 2012. https://doi.org/10.4028/www.scientific.net/amr.413.154

[18] Vavrik, William R., William J. Pine, and Samuel H. Carpenter. "Aggregate blending for asphalt mix design: Bailey method." *Transportation Research Record* 1789, no. 1 (2002): 146-153. https://doi.org/10.3141/1789-16

[19] MS-2: Asphalt Mix Design Methods, 7th edition, Asphalt Institute, Manual Series No. 02 (2014).

[20] ASTM, D., 4402/4402M-15. Standard Test Method for Viscosity Determination of Asphalt at Elevated Temperatures Using a Rotational Viscometer, ASTM International, West Conshohocken, PA (2015). https://doi.org/10.1520/d4402_d4402m-12

[21] ASTM, D., 7175-15. Standard Test Method for Determining the Rheological Properties of Asphalt Binder Using a Dynamic Shear Rheometer, ASTM International, West Conshohocken, PA (2015). https://doi.org/10.1520/d7175

[22] ASTM, D., 2872-12. Standard Test Method for Effect of Heat and Air on a Moving Film of Asphalt (Rolling Thin-Film Oven Test), ASTM International, West Conshohocken, PA (2012). https://doi.org/10.1520/d2872-04

[23] ASTM, D., 6521-18. Standard Practice for Accelerated Aging of Asphalt Binder Using a Pressurized Aging Vessel (PAV), ASTM International, West Conshohocken, PA (2018). https://doi.org/10.1520/d6521-04

[24] ASTM, D., 6816-11, Standard Practice for Determining Low-Temperature Performance Grade (PG) of Asphalt Binders, ASTM International, West Conshohocken, PA (2011). https://doi.org/10.1520/d6816-11

[25] ASTM, D., 4791-10. Standard Test Method for Flat Particles, Elongated Particles, or Flat and Elongated Particles in Coarse Aggregate, ASTM International, West Conshohocken, PA (2010). https://doi.org/10.1520/d4791-05e01

[26] ASTM, D., 5821-13, 2017.Standard Test Method for Determining the Percentage of Fractured Particles in Coarse Aggregate, ASTM International, West Conshohocken, PA. https://doi.org/10.1520/jte12077j

[27] ASTM, C., 1252-17. Standard Test Methods for Uncompacted Void Content of Fine Aggregate (as Influenced by Particle Shape, Surface Texture, and Grading), West Conshohocken, PA: ASTM International (2017). https://doi.org/10.1520/c1252-06

[28] AASHTO, T 84/ 85. Specific gravity and absorption of fine aggregate. Washington, DC. n.d (2002).

[29] AASHTO, M 323. Standard Specification for Superpave Volumetric Mix Design Washington, DC. n.d (2017).

[30] Khasawneh, Mohammad Ali, and Mohammad Ahmad Alsheyab. "Effect of nominal maximum aggregate size and aggregate gradation on the surface frictional properties of hot mix asphalt mixtures." *Construction and Building Materials* 244 (2020): 118355. https://doi.org/10.1016/j.conbuildmat.2020.118355

[31] ASTM, C., 29/29M-07. Standard Test Method for Unit Weight and Voids in Aggregate, West Conshohocken, PA: ASTM International (2007). https://doi.org/10.1520/c0029_c0029m-16

[32] Alsheyab, Mohammad Ahmad, and Mohammad Ali Khasawneh. "Quantifying the effect of modified mixture volumetrics and compaction effort on skid resistance of asphalt pavements." *International Journal of Pavement Engineering* 23, no. 5 (2022): 1552-1560. https://doi.org/10.1080/10298436.2020.1810688

[33] AASHTO, T 321. Standard Method of Test for Determining the Fatigue Life of Compacted Asphalt Mixtures Subjected to Repeated Flexural Bending. Washington, DC. n.d (2022).

[34] Khasawneh, Mohammad Ali, Ansam Adnan Sawalha, Mohannad Theeb Aljarrah, and Mohammad Ahmad Alsheyab. "Effect of aggregate gradation and asphalt mix volumetrics on the thermal properties of asphalt concrete." *Case Studies in Construction Materials* 18 (2023): e01725. https://doi.org/10.1016/j.cscm.2022.e01725

Advanced Topics in Mechanics of Materials, Structures and Construction - AToMech1-2023 Materials Research Forum LLC
Materials Research Proceedings 31 (2023) 262-268 https://doi.org/10.21741/9781644902592-27

Stone dust as replacement for fine aggregate in cellular lightweight concrete (CLC): Volume weight and compressive strength

PAREA R. RANGAN[1,a*], M. TUMPU[2,b], Y. SUNARNO[1,c] and MANSYUR[3,d]

[1]Civil Engineering Department, Faculty of Engineering, Christian University of Indonesia, Toraja, Indonesia

[2]Disaster Management Study Program, The Graduate School, Hasanuddin University, Makassar, Indonesia

[3]Civil Engineering Department, Faculty of Engineering, SembilanBelas November University, Kolaka, Indonesia

[a]pareausanrangan68@gmail.com, [b]tumpumiswar@gmail.com, [c]yohanssunarno@gmail.com, [d]mansyurusn14@gmail.com

Keywords: Stone dust, Fine Aggregate, CLC, Volume Weight, Compressive Strength

Abstract. A type of lightweight concrete called cellular lightweight concrete (CLC) has a lower volume weight than ordinary concrete and is made of cement, sand, water, and a foaming agent. In this investigation, CLC lightweight concrete was utilized as a lightweight brick. The purpose of this study was to ascertain how the compressive strength and volume weight of CLC lightweight bricks would change if stone dust were used in place of fine aggregate. Making lightweight bricks with stone dust substitutions of 0%, 50%, and 100% of the weight of the fine aggregate was the experimental process used in this study. The compressive strength and volume weight of the light-weight bricks that had been cured for 3, 7, 14, and 28 days were then measured. Because stone dust has a good binding capacity, the results showed that using it as a fine aggregate replacement in a mixture of lightweight bricks increased the compressive strength; the highest compressive strength value was obtained at a substitution of 100% stone dust at 28 days, which was 24.62 kg/cm^2. The volume weight of a mixture of lightweight bricks increased by 0.66 gr/cm^3 when stone dust was used in place of fine aggregate. In place of 50% stone dust, the volume weight value increased by 2% to 0.67 gr/cm^3, and by 4% to 0.68 gr/cm^3 for 100% stone dust. Due to its volume weight range of 0.6-1.8 gr/cm^3, this lightweight brick can still be designated as lightweight concrete.

Introduction

A porous brick that is lighter in density than regular brick is said to be lightweight. Cellular lightweight concrete (CLC) is one of various forms of lightweight bricks available on the market [1-3]. In general, Portland cement, fine aggregate, water, and a foam agent were combined to make CLC lightweight bricks [4-6]. By adding foam to a concrete mortar mixture without generating a chemical reaction, CLC lightweight bricks are created. Air bubbles were kept stable during rapid mixing using a foaming agent. Both natural and artificial materials can be used to generate foam agents [7–9].

The CLC lightweight brick business had to grapple with the problem of how to boost the compressive strength of CLC lightweight bricks because the strength produced was still quite low while maintaining the density value, keeping the bricks light. As a result, both in terms of the mixed materials and the composition of the mixture employed in the manufacturing process, new developments in the production of CLC lightweight bricks continue to be developed. lightweight CLC bricks Lesovik et al., 2020 [10], who did experimental study on the use of silica sand as a substitute for fine aggregate to improve the performance of foam concrete, carried out some of these investigations. It was shown that the compressive strength and indirect tensile strength of foam concrete could be maximized by utilizing silica sand and adding calcium. White dirt was found to boost the compressive strength of lightweight bricks in a study by Amran et al. (2015 [11], which looked at the experimental examination of the compressive strength and water absorption of CLC with that material as an aggregate.

Figure 1. *Cellular Lightweight Concrete (CLC)*

It was vital to find other options by utilizing the abundant natural resources since as development expanded, so did the demand for fundamental building construction materials. For instance, stone dust from the waste processing of coarse aggregate, which granules pass through perforated sieve by 4.75 mm and kept sieved by 0.075 mm, can be utilized as a light-weight brick-making material in Toraja, South Sulawesi Province.

On the other hand, the local population exclusively utilized stone dust for straightforward building tasks like printing stone and other such tasks. The stone dust employed in this study was created utilizing a used stone crusher in Buakayu, Bonggakaradeng District, Toraja. The findings of a study by A. Haris Ratih in 2016 [12] that looked at how using stone ash affected the compressive strength of concrete of the K-350 quality indicated that the stone ash-containing mixture could have an impact on the compressive strength of concrete.

The compressive strength decreased with the amount of stone ash utilized, but the 350 kg/cm^2 limit could still be reached. A combination containing 40% stone ash had the highest average compressive strength. In this study, it is envisaged that the usage of stone dust can be employed as a substitute for fine aggregate in the composition of lightweight bricks and can enhance the quality of lightweight bricks with targeted compressive strength.

Materials and Methodology

Lightweight Brick Physical Requirement

The Indonesian National Standard (SNI) requirement has not been found to address the viability of lightweight bricks. Based on this, SNI 03-3449-2002 [13] was utilized as a criteria to be employed for lightweight bricks regarding concrete bricks for masonry walls [14]. The lightweight brick reference C will be used to determine the physical requirements for the feasibility of concrete bricks.

Table 1. Lightweight brick physical requirement [14]

Physical requirement	Unit	Lightweight brick quality level			
		I	II	III	IV
Minimum average of compressive strength	kg/cm^2	100	70	40	25
Bruto compressive strength of specimen	kg/cm^2	90	65	35	21
Minimum average of water absorption	%	25	35	-	-

Materials Research Proceedings 31 (2023) 262-268 https://doi.org/10.21741/9781644902592-27

Stone Dust

Stone dust was an artificial aggregate-based building material (minerals or fillers with a particle size generally less than 0.075 mm, which is a by-product or processed crushed stone used in stone crushers). Stone ash was sharp and gray in color, with a fine-grained texture. It has pozzolanic components, is tough and long-lasting (it contains silica and alumina compounds that are not cementitious, but their smooth form when mixed with water can turn into a solid mass).

Stone dust's water content, volume weight in a loose state, volume weight in a solid state, and sludge content were all examined for their physical properties. The results were 3.1%, 1.3, 1.5, and 2.4%, respectively. 2.3, 2.4, 2.5, and 4.2%, respectively, were the results of the bulk specific gravity, saturated surface dry specific gravity, apparent specific gravity, and water absorption.

Fine Aggregate

River sand, the fine aggregate employed in this study, had water content values of 4.2%, 1.2, 1.4, and 3.4%, respectively, as well as volume weights in loose, solid, and sludge conditions. 2.4, 2.5, 2.7, and 4.6%, respectively, were the results of the bulk specific gravity, saturated surface dry specific gravity, apparent specific gravity, and water absorption.

Foam Agent

Foam agent is one of the foam agents made from hydrolyzed protein-based ingredients. Natural ingredients in the form of proteins with a density of 80 grams per liter can be used as foam agent-forming substances. In order to generate lightweight concrete, this foam ingredient had to stabilize air bubbles during quick mixing.

Research Design

Making lightweight bricks with stone dust substitutions of 0%, 50%, and 100% of the weight of the sand was the experimental procedure employed in this investigation. We examined lightweight bricks to determine their compressive strength and volume weight value after curing them for 3, 7, 14, and 28 days. Table 2 lists each of the 36 test objects that were used in this investigation.

Table 2. The number of test object

Age (Day)	Normal lightweight briks	Subtitusion of stone dust	
		50%	100%
3	3	3	3
7	3	3	3
14	3	3	3
28	3	3	3
Σ	12	12	12
Total	36		

Results and Discussion

Volume Weight of CLC Lightweight Bricks

Figure 2 displays the volume weight calculations for standard CLC lightweight bricks and CLC lightweight bricks with 50% and 100% substitutes of stone dust. As can be seen, the typical CLC lightweight bricks had an average volume weight of 0.75 grams per cubic centimeter at 3 days, 0.75 gram/cm^3, at 7 days, 0.75 gram/cm^3, at 14 days, 0.72 gram/cm^3 and 0.66 gram/cm^3 at 28 days. The average volume weight of lightweight bricks with CLC substitution of 50% stone dust at the age of 3 days was 0.81 gram/cm^3, at the age of 7 days it was 0.77 gram/cm^3, at the age of 14 days it was 0.76 gram/cm^3, and at the age of 28 days it was 0.67 gram/cm^3. The average volume weight of CLC lightweight substitution of 100% stone dust at the age of 3 days was 0.83 gram/cm^3, at the age of 7 days it was 0.78 gram/cm^3, at the age of 14 days it was 0.77 gram/cm^3, and at the age of 28 days it was 0.68 gram/cm^3.

The results show that all the specimens were lightweight concrete in accordance with SNI 03:2847:2013, namely the volume weight for lightweight concrete is 650–1840 kg/m^3, while the volume weight of foam concrete produced is an average of 770 kg/m^3 [15]. Foam usage had a significant impact on the weight and volume of foam concrete. The foam concrete became lighter and had a decreased compressive strength as more foam was added.

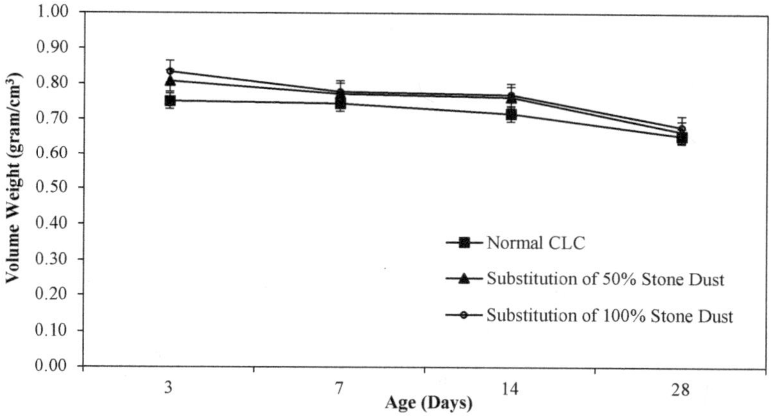

Figure 2*. Volume weight of Cellular Lightweight Concrete (CLC)*

Compressive Strength of CLC Lightweight Bricks

Figure 3 displays the outcomes of the lightweight bricks' compressive strength test at ages 3, 7, 14, and 28 days. As can be shown, the typical CLC lightweight bricks had an average compressive strength of 8.54 kg/cm^2 at three days, 13.44 kg/cm^2 at seven days, 21.90 kg/cm^2 at fourteen days, and 22.66 kg/cm^2 at twenty-eight days. Lightweight bricks with a CLC substitution of 50% stone dust had an average compressive strength of 9.67 kg/cm^2 after three days, 13.60 kg/cm^2 after seven days, 22.21 kg/cm^2 after fourteen days, and 23.26 kg/cm^2 after 28 days. At the ages of 3, 7, 14, and 28, the compressive strength of lightweight bricks with 100% stone dust CLC substitution was, on average, 10.12 kg/cm^2, 15.11 kg/cm^2, 23.42 kg/cm^2, and 24.62 kg/cm^2.

All mixes' compressive strengths rose as the specimens aged, which is consistent with the behavior of concrete made with Portland cement. As the material aged, the compressive strength of foam concrete kept rising. The stabilization of the bubbles that had developed as a result of the response of composite portland cement when mixing foam into concrete led to the rise in compressive strength, which subsequently persisted with the hydration process lasting until the age of 28 days. The compressive strength of foam concrete can be impacted by the addition of foam to the mixture [16,17].

The amount of load that the lightweight bricks can support depends on the amount of foam added to the lightweight concrete mixture. According to the test results, the amount of foam agent added to the foam concrete mixture significantly affected the load that the lightweight bricks could support. It is evident from the resulting load value that as the volume of foam utilized rose, the load value that the lightweight bricks could support declined. In all combinations of lightweight concrete, the presence of gas results in the development of a deflection value that contrasts with a considerable difference; frequently, this physical attribute is utilized to identify the zone of gas accumulation [18-21].

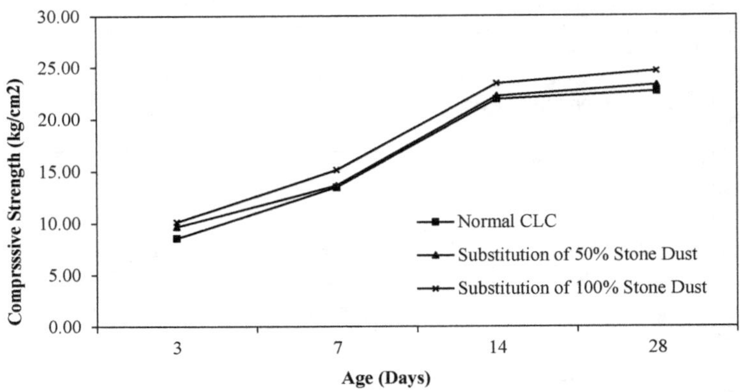

Figure 3. *Compressive strength of Cellular Lightweight Concrete (CLC)*

Conclusion

It was possible to combine PCC cement, sand, and stone dust to create a mortar that, when combined with foam, produced lightweight concrete that demonstrated good compaction. The rise in the compressive strength of lightweight concrete bricks from the age of 3 to 28 days demonstrates that the good compatibility of all the ingredients employed allows the hardening of the concrete to proceed smoothly during the binding process.

The volume weight of cellular lightweight concrete (CLC) increased by 0.66 gram/cm^3 over time when stone dust was used as a fine aggregate replacement. In place of 50% stone dust, the volume weight value increased by 2% to 0.67 gram/cm^3, and by 4% to 0.68 gram/cm^3 for 100% stone dust. Due to its volume weight range of 0.600 to 1,800 gram/cm^3, this light brick can still be referred to as being light.

Because stone dust has strong bonding power, it can be used as a fine aggregate replacement in cellular lightweight concrete (CLC) mixtures to boost compressive strength. The highest compressive strength value is obtained by substituting 100% stone dust aged for 28 days at 24,62 kg/cm2, which is in accordance with Indonesian requirements for lightweight concrete bricks for infilled walls. However, due to the influence of foam agents, which cause pores in lightweight bricks, the compressive strength does not meet the requirements to be used as wall pairs.

References

[1] Sunarno. Y, Tjaronge. MW. and Irmawaty. R., "Preliminary Study on Early Compressive Strength of Foam Concrete Using Ordinary Portland Cement (OPC) and Portland Composite Cement (PCC)," IOP Conf. Ser.: Earth Environ. Sci, vol. 419 (2020), 012033.https://doi.org/10.1088/1755-1315/419/1/012033

[2] Tumpu. M, Parung. H, Tjaronge. MW, and Amiruddin. A, A., "Failure Pattern of Prefabricated Foam Concrete as Infill Wall Under In-Plane Lateral Loading," Design Engineering, issue 7 (2021), 7168-7178. ISSN: 0011-9342.

[3] Amirudin A.A, Parung H., Tjaronge M.W., Mansyur M., Tumpu M. 2022. Influence of prefabricated foam concrete as infilled wall on the strength due to cyclic loading. International Journal of GEOMATE. 22, pp 114-121. https://doi.org/10.21660/2022.93.j2343

[4] Syahrul, Tjaronge. MW, Djamaluddin. R, and Amiruddin. AA., "Flexural Behavior of Normal and Lighweight Concrete Composite Beams," Civil Engineering Journal. vol. 07:03 (2021), 549-559. https://doi.org/10.28991/cej-2021-03091673

[5] Rangan, PR., and Tumpu M., "Effect Calcium Hydroxide (Traditionally Called Slaked Lime) to Stabilization of Laterite Soil," IOP Conf. Series: Materials Science and Engineering 1088 (2021). https://doi.org/10.1088/1757-899X/1088/1/012105

[6] Bindiganavile. V, and Hoseini. M, "Foamed Concrete," Developments in the Formulation and Reinforcement of Concrete," Woodhead Publishing Series in Civil and Structural Engineering, (2019), 365-390. https://doi.org/10.1016/B978-0-08-102616-8.00016-2

[7] Narayanan. N, Ramamurthy. K, "Structure and Properties of Aerated Concrete: Review, "Cem. Concr. Compos, vol. 22:5 (2000), 321-329. https://doi.org/10.1016/S0958-9465(00)00016-0

[8] Rangan, PR., Tumpu M., Caroles L., and Mansyur, "Compressive Strength of High-Strength Concrete With Cornice Adhesive as a Partial Replacement for Cement," IOP Conf. Series: Earth and Environmental Science 871 (2021) 012006. https://doi.org/10.1088/1755-1315/871/1/012006

[9] Hajimohammadi A, Ngo T, Mendis. P, "Enhancing the Strength of Pre-Made Foams for Foam Concrete Applications," Cem. Concr. Compos, vol. 87 (2018), 164-171. https://doi.org/10.1016/j.cemconcomp.2017.12.014

[10] Lesovik. V, Voronov. V, Glagolev. E, Fediuk. R, Alaskhanov. A, Amran. YHM, Murali. G, and Baranov. A., "Improving the Behaviors of Foam Concrete Through The Use of Composite Binder," Journal of Building Engineering, vol. 31 (2020), 101414. https://doi.org/10.1016/j.jobe.2020.101414

[11] Amran. YHM, Farzadnia. N, Ali AA., "Properties and Applications of Foamed Concrete; A Review," Constr. Build. Mater, vol. 101 (2015), 990-1005. https://doi.org/10.1016/j.conbuildmat.2015.10.112

[12] A. Haris HA. The Effect of Using Stone Ash on the Compressive Strength of K-350 Quality Concrete. Civil Engineering Study Program, Adhi Tama Institute of Technology Surabaya (in Indonesian).

[13] Standard National of Indonesia. Standard Test Spesification for Lightweight Aggregates for Structural Concrete. SNI 03-3449-2002.

[14] Valore RC. Cellular concrete part 2 physical properties. ACI J 1954; 50:817-36. https://doi.org/10.14359/11795

[15] Tumpu M. and Mabui D. S. 2022. Effect of Hydrated Lime (Ca(OH)2) to Compressive Strength of Geopolymer Concrete. AIP Conference Proceedings 2391, 070011 (2022). https://doi.org/10.1063/5.0086702

[16] Mansyur and Tumpu M. 2022. Compressive Strength of Normal Concrete Using Local Fine Aggregate from Binang River in Bombana district, Indonesia. AIP Conference Proceedings 2391, 070010 (2022). https://doi.org/10.1063/5.0072888

[17] Mansyur, Amiruddin A. A., Parung H., Tjaronge M. W. and Tumpu M. 2021. Utilization of Sea Water to Production of Concrete in Terms of Mechanical Behavior. IOP Conf. Series: Earth and Environmental Science 921 (2021) 012068. https://doi.org/10.1088/1755-1315/921/1/012068

[18] Tumpu, M., Parung H., Tjaronge, M. W., Amiruddin, A. A. 2021. Failure Pattern of Prefabricated Foam Concrete as Infill Wall Under In-Plane Lateral Loading. Design Engineering, Issue: 7, Pages: 7168-7178.

[19] Mansyur and Tumpu M. 2022. Compressive Strength of Non-Sand Concrete with Coarse Aggregate in Kolaka District as Yard Pavement. AIP Conference Proceedings 2391, 070022 (2022). https://doi.org/10.1063/5.0072889

[20] Rangan, Parea Rusan and Tumpu, M. 2022. The Potential Utilization of Candlenut Shell Waste as Coarse Aggregate Replacement in Concrete. Design Engineering, Issue: 1, Pages:458-463.

[21] Rangan, Parea Rusan, Tumpu, M., Caroles, L., Mansyur. Compressive strength of high-strength concrete with cornice adhesive as a partial replacement for cement. IOP Conference Series: Materials Science and Engineering, https://iopscience.iop.org/issue/1755-1315/871/1. https://doi.org/10.1088/1755-1315/871/1/012006

Advanced Topics in Mechanics of Materials, Structures and Construction - AToMech1-2023 Materials Research Forum LLC
Materials Research Proceedings 31 (2023) 269-280 https://doi.org/10.21741/9781644902592-28

Contemporary conceptual models for predicting carrying capacity of multi helices screw piles

Tahar AYADAT[1,a *], Hocine LEFKIR[2,b], Zahar SAID HADDAD[3,c], Danish AHMED[1,d] and Andi ASIZ [1,e]

[1]Prince Mohammad Bin Fahd University, Saudi Arabia

[2]Civil Protection, BBA, Algeria

[3]Municipality, BBA, Algeria

[a]tayadat@pmu.edu.sa, [b]hocine.34@live.fr, [c]saidhaddadzahar@gmail.com, [d]dahmed@pmu.edu.sa, [e]aasiz@pmu.edu.sa

Keywords: Screw Pile, Multi Helices, Axial Loading, Theoretical Model, Carrying Capacity

Abstract. Contemporary civil constructions are based on foundation systems consisting mainly of shallow, semi deep and deep foundations. The screw pile resembles essentially to a large screw that is inserted into the ground by a special machine. Basically, it acts as a foundation or an anchor. In general, it is composed of a central shaft and one or more graded steel helices. Screw piles have made great progress in recent decades, particularly when it comes to their axial capabilities and installation processes. Using screw piles for anchoring and supporting structures is advancing rapidly. Despite this, screw piles are still lacking design methods that require further research. This study consists on developing three conceptual models to predict the carrying capacity of a screw pile of multi helices. The first analytical model takes into account shaft friction, grooves' geometry, and configuration of helices' assembly. The second analytical model is based on the observed failure surface of screw piles proclaimed in literature. Whereas, the third empirical model considers the existing correlation between the torsional resistance generated during screw pile installation and its carrying capacity. The three models were validated against some actual testing results and some experimental data reported in literature. It was noted that the first analytical model underestimate the compressive carrying capacity of screw piles by about 5 to 15%. For the second model, a reasonable agreement was noted between the calculated and the measured results when the ratio $L/B \geq 10$. However, in the opposite, the model overestimates the ultimate capacity of screw piles by about 15.3%. Likewise, the empirical model (third model) overestimates the tensile carrying capacity of screw piles by 10 to 20% with an average of 12.4%.

Introduction and Background

Foundations that are deep (such as piles and shafts) play an important role in transmitting loads or stresses from superstructures to the ground beneath them. When the soils are problematic or have mediocre properties, their use can be crucial. Construction of any type relies heavily on the foundations because they support the entire structure. It is therefore essential to choose a foundation that best suits the requirements and the various essential parameters.

In the construction industry, screw piles (Fig. 1) are one of the most commonly used types of piled foundations. They are made from circular hollow sections of steel that have one or more helices welded to the shaft to provide self-tapping properties during installation. Upon installation, the hollow stem may be filled with reinforced concrete, and it is structurally connected to the building substructure. Depending on capacity requirements, shaft diameters range from 50 mm to 600 mm, and helices' diameters range from 150 mm to 1200 mm. Compared to other pile types, screw piles can provide significant advantages in terms of speed and ease of installation. As well as providing a high bearing capacity and a wide range of applications, they are also quick to form piles, have a flexible size, do not vibrate or make noise, and are of low unit cost. They have been widely used in foundations of high-rise buildings [1, 2, 3]. A multitude of constraints can be overcome with screw piles, including the following: excavation is not possible, unstable ground, underpinning (cracked and collapsing foundations), work schedules are limited by weather or climate, soil is contaminated, soil has low bearing capacity, access to the work site is restricted, and there are environmental restrictions When screw piles are installed, they can be loaded immediately after installation, making them a unique type of foundation. A screw pile does not require waiting for excess pore water pressures to subside, nor does concrete or grout need to harden. This can be important, for example, emergency response projects may require fast foundation installation if the construction schedule is short and the rest of the project is dependent on it. Screw piles are galvanized steel supports or anchors that are highly corrosion-resistant. In comparison to formwork tubes or poured concrete foundations, they are much more cost-effective.

A screw pile's capacity is affected by ground conditions and its structural design. The capacity of piles depends on many factors, such as the type of soil, shaft size, number of helices, helical size and spacing, concrete in-filling, helix-to-shaft weld, pile spacing and installation angle, and installation torque. The bearing properties and failure mechanism of screw piles under compressive, tensile and lateral loading conditions were examined through numerical analyses, laboratory tests, and field load tests [4,5,6,7,8,9,10,11, 12]. A screwed pile is shown to have a higher ultimate carrying capacity than a circular pile due to occlusion effects between the screw or helices and soil.

Recent advances have been made in enhancing the axial capacities of screw piles and their installation processes [13,14,15]. The design methods for screw piles, however, are still lacking, which calls for further investigation [16]. As a result, it is evident that screw piling requires further research to continue to evolve and become a common engineering practice. Different design methods related to screw piles are reported in the literature. They consist of the cylindrical shear method, the individual bearing method and a relationship between the installation torque and the bearing capacity [17]. In spite of this, screw piles are designed so most of the capacity of the pile is generated through the bearing of the helix plates against the soil rather than shaft friction.

This study consists on developing three conceptual models to predict the carrying capacity of a screw pile of multi helices. The first analytical model takes into account shaft friction, grooves' geometry, and configuration of helices' assembly. The second analytical model is based on the observed failure surface of screw piles proclaimed in literature. Whereas, the third empirical model consider the existing correlation between the torsional resistance generated during screw pile installation and its carrying capacity. The three models are validated against some actual testing results and some experimental data reported in literature.

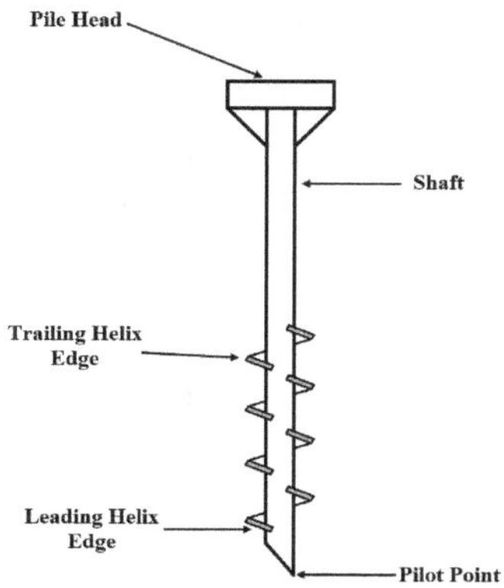

Figure 1. Structural elements of a screw pile

Theoretical Developments

The lack of knowledge about screw piles and the minute data that supports their capabilities has led foundation designers to follow conventional foundation techniques [18]. Despite improvements in screw pile practice, guidelines for helical piles have not been widely adopted yet, though further efforts are needed [15]. As such, this section aims to further enlighten the design of screw piles in a contemporary context, thereby raising awareness and improving the effectiveness of screw piles during construction.

Prediction of Compressive Carrying Capacity of Screw Piles:

A screw pile of multi helices derives its load-carrying capacity by friction or adhesion along the pile shaft with surrounding soil, by compressive resistance at the contact of the pile base with underlying soil, and by passive resistance of soil at the contact of pile grooves (Fig. 2).

The load Q_u at the head of a single screw pile can be expressed as the sum of the bearing load (Q_p) carried at the base, the friction load (Q_s) carried by the shaft of the pile, and the bearing load carried by the helices (Q_g), as follows:

$$Q_u = Q_p + Q_s + Q_g \tag{1}$$

Where:
Q_u = Ultimate carrying capacity.
Q_p = Pointe bearing capacity of screw pile.
Q_s = Skin resistance (friction) of screw pile.
Q_g = Resistance of helices' grooves.

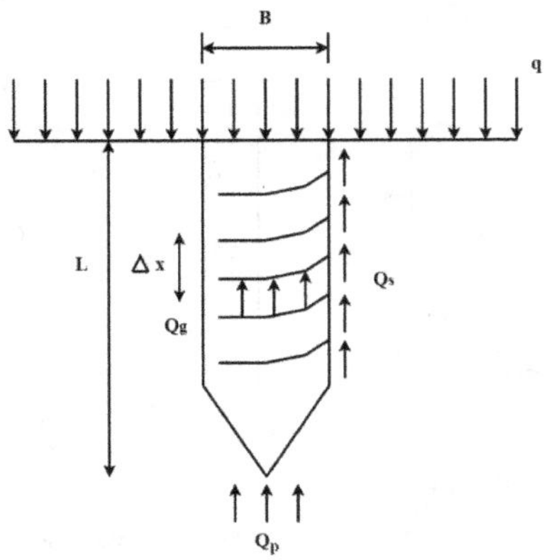

Figure 2. Components of pile's resistances

For an embedment depth L greater than L_o, the point bearing capacity (Q_P) of the screw pile can be determined by the following expression:

$$Q_p = q_p \times A_p \tag{2}$$

$$q_p = (\gamma L + q)N_q^* + 1.3cN_c^* \tag{3}$$

$$A_p = \frac{\pi B^2}{4\cos\beta} \tag{4}$$

$$L_o = \frac{B}{4} N_q^{2/3} \tag{5}$$

$$N_q = \tan^2\left(\frac{\pi}{4} + \frac{\varphi}{2}\right) e^{\pi \tan\varphi} \tag{6}$$

Where:
B = Diameter of screw pile
L = Embedment depth (i.e. length) of the screw pile
β= Angle of inclination of screw pile's tip
A_p = Cross-section of screw pile's shaft
γ= Soil unit weight
c = Soil cohesion
φ= Soil shearing resistance
q = Applied surcharge (if any)
N_q^* and N_q^* = Bearing capacity factor determined from Meyerhof chart [19]

Likewise, the shaft friction or the positive skin friction on shaft of a screw pile can be determined by the following expression:

$$Q_s = \pi B\left(cD + k_o\left(\frac{1}{2}\gamma D^2 + qD\right)\tan\varphi \right) \tag{7}$$

Where:
k_o = Coefficient of earth pressure at rest $\left(k_o = 1 - \sin\varphi\right)$
D = length of the pile's shaft (i.e. the distance between the pile's head and the helices).

Concerning the resistance of the helices' grooves (i.e. the third term of Eq. 1), it can be assumed that all the grooves form circular ring foundations. Then, the helices' grooves resistance can be determined by the following expression:

$$q_g = \sum_{i=1}^{n} q_{gi} = \left[1.3cN_c + 0.3\gamma(B_{ext} - B_{\text{int}})N_\gamma\right]n + \sum_{i=1}^{n}(\gamma D_i + q)N_q \tag{8}$$

Where:

$$\begin{aligned}
\sum_{i=1}^{n}(\gamma D_i + q)N_q &= (\gamma D_1 + q)N_q + (\gamma D_2 + q)N_q + \dots\dots\dots + (\gamma D_n + q)N_q \\
\sum_{i=1}^{n}(\gamma D_i + q)N_q &= N_q\left[(\gamma D_1 + q) + (\gamma D_2 + q) + \dots\dots + (\gamma D_n + q)\right] \\
\sum_{i=1}^{n}(\gamma D_i + q)N_q &= N_q\left[\gamma(D_1 + D_2 + \dots\dots + D_n) + nq)\right] \\
\sum_{i=1}^{n}(\gamma D_i + q)N_q &= N_q\left[\gamma\sum_{i=1}^{n} Di + n.q\right] \\
\sum_{i=1}^{n} D_i &= D_1 + D_2 + \dots\dots\dots + D_n \\
\sum_{i-1}^{n} D_i &= (\Delta x_1 + e) + (\Delta x_2 + 2e) + \dots\dots\dots\dots + (\Delta x_n + ne)
\end{aligned} \tag{9}$$

But,

$$\Delta x_{i+1} = 2\Delta x_i \tag{10}$$

Therefore,

$$\sum_{i=1}^{n} D_i = (\Delta x_1 + e) + 2(\Delta x_1 + e) + \dots . + n(\Delta x_1 - e)$$

$$\sum_{i=1}^{n} D_i = (\Delta x_1 + e)(1 + 2 + \dots\dots . + n)$$

This implies:

$$\sum_{i=1}^{n} D_i = (\Delta x_1 + e)\frac{n(n+1)}{2} \tag{11}$$

It can be written that:

$$Q_g = q_g.\pi.\frac{(B_{ext} - B_{\text{int}})^2}{4} \tag{12}$$

Therefore,

$$Q_g = \left[\left(1.3cN_c + 0.3\gamma(B_{ext} - B_{int})N_\gamma\right)n + \left(\gamma(\Delta x_1 + e)\frac{n(n+1)}{2} + nq \right)N_q \right] . \frac{\pi(B_{ext} - B_{int})^2}{4} \quad (13)$$

Finally, the ultimate carrying capacity of screw piles can be determined as follows:

$$Q_u = \left[(\gamma L + q)N_q^* + 1.3cN_c^*\right]A_p + \left[\pi B\left(cD + k_o\left(\frac{1}{2}\gamma D^2 + qD\right)\tan\varphi \right)\right] + \left[\left(1.3cN_c + 0.3\gamma(B_{ext} - B_{int})N_\gamma\right)n + \left(\gamma(\Delta x_1 + e)\frac{n(n+1)}{2} + nq \right)N_q \right] . \frac{\pi(B_{ext} - B_{int})^2}{4} \quad (14)$$

It is worthy to note that, this equation is applicable in case: $L \geq L_o$. However, in case of $L < L_o$, the parameters N_q^* and N_q^* should be replaced by the N_q and N_c.

Where:
D_i = Distance between pile's head and helice "i".
B_{ext} = External diameter of helices.
B_{int} = Internal diameter of helices.
Δx = Distance between two successive grooves.
e =Thickness of grooves.
n = Number of grooves.

Prediction of Tensile Carrying Capacity of the Screw Piles:
In this section, an analytical model to predict the tensile carrying capacity of a screw pile of multi helices is proposed. The model was developed based on the failure mechanism (i.e. the failure surface) reported in literature [20]. A conical failure surface was observed during the pull-out of a screw pile in a homogeneous soil. Cone pull-out failure occurs when shear forces are reduced to zero along the failure plane. At this critical point the pile and some surrounding soil is pulled out in the shape of a truncated inverted cone or pyramid (Fig. 3). The angle of pull-out will vary depending on soil type and is typically between 0.45φ to 0.55φ (φ is the soil angle of shearing resistance). The ultimate tensile carrying capacity mobilized along the truncated inverted cone failure surface can be computed as follows:

$$Q_u = \int_0^L pdz\left[k_p(c + \gamma z \tan\varphi)\right] = p\int_0^L \left[k_p(c + \gamma z \tan\varphi)\right]dz \quad (15)$$

$$Q_u = pk_p\left(cL + \frac{\gamma L^2}{2}\tan\varphi \right) \quad (16)$$

$$Q_u = k_p\left(\frac{\pi B + \pi(B + 2L\tan\alpha)}{2} \right)\left(cL + \frac{\gamma L^2}{2}\tan\varphi \right) \quad (17)$$

$$Q_u = \tan^2\left(\frac{\pi}{4} + \frac{\varphi}{2}\right)(\pi B + \pi L\tan\alpha)\left(cL + \frac{\gamma L^2}{2}\tan\varphi \right) \quad (18)$$

The failure plane angle (α) depends mainly on the soil shearing resistance angle (φ) and the number of the helices (n). It can be apparoximated by the following expression:

$$\tan\alpha = \tan(0.01n + 0.44\varphi) \tag{19}$$

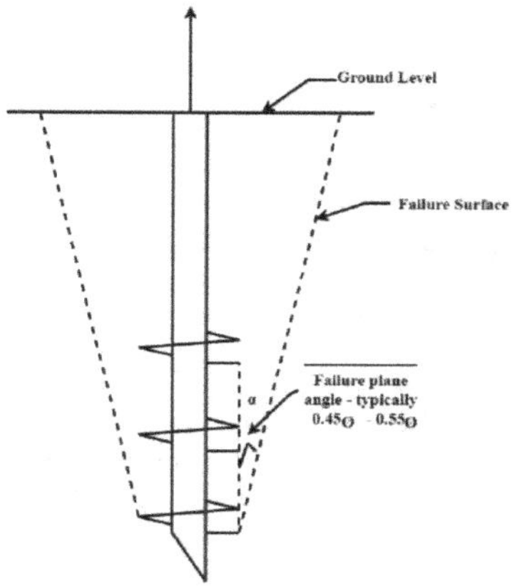

Figure 3. Cone Pull out Failure Surface of Screw Pile [20]

Prediction of Compressive Carrying Capacity of Screw Piles Based on Installation Torque: An attempt was conducted to correlate the compressive carrying capacity of screw piles and the installation torque. It was reported in literature that a straight line could approximate the relationship between the ultimate capacity of screw piles and the installation torque [20, 21]. The torsional resistance generated during screw pile installation is an indication of soil shear strength and provides an alternative (empirical) way to find the capacity of a screw pile. According to the proclaimed observation in literature, the following expression can be used to predict the compressive carrying capacity of screw piles:

$$Q_u = aT \tag{20}$$

Where:
a = Parameter which depend mainly on soil type and pile shaft diameter (1/m)
T = Installation torque (kN.m)
Q_u = Ultimate capacity of screw piles (kN)

In order to calibrate the parameter (a), an analogy between a screw pile in soil and a bolt and its nut was adopted in this investigation. Similarly, a torque is simply a measure of the twisting force required to spin the nut up along the threads of a bolt. In this case, the relationship between the torque and the desired clamp load tension is given by the following formula [22]:

$$T = \frac{KBP}{12} \tag{21}$$

Therefore, for a screw pile in the ground, it can be written that:

$$Q_u = \frac{12}{KB} T \tag{22}$$

So,

$$a = \frac{12}{KB} \tag{23}$$

Where:
T = Torque (lbs.ft)
B = Nominal diameter (in.)
Q_u = Ultimate carrying of screw piles (lbs)
K = Torque coefficient (dimensionless)
a = Parameter depending on soil type (1/in.)

The K coefficient can vary depending on the condition of the soil. The following coefficient range can be adopted (the different values of K were deduced by an analogy applied between pile/soil and bolt/nut): K = 0.10 for soft clay or loose sand, K = 0.20 for medium stiff clay or medium-to-dense sand, and K = 0.30 for firm clay or dense/very dense sand.

Validation of the Analytical Model

To validate the analytical models described previously in sections 2.1 and 2.2 (i.e. Eqs. 14 and 18), a small scale testing program was conducted in laboratory. Whereas, the third model, purported in section 2.3 (Eq. 22), was endorsed against some experimental data reported in literature. The laboratory testing program consisted on loading in a test rig three types of steel screw piles of different lengths, diameters and distances between grooves. The diameters adopted were 8, 10 and 12 mm; while, the piles' lengths used were 50, 100 and 150 mm. The soil bed was dense uniform sand with an angle of shearing resistance φ = 38°.

The results of the different comparisons are summarized in Tables 1, 2 and 3. According to the data grouped in Table 1, it can be deduced that the first analytical model, represented by Eq. 14, underestimates the compressive carrying capacity of screw piles by about 5 to 15%. For the second model, a reasonable agreement was noted between the calculated and the measured results when the ratio $L/B \geq 10$. However, in the opposite, the model overestimates the ultimate capacity of screw piles by about 15.3%. Likewise, the empirical model (Eq. 22), overestimates the tensile carrying capacity of screw piles by 10 to 20% with an average of 12.4%. In consequence, Eq. 22 was revised in order to reduce or eliminate this discrepancy. The revised equation was obtained by multiplying the parameter (a) by 0.847. Accordingly, the current equation representing the empirical model is given as follows:

$$Q_u = \frac{10}{KB} T \tag{24}$$

Materials Research Proceedings 31 (2023) 269-280 https://doi.org/10.21741/9781644902592-28

Table 1. *Comparison between Theoretical (Eq. 14) and Present Laboratory Results*

Screw Piles Characteristics			Compressive Carrying Capacity, Q_u(N)	
B **(mm)**	***L*** **(mm)**	**Δx(mm)**	**Theoretical (Eq. 14)**	**Experimental**
8	50	0	22.97	25.18
	100		34.97	37.57
	150		40.29	43.33
10	50	0	23.82	26.67
		2	27.22	29.10
		3.5	32.14	36.67
		5	33.71	37.95
	100	0	34.72	37.39
		2	36.73	41.48
		3.5	37.93	42.50
		5	37.79	44.00
	150	0	47.03	55.00
		2	50.83	59.60
		3.5	57.64	64.35
		5	59.29	67.72
12	50	0	39.72	45.33
		2	41.02	47.38
		3.5	44.77	50.00
		5	49.62	57.87
	100	0	39.94	46.90
		2	42.25	48.73
		3.5	48.12	50.66
		5	53.28	62.07
	150	0	65.34	76.08
		2	68.05	78.67
		3.5	78.97	84.61
		5	75.53	88.31

Table 2. *Comparison between Theoretical (Eq. 18) and Present Laboratory Results*

Screw Piles Characteristics			Compressive Carrying Capacity, Q_u(N)	
B **(mm)**	***L*** **(mm)**	**φ(°)**	**Theoretical (Eq. 18)**	**Experimental**
10	100	38	51.31	44.00
	150		73.42	64.35
12	50	38	48.85	50.00
	100		46.46	48.73
	150		85.71	88.31

Table 3. *Comparison between Theoretical Results (Eq. 22) and Experimental Data Reported in Literature*

Screw Piles Characteristics			**Tensile Carrying Capacity, Q_u, kN**(lbs)	
***B*, mm** (in.)	***Soil State*** ***(K)***	***T*, kN.m** (lbs.ft×10^3)	**Theoretical (Eq. 22)**	**Experimental [23]**
114 (4.49)	0.1	**15** (11.06)	**179.42** (40.33×10^3)	**150** (33.72×10^3)
	0.2	**20** (14.75)	**222.14** (49.94×10^3)	**200** (44.96×10^3)
140 (5.51)	0.1	**25** (18.44)	**281.61** (63.31×10^3)	**255** (57.32×10^3)
	0.2	**35** (25.81)	**342.58** (77.01×10^3)	**310** (69.69×10^3)
	0.3	**50** (36.88)	**468.37** (105.29×10^3)	**425** (95.54×10^3)

Definitely, in geomechanical and geomaterial design a variability of the ultimate capacity parameters of 5 – 20% is fairly frequent and acceptable [24]. Furthermore, the calculated values for the mean and the standard deviation of the results grouped in Tables 1, 2 and 3 are 11.3, 12.4 and 15.3% (for the mean), and 3.03, 4.06 and 1.7 (for the standard deviation), respectively. In addition, the coefficient of variation, defined as the ratio of the standard deviation to the mean (which is a measure of the dispersion of a data set), is less than one, indicating a relatively low variation in the results.

Conclusion

Significant advances are being made toward the utilisation of screw piles as anchors and foundations of structures. Nonetheless, there is still a lack of design methods related to screw piles which require continued investigation. The main conclusions drawn from this investigation are summarized as follows:

1- Three conceptual models to predict the carrying capacity of a screw pile of multi helices were developed.
2- The first analytical model takes into account shaft friction, grooves' geometry, and configuration of helices' assembly (Eq. 14). It was noted that this analytical model underestimate the compressive carrying capacity of screw piles by about 5 to 15%.
3- The second analytical model is based on the observed failure surface of screw piles proclaimed in literature (Eq. 18). For this model, a reasonable agreement was noted between the calculated and the measured results when the ratio $L/B \geq 10$. However, in the opposite, the model overestimates the ultimate capacity of screw piles by less than 15.3%.
4- The third empirical model considered the existing correlation between the torsional resistance generated during screw pile installation and its carrying capacity (Eq. 22). Likewise, the empirical model overestimates the tensile carrying capacity of screw piles by 10 to 20% with an average of 12.4%.

References

[1] Deng, R. F. Construction technique of half-screwed filling pile and its application in railway engineering. Railway Construction Technology 33(4): 84-87. (2016).

[2] Liu, Z. Application research of screw pile in deep fill foundation reinforcement. Resources Environment& Engineering 29(3): 323-326. (2015).

[3] Wan, P. Z. Key construction techniques and bearing behaviors of railway foundation treatment with screw pile in saline soil area. Railway Construction Technology 33(5): 82-86. (2016).

[4] Rao, S. N., Prasad Y.V.S.N. and Shetty M. D. The behavior of model screw piles in cohesive soils. Soils & Foundations 31(2): 35-50. (2008). https://doi.org/10.3208/sandf1972.31.2_35

[5] Kurian, N.P., Shah, S.J. Studies on the behaviour of screw piles by the finite element method. Can. Geotech. J. 46 (6), 627-638. (2009). https://doi.org/10.1139/T09-008

[6] Sakr, M. Performance of helical piles in oil sand. Can. Geotech. J. 46 (9), 1046-1061. (2009). https://doi.org/10.1139/T09-044

[7] Meng Z., Chen J.J., Wang J.H. and Yin Z.Y. Study of model test on bearing capacity of screw piles in sand. Rock and Soil Mechanics 33(s1): 141-145. (2012).

[8] Stanier, S.A., Black, J.A. and Hird, C.C. Modelling helical screw piles in soft clay and design implications. Geotechnical Engineering 167(5): 447-460. (2014). https://doi.org/10.1680/geng.13.00021

[9] Qian, J., Wang, B., Chen, H., and Huang, M. Model test and bearing mechanism study on grouting-screw uplift piles. Journal of Building Structures 36(10): 146-152. (2015).

[10] Lee C.W., Kim Y.S. and Park S.Y. Development of pre-bored screw pile method and evaluation of its bearing characteristics. Marine Georesources & Geotechnology 34(1): 42-56. (2016). https://doi.org/10.1080/1064119X.2014.954682

[11] Zahedi, P. and Soltani-Jigheh, H. A study of the bearing capacity of screw piles in sand. International Geotechnical Symposium, Istanbul, turkey, November. (2019).

[12] Karkush, M.O. and Hussein, A.A. Experimental Investigation of Bearing Capacity of Screw Piles and Excess Pore water Pressure in Soft Clay under Static Axial Loading, Second International Conference on Geotechnical Engineering, Iraq, Volume 318, (2021). https://doi.org/10.1051/e3sconf/202131801001. https://doi.org/10.1051/e3sconf/202131801001

[13] Pack, J.S. Design of helical piles for heavily loaded structures. In: New Technological and Design Developments in Deep Foundations. American Society of Civil Engineers, USA, pp. 353-367. (2000). https://doi.org/10.1061/40511(288)25

[14] Sakr, M. Lateral resistance of high capacity helical piles: case study. In: Proceedings of the 63rd Canadian Geotechnical and 6th Canadian Permafrost Conference. Calgary, Alberta, 12-16 September, pp. 402-412. (2010).

[15] Perlow Jr., M. Helical pile acceptance criteria, design guidelines and load test verification. In: Geo-Frontiers 2011. American Society of Civil Engineers, USA, pp. 94-102. (2011). https://doi.org/10.1061/41165(397)11

[16] Livneh, B., Naggar, M.H.M. Axial testing and numerical modelling of square shaft helical piles under compressive and tensile loading. Can. Geotech. J. 45 (8), 1142-1155. (2008). https://doi.org/10.1139/T08-044

[17] Mohajeranin, A., Bosnjak, D., Bromwich, D. Analysis and design methods of screw piles: A review, Soils and Foundations, 56(1):115-128. (2016). https://doi.org/10.1016/j.sandf.2016.01.009

[18] Vito, D., Cook, T. Highly loaded helical piles in compression and tension applications: a case study of two projects. In: Proceedings of the Pan-Am CGS Geotechnical Conference, pp. 1-4.(2011).⟨http://geoserver.ing.puc.cl/info/conferences/PanAm2011/panam2011/pdfs/GEO11Paper902.pdf⟩.

[19] Das, B. M. and Sivakugan, N. Principles of Foundation Engineering, 9th Edition, Cengage Learning, ISBN-13: 978-1-337-70503-5. (2019).

[20] IPENZ. Practice Note 28 Screw Piles: Guidelines for Design, Construction & Installation, e IPENZ Engineering Practice Advisory Committee, Version 1, October, (2015).

[21] Harnish, J.L. (2015). Helical Pile Installation Torque and Capacity Correlations, Electronic Thesis and Dissertation Repository. 2855. The University of Western Ontario, https://ir.lib.uwo.ca/etd/2855.

[22] Mudge Fasteners, Inc. Determining Torque: The Facts About Required Torque, Tension and Clamp Loads, February, (2018).

[23] Tappenden, K. M. Predicting the Axial Capacity of Screw Piles Installed in Western Canadian Soils. Edmonton: The University of Alberta. (2007).

[24] European Standard (EN1926), European Committee for Standardization, Technical Committee CEN/TC 246Brussels, December (2006).

Size-effects due to Burgers tensor in classical deformation of isotropic thermoplastic materials

A.S. Borokinni[a]*, A.P. Akinola[b], O.P. Layeni[c], O.O. Fadodun[d], and B.A. Olokuntoye[e]

[1]Department of Mathematics, Obafemi Awolowo University, Ile-Ife, Nigeria

[a]aborokinni@oauife.edu.ng, [b]aakinola@oauife.edu.ng, [c]olayeni@oauife.edu.ng, [d]ofadudun@oauife.edu.ng, [e]toyeolokun@oauife.edu.ng

Keywords: Thermoplasticity, Size-Effect, Distortion Gradient Plasticity, Plastic Flow Rule, Burgers Tensor

Abstract. This work considers size-effects phenomena associated with Burgers tensor and heat transfer in isotropic plastic materials under thermal loading. The virtual power principle, the first and second laws of thermodynamics are used to obtain the balance of forces, the balance of energy, and the free-energy imbalance in local forms. Also, the constitutive relations for microscopic stresses associated with the Burgers tensor are obtained. The balance of microscopic forces is supplemented with the constitutive relations for the stresses to form the plastic flow rule. The presence of material length scale in the flow rule shows that it is possible to study size effects through the Burgers tensor.

Introduction

Investigations have shown that within the micron range of about 500 nanometers to 50 micrometers, most metals generally exhibit size-dependent behaviors. These size-effects are not known within the context of the classical plasticity because of its inability to accommodate intrinsic material length scales. The theories of strain gradient plasticity have been developed to address this shortcoming.

There are a number of phenomenological gradient theories in the literature, with the earliest attempt by Aifantis [1] whose theory incorporates and energetic length scale via the Laplacian of the accumulated plastic strain within a modified von Mises yield criterion for the stress. Other well established theories oinclude the works of Gurtin [2], Gudmundson [3] and Gurtin and Anand [4].

Recently, Borokinni *et.al.,* [5] investigated size-effects in isotropic materials associated with the divergence of plastic distortion. The work shows that the divergence of the transpose of plastic distortion is a measure of the skew part of the Burgers tensor. However, it is observed that such size-effect does not include the symmetric part of the Burgers tensor which also plays a role in investigating size-effects.

Furthermore, many theories of gradient plasticity are purely mechanical, and so not much attention is given to thermoplastic materials exhibiting size-dependent behaviors.

This paper present a coupled thermo-mechanical theory of distortion gradient plasticity that accounts for the Burgers tensor.

Basic Kinematics Relations

Suppose a point $\boldsymbol{X}$ of a body B in a region of space E has a displacement $\boldsymbol{u}$ at time t. The small deformation theory of continuum plasticity allows the displacement gradient $\nabla\boldsymbol{u}$ to be additively decomposed

$$\nabla\boldsymbol{u} = \boldsymbol{H}^e + \boldsymbol{H}^p \tag{1}$$

into elastic distortion $\boldsymbol{H^e}$ and plastic distortion $\boldsymbol{H^p}$. The second order tensor $\boldsymbol{H^e}$ measures the stretch and rotation of the underlying material structure, in this case, a lattice. The second order tensor $\boldsymbol{H^p}$ measures defects in that material structure arising from the motion of dislocations through the lattice. The Burgers tensor $\boldsymbol{G}$ is defined as the second order tensor

$$\boldsymbol{G} = \nabla \times \boldsymbol{H^p}, \tag{2}$$

which is also a measure of defect in the material associated with geometrically necessary dislocation densities (GNDs). For most metals, the volumetric change is not accompanied by plastic deformation, and so the trace of $\boldsymbol{H^p}$ is zero. That is,

$$tr\boldsymbol{H^p} = 0. \tag{4}$$

Macroscopic and Microscopic Forces

Here, we introduce power expended through the Burgers tensor rate $\dot{\boldsymbol{G}}$ by a second order microscopic stress denoted as $\boldsymbol{S}$. Thus, we shall assume the following:

- The Cauchy stress $\boldsymbol{T}$ is work-conjugate to the elastic distortion $\boldsymbol{H^e}$;
- The symmetric and deviatoric plastic microscopic stress $\boldsymbol{T^p}$ is work-conjugate to the plastic distortion $\boldsymbol{H^p}$; and
- The second order microscopic stress $\boldsymbol{S}$ is work-conjugate to the Burgers tensor $\boldsymbol{G}$.

In addition to work done by the body force $\boldsymbol{b}$ and macrotraction force $\boldsymbol{t}$, we assume that there exist a microtraction stress tensor $\boldsymbol{K}$ which is work-conjugate to the plastic distortion rate.
The macroscopic force balance is the classical momentum equation in local form given by

$$\text{div}\boldsymbol{T} + \boldsymbol{b} = \boldsymbol{0} \text{ in } P, \text{ and } \boldsymbol{Tn} = \boldsymbol{t} \text{ in } \partial P, \tag{5}$$

where P is an arbitrary small portion of the body B and ∂P is the boundary of P.
To obtain the microscopic force balance, we note that the rate-like kinematic relation is given by

$$\nabla\dot{\boldsymbol{u}} = \dot{\boldsymbol{H}}^e + \dot{\boldsymbol{H}}^p. \tag{6}$$

If the motion is microscopic then $\dot{\boldsymbol{u}} = \boldsymbol{0}$, so that $\dot{\boldsymbol{H}}^e = -\dot{\boldsymbol{H}}^p$. The power balance is given by

$$\int_P \left(\boldsymbol{T}:\dot{\boldsymbol{H}}^e + \boldsymbol{T^p}:\dot{\boldsymbol{H}}^p + \boldsymbol{S}:\dot{\boldsymbol{G}}\right)dV = \int_P \boldsymbol{b}\cdot\dot{\boldsymbol{u}}\,dv + \int_{\partial P} \left(\boldsymbol{t}\cdot\dot{\boldsymbol{u}} + \boldsymbol{K}:\dot{\boldsymbol{H}}^p\right)dA. \tag{7}$$

For microscopic motion, we have

$$\int_P \left((\boldsymbol{T^p} - \boldsymbol{T_o}):\dot{\boldsymbol{H}}^p + \boldsymbol{S}:\dot{\boldsymbol{G}}\right)dV = \int_{\partial P} \boldsymbol{K}:\dot{\boldsymbol{H}}^p dA, \tag{8}$$

where the quantity $\boldsymbol{T_o}$ is the deviatoric part of the Cauchy stress tensor $\boldsymbol{T}$.
We note that

$$\boldsymbol{S}:\dot{\boldsymbol{G}} = S_{ij}\mathcal{E}_{irs}H^p_{js,r}.$$

Define the component M_{jsr} of the hyperstress $\mathbb{M}$ as

$$M_{jsr} = S_{ij}\mathcal{E}_{irs}. \tag{9}$$

The power balance in Eq. 8 can be written as

$$\int_P \left((\boldsymbol{T^p} - \boldsymbol{T_o}) : \dot{\boldsymbol{H}}^p + \mathbb{M} \vdots \nabla\dot{\boldsymbol{H}}^p\right)dV = \int_{\partial P} \boldsymbol{K} : \dot{\boldsymbol{H}}^p dA. \tag{10}$$

By Gauss divergence theorem, the microscopic force balance in local form is given as

$$\boldsymbol{T_o} = \boldsymbol{T^p} - \mathrm{div}\mathbb{M}, \tag{11}$$

with microtraction condition given as

$$\mathbb{M}\boldsymbol{n} = \boldsymbol{K}. \tag{12}$$

Balance of Energy

The balance of energy is essentially the first law of thermodynamics which is mathematically written as

$$\int_P \dot{\mathcal{E}}\, dV = \int_P \boldsymbol{T} : \dot{\boldsymbol{H}}^e\, dV + \int_P \boldsymbol{T^p} : \dot{\boldsymbol{H}}^p\, dV + \int_P \mathbb{M} \vdots \nabla\dot{\boldsymbol{H}}^p\, dV - \int_{\partial P} \boldsymbol{q} \cdot \boldsymbol{n}\, dV + \int_P Q\, dV, \tag{13}$$

where $\mathcal{E}$ is the internal energy measured per unit volume, $\boldsymbol{q}$ is the heat flux measured per unit area, and Q is the heat supply measures per unit volume.
By using the Gauss divergence theorem and noting that P is arbitrary, then the balance of energy in local form is given by

$$\dot{\mathcal{E}} = \boldsymbol{T} : \dot{\boldsymbol{H}}^e + \boldsymbol{T^p} : \dot{\boldsymbol{H}}^p + \mathbb{M} \vdots \nabla\dot{\boldsymbol{H}}^p - \mathrm{div}\boldsymbol{q} + Q. \tag{14}$$

Entropy Imbalance

Let η be the entropy at an arbitrary point of P. The second law of thermodynamics in local form is the given by the Clausius-Duhem inequality

$$\dot{\eta} \geq -\mathrm{div}\left(\frac{\boldsymbol{q}}{\vartheta}\right) + \frac{Q}{\vartheta}, \tag{15}$$

where $\vartheta > 0$ is the absolute temperature.
The free-energy φ is defined via the relation

$$\mathcal{E} = \varphi + \vartheta\eta,$$

so that we have

$$\dot{\mathcal{E}} = \dot{\varphi} + \vartheta\dot{\eta} + \dot{\vartheta}\eta. \tag{16}$$

The balance of energy in term of free-energy and entropy is given by

$$\dot{\varphi} + \vartheta\dot{\eta} + \dot{\vartheta}\eta = \boldsymbol{T}:\dot{\boldsymbol{H}}^e + \boldsymbol{T}^p:\dot{\boldsymbol{H}}^p + \mathbb{M} \vdots \nabla\dot{\boldsymbol{H}}^p - \mathrm{div}\boldsymbol{q} + Q$$

$$\Longrightarrow \quad \boldsymbol{T}:\dot{\boldsymbol{H}}^e + \boldsymbol{T}^p:\dot{\boldsymbol{H}}^p + \mathbb{M} \vdots \nabla\dot{\boldsymbol{H}}^p - \mathrm{div}\boldsymbol{q} + Q \geq \dot{\varphi} + \eta\dot{\vartheta} + \frac{\boldsymbol{q}}{\vartheta}\cdot\nabla\vartheta - \mathrm{div}\boldsymbol{q} + Q. \tag{17}$$

The free-energy imbalance essential for the development of thermodynamically consistent constitutive equations is given by

$$\dot{\varphi} + \eta\dot{\vartheta} - \boldsymbol{T}:\dot{\boldsymbol{H}}^e - \boldsymbol{T}^p:\dot{\boldsymbol{H}}^p - \mathbb{M} \vdots \nabla\dot{\boldsymbol{H}}^p + \frac{\boldsymbol{q}}{\vartheta}\cdot\nabla\vartheta \leq 0. \tag{18}$$

Constitutive Relations
We shall assume that the free-energy φ is additively decomposed into

$$\varphi = \varphi^e + \varphi^p \tag{19}$$

into elastic and plastic free-energies, and it is assumed that

$$\varphi^e = \varphi^e(\boldsymbol{E}^e, \vartheta) \text{ and } \varphi^p = \varphi^p(\boldsymbol{G}, \vartheta) \text{ so that we have } \varphi = \varphi(\boldsymbol{E}^e, \boldsymbol{G}, \vartheta). \tag{20}$$

Furthermore, let the plastic stress $\boldsymbol{T}^p$ be purely dissipative and rate-independent, while $\mathbb{M}$ is purely energetic. By chain rule, we have

$$\dot{\varphi} = \frac{\partial\varphi}{\partial\vartheta}\dot{\vartheta} + \frac{\partial\varphi^e}{\partial\boldsymbol{E}^e}:\dot{\boldsymbol{E}}^e + \frac{\partial\varphi^p}{\partial\boldsymbol{G}}:\dot{\boldsymbol{G}}. \tag{21}$$

Observe that

$$\frac{\partial\varphi^p}{\partial\boldsymbol{G}}:\boldsymbol{G}^{\cdot} = \frac{\partial\varphi^p}{\partial G_{ij}}\,\varepsilon_{irs}\,\dot{H}^p_{js,r} \text{ and } \mathbb{M} \vdots \nabla\dot{\boldsymbol{H}}^p = \boldsymbol{S}:\dot{\boldsymbol{G}}.$$

The free energy imbalance becomes

$$\left(\frac{\partial\varphi}{\partial\vartheta} + \eta\right)\dot{\vartheta} + \left(\frac{\partial\varphi^e}{\partial\boldsymbol{E}^e} - \boldsymbol{T}\right):\dot{\boldsymbol{E}}^e + \left(\frac{\partial\varphi^p}{\partial\boldsymbol{G}} - \boldsymbol{S}\right):\dot{\boldsymbol{G}} - \boldsymbol{T}^p:\dot{\boldsymbol{H}}^p + \frac{\boldsymbol{q}}{\vartheta}\cdot\nabla\vartheta \leq 0. \tag{22}$$

By the Coleman-Noll procedure (Gurtin *et al.*, 2010), the constitutive relations for the entropy, Cauchy stress, plastic stress and polar microstresses are

$$\eta = -\frac{\partial\varphi}{\partial\vartheta}, \quad \boldsymbol{T} = \frac{\partial\varphi^e}{\partial\boldsymbol{E}^e}, \quad \boldsymbol{S} = \frac{\partial\varphi^p}{\partial\boldsymbol{G}}. \tag{23}$$

Clearly, we have

$$M_{jsr} = \mathcal{E}_{irs}\frac{\partial\varphi^p}{\partial G_{ij}}. \tag{24}$$

The mechanical dissipation and heat conduction inequalities are thus given as

$$\boldsymbol{T}^p : \dot{\boldsymbol{H}}^p \geq 0 \text{ and } \boldsymbol{q} \cdot \nabla\vartheta \leq 0. \tag{25}$$

Isotropic Thermoplastic Solids

Assume the quadratic form of the free energy

$$\varphi = \mu|\boldsymbol{E}^e|^2 + \frac{1}{2}\lambda|\mathrm{tr}\boldsymbol{E}^e|^2 + \frac{1}{2}\mu L^2|\boldsymbol{G}|^2 + (\boldsymbol{M}^e : \boldsymbol{E}^e)(\vartheta - \vartheta_o) + (\boldsymbol{A} : \boldsymbol{G})(\vartheta - \vartheta_o) - \frac{c_o(\vartheta - \vartheta_o)^2}{2\vartheta_o}, \tag{26}$$

where $\boldsymbol{M}^e$ and $\boldsymbol{A}$ are the macroscopic and microscopic stress-temperature moduli respectively. Following Eqs. 24 and 26, the constitutive relations for isotropic materials are given as

$$\eta = -\boldsymbol{M}^e : \boldsymbol{E}^e - \boldsymbol{A} : \boldsymbol{G} + \frac{c_o(\vartheta - \vartheta_o)}{\vartheta_o} \tag{27a}$$
$$\boldsymbol{T} = 2\boldsymbol{E}^e + \lambda(\mathrm{tr}\boldsymbol{E}^e)\boldsymbol{I} + \boldsymbol{M}^e(\vartheta - \vartheta_o) \tag{27b}$$
$$\boldsymbol{S} = \mu L^2\boldsymbol{G} + \boldsymbol{A}(\vartheta - \vartheta_o). \tag{27c}$$

We shall assume that the plastic stress $\boldsymbol{T}^p$ obeys the codirectionality constraint and the heat flux obeys the Fourier law, given as (Gurtin *et al.*, 2010; Borokinni *et al.*, 2020):

$$\boldsymbol{T}^p = Y_o \frac{\dot{\boldsymbol{H}}^p}{|\dot{\boldsymbol{H}}^p|} \quad \text{for } \dot{\boldsymbol{H}}^p \neq \boldsymbol{0}, \text{ and } \boldsymbol{q} = -k\nabla\vartheta, \tag{28}$$

where $\boldsymbol{Y_o}$ and $\boldsymbol{k}$ are the flow resistance and coefficient of thermal conductivity respectively. To obtain the non local plastic flow rule that account for size effect due to Burgers tensor, the following relations between the partial derivative of the Burgers tensor and partial derivatives of the plastic distortion tensor will be useful

$$\varepsilon_{irs}G_{ij,r} = H^p_{js,rr} - H^p_{jr,rs}, \tag{29}$$

where

$$\boldsymbol{G}_{ij,r} = \frac{\partial \boldsymbol{G}_{ij}}{\partial \boldsymbol{X}_r}, \qquad \boldsymbol{H}^p_{js,rr} = \frac{\partial^2 \boldsymbol{H}^p_{js}}{\partial \boldsymbol{X}_r \partial \boldsymbol{X}_r} \text{ and } \boldsymbol{H}^p_{js,rr} = \frac{\partial^2 \boldsymbol{H}^p_{jr}}{\partial \boldsymbol{X}_r \partial \boldsymbol{X}_s}.$$

Also, observe that

$$\mathrm{div}\,\mathbb{M} = \mu L^2[\Delta \boldsymbol{H}^p - \nabla\mathrm{div}\boldsymbol{H}^p] + \boldsymbol{A}^{\mathrm{T}}(\nabla\vartheta \times).$$

Thus by substituting the relevant constitutive relations for the microscopic stresses as deduced from Eqs. 27c and 28, the non-local plastic flow rule accounting for size effect due to Burgers tensor is

$$\boldsymbol{T_o} + \boldsymbol{\mu L^2[\Delta H^p - \nabla\mathrm{div}H^p]} + \boldsymbol{A}^{\mathrm{T}}(\boldsymbol{\nabla\vartheta} \times) = \boldsymbol{Y_o}\frac{\dot{\boldsymbol{H}}^p}{|\dot{\boldsymbol{H}}^p|} \quad \textbf{provided } \dot{\boldsymbol{H}}^p \neq \boldsymbol{0}.$$

Conclusion

This paper has only presented plastic flow rule associated with the Burgers tensor. It has shown that the flow rule is non-local as it involves system of second-order partial differential equations in plastic distortion. Consequently, it is required that the plastic flow law be supplemented by appropriate initial-boundary conditions, and further, the system of equations obtained from balance of macroscopic forces, balance of energy and microscopic force balance be translated to variational problem, where well-posedness of the initial-boundary value problem could be investigated, and a finite element model can be provided. A report of the well-posedness and numerical implementations of this problem will be considered as a future work.

References

[1] E.C. Aifantis, On the microstructural origin of certain inelastic models. Trans. ASME J. Eng. Mater. Tech. 106, 326-330, 1984. https://doi.org/10.1115/1.3225725

[2] M.E. Gurtin, A gradient theory of small- deformation isotropic plasticity that accounts for Burgers vector and dissipation due to plastic spin. J. Mech. Phys. Solids , 52, 2545-2568, 2004. https://doi.org/10.1016/j.jmps.2004.04.010

[3] P. Gudmundson, A unified treatment of strain gradient plasticity. J. Mech. Phys. Solids 52, 1379-1406, 2004. https://doi.org/10.1016/j.jmps.2003.11.002

[4] M.E. Gurtin and L. Anand, Thermodynamics applied to gradient theories involving the accumulated plastic strain: the theories of Aifantis and Fleck and Hutchinson and their generalization. J. Mech. Phys. Solids 57, 405-421, 2009. https://doi.org/10.1016/j.jmps.2008.12.002

[5] A.S. Borokinni, O.O. Fadodun, O.P. Layeni, A.P. Akinola and B.A. Olokuntoye, Size-effect associated with skew-symmetric Burgers tensor. Theoretical and Applied Mechanics 47, 1, 99-112, Belgrade, 2020. https://doi.org/10.2298/TAM191125001B

[6] M. Gurtin, E. Fried and L. Anand, The mechanics and thermodynamics of continua. Cambridge University Press, Cambridge, 2010. https://doi.org/10.1017/CBO9780511762956

Advanced Topics in Mechanics of Materials, Structures and Construction - AToMech1-2023 Materials Research Forum LLC
Materials Research Proceedings 31 (2023) 287-298 https://doi.org/10.21741/9781644902592-30

Aeroponic tower garden solar powered vertical farm

Omar Yahya[1,a *], Gharam Yahya[2,b], Ahmad Al-Omair[1,c], Emad Yahya[3,d], Esam Jassim[1,e], and Faramarz Djavanroodi[1,f]

[1]Mechanical Engineering Department, College of Engineering, Prince Muhammad Bin Fahd University, 31952 Al-Khobar, Kingdom of Saudi Arabia

[2]Physics Department, College of Science, Imam Abdulrahman Bin Faisal University, 31952 Dammam, Kingdom of Saudi Arabia

[3]Electrical Engineering Department, College of Engineering, Prince Muhammad Bin Fahd University, 31952 Al-Khobar, Kingdom of Saudi Arabia

[a]201502345@pmu.edu.sa, [b]miss-gharamy@mail.ru, [c]ahmadaalomair@gmail.com, [d]201601149@pmu.edu.sa, [e]ejassim@pmu.edu.sa, [f]fdjavanroodi@pmu.edu.sa

Keywords: Vertical Farming, Vertical Garden, VF Garden, Tower Garden

Abstract. This review paper is written in order to specifically examine, explore, and evaluate vertical farming (VF) with the application of solar power as the powering sources for the entire farm. Nevertheless, a huge number of 2 billion individual suffers from micronutrient deficiencies and other related major health problems led by it. Moreover, with vertical farming processes we can offer a massive key sustainable food resource along with many advantages compared with horizontal farming (HF) processes, such cutting the need of water up to 95% while providing an efficient use of almost any space and many much more positives that will be discussed. Some positives are eliminating risks of biotic, abiotic, and climate issues. On the other hand, VF known to consume a huge amount annually of power per square meter in farming field of 3500 kW h, which is considered as a major setback for this process. However, with efficient use of green renewable power resources alternatives such solar panels, we are able to cut-down the power consumption dramatically to offer a true sustainable food resource for us and for next generations. Therefore, the purpose of the paper is to discuss vertical farming VF using solar power to create a sustainable food resource.

Introduction

We all familiar with conventional farming, however when it comes to vertical farming, a limited number of people heard about this term. Vertical farming (VF) is a process to produce food by the use of vertical dimension for hydroponic growing of crops with indoor controlled-environment agriculture (CEA). In addition, these methods such vertical farming designed to play a role in facing these major crises such the scarcity of water and the major growing in our population [1,2,3]. Moreover, in compact cities, the pressure on food resources rises, since typical farming lands are limited. Nevertheless, researchers predict an increasing need of 25%-70% in the crops demand by 2050 which we have to supply. However, with the decreasing availability of growing lands while the condition of climate getting worse, the demand in VF a rises [4]. VF uses the application of LED lights, heaters, ventilation and air conditioning (HVAC), sensors, software, internet of things (IOT), drones, applications and many more factors to create an efficient growing environment for crops. All this to plant in a vertical stack of crops layered up each other in order to save space, energy, and water. VF Systems generally categorized onto two main systems, one is green walls and cylinder shaped vertical growing space. A Hydroponics branch known as hydroponics uses water as the solvent to produce vegetables without using soil. As a result, this technique offers significant advantages that enable farming everywhere without the need for soil.

However, it is projected that there would be a larger lack of areas and soils suited for agricultural. Because there is no requirement for soil, this method has several advantages that make it possible to farm everywhere. On the other hand, the Aquaponic farming technique, mixes farming with aquaculture. It is also utilized in hydroponic systems that incorporate fish tanks. Thus, the nutrients required by plants in a growing tray are provided by fish waste. Lastly, using only a little amount of sodium water, the farming technique known as Aeroponic VF grows plants in a misty atmosphere without the use of soil or any growth medium has considered to be the best option in our case.

Vertical farming as the name refers uses multiple vertical layers of crops laid indoor a warehouse where crops will find an artificial growing environment simulating horizontal farming. Additionally, this simulated indoor environment includes factors such light needed, temperature required, humidity level, concentration of carbon dioxide, water amount, and nutrients. Moreover, by keeping in mind every playing role result in producing a large number of crops produced precisely with high quality and freshness sustainably 100% indoor [5]. Prices of food keep increasing due to multiple reasons and the increase in food scarcity. Though, the cost of traditional HF less than up keeping and installation of greenhouses, still they widely increasing due to the extension of the growing season they provide while offering various products to be possibly grown without worrying of the external factors. VF known to consume a huge amount of power. In general, buildings consume about 40% of the total power consumed, while 40% - 50% of that consumed by lighting. As result, the application of renewable started to attract more attention in this field. In addition, the study shows 56% - 89% of potential savings [6].

Hydroponics Systems

Facts shows that by 2050, our world population expected to grow up to 9.7 billion people, while almost 70% of this population expected to live in cities. With that being said, the scarcity in lands and appropriate farming soil expected to increase. Hydroponics is a hydro cultural branch that grows crops eliminating soil with solutions of mineral nutrient in water solvent. As a result, this method offers major positive impacts which allow farming in any place without the need for soil [7]. That being said, it is anticipated that there would be a greater shortage of suitable farming soil and lands. A hydro-cultural branch known as hydroponics uses water as the solvent to produce vegetables without using soil [8]. As a result, this technique offers significant advantages that enable farming everywhere without the need for soil. Additionally, the most popular VF system in the world is hydroponic. Fig 1. is demonstrating Hydroponic System.

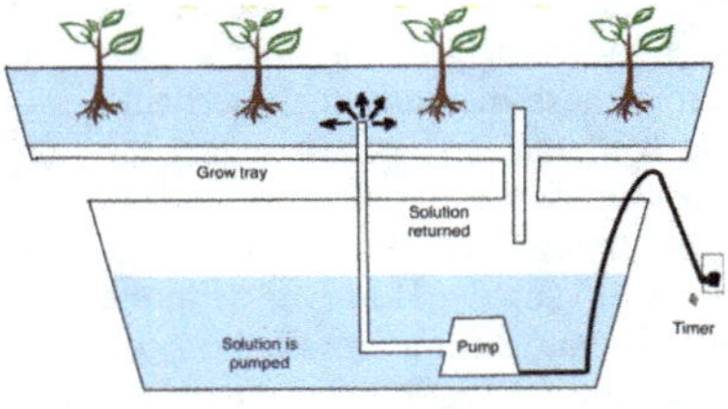

Fig 1. *Hydroponic System*

Aeroponic Systems

The farming method known as aeroponic VF produces plants in a misty environment without the need of soil or a growing medium other than a tiny quantity of nutrient-rich water. This kind of farming differs from hydroponics and conventional agricultural methods. Aeroponic systems use 90% less water than hydroponic systems, according to a comparison of the two. Furthermore, because this VF technique contains vitamins and minerals, it's claimed to be healthier [7]. Fig 2. is demonstrating Aeroponic System.

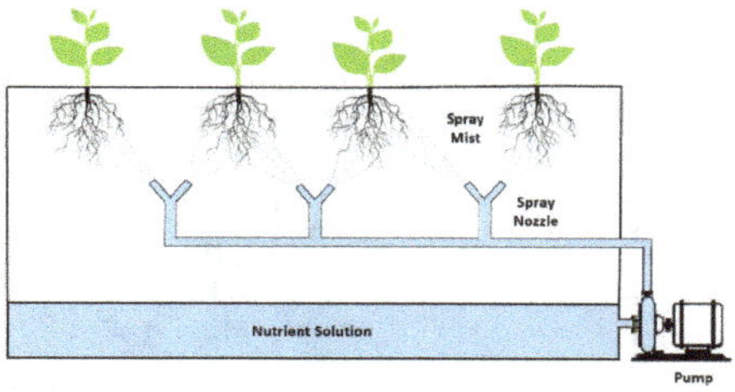

Fig 2. Aeroponic System

Aquaponic Systems

This agricultural method combines agriculture and aquaculture. It is also used in hydroponic systems that include fish tanks into one system [2]. Having said that, fish excrement is utilized to supply the nutrients that plants in a growing tray need. Remember that the water is then recycled after being filtered by crops to remove fish excrement. One major feature of this method is that, after the initial month of intense monitoring, pH and ammonia levels only need to be checked once a week when the system is up and running effectively. Fig 3. is demonstrating Aquaponic System.

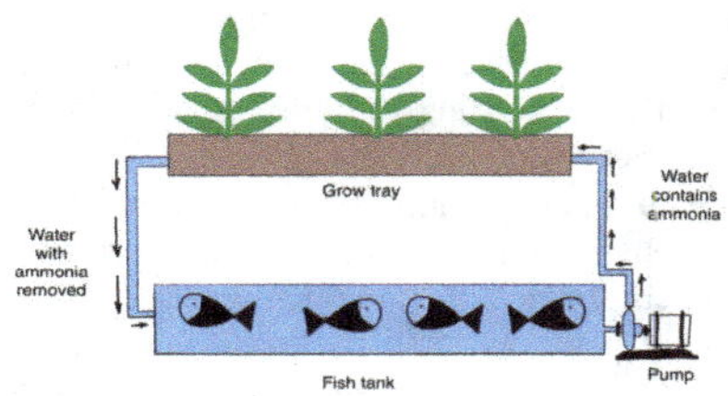

Fig 3. Aquaponic System

An aeroponic system is ideal since it enables you to cultivate your garden anywhere you like, indoors or outside. They are better for the environment than traditional gardens since they consume less water and electricity. Although not every plant can grow in an aeroponics system, you still have a huge selection to pick from!

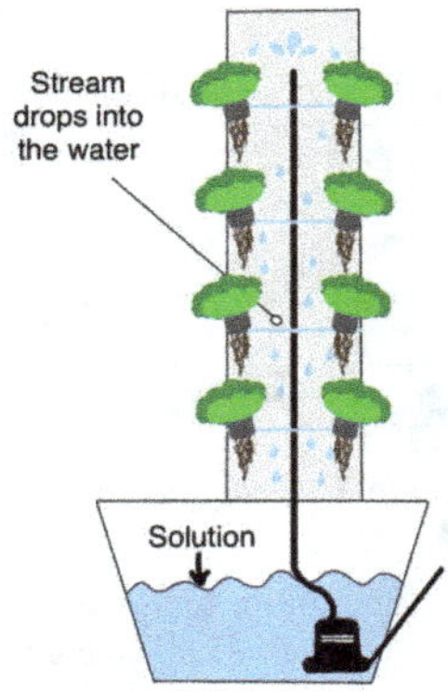

Fig 4. *Vertical Garden Column Type*

Tower gardens column type are the latest agricultural revolutions in the Western world, and they spread after cultivating roofs with vegetables and fruits and after cultivating balconies, which are agricultural methods such as plastic pipes, wooden cylinders, or wooden pyramids. Their processing and cultivation differ according to the crop, or the type of leafy vegetables or fruits such as strawberries. And parsley, mint, basil and coriander. Note that the plastic cylinders are also carried in plastic pockets with one hole, and the pockets are filled with seedlings of plants to be cultivated. The tower may consist of a number of cylinders connected from the bottom to collect irrigation filtrate and fertilizers. Basically, side openings are opened in the cylinder, by sawing it with a straight line, exposing it to heat, and pulling it out, forming a pocket. The pockets are separated from each other horizontally and vertically according to the type of crop and the length of its plants. The cylinder is filled with an agricultural medium instead of soil, and a perforated hose is inserted into it passing through the middle of the cylinder. To provide plants with irrigation and fertilizers, these towers are usually placed in kitchens, near and far from windows, according to their need for light. They are characterized by high productivity per unit area. A 4-inch cylinder with a length of two meters can carry 40-50 plants.

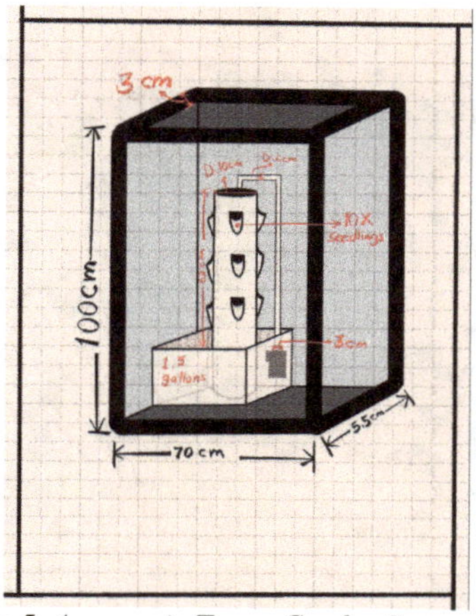

Fig 5. Aquaponic Tower Garden System

Simulation
The measurements and final building design are provided. Because of its high yield strength and stiffness, carbon steel was chosen as the primary structural material for the initial prototype. Carbon steel was the best choice since the structure was going to support over 140 kg of dead weight and a sizable amount of living weight from the atmosphere [8,9,10,11,12]. With dimensions of a length: 1.5 m, width: 1 m and height: 1.56 m [13,14,15,16,17]. By providing several degrees of service for checking and validating the application, Ansys makes it simple to check that it is functioning correctly on the computer and operating system [18]. When your system changes, such as when the operating system is upgraded or new math or vector libraries are installed, we can benefit from this testing. With reference to the project boundaries and data, ANSYS-FLUENT was used for the simulation of the tower garden system [19].

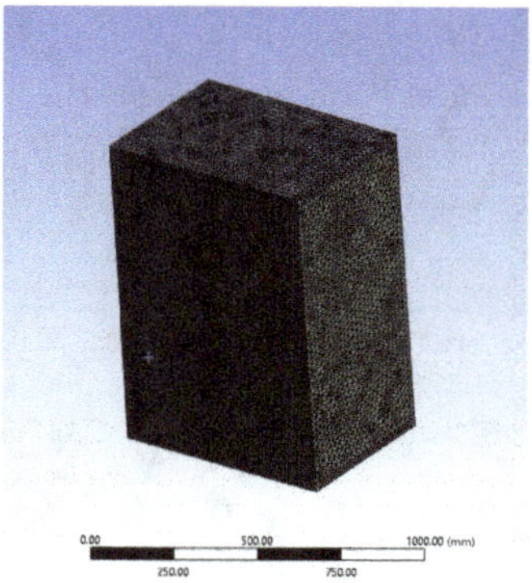

Fig 6. 512K Elements Mesh for The System

As fig 6. shown, the type of elements used is the tetrahedral, and the number of elements is 329723, and the number of nodes is 64762.

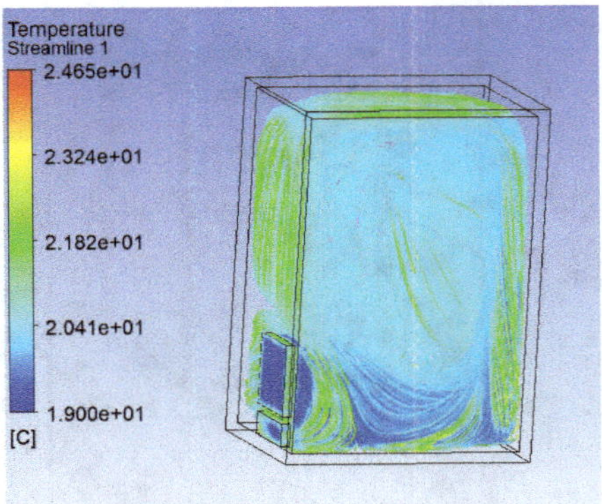

Fig 7. *Air Temperature Streamline*

Fig 7. is showing how the air is going in to the system, and how temperature is changing while air is flowing & distributed.

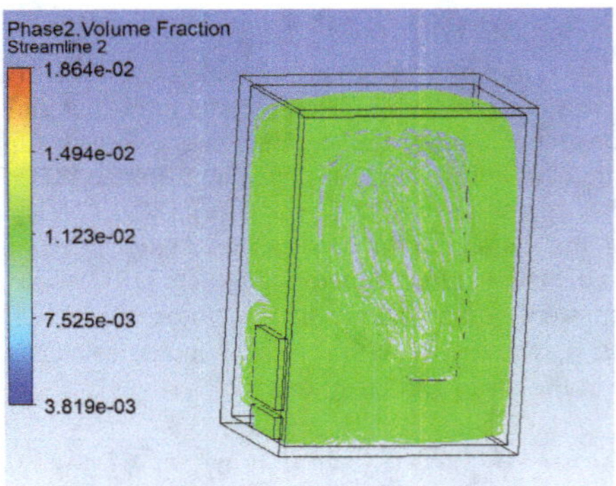

Fig 8. *Water Vapor Streamline*

Fig 8. is showing how the water vapor is going in to the system, and how temperature is changing while vapor is flowing & distributed.

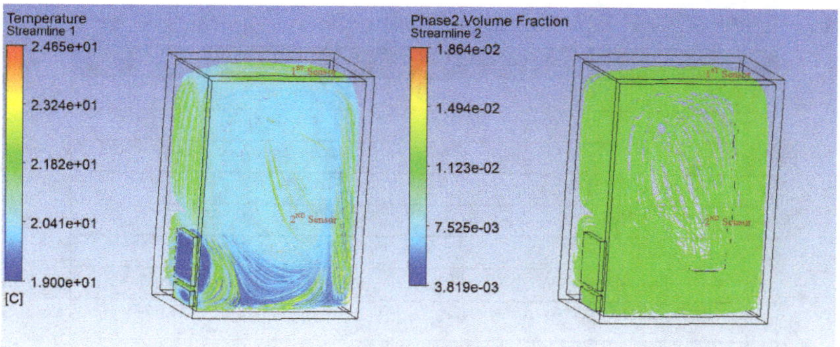

Fig 9. *Humidity & Temperature Sensors Placing*

However, for the prototype we are going to use to place one sensor only since it is for demonstrating purposes. In other words, the control system will be programmed to sense and check the humidity and temperature of the environment inside the prototype using one sensor placed inside at the top.

System Design and Control System

Polystyrene is a waterproof and thermoplastic substance that acts as a sound and temperature insulator, sometimes referred to as Styrofoam. The two variations have differing cost and performance ratings. The R-value of the more costly XPS is R-5.5. It makes up around 7% of the whole thermoplastic industry, making it one of the most popular commodity plastics. [20,21,22] Furthermore, the suitable environment for the lettuce to grow is that for vegetables require high relative humidity between 75 and 95 percent, with an average of 85 percent, and that the majority of vegetables grow best between 18 and 24 degrees Celsius on average lights [23]. It makes up around 7% of the whole thermoplastic industry, making it one of the most popular commodity plastics. Undoubtedly, it is one of the most significant styrene copolymers (PSAN) is poly(styrene co-acrylonitrile).

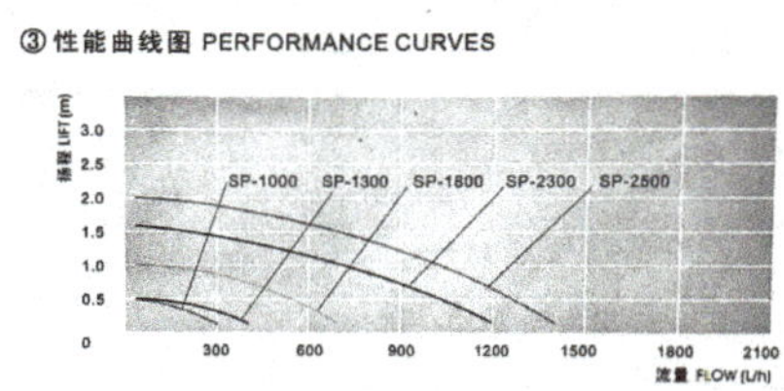

Fig 10. *The Chart of Pump*

And the amount of water to pump and the irrigation rate is 1.5 gallons per hour, or GPH = 5 liters per hour. In reality, the process of vertical farming depends on understanding the heat load. For instance, it was shown that a conditioning procedure of short days followed by 7- 9 hours at 21/12 °C (day/night) will increase the amount of fruit produced (41%–152%) annually compared to non-conditioned controls [22,24]. Thus, the suitable option for the submersible pump that could find in the stores was with a voltage of 230V and 13W with a maximum flow rate (Qmax) is 700L/h and a max head (Hmax) is 1.0m.

Tab 1 Engineering standards

BATTERY SIZING DATA	Solar System	Project load
Load Power 50W + Future 5W	55	200
Voltage (V) 24	24	12
Load Current (A)	2.291666667	16.6666667
Backup Time hours(BT) 120	120	120
Aging Factor (AF) 1.25	1.25	1.25
Design Factor (DF) 1.1	1.1	1.1
Temperature Compensation factor @ 25 C 1.05	1.05	1.05
Battery Ah @ 120 (A8= A3*A4*A5*A6*A7)	397.03125	2887.5
Selected battery Sizing	405	3000
Battery Bank C120 In Wh 9840	9720	36000
B PV SIZING DATA		
Total Days in month	30	30
Daily Peak hours in worst month	5	5
Recharge Time	30 Days	30 Days
Recharge Hours= Total days in monthX Daily peak hours	150	150
Peak PV output as per vendor (A)	9.1	9.1
Daily PV peak output (Ah/ Day)	45.5	45.5
Load current (A)	2.291666667	16.6666667
Battery inefficiency factor (BIF)	1.15	1.15
Battery charging current= (BATTERY AH*BIF)/RECHARGE HOURS	3.105	23
Daily Load Power (Ah/Day)	55	200
Daily Battery charging Power = Battery charging currentXDaily Peak hours	15.525	115
Total daily Power (Ah/ Day)= Daily Battery charging Power+ Daily Load Power	70.525	315
Adjusted Daily Power (Ah/ Day)= Total daily PowerX Aging(1.1)X Dust (1.1)+ Future (20%)	99.44025	444.15
No. of Parrallel Modules	2.1855	9.76153846
Selected No. of Parallel Modules	3	10

System Testing and Analysis

Fig 10. is highlighting the Connection of the control system. The control system designed using four relays to control the powering of each component of the prototype. On the other hand, the humidity & temperature sensor is the key element for controlling the system as shown in previous shown code. We added the screen to show the current readings of sensor in order to have observation of current environment inside the system.

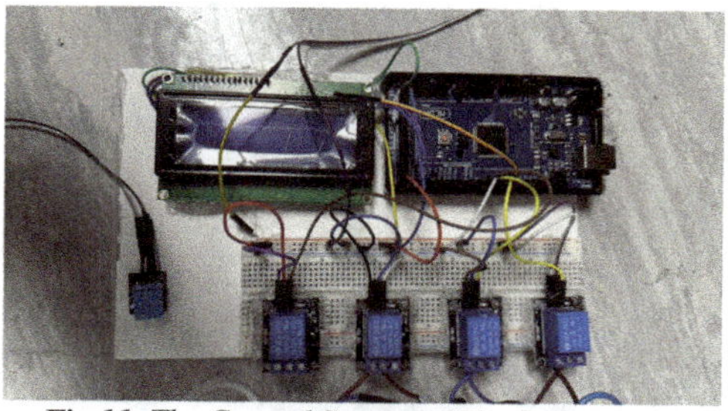

Fig 11. *The Control System of The Prototype*

Once the prototype was completed, I started at the main project. Choosing the Styrofoam for insulation to provide the suitable environment for the lettuce to grow in within the average humidity of 75-85%, and temperature between 18-24 °C on average lights. Moreover, the voltage provided by the growth lights is and 220-240V with wattage of 16W and it is the same voltage for the evaporative cooler AC with 65W. And for the submersible pump, the suitable option that could find in the stores was with a voltage of 230V and 13W with a maximum flow rate (Qmax) is 700L/h and a vertical lift or max head (Hmax) is 1.0m. Also, the humidifier has the voltage of 240V and 25W. And the dehumidifier that has been chosen is breathable design for moisture obsorption and moisture drainages are efficient, high usable material, economical and environmentally friendly. And its power is up to 25W and it recovers about 0.4kWh to dry. And the humidifier designed with a filter effectively removes impurities from water with a capacity of 3 liters. Furthermore, its ultrasonic system efficiently produces a high-blur effect. With a knob, this humidifier allows you to set the fog intensity. Once the setup finished, through the central conduit, the low-wattage submersible pump in the reservoir forces the nutrients solution to the peak of the tower garden. The fertilizer solution then uniformly cascades over the exposed plant roots as it descends down the interior of the tower garden. A timer makes sure that this procedure is repeated continually so that plants receive the right dosage of oxygen, water, and nutrients at the ideal moment.

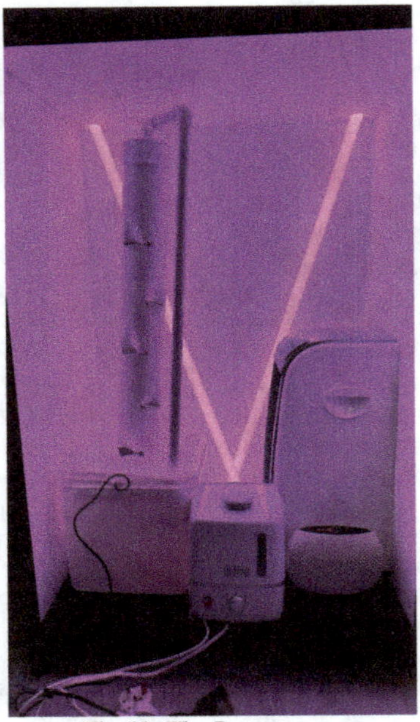

Fig 12. *The Prototype*

Conclusions

In conclusion, vertical farming innovates the agricultural industry by enhancing food production in an optimum space inside a regulated environment without wasting natural resources. Furthermore, sustainable food sources are not readily available for a number of reasons. One practical solution to the food shortage may be vertical farming operations. With this project, we aimed to build a scalable industrial-grade vertical farming unit that has been accessible to everyone, balanced, and sustainable. Therefore, we begin by selecting the kind of vertical farm (Aeroponic) since it requires fewer soil, uses less water, and produces crops with a better yield. Then, when we build the structure, we take insulation and inside systems like the HVAC, lighting, and irrigation systems into account. Additionally, we choose Polystyrene (XPS), for waterproofing substance and insulating sound and temperature. Moreover, we look for artificial lighting, most often electric lighting, therefore we utilize LED grow lights since they provide the necessary light intensity while using less electricity. We use an evaporative cooler as part of our heating, ventilation, and air conditioning (HVAC) system to provide the proper humidity level and temperature for the plants inside the farm. Additionally, to meet our power needs, we combined solar PV panels with batteries and a hybrid electrical system. To offer a proper environment for growing and harvest, we were able to monitor and assess temperature, relative humidity, CO2 concentration, light intensity, and power usage through this research. This project has the potential to develop a system for monitoring and controlling. As well as an alternative to structure and insulation, think about adopting stiff composite insulators. Also, improving the design of the grow towers inside the unit. Include HVAC ducting in the device.

References

[1] Pinstrup-Andersen, P. (2018). Is it time to take vertical indoor farming seriously? Global Food Security, 17(September 2017), 233–235. https://doi.org/10.1016/j.gfs.2017.09.002Plant characteristics and the potential for living walls to reduce temperatures and sequester carbon 2020, Energy and Buildings

[2] Teo, Y. L., & Go, Y. I. (2021). Techno-economic-environmental analysis of solar/hybrid/storage for vertical farming system: A case study, Malaysia. Renewable Energy Focus, 37(June), 50–67. https://doi.org/10.1016/j.ref.2021.02.005

[3] De Oliveira, F. B., Forbes, H., Schaefer, D., & Syed, J. M. (2020). Lean principles in vertical farming: A case study. Procedia CIRP, 93, 712–717. https://doi.org/10.1016/j.procir.2020.03.017

[4] Tuomisto, H. L. (2019). Vertical Farming and Cultured Meat: Immature Technologies for Urgent Problems. One Earth, 1(3), 275–277. https://doi.org/10.1016/j.oneear.2019.10.024

[5] SharathKumar, M., Heuvelink, E., & Marcelis, L. F. M. (2020). Vertical Farming: Moving from Genetic to Environmental Modification. Trends in Plant Science, 25(8), 724–727. https://doi.org/10.1016/j.tplants.2020.05.012

[6] Yalçın, R. A., & Ertürk, H. (2020). Improving crop production in solar illuminated vertical farms using fluorescence coatings. Biosystems Engineering, 193(2006), 25–36. https://doi.org/10.1016/j.biosystemseng.2020.02.07

[7] Mishra, A. P., & Sahoo, J. P. (2021). Agriculture Letters (ISSN : 2582-6522) Vertical Farming - The Foreseeable Future of Agriculture. May. https://doi.org/10.13140/RG.2.2.32900.45445

[8] Le Blanc, R. (2019, February 27). Grow light options for indoor and vertical farming. The Balance Small Business. Retrieved February 16, 2022, from https://www.thebalancesmb.com/grow-light- options-for-indoor-and-vertical-farming-4147429

[9] Khan, S. A., Devi, T. P., & Sharma, P. P. (2020). Vertical farming: Why it matters for Bhutan. Studies in Indian Place Names, 40(1), 1323–1329.

[10] Lomax, R. (2021). Endogenous Clocks and Vertical Farming From a Commercial Viewpoint. June. https://doi.org/10.13140/RG.2.2.28918.32328

[11] M. Manso et al. Green roof and green wall benefits and costs: a review of the quantitative evidence

[12] Osmond, C. (2021, July 15). Massive guide for growing plants indoors under lights. Backyard Boss. Retrieved February 16, 2022, from https://www.backyardboss.net/lights-for-growing-plants- indoors/

[13] Perceived sensory dimensions: an evidence-based approach to greenspace aesthetics. Urban For. Urban Green. (2021)

[14] Texas A&M Agrilife Extention. (0000, 00 00). Greenhouse Structures . Retrieved 2 19, 2022, from Texas A&M Agrilife Extention: https://aggie-horticulture.tamu.edu/ornamental/greenhouse- management/greenhouse-structures/

[15] Thermal and energy performance of algae bioreactive façades: A review 2020, Journal of Building Engineerin J. Stoltz et al.

[16] Vertical Farming Perspectives in Support of Precision Agriculture Using Artificial Intelligence: A Review 2022, Computers

Materials Research Proceedings 31 (2023) 287-298 https://doi.org/10.21741/9781644902592-30

[17] Well-being, health and urban coherence-advancing vertical greening approach toward resilience: a design practice consideration

[18] Ansys for Students on https://www.ansys.com/academic/students

[19] Fatimah H. Naser, Ali Hameed Naser Al Mamoori, Mohammed K. Dhahirc. (3,2021) Effect of using different types of reinforcement on the flexural behavior of ferrocement hollow core slabs embedding PVC pipes

[20] CB Insights. (2021, 05 20). How Vertical Farming Is Impacting The Food Supply Chain And Enabling Taste Innovation. Retrieved (2 19, 2022) from CB Insights: https://www.cbinsights.com/research/what- is-vertical-farming/

[21] Comments, Nicole, Jahan, N., Moody, B., Kerr, J., Cynthia, D, & Lawson, C. (2021, April 9). Grow lights for your vertical garden: A complete guide. Garden Tabs. Retrieved February 16, 2022, from https://gardentabs.com/grow-lights/

[22] Energy cost reduction by shifting electricity demand in indoor vertical farms with artificial lighting. Biosystems Engineering, 211, 219–229. https://doi.org/10.1016/j.biosystemseng.2021.09.06

[23] Avgoustaki, D. D., & Xydis, G. (2021). Energy cost reduction by shifting electricity demand in indoor vertical farms with artificial lighting. Biosystems Engineering, 211, 219–229. https://doi.org/10.1016/j.biosystemseng.2021.09.06

[24] Xiaoying Dingabc, Zhiyuan Zhao, Jie Zhengd, Xiaopeng Yue, Han Jina, Yukun Zhangd. (9, 2020) Community Gardens in China: Spatial distribution, patterns, perceived benefits and barriers.

Using powder metal gears in industrial applications- A review

Omar D. Mohammed[1,a *], QatrAlnada Mahmood[2,b]

[1]Mechanical Engineering Department, Prince Mohammad Bin Fahd University, Al-Khobar, KSA

[2]College of Engineering, University of Mosul, Mosul, Iraq

[a] osily@pmu.edu.sa, [b]qatralnada.mahmood@gmail.com

Keywords: Gears, Powder Metal (PM) Technology, Sintered PM, PM Gears

Abstract. Due to the high performance, lightweight, competitiveness and net shape of the parts produced using sintered Powder Metallurgy (PM), PM-based products are used in different industrial applications. PM method offers high precision and repeatability, maximum material utilization, tailored shape properties, freedom of design, cost efficiency, and sustainability. These advantages are driving the ongoing demand for PM products and for the PM method as a good alternative to the conventional gear production methods. The use of PM is confirmed as an applicable technology for gear production. However, the strength and NVH characteristics of PM-products still require more investigation. In the current paper, the mechanical properties, strength and NVH characteristics are presented and discussed. The paper concludes based on the presented overview of the applicability of PM gears in the industry, further research investigation is needed on both the system and component levels to improve the durability and NVH behavior which are important for extending the PM parts' application. Moreover, the sustainability aspects of powder metallurgy technology should be emphasized.

Introduction

In the industry, optimizing fuel efficiency and minimizing weight while increasing power density is challenging. However, the industry still needs to optimize the applied manufacturing methods and reduce investment and manufacturing costs for the upcoming powertrain operations.

Each method that can be applied for structural parts production has certain limitations and offers certain advantages. The Powder Metal (PM) method as compared to other manufacturing methods has the ability to produce complicated structural shapes of parts of high dimensional accuracy in large series without chip formation at reasonably low costs. PM method offers a unique and important facility to generate shapes which are not achievable with other methods or can only be achieved with difficulties and at high costs [1]. Different research works were presented to review the use and application of powder metal gears.

During the last two decades, the use of PM technology has gained a lot of attention in the automotive industry to investigate the possibility of producing gears with the attributes of PM. Transmission synchronizer hubs were produced using PM and used at different gear speeds. More recently, PM gears were introduced for the low-load gears to prove the applicability of this technology to survive under the applied transmission load and meet the durability requirements [2].

The recent developments in the manufacturing of gears were presented in [3]. Due to international competitiveness and environmental regulations, industries are forced to adopt economical, ecological and efficient means of producing products. The availability of financial and technical data which is needed for economic analysis of PM gear production was reviewed in [4]. The author concluded that the operational cost of PM gear manufacturing is comparable to that of conventional gear manufacturing methods. However, further research is needed for proving the benefits of adopting PM gear manufacturing including the economic benefits as compared with the conventional methods.

In the literature discussed in the above-mentioned references, the use of PM is confirmed as an alternative method for gear production. However, the strength and NVH characteristics of PM still require more investigation. In this paper, the mechanical properties, strength and NVH characteristics are discussed to give a good overview of the applicability of PM gears.

Benefits and limitations of PM gears

In general, the main advantage of applying PM technology in production is the possibility of producing complicated parts of high dimensional accuracy in large series at reasonably low costs [1]. PM technology has been applied for gear production to use the advantages of optimizing the tooth root shape for better stress distribution, reducing gear weights, adding additional design features, and reducing structural-borne vibrations, which means reducing gear noise, due to the damping effect [5]. Fig.1 [5] shows how the use of PM can change the stress distribution due to the possibility of optimizing the tooth root shape, in contrast to the conventional hobbed gear. Moreover, the product weight is relatively lighter than if it is produced from wrought steel [1]. New techniques of adding alloys to the steel PM for having lighter weight were discussed in [6,7].

The strong ongoing trend of electrification in the automotive industry introduced the challenges of increasing the input speed of the transmission and increasing the required power density. Furthermore, the generated gearbox noise is of high importance, since the noise of the gearbox is no longer masked by the combustion engine sound. PM could be a good alternative to meet those requirements. Due to the internal porosity PM gears can be of lighter overall weight and can have a dampening effect on the noise and vibration [8].

However, the current manufacturing of powder metallurgical steel has also limitations. In general, PM gears have limitations in strength and size. PM gear teeth as compared to wrought steel ones have approximately 50% lower impact resistance and 33% lower contact fatigue strength due to the porosity of the PM structure. In manufacturing, this limitation can be partially made up by increasing the density of the produced PM gear teeth. This can be achieved using double pressing and double sintering. Also, high-temperature sintering or case hardening can be used. Another important limitation of PM technology is the gear face width. The amount of the used powder in most of the compaction presses determines the gear face width to about 3 inches. Furthermore, frictional losses between the die and the powder cause decreased density along the gear face width, where the lowest density is at the mid-point. The wider the gear face width, the more density reduction. The density variation along the face width can result in dimensional variations during sintering and also when it is subjected to heat treatment. High-density variation leads to distortion, especially in big gears [9].

One new technology of Nanotechnology Enhanced Sintered Steel Processing works on enhancing the PM gear strength to reduce the applicability difference between conventional steel and powder metallurgical steel. One of the significant benefits of this technology is that the density can be increased with the inclusion of nano-powder particles [8].

Applying PM technology involves sustainability aspects which can be of great environmental benefit if the approach of recycling materials can be applied. Höganäs AB in Sweden presented its new sustainable Cr pre-alloyed metal powder. The new powder (Astaloy® CrS) is recyclable and also it can be used as a raw material for the next product when the first product has reached its end of life [10]. PM technology can be applied for converting oxidized ferrous metal scrap into usable parts [11]. An optimized PM-based recycling process is studied in [11] to obtain significantly higher mechanical and physical properties of PM products. Recycling or reusing materials involves environmental and also economical benefits, due to the increasing material disposal costs and the lack of land fields, in addition to the raw material extraction costs [12].

Further research on the sustainability aspects of gear manufacturing processes including powder metallurgy is needed to enhance this approach.

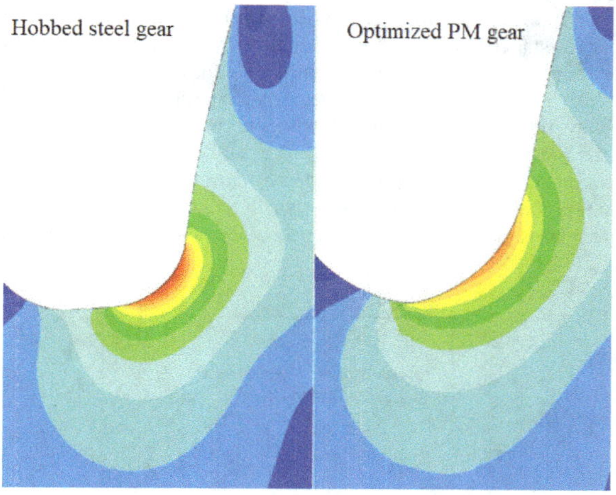

Fig.1 Tooth root stress distribution in a PM gear as compared to a hobbed gear [5].

Mechanical properties and strength of PM gears

In general, the flank load-carrying capacity of the PM gears is lower than the wrought steel gears because of the porous structure. In the PM parts of the porous structure, every pore can be considered a weak spot where a crack can be initiated. Therefore, the tooth flank surface of the highly loaded PM gears is usually densified to reduce the porosity in the surface region to ensure a sufficient load-carrying capacity. The material properties such as Poisson´s ratio and Young´s modulus are mainly affected by the local porosity [13]. The material properties were typically reported to be, for example, for copper powder (Young´s modulus =110-120 GPa, Poisson´s ratio = 0.35) [14,15] and for stainless steel powder (Young´s modulus =190-200 GPa, Poisson´s ratio = 0.29) [14,16]. The normalized Young´s modulus magnitude for spherical copper and stainless-steel powders shows similar dependence on the normalized density, Fig. 2 [14].

Efforts made by tooling manufacturers and PM parts producers have shown significant improvement in helical gear compaction. Although this technology is expensive and limited to a few specialist companies, it is applied in the powder metallurgy industry. The possibility of compacting a helical gear with a 33° helix angle using conventional alloy steel powders is proofed at a density close to 7.2 g/cm3. In addition to compaction complexity in PM production and the difficulties of high core density, PM gear blanks should get deep surface densification in order to maximize the strength in the root fillet and on the active tooth surfaces [17].

A comparison of the contact fatigue of hobbed wrought steel gears and surface-densified gears using a powder metallurgy process is presented in Fig.3 [17]. As shown in the figure, surface-densified PM gears have an endurance limit of around 1764 MPa as compared to 1839 MPa for the hobbed gears. The difference in fatigue life, below 5 per cent in this case, makes the surface-densified gears a good and applicable alternative to the ordinary hobbed gears with respect to performance in applications of high contact stress [17].

In the production of PM parts, the rolling densification process was processed using a rolling machine provided by Escofier as illustrated in Fig. 4 [18]. In the densification rolling process, two rolling wheels of the same speed are spinning in the same direction [18]. However, the Fe-base PM parts that are produced using the conventional PM production methods have a porosity of about 5% to 15%, and their application range in the automotive industry is then limited due to the relatively low wear resistance, strength, and hardness caused by the existence of pores [19]. The mechanical properties of Fe-base parts can be comparable to wrought steel parts when the porosity is made close to 0%, which can be obtained using new technologies to produce PM parts with low porosities. In this way, the applicability of PM-products can be expanded in the industry [18].

The impact of the precipitation-hardened structures and the heterogeneous grain structures are combined by processing a certain dual-phase microstructure based on the Ti-2448 alloy. Actually, several microstructural features have been investigated and modified by changing the processing parameters, e.g. grain sizes in both shell and core, chemical concentrations, and phase precipitation [20].

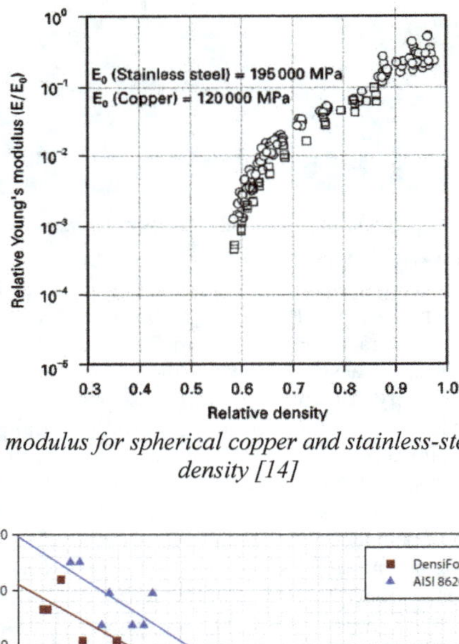

Fig.2 Relative Young's modulus for spherical copper and stainless-steel powders with relative density [14]

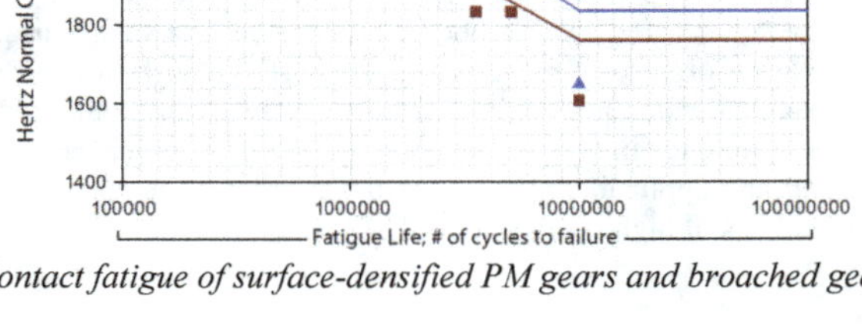

Fig.3 Contact fatigue of surface-densified PM gears and broached gears [17]

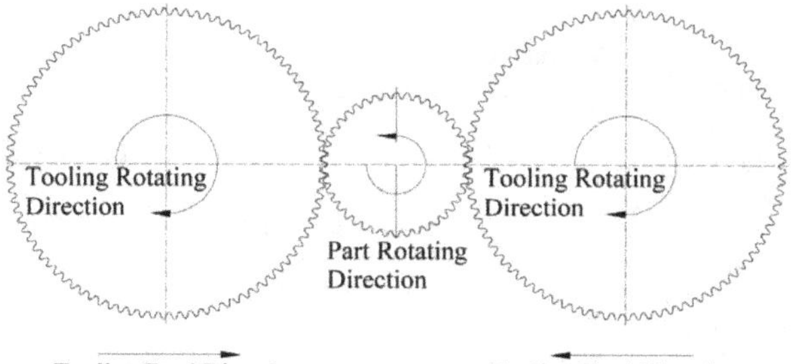

Fig.4 Schematic view of the experimental apparatus [18]

Noise and vibration NVH characteristics

In the new era of electrification in the automotive industry, reducing the generated noise from gearboxes is of high importance since the noise of the gearbox is no longer masked by the combustion engine. Hybrid electric vehicles (HEV) and electric vehicles (EV) are facing different challenges, like the need for performance enhancement to meet the durability requirements and also the increasing NVH requirements for minimizing the generated noise and vibration. PM gears can be a good alternative to be investigated for meeting those requirements.

Examining different PM densities, it is observed the damping property increases while decreasing the gear density and a more increase can be obtained in damping when further decreasing in PM density [21,22].

Using the original geometry of a steel gear to a PM gear leads to a higher excitation, but the optimized PM gear geometry results in a lower excitation at the applied torque range. As a result of this, gear micro-geometry optimization, in addition to tooth root area optimization, should be included in the design process for PM gear surface densification [21].

In the current stage, testing is important to further investigate the improvement in the NVH by using the PM. Damping potential is examined on the component level, but the impact on NVH is still not obviously evaluated, especially on the system level. On both the system and component levels testing is needed to improve the NVH behavior which facilitates further potential to apply PM Gears in the current and future gear transmissions and machines [21].

PM gear parts with an overall density of 6.8g/cm^3 give a significant change in the NVH behaviour, where the change in material density and porosity has an impact on the obtained Eigen frequencies. Depending on the applied torque and rotational speed, the structure-borne noise could be reduced by about 3dB [23].

The porosity due to the process leads to difficulties and challenges in the manufacturing of PM gears but also can be good in terms of offering the potential for higher damping and then an improvement of the noise, vibration, and harshness (NVH) behavior can be obtained. PM gears offer the feasibility of an optimized gear design for better NVH behavior. To evaluate this feasibility, tests and experimental investigations were conducted, and both the excitation and the structure-borne noise emission were measured and evaluated. In terms of gear damping and noise reduction properties of the materials, the structure-borne noise emission of the produced PM gears is proportional to the density of the gear structure. By reducing the density of the gear structure, the NVH behaviour is improved. [24].

Conclusions

- Powder Metallurgy (PM) method offers lightweight, high precision and repeatability, maximum material utilization, tailored shape properties, freedom of design, cost efficiency, and sustainability.
- PM gears have limitations in strength and size, so further work is needed to improve the applicability of PM gears in terms of strength and size.
- Highly loaded PM gears are usually surface-densified to reduce the porosity in the surface region to ensure a sufficient load-carrying capacity. So, improving the surface densification process is important to close the gap, in terms of durability, between the PM gears and those from wrought steel.
- The NVH behavior is affected by the change in material density and porosity and thus system Eigen frequencies are affected. In general, the higher porosity will result in higher damping effects.
- Further investigation is needed on both the system and component levels to prove the applicability of PM gears for more industrial applications in terms of durability and NVH behavior.
- Further research on the environmental benefits of applying PM technology is needed to enhance the sustainability aspects.

References

[1] Höganäs Handbook for Sintered Components, Design and Mechanical Properties, Höganäs AB, Sweden, 2015.

[2] A. Flodin, Powder metal gear technology: a review of the state of the art, Power Transmission Engineering. March (2016) 38-43.

[3] K. Gupta, R.F. Laubscher, J. P. Davim, N. K. Jain, Recent developments in sustainable manufacturing of gears: a review, Journal of Cleaner Production. 112 (2016) 3320-3330. https://doi.org/10.1016/j.jclepro.2015.09.133

[4] B. Kianian, Comparing acquisition and operation life cycle costs of powder metallurgy and conventional wrought steel gear manufacturing techniques, Procedia CIRP. 81 (2017) 1101-1106. https://doi.org/10.1016/j.procir.2019.03.260

[5] North American Höganäs Handbook, Gear up for new PM business opportunities, Höganäs AB, Sweden, 2015.

[6] B. Kieback, G. Stephani, T. WeiBgarber, T. Schubert, U. Waag, A. Bohm, O. Andersson, H. Gohler, M. Reinfried, Powder metallurgy for light weight and Ultra-Light weight materials, Journal of Korean Powder Metallurgy Institute. 10 (2003) 383-389. https://doi.org/10.4150/KPMI.2003.10.6.383

[7] F. H. Froes, J. R. Pickens, Powder metallurgy of light metal alloys for demanding applications, The Journal of The Minerals, Metals & Materials Society. 36 (1984)14-28. https://doi.org/10.1007/BF03339911

[8] E. Bergstedt, A Comparative Investigation of Gear Performance Between Wrought and Sintered Powder Metallurgical Steel, PhD thesis, KTH Royal Institute of Technology, Sweden, 2021.

[9] H. Sanderow, Powder metallurgy gears gain strength and viability, Machine Design. March 2000. https://www.machinedesign.com/motors-drives/article/21828990/powder-metallurgy-gears-gain-strength-and-viability

[10] Höganäs Powder Metallurgy Review, Improved sustainability enhanced performance, Inovar Communications Ltd. 11 (10) (winter-2022) 1-124.

[11] K. Rane, P. Date, Sustainable recycling of ferrous metallic scrap using powder metallurgy process, Journal of Sustainable Metallurgy. 3 (2017) 251-264. https://doi.org/10.1007/s40831-016-0075-3

[12] J. Mascarenhas, Powder Metallurgy: A Major partner of the sustainable development, Materials Science Forum. 455-456 (2004) 857-860 https://doi.org/10.4028/www.scientific.net/MSF.455-456.857

[13] T. Frech, P. Scholzen, P. Schaflein, C. Lopenhaus, P. Kauffmann, F. Klocke, Design for PM, Procedia CIRP. 70 (2018) 186-191. https://doi.org/10.1016/j.procir.2018.03.267

[14] P.C. Carnavas, N.W. Page, Elastic Properties of Compacted Metal Powders, Journal of Materials Science. 33 (1998) 4647-4655 https://doi.org/10.1023/A:1004445527430

[15] E. Dean, Elastic moduli of porous sintered materials as modeled by a variable-aspect-ratio self-consistent oblate-spheroidal-inclusion theory, Journal of American Ceramic Society. 66 (1983) 847-854. https://doi.org/10.1111/j.1151-2916.1983.tb10999.x

[16] J. Wang, Young's modulus of porous materials, Journal of Materials Science. 19 (1984) 801-808. https://doi.org/10.1007/BF00540451

[17] S. Nigarura, R. Parameswaran, M. Scott, C. Dennert, A Comparison of wrought steel gears and surface-densified powder metallurgy gears, Gear Solutions. August (2015) 33-38. https://doi.org/10.31399/asm.hb.v07.a0006114

[18] J. Peng, Y. Zhao, D. Chen, K. Li, W. Lu, B. Yan, Effect of surface densification on the microstructure and mechanical properties of powder metallurgical gears by using a surface rolling process, Materials. 9 (2016) 846. https://doi.org/10.3390/ma9100846

[19] L.S. Sig, G. Rau, C. Dennert, Selective surface densification for high performance P/M components, International Conference on Powder Metallurgy and Particulate Materials. Denver, U.S.A, 2007.

[20] B. Fer, D. Tingaud, A. Hocini, Y. Hao, E. Leroy, F. Prima, G. Dirras, Powder metallurgy processing and mechanical properties of controlled Ti-24Nb-4Zr-8Sn heterogeneous microstructures, Metals. 10 (2020) 1626. https://doi.org/10.3390/met10121626

[21] G. Kotthoff, NVH potential of PM gears for electrified drivetrains, GEAR TECHNOLOGY. September/October (2018) 40-43.

[22] B. Kianian, A comparative cost analysis of conventional wrought steel and powder metallurgy (PM) gear manufacturing technologies, Licentiate Thesis, Lund University, Sweden, 2019.

[23] B. Leupold, V. Janzen, G. Kotthoff, D. Eichholz, Validation approach of PM gears for e-drive applications, Power Transmission Engineering. October (2019) 49-55.

[24] T. Frech, P. Scholzen, C. Loepenhaus, F. Klocke, Powder metal gears for highly loaded powertrains: How powder metallurgy supports current trends in transmission technology, SAE International Journal of Materials and Manufacturing. 11(2018) 431-440. https://doi.org/10.4271/2018-01-0989

Preliminary investigation and SWOT analysis of sustainable mortar utilizing Arabic Gum

Muhammad Nasir, Walid A. Al-Kutti*, A.B.M. Saiful Islam, Khalid Saqer Alotaibi, Ammar Al Eid, Yousef Khalid Algatam, Ali Alqahtani

Department of Civil and Construction Engineering, College of Engineering, Imam Abdulrahman Bin Faisal University, Dammam 31451, Saudi Arab

* walsalem@iau.edu.sa

Keywords: Sustainable Mortar, Portland Cement, Arabic Gum, Admixture, Bleeding, Flow, Compressive Strength, Microscopy, SWOT Analysis

Abstract. Modern concrete technology requires commercial chemical additives or admixtures - synthetic in nature which are considered as non-ecological construction practice. Conversely, Arabic Gum (AG) is also known to alter the properties of the concrete without compromising the techno-economic and environmental merits when adequately incorporated. This research explores the potency of AG in the development of ordinary Portland cement (OPC) mortar. Four types of mixes were prepared after admixing the dosage of AG in the range of 0% to 1.5% at an interval of 0.5% by weight of the OPC. The specimens were tested by evaluating representative fresh and mechanical properties, such as visual inspection, flow, and compressive strength. The findings were also correlated with the scanning electron microscopy. The results indicated that workability was linearly enhanced by the inclusion of the AG. On the other hand, the strength progressed up to 1% dosage of AG afterwards insignificant development was noticed attributed to bleeding phenomenon leading to a non-cohesive matrix. Therefore, 1%-admixed AG mixture was considered as optimum mix yielding a maximum flow of 250 mm and a 28-day compressive strength of 47.4 MPa. These values were 47% and 22% higher than that in the AG-free mix, respectively. It is postulated that the AG acts like a superplasticizer and contribute to enhance the workability of the mixture, whereas optimum dose provide microstructural densification resulting in higher strength. Hence, up to 1% of AG can be beneficially admixed depending on the flow and strength requirements. The SWOT analysis indicated that the usage of AG in production of construction products might expand the market for the farmers, cut-short the material cost and contribute to the circular economy.

1. Introduction

Chemical admixture plays vital role in modern construction applications and help control the water demand, setting times, air entrainment, corrosion, and shrinkage characteristics when admixed at suitable dosage to the fresh mix. The worldly demand of the admixtures has grown to approximately $18 billion in 2020 [1] and continuously escalating due to increase in rate of population and urbanization. However, the existing commercial admixtures are synthesized using oil-based polymers which are known to be non-renewable, the direct exposure of them to human and the environment is considered as hazardous, they may also behave unpredictable sometimes resulting in abnormal fresh and hardened properties of concrete, they may also require more than 10% of the total construction cost, and need to be stored at favorable temperature having a shelf life of not more than one year [2].

The problems associated with the chemical admixtures have directed the attention of researchers to propose sustainable and viable solutions. One of the options to replace the traditional admixtures is to incorporate the supplementary cementitious materials (SCMs) which includes the by-products of various agricultural (rice husk ash, date palm ash, palm oil fuel ash, etc.), industrial (fly ash, slag, silica fume, etc.), and natural minerals (natural pozzolan, metakaolin, clay, etc.) [3,4]. A few studies reported that Arabic gum (AG) – a water-soluble material obtained from the tall Acacia trees exist in Arabian Peninsula – has potential to be used as an admixture. The typical AG tree is depicted in Figure 1.

Fig. 1 Typical Arabic gum tree, its secretion from branches, and collected fruit.

According to the review of literature, Satti and Ahmed [5] observed that a 0.8% dosage of the AG by weight of cement was beneficial in minimizing the water content of up to 11.5% without impairing the workability but slightly extending the setting times of the concrete composites. The dispensing effect imparted by the AG tends to exhibit a good fluidity as well [6]. A study in hot weather conditions of Africa suggested that AG can be successfully deployed into the concrete production for resisting the chloride permeability and enhancing the compressive strength [7]. It was posited that the increased strength is attributed to the precipitation of the minerals during the hydration reaction and the formation of dense CSH gel [8].

The former studies on AG were mainly sourced from African region, they were sparse, and non-systematic. Since Saudi Arabia is planting the Acacia trees at thousands of hectares of land to cope with desertification and climate change, beside commercial utilization of the AG in various industries, AG can be beneficially utilized in the construction industry of Saudi Arabia. It is expected that this preliminary study will encourage the farmers to increase plantation, cut-short the construction cost, and contribute to circular economy without compromising the technical merits of the construction products.

2. Materials and methods

2.1. Raw material

Type I cement (i.e., ordinary Portland cement (OPC)) was used as a main binding material in this study. The Arabic gum (AG) was procured from local market and admixed at varying content of 0 - 1.5% at an interval of 0.5% by weight of OPC. The AG was powdered and sieved passing # 100 size prior to usage. The fine aggregate (FA) used was the dune sand which was procured from a local Ready-mix company having a maximum grain size of 600 microns. The FA to OPC and water to OPC ratios were kept invariant as 1.5 and 0.4 in mixes, respectively.

2.2. Synthesis of aqueous AG and fabrication of specimens

Firstly, aqueous AG was synthesized by mixing water content of a mix with the dosage of powdered AG powder using a stirrer. Next, the mortar specimens were fabricated using a Hobart mixer. All ingredients were added to the bowl and mixed for a duration of approximately 5 min. The mix proportion is presented in Table 1.

Table 1 Mix design (All quantities in kg/m³)

<table>
<tr><th>Mix IDs</th><th>OPC</th><th>Sand</th><th>Water</th><th>AG</th></tr>
<tr><td>M1-0% AG</td><td rowspan="4">768</td><td rowspan="4">1151</td><td rowspan="4">307</td><td>0</td></tr>
<tr><td>M2-0.5%AG</td><td>3.83</td></tr>
<tr><td>M3-1%AG</td><td>7.67</td></tr>
<tr><td>M4-1.5%AG</td><td>11.52</td></tr>
</table>

2.3. Evaluation

The workability or flow of mortar was measured in fresh state using an ELE brand flow table, as per the specifications of ASTM C230. The spread of mortar was measured in three directions and average value was reported. Next, a visual inspection was carried out on the freshly prepared specimens to qualitatively assess the rate of bleeding and segregation. The compressive strength was evaluated after 7-, 14-, and 28-days of curing on 50 x 50 x 50 mm cubes, as per the specifications of ASTM C109. Thereafter, scanning electron microscope (SEM) was employed to investigate the microstructure of the raw ingredients and OPC-AG paste matrix. The picture gallery of the experimental program is shown in Fig. 2. Finally, a strength, weakness, opportunities, and threats (SWOT) analysis was presented.

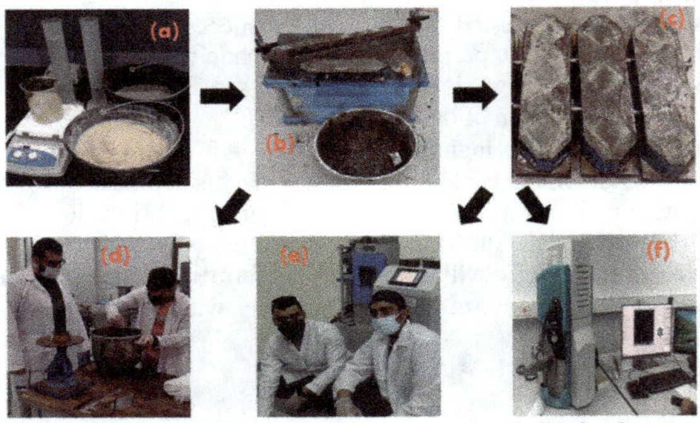

Fig. 2 Pictorial view of the experimentation: (a) raw material, (b) fresh mortar and compaction set-up, (c) fresh cubic specimens, (d) testing of flow, (e) testing of compressive strength, (f) microscopic analysis.

3. Results and Discussions

3.1. Flow

The variation of the flow of fresh mortar is depicted in Figure 3. It was clearly noticed that an increment in flow is likely when the amount of the AG in increased to 1.5%. The control specimen registered a flow of 170 mm, whereas the flow in M2-0.5%AG, M3-1%AG, and M4-1.5%AG were 195, 242, and 250 mm, respectively. These flow values in AG-based mixes were 14, 42, and 47% higher than that recorded in AG-free mix. An increase in flow values due to an increase in the AG content is attributable to its superplasticizing nature or lubricating feature which provides ball bearing effect and consequently enhancing the flow mobility. Thus, AG can be used as an alternative to the commercial superplasticizer or as a pumping aid. The trend of flow results follows Athman et al. [9] who found AG as a superplasticizer during the manufacture of a self-consolidating concrete. Elsewhere [10] also reported that the flow of AG-free mortar was 120 mm which was linearly or exponentially increased with the addition of AG content of up to 1.1%. The authors found that the increased flow due to an increase in the dosage of the AG is basically due to the emulsifying effect provided by the AG itself. Mbugua et al. [11] observed a 19% increase in the flow when AG was admixed between 0.4 and 0.7%, whilst any further addition led to a flow of more than 70% of AG-free mortar.

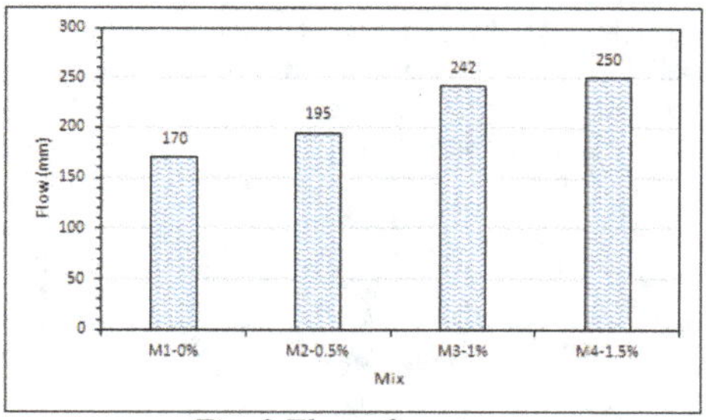

Fig. 3 Flow of mortars.

3.2. Visual inspection

Once the specimens were cast, a visual inspection was carried out to qualitatively assess the rate of bleeding and segregation in each mixture. Generally, the rate of bleeding appeared to be increased due to increment in the amount of the used AG, as the trend noticed in the flow data. However, the bleeding was significantly increased in M4 specimens prepared with 1.5% dosage of the AG. This is evident from Figure 4(a) which shows the sheen of water appeared on the surface of the fresh mortar, as compared to that prepared with ≤ 1% of AG. This indicates that the content of AG should be limited to 1% to avoid significant bleeding and segregation. Similar bleeding phenomenon was confirmed by [11] by measuring the rate of bleeding in AG-based mixes. They reported a marginal bleeding in specimens prepared with up to 0.8% AG content, whereas excessive bleeding was observed in specimens synthesized with a threshold limit of 1% of AG mainly due to occurrence of de-flocculation that converts the mix in to waterier.

Fig. 4 Visual inspection of fresh mortar: (a) bleeding sign in M4-1.5% and (b) non-bleeding in typical mortar prepared with ≤ 1% of AG.

3.3. Compressive strength

The variation in the development of compressive strength of mortar is depicted in Figure 5. In general, the strength tended to increase as the curing progressed, regardless of the dosage of the AG. However, the strength values tended to slightly drop when more that 1% dosage of AG was admixed. It was also observed that the rate of strength development was rather slow at early ages of 7 and 14 days in the AG-based specimens, whereas the degree of strength development was more pronounced after a curing period of 14 days. This trend was expected as AG was already found to be act like a retarder in the investigation carried out by Mohamed et al. [10] and Mbugua et al. [11]. According to the results obtained in this study, the compressive strength in AG-free specimens, M2-0.5%, M3-1%, and M4-1.5% was recorded to be 27.4, 29.8, 31.7, and 31.1 MPa at the age of 7 days, respectively. Likewise, these strength values were recorded to be 38.5, 40.4, 47.4, and 47.1 MPa after 28 days of curing, respectively. This indicates that the optimum mix is M3 prepared with admixing 1% of AG for enhancing both the early-age and later-age strength of mortar. Numerically, up to 8.8%, 22.9%, and 22.2% increment in strength values were noticed in AG-based mixes prepared with 0.5, 1%, and 1.5% dosage, respectively. These findings are complying to Elinwa et al. [8] who reported an optimum dosage of AG of about 0.75% which results in increment of the strength of up to 39.5% due to the extra formation of the minerals.

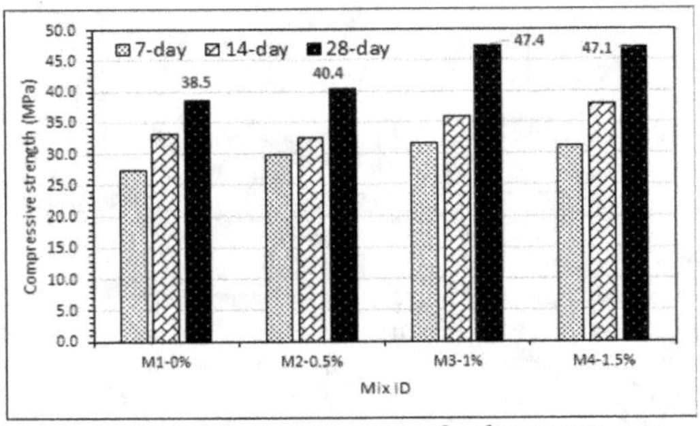

Fig. 5 Compressive strength of mortars.

3.4. Scanning electron microscopy

The variation in the morphology of the neat OPC and neat AG alongside the evolved microstructure of OPC-alone and OPC-AG paste matrix is depicted in Figure 6. It can be noticed that the texture of both raw materials was heterogeneous, and the grain sizes were also widely varied. Obviously, the morphology of the paste was significantly changed due to the hydraulic reaction. Among the paste mixes the microstructure of the M3 specimens was more compact with fewer unreacted particles and micro-cracks, whereas the microstructure of the excessive AG mix exhibited high porosity and non-uniformity of reaction products. It can be hypothesized that the increased compressive strength with an increment of the used AG dosage can be attributed to the microstructural densification or the more formation of the gel products in contrast to AG-free specimens; however, excessive AG content does not necessarily contribute to the microstructural densification and may only act as pore-filling agent. This argument is in compliance with the previous findings. For instance, Rustum and Oweed [12] carried out SEM analysis to investigate the microstructure of the AG-free and optimum AG-based specimens. They revealed a significant reduction in the porosity of the AG-based specimens compared to the former. Nonetheless, upon visual inspection of the freshly cast specimens, it was confirmed that there was remarkable bleeding in the specimens prepared with more than 1% AG compared to other counterparts (Fig. 4). This bleeding caused non-uniform or non-cohesive mixture resulting in strength retrogression at higher dosage of AG.

Fig. 6 Morphology of the raw materials and OPC-AG pastes.

3.5. SWOT analysis

The strength, weakness, opportunities, and threats (SWOT) analysis is summarized in Fig. 7.

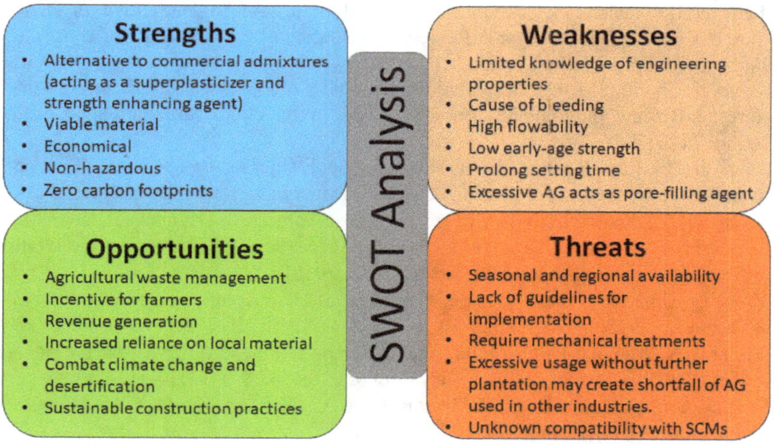

Fig. 7 SWOT analysis of the AG as an alternative admixture.

4. Summary

Following are the conclusions based on this experimental study:

1. A flow of 195, 242, and 250 mm was recorded in M2-0.5%AG, M3-1%AG, and M4-1.5%AG, respectively. These flow values were 14, 42, and 47% higher than that recorded in AG-free mix. The increased in flow values has direct relationship to the dosage of AG attributable to its superplasticizing nature or lubricating feature which provides ball bearing effect and consequently enhancing the flow mobility.
2. The compressive strength in AG-free specimens, M2-0.5%, M3-1%, and M4-1.5% was recorded to be 27.4 to 38.5 MPa, 29.8 to 40.4 MPa, 31.7 to 47.4 MPa, and 31.1 to 47.1 MPa at the age of 7 to 28 days, respectively. Numerically, up to 8.8%, 22.9%, and 22.2% increment in strength values were noticed in AG-based mixes of 0.5, 1%, and 1.5% dosage, respectively.

3. The visual inspection of freshly cast specimens revealed that the increased compressive strength up to 1% dosage of AG enabled a homogeneous mix leading to a dense mortar matrix, whilst strength retrogression above 1% dosage of AG was ascribed to excessive bleeding and the formation of a non-cohesive mixture.
4. The incorporation of AG significantly improved the workability and strength characteristics. The SEM analysis confirmed that the optimum mix is M3 prepared with admixing 1% of AG for maximizing both the early-age and later-age strength of mortar.
5. The SWOT analysis of the AG as an additive in cementitious composites indicated a high potential for its usage from techno-economic-ecological point of views. However, certain weaknesses and threats associated with the use of AG are imperative to be resolved.

References

[1] Global Market Insight, Concrete Admixture Market, Industry Analysis Report, Regional Outlook, Growth Potential, Competitive Market Share & Forecast, 2021 - 2027, Report ID: GMI1459. (2021). https://www.gminsights.com/industry-analysis/concrete-admixtures-market (accessed July 20, 2022).
[2] H. Faqe, H. Dabaghh, A. Mohammed, Natural Admixture As An Alternative for Chemical Admixture in Concrete Technology: A Review, The Journal of the University of Duhok. (2020) 301-308. https://doi.org/10.26682/csjuod.2020.23.2.24
[3] T. Ayub, N. Shafiq, S. Khan, M. Nuruddin, Durability of Concrete with Different Mineral Admixtures: A Review, International Journal of Civil, Architectural, Structural and Construction Engineering. 7 (2013).
[4] V. Mahesh, H. Chemalapati, A. Vani, S. Dandagala, Evaluation of admixture effect on strength and durability properties of concrete - A critical review, International Journal of Engineering and Technology(UAE). 7 (2018). https://doi.org/10.14419/ijet.v7i2.1.9878
[5] S.A.A. Satti, Y.H. Ahmed, Use of Gum Arabic (Acacia Seyal) As Concrete Admixture, in: 2nd Conference Proceedings Civil Engineering 2018, 2018.
[6] C. Zhao, Q. Zhao, Y. Zhang, M. Zhou, The Effect of Gum Arabic on the Dispersion of Cement Pastes, in: 2015. https://doi.org/10.1007/978-3-319-13948-7_47
[7] R. Mbugua, S. Wanjala, J. Ndambuki, Acacia Karroo as Potential Admixture for Hot African Weather, in: RILEM Bookseries, 2020. https://doi.org/10.1007/978-3-030-33342-3_19
[8] A.U. Elinwa, G. Abdulbasir, G. Abdulkadir, Gum Arabic as an admixture for cement concrete production, Constr Build Mater. 176 (2018). https://doi.org/10.1016/j.conbuildmat.2018.04.160
[9] C.M. Athman, S.O. Abuodha, T. Nyomboi, Use of gum Arabic as a superplasticizer in self-compacting concrete, International Journal of Innovative Science and Modern Engineering. 5 (2018).
[10] A.M. Mohamed, M.H. Osman, H. Smaoui, M.A. Mohd Ariffin, Durability and microstructure properties of concrete with arabic gum biopolymer admixture, Advances in Civil Engineering. 2018 (2018). https://doi.org/10.1155/2018/1962832
[11] R. Mbugua, R. Salim, J. Ndambuki, Effect of Gum Arabic Karroo as a water-reducing admixture in cement mortar, Case Studies in Construction Materials. 5 (2016). https://doi.org/10.1016/j.cscm.2016.09.002
[12] M.K. Rustum, K.M. Oweed, Investigation the effect of arabic gum on the physical and mechanical properties of ordinary cement mortar, Journal of Engineering and Sustainable Development. 24 (2020). https://doi.org/10.31272/jeasd.conf.1.46

Advanced Topics in Mechanics of Materials, Structures and Construction - AToMech1-2023 Materials Research Forum LLC
Materials Research Proceedings 31 (2023) 313-323 https://doi.org/10.21741/9781644902592-33

A numerical study on the effect of basement excavation and fibre reinforced concrete as tunnel lining material

Farzana Iqbal[1,a *] and Malavika Varma[1,b]

[1] Cochin University of Science and Technology, India

[a]farzapallath@gmail.com, [b]malavikavarmamail@gmail.com

Keywords: Tunnel, Tunnel Lining, Basement Excavation, Finite Element Modelling (FEM), Fibre Reinforced Concrete (FRC)

Abstract. Metros are one of the important transportation systems in urban areas. Due to rapid development in urban areas, there is huge commercial demand resulting in the development of underground spaces are rapid. Some of the developments may have deep excavations for basements in close proximity to existing tunnels. If the induced tunnel deformation and internal forces exceed the capacity of the tunnel structures then damages such as segmental cracking, leakage and longitudinal distortion of the tracks occur and it threatens the safety of the passengers and hence is a major concern. A parametric study on the effect of basement excavation on the underground metro in soil gives an idea about the influence of various factors like lining material and lining thickness. The crown and right spring line undergo more displacements in both the excavation and loading stage. From the study it was observed that, providing FRC as tunnel lining material has significant effect on reducing foundation settlement.

Introduction

Infrastructure is rapidly evolving, leading to the development of additional buildings in major cities and urban areas. Due to the scarcity of space in urban areas, especially metropolitan cities, the use of underground space for transit as metros or multiple levels of basements in buildings becomes a necessity. The ground movement towards the excavation will certainly occur during the construction of such subterranean metro tubes, resulting in ground surface settlement. This ground settlement is likely to have an impact on nearby structures and constitute a risk to the environment. Construction of the basement inevitably causes stress changes in the ground leading to soil movements which may cause potential damage to adjacent tunnels. Thus, it is essential for designers and engineers to assess excavation-induced movements occurring in the tunnel as well as nearby structures. Ground movements that occur during the construction of metro tunnels result in ground settlements and these settlements induce high risk to the nearby existing structures. There are also a number of numerical studies that explored the influence of tunnelling on foundations, in addition to studies on green-field settlement prediction and control owing to tunnelling.

Gang, (2008) studied the response of existing tunnel due to overlying excavation using 2D FEM. The study concluded that this deformation is mainly due to the uneven changes in ground contact pressure on tunnel linings. Also, as the tunnel embedded depth beneath the excavation increased the vertical and horizontal displacement of the tunnel decreased. Shi et al., (2013,2015) conducted a three-dimensional numerical parametric study to investigate the influence of excavation geometry, sand density, tunnel stiffness and joint stiffness on tunnel responses by overlying basement excavation. The heave and tensile strain induced in the tunnel reduced rapidly as the tunnel stiffness increased. Stiffening a tunnel can be an effective way to reduce the adverse effects on the tunnel due to basement excavation. Huang et al., (2014) investigated the effect of deep excavation above an existing tunnel by conducting a series of centrifuge model tests. When the tunnel cross-section was 1.5 times the excavation depth away from the center of the excavation, the additional moments decreased rapidly to negligible. Mahajan et al., (2016, 2019) carried out a numerical analysis to examine the effect of basement raft loading on the existing tunnel in sand using PLAXIS 3D. The results indicated that the displacement and moment at the crown, invert, springing lines and deformations are significantly affected by various stages of construction of the raft and loading. The variation in the total displacement is more when excavation is closer to tunnel location, i.e., X/D ratio is zero. This study concluded that the critical distance between the tunnel edge and the basement foundation is about 2.5 times the tunnel diameter to avoid the detrimental effects on tunnel lining.

Marara et al. (2011) studied the toughness of normal strength steel fibre reinforced concrete (NSSFRC). Concrete cylinders reinforced with three different aspect ratios of hooked-end steel fibres 60, 75, and 83 and six different percentages of steel fibres L/D (length/diameter of fibre) = 0.5, 1.0, 1.25, 1.5, 1.75, and 2.0% were tested. And a compressive strength of 83 MPa was obtained for the ratio of L/D=1 and 83% fibre volume fraction.

Although there have been a few studies on the impact of pile loading on existing tunnels, there hasn't been enough focus on understanding the interplay of basements and tunnels. Also, very few studies have been carried out regarding the use of fibre-reinforced concrete (FRP) as tunnel lining material for reducing the deformation in tunnels and foundations. The scopes of the present study are to conduct a numerical analysis on the influence of basement excavation near existing tunnels and understand the deformations of the foundation and basement. In this study, the tunnel is considered to be constructed before the basement excavation, and the excavation takes place at close proximity to the tunnel i.e., at X/D=0 (where X is the clear distance between the foundation and tunnel and D is the diameter of the tunnel). In a later stage parametric studies were conducted to study the influence of different lining materials (PCC and FRC) for different tunnel lining thicknesses on the tunnel and foundation. Foundation settlement and tunnel deformations mainly in the crown, invert, left and right spring lines are studied in detail.

Effect of basement excavation on tunnel

In this numerical study, the effect of a three-floor basement excavation and foundation loading on an existing underground metro tunnel at X/D=0 is analysed using PLAXIS 2D. A parametric study is then conducted to understand the effect of different tunnel lining thicknesses (0.25 m,0.3 m, 0.35 m, 0.4 m) on tunnel deformation mainly at the crown, invert, left and right spring line. The effect on foundation settlement with existing structures are also considered. Similarly for different tunnel lining materials Reinforced cement concrete (RCC) and fibre reinforced concrete (FRC) and their effect on tunnel deformation and foundation settlement is also studied.

In this proposed problem there exists a tunnel with a diameter of 6 m considered similar to that of the Delhi Metro Tunnel. In this study, the pressure simulated is 270 kPa, which corresponds to a 15-storey and three basement structure. The tunnel crown is located 18 m below the ground level. Soil properties are taken as that of Delhi Yamuna sand. The basement foundation of 15m width is considered to be made of reinforced cement concrete (RCC) and is designed for 15-storey loading and 3 basement floors. Retaining wall up to the height of 11 m i.e., 9 m height of three basements and 2 m thick foundation. Counterforts with high stiffness have been provided to control the deflection at the top of the retaining wall.

The properties of the soil, tunnel, basement foundation and retaining wall are provided in Tables 1 and 2 [21]. Temporary sheet pile walls and struts are provided to stabilize the soil before excavation and to limit sheet pile deflection as well as the embedment depth respectively. The total length of sheet piles is taken as 15 m. The sheet piles and retaining walls are kept at a clear distance of 0.5 m in the numerical model to avoid numerical discrepancies. Struts with normal stiffness EA =2 x 10^6 kN as elastic members with an out-of-plane spacing of 5 m. The property of temporary sheet pile walls is provided in Table 2. In PLAXIS 2D, the staged excavation and foundation loading can be numerically simulated as it occurs in the field. This study tries to understand the most critical case where the deformations are expected to be maximum. The schematic diagram of the model is presented in Fig. 1. The numerical model schematic diagram is given in Fig. 2, which is considered for all the presented in Fig. 1.

Table 1. Soil Properties [21]

Property	Soil (Yamuna sand)
Dry unit weight	15 [kN/m^3]
Young's modulus	25000 [kPa]
Poisson's ratio	0.3
Friction angle	34°
Cohesion	1 [kPa]

Numerical Modelling PLAXIS 2D – Basement Excavation and Foundation Loading
The plane strain model with 15-noded elements was used for the numerical modelling. Mohr-Coulomb model was considered for modelling the soil. Tunnel lining was modelled using plate elements as an elastic material. Model geometry was considered as 220 m x 50 m, where the sides were laterally restrained, and the bottom is fixed. Fig 2 shows the numerical model created and Fig 3 shows the finite element mesh generated. The meshing adopted is very fine and had a total of 1564 elements. The total number of nodes are 12919 and the average element size is 2.67m. The initial stresses are generated using K_o procedure (K_o=1) and the initial effective principal stress generated is -748.29 kN/m^2. In the calculation stage total of 9 phases are defined. These are defined in such

Table 2. Material properties [21]

Parameter	Tunnel Lining (D=6.0m)	Basement Foundation	Retaining wall	Sheet Pile
Material (model behaviour)	Concrete (Linear elastic)	Concrete (Linear elastic)	Concrete (Linear elastic)	Steel (Linear elastic)
Equivalent thickness (m)	0.25	2.0	1.5	0.25
Flexural rigidity (EI) kNm2/m	3.2x10^4	1.7x10^7	7 03x10^6	2.6x10^6
Normal stiffness (EA) kN/m	6.25x10^6	5x10^7	3 75x10^7	5x10^7
Poisson's ratio	0.15	0.15	0.15	0.2
Weight (kN/m/m)	6.0	48	36	19.7

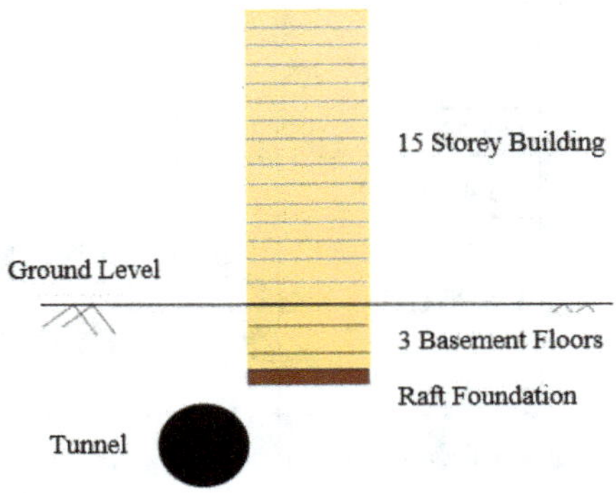

Fig 1. Schematic diagram – Storeys + Basements + Tunnel

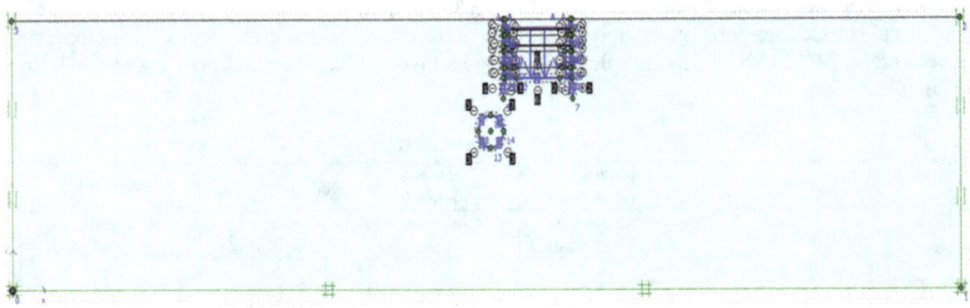

Fig 2. PLAXIS 2D Numerical model

These are defined in such a way that it represents how the construction is stimulated in the field. In the first stage, the tunnel was excavated up to a cover depth of 18 m from the ground level. In phase 2, temporary sheet pile walls were driven. The first phase of excavation and struts were provided in phase 3. Similarly, the second and the third phase of excavations are defined in phases 4 and 5. In phase 6, after removing the struts, foundation was being constructed. In phase 7, retaining wall and counterfort walls were numerically stimulated.
In phase 8 the sheet pile walls were removed and finally,
in phase 9 the loading was provided. In this first study, the analysis is carried out for three cases:
1. Effect on tunnel points after excavation
2. Effect on tunnel points after loading
3. Effect on foundation settlement

The effect on tunnel lining due to the three-floor basement excavation is represented by the deformed mesh in Fig 4 and tunnel deformation in Fig 5. Due to the basement excavation, stress relief is induced at the base of the excavation pit which results the tunnel lining to move upwards, i.e., towards the excavation. Table 4 shows the deformations at different tunnel points after excavation.

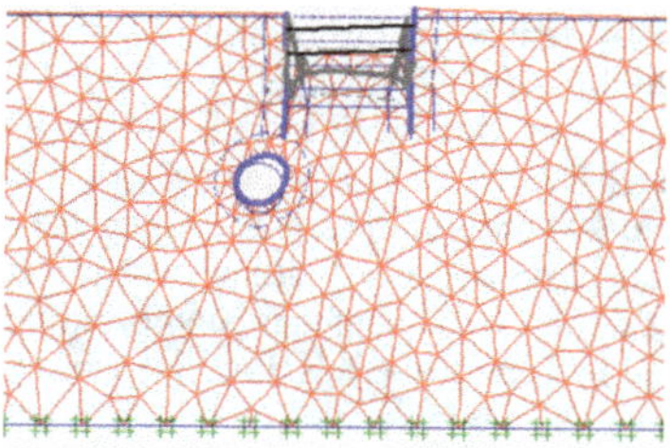

Fig 4. Deformed mesh- After excavation

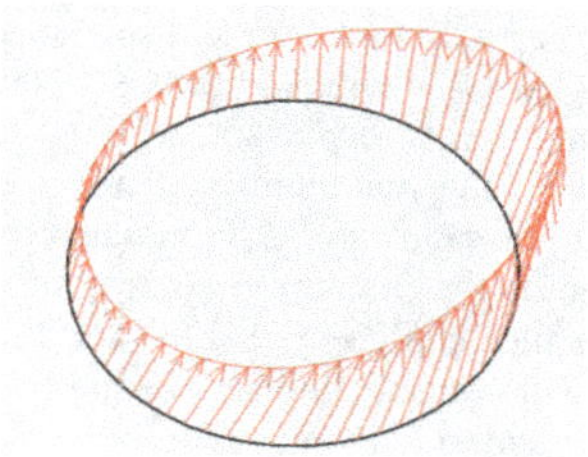

Fig 5. Tunnel lining deformation-After Excavation

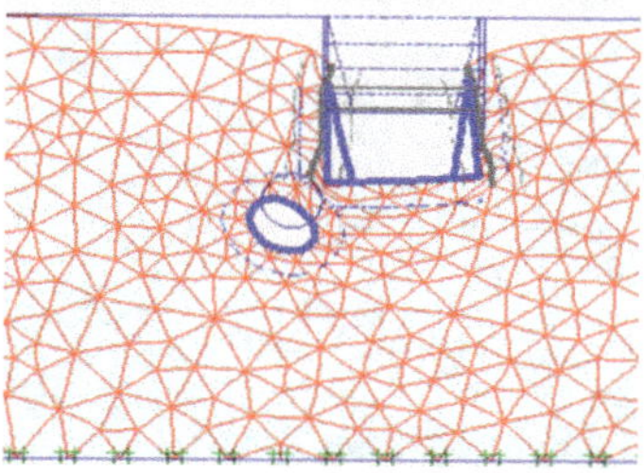

Fig 6. Deformed mesh- After loading

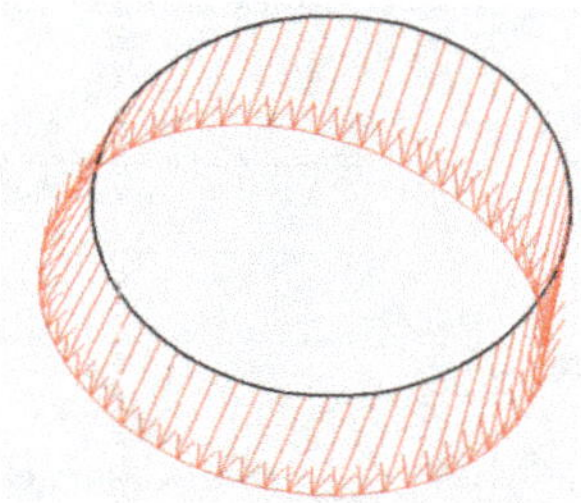

Fig 7. Tunnel lining deformation-After Loading

Effect on Tunnel Points and Foundation after Loading

The effect on tunnel lining due to the three-floor basement loading is represented by the deformed mesh in Fig 6 and tunnel deformation in Fig 7. Due to the basement foundation loading the tunnel lining moves away from the pit this is represented by displacement of tunnel lining in the opposite direction as shown in Fig 7. The total displacements at tunnel points after loading are represented in Table 4. For the given problem, the maximum displacement of the foundation after loading was found to be 158.41 mm. From Table 4, it can be analysed that the crown and right spring line undergoes more displacement in both the excavation and loading stages compared to invert and left spring line. Among them the right spring line undergoes maximum displacement because it is in close vicinity to the basement excavation compared to other tunnel points.

Table 4. Tunnel deformations – After Loading

Stage	Crown [mm]	Invert [mm]	Right spring line [mm]	Left spring line [mm]
After excavation	14.084	21.349	27.446	10.519
After loading	61.986	53.110	65.934	50.097

Materials Research Proceedings 31 (2023) 313-323 https://doi.org/10.21741/9781644902592-33

The tunnel crown is affected more by foundation loading than excavation of basement and foundation compared to other tunnel positions. After loading the total displacement at the crown increased by about 48 mm. Allowable limit of total displacement (in any direction) of underground, transition and subaqueous rapid transit system structures is specified as 15 mm by Land Development Authority [15] and 20 mm by Buildings Department [4].

The maximum displacement of the foundation after loading was found to be 158.41 mm. As per IS 1904, the maximum settlement for the RCC/Steel framed buildings over a raft foundation resting on sandy soil is 75 mm and can be increased to 100 mm for the structure like water towers and silos the maximum settlement under a raft foundation of RCC building and towers/water tanks/silos can go up to 125 mm. In some cases, maximum settlement can be allowed up to 150 mm without affecting serviceability of the building [11]. Hence, there is a need to limit the total settlement to 150 mm.

Effect of tunnel lining material and tunnel lining thickness

For this study, two types of tunnel lining materials were considered: Plain cement concrete (PCC) and Fibre reinforced concrete (FRC). The properties of PCC lining material are given in Table 5. The 28-day compressive strength of fibre reinforced concrete cylinder sample is higher than plain concrete. The properties of fibre-reinforced concrete are given in Table 5 [24].

Table 5. Fibre concrete material properties[24]

f_c [MPa]	EA [kN/m]	EI [kNm^2/m]	T[mm]	ν
83	9.2×10^6	6.2×10^4	300	0.2

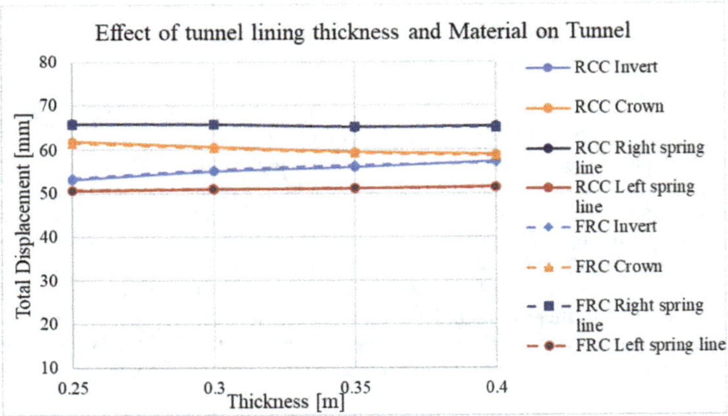

Fig 8. Effect of tunnel lining thickness for PCC and FRC on tunnel positions

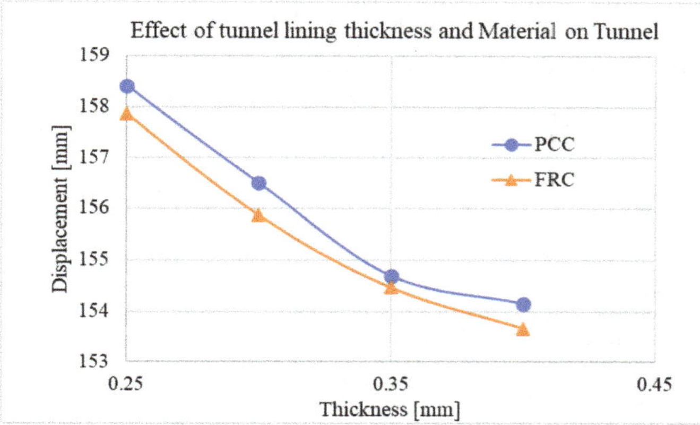

Fig 9. Effect of tunnel lining thickness for PCC and FRC on foundation settlement

The total displacements of tunnel crown, invert, left and right spring line for PCC as tunnel lining for different tunnel lining thicknesses (0.25 m, 0.3 m, 0.35 m, 0.4 m) and total displacements for FRC as lining material for different tunnel lining thicknesses are plotted in Fig 8. The foundation settlement shows a similar pattern (Fig. 9). From the results obtained it can be observed the effect of using FRC instead of PCC as a tunnel lining material in reducing tunnel deformations is very small. Both PCC and FRC show similar behaviour. However, the total displacement increases at the invert compared to the crown and either of the two spring levels. This may be due to the fact that as the tunnel lining thickness increases the total unit weight of concrete concentrating at the invert position which leads to an increase in the total displacement.

Unlike tunnels, the effect of using FRC in reducing foundation settlement is significant. By using FRC as tunnel lining material and with a tunnel lining thickness of 0.4 m the foundation settlement is reduced from 158.41 mm to 153.68 mm.

Conclusion

Construction of any structure (pile, super-structure) or any excavation activity near an existing tunnel gives rise to a complicated and sensitive condition in terms of deformations and stresses generated. Tunnelling near pile foundations results in an increase in the lateral displacement of the pile when the lateral distance between the pile and tunnel is minimum.

1. The crown and right spring line undergo more displacement in both the excavation and loading stage compared to the invert and left spring line. Among them, the right spring line undergoes maximum displacement because it is in close vicinity to the basement excavation compared to other tunnel points. The tunnel crown is affected more by building loading than the basement excavation than other tunnel positions.
2. Both PCC and FRC show similar behaviour. From the results obtained it can be observed the effect of using FRC instead of PCC as tunnel lining material in reducing tunnel deformations is very small. Unlike tunnel, the effect of using FRC in reducing foundation settlement is significant. Foundation settlement was reduced from 158.41 mm to 153.68 mm by using FRC as tunnel lining material with 0.4m thickness.

From the results obtained from this study, the total displacements at the tunnel positions were not reduced to the permissible limit as per Land Development Authority [15] and Buildings Department [4]. However, the foundation settlement was considerably reduced to within the permissible limit. Since the tunnel deformations were not reduced within the permissible limit in this study where X/D =0, more studies are required to determine the best way to reduce the tunnel deformations more effectively.

References

[1] Ayothiraman,R., & Arunkumar,S., (2011). Influence of vertical pile loading on existing tunnel lining in soft clay. In: 14th Australasian tunnelling conference Auckland: New Zealand, pp 1-13.

[2] Boldini, D., Losacco, N., Bertolin, S., & Amorosi, A. (2016). Modelling of reinforced concrete framed structures interacting with a shallow tunnel. Procedia Engineering. 158,176-181. https://doi.org/10.1016/j.proeng.2016.08.425

[3] Boushehrian, A., (2020). Machine Foundations on Fiber Concrete Tunnel in Reinforced Sand. Journal of Geotechnical Geology 16 (1), 359-363.

[4] Buildings Department. Practice notes for authorized persons APP-24 (2009) Technical notes for guidance in assessing the effects of civil engineering construction/building development on railway structures and operations. Buildings department of the government of HKSAR (BD)

[5] Chen, L.T., Poulos, H.G., & Loganathan, N., (1999). Pile responses caused by tunnelling. Journal for Geotechnical & Geoenvironmental Eng. 125 (3), pp 207-215. https://doi.org/10.1061/(ASCE)1090-0241(1999)125:3(207)

[6] Franza, A.,Marshall, A.M.,Haji, T., et al.,(2016).A simplified elastic analysis of tunnel-piled structure interaction. Tunnelling and Underground Space Technology, Volume 61, pp 104-121, ISSN 0886-7798. https://doi.org/10.1016/j.tust.2016.09.008

[7] Gang, W.S., (2008). Numerical analyses of influence of overlying pit excavation on existing tunnels. J. Cent. S. Univ. Technol. 15(2), pp 069-075. https://doi.org/10.1007/s11771-008-0438-4

[8] Haema, N., Jongpradist, P., Lueprasert, P., & Suwansawat, S., (2017). Investigation on tunnel responses due to adjacent loaded pile by 3D finite element analysis, International Journal for Geomaterials 12(31): 51-57. https://doi.org/10.21660/2017.31.6542

[9] Haibin, H.,Hao, C.H., & Jubo,Z.,(2014) .The Influence of foundation excavation on the existing metro tunnel in complicated environment. Electron. J. Geotech. Eng., vol. 19 N, pp. 3377-3385, 2014

[10] Huang, X., Huang, H., & Zhang, D., (2014). Centrifuge modelling of deep excavation over existing tunnels. Proc. Inst. Civ. Eng. - Geotech. Eng. 167(1), 3-18. https://doi.org/10.1680/geng.11.00045

[11] Indian Standard IS: 456 (2000). Plain and reinforced concrete code of practice

[12] Kitiyodom, P., Matsumoto, T., & Kawaguchi, K., (2005). A simplified analysis method for piled raft foundations subjected to ground movements induced by tunnelling. International Journal for Numerical and Analytical Methods in Geomechanics. 26(13), 2002, 1349-1369.[13] Kong, S.M., et al. (2021). Effects of Pile Installation on Existing Tunnels Using Model Test and Numerical Analysis with Medium Density Sand. Journal of Appl. Sci. 2021, 11(15), 6904. https://doi.org/10.3390/app11156904

[14] Buildings Department. Practice notes for authorized persons APP-24 (2009) Technical notes for guidance in assessing the effects of civil engineering construction/building development on railway structures and operations. Buildings department of the government of HKSAR (BD)

[15] Chen, L.T., Poulos, H.G., & Loganathan, N., (1999). Pile responses caused by tunnelling. Journal for Geotechnical & Geoenvironmental Eng. 125 (3), pp 207-215. https://doi.org/10.1061/(ASCE)1090-0241(1999)125:3(207)

[16] Franza, A.,Marshall, A.M.,Haji, T., et al.,(2016).A simplified elastic analysis of tunnel-piled structure interaction. Tunnelling and Underground Space Technology, Volume 61, pp 104-121, ISSN 0886-7798. https://doi.org/10.1016/j.tust.2016.09.008

[17] Gang, W.S., (2008). Numerical analyses of influence of overlying pit excavation on existing tunnels. J. Cent. S. Univ. Technol. 15(2), pp 069-075. https://doi.org/10.1007/s11771-008-0438-4

[18] Haema, N., Jongpradist, P., Lueprasert, P., & Suwansawat, S., (2017). Investigation on tunnel responses due to adjacent loaded pile by 3D finite element analysis, International Journal for Geomaterials 12(31): 51-57. https://doi.org/10.21660/2017.31.6542

[19] Haibin, H.,Hao, C.H., & Jubo,Z.,(2014) .The Influence of foundation excavation on the existing metro tunnel in complicated environment. Electron. J. Geotech. Eng., vol. 19 N, pp. 3377-3385, 2014

[20] Huang, X., Huang, H., & Zhang, D., (2014). Centrifuge modelling of deep excavation over existing tunnels. Proc. Inst. Civ. Eng. - Geotech. Eng. 167(1), 3-18. https://doi.org/10.1680/geng.11.00045

[21] Indian Standard IS: 456 (2000). Plain and reinforced concrete code of practice

[22] Kitiyodom, P., Matsumoto, T., & Kawaguchi, K., (2005). A simplified analysis method for piled raft foundations subjected to ground movements induced by tunnelling. International Journal for Numerical and Analytical Methods in Geomechanics. 26(13), 2002, 1349-1369. doi: 10.1002/nag.24

[23] Kong, S.M., et al. (2021). Effects of Pile Installation on Existing Tunnels Using Model Test and Numerical Analysis with Medium Density Sand. Journal of Appl. Sci. 2021, 11(15), 6904; https://doi.org/10.3390/app11156904 https://doi.org/10.3390/app11156904

[24] Kou,Y., Shukla, S., & Mohyeddin, A., (2019). Effect of Width of Geosynthetic Reinforcement within the Granular Cover on the Load Distribution over the Tunnel Lining: Proceedings of the 5th GeoChina International Conference 2018 - Civil Infrastructures Confronting Severe Weathers and Climate Changes: From Failure to Sustainability, held on July 23 to 25, 2018 in HangZhou, China. 10.1007/978-3-319-95645-9_6. https://doi.org/10.1007/978-3-319-95645-9_6

[25] Land Transport Authority (2000) Code of practice for railway protection. Singapore: Development and Building Control Department, Land Transport Authority (LTA)

[26] Liang,R., Xia, W., Huang, M., & Lin, C., (2017). Simplified analytical method for evaluating the effects of adjacent excavation on shield tunnel considering the shearing effect. Computational Geotechnology, 81:167-187 https://doi.org/10.1016/j.compgeo.2016.08.017

[27] Liang, R., Wu, W., Yu, F., Jiang, G., & Liu, J.,(2018).Simplified method for evaluating shield tunnel deformation due to adjacent excavation. Tunnelling and Underground Space Technology. 71(August 2015), 94-105 (2018). https://doi.org/10.1016/j.tust.2017.08.010

[28] Liu, Q., Guo, J., et al., (2020). Optimization Analysis of Smart Steel-Plastic Geogrid Support for Tunnel. Advances in Civil Engineering. 1-11.https://doi.org/10.1155/2020/6661807

[29] Loganathan, N., & Poulos, H.G., (1998). Analytical prediction for tunnelling-induced ground movement in clays. Journal for Geotechnical & Geoenvironmental Eng ASCE 124(9),pp 846-856. https://doi.org/10.1061/(ASCE)1090-0241(1998)124:9(846)

[30] Loganathan, N., Poulos, H.G., & Xu, K.J., (2001). Ground and Pile-Group Responses Due to Tunnelling. Soils and Foundations Vol 41, Issue 1,pp 57-67,ISSN 0038-0806. https://doi.org/10.3208/sandf.41.57

[31] Mahajan, S., Ayothiraman, R., & Sharma K.g.,(2019). A Parametric Study on Effects of Basement Excavation and Foundation Loading on Underground Metro Tunnel in Soil. Indian Geotech J (December 2019) 49(6):667-686. https://doi.org/10.1007/s40098-019-00361-x

[32] Mahajan, S., et al. (2016). Numerical analysis on effect of basement raft loading on existing urban tunnel in soil. Indian Geotechnical Conference IGC2016,15-17 December 2016, IIT Madras, Chennai, India.

[33] Malhotra, M., Sahu, V., Srivastava, A., & Misra A.M., (2019). Impact of pile foundations adjacent to tunnels in sandy strata. Indian Academy of Sciences (2019) 44:184 Indian Academy of Sciences. https://doi.org/10.1007/s12046-019-1169-y

[34] Marara, K., Eren, Ö., & Yitmena, İ., (2010). Compression Specific Toughness of Normal Strength Steel Fiber Reinforced Concrete (NSSFRC) and High Strength Steel Fiber Reinforced Concrete (HSSFRC). Materials Research. 14. 239-247.https://doi.org/10.1590/S1516-14392011005000042

[35] Marshall, A.M., 2013. Closure to tunnel-pile interaction analysis using cavity expansion methods by Alec M. Marshall. ASCE J. Geotech. Geoenviron. Eng. 139(11), 2002-2004. https://doi.org/10.1061/(ASCE)GT.1943-5606.0000975

[36] Moffat, R., Jadue, C., et al., (2017). Experimental evaluation of geosynthetics as reinforcement for shotcrete. Geotextiles and Geomembranes 45 (3), 161-168. https://doi.org/10.1016/j.geotexmem.2017.01.007

[37] Naqvi, M. W., Akhtar, M. F., et al. (2021). Effect of Superstructure on the Stability of Underground Tunnels. Transportation Infrastructure Geotechnology. 8. 10.1007/s40515-020-00119-6 https://doi.org/10.1007/s40515-020-00119-6

[38] Ng, C.W.W. & Shi, Jiangwei & Hong, Yi. (2013). Three-dimensional centrifuge modelling of basement excavation effects on an existing tunnel in dry sand. Canadian Geotechnical Journal. 50. 10.1139/cgj-2012-0423. https://doi.org/10.1139/cgj-2012-0423

[39] Ng, C.W.W., Shi, J., Mašín, D., & Sun, et al. (2015). Influence of sand density and retaining wall stiffness on three dimensional responses of tunnel to basement excavation. Canadian Geotechnical Journal. 52. 150427143611003. 10.1139/cgj-2014-0150. https://doi.org/10.1139/cgj-2014-0150

[40] Panji, M., Koohsari, H., Adampira, M., Alielahi, H., & Asgari Marnani, J., (2016). Stability analysis of shallow tunnels subjected to eccentric loads by a boundary element method. J. Rock Mech. Geotech. Eng. 8(4), 480-488. https://doi.org/10.1016/j.jrmge.2016.01.006

[41] Shi, J., Ng, C.W.W., Chen, Y., (2017). A simplified method to estimate three-dimensional tunnel responses to basement excavation. Tunnelling and Underground Space Technology, (62):53-63. https://doi.org/10.1016/j.tust.2016.11.007

[42] Sun, H., & Sun, W., (2019). Effect of Soil Reinforcement on Tunnel Deformation as a Result of Stress Relief. Journal of Appl. Sci. 2019, 9, 1420; doi:10.3390/app9071420. https://doi.org/10.3390/app9071420

[43] Sun,H.S., Wang,L.W.,et al.,(2018). Nonlinear analysis of tunnel deformation and mechanisms of stress transfer caused by basement excavation.Soil Mechanics and Foundation Engineering, vol. 55, no. 5, pp. 298-304. https://doi.org/10.1007/s11204-018-9540-2

[44] Sun,H.S., Wang,L.W.,et al.,(2021). Tunnel Behaviour Caused by Basement Excavation in Clay. Geofluids, vol. 2021, Article ID 5570846, 11 pages, 2021. https://doi.org/10.1155/2021/5570846. https://doi.org/10.1155/2021/5570846

[45] Yao, J., Taylor, R.N., & McNamara, A.M., (2006). The effects of loaded bored piles on existing tunnels. Geotechnical Aspects of Underground Construction in Soft Ground,Taylor & Francis Group, London, ISBN 978-0- 415-48475-6.

[46] Yao, J., Taylor, R.N., & McNamara, A.M., (2009). The effects of loaded bored piles on existing tunnels. In: Proceedings of the 6th International Symposium on Geotechnical Aspects of Underground Construction in Soft Ground, Balkema, Leiden, Netherlands, pp. 735-741. https://doi.org/10.1201/9780203879986.ch102

[47] Zidan, A.F., & Ramadan, O., (2015). Three-dimensional numerical analysis of the effects of tunnelling near piled structures. KSCE J. Civ. Eng. 19(4), 917-928. https://doi.org/10.1007/s12205-014-0741-6

Advanced Topics in Mechanics of Materials, Structures and Construction - AToMech1-2023 Materials Research Forum LLC
Materials Research Proceedings 31 (2023) 324-335 https://doi.org/10.21741/9781644902592-34

Estimation of failure load of adhesively bonded composite joints with embedded crack in adherends: bond EM

JAMES Polagangu[1,a*], BYJI Varughese[1,b]

[1]Advanced Composites Division, CSIR National Aerospace Laboratories, Bangalore, India

[a]james@nal.res.in*, [b]byji@nal.res.in

Keywords: Composite Structures, Failure load, Bonded Joints, Delamination, Crack, Bond Energy Method

Abstract. The authors developed Bond Energy Method (Bond EM), a novel quasi-numerical-analytical failure estimation method based on the energy conservation principle. This method previously estimated the ultimate failure load of adhesively bonded composite (ABC) repaired joints of third-party experiments without crack. In this present study, authors used the same method to estimate the ultimate failure load of third-party ABC joint test specimens with an embedded crack in the adherends. The adhesive layer and adherends are modeled using one and two-dimensional finite elements in such a way that *one plus two is three* to capture three-dimensional stresses. The crack in adherend is modeled by adding a Teflon layer of 0.02 mm between its composite stacking sequence. Static stress analysis is carried out to obtain the precise force and stress values in the adhesive layer and adherends per unit load (1000 N). These values are appropriately substituted in mathematical equations of Bond EM and estimated the ultimate failure load of L. Tong's test specimens. The difference in mathematical estimation is found in the range of (+4.29, +18.15)% for higher side estimation; (-4.80, -34.50)% for lower side estimation. The study compared the estimated failure load by Bond EM with that of other popular third-party methods, and Bond EM is found superior to all other methods considered in this study.

1. Introduction

Council of Scientific and Industrial Research-National Aerospace Laboratories (CSIR-NAL) Bangalore is a premier scientific research institute funded by the Government of India, engaged in the design, development, manufacturing, and certification of both military and civil transportation category aircraft. Advanced Composites Division of CSIR-NAL carries out research on the design and development of lightweight composite structures. The composite parts may show the deviations such as delamination, damage, resin rich, resin starved areas, and uninvited foreign debris in the composite laminate region. These deviations pose a serious technical challenge to the designer for acceptance of the part. Understanding the effect of such deviations on the overall performance and structural integrity of composite parts became the subject of interest for composite structural designers. Several analytical models are available for understanding the failure of a composite laminate of test coupons, however, those methods cannot be implemented on a global two-dimensional finite element model with induced delamination, or crack. Therefore, the authors developed an innovative and simplified finite element analysis procedure for modeling ABC joints with an embedded crack in composite laminate; and mathematically estimated the ultimate failure load of such joints using Bond EM [1]. Authors developed this method to understand analytically the effect of a manufacturing defect in bigger composite parts. Authors estimated the ultimate failure load of third-party ABC joints without crack; which was designed using various material distribution, size, and shape subjected to both uniaxial tension and compression loading. In the present study, the same method is used in its original form to estimate the ultimate failure load of ABC joints with the embedded crack in the adherends. The details of

ABC joints considered in the present study have reference to the experimental models tested by L. Tong in 1998 [2] with the embedded crack in adherends.

2. Literature

Without conducting any experiments, it is difficult to estimate the ultimate failure load of ABC joints using allowable material strength properties of the adhesive layer and adherends. Prediction of progressive failure of bonded joints through a conventional material strength approach is not possible. James Polagangu *et al* [1] made it simple to estimate the ultimate failure load of such ABC joints through a novel FE modeling approach and Bond EM; validated the method using *third-party* experimental results referenced by various researchers [3-5]. Phil Yarrington *et al.* [6] carried out the analytical study to validate L. Tong's experimental models [2] using Linear and Nonlinear HyperSizer finite element analysis tools. But, this method could predict the first failure load of L. Tong's experimental specimen of group A without a crack in adherend. He reported that these HyperSize analyses ended the solutions due to the singularity problems encountered because of the modeling of cracks in three-dimensional finite element models; therefore, not reported the estimated failure loads of other groups of specimens from B to F.

2.1. Status of failure theories

M. J. Hinton *et al.* [7] carried out a worldwide review exercise with the title *World Wide Failure Exercise* on 12 leading failure theories popularly known to predict the failure of composite laminates. The team carefully selected 14 test cases, contacted the originators of failure theories, and collected the necessary information on test data to predict the failure. The exercise avoided the information on theoretical and test data from third-party researchers. This exercise summarized the weakness and strengths of all leading failure theories by comparing the theoretically predicted failure load and test data. This worldwide exercise concluded that *a huge gap* existed between theoretical predictions and experimental data. Therefore, the mechanics of composite laminates, composite structures, composite co-cured T-joint, and ABC joints is still an open-ended research area. James Polagangu *et al.* [8] developed a simplified finite element modeling approach to understand the stress distribution around Bermuda Triangle (BT) region in the composite co-cured T-joint intersection. The load-carrying capacity of such a complex composite joint was explained through a novel failure criterion, and showed compliance with airworthiness certification requirements.

2.2. Material science and engineering at a glance

The concept of stress, strain, and elasticity explains the overall material strength, behavior, and structural mechanics of small prismatic member to large structures. The name of Scientists and year of discovery [9,10, 11] of various basic material engineering parameters is briefly summarised for ready reference. Understanding the subject of material science and engineering, mechanics of structures, and failure mechanics took centuries of the period to add a new engineering parameters like stress, strength, stiffness, and mechanical energy terms.

Sir Robert Hooke proposed the law of linear elasticity of material within the proportional limit, named Hooke's law in 1660, and published in 1678 (*17th Century*). Gottfried Wilhelm Leibniz in 1684, and Jakob Bernoulli in 1691 explained the idea of internal tension acting across the surface in a deformable solid. In 1705 (*18th Century*), Jakob Bernoulli observed and described the deformation through force per unit area or stress. In 1727, Leonhard Euler proposed a linear relationship between stress σ and strain ε. In 1752, Euler introduced the idea of compressive normal stress as the pressure in a fluid. For the first time, Charles-Augustin Coulomb introduced the theory of a beam as a bent elastic line; and related the bending to stress and strain in an actual beam, he developed the famous expression $\sigma = (M/I)y$ for the stress due to the pure bending of a homogenous linear elastic beam. In 1782, Giordano Riccati performed the first experiments similar to the concept of Young's modulus. Charles Augustin Coulomb correctly formulated the entire problem of cantilever beam bending in a paper published in 1773. But, In 1807 (*19th Century*), Thomas Young related the stress and strain in the form σ=Eε. There appears to be a timeline disparity between the above two time periods, however, it is considered to be insignificant from the present contest. Euler–Bernoulli published the beam theory for the first time in 1750 but it became popular only in the 19th century during erection of the Eiffel tower in 1887 due to its validation. Therefore, it is understood that it took almost two centuries from the time of Hooke's law to erection of the Eiffel Tower, 1660 to 1887, to understand, formulate and validate the simple theory of bending equation M/I=f/y=E/R. In 1713, Antoine Parent introduced the concept of shear stress. In 1773, Coulomb developed this concept further, explained the failure of solids in connection with the stressing beams; and studied frictional slips in 1779. This literature review convinces us that any mathematical, analytical /numerical method developed by one scientist should be validated by the work of another scientist/ experimentalist /designer. Therefore that particular method can be used by other designers for designing large structures. The present paper also aimed at validating the Bond EM through third-party experimental results and proving its adaptivity for any type of ABC joints.

2.3. Bond energy method

James Polagangu *et al.* mathematically formulated Bond EM [1], a novel failure estimation formulation developed based on mass and energy conservation laws.

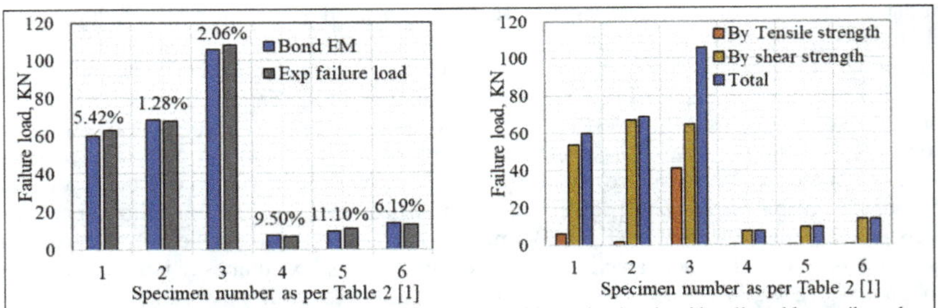

(a) Percentage difference between the ultimate failure load estimated by Bond EM and the experimental failure load of third-party test specimens

(b) Ultimate load resisted by allowable tensile and shear strength properties of the adhesive layer estimated by Bond EM

Figure 1. Failure load estimated by Bond EM and third-party test results. Refer Table 2[1]

The mathematical form of this method consists of widely known basic mechanical strain energy terms such as $\sigma^2/2E$ and $\tau^2/2G$ with an additional new term named Bond energy $\tau^2/2E$. The original work of this method explained the structural behavior of ABC joints of various popular configurations through a simplified one and two-dimensional finite element modeling approach. For the first time, this method estimated the failure load of ABC repaired joints [3-5] of different sizes, shapes, and materials subjected to both tensile and compressive loads. The lowest difference in mathematical estimation was 1.28%, as shown in Fig.1(a). Figure. 1(b) shows the uniqueness of Bond EM. The method quantifies the load carried by the allowable tensile and shear strength properties of the adhesive layer separately. Most of the specimens predominantly failed in shear except the 3rd specimen, which resisted the applied load considerably by tensile strength property of the adhesive layer. The vertical face of the stepped bonded joint configuration consumed the internally stored tensile strain energy from the adhesive layer, which satisfies the general understanding of the structural behavior of ABC joints of that configuration.

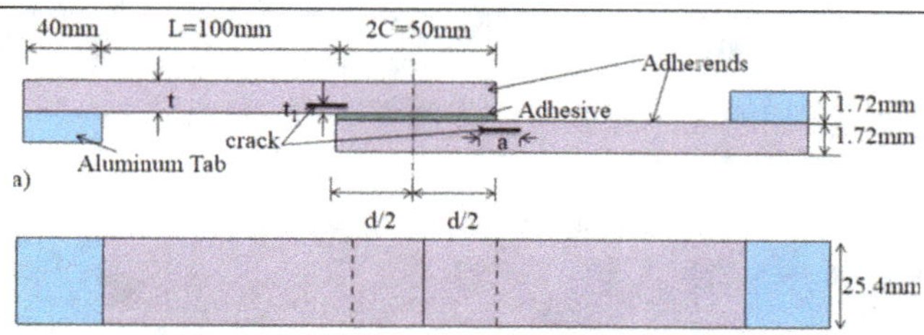

(a) Geometrical configuration of L. Tong's specimens

Summary of geometrical dimensions of six specimens

Experimental group	A	B	C	D	E	F
a, crack width in mm	0	4	6	8	6	6
d, offset distance in mm	50	50	50	50	60	40

(b)Left hand side adherend and tab

(c) Right hand side adherend and tab

(d) Adhesive layer and crack in upper adherend

(e) Adhesive layer and crack in lower adherend

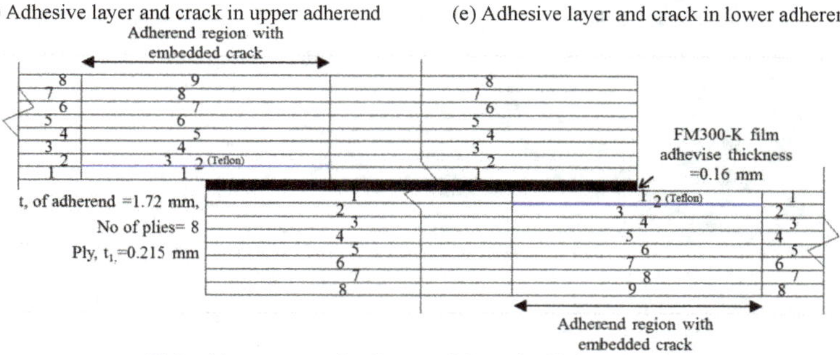

(f) Stacking sequence details around the embedded crack region

Figure 2. Geometrical details of L. Tong's experimental models [2]

3. ABC Joints with the embedded crack in Adherends

The present study aims at estimation of the failure load of ABC joints with the embedded crack in the adherends, however, determination of deflection, rotation, and conducting nonlinear progressive failure analysis of such joints is beyond the scope. The experimental studies carried out by L. Tong in 1998 [2] with the embedded crack in both adherends are the reference models considered in the study. Figure 2 shows the geometrical dimensions of L. Tong's specimens, with the crack embedded at the second layer of each adherends with different sizes and positions.

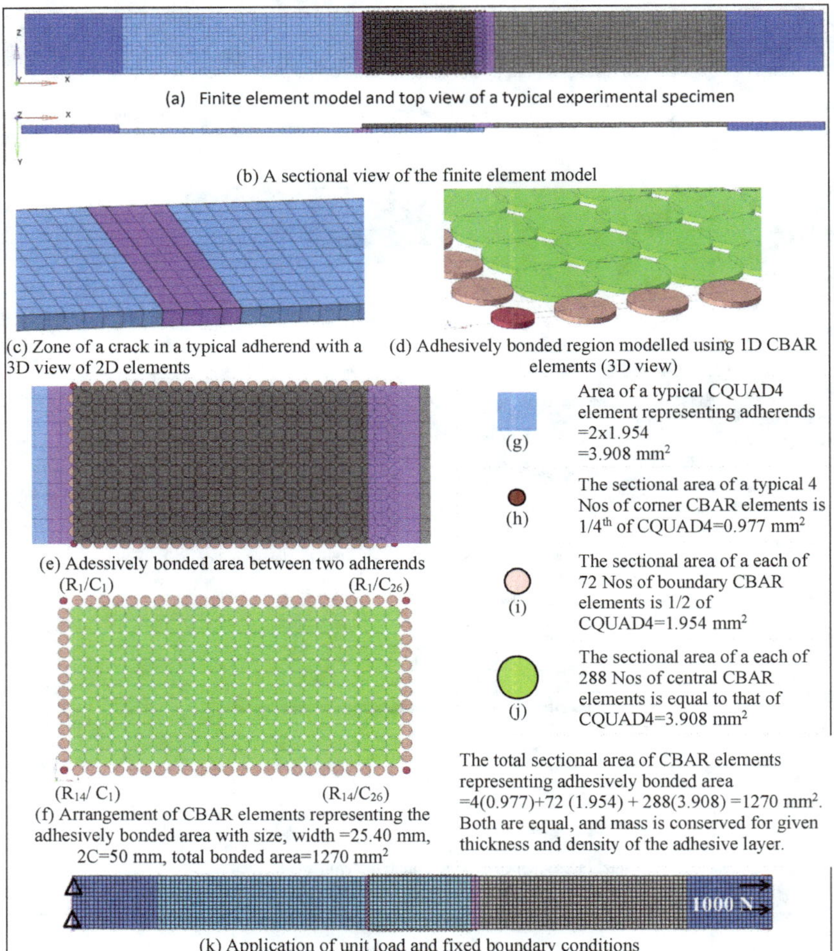

Figure 3. Details of the FE modeling approach of ABC joints

3.1. Finite element model

The authors explained the procedure for creating one and two-dimensional finite element models of ABC joint [1]. The present study adopted the same approach for creating such models with the embedded crack using Hypermesh® Pre- and Post Processor. The top view of a typical specimen model is shown in Fig.3(a). The adherend region with and without crack is shown in Fig.3(b)&(c). The adherends are modeled using 2D-CQUAD4 elements with PCOMP property, MAT8 material cards; and the adhesive layer is modeled using ID-CBAR beam elements of NASTRAN® element library defined by PBARL property card with the circular sectional area as shown in Fig.3(d). The arrangement and sectional area of elements are shown in Fig.3(f) to (j) respectively. The adhesive layer surface area is 1270 mm^2, and that of CBAR elements is 1270 mm^2, which confirms the conservation of mass from a single surface to multiple CBAR sectional areas. These elements have all six degrees of freedom to capture the force components along three axes, and bending moment about three axes that arise due to the eccentricity of a single lap joint. The composite stacking sequence of both adherends consists of eight plies of $(0°)_8$ orientation as per the layup details marked in Fig. 2(f). The composite stacking sequence of adherend includes a Teflon film layer of 0.02 mm (t_1) embedded by adding a layer at the position shown in Fig. 2(a),(d) to (f). The Teflon layer defined by the fictitious isotropic material properties (E=0.55 GPa, and υ=0.3) virtually represents and simulate the condition of the crack in adherend. The finite element model for six specimens is created with appropriate crack dimension and position as per the summary given in Fig.2. Figure 3(k) shows loading conditions applied to the finite element models. The top adherend is constrained against all rotation and translations; a unit load of 1000 N is applied at another end of the bottom adherend that pulls the joint away from the bonded region.

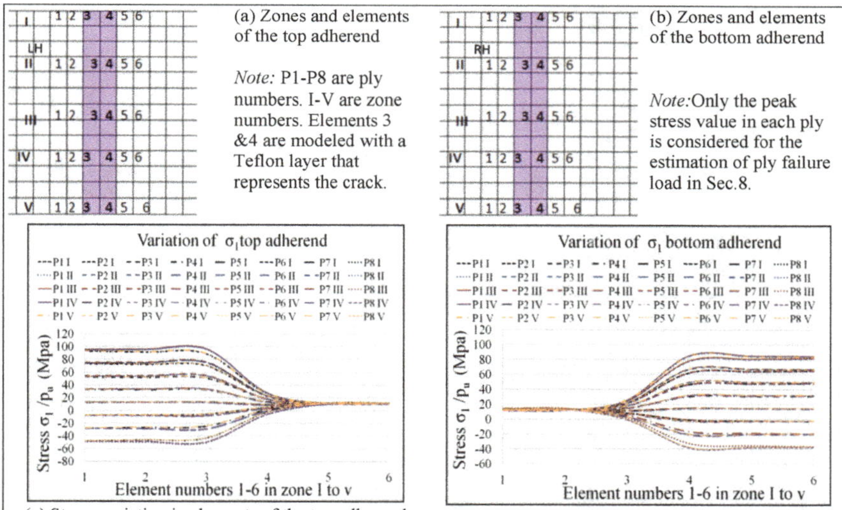

(c) Stress variation in elements of the top adherend (d) Stress variation in elements of the bottom adherend

Figure 4. Layer-wise stress in the adherends per 1000 N for group A specimen

4. Analysis and Results

4.1. Stress in the composite plies of adherends

The static stress analysis is carried out using MD Nastran® solver for all six models. The analysis called for output data of composite ply-wise stresses and force components for all CBAR elements. The output data is post-processed outside the analysis domain using a Microsoft® Excel spreadsheet inline with the CBAR element position marked in rows R_i (*i=1 to 14*) and columns C_j (j=1 to 26) as shown in Fig.3(f); zones I to V and element numbers 1 to 6 marked for both adherends as shown in Fig.4 (a)&(b). The ply-wise stress distribution in those elements located near to the crack region is shown in Fig. 4(c)&(d) for both top and bottom adherends respectively w.r.t the zones and number of elements marked in the same figure. The ply-wise stress distribution shows that the 1st layer of 3rd element in zone IV attains maximum tensile stress of 95.51 MPa and the 8th layer of 3rd element in zone IV linearly attains the maximum compressive stress of -49.59 MPa due to known eccentricity in single lap joint configuration. Similar observations are made for another group of specimens with a significant changes in the stress values.

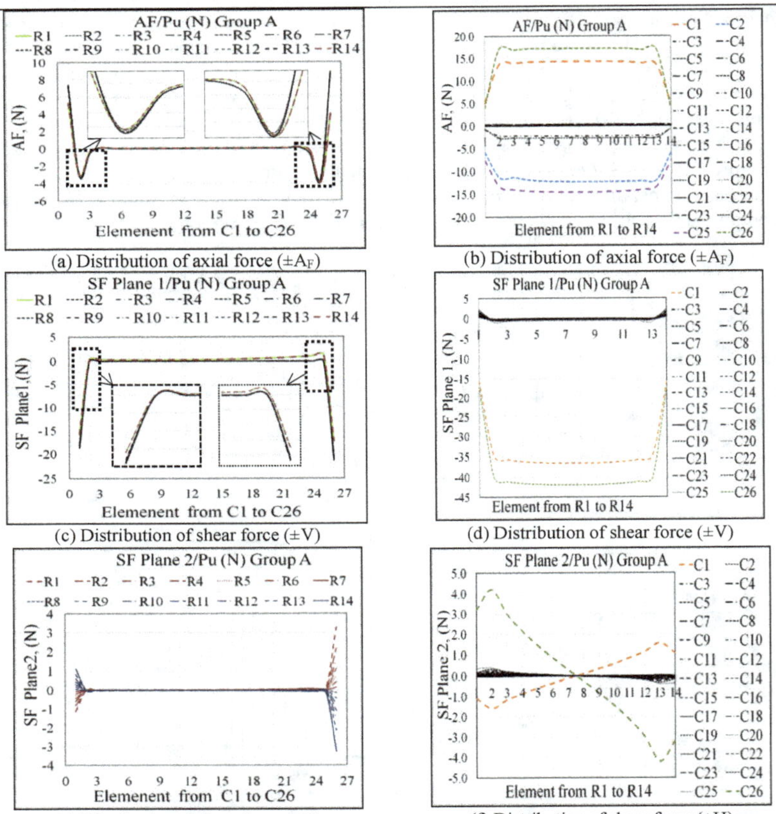

(a) Distribution of axial force (±A_F) (b) Distribution of axial force (±A_F)

(c) Distribution of shear force (±V) (d) Distribution of shear force (±V)

(e) Distribution of shear force (±H) (f) Distribution of shear force (±H)

Figure 5. Force distribution in CBAR elements per 1000 N for group A FE specimen

4.2. Force in the adhesive layer

Figure 5(a) to (f) shows the force distribution in CBAR elements per 1000 N for a typical group A experimental specimen. The force component divided by the respective sectional area of that element gives the stress value in the respective direction. The sense of axial force (AF) is denoted by ±AF, where +AF is axial tension and –AF is axial compression force in the adhesive layer which is acting normal to the adherend. Force components acting in the direction of applied load or material orientation of the adherends is denoted by ±V w.r.t the direction of applied force or vector *v1* of CBAR element. Force components acting in the plane of the adhesive layer and transverse direction to the applied load is denoted by ±H or w.r.t the vector *v2* of CBAR element.

Force distribution shows that those elements located around the boundary of the adhesive layer experience high out-of-plane force or stress for all models.

5. Ultimate failure load estimation by Bond EM

5.1. Ultimate failure load of adhesive layer

Force components are converted into equivalent stress values in respective CBAR elements. The same stress values are appropriately substituted in Eq.1 to Eq.3. The allowable tensile strength property of the adhesive layer stores the tensile strain energy F_T in the form of $(\sigma^2/E)_{ij}$, in all elements that are exclusively subjected to tensile stress as given in Eq.2. The allowable shear strength property of the adhesive layer stores the shear strain energy F_S in the form of $(\tau^2/G)_{ij}$, in all elements that are exclusively subjected to shear stress in both longitudinal and transverse directions as given in Eq.3. While those CBAR elements that are exclusively subjected to compressive load stores the Bond energy in the form of $(\tau^2/E)_{ij}$ as given in Eq.3.

The ultimate failure load of ABC joints is defined by $F_{ufl}^{Ad} = \frac{P_U^2}{n_b}\left[F_{ufl}^{T} + F_{ufl}^{Sh}\right]$ (1)

Where,

$$F_T = F_{ufl}^{T} = \left(\frac{\int_{i=1}^{n}\int_{j=1}^{m}\left(\sigma_{z,al.}^2 a_a.t_a / E_a\right)_{ij}^{+A_F}}{\int_{i=1}^{n}\int_{j=1}^{m}\left(\sigma_{y,ap}^2 a_a.t_a / E_a\right)_{ij}^{+V} + \int_{i=1}^{n}\int_{j=1}^{m}\left(\sigma_{y,ap}^2 a_a.t_a / E_a\right)_{ij}^{-V} + \int_{i=1}^{n}\int_{j=1}^{m}\left(\sigma_{z,ap}^2 a_a.t_a / E_a\right)_{ij}^{+A_F} + \int_{i=1}^{n}\int_{j=1}^{m}\left(\sigma_{z,al}^2 a_a.t_a / E_a\right)_{ij}^{-A_F}} \right) \quad (2)$$

$$F_S = F_{ufl}^{Sh} = \left(\frac{\int_{i=1}^{n}\int_{j=1}^{m}\left(\tau_{xy,al}^2 a_a.t_a / E_a\right)_{ij}^{-A_F} + \int_{i=1}^{n}\int_{j=1}^{m}\left(\tau_{yz,al}^2 a_a.t_a / G_a\right)_{ij}^{+H} + \int_{i=1}^{n}\int_{j=1}^{m}\left(\tau_{yz,al}^2 a_a.t_a / G_a\right)_{ij}^{-H}}{\int_{i=1}^{n}\int_{j=1}^{m}\left(\tau_{yz,ap}^2 a_a.t_a / G_a\right)_{ij}^{+H} + \int_{i=1}^{n}\int_{j=1}^{m}\left(\tau_{yz,ap}^2 a_a.t_a / G_c\right)_{ij}^{-H}} \right) \quad (3)$$

NOTE: Wherein P_u is the number of unit loads applied. If the applied load is 1C00 N, then P_u=1. If the applied load is 10,000 N, then P_u =10. The numerical value of n_b is equal to the number of bonded faces, n_b =1 for one side bonded joint, n_b =2 for two sides bonded joint. *Nomenclature*: σ tensile stress, τ shear stress, a_a cross sectional area, t_a thickness, E_a Young's modulus, and G_a shear modulus of respective CBAR element.

Ultimate failure load is the load required to fail the adhesive layer or composite plies completely so that it separates both the adherends. This condition occurs when two criteria are satisfied. The first criterion is satisfied when externally applied load or tensile stress consumes the internally stored tensile strain energy F_T of the adhesive layer. The second criterion is satisfied when externally applied load in the form of shear stress consumes the internally stored shear strain energy F_S. The ABC joint sees both tensile and shear stress regions due to complex structural details and loading conditions. Therefore, the total ultimate failure load is the sum of two strain energy terms F_T+ F_S respectively. Figure 6 shows the comparison of the ultimate failure load of Tong's experimental specimens with that of Bond EM. L. Tong had not reported the ultimate failure load of all five specimens of groups A and F due to large variations in failure load [2].

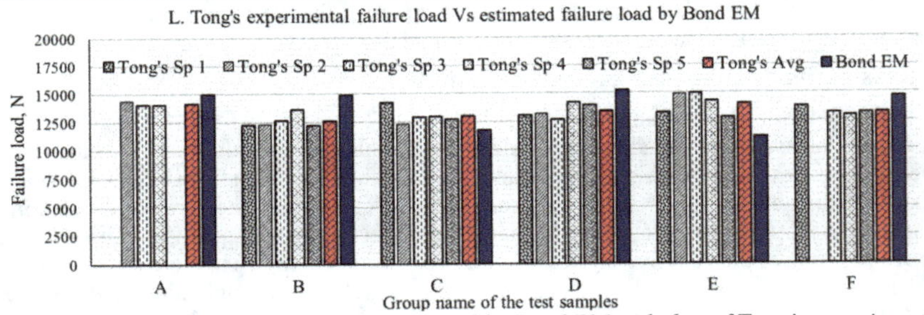

Figure 6. Comparison of failure load estimated by Bond EM with that of Tong's experimental models [2]

5.2. Comparison of failure load

Table 1 gives the summary of the ultimate load estimated by Bond EM. The percentage difference is estimated w.r.t maximum, minimum, and average ultimate failure load value of experiment models. It shows that the load estimated by Bond EM differs by +4.29%, 18.15% for group A and B specimens; -4.80%, and -34.50% for group C and E specimens respectively. The positive difference indicates the estimated failure load is higher than experimental values and Vice-Versa. The higher difference between the minimum and maximum values indicates the larger variation in the experimental failure load of specimens of that group.

Table 1. Summary of ultimate failure load estimated by Bond EM

S. No [1]	Name of specimen group [2]	A [3]	B [4]	C [5]	D [6]	E [7]	F [8]
1	Failure load by Bond EM, N	15018	14917	11718	15242	11114	14674
2	The difference with minimum load (%)	*4.29*	8.64	*-4.80*	7.01	-15.28	6.14
3	The difference with maximum load (%)	6.35	*18.15*	-21.32	16.89	*-34.50*	11.93
4	The difference with the average load (%)	5.64	15.55	-10.97	12.11	-26.15	9.62

5.3. Reason for higher difference

The higher difference between the experimental and estimated failure loads attribute to various parameters that influence the manufacturing and testing procedure. The temperature, pressure, and duration of curing affect the quality of the specimen. The variation in thickness of the adhesive layer and the inclusion of foreign debris sometimes may reduce the failure load. The present study considered the uniform thickness of the adhesive layer throughout the bonded region as defined[2]. The study assumed perfect compactness of the joint with no manufacturing defects induced in both the adhesive layer and adherends. These assumptions have resulted in the higher prediction of failure load as seen in groups A, B, D, and F; for the other two groups C and E, the prediction is on the lower side. The lower prediction of failure load is attributed to the possible deviation of experimental specimens[2] from the standard dimensions. Authors report based on experience that the specimens with deviation were found to take a higher load than the pristine specimens. A good practice is to report any deviation/defect in the experimental specimens, as the Bond EM understands the structural behavior of ABC joints with such deviations.

7. Bond EM versus other methods

L. Tong [2] predicted the failure load of test specimens using maximum stress criterion FEA-MSC, energy release criterion ERC, and other theoretical methods. Phil Yarrington [6] also predicted the failure load of L. Tong's test specimens using initial failure load, HyperSizer Linear, and Nonlinear methods. The ultimate failure load predicted by various analytical models, including Bond EM, with that of the average failure load of respective specimen groups is shown in Fig.7.

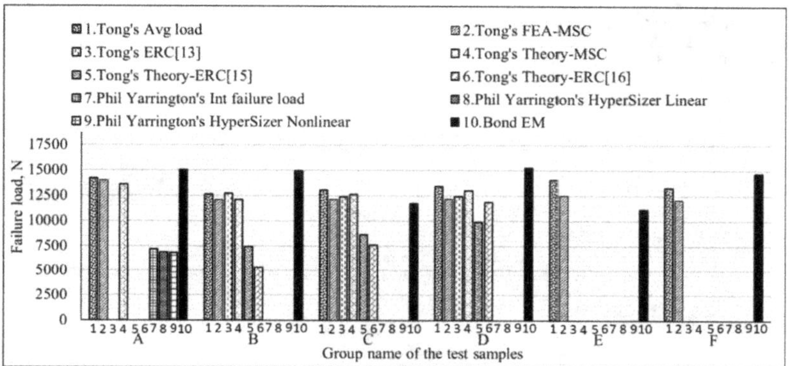

Note: References [13,15,16] from [2,6]

Figure 7. Comparison of Tong's failure load estimated by various analytical models and Bond EM

The failure load predicted by L. Tong using FEA-MSC closely correlated to the own experimental specimens of all groups, but the failure load predicted by other third-party methods is comparable with a few groups of specimens with a large gap. Phil Yarrington [6] estimated the first ply failure load of group A-specimens with a considerable variation but not reported for other groups of specimens as the singularity error encountered while solving ABC joint with crack. The present study also predicted the ultimate failure load of all groups of specimens with the variations discussed in *Sec 5.3.* The one and two-dimensional finite element modeling approach adopted in this study never encounter reported *singularity* error. This comparison reveals that Bond EM is *superior* to all analytical and theoretical methods discussed in this paper as it also predicted the failure load of all ABC joints with cracks.

8. Linear progressive failure analysis

Section 5.1 discussed the method of estimating the ultimate failure load of an adhesive layer of all groups of test specimens. In this section, a simple mathematical linear progressive analysis is carried out to understand the load at which a particular composite ply of adherend failed. The in-plane tensile stress value per unit load along with the allowable tensile strength property is sufficient for estimating the ultimate load at which the composite plies failed in sequence / progressively. For the demonstration purpose of this approach, the study considered the composite ply in-plane stress distribution in the top adherend of group A specimen as shown in Fig.4(c). Table 2 gives the sequence of simple mathematical calculations performed to estimate the ultimate failure load of each composite ply. L. Tong's [2] experimental models showed that composite plies simultaneously failed along with the adhesive layer. Phil Yarrington *et al* [6] determined the margin of safety values available in the composite plies of L. Tong's experimental models using HyperSizer 3D finite element analysis. The present study estimated and explained the sequence of failure of composite plies and the adhesive layer, and a numerical value of the failure load of each ply is reported in Table 2. Similarly, the peak in-plane tensile stress values in the composite plies of other specimen groups govern the failure. The first failure occurred in the composite ply adhered to the adhesive layer, and the failure progressed towards the free surface until the adhesive layer ultimately failed. Both top and bottom adherends exhibited this phenomenon simultaneously. Figure 8 shows the path of failure across the thickness of adherends and the adhesive layer, which is in agreement with the reported failure mode[2].

Table 2. Estimation of failure load and mode of failure (for group A specimen)

Ply No	Ply stress per 1000 N, MPa *Refer Fig. 4(c)*		Ply failure load, N		Description of failure
	Top Adh	Bot. Adh	Top Adh	Bot. Adh	
1	*2*	*3*	*4*	*5*	*6*
Ply 1	95.51	84.15	5423	6156	Tensile failure*
Ply 2	75.35	66.66	6875	7771	Tensile failure*
Ply 3	54.41	49.18	9520	10533	Tensile failure*
Ply 4	33.46	31.69	**15481**	16346	Adhesive failure**
Ply 5	12.73	14.38	-	-	No failure++
Ply 6	-8.13	-3.28	-	-	No failure ++
Ply 7	-28.86	-20.76	-	-	No failure ++
Ply 8	-49.59	-38.24	-	-	No failure++

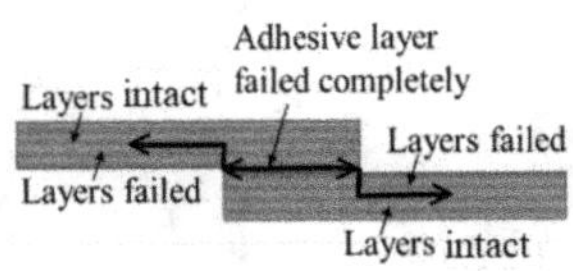

Figure 8. Schematic representation of failure mode in both adhesive layer and adherends

NOTE: (1) In-plane tensile strength of composite ply is 518 MPa [2,3]; (2) Value in column 4 =1000x 518 / Value in column 2; (3) Value in column 5 =1000x 518 / Value in column 3; (*) Ply failure initiated only at the highly stressed region, but plies continued to taking load along with other plies and adhesively layer; (**) The joint failed due to the separation of the adhesive layer; (++) the failure was not initiated in these plies but continued taking load along with other plies and the adhesive layer. Further investigation is needed to work out the load at which the individual ply fully failed across the width. Ply 4 in the top adherend and adhesive layer failed simultaneously. Difference in estimation of the ultimate failure load by this mehod is only 3.08% (=15481/15018-1)%.

9. Conclusions

This paper discussed the Bond Energy Method, finite element modeling approach, and method of estimating the ultimate failure load of adhesively bonded composite joints with embedded cracks. The method estimated and validated the ultimate failure load of L. Tong's experimental specimens of various groups with a crack embedded in both the adherends with a close correlation within the range of (+4.29, +18.15)% for higher side estimation; (-4.80, -34.50)% for lower side estimation, which is the closest estimation of all other methods. The comparison showed that Bond EM is superior to all other methods considered in this study because it predicted the ultimate failure load of all specimen groups with cracks embedded in the adherends without encountering singularity error. The mathematical linear progressive failure analysis is also carried out for understanding the failure mode in both adherends and adhesive layers, which is in agreement with the reported experimental findings. Therefore, Bond EM is suitable for estimating the ultimate failure load of ABC joints in a global or large size one and two-dimensional finite element model with and without composite ply delamination/crack induced due to various reasons. This method is practically adaptable by the composite structural designer for understanding the structural behavior of large composite aircraft parts with manufacturing defects and damages. So far, Bond EM predicted the failure load of *12* experiments of third-party specimens and compared it with *eight* other theoretical and numerical failure load estimation methods. The authors have formulated mathematical equations and carried out analytical studies. All test results are considered from other references.

Acknowledgments

The authors have carried out this analytical study using the available infrastructures of CSIR-NAL Bengaluru, an aerospace research and design organization funded by the Government of India.

Therefore, no claims and conflicts from others involved in this study. The contribution of Ms. Dixi Patel is acknowledged.

References

[1] James Polagangu, K. Mahesh, Varghese Byji, "Estimation of failure load of composite bonded joints using 1D and 2D FE analysis and the mathematical equation of strain and bond energy in the adhesive layer", Mechanics of Advanced Materials and Structures, 2019, Vol. 26, Issue 1, pp: 20-28. https://doi.org/10.1080/15376494.2018.1516254

[2] L. Tong. Failure of Adhesive-Bonded Composite Single Lap Joints with Embedded Cracks. J. AIAA, Vol. 36, No 3, March 1998. https://doi.org/10.2514/2.385

[3] Mohammad Kashfuddoja., M, Ramji. Whole-field strain analysis and damage assessment of adhesively bonded patch repaired CFRP laminates using 3D-DIC and FEA. J. Composite. Part B, 53 (2013), 46-61. https://doi.org/10.1016/j.compositesb.2013.04.030 https://doi.org/10.1016/j.compositesb.2013.04.030

[4] Kumari Asha, Byji Varughese, D. Saji. Stress analysis of stepped lap bonded repair joint under compressive loading. Proceedings of ISAMPE National Conference on Composites INCCOM-12, CP160 (2013). Bangalore, India.

[5] Zong-Hong Xie., Xiang Li., Sui-An Wang. Parametrical study on stepped-lap repair of composite laminates. The 2016 Structures Congress. Structures16. Jeju Island, Korea.

[6] Phil Yarrington, James Zhang and Craig Callier. Failure Analysis of Adhesively Bonded Composite Joints. 46th AIAA/ASME/ASCE/AHS/ASC/ Structures, Structural Dynamics & Material Conference, 18-21 April 2005, Austin, Texas. https://doi.org/10.2514/6.2005-2376

[7] M. J. Hinton, A. S. Kaddour, P. D. Soden. A Comparison of the Predictive capabilities of current Failure Theories for Composite Laminates, Judged against Experimental Evidence. Composites Science and Technology 62 (2002) pp: 1725-1797. https://doi.org/10.1016/S0266-3538(02)00125-2

[8] Polagangu James, Byji Varughese, Ramesh Sundaram. Analytical qualification of composite co-cured T-Joint for the internal fuel pressure requirement using a novel FE analytical approach: Bubbles in Bermuda Triangle. CPID-8, SL. No 19, INCCOM-16: ISAMPE-National Conference on Composites.September 20-21, 2019, Mascot Hotel, Thiruvananthapuram, Kerala, India.

[9] https://www.britannica.com/science/mechanics-of-solids#ref611523.

[10] https://www.britannica.com/science/mechanics-of-solids/History.

[11] S P Timoshenko. History of strength of materials. McGraw-Hill New York,1953.

Finite electroelastic deformation of dielectric semilinear hyperelastic tubes

Odunayo O. Fadodun*, Bolanle A. Olokuntoye, Patrick O. Layeni and Adegbola P. Akinola

Department of Mathematics, Obafemi Awolowo University, Ile-Ife 220005, Nigeria

ofadodun@oauife.edu.ng, toyeolokun@oauife.edu.ng, olawanle.layen@gmail.com, aakinola@oauife.edu.ng

Keywords: Dielectric, Finite Deformation, Electroelasticity, Cylindrical Tube, Hyperelasticity

Abstract. This study examines the finite electroelastic deformation problem of extension and inflation coupling of dielectric semilinear hyperelastic tubes with closed ends under the influence of internal pressure, axial loads and radial electric field. The laws of thermodynamics and Coleman-Noll procedure are used to derive the electroelastic constitutive model of the tube. The solution of the consequent electromechanical field equations shows that the applied radial electric field associated with the equal and opposite charges on the electrode coated surfaces contributes to both internal pressure and axial loads of the closed tube. Furthermore, it is obtained that the stress propagation in dielectric semilinear hyperelastic solids is sensitive to the electric displacement field generated within the solids while the accompanying electric field interacts with the deformation of the solids. Finally, and among other things, the graphical illustration shows that the radial electric field generated within the tube increases with the increasing azimuthal stretch.

Introduction

Dielectric elastomers (DEs) are smart materials that change their mechanical behaviour in response to the application of electric field. These materials belong to a class of electroactive polymers, which generate finite deformation under the action of external electric field [1-9]. The high elastic strain energy density, electromechanical coupling and fast actuation speeds make DEs extremely attractive for widespread applications. For instance, the out-of-plane deformation response of an edge-clamped DE membrane has been investigated as a potentials replacement for the passive diaphragm in a left ventricular assist device; and using a DE membrane eliminates the need for a separate actuation source in the prosthetic pump, this enables a simpler, lighter, and more compact device that mimics the behaviour of the natural heart [2]. Other applications of DEs include but not limited to soft robotics, actuators, sensors, pvalves. energy harvesting, adaptive optics and haptic feedback.

In respect of the widespread applications of dielectric elastomers, and since the pioneering work of Toupin [10], the studies of electroelastic deformation of these materials have continued to attract considerable interest of reserachs in the fields of applied mathematics, physics, chemistry, and engineering. Melnikov and Ogden [11] used the reduced energy procedure and solved the problem of extension and inflation of a circular cylindrical incompresible tube made of dielectric elastomer based on neo-Hookean, Ogden and Gent material models; Saxena et al. [12] obtained the solution of combined extension, torsion and inflation in compressible electroelastomeric thin tubes; Dorfmann and Ogden [13] analysed the influence of deformation dependent permitivity on the elastic response of a finitely deformed dielectric incompressible thick-walled tube made of electro-sensitive neo-Hookean and Gent materials; Fu et al. [14] derived a reduced electroelastic plate model which describes the incremental formulation of an electrodes-coated dielectric plate that takes the leading-order thickness effect into account; and Abd-alla et al. [15] formulated and solved a mathematical model for longitudinal wave propagation in a magnetoelastic hollow circular cylinder of anisotropic material under the influence of initial hydrostatic stress to mention a few. The present study incorporates material incompressibility constraint into the rencent work of Fadodun et al. [1] and solve the problem of extension-inflation coupling of dielectric semilinear hyperelastic tube subject to internal pressure, axial load, and a radial electric field. Electroelasticty field equations, together with the consequent constitutive model and accompanying boundary conditions are used to obtain the solution of the problem. The rest of the paper is as follow: scetion two highlights the notaion, section three gives the basic equations of electroelasticity, section four details the application of electroelasticity theory presented in the previous sections to the problem of extension and inflation coupling of a thick circular cylindrical tube, section five gives the solution of the resulting electroelastic field equations, while section six concludes the study.

Notation

In this study, we employ upper-case boldface such as **A**, **B** for second-rank tensors. The tensors $\boldsymbol{A}^T$ and A^{-1} denote the transpose and inverse of tensor $\boldsymbol{A}$ respectively. Both upper- and lower-cases with overhead arrow such as $\vec{A}$ and $\vec{n}$ denote vectors, and parameters such as $p, \eta, F, F_{\text{red}}$ denote scalars.

Kinematics

Let a stress-free dielectric semilinear hyperelastic solid occupying the reference configuration $\Omega_0 \subset R^3$,with smooth boundary $\partial\Omega_0$ deform onto current/deformed configuration Ω, with smooth boundary $\partial\Omega$, when subject to electric field and/ or mechanical loads. Let $\vec{N}$ and $\vec{n}$ denote the outward unit normal vectors on boundaries $\partial\Omega_0$ and $\partial\Omega$ respectively, and let the vector function $\vec{\varphi}$

$$\vec{\varphi}: \overline{\Omega_0} \to \overline{\Omega}, \tag{1}$$

give the deformation of the body such that $\vec{x} = \vec{\varphi}(\vec{X})$, where $\vec{X}$ is the position vector of an arbitrary point in Ω_0 prior to the deformation and $\vec{x}$ is the position vector of the corresponding point in Ω. The closures $\overline{\Omega_0}$ and $\overline{\Omega}$ in Eq. (1) are defined by

$$\overline{\Omega_0} = \Omega_0 \cup \partial\Omega_0 \quad \text{and} \quad \overline{\Omega} = \Omega \cup \partial\Omega.$$

The deformation gradient **F** is defined by

$$\boldsymbol{F} = \text{Grad}\,\vec{x} = \text{Grad}\,\vec{\varphi}(\vec{X}), \tag{2}$$

where Grad is the gradient operator with respect to Ω_0. At an arbitrary point $\vec{X}$, the determinant of deformation gradient $\det(\boldsymbol{F}) > 0$ measures the local volume change. The left polar decomposition of **F** gives

$$\boldsymbol{F} = \boldsymbol{O}^D\,\boldsymbol{U}, \tag{3}$$

where $\boldsymbol{U}$ is the right stretch symmetric tensor and $\boldsymbol{O}^D$ is the orthogonal rotation tensor. The tensors $\boldsymbol{U}$ and $\boldsymbol{O}^D$ are obtained from the relations

$$\boldsymbol{U} = \sqrt{\boldsymbol{F}^T\boldsymbol{F}} = \sqrt{\boldsymbol{C}} \quad and \quad \boldsymbol{O}^D = \boldsymbol{F}\,\boldsymbol{U}^{-1} \tag{4}$$

where $\boldsymbol{F}^T$ is the transpose of F, $\boldsymbol{U}^{-1}$ is the inverse of $\boldsymbol{U}$ and $\boldsymbol{C} = \boldsymbol{F}^T\boldsymbol{F}$ is the right Cauchy-Green deformation tensor [1].

Eulerian Electroelasticity Field Equations

The problem of electroelasticity being one of electromechanical problems, the basic equations are those of electrostatic and elasticity. Let $\boldsymbol{T}$ denote the total stress tensor and suppose $\vec{E}$ and $\vec{D}$ denote the Eulerian electric field and electric displacement vectors respectively. For a purely electrostatic situation in the absence of mechanical body force, magnetic field, free electric current and free volumetric charges, the electroelasticity field equations are

$$\operatorname{div}\boldsymbol{T} = \vec{0}, \ \operatorname{curl}\vec{E} = \vec{0} \ \text{ and } \operatorname{div}\vec{D} = 0, \tag{5}$$

where div and curl are the divergence and rotor operators with respect to the deformed configuration Ω. Note that the electric body force has been incorporated through $\boldsymbol{T}$, which by virtue of angular momentum balance is symmetric.

Let $\overrightarrow{E^*}$ and $\overrightarrow{D^*} = \epsilon_0\overrightarrow{E^*}$ denote the corresponding fields in the free space outside the body, where ϵ_0 is the permitivity of free space, both $\overrightarrow{E^*}$ and $\overrightarrow{D^*}$ satisfy Eqs. 5_2 and 5_3 respectively. The standard boundary conditions accompanying Eq. (5) are

$$\boldsymbol{T}\vec{n} = \overrightarrow{t_a} + \overrightarrow{t_m^*}, \qquad \vec{n}\times\left(\overrightarrow{E^*} - \vec{E}\right) = \vec{0}, \quad \vec{n}\cdot\left(\overrightarrow{D^*} - \vec{D}\right) = \sigma_f \quad \text{on} \quad \partial\Omega \tag{6}$$

where σ_f is the free surface charge on $\partial\Omega$ per unit area, $\overrightarrow{t_a}$ is the surface mechanical traction on $\partial\Omega$ per unit area and $\overrightarrow{t_m^*} = \boldsymbol{T}^*\vec{n}$ is the load due to the Maxwell stress [1, 11]

$$\boldsymbol{T}^* = \epsilon_0\overrightarrow{E^*}\otimes\overrightarrow{E^*} - \frac{1}{2}\epsilon_0\left(\overrightarrow{E^*}\cdot\overrightarrow{E^*}\right)\mathbf{I}\,. \tag{7}$$

Lagrangian Electroelasticity Field Equations

Let $\boldsymbol{P}$ denote the total first Piola-Kirchhoff stress tensor and suppose $\overrightarrow{E_L}$ and $\overrightarrow{D_L}$ denote the Lagrangian electric field and electric displacement vectors respectively. The electromechanical field variables $\boldsymbol{P}, \overrightarrow{E_L}, \overrightarrow{D_L}$ and $\boldsymbol{T}, \vec{E}, \vec{D}$ are related by

$$\boldsymbol{P} = \det(\boldsymbol{F})\,\boldsymbol{F}^{-1}\boldsymbol{T}\,, \ \overrightarrow{E_L} = \boldsymbol{F}^T\vec{E}, \ \overrightarrow{D_L} = \det(\boldsymbol{F})\,\boldsymbol{F}^{-1}\vec{D}, \tag{8}$$

where the tensors $\boldsymbol{F}^{-1}$ is the inverses of the tensors $\boldsymbol{F}$.

The corresponding electroelasticity field equations in Lagrangian forms are

$$\operatorname{Div}\boldsymbol{P} = \vec{0}, \ \operatorname{Curl}\overrightarrow{E_L} = \vec{0}, \ \operatorname{Div}\overrightarrow{D_L} = 0 \tag{9}$$

where Div and Curl are the divergence and rotor operators with respect to the reference configuration Ω_0.

Similarly, the standard boundary conditions associated with Eq. 9 are

$$\boldsymbol{P}\vec{N} = \overrightarrow{t_F} + \overrightarrow{t_{mF}^*}, \ \vec{N}\times\left(\boldsymbol{F}^T\overrightarrow{E^*} - \overrightarrow{E_L}\right) = \vec{0}, \ \vec{N}\cdot\left(\det(\boldsymbol{F})\boldsymbol{F}^{-1}\overrightarrow{D^*} - \overrightarrow{D_L}\right) = \sigma_F \text{ on } \partial\Omega_0 \tag{10}$$

where σ_F is the free surface charge on $\partial\Omega_0$ per unit area, $\overrightarrow{t_F}$ is the surface mechanical traction on Ω_0 per unit area, $\overrightarrow{t_{mF}^*} = \boldsymbol{P}^*\vec{N}$ and

$$\boldsymbol{P}^* = \det(\boldsymbol{F})\,\boldsymbol{F}^{-1}\boldsymbol{T}^*,$$

is the pull back version of the Maxwell stress T^* [1, 11].

Electroelastic Constitutive Model for Incompressible Dielectric Hyperelastic Solids

In view of Fadodun et al. [1], Mehnikov and Ogden [11] and Dorfmann and Ogden [13], let $\Phi^*(\boldsymbol{F}, \overrightarrow{D_L})$ denote the electroelastic Helmholtz's free-energy function per unit volume, where $\boldsymbol{F}$ and $\overrightarrow{D_L}$ are the independent variables. Using the Coleman-Noll procedure, the total first Piola-Kirchhoff stress tensor **P** and Lagrangian electric field vector $\overrightarrow{E_L}$ are obtained by

$$\boldsymbol{P} = \frac{\partial\Phi^*(F,\overrightarrow{D_L})}{\partial F} \text{ and } \overrightarrow{E_L} = \frac{\partial\Phi^*(F,\overrightarrow{D_L})}{\partial\overrightarrow{D_L}}. \tag{11}$$

Let $\Phi(F)$ denote the strain energy per unit volume for purely elastic materials. In the case of an isotropic semilinear hyperelastic solids, the function $\Phi(F)$ reads

$$\Phi(F) = \mu_e I_1^2(\boldsymbol{U} - \boldsymbol{I}_0) + \frac{1}{2}\lambda_e I_1(\boldsymbol{U} - \boldsymbol{I}_0)^2, \tag{12}$$

where $I_1(\boldsymbol{U} - \boldsymbol{I}_0)$ is the first invariant of the tensor $(\boldsymbol{U} - \boldsymbol{I}_0)$, $\boldsymbol{I}_0$ is the unit tensor in the reference configuration Ω_0 and λ_e, μ_e are the material Lame constants [1, 16, 17].
On the basis of free-energy function in Eq. 12, and using the first and second laws of thermodynamics, Fadodun et al. [1] proposed electroelastic Helmholtz's free energy function $\Phi^*(\boldsymbol{F}, \overrightarrow{D_L})$

$$\Phi^*(\boldsymbol{F}, \overrightarrow{D_L}) = \mu_e I_1^2(\boldsymbol{U} - \boldsymbol{I}_0) + \frac{1}{2}\lambda_e I_1(\boldsymbol{U} - \boldsymbol{I}_0)^2 + \frac{1}{2\epsilon}\overrightarrow{D_L} \cdot \boldsymbol{U} \cdot \overrightarrow{D_L}, \tag{13}$$

for the dielectric semilinear hyperelastic solid under consideration, where ϵ is the electric permitivity of the material.
Substituting Eq. 13 into Eq. 11 gives the electro-sensitive stress tensor

$$\boldsymbol{P} = \frac{\partial\Phi^*(F,\overrightarrow{D_L})}{\partial F} = 2\mu_e\boldsymbol{F}^T + \left((\lambda_e I_1(\boldsymbol{U} - \boldsymbol{I}_0) - 2\mu_e)\boldsymbol{I}_0 + \frac{1}{2\epsilon}(\overrightarrow{D_L} \otimes \overrightarrow{D_L})\right)\boldsymbol{O}^{DT}, \tag{14}$$

and deformation-dependent electric field vector

$$\overrightarrow{E_L} = \frac{\partial\Phi^*(F,\overrightarrow{D_L})}{\partial\overrightarrow{D_L}} = \frac{1}{\epsilon}\boldsymbol{U}\overrightarrow{D_L}, \tag{15}$$

as the electroelastic constitutive model for the solids, where $\otimes$ is the tensor product [1].
In the case of an incompressible dielectric semilinear hyperelastic material, the first Piola-Kirchhoff stress tensor $\boldsymbol{P}$ in Eq. 14 assumes the form

$$\boldsymbol{P} = \frac{\partial\Phi^*(F,\overrightarrow{D_L})}{\partial F} = 2\mu_e\boldsymbol{F}^T + \left((\lambda_e I_1(\boldsymbol{U} - \boldsymbol{I}_0) - 2\mu_e)\boldsymbol{I}_0 + \frac{1}{2\epsilon}(\overrightarrow{D_L} \otimes \overrightarrow{D_L})\right)\boldsymbol{O}^{DT} - \eta\boldsymbol{F}^{-T}, \tag{16}$$

where η is the Lagrange multiplier associated with the incompressibility constraint $\det(\mathbf{F}) = 1$.

Application: Extension and Inflation Coupling of a Thick Electroelastic Tube

The theory of electroelasticity summarized in the previous sections is now specialized for application to the problem of combined extension and inflation of a relatively thick-walled circular cylindrical tube. The tube under consideration has closed ends and is subject to internal pressure, axial load and a radial electric field generated by a potential difference between flexible electrodes coated on its inner and outer radial surfaces.

Let $\vec{R}(R,\Theta,Z)$ be the position vector of an arbitrary point in the electroelastic tube Ω_0 prior to the deformation and let $\vec{r}(r,\theta,z)$ denote the position vector of the correspoonding point in the deformed tube Ω. The reference configuration Ω_0 of the tube is described by

$$A \leq R \leq B, \qquad 0 \leq \Theta \leq 2\pi, \qquad 0 \leq Z \leq L, \tag{17}$$

where A and B are the inner and outer radii and L is the length of the tube in Ω_0.
Invoking the constraint of circular symmetric, the deformed configuration Ω is defined by

$$a \leq r \leq b, \qquad 0 \leq \theta \leq 2\pi, \qquad 0 \leq z \leq l, \tag{18}$$

where a and b are the inner and outer radii and l is the length of the tube in Ω. Note that (R,Θ,Z) and (r,θ,z) are the cylindrical polar coordinates in the reference configuration Ω_0 and deformed configuration Ω respectively.
Using the incompressibility constraint of the tube material, the deformation is defined by the relations [11]

$$r = r(R) = \sqrt{a^2 + \lambda_z^{-1}(R^2 - A^2)}, \qquad \theta = \Theta, \qquad z = \lambda_z Z, \tag{19}$$

where λ_z is the uniform axial stretch of the tube. In view of Eq. (2), the deformation gradient $\boldsymbol{F}$ is

$$\boldsymbol{F} = \mathrm{Grad}\,\vec{r} = \frac{\partial r}{\partial R}\,\overrightarrow{e_r^*} \otimes \overrightarrow{E_R^*} + \frac{r}{R}\,\overrightarrow{e_\theta^*} \otimes \overrightarrow{E_\Theta^*} + \frac{\partial z}{\partial Z}\,\overrightarrow{e_z^*} \otimes \overrightarrow{E_Z^*}, \tag{20}$$

where $\overrightarrow{e_r^*}, \overrightarrow{e_\theta^*}, \overrightarrow{e_z^*}$ and $\overrightarrow{E_R^*}, \overrightarrow{E_\Theta^*}, \overrightarrow{E_Z^*}$ are the orthonormal basis vectors in Ω and Ω_0 respectively.
Substituting Eq. 19 into Eq. 20 and introducing $\lambda_\theta = \frac{r}{R}$ as the azimuthal stretch give

$$\boldsymbol{F} = \frac{1}{\lambda_z\lambda_\theta}\,\overrightarrow{e_r^*} \otimes \overrightarrow{E_R^*} + \lambda_\theta\,\overrightarrow{e_\theta^*} \otimes \overrightarrow{E_\Theta^*} + \lambda_z\,\overrightarrow{e_z^*} \otimes \overrightarrow{E_Z^*}. \tag{21}$$

Using Eqs. 3, 4 and 21 gives

$$\boldsymbol{U} = \boldsymbol{F} \quad \text{and} \quad \boldsymbol{O}^D = \boldsymbol{I}_0. \tag{22}$$

Let the parameters λ_a and λ_b be defined by

$$\lambda_a = \frac{a}{A}, \quad \lambda_b = \frac{b}{B}, \quad \text{and} \quad b = r(B). \tag{23}$$

Using Eq. 19_1 and Eq. 23 give the relations

$$\frac{\lambda_a^2\lambda_z-1}{\lambda_\theta^2\lambda_z-1}=\frac{B^2}{A^2}\left(\frac{\lambda_b^2\lambda_z-1}{\lambda_\theta^2\lambda_z-1}\right)=\frac{R^2}{A^2}. \tag{24}$$

Equation 24 shows that the term $R^2(\lambda_\theta^2\lambda_z-1)$ defined by

$$R^2(\lambda_\theta^2\lambda_z-1)=B^2(\lambda_b^2\lambda_z-1)=A^2(\lambda_a^2\lambda_z-1)=\tau^2,\quad \tau\in R \tag{25}$$

is independent of both R and r; and for the case of inflation of cylindrical tube at fixed length $\lambda_\theta^2\lambda_z-1>0$ for $\lambda_b\le\lambda_\theta\le\lambda_a$.

Components of Stress Tensor and Electric Field Vector

In view of Melnikov and Ogden [11], an applied potential difference between the electrode coated surfaces $R = A$ and $R = B$ generates a radial electric field within the tube and is associated with equal and opposite charges on the electrode coated surfaces. By Gauss's law and neglecting the edge effect, the electric field vector vanishes outside the tube. The components of electric displacement vector $\overrightarrow{D_L}$ within the tube assume the form

$$D_R=D(R)\neq 0,\ D_\Theta=0,\ D_Z=0, \tag{26}$$

where D_R, D_Θ and D_Z are the radial, azimuthal and axial components of $\overrightarrow{D_L}$ respectively and $D(R)$ is a function of radius R only.
Using Eqs. 15, 21, 22 and 26 gives

$$E_R=\frac{1}{\lambda_\theta\lambda_z\epsilon}D_R,\qquad E_\Theta=0,\qquad E_Z=0, \tag{27}$$

where E_R, E_Θ and E_Z are the radial, azimuthal and axial components of $\overrightarrow{E_L}$ respectively.
Substituting Eqs. 21, 22, and 26 into Eq. 16 gives

$$P_{RR}=\frac{2\mu_e}{\lambda_\theta\lambda_z}+\lambda_e\left(\frac{1}{\lambda_\theta\lambda_z}+\lambda_\theta+\lambda_z-3\right)-2\mu_e+\frac{1}{2\epsilon}D_R^2-\eta\lambda_\theta\lambda_z,\quad P_{R\Theta}=0,\quad P_{RZ}=0, \tag{28}$$

$$P_{\Theta\Theta}=2\mu_e\lambda_\theta+\lambda_e\left(\frac{1}{\lambda_\theta\lambda_z}+\lambda_\theta+\lambda_z-3\right)-2\mu_e-\frac{\eta}{\lambda_\theta},\quad P_{\Theta R}=0,\quad P_{\Theta Z}=0, \tag{29}$$

$$P_{ZZ}=2\mu_e\lambda_z+\lambda_e\left(\frac{1}{\lambda_\theta\lambda_z}+\lambda_\theta+\lambda_z-3\right)-2\mu_e-\frac{\eta}{\lambda_z},\quad P_{ZR}=0,\quad P_{Z\Theta}=0, \tag{30}$$

where P_{ij}, $i,j=R,\Theta,Z$ are the components of total first Piola-Kirchhoff stress tensor **P**.

Solution of Reduced Electroelasticity Field Equations for Extension-Inflation Coupling of Dielectric Semilinear Hyperelastic Tube

In view of Eqs. 28-30, Eq. 9_1 reduces to

$$\frac{d}{dR}P_{RR}+\frac{1}{R}(P_{RR}-P_{\Theta\Theta})=0. \tag{31}$$

Substituting Eqs. 28_1 and 29_1 into Eq. 31 gives

$$R\frac{d\eta}{dR} = -\left(\frac{2\mu_e+\lambda_e}{\lambda_\theta^4\lambda_z^3}\right)\left(\lambda_\theta^2\lambda_z - 1\right)^2 + \frac{1}{\epsilon\lambda_\theta\lambda_z}\left(\frac{1}{2}D_R^2 + RD_R\frac{dD_R}{dR}\right). \quad (32)$$

The form of electric field vector $\overrightarrow{E_L}$ in Eq. 27 shows that Eq. 9_2 is satisfied identically; and substituting Eq. 26 into Eq. 9_3 gives

$$\frac{d}{dR}(RD_R) = \frac{d}{dR}\left(RD(R)\right) = 0. \quad (33)$$

The integration of Eq. 33 gives

$$RD(R) = const. \quad (34)$$

At the coated surfaces $R = A$ and $R = B$, we have

$$R\,D(R) = A\,D(A) = B\,D(B). \quad (35)$$

Let the total charge at the coated surface $R = A$ be Q, then the total charge at the coated surface $R = B$ is -Q. The free surface charge densities per unit area are

$$\sigma_F = \frac{Q}{2\pi AL} \text{ on } R = A \text{ and } \sigma_F = -\frac{Q}{2\pi BL} \text{ on } R = B. \quad (36)$$

In view of the boundary condition Eq. 10_3 and setting $\overrightarrow{D^*} = \vec{0}$ yield

$$D(A) = \frac{Q}{2\pi AL} \quad \text{and} \quad D(B) = \frac{Q}{2\pi BL}. \quad (37)$$

Using Eqs. 35 and 37 gives

$$D_R = D(R) = \frac{Q}{2\pi RL}. \quad (38)$$

The combination of Eq. 38 and Eq. 27_1 yields [1]

$$E_R = \frac{1}{\lambda_\theta\lambda_z}\left(\frac{Q}{2\epsilon\pi RL}\right). \quad (39)$$

Substituting Eq. 38 into Eq.32 gives

$$d\eta = -\left(\frac{2\mu_e+\lambda_e}{\lambda_\theta^4\lambda_z^3}\right)\left(\lambda_\theta^2\lambda_z - 1\right)^2\frac{dR}{R} - \frac{1}{2\epsilon\lambda_\theta\lambda_z}\left(\frac{Q}{2\pi L}\right)^2\frac{1}{R^2}\frac{dR}{R}. \quad (40)$$

The relation $\lambda_\theta = \frac{r}{R}$ and knowing that $\frac{dr}{dR} = (\lambda_\theta\lambda_z)^{-1}$ gives

$$\frac{dR}{R} = \left(\frac{\lambda_\theta\lambda_z}{1-\lambda_\theta^2\lambda_z}\right)d\lambda_\theta. \quad (41)$$

Substituting Eq. 41 into Eq. 40 gives

$$d\eta = \left(\frac{2\mu_e+\lambda_e}{\lambda_\theta^3\lambda_z^2}\right)(\lambda_\theta^2\lambda_z - 1)d\lambda_\theta + \frac{1}{2\epsilon}\left(\frac{Q}{2\pi L}\right)^2 \frac{d\lambda_\theta}{R^2(\lambda_\theta^2\lambda_z-1)}. \tag{42}$$

Substituting Eq. 25 into Eq. 42 gives

$$d\eta = \left(\frac{2\mu_e+\lambda_e}{\lambda_\theta^3\lambda_z^2}\right)(\lambda_\theta^2\lambda_z - 1)d\lambda_\theta + \frac{1}{2\epsilon}\left(\frac{Q}{2\pi\tau L}\right)^2 d\lambda_\theta. \tag{43}$$

The integration of Eq. 43 gives

$$\eta = (2\mu_e + \lambda_e)\left(\frac{1}{\lambda_z}\ln\lambda_\theta + \frac{1}{2\lambda_z^2\lambda_\theta^2}\right) + \frac{1}{2\epsilon}\left(\frac{Q}{2\pi\tau L}\right)^2 \lambda_\theta + C, \tag{44}$$

where C is the constant of integration.
Substituting Eq. 44 into Eq. 28_1 gives

$$P_{RR} = (2\mu_e + \lambda_e)\left(\frac{1}{2\lambda_\theta\lambda_z} - \lambda_\theta\ln\lambda_\theta\right) + \lambda(\lambda_\theta + \lambda_z)$$

$$+\frac{1}{2\epsilon}\left(\frac{Q}{2\pi L}\right)^2\left(\frac{1}{R^2} - \frac{\lambda_\theta^2\lambda_z}{\tau^2}\right) - (2\mu_e + 3\lambda_e) - \lambda_\theta\lambda_z C. \tag{45}$$

The mechanical boundary conditions on the surfaces $R = A$ and $R = B$ are

$$P_{RR} = -p \quad \text{on} \quad R = A \quad \text{and} \quad P_{RR} = 0 \quad \text{on} \quad R = B, \tag{46}$$

where p is the internal pressure.
Substituting Eq. 45 into Eq. 46_2 gives the integration constant C

$$C = \left(\frac{2\mu_e+\lambda_e}{\lambda_b\lambda_z}\right)\left(\frac{1}{2\lambda_b\lambda_z} - \lambda_b\ln\lambda_b\right) + \frac{1}{2\epsilon\lambda_b\lambda_z}\left(\frac{Q}{2\pi L}\right)^2\left(\frac{1}{B^2} - \frac{\lambda_b^2\lambda_z}{\tau^2}\right) + \frac{\lambda(\lambda_b+\lambda_z)}{\lambda_b\lambda_z} - \left(\frac{2\mu_e+3\lambda_e}{\lambda_b\lambda_z}\right). \tag{47}$$

Substituting Eq. 47 into Eq. 45 gives

$$P_{RR} = (2\mu_e + \lambda_e)\left(\frac{1}{2\lambda_\theta\lambda_z} - \lambda_\theta\ln\lambda_\theta + \frac{\lambda_\theta}{\lambda_b}\left(\lambda_b\ln\lambda_b - \frac{1}{2\lambda_b\lambda_z}\right)\right) + (2\mu_e + 3\lambda_e)\left(\frac{\lambda_\theta}{\lambda_b} - 1\right)$$

$$\frac{1}{2\epsilon}\left(\frac{Q}{2\pi L}\right)^2\left(\frac{1}{R^2} - \frac{\lambda_\theta^2\lambda_z}{\tau^2} + \left(\frac{\lambda_b^2\lambda_z}{\tau^2} - \frac{1}{B^2}\right)\frac{\lambda_\theta}{\lambda_b}\right) + \lambda\left((\lambda_\theta + \lambda_z) - (\lambda_b + \lambda_z)\frac{\lambda_\theta}{\lambda_b}\right). \tag{48}$$

Substituting Eq. 48 into Eq. 46_1 gives the expression for the internal pressure p

$$p = (2\mu_e + \lambda_e)\left(\frac{1}{2\lambda_z}\left(\frac{\lambda_a}{\lambda_b^2} - \frac{1}{\lambda_a}\right) + \lambda_a\ln\left(\frac{\lambda_a}{\lambda_b}\right)\right) + \frac{1}{2\epsilon}\left(\frac{Q}{2\pi L}\right)^2\left(\frac{\lambda_a\lambda_z}{\tau^2}(\lambda_a - \lambda_b) - \frac{1}{A^2} + \frac{1}{B^2}\frac{\lambda_a}{\lambda_b}\right)$$

$$+ \lambda_e\lambda_z\left(\frac{\lambda_a}{\lambda_b} - 1\right) + (2\mu_e + 3\lambda_e)\left(1 - \frac{\lambda_a}{\lambda_b}\right). \tag{49}$$

In addition to the internal pressure obtained in Eq. 49, we proceed to find the axial load F applied to the ends of the closed tube. The applied axial load F is defined by

$$F = 2\pi \int_A^B P_{ZZ}\ RdR. \tag{50}$$

Let the stresses $P_{\Theta^*R^*}$ and $P_{Z^*R^*}$ be defined by

$$P_{\Theta^*R^*} = P_{\Theta\Theta} - P_{RR} \quad \text{and} \quad P_{Z^*R^*} = P_{ZZ} - P_{RR}. \tag{51}$$

The axial stress P_{ZZ} is related to the radial stress P_{RR} and the stresses $P_{\Theta^*R^*}, P_{Z^*R^*}$ by

$$P_{ZZ} = \frac{1}{2}\left(\frac{1}{R}\frac{d}{dR}(R^2 P_{RR})\right) - \frac{1}{2}P_{\Theta^*R^*} + P_{Z^*R^*}. \tag{52}$$

Using Eqs. 52 and 46 in Eq. 50 gives

$$F = \pi A^2 p + \pi \int_A^B (2P_{Z^*R^*} - P_{\Theta^*R^*})R\,dR. \tag{53}$$

It is obvious from Eq. 53 that the internal pressure p contributes to the axial load F; consequently the reduced axial load F_{red} is defined by

$$F_{red} = F - \pi A^2 p = \pi \int_A^B (2P_{Z^*R^*} - P_{\Theta^*R^*})R\,dR. \tag{54}$$

Using Eqs. 28, 29 and 51 in Eq. 54 gives

$$F_{red} = \frac{(2\mu_e+\lambda_e)\tau^2}{\lambda_z}\ln\left(\frac{\lambda_a}{\lambda_b}\right) + (6\mu_e + \lambda_e)\tau^2\left(\frac{\lambda_a}{\lambda_a^2\lambda_z-1} - \frac{\lambda_b}{\lambda_b^2\lambda_z-1}\right)$$

$$-\frac{1}{2\epsilon\sqrt{\lambda_z}}\left(\frac{Q}{2\pi L}\right)^2\left(\tanh^{-1}(\sqrt{\lambda_z}\lambda_a) - \tanh^{-1}(\sqrt{\lambda_z}\lambda_b)\right)$$

$$-\left(\frac{4\mu_e+\lambda_e}{\lambda_z} + \lambda_e\lambda_z(\lambda_z - 1)\right)\tau^2\left(\frac{1}{\lambda_a^2\lambda_z-1} - \frac{1}{\lambda_b^2\lambda_z-1}\right) + 2\mu\lambda_z\tau^2\left(\frac{\lambda_a+\lambda_z-3}{\lambda_a^2\lambda_z-1} - \frac{\lambda_b+\lambda_z-3}{\lambda_b^2\lambda_z-1}\right). \tag{55}$$

Remark: It is obvious from Eqs. 49 and 55 that both internal pressure and axial load interact with the applied radial electric field associated with the equal and opposite charges on the electrode coated surfaces of the tube. These results are in agreement with study in literature.

Finally, in view of Eq. 39, the electric field generated within tube in now expressed in terms of azimutal stretch λ_θ

$$E_R = \frac{1}{\epsilon A\lambda_\theta\lambda_z}\left(\frac{Q}{2\pi L}\right)\sqrt{\frac{\lambda_\theta^2\lambda_z-1}{\lambda_a^2\lambda_z-1}}. \tag{56}$$

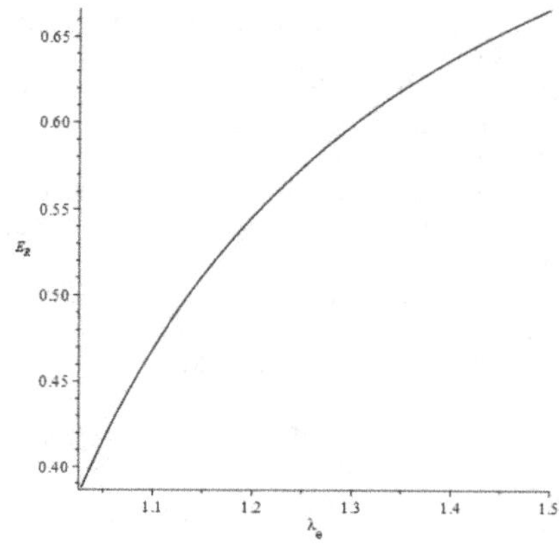

Fig.: This plot shows that the radial electric field generated within the tube increases with the increasing azimuthal stretch of the tube

Conclusion

The study formulated electroelastic constitutive model for incompressible dielectric semmilinear hyperelastic materials. The consequent electroelastic equations are used to solve the problem of extension and inflation coupling of circular cylindrical tube. It is obtained that the applied radial electric field associated with the equal and opposite charges on the electrode coated surfaces contributes to both internal pressure and axial loads of the cylindrical tube. The aplications of the results in this study include but not limited to actutors, sensors, and soft robotic.

References

[1] O. O. Fadodun, O. G. Fadodun, A. S. Borokinni, B. A. Olokuntoye, O. P. Layeni and A. P. Akinola, Electroelastic constitutive model for dielectric semilinear hyperelastic solids with application to radial deformation of a rotating tube, Meccanica, 57 (2022) 2355-2363.

[2] N. C. Goulbourne, E. Mockensturm and M. I. Frecker, Modelling dielectric elastomer membrane. Dielectric Elastomers as Electromechanical Transducers, (2008) doi: 10.1016/B978-0-08-047488-500018-6.

[3] M. Moscardo, X. Zhao, Z. Suo and Y. Lapusta, On designing dielectric elastomer actuators, J. Appl. Phys. 104 (2008), doi.org/10.1063/1.3000440.

[4] A. Melnikov, L. Dorfmann and R. W. Ogden, Bifurcation of finitely deformed thickwalled electroelastic spherical shell subject to radial electric field, Int. J. Nonlinear Mech., (2020) doi.org/10.1016/j.ijnonlinmec.2020.103429.

[5] X. Y. Xie, Y. C. Liu and Y. B. Fu, Bifurcation of a dielectric elastomer balloon under pressurized inflation and electric actuation, Int. J. Solids Struct. 78 (2016) 182-188.

[6] L. Dorfmann and R. W. Ogden, Waves and vibrations in a finitely deformed electroelastic circular cylindrical tube, Proc R Soc A, (2020) doi.org/10.1098/rspa.20190701.

[7] R. D´ıaz-Calleja, M. J. Sanchis and E. Riande, Effect of an electric field on the bifurcation of a biaxially stretched incompressible slab rubber, Eur. Phys. J. E. 30 (2009) 417-426.

[8] R. D´ıaz-Calleja, E. Riande and M. J. Sanchis, On electromechanical stability of dielectric elastomers, Appl. Phys. Lett. 94 (2008), doi.org/10.1063/1.3138153.

[9] A. Melnikov and R. W. Ogden, Bifurcation of finitely deformed thick-walled electroelastic cylindrical tubes subject to a radial electric field, Z. Angew. Math. Phys. 67 (2016) 140.

[10] R. A. Toupin, The dielectric, J. Ration. Mech. Anal. 5 (1958) 849-915.

[11] A. Melnikov and R. W. Ogden, Finite deformations of an electroelastic circular cylindrical tube, Z. Angew. Math. Phys. 69 (2016) 60.
[12] S. Sazena, D. D. Barreto and A. Kumar, Extension-torsion-inflation coupling incompressible electroelastomeric thintubes, Math. And Mech. Solids, (2019) DOI:10.1177/1081286519886901.
[13] L. Dorfmann and R. W. Ogden, The effect of deformation dependent permittivity on the elastic response of a finitely deformed dielectric tube, Mech Res Commun 93 (2018) 47-57.
[14] Y. Fu, Y. X. Xie and L. Dorfmann, A reduced model for electrodes-coated dielectric plates, Int. J. Nonlinear Mech. 106 (2018) 60-69.
[15] A. N. Abd-alla, F. Alshaikh and I. Giorgio, (2015). A mathematical model for longitudinal wave propagation in a magnetoelastic hollow circular cylinder of anisotropic material under the influence of initial hydrostatic stress, Math. And Mech. Of Solids, Doi.org/10.1177/1081286515582883.
[16] O. O. Fadodun, Analysis of axisymmetric fractional vibration of an isotropic thin disc in finite deformation, Computer and Concrete, 23 (2019) 303-309.
[17] A. P, Akinola, An application of nonlinear fundamental problems of a transversely isotropic layer in finite elastic deformation, Int. J. Nonlin. Mech. 36 (2001) 307-321.

Advanced Topics in Mechanics of Materials, Structures and Construction - AToMech1-2023 Materials Research Forum LLC
Materials Research Proceedings 31 (2023) 347-355 https://doi.org/10.21741/9781644902592-36

Design of a solar PV power plant at KFU premises

Mahdi Alnasser[1a], Hassan Alghazal[1b], Amjad Almeihan[1c], Mounir Bouzguenda[1d*], Mohammed Abdulaziz Alaqil[1e]

[1]Department of Electrical Engineering, King Faisal University, Al Ahasa, PC 31982, Kingdom of Saudi Arabia

[a]mahdialnasser10@gmail.com, [b]HassanAdelAlghazal@gmail.com, [c]almeihanamjad@gmail.com, [d]mbuzganda@kfu.edu.sa, [e]malaqil@kfu.edu.sa

Keywords: Grid-Connected Solar PV System, KSA 2030 Vision, Two-Way Meters, Life Cycle Cost Analysis

Abstract. This research project involves the design of an on-grid solar photovoltaic system on a $6500m^2$ triangular-shaped land at the King Faisal University (KFU) campus. This project consists of two parts: general knowledge of Solar PV systems and design aspects of the solar PV system. The first part highlights key theoretical knowledge about PV systems, i.e., the definition of a PV system, the types, current technologies, and the components of the system. The second part presents the design aspects of an on-grid PV system and shows some standards used in designing, and sizing PV components including the distance between arrays and comparison between vertical and horizontal mounted PV panels. The proposed modeling approach utilizes PVsyst solar simulation program to help optimize and design the proposed solar PV system. Based on the performed numerical calculations, the final design consisted of 8 inverters rated at 110kVA each and 1200 panels rated at 540W to get the maximum output power to feed the 1MW water treatment plant. Several scenarios were carried out using different economic parameters. For the optimal scenario, the total operation cost using the 650 kW PV system is estimated to be 451,215 SAR with savings of 109,425 compared to the grid alone scenario. This translated to 0.254 SAR/kWh. In terms of oil and pollution and oil savings, the optimal results for adding the 650 kW PV system yield he life cylce cost anaylsis yield savings of more than 17,000 oil barrels and more than 23 thousand tons of CO_2. The corresponding monetary revenues are about 13.4 million SAR.

Introduction

The Kingdom of Saudi Arabia launched the National Renewable Energy Program, aiming to activate the local resources to produce and benefit from renewable energy opportunities. It is planned to produce at least 9.5 GW of renewable energy by 2023, which is anticipated to increase by 2030. This program will contribute to supporting the Kingdom's economy and providing jobs. The proposed project will be an opportunity to contribute to this program [1]. The project will be based on solar energy, which is one of the most important sources of renewable energy. It involves designing a photovoltaic (PV) solar power plant at King Faisal University premises. The university is in Al-Ahasa, which is one of the best areas in the Kingdom to benefit from solar energy. The design will be an on-grid solar PV power plant on a triangular land of $6{,}500m^2$ granted for energy projects at the university. It will supply the university's electricity network (Additional to the existing network from Saudi Electric Company).

The paper is organized as follows: In section 2, the authors compare (a) the existing solar PV systems and (b) water treatment power supplies. Section 3 is devoted to describing the design methodology and the proposed grid-connected PV at KFU. Section 4 describes the performance of the designed PV system using the PVGIS software tool and the economic analysis of the proposed system. The paper is concluded with Section 5 where the conclusions, recommendations, and future improvement of the design are discussed.

Materials Research Proceedings 31 (2023) 347-355 https://doi.org/10.21741/9781644902592-36

Literature Review and Problem Statement.
The three main existing solar PV types that are commercially applied are off-grid, on-grid, and hybrid PV solar grids. Each type has its own advantages and disadvantages, and it actually comes down to what the customer demands to gain from solar PV panel installation. The three types are compared in Table 1 below.

TABLE 1. Comparison of: off-grid, on-grid, and hybrid PV solar grids

Type	Advantages	Disadvantages	Notes
Off-grid	No need for transmission and distributions lines	Expensive energy storage devices	Valid for remote areas (mountains, far from the grid)
Grid-connected	Energy exchange with the grid.	If the grid is down, the load is not served	
Hybrid	Available 24 hours, Grid energy exchange	Expensive energy storage devices	areas with high wind potential

Water treatment power supplies
Water treatment plants (WTPs) are used to treat and recycle used water from industries, residential and commercial customers, and agriculture. The water treatment power supplies can use either the local power grid or a renewable energy source. Typical water treatment plants are supplied through conventional power plants. With recent improvements in renewable energy and increasing pollution issues, municipalities and owners start using renewable energy sources such as solar PV, Wind, Hydro, fuel cells, and storage. In fact, several WTPs are supplied with solar energy. For instance, the Greenskies Renewable Energy project shown in Figure 1 consists of a 137-kW solar array in Sprague, Connecticut, generating solar electricity for the Sprague wastewater treatment plant [1]. It is located on adjacent property and provides 80% of the power needed at the plant. Meanwhile, Figure 2 shows a sewage treatment with solar-produced green hydrogen [2].

Figure 1. Greenskies 137-kW Solar Array at the Water Treatment Plant [1].

Figure 2. A Water Treatment Facility Operated by Aqualia [2]

Potential Resources in Saudi Arabia

1) Solar PV potential resources in Saudi Arabia: The expansion of power generation in Saudi Arabia is necessary to meet the expected growth of power demand. Solar energy resources can be used effectively in Saudi Arabia due to the high solar irradiation, vast rainless area, and long-term sunshine [3].
2) Wind resources in Saudi Arabia are moderate and are high in coastal areas: Saudi Arabia has the potential to produce more than 200GW of onshore wind energy with an average capacity factor of 35.2%, higher than most countries paving the way in wind energy generation including the US (33.9%), UK (27.8%), Denmark (28.4%) and Germany (19%). By 2025, Saudi Arabia aims to generate 10 GW of wind energy, creating more than 7,500 jobs and

contributing more than $15 billion to Saudi Arabia's gross domestic product [5]. Potential renewable energy sites in Saudi Arabia including Al Jawf, Tabuk, Najran, Taif, and Yadamah have high wind speeds and promising capacity factors to make wind energy projects a reality.

3) Hydro resources in Saudi Arabia: The electricity production from hydroelectric sources (kWh) in Saudi Arabia was reported at 0 kWh in 2014. However, in 2016, the desalination plant located in the city of Ras Al-Khair is one of the highest hydropower-producing cities, with an average production of 18 million and 977 thousand megawatts per hour. This was followed by the desalination plant in the city of Jubail, which produced about 10 million and 366 thousand megawatts per hour, followed by a desalination plant in the city of Ras Al Khair, which produced about 5 million and 453 thousand megawatts per hour as shown in Figure 5 [7].

4) Fuel cells in Saudi Arabia: Saudi Arabia is seeking to develop and utilize all the various energy resources, with which the Kingdom is endowed. And to get all possible advantages of fuel cells in the kingdom, the Ministry of Energy has signed many memoranda of understanding (MoUs) with several entities for implementing pilot projects for hydrogen fuel-cell-based vehicles, buses, trains, and transportation applications and sustainable jet fuel production in selected areas and roads in various cities in the Kingdom [8].

5) Energy storage is being considered in Saudi Arabia in the last few years. The Red Sea project has the largest energy storage facility in the world. The battery storage facility will supply the Red Sea project with renewable energy during the night when there is no possibility of solar power generation. It will also ensure supply in the case of outages when shutdowns occur due to potential faults or sandstorms affecting production. The blend of solar and wind power generation will also guarantee a reliable supply of energy to this tourism destination [9].

Because of the high potential and focus on solar PV resources in the KFU site, the authors have recommended designing a power system for the 1 MW water treatment plant.

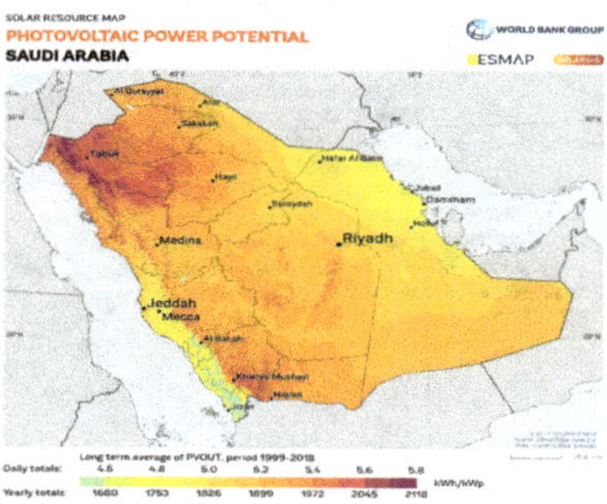

Fig 3. Photovoltaic Power Potential in Saudi Arabia [4]

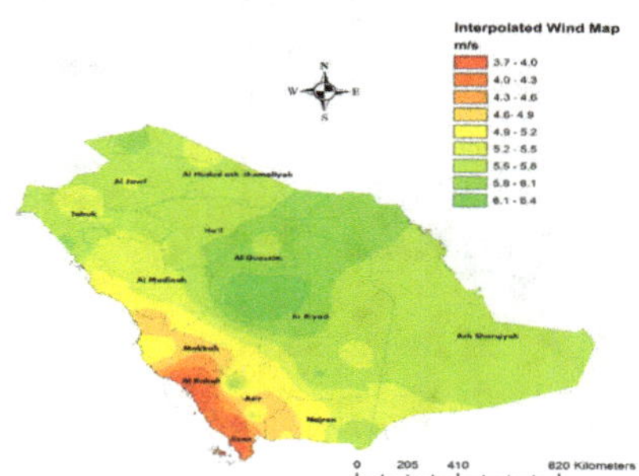

Fig 4. Wind Energy Map of Saudi Arabia [6]

Design methodology of the proposed solar PV system

The main objective of this paper is to present the mechanical design of a grid-connected solar PV system as well as calculate its power production capabilities. The designed system feeds the water treatment plant which is subjected to the following constraints:

a. Water treatment plant power requirements.
The existing WTP at KFU is operated through the Saudi Electricity Company (SEC) grid. It is supplied using a 3-phase transformer rated at 1 MVA (13.8kV/380 V). Therefore, the size of the PV plant must not exceed 1 MVA since it is expected that excess energy from the solar PV system could be sold to the local electricity company.

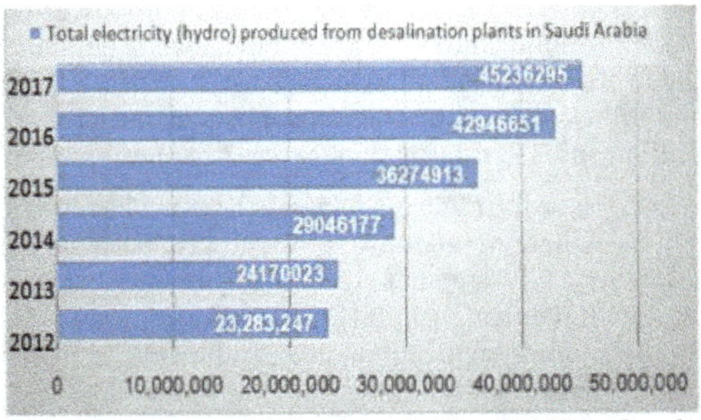

Figure 5. Hydroelectric Power between 2012 to 2017 in Saudi Arabia [7]

b. Limited land area.
The area of the allotted piece of land for this project is approximately 6500 m^2, with a useful land area to accommodate the solar PV system anticipated at 5850 m^2. The schematic drawing of the allotted land is described in the following Figure 6. As shown in the figure, the land is triangular with 162 m as a base, 82 m as height, and 115 m from the right and left sides.

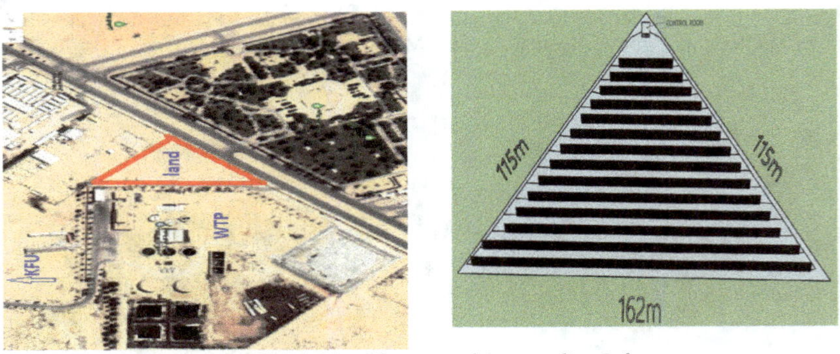

Figure 6. Solar PV land layout and surrounding Infrastructure.

c. The solar array spacing.
In this project, the best dimensions of solar panels to use are (2.3m×1.1 m) with 540 W per panel and 3.3 m array spacing. That was considered based on the shape of the land.

d. Solar PV Standards and Design.
The applied PV standards in the project are listed in Table 2 and the final single-line diagram of the solar PV system is shown in Figure 7. The final Single line diagram of the solar PV system connected to the WTP is shown in Figure 8. Table 3 lists the final design components. The 1 MW transformer is already installed.

Table 2. Engineering Standards [10-14]

Standard Code	Standard details
IEC 62548:2016	Photovoltaic (PV) arrays - Design requirements [10].
IEC TS 62738:2018	Ground-mounted photovoltaic (PV) power plants - Design guidelines and recommendations [11].
IEC 62446-1:2016	Photovoltaic (PV) systems - Requirements for testing, documentation, and maintenance - Part 1: Grid-connected systems - Documentation, commissioning tests, and inspection [12].
IEC 62109-2:2011	Safety of power converters for use in photovoltaic power systems - Part 2: Particular requirements for inverters [13].
UL 1741:2021	Inverters, Converters, Controllers, and Interconnection System Equipment for Use With Distributed Energy Resources [14].

Table 3. Final Design Solar PV System Components

Item	Description	Size/Number
PV system size	5850 m^2	648 kW
Solar panels	P_{rated}=540 W, V_m=41.76V, I_m=12.94A, η=20.89%	1200
DC-AC inverters	110 kVA	8
AC Combiner boxes	400A 1000V	4
DC Combiner boxes	160A 1500V	8
Disc. DC Switch	250A	8

Performance and Economic Analysis of the Grid PV System at KFU.
The PV system energy performance is carried out using PVGIS. The monthly energy production is listed in Figure 10.

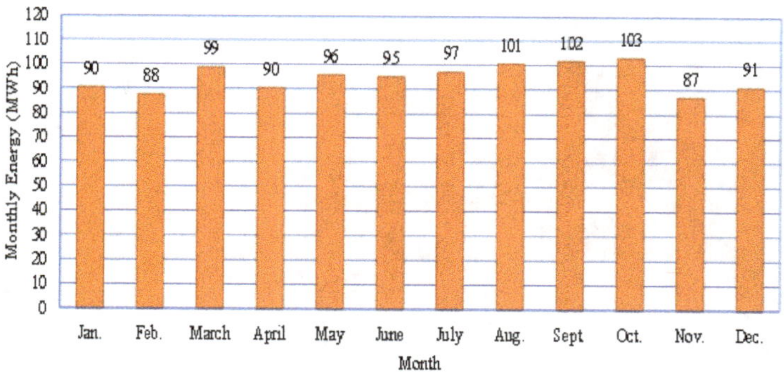

Figure 10. Monthly Output of the Solar PV Power Plant [15].

The life cycle cost analysis results for four different scenarios are listed in Table 4.

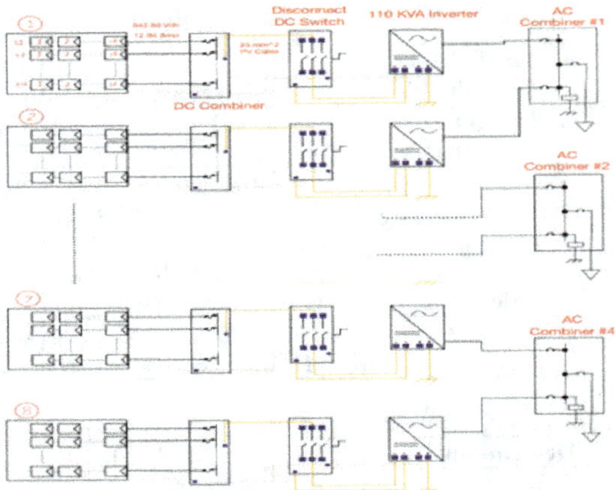

Figure 7. Single Line Diagram of the Solar PV System [Ref: own elaboration].

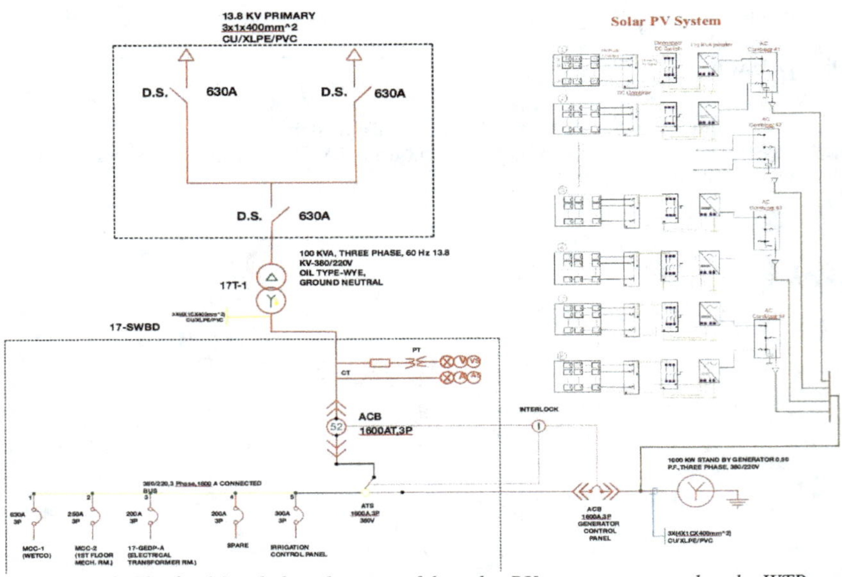

Figure 8. The final Single line diagram of the solar PV system connected to the WTP

In this study, the economic analysis is carried out using the following model described by equations (1) through (5) [16].

The total cost of the PV array can be calculated by using the following equation:

$$C_{pv} = Unit\ Cos\ of\ PV \times No.of\ Modules \times Peak\ Module\ Power \tag{1}$$

The present worth of maintenance cost (C_m), is estimated using the equation [16]:

$$C_m = M_{\frac{c}{yr}} \times \left(\frac{1+i}{1+d}\right) \times \left[\frac{1-\left(\frac{1+i}{1+d}\right)^N}{1-\left(\frac{1+i}{1+d}\right)}\right] \tag{2}$$

Where Mc is used to represent the annual maintenance cost and N is the Lifetime of the system. Life Cycle Cost (LCC) is computed as follows [16]:

$$LCC = C_{PV} + C_{inv} + C_{AC.com} + C_{DC.com} + C_{inst} + C_m + \cdots \tag{3}$$

The Annualized Life Cycle Cost (ALCC) can be calculated using the following equation [16]:

$$ALCC = LCC\left[\frac{1-\left(\frac{1+i}{1+d}\right)}{1-\left(\frac{1+i}{1+d}\right)^N}\right] \tag{4}$$

The unit electrical cost can be calculated using the equation below [16]:

$$UC_{el} = \frac{ALCC}{365 \times E_{load}} \tag{5}$$

Where E_{Load} is the necessary amount of daily energy in kWh/day.

Table 4. Energy cost for the four scenarios

Scenario	1	2	3	4
Solar Panels Cost (SAR)	576,243	2,352,760	576,243	576,243
Inverters Cost (SAR)	153,073	153,073	153,073	145,899
DC/AC String Boxes Cost(SAR)	15,040	15,040	18,800	15,040
Switch Disconnects Cost (SAR)	1,263	1,263	1,639	1,263
Cable Cost (SAR)	112,800	112,800	150,400	112,800
Installation Cost (SAR)	57,626	57,626	57,626	86,435
Inflation Rate (%)	7	7	6	7
O&M Cost Rate (%)	3	3	3	5
Annual O&M Cost (SAR)	34,224	87,522	34,637	11,977
ALCC (SAR)	270,175	707,594	437,871	347,044
WTP Energy Consumption (kWh)	1,752,000	1,752,000	1,752,000	1,752,000
Energy Cost (SAR)	560,640	560,640	560,640	560,640
PV Energy (kWh)	1,186,250	1,186,250	1,186,250	1,186,250
Energy Purchased from SEC (SAR)	565,750	565,750	565,750	565,750
Energy Purchased Cost (SAR)	181,040	181,040	181,040	181,040
Total Cost (SAR)	451,215	888,634	618,911	528,084
Cost Savings (SAR)	109,425	-327,994	-58,271	32,556
Cost of Energy (SAR/kWh)	0.258	0.507	0.353	0.301

The results in Table 4 are obtained by considering alternative component cost and O&M cost ratio. From Table 4, the first scenario is considered to be optimal since it provides the highest profits of all four cases. The profits are estimated at 109,425 SAR. Moreover, 67% of the WTP energy is provided by the designed solar PV system. The final design economic parameters are listed in Table 5.

In addition to the energy cost savings, environmental benefits occur because of the design PV system. Table 6 displays the annual savings in oil consumption and CO_2 tax savings.

Table 5. Final design size and component cost

Component	Size	Number of items	Unit Cost
PV Panels	540 W	1200	480 SAR/unit
Inverters	110 kW	8	19,130 SAR/unit
DC Combiners	160 A /1500 V	8	1,200 SAR/unit
AC Combiners	400 A / 1000V	4	1,350 SAR/unit
Disconnect DC Switches	250 A	8	160 SAR/unit
Cables	1000 V	-	30,000 SAR
Installation Cost Rate	-	-	10%-20%
O&M/year Rate	-	-	10%

Table 6. Oil Production and CO_2 tax savings

	PV System Life Cycle Energy (MWh)	29,656
Life Cycle Results	Life Cycle Oil Saving (Barrels)	17,445
	Life Cycle Oil Savings (Million SAR)	6.673
	Life Cycle CO_2 Savings (ton)	23,725
	Life Cycle CO_2 Tax Savings (Million SAR)	6.673
Economic Assumptions	Barrel Energy content (kWh)	1,700
	Oil Price ($/Barrel)	102
	CO_2/kWh (kg)	0.8
	CO_2 Tax ($/ton) [17]	75

Conclusion

To meet KSA 2030 Vision which aims to produce electricity from clean energy resources and reduce carbon emissions that cause harm to the environment, and because of the high solar PV resources in Saudi Arabia, the authors recommend designing a solar PV power system for the 1 MW water treatment plant at KFU. This system is an on-grid solar PV system which gives an advantage to KFU in that if there is any excess of the generated electricity from the PV system, it is sold to the electricity company. One important fact about this project is that the authors have taken into consideration the environmental and societal impacts of this project.

For the optimal scenario, the total operation cost using the 650 kW PV system is estimated to be 451,215 SAR with savings of 109,425 compared to the grid alone scenario. This translated to 0.254 SAR/kWh.

In terms of oil and pollution and oil savings, the optimal results for adding the 650 kW PV system yield he life cylce cost anaylsis yield savings of more than 17,000 oil barrels and more than 23 thousand tons of CO_2. The corresponding monetary revenues are about 13.4 million SAR.

And finally found that the proposed design saves the earth's planet from 23,750 tons of CO_2 in the next 25 years.

Acknowledgment

The authors would like to thank King Faisal University for funding this research. The grant number will be provided in due time.

References

[1] "Greenskies Finishes 137-KW Solar Array at Water Treatment Plant." Solar Power World, 23 May 2019, https://www.solarpowerworldonline.com/2019/05/greenskies-finishes-137-kw-solar-array-at-water-treatment-plant/

[2] Cabello, Luisa. "Spanish Consortium Wants to Produce Green Hydrogen with Sewage." PV Magazine International, 16 Mar. 2022, https://www.pv-magazine.com/2022/03/16/spanish-consortium-wants-to-produce-green-hydrogen-with-sewage

[3] (PDF) Solar Power Potential in Saudi Arabia - Researchgate.https://www.researchgate.net/publication/271722313_Solar_Power_Potential_In_Saudi_Arabia

[4] "Solar Resource Maps of Saudi Arabia." Solargis, https://solargis.com/maps-and-gis-data/download/saudi-arabia.

[5] ACWA Power. Saudi Arabia Is Unlocking the Potential of Wind Energy, ACWA Power, 8 May 2019, https://acwapower.com/news/saudi-arabia-is-unlocking-the-potential-of-wind-energy/

[6] Khalid A. Althobiti, and Ahmed M. Nahhas, "Recent Advances of Hydropower Energy in Saudi Arabia." American Journal of Energy Research, vol. 8, no. 1 (2020): 1-6. doi: 10.12691/ajer-8-1-1.

[7] Information on https://www.moenergy.gov.sa/

[8] Information on https://www.redseaglobal.com/

[9] S., Rehman, et al. "Directory of Open Access Journals." FME Transactions, University of Belgrade - Faculty of Mechanical Engineering, Belgrade, 1 Jan. 2020, doaj.org/article/44039c24f9b44daf9659171b331e71bd.

[10] "IEC." IEC 62548:2016 | IEC Webstore | Solar Panel, Photovoltaic, PV, Solar Power, Rural Electrification, LVDC, https://webstore.iec.ch/publication/25949

[11] "IEC." IEC TS 62738:2018 | IEC Webstore, https://webstore.iec.ch/publication/26942.

[12] "IEC." IEC 62446-1:2016 | IEC Webstore | Rural Electrification, Solar Power, LVDC, https://webstore.iec.ch/publication/24057.

[13] "IEC." IEC 62109-2:2011 | IEC Webstore | Rural Electrification, Solar Panel, Photovoltaic, PV, Solar Power, Electricity, LVDC, https://webstore.iec.ch/publication/6471.

[14] "UL Standards Sales Site." Buy UL Standards Online | UL Standards Catalog | Comm 2000, https://standardscatalog.ul.com/ProductDetail.aspx?productId=UL1741.

[15] K. Basaran and H. Sırdaş, "Economic analysis of an on-grid photovoltaic system," international congress on energy research, 2018.

[16] JRC Photovoltaic Geographical Information System (PVGIS) - European Commission, 11 Jan. 2016, https://re.jrc.ec.europa.eu/pvg_tools/en/#PVP.

[17] Minh Ha Duong, "What is the Price of Carbon? Five definitions", S.A.P.I.EN.S [Online], 2.1 | 2009, Online since 03 June 2009, connection on 30 January 2023. URL: http://journals.openedition.org/sapiens/793

Experimental investigations on the thermo-mechanical properties of carbon-basalt-aramid/epoxy and glass-basalt-aramid/epoxy hybrid interply composites under different aging environments

Zekiye ERDOĞAN[1,a] and Munise Didem DEMİRBAŞ[2,b*]

[1] Erciyes University,Graduate School of Natural and Applied Sciences, Turkey

[2] Erciyes University, Engineering Faculty, Turkey

[a]zekiyeerdogan4@gmail.com [b]mddemirbas@erciyes.edu.tr

Keywords: Carbon-Basalt-Aramid Hybrid, Glass-Basalt-Aramid Hybrid, Thermal Expansion Coefficient, Tg

Abstract. In this study, Carbon-Basalt-Aramid (C-B-A) and Glass-Basalt-Aramid (G-B-A) samples were produced, and their thermal expansion coefficient (α), glass-transition temperature (Tg), and dimension change properties were investigated. In the study, thermo-mechanical properties of hybrid composites with and without 12 h and 48 h aging processes were determined. As a result, the properties obtained according to the results of TMA analysis of both hybrid composites with the aging process changed significantly.

1. Introduction

High-performance fiber-reinforced composites offer excellent strength and stiffness properties, but their relatively high material and manufacturing cost, and their brittle, catastrophic failure without sufficient warning, limit their use in high-volume applications such as mass-produced automotive and construction. To expand their use, the development of high-performance ductile or pseudo-ductile composites with safe failure mechanisms similar to metals, with detectable warning and a wide margin before final failure, is of significant interest. However, adding ductility to composite materials is challenging as both traditional constituents of high-performance long fiber-reinforced thermoset polymer matrix composites are brittle [1-3]. Researchers have investigated various approaches to improve the ductility of composites such as modified matrix systems, new ductile fibers, and modified composite architecture [4-5].

A common one of these approaches is with out-of-plane fluctuation, which can create extra stress before breaking by creating additional tension through the realignment of conventional laminated composites or allowing the fibers to be reorientated [6-9]. Thus, the fracture is delayed and nonlinearity can be created through controlled damage before eventual failure. The other can be accomplished by hybridizing commercially available unidirectional plies so that they can be used to maintain a high initial modulus and potentially cause a cascading failure, but this can often produce undesirable large load drops when lower strain fibers break [10-11]. Layered hybrid [12-13] and mixed hybrid [14-15] structures have been investigated to improve the brittleness of single fiber composites. Then, hybrid composites with more than one type of single-phase reinforcement gained importance due to the freedom of adaptation of single fiber type composites according to needs. Thus, high strength can be achieved with less weight and cost for extensive engineering applications.

Some of the recent research on these composites, called hybrids, is as follows:

Czel et al. [16] studied the optimum hybrid configuration of carbon/glass reinforced hybrid composites and emphasized that the increase in the number of carbon layers increases the ductility of the material. Wu et al. [17] studied the effects of hybridization on the bending performance of carbon/glass interlayer and interlayer composites and emphasized that interlayer hybrid composites containing less carbon fiber exhibit excellent bending behavior. Bhagwat et al. [18] obtained Young's modulus, tensile strength and compressive strength, and Poisson's ratio from tensile and compression tests of hybrid glass/carbon composites. Chelliah et al. [19] characterized various failure modes in glass, carbon hybrid composite laminates under uniaxial tensile loading with acoustic emission (AE) monitoring and presented individual failure modes and failure times. Aslan et al. [20] studied the tribological and mechanical performance of sisal-filled waste carbon and glass fiber hybrid composites. Melo et al. [21] studied the effect of Portland cement admixture on hybrid glass fiber-reinforced composites based on full factorial design. Ramlee et al. [22] studied the physical and morphological and mechanical properties of natural fiber-reinforced phenolic hybrid composites.

In recent studies, the investigation of mechanical and thermo-mechanical properties under the static and dynamic influence, including fibers other than carbon and glass, has also been the subject of many studies [23-25]. Supian et al. [26] investigated the high impact strength of a filament wound kenaf/glass hybrid composite energy absorption tube under medium velocity impact load conditions. They emphasized that the energy absorption value of the hybrid sample with high kenaf fiber fraction mass increased significantly. Wahab et al. [27] discussed the conceptual design of a glass/renewable natural fiber reinforced polymer hybrid composite motorcycle side cladding. They stated that the glass/coconut fiber reinforced polypropylene hybrid composite increases weight but reduces thickness due to higher density.

Sahu et al. [28] numerically investigated the time-dependent displacement behavior of glass-Carbon-Kevlar fiber-reinforced hybrid shell panels under thermo-mechanical loading. Charvani et al. [29] performed impact and Thermogravimetric analysis of the hybrid composite produced using carbon and Aramid fibers by adding alumina nanoparticles to the binder resin matrix. They observed that the thermal stability increased with nanoparticle loading and stated that the use of both fibers together increased the impact strength. Wu et al. [30] investigated the longitudinal and transverse shear properties of 3D braided hybrid composites using aramid and carbon fibers. They stated that the transverse shear property of the hybrid braided structure is superior to the longitudinal shear and showed the failure modes. Protchenko et al. [31] investigated the tensile and shear strength of basalt/carbon fibers and hybrid composite rods. They emphasized that hybrid rods have better tensile and shear strength than basalt fiber rods.

Yashas Gowda et al. [32] studied the effect of stacking order on the mechanical and thermal properties of flax, kevlar, carbon, and carbon-kevlar hybrid fiber composites. They investigated the thermal behavior by performing tensile, bending, interlayer shear, and impact tests for four-layer composite laminates with different stacking orders. Shishevan et al. [33] studied the mechanical and thermal properties of carbon fiber composites. They performed the different test to determine the tensile and bending behaviors. They determined that the viscoelastic properties of carbon fibers were significantly affected by the increase in thermal energy.

Grimurugan et al. [34] studied the thermal properties of carbon and kevlar-reinforced hybrid material. They determined the materials' thermomechanical properties, hardness, impact strength, bending strength, and thermal conductivity. Mishra et al. [35] investigated the mechanical and thermomechanical properties of basalt nanoparticle basalt/basalt reinforced hybrid composite material and basalt/jute reinforced hybrid composite material. For thermal stability, they performed different tests and emphasized that the basalt/basalt material with basalt nanoparticles had better mechanical and thermal properties.

Karacor and Ozcanlı [36] investigated the thermal and mechanical properties of basalt/carbon hybrid and basalt/aramid hybrid structures. They emphasized that the basalt/carbon hybrid is good in terms of tensile and microhardness. They also stated that the carbon fiber hybridization to basalt fabric greatly improves the deformation resistance and heat resistance of the materials compared to the aramid fabric hybridization process.

Raj et al. [37] fabricated the glass, basalt, aramid, and carbon hybrid composite with different interlayer configurations and investigated its natural frequency and damping behavior and stated that hybrid composites with carbon-glass-aramid fibers have the best damping properties.

Pai et al. [38] investigated the mechanical properties of basalt and aramid fiber hybrid composites under three different aging conditions and emphasized that moisture absorption has a negative effect. They showed that matrix segregation, matrix cracks, and interfacial bond rupture occur under the aging effect.

In this study, thermo-mechanical properties of carbon-basalt-aramid/epoxy and glass-basalt-aramid/epoxy hybrid composite materials such as α, Tg, and dimensional change properties under different aging conditions will be investigated. These hybrid composites to be produced with three different reinforcement elements have not been found in the detailed literature, and in this respect, our study will create an innovation.

2. Experimental process

In this study, the selection of reinforcement elements was made by considering the properties we want to achieve. Glass fiber (200 gr/m^2) with high strength and thermal insulation properties; carbon fiber (200 gr/m^2) for lightness, good thermal and electrical conductivity; basalt fiber (210 gr/m^2) for thermal strength; Aramid fiber (200 gr/m^2) was chosen for high friction and abrasion resistance. Hexion MGS® L 326 high-temperature system was used as resin and 770-NC was used as a mold release agent.

2.1 Hybrid interply composite sample production

Two hybrid composites were produced for this study. The first one is reinforced with G-B-A fiber and the second one is reinforced with C-B-A fiber. The samples are 3mm in total thickness and 500x500 mm in size. Each of the samples was produced using 16 fibers in 9 hours, including curing, and these fibers were sorted according to their manufacturability. The fibers placed in each layer of the plates in production were arranged by hand laying method and epoxy was applied and laminated with a deaeration roller. Between each fiber stack, 10 minutes was allowed for the gel time of the epoxy. After the epoxy process applied to each floor was finished, it was pressed between 120-150°C for 6 hours in order for the material to become a plate.

In this study, the thermo-mechanical behavior of the material was investigated by determining the α, Tg, and dimensional change properties under different temperatures by applying the aging process to the samples produced. The fiber array of two different hybrid composite samples with three reinforcement elements is shown in Figure 1.

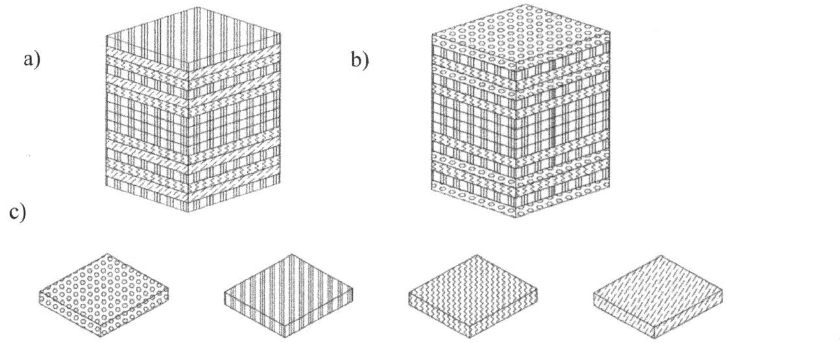

Figure 1. Fiber sequence of hybrid composite material a) G-B-A hybrid composite, b) C-B-A hybrid interply composite, c) fibers in samples (carbon, basalt, aramid, glass fibers, from left to right, respectively)

2.2 Aging process

12 and 48 h UV aging of hybrid composite samples were performed with PN-EN-ISO 4892-3 standard / fluorescent UV lamps and ATLAS ultraviolet TEST device. The aging process was carried out in an aging chamber with 8 fluorescent UV lamps (UVA 340, UVB 313, UVA 351). This instrument is designed to test resistance to photo-oxidation, i.e., ultraviolet A and ultraviolet B, the most destructive range to UV radiation. According to this test standard, it is designed to simulate the wear effects that occur when materials are exposed to global solar radiation or solar radiation in real end-user environments. The samples were cut in dimensions of 10mmx10mmx3mm before being tested. Two different hybrid samples were aged for 12 and 48 h (Figure 2). In the aging process, the samples were exposed to UVA -340 lamps and a radiation intensity of 0.75W/m^2. In the 12 h aging process, the samples were exposed to irradiation at 8 h and 50 °C, then left to rest for 3 h, 45 min, and below 50 degrees. After the listening process, 15 minutes of spraying was done, and 12 hours were completed. For 48 h aging, similar processes to the hybrid composite material were applied at different times.

Figure 2. C-B-A reinforced hybrid composite specimens from left to right, respectively; no aging process was applied, 12 h aging process was applied and 48 h aging process was applied.

2.3 Thermo-mechanical analysis (TMA)

TMA test in expansion mode with TA Instruments TMA Q400 device was performed by increasing the temperature with 5ºC/min heating rate under 0.02 N load in the range of -30 ºC to 120 ºC according to ASTM E831 standard. This device allows us to determine the α, Tg, and dimensional change properties that is, the thermo-mechanical behavior of the samples under a certain load, according to the parameters determined at different temperatures.

In these tests, the determination of α and Tg. TMA test was performed for 6 samples (no aging process, 12 h aging process, and 48 h aging process) for 2 different hybrid composites.

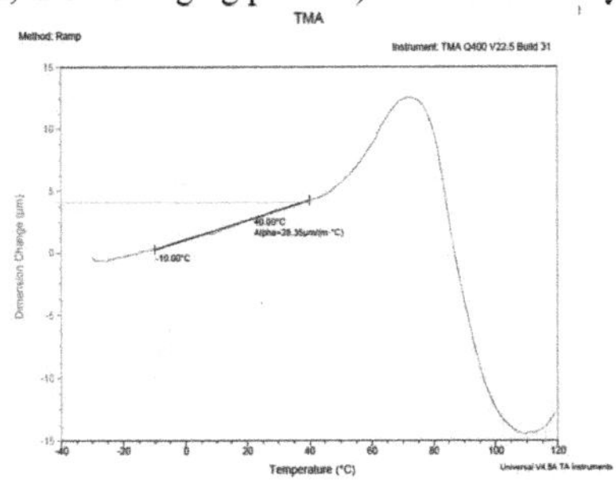

Figure 3. TMA result of C-B-A reinforced hybrid composite material without aging process.

The TMA results for the C-B-A hybrid composite are shown in Figure 3 for the sample without the aging process. Here the dimensional change was measured as 4.2 µm, α=28.35 µm/(m°C), and Tg=115 °C.

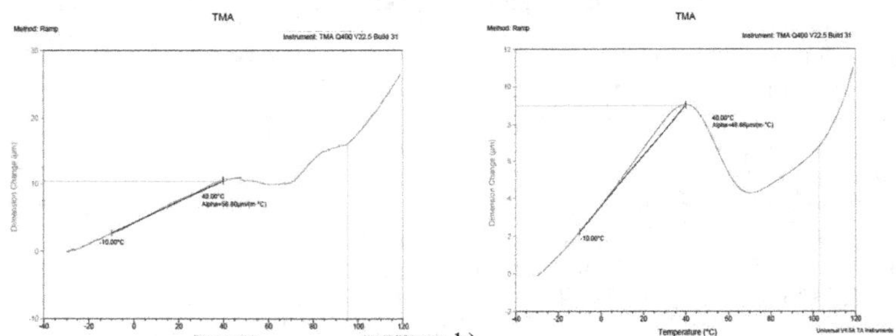

a) b)

Figure 4. TMA result of C-B-A reinforced hybrid composite material that has been aged for a) 12 h b) 48 h.

The TMA results for the C-B-A hybrid composite sample with the aging process applied for 12 h are shown in Figure 4-a). Here the dimensional change was measured as 10.3 µm, α=56.80 µm/(m°C), and Tg= 95°C. TMA result of C-B-A reinforced hybrid composite material that has been aged for 48 h. The TMA results for the C-B-A hybrid composite sample with the aging process applied for 48 h are shown in Figure 4-b). Here the dimensional change was measured as 9 µm, α=48.66 µm/(m°C), and Tg=102.5°C. The Tg for the C-B-A hybrid composite without the aging process, 12 h aging process, and 48 h aging process applied for the samples, respectively, is 115 °C, 95 °C, and 102.5 °C, this value decreases when the aging process is performed and decreases when the processing time is increased. Similarly, the expansion coefficients of the samples vary as α=28.35 µm/(m°C), α=56.80 µm/(m°C), and α=48.66 µm/(m°C), respectively. Dimensional change amounts are respectively; 4.2 µm, 10.3 µm, and 9 µm. In both values, depending on the aging application time, first an increase and then a decrease occurs. This shows that the thermo-mechanical properties of the C-B-A reinforced hybrid composite are significantly affected by aging.

Figure 5. G-B-A reinforced hybrid composite specimens from left to right, respectively; no aging process was applied, 12 hours aging process was applied and 48 hours aging process was applied.

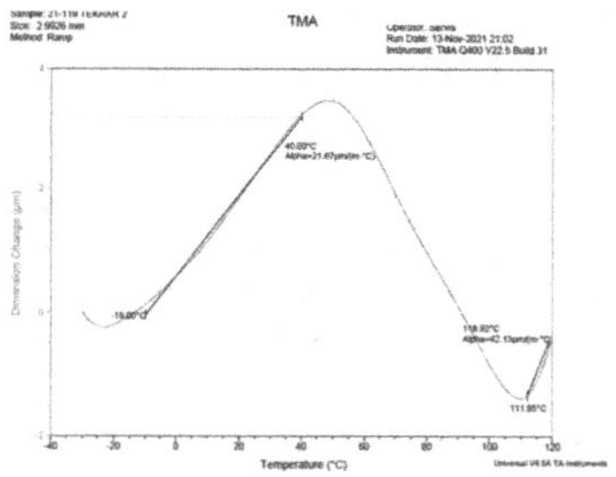

Figure 6. TMA result of G-B-A reinforced hybrid composite material without aging process.

The TMA results for the G-B-A hybrid composite are shown in Figure 6 for the sample without the aging process. Here the dimensional change was measured as 3.2 µm, α=21.67 µm/(m°C), and Tg=111.95 °C.

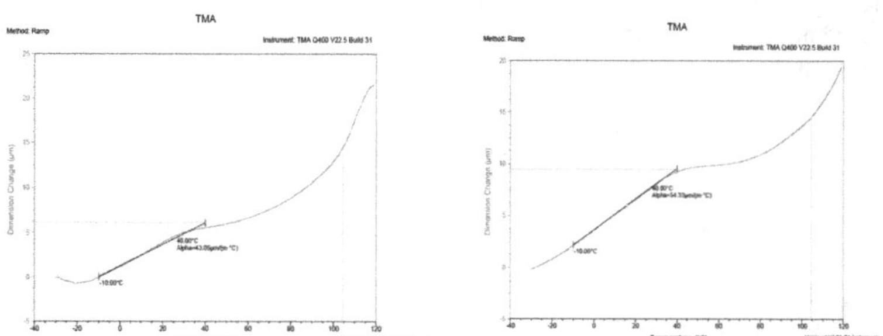

a) b)

Figure 7. TMA result of G-B-A reinforced hybrid composite material that has been aged for a) 12 h b) 48 h.

The TMA results for the G-B-A hybrid composite sample with the aging process applied for 12 h are shown in Figure 7-a. Here the dimensional change was measured as 6.25 µm, α=43.05 µm/(m°C), and the Tg=104.85°C. The TMA results for the G-B-A hybrid composite sample with the aging process applied for 48 h are shown in Figure 7-b. Here the dimensional change was measured as 9.4 µm, α=54.33 µm/(m°C), and Tg=105°C. The Tg for the G-B-A hybrid composite without the aging process, 12 h aging process, and 48 h aging process applied for the samples, respectively, is 111.95 °C, 104.85 °C, and 105 °C, this value increases when the aging process is performed. Similarly, the expansion coefficients of the samples vary as α=21.67 μm/(m°C), α=43.05 μm/(m°C), and α=54.33 μm/(m°C), respectively. Dimensional change amounts are respectively; 3.2 μm, 6.25 μm, and 9.4 μm. Both values increase depending on the aging application time. This shows that the thermo-mechanical properties of the G-B-A reinforced hybrid composite are significantly affected by aging.

Table 1. TMA test results for two hybrid composites

Aging process time [h]	Sample	Dimensional change [μm]	α [μm/(m°C),]	Tg [°C]
-	C-B-A	4.2	28.35	115
-	G-B-A	3.2	21.67	111.95
12	C-B-A	10.3	56.80	95
48	C-B-A	9	48.66	102.5
12	G-B-A	6.25	43.05	104.85
48	G-B-A	9.4	54.33	105

It is seen that the thermal properties of the G-B-A reinforced hybrid composite material with and without the aging process are more affected than the C-B-A reinforced hybrid composite material (Table 1). According to these results, the high-density fragility of glass fiber in G-B-A reinforced composite, combined with low thermal properties, high strength of aramid fiber, and high thermal resistance properties of basalt fiber, strengthened the weaknesses of glass fiber.

3. Result and conclusion

In this study, two different hybrid composite samples with three reinforcements were produced and the α, Tg, and dimensional change amounts of the composites were determined by TMA. In addition, the change in these properties was investigated by applying the aging process to both composite samples.

The dimensional change and α of the aged C-B-A hybrid composite material increased by an average of 2.3 times and 1.85 times, respectively, while the Tg decreased by an average of 0.855 times compared to the unaged condition. The dimensional change and α of the aged G-B-A hybrid composite material increased by an average of 2.44 times and 2.24 times, respectively, while the Tg decreased by an average of 0.93 times compared to the unaged condition. According to the results of this study, it has been shown that the aging process has a significant effect on the thermo-mechanical results presented and the effectiveness of the selection of each reinforcement element of the hybrid composite on these properties.

Acknowledgment

This work was supported by the Erciyes University Scientific Research Projects Foundation with Project No: FYL-2021-11082.

References

[1] K.K. Chawla. Composite materials:science and engineering, Springer-Verlag, New York, 2012. https://link.springer.com/book/10.1007/978-3-030-28983-6

[2] S. Boncel, M. Rajyashree, R.M. Sundaram, A.H. Windle, K.K.K. Koziol, Enhancement of the mechanical properties of directly spun CNT fibers by chemical treatment, ACS Nano. 5 (2011) 9339-9344. https://pubs.acs.org/doi/10.1021/nn202685x

[3] L. Mészáros, T. Turcsán, Development and mechanical properties of carbon fibre reinforced EP/VE hybrid composite systems, Periodica Polytech Mech Eng. 58 (2014) 127-133. https://doi.org/10.3311/PPme.7237

[4] M.G. Callens, L. Gorbatikh, I. Verpoest, Ductile steel fibre composites with brittle and ductile matrices, Composites Part A: Applied Science and Manufacturing. 61 (2014) 235-244. https://doi.org/10.1016/j.compositesa.2014.02.006

[5] M.G. Callens, P. De Cuyper, L. Gorbatikh, I. Verpoest, Effect of fibre architecture on the tensile and impact behaviour of ductile stainless steel fibre polypropylene composites, Composite Structures. 119 (2015), 528-533. https://doi.org/10.1016/j.compstruct.2014.09.028

[6] J.D. Fuller, M.R. Wisnom, Pseudo-ductility and damage suppression in thin ply CFRP angle-ply laminates, Composites Part A: Applied Science and Manufacturing. 69 (2015) 64-71. https://doi.org/10.1016/j.compositesa.2014.11.004

[7] J.D. Fuller, M. Jalalvand, M.R. Wisnom, Combining fibre rotation and fragmentation to achieve pseudo-ductile CFRP laminates, Composite Structures. 142 (2016) 155-166. https://doi.org/10.1016/j.compstruct.2016.01.073

[8] G. Grail, S. Pimenta, S.T. Pinho, P. Robinson, Exploring the potential of interleaving to delay catastrophic failure in unidirectional composites under tensile loading, Composites Science and Technology. 106 (2015) 100-109. https://doi.org/10.1016/j.compscitech.2014.11.006

[9] G. Czel, S. Pimenta, M.R. Wisnom, P. Robinson Demonstration of pseudo-ductility in unidirectional discontinuous carbon fibre/epoxy prepreg composites Composites Science and Technology. 106 (2015) 110-119. https://doi.org/10.1016/j.compscitech.2014.10.022

[10] N.L. Hancox. Fibre composite hybrid materials, Applied Science Publishers Ltd., London, 1981.

[11] A.R. Bunsell, B. Harris, Hybrid carbon and glass fibre composites, Composites. 5(4) (1974), 157-164. https://doi.org/10.1016/0010-4361(74)90107-4

[12] Hayashi T. Development of new material properties by hybrid composition. 1st report. Fukugo Zairyo (Composite Materials). (1972) 1 18-20. https://cir.nii.ac.jp/crid/1573950398981322752

[13] Hayashi T, Koyama K, Yamazaki A, Kihira M, Development of new material properties by hybrid composition. 2nd report. Fukugo Zairyo (Composite Materials). (1972) 1 21-25. https://cir.nii.ac.jp/crid/1570854174237505408

[14] G. Marom, S. Fischer, F.R. Tuler, H.D. Wagner, Hybrid effects in composites: conditions for positive or negative effects versus rule-of-mixtures behaviour, Journal of Materials Science. (1978) 13 (7) 1419-1426. https://link.springer.com/article/10.1007/BF00553194

[15] N. Svensson, Manufacturing of thermoplastic composites from commingled yarns – a review, Journal of Thermoplastic Composite Materials. (1998) 11 (1) 22-56. https://doi.org/10.1177/089270579801100101

[16] G. Czél, M. Jalalvand, M. Wisnom, Design and characterisation of advanced pseudo-ductile unidirectional thin-ply carbon/epoxy glass/epoxy hybrid composites, Composite Structures. (2016) 143 362-370. https://doi.org/10.1016/j.compstruct.2016.02.010

[17] W. Wu, Q. Wang, A. Ichenihi, Y. Shen, W. Li, The effects of hybridization on the flexural performances of carbon/glass interlayer and intralayer composites, Polymers. (2018) 10(5) 549. https://doi.org/10.3390/polym10050549

[18] P.M. Bhagwat, M. Ramachandran, P. Raichurkar, Mechanical properties of hybrid glass/carbon fiber reinforced epoxy composites. Materials Today: Proceedings. (2017) 4(8) 7375-7380. https://doi.org/10.1016/j.matpr.2017.07.067

[19] S.K. Chelliah, S.K. Kannivel, A. Vellayaraj, Characterization of failure mechanism in glass, carbon and their hybrid composite laminates in epoxy resin by acoustic emission monitoring, Nondestructive Testing and Evaluation. (2019) 34(3), 254-266. https://doi.org/10.1080/10589759.2019.1590829

[20] M. Aslan, M. Tufan, T. Küçükömeroğlu, Tribological and mechanical performance of sisal-filled waste carbon and glass fibre hybrid composites, Composites Part B: Engineering. (2018) 140, 241-249. https://doi.org/10.1016/j.compositesb.2017.12.039

[21] A.B.L. Melo, T.H. Panzera, R.T.S. Freire, F. Scarpa, The effect of Portland cement inclusions in hybrid glass fibre reinforced composites based on a full factorial design, Composite Structures. (2018) 202, 233-240. https://doi.org/10.1016/j.compstruct.2018.01.069

[22] N.A. Ramlee, M. Jawaid, E.S. Zainudin, S.A.K. Yamani, Tensile, physical and morphological properties of oil palm empty fruit bunch/sugarcane bagasse fibre reinforced phenolic hybrid, composites, Journal of Materials Research and Technology. (2019) 8(4), 3466-3474. https://doi.org/10.1016/j.jmrt.2019.06.016

[23] N.M. Nurazzi, M.R.M. Asyraf, S. Fatimah Athiyah, S.S. Shazleen, S.A. Rafiqah, M.M. Harussani, S.H. Kamarudin, M.R. Razman, M. Rahmah, E.S. Zainudin et al., A Review on Mechanical Performance of Hybrid Natural Fiber Polymer Composites for Structural Applications, Polymers. (2021) 13 2170. https://doi.org/10.3390/polym13132170

[24] M.J. Suriani, R.A. Ilyas, M.Y.M. Zuhri, A. Khalina, M.T.H. Sultan, S.M. Sapuan, C.M. Ruzaidi, F.N. Wan, F. Zulkifli, M.M. Harussani, M.A. Azman, F.S.M. Radzi, S. Sharma, Critical Review of Natural Fiber Reinforced Hybrid Composites: Processing, Properties, Applications and Cost, Polymers. (2021) 13 3514. https://doi.org/10.3390/polym13203514

[25] L. Prabhu, V. Krishnaraj, S. Sathish, S. Gokulkumar, N. Karthi, L. Rajeshkumar, D. Balaji, N. Vigneshkumar, K.S. Elango, A review on natural fiber reinforced hybrid composites: Chemical treatments, manufacturing methods and potential applications, Materials Today Proceedings. (2021) 45 8080-8085. https://doi.org/10.1016/j.matpr.2021.01.280

[26] A.B.M. Supian, S.M. Sapuan, M. Jawaid et al. Crashworthiness Response of Filament Wound Kenaf/Glass Fibre-reinforced Epoxy Composite Tubes with Influence of Stacking Sequence under Intermediate-velocity Impact Load, Fibers and Polymers. (2022) 23 222-233. https://doi.org/10.1007/s12221-021-0169-9

[27] M. A. F. A. Wahab, S. M. Sapuan, M. M. Harussani, M. Y. M. Zuhri, A. A. Saleh, Conceptual Design of Glass/Renewable Natural Fibre-Reinforced Polymer Hybrid Composite Motorcycle Side Cover, Natural Fibre Composites: Design, Materials Selection and Fabrication, Journal of Renewable Materials. (2021) 9(11) 1973-1989. https://doi.org/10.32604/jrm.2021.016221

[28] P. Sahu, N. Sharma, H.C. Dewangan, S.K. Panda, Thermo-Mechanical Transient Flexure of Glass-Carbon-Kevlar-Reinforced Hybrid Curved Composite Shell Panels: An Experimental Verification, International Journal of Applied Mechanics. (2022) 14(1) 2150120. https://doi.org/10.1142/S1758825121501209

[29] S. Charvani, Ch. Gopal Reddy, G. Narendar, Investigation on Impact and Thermogravimetric Analysis of Aramid and Carbon based Nanocomposites, IOP Conference Series: Materials Science and Engineering. (2021) 1126 012042. https://doi.org/10.1088/1757-899X/1126/1/012042

[30] L. Wu, X. Sun, C. Xiang, W. Wang, F. Zhang, Q. Jiang, Y. Tang, J.-H. Lin, Short Beam Shear Behavior and Failure Characterization of Hybrid 3D Braided Composites Structure with X-ray Micro-Computed Tomography, Polymers. (2020) 12 1931. https://doi.org/10.3390/polym12091931

[31] K. Protchenko, F. Zayoud, M. Urbański, E. Szmigiera, Tensile and Shear Testing of Basalt Fiber Reinforced Polymer (BFRP) and Hybrid Basalt/Carbon Fiber Reinforced Polymer (HFRP) Bars, Materials (Basel). (2020) 13(24) 5839. https://doi.org/10.3390/ma13245839

[32] T.G. Yashas Gowda A. Vinod, P. Madhu, S. Mavinkere Rangappa, S. Siengchin, M. Jawaid, Mechanical and thermal properties of flax/carbon/kevlar based epoxy hybrid composites, Polymer Composite. (2022) 43(8) 5649. https://doi.org/10.1002/pc.26880

[33] F.A. Shishevan, H. Akbulut, M.A. Mohtadi-Bonab, Thermal shock behavior of twill woven carbon fiber reinforced polymer composites, Journal of Composites Science. (2021) 5(1) 33. https://doi.org/10.3390/jcs5010033

[34] R. Grimurugan, N. Senniangiri, B. Pitchia Krishnan, Dr.S. Kavitha, M. Vairavel, Tensile behaviour of hybrid polymer composites, IOP Conference Series. Materials Science and Engineering. (2021) 1059 012033. https://doi.org/10.1088/1757-899X/1059/1/012033

[35] R. Mishra, H. Jamshaid, J. Militky, Basalt nanoparticle reinforced hybrid woven composites mechanical and thermo-mechanical performance, Fibers and Polymers. (2017) 18(12) 2433-2442. https://doi.org/10.1007/s12221-017-7545-5

[36] B. Karacor, M. Ozcanli, Thermal and mechanical characteristic investigation of the hybridization of basalt fiber with aramid fiber and carbon fiber, Polymer Composites. (2022) 43(11) 8529. https://doi.org/10.1002/pc.27022

[37] K. Raj, A. Vasudevan, V. Mohanavel, Fast Fourier Transform (FFT) analysis on different inter-ply configurations of glass/basalt/aramid/carbon hybridized composites for aircraft structures, Materials Today: Proceedings. (2021) 37(2) 2375-2381. https://doi.org/10.1016/j.matpr.2020.08.079

[38] Y. Pai, D. Pai K M. Vijaya Kini, E. Wong, Experimental investigations on the moisture absorption and mechanical behaviour of basalt-aramid/epoxy hybrid interply composites under different ageing environments, Cogent Engineering. (2022) 9 2080354. https://doi.org/10.1080/23311916.2022.2080354

Critical review on ethanol producing feedstock and methods

Jaanvi Garg[1,a], Akshay Jain[1,b], Avadhoot Abaso Mohite[1,c], Bhaskor Jyoti Bora[1*,d], Prabhakar Sharma[2,e], Bhaskar Jyoti Medhi[3,f], Debrata Barik[4,g]

[1]Energy Institute Bengaluru, Centre of Rajiv Gandhi Institute of Petroleum Technology, Bengaluru, Karnataka-562157, India

[2]Delhi Skill and Entrepreneurship University, Delhi-110089, India

[3]Assam Energy Institute, Centre of Rajiv Gandhi Institute of Petroleum Technology, Sivsagar, Assam-785697, India

[4]Department of Mechanical Engineering, Karpagam Academy of Higher Education, Coimbatore-641021, India

[a]garg27jaanvi@gmail.com, [b]ajainmech@gmail.com, [c]aamandindia@gmail.com, [d]bhaskorb3@gmail.com, [e]psharmahal@gmail.com, [f]m.bhaskarjyoti@gmail.com, [g]debabrata93@gmail.com

Keywords: Ethanol production, Biomass, Bioethanol

Abstract. Dwindling of fossil fuels and the change in global climate have led humankind to explore alternative energy resources. In the force to explore inexhaustible and sustainable resources, research in the domain of solar, wind, tidal, biomass and geothermal are delved into. The energy derived from these various domains proved to be the solution to the greater cause of climate change. Among all these various sources of energy, biomass have tremendous potential. Bioethanol produced from biomass is an attractive biofuel having great prospect for energy security and environmental safety over fossil fuels. The conversion process of biomass to ethanol have improved significantly over the years to make it a viable option in transportation sector thereby reducing dependency on petrol and diesel. The major steps in typical conversion process include pre-treatment, enzymatic hydrolysis, and fermentation of sugars, greatly influenced by microbial strains. Yeast saccharomyces cerevisiae and a bacterial species, zymomonas mobilis, are the two most widely used in ethanol fermentation technology. The final yield of ethanol is directly dependent on the various factors affecting the production process. This comprehensive review study gives an overview of second-generation bioethanol production in regard with the various factors affecting its overall production process.

Introduction

The two significant challenges currently present in the world are (1) providing enough energy sources and (2) protecting the environment. In the past few decades, humankind's intense use of difficult-to-recover petroleum reserves for commercial and personal use has heightened environmental and energy security issues. The dependency on fossil fuels by the world economy is because it is the primary source of energy (about 80%) [1]. Uncontrolled use of non-renewable fuels harms nature because of the greenhouse emissions and environmental pollution resulting in global warming [2]. In 2021 globally coal accounted for over 40% of the overall growth in CO2 emissions. The highest increase in CO_2 emissions in 2021 occurred in heat production and electricity sector since the utilization of fossil fuels shoot up to help meet the growth of electricity demand [3]. It is imperative to slow down fossil fuel dependency as it is a limited resource and will get exhausted soon. Thus, increasing demands of fuel and extensive climate change issues are the main driving factors that is shifting the world's attention towards finding alternative energy resources. These alternative energy resources can result in reduced dependency on fossil fuels and thereby pave the way towards sustainability. Regarding this, The Sustainable Development Goals (SDGs) were adopted by United Nations in 2015 to end hunger, make energy affordable and clean, form sustainable cities and communities, and ensure that by 2030 people all around the globe enjoy peace and prosperity [4]. Also, governments are promoting the concept of circular economy, in which the primary motive is to ensure resource efficiency and less waste. It is a new way of producing value where resources are used more efficiently by using them multiple times instead of just once [5].

Alternative energy sources are prime solution to tackle the increasing prices of fossil fuel supplies. This is the reason production of biofuel is advocated as a resolution to escalating crude oil costs and viable choice to non-renewable resources. To enhance India's energy security, Government of India aim to reduce import dependency on fuel by extensively encouraging the Ethanol Blended Petrol (EBP) Programme. India has achieved the target of 10% ethanol blending in petrol in June 2022, five months ahead of the tentative schedule [6]. This policy further is amended in last year, namely National Policy on Biofuels-2018 Amendment, 2022 as per which oil companies shall sell Ethanol Blended Petrol (EBP) with ethanol percentage being 20% throughout the country from 1st April 2023 [7]. As per Pradhan Mantri Ji-Van Yojana, lignocellulosic surplus biomass availability in India is about 12-16 crore tons per annum which if utilized properly, there's possibility to yield 2500 to 3000 crore litres of ethanol annually [8]. This can further bring down India's dependency on imported crude oil significantly. In the effort to combat increasing fuel prices and pollution, flex fuels have been graining footing as alternative and environment friendly fuels. Usage of ethanol as flex-fuel will curb the dependency on imported crude oil as well as carbon footprints. In 2022, Toyota launched first of its kind project on Flexi-Fuel Strong Hybrid Electric Vehicles (FFV-SHEV) which would function on 100% petrol as well as 20% to 100% blended ethanol and electric power [9].

Biofuels are generally divided into first generation, second generation, third generation and fourth generation. Bioethanol obtained from first-generation biofuel comes from starchy food crops, namely sugar cane, wheat, maize. Thus, its production disputes with the accessibility of food and cultivable land [10]. Bioethanol obtained from lignocelluloses are biofuels belonging to the second generation, namely sugarcane bagasse, corn husks, and rice husks. Third generation biofuels are obtained from various species of algae biomass. Algae has a high lipid content and requires very less space for production, which makes them a viable option for biofuel production. Also, the fourth-generation biofuels are algae which are modified genetically [11]. The availability of lignocellulosic is in large quantities and is also easily accessible which is why biofuels produced from these biomasses by means of the alcoholic fermentation of sugar is getting more attention [12,13]. Lignocellulosic biomass (LCM) mainly consists of cellulose (~50%), hemicelluloses (~30%) and lignin (~20%). These components together constitute about 90% of the dry matter in lignocelluloses, and the remaining percentage is composed of extract and ash. These energy sources have tremendous potential in the shape of biofuel, more specifically bioethanol [14]. However, the conversion of sugars, starch and lignocellulosic biomass into bioethanol is different, regarding obtaining sugar solutions. To get fermentable sugars, raw materials which are sugar based require an extraction process, whereas starchy crops must undergo hydrolysis for the conversion of starch into glucose. In the case of lignocellulosic biomass, pre-treatment before hydrolysis is conducted to alter the cellulose structures for smooth enzyme accessibility [15].

Production process of ethanol from lignocellulosic biomass

Lignocellulosic are formed of three main components namely cellulose, hemicellulose and lignin. Except lignin, composition of cellulose and hemicellulose is of chains of sugar molecules. These chains when hydrolyzed leads to the production of monomeric sugars. Some part of monomeric sugars can be further fermented using baker's yeast. To make the production economically successful, high ethanol yield is a prime requirement. Lignocellulosic biomasses are abundant and easily available and is a great alternative to petroleum and other natural resources. These are primarily comprised of non-food crops (namely switchgrass and miscanthus), agricultural wastes (namely rice husks, sugarcane bagasse, coconut husks), forest residues, energy crops. The chemical composition of lignocellulosic biomass consists of cellulose (30–60%), hemicellulose (20–40%) and lignin (10–25%), along with tiny portion of ash, protein, and extractives [16]. Biological transformation of lignocellulosic biomass to bioethanol has attracted more and more attention over the period of years. The procurement of bioethanol involves all aspects of the key processing steps viz., pre-treatment, enzymatic hydrolysis, and fermentation [17]. However, its conversion to second generation ethanol is very demanding in comparison to first generation ethanol essentially owning to impenetrable characteristics of biomass matrix, ineffective disintegration of carbohydrate-lignin structure composite, high lignin content. For the conversion of lignocellulosic biomass, most commercial second-generation production units involves downstream and upstream technologies. The pathways of production initiate with upstream pre-processing of biomass feedstocks (e.g., sieving, cutting, milling, size reduction, grinding); pre-treatment for the eradication of the legitimate roadblock and further making it attainable to enzymes. After this enzymatic saccharification is done which releases free monomeric sugars. In case of downstream technologies, the fermentation of monomeric sugars is conducted via microorganisms. The obtained fermented broth is then distilled and further dried to obtain complete ethanol for blending motives [18].

Pre-treatment process used for Bioethanol process

Pre-treatment step alters the macroscopic, sub–microscopic and microscopic structures of the biomass. This step is costly and critical. Pre-treatment process remove hemicellulose along with lignin from the lignocellulosic matrix. It further decreases the crystallinity of cellulose, at the same time increasing the surface area of biomass [19]. To facilitate the hydrolysis of cellulose for final production of ethanol, various pre-treatment techniques have been developed namely physical, physiochemical, and chemical pre-treatments. Physical treatment includes hydrolysis and comminution whereas chemical pre-treatment is conducted using alkali, acid, ozone, and solvent based procedures. Another type of pre-treatment namely physiochemical exists which is basically the amalgamation of chemical and physical methods such as ammonia fibre expansion and steam explosion. The main purpose of this procedure is to disarray the complex structures of lignocellulosic biomass and its major components into simple sugars, which will ensure fruitful bioconversion process. The various pre-treatment methods namely include steam explosion, acid pre-treatment, alkaline pre-treatment, thermal pre-treatment, ammonia fibre explosion (AFEX), wet oxidation pre-treatment, carbon-dioxide explosion pre-treatment, ozonolysis pre-treatment, organosolvent pre-treatment and aqueous ammonia solution pre-treatment. In recent times new pre-treatment methods have been developed to value lignocellulosic biomass. For the separation of lignocellulosic feedstock to produce pure cellulose and lignin portions in a feasible biorefinery, deep eutectic solvent (DES) is a prominent alternative proposition. This approach reveals the ionic characteristics which are tough and being ideal like ionic liquids for recovery and reuse [17].

Cellulose Hydrolysis

Hydrolysis process is conducted after pre-treatment to crack down the biomass into fermentable sugars. The two hydrolysis methods which are most used are acidic and enzymatic. Acid hydrolysis is categorized into dilute and concentrated hydrolysis. Dilute acid hydrolysis is conducted using lower acid concentration at elevated temperature. On the other hand, concentrated acid hydrolysis is done in high acid concentration at lower temperature. Enzymatic hydrolysis needs enzymes to hydrolyze the feedstocks into fermentable sugars. The three commonly used types of enzymes are endo-β—1,4-glucanases, cello-biohydrolases and β-glucosidases. The cellulase enzyme activity is impacted by the source of enzyme and its concentration. The degradation of cellulose into reducing sugars will be at mild conditions of pH: 4.8-5.0, temperature: 45-50 °C. The enzymatic hydrolysis efficiency is greatly influenced by optimized conditions such as time, temperature, pH, substrate concentration and enzyme loading [20]. Enzymatic hydrolysis is a relatively slow procedure; however, it has the upper hand of no by-product being throughout the entire procedure. The use of cellobiase of Trichoderma reesei and cellulase has been reported to hydrolyze sugarcane after alkaline delignification [21]. However, to exhibit the fibers for enzyme accessibility pre-treatment is necessary before enzymatic hydrolysis.

Fermentation

To produce alcohol, fermentation of hydrolysate from lignocellulosic biomass is prime necessity. Regarding this, pentose-utilizing negative yeast strains and pentose-utilizing yeast strains have been used. Zymomonas mobilis, Escherichia coli, Kluyveromyces marxianus, Z. bailii, B. claussenii, Scheffersomyces stipitis and S. cerevisiae, microbial strains have been build using recombinant DNA technology as they are competent of utilizing hexose and pentose sugars making the entire process efficient [22–24]. Fermentation of biomass is conducted mostly via separate hydrolysis and fermentation (SHF) and simultaneous hydrolysis and fermentation (SSF). In the separate hydrolysis and fermentation (SHF) process two steps are involved: hydrolysis and fermentation. In hydrolysis method, sugars are released from cellulose after pre-treatment. In fermentation, microbial strain used for conversion, converts released sugar into ethanol. Each and all steps are conducted separately at optimal conditions. In case of simultaneous hydrolysis and fermentation (SSF), hydrolysis and fermentation processes are combined and performed in one vessel simultaneously. The inhibition caused by the accumulation of sugars are eradicated in this process as the sugars released are transformed into bioethanol instantly with the help of appropriate microbial strains. The eradication of sugar accumulation results in increasing the extent of hydrolysis process and overall decrease the time taken in the step [25].

Literature review on bioethanol production

This section discusses the process and quantity of bioethanol production utilizing feedstocks from various agricultural and industrial fields, refer Table 1. Table 1 also discusses the process of pretreatments at varying conditions while using enzymatic hydrolysis and fermentation process for maximizing the bioethanol yield.

Table 1. Literature Review on Bioethanol Production

Feedstocks	Descriptions	References
Wheat straw	Subcritical water pre-treatment is conducted (220.5 °C; 22.0 min). Further feedstock undergoes separate high solid (15%) hydrolysis and fermentation (SHF) is done. Bioethanol obtained 37.0 g/L	Chen et al. [26]
Bamboo	Dendrocalamus sinicus species of bamboo is subjected to sulfomethylation-aided phosphoric acid pre-treatment. Cellic Ctec2 and baker's yeast is used for fermentation. Ethanol obtained is 13.26 g/L	Jin et al. [27]
Whole plant cassava	Hydrothermal pre-treatment is conducted (180 °C; 2 MPa; 60 min). Fermentation of C_5 integrated cellulosic sugar and starch in simultaneous saccharification and fermentation method (SSF).	Lyu et al. [28]

Polyploidy potato straw	Two pre-treatments (8 min liquid hot water; 5% CaO) at minimized conditions. Maximum bioethanol yield of 24 % (% dry matter)	Madadi et al. [29]
Wheat and rye stillages	Pre-treatment method is microwave-assisted with dilute acid. Produced >156 mg/g glucose, rated microwave power 300 W (15 min, 54 PSI in 24 h process), 48 h of fermentation with *S. cerevisiae*. Bioethanol obtained is 20 g/L.	Mikulski and Kłosowski [30]
Cotton stalk	Pre-treatment is done in hydrothermal and organosolv processes, thereafter pre-hydrolysis with use of 80 FPU/g cellulose (at 50 ◦C, at pH 5.0, for 6 h). After pre-treatment 15 mg yeast per gram undergoes fermentation at 30 °C with pH level of 5.0. Bioethanol obtained is 47.0 g/ L	Dimos et al. [31]
Sugarcane Bagasse	Aqueous ammonia soaking pre-treatment is done (50 °C for 48 hr). Separate hydrolysis and fermentation: hydrolysed via fed batch at 40% high solids loading. Microorganism used is Cellic Ctec2. Glucose and xylose (96 hr hydrolysis) obtained is 157.65 g/l and 57.12 g/l respectively.	Raj and Krishnan [32]
Sugarcane Bagasse	Pre-treated with hydrodynamic cavitation, further assisted alkaline hydrogen at optimum conditions. Enzymes used for hydrolysis is cellulase. For fermentation Scheffersomyces stipitis NRRL-Y7124 is used. Bioethanol produced is 31.50 g/L.	Hilares et al. [33]
Eucalyptus biomass	At 150 °C, alkaline extrusion pre-treatment is conducted to achieve highest xylan and glucan conversion through enzymatic hydrolysis, about attaining 40% and 70% yields.	Duque et al. [34]
Sugarcane Bagasse	During pre-treatment process at 140 °C for 1 hr, 5% Na2CO3 solution was used. In the hydrolysis step glucose obtained was 97.6%. At 37 °C for 72 h fermentation is done thereby bioethanol obtained is 7.27 g/L.	Nosratpour et al. [35]
Corncob	Spathaspora passalidarum U1-58, is used to ferment the pre-treated corncob. It uses pentoses and hexoses. Here, SSF gave higher yield value than SHF. After 96 hr of fermentation values obtained were 42.46 g/L and 53.24 g/L, respectively	Yu et al. [36]
Water Hyacinth	Pre-treatment is done with dilute H_2SO_4 separate hydrolysis and fermentation. Fermentation is done by the combination of S.cerevisiae (MTCC 173) and Z. mobilis (MTCC 2428). The obtained ethanol was 13.6 mg/ml.	Das et al. [37]
Green coconut husk fibers	The pre-treatment method used is alkaline method, followed by enzymatic hydrolysis. In fermentation process *S. cerevisiae* were used. For fermentable sugars the bioethanol conversion efficiency was high as 59.6%.	Cabral et al. [38]

Water Hyacinth	NaOH pre-treatment is done, followed by separate hydrolysis and fermentation and simultaneous hydrolysis and fermentation. Strain used is control S. cerevisiae and Kluyveromyces marxianu K213. Maximum ethanol yield obtained is 7.34 g/L.	Yan et al. [39]
Rice straw	CO_2 gets incorporated in ammonia explosion pre-treatment at conditions: 14.3% ammonia, 2.2MPaCO_2, 165.1°C for 69.8 min. The obtained glucose yield is 93.6%. Bioethanol obtained is 97% reportedly after simultaneous saccharification and fermentation method (SSF)	Cha et al. [40]

Conclusion

Lignocellulosic wastes are a great source to produce bioethanol. However, the main challenge of researchers is to attain greater yield of bioethanol via economically achievable pre-treatment/hydrolysis processes. Pre-treatment process of the feedstocks is a very crucial footstep in bioethanol generation. These methods in most cases are costly and high energy consuming, which acts as a vital obstacle in the application of these propitious feedstocks on a commercially grand measure. To make it more effective, process integrations are conducted, however in practical scenario to make it more economically viable implementation of these advances in small scale or commercial scale would be more suitable regarding growing a sustainable lignocellulosic ethanol industry.

References

[1] Tyagi S, Lee K-J, Mulla SI, Garg N, Chae J-C. Production of Bioethanol from Sugarcane Bagasse: Current Approaches and Perspectives. Appl. Microbiol. Bioeng., Elsevier; 2019, p. 21–42. https://doi.org/10.1016/B978-0-12-815407-6.00002-2

[2] Sánchez ÓJ, Cardona CA. Trends in biotechnological production of fuel ethanol from different feedstocks. Bioresour Technol 2008;99:5270–95. https://doi.org/10.1016/j.biortech.2007.11.013

[3] Global Energy Review: CO2 Emissions in 2021. IEA 2022.

[4] The SDGS in action. Sustainable development goals. Accessed on February 06, 2023. UNDP 2022. Information on https://www.undp.org

[5] Sohal A, Nand AA, Goyal P, Bhattacharya A. Developing a circular economy: An examination of SME's role in India. J Bus Res 2022;142:435–47. https://doi.org/10.1016/j.jbusres.2021.12.072

[6] India has achieved the target of 10 percent ethanol blending, 5 months ahead of schedule. Ministry of Petroleum & Natural Gas. PIB 2022.

[7] National Policy on Biofuels-2018 Amendment, 2022. The Gazette of India. Minist Pet Nat GAS Notif 2022.

[8] "Pradhan Mantri Jaiv Indhan- Vatavaran Anukool fasal awashesh Nivaran (JI-VAN) Yojana. Ministry of Petroleum and Natural Gas notification. New Delhi: The Gazette of India; 2019.

[9] Shri Nitin Gadkari Launches first of its kind pilot project on Flexi-Fuel Strong Hybrid Electric Vehicles (FFV-SHEV) in India. Ministry of Road Transport & Highways. Delhi: PIB; 2022.

[10] Zhao W, Zhao F, Zhang S, Gong Q, Chen G. Ethanol production by simultaneous saccharification and cofermentation of pretreated corn stalk. J Basic Microbiol 2019;59:744–53. https://doi.org/10.1002/jobm.201900117

[11] Sivagurunathan P, Kumar G, Mudhoo A, Rene ER, Saratale GD, Kobayashi T. Fermentative hydrogen production using lignocellulose biomass: An overview of pre-treatment methods, inhibitor effects and detoxification experiences. Renew Sustain Energy Rev 2017;77:28–42. https://doi.org/10.1016/j.rser.2017.03.091

[12] Tiwari R, Nain L, Labrou NE, Shukla P. Bioprospecting of functional cellulases from metagenome for second generation biofuel production: a review. Crit Rev Microbiol 2018;44:244–57. https://doi.org/10.1080/1040841X.2017.1337713

[13] Zhang M, Shukla P, Ayyachamy M, Permaul K, Singh S. Improved bioethanol production through simultaneous saccharification and fermentation of lignocellulosic agricultural wastes by Kluyveromyces marxianus 6556. World J Microbiol Biotechnol 2010;26:1041–6. https://doi.org/10.1007/s11274-009-0267-0

[14] Bagewadi ZK, Mulla SI, Ninnekar HZ. Optimization of endoglucanase production from Trichoderma harzianum strain HZN11 by central composite design under response surface methodology. Biomass Convers Biorefinery 2018;8:305–16. https://doi.org/10.1007/s13399-017-0285-3.

[15] Bušić A, Marđetko N, University of Zagreb, Faculty of Food Technology and Biotechnology, Pierottijeva 6, 10000 Zagreb, Croatia, Kundas S, Belarussian National Technical University, Power Plant Construction and Engineering Services Faculty, Nezavisimosti Ave. 150, 220013 Minsk, Belarus, Morzak G, et al. Bioethanol Production from Renewable Raw Materials and its Separation and Purification: a Review. Food Technol Biotechnol 2018;56. https://doi.org/10.17113/ftb.56.03.18.5546

[16] Su T, Zhao D, Khodadadi M, Len C. Lignocellulosic biomass for bioethanol: Recent advances, technology trends, and barriers to industrial development. Curr Opin Green Sustain Chem 2020;24:56–60. https://doi.org/10.1016/j.cogsc.2020.04.005.

[17] Chandra MS, Srinivasulu M, Yadav PS, Ramesh B, Kameswaran S, Madhavi A. New insights of cellulosic ethanol production from lignocellulosic feedstocks. Microbes Microb. Biotechnol. Green Remediat., Elsevier; 2022, p. 749–79. https://doi.org/10.1016/B978-0-323-90452-0.00045-1

[18] Raj T, Chandrasekhar K, Naresh Kumar A, Rajesh Banu J, Yoon J-J, Kant Bhatia S, et al. Recent advances in commercial biorefineries for lignocellulosic ethanol production: Current status, challenges and future perspectives. Bioresour Technol 2022;344:126292. https://doi.org/10.1016/j.biortech.2021.126292

[19] Zabed H, Sahu JN, Boyce AN, Faruq G. Fuel ethanol production from lignocellulosic biomass: An overview on feedstocks and technological approaches. Renew Sustain Energy Rev 2016;66:751–74. https://doi.org/10.1016/j.rser.2016.08.038

[20] Canilha L, Chandel AK, Suzane dos Santos Milessi T, Antunes FAF, Luiz da Costa Freitas W, das Graças Almeida Felipe M, et al. Bioconversion of Sugarcane Biomass into Ethanol: An Overview about Composition, Pretreatment Methods, Detoxification of Hydrolysates, Enzymatic

Saccharification, and Ethanol Fermentation. J Biomed Biotechnol 2012;2012:1–15. https://doi.org/10.1155/2012/989572

[21] Krishnan C, Sousa L da C, Jin M, Chang L, Dale BE, Balan V. Alkali-based AFEX pretreatment for the conversion of sugarcane bagasse and cane leaf residues to ethanol. Biotechnol Bioeng 2010;107:441–50. https://doi.org/10.1002/bit.22824

[22] Indira D, Das B, Bhawsar H, Moumita S, Johnson EM, Balasubramanian P, et al. Investigation on the production of bioethanol from black tea waste biomass in the seawater-based system. Bioresour Technol Rep 2018;4:209–13. https://doi.org/10.1016/j.biteb.2018.11.003

[23] Singh A, Bishnoi NR. Comparative study of various pretreatment techniques for ethanol production from water hyacinth. Ind Crops Prod 2013;44:283–9. https://doi.org/10.1016/j.indcrop.2012.11.026

[24] Tinôco D, Genier HLA, da Silveira WB. Technology valuation of cellulosic ethanol production by Kluyveromyces marxianus CCT 7735 from sweet sorghum bagasse at elevated temperatures. Renew Energy 2021;173:188–96. https://doi.org/10.1016/j.renene.2021.03.132

[25] Devi A, Singh A, Bajar S, Pant D, Din ZU. Ethanol from lignocellulosic biomass: An in-depth analysis of pre-treatment methods, fermentation approaches and detoxification processes. J Environ Chem Eng 2021;9:105798. https://doi.org/10.1016/j.jece.2021.105798

[26] Chen J, Wang X, Zhang B, Yang Y, Song Y, Zhang F, et al. Integrating enzymatic hydrolysis into subcritical water pretreatment optimization for bioethanol production from wheat straw. Sci Total Environ 2021;770:145321. https://doi.org/10.1016/j.scitotenv.2021.145321

[27] Jin Y, Liu J, Yang H, Shi Z, Zhao P, Yang J. Improving enzymatic saccharification and ethanol production of bamboo residues with sulfomethylation-aided phosphoric acid pretreatment. Ind Crops Prod 2021;170:113733. https://doi.org/10.1016/j.indcrop.2021.113733

[28] Lyu H, Yang S, Zhang J, Feng Y, Geng Z. Impacts of utilization patterns of cellulosic C5 sugar from cassava straw on bioethanol production through life cycle assessment. Bioresour Technol 2021;323:124586. https://doi.org/10.1016/j.biortech.2020.124586

[29] Madadi M, Zhao K, Wang Y, Wang Y, Tang S, Xia T, et al. Modified lignocellulose and rich starch for complete saccharification to maximize bioethanol in distinct polyploidy potato straw. Carbohydr Polym 2021;265:118070. https://doi.org/10.1016/j.carbpol.2021.118070

[30] Mikulski D, Kłosowski G. Microwave-assisted dilute acid pretreatment in bioethanol production from wheat and rye stillages. Biomass Bioenergy 2020;136:105528. https://doi.org/10.1016/j.biombioe.2020.105528

[31] Dimos K, Paschos T, Louloudi A, Kalogiannis KG, Lappas AA, Papayannakos N, et al. Effect of Various Pretreatment Methods on Bioethanol Production from Cotton Stalks. Fermentation 2019;5:5. https://doi.org/10.3390/fermentation5010005

[32] Raj K, Krishnan C. Improved high solid loading enzymatic hydrolysis of low-temperature aqueous ammonia soaked sugarcane bagasse using laccase-mediator system and high concentration ethanol production. Ind Crops Prod 2019;131:32–40. https://doi.org/10.1016/j.indcrop.2019.01.032

[33] Terán Hilares R, Kamoei DV, Ahmed MA, da Silva SS, Han J-I, Santos JC dos. A new approach for bioethanol production from sugarcane bagasse using hydrodynamic cavitation

assisted-pretreatment and column reactors. Ultrason Sonochem 2018;43:219–26. https://doi.org/10.1016/j.ultsonch.2018.01.016

[34] Duque A, Manzanares P, González A, Ballesteros M. Study of the Application of Alkaline Extrusion to the Pretreatment of Eucalyptus Biomass as First Step in a Bioethanol Production Process. Energies 2018;11:2961. https://doi.org/10.3390/en11112961

[35] Nosratpour MJ, Karimi K, Sadeghi M. Improvement of ethanol and biogas production from sugarcane bagasse using sodium alkaline pretreatments. J Environ Manage 2018;226:329–39. https://doi.org/10.1016/j.jenvman.2018.08.058

[36] Yu H, Guo J, Chen Y, Fu G, Li B, Guo X, et al. Efficient utilization of hemicellulose and cellulose in alkali liquor-pretreated corncob for bioethanol production at high solid loading by Spathaspora passalidarum U1-58. Bioresour Technol 2017;232:168–75. https://doi.org/10.1016/j.biortech.2017.01.077

[37] Das A, Ghosh P, Paul T, Ghosh U, Pati BR, Mondal KC. Production of bioethanol as useful biofuel through the bioconversion of water hyacinth (Eichhornia crassipes). 3 Biotech 2016;6:70. https://doi.org/10.1007/s13205-016-0385-y

[38] Cabral MMS, Abud AK de S, Silva CE de F, Almeida RMRG. Bioethanol production from coconut husk fiber. Ciênc Rural 2016;46:1872–7. https://doi.org/10.1590/0103-8478cr20151331

[39] Yan J, Wei Z, Wang Q, He M, Li S, Irbis C. Bioethanol production from sodium hydroxide/hydrogen peroxide-pretreated water hyacinth via simultaneous saccharification and fermentation with a newly isolated thermotolerant Kluyveromyces marxianu strain. Bioresour Technol 2015;193:103–9. https://doi.org/10.1016/j.biortech.2015.06.069

[40] Cha Y-L, Yang J, Ahn J-W, Moon Y-H, Yoon Y-M, Yu G-D, et al. The optimized CO2-added ammonia explosion pretreatment for bioethanol production from rice straw. Bioprocess Biosyst Eng 2014;37:1907–15. https://doi.org/10.1007/s00449-014-1165-x

Advanced Topics in Mechanics of Materials, Structures and Construction - AToMech1-2023 Materials Research Forum LLC
Materials Research Proceedings 31 (2023) 375-384 https://doi.org/10.21741/9781644902592-39

Investigation of blast resistance of cladding with square dome-shaped Kirigami folded structures as core

Vishal M Birari[1,a *], B N Rao[1,b]

[1]Structural Engineering Division, Department of Civil Engineering, Indian Institute of Technology Madras, Chennai, India - 600036

[a]vishalbirari09@gmail.com, [b]bnrao@iitm.ac.in

Keywords: Kirigami, Foldcore, Sacrificial Cladding, Quasi-Static Compression, Blast Load

Abstract. This study examines the response of Square Dome-shaped Kirigami (SDK) structures when used as a core in cladding systems under blast loading conditions. A numerical model of the SDK foldcore is developed on the commercially available software ABAQUS. The SDK foldcore made of an aluminium sheet is placed between two rigid plates. The model is put under a quasi-static compression test to simulate the crushing effect and the results are used to calibrate the simulation with the experimental data from the literature. To evaluate the blast resistance of the system, four different levels of blast loading conditions are applied to the top of the top plate using Trinitrotoluene (TNT) explosive, with a distance of 1500 mm from the centre of the top plate. The structural response of the SDK foldcore is then compared with that of a traditional Square Honey Comb (SHC) core under the same blast loading conditions. This study aims to evaluate the relative performance of the two different cores in terms of their ability to mitigate the effects of blast loading. The SDK foldcore demonstrated the capability to disperse blast energy over a wider area, thus decreasing the stress the cladding system experienced. The results of the study show that SDK foldcore provides a significant improvement in energy absorption, with a maximum reduction of 70% in the peak load transmitted compared to the case with no cladding. The peak load transmitted by the SDK foldcore is much more consistent than the SHC core, even under different blast loading conditions. This is due to its favourable plastic deformation, which prevents complete densification. These results suggest that the SDK folded structure has better performance in mitigating the effects of blast loading.

Introduction

Sandwich structures are commonly used in various fields, including transportation (such as vehicles, aircraft, and ships), packaging, and protective structures, due to their light weight and ability to absorb a large amount of energy [1]. Sacrificial cladding is used to withstand significant blast and impact loads, resulting in a large change in the geometry of the structure. Its failure is primarily caused by deformation, which is brought on by a combination of bending and stretching of the materials. Therefore, most of the energy absorbers are made of ductile material [1]. It is essential that the cladding should convert the input kinetic energy from the blast into inelastic energy. This should occur through plastic deformation or other dissipation processes, rather than simply storing it elastically. The maximum elastic deformation is reached when the initial kinetic energy is converted into elastic strain energy. The elastic strain energy then is released, potentially causing subsequent damage to the structure and those it is designed to protect.

Sacrificial cladding typically consists of two layers: an inner core and an outer skin panel. Most of the energy from the blast load is absorbed by the inner core, which deforms gradually. The outer skin panel distributes blast pressure uniformly over the inner core. The sacrificial layer absorbs energy from a nearby explosion and as a result, it experiences a significant deformation in a very short time interval. The strength of sacrificial cladding depends on various factors, such as the material properties, thickness, design of the cladding, and type of core. Various types of cores have been developed which include lattices [2], polymeric foams [3], aluminium honeycombs [4] and metallic foams [5].

A folded structure can serve as the core for sacrificial cladding. It has several advantages such as being lightweight, having a high strength-to-weight ratio, and having the ability to fold into a compact configuration. Miura introduced the folded energy-absorbing structures [6]. It is constructed from a single, unbroken sheet of material that is folded along predetermined creases without stretching or twisting the faces. It makes them suitable for use in solar panels in satellites [7]. It is relatively simple to fabricate, as it requires just a single sheet of material and a folding process, making them cost-effective. In foldcore construction, the open design of the cells allows moisture to escape through the channels, preventing it from being trapped. However, the Miura-type origami core is not as efficient in terms of crushing resistance and energy absorption capacity when compared to other core materials with similar densities [8, 9].

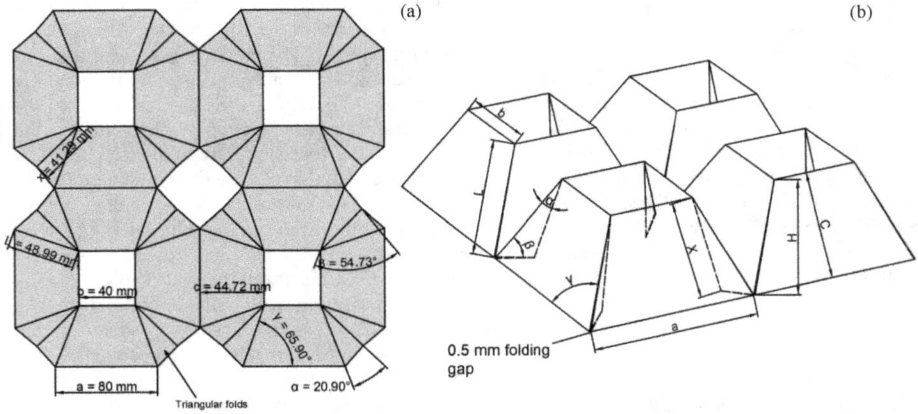

Figure 1: *(a) crease patterns and geometric parameters and (b) isometric view of folding configuration.*

Kirigami foldcore provides a way to create more efficient energy-absorbing structures. It is a variation of origami which includes folding as well as cutting to create a complex 3D structure from a single sheet of material. A study comparing the performance of the diamond strip core (kirigami foldcore) with that of the Miura-type foldcore revealed that the diamond strip core exhibited an increase in peak and average stresses of 74% and 92%, respectively [9]. Previous studies have demonstrated the potential use of a kirigami structure as a core in sacrificial cladding. However, further research is required to fully understand the various kirigami fold cores and their applications.

This study focuses on the performance of the kirigami foldcore under blast loading condition and how it compares to that of the honeycomb core. Square Dome-shaped Kirigami (SDK) foldcore is used as a core for sacrificial cladding using a single piece of aluminium sheet. Fig. 1 shows the crease patterns and the geometric properties of the foldcore. The energy absorption and crushing resistance of the foldcore increase as the vertical edges are connected since it is made from a single sheet of material. This connection provides more constraint against the out-of-plane crushing of the foldcore. This research examines the ability of the core to reduce the impact of a blast, the energy absorption capacity, and the peak load transmitted to the protected structure.

Geometry and Material Properties

The examined sandwich panels consist of top and bottom plates with a thickness of 5 mm and an aluminium SDK foldcore of 0.94 mm, giving the core a 5% relative density. The core's height is considered to be 40 mm. The material properties and true stress-strain data of Aluminium 1060 are shown in Table 1 and Table 2 respectively. Both the top and bottom plates have a density of 2400 kg/m^3 and Young's Modulus of 200 GPa.

***Table 1:** Material properties of Aluminium 1060*

Parameter	Young's Modulus (GPa)	Poisson's Ratio	Yield Stress (MPa)	Density (kg/m^3)
Value	69	0.33	67.7	2710

Table 2: True stress-strain data of Aluminium 1060

Strain	0	0.002	0.005	0.013	0.063	0.121
Stress (MPa)	0	67.7	112.3	120.1	125.8	130.6

The SDK foldcore has a dimension of 80×80×40 mm. Fig. 1 shows the geometric parameters and crease pattern of foldcore. A 2 mm high boundary strip is considered on the bottom plate to constrain the outer edges of the foldcore under out-of-plane crushing and to restrict the horizontal movement of the foldcore outer edges. The foldcore and top plate are simply supported and no tie constraint is defined between the foldcore and the plates.

The geometry of the foldcore is determined by three parameters only, i.e., the length of the bottom edge (*a*), the length of the top edges (*b*), and the height of the core (*H*). The parameters are illustrated in Fig. 1(a). Other parameters can be expressed in terms of a, b and H as follows:

$$c = \sqrt{\left(\frac{a-b}{2}\right)^2 + H^2} \tag{1}$$

$$L = \sqrt{\left(\frac{a-b}{2}\right)^2 + c^2} \tag{2}$$

$$\gamma = tan^{-1}\left(\frac{2c}{a-b}\right) \tag{3}$$

$$\alpha = \gamma - \frac{\pi}{4} \tag{4}$$

$$\beta = cos^{-1}\left(\frac{\sqrt{2}a-\sqrt{2}b}{2l}\right) \tag{5}$$

$$X = \frac{sin\beta.l}{\sin(\pi-\alpha-\beta)} \tag{7}$$

The total surface area of each SDK cell,

$$A_{surf} = 4.\frac{1}{2}c(a+b) + 8.\frac{1}{2}sin\alpha.Xl \tag{8}$$

The relative density,

$$\rho_{\vartheta} = \frac{A_{surf}.T}{a^2H} \tag{9}$$

The performance of SDK foldcore is compared to that of a Square Honey Comb (SHC) Core structure made of the same material (Aluminium 1060). The SHC core has a thickness of 0.87 mm. Both cores have the same height (40 mm) and the relative density of the core is kept the same for a fair comparison.

Numerical Modelling

A finite element analysis was performed on sacrificial cladding with SDK foldcore and SHC core as the inner core on a commercially available software ABAQUS/Explicit. A quasi-static compression test was conducted in an experiment, and it was determined that the core did not tear [10]. Therefore, an isotropic-hardening plasticity material model was used with no damage initiation or evolution behaviour. The aluminium sheet is modelled as an S4R 4-noded shell element having a linear geometric order, while the top and bottom plate are modelled as a C3D8R 8-noded solid element with linear geometric order as shown in Fig. 2. The inclusion of geometric imperfections in the model improves the accuracy of predictions for the performance of the core [8, 10, 11]. Therefore, a small folding gap of 0.5 mm is considered near the corner of the unit cell for the numerical modelling to account for any imperfections that may occur during the manufacturing of the SDK foldcore.

The top and bottom plates are modelled as rigid parts. The top plate is only allowed to move in the z-direction (towards the bottom plate) while the bottom plate is fixed. The mesh size of the outer plates is not critical, as they are considered to be rigid, i.e. no stress will be developed on them. The coefficient of friction for tangential contact behaviour is set at 0.3, while the normal contact behaviour is set as hard.

Table 3: Mesh convergence study

Mesh Size (mm)	Peak Load Transmitted (kN)	Average Force Transmitted (kN)
4	127.07	12.48
2	104.97	12.62
1	100.90	12.62
0.75	100.10	12.62
0.5	103.27	12.62

A mesh convergence study was conducted using the SDK foldcore cladding model to evaluate the effects of a 1 kg Trinitrotoluene (TNT) explosion at a stand-off distance of 1.5 m. The results showed that the peak and average transmitted force on the protected structure were similar when using mesh sizes of 0.5 mm and 2 mm, as shown in Table 3. As a result, a mesh size of 2 mm, which results in 66,000 elements, was deemed sufficient for further numerical studies. The same mesh size was used for the SHC core.

Figure 2: *Numerical model of the SDK foldcore and its base plate.*

Quasi-Static Compression Test

The numerical model considering a four-unit cell of the SDK foldcore is crushed under a quasi-static compression test. A boundary condition is applied to the top plate in which a constant loading rate of 50 mm/s is applied until 0.8H is reached. As shown in Fig. 3, the critical including plateau stress and densification strain is close for the two curves from the numerical simulation obtained in this study and the curve given by Li et al. (2018) [10]. In this study, the maximum stress achieved before complete densification is significantly higher than what was observed in experiments. This discrepancy is attributed to flaws in folding the sheet, such as not taking into account the appropriate folding gap and the initial stiffness of the foldcore. The high-pressure loading incident on the structure should be converted into a lower magnitude load with a much longer duration (due to conservation of momentum) and hence the core should provide predictable and constant load transfer up to densification. Once densification occurs, the load increase and the advantages of the cellular material are lost.

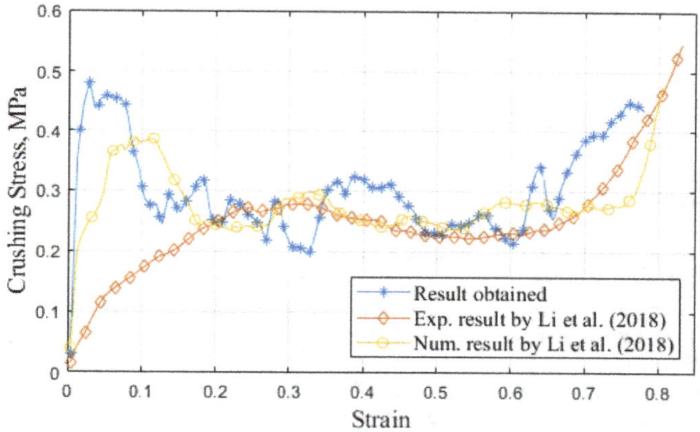

Figure 3: *Validation of Stress-Strain curve of numerical result with experimental results*

During the analysis, three distinct states of deformation can be observed. The first stage is when the peak stress is reached. This is the point at which the maximum stress is experienced by the core before it begins to deform. The second stage is when the faces of the SDK foldcore begin to bend. In the third stage, the load is consistently transferred throughout the core, resulting in a constant load transfer. In the final stage, densification of the core occurs. This is when the core becomes more compact, its relative density increases and it becomes more resistant to further deformation.

Ideal Blast Wave Profile

After an explosion occurs, the pressure increases instantaneously and then decays exponentially. The blast pressure wave depends on the mass of charge, the distance of the centre of charge, and time. Fig. 4 shows an ideal blast wave resulting from an explosion in the air. The maximum pressure reached is known as peak positive pressure, P_{pos}. The time required for the pressure to become P_{pos} after the blast is called arrival time, t_a. In the decay phase, the pressure becomes lower than the ambient pressure (P_o), known as under pressure, P_{neg}. The blast wave profile is described by the Modified Friedlander Equation [12]. It depends on time, t which starts at the arrival of the pressure wave at this point, i.e. $t = t_o - t_a$ as,

$$P(t) = P_o + P_{pos}\left(1 - \frac{t}{t_{pos}}\right)e^{-b\frac{t}{t_{pos}}} \tag{10}$$

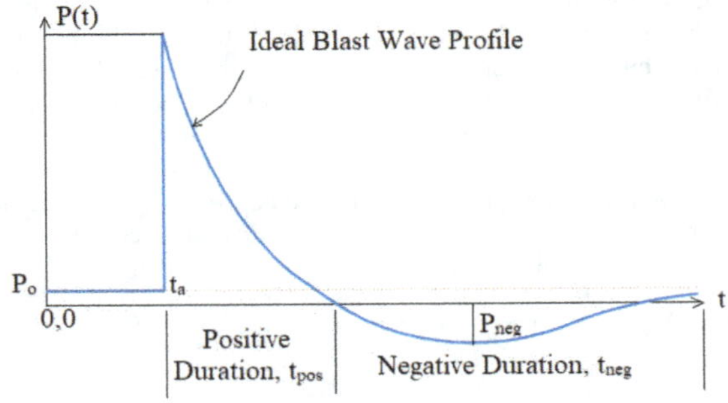

Figure 4: *Ideal Blast Wave resulting from an explosion in the air.*

Blast Loading

Sixteen SDK unit cells of dimension 80×80×40 mm are sandwiched between two rigid plates making the core 320×320×40 mm, as shown in Fig. 5(a). The unit cell size of the SHC has been established as 40×40×40 mm, resulting in a top-opening dimension that matches that of the SDK foldcore as shown in Fig. 5(b). The top and bottom plates are of dimension 330×330×5 mm with a 2 mm high boundary strip at the edges of the bottom plate. TNT of 1kg, 2kg, 4kg and 6kg is used as a spherical free airblast source at a distance of 1500 mm (scaled distance of 1500 mm/kg$^{1/3}$, 1190.55 mm/kg$^{1/3}$, 944.94 mm/kg$^{1/3}$ and 825.48 mm/kg$^{1/3}$ respectively referring to Eq. 11) from the centre of mass of the top plate. The expression for scaled distance is as follows:

$$Z = \frac{R}{M^{1/3}} \tag{11}$$

Where Z = scaled distance (m/kg$^{1/3}$)
R = distance to the structure of concern (m)
M = mass of blast charge (kg)

CONWEP [13] is used to calculate the pressure wave and the arrival time and is directly applied on the top surface of the rigid top plate. The force transmitted to the top of the bottom plate is observed for 3 ms. The transmitted force-time history curve of the protected structure under various blast loads is shown in Fig. 7. The peak load transmitted to the protected structure for the no-cladding case is 141.51 kN. The deformation of the cladding can be divided into three states i.e. (i) elastic state, (ii) plastic state and (iii) fully densified state. The elastic state is the temporary deformation of a structure that returns to its original shape when the stress or force causing the deformation is removed. This type of deformation can be observed when the transmitted force fluctuates multiple times. The SHC core is observed to be in an elastic state for initial cases of loading considered in this study. The maximum displacement of the top plate is 0.88 mm for the SHC core (2.2% of the depth of the core). As the strength of the blast wave increases while maintaining a constant distance from the structure, the arrival time decreases and the peak transmitted force increases.

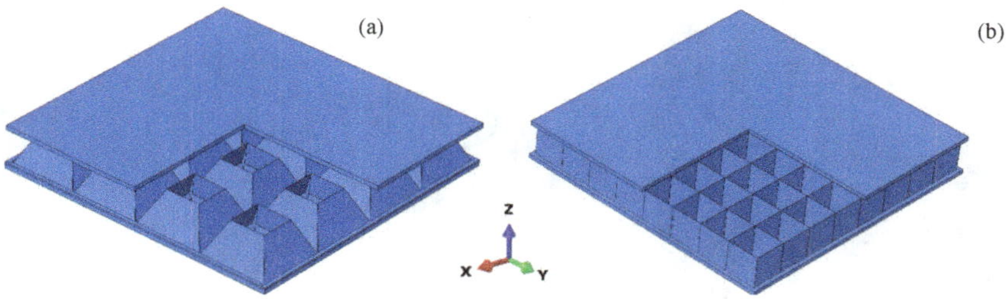

Figure 5: *Numerical model of (a) SDK foldcore for blast analysis and (b) square honeycomb core for blast analysis.*

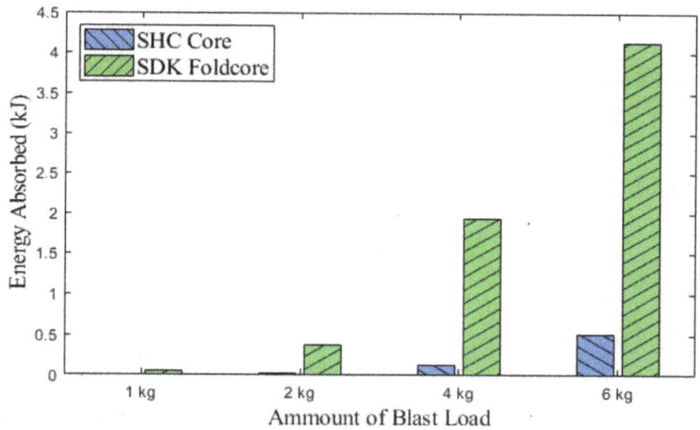

Figure 6: *Energy absorption by core for various blast loading*

The SDK foldcore exhibits a higher average transmitted force for all the loads considered in this study, which supports the presence of plastic deformation across all loading scenarios (as listed in Table 4). In contrast, the SHC core transmits more load than the no-cladding case for 1kg, 2kg, and 4kg of TNT blast load, which may result in greater damage to the protected structure. The SDK foldcore can reduce a maximum of 68.7% of the peak load in the no-cladding case. It demonstrates favourable energy absorption properties under both quasi-static and dynamic loading. This is evident through the higher plastic strain, avoiding full densification, low initial peak stress, and a small ratio of average stress to peak stress. The comparison of energy absorption by the two cores is shown in Fig. 6.

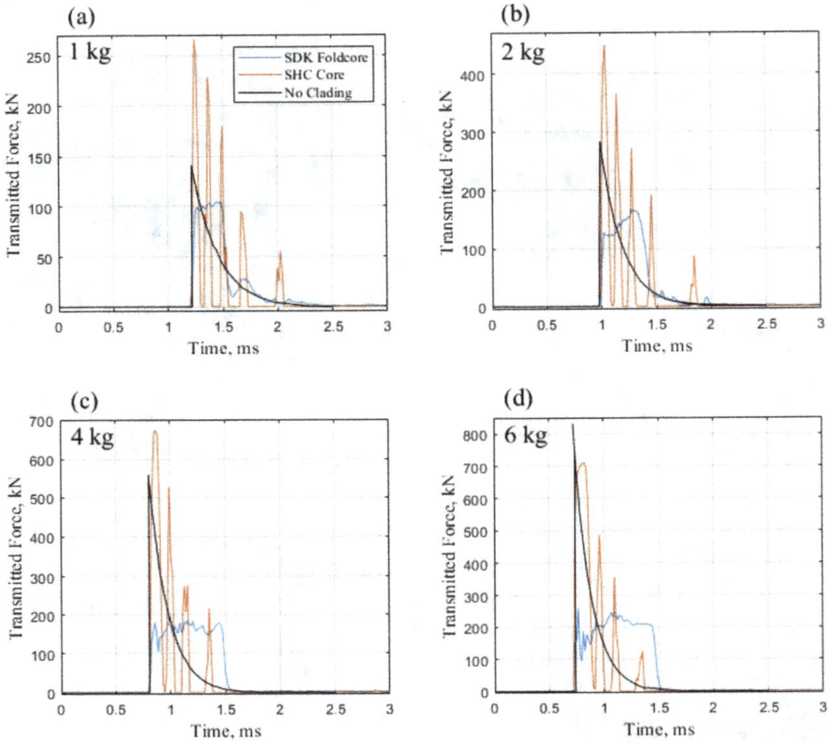

Figure 7: *Comparison of transmitted force-time history curves under different blast loads; (a) 1 kg TNT; (b) 2 kg TNT; (c) 4 kg TNT; (d) 6 kg TNT.*

As the blast load increases, the capacity for energy absorption also increases when subjected to higher blast loads. Significant improvement in energy absorption is observed for SDK foldcore. The low value of energy absorbed by the SHC core when subjected to 1 kg and 2 kg of blast load suggests that it remains in an elastic state. An energy-absorbing structure or material under impact must provide enough total energy-absorption capacity during the large deformation process, while also keeping the peak force (and thus the peak deceleration) below the level that would cause harm or injury. Additionally, the reactive force should remain stable or almost consistent to prevent an excessively high rate of deceleration, which is demonstrated by SDK foldcore.

When the core is completely densified, it should be able to absorb some of the energy from the blast. However, the peak transmitted force may increase to a value that is higher than it would be without cladding in some cases. The peak transmitted force in an elastic state will be observed immediately after the blast, but for a densified state, it will be observed near the end of the test after the material has undergone plastic deformation. As a result, the plastic state is the most desirable, which absorbs more energy and transmits less load to the protected structure.

Table 4: Peak transmitted load, duration and the crushed distance at the cladding centre by a core of different cladding configurations under various TNT blast loads.

Amount of TNT	Cladding type	P_{peak} (kN)	$P_{average}$ (kN)	Duration (ms)	Peak crushed distance at centre δ (mm)	Energy Absorbed by core (J)
1 kg TNT 1.5 m/kg$^{1/3}$	No-cladding	141.51	-	0.78	-	-
	SHC core	265.33	11.76	0.79	0.05	3
	SDK foldcore	104.97	12.62	0.75	0.74	51.45
2 kg TNT 1.19 m/kg$^{1/3}$	No-cladding	284.05	-	0.79	-	-
	SHC core	448.21	19.43	0.90	0.11	16.21
	SDK foldcore	167.07	21.62	0.84	3.21	368.12
4 kg TNT 0.94 m/kg$^{1/3}$	No-cladding	560.59	-	0.80	-	-
	SHC core	672.81	33.50	0.59	0.32	121.25
	SDK foldcore	184.90	37.82	0.81	13.16	1934.60
6 kg TNT 0.83 m/kg$^{1/3}$	No-cladding	832.29	-	0.79	-	-
	SHC core	711.67	45.39	0.62	0.88	503.34
	SDK foldcore	260.74	50.98	0.84	21.57	4128.50

Summary and Conclusions

The study presented in the paper focused on evaluating the performance of an SDK foldcore as the core of a sacrificial cladding system. The results were compared to those of an SHC core with the same relative density. SDK foldcore can be made from a single sheet of material, making it different from other high-performing cores. SDK foldcore can be made through stamping, but the thickness of the core needs to be reduced to minimize folding gaps and join the vertical edges more accurately. The SDK foldcore showed plastic deformation and prevented complete densification, which is a desirable type of failure, while the SHC core showed elastic deformation, leading to a higher peak transmitted load than without cladding. The other two states, elastic and fully densified, occur when the cladding is too weak or too strong with respect to the reflected blast pressure, resulting in reduced effectiveness of the cladding.

The square dome kirigami foldcore design shows effective energy absorption properties in both static and dynamic conditions, due to its high-density strain, low maximum stress, and a low ratio between average stress and peak stress. The unique blend of their strength and energy-absorbing abilities, combined with the capacity to fold and unfold, makes kirigami structures an optimal choice for situations where maintaining structural stability and resilience to impact are crucial. Further studies are needed to determine the optimal SDK geometry for various applications and to explore its potential use in sacrificial cladding, due to its uniform crushing resistance and insensitivity to strain rate.

References

[1] Lu, G., and T. Yu. 2003. Energy Absorption of Structures and Materials. Energy Absorpt. Struct. Mater.

[2] Xue, Z., and J. W. Hutchinson. 2004. "A comparative study of impulse-resistant metal sandwich plates." Int. J. Impact Eng., 30 (10): 1283–1305. https://doi.org/10.1016/j.ijimpeng.2003.08.007

[3] Ousji, H., B. Belkassem, M. A. Louar, B. Reymen, J. Martino, D. Lecompte, L. Pyl, and J. Vantomme. 2017. "Air-blast response of sacrificial cladding using low density foams: Experimental and analytical approach." Int. J. Mech. Sci., 128–129: 459–474.

https://doi.org/10.1016/j.ijmecsci.2017.05.024

[4] Nurick, G. N., G. S. Langdon, Y. Chi, and N. Jacob. 2009. "Behaviour of sandwich panels subjected to intense air blast - Part 1: Experiments." Compos. Struct., 91 (4): 433–441. Elsevier Ltd. https://doi.org/10.1016/j.compstruct.2009.04.009

[5] Wu, C., and Y. Zhou. 2011. "Simplified Analysis of Foam Cladding Protected Reinforced Concrete Slabs against Blast Loadings." Int. J. Prot. Struct., 2 (3): 351–365. SAGE Publications. https://doi.org/10.1260/2041-4196.2.3.351

[6] Miura, K. 1972. Zeta-Core Sandwich- Its Concept and Realization. Institute of Space and Aeronautical Science, University of Tokyo.

[7] Miura, K. 1985. "Method of Packaging and Deployment of Large Membranes in Space." Inst. Sp. Astronaut. Sci. Rep., 618: 1–9. 宇宙科学研究所.

[8] Gattas, J. M., and Z. You. 2014. "Quasi-static impact of indented foldcores." Int. J. Impact Eng., 73: 15–29. https://doi.org/10.1016/j.ijimpeng.2014.06.001

[9] Fathers, R. K., J. M. Gattas, and Z. You. 2015. "Quasi-static crushing of eggbox, cube, and modified cube foldcore sandwich structures." Int. J. Mech. Sci., 101–102: 421–428. Elsevier. https://doi.org/10.1016/j.ijmecsci.2015.08.013

[10] Li, Z., W. Chen, and H. Hao. 2018. "Blast mitigation performance of cladding using square dome-shape kirigami folded structure as core." Int. J. Mech. Sci., 145 (July): 83–95. Elsevier Ltd. https://doi.org/10.1016/j.ijmecsci.2018.06.035

[11] Gattas, J. M., and Z. You. 2015. "The behaviour of curved-crease foldcores under low-velocity impact loads." Int. J. Solids Struct.

[12] Dewey, J. M. 2018. "The Friedlander Equations." Blast Eff. Phys. Prop. Shock Waves, I. Sochet, ed., 37–55. Cham: Springer International Publishing.

[13]"Conventional Weapons Effects Program (CONWEP)." 1986. Tech. Man. TM5-855-1, Fundam. Prot. Des. Conv. Weapons.

Advanced Topics in Mechanics of Materials, Structures and Construction - AToMech1-2023 Materials Research Forum LLC
Materials Research Proceedings 31 (2023) 385-397 https://doi.org/10.21741/9781644902592-40

Von mises stress distribution along the cruciform specimen under biaxial loading: geometric variation effect

Abdul Mateen Mohammed[1,a*], Jaya KandaSamy[2,b], Bhavani Venkata Yashwanth Dasari [3,c] Lingala Syam Sundar[4,d]

[1]Associate Professor, School of Engineering and Technology, CMR University, Bengaluru - 562149, INDIA

[2]Professor, Department of Mechanical Engineering, Maturi Venkata Subba Rao Engineering College, Hyderabad – 501510, INDIA

[3]Junior Modeler, CADMAXX Solutions Pvt. Ltd, Vijayanagar, Bengaluru, Karnataka 560040 INDIA

[4] Professor, Prince Mohammad Bin Fahd University, Al Jawharah, Khobar, Dhahran 34754, Saudi Arabia

[a*]abdulmateen7@gmail.com, [b] kandan_kandan@yahoo.co.in, [c] dasari368@gmail.com, [d]sslingala@gmail.com

Keywords: Biaxial Stress, Cruciform Specimen, Taguchi, Stress-Strain

Abstract. The aim of this work is to optimize the geometry of the cruciform specimen required for in-plane biaxial testing where the failure has to occur in the central gauge region. The specimen has three different zones with three different thickness, i.e., (i) along the arm, (ii) Milled region and (iii) along the central gauge region. The geometric variables in the specimen are the radii at four different locations in the geometry (i) central gauge (ii) radius of the slots milled along the arm, corner radius at (iii) Milled region and (iv) at the intersection of arms. The Taguchi method was realized for the design of experiments considering these four factors varied at three level. However, the effect of the variation in the thickness was studied independently. During which the thickness at the central region was maintained constant to 0.7mm, whereas the thickness at the other two positions were varied from 3 mm to 3.5mm and 1 to 1.5 mm respectively. In the first set of experiments the Von Mises stress was maximum at the gauge section along with a portion of it also being appearing along the milled region on the arm of the specimen. However, for the second set of experiments the maximum stress occurred along the gauge region for the three geometries. Later, Taguchi analysis indicated the radius of the slots milled along the arm is the prime statistically influencing factor and the radius at the interface of the arm section is the second statistically influencing factor. Whereas the other two radii are found to be statistically non-influencing factor when considered individually. However, a strong interaction has been observed between all the four parameters.

Introduction

The sheet metal operation is a process where in the materials being processed are generally subjected to most complex stress state. Customarily, these process are basically designed based on the uniaxial test data which is found to be insufficient. Because the test data resulting from the regular tensile test would be able to determine the properties solitary along the direction of the load being applied which, possibly will not be suitable to use for these processing as they experience a multi-directional stress state [1]. Subsequently, the biaxial testing of materials have become a subject of great significance in order to establish the material characteristics under such loading conditions [2]. Typically, a thin-walled tubular specimen or a bar of circular cross-section which is subjected to either an axial or a torsional loading conditions are regularly used for the material evaluation for biaxial state of stress. However, these geometries experience few challenges such as (i) the limitation of producing a pure thin cylinder, (ii) eliminating the edge effect and (iii) these test are not appropriate for materials which are produced by rolling [3-4]. As a result the flat cruciform specimen subjected to the in-plane biaxial loading is gaining importance.

However, the key for the success of these samples is the cruciform specimen design. Typically, one can identify three precincts in such geometries which are sensitive. These areas are (1) the specimen arms section, (2) The center region of the geometry, and (3) the area about the transition zone between two adjacent arms. While designing such geometry it is significant to confirm that a huge extent of the deformation occur at the central region of the specimen, with no stress concentrations zones in the other regions, principally while the large strains are essential [5]. In the recent past, different cruciform shaped specimen geometries made of metallic materials were subjected for analysis either under a quasi-static biaxial pull or under a fatigue loading [6-12]. The international standard ISO 16842: 2014 (E) [13] proposed a cruciform specimen with slots in the arm along with a fillet at the transition zone of the. Nevertheless, owing to a low strain level at the specimen central region, it is challenging to realize either a fracture or necking state at that region. Furthermore, if due to the slots or slits that are made about the arm, strength at may be decreased when compared with the central zone. Thus, creating a possibility of failure being occurring at the slits under a fatigue loading situation making the geometry to be inappropriate for testing under such conditions. Thus, it is not applicable to the biaxial fatigue test. With the intention to get over the above said limitations a reduction of thickness and designing the fillet radius or fillet geometry was suggested. The base geometry was the optimal geometry presented in the study [14] were the effect of the geometric variations were studied. Three geometries were subject for study of which cruciform geometry with round corner at the intersection of arms and tapered along the length was proposed to be optimal. However, it was not presented if the central region is the only place where the maximum stress is occurring or there is a possibility of the same to occur at the interface of the two arms.

The present paper is focused on selecting a suitable geometry from the modified base geometry which could be effectively implemented to understand the biaxial behavior of metallic materials. Taguchi design of experiments was implement to in order to understand the outcome of the geometric variation on the maximum Von Mises stress and its occurrence along the sample. An L9 orthogonal array was used as the factors considered were the geometric radii at four different positions under three levels.

Optimization of the specimen geometry:

A complex stress situation along the central region in a cruciform shape specimen is observed when it is subjected to a biaxial tension. Where a non-uniform direct stress and a shearing stress are observed along the central region of the specimen. Moreover, the arms of the specimen experiences a uniaxial loading and the central region experiences a biaxial loading causing a failure to occur along the arm of the specimen. This failure is due to the fact that under the uniaxial loading the deformation capacity is less when capered to that in a biaxial loading case [15]. Thus the main objective of the optimization is to determine an appropriate specimen geometry which generates a uniform and maximum stress along the center region. The basic geometry for the analysis as mentioned above is the optimal geometry proposed by Mohammed et al [14]. Where in the geometry of the specimen had two circular slots of 1 mm depth along the arm length and a 1.15 mm deep circular slot at the center. These slots were symmetric about the thickness of the specimen which was 3mm and the elementary geometry is represented in fig 1. However, in the present study the radius for the milled region and the curvature R1, R2, R3 and R4 are varied in order to optimize the geometry to produce a uniform maximum stress about the circular milled region and the variation in the radii are presented in table 1. This circular milled region in the specimen geometry is the gauge section for the present study. Moreover, the thickness of the specimen at different sections i.e., the thickness along the arm length and the center milled region are as so altered, later in the study. Two different categories of specimens were observed, the first set which is the case –I had the same the thickness variation as suggested earlier [14]. Whereas for the second category i.e, the case – II the thickness of the overall sample and about the circular slots was increased to 3.5mm 1.5 mm respectively. However, for optimization this variation in thickness is not considered as a parameter for the analysis. Rather than going with all the variations an orthogonal array L9 was used for designing the experiments which is presented in table 2 for both the category.

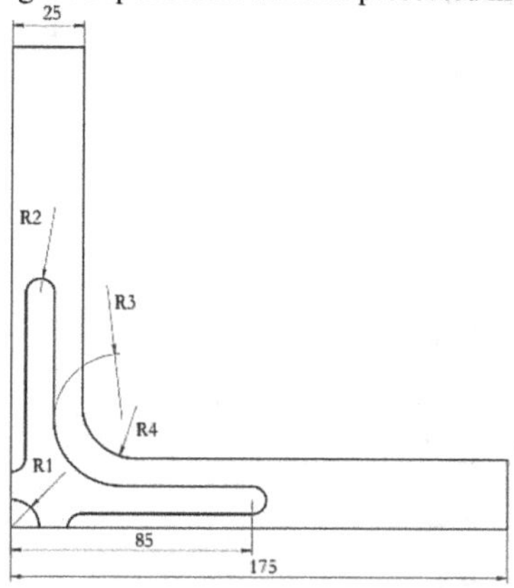

Fig1: Elementary specimen geometry.

Table1: Variations in the radii

	1	2	3
R1	10	12	14
R2	5	6	7
R3	18	20	22
R4	36	38	40

Table2: Design of experiments

Specimen Name	*R1*	*R2*	*R3*	*R4*
CA - I	10	5	18	36
CA - II	10	6	20	38
CA - III	10	7	22	40
CA - IV	12	5	20	40
CA – V	12	6	22	36
CA - VI	12	7	18	38
CA - VII	14	5	22	38
CA - VIII	14	6	18	40
CA - IX	14	7	20	36

Finite Element Analysis:

With the intention of simulating the effect of biaxial stress on the mentioned nine different geometries, ANSYS® a commercially accessible finite element analysis package was implemented as a numerical tool. The material for the specimens was A1050-H14 a ductile aluminium alloy which finds its application in the electrical and chemical industries whose properties are tabulated in table 3. In order to utilize the advantage of geometric symmetricity a quarter geometry was modelled during the analysis and a layered eight nodded linear three-dimensional shell element was selected for discretisation. At each node, the shell element possess six degrees of freedom with three translational and three rotational degree about the nodal coordinates (x, y and z-axis). A biaxial stress ratio of 1:1 was applied along the two perpendicular axes of the arms of the sample and the boundary conditions employed are presented in Figure 2. Where, the two symmetric edges were constrained by symmetricity and a distributed force of 1kN was applied on the arms of the sample.

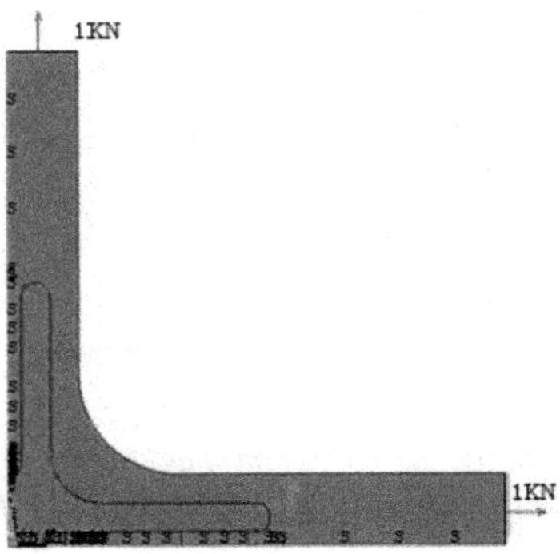

Fig 2: Boundary conditions

Results

The von misses stress dissemination over the nine different geometries under the first category under study is presented in fig 3. And it is clear that, the stress for all the nine geometries is uniform at the gauge section but the existence of maximum stress is at different locations. The maximum Von Mises stress for the geometries CA – I, CA – II, and CA – V, is along the gauge region. However, a small portion of it is also observed along the milled region (i.e., at the slot section) Whereas, for the other specimens the maximum Von Mises stress is either at the milled region or at the interface/intersection of two arms.

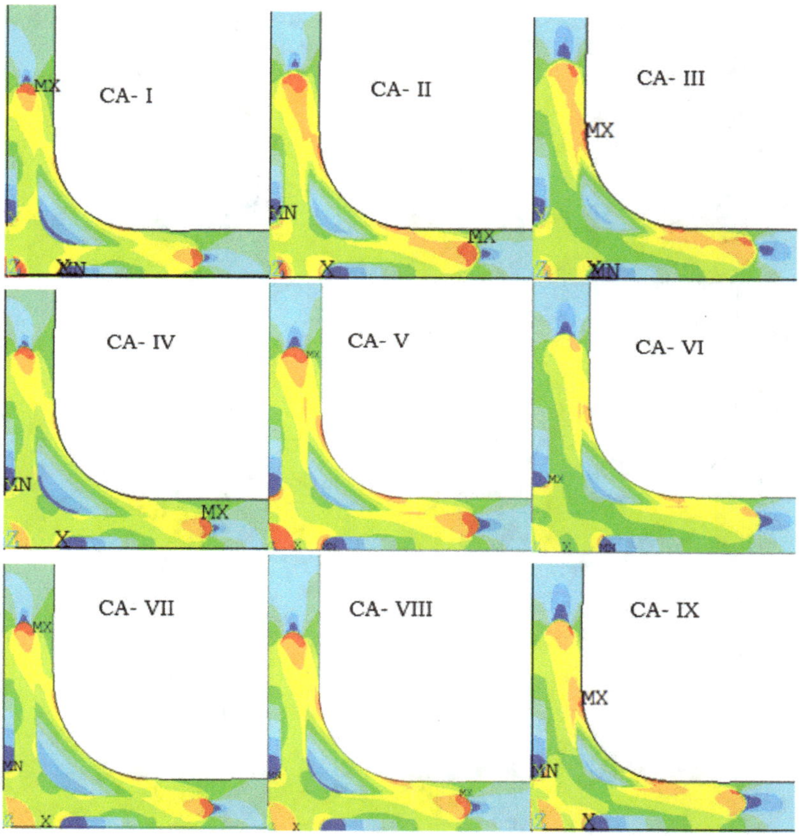

Fig 3: Von Mises Stress distribution on various geometries of first category.

Thus, the thickness of the sample about the milled region were varied as mentioned in the section 2 in order to study the variation of Von Mises stress at the region of interest. Figure 4 signifies the deviation of the Von Mises stress over the nine geometries belonging to the second category. It is clearly evident from the analysis that for all the samples the Von Mises stress was maximum at the central gauge region.

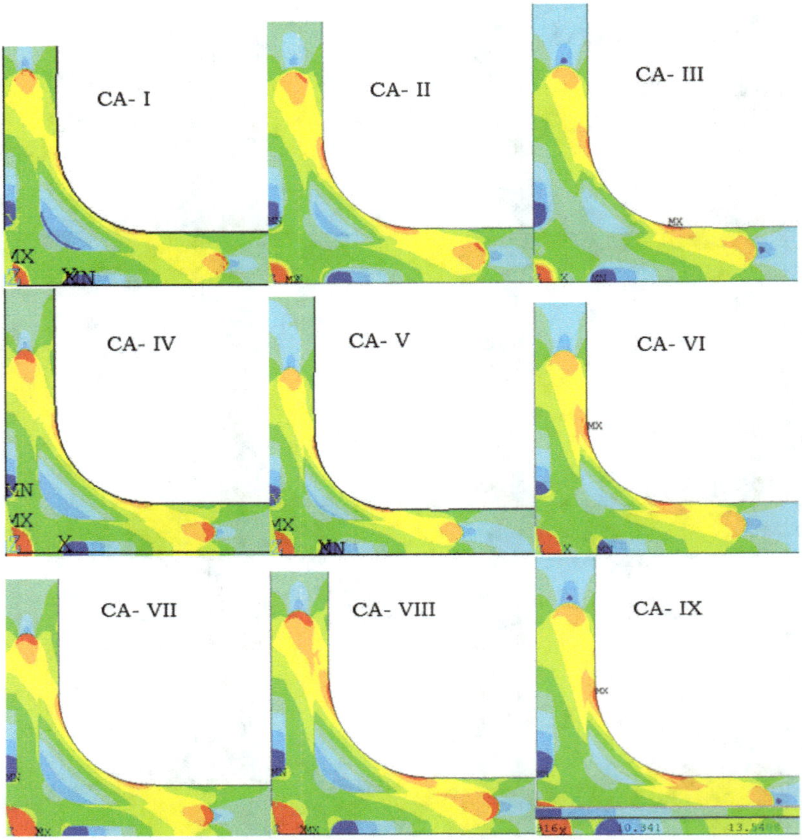

Fig 4: Von Mises Stress distribution on various geometries of Second category.

However, there is a small portion of maximum Von Mises stress even at (i) either the milled region i.e., along the slots or (ii) at the interface region for all the geometries except CA- V. Moreover, to confirm the better geometry the stress distribution along the central region equidistant nodal points along the diagonal and along the arm length were selected as represented fig 5. The path 1 in the fig 5 indicates the selection of the nodal points along the diagonal of the specimen and sequence 2 represents the selection of the nodal points along the arm length.

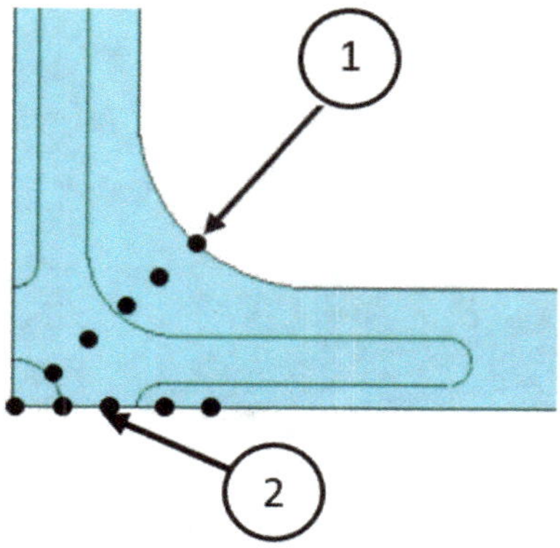

Fig 5: Selection of nodal points along the (1) diagonal and (2) along the length of the sample.

The Von Mises stress distribution was approximately linear for most of the geometries along both the paths consider and the variation is presented in fig 6 and fig 8. However, for the samples with geometries CA I, CA II, CA III the stress distributions is uniform only for about 1 mm or less along the path 1 beyond which the stress decreases. Whereas, for all the other sample geometries the Von Mises stress is uniform until the gauge section. The sample geometry CA IX indicated a high stress value beyond the gauge section and the variation could be seen as a nonlinear with a very low degree of nonlinearity. The fig 8 represents the strain distribution along the diagonal of the sample (i.e., along path 1). A negative strain is observed throughout the path selected with an increasing strain value until the gauge section, beyond which the strain started reducing until 25 mm from the center of the specimen. The stress distribution to be more specific the Von Mises stress along the path 2 is presented in fig 8. Where a similar pattern of stress distribution as that of path 1 is observed. However, the stress uniformity for the samples with geometries CA I, CA II, CA III was linear until 2 mm beyond which it decreased in stepped form. The shear strain distribution along the path 2 is presented in fig 9 and it is clearly evident that the variation in the shear strain with respect to the distance was uniform with a zero strain for all the samples except CA I, CA II and CA III.

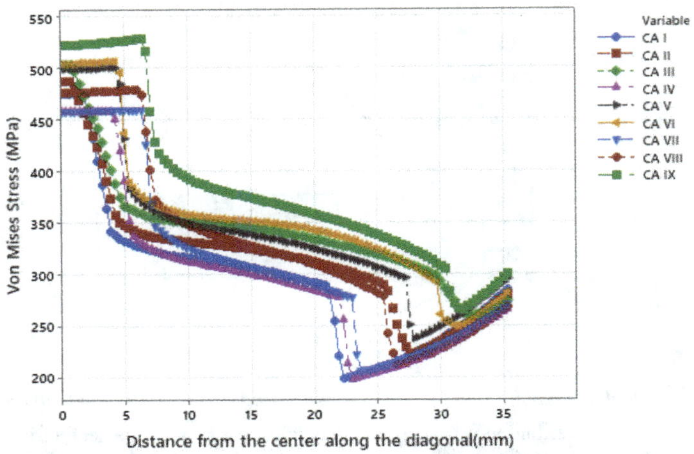

Fig 6: Von Mises Stress distribution along the diagonal of the sample.

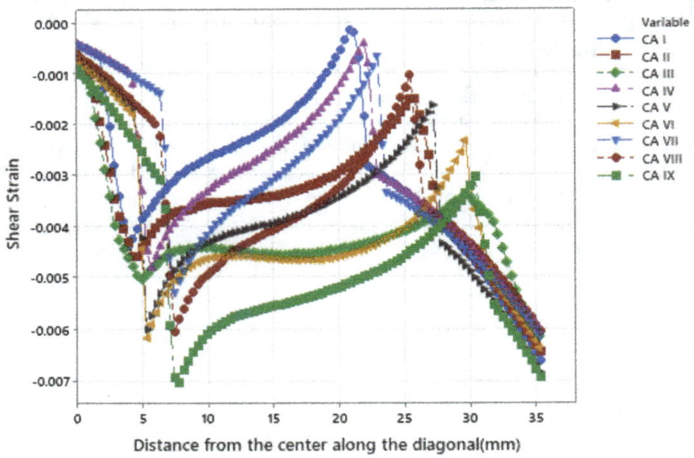

Fig 7: Maximum Shear Strain distribution along the diagonal of the sample.

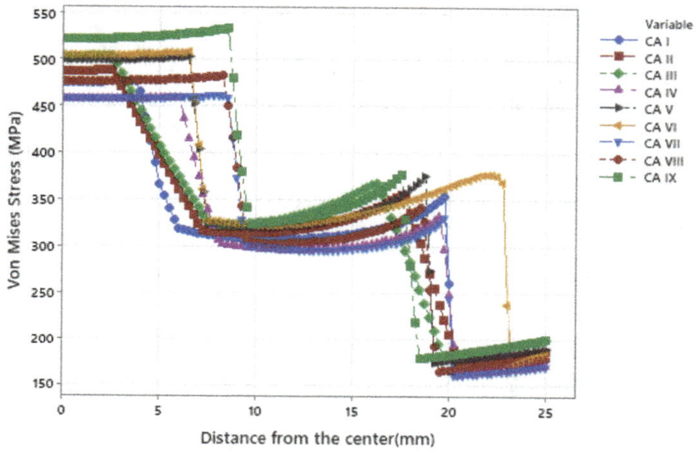

Fig 8: Von Mises Stress distribution along the arm length of the sample.

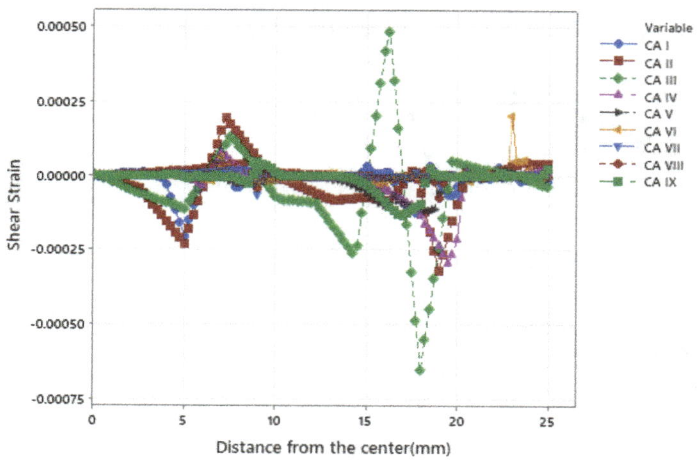

Fig 9: Shear Strain distribution along the arm length of the sample.

Taguchi analysis was carried out to determine the effect of individual variables on the Von Mises stress at the central region, to be more specific at the centre of the specimen. The table 3 and 4 presents the response table with respect to the signal to noise ratio and means. It is clear that the radius of the slot section, R2 has a rank 1 indicating that this variable is most influencing factor and the second being the R4 with rank 2 which is the radius at the interface of the arm section. However, the other two factors i.e., R1 and R3 have a rank of 3 and 4 respectively for the signal to noise ratio and rank 4 and 3 respectively in the mean table. The fig 10 represents the main effect plots of means for the maximum Von Mises stress at the centre. It is clear from the figure that the variables R1 and R3 have no significant effect on the occurrence of the maximum Von Mises stress individually as they are approximately linear and are very near to the mean line. However, as mentioned earlier the other variables R2 and R4 have a considerable effect on the occurrence of maximum Von Mises stress. It is also evident that as the radius R2 increases an increase in the Von Mises stress is observed. However, a converse situation is witnessed with respect to the radius R4 i.e., as the radius R4 increases a reduction in Von Mises stress is observed. Furthermore, the optimal variable sequence suggested by the analysis is R1, R2, R3 and R4 being 10, 7, 20 and 36 mm respectively.

Table 3: Response Table for Signal to Noise Ratios (Larger is better)

Level	R1	R2	R3	R4
1	53.77	53.32	53.71	53.95
2	53.74	53.75	53.77	53.68
3	53.71	54.14	53.74	53.59
Delta	0.07	0.83	0.06	0.36
Rank	3	1	4	2

Table 4: Response Table for Means (Larger is better)

Level	R1	R2	R3	R4
1	488.3	463.4	485.1	498.5
2	486.8	487.3	488.7	483.2
3	485.2	509.6	486.6	478.6
Delta	3.1	46.2	3.6	20.0
Rank	4	1	3	2

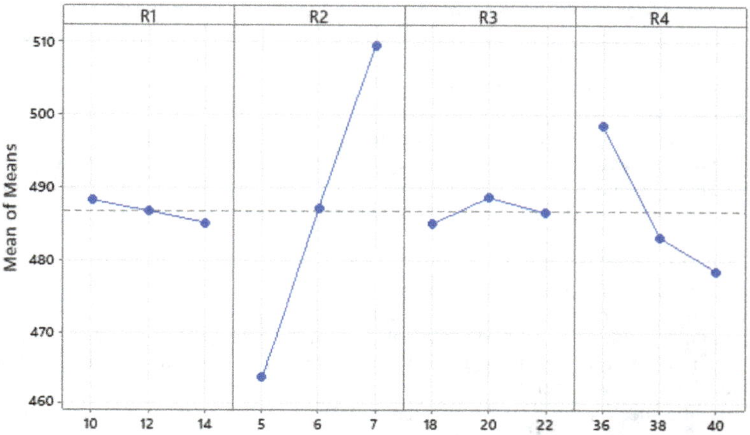

Fig 10: Main effect plot for means.

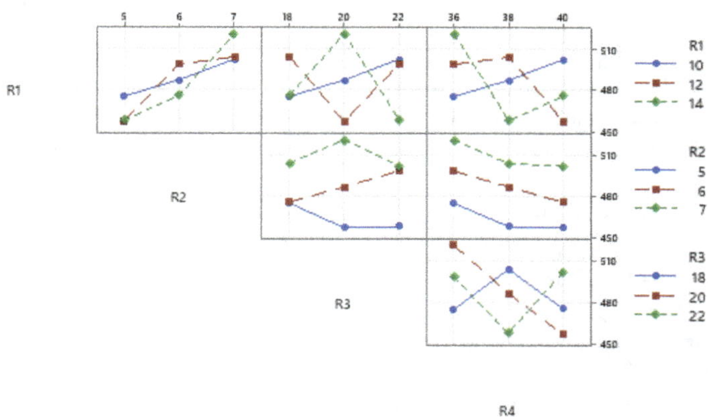

Fig 11: Interaction plot for Von Mises stress.

Nevertheless, the sequence suggested by the analysis is by considering the individual effect which cannot be considered without determining the effect of interactions. The fig 11 represents the interaction plot where a good interaction between all the variables is observed. Yet, a distinctive relation between the main effecting variables i.e., the radius R2 and R4 indicated no interaction.

Conclusions

In the prime aim of the present study was to determine the suitable geometry required for the in-plane biaxial testing of aluminium A1050-H14 material. A basic geometry suggested by Mohammed et al.[14] was selected and the geometric variant in the radius at the gauge section, milled region along the arms and the interception of the arms were varied for the current study. As there were four variables (factors) which were varied at three levels Taguchi design of experiments was implemented to design 9 experimental sequence rather than full experimental sequence. The effect of the thickness was not considered while designing the experiments in the first stage i.e., in case - I rather another set of 9 experiments with same variable sequence was carried out in the second stage in case –II. Later, the following conclusions are drawn from the study.

1. In the case – I, the specimen geometries CA – I, CA – II, and CA – V indicated the occurrence of the maximum Von Mises stress at the gauge zone with a small portion of it is also along the milled region (i.e., at the slot section). While, for the other specimens' occurrence of maximum Von Mises stress was either along the milled region or at the interface/intersection of two arms.
2. In the case – II, for the entire sample geometries range the maximum Von Mises stress occurred at the central gauge region, except the geometries CA- III, CA – VI and CA – IX. In these three geometries the maximum stress has occurred along the interface of the arm yet a portion which is not maximum but almost impending the maximum value is observed at the gauge region.
3. A uniform Von Mises stress distribution was observed in all the geometries of case – II when the elements were selected along the specimen diagonal and along the arm length. Conversely, the shear strain along the diagonal elements indicated an increment in shear strain along the negative direction. While a constant shear strain was observed for elements along the arm length at gauge length except for the geometries CA – I, CA – II and CA – III. For these geometries the strain has changed its orientation from negative to positive.
4. Taguchi analysis indicated that the radius of the slot section, i.e., R2 has a rank 1 indicating itself being a statically most influencing factor and the radius at the interface of the arm section i.e., R4 is the second statically influencing factor. However, the other two factors i.e., R1 and R3 were found to be non-influencing factors when analysed independently. The optimal variable sequence suggested by the analysis is R1, R2, R3 and R4 being 10, 7, 20 and 36 mm respectively. However, a strong interaction between all the factors with each other except R2 and R4 is observed.

Reference

[1]. Jones, C., 2001. "Biaxail testing of polymer composites". Mater. World 9 (11), 19-21.

[2]. Demmerle S, Boehler JP (1993), "Optimal design of biaxial tensile cruciform specimens". J Mech Phys Solid 41(1):143- 181 https://doi.org/10.1016/0022-5096(93)90067-P

[3]. Makinde A, Thibodeau L, Neal KW, Lefebvre D (1992), "Design of a biaxial extensiometer for measuring strains in cruciform specimens". Exp Mech 32(6):132-137 https://doi.org/10.1007/BF02324724

[4]. Tiernan, P.; Hannon, A. Design optimization of biaxial tensile test specimen using finite element analysis. Int. J. Mater. Form. 2014, 7, 117-123. https://doi.org/10.1007/s12289-012-1105-8

[5]. Hannon, A.; Tiernan, P. A review of planar biaxial tensile test systems for sheet metal. J. Mater. Process. Tech. 2008, 198, 1-13. https://doi.org/10.1016/j.jmatprotec.2007.10.015

[6]. Zhutao Shaoa, Nan Lia, Jianguo Lin, The optimisation of cruciform specimen for the formability evaluation of AA6082 under hot stamping conditions, Procedia Engineering 207 (2017) 735-740. https://doi.org/10.1016/j.proeng.2017.10.821

[7]. Zheng Lu, Jia-Yu Zhao, Chang-Yu Zhou, and Xiao-Hua He, Optimization Design of a Small-Sized Cruciform Specimen for Biaxial Fatigue Testing, Metals 2020, 10, 1148; https://doi.org/10.3390/met10091148

[8]. Liu,W.; Guines, D.; Leotoing, L.; Ragneau, E. Identification of sheet metal hardening for large strains with an in-plane biaxial tensile test and a dedicated cross specimen. Int. J. Mech. Sci. 2015, 101, 387-398. https://doi.org/10.1016/j.ijmecsci.2015.08.022

[9]. Kwon, H.J.; Jar, P.-Y.B.; Xia, Z. Characterization of biaxial fatigue resistance of polymer plates. J. Mater. Sci. 2005, 40, 965-972. https://doi.org/10.1007/s10853-005-6515-2

[10]. Kulawinski, D.; Nagel, K.; Henkel, S.; Hübner, P.; Fischer, H.; Kuna, M.; Biermann, H. Characterization of stress-Strain behavior of a cast trip steel under different biaxial planar load ratios. Eng. Fract. Mech. 2011, 78, 1684-1695. https://doi.org/10.1016/j.engfracmech.2011.02.021

[11]. Kuwabara, T. Biaxial stress testing methods for sheet metals. In Comprehensive Materials Processing; Elsevier: Amsterdam, The Netherlands, 2014; Volume 1, pp. 95-111. https://doi.org/10.1016/B978-0-08-096532-1.00106-0

[12]. Ackermann, S.; Lippmann, T.; Kulawinski, D.; Henkel, S.; Biermann, H. Biaxial fatigue behavior of a powder metallurgical trip steel. Frat. Ed Integrita Strutt. 2015, 34, 580-589. https://doi.org/10.3221/IGF-ESIS.34.64

[13]. ISO/CD 16842. Metallic Materials-Sheet and Strip-Biaxial Tensile Testing Method Using Cruciform Specimen; ISO: Geneva, Switzerland, 2014.

[14]. Mohammed A M, Sayyadh R, Dasari V R S, Mohammed M H, Stress distribution along the cruciform geometry under pure in plane biaxial loading condition. Journal of the Brazilian Society of Mechanical Sciences and Engineering (2019) 41:416. https://doi.org/10.1007/s40430-019-1928-z

[15]. Yu Y, Wan M, Wu XD, Zhou XB (2002) Design of a cruciform biaxial tensile specimen for limit strain analysis by FEM. J Mater Process Technol 123:67-70. https://doi.org/10.1016/S0924-0136(02)00062-6

Effect of steel fibres on the mechanical strengths of fly ash/GGBS based geopolymer concrete under ambient curing condition

Musa Adamu[1,a*], Tejas D Doshi[2,b], Veerendrakumar C. Khed[2,c*], Yasser E. Ibrahim[1,d]

[1]Engineering Management Department, College of Engineering, Prince Sultan University, Riyadh 11586, Saudi Arabia

[2]Department of Civil Engineering, KLE Technological University's Dr. M.S. Sheshgiri College of Engineering & Technology, Udyambag, Belagavi 590008, Karnataka, India

[a]madamu@psu.edu.sa, [b]tejas019@gmail.com, [c]veerendra.khed@klescet.ac.in, [d]ymansour@psu.edu.sa

Keywords: Geopolymer Concrete, Fly Ash, Ground Granulated Blast Furnace Slag, Steel Fiber, Mechanical Strengths

Abstract. The production of concrete involves the use of huge quantity of cement which is the main binder material. However, cement production involves the use of high amount of embodied energy and reduce of embodied CO_2 emission. Hence one of the alternatives is the use of concrete with zero cement called geopolymer. However, the main shortcoming of geopolymer is its higher brittleness and low modulus of elasticity compared to conventional cement concrete. This led to the addition of fibres to geopolymer. In this study, the effect of steel fibres addition on the mechanical strengths of ambient cured geopolymer concrete was investigated. 60% Fly ash class F and 40% GGBFS are used as binder materials to produce ambient geopolymer concrete of grade 30 MPa, which was activated by alkaline solution (mixture of sodium Hydroxide and sodium silicate) with a constant Molarity of Sodium Hydroxide as 10M without any cement. End hooked steel fibres of aspect ratio 35, were added to the geopolymer at 0%, 0.25% and 0.5% by volume fraction. The results findings showed that the workability of the geopolymer decreased with increase in addition of steel fibres. Ambient cured specimens yield good results and higher strength is observed due to high polymerization process. Furthermore, the compressive, split tensile and flexural strengths all improved significantly with increase in percentage of volume of steel fibres at any ambient curing period.

Introduction

Due to easy preparation, availability and fabrication, concrete is very widely used construction material. To produce one tonne Ordinary Portland Cement (OPC) concrete requires 150 to 250 kWh Embodied Energy (EE), produces 75–175 kg embodied carbon dioxide (ECO_2) [1, 2]. Concrete could not be replaced with other materials due to its effectiveness, performance and cost. An alternative solution is required to reduce or replace OPC as its manufacturing process is highly energy intensive and also should be eco-friendly. Use of SCM's appears to be most promising sources to manufacture Geo Polymer Concrete (GPC) because of its lower water demand, alkaline solutions and reliable rheological properties.

In above context, for a complete replacement of cement with a new binder material "Geopolymer" was introduced and it had Geopolymeric alumina silicate gel performing as binder. Geopolymer is an inorganic polymer and was developed by Davidovits in the year 1978 [3, 4]. An alkaline liquid is added to react with the silicon, aluminium present in the source materials. Geopolymer concrete mainly consists of alkaline liquid and source materials. Alkali activation of alumina and silica including blast furnace slag powder, known as GGBFS, has been found out since long back. Due to the activation of FA and GGBFS, alkali hydroxides and silicates were observed. Processing conditions for Geopolymer cement concretes (GPCC) are almost similar to cement concretes, except that during mixing operation of concrete, a premixed alkaline solution is added instead of water [3, 4]. The major shortcomings of geopolymer concrete in comparison to conventional cement concrete is its higher brittleness and lower modulus of elasticity [5]. To address this shortcoming of geopolymer concrete, fibres are mostly added. The fibres play a role of reducing the brittleness and increasing the ductility and elastic modulus of the geopolymer. Many types of fibres such as polyvinyl alcohol, polypropylene, glass and carbon fibres, natural fibres etc have been added to geopolymer to enhance its ductility and modulus of elasticity. Steel fibre have been reported to have give more improvement to the ductility and elasticity of geopolymer due to its high tensile strength, modulus of elasticity and fracture toughness

Therefore, the main objective of this study was to investigate the influence of steel fibres on the mechanical strengths of geopolymer concrete produced using hybrid of fly ash and GGBS as the main binder materials under ambient curing condition.

Materials and Methods

Materials

Class F fly ash was used as one of the binder materials. The properties of the fly ash are presented in Table 1. GGBFS is glassy, granular, nonmetallic material with silicates and aluminates of calcium and other bases that are used in geopolymer typically. In the present investigation GGBFS was used as another source binder material. The properties of the GGBS are also presented in Table 2.

Manufactured sand (M-sand) was used as fine aggregates instead of natural sand for sustainable development and as per available conditions. The M-sand has a specific gravity of 2.9, loose and bulk densities of 1664 kg/m^3 and 1894 kg/m^3 respectively, and water absorption of 1.2%. Crushed gravel with maximum size of 20 mm was used as coarse aggregate. The aggregate has a specific gravity value of 2.87, bulk density of 1614 kg/m^3 and water absorption of 1.96%. Hooked end type steel fibres were used in this study. The fiber has an aspect ratio (L/D) of 35, tensile strength of 1123 MPa.

Table 1. Properties of Fly Ash

Property	Requirements As Per IS 3812 Part 1 [6]	Observed Value
Fineness (m^2/Kg)	320	355
Soundness Auto Clave Expansion (%)	0.8 Max	0.06
Specific Gravity	------	2.19
Particle Retained On 45μ IS Sieve, % By Mass	34 Max	30
$SiO_2+Al_2O_3+Fe_2O_3$	70.0min	92.33
SiO_2	35.0min	61.24
Reactive Silica	20.0min	31.11
MgO	5.0max	1.01
SO_3	3.0max	0.1
Cl	0.05max	≤.01
Loss on Ignition (LOI)	5.0max	0.63

Table 2. Properties of GGBS

Property	**Requirement as per IS 16714 [7]**	**Test Result**
Fineness(m^2/Kg)	320 (Min)	381
Specific Gravity	–	2.89
Residue by wet basis on 45μ (%)	–	5.00
Manganese oxide	5.50(Max)	0.21
Magnesium oxide	17.00(Max)	8.02
Sulphur Sulphide	2.00(Max)	0.52
Sulphate	3.00(Max)	0.16
Insoluble residue	3.00(Max)	0.14
Chloride content	0.10(Max)	0.005
Loss on ignition	3.00(Max)	0.39
Moisture	1.00 (Max)	0.022
Glass	85 (Min)	97.75
Initial Setting Time	more than OPC	180min
Slag Activity Index		
7 days	Not less than 60 % of control OPC 43 Grade cement mortar cube	69.78
28 days	Not less than 75 % of control OPC 43 Grade cement mortar cube	89.74
Chemical Moduli		
$(CaO+MgO+ 1/3Al_2O_3)$ / $SiO_2+2/3Al_2O_3$	1.00 (Min)	1.11
$CaO + MgO + Al_2O_3 / SiO_2$	1.00 (Min)	1.84
$(CaO + MgO + SiO_2)$	66.66 (Min)	79.96
$(CaO + MgO/SiO_2)$	>1.0	1.26
(CaO/SiO_2)	<1.40	1.03

Alkaline Solutions

10M Sodium hydroxide (NaOH) solution was prepared by dissolving pellets in the water. Sodium hydroxide (NaOH) solution must be prepared before 24 hours prior to casting and also strongly recommended that the prepared solution should not be exceeding 36 hours because it terminates to semi solid-liquid state. 10M NaOH solution consists of 400grams of solid NaOH per liter solution

The sodium silicate (Na_2SiO_3) (water glass or liquid glass) which is liquid or gel was used. In the present investigation the ratio of sodium silicate to sodium hydroxide is kept to be 1.23. Alkaline liquid was prepared by mixing of Na_2SiO_3 and NaOH at room temperature according to mix ratio considered for design. When both the solutions are mixed together they start reacting, it means polymerization starts taking place. It is to be noted that, as it liberates large amount of heat, it should be kept for about 24 hours before using it for mixing. The alkaline liquid to solid binder ratio was chosen to be 0.36.

Mix Design

Grade M30 (30 MPa compressive strength) geopolymer mix was designed based on trial and error methods. 60% Fly ash class F and 40% GGBFS were used as binder materials to produce ambient geopolymer concrete of grade 30 MPa, which is activated by alkaline solution (mixture of sodium Hydroxide and sodium silicate) with a constant Molarity of Sodium Hydroxide as 10M without any cement. The mix proportions of the conventional geopolymer is given in Table 3. The proportion of steel fibres added were 0.25% and 0.5% by volume of the geopolymer. Each mix was assigned a unique ID. Mix G-S-0, G-S-0.25 and G-S-0.5 are the geopolymer mixes with 0%, 0.25% and 0.5% steel fibres respectively.

Table 3. Mix Proportions

Mix id	Steel fibre	Fly ash	GGBFS	Sodium silicate solution	Sodium hydroxide solution	CA	FA	Water
	(%)	(kg/m^3)						
G-S-0	0.00	239	159	96	77	1111	721	52
G-S-0.25	0.25							
G-S-0.5	0.50							

Manufacturing of Fresh Geopolymer Concrete

Geopolymer concrete specimens were prepared based on the methods adopted by Rafeet, et al. [8]. As per mix proportions the Alkaline Activator Solution (AAS) was prepared one day before to the casting of fresh geopolymer concrete. The coarse aggregate and M-sand were in saturated surface dry condition. The fly ash, ground granulated blast furnace slag and the aggregates were first mixed together in pan mixer with rotation for about 3 minutes. The liquid component Alkaline Activator Solution (AAS) was then added to the dry materials gradually with addition of steel fibres and calculated free water to the mix and the mixing continued by rotating pan mixer for further about 4 minutes to manufacture the fresh geopolymer concrete. The fresh Geopolymer concrete is tested for workability tests. After the tests are performed the concrete is cast into the specimen moulds. The demoulding is done carefully to take out specimens without damaging from moulds. The weight of all specimens along with specific names was noted. After demoulding specimens were kept for ambient curing that is at room temperature till they are tested.

Experimental Methods

Workability tests namely slump cone and compaction factor tests are conducted on freshly prepared geopolymer concrete mixes before casting the specimens. Compressive strength test is carried out on cubes of size 150mm*150mm*150mm after 28 days, 56 days and 90 days of ambient curing. Following are the results for different mixes of geopolymer concrete. Split tensile strength test is conducted on cylinders of diameter 150mm and Length 300mm after 28 days, 56 days and 90 days of ambient curing. Following are the results for different mixes of geopolymer concrete. Flexural strength test was conducted on beams of size 100mm*100mm*500mm after 28 days, 56 days and 90 days of ambient curing. Following are the results for different mixes of geopolymer concrete.

Results and Discussion

Workability

Slump cone and compaction factor test results for different steel fibre reinforced geopolymer concrete (SFRGC) mixes are given in Table 4. It can be observed that that as steel fiber content in

Materials Research Proceedings 31 (2023) 398-407 https://doi.org/10.21741/9781644902592-41

geopolymer mix increases the slump and compaction factor values decreases. The slump and compaction factor values are higher for G-S-0 compared to G-S-0.25 and G-S-0.5. Even though the slump and compaction factor values decrease due to addition of steel fibres to concrete mixes, all mixes are in good workable stage. Due to high viscosity of geopolymer it is necessary to perform workability of fresh geopolymer. From the results it is seen that there is no balling effect of steel fibres in concrete and fibres are dispersed uniformly throughout the mix. It is also seen that for all geopolymer concrete mixes compaction factor value is less than unity. An increase in percentage of steel fiber content decreases workability of GPC due to more stiffness of steel fiber compared to 0% SFRGPC.

Table 4. Workability results of SFRGPC

Mix ID	Slump (mm)	Compaction Factor
G-S-0	198	0.90
G-S-0.25	187	0.87
G-S-0.5	178	0.84

Compressive Strength

The results of the compressive strength of SFRGPC mixes is presented in Fig. 1. It is noticed that as the age of concrete increases the compressive strength also gets increased due to continuous polymerization reaction. The addition of steel fiber resulted to enhancement of the compressive strength of the geopolymer concrete. The compressive strength of G-S-0.25 and G-S-0.5 is observed to be higher than G-S-0 at all ages. At 28 days, the compressive strength of mixes G-S-0.25 and G-S-0.5 improved by 5.8% and 8.2% respectively compared to mix G-S-0. Similarly, at 90 days the compressive strength of mixes G-S-0.25 and G-S-0.5 were superior by 7.6% and 15.6% respectively compared to mix G-S-0. The improvement in strength of the geopolymer was more pronounced at higher age of curing. This is due to the fact that the continuous polymerization led to stronger bonding between the fiber and binder matrix, hence higher strength. The improvement in compressive strength with addition of steel fiber can be ascribed to the binder mechanism as well as bond mechanism of geopolymer. As hooked end steel fiber is used, the bonding between the fiber and binder matrix is very strong which contributed to the strength improvement. Steel fibre enhanced the structural integrity of the cement matrix, delay the beginning, growth and spread of macro and micro crack, which led to significant rise in post-cracking load resistance and consequently improvement in compressive strength [9, 10].

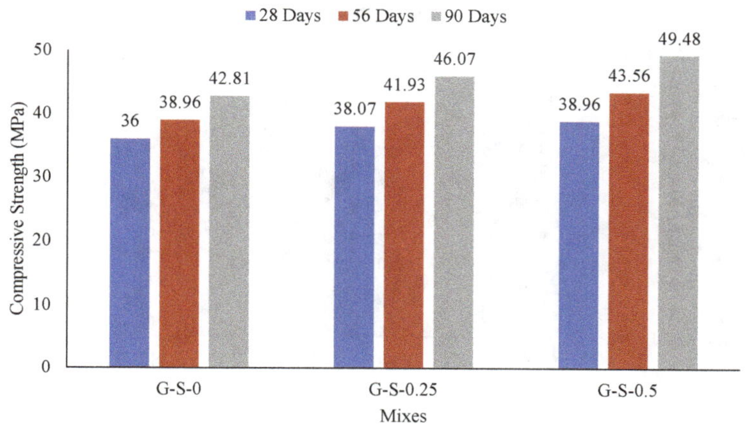

Fig. 1. Compressive Strength of SFRGPC

Split Tensile Test

The results of the splitting tensile strength of the SFRGPC mixes is presented in Fig. 2. The split tensile strength increased with age due to continues polymerization reaction of the geopolymer. The addition of steel fiber to the geopolymer concrete under ambient curing conditions led to improvement in tensile strength. The split tensile strength of G-S-0.25 and G-S-0.5 is observed to be higher than G-S-0 at all ages. At 28 days, the compressive strength of mixes G-S-0.25 and G-S-0.5 were higher by 26.9% and 44.2% respectively compared to mix G-S-0. Similarly, at 90 days the compressive strength of mixes G-S-0.25 and G-S-0.5 were superior by 17.7% and 22.9% respectively compared to mix G-S-0. The increase in split tensile strength with addition of steel fiber to the geopolymer is due to the high tensile strength of the fiber which invariable increased the tensile strength of the concrete. Additionally, the steel fiber due to its ability to prevent the occurrence, growth and propagation of cracks in the geopolymer when subjected to tensile load, enhanced the post-cracking load resistance and ductility of the concrete and hence improvement in tensile strength [11].

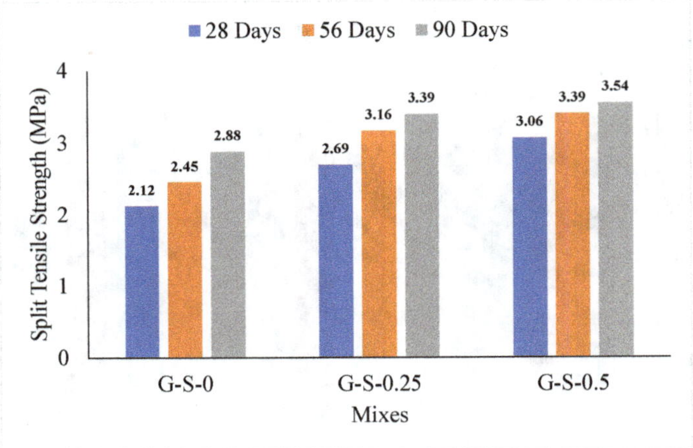

Fig. 2. Split Tensile Strength of SFRGPC

Flexural Strength

The results of the SFRGPC mixes is presented in Fig. 3. It is noticed that as the age of the geopolymer increases the flexural strength also gets increased due to continuous polymerization process. Similar to compressive and split tensile strengths, the flexural strength of the geopolymer concrete increased with increase in streel fiber content at all ages. In comparison to the control mix (G-S-0), the flexural strength of mixes G-S-0.25 and G-S-0.5 were higher by 21.5% and 42.9% respectively at 28 days, and by 19.4% and 42.1% respectively at 56 days, and 14.8% and 35.2% respectively at 90 days. One of the main reason for the enhancement in flexural strength with addition of steel fibers to geopolymer is the strong adhesion between the binder matrix and steel fiber which is better for SFRGPC compared to steel fiber reinforced concrete [12]. Additionally, steel fiber due to its high tensile loads when added to geopolymer concrete significantly improved its bending resistance. Furthermore, steel fiber due to its crack bridging through strong bonding, delay failure of the geopolymer when subjected to bending loads through redistribution and transfer of stresses across the matrix and improving the post cracking failure resistance [11, 13]

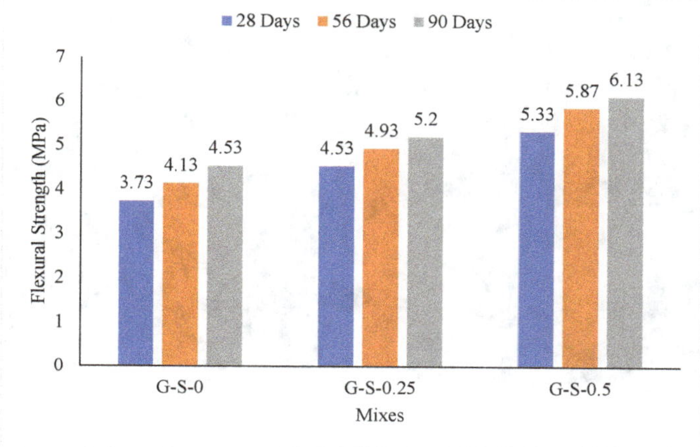

Fig. 3. Flexural Strength of SFRGPC

Microstructural Evaluation

The microstructural morphology of mix G-S-0 as obtained using scanning electron microscopy (SEM) is presented in Fig. 4. It can be observed that pores and formation of geopolymer matrix are present. The aggregate surface, reacted fly ash, reacted GGBFS, partially reacted fly ash and unreacted GGBFS can also be seen on the surface. Crystal needles and N-A-S-H gel matrix and microcracks can also be seen. Fig. 5 presents the microstructural morphology of mix G-S-0.5. In the microstructure pores, cracks, reacted GGBFS, partially GGBFS, N-A-S-H gel matrix, unreacted GGBFS, unreacted fly ash, partially reacted fly ash, reacted fly ash are observed. By comparing the morphology of mixes G-S-0 with G-S-0.5, it can be seen that the latter is more densified with less microcracks and pores compared to the former.

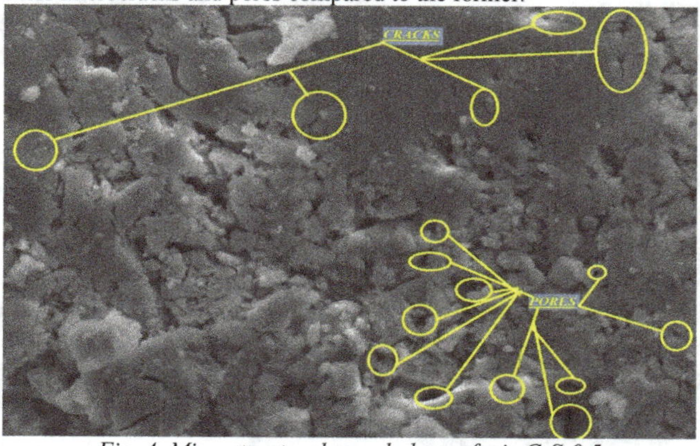

Fig. 4. Microstructural morphology of mix G-S-0.5

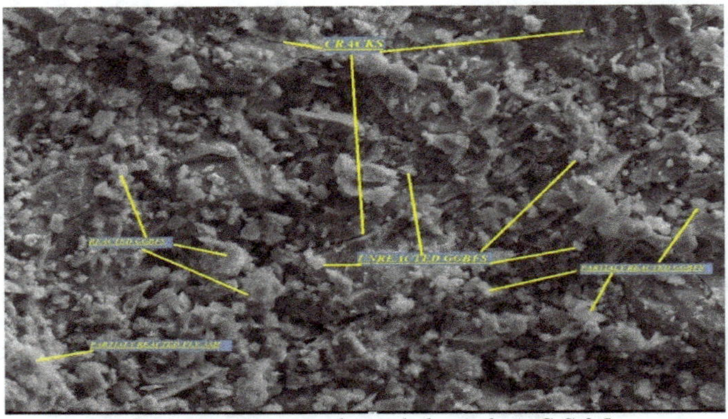

Fig. 5. Microstructural morphology of mix G-S-0.5

Conclusions

The following conclusions were drawn.

1. Workability of GPC decreases with increase in percentage of steel fibres.
2. Ambient cured specimens yield good results and higher strength is observed due to high polymerization process.
3. Compressive, split tensile and Flexure strength increases with increase in percentage of volume of steel fibres.
4. Strength of geopolymer concrete depends on binder mechanism as well as bond mechanism of concrete.
5. The addition of steel fiber densified the microstructure of the geopolymer concrete by preventing the occurrence and growth of microcracks.

Acknowledgement

The authors wish to acknowledge the support of the Structures and Materials Research Laboratory of the College of Engineering, Prince Sultan University, Riyadh Saudi Arabia, for their valuable support.

References

[1] A. Bouaissi, L.-y. Li, M. M. A. B. Abdullah, Q.-B. Bui, Mechanical properties and microstructure analysis of FA-GGBS-HMNS based geopolymer concrete, Constr. Build. Mater, 210 (2019) 198-209. https://doi.org/10.1016/j.conbuildmat.2019.03.202

[2] V. Malhotra, Introduction: sustainable development and concrete technology, Concr. Int. 24 (2022), no. 7.

[3] J. Davidovits, Geopolymers and geopolymeric materials, J. therm. Anal. 35 (1989) 429-441 https://doi.org/10.1007/BF01904446

[4] J. Davidovits, Properties of geopolymer cements," in First international conference on alkaline cements and concretes, 1 (1994) 131-149.

[5] Y. Ding, Y.-L. Bai, Fracture properties and softening curves of steel fiber-reinforced slag-based geopolymer mortar and concrete, Mater. 11-8 (2018), 1445. https://doi.org/10.3390/ma11081445.

[6] IS 3812 Part 1, Pulverized Fuel Ash Specification, New Delhi, India, (2013).

[7] IS 16714, Ground Granulated Blast Furnace Slag for Use in Cement, Mortar and Concrete - Specifications, New Delhi, India, (2018).

[8] A. Rafeet, R. Vinai, M. Soutsos, W. Sha, Guidelines for mix proportioning of fly ash/GGBS based alkali activated concretes, Constr. Build. Mater 147(2017) 130-142. https://doi.org/10.1016/j.conbuildmat.2017.04.036

[9] A. C. Ganesh, K. Sowmiya, M. Muthukannan, Investigation on the effect of steel fibers in geopolymer concrete, in IOP Conference Series: Materials Science and Engineering 872-1 (2020) 012156. https://doi.org/10.1088/1757-899X/872/1/012156

[10] Y. Liu, Z. Zhang, C. Shi, D. Zhu, N. Li, Y. Deng, Development of ultra-high performance geopolymer concrete (UHPGC): Influence of steel fiber on mechanical properties, Cem. Concr. Comp. 112 (2020) 103670. https://doi.org/10.1016/j.cemconcomp.2020.103670

[11] P. Zhang, J. Wang, Q. Li, J. Wan, Y. Ling, Mechanical and fracture properties of steel fiber-reinforced geopolymer concrete, Sci. Eng. Comp. Mater. 28-1(2021) 299-313. https://doi.org/10.1515/secm-2021-0030

[12] K. H. Mo, K. H. Yeoh, I. I. Bashar, U. J. Alengaram, M. Z. Jumaat, Shear behaviour and mechanical properties of steel fibre-reinforced cement-based and geopolymer oil palm shell lightweight aggregate concrete, Constr. Build. Mater. 148(2017) 369-375. https://doi.org/10.1016/j.conbuildmat.2017.05.017.

[13] G. L. Golewski, Changes in the fracture toughness under mode II loading of low calcium fly ash (LCFA) concrete depending on ages, Mater. 13-22 5241. https://doi.org/10.3390/ma13225241

Advanced Topics in Mechanics of Materials, Structures and Construction - AToMech1-2023 Materials Research Forum LLC
Materials Research Proceedings 31 (2023) 408-414 https://doi.org/10.21741/9781644902592-42

Design of an automated cleaning system for solar panels

Mounir Bouzguenda[1a*], Bader Bashmail[1b], Mohammed Ben Youssef Al-Khaldi[1c], Nasser Al-Jafar[1d], Saad Al-Jasim[1e], and Mohammed A. Al Aqil[1f]

[1]College of Engineering, King Faisal University, Eastern Province, AlAhsa, Saudi Arabia

[a]mbuzganda@kfu.edu.sa, [b]bader.bashmail@gmail.com, [c]218010825@student.kfu.edu.sa, [d]NasserAdel22@hotmail.com, [e]218038109@student.kfu.edu.sa, [f]malaqil@kfu.edu.sa

Keywords: Maximum Power Point, KSA 2030 Vision, Automated Cleaning System, Arduino-based Controller, Sensors

Abstract. This project involves the design of an automated system for solar panel cleaning. The purpose of the design is to reduce the effect of dust on solar PV efficiency. The result is an increase in the efficiency of panel power output by 6.89%. The system is useful in the solar panel field where the system can be operated for a long time without human intervention. It measures and records current, voltage, temperature, radiation, and humidity. The measured data are sent to the designated supervisor.

Introduction

Fossil fuel is commonly used in Saudi Arabia, mostly in power plants, which causes lots of emissions into our air. So, one of the solutions to generating electrical energy and reducing its usage is using renewable energy sources. Solar energy is one of the most important renewable energy sources that use the sun's radiation. It is clean and environmentally friendly. The power produced by photovoltaic cells is mostly determined by the amount of sunlight that reaches them. In fact, many studies concentrate on the design and economic analysis of PV solar systems [1]. Such as using a maximum power point tracker (MPPT) to improve the efficiency of a PV system [2,3], or hybrid systems to improve the performance of a PV system [4]. All of these studies are being conducted in order to increase the efficiency of solar cells. On the other hand, there are many reasons that affect the efficiency and the output power of solar cells, such as solar radiation, Humidity, high temperatures, tilted angle, and the accumulation of dust on solar panels, particularly in desert areas.

Saudi Arabia is a desert area, with high temperatures reaching their peak in summer. The temperature ranges from 40°C to 50°C. Drought and sandstorms are common, especially in Al-HUFUF due to its geographical location near the desert called the Empty Quarter. The issue at hand is the impact of external factors like high temperatures and dust accumulation on the PV panels' surfaces. The efficiency of solar panels is greatly reduced by these two characteristics. As a result, we anticipate that a study of the impact of dust accumulation on PV module output power and efficiency will be required. Dust can also refer to dust storms, which is relevant to this study. To solve this problem, there was a solution to get rid of accumulated dust on the photovoltaic panels using cleaning.

Cleaning solar panels adopted through many stages, starting from traditional cleaning methods are ineffective and detrimental. New techniques are being evolved. Natural forces inclusive of wind and rain will eliminate dust. Mechanical methods, self-cleaning nano-film, and electrostatic gear are employed [5]. Those are referred to as energetic cleaning methods and are applied directly or in a roundabout way to the solar panel [6,7].

The paper is organized as follows. In Section 2, the authors compare the existing cleaning methods of solar panels. Section 3 is devoted to describing the design methodology and the proposed automatic cleaning systems. While Section 4 describes the performance of the designed cleaning system. Finally, section 5 involves the conclusion, recommendations, and future improvement of the design.

Review of Solar Panel Cleaning Techniques

The operation and maintenance of photovoltaic panels should be carried out frequently, not only to improve their efficiency but also to prevent damage caused by dust accumulation due to environmental pollution. A study on several dust removal techniques has been carried out in this work. These methods can be classified as shown in Figure 1. Some of these techniques are discussed below. Their advantages and disadvantages are summarized in Table 1.

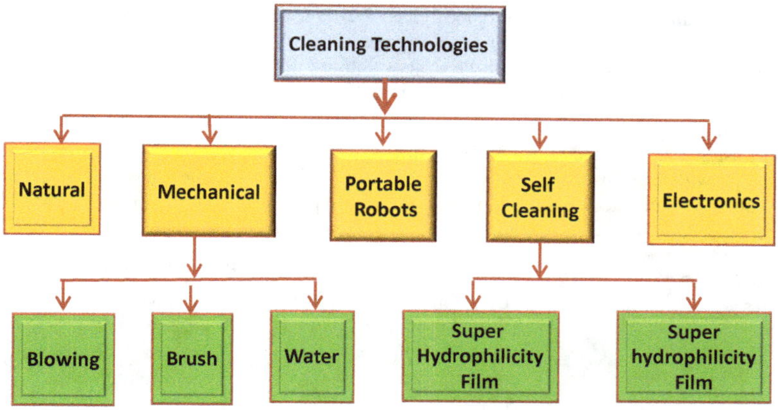

Figure 1. Common Solar PV panel cleaning methods

Dry Robotic Cleaning

Robots can clean solar panels without water. Robots use dry brushes and air pressure to remove dirt. Dry cleaning is generally less effective than wet cleaning. The water involved in wet cleaning removes dust particles since it acts as a medium for the particles to disperse. Despite this, water is not always available everywhere, and this makes it difficult to clean the modules. Therefore, dry cleaning is a great alternative.

Electrostatic Cleaning

Solar panels can be cleaned by means of moving wave electric charge on small particles suspended in liquid [8], allowing dust and similar dirt; except algae to be removed. Dust particles are emitted from the panels when a negative voltage is applied to one electrode and a positive voltage is applied to the other electrode [9,10]. A different mechanism for removing dust involved the distribution of the electric field density to create an electric curtain. By varying the electric field density over the plate, this system forces the dust off the plate [11].

Natural Method

Rain is considered a natural way that cleans PV panels. The amount of rain limits the effectiveness where some countries get rainy less often than others. According to an experimental study [12], accumulated soiling on PV panels for 5 years reduces the generated power by 3%, and the rain could improve the efficiency by only 1% of the full power rating [13]. The results of dry months hint that the output of the PV panel could be reduced by 6%.[14] It is worth noting that light rain might also reduce the efficiency of PV panels and that is due to the dust slides by the rainwater from the upper cells of PV panels onto the lower ones.

Coating Vibration method
Coating Vibration is another method used to prevent dust accumulation by applying a special spray on the surface of PV solar panels and shaking them with a mechanical vibration system. An experimental study was applied to compare the power efficiency of PV panels with different scenarios. According to the experiment PV panel with coating and vibration technique had been reduced by 10% of power efficiency in 6 weeks, while the PV panel with no coating or vibration had been reduced by 10% of power efficiency in 1 week and goes along 33% in 6 weeks [15].

Mechanical Methods
Definition: The mechanical methods remove the dust by brushing, blowing, vibrating, and ultrasonic driving. ` The brushing methods clean the solar cell with something like the broom or brush that was driven by the machine was designed just like a windscreen- wiper. A mechanical method has four techniques to expel the dust which are a robotic method, air-blowing method, water-blowing method, and ultrasonic vibration method.

Wet cleaning
Wet cleansing is completed with ionized distilled water, which is desired for this task. Cleaning PV panels with pure water has massive benefits; it entirely eliminates the contaminating particulates on the surfaces and leaves no residue, which increases the panel output voltage [16,17].

Table 1 Comparison of the forthmentioned cleaning methods

Technique	Approach	Advantages	Demerits
Coating	Water-Free	•High efficiency in humid regions •No electrical power consumption. •Availability of providing other features such as anti-icing, more stability, anti-reflecting, Photocatalysis reaction, and anti-fogging	•Recoating requirement •Treats of realized chemical materials for the environment
Electrostatic charge system	electric curtain	•No water is required •No heavy gears are required	•Needs recharging
Linear Piezoelectric System	Vibration	•Minimum gears required •Lightweight on the surface of the panel	•Comparatively complex system
Manual Cleaning	Water-Based	•Simple with a low cleaning cost •No need for a power supply	•Cleaning efficiency is dependent on the technician's expertise •High water consumption •Restrictions of the floating structure and weight-bearing
Robots	Water-Based and/or Water-free	•Modules can be cleaned every single day with robotic cleaning systems, which is considerably more convenient than manual cleaning.	•This type of cleaning requires a significant investment. This makes it unsuitable for residential use.

Materials Research Proceedings 31 (2023) 408-414 https://doi.org/10.21741/9781644902592-42

		•Compared to manual cleaning, dry cleaning produces more energy because modules are cleaned daily	•Electricity is required to operate these robotic cleaners. That adds to the cost.
Self-Cleaning	Water-Based /Water-Free	•Fully automated •Cooling effect (water-based approach)	•Low efficiency •High consumption of water •High initial power cost
Wet cleaning	Distilled water	•Does not leave a residue •Does not cause corrosion	•High consumption of distilled water

Design methodology and the proposed automatic cleaning systems

The proposed system design will be environmentally friendly and fully automated to clean dust accumulation. The system consists of these main components: Solar panel, Arduino Mega controller, sensors, DC motors, water pump, water tank, and wiper shown in Figure 2. The process of cleaning starts when the system detects a power drop by comparing the PV-measured power P_M and the expected power P_{th} given by:

$$P_M = V_M \times I_M \tag{1}$$

$$P_{th} = E \times A_c \times \%\eta \tag{2}$$

Where

V_M is the measured voltage,
I_M is the measured current
E is the solar radiation at the time of the test
Ac is the area of the solar panel.
η is the solar panel efficiency.

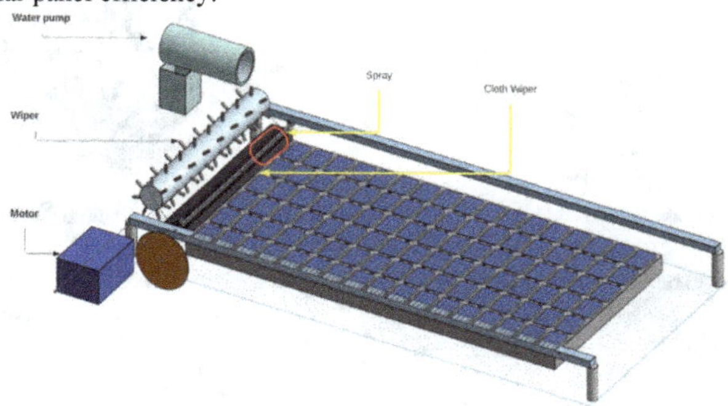

Figure 2: 3D model of the automatic cleaning system.

The control unit is the heart of the cleaning system. It consists of an Arduino Mega unit. It processes all sensor values; it compares the expected power output with the actual output and calls for cleaning. In addition, it sends the data to the transmitter and the display unit. The cleaning unit consists of the DC Motor to move the cleaning mop back and forth. The motor operates at a low speed for effective cleaning. The chosen DC motor has a torque of 2.88 kg.cm at 30.07 rpm. The wiper is chosen in a way such that the weight of the rod and the electric motor are developing a

Materials Research Proceedings 31 (2023) 408-414 https://doi.org/10.21741/9781644902592-42

normal force at the PV panel. This normal force creates a frictional one against the movement of the wiper, which will help to clean the cell very effectively.

The water pumping system's main function is to supply the system with adequate water. The principle of operation of this system is to pump water from a low-pressure head to a high pressure to increase the flow rate of the liquid. In addition, a solar radiation sensor is needed to measure the radiation level hitting the solar panel. The output of the sensor is fed to the control unit to decide if the panel power output is as expected or below expected in case of dust accumulation.

The sensor system includes temperature and humidity sensors, a voltage sensor, and a current sensor. The voltage and current sensors are used to calculate the panel power output. The control unit compares it with the expected power output and activates the cleaning process. The following flowchart shows the principles of the automatic cleaning system. The final cleaning system is shown in Figure 3. The prototype performance and benefits are evaluated through field measurements.

Results

The main outcome of the proposed design is improving the solar panel performance by removing the dust settling on the panel. Many cleaning tests have been made. Figures 4 and 5 show the I-V and P_V curves for a 30-Watt solar panel before and after cleaning. The two figures depict that the current and power have increased after removing the dust. The current increased by 9.47% and the maximum power increased by 6.87%.

Conclusion and Future Work

In conclusion, we have looked up factors that affect PV efficiency, and depending on the previous methods and techniques, the combination of water cleaning and nano-spray automated cleaning system is the best method for cleaning dust accumulation.

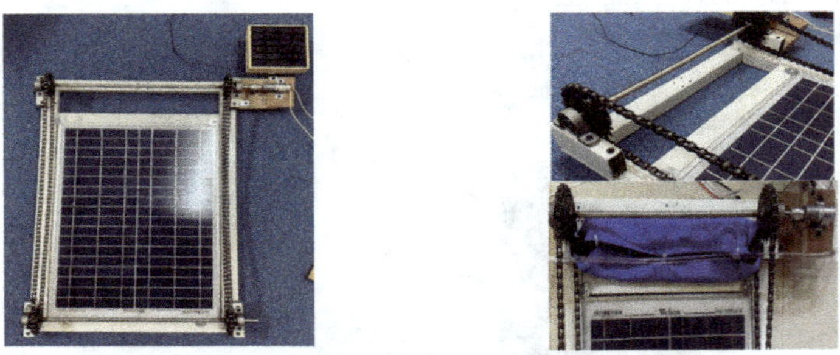

Figure 3: The automatic cleaning system final design

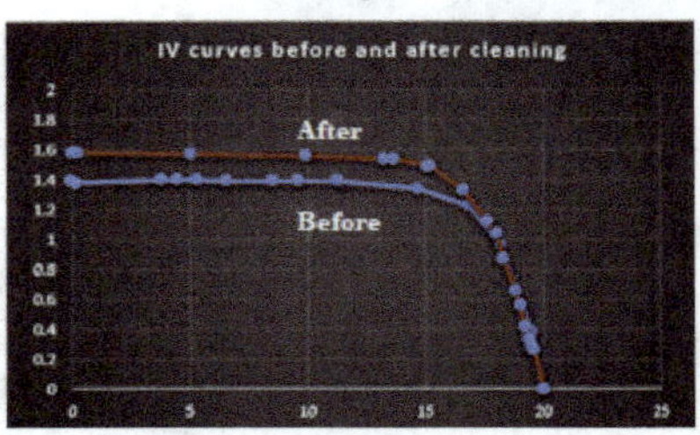

Figure 4 I-V curve for a 30-Watt solar panel before and after cleaning

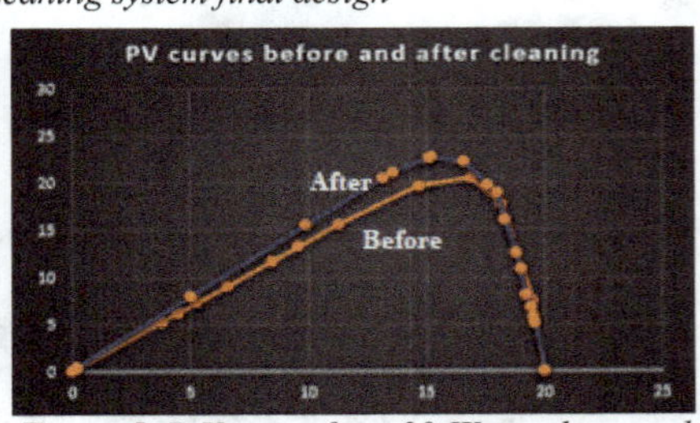

Figure 5 P-V curve for a 30-Watt solar panel before and after cleaning

Based on the experiments conducted at KFU, the dust effect on PV efficiency was recorded, and cleaning the tested solar panel raises the power efficiency by 6.86%. Our system can be quite useful in the solar panel field where the system can be operated for a long time without human intervention. Also, it is considered green energy and in the line with the future vision of Saudi Arabia. However, power decrease is caused by different natural factors such as (temperature, shadow, dust, etc.) which can be recognized by the control unit.

On the other hand, the system's total energy consumption during the cleaning process is about 66 mWh. The cleaning system will definitely improve the power efficiency by reducing dust accumulation by nano-spray and cleaning the solar panel when the dust starts to affect the power efficiency. Even though a test bed for cleaning solar panels is assembled, the following improvements are needed. These include an automated nano spray, a higher cleaning range (suitable for many solar arrays), and a WiFi-based remote monitoring system.

Acknowledgment
The authors would like to thank King Faisal University for funding this project.

References

[1] H. Shahinzadeh et al., "Design and Economic Study for Use the Photovoltaic Systems for Electricity Supply in Isfahan Museum Park," International Journal of Power Electronics and Drive System (IJPEDS), 3 (1), 83-94 (2013). https://doi.org/10.11591/ijpeds.v3i1.1797

[2] A. Hassani et al., "A New High Speed and Accurate FPGA-based Maximum Power Point Tracking Method for Photovoltaic Systems," International Journal of Power Electronics and Drive System (IJPEDS), 8 (1), 1335-1344 (2017). https://doi.org/10.11591/ijpeds.v8.i3.pp1335-1344

[3] H. Abouobaida and E. B. Said, "Practical Performance Evaluation of Maximum Power Point Tracking Algorithms in a Photovoltaic System," International Journal of Power Electronics and Drive System (IJPEDS), 8 (4), 1744-1755 (2017). https://doi.org/10.11591/ijpeds.v8.i4.pp1744-1755

[4] A. Lekbir et al., "The Recovery of Energy from a Hybrid System to Improve the Performance of a Photovoltaic Cell," International Journal of Power Electronics and Drive System (IJPEDS), 9 (3), 957-964 (2018). https://doi.org/10.11591/ijpeds.v9.i3.pp957-964

[5] He, G.; Zhou, C.; Li, Z. Review of self-cleaning method for solar cell array. Proc. Eng. 2011, 16, 640-645 https://doi.org/10.1016/j.proeng.2011.08.1135

[6] Chanchangi, Y.N.; Ghosh, A.; Sundaram, S.; Mallick, T.K. Dust and PV performance in Nigeria: A review. Renew. Sustain. Energy Rev. 2020, 21, 1-14 https://doi.org/10.1016/j.rser.2020.109704

[7] Maghami, M.R.; Hizam, H.; Gomes, C.; Radzi, M.A.; Rezadad, M.I.; Hajighorbani, S. Power loss due to soiling on solar panel: A review. Renew. Sustain. Energy Rev. 2016, 59, 1307-1316. https://doi.org/10.1016/j.rser.2016.01.044

[8] Masuda, S.; Washizu, M.; Iwadare, M. Separation of small particles suspended in liquid by nonuniform traveling field. IEEE Trans. Ind. Appl. 1987, 23, 474. https://doi.org/10.1109/TIA.1987.4504934

[9] Biris, A.S.; Sanini, D.; Srirama, P.K.; Mazumder, P.K.; Sims, R.A.; Calle, C.I. Electrodynamic removal of contaminant particles and its applications. In Proceedings of the Conference Record of the 2004 IEEE Industry Applications Conference, 2004, 39th IAS Annual Meeting, Seattle, WA, USA; 2004; pp. 1283-1286.

[10] Hudedmani, M.G.; Joshi, G.; Umayal, R.M.; Revankar, A. A comparative study of dust cleaning methods for the solar PV panels. Adv. J. Grad. Res. 2017, 1, 24-29. https://doi.org/10.21467/ajgr.1.1.24-29

[11] Guangming, W.; Dan, L.; Guangjian, X.; Tianlan, Y. The mechanism study of dust removal with transparent interdigitated electrodes. Integr. Ferroelec. 2016, 171, 1-7. https://doi.org/10.1080/10584587.2016.1162585

[12] Hammond, R.; Srinivasan, D.; Harris, A.; Whitfield, K.; Wohlgemuth, J. Effects of soiling on PV module and radiometer performance. In Proceedings of the Conference Record of the Twenty Sixth IEEE Photovoltaic Specialists Conference-1997, Anaheim, CA, USA, 29 September-3 October 1997; pp. 1121-1124.

[13] Vignola, F.E.; Krumsick, J.; Mavromatakis, F.; Walwyn, R. Measuring degradation of photovoltaic module performance in the field. In Proceedings of the 38th ASES National Solar Conference (SOLAR 2009), Buffalo, NY, USA, 16 May 2009; pp. 11-16.

[14] Khonkar, H.; Alyahya, A.; Aljuwaied, M.; Halawani, M.; Al Saferan, A.; Al-khaldi, F.; Alhadlaq, F.; Wacaser, B.A. Importance of cleaning concentrated photovoltaic arrays in a desert environment. Sol. Energy 2014, 110, 268-275. https://doi.org/10.1016/j.solener.2014.08.001

[15] Moharram, K.A., Abd-Elhady, M.S., Kandil, H.A., El-Sherif, H., 2013. Influence of cleaning using water and surfactants on the performance of photovoltaic panels. Energy Covers. 68,266-272. https://doi.org/10.1016/j.enconman.2013.01.022

[16] Myenerjisolar, Available online: https://www.myenerjisolar.com/gunes-enerjisi-panel-temizligi/ (accessed on 1 November 2020).

[17] Kursun, M. The effect of natural dust accumulation on the efficiency of solar panels in Samsun climate conditions. Master's Thesis, University of Ondokuz Mayıs, Samsun, Turkey, 2019. Available online: http://libra.omu.edu.tr/tezler/131186.pdf (accessed on 1 November 2022).

Strength properties of self-compacting concrete incorporating iron ore tailings

AJAGBE Wasiu Olabamiji[1,a], OKPANACHI Robert Ojonugwa[1,b], GANIYU Abideen Adekunle[2,c*] and ABDULKAREEM Muyideen Oladimeji[3,d]

[1]Department of Civil Engineering, Faculty of Technology, University of Ibadan, Ibadan, Nigeria

[2]Department of Civil Engineering, British University of Bahrain, Saar, Kingdom of Bahrain

[3]Department of Civil & Environmental Engineering, UCSI University, Kuala Lumpur, Malaysia

[a]ajagbewas@gmail.com, [b]robert4stephen@gmail.com, *[c]a.ganiyu@bub.bh, [d]muyikareem@gmail.com

Keywords: Industrial Wastes, Iron Ore Tailings, Fine Aggregates, Compressive Strength, Flexural Strength, Self-Compacting Concrete

Abstract. The quest for sustainable production of concrete is a major step towards achieving net-zero Carbon. The utilisation of industrial waste in the production of concrete entrenches both sustainability and economic viability. Iron Ore Tailings (IOT) are a byproduct of the beneficiation process of iron ore. The unbridled disposal of Iron Ore Tailings (IOT) has proven to be a great threat to the natural environment. Self-compacting concrete (SCC) has gained prominence in the construction industry based on its unique flowability quality. This study investigated the strength characteristics of SCC incorporating IOT as replacement for fine aggregate. Concrete containing 5%, 10%, 15%, 20%, 30%, 40%, and 100% IOT as fine aggregate replacement were prepared and subjected to compressive and flexural strengths tests. The results reveal a progressive increase in strength with age for compressive strength, and progressive increase in strength with increasing percentage contents of IOT for both compressive and flexural strength tests. The inclusion of IOT as partial replacement of fine aggregates in SCC has a potential to mitigate the environmental degradation caused by its disposal and excessive depletion of natural sand. Besides, it is a viable measure towards reduction of carbon footprint of concrete production with evident economical benefits.

Introduction

Aggregates are the key constituents of concrete accountable for the precipitous exhaustion of natural resources. Sand mining has a multitude of negative environmental effects, including depletion of virgin resources, reduction of water tables, riverbank collapse, and water contamination [1]. A novel approach to sustainable development of concrete is the partial replacement of aggregates with suitable industrial waste material thereby preserving the naturally available aggregates, reducing the cost, and preventing indiscreet dumping of industrial wastes into the environment and all its associated complications [2; 3]. In this vein, the adoption of alternative fine aggregates such as crushed rock sands, recycled fine aggregates, and industrial by-products has been put forward [4; 5].

Iron Ore Tailings (IOT) are a by-product of the beneficiation process of iron ore. The unbridled disposal of several tons of IOT in landfills and quarries has proven to be a great threat to the natural environment. In Nigeria, the National Iron Ore Mining Company (NIOMCO) Itakpe, produces 3,072 tonnes of IOT per day, a significant proportion of which are not productively utilised [6]. The high specific surface area, high density, and irregular shape of IOT have a detrimental effect on the workability and shrinkage of concrete [7; 8]. However, the chemical composition and

micro-structural analysis of IOT reveal that it is suitable as a binding material in concrete either by micro filling of voids or in pozzolanic reaction [9].
Self-compacting concrete (SCC) is a concrete type that can flow, fill formwork, encapsulate the reinforcement, and maintain its homogeneity under its own weight. The adoption of SCC in construction reduces the labour cost, accelerates project schedule, and improves structural integrity and sustainability [10]. The rheology, strength, shrinkage, and durability of SCC are affected by the mix design process, raw material properties, chemical and mineral admixtures, aggregate packing density, and the water to cement ratio [11; 12]. SCC does not require compaction and has different strength properties as compared to normal concrete, although increase in strengths is observed with decrease in water cement ratio as observed for normal concrete [13].

Previous research on the utilisation of IOT in concrete has been limited to the normal concrete and/or high-density concrete as partial replacement for cement, fine and/or coarse aggregates, engineered cementitious composites (ECC), ultra-high performance concrete (UHPC) [14-16], but the adoption of IOT in SCC is less prominent. Ferreira *et al.* [17] utilised IOT as aggregate in the production of metakaolin based geopolymer and reported that IOT acts satisfactorily at improving the quality of the geopolymer. Siamardi [18] modelled the fresh and hardened properties of powder type light weight SCC produced with coarse grained light expanded clay aggregates (LECA) as partial replacement of normal weight aggregates. Tang *et al.* [19] utilised red mud to partially replace fly ash in SCC and observed that samples containing 50% red mud have the best performance in compressive strength and elastic modulus. This research investigated the strength characteristics of self-compacting concrete incorporating IOT as partial replacement for fine aggregates by conducting compressive and flexural strength tests on samples of different replacement percentages of IOT.

Materials and Methods
Natural river sand obtained from Osogbo, Nigeria was used in this research. Locally available granite, also sourced from Osogbo with a maximum size of 20mm was used. The coarse aggregate used is the locally available granite, also sourced from Osogbo, the maximum size of which was 20mm. Figure 1 (a and b) show the aerial view of NIOMCO Itakpe, Nigeria, and the IOT sample obtained therefrom.

Figure1: (a) Aerial view of NIOMCO, Itakpe, Nigeria (b) IOT sample

Particle size distribution, specific gravity and bulk density tests were conducted on all aggregates. Portland Limestone cement, of Grade 32.5R (locally available Dangote brand) which complied with BS EN 197-1 was the binder used. MasterRheobuild 858 superplasticizer, a ready-to-use, high range water-reducing admixture designed to produce high slump concrete was used as chemical admixture to aid the workability of the concrete. It was added at an optimum dosage not exceeding 1.0 % by weight of cement. Table 1 gives the properties of the superplasticizer as obtained from the manufacturer.

Table 1: Properties of the Superplasticizer

Colour	Specific gravity @ 25°C	Chloride content	Freezing point	Flashpoint
Dark Brown	1.240	Chloride-free to EN 934	0°C	N/A

Potable water conforming to the requirements of water for concreting was utilised for mixing the concrete. A mix ratio of 1:2.2:2 for cement, fine and coarse aggregates, respectively with a water/cement ratio of 0.5 was used. The mould of 100 mm×100 mm × 100 mm size was used to produce the test samples for the concrete cubes and 100 mm× 100 mm × 500 mm was used to produce the test samples for the beams in accordance with BS EN 12390. Samples containing IOT as replacement of fine aggregates at 0, 5, 10, 15, 20, 30, 40 & 100 % were produced. The concrete was poured into the mould and compacted thoroughly, its surface was levelled, after which it was kept for 24 hours before being demoulded and immersed in a water curing tank containing clean water until the day of testing (7, 14 & 28 days for cubes, and 28 days for beams). Three samples were made and tested for each mix, and the results reported for each mix are the average of three samples, the cube samples were tested for compressive strength and the beams for flexural strength. Figure 2 (a and b) show the concrete samples in compressive and flexural strength tests, respectively.

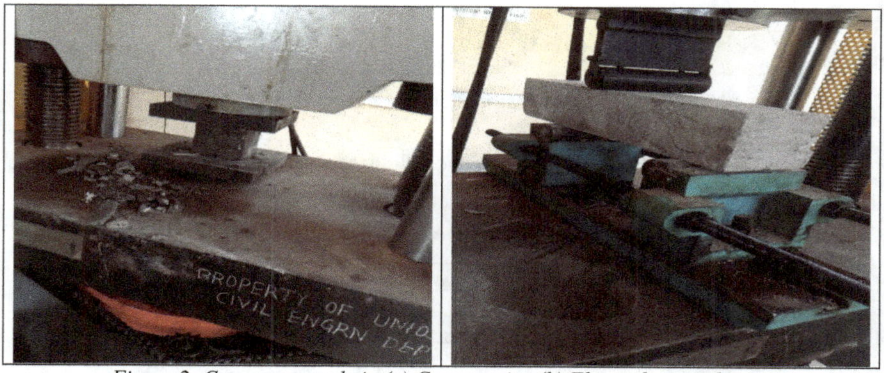

Figure 2: Concrete sample in (a) Compressive (b) Flexural strength tests

Results and Discussions

Figure 1 presents the particle size distribution obtained from the sieve analysis for coarse aggregates (gravel) and fine aggregates (sand and IOT), the curve indicates that the size of fine aggregates used was between 0.063 mm and 4.75 mm, while the size of coarse aggregates was between 4.75 mm and 19.5 mm. The bulk density, specific gravity, compacted and uncompacted bulk density of the aggregates are presented in Table 2. The void in IOT has a greater tendency to absorb water, hence the high moisture content, also IOT is denser than both fine (sand) and coarse aggregates used, thus a rise in the density of concrete is expected with the addition of IOT. The ratio of uncompacted to compacted bulk densities are 0.81, 0.88 & 0.89 for the gravel, sand and IOT, respectively.

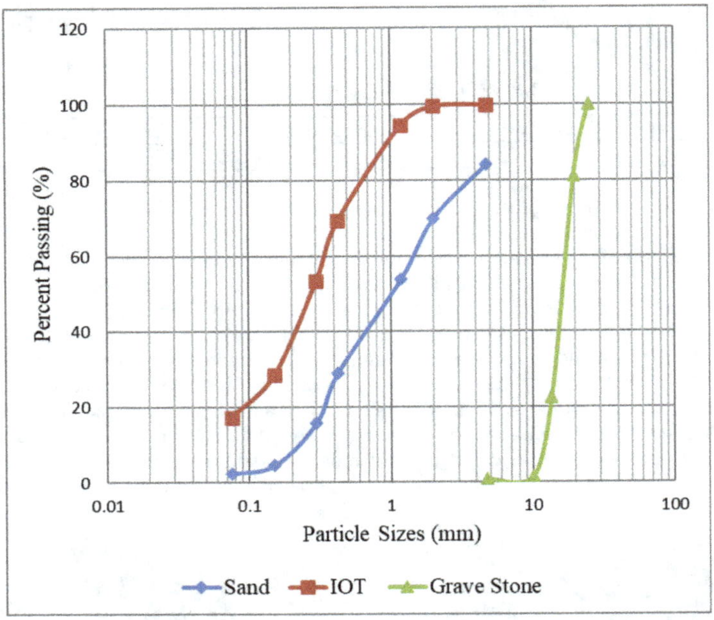

Figure 3: Particle size distribution curve for gravel, sand and IOT

Table 2: Physical and Mechanical Properties of the Aggregates

	Moisture Content (%)	Specific gravity	Bulk Density Uncompacted (kg/m3)	Bulk Density Compacted (kg/m3)
Gravel	0.92	2.63	834.8	1028.4
Sand	4.51	2.53	1052.9	1185.4
IOT	6.2	3.26	1747.5	1953.6

Figure 4 presents the compressive strengths at 7, 14 & 28 days of curing for the different percentages of IOT replacement of fine aggregates. The compressive strength of all the concrete samples increased with the increasing curing age, the compressive strength also increased with increase in IOT percentage, with 100 % replacement having the highest value and the control sample (0% IOT) the least strength for all curing days. The finer sizes of IOT particles may have contributed to the observed increase in strength by filling the pore and optimizing the pore structure of the mixes. Also, IOT particles have a rough and angular texture that strengthens the bond between the binding medium (cement) and the aggregate interface, resulting in increased strength. Likewise, the presence of iron at slightly higher quantities due to the inclusion of IOT in the mixes has a good effect on strength development similar to the incorporation of iron fibers [20-22].

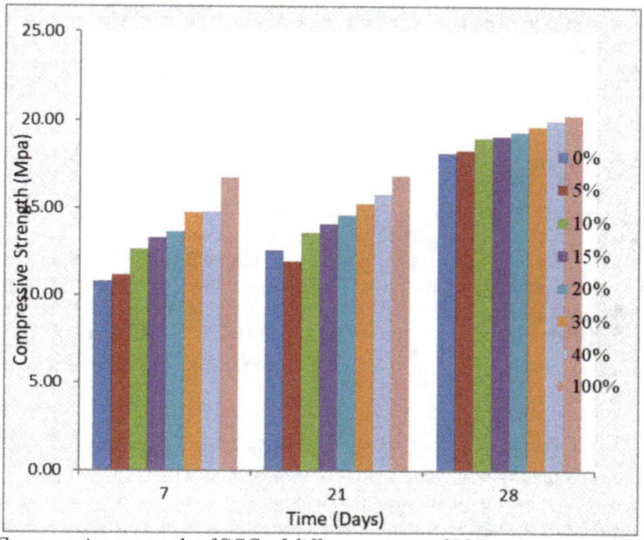

Figure 4: Compressive strength of SCC of different mixes of IOT at 7-, 21- & 28-days curing

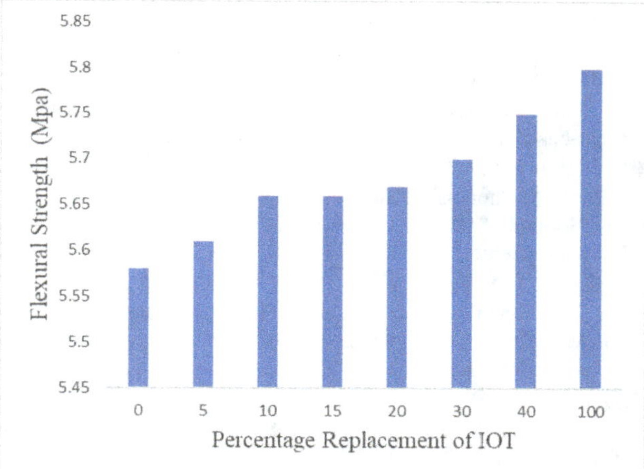

Figure 5: Flexural of SCC of different mixes of IOT at 28-days curing

The results of the 28-day flexural strength of the concretes are presented in Figure 5. From the Figure, the flexural strength of the different mixes behaved in a similar way to the compressive strength, as it increased with increase of IOT percentages, while the lowest and highest value of flexural strength are at 0% and 100% IOT replacements, respectively. Similar observation was reported by Zhao et al. [23] where IOT was used as partial replacement of aggregates in ultra-high-performance concrete (UHPC). The observed increase could be attributed to similar phenomena of finer particles, rough and angular texture of particles and increment of iron contents as earlier alluded to in the compressive strength. Thus, the partial replacement of fine aggregates with IOT increases the modulus of rupture of the concrete.

Conclusion

This study investigated the strength characteristics of SCC incorporating IOT as partial replacement of fine aggregate. The intent is to create large scale alternative uses for IOT, a by-product of a major industrial process which is being dumped recklessly and has already constituted a great nuisance to the environment. The results obtained revealed that the incorporation of IOT in SCC is viable and yielded improved compressive and flexural strength values. The adoption of IOT as replacement of fine aggregates in IOT will also reduce the wanton extraction of natural sand for concrete production, another positive step towards the reduction of carbon footprint of concrete. Already being a waste, any economical use of IOT is beneficial to the cost implications of its handling and disposal, and equally implies a reduction in the financial implications of concrete production.

References

[1] E.S. Rentier, L.H. Cammeraat, The environmental impacts of river sand mining. Sci Total Environ 838 (2022) 155877. https://doi.org/10.1016/j.scitotenv.2022.155877

[2] G. Bangalore Chinnappa, R.C. Karra, Experimental and Statistical Evaluations of Strength Properties of Concrete with Iron Ore Tailings as Fine Aggregate. Journal of Hazardous, Toxic, and Radioactive Waste 24 (2020). https://doi.org/10.1061/(ASCE)HZ.2153-5515.0000480

[3] A.M. Mhaya, S. Shahidan, H.A. Algaifi, S.S.M. Zuki, O. Benjeddou, M.H.W. Ibrahim, G.F. Huseien, Thermal Conductivity of Coconut Shell-Incorporated Concrete: A Systematic Assessment via Theory and Experiment. Sustainability 14 (2022). https://doi.org/10.3390/su142316167

[4] J. De Brito, R. Kurda, The past and future of sustainable concrete: A critical review and new strategies on cement-based materials. Journal of Cleaner Production 281 (2021). https://doi.org/10.1016/j.jclepro.2020.123558

[5] K. S.K, S.K. Singh, A. Chourasia, Alternative fine aggregates in production of sustainable concrete- A review. Journal of Cleaner Production 268 (2020). https://doi.org/10.1016/j.jclepro.2020.122089

[6] A.Q. Babatunde, E.A. Oshioname, Y. Paul, O.K. Junwolo, Effect of Elapsed Time after Mixing on the Strength Properties of Lime-Iron Ore Tailings Treated Black Cotton Soil as a Road Construction Material. Infrastructures 5 (2020). https://doi.org/10.3390/infrastructures5110089

[7] Z. Zhang, Z. Zhang, S. Yin, L. Yu, Utilization of Iron Tailings Sand as an Environmentally Friendly Alternative to Natural River Sand in High-Strength Concrete: Shrinkage Characterization and Mitigation Strategies. Materials (Basel) 13 (2020). https://doi.org/10.3390/ma13245614

[8] J. Zhao, K. Ni, Y. Su, Y. Shi, An evaluation of iron ore tailings characteristics and iron ore tailings concrete properties. Construction and Building Materials 286 (2021). https://doi.org/10.1016/j.conbuildmat.2021.122968

[9] M.M. Arbili, M. Alqurashi, A. Majdi, J. Ahmad, A.F. Deifalla, Concrete Made with Iron Ore Tailings as a Fine Aggregate: A Step towards Sustainable Concrete. Materials (Basel) 15 (2022). https://doi.org/10.3390/ma15186236

[10] W.O. Ajagbe, J.I. Braimah, A.A. Ganiyu, Shear Performance of Bamboo Reinforced Self Compacting Concrete Beams with Stirrups, ISPEC 10th International Conference on Agriculture, Animal Sciences and Rural Development, Sivas, Turkiye, 2022.

[11] A. Cwirzen, Properties of SCC with industrial by-products as aggregates, Self-Compacting Concrete: Materials, Properties and Applications, 2020, pp. 249-281. https://doi.org/10.1016/B978-0-12-817369-5.00010-6

[12] K. Nandhini, J. Karthikeyan, A review on sustainable production of self-compacting concrete utilizing industrial by-products as cementitious materials. Innovative Infrastructure Solutions 7 (2022) 199. https://doi.org/10.1007/s41062-022-00792-1

[13] P. Aggarwal, Y. Aggarwal, Strength properties of SCC, Self-Compacting Concrete: Materials, Properties and Applications, 2020, pp. 83-115. https://doi.org/10.1016/B978-0-12-817369-5.00004-0

[14] M. Gou, L. Zhou, N.W.Y. Then, Utilization of tailings in cement and concrete: A review. Science and Engineering of Composite Materials 26 (2019) 449-464. https://doi.org/10.1515/secm-2019-0029

[15] X. Huang, R. Ranade, V.C. Li, Feasibility Study of Developing Green ECC Using Iron Ore Tailings Powder as Cement Replacement. Journal of Materials in Civil Engineering 25 (2013) 923-931. https://doi.org/10.1061/(ASCE)MT.1943-5533.0000674

[16] W. Zhang, X. Gu, J. Qiu, J. Liu, Y. Zhao, X. Li, Effects of iron ore tailings on the compressive strength and permeability of ultra-high performance concrete. Construction and Building Materials 260 (2020). https://doi.org/10.1016/j.conbuildmat.2020.119917

[17] I.C. Ferreira, R. Galéry, A.B. Henriques, A. Paula de Carvalho Teixeira, C.D. Prates, A.S. Lima, I.R. Souza Filho, Reuse of iron ore tailings for production of metakaolin-based geopolymers. Journal of Materials Research and Technology 18 (2022) 4194-4200. https://doi.org/10.1016/j.jmrt.2022.03.192

[18] K. Siamardi, Optimization of fresh and hardened properties of structural light weight self-compacting concrete mix design using response surface methodology. Construction and Building Materials 317 (2022). https://doi.org/10.1016/j.conbuildmat.2021.125928

[19] W.C. Tang, Z. Wang, Y. Liu, H.Z. Cui, Influence of red mud on fresh and hardened properties of self-compacting concrete. Construction and Building Materials 178 (2018) 288-300. https://doi.org/10.1016/j.conbuildmat.2018.05.171

[20] S. Grünewald, J.C. Walraven, Properties of fibre reinforced SCC, Self-Compacting Concrete: Materials, Properties and Applications, 2020, pp. 309-370. https://doi.org/10.1016/B978-0-12-817369-5.00012-X

[21] A.U. Shettima, M.W. Hussin, Y. Ahmad, J. Mirza, Evaluation of iron ore tailings as replacement for fine aggregate in concrete. Construction and Building Materials 120 (2016) 72-79. https://doi.org/10.1016/j.conbuildmat.2016.05.095

[22] M. Yellishetty, V. Karpe, E.H. Reddy, K.N. Subhash, P.G. Ranjith, Reuse of iron ore mineral wastes in civil engineering constructions: A case study. Resources, Conservation and Recycling 52 (2008) 1283-1289. https://doi.org/10.1016/j.resconrec.2008.07.007

[23] S. Zhao, J. Fan, W. Sun, Utilization of iron ore tailings as fine aggregate in ultra-high performance concrete. Construction and Building Materials 50 (2014) 540-548. https://doi.org/10.1016/j.conbuildmat.2013.10.019

Advanced Topics in Mechanics of Materials, Structures and Construction - AToMech1-2023 Materials Research Forum LLC
Materials Research Proceedings 31 (2023) 422-436 https://doi.org/10.21741/9781644902592-44

Tensor as a tool in engineering analysis

A.P. Akinola[1,a *], A.S. Borokinni[1,b] and O.O. Fadodun[1,c]

[1]Department of Mathematics, Obafemi Awolowo University, 220005 Ile-Ife, Nigeria

[a]aakinola@oauife.edu.ng, [b]aborokinni@oauife.edu.ng, [c]ofadodun@oauife.edu.ng

Keywords: Invariants, Christofell Symbols, Exterior Differential Forms, Hyperelasticity, Strain Gradient Plasticity

Abstract. This paper underscores the potency of the invariant character of tensor and its derivative concepts and accentuate the synergy between isotropic tensor and other tensors and the corresponding vector operations. The equivalence of *covariant derivative* in a curvilinear coordinates system embedded with a non-constant vector field and the *partial derivative* in an affine coordinates system ingrained with a constant vector field is interrogated. The corresponding role of the Christofell symbols as the affine connector of vectors with their derivatives in a variable field are compared to the *Frenet-Seret* skew-matrix connecting the *trihedrons* (i.e. *tangent*, *normal* and *binormal*) of a moving space curve with their derivatives. The nexus of the Christofell symbols with the *geodesics* is also shown. The structure of the metric tensor $\tilde{I}$ and the Levi-Chivita skew-symmetric tensor $\tilde{\boldsymbol{\varepsilon}}$, as isotropic tensor rank-2 and rank-3 respectively is highlighted, such that the usual operations of *dot product* (or *scalar product* or *inner product*) and *cross product* or (*vector product* or *spin/rotation operation*}) are now expressed through the isotropic tensors. Recalling the theory of *exterior differential form* and invoking the Poincare's theorem we show the application of the exterior product in establishing *exact differential* (or *total differential*) in calculus in relation to *plane problem* of Elasticity. The invariant nature of the tensor objects and operations therefrom are then copiously invoked and deployed to establish constitutive relation for materials: in finite elasticity, within the context of *hyperelasticity*; composites, where there is a trade-off between heterogeneity and anisotropy through homogenisation process whereby differential equations with variable coefficients are converted to differential equations with constant coefficients; and plasticity, where application of tensor is exhibited with strain gradient plasticity, and shown how the concepts provide balance of microscopic forces, balance of macroscopic forces, and plastic flow laws as concise mathematical equations.

Introduction

This work, *Tensor as a Tool in Engineering Analysis* provides a brief survey of what tensor is, its relevance and application to continuum mechanics and engineering investigations and analysis. *Tensor theory*, simply and succinctly put, is the ***the theory of invariants***; which encompasses familiar physical objects such as *scalars, vectors, stress, strain*, *isotropes* and similar objects of various ranks and pertinent operations on them. It is even *a shorthand tool* in underpinning or driving home fundamental physical concepts, concisely [1-3]. The richness in the invariant nature of tensor, for example, enables us realize the equivalence in the actions of *covariant derivative* operator in a varying field, with curvilinear coordinates system, and the *partial derivative* operator in *affine*/rectilinear coordinates system. What is more, the effect of covariant derivative of metric tensor corresponds to the partial derivative of the Kronecker delta tensor, as a vanishing quantity.

Tensor as a discipline provides a potent and convenient tool to interrogate and navigate the complex labyrinth of the world of continuum mechanics and by implication, the modeling of real life phenomenon and engineering designs and analysis [4-5]. So, it is often pertinent and profitable to possess good and deep knowledge and understanding of this tool and its deployment in scientific, engineering and technological endeavours. This, we would demonstrate in a number of papers that would be presented in this session of the Conference.

In the setting of engineering mathematics/continuum mechanics as related to engineering design, underlining governing rules are deduced from concepts similar to those used in theoretical mechanics or dynamics [6]. Fundamental amongst these principles are the: *conservation of mass, conservation of linear momentum, conservation of moment of linear momentum, conservation of energy*. These constitute the so-called conservation laws. They are complimented by the so-called constitutive laws, in solving problems for specific material of a continuum; be it elastic, plastic, viscoelastic, viscoplastic, fliud - linear or nonlinear [7-8].

Here, a formal definition of tensor is given, and the relevance of the Christofell symbols as a coefficient of affine connectivity or coefficient of proportionality of a vector to its derivative in a varying space in analogy to the Frenet-Seret Matrix for a space curve is highlighted; it also serves as a measure of geodesics in an Euclidean space. The theory of exterior form is invoked, through the Poincare's Theorem for a differential form, to establish the Airy's stress function for plane elasticity [9-12]. The concept of *Homogenisation* [13-15] for periodic composites is illuminated, giving its implication as a process converting a system of partial differential equations with variable coefficients to that of constant coefficients, but with an incurred anisotropy as penalty. The property of isotropic tensor is explored to navigate the process of tensor derivative of functions of tensor argument. The deep endowment of tensor operations is richly displayed with the interrogation of strain-gradient-divergence plasticity [16-17].

Tensor

An object $\tilde{\mathbf{T}} \in \mathsf{T} \subset \Omega(\mathsf{E}^n)$ is a tensor in an n-dimensional sub-space of Euclidean space E^n if it is invariant under the transformation of coordinates system $\mathbf{q}(q^i)$, however, such that the components vary linearly and homogeneously, under the said transformation:

$\mathsf{O}: q^i \to q'^i,\ i = 1,2,3,\ldots,n$.

Let $\tilde{\mathbf{T}}$ be a tensor in a three-dimensional euclidean space E^n in which we have introduced an arbitrary curvilinear system of coordinates $\mathbf{q} = (q^1, q^2, q^3, \ldots)$ with the accompanying orthogonal covariant basis $\mathbf{e}_1, \mathbf{e}_2, \mathbf{e}_3, \ldots$ and the corresponding contravariant basis $\mathbf{e}^1, \mathbf{e}^2, \mathbf{e}^3, \ldots$. Then, either through the contravariant base vectors (basis) $\mathbf{e}^i$ or the covariant base vectors (basis) $\mathbf{e}_i$ we can express an *n-ranked* tensor $\tilde{\mathbf{T}}$ as

$$\tilde{\mathbf{T}} = T^{i_1 i_2 \cdots i_n} \mathbf{e}_{i_1} \mathbf{e}_{i_2} \cdots \mathbf{e}_{i_n} \text{ or } \tilde{\mathbf{T}} = T_{i_1 i_2 \cdots i_n} \mathbf{e}^{i_1} \mathbf{e}^{i_2} \cdots \mathbf{e}^{i_n}. \tag{1.1.1}$$

Now, if in E^n in place of $\mathbf{q}$ we introduce another coordinate system $\mathbf{q}'$, such that these two systems relate to one another by the non-singular matrix of transformation (or the so-called fundamental matrix)

$$\frac{\partial q^i}{\partial q^{i'}} \quad \text{and its inverse} \quad \frac{\partial q^{i'}}{\partial q^i}; \quad \left|\frac{\partial q^i}{\partial q^{i'}}\right| \neq 0, \tag{1.1.2}$$

then, the invariant character of $\tilde{\mathbf{T}}$ is expressed in the following.

(i) $\tilde{\mathbf{T}}'(\mathbf{q}') = T^{i_1' i_2' \cdots i_n'}(\mathbf{q}') \mathbf{e}_{i_1'} \mathbf{e}_{i_2'} \cdots \mathbf{e}_{i_n'} = T_{i_1' i_2' \cdots i_n'}(\mathbf{q}') \mathbf{e}^{i_1'} \mathbf{e}_{i'2} \cdots \mathbf{e}^{i_n'}$

$$= T_{i_1 i_2 \cdots i_n}(\mathbf{q}) \mathbf{e}^{i_1} \mathbf{e}^{i_2} \cdots \mathbf{e}^{i_n} = T^{i_1 i_2 \cdots i_n}(\mathbf{q}) \mathbf{e}_{i_1} \mathbf{e}_{i_2} \cdots \mathbf{e}_{i_r} = \tilde{\mathbf{T}}(\mathbf{q}). \tag{1.1.3}$$

(ii) But, with respect to the components, they transform accordingly and we have

$$T_{i_1' i_2' \cdots i_n'}(\mathbf{q}') = T_{i_1 i_2 \cdots i_n}(\mathbf{q}) \frac{\partial q^{i_1}}{\partial q^{i_1'}} \frac{\partial q^{i_2}}{\partial q^{i_2'}} \cdots \frac{\partial q^{i_n}}{\partial q^{i_n'}},\ T^{i_1' i_2' \cdots i_n'}(\mathbf{q}') = T(\mathbf{q})^{i_1 i_2 \cdots i_n} \frac{\partial q^{i_1'}}{\partial q^{i_1}} \frac{\partial q^{i_2'}}{\partial q^{i_2}} \cdots \frac{\partial q^{i_n'}}{\partial q^{i_n}}; \tag{1.1.4}$$

$$T_{i_1 i_2 \cdots i_n}(\mathbf{q}) = T_{i_1' i_2' \cdots i_n'}(\mathbf{q}') \frac{\partial q^{i_1'}}{\partial q^{i_1}} \frac{\partial q^{i_2'}}{\partial q^{i_2}} \cdots \frac{\partial q^{i_n'}}{\partial q^{i_n}} \quad T^{i_1 i_2 \cdots i_n}(\mathbf{q}) = T^{i_1' i_2' \cdots i_n'}(\mathbf{q}') \frac{\partial q^{i_1}}{\partial q^{i_1'}} \frac{\partial q^{i_2}}{\partial q^{i_2'}} \cdots \frac{\partial q^{i_n}}{\partial q^{i_n'}}. \quad (1.1.4)'$$

Here, $T^{i_1 i_2 \cdots i_n}(\mathbf{q})$ and $T^{i_1' i_2' \cdots i_n'}(\mathbf{q}')$ are the *contravariant components* of the tensor $\tilde{\mathbf{T}}$ in the coordinates systems $\mathbf{q}$ and $\mathbf{q}'$ respectively. Similarly, $T_{i_1 i_2 \cdots i_n}(\mathbf{q})$ and $T_{i_1' i_2' \cdots i_n'}(\mathbf{q}')$ are referred to as the *covariant components* of the tensor $\tilde{\mathbf{T}}$ in the coordinates systems $\mathbf{q}$ and $\mathbf{q}'$ respectively. For example, g_{ij}, g^{ij} the *covariant metric tensor, contravariant metric tensor* are respectively $g_{ij} \equiv \mathbf{e}_i \cdot \mathbf{e}_j$ and $g^{ij} \equiv \mathbf{e}^i \cdot \mathbf{e}^j$.

Christofell Symbols and Relevance

The objects Γ_{ijk} and Γ_{ij}^k are known as the *Christofell symbols* of *first kind* and *second kind* or simply, *1st kind* and *2nd kind* respectively. The 2nd kind is defined as

$$\frac{\partial \mathbf{e}_m}{\partial q^i} \equiv \Gamma_{im}^k \mathbf{e}_k, \ \frac{\partial \mathbf{e}^m}{\partial q^i} \equiv -\Gamma_{ik}^m \mathbf{e}^k; \Gamma_{ijm} = g_{mk} \Gamma_{ij}^k, \ \Gamma_{ij}^m = g^{mk} \Gamma_{ijk}, \ i,j,k,m = 1,2,3. \quad (1.2.1)$$

The Christofell symbols, related to one another as above, and to the metric tensor through the expression:

$$\Gamma_{ijk} = \frac{1}{2}(g_{kj,i} + g_{ik,j} - g_{ij,k}), \Gamma_{ij}^m = \frac{1}{2} g^{mk}(g_{kj,i} + g_{ik,j} - g_{ij,k}), \ i,j,k,m = 1,2,3. \quad (1.2.2)$$

They constitute what is called Symbols of Space Connectivity or Coefficients of Affine Connectivity.

Frenet-Seret Matrix analogous to Christofell Symbols

It could be noted that in this, the *2nd kind Christofell Symbols* plays an anologuous role as the *Frenet-Seret matrix* in the case of space curves, where the matrix relates the non-constant unit orthogonal *thriedral* $(\mathbf{t}, \mathbf{n}, \mathbf{b})$ (i.e. *tangent, normal* and *bi-normal*) to their derivatives $(\frac{d\mathbf{t}}{ds}, \frac{d\mathbf{n}}{ds}, \frac{d\mathbf{b}}{ds})$, on the space curve, parametrised by s, the *arclength*; $\mathbf{r} = \mathbf{r}(s)$:

$$\begin{pmatrix} \frac{d\mathbf{t}}{ds} \\ \frac{d\mathbf{n}}{ds} \\ \frac{d\mathbf{b}}{ds} \end{pmatrix} = \begin{pmatrix} 0 & \kappa & 0 \\ -\kappa & 0 & \tau \\ 0 & -\tau & 0 \end{pmatrix} \begin{pmatrix} \mathbf{t} \\ \mathbf{n} \\ \mathbf{b} \end{pmatrix} \text{.Frenet-Seret } \textit{matrix} \text{ and } \textit{rule} \quad (1.2.3)$$

The covariant derivative

The covariant derivative of any mixed tensor $T_{j_1 j_2 \cdots j_n}^{i_1 i_2 \cdots i_m}$ is denoted and given by the expression

$$\nabla_k T_{j_1 j_2 \cdots j_n}^{i_1 i_2 \cdots i_m} \equiv \frac{\partial}{\partial q^k} T_{j_1 j_2 \cdots j_n}^{i_1 i_2 \cdots i_m} - T_{p j_2 \cdots j_n}^{i_1 i_2 \cdots i_m} \Gamma_{j_1 k}^p - T_{j_1 p j_3 \cdots j_n}^{i_1 i_2 \cdots i_m} \Gamma_{j_2 k}^p \cdots - T_{j_1 j_2 \cdots j_{n-2} p j_n}^{i_1 i_2 \cdots i_m} \Gamma_{j_{n-1} k}^p - T_{j_1 j_2 \cdots j_{n-1} p}^{i_1 i_2 \cdots i_m} \Gamma_{j_n k}^p$$

$$+ T_{j_1 j_2 \cdots j_n}^{p i_2 \cdots i_m} \Gamma_{pk}^{i_1} + T_{j_1 j_2 \cdots j_n}^{i_1 p i_3 \cdots i_m} \Gamma_{pk}^{i_2} \cdots + T_{j_1 j_2 \cdots j_n}^{i_1 i_2 \cdots i_{m-2} p i_m} \Gamma_{pk}^{i_{m-1}} + T_{j_1 j_2 \cdots j_n}^{i_1 i_2 \cdots i_{m-2} i_{m-1} p} \Gamma_{pk}^{i_m}. \quad (1.2.4)$$

We emphasize, the partial derivative of the Kronecker delta tensor is to the covariant derivative of the metric tensor as the rectilinear coordinates (affine coordinates) system is to the curvilinear coordinates system. In fact, the partial derivative of Kronecker delta δ_{ij} and the covariant derivative of the metric tensor g_{ij} correspondingly vanish:

$$\partial\delta_{ij} = 0 \to \nabla_k g_{ij} = \frac{\partial g_{ij}}{\partial q^k} - g_{mj}\Gamma^m_{ki} - g_{im}\Gamma^m_{jk} = \frac{\partial g_{ij}}{\partial q^k} - \Gamma_{ki,j} - \Gamma_{jk,i} \overset{(1.2.2)}{=} 0. \quad (1.2.4)$$

Measure of Geodesics

The concept of *Geodesics* derives from the notion of *absolute derivative*. In fact, let a space curve be parametrized on interval $I \subset \mathrm{R}$ with parameter $t \in I$, such that its position vector is

$$\mathbf{r} = \mathbf{r}(u^1(t), u^2(t), u^3(t)) = \mathbf{r}(t). \quad (1.2.5)$$

Note that

$$\mathbf{e}_i = \mathbf{e}_i(u^k(t)) \equiv \frac{\partial \mathbf{r}}{\partial u^i}; \; \frac{d\mathbf{e}_i(u^k(t))}{dt} = \frac{\partial \mathbf{e}_i}{\partial u^k}\frac{du^k}{dt} \overset{(1.2.1)}{=} \Gamma^m_{ik}\mathbf{e}_m \frac{du^k}{dt}. \quad (1.2.6)$$

Now, we can consider the derivative of a vector $\mathbf{v} = v^i \mathbf{e}_i$ along this curve in the domain with arbitrary coordinates u^i and the orthogonal basis $\mathbf{e}_i$. Its derivative with respect to parameter t, taking cognizance of the basis $\mathbf{e}_i$ not being constant, in a fashion similar to the process of covariant derivative, is

$$\frac{d\mathbf{v}}{dt} = \frac{dv^i}{dt}\mathbf{e}_i + v^i \frac{d\mathbf{e}_i}{dt} = \frac{dv^i}{dt}\mathbf{e}_i + v^i\Gamma^m_{ij}\frac{dv^j}{dt}\mathbf{e}_m = (\frac{dv^i}{dt} + v^m\Gamma^i_{mj}\frac{dv^j}{dt})\mathbf{e}_i; \; \nabla_t \equiv \frac{dv^i}{dt} + v^m\Gamma^i_{mj}\frac{dv^j}{dt}, \quad (1.2.7)$$

where $\nabla_t \equiv \nabla_i$ is referred to as the *absolute derivative* of the component v^i of vector $\mathbf{v}$. This is at times also referred to as *intrinsic derivative*. Note that this notion can be extended to invariants of higher rank.

We can similarly deduce the absolute derivative of the (i) covariant vector $v_i(t)$, (ii) covariant tensor $T_{ij}(t)$, (iii) mixed tensor T_i^j and contravariant tensor T^{ij}.

A geodesic is *a curve* $\mathbf{r} = \mathbf{r}(s)$ *parametrized with the arclength* s *such that the absolute derivative of its tangent vanishes*. (i.e. *a curve of constant tangent* or *vanishing acceleration*). This is given by the expression

$$\frac{d\mathbf{t}}{ds} = 0 \text{ i.e. } \frac{dt^i}{ds} + \Gamma^i_{mj}t^m\frac{du^j}{ds} = 0, \to \frac{d^2u^i}{ds^2} + \Gamma^i_{mj}\frac{du^j}{ds}\frac{du^m}{ds} = 0, \quad (1.2.8)$$

noting that $t^i = du^i / ds$, $\mathbf{t} \equiv \dfrac{d\mathbf{r}}{ds}$.

Example

The Gauss curvature and the geodesic line for the Poincare's half plane,

$$ds^2 = \frac{du^2 + dv^2}{v^2}, \; v > 0 \quad (1.2.9)$$

are obtained respectively as

$$K = -\frac{R_{1212}}{g} = \frac{b_{11}b_{22} - b_{12}^2}{g_{11}g_{22} - g_{12}^2} = -1; \; (u - u_0)^2 - v^2 = c^2.$$

Indeed, taking into recognition (1.2.8) and (1.2.9), the required geodesic line is the family of curves $(u - u_0)^2 + v^2 = c^2$ (i.e. shifted circles); noting that for the half plane (1.2.9),

$$g_{11} = \frac{1}{v^2}, g_{12} = g_{21} = 0, \; g_{22} = \frac{1}{v^2} \;\; g^{11} = v^2, g^{12} = g^{21} = 0, \; g^{22} = v^2.$$

$$\Gamma^1_{11} = 0, \; \Gamma^1_{12} = -\frac{1}{v}, \; \Gamma^1_{22} = 0, \; \Gamma^2_{11} = \frac{1}{v}, \; \Gamma^2_{12} = 0, \; \Gamma^2_{22} = -\frac{1}{v}.$$

The Gauss curvature is deduced from the expression relating the Riemann curvature R_{1212} and the Gauss curvature K such that $R_{1212} = -gK = b_{11}b_{22} - b_{12}^2$, while the determinant of the 1st fundamental matrix of the surface is $g = g_{11}g_{22} - g_{12}^2$ and the determinant of the 2nd fundamental matrix of a surface $b_{11}b_{22} - b_{12}^2$ is $-(\frac{1}{v^2})^2$..

Exterior Form and Application

Definition of Exterior Form

Consider Q, the space of m-$tuple$ tensor product of the conjugate space $\Re^*$ on itself. That is $Q = \Re^* \otimes \Re^* \otimes \cdots \otimes \Re^* \otimes \Re^*$. Let Q_o be a space of skew-symmetric/anti-symmetric tensor, a subspace of Q; $Q_o \subset Q$. Then, the element $\Phi \in Q_o$ is called the *Exterior Form*, rank-m and presented as

$$\Phi = \varphi_{i_1 i_2 \ldots i_{m-1} i_m} \mathbf{e}^{i_1} \Lambda \mathbf{e}^{i_2} \cdots \Lambda \mathbf{e}^{i_{m-1}} \Lambda \mathbf{e}^{i_m}, \setminus m = 1,2,3,\ldots, \tag{2.1.1}$$

where $\mathbf{e}^{i_k}, k = 1,2,3,\ldots$ are referred to as *covectors.*

Differential Form

An *Exterior Differential Form* ω is any exterior form constructed on differential basis; when the covectors are now the *differential,* dx^{i_k} .

$$\omega = a_{i_1 i_2 \ldots i_{m-1} i_m}(x^1, x^2, \ldots, x^{n-1}, x^n) dx^{i_1} \Lambda dx^{i_2} \ldots \Lambda dx^{i_{m-1}} \Lambda dx^{i_m}, \tag{2.1.2}$$

where in any $n-$ dimensional domain Ω, the ring of coefficients

$$a(x^1, x^2, \ldots, x^{n-1}, x^n), \ a \in C^q(\Omega); \ q \geq 1.$$

i.e. a is $q-$times continuously differentiable or possesses partial derivatives of order q inclusively.

The base elements (covectors) are the differentials of the variables $x^1, x^2, \ldots, x^{n-1}, x^n$, $dx^1, dx^2, \ldots, dx^{n-1}, dx^n$ and view them as abstract unit. It is obvious that all the properties of an exterior form are endowed any differential forms [11,18].

The Exterior differentiation of an exterior differential form (2.1.2) is defined as

$$\mathsf{D}\omega \equiv da_{i_1 i_2 \ldots i_{m-1} i_m}(x^1, x^2, \ldots, x^{n-1}, x^n) dx^{i_1} \Lambda dx^{i_2} \ldots \Lambda dx^{i_{m-1}} \Lambda dx^{i_m}$$

$$= \frac{\partial a_{i_1 i_2 \ldots i_{m-1} i_m}}{\partial x^i} dx^j \Lambda dx^{i_1} \Lambda dx^{i_2} \ldots \Lambda dx^{i_{m-1}} \Lambda dx^{i_m}, \ i = 1,2,3,\ldots,n-1,n. \tag{2.1.3}$$

We note that *The operation of exterior diferential form (like that of covariant differentiation) increases the rank of a form by one*. And when a change of coordinates is executed we have

$$dx^{i_k} \to dx^{i'_k} \to dx^{i'_1} \Lambda dx^{i'_2} \ldots \Lambda dx^{i'_{n-1}} \Lambda dx^{i'_n} = det \left| \frac{\partial x^{i'}}{\partial x^i} \right| dx^{i_1} \Lambda dx^{i_2} \ldots \Lambda dx^{i_{n-1}} \Lambda dx^{i_n}.$$

Poincare Theorem

The second exterior differential of an exterior form is zero,

$$\mathsf{DD}\omega = 0. \tag{2.2.1}$$

Proof:

The second exterior differential is

$$\mathsf{DD}\omega \overset{(2.1.3)}{=} \frac{\partial^2 a_{i_1 i_2 \ldots i_{m-1} i_m}}{\partial x^j \partial x^k} dx^k \Lambda dx^j \Lambda dx^{i_1} \Lambda dx^{i_2} \ldots \Lambda dx^{i_{m-1}} \Lambda dx^{i_m}.$$

$$\text{But } a_{i_1 \ldots i_m}(x^1, \ldots, x^n) \in C^m(\Omega) \to \frac{\partial^2 a_{i_1 i_2 \ldots i_{m-1} i_m}}{\partial x^j \partial x^k} = \frac{\partial^2 a_{i_1 i_2 \ldots i_{m-1} i_m}}{\partial x^k \partial x^j}.$$

At the same time, the form $dx^k \Lambda dx^j \Lambda dx^{i_1} \Lambda dx^{i_2} \ldots \Lambda dx^{i_{m-1}} \Lambda dx^{i_m}$ is skew-symmetric, including in the indices k and j. For this, we have $-\mathsf{DD}\omega = \mathsf{DD}\omega$. Hence, the required proof, $\mathsf{DD}\omega = 0$.

Example

Let a linear differential form be $\omega = Pdx + Qdy$. By (2.1.3), the exterior differential form is

$$\mathsf{D}\omega \overset{(2.1.3)}{=} dP\Lambda dx + dQ\Lambda dy \overset{(2.1.3)}{=} \frac{\partial P}{\partial y} dy\Lambda dx + \frac{\partial Q}{\partial x} dx\Lambda dy = (\frac{\partial Q}{\partial x} - \frac{\partial P}{\partial y})dx\Lambda dy;\ dx\Lambda dy = -dx\Lambda dy. \quad (2.2.2)$$

We note that (i.) ω is a total differential if $\frac{\partial P}{\partial y} = \frac{\partial Q}{\partial x}$, since $d\omega = \frac{\partial \omega}{\partial x}dx + \frac{\partial \omega}{\partial y}dy$ and $\frac{\partial^2 \omega}{\partial x \partial y} = \frac{\partial^2 \omega}{\partial y \partial x}$. (ii.) if ω is a total differential then, $\frac{\partial P}{\partial y} = \frac{\partial Q}{\partial x}$ and we shall obtain from (2.2.2) that $\mathsf{D}\omega = 0$. And by the corollary to the Poincare's theorem, ω is a differential form. (iii.) the concept of differential form can be used to establish whether or not an expression constitutes a *total differential*.

Application of Poincare's Theorem in Elasticity

The boundary value problem of *plane elasticity* can be formulated in terms of the Airy's function, $\varphi(x_1, x_2)$. At the root of this formulation is the Poincare's theorem, deriving from the theory of *Exterior Form* or *Exterior Differential Form*. This fundamental background, often, is omitted in textbooks of elasticity.

Theorem: $\frac{\partial \sigma_{11}}{\partial x_1} + \frac{\partial \sigma_{12}}{\partial x_2} = 0;\ \frac{\partial \sigma_{12}}{\partial x_1} + \frac{\partial \sigma_{22}}{\partial x_2} = 0$ if only and only if $\sigma_{11}dx_2 - \sigma_{12}dx_1$ and $\sigma_{22}dx_1 - \sigma_{12}dx_2$ are respectively an exact differentials.

Proof: By Poincare's theorem, ω is an exact differential if $d\omega = 0$. That is, we can find φ such that $\omega = d\varphi$ and consequently, $\omega = dd\varphi = 0$. Thus, $\sigma_{11}dx_2 - \sigma_{12}dx_1$ is an exact differential $\rightarrow$ $d(\sigma_{11}dx_2 - \sigma_{12}dx_1) = 0$. As a differential form, we have

$$d(\sigma_{11}dx_2 - \sigma_{12}dx_1) = \frac{\partial}{\partial x_1}(\sigma_{11}dx_2 - \sigma_{12}dx_1)dx_1 + \frac{\partial}{\partial x_2}(\sigma_{11}dx_2 - \sigma_{12}dx_1)dx_2$$

$$= \frac{\partial \sigma_{11}}{\partial x_1} dx_2 \Lambda dx_1 - \frac{\partial \sigma_{12}}{\partial x_1} dx_1 \Lambda dx_1 + \frac{\partial \sigma_{11}}{\partial x_2} dx_2 \Lambda dx_2 - \frac{\partial \sigma_{12}}{\partial x_2} dx_1 \Lambda dx_1$$

$$= (\frac{\partial \sigma_{11}}{\partial x_1} + \frac{\partial \sigma_{12}}{\partial x_2}) dx_2 \Lambda dx_1 = 0;\ \text{since } dx_1 \Lambda dx_1 = dx_2 \Lambda dx_2 = 0. \Rightarrow (\frac{\partial \sigma_{11}}{\partial x_1} + \frac{\partial \sigma_{12}}{\partial x_2}) = 0.$$

Conversely, Let $(\frac{\partial \sigma_{11}}{\partial x_1} + \frac{\partial \sigma_{12}}{\partial x_2}) = 0 \Rightarrow d(\sigma_{11}dx_2 - \sigma_{12}dx_1) = 0$. Then, $\sigma_{11}dx_2 - \sigma_{12}dx_1$ is an exact differential. $\Rightarrow \exists$ a function $P(x_1, x_2)$;

$$\sigma_{11}dx_2 - \sigma_{12}dx_1 = dP(x_1, x_2) = \frac{\partial P}{\partial x_1}dx_1 + \frac{\partial P}{\partial x_2}dx_2;\ \Rightarrow \sigma_{11} = \frac{\partial P}{\partial x_2},\ \sigma_{12} = -\frac{\partial P}{\partial x_1}. \quad (*)$$

Similarly: There exists a function $Q(x_1, x_2)$;

$$\sigma_{22}dx_1 - \sigma_{12}dx_2 = dQ(x_1, x_2) = \frac{\partial Q}{\partial x_1}dx_1 + \frac{\partial Q}{\partial x_2}dx_2;\ \Rightarrow \sigma_{22} = \frac{\partial Q}{\partial x_i},\ \sigma_{12} = -\frac{\partial Q}{\partial x_2}. \quad (**)$$

By (*) and (**) $\Rightarrow \exists$ the *form* $\varphi(x_1, x_2)$ (i.e. *Airy's stress function*) such that

$$\sigma_{11} = \frac{\partial^2 \varphi}{\partial x_2^2},\ \sigma_{22} = \frac{\partial^2 \varphi}{\partial x_1^2},\ \sigma_{12} = \frac{\partial^2 \varphi}{\partial x_2 \partial x_1} = \frac{\partial^2 \varphi}{\partial x_1 \partial x_2} = \sigma_{21}.$$

Homogenisation of Heterogeneous Medium

Linear Elasticity: A Sample Elastostatic Problem

Consider, in a three-dimensional Euclidean space E^3, the already familiar boundary value problem of classical elastostatics in terms of displacement $\mathbf{u} \in \Omega \subset \mathsf{E}^3$,

$$[C_{ijkl}u_{k,l}]_{,j} + \rho X_i = 0, \text{ in } \Omega, \tag{3.1}$$

$$\varepsilon_{ij} = \frac{1}{2}(u_{i,j} + u_{j,i}), \text{ in } \Omega, \tag{3.2}$$

$$\sigma_{ij} = C_{ijkl}\varepsilon_{kl}, \text{ in } \Omega, \tag{3.3}$$

$$u_i |_{\Sigma_u} = u_i^0; \;\; C_{ijkl}u_{k,l}n_j |_{\Sigma_\sigma} = S_i^0, \text{ on } \Sigma = \Sigma_u \bigcup \Sigma_\sigma, \tag{3.4}$$

where C_{ijkl} are the generalized Hooke's material characteristics/constants, X_i is the mass or volume body force, u_i is the displacement, σ_{ij} is the stress, ρ is the density, ε_{ij} is the strain, u_i^0 is the displacement specified on $\Sigma_u = \partial\Omega_u$, a part of the boundary of Ω, while S_i^0 is the load/force specified on $\Sigma_\sigma = \partial\Omega_\sigma$, the other part of the boundary Ω: $\Sigma_u \bigcup \Sigma_\sigma = \partial\Omega_u \bigcup \partial\Omega_\sigma = \partial\Omega$. Here and elsewhere, $f_{m,n} \equiv \dfrac{\partial f_m}{\partial x_n}$, $i, j, k, l = 1,2,3$.

Alternate Compatibility Equation - Chezaro's Formular:

Taking advantage of tensor notation and its invariance property, an alternate presentation of the compatibility equation (or Cauchy equations) (3.2) is provided by Chezaro [9-10]. This is just the solution to the differential equations (3.2); obtained through integration of (3.2):

$$u_i = u_i^0 + \omega_{ij}^0(x_j' - x_j^0) + \int_M^{M'} [\varepsilon_{ij} + (x_n' - x_n)(\frac{\partial \varepsilon_{in}}{\partial x_j} - \frac{\partial \varepsilon_{jn}}{\partial x_i})]dx_j, \tag{3.2*}$$

where the body has been fixed at point M^0, with displacement u_i^0 and rotation tensor $\omega_{i,j}^0 = \frac{1}{2}(u_{i,j}^0 - u_{j,i}^0)$. Now, let Ω be an heterogeneous medium, say a *lamina composite*, of isotropic elastic layers.

The tensor rank-4 (i.e. the elastic characteristics) C_{ijkl} will be non-constant, but a periodic function of the coordinates

$$C_{ijkl}(\mathbf{x}) = C_{ijkl}(\mathbf{x}, \mathbf{l}), \tag{3.5}$$

where l is the periodic geometric length. We introduce second argument (or the so-called *local coordinate* or the *fast coordinate*) ξ in which case $\mathbf{x}$ is referred to as the *global coordinate*, and then take the asymptotic expansion [13,15]

$$u_i(\mathbf{x}, \xi) = \sum_{m=0} \alpha^m N_{ipq_1 \cdots q_m}^{(m)}(\xi) v_{p,q_1 \cdots q_m}(\mathbf{x}); \;\; \xi = \frac{\mathbf{x}}{\alpha}, \;\; |\alpha| < 1, \tag{3.6}$$

where $\alpha \equiv \dfrac{l}{L}$ is the so-called *small parameter*, which is the ratio of characteristic lengths l and L of the periodic cell ω and the whole body Ω respectively.

We then insert (3.6) in (3.1), (3.2) to obtain ensuing boundary value problems, now for the *homogenised* medium, howbeit now with acquired anisotropy:

$$\sum_{m=0} \alpha^m h_{ipq_1 \cdots q_{m+2}}^{(m)} v_{p,q_1 \cdots q_{m+1}}(\mathbf{x}) + X_i = 0, \tag{3.1'}$$

$$\sum_{m=0}\alpha^m N^{(m)}_{ipq_1\ldots q_m}(\xi)v_{p,q_1\ldots q_m}(\mathbf{x})\,|_{\partial\Omega_1} = u_i^0;\ \sum_{m=0}\alpha^m h^{(m)}_{ipq_1\ldots q_{m+2}}v_{p,q_1\ldots q_{m+1}}(\mathbf{x})n_{q_{m+2}}\,|_{\partial\Omega_1} = S_i^0, \qquad (3.2)'$$

where

$$\mathbf{N^m} = \mathbf{N^{(m)}_{ipq_1\ldots q_m}}(\xi) \text{ and } \mathbf{h^m} = \mathbf{h^{(m)}_{ipq_1\ldots q_{m+2}}}$$

are the periodic *structural/local function* and
the *effective material elastic constants*, the latter which is independent of both $\mathbf{x}$ and ξ.

$$\mathbf{v}(\mathbf{x}) \equiv \mathbf{v_p}(\mathbf{x}) \equiv < \mathbf{u}_0 >$$

is the *average* displacement vector, independent of ξ,

$$N^p \equiv \delta_{ij},\ i,j = 1,2,3,\ldots; N^m = 0 \text{ if } m<0,\ \delta_{ij} = \begin{cases}1 & \text{if } i=j,\\ 0 & \text{if } i\neq j,\end{cases} i,j = 1,2,3, \qquad (3.7)$$

where δ_{ij} is the so-called *Kronecker delta*, unit tensor.

Mechanical Characteristics $\tilde{h}$ and Conjugate Local Functions $\mathrm{N}(\xi)$

To obtain $\tilde{h}$, the mechanical characteristics, and the corresponding $\tilde{N}(\xi)$, the *structural functions* (or *local functions*) we solve the so-called *non-boundary valued periodic problems* $P^{(m+1,m)}$, sequentially:

$$h^{(m)}_{ipq_1\ldots q_{m+2}} = (C_{ijkl}N^{(m+2)}_{ipq_1\ldots q_{m+2}}(\xi)|_l)|_j + (C_{ijq_{m+1}k}N^{(m+2)}_{ipq_1\ldots q_{m+1}}(\xi))|_j$$
$$+C_{iq_{m+2}kl}N^{(m+1)}_{kpq_1\ldots q_{m+1}}(\xi)|_l + C_{iq_{m+2}kq_{m+1}}N^{(m)}_{kpq_1\ldots q_m}(\xi)|_l,\ m=-1,0,1,\ldots$$
$$h^{(m)}_{ipq_1\ldots q_{m+2}} = \langle C_{iq_{m+2}kl}N^{(m+1)}_{kpq_1\ldots q_{m+1}}(\xi)|_l + C_{iq_{m+2}kq_{m+1}}N^{(m)}_{kpq_1\ldots q_m}(\xi)\rangle, \qquad (3.9)$$

where for any function $f(\mathbf{x},\xi)\in\Omega$, $f(\mathbf{x},\xi)_{,m} \equiv \dfrac{\partial f}{\partial x_m}$, while $f(\mathbf{x},\xi)|_m \equiv \dfrac{\partial f}{\partial \xi_m}$.

Thus, at the zeroth level, we have the periodic problem $P^{(1,0)}$. This implies that the boundary value problem (3.1)-(3.4), which is a set of equations with variable coefficients, has now become a system of differential equations now with constant coefficients; howbeit, with an incurred anisotropy:

$$[h_{ijkl}v_{k,l}]_{,j} + \rho X_i = 0, \text{ in } \Omega, \qquad (3.1)''$$

$$\varepsilon^{(0)}_{ij} = \frac{1}{2}(v_{i,j}+v_{j,i}), \text{ in } \Omega, \qquad (3.2)''$$

$$\sigma_{ij} = h_{ijkl}\varepsilon^{(0)}_{kl}, \text{ in } \Omega, \qquad (3.3)''$$

$$v_i|_{\Sigma_u} = u_i^0;\ h_{ijkl}v_{k,l}n_j|_{\Sigma_\sigma} = S_i^0, \text{on } \Sigma = \Sigma_u \bigcup \Sigma_\sigma, \qquad (3.4)''$$

Isotropic Tensor and Invariant Operations

Isotropy

We recall that *isotropy* is the property of a material such that every direction in it constitutes an axis of rotational symmetry.

Isotropic Scalar Function:

A *scalar function* $\phi(a^{ij})$ of a tensor $\mathbf{A} = (a^{ij})$ is isotropic if it preserves its value and form of dependence on those components in any orthogonal mapping H of the coordinates system

$$\phi(a'^{ij}) = \phi(a^{ij}),\ \mathsf{H}: a^{ij}\to a'^{ij},\ \mathbf{r}' = \mathbf{r}\cdot\mathsf{H}.$$

Isotropic Tensor:
A tensor is *isotropic* if its components are invariant under any orthogonal mapping of the base vectors in which it is defined. (That is, both the tensor and its components are invariant.)
Note: Any would-be tensor, by definition, is meant to be invariant under mappings of coordinates system, while only its components would admit appropriate changes by the imposed rule of transformation. But for an isotropic tensor, even its components are required to be unchanged.
Using these properties of isotropic function and tensor, energy functions can be constructed and continuum mechanics problems solved [10,12].

Example
Let $\tilde{O}$ be an orthogonal tensor that maps base vector $\mathbf{e}_i$ into another base vector $\mathbf{e}'_i = \mathbf{e}_i \cdot \tilde{O}^t$. Consider the unit metric tensor $\tilde{I} = g_{ij}\mathbf{e}^i\mathbf{e}^j$, which is known to be an isotropic tensor. Under transformation of its components we have

$$g_{ij} = \mathbf{e}_i \cdot \mathbf{e}_j = \mathbf{e}'_i \cdot \tilde{O} \cdot \tilde{O}^t \cdot \mathbf{e}'_j = \mathbf{e}'_i \cdot \mathbf{e}'_j = g'_{ij} \Rightarrow \tilde{I} = g_{ij}\mathbf{e}^i\mathbf{e}^j = g^{ij}\mathbf{e}_i\mathbf{e}_j = \mathbf{e}^i\mathbf{e}_i = \mathbf{e}_i\mathbf{e}^i$$

is isotropic.

Basic Isotropic Tensor
The basic isotropic tensors include [10]:
(i.) The *rank-0 isotropic tensor*, (the *scalar*), (ii.) The *rank-2 isotropic tensor*, (the unit tensor $\tilde{I}$); it is the only rank-*2* isotropic tensor, and (iii.) The *rank-3 isotropic tensor*, (the Levi-Chivita tensor $\tilde{\epsilon}$).
Any other isotropic rank-3 tensor $\tilde{\mathbf{T}}$ at most is a constant multiple of it, $\lambda\tilde{\epsilon}$, where λ is a constant number (hence, $\tilde{\epsilon}$ is *pseudoisotropic tensor*):

$$\tilde{\epsilon} = -\tilde{I} \times \tilde{I}; \epsilon^{ijk}\epsilon_{lmn} = \begin{vmatrix} \delta_{il} & \delta_{im} & \delta_{in} \\ \delta_{jl} & \delta_{jm} & \delta_{jn} \\ \delta_{kl} & \delta_{km} & \delta_{kn} \end{vmatrix}; \delta_{ij} = \begin{cases} 1 & \text{if} \quad i = j, \\ 0 & \text{if} \quad i \neq j, \end{cases} i, j, k, m, n = 1, 2, 3, \qquad (4.1.1)$$

where δ_{ij} is the so-called *Kronecker delta.*

(iv.) The rank-4 isotropic tensor ${}^4\tilde{C}$ consists of the basic isomers

$$\tilde{C}_I \equiv \tilde{I}\tilde{I} = \mathbf{e}_i\mathbf{e}^i\mathbf{e}_j\mathbf{e}^j, \ \tilde{C}_{II} \equiv \mathbf{e}_i\mathbf{e}_j\mathbf{e}^i\mathbf{e}^j, \ \tilde{C}_{III} \equiv \mathbf{e}_i\tilde{I}\mathbf{e}^i = \mathbf{e}_i\mathbf{e}_j\mathbf{e}^j\mathbf{e}^i. \qquad (4.1.2)$$

Isotropic Tensor in Product Operation with Rank-2 Tensor
$\forall$ vectors $\mathbf{a}, \mathbf{b}, \mathbf{c}$ and rank-2 tensors $\tilde{P}, \tilde{Q}$ we highlight interaction of the isotropic tensors on them via indicated operations.

$$\tilde{I} \cdot\cdot \mathbf{ab} = \mathbf{a} \cdot \mathbf{b} = a_i b_j \delta_{ij}, \tilde{\epsilon} \cdot\cdot \mathbf{ab} = \mathbf{a} \times \mathbf{b} = \epsilon^{ijk} a_i b_j \mathbf{e}_k = \mathbf{c},$$

$$\tilde{\epsilon} \cdot \tilde{\epsilon} = (\tilde{I} \times \tilde{I}) \cdot (\tilde{I} \times \tilde{I}) = \tilde{I} \times \tilde{I} \times \tilde{I} = \tilde{C}_{II} - \tilde{C}_{III} <-> (\tilde{C}_{III} - \tilde{C}_{II}) \cdot\cdot \tilde{\mathbf{T}} = -(\tilde{\mathbf{T}} - \tilde{\mathbf{T}}^t), \qquad (4.1.3)$$

$$\tilde{\epsilon} \cdot\cdot \tilde{\epsilon} = \mathbf{e}_s\mathbf{e}_k \cdot \mathbf{e}^s\mathbf{e}^k - \mathbf{e}_s\mathbf{e}_m \cdot \mathbf{e}^m\mathbf{e}^s = \tilde{I} - 3\tilde{I} = -2\tilde{I},$$

$$\tilde{C}_I \cdot \tilde{P} = \tilde{I} \cdot \tilde{P}; \ \tilde{C}_{III} \cdot \tilde{P} = \tilde{P} \cdot \tilde{C}_{III}; \tilde{C}_I \cdot\cdot \tilde{P} = I_1(\tilde{P})\tilde{I} = \tilde{P} \cdot\cdot \tilde{C}_I; \tilde{C}_{II} \cdot\cdot \tilde{P} = \tilde{P} \cdot\cdot \tilde{C}_{II} = \tilde{P}^t; \ \tilde{C}_{III} \cdot\cdot \tilde{P} = \tilde{P} \cdot\cdot \tilde{C}_{III} = \tilde{P},$$

$$(\tilde{C}_I \cdot\cdot \tilde{P}) \cdot \tilde{Q} = I_1(\tilde{P})\tilde{Q}; \ (\tilde{C}_{II} \cdot\cdot \tilde{P})\tilde{Q} = \tilde{P}^t \cdot \tilde{Q}; \ (\tilde{C}_{III} \cdot\cdot \tilde{P}) \cdot \tilde{Q} = \tilde{P} \cdot \tilde{Q}, \qquad (4.1.3)'$$

$$\tilde{C}_I \cdot\cdot \tilde{P} \cdot \tilde{Q} = I_1(\tilde{P} \cdot \tilde{Q})\tilde{I}; \ \tilde{C}_{II} \cdot\cdot \tilde{P} \cdot \tilde{Q} = \tilde{P}^t \cdot \tilde{Q}^t; \ \tilde{C}_{III} \cdot\cdot \tilde{P} \cdot \tilde{Q} = \tilde{P} \cdot \tilde{Q}.$$

Lemma

The space of *rank*- 4 isotropic tensors ${}^4\tilde{C}$ is spanned by its *isomers*, $\{\tilde{C}_I, \tilde{C}_{II}, \tilde{C}_{III}\}$.

Proof

Since the set $\{\tilde{C}_I, \tilde{C}_{II}, \tilde{C}_{III}\}$ (4.1.2), forms the basis for the rank-4 isotropic tensor, then given constants λ, μ, ν we have:

$$ {}^4\tilde{C} = \lambda\tilde{C}_I + \mu(\tilde{C}_{III} + \tilde{C}_{II}) + \nu(\tilde{C}_{III} - \tilde{C}_{II}). \tag{4.1.4}$$

It could be seen that each of the tensors $\tilde{C}_\alpha, \alpha{=}I, II, III$ is isotropic. In fact, for $\tilde{C}_{III}$, $\mathbf{e}'^i = \mathbf{e}^i \cdot O^k_{.m}\mathbf{e}_k\mathbf{e}^m = O^i_{.m}\mathbf{e}^m, \mathbf{e}'_i = \mathbf{e}_i \cdot O^n_m\mathbf{e}^m\mathbf{e}_n = O_i^{\ n}\mathbf{e}_n$. Then, for $\tilde{C}_{III}$, $\tilde{C}'_{III} = \mathbf{e}'^i\tilde{I}\mathbf{e}'_i = O^i_{.m}O_i^{\ n}\mathbf{e}^m\tilde{I}\mathbf{e}_n = \delta^n_m\mathbf{e}^m\tilde{I}\mathbf{e}_n = \mathbf{e}^n\tilde{I}\mathbf{e}_n = \tilde{C}_{III}$. This confirms isotropy of $\tilde{C}_{III}$, which is similarly true for the other two isomers. Further, it would be seen that $\tilde{C}_\alpha \cdot\cdot \tilde{C}_\beta = \tilde{C}_\gamma, \alpha, \beta, \gamma = I, II, III$. Hence, (4.1.4).

Remark

[a] Any isotropic tensor *rank*- n ($n > 2$) is expressed through the rank- 2 isotropic tensor, $\tilde{I}$.
[b] We further note (i.) that, in consonance with the concept of *isometry*, there could be other ways of writing the basic isotropic tensors; (ii.) any n-ranked tensor has $n!$ isomers. $\tilde{\mathbf{T}}^t$ is the isomer of $\tilde{\mathbf{T}}$. Likewise, $C_{ijk}\mathbf{e}^i\mathbf{e}^j\mathbf{e}^k$ has one of its isomers as $C_{ijk}\mathbf{e}^k\mathbf{e}^j\mathbf{e}^i$.; (iv.) the structure of a given tensor determines the number of independent isomers it will possess, such that the existence of symmetry in its internal structure reduces the number of independent isomers.
For this, a rank-4 tensor will possess $4! = 24$ isomers. But if this tensor is isotropic, then only 3 of its isomers are independent. Hence, for any constants λ, μ and ν any rank-4 isotropic tensor is expressed through the 3 isomers:
[c] In the orthonormal system of coordinates, (4.1.4) reduces to the known *Hooke's elastic tensor* in the case of isotropic material. It is constructed on the rank-4 isotropic tensor $\delta_{ij}\delta_{kl}$, which has the three *isomers* ($\delta_{ij}\delta_{kl}$, $\delta_{ik}\delta_{jl}$, $\delta_{il}\delta_{jk}$), for which the elastic tensor C_{ijkl} for isotropic material takes the expression $C_{ijkl} = \lambda\delta_{ij}\delta_{kl} + \mu(\delta_{ik}\delta_{jl} + \delta_{il}\delta_{jk})$, λ and μ are the Lame constants.

Isotropy

Further, given any rank-2 tensor $\tilde{\mathbf{T}}$, its *cofactor tensor* $\tilde{\mathbf{T}}^c$ can be obtained simply as the *dyad* of vector $\mathbf{e}_i \times \mathbf{e}_j$ and vector $(\tilde{\mathbf{T}}\cdot\mathbf{e}^i) \times (\tilde{\mathbf{T}}\cdot\mathbf{e}^i)$ such that

$$\tilde{\mathbf{T}}^c = [(\tilde{\mathbf{T}}\cdot\mathbf{e}^i) \times (\tilde{\mathbf{T}}\cdot\mathbf{e}^i)][\mathbf{e}_i \times \mathbf{e}_j] = \frac{1}{2}\epsilon_{skq}\epsilon^{mnp}T_n^{\ k}T_p^{\ q}\mathbf{e}^s\mathbf{e}_m. \tag{4.2.1}$$

Thus, the *1st invariant* $I_1(\tilde{\mathbf{T}})$ and *2nd invariant* $I_2(\tilde{\mathbf{T}})$ of tensor $\tilde{\mathbf{T}}$ can be given through the double dot product:

$$I_1(\tilde{\mathbf{T}}) = \tilde{I}\cdot\cdot\tilde{\mathbf{T}}, I_2(\tilde{\mathbf{T}}) = I_1(\tilde{\mathbf{T}}^c) = \tilde{I}\cdot\cdot\tilde{\mathbf{T}}^c = \frac{1}{2}[I_1^2(\tilde{\mathbf{T}}) - I_1(\tilde{\mathbf{T}}^2)]. \tag{4.2.2}$$

This process can be followed to, compute other invariants, obtain the Hamilton-Cayley relations, take tensor derivative of scalar and tensor functions and thus deduce constitutive relations from pertinent energy; essentially, using already known operations of dot and cross products on invariant objects, vectors inclusive, and many more operations.

Hamilton-Kelly's Theorem
The set $\tilde{\mathbf{T}}^2, \tilde{\mathbf{T}}, \tilde{I}$, where $\tilde{\mathbf{T}}^0 \equiv \tilde{I}$, forms the basis for any tensor. $\tilde{\mathbf{T}}^m$ in the space $\tilde{\mathbf{T}} \in \mathsf{T}^m \subset \Omega(\mathsf{E}^n)$, $m \geq 3$ such that

$$\tilde{\mathbf{T}}^3 = I_1(\tilde{\mathbf{T}})\tilde{\mathbf{T}}^2 - I_2(\tilde{\mathbf{T}})\tilde{\mathbf{T}} + I_1(\tilde{\mathbf{T}})\tilde{I}. \tag{4.2.3}$$

i.e. the tensor $\tilde{\mathbf{T}}^m$, $m \geq 3$ is expressible linearly through tensors of lower degree $\tilde{\mathbf{T}}^2, \tilde{\mathbf{T}}, \tilde{I}$.
We further recall, in addition to (4.2.2), the following relations for nonsingular tensor $\tilde{\mathbf{T}}$

$$\tilde{\mathbf{T}}^{-1} = \frac{1}{I_1(\tilde{\mathbf{T}})}[\tilde{\mathbf{T}}^2 - I_1(\tilde{\mathbf{T}})\tilde{\mathbf{T}} + I_2(\tilde{\mathbf{T}})\tilde{I}], \quad I_3(\tilde{\mathbf{T}}) = \frac{1}{6}[I_1^3(\tilde{\mathbf{T}}) - 3I_1(\tilde{\mathbf{T}})I_1(\tilde{\mathbf{T}}^2) + 2I_1(\tilde{\mathbf{T}}^3)],$$

$$I_1(\tilde{\mathbf{T}}^{-1}) = \frac{I_2(\tilde{\mathbf{T}})}{I_3(\tilde{\mathbf{T}})}, \; I_2(\tilde{\mathbf{T}}^{-1}) = \frac{I_1(\tilde{\mathbf{T}})}{I_3(\tilde{\mathbf{T}})}, \; I_3(\tilde{\mathbf{T}}^{-1}) = \frac{1}{I_3(\tilde{\mathbf{T}})} = det(\tilde{\mathbf{T}}^{-1}).$$

We also note the following relations, including inequalities,

$$I_k(\tilde{\mathbf{T}}) = I_k(\tilde{\mathbf{T}}^t)\; k = 1,2,3; \quad I_1(\tilde{\mathbf{T}}) \geq 3I_3^{\frac{1}{3}}(\tilde{\mathbf{T}}), \; I_2(\tilde{\mathbf{T}}) \geq 3I_3^{\frac{2}{3}}(\tilde{\mathbf{T}}), \; I_1^2(\tilde{\mathbf{T}}) \geq 3I_2(\tilde{\mathbf{T}}). \tag{4.2.4}$$

Tensor Derivative of Invariants - Energy Function

Differentiation in Tensor Argument: Frechet Derivative}

Frechet Derivative
Here, the tensor derivative of an invariant is taken from the first principle of variation with respect to the {argument} in the sense of *Frechet*'s derivative.
Thus, we define derivative of a differentiable function $F \in C(\Omega)$ as the *linear coefficient of the variation of the argument due to the variation of the function*:

$$\delta F = F(x + \delta x) - F(h) = F'(x)\delta x, \tag{5.1.1}$$

where δF is the variation of the function $F(x)$ due to variation of the argument δx. Here, $F'(x)$ as the coefficient of δx in its linear form, constitutes the derivative of the function F. It is in this form that the notion of *Frechet Derivative* is invoked in respect of invariants/tensors [10].

Computation of Derivative of Scalars
Lemma: The derivative with respect to tensor $\tilde{\mathbf{T}}$ of the 1st, 2nd and 3rd invariants, $I_1(\tilde{\mathbf{T}})$, $I_2(\tilde{\mathbf{T}})$, $I_3(\tilde{\mathbf{T}})$, is given respectively as

$$I_1(\tilde{\mathbf{T}})_{\tilde{\mathbf{T}}} = \tilde{I}, \; I_1(\tilde{\mathbf{T}}^2)_{\tilde{\mathbf{T}}} = 2\tilde{\mathbf{T}}^t, \; I_1(\tilde{\mathbf{T}}^3)_{\tilde{\mathbf{T}}} = 3\tilde{\mathbf{T}}^{2t}.$$

Proof

Derivative of Tensor Invariants.

a. We note that the first invariant of a tensor is a linear scalar function of its argument and it is given by the double dot product between the tensor and the unit tensor:

$$I_1(\tilde{\mathbf{T}}) \overset{(4.2.2)}{=} \tilde{I} \cdot\cdot \tilde{\mathbf{T}}.$$

Then, from the first principle we have

(i) $\delta I_1(\tilde{\mathbf{T}}) \overset{(5.1.1)}{=} I_1(\tilde{\mathbf{T}} + \delta\tilde{\mathbf{T}}) - I_1(\tilde{\mathbf{T}}) = \tilde{I} \cdot\cdot (\tilde{\mathbf{T}} + \delta\tilde{\mathbf{T}}) - \tilde{I} \cdot\cdot \tilde{\mathbf{T}} = \tilde{I} \cdot\cdot \delta\tilde{\mathbf{T}} = \tilde{I} \cdot\cdot \delta\tilde{\mathbf{T}}^t. \Rightarrow I_1(\tilde{\mathbf{T}})_{\tilde{\mathbf{T}}} = \tilde{I}.$

(ii) Likewise we have

$$\delta I_1(\tilde{\mathbf{T}}^2) = I_1[(\tilde{\mathbf{T}} + \delta\tilde{\mathbf{T}}) \cdot (\tilde{\mathbf{T}} + \delta\tilde{\mathbf{T}})] - I_1(\tilde{\mathbf{T}}^2) = I_1(\tilde{\mathbf{T}} \cdot \delta\tilde{\mathbf{T}} + \delta\tilde{\mathbf{T}} \cdot \tilde{\mathbf{T}}),$$

and ignoring the nonlinear term $\delta\tilde{\mathbf{T}}\cdot\delta\tilde{\mathbf{T}}$ due definition,

$$=\tilde{I}\cdot\cdot\tilde{\mathbf{T}}\cdot\delta\tilde{\mathbf{T}}+\tilde{I}\cdot\cdot\delta\tilde{\mathbf{T}}\cdot\tilde{\mathbf{T}}=\tilde{\mathbf{T}}\cdot\cdot\delta\tilde{\mathbf{T}}+\tilde{\mathbf{T}}\cdot\cdot\delta\tilde{\mathbf{T}}=2\tilde{\mathbf{T}}'\cdot\cdot\delta\tilde{\mathbf{T}}'.\ \Rightarrow\ I_1(\tilde{\mathbf{T}}^2)_{\tilde{\mathbf{T}}}=2\tilde{\mathbf{T}}'.$$

(iii) $\delta I_1(\tilde{\mathbf{T}}^3)=I_1(\tilde{\mathbf{T}}^2\cdot\delta\tilde{\mathbf{T}}+\tilde{\mathbf{T}}\cdot\delta\tilde{\mathbf{T}}\cdot\tilde{\mathbf{T}}+\delta\tilde{\mathbf{T}}\cdot\tilde{\mathbf{T}}^2)=3\tilde{\mathbf{T}}^2\cdot\cdot\delta\tilde{\mathbf{T}}=3\tilde{\mathbf{T}}^{2t}\cdot\cdot\delta\tilde{\mathbf{T}}'.\ \Rightarrow I_1(\tilde{\mathbf{T}}^3)_{\tilde{\mathbf{T}}}=3\tilde{\mathbf{T}}^{2t}.$

Hence,

$$I_1(\tilde{\mathbf{T}})_{\tilde{\mathbf{T}}}=\tilde{I},\ I_1(\tilde{\mathbf{T}}^2)_{\tilde{\mathbf{T}}}=2\tilde{\mathbf{T}}',\ I_1(\tilde{\mathbf{T}}^3)_{\tilde{\mathbf{T}}}=3\tilde{\mathbf{T}}^{2t}.$$

b. **Derivative of 2nd and 3rd Invariants**

By invoking the Hamilton-Kelly theorem relating tensor of any degree through degree not more than two *via-a-vis* the invariants of that tensor, we deduce the pertinent expressions for the derivative of higher invariants.

(iv) Indeed, in cognizance of (4.2.4) and the derivatives of first invariants indicated above, we obtain the required derivatives of the 2nd and 3rd invariants as

$$I_2(\tilde{\mathbf{T}})_{\tilde{\mathbf{T}}}=I_1(\tilde{\mathbf{T}})\tilde{I}-\tilde{\mathbf{T}}',\quad I_3(\tilde{\mathbf{T}})_{\tilde{\mathbf{T}}}=\tilde{\mathbf{T}}'^2-I_1(\tilde{\mathbf{T}})\tilde{\mathbf{T}}'+I_2(\tilde{\mathbf{T}})\tilde{I}=I_3(\tilde{\mathbf{T}})(\tilde{\mathbf{T}}')^{-1}.$$

Theorem: The derivative of any scalar function of tensor argument $\varphi(\tilde{\mathbf{T}})$ is

$$\varphi(I_1(\tilde{\mathbf{T}}),I_2(\tilde{\mathbf{T}}),I_3(\tilde{\mathbf{T}}))_{\tilde{\mathbf{T}}}=(\frac{\partial\varphi}{\partial I_1}+I_1(\tilde{\mathbf{T}})\frac{\partial\varphi}{\partial I_2}+I_2\frac{\partial\varphi}{\partial I_3})\tilde{I}-(\frac{\partial\varphi}{\partial I_2}+I_1\frac{\partial\varphi}{\partial I_3})\tilde{\mathbf{T}}'+\frac{\partial\varphi}{\partial I_3}\tilde{\mathbf{T}}'^2.$$

Proof: Any scalar function of a tensor argument is ultimately expressed through the invariants of the tensor $\varphi(\tilde{\mathbf{T}})=\varphi(I_1(\tilde{\mathbf{T}}),I_2(\tilde{\mathbf{T}}),I_3(\tilde{\mathbf{T}}))$. Invoking the last lemma, the Hamilton-Kelly theorem and the established derivatives of invariants, the theorem is proved:

$$\varphi(I_1(\tilde{\mathbf{T}}),I_2(\tilde{\mathbf{T}}),I_3(\tilde{\mathbf{T}}))_{\tilde{\mathbf{T}}}=[\frac{\partial\varphi}{\partial I_1}+I_1(\tilde{\mathbf{T}})\frac{\partial\varphi}{\partial I_2}]\tilde{I}-\frac{\partial\varphi}{\partial I_2}\tilde{\mathbf{T}}'+\frac{\partial\varphi}{\partial I_3}I_3(\tilde{\mathbf{T}})(\tilde{\mathbf{T}}')^{-1}$$

$$=(\frac{\partial\varphi}{\partial I_1}+I_1(\tilde{\mathbf{T}})\frac{\partial\varphi}{\partial I_2}+I_2\frac{\partial\varphi}{\partial I_3})\tilde{I}-(\frac{\partial\varphi}{\partial I_2}+I_1\frac{\partial\varphi}{\partial I_3})\tilde{\mathbf{T}}'+\frac{\partial\varphi}{\partial I_3}\tilde{\mathbf{T}}'^2.$$

Theorem: Suppose the invariants of tensor, arising from geometry of deformation in elasticity, are

$$s_0=\mathbf{c}\cdot(\tilde{U}-\tilde{I})\cdot\mathbf{c},\quad s_4=\mathbf{c}\cdot(\tilde{U}-\tilde{I})^2\cdot\mathbf{c},$$

where, $\tilde{U}$ such that $\tilde{U}^2=\overset{o}{\nabla}\mathbf{R}\cdot\overset{o}{\nabla}\mathbf{R}'$ is the symmetric *stretch tensor*, $\overset{c}{\nabla}\mathbf{R}$ is a non-symmetric rank-2 tensor of *deformation gradient* such that $\overset{o}{\nabla}\mathbf{R}=\tilde{U}\cdot\tilde{O}^D$, $\tilde{O}^D=\tilde{U}^{-1}\cdot\overset{o}{\nabla}\mathbf{R}$ is the *deformative rotation tensor*, $\tilde{I}$ is the unit tensor, $\mathbf{R}$ is the position vector and $\mathbf{c}$ is a unit directional vector. Then, the tensor derivative with respect to $\overset{o}{\nabla}\mathbf{R}$ of s_0 and s_4 are

$$s_{o_{\overset{o}{\nabla}\mathbf{R}}}=\frac{\partial s_o}{\partial\overset{o}{\nabla}\mathbf{R}}=\frac{1}{2}(\mathbf{cc}\cdot\tilde{O}^D+\tilde{U}^{-1}\cdot\mathbf{cc}\cdot)\overset{o}{\nabla}\mathbf{R}\quad\text{and}$$

$$s_{4_{\overset{o}{\nabla}\mathbf{R}}}=\frac{\partial s_4}{\partial\overset{o}{\nabla}\mathbf{R}}=2\mathbf{cc}\cdot\overset{o}{\nabla}\mathbf{R}-\mathbf{cc}\cdot\tilde{O}^D-\tilde{U}^{-1}\cdot\mathbf{cc}\cdot\overset{o}{\nabla}\mathbf{R}.$$

Proof: The proof of this follows directly from the application of the previously enunciated procedure for Frechet derivative of a tensor, on any $\tilde{I}$, $\tilde{\mathbf{T}}$ and $\tilde{\mathbf{T}}^2$.

Theorem:

Suppose in addition to the invariants given in the previous theorem we have also

$$s_1 = I_1(\tilde{U}-\tilde{I}) = \tilde{I}\cdot\cdot(\tilde{U}-\tilde{I}) \text{ and } s_2 = I_1(\tilde{U}-\tilde{I})^2 = \tilde{I}\cdot\cdot[(\tilde{U}-\tilde{I})\cdot(\tilde{U}-\tilde{I})).$$

Now let the scalar energy function W be given as

$$(i) W = \mu s_2 + \frac{1}{2}\lambda s_1^2, \quad (ii) W = \lambda s_2 + \frac{1}{2}\lambda_1 s_1^2 + \lambda_0 s_0,$$

respectively for an isotropic elastic material, for a plane transversely isotropic material [19-20].

Then the *Piola stress tensor* resulting from the derivation of W with respect to $\overset{o}{\nabla}\mathbf{R}$, is respectively

$$(i)\ \tilde{P} \equiv \frac{\partial W}{\partial \overset{o}{\nabla}\mathbf{R}} = 2\mu\overset{o}{\nabla}\mathbf{R} + [\lambda s_1 + 2\mu]\tilde{O}^D = [(\lambda s_1 - 2\mu)\tilde{U}^{-1} + 2\mu]\cdot\overset{o}{\nabla}\mathbf{R},$$

$$(ii)\ \tilde{P} \equiv \frac{\partial W}{\partial \overset{o}{\nabla}\mathbf{R}} = 2\lambda_2\overset{o}{\nabla}\mathbf{R} + (\lambda_1 s_1 - 2\lambda_2)\tilde{O}^D + \lambda_0\mathbf{cc}\cdot\overset{o}{\nabla}\mathbf{R},$$

where λ and μ are the so-called Lame's constants from elasticity, λ_0, λ_1 and λ_2 are pertinent constants associated with the anisotropy of the elastic composite materials [20].

Proof: The proof of this follows directly from the application of the procedure for Frechet derivative of a tensor on $\tilde{\mathbf{T}}$, $\tilde{\mathbf{T}}^2$ and $\tilde{I}$ and the previous theorem, straightforwardly.

Theory of Strain-Gradient-Divergence Plasticity

Plasticity is an aspect of elasticity, when irreversible process takes place. Here, the robustness of tensor operations and the concepts therein, such as: scalar functions and invariants and their tensor derivatives; multiple scalar products and conjugacy between geometric and mechanical characteristics have been deftly applied to upgrade the theory of strain-gradient plasticity of Gurtin and Anand [17] to obtain the theory of strain-gradient-divergence plasticity [16].

Now, consider a body Ω undergoing plastic deformation. Suppose $\mathbf{u}(\mathbf{x},t)$ denotes the displacement vector of an arbitrary point $\mathbf{x}$ in a region Ω describing a body composed of manifolds of particles. The classical theory of isotropic plastic solids undergoing small deformations is based on the kinematic relations given by the decomposition of the displacement gradient;

$$\nabla\mathbf{u} = \tilde{H}^e + \tilde{H}^p; \quad tr\tilde{H}^p = I_1(\tilde{H}^p) = \tilde{H}^p\cdot\cdot\tilde{I} = 0. \tag{6.1}$$

$$\tilde{E}^e = \frac{1}{2}(\tilde{H}^e + \tilde{H}^{et}); \quad \tilde{E}^p = \frac{1}{2}(\tilde{H}^p + \tilde{H}^{pt}), \tag{6.2}$$

where $\tilde{E}^e$ represents rotation and stretching while $\tilde{E}^p$ denotes the plastic distortion characterizing the evolution of dislocations and other defects through the structure. $I_1(\tilde{H}^p) = 0$ defines the condition of plastic incompressibility. The elastic and plastic strains are defined by (6.2), while $\tilde{W}^e = skw\tilde{H}^e$ and $\tilde{W}^p = skw\tilde{H}^p$ are the elastic and plastic rotation tensor respectively.

Let the internal and external virtual power expenditure over a micro-region $P \subset \Omega$ be

$$W_{int} = \int_P [\tilde{\mathbf{T}}\cdot\cdot\tilde{E}^e + \boldsymbol{\chi}\cdot\nabla\cdot\tilde{\mathbf{E}}^p + \tilde{\mathbf{T}}^p\cdot\cdot\tilde{\mathbf{E}}^p + \tilde{\mathbf{K}}^p \vdots \nabla\tilde{\mathbf{E}}^p]dV;$$

$$W_{ext}(P,V) = \int_P [\mathbf{t}(\mathbf{n})\cdot\mathbf{u} + \mathbf{K}(\mathbf{n})\cdot\cdot\tilde{E}^p]dA + \int_{\partial P} \mathbf{b}\cdot\mathbf{u}dV,$$

for which, given the set of virtual velocities $\vartheta = (\delta\mathbf{u}, \delta\dot{\tilde{H}}^e, \delta\dot{\tilde{E}}^e)$, by the principle of virtual power we have $W_{int}(P,\vartheta) = W_{ext}(P,\vartheta)$, which results in the set of *macrobalance force* with the corresponding *macrotraction* and *microbalance force* with the corresponding *microtraction*:

$$\nabla\cdot\tilde{\mathbf{T}} + \mathbf{b} \text{ and } \tilde{\mathbf{T}}\cdot\mathbf{n} = \mathbf{t}, \tag{6.3}$$

$$\tilde{\mathbf{T}}_o = \tilde{\mathbf{T}}^p - sym_o(\nabla\boldsymbol{\chi}) - \nabla\cdot\tilde{K}^p \text{ and } \tilde{K} = sym_o(\boldsymbol{\chi}\otimes\mathbf{n}) + \tilde{K}^p\cdot\mathbf{n}, \tag{6.3'}$$

where $\tilde{\mathbf{T}}_o = \tilde{\mathbf{T}} - \frac{1}{3}(\mathrm{tr}\tilde{\mathbf{T}})\tilde{I}$ is the deviatoric part of the macrostress $\tilde{\mathbf{T}}$, $sym_o(\nabla\boldsymbol{\chi})$ is the symmetric deviatoric part of the gradient of the microforce $\boldsymbol{\chi}$, while $\boldsymbol{\chi}\otimes\mathbf{n}$ is a dyad. Note that symmetric-deviatoric part of tensor $\tilde{\mathbf{T}}$ is $sym_o\tilde{\mathbf{T}} \equiv [\frac{1}{2}(\tilde{\mathbf{T}}+\tilde{\mathbf{T}}') - \frac{1}{3}(\mathrm{tr}\tilde{\mathbf{T}})\tilde{I}]$.

Here, the internal microforce $\boldsymbol{\chi}$ is energy conjugate to $\nabla\cdot\tilde{E}^p$, macrostress $\tilde{\mathbf{T}}$ is energy conjugate to $\tilde{E}^e$, microstress $\tilde{\mathbf{T}}^p$ is energy conjugate to $\tilde{E}^p$, the polar microstress rank3 tensor $\tilde{K}^p$ is energy conjugate to the rate of plastic strain gradient $\nabla\tilde{E}^p$. $\mathbf{b}$ is a body force in P, $\mathbf{t}$ is a microtraction vector and $\mathbf{K}$ a rank2 microtraction tensor both on ∂P, $\mathbf{n}$ is the outward unit normal vector, orientating the surface ∂P.

By the second law of thermodynamics, the free energy imbalance for this plastic process is taken, due (6.3), as

$$\psi \le \tilde{\mathbf{T}}\cdot\tilde{E}^e + \boldsymbol{\chi}\nabla\cdot\tilde{E}^p + \tilde{\mathbf{T}}^p\cdot\cdot\tilde{E}^p + \tilde{K}^p\vdots\nabla\tilde{E}^p,$$

which nudges us to take ψ as the Helmholtz free energy in the form

$$\psi = \psi(\tilde{E}^e, \tilde{E}^p, \nabla\cdot\tilde{E}^p, \nabla\times\tilde{E}^p). \tag{6.4}$$

Following from the previous theory, we then take the tensor derivative of the scalar (i.e. the free energy) with respect to each argument to obtain the corresponding conjugate mechanical parameter:

$$\tilde{\mathbf{T}} = \frac{\partial\psi}{\partial\tilde{E}^e},\quad \tilde{\mathbf{T}}^p = \frac{\partial\psi}{\partial\tilde{E}^p}\quad \boldsymbol{\chi}_{en} = \frac{\partial\psi}{\partial\nabla\cdot\tilde{E}^p},\quad (\tilde{K}_{en}{}^p)_{jqp} = (sym_o P)_{jqp}, \tag{6.5}$$

where $(P)_{jqp} = \frac{\partial\psi}{\partial G_{ij}}\grave{o}_{ipq}$ and $G_{ij} = \grave{o}_{irs}E^p_{js,r}$.

Consequently, the energy is obtained in the quadratic form as

$$\frac{1}{2}\kappa I_1^2(\tilde{E}^e) + \mu|\tilde{E}^e_o|^2 + \frac{1}{2}\mu|\tilde{E}^p|^2 + \frac{1}{2}\mu Q^2|\nabla\cdot\tilde{E}^p|^2 + \frac{1}{2}\mu L^2|\nabla\times\tilde{E}^p|^2, \tag{6.6}$$

where Q and L are called energetic length scales associated with $\nabla\cdot\tilde{E}^p$ and $\nabla\times\tilde{E}^p$ respectively. The dissipative microstresses based on the von Mises yield criterion are given in terms of their corresponding power conjugates as follows

$$\boldsymbol{\chi}_{\mathrm{dis}} = q^2 S_o(\frac{d^p}{d_o})^m\frac{\nabla\cdot\tilde{E}^p}{d^p};\ \tilde{\mathbf{T}}^p_{\mathrm{dis}} = S_o(\frac{d^p}{d_o})^m\frac{\tilde{E}^p}{d^p};\ \tilde{K}^p_{\mathrm{en}} = l^2 S_o(\frac{d^p}{d_o})^m\frac{\nabla\tilde{E}^p}{d^p}, \tag{6.7}$$

S_o is the initial yield strength, d_o is the initial flow rate, m is the rate sensitivity parameter and d^p is the effective flow rate defined by

$$d^p = \sqrt{|\tilde{E}^p|^2 + q^2|\nabla\cdot\tilde{E}^p|^2 + q^2|\nabla\tilde{E}^p|^2},$$

where q and l are called dissipative length scales associated with $\nabla\cdot\tilde{E}^p$ and $\nabla\tilde{E}^p$ respectively. Putting (6.5) and (6.7) into the microforce balance and microtraction condition $(6.3)'$, we obtain the interested viscoplasticity flow rule, which is a generalization of the Gurtin-Anand model:

$$\tilde{\mathbf{T}}_o + \mu(L^2\Delta\tilde{E}^p + (Q^2 - L^2)sym_o(\nabla\nabla\tilde{E}^p) - \tilde{E}^p) = S_o(\frac{d^p}{d_o})^m\frac{\tilde{E}^p}{d^p} - q^2 S_o(sym_o(\nabla[(\frac{d^p}{d_o})^m\frac{\nabla\cdot\tilde{E}^p}{d^p}])) - l^2 S_o\nabla[(\frac{d^p}{d_o})^m\frac{\nabla\tilde{E}^p}{d^p}].$$

References

[1] I.S. Sokolnikoff, Tensor Analysis with Application to Geometry and the Mechanics of Continua. John Wiley and Son, Inc. (1964) New York.

[2] L. Brillouin, Tensors in Mechanics and Elasticity. Paris, Masson (1964).

[3] V.W. Eduardo, Notes on Continuum Mechanics. Springer Publishers (2013).

[4] R. Ogden, Application of Nonlinear Elasticity to Soft Tissue Biomechanics. 24th International Congress of Theoretical and Applied Mechanics (ICTAM2016), (2016) Montreal, Canada.

[5] A. Akinola, B. Olokuntoye, A. Borokinni, O. Fadodun, O. Layeni, Effective teaching of vector to mitigate tensor-phobia in mechanics and application of exterior form. Proceeding of the 25th International Congress of Theoretical and Applied Mechanics (2021).

[6] C.A Truesdell, The Mechanical Foundation of Elasticity and Fluid Dynamics International Science Review Series Vol III, Gordon and Breach Science Publishers, (1966) New York.

[7] A.A. Ilyushin Mechanics of Continuum Mechanics. Moscow State University Press, (1978) Moscow. (2nd Ed. in Russian).

[8] G.A. Maugin, Continuum Mechanics through the Ages - From the Renaissance to the Twentieth Century. From Hydraulic to Plasticity, Solid Mechanic and Applications, (2016) Vol. 223. Springer International Publishing, Switzerland. https://doi.org/10.1007/978-3-319-26593-3

[9] Y.A. Amenzade, Theory of Elasticity. Mir. Publishers, (1979) Moscow.

[10] A.I. Lurie, Nonlinear Theory of Elasticity. Nauka Publishers, (1980) Moscow. (in Russian).

[11] D. Bachman, A Geometric Approach to Differential Forms, Birkhauser (2006).

[12] A.E. Green, W. Zerna, Theoretical Elasticity, Oxford University Press, (1968) London.

[13] E. Sanchez-Palencia, Non-homogeneous media and vibration theory. Springer-Verlag (1980).

[14] N.S. Bakhvalov, Averaging of partial differential equations with oscillating coefficients. Doklad Akademik Nauk, USSR, (1975) Vol. 225, No 2, pp245-252.

[15] B.E. Pobedria, Mechanics of Composite Materials. Moscow State University Publishers, Moscow (1984). (in Russian).

[16] A.P. Akinola, A.S. Borokinni, O.P. Layeni, A theory of strain-gradient-divergence plasticity. Proceeding of the 24th International Congress of Theoretical and Applied Mechanics (2016).

[17] M.E. Gurtin, L. Anand, A theory of strain gradient plasticity for istropic, plastically irrotational materials. Part I: Small deformations. J. Mech. Phys. Solids (2005) 53: 1624-1649. https://doi.org/10.1016/j.jmps.2004.12.008

[18] J. Milnor, Towards the Poincare Conjecture and the Classification of 3-manifolds. Notices of the American Mathematical Society, (2003) 50, no. 10, pp1226-1233.

[19] A. Akinola On Interacting Longitudinal-Shear Waves in Large Elastic Deformation of a Composite Laminate. International Journal of Nonlinear Mechanics, (1999) 34, pp405-414. https://doi.org/10.1016/S0020-7462(98)00019-5

[20] A. Akinola, On Complex Potentials for Finite Plane Deformation of a Transversely Isotropic Material. Mechanic Research Communication (MRC), (2005) 32, pp99-114. https://doi.org/10.1016/j.mechrescom.2004.04.006

Advanced Topics in Mechanics of Materials, Structures and Construction - AToMech1-2023 Materials Research Forum LLC
Materials Research Proceedings 31 (2023) 437-447 https://doi.org/10.21741/9781644902592-45

Bayesian regularization optimization algorithm for the experimental thermophysical property for 80:20% water and ethylene glycol based ZrO_2 nanofluids

M. Manzoor Hussain[1,*], L. Syam Sundar[2], Feroz Shaik[2]

[1]Department of Mechanical Engineering, JNTUH College of Engineering, Jawaharlal Nehru Technological University Hyderabad, Kukatpally 500085, Hyderabad, India

[2]Department of Mechanical Engineering, College of Engineering, Prince Mohammad Bin Fahd University, P.O. Box 1664, Al-khobar 31952, Saudi Arabia

* manzoorjntu@jntuh.ac.in

Keywords: Water and Ethylene Glycol, Thermophysical Properties, Bayesian Regularization Approach, Correlations

Abstract. In the current study, water and ethylene glycol (W/EG 80:20%) are used as the base fluid, and sodium dodecyl benzene sulfonate is used as a surfactant to create nanofluids using ZrO_2 nanoparticles prepared using the sol-gel technique. For temperatures ranging from 20 °C to 60 °C and various volume loadings of nanoparticles, 0.2, 0.4, 0.6, 0.8, and 1.0%, respectively, the thermal conductivity, dynamic viscosity, density, and viscosity of these ZrO_2 nanofluids are experimentally evaluated. Artificial neural network based Bayesian regularization algorithm was used to find the correlation coefficient R^2 and root-mean square error. New correlations were also suggested for each of the thermophysical properties. Experiments show that temperatures and concentrations of nanoparticles have a significant impact on the thermophysical properties of nanofluids. In fact, it is shown that, at 20 °C and 60 °C, respectively, increasing the thermal conductivity of nanofluids by 1.0 vol% leads to increases of almost 10.16% and 24.53%. Additionally, at 1.0 vol and 20 °C to 60 °C, the dynamic viscosity is reduced from 61.94% to 50.79%. The correlations and outcomes of the developed artificial neural network are in perfect agreement with the experimental data.

Introduction

The majority of engineering fields have combined their heat transfer methods, and in recent years, research has focused heavily on creating smaller and more effective heat exchangers. Many research activities are now concentrated on increasing the low heat transfer capacities of conventional liquids like water (W), ethylene glycol (EG), or engine oils after extensive use of various approaches, such as modifying materials, using extended surfaces, or improving process standards. In this regard dispersing high conductivity nanoparticles, also referred to as nanofluids [1] has improved the thermal transport properties of heat transfer fluids, making them important research tools [2].

A wide range of nano-additives were used to create the nanofluids including metallic oxides, organic materials, and inorganic materials. Numerous studies have demonstrated that using single phase nanofluids may enhance the heat transfer capabilities of the thermal devices [3-4]. Sundar and Sharma [5] have observed an enhanced thermal conductivity (k_{nf}) with the use of water based Al_2O_3 nanofluids. Wang et al. [6] obtained an enhanced thermal conductivity by using Al_2O_3 and CuO nanoparticles into water, vacuum pump fluid, engine oil, and ethylene glycol. Murshed et al. [7] also found an increased thermal conductivity for TiO_2/water nanofluid. Liu et al. [8] have seen 24% augment in thermal conductivity of 0.1% water mixed Cu nanofluids. Mintsa et al. [9] revealed an augmented thermal conductivity with Al_2O_3/water, and CuO/water nanofluids.

Apart from the water, the mixture of water and ethylene glycol is used as a base fluid for the preparation of nanofluids. The freezing temperature of water can be enhanced by adding small quantity of ethylene glycol [10]. The water and ethylene glycol mixture fluids can be used as engine coolant in automobile radiators in the cold region countries. Vajjha and Das [11] found an increased thermal conductivity of CuO and Al_2O_3 nanofluids, but they used 60:40% of ethylene glycol and water (EG/W) mixture as a base fluid instead of water, because this water and ethylene glycol is used as engine coolant. Sundar et al. [12] have seen an augmented k_{nf} of 50:50% W/EG Al_2O_3 and CuO nanofluids. They determined thermal conductivity in the temperature range from 15 °C to 50 °C and in the volume loadings from 0% to 0.8%. Banisharif et al. [13] observed thermal conductivity enhancement of 9.5% and 14.3% at 0.1% vol. of 50:50% W/EG Fe_3O_4 nanofluid at temperatures of 263.15K and 293.15K respectively. Usri et al. [14] noticed an augmented thermal conductivity for 40:60%, 50:50% and 60:40% W/EG Al_2O_3 nanofluid in the temperatures of 30 to 70 °C and over ϕ of 0.5% to 2.0 %. Alawi et al. [15] found raised k_{nf} for 70:30% and 50:50% W/EG CuO nanofluids. Sundar et al. [16] observed higher thermal conductivity enhancement for 20:80% EG/W Al_2O_3 nanofluid among 20:80%, 40:60% and 60:40% EG/W nanofluids. Sundar et al. [17] also observed an increased thermal conductivity by using 20:80%, 40:60% and 60:40% W/EG Fe_3O_4 nanofluids in the temperature range from 20 °C to 60 °C and in the volume concentration range from 0.2% to 2.0%.

This paper deals with the experimental determination of thermophysical properties of 20:80% EG/W mixture based ZrO_2 nanofluids and validated with Bayesian regularization algorithm approach. The experiments were performed in the volume concentration ranging from 0.2–1.0% and temperature ranging from 20–60°C. From the sol-gel technique, the ZrO_2 nanoparticles were synthesized. The measured thermophysical properties were validated through the literature data. New equations were developed for the thermophysical properties. The regression coefficients were developed through the Bayesian Regularization algorithm approach.

Experimental study

Development of ZrO_2 nanoparticles

ZrO_2 nanoparticles were made through the sol-gel technique. The purified chemicals such as zirconium (IV) propoxide, propanol, NH_3, ethanol and ethylene glycol were procured through the Sigma-Aldrich Chemicals, USA. In a large beaker, the propanol, water and ammonium were taken and then agitated for 10 minute. The zirconium propoxide was added to the above solution and the whole mixture was stirred for 1 hour and observe the formation of while sol. The mole ratio of zirconium propoxide, water and propanol is fixed at 1:8:20. The formed sol is heated around 80 °C to remove the impurities and it is dried in a vacuum. Further the powder is dried in a furnace at a heating rate of 1°C per/min and kept at a temperature of 500 °C for 2.

Preparation of 80:20% W/EG ZrO_2/EG nanofluids

The base liquid is considered as 80:20% W/EG mixture. The stable 80:20% W/EG ZrO_2 nanofluids were prepared by adding SDBS surfactant. The ZrO_2 nanoparticles required for known particle loadings of 20 g of base liquid was calculated from Eq. (1).

$$W_{ZrO_2} = \left(\frac{\phi}{(1-\phi)}\right) \times \left(\frac{W}{\rho}\right)_{bf} \times \rho_{ZrO_2} \tag{1}$$

Where, the ρ_{ZrO_2} and ρ_w is 5680, and 1029.72 kg/m^3, the W_{bf} is 20 g, and W_{ZrO_2} weight of nanoparticles (g). The dry ZrO_2 nanoparticles of 0.22, 0.44, 0.66, 0.88, and 1.11g were used for 0.2, 0.4, 0.6, 0.8 and 1.0% vol. loadings of nanofluids.

Estimation of thermophysical properties

Thermal conductivity of nanofluids

The k_{nf} was evaluated through KD2 Pro (Decagon Devices Inc., USA) instrument. The KD2 Pro works under the principle of transient hot-wire technique. The instrument contains microcontroller and KS-1 sensor and its length and diameters of 60 and 1.3 mm was used. Accuracy of KS-1 sensor was ±2.5% and it measures the thermal conductivity in the ranging from 0.2 to 2 W/mK. The temperature of the nanofluids sample was controlled by Julabo temperature controller, Germany with an accuracy of ±0.1 °C.

Dynamic viscosity of nanofluids

The A&D vibro viscometer, Japan considered for determine the dynamic viscosity of ZrO_2/water nanofluids, equipment consists of electric driven two gold coated vibrating plate sensors and it also consists of a temperature sensor for checking the temperature of the nanofluid sample. The accuracy of the instrument is ±1% and it measures the viscosity over the range from 0.3 to 10,000 mPa.s. Initially the equipment was calibrated with the know viscosity of the fluid (i.e. water) and then used for base fluid (80:20% W/EG) and nanofluids. The viscosity measuring fluid was poured in a cup and it is located on the table. Slowly adjust the table so that, the gold plates are partially immersed into the fluid. There is a mark on the gold plates, up the mark the gold plates should immerse into the fluid.

Density of nanofluids

The density of ZrO_2 nanofluids were measured by utilizing the Archimedes principle. The definition of density says, it is ratio between mass to volume. If we know the volume, then we measure the weight of the fluid by using the precision weighing machine. Initially 50 ml weight was measured after that 20 ml nanofluids and then measure the weight of the beaker. Then calculate the density values. The accuracy of the weighing machine is ±0.001 mg. The same procedure is adopted for measuring the density of other nanofluids concentrations. The law of mixtures can be used to determine the density of nanofluid, and which is given below:

$$\rho_{nf} = \rho_p \phi_p + \rho_{bf}(1 - \phi_p) \tag{15}$$

Specific heat of nanofluids

The nanofluids C_p have been measured by using DSC 2920 model of TA instruments. The cell is first validated with indium, water and then it is used for nanofluids. A nanofluid sample of 10 mg was placed in the instrument. The specific heat of nanofluids was measured over 20 to 60 °C.

Bayesian regularization algorithm

As opposed to traditional backpropagation networks, Bayesian regularization neural networks are thought to be more dependable, robust, and efficient and may reduce or even do away with the need for cross-validation during the learning process. The Bayesian regularization technique makes use of a mathematical technique called ridge regression, which converted a nonlinear regression problem into a statistical task that was equally well-posed provide a more thorough explanation of Bayesian regularization. As part of the Levenberg-Marquardt approach, the backpropagation is typically used to compute the Jacobian $'jX'$ of the performance taking into account the weight and bias variables X. Following the Levenberg-Marquardt algorithm's basic tenets, each variable is adjusted as follows:

$$\left.\begin{aligned} jj &= jX \times jX \\ je &= jX \times E \\ dX &= \frac{-(jj + I \times mu)}{je} \end{aligned}\right\} \tag{18}$$

Where, E indicates all errors, while I reflects identity matrix. The adaptive controlling parameter mu is raised by the factor of mu_inc until the change lowered the performance.

A volumetric concentration (ϕ) and temperature (T) are the input in the proposed NN. The proposed NN is trained individually for each property, i.e., thermal conductivity (k), viscosity (μ), density (ρ), and specific heat (Cp). The schematic diagram is shown in **Fig. 1.**

Getting an optimum NN network is one of the critical tasks. Appropriate selection of hidden layers and the number of neurons in hidden layer determines the accuracy of the prediction. Therefore, experimental data is prepared with six volumetric concentrations ranging from 20 °C to 60 °C. Then, the training of the network is analyzed with minimum mean square error (MSE) over 1000 epochs and R-Value as expressed in the following equations.

$$MSE = \frac{1}{N}\sum_{i=1}^{N}\left(y_{\exp(i)} - y_{\mathrm{ANN}(i)}\right)^2 \tag{19}$$

$$R = \sqrt{1 - \frac{\sum_{i=1}^{N}\left(y_{\exp(i)} - y_{\mathrm{ANN}(i)}\right)^2}{\sum_{i=1}^{N}\left(y_{\exp(i)}\right)^2}} \tag{20}$$

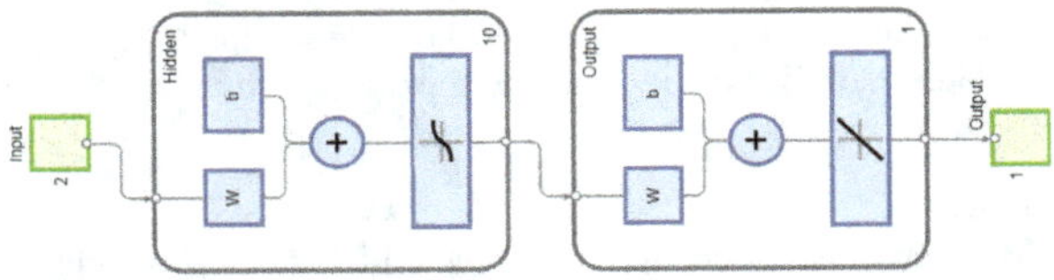

Fig. 1: *The proposed NN network structure with ten neurons in the hidden layer*

Results and discussion

Thermophysical properties

Fig. 2(a) is plotted for the k_{nf} of ZrO_2 nanofluids at dissimilar particle volume loadings and temperatures. As it is observed that the thermal conductivity of increased with an increase of particle volume loadings and temperatures. At particle loadings of 0.2%, 0.4%, 0.6%, 0.8%, and 1.0% vol. of nanofluids at 20 °C, the k_{nf} is raised by 3.86%, 5.69%, 7.72%, 8.13% and 10.16%, respectively, whereas at 60 °C, the k_{nf} is augmented by 7.17%, 12.45%, 15.28%, 19.06% and 24.53% against the base fluid. The base liquid considered in the current analysis is 80:20% W/EG mixture and the thermal conductivity ratio of ZrO_2/80:20% W/EG nanofluid data is compared with Sundar et al. [17] of Fe_3O_4/80:20% W/EG nanofluid and it is shown in **Fig. 2(b)**. The thermal conductivity ratio, $k_r = k_{nf}/k_{bf}$ of the present ZrO_2/80:20% W/EG nanofluid is 1.245, whereas, the thermal conductivity ratio of the Fe_3O_4/80:20% W/EG nanofluid is 1.34 at ϕ = 1.0% and 60°C.

The measured dynamic viscosity of ZrO_2 nanofluid is plotted in **Fig. 2(c)** at different particle loadings and temperatures. Interestingly at higher particle loadings the viscosity is higher, but at the same time measured between 20 °C to 60 °C the μ_{nf} is gradually decreased. The increased μ_{nf} may directly impact on the friction factor. The μ_{nf} is raised by 25.16% to 17.46% (ϕ = 0.2%), 36.47% to 25.40% (ϕ = 0.4%), 44.52% to 31.75% (ϕ = 0.6%), 55.48% to 38.10% (ϕ = 0.8%), and 61.94% to 50.79% (ϕ = 1.0%) from 20 °C to 60 °C, in comparison with base fluid data. The larger resistance between the fluid layers leads to a larger μ_{nf} values. The similar nature of viscosity enhancement has noticed by Minakov et al. [30] for Al_2O_3, TiO_2, ZrO_2, CuO, Fe_2O_3, and Fe_3O_4, as well as nanodiamonds nanofluids. The present study 80:20% W/EG mixture ZrO_2 nanofluids are validated through the Sundar et al. [22] of Fe_3O_4/80:20% W/EG nanofluid and it is provided in **Fig. 2(d)**. The viscosity ratio, $\mu_r = \mu_{nf}/\mu_{bf}$ of the present ZrO_2/80:20% W/EG nanofluid is

1.619, whereas, the viscosity ratio of Fe_3O_4/80:20% W/EG nanofluid is 1.606 at ϕ = 1.0% and 20°C.

The density of ZrO_2 nanofluids were plotted in **Fig. 3(a)**. The density is increased for nanofluid at higher particle volume loadings and those are lowered at higher temperatures. The density of base fluid is 1029.72 kg/m^3, whereas the density is increased to 1076.22 kg/m^3 at 1.0% vol. loadings at 20°C. Moreover at 60°C, the density of the base fluid is 1011.99 kg/m^3 and the density of the 1.0% vol. of nanofluid is increased to 1058.67 kg/m^3. Similar kind of an enhanced density with nanofluids have been presented by Sharifpur et al. [27] and Shoghl et al. [28] by using water -CuO, -MgO, -CNT, -TiO_2, -Al_2O_3 and -ZnO nanofluids.

Measured C_p values were presented in **Fig. 3(b).** As it is seen from the figure, with respect to increase of temperature, the C_p is increases, but the with respect to increase of particle volume loadings, the C_p is decreases. The C_p of base fluid (80:20% W/EG) is 3815 J/kg K, whereas the C_p of 1.0% nanofluid is 3781.05 J/kg K at 20 °C. Similarly, the C_p of base fluid is 3907 J/kg K, but the C_p of 1.0% nanofluid is 3872.13 J/kg K at 60 °C.

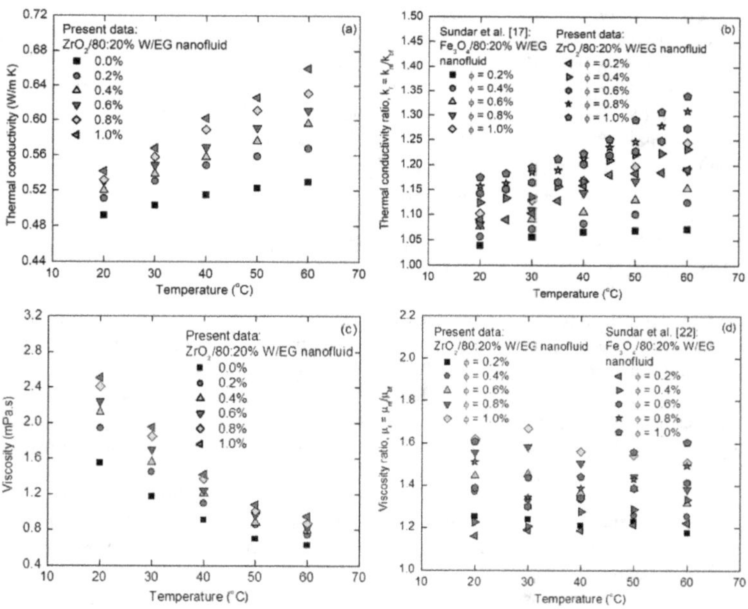

Fig. 2: *(a) Thermal conductivity with respect to temperature, (b) thermal conductivity ratio, (c) viscosity with respect to temperature, and (d) viscosity ratio*

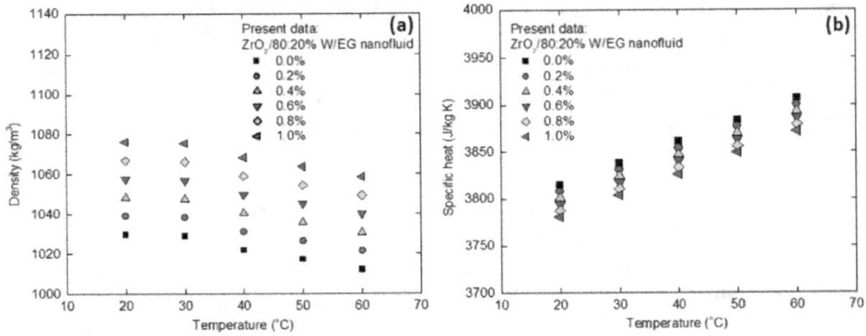

*Fig. 3(a): Density of ZrO_2/80:20% W/EG nanofluid, **(b)** Specific heat of ZrO_2/EG nanofluids*

Bayesian regularization algorithm approach

The proposed network is trained separately for four properties. To avoid the over fitting problem of artificial neural network (ANN), the data is divided into 86% training, 7% testing, and 7% validation dataset. During the training phase, 1000 epochs are used. The training stops if the target mean square error (MSE) is achieved or after completing a number of the epoch. **Figure 4** presents the proposed R^2 analysis for k_{nf}, (a) train values, (b) confirmation of train values, (c) test values, and (d) total values. It is observed from the figure, the training data, R^2 is equal to 0.99966, whereas, for all the data (both training and test data), the R^2 is equal to 0.99938. A good performance can be expressed with the closeness of sample data towards the equality line. The R^2 values obtained for all datasets are close to one, which shows that the developed model is well trained, giving the best performance for predicting the data.

Figure 5 presents the proposed R^2 examination for viscosity, (a) train values, (b) confirmation of train values, (c) test values, and (d) total values. As seen from the figure, the training data, R^2 is equal to 0.99772, but whereas, for all the data (both training and test data), the R^2 is equal to 0.99747. A good matching of the data can be expressed with the closeness of sample data towards the equality line. The R^2 values obtained for all the viscosity datasets are nearly equals to one, which indicates the proposed model is well trained and providing the best results. **Figure 6** presents the proposed correlation coefficient R^2 analysis for density, (a) train values, (b) confirmation of train values, (c) test values, and (d) total values. As it is seen that, the training data, R^2 is equal to 0.99999, but whereas, for all the data (both training and test data), the R^2 is equal to 0.99998. Perfect matching of the experimental and optimized data was observed. The R^2 values obtained for all the density data points are almost equals to one, which provides the proposed model data is well predicted the experimental data.

Figure 7 presents the proposed R^2 analysis for specific heat, (a) train values, (b) confirmation of train values, (c) test values, and (d) total values. As it is seen that, the training data, R^2 is equal to 1, and also all the data (both training and test data) R^2 equal to 1. The R^2 value of specific heat is 1 it means that the used model data is well predicted the experimental data.

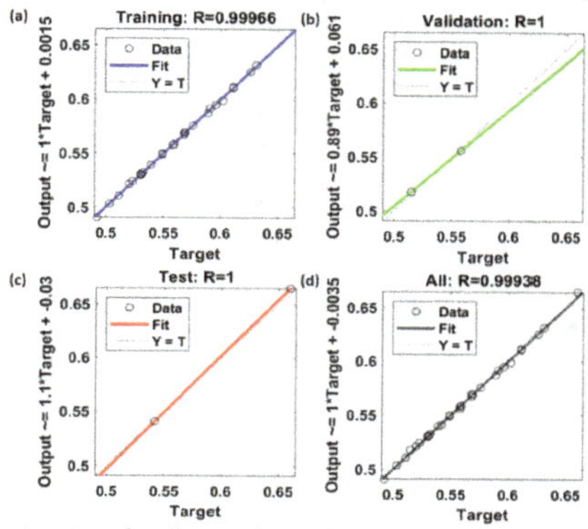

Fig. 4: *Proposed R^2 examination for thermal conductivity: (a) train values, (b) confirmation of train values, (c) test values, and (d) total values.*

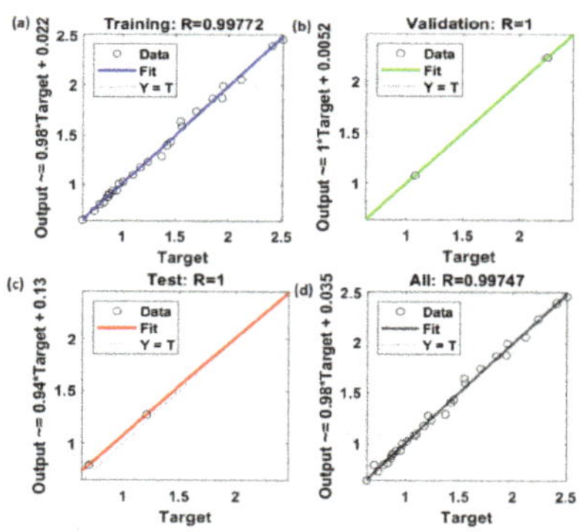

Fig. 5: *Proposed R^2 examination for viscosity: (a) train values, (b) confirmation of train values, (c) test values, and (d) total values*

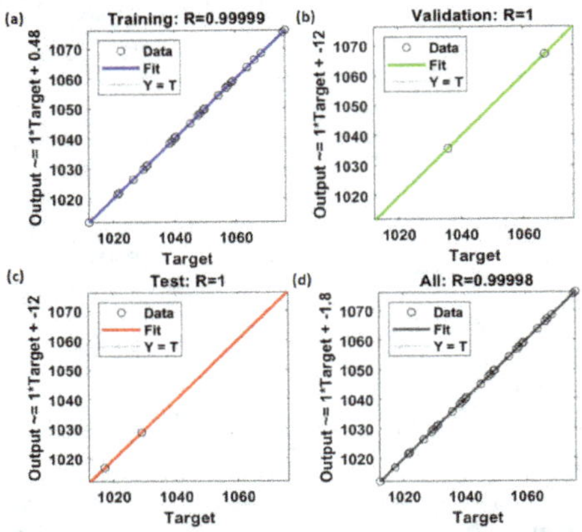

***Fig. 6:** Proposed R^2 examination for density: (a) train values, (b) confirmation of train values, (c) test values, and (d) total values*

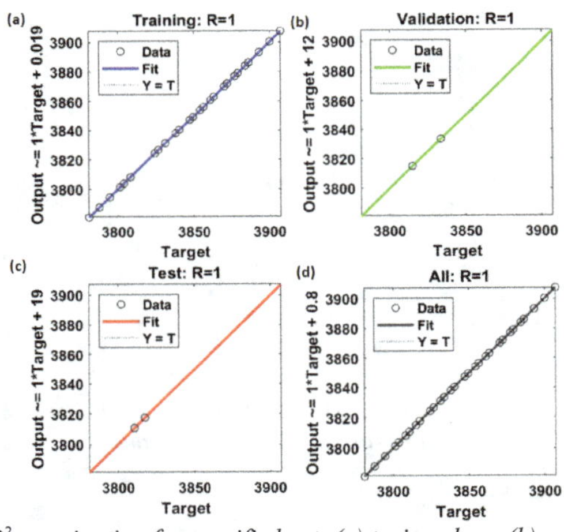

***Fig. 7:** Proposed R^2 examination for specific heat: (a) train values, (b) confirmation of train values, (c) test values, and (d) total values.*

Conclusions

Experiments were performed for the analysis of 80:20% W/EG ZrO_2 nanofluids at dissimilar loading concentrations and temperatures. The ZrO_2 were developed through sol gel procedure. The developed nanofluids were offered ±30 mV of zeta potential which states that the nanofluids are stable. With increased particle volume loadings and temperatures, the thermal conductivity of nanofluids is enhanced. At higher particle loading of 1.0%, the thermal conductivity enhanced is 24.53% at 60°C. With the increased temperature, the dynamic viscosity of the nanofluid is decreased, but oppositely, with an increased particle loading, the viscosity is enhanced. Higher particle concentration of 1.0%, the viscosity increased is 61.94% at 20 °C against base fluid. Other side density is larger and specific heat is lowered to an increase of particle volume loadings.

Correlation coefficient R^2 and mean square error was analyzed using the neural network of Bayesian regularization algorithm approach for all the properties. The correlation coefficient R^2 of specific heat is 1, whereas, correlation coefficient R^2 for thermal conductivity, dynamic viscosity, and density is 0.99966, 0.99772, and 0.9999, respectively. The measurements were appropriately augmented for all the data points used by the neural network approach. Bayesian regularization algorithm utilizes the more accurate tool for modeling the experimental data of nanofluids.

References

[1] S.U.S. Choi, Enhancing thermal conductivity of fluids with nanoparticles. Developments and Applications of Non-Newtonian Flows FED-vol. 231/MD-vol. 66, (1995) 99-105.

[2] R. Saidur, K.Y. Leong, H.A. Mohammad, A review on applications and challenges of nanofluids, Renewable and Sustainable Energy Reviews, 15(3) (2011) 1646 -1668. https://doi.org/10.1016/j.rser.2010.11.035

[3] L.S. Sundar, M.T. Naik, K.V. Sharma, M.K. Singh, T.Ch. Siva Reddy, Experimental investigation of forced convection heat transfer and friction factor in a tube with Fe3O4 magnetic nanofluid, Exp. Therm. Fluid Sci. 37 (2012) 65-71. https://doi.org/10.1016/j.expthermflusci.2011.10.004

[4] L.S. Sundar, M.K. Singh, I. Bidkin, A.C.M. Sousa, Experimental investigations in heat transfer and friction factor of magnetic Ni nanofluid flowing in a tube, Int. J. Heat and Mass Transfer 70 (2014) 224-234. https://doi.org/10.1016/j.ijheatmasstransfer.2013.11.004

[5] L.S. Sundar, K.V. Sharma, Thermal conductivity enhancement of nanoparticles in distilled water, Int. J. Nanoparticles 1 (2008) 66-77. https://doi.org/10.1504/IJNP.2008.017619

[6] X. Wang, X. Xu, S.U.S. Choi, Thermal conductivity of nanoparticle-fluid mixture, J. Thermophysics and Heat Transfer 13 (1999) 474-480. https://doi.org/10.2514/2.6486

[7] S.M.S. Murshed, K.C. Leong, C. Yang, Investigations of thermal conductivity and viscosity of nanofluids, Int. J. Thermal Science 47 (2008) 560-568. https://doi.org/10.1016/j.ijthermalsci.2007.05.004

[8] M.S. Liu, M.C.C. Lin, I.T. Huang, C.C. Wang, Enhanced thermal conductivity with CuO for nanofluids, Chemical Engineering and Technology 29 (2006) 72-77. https://doi.org/10.1002/ceat.200500184

[9] H.A. Mintsa, G. Roy, C.T. Nguyen, D. Doucet, New temperature dependent thermal conductivity data for water-based nanofluids, Int. J. Thermal Science 48 (2009) 363-371. https://doi.org/10.1016/j.ijthermalsci.2008.03.009

[10] ASHRAE, in: Handbook-Fundamentals, 2009, pp. 31.1-31.13. https://doi.org/10.1049/SBRA018E_ch1

[11] R.S. Vajjha, D.K. Das, Experimental determination of thermal conductivity of three nanofluids and development of new correlations, Int. J. Heat and Mass Transfer 52 (2009) 4675-4682. https://doi.org/10.1016/j.ijheatmasstransfer.2009.06.027

[12] L.S. Sundar, Md. Hashim Farooky, S.N. Sarada, M.K. Singh, Experimental thermal conductivity of ethylene glycol and water mixture based low volume concentration of Al2O3 and CuO nanofluids, Int. Comm. Heat and Mass Transfer 41 (2013) 41-46. https://doi.org/10.1016/j.icheatmasstransfer.2012.11.004

[13] A. Banisharif, M. Aghajani, S.V. Vaerenbergh, P. Estellé, A. Rashidi, Thermophysical properties of water ethylene glycol (WEG) mixture-based Fe3O4 nanofluids at low concentration and temperature, J. Molecular Liquids, 302 (2020) 112606. https://doi.org/10.1016/j.molliq.2020.112606

[14] N.A. Usri, W.H. Azmi, R. Mamat, K. Abdul Hamid, G. Najafi, Thermal conductivity enhancement of Al2O3 nanofluid in ethylene glycol and water mixture, Energy Procedia 79 (2015) 397-402. https://doi.org/10.1016/j.egypro.2015.11.509

[15] O.A. Alawi, A.R. Mallah, S.N. Kazi, M.N.M. Zubir, C.S. Oon, Thermal transport feasibility of (water + ethylene glycol)- based nanofluids containing metallic oxides: Mathematical approach, IOP Conf. Series: Materials Science and Engineering 854 (2020) 012023. https://doi.org/10.1088/1757-899X/854/1/012023

[16] L.S. Sundar, E.V. Ramana, M.K. Singh, A.C.M. Sousa, Thermal conductivity and viscosity of stabilized ethylene glycol and water mixture Al2O3 nanofluids for heat transfer applications: An experimental study, Int. Comm. Heat and Mass Transfer 56 (2014) 86-95. https://doi.org/10.1016/j.icheatmasstransfer.2014.06.009

[17] L.S. Sundar, M.K. Singh, A.C.M. Sousa, Thermal conductivity of ethylene glycol and water mixture based Fe3O4 nanofluid, Int. Comm. Heat and Mass Transfer 49 (2013) 17-24. https://doi.org/10.1016/j.icheatmasstransfer.2013.08.026

[18] J.C. Maxwell, A treatise on electricity and magnetism, 2nd Edition, Oxford University Press, Cambridge, UK, 1904.

[19] R.L. Hamilton, O.K. Crosser, Thermal conductivity of heterogeneous two component systems, Ind. Eng. Chem. Fundamental. 1 (3) (1962) 187-191. https://doi.org/10.1021/i160003a005

[20] F.J. Wasp, Solid-Liquid Slurry Pipeline Transportation, Trans, Tech, Berlin, 1977.

[21] W. Yu, S.U.S. Choi, The role of interfacial layers in the enhanced thermal conductivity of nanofluids: a renovated maxwell model, J. Nanoparticle Research 5 (2003) 167-171. https://doi.org/10.1023/A:1024438603801

[22] Y. Xuan, Q. Li, W. Hu, Aggregation structure and thermal conductivity of nanofluids, AIChE J. 49 (4) (2003) 1038-1043. https://doi.org/10.1002/aic.690490420

[23] C.H. Chon, K.D. Kihm, S.P. Lee, S.U.S. Choi, Empirical correlation finding the role of temperature and particle size for nanofluid (Al2O3) thermal conductivity enhancement, Appl. Phys. Lett. 87 (15) (2005) 153107. https://doi.org/10.1063/1.2093936

[24] E.V. Timofeeva, A.N. Gavrilov, J.M. McCloskey, Y.V. Tolmachev, Thermal conductivity and particle agglomeration in alumina nanofluids: Experiment and theory, Physical Review 76 (2007) 061203. https://doi.org/10.1103/PhysRevE.76.061203

[25] G.K. Batchelor, The effect of Brownian motion on the bulk stress in a suspension of spherical particles, J. Fluid Mechanics 83 (1977) 97-117. https://doi.org/10.1017/S0022112077001062

[26] A. Einstein, A new determination of molecular dimensions, Annalen der Physik 19 (1906) 289-306. https://doi.org/10.1002/andp.19063240204

[27] H.C. Brinkman, The viscosity of concentrated suspensions and solution, J. Chemical Physics 20 (1952) 571-581. https://doi.org/10.1063/1.1700493

[28] X. Wang, X. Xu, S.U.S Choi, Thermal conductivity of nanoparticles-fluid mixture, J. Thermophysics Heat Transfer 13 (4) (1999) 474-480. https://doi.org/10.2514/2.6486

[29] P.K. Namburu, D.K. Das, K.M. Tanguturi, R.K. Vajjha, Numerical study of turbulent flow and heat transfer characteristics of nanofluids considering variable properties, Int. J. Thermal Sciences 48 (2009) 290-302. https://doi.org/10.1016/j.ijthermalsci.2008.01.001

[30] A.V. Minakov, Valery Ya. Rudyak, Maxim I. Pryazhnikov, Systematic experimental study of the viscosity of nanofluids, Heat Transfer Engineering 42 (2021) 1024-1040. https://doi.org/10.1080/01457632.2020.1766250

Modelling the effect of tool material on material removal rate in electric discharge machining

Umair Arif[1,a*], Imtiaz Ali Khan[2,b], Faisal Hasan[2,c]

[1]Research Scholar, Department of Mechanical Engineering, Z.H. College of Engineering & Technology, Aligarh Muslim University

[2]Professor, Department of Mechanical Engineering, Z.H. College of Engineering & Technology, Aligarh Muslim University

[a]umairarif@zhcet.ac.in, [b]iakhan.me@amu.ac.in, [c]Faisalhasan@zhcet.ac.in

Keywords: Thermal Model, Electric Discharge Machining, Tool Material, Gaussian Heat Flux

Abstract. Present study aims at modelling the impact of tool materials such as copper, tungsten carbide and brass tool on the electric discharge machining of AISI 202 stainless steel. It is well known that the electrical conductivity of tool material has an influence on the current density passed through interelectrode gap and hence sparking process and MRR are affected accordingly. A finite element model was made using gaussian heat flux equation, spark radius and fraction of heat transferred to workpiece as a function of pulse on time and pulse current, latent heat in specific heat values and thermal conductivity properties. However, for the above reasons, current density used in gaussian heat flux equation was modified and electrical resistivity (which is inverse of electrical conductivity) of tool and workpiece were incorporated in it. This theorized heat flux formulae were then tested with the literature and found to give MRR similar to the literature.

Nomenclature

EDM	Electric discharge machining	Kt	Thermal conductivity
MRR	Material Removal Rate	ρ	Density
TWR	Tool Wear Rate	r_p	Spark radius
SR	Surface Roughness	R	Radius under consideration for heat flux
F_c	Energy distribution factor or fraction of heat transferred to workpiece	Qw	Heat flux applied
V	Voltage	delta T	Melting point temperature of work material - room temperature
I	Pulse current	ρ_t	Electrical resistivity of tool
T_{on}	Pulse on time	R_t	Resistance of tool
T	Temperature	R_w	Resistance of workpiece
ρ_W	Electrical resistivity of workpiece		

Introduction

Electric discharge machining (EDM) is an unconventional machining process where 2000-500,000 sparks/minute are generated during pulse on time. These sparks ionize the dielectric fluid at a point and impinges the workpiece resulting in an intense localized heat generation. This causes vaporization of workpiece followed by crater formation. The MRR depends on the electrical conductivity of the tool material that allows certain current density. The tool with high electrical conductivity facilitates the sparking process and increases effective discharge energy which increase MRR. [1] [2][3].

Researchers have modeled EDM process by used gaussian heat flux and spark radius as a function of process parameters [4]–[12]. Kalajahi et al. [6] and Tang et al. [9] have used spark radius as a function of pulse on time and pulse current. Joshi et al. [12] used F_c as 18.3%, PFE as

Materials Research Proceedings 31 (2023) 448-455 https://doi.org/10.21741/9781644902592-46

100% and spark radius as a function of pulse on time and pulse current in their 2D FEM model of a single-spark EDM process using AISI W1 tool steel workpiece. They compared their results from Dibitonto et al [13] and discovered that theirs result, better matched the experimental reading in terms of the shape of crater cavity, MRR and TWR. Further Shabgard et al.[14] used gaussian heat flux, current, and pulse on time based spark radius to develop a model that demonstrated that F_c is a function of pulse on time and current. Harminder singh also showed similar results [15]. Ming et al [16] used Harminder data in their model to develop a relationship between F_c and pulse current, pulse on time. They used that F_c in their model along with gaussian heat distribution, spark radius as a function of pulse current and pulse on time, the latent heat, and PFE and compared their results with Dibitonto et al [13] and Joshi et al.[12]. They found that their results were more comparable to experimental results. Ahmed et al [17] presented a thermal model of EDM using F_c and spark radius as a function of current, temperature dependent thermal properties of the material, latent heat of fusion and found better accuracy of the obtained results for inconel 718. M. Kliuev, et al.[18] also concluded in their model that F_c is strongly influenced by discharge current and then by pulse on time. However, authors did not find any literature that incorporates the effect of using different tool material on performance parameters in their model. Experimentally it was observed that the tools with better electrical conductivity had a major impact on discharge channel and discharge energy to workpiece resulting in better MRR[1] [2]. Hence this study models the effect of using different tool material on MRR by incorporating electrical resistivity in the heat flux formulae.

Thermal analysis model

Here, Fourier heat conduction equation is taken as governing equations, with necessary boundary conditions. Assuming that all the sparks have the same nature, a transient nonlinear analysis of single discharge EDM is performed on A1S1 202 steel. The model is then extended to all the sparks occurring during the process. During the analysis, the following presumptions were made:

1) The material of the workpiece was assumed to be isotropic and homogeneous.
2) The work domain is axis symmetric about X and Y axis.
3) Heat is transferred to the workpiece via conduction and radiation; convective heat losses are negligible.
4) Gaussian heat flux is utilized as the heat source, and only a portion of the spark energy is lost as heat in the workpiece.
5) It is assumed that the spark radius is a function of pulse current and pulse on time.
6) The fraction of heat transferred to workpiece is taken as a function of pulse on time and pulse current.

Governing equation

The Fourier heat conduction equation is used as the governing equation for the thermal analysis of the EDM process.

$$\frac{\partial T}{\alpha \partial t} = \frac{\partial T}{r \partial r} + \frac{\partial \partial T}{\partial Z^2} + \frac{\partial \partial T}{\partial r^2} \qquad 1$$

Where $\alpha = \frac{Kt}{\rho(Cpeff+\frac{m}{Tm})}$, r and z are coordinates.

Boundary conditions

Figure 1 depicts the model's assumed boundary conditions. A Gaussian heat flux distribution represents the heat transferred to the workpiece during the pulse on-time on the top surface. Surface B1 shows convective heat transfer to the dielectric for $R \geq r_p$. Surfaces B2, B3, and B4 are considered as insulated and no heat transfer occur across them. The initial temperature is considered equal to dielectric temperature. The applied boundary and initial conditions are represented mathematically as

$$k\frac{\partial T}{\partial z} = Qw \text{ for } R \leq r_p \text{ on } B1 \quad (2)$$

$$\frac{k\partial T}{\partial z} = hc\,(T - To) \text{ for } R \geq r_p \text{ on } B1$$

$$\frac{k\partial T}{\partial z} = 0 \text{ for } T_{off} \text{ (Pulse off time)}$$

$$\frac{\partial T}{\partial z} = 0 \text{ for } B2, B3, B4.$$

Where h_c is the dielectric fluid's heat transfer coefficient and T_0 (K) is the initial temperature of dielectric before starting of the EDM process.

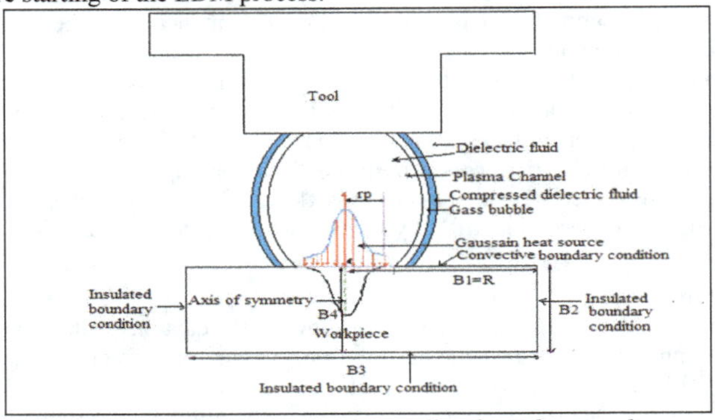

Figure 1 Axis symmetric model as well as boundary conditions

Heat flux

Many researchers have demonstrated that the gaussian heat flux distribution is more realistic and produces better results [12] [16]. Patel at el. [19] proposed a Gaussian-based heat input formulation for thermal modelling of the EDM process, which has been used by many researchers [12], [20].

$$Q = 4.57\frac{FcVI}{\pi rp^2} EXP\{-4.5\left(\frac{R}{rp}\right)^2 \quad (3)$$

Total resistance offered to flow of current in EDM can be written= $R_t + R_w$

$$Resistance = \frac{\rho * length}{Area} \quad (4)$$

The fraction of the total current that flows through the tool is given by:

$$\frac{Current\ tool}{Current\ total} = \frac{Rt}{(Rt + Rw)} = \frac{\rho t}{\rho t + \rho w} \quad (5)$$

To incorporate the effect of tool material on heat flux we replaced the energy density $(\frac{VI}{\pi rp^2})$ in the equation 3 by $(\frac{VI\left(\frac{\rho t}{\rho t+\rho w}\right)}{\pi rp^2})$. However still the simulation reading was not the same as in the experiment[1], so by using the initially obtained simulation readings and comparing it with experiment readings in literature[1], the exponential power was calculated. The final heat flux equation became as shown in equation 6. This heat flux was then validated by using the experimental reading in the literature[1].

Materials Research Proceedings 31 (2023) 448-455 https://doi.org/10.21741/9781644902592-46

$$Q = 4.57\frac{FcVI\left(\frac{ft}{ft+fw}\right)^{0.299}}{\pi rp^2}EXP\{-4.5\left(\frac{R}{rp}\right)^2 \qquad (6)$$

Spark radius is an important parameter to consider when modelling the EDM process. In the literature, various researchers have proposed spark radius as a point source, constant radius, or as a function of pulse current and pulse on time. However, Ikai and Hashiguchi's [21] showed that EDM radius calculations are more accurate when based on a semiempirical equation in equation 7.

$$rp = (2.04e - 3)I^{0.43}\,Ton^{0.44} \qquad (7)$$

Energy distribution factor

Another important factor in modelling EDM process is the energy distribution factor or fraction of heat transferred to workpiece (F_c). Shabgard et al.[14] demonstrated that the fraction of heat absorbed by the electrodes varied with pulse current (I) and Pulse on time (T_{on}). Shabgard defined F_c as $F_c = 5.5998xI^{-0.3401}xT_{on}^{0.2989.}$ As a result, it was decided to model the results by using F_c as the function provided by Shabgard et al.[14].

Plasma flushing efficiency

Plasma flushing efficiency is the ratio of the actual volume of material eliminated per pulse to the theoretical volume of material melted per pulse. It is determined as

$$PFE = \frac{Volume\ Exp}{Volume\ FEM} * 100 \qquad (8)$$

Table 1 Mechanical properties of AISI 202 stainless steel and tools used

AISI 202 stainless steel								
Electrical resistivity (ohm-cm)	Tensile yield strength (MPa)	Specific heat capacity (C_p)(J/kg°c)	Melting point (°C)	Density (kg/m³)	Young 'modulus (Gpa)	Poisson's ratio	Thermal conductivity (W/m-k)	Thermal expansion coefficient (1/°c)
6.9E-7	275	500	1450	7800	200	0.28	16.2	$2.06x10^{-5}$

Copper tool	Brass tool	Tungsten tool
Electrical resistivity – 1.67E-8	Electrical resistivity - 5.98E-7	Electrical resistivity – 1.56E--7

Results

Table 2 shows simulated MRR and SR readings for different tool material at 9A current and 0.4 duty cycle.

Tool material	Current 9 A				
	Gap voltage	spark radius	Radius (mm)	Height (mm)	MRR (mm³/min)
Not considering any tool material	40	6.09E-05	0.037	0.017	23.46585
Copper tool	40	6.09E-05	0.018	0.005	1.56529
Brass tool	40	6.09E-05	0.026	0.009	5.960034
Tungsten carbide tool	40	6.09E-05	0.03	0.012	10.71619

Copper tool	60	6.09E-05	0.026	0.01	6.68192
Brass tool	60	6.09E-05	0.032	0.013	13.22976
Tungsten carbide tool	60	6.09E-05	0.036	0.015	19.37223
Copper tool	80	6.09E-05	0.03	0.012	10.71619
Brass tool	80	6.09E-05	0.035	0.015	18.369
Tungsten carbide tool	80	6.09E-05	0.039	0.018	27.62132

*Fraction of heat transferred to the workpiece was calculated as 7.989 % using Shabgard et al.[14] equation, Pulse on time was taken as 40μs and Pulse off time was taken as 60μ.

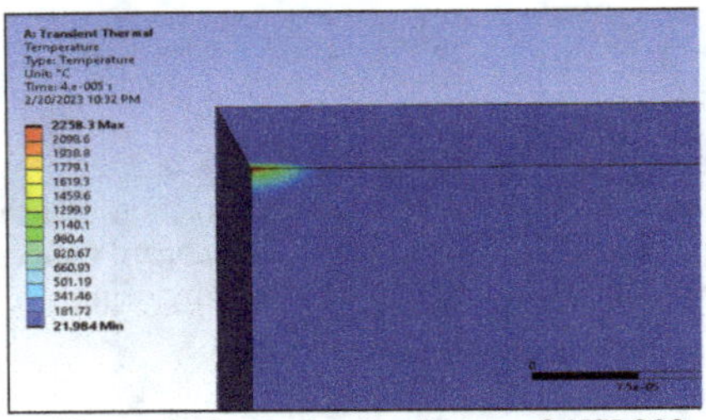

Figure 2 shows the simulation of AISI 202 Stainless steel using Copper tool at 9A.

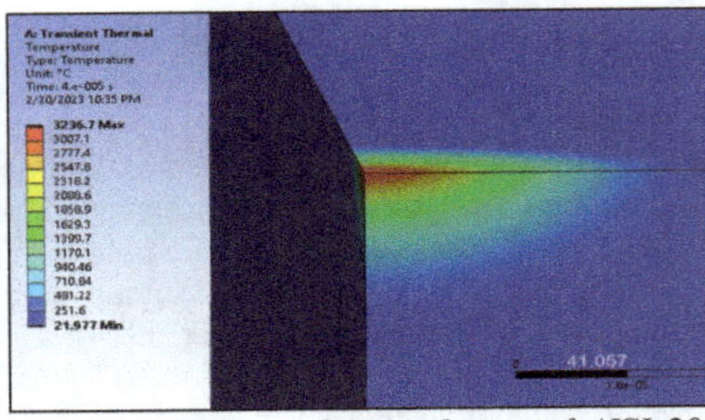

Figure 3 shows the simulation of AISI 202 Stainless steel using Brass tool at 9A.

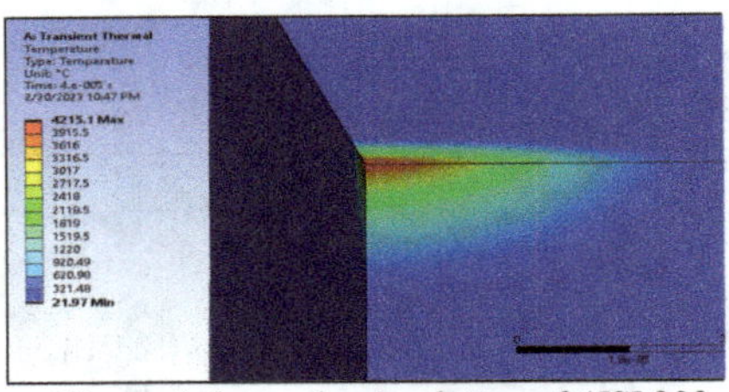

Figure 4 shows the simulation of AISI 202 Stainless steel using Tungsten carbide tool at 9A.

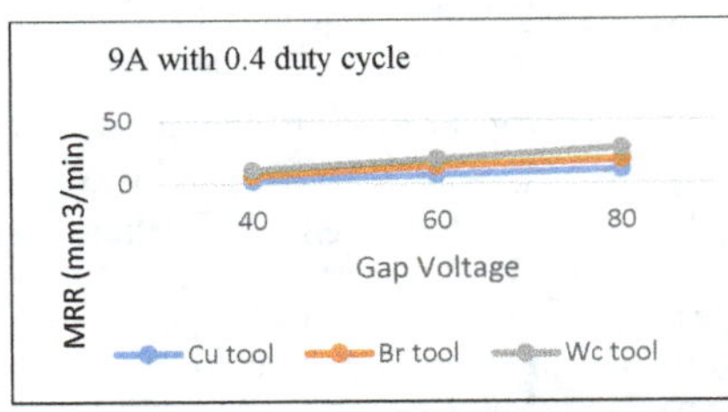

Figure 5 shows the predicted MRR results from the model for different tool materials at 9A.

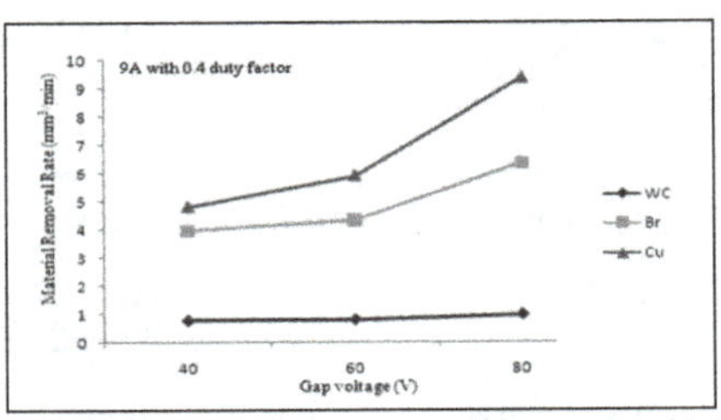

Figure 6 show the MRR results of Muthuramalingam and Mohan[1] at 9A.

Discussion

In this research, a thermal model was developed for the EDM process using a Gaussian heat flux, spark radius, and F_c, as functions of pulse on time and pulse current, specific heat values and thermal conductivity properties. The Finite Element Modelling method was used to solve the differential equation with the above-described boundary conditions using Ansys transient thermal. The axisymmetric workpiece model was used. The material properties from Table 1 were given to Ansys. At the spark position, the mesh size was refined for improved convergence of results. The gaussian heat fluxes was calculated incorporating the F_c values calculated using Shabgard et al.[14] equation. The gap voltage was taken as 40, 60, 80 V. The spark radius obtained from the above equation was divided into ten parts and accordingly the Gaussian heat flux was calculated and applied to each part. On the remaining area, convective boundary conditions were used to generate temperature profiles. The convective heat transfer coefficient was taken into account to be 10,000 W/m^2K.The melting temperature of AISI 202 stainless steel ie 1450°C was chosen as the temperature for crater formation, as shown in Figure *2*. Nodes with temperatures higher than the melting point of AISI 202 stainless steel were eliminated.

It is experimentally proven that tool material with better electrical conductivity gave more MRR[1] [2]. Using this fact, the same has been incorporated in the gaussian heat flux formulae. As a consequence, the energy density in gaussian heat flux ie VI/ πr^2 was modified as $\left(\frac{VI\left(\frac{ʃt}{ʃt+ʃw}\right)}{\pi rp^2}\right)$. However, modifying the current by multiplying it by simple ratio $\left(\frac{ʃt}{ʃt+ʃw}\right)$did not gave good results for all materials. MRR reading was obtained from the model and compared to the obtained from literature[1]. It was observed that the model reading was much larger than the experimental values[1]. Hence by comparing the simulation reading and the experimental reading[1], the exponential power was calculated and the ratio become $\left(\frac{ʃt}{ʃt+ʃw}\right)^{0.299}$ which gave much closer results to experimental values.

The resulting isotherms obtained from Ansys are shown in Figure *2* to Figure 4. Table 2 shows the coordinates of the crater formed. Crater volume is determined by using formulae in equation 9. MRR is calculated by multiplying the crater volume by the number of pulses per minute as shown in equation 10 and 11 [22]. Table 2 displays the final results.

$$\text{Crater volume} = 3.14 * \text{height of crater} * \frac{3*(\text{radius of crater})^2+(\text{height of crater})^2}{6} \tag{9}$$

$$\text{Total no of pulses per min} = \frac{60}{\text{pulse on time+pulse off time}} \tag{10}$$

$$\text{MRR} = \text{Crater volume} * \text{Number of pulses per min} \tag{11}$$

The proposed finite element model gave MRR readings for three tool materials namely copper, brass and tungsten carbide by incorporating their electrical resistivity in the gaussian heat flux formulae. These readings were validated from literature [1] and graphs were plotted as depicted in Figure 5. It was observed that the modified heat flux formula readings as shown in Figure 5, gave similar trends for all three tool materials as obtained in literature [1] also depicted in Figure 6. Hence this model shows better agreement with the experimental reading for different tool materials. It was observed from the model results that for all three tools, tool material having larger electrical conductivity gave more the MRR for same process parameters. Besides it was also observed that increasing the gap voltage results in improved MRR for same current values.

However, predicted values differed from the literature values which can be explained by the fact that the simulation was modelled for a single spark as well as plasma flushing efficiency was not incorporated in the model. This also means that above gaussian heat source model has a potential to be further developed by incorporating a better regression analysis and validating it with more experiments.

Conclusion

The present study aims to investigate the impact of tool materials, such as, tungsten carbide, copper and brass on electric discharge machining of AISI 202 stainless steel. It is well known that the electrical conductivity of the tool material plays a crucial role in determining the current density passing through the interelectrode gap, thereby affecting the discharging process and MRR. To develop a finite element model of EDM, a Gaussian heat flux equation was employed, which incorporated spark radius, fraction of heat transferred to the workpiece as a function of pulse current and pulse on time, specific heat values, and thermal conductivity properties. However, to account for the effect of electrical conductivity, modifications were made to the energy density used in the heat flux equation, and the electrical resistivity of both the tool and workpiece was incorporated. The proposed heat flux equation was validated against existing literature, and the results showed that the MRR obtained was in good agreement. Besides it was observed that increasing tool materials electrical conductivity or gap voltages resulted in improved MRR. However, the predicted values obtained from the simulation were found to be higher than the experimental values. This discrepancy can be explained by the fact that the simulation model was designed for a single spark and did not take into account the plasma flushing efficiency. Such improvements would enhance the accuracy and applicability of the model.

Reference

[1] T. Muthuramalingam and B. Mohan, "Influence of tool electrode properties on machinability in spark erosion machining," *Mater. Manuf. Process.*, vol. 28, no. 8, pp. 939–943, 2013. http://doi.org/10.1080/10426914.2013.763973

[2] M. P. Jahan, Y. S. Wong, and M. Rahman, "A study on the fine-finish die-sinking micro-EDM of tungsten carbide using different electrode materials," vol. 9, pp. 3956–3967, 2008. http://doi.org/10.1016/j.jmatprotec.2008.09.015

[3] U. Arif, I. Ali Khan, and F. Hassan, "Green and sustainable electric discharge machining: a review," *Adv. Mater. Process. Technol.*, vol. 00, no. 00, pp. 1–75, 2022. http://doi.org/10.1080/2374068X.2022.2108599

[4] V. Yadav, V. K. Jain, and P. M. Dixit, "Thermal stresses due to electrical discharge machining," *Int. J. Mach. Tools Manuf.*, vol. 42, no. 8, pp. 877–888, 2002. http://doi.org/10.1016/S0890-6955(02)00029-9

[5] S. H. Yeo, W. Kurnia, and P. C. Tan, "Electro-thermal modelling of anode and cathode in micro-EDM," *J. Phys. D. Appl. Phys.*, vol. 40, no. 8, pp. 2513–2521, 2007. http://doi.org/10.1088/0022-3727/40/8/015

[6] M. Hosseini Kalajahi, S. Rash Ahmadi, and S. Nadimi Bavil Oliaei, "Experimental and finite element analysis of EDM process and investigation of material removal rate by response surface methodology," *Int. J. Adv. Manuf. Technol.*, vol. 69, no. 1–4, pp. 687–704, 2013. http://doi.org/10.1007/s00170-013-5059-x

[7] S. Plaza, I. Pombo, N. Ortega, B. Ã. Izquierdo, and J. A. Sa, "International Journal of Machine Tools & Manufacture A numerical model of the EDM process considering the effect of multiple discharges," vol. 49, pp. 220–229, 2009. http://doi.org/10.1016/j.ijmachtools.2008.11.003

[8] H. K. Kansal, S. Singh, and P. Kumar, "Numerical simulation of powder mixed electric discharge machining (PMEDM) using finite element method," *Math. Comput. Model.*, vol. 47, no. 11–12, pp. 1217–1237, 2008. http://doi.org/10.1016/j.mcm.2007.05.016

[9] L. Tang, L. Ren, and Q. L. Zhu, "EDM multi-pulse temperature field simulation of SiC/Al functionally graded materials," *Int. J. Adv. Manuf. Technol.*, vol. 97, no. 5–8, pp. 2501–2508, 2018. http://doi.org/10.1007/s00170-018-2136-1

[10] M. Singh, P. Saxena, J. Ramkumar, and R. V. Rao, "Multi-spark numerical simulation of the micro-EDM process: an extension of a single-spark numerical study," *Int. J. Adv. Manuf. Technol.*, vol. 108, no. 9–10, pp. 2701–2715, 2020. http://doi.org/10.1007/s00170-020-05566-6

[11] K. P. Somashekhar, S. Panda, J. Mathew, and N. Ramachandran, "Numerical simulation of micro-EDM model with multi-spark," *Int. J. Adv. Manuf. Technol.*, vol. 76, no. 1–4, pp. 83–90, 2015. http://doi.org/10.1007/s00170-013-5319-9

[12] S. N. Joshi and S. S. Pande, "Development of an intelligent process model for EDM," *Int. J. Adv. Manuf. Technol.*, vol. 45, no. 3–4, pp. 300–317, 2009. http://doi.org/10.1007/s00170-009-1972-4

[13] D. D. Dibitonto, P. T. Eubank, M. R. Patel, M. A. Barrufet, and D. D. Diitonto, "Theoretical models of the electrical discharge machining process . I . A simple cathode erosion model Theoretical models of the electrical discharge machining process . I . A simple cathode erosion model," vol. 4095, no. 1989, 1989. http://doi.org/10.1063/1.343994.

[14] M. Shabgard, R. Ahmadi, M. Seyedzavvar, and S. N. B. Oliaei, "Mathematical and numerical modeling of the effect of input-parameters on the flushing efficiency of plasma channel in EDM process," *Int. J. Mach. Tools Manuf.*, vol. 65, pp. 79–87, 2013. http://doi.org/10.1016/j.ijmachtools.2012.10.004

[15] H. Singh, "Experimental study of distribution of energy during EDM process for utilization in thermal models," *Int. J. Heat Mass Transf.*, vol. 55, no. 19–20, pp. 5053–5064, 2012. http://doi.org/10.1016/j.ijheatmasstransfer.2012.05.004

[16] W. Ming *et al.*, "A hybrid process model for EDM based on finite-element method and Gaussian process regression," *Int. J. Adv. Manuf. Technol.*, vol. 74, no. 9–12, pp. 1197–1211, 2014. http://doi.org/10.1007/s00170-014-5989-y

[17] A. Ahmed, A. Fardin, M. Tanjilul, Y. S. Wong, M. Rahman, and A. Senthil Kumar, "A comparative study on the modelling of EDM and hybrid electrical discharge and arc machining considering latent heat and temperature-dependent properties of Inconel 718," *Int. J. Adv. Manuf. Technol.*, vol. 94, no. 5–8, pp. 2729–2737, 2018. http://doi.org/10.1007/s00170-017-1100-9

[18] M. Kliuev, K. Florio, M. Akbari, and K. Wegener, "In fl uence of energy fraction in EDM drilling of Inconel 718 by statistical analysis and fi nite element crater-modelling," *J. Manuf. Process.*, vol. 40, no. February, pp. 84–93, 2019. http://doi.org/10.1016/j.jmapro.2019.03.002

[19] M. R. Patel, M. A. Barrufet, P. T. Eubank, and D. D. Dibitonto, "Theoretical models of the electrical discharge machining process . II . The anode erosion model," vol. 4104, 1989. http://doi.org/10.1063/1.343995

[20] S. N. Joshi and S. S. Pande, "Thermo-physical modeling of die-sinking EDM process," *J. Manuf. Process.*, vol. 12, no. 1, pp. 45–56, 2010. http://doi.org/10.1016/j.jmapro.2010.02.001

[21] T. Ikai and K. Hashigushi, "Heat input for crater formation in EDM," in *In Proceedings of the International Symposium for Electro-Machining-ISEM XI, EPFL*, 1995, pp. 163–170.

[22] Vishwakarma, A. Dvivedi, and F. E. Analysis, "FINITE ELEMENT MODELING OF MATERIAL REM OVAL RATE IN POWDER MIXED ELECTRIC DISCHARGE MACHINING OF AL-SIC METAL MATRIX COMPOSITES," no. iii, pp. 151–158, 2013.

[23] M. A. Younis, M. S. Abbas, M. A. Gouda, F. H. Mahmoud, and S. A. Abd, "Effect of electrode material on electrical discharge machining of tool steel surface," pp. 977–986, 2015.

Energy harvesting and energy conversion in an electromechanical coupling acoustic black hole beam

ZHANG Linli[1,2,a], SHENG Xiaozhen[1,3,b*] and LI Meng[4,c]

[1]School of Urban Railway Transportation, Shanghai University of Engineering Science, Shanghai, China

[2]Department of Mechanical Engineering, The Hong Kong Polytechnic University, Hong Kong SAR, China

[3]Shanghai Engineering Research Centre of Vibration and Noise Control Technologies for Rail Transit, Shanghai, China

[4]School of Mechanics and Aerospace Engineering, Southwest Jiaotong University, Chengdu, China

[a]linli.zhang@sues.edu.cn, [b]shengxiaozhen@hotmail.com, [c]meng.li@swjtu.edu.cn

Keywords: Acoustic Black Hole, Energy Harvesting, Energy Conversion, Electromechanical Coupling

Abstract. Acoustic black hole (ABH) shows unique and attractive features of energy focusing when the flexural wave propagates along a structure with a variable power-law thickness profile, which are found to be conducive to effective energy harvesting. In this paper, an improved electromechanical ABH model is established based on Timoshenko beam theory, which considers the high-frequency shear and rotational effects of the beam, as well as its coupling with PZT coating and other additional elements like damping layers. External electrical modules including both linear and nonlinear circuits can also be easily integrated into the system to form a fully coupled electromechanical model. The proposed model is then used to analyze typical ABH-specific features such as energy focusing and energy harvesting. Numerical results demonstrate the excellent energy harvesting performance and confirm that installing PZT on the ABH beam warrants higher electrical power than the uniform beam. Moreover, studies explore the relationship between the electromechanical coupling and the energy harvesting efficiency, and different methods to enhance the electromechanical coupling are also investigated. Finally, experimental results are presented to demonstrate the feasibility of ABH beam in energy harvesting.

Introduction

Flexural waves propagate along a thin-walled structure and interact with the surrounding medium, causing structural vibrations and noise. Manipulating waves by tailoring the physical or geometrical properties of a structure is the foundation for the vibration and noise control and has drawn increasing attention in the research community [1-3]. The concept of Acoustic Black Hole (ABH) was proposed based on this principle, which is a new wave manipulation method and has aroused intense interest since its inception [4, 5].

ABH shows unique and attractive features of energy focusing when flexural waves propagate along a structure with a tailed thickness profile following a reduced power law [6-8]. The local phase velocity and the group velocity of the flexural wave gradually decrease as the thickness decreases [9, 10]. In an ideal scenario, wave reflection is annulled when the thickness becomes zero, and this causes a high energy concentration. In contrast to traditional approaches, in ABH only a small amount of damping material can effectively dissipate concentrated energy [11, 12]. In addition to energy dissipation, strong energy concentration in ABH portion also offers new opportunities for effective energy harvesting [13-15].

However, existing studies on ABH-based energy harvesting are still limited to focus on phenomenon analyses of specific structures through FEM simulations. There is a lack of basic theoretical investigations and flexible simulation tools to guide system design and optimization. Moreover, the physical process of energy harvesting involves numerous parameters of the electromechanical coupling system, including those of the host structures, piezoelectric materials and external circuits, posing a challenge to parameter optimization design. To sum up, a flexible simulation tool is needed 1) to solve the specific requirements of ABH-based energy harvesting to guide the design and optimization of the coupled system; and 2) to better understand and quantify the energy conversion and electromechanical coupling in the system.

Motivated by the above, this paper proposes an electromechanical semi-analytical ABH model based on Timoshenko beam theory, which considers the high-frequency shear and rotational effects. PZT patches and other additional elements like damping layers can be coated in this model, and further, external circuit modules can also be easily integrated into the system to form a fully coupled electromechanical model. The proposed model is then used to investigate a few important issues pertinent to energy harvesting. Numerical results prove the better energy harvesting performance of ABH beam than the uniform beam. Studies explore the relationship between electromechanical coupling and energy harvesting efficiency, and different methods to enhance the electromechanical coupling are also investigated. Finally, experimental results are presented to demonstrate the feasibility of ABH beam in energy harvesting.

Theoretical Model and Formulation

Figure 1 shows a beam with flexural vibration while a point force $f_{ext}(t)$ is excited at x_f. It is symmetrical with respect to the mid-line, the width is constant b. This beam is composed of an uniform part from 0 to x_u (the thickness is constant $2h_u$) and an ABH part from x_u to l (the thickness $2h_b$ is variable following a power-law profile), $h_b(x)=\beta(L-x)^m$. L is the total length when the thickness can ideally be decreased to 0 without truncation. Besides, single or multiple piezoelectric patches with constant thickness h_p can be symmetrically laid on the two sides of the beam. Fig.1 shows an example using a resistance (R) resonant circuit for the purpose of energy harvesting, while other circuits can also be connected. Both ends of the beam are elastically supported by two springs, the stiffness can be adjusted to simulate different boundary conditions [16, 17].

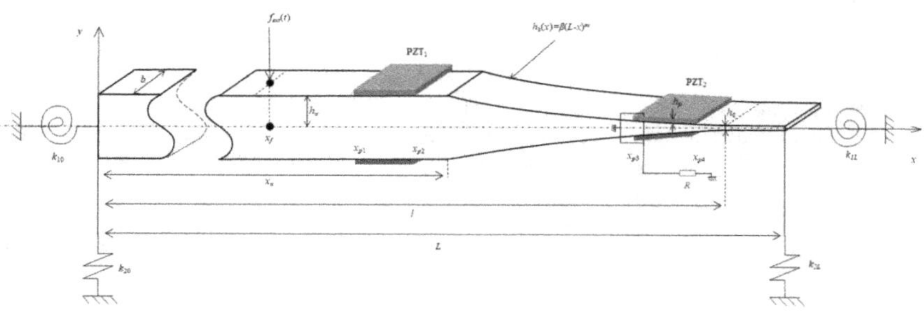

Figure 1. An ABH beam connected with multiple piezoelectric patches and a resistance resonant circuit.

In this paper, an electromechanical coupling ABH model based on Timoshenko beam with PZT patches and the external circuit is proposed through Rayleigh-Ritz method [18]. The out-of-plane displacement $w(x, t)$ and the rotation angle $\theta(x, t)$ of the beam are decomposed into a set of assumed shape functions [19], whose corresponding temporal coordinates (in two unknown vectors $\mathbf{a}$(t) and $\mathbf{b}$(t)). And the kinetic energy, potential energy, electrical load and work done by $\mathbf{f}_{ext}$(t) can be mathematically expressed to form the Lagrangian of the coupled system. This electromechanical ABH beam with a pure resistance circuit can be written as the matrix form:

$$\begin{bmatrix} \boldsymbol{M}_{a1}+\boldsymbol{M}_{a2} & \boldsymbol{M}_{b1} & 0 \\ \boldsymbol{M}_{b1} & \boldsymbol{M}_{b2} & 0 \\ 0 & 0 & 0 \end{bmatrix} \cdot \begin{bmatrix} \ddot{\boldsymbol{a}}(t) \\ \ddot{\boldsymbol{b}}(t) \\ \ddot{q}(t) \end{bmatrix} + \begin{bmatrix} 0 & 0 & 0 \\ 0 & 0 & 0 \\ 0 & 0 & R \end{bmatrix} \cdot \begin{bmatrix} \dot{\boldsymbol{a}}(t) \\ \dot{\boldsymbol{b}}(t) \\ \dot{q}(t) \end{bmatrix} + \begin{bmatrix} \boldsymbol{K}_{a1}+\boldsymbol{K}_{a2}+\boldsymbol{K}_{a3} & \boldsymbol{K}_{b1}+\boldsymbol{K}_{b2} & -C_{eq}^{-1}\boldsymbol{\Theta}_1 \\ \boldsymbol{K}_{b1}+\boldsymbol{K}_{b2} & \boldsymbol{K}_{b3}+\boldsymbol{K}_{b4}+\boldsymbol{K}_{b5} & -C_{eq}^{-1}\boldsymbol{\Theta}_2 \\ -C_{eq}^{-1}\boldsymbol{\Theta}_1^T & -C_{eq}^{-1}\boldsymbol{\Theta}_2^T & C_{eq}^{-1} \end{bmatrix} \cdot \begin{bmatrix} \boldsymbol{a}(t) \\ \boldsymbol{b}(t) \\ q(t) \end{bmatrix} = \begin{bmatrix} \boldsymbol{f}_{ext}(t) \\ \boldsymbol{0} \\ 0 \end{bmatrix}. \quad (1)$$

The linear frequency response curve (FRC) of this forced vibration system and the harvested electrical charge can be obtained by solving Eq. (1) directly. For free vibration, setting the force vector $f_{ext}(t)$ to zero leads to the corresponding eigenvalue equation.

Numerical Analyses

An electromechanical cantilever ABH beam is numerically investigated, its geometrical and material parameters are shown in Table 1. The cantilever beam can be simulated by setting k_{10} and k_{20} to 10^{12} for the uniform end, and k_{1L} and k_{2L} assigning to 0 at the ABH portion as detailed in [20]. The ABH beam is excited by a harmonic point force with amplitude of 1N at x_f=100mm on the uniform portion.

Table 1. Geometrical and material parameters of the beam, PZT and external circuit.

Geometrical parameters	Material parameters
Beam $m = 2$ $\beta = 0.1$ b = 50mm x_u = 250mm l = 450mm L = 500mm h_u = 6.25mm h_0 = 0.5mm PZT x_{p1} = 400mm x_{p2} = 450mm h_p = 0.5mm	Beam Density: ρ_b = 7800kg/m^3 Damping loss factor: η_b = 0.005 Elasticity modulus: E_b = 210GPa PZT Density: ρ_p = 7600kg/m^3 Damping loss factor: η_p = 0 Elasticity modulus: E_p = 132GPa Piezoelectric stress constant: e = -3C/m^3 Dielectric constant: $\varepsilon^s = 2.8\times10^{-9}$F/m External circuit Resistance: R = 1000Ω

(1) Energy focusing of ABH

Firstly, in order to illustrate the potential advantages from energy harvesting perspective, the modal shapes of some modes of the ABH beam without any external module are shown in Figure 2. They are beneficial to understand the baseline system dynamics and the subsequent ABH phenomenon and the corresponding energy focusing feature. It can be seen that the ABH portion (the grey regions), particularly near the tip, oscillates very strongly, the vibration amplitudes are much greater than the uniform part, especially the higher order modes. This indicates a strong concentration of vibration energy around the tip, which is a typical ABH feature and facilitates energy harvesting.

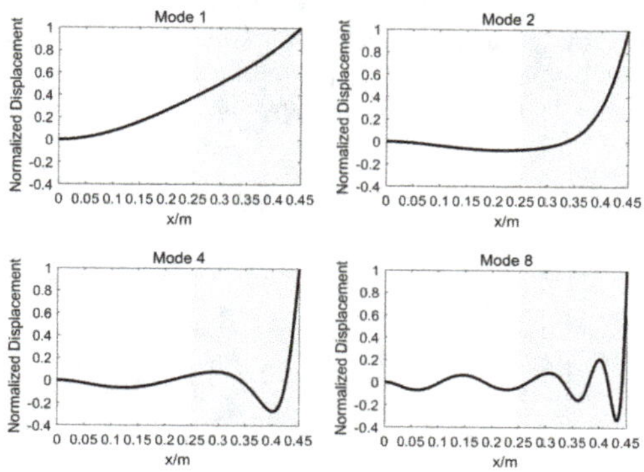

Figure 2. Mode shapes of modes 1, 2, 4 and 8 without PZT.

(2) Energy harvesting and energy conversion
Having demonstrated the unique advantages of ABH beam on energy focusing, PZT patches are added to the beam to harvest the electrical power. The harvested output power of each PZT patch can be calculated as $P = V^2/R$, so the total harvested output power is $P_{out} = 2P$. The harvested electrical power has also been compared in Figure 3 with that when the beam is uniform all the way. From the comparison it can be seen that the energy harvesting efficiency of ABH beam is much higher than the uniform beam in a wide band range, demonstrating the advantages of ABH beam in energy harvesting.

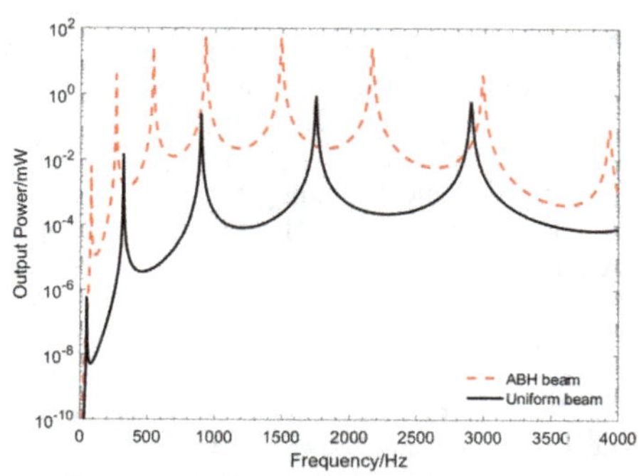

Figure 3. Comparison of harvested electrical power between ABH beam and uniform beam.

As an important indicator of energy recovery efficiency, in addition to the harvested output power, we also introduce an energy conversion rate, δ, which is defined as the ratio of the harvested output power from PZTs to the input mechanical power [20]:

$$\delta = \frac{P_{out}}{P_{in}} = \frac{P_{out}}{1/2\, Re\{F \cdot v^*\}}. \tag{2}$$

Note that, due to energy conservation, P_{in} includes both the harvested output power from PZTs and the dissipated power by the mechanical system, which is related to the amplitude of the external force and the velocity of the excitation point. The comparison of energy conversion rate between the ABH beam and the uniform beam is shown in Figure 4.

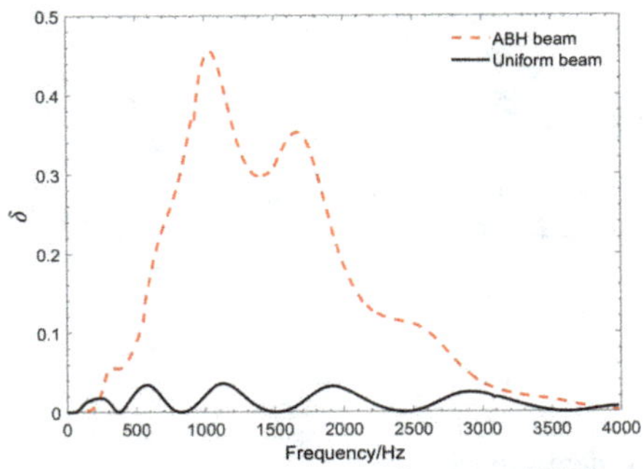

Figure 4. Comparison of energy conversion rate between ABH beam and uniform beam.

As shown in Figure 4, by comparing the energy conversion rate between the two different beams, the obvious advantages of the ABH beam in energy harvesting can be better seen. We can immediately conclude that the uniform beam shows no advantages over ABH beam in a fairly wide frequency range, when PZT patches are placed on the same position for the uniform beam and ABH beam, highly consistent with the conclusion obtained from the output power. It is precisely because of the energy focusing effect of ABH structures that energy from the uniform portion is trapped by the ABH portion, resulting in a high energy focusing around the ABH tip, and then harvest by the PZTs. It means that when the ABH effect occurs, broadband energy harvesting based on ABH structure shows great potential, including energy conversion in the high frequency range.

(3) Energy conversion and electromechanical coupling

There are many parameters in the system affecting the energy harvesting efficiency and the most fundamental factor is the strength of electromechanical coupling. Therefore, designing an efficient system is a complex problem that requires a clear understanding the role of electromechanical coupling in the energy conversion process, and to take a good quantification of the electromechanical coupling into consideration. In order to establish its possible relationship between the energy conversion rate and electromechanical coupling, the electromechanical coupling factor k is introduced, which is defined as [21, 22]:

$$k^2 = \frac{\omega^2{}_{oc} - \omega^2{}_{sc}}{\omega^2{}_{sc}}. \tag{3}$$

where ω_{oc} and ω_{sc} are the angular natural frequencies of the coupled system when PZT is open-circuited and short-circuited, respectively. They can be obtained easily, including by solving corresponding eigenvalue problems through this established model and by experimental tests. The relationship between the electromechanical coupling factor k and the energy conversion rate δ corresponding to the ABH beam are shown in Figure 5.

Advanced Topics in Mechanics of Materials, Structures and Construction - AToMech1-2023 Materials Research Forum LLC
Materials Research Proceedings 31 (2023) 456-465 https://doi.org/10.21741/9781644902592-47

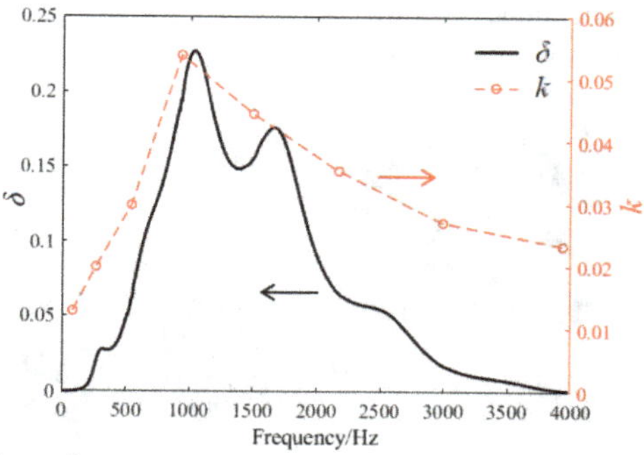

Figure 5. Comparison of energy conversion rate and electromechanical coupling factor.

Figure 5 shows that electromechanical coupling factor and energy conversion rate have very consistent trends with respect to the frequency. It means that the system with stronger electromechanical coupling can obtain the higher energy harvesting efficiency. Therefore, the electromechanical coupling factor k can be used as an intrinsic and simple indicator of the energy harvesting efficiency to guide the system design.

There are many different methods to enhance the electromechanical coupling effect, including designing external circuits, laying multiple sets of piezoelectric patches and so on. Figure 6(a) and 6(b) show the effect of adding negative capacitance in external circuit and using two sets of PZTs on the enhancement of electromechanical coupling. The results show the effectiveness of these methods, especially by adding the negative capacitance (Figure 6(a)), the electromechanical coupling effect can be greatly increased, and the design of the external circuit is relatively more flexible and convenient.

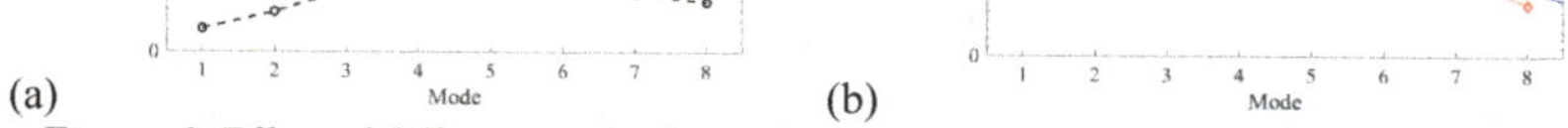

Figure 6. Effect of different methods on electromechanical coupling factor: (a) negative capacitance; (b) two sets of PZTs.

Experimental Results

The phenomena of ABH phenomenon and corresponding energy harvesting effect have been analyzed and discussed in the previous section. Here, experiments were conducted to further validate the energy focusing phenomenon of ABH effect and the advantage of energy harvesting. The test sample is shown in Figure 7. The uniform end of the beam is clamped by a fixed bracket and the ABH end is set free. The beam is excited at x_f=0.1m, measured from the clamped end on the uniform portion, by an electromagnetic shaker using a sweep signal from 5Hz to 1000Hz. A laser vibrometer (NLV-2500) is used to measure the point velocity of the beam at different positions (every 1cm) on the ABH beam for the modal shape test. Figure 8 shows the mode shapes for the first four structural modes of the ABH beam without PZTs, the energy focusing phenomenon has been validated in experiments.

Figure 7. Experimental setup of mechanical ABH system.

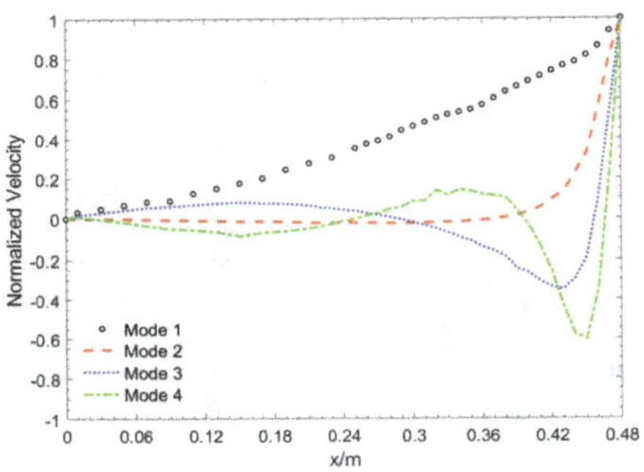

Figure 8. Experimentally measured modal shapes of the first four modes without PZTs.

The next step is to verify the advantages of ABH beam in terms of energy harvesting. An essential part in an energy harvesting circuit is rectification, which is also called AC/DC conversion. We use the most common rectifier circuit, the bridge rectifier circuit, including four rectifier diodes, to convert the AC voltage into a positive voltage that can be directly applied to the load. Although the voltage rectified by the bridge rectifier circuit can continuously supply power to the load, it remains unstable. Therefore, a smoothing capacitor is added to the rectifier circuit to form a full-wave rectifier circuit, as shown in Figure 6.9, wherein the smoothing capacitor is used to convert the rectified positive voltage signal into DC to better supply power to the load. Here, a LED is connected into the circuit to visually demonstrate the energy harvesting effect.

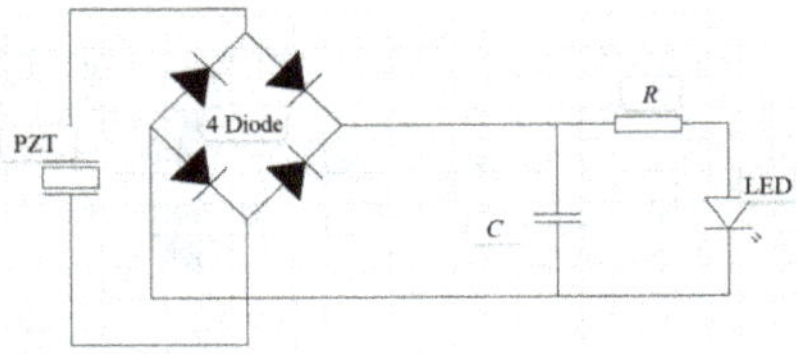

Figure 9. Schematic diagram of energy harvesting circuit in the experiment.

Figure 10 shows the working process of the energy harvesting circuit. Firstly, the alternating current generated by the piezoelectric patches is rectified by the bridge rectifier circuit to charge the capacitance, as shown in Figure 10(a). The capacitance then starts to discharge, which can supply power to the load to light up a LED (Figure 10(b)). Finally, the capacitance will finish discharge and the process repeats (shown as Figure 10(c)). The corresponding prototype of energy harvesting circuit used in the experiment is shown in Figure 11, which demonstrates the effect of energy harvesting: the LED light can be always on.

(a) (b) (c)

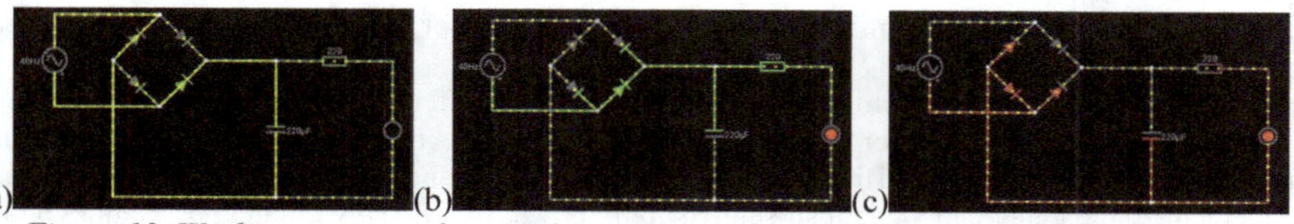

Figure 10. Working process of energy harvesting circuit: (a) capacitance is charging; (b) capacitance starts to discharge; (c) capacitance will finish discharge.

Figure 11. Experimentally measured effectiveness of energy harvesting: LED can be always on.

An oscilloscope is used to measure the voltage produced by the piezoelectric patches. Fast Fourier transform is performed on the entire voltage signal with the resultant spectrum shown in Figure 12. The output voltage can reach up to 6V, exceeding the 2V required for lighting up the LED. The electricity generated by the electromechanical ABH beam is deemed feasible as energy harvesting devices to not only power the LED light, but also for other microelectronic devices.

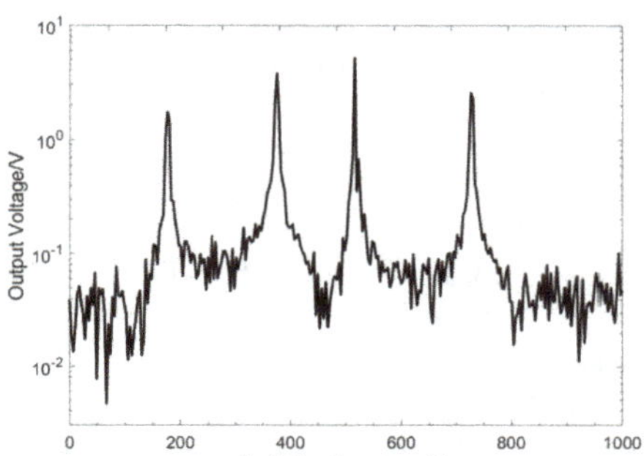

Figure 12. Experimentally measured output voltage from PZTs.

Conclusion

In this paper, a semi-analytical electromechanical ABH model is established to analyze the electromechanical coupling of Timoshenko beam coated with PZT patches, and its effects on energy harvesting and energy conversion. The unique energy focusing feature of the ABH effect makes the energy density near the ABH tip higher, favoring the PZT installation in the vicinity of its tip area and warranting better energy harvesting performance than the uniform beam. Moreover, the electromechanical coupling is shown to exhibit consistent variations with the energy harvesting performance of the system and thus is a good indicator for either evaluating the energy harvesting performance or tuning/optimizing system parameters for enhanced energy harvesting performance.

Experimental results confirm the excellent energy focus effect in the mechanical ABH beam, which offers the possibility for effective energy harvesting. After the voltage generated by the piezoelectric patches is rectified by the rectifier circuit and converted by the smoothing capacitance, the generated signal can stably supply power to a LED, with an appreciable voltage level, which holds promises for powering other microelectronic devices. As a final remark, it is relevant to note that theoretical analyses and experimental validation were used to emphasize some fundamental issues pertinent to ABH-based energy harvesting, but no effort has been made to achieve optimal energy harvesting performance. If necessary, the optimal design can be realized using the proposed simulation model.

Acknowledgements

This work is funded by the National Natural Science Foundation of China (52272352).

References

[1] D.J. Mead, Passive vibration control, John Wiley & Sons, 1999.

[2] A.D. Nashif, D. Jones, J.P. Henderson, Vibration damping, John Wiley & Sons, 1985.

[3] D.J. Inman, Vibration with Control, John Wiley & Sons, 2006. https://doi.org/10.1002/0470010533

[4] C. Pekeris, Theory of propagation of sound in a half-space of variable sound velocity under conditions of formation of a shadow zone, Journal of the Acoustical Society of America, 18: 295-315, 1946. https://doi.org/10.1121/1.1916366

[5] M. Mironov, Propagation of a flexural wave in a plate whose thickness decreases smoothly to zero in a finite interval, Soviet Physics Acoustics, 34: 318-319, 1988.

[6] V.V. Krylov, Acoustic black holes: Recent developments in the theory and applications, IEEE Transactions on Ultrasonics Ferroelectrics and Frequency Control, 61: 1296-1306, 2014. https://doi.org/10.1109/TUFFC.2014.3036

[7] A. Pelat, F. Gautier, S.C. Conlon and F. Semperlotti, The acoustic black hole: A review of theory and applications, Journal of Sound and Vibration, 476: 115316, 2020. https://doi.org/10.1016/j.jsv.2020.115316

[8] V.V. Krylov and A. Shuvalov, Propagation of Localised Flexural Vibrations along Plate Edges Described by a Power Law, Proceedings of the Institute of Acoustics, 22: 263-70, 2000.

[9] H. Ji, W. Huang, J. Qiu and L. Cheng, Mechanics problems in application of acoustic black hole structures, Advances in Mechanics, 47: 333, 2017.

[10] V.V. Krylov, On the velocities of localized vibration modes in immersed solid wedges, Journal of Acoustical Society of America, 103: 767-770, 1998. https://doi.org/10.1121/1.421240

[11] D.J. O'Boy and V.V. Krylov, Damping of flexural vibrations in circular plates with tapered central holes, Journal of Sound and Vibration, 330: 2220-2236, 2011. https://doi.org/10.1016/j.jsv.2010.11.017

[12] V.V. Krylov and F. Tilman, 'Acoustic 'Black Holes' for Flexural Waves as Effective Vibration Dampers', Journal of Sound and Vibration, 274: 605-19, 2004. https://doi.org/10.1016/j.jsv.2003.05.010

[13] L. Zhao, S.C. Conlon and F. Semperlotti, Broadband energy harvesting using acoustic black hole structural tailoring, Smart Materials and Structures, 23: 065021, 2014. https://doi.org/10.1088/0964-1726/23/6/065021

[14] H. Ji, Y. Liang, J. Qiu, L. Cheng, et al., Enhancement of Vibration Based Energy Harvesting Using Compound Acoustic Black Holes, Mechanical Systems and Signal Processing, 132: 441-56, 2019. https://doi.org/10.1016/j.ymssp.2019.06.034

[15] L. Zhang and X. Sheng, A review on the research progress of mechanical meta-structures and their applications in rail transit, Intelligent Transportation Infrastructure, 1: 1-22, 2022. https://doi.org/10.1093/iti/liac010

[16] L. Tang, L. Cheng, H. Ji and J. Qiu, Characterization of acoustic black hole effect using a one-dimensional fully-coupled and wavelet-decomposed semi-analytical model, Journal of Sound Vibration, 374: 172-184, 2016. https://doi.org/10.1016/j.jsv.2016.03.031

[17] L. Cheng, Vibroacoustic modeling of mechanically coupled structures: artificial spring technique applied to light and heavy medium, Shock and Vibration 3: 193-200, 1996. https://doi.org/10.1155/1996/343429

[18] J. Deng, O. Guasch and L. Zheng, A semi-analytical method for characterizing vibrations in circular beams with embedded acoustic black holes, Journal of Sound and Vibration 476: 115307, 2020. https://doi.org/10.1016/j.jsv.2020.115307

[18] Y. Wang, J. Du and L. Cheng, Power flow and structural intensity analyses of acoustic black hole beams, Mechanical Systems and Signal Processing, 131: 538-553, 2019. https://doi.org/10.1016/j.ymssp.2019.06.004

[19] L. Zhang, G. Kerschen and L. Cheng, Electromechanical Coupling and Energy Conversion in a PZT-Coated Acoustic Black Hole Beam, International Journal of Applied Mechanics, 12: 2050095, 2020. https://doi.org/10.1142/S1758825120500957

[20] G. Raze, A. Jadoul, S. Guichaux, V. Broun, V. and G. Kerschen, A digital nonlinear piezoelectric tuned vibration absorber, Smart Materials and Structures, 29: 015007, 2019. https://doi.org/10.1088/1361-665X/ab5176

[21] Linli Zhang, Gaetan Kerschen and Li Cheng, Nonlinear Features and Energy Transfer in an Acoustic Black Hole Beam through Intentional Electromechanical Coupling, Mechanical Systems and Signal Processing, 177: 109244, 2022. https://doi.org/10.1016/j.ymssp.2022.109244

Bi-axial behavior of glass/epoxy pipe subjected to internal pressure

Abdul Mateen Mohammed[1,a*], Tajuddin Mohammed[2,b], Ravi Shankar D V[3,c] Manzoor Hussain M[4,d] and Puneetha[5,e]

[1]Associate Professor, School of Engineering and Technology, CMR University, Bengaluru - 562149, INDIA

[2]Department of Non-Destructive Engg & Tech (NDET), University of Hafr Al Batin, Hafr Al Batin, 39524, Kingdom of Saudi Aribia

[3]Department of Mechanical Engineering, TKR College of Engineering and Technology, Hyderabad, Telangana - 500 097, INDIA

[4]Department of Mechanical Engineering, Jawaharlal Nehru Technological University Hyderabad, Hyderabad, Telangana - 500 085, INDIA

[5] Department of Mathematics, School of Engineering and Technology, CMR University, Bengaluru, Karnataka, 562149, INDIA

[a*]abdulmateen7@gmail.com, [b] tajmohammed@uhb.edu.sa, [c] shankardasari64@gmail.com, [d]manzoorjntu@gmail.com, [e] puneethmath@gmail.com

Keywords: Open End Burst Test, GFRP, Bi-Axial Behaviour, Stress and Strain

Abstract. Composite structures are considered as the auxiliary to the conventional metallic materials. However, a thorough understanding of the material behaviour is required to replace the materials in use with scientific evidence for design rather than the designing based on high safety factor. The behaviour of filament wound E-glass/epoxy pipes of different helix angles exposed to internal pressure are studied. The sample GFRP pipe of different helix angles namely [±45o], [±55o] & [±70o] were subjected to open end pressure until the sample failed due to burst(pressure) and the corresponding strains were recorded using the strain smart 5000 data acquisition system (DAQ). The theoretical first-ply failure and burst pressure were determined using (i) Maximum stress theory, (ii) Maximum strain theory and (iii) Tsai Hill theory and were later compared with the experimental results. The first ply failure of the samples was observed by the whitening phenomenon. Later, the damage initiated by the formation of pine holes that act as a micro crack and which later grows to a macro crack in the fiber direction until final failure. A nonlinear relation between the longitudinal stress and strain was observed.

1. Introduction

Fiber reinforced polymeric (FRP) composites are considered analogues to the conventional materials such as metals for a specific advantage as these materials possess high strength to weight ratio. Among the few automated processing techniques used in the production of FRP components, filament winding method is the forthwith methodology that is available for production units where in a continuous fiber is carefully arranged in controlled direction. Predominantly, the filament-wound structures are axisymmetric shells, where the fibers are wound with $\pm\theta$ helix angle with respect to the symmetric axis. The properties like High strength to low weight ratio, provision for tailoring the material properties and corrosion resistance, has led to the use of these fiber reinforced structural components in many engineering applications. In general, these structures are subject to complex loading conditions and the mechanical performance of these structures depends on the helix angle used to develop the structures. The Contemporary advancements in the computer-control system and the design software has facilitated the designer to generate multiple winding sequences depending on specific requirement. Generally, to improve the productivity intermediate angles between 0° and 90° are selected. However, the design procedure in specifying the lay-up depends on (a) the selection of fiber and matrix, (b) fiber layup in each layer, (c) the thickness of each lamina and (d) overall laminate thickness. While FRP pipes are tested under different mechanical loadings and pressures, the designer should note that the mechanical properties measured are affected by the internal pressure applied [1]. In general, the stress and failure analysis are normally carried out deprived of considering (i) the internal liner and (ii) the outer cover. It is assumed that these two parts, the inner liner and the outer cover would not contribute in resistance to deformation [2]. Composite pipes under internal pressure are exposed to both hoop and axial stresses with hoop-to-axial stress of 2:1. In the case of open ends tubes the axial stress produced in the structure due to the applied pressure is zero. Furthermore, auxiliary stresses may upshot due to installation, weight, external pressure, etc.

A preliminary structure to regulate the failure of GRE pipes in its function, which was imperiled to hydrostatic internal pressure was considered by Roham and Farshid [3] in order to inquire about the factures such as (i) winding angle and (ii) fibre volume fraction on the failure. Five distinct fiber volume fractions ranging from 50%-60% with and increment of 2.5% along with three specific winding angles, (i) ±52.5°, (ii) ±57.5°, and (iii) ±60.9° were considered to resolve the first ply failure (FPF) and burst pressures. Failure test of a composite tube at normal ambient temperature and at the higher temperature of 100 °C was conducted by P. Mertiny et al [4]. This failure test basically a leakage was done by means of a permeability-based and fluid volume loss methodology. In the course of the analysis, the leakage in the tube was found based on two phenomenon, (i) the aggregate loss in the volume of the fluid and (ii) the active permeability established with reference to the average fluid flux. A composite tubes with the ±60° winding angle was pressurization by a hydraulic oil in a closed ended condition. The pressure was perpetually augmented under three steps of the loading, and later it was reported that the leakage stress had a direct relation with the loading rate.

Rosenow [5] tested the pipes made of GRP individually under different loading cases (i) biaxial pressure (ii) tensile and (iii) hoop pressure. The samples with a diameter of 50.8 mm were made of six different winding angles ranging between 15° - 85° with an increment of 15 degrees. The axial and hoop strains were measured using strain gauges which were later contrasted with the strain values calculated based on classical lamination theory (CLT) and a good similitude was reported. The optimum winding angle for the pipe under the biaxial pressure loading was 54.75° and 75° for hoop pressure loading was concluded. It was quantified that CLT could be reliable until the nonlinearity point is obtained.

Spencer and Hull [6] has considered GRP with four winding patterns namely ±35°, ±45°, ±65° and ±75° with an internal diameter of 50 mm for the study under internal pressure. And the study strongly demonstrated that the failure mechanism and the deformation are intensely influenced by the winding angles. The failure structure from weepage to the burst was along with the deformation characteristic were decided using strain gauges which were mounted along the hoop and axial directions of pipes.

Arikan [7.] evaluated the failure behaviour of filament wound pipes which possess an inclined crack along the surface under internal pressure. The helix angle was maintained at [±55°] for all the specimens was as the orientation of surface crack angles was varied from 0° to 90°, with an increment of 15° along the longitudinal axis of the specimen. The failure was under three stages, however, the pattern of failure that occurred is not reported. Hugo [8] implemented the European standards, EN1447 and EN1227 to investigate the extended properties on a scale down test pool of GFRP pipe. The Ultimate Elastic Wall Stress (UEWS) along with the strain to failure tests, ascertains the first non-elastic behaviour, which furnishes a surrogate explanation to anticipate the abiding hydrostatic pressure strength [9, 10]. Under the biaxial and hydrostatic loading conditions filament wound glass fiber reinforced epoxy (GRE) composite pipes at varying temperatures up to

95 °C were subjected to test by Majid et al [11.–15.]. Wherein they have used a UEWS test that is a momentary test operated by the manufacturing companies in-order to assure the combat protracted functioning of pipes under regression test as per the ISO 14692 through ASTM D2992. However, the failure behaviour and the failure pattern of these materials are further needed to be understood so that the designers could effectively utilizes these materials for further more applications. Thus, the present paper aims at determining the internal burst pressure capacity of glass/epoxy pipes for different winding angles required for pressure vessel applications. The theoretical first-ply failure & burst pressure based on Tsai Hill theory, Maximum strain theory and Maximum stress theory were used as baseline and compared with experimental results. The filament wound GFRP pipes of different helix angles namely [±45°], [±55°] & [±70°] are pressurized until they burst with water as medium using hand operated pump and strains were recorded using the DAQ.

2. Experimentation:

The experiments were conducted using a open-end burst test on the samples made from E-glass/epoxy and the procedure of sample preparation and the testing is detailed in the sub section below.

2.1. Sample preparation

The E-glass/epoxy pipe samples required for the present study were produced using the filament winding facility available at CNC Techniques, Hyderabad, India. The pipes were of two-meter length, 50mm internal diameter and 3mm thickness, with three different fiber orientations, [±45°], [±55°] and [±70°]. The pipes after winding on the mandrel were allowed to cure under an IR heater, which was maintained at a temperature of 1300C while the mandrel was rotating at very low speed. After the samples were cured, a 600 mm length tubes were cut using a diamond dust cutter and the ends were prepared for the open-end burst test. A small piece from each sample was take in-order to determine the volume fraction using burn test and was found to be 58%. At the Centre of the pipe sample, a rosette type strain gauge was bonded and care was taken during bonding so as to form a perfect bond between the sample and the strain gauge.

2.2. Test Equipment

The experiments were conducted on the test set-up similar to that of the open- end burst test by P. D. Soden, et al. [13], however instead of the hydraulic power pack a hand operated pump was used to pressurize the system with trivial modifications. Figure 1 describes the test set-up used in the nonce study for testing GFRP pipes under internal pressure. The test set-up comprises of (i) Mechanical Structure and (ii) Strain measuring unit. The mechanical structure constitutes of a hand operated pump used to pressurize the system that is connected to a non-return valve. The other side of the non-return valve is connected to one of the two end caps that are used to firmly hold the sample. The end caps are provided with a circular grove in them, so as to accommodate the pipe and rubber seal. The sample is fixed using four tie rods which are of sufficient strength to restrict the leakage at the plates. The strain measuring system consists of strain smart 5000 data acquisition system (DAQ) from Micro Measurements which has an in-built compensation circuit required for completing the bridge circuit, depending on the resistance of the strain gauge.

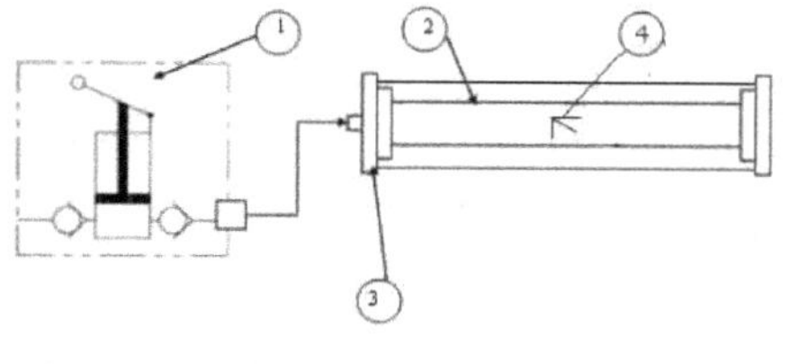

Fig 1: Schematic diagram for pipe burst test.

3. Results and Discussion

The pipe samples with orientation mentioned earlier were subjected to burst test using the test equipment described above. The system was pressurized until the sample fail due to burst and the equivalent strains were logged using the DAQ system. The theoretical first-ply failure and burst pressure were anticipated using (i) Tsai Hill theory, (ii) Maximum strain theory and (iii) Maximum stress theory which were later compared with the experimental results. The results obtained from the experimentation and the theoretical prediction of first-ply failure and the burst failure are tabulated in table 1. The experimental bust pressures for the sample with [±45°], [±55°] and [±70°] fiber orientation is found to be 4.383 MPa, 7.71 MPa and 4.2762 MPa respectively. However, in the present study the first ply failure of the samples was observed in the form of whitening phenomenon which occurred on the sample during the test.

The stress – strain relation for the sample with [±45°], [±55°] and [±70°] fiber orientation is represented in fig 2-4. A nonlinear stress – strain behaviour was witnessed intended for the samples in the longitudinal direction, however for the sample with [±70°] fiber orientation the stress –strain relation was in the positive side indicating a tensile stress state induced in the sample. A linear stress strain relation is observed till the theoretical first ply failure for the sample with [±45°] and [±55°] fiber orientations which later changes to nonlinear until the final failure. Moreover, the nonlinear behaviour for the sample with [±45°] fiber orientation due to the sliding of the fiber the strain changes its sign.

Table 1. Comparison between the experimental results and the failure theories.

Winding Angle	First Ply Failu e			Burst Pressure			
Deg	in MPa						
	Maximum Stress theory	Maximum Strain theory	Tsai Hill theory	Experimental	Maximum Stress theory	Maximum Strain theory	Tsai Hill theory
[±45º]	2.84402	2.36137	2.42798	4.38311	4.66024	4.66024	3.58222
[±55º]	5.4781	3.31578	3.97099	7.70968	7.25047	6.26064	6.87799
[±70º]	2.63801	2.11788	2.41895	4.27618	4.46159	4.17592	4.27333

Materials Research Proceedings 31 (2023) 466-474 https://doi.org/10.21741/9781644902592-48

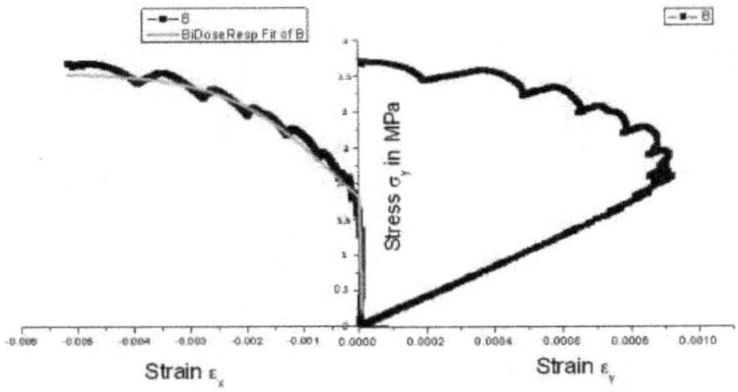

Fig 2: Stress – Strain distribution for [±45°] sample under S.R=1:0.

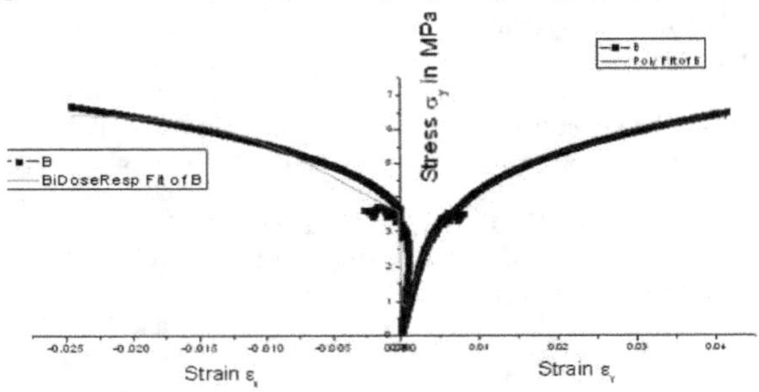

Fig 3: Stress – Strain distribution for [±55°] sample under S.R=1:0.

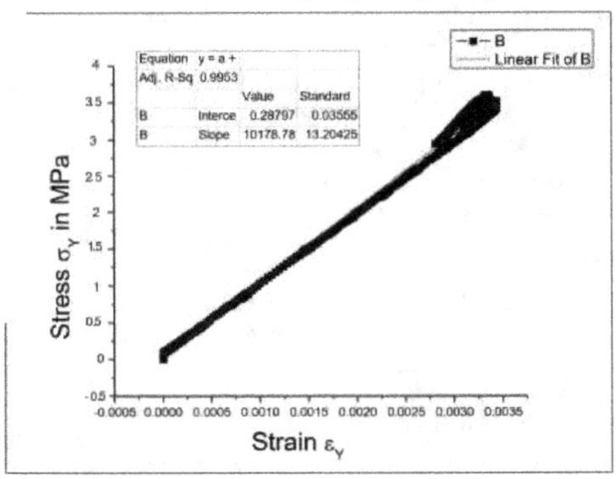

Fig 4: Longitudinal Strain distribution for [±70°] sample under S.R=1:0.

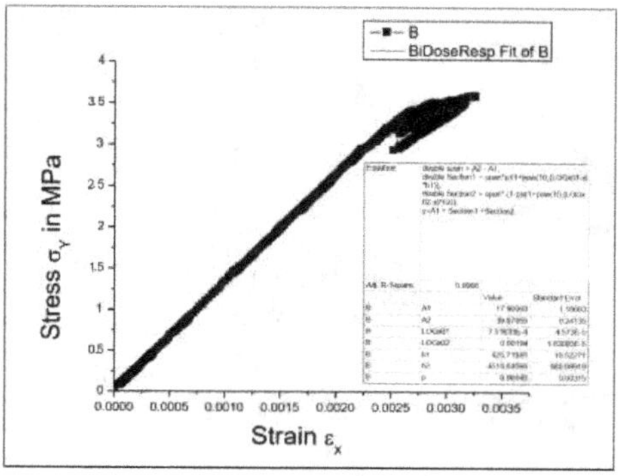

Fig 5: Hoop Strain Distribution for [±70°] sample under S.R=1:0.

The failure pattern which is represented in the fig 6 indicates an outflow along the pipe body. The close view at the middle of the pipe sample indicates the appearance of crack between fiber and matrix when attitude to the pressure being applied. Furthermore, there are numerous cracks along the axial direction at final failure position indicated in Figure 6(c), representing a larger stress along the circumferential direction is being experienced by the outer wall of the sample which is a resin-rich layer. These axial cracks are the typical mode of failure that the GFRP pipe experiences due to the internal pressure. Two different types of failure patterns were typically found when the crack morphology at the failure was observed with naked eye. The first one being the appearance of fiberglass whitening along the area of failure and the second being a great number of cracks along the circumferential direction. Besides the presence of the cracks along the circumferential direction there are small pin holes which are also the axial cracks which can be seen in Fig 6(a). These pin holes are found in large number on the sample which are caused owing to the fiber fracture. Though these pin holes could be viewed as the part of the circumferential cracks that are appeared along the sample which are the damage initiation sites. These damage initiation i.e., the pine holes which form a micro crack, later grows into a macro crack to form a thin white line parallel to the fiber direction, which is highlighted in fig 6(a&c). This whitening, originates to a matrix crack fallowed by fiber–matrix de-bonding which are produced due to the shearing of matrix which occurs parallel to the fiber. The whitening physical process has been even reported by other investigators and was ascribe mainly due to the matrix cracking and local interface debonding. A careful examination indicated that the damage or microcracked in a band form proliferates about the fibres without effecting the interface between the fibre and matrix which is represented in fig 7.

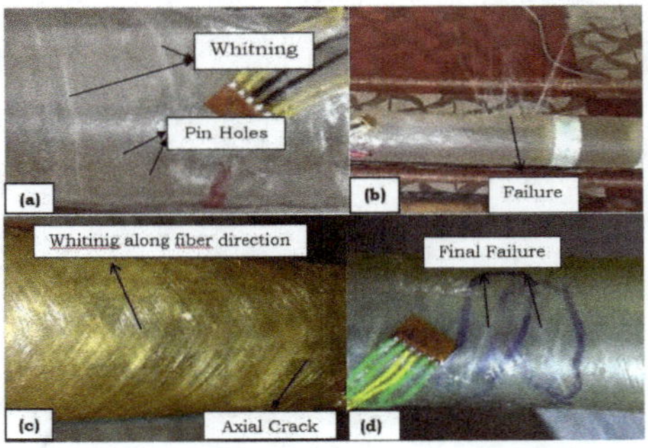

Fig 6: Macroscopic Failure Morphology of the failed sample.

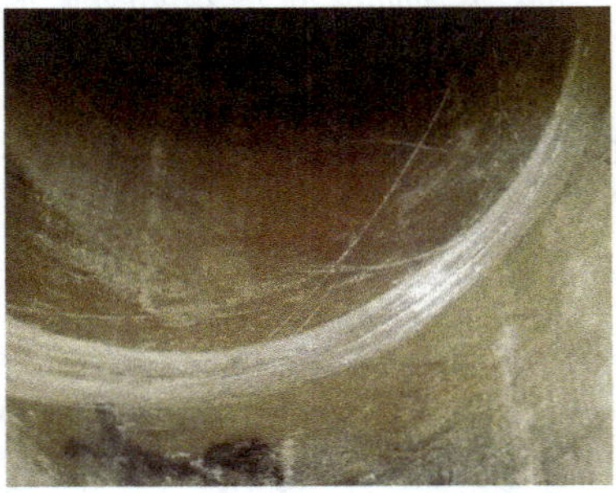

Fig 7: Morphology of failure along the fiber direction.

Conclusions

GFRP Tubular samples with three different fiber lay-up configurations, i.e. [±45°], [±55°] and [±70°], were tested under internal pressure with open end burst test arrangement. Data in the form of strains with respect to the applied stress were recorded using a strain smart data acquisition system. Later the stresses at both the functional and structural breakdown, along with the stress–strain relation were computed and following conclusions are drawn:

1. The experimental results signify that the samples with different winding angles failed at different burst pressures. The sample with [±45°] fiber orientation failed under burst at 4.38311 MPa whereas the sample with [±55°] and [±70°] fiber orientation failed at 7.70968 MPa, 4.27618 MPa respectively.
2. The sample with [±45°] and [±55°] fiber orientation showed a nonlinear stress – strain behaviour. However, the sample with [±70°] indicated an approximate linear relation between the applied stress and the longitudinal strain.
3. The samples with [±45°] and [±55°] fiber orientation indicated a linear stress strain relation until the theoretical first ply failure (both in longitudinal and hoops direction); later the behaviour changes to non-linear. Moreover, the strain in the hoops direction decreases after the first ply failure for the [±45°] sample indicating a strain reliving phenomena.
4. A micro crack is observed as pin holes on the sample just before the first fly failure which later increases in number with an increase in pressure. These pin holes become unstable and propagates along the fiber direction after the theoretical first ply failure. Theses micro cracks

accumulate and grow in to macro cracks along the fiber direction and appear as white lines along the fiber direction.

Acknowledgments
The authors would like to acknowledge the support rendered by CNC Technology, Hyderabad in preparation of samples. The authors would like to thank Mr. K. Vidya Sagar, Founder Vijaya Duraga Associates, Hyderabad for providing the strain smart data acquisition system during the test schedule.

Reference

[1] P. A. Zinoviev, S. V. Tsvetkov, Mechanical properties of unidirectional organic-fiber reinforced plastics under hydrostatic pressure, Composites Science and Technology 58 (1998) 31-9. https://doi.org/10.1016/S0266-3538(97)00079-1

[2] DNVGL-RP-F119, Recommended Practice for Thermoplastic composite pipes, Technical Report, DNV GL, 2015.

[3] R. Rafiee, F. Reshadi, Simulation of functional failure in GRP mortar pipes, Compos. Struct. 113 (2014) 155-163. https://doi.org/10.1016/j.compstruct.2014.03.024

[4] P. Mertiny, A. Gold, Quantification of leakage damage in high-pressure fibre reinforced polymer composite tubular vessels, Polym. Test. 26 (2) (2007) 172-179. https://doi.org/10.1016/j.polymertesting.2006.09.009

[5] Rosenow MWK. Winding angle effects in glass fibre reinforced polyester filament wound pipes. Composites 1984;15(2):144-52. https://doi.org/10.1016/0010-4361(84)90727-4

[6] Spencer B, Hull D. Effect of winding angle on the failure of filament wound pipes. Composites 1978:263- 71. https://doi.org/10.1016/0010-4361(78)90180-5

[7] Arikan H. Failure analysis of (±55_) filament wound composite pipes with an inclined surface crack under static internal pressure. Compo Struct 2010;92 (1):182-7. https://doi.org/10.1016/j.compstruct.2009.07.027

[8] H.F. Schwencke, The use of the ultimate elastic wall stress (UEWS) in evaluating PVC polymer blends, J. Vinyl Technol. 11 (1) (1989) 28-32. https://doi.org/10.1002/vnl.730110108

[9] H. Faria, R.M. Guedes, Long-term behaviour of GFRP pipes: reducing the prediction test duration, Polym. Test. 29 (3) (2010) 337-345. https://doi.org/10.1016/j.polymertesting.2009.12.008

[10] T.A. Assaleh, L.A. Almagguz, Ultimate elastic wall stress (UEWS) test under biaxial loading for glass-fibre reinforced epoxy (GRE) pipes, Adv. Mater. Res. 974 (2014) 188-194. https://doi.org/10.4028/www.scientific.net/AMR.974.188

[11] M.S. Abdul Majid, M. Afendi, R. Daud, N.A.M. Amin, A. Mohamad, E.M. Cheng, M. Hekman, General lifetime damage model for glass fibre reinforced epoxy (GRE) composite pipes under multiaxial loading, Key Eng. Mater. 594 (2014) 624-628. https://doi.org/10.4028/www.scientific.net/KEM.594-595.624

[12] M.S. Abdul Majid, A.G. Gibson, M. Hekman, M. Afendi, N.A.M. Amin, Strain response and damage modelling of glass/epoxy pipes under various stress ratios. Plastics, (2014) 290-Rubber and Composites, 299. https://doi.org/10.1179/1743289814Y.0000000101

[13] M.S. Abdul Majid, M. Afendi, R. Daud, A.G. Gibson, T.A. Assaleh, J.M. Hale, M. Hekman, Acoustic emission monitoring of multiaxial ultimate elastic wall stress tests of glass

fibre-reinforced epoxy composite pipes, Adv. Compos. Mater. (2014) 1-17. https://doi.org/10.1080/09243046.2013.871175

[14] M.S. Abdul Majid, M. Afendi, R. Daud, A.G. Gibson, M. Hekman, Effects of winding angles in biaxial ultimate elastic wall stress (UEWS) tests of glass fibre reinforced epoxy (GRE) composite pipes, Adv. Mater. Res. 795 (2013) 424-428. https://doi.org/10.4028/www.scientific.net/AMR.795.424

[15] M.S. Abdul Majid, T.A. Assaleh, A.G. Gibson, J.M. Hale, A. Fahrer, C.A.P. Rookus, M. Hekman, Ultimate elastic wall stress (UEWS) test of glass fibre reinforced epoxy (GRE) pipe, Compos. A: Appl. Sci. Manuf. 42 (10) (2011) 1500-1508. https://doi.org/10.1016/j.compositesa.2011.07.001

[16] Qi Guoquan1, Qi Dongtao1, Bai Qiang1, Li Houbu1, Wei Bin1, Ding Nan1, Zhang Dongna1, and Shao Xiaodong, Failure Analysis on Pressure Leakage of FRP, Fibers and Polymers 2019, Vol.20, No.3, 595-60. https://doi.org/10.1007/s12221-019-8617-5

Manufacturing strategy for additive manufacturing of a piping component for oil and gas application

Meshal Alsaiari[1, a *], Mushtaq Khan [2, b] Faramarz Djavanroodi[2, c]

[1]Saudi Aramco, Abqaiq, Kingdome of Saudi Arabia

[2] Prince Mohammed Bin Fahd University, Kingdome of Saudi Arabia

[3]List all distinct addresses in the same way

[a]mishal.saiari@aramco.com, [b]mkhan7@pmu.edu.sa, [c]fdjavanroodi@pmu.edu.sa

Keywords: Additive Manufacturing, DfAM, Polymers, FDM, HDPE

Abstract. Additive manufacturing, or 3D printing, has been the most developing, desired manufacturing process in the manufacturing industry for the last three decades. Ease of application, design of freedom, and variety of materials application attracted all industries manufacturers to shift their dependency of the traditional manufacturing processes to acquire an innovative state-of-the-art solution at the most affordable cost. AM manufacturing provides parts with the optimum processing factors and the least material wastage. Strength and final dimensional accuracy are vital in AM parts creation. This paper will demonstrate the DfAM to fabricate a piping spacer for the oil and gas piping system. Also, it will shed light on the main elements that impact DfAM's strategy. All related manufacturing parameters: including infill, printing orientation, and material selection analyzed, to acquire a robust part created using the FDM process to ensure the final product is safely utilized for the oil and gas industry application.

Introduction

Additive manufacturing is an evolving state-of-the-art fabrication process wherein all industries seek to adopt as an alternative to traditional fabrication processes. 3DP or additive manufacturing (AM) is a process driven by manufacturing using three-dimensional model data and building the materials to develop the final product. This fabrication process can produce parts with various complex geometries and design freedom from constraints. Initially, 3DP was only used for prototyping purposes. In recent years the convergence of 3DP from prototyping to products was increased due to ease of creation and diverse final product sizes with affordable cost regardless of the product quantities. AM is one of the leading enablers transforming the manufacturing industries to a new perspective of sustainability and recyclability. AM has become an attractive market for manufacturers, leading to tremendous acceleration in the development of AM methodologies. The predicted global economic growth for 3DP was estimated to approach USD 23.33 billion by 2026 [1]. AM is generated from a physical model that relies on a computer-aided design (CAD) model. The AM starts by transforming the CAD model format into a Standard Tessellation Language (STL) type file, entailing the part geometry and layout. Then, the STL files are sliced into the 2D cross-sectional layers with nearly 0.01–0.7 mm of thickness for the component and printed using one of AM methodologies [2]. The overall AM essential manufacturing steps are illustrated in Fig 1. All AM methods use the same processing steps until the construction of all layers one above each other, which is finally completed by post-processing or curing as required [3]. ISO/ASTM 52900 classified AM processes into the following categories: material extrusion, powder bed fusion, vat polymerization, directed energy deposition, material jetting, binder jetting, and sheet lamination [3].

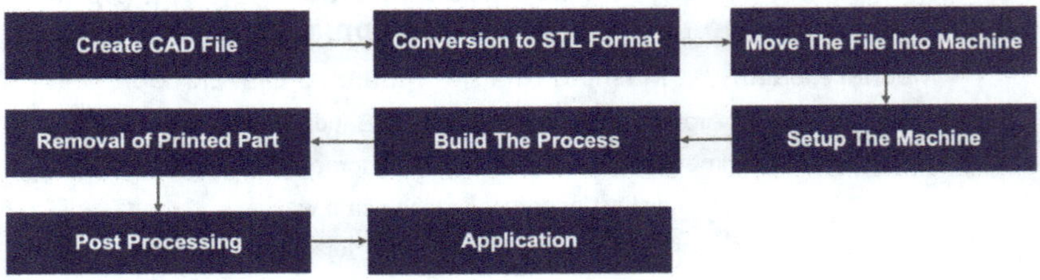

Fig 1 AM Overall essential Manufacturing Steps

Process Flowchart
AM part creation process for oil and gas applications starts with the potential component selection by evaluating: needs, risk, and experience for fabrication. The main target is to assess the overall feasibility of part fabrication to move to the next step. DFAM framework is the second step that is developed based on the performance specification of the part. This step is crucial since the component material and design are created, guiding the AM process selection.

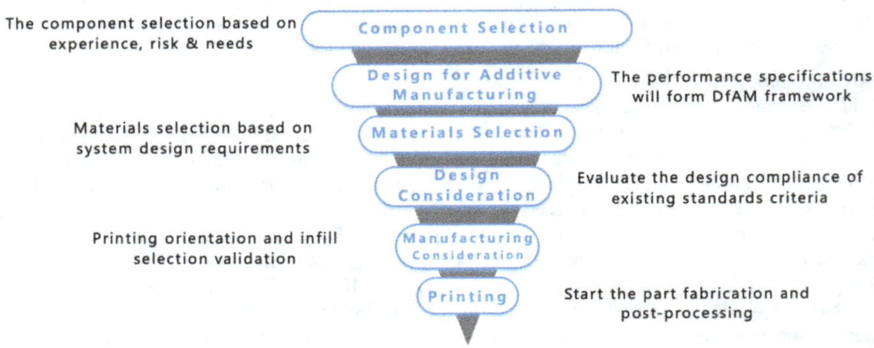

Figure 2 AM of Oil & Gas Component Creation Framework

Component Selection
Spacers and spectacle blinds are piping components used in any processing industry to retain a pressured piping loop. The primary function of these blinds is to isolate and shut down a flow in the case of spectacle blinds or to fill a gap between two piping flanges for piping spacers. Usually, blinds applications temporarily or permanently based on the piping system configuration and process requirements. ASME B16.48 is the governor standard to fabricate and manufacture the piping blinds/spacers for the installation between two pressurized piping flanges [4]. The fabrication for these blinds is either forged or casted steel for metallic piping. For nonmetallic piping, the current industry practice of nonmetallic piping systems is to use a steel-coated piping spacer between the piping flanges; if the coating fails, the part corrosion vulnerability is present. Another concern is over-torqueing leads to damage to the piping flanges. Based on this, the 3DP

method is selected to fabricate the piping spacer with nonmetallic material that can suit the subject piping network. A specific case of study for a High-density Polyethylene piping loop is determined for an existing piping circuit to fabricate and manufacture a piping spacer. Table 1 shows the selected pipe specification and processing details.

Table 1 The Case of Study Piping Details

Pipe Specification and Details	
Pipe Material	High-density Polyethylene
Pipe Class Rating	150
Pipe Diameter	4"
Reference Standards	ASME B16.5, ASME B16.48
Fluid Type	Raw Water
Fluid Temperature	115F
Operating Pressure	175 PSI

Materials Selection

Materials selection involves various factors to make the final decision in manufacturing, such as application criticality, temperature, applied stresses, chemical compatibility, and anticipated failure modes. Polymers are the most commonly used materials in the 3DP as filaments to produce parts due to their ease of application and lower cost. 3DP uses filaments made of polymer: polylactic acid (PLA), acrylonitrile butadiene styrene (ABS), poly-ether-ketone (PEEK), etc. Some materials, such as ABS and PLA, have limited applications due to their lower mechanical properties. For instance, these materials can't be utilized in aerospace applications. Instead, PEEK has superior mechanical and thermal properties to fit high-temperature surfaces [6].The material selection has its own cost. Thus, the materials are selected considering the part's functionality and criticality. In general, pure polymers have lower mechanical performance. As a result, manufacturers tend to use composite polymeric materials (reinforced). In the context of fiber-reinforced polymers, there are two forms of fiber reinforcement: short fibers, and continuous fibers, impeded during printing to attain superior mechanical properties [7]. Continuous fibers are available with a longer length and control fiber orientation, while short fibers have shorter aspect ratios and generally are not fully aligned. Fiber orientation is vital in manufacturing since aligned fibers deliver higher strength than unaligned ones [8]. For this reason, continuous fibers are preferred from a strength standpoint. From a manufacturability aspect, the short fibers are much more appealing due to ease of production and lower cost. The common fiber used in the 3DP field are: carbon, glass, and Kevlar fibers. These fibers deliver higher stiffness in manufacturing with the matrix material after bonding [8]. Various polymeric materials with their composite mechanical properties were reviewed and analyzed from multiple resources with the aim of identifying the highest composite material performance. As indicated in table 2, it is clear that the 3DP composite filament tensile strength performance is much higher than pure polymers. The highest tensile value was on polymers reinforced with continuous fibers. Although continuous fiber impregnation delivers better mechanical properties, poor bonding is generated when the fiber content increases [9]. For this reason, the fabrication of continuous fiber reinforcement filament is very complex since it requires thorough control during processing. On the other side, the short fiber reinforcement avails a significant enhancement in the tensile strength by almost 45% versus pure polymers. In this case of study, the selected material was HPDE to fit the existing piping materials. As known HDPE crystallization and solidification process is very critical since failures such as warping, distortion and voids are present during fabrication. Due to the challenges identified in HDPE, the piping spacer materials will be utilized will pure and reinforced HDPE to evaluate the overall product cost and performance.

Table 2 Summary of Different Filaments with Their Reinforced Composite Materials

Materials	Pure polymer Tensile Strength (MPa)	Composite polymer Tensile Strength (MPa)	Fiber Type	Fiber Form	Ref
ABS	**32-43**	**63**	**Carbon**	**Short**	[11] [12] [13]
HDPE	**26.2**	**300.2**	**Carbon**	**Continuous**	[14] [15]
PLA	**46.61-65**	**241**	**Glass**	**Continuous**	[16] [7] [12] [9]
NYLON	**35.25**	**50.6**	**Carbon**	**Short**	[17]

Design for Additive Manufacturing (DfAM)
AM design consideration is fundamental while creating the desired printing model sliced to produce the final part. Not only the product's geometrical complexity and functionality are considered, but other factors such as quality, time, and the final cost are vital in designing for manufacturing. The design stage is a very critical and challenging task for AM producers because each new part will be created and evaluated based on its complexity, design criteria, performance specification, and application, making each manufactured part a unique product. (Wiberg et al.,2019) Defined and showed the design of AM automation application to be with three main stages: system, part, and process design [18]. The 3DP design process starts with the component design, where the component deficiency or problem, material, and anticipated load are considered while selecting the coveted printing method. Then, the part design follows the previous step by creating part initial design with interpretation and evaluating the need for the new part support structure. In this stage, verification is essential before proceeding with the processing to validate the created part design and ensure the design variables are examined. As shown in Fig 2, the overall design phases with an iterative and assessment process to optimize the final product design.

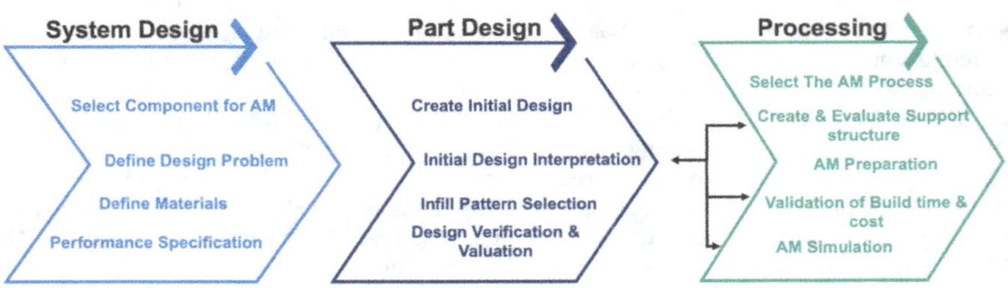

Fig 3 AM Design Phases Today Manufacturers Use.

AM preliminary design feature option is the overhang support structure. Generally, overhung structure design characteristics can only be achieved if the printing platform's inclination angle is above 45 degrees [19]. This is true because the self-supporting and overhanging are linked together with the difference in manufacturing platform angles. The supporting mechanism is defined as when the printing baseplate angle is less than 45° as overhang and self-supported if the angle is greater than 45 [19].

Design Consideration for FDM
Fused Deposition Modeling (FDM), or fused filament fabrication (FFF), is one of the most widely AM technologies currently used in the 3DP industry due to its simplicity, lower cost, and flexibility to fabricate parts with complex geometries. The FDM is a material extrusion-based process. In this process, the filaments are injected into a liquefier head and deposited with ultra-thin layers in a semi-solid state and solidify immediately to the previous layer formation until the final product is built [6]. The FDM 3D printed components are usually supported using a weaker material, and upon process completion, the support materials are removed, providing the final finished part [7]. Adhesion between printed layer elements and bonding of the deposited filament without facile detachment are the main factors that impose objects free of imperfections [8]. Without proper control of the printing process parameters, the tendency of voids formation is high, affecting the produced part strength compared to other traditional manufacturing methods. In this case study, the FDM process was selected to fabricate the part since the part materials that will be used is HDPE. The part design was started by selecting the required part size and dimensions. The piping component chosen was designed as per ASME B16.48 for steel piping components since no standards are available for nonmetallic. As shown in Fig 3, the selected piping component CAD design was developed and verified for any design uncertainties.

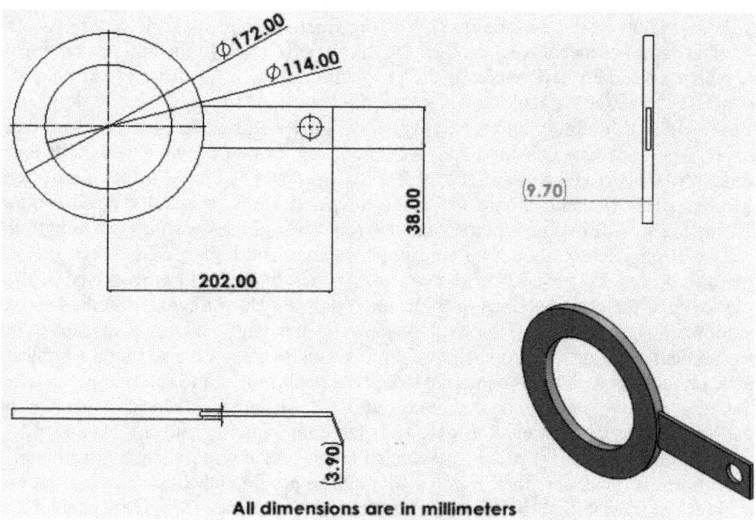

Figure 4 Isometric Drawing of The Piping Spacer

Manufacturing Consideration

Defects and imperfections are formed in manufacturing due to several causes, including but not limited to process parameter fluctuation, poor design, low-quality materials, and improper operation while producing new components. Thus, it is vital in manufacturing to understand and predict the product failure modes and defect formation to be addressed in advance before processing. In the FDM process, warping or distortion are the main issues that disturb FDM parts dimensions accuracy caused by the produced internal stresses in the manufacturing stage. These internal stresses rely essentially on the amount of volume reduction during the cooling phase from the glass transition temperature, precisely due to the discrepancy in cooling rates between different printed layers [20], [21]. Consequently, if the adhesion between the printed layers and the fabrication baseplate is enhanced using a suitable melting temperature, such defects can be avoided [21].Another quality concern impacting the part fabrication is improper feeding or extrusion. Basically, unsuitable extrusion, either over-run or under-run during the extrusion process, can cause a lack of adhesion, delamination, and debonding between the filament's layer, providing an uneven surface profile with dimensional deviations, porosity, and cracks [22], [23]. All these defects are profoundly affected by the printing process quality envelope. In-depth analysis of the main contributing elements for the FDM process are categorized, analyzed, and divided into two main domains: Printing Orientation and Infills.

Printing orientation and infills

Printing or building orientation is essential while producing AM components, particularly in the FDM process. The printing orientation is defined as the position of created part concerning the manufacturing machine coordinates within its printing platform. Build orientation is one of the main parameters that influences the staircase effect and the amount of support structure production. The stair-stepping effect is formed due to thickness differences between produced layers, which imposes deterioration in the surface quality of the final component [24]. As known, the stair-stepping effect can be reduced by controlling the thickness of the layer, leading to a more extended printing, which eventually will increase the production time. Infill density and pattern are the fundamental keys in 3DP, provoking designers to advance AM parts design seeking the lowest cost with the highest durability and reliability. This is not a straightforward equation that can be applied everywhere; several considerations shall be taken for part design based on its application for the final infill density and pattern. The infill density is described as the infill volume percentage using the deposited filament, where 0% is hollow, and 100% is a fully filled product [21]. Basically, the higher infill volume percentage delivers filled objects with upper strength limits; of course, this will impose extra material consumption, cost, and higher weight. For this reason, a well-organized design and production strategy is essential. In the FDM process, the infill strategy can be deployed at different sections of the created parts, such as the exterior shell or walls and upper and lower layers, by modifying each element independently, having a significant influence on the mechanical and physical properties. AM parts usually are not created with completely solid infill states, rather than with hollow internal structures to optimize time and cost in manufacturing. These internal hollow structures are produced with different structure geometry patterns: rectilinear, concentric, grid, triangular, gyroid, octet, cubic, quarter cubic, tri-hexagon, and stars patterns [21]. The desired infill pattern is selected based on the required strength, flexibility, and time. The concentric, grid, and honeycomb infill patterns provided the highest tensile strength due to their higher susceptibility to holding intermolecular deposits [25]. The part's mechanical characteristics heavily impact the infill density; as infill density increased, higher tensile strength and young's modulus were attained.

Conclusion

This paper outlines a case study of designing a piping component for process piping in the oil and gas industry. The DfAM framework demonstrates its effectiveness by minimizing iteration work at the processing stage. This design has the potential for deployment with any processing equipment. Utilizing this DfAM structure requires an investigation of the probability of risk failure for the selected part. The material selection and the part performance specification should integrate and automate at one stage to have an efficient and optimal design, avoid any constraints, and reduce the overall cost. The future work of this case study will focus on creating a piping spacer made of nonmetallic material suitable for outdoor applications that is compatible with the selected piping system specifications.

References

[1] F. A. Cruz Sanchez, H. Boudaoud, M. Camargo, and J. M. Pearce, "Plastic recycling in additive manufacturing: A systematic literature review and opportunities for the circular economy," *J. Clean. Prod.*, vol. 264, p. 121602, 2020. https://doi.org/10.1016/j.jclepro.2020.121602

[2] A. V. D. & R. M. Metkar, *Reconstruction of Damaged Parts by Integration Reverse Engineering (RE) and Rapid Prototyping (RP)*. 2018.

[3] N. Yaragatti and A. Patnaik, "A review on additive manufacturing of polymers composites," *Mater. Today Proc.*, vol. 44, pp. 4150–4157, 2020. https://doi.org/10.1016/j.matpr.2020.10.490

[4] T. A. S. of M. E. ASME, "ASME B16.48 Line Blanks," 2005.

[5] T. D. Ngo, A. Kashani, G. Imbalzano, K. T. Q. Nguyen, and D. Hui, "Additive manufacturing (3D printing): A review of materials, methods, applications and challenges," *Compos. Part B*, vol. 143, no. 13 February 2018, pp. 172–196, 2018. https://doi.org/10.1016/j.compositesb.2018.02.012

[6] K. Rajan, M. Samykano, K. Kadirgama, W. S. W. Harun, and M. M. Rahman, *Fused deposition modeling: process, materials, parameters, properties, and applications*, vol. 120, no. 3–4. Springer London, 2022.

[7] Y. Yao, M. Li, M. Lackner, and L. Herfried, "A continuous fiber-reinforced additive manufacturing processing based on PET fiber and PLA," *Materials (Basel).*, vol. 13, no. 14, 2020. https://doi.org/10.3390/ma13143044

[8] J. Yan, E. Demirci, A. Ganesan, and A. Gleadall, "Extrusion width critically affects fibre orientation in short fibre reinforced material extrusion additive manufacturing," *Addit. Manuf.*, vol. 49, no. October 2021, p. 102496, 2022. https://doi.org/10.1016/j.addma.2021.102496

[9] K. Chen, L. Yu, Y. Cui, M. Jia, and K. Pan, "Optimization of printing parameters of 3D-printed continuous glass fiber reinforced polylactic acid composites," *Thin-Walled Struct.*, vol. 164, no. November 2020, p. 107717, 2021. https://doi.org/10.1016/j.tws.2021.107717

[10] A. Borkar, A. Hendlmeier, Z. Simon, J. D. Randall, F. Stojcevski, and L. C. Henderson, "A comparison of mechanical properties of recycled high-density polyethylene/waste carbon fiber via injection molding and 3D printing," *Polym. Compos.*, vol. 43, no. 4, pp. 2408–2418, 2022. https://doi.org/10.1002/pc.26550

[11] C. Casavola, A. Cazzato, V. Moramarco, and G. Renna, "Mechanical behaviour of ABS-Fused Filament Fabrication compounds under impact tensile loadings," *Materials (Basel).*,

vol. 12, no. 8, 2019. https://doi.org/10.3390/ma12081295

[12] M. Algarni and S. Ghazali, “Comparative Study of the Sensitivity of PLA, ABS, PEEK, and PETG’s Mechanical Properties to FDM Printing Process Parameters,” *Crystals*, p. 21, 2021. https://doi.org/10.1016/b978-075065129-5/50007-6

[13] H. L. Tekinalp *et al.*, “Highly oriented carbon fiber-polymer composites via additive manufacturing,” *Compos. Sci. Technol.*, vol. 105, pp. 144–150, 2014. https://doi.org/10.1016/j.compscitech.2014.10.009

[14] C. G. Schirmeister, T. Hees, E. H. Licht, and R. Mülhaupt, “3D printing of high density polyethylene by fused filament fabrication,” *Addit. Manuf.*, vol. 28, no. April, pp. 152–159, 2019. https://doi.org/10.1016/j.addma.2019.05.003

[15] M. Zhang, X. Tian, and D. Li, “Interfacial transcrystallization and mechanical performance of 3d-printed fully recyclable continuous fiber self-reinforced composites,” *Polymers (Basel).*, vol. 13, no. 18, 2021. https://doi.org/10.3390/polym13183176

[16] N. Lokesh, B. A. Praveena, J. Sudheer Reddy, V. K. Vasu, and S. Vijaykumar, “Evaluation on effect of printing process parameter through Taguchi approach on mechanical properties of 3D printed PLA specimens using FDM at constant printing temperature,” *Mater. Today Proc.*, vol. 52, pp. 1288–1293, 2022. https://doi.org/10.1016/j.matpr.2021.11.054

[17] M. Mohammadizadeh, A. Gupta, and I. Fidan, “Mechanical benchmarking of additively manufactured continuous and short carbon fiber reinforced nylon,” *J. Compos. Mater.*, vol. 55, no. 25, pp. 3629–3638, 2021. https://doi.org/10.1177/00219983211020070

[18] A. Wiberg, J. Persson, and J. Ölvander, “Design for additive manufacturing – a review of available design methods and software,” *Rapid Prototyp. J.*, vol. 25, no. 6, pp. 1080–1094, 2019. https://doi.org/10.1108/RPJ-10-2018-0262

[19] Y. H. Kuo and C. C. Cheng, “Self-supporting structure design for additive manufacturing by using a logistic aggregate function,” *Struct. Multidiscip. Optim.*, vol. 60, no. 3, pp. 1109–1121, 2019. https://doi.org/10.1007/s00158-019-02261-3

[20] H. I. Medellin-Castillo and J. Zaragoza-Siqueiros, “Design and Manufacturing Strategies for Fused Deposition Modelling in Additive Manufacturing: A Review,” *Chinese J. Mech. Eng. (English Ed.*, vol. 32, no. 1, 2019. https://doi.org/10.1186/s10033-019-0368-0

[21] H. K. Dave and S. T. Patel, “Introduction to Fused Deposition Modeling Based 3D Printing Process,” 2021, pp. 1–21.

[22] T. Ravi, R. Ranganathan, S. P. Ramesh, and D. S. Dandotiya, “3D Printed Personalized Orthotic Inserts Using Photogrammetry and FDM Technology,” in *Fused Deposition Modeling Based 3D Printing,Materials Forming, Machining and Tribology*, 2021, pp. 349–361.

[23] E. G. Gordeev, A. S. Galushko, and V. P. Ananikov, “Improvement of quality of 3D printed objects by elimination of microscopic structural defects in fused deposition modeling,” *PLoS One*, vol. 13, no. 6, 2018. https://doi.org/10.1371/journal.pone.0198370

[24] L. Di Angelo, P. Di Stefano, and E. Guardiani, “Search for the optimal build direction in additive manufacturing technologies: A review,” *J. Manuf. Mater. Process.*, vol. 4, no. 3, 2020. https://doi.org/10.3390/JMMP4030071

[25] M. Rismalia, S. C. Hidajat, I. G. R. Permana, B. Hadisujoto, M. Muslimin, and F.

Triawan, "Infill pattern and density effects on the tensile properties of 3D printed PLA material," *J. Phys. Conf. Ser.*, vol. 1402, no. 4, 2019. https://doi.org/10.1088/1742-6596/1402/4/044041

[26] H. Gonabadi, A. Yadav, and S. J. Bull, "The effect of processing parameters on the mechanical characteristics of PLA produced by a 3D FFF printer," *Int. J. Adv. Manuf. Technol.*, vol. 111, no. 3–4, pp. 695–709, 2020. https://doi.org/10.1007/s00170-020-06138-4

[27] P. K. Penumakala, J. Santo, and A. Thomas, "A critical review on the fused deposition modeling of thermoplastic polymer composites," *Compos. Part B*, vol. 201, no. July, p. 108336, 2020. https://doi.org/10.1016/j.compositesb.2020.108336

A homogenized finite element analysis of the deformation of axially-loaded thin-walled epoxy/coir fibre-reinforced aluminum 6063 composite tubes

*B.O. Malomo

Department of Mechanical Engineering, Obafemi Awolowo University, Ile-Ife, Nigeria

bobmalom@oauife.edu.ng

Keywords: Tube Thickness, Coir-Fibre, Deformation, Aluminum 6063, Energy Absorption

Abstract. Metal-fibre hybridization is strategic to the development of efficient lightweight energy absorbing structures. In this study, the deformation of variable coir-fibre reinforcement volumes (10%-90%) hybridized at reinforcement thicknesses (10T, 15T and 20T) with aluminum alloy 6063 tubes was investigated from a continuum standpoint to unravel the underlying mechanisms of the composite's performance potential. A representative volume element (RVE) was developed to determine the effective properties of coir-fibers based on Mori-Tanaka/Benveniste theoretical model in DIGIMAT 2017.0, while the aluminum alloy properties were defined by Johnson-Cook's plasticity. Finite element (FE) simulations were implemented in ABAQUS Explicit Dynamics by deformable continuum elements to capture the damage initiation and evolution response based on the Hashin's stiffness degradation failure criteria and the strength-based cohesive zone, traction-separation law. Incidental high peaks and unstable buckling loads at the onset of crushing was significant for 10T and 15T tubes with 10-30% and 80-90% reinforcement concentrations (V_f) relative to the low frequency of fluctuations at 40-70%, V_f. Axisymmetric to non-asymmetric transformations occurred with relative significance for the tubes as the resistance to global bending increased with reinforcement thickness. The interaction between the tube walls and reinforcement layers in preventing the formation of complete folds notably increased with 15T and 20T tubes at 60% fiber concentration, where the extensional collapse by progressive delamination of the 20T tubes improved the buckling resistance at densification. In agreement with FE simulations, experimental stress-strain plots at 60% V_f, confirmed that 20T tubes indicated wider plateau region of stable crushing to underscore observed high capacities for energy absorption.

1. Introduction

As one of the most compelling performance metrics in the automotive industry to achieve a much-required competitive edge is enhanced occupant safety, automakers have continued to deploy scientific novelties that advance superior crashworthiness through the incorporation of effective structural mechanisms in state-of-the-art material forms geared towards mitigating catastrophic effects from crash conditions. In recent times, a flurry of research efforts has intensified its focus on optimizing the energy absorption efficiencies of advanced structures evolving from the profound pioneering work of Weirzbicki and Abramowicz, [1] who jointly postulated the super folding element (SFE) theory to describe the relationships between deformation mechanisms and energy absorption criteria. Their study has facilitated the development of thin-walled metallic structures that have found wide applicability in safety designs over the decades. Based on the premise that mechanisms for the dissipation of kinetic energy in crash situations in high-speed transportation can easily follow a predictable and controllable scenario, tailorable aluminum composites have evolved for best performances. However, one of the major constraints associated with the arbitrary incorporation of metallic structures is the density effect, because it is somewhat challenging to have a lightweight absorber with a correspondingly high specific energy absorption (SEA) index which is the major requirement in many applications [2]. To circumvent this limitation in view of the sophisticated requirements for safety in modern transportation, the quest for light-weight structures in certain automotive applications continues to expand with growing interest in fiber-reinforced composites [2]. To this end, hybrid composites and other carefully-engineered, synthesized material forms are candidates for superior performance in the automotive and aerospace applications.

The hybridization of fibers and metals has been shown to achieve a good compromise in specific strength and modulus for shell structures and thin tubular sections with profound energy absorption performances. For instance, the authors [3] investigated the quasi-static axial crushing behavior of hybrid wrapped glass fibre reinforced polymer (GFRP) and square aluminum tubes in order to evaluate the inference of ply orientations, and they found that the performances of the hybrid tubes were superior to those of the aluminum tubes. The crashworthiness of a vehicle energy absorber has been studied based on a hybrid square hollow steel tube filled with glass-fiber reinforced 'polyamide 66' honeycomb structure [4]. Furthermore, to address the scenarios of oblique collisions in crashes, the authors investigated the ultimate bending moments and energy-absorption performance of aluminum/GFRP hybrid tube beams with emphasis on the effects of thin GFRP skin layer in relation to bending deformation behavior and fracture characteristics and they found the optimal thickness and layup of the composite skin layer consistent with the best performance of the hybrid tubes [5]. Metal-hybrids with synthetic fibers have also been achieved for strategic purposes. The effect of geometry and energy absorption of square cross section, graphite/epoxy and Kevlar/epoxy tubes under static crushing has been investigated in which the effect of decreasing width-wall thickness (b/t) ratio was correlated with increasing energy absorption [6].

As the requirements for energy absorption performance have heightened particularly in motorsport applications where fiber-reinforced polymers are often the primal choice, the prohibitive costs of synthetic fibres (carbon-fibre, kevlar, aramid, etc.,) along with growing environmental concerns on recyclability, has inspired the need for technologically-viable alternatives. In this purview, incorporating natural fibres as second phase materials in developing hybrid composites has indicated promising and desirable results [7]. Some natural fibers such as jute, hemp, sisal, kenaf, coir, flax, banana etc., have been shown to exhibit mechanical properties comparable to glass fibers [7]. This quality, coupled with their low densities, biodegradability, renewability, availability and low costs, make them directly applicable in lightweight, cost-effective automotive applications. Therefore, in line with the requirements of crash safety, hybrid natural fibre based composites have been investigated on their impact performances. For instance, the authors [8] have documented that a hybrid composite laminate of banana/E-glass fabrics reinforced polyester could possess a considerable impact resistance; also a hybridization mechanism to enhance the impact strength of jute/glass and jute/carbon-reinforced composites has been postulated [8], similarly, the post-impact behavior and damage tolerance capability of hemp-based fiber-reinforced polymers have been studied to determine the hybridization effect of basalt fiber additions [8].

The hybridization concept is very significant in improving the overall performance of composites and on this premise, this study focuses on investigating the hybridization effect of the inexpensive, naturally-occurring coir-fibres with high strength aluminum alloy 6063 taking into consideration variations in the concentrations of the coir-fibre and thickness of cylindrical aluminum tubes towards developing an efficient, low-cost, high-performance energy absorbing structures.

2. Materials and Methods

2.1 Composite material specifications

The hybrid composite structure consists of a primary (natural coir fibre) and secondary aluminum alloy 6063 in combination with an epoxy-resin matrix. The coir-fibre is a biodegradable natural lignocellulosic filler and the aluminum alloy (AA 6063) is an age-hardenable alloy with excellent dynamic deformation characteristics. The physical properties of the coir fibre, aluminum alloy and epoxy resin (LY556) are shown in Table 1.

Table 1. Baseline physical properties of coir fibre, AA 6063 and epoxy

Material	Tensile strength (MPa)	Young's modulus (GPa)	Elongation at break (%)	Density (g/cm^3)
Coir fibre [30]	220	6	15-25	1.25
AA 6063	145-186	68.3	18-33	2.69
Epoxy	31	3.7	-	1.15

2.2 Material constitutive modelling

The coir-fiber is constituted by continuous fibres with high aspect ratio, corresponding to infinitely large length-to-diameter (L/D) ratio, an approximation consistent for stiff and strong fibers. The failure initiation of the hybrid composite can be described according to the Hashin's damage law when any of the following four failure modes of fiber tension (F_t), fibre compression (F_c), matrix tension (F_{mt}) and matrix compression (F_{mc}) has a value that is equal or greater than unity.

$$F_t = \left[\frac{\widehat{\sigma_{11}}}{X_T}\right]^2 + \left[\frac{\widehat{\sigma_{12}}}{S_L}\right]^2 = 1 \quad (1)$$

$$F_C = \left[\frac{\widehat{\sigma_{11}}}{X_C}\right]^2 = 1 \quad (2)$$

$$F_{tm} = \left[\frac{\widehat{\sigma_{22}}}{Y_T}\right]^2 + \left[\frac{\widehat{\sigma_{12}}}{S_L}\right]^2 = 1 \quad (3)$$

$$F_{cm} = \left[\frac{\widehat{\sigma_{22}}}{2S_T}\right]^2 + \left(\left[\frac{Y_c}{2S_T}\right]^2 - 1\right)\left[\frac{\widehat{\sigma_{22}}}{Y_C}\right]^2 + \left[\frac{\widehat{\sigma_{12}}}{S_L}\right]^2 = 1 \quad (4)$$

Where $\widehat{\sigma_{11}}$ and $\widehat{\sigma_{22}}$ are the effective stress tensor components, X_T and X_C, are the fibre tensile and fibre compressive stresses; Y_T, and Y_C are the matrix tensile and matrix compressive stresses; while S_L and S_T are the longitudinal and transverse shear strengths respectively. The constitutive equations for interlaminar failure is of the form Eq. 5.

$$\sigma = C_{dm}\varepsilon \quad (5)$$

Where σ, ε are the stress and strain components respectively. The C_{dm} is the stiffness matrix indicating damage which is further expressed according to the elastic constants by;

$$C_{dm} = \frac{1}{D}\begin{bmatrix} (1-d_f)E_{11} & (1-d_f)(1-d_m)\mathrm{v}_{21}E_{11} & 0 \\ (1-d_f)(1-d_m)\mathrm{v}_{12}E_{22} & (1-d_m)E_{22} & 0 \\ 0 & 0 & D(1-d_s)G_{12} \end{bmatrix} \quad (6)$$

Where the damage variables for tension, compression and shear are given by d_f, d_m and d_s respectively. The scalar damage parameter D is associated with plastic strain and defines the degradation of material stiffness according to the expression;

$$\boldsymbol{D} = 1 - (1 - d_f)(1 - d_m)\mathrm{v}_{12}\mathrm{v}_{21} \tag{7}$$

Delamination between coir fibres and epoxy resin is analyzed by the cohesive zone (CZM) model approach according to the quadratic loss function or traction-separation criterion;

$$\left\{\frac{\langle\tau_n\rangle}{\tau_n^0}\right\}^2 + \left\{\frac{\langle\tau_s\rangle}{\tau_s^0}\right\}^2 + \left\{\frac{\langle\tau_t\rangle}{\tau_t^0}\right\}^2 = 1 \tag{8}$$

Where, t_n^0, t_s^0, t_t^0 are the peak values of the nominal stress components across the interface. The failure initiation displacements U_n^0, and U_s^0, can be expressed as follows:

$$U_n^0 = {\sigma_n}/{E_p}; \qquad U_s^0 = {\sigma_s}/{E_p} \tag{9}$$

Where, σ_n, σ_s, E_p, are the normal compressive strength, shear strengths and stiffness at the interface respectively. The final displacements U_n^1 and U_s^1 are obtained from the fracture toughness G_n for the normal mode and critical energy release G_s for the shear mode according to the relations:

$$U_n^1 = {2G_n}/{\sigma_n} \qquad U_s^1 = {2G_s}/{\sigma_s} \tag{10}$$

The effective separation displacement for the normal and shear separations is determined using;

$$\delta_m = \sqrt{\delta_n^2 + \delta_s^2} \tag{11}$$

The mixed-mode damage initiation criteria can then be expressed by;

$$\delta_m^0 = U_n^0 \times U_s^0 \times \sqrt{\frac{1+\beta^2}{{U_s^0}^2 + (\beta \times U_n^0)^2}} \tag{12}$$

Where $\beta = \frac{U_s}{U_n}$, the total separation by mixed-mode failure is then obtained through the Benzeggagh-Kenane (B-K) criterion with γ is the (B-K) criterion exponent relations;

$$\delta_m^f = \frac{2}{E_p \times \delta_m^0}[G_n + (G_s - G_n)] \times \left(\frac{\beta^2}{1+\beta^2}\right)^\gamma \tag{13}$$

The damage evolution variable $\boldsymbol{D}$ is expressed in terms of the maximum mixed-mode δ_{max}, by

$$D = \frac{\delta_m^f \times (\delta_{max} - \delta_m^0}{\delta_{max} \times (\delta_m^f - \delta_m^0)} \tag{14}$$

2.3 Numerical implementation

2.3.1 RVE homogenization

To simulate composite structures, effective material properties are required, and in most cases, these properties are difficult to measure prompting the need for numerical homogenization. The elastic properties and strength parameters of coir fibre/epoxy are required for the Hashin's progressive damage model and analysis of the hybrid composite, but since this is not readily available in the open literature it was found by implementing the Mori-Tanaka/Eshelby mean field model as a homogenization scheme in DIGIMAT 2017.0 software. In this special case, the elastic moduli viz; longitudinal modulus E_{11}, transverse modulus E_{22}, longitudinal shear modulus G_{12}, transverse shear modulus G_{23} and major Poisson's ratio v_{12}, were determined. The governing mathematical formulation of Mori-Tanaka and Benveniste [9] are given by the following relations that express the stress in the reinforcing fibres and the matrix by Eq. 15 and Eq. 16 respectively.

$$\sigma_f = \sigma_m + C_m(e^0 + e_f^C - e_f^T) + C_m(\bar{e} + e_f^C - e_f^T) \quad (15)$$
$$\sigma_m = \sigma_0 + C_m\bar{e} = C_m e^0 + C_m\bar{e} \quad (16)$$

Where, σ_0 is the external stress applied to the composite, C_m, e^0 and $\bar{e}$ are the stiffness tensor, the mean strain of the matrix and the eigenstrain or disturbance strain in the matrix due to the fibers. The constrained strain in the fibres and the mismatch strain between the matrix and the fibres are given e_f^C and e_f^T respectively. The stiffness tensor of the composite C_s is expressed via the relations;

$$C_s^{-1} = [I + fB(I + Q)]C_m^{-1} \quad (17)$$

Where, $B = \left((C_f - C_m)S + C_m^{-1}\right)(C_f - C_m)$; $Q = (I + f(S - I)B)^{-1}(-f(S - I)B)$ (18)

C_f, f, S and I are the stiffness tensor of fiber, fiber volume fraction, Eshelby tensor and identity tensor, respectively. In DIGIMAT 2017.0, the boundary value problem on the RVE of the epoxy/coir fiber composite (Fig. 1) is solved by substituting the elastic modulus and Poisson's ratio of the epoxy matrix and coir fibres into Eq. 17, to generate a homogenized and stress-strain response which determines the effective elastic moduli of the composite.

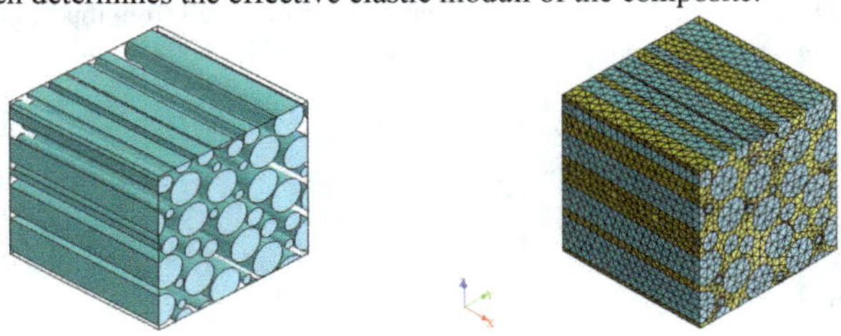

Fig. 1. Representative volume element (a) fibre-matrix (b) meshed structure.

2.3.2 Structural modelling

Numerical analysis was performed using ABAQUS/Explicit [41] to predict the load-deformation behavior and stress-contour profiles of the tubes, incorporating the interlaminar damage initiation and evolution criteria. The model as shown in Fig. 2, consists of a rigid top and bottom plates in counter-acting positions with the deformable coir-fibre/aluminum structure. The motion of the top plate is restricted to an axial vertical direction while the bottom plate is fixed. To prevent the initial penetration of the top plate and the deformable structure, a minimum clearance 0f 0.001mm was applied between the assemblies. A tangential behavior, 'surface-to-surface' contact condition was used to describe the interacting contacting surfaces with a penalty frictional formulation of 0.2. The top surface of the rigid bottom plate and the bottom face of the deformable structure are coupled by the 'node-to-surface' interaction criterion and the lateral motion of the contact was prevented by the 'rough' frictional contact condition. The top plate and rigid bottom were discretized with the 4-node 3D bilinear quadrilateral element 'R3D4' with an element size of 4.8mm. The deformable structure was discretized by the shell element 'S4R' with three integration points. The material properties for the homogenized coir-fiber/epoxy systems were incorporated to the deformable shell structure using the composite lay-up function. Axial load of *100N* was imposed from the top plate with a velocity of 15.6 m/s. The effective crushing distance is set at 20mm which is about 67% of the total length of the overall assembly.

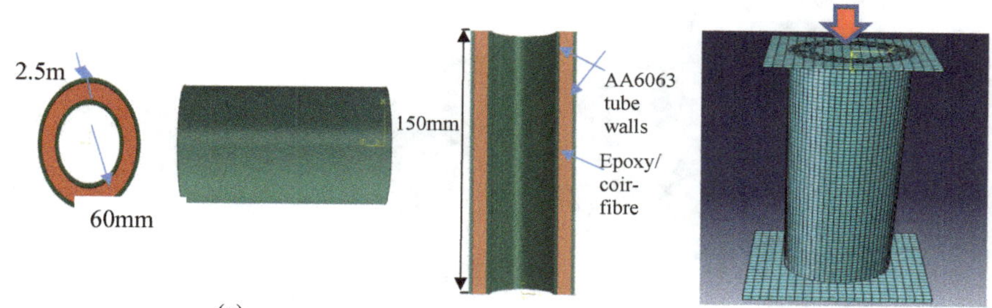

(a) (b)

Fig.2. (a) Geometry of aluminum/coir-fibre tubes (b) meshed geometry

3. Results and Discussions

3.1 Deformation profiles of composite tubes

The load-displacement curves in Fig. 3 describe the coupled elastic-plastic response based on the von Mises J_2 plasticity, and the damage initiation to evolution behavior of Hashin. The stress-strain plots suggest that a resistance to crush loads is much higher for the 20T samples over a wide range of displacement as compared with the 10T samples, and the variations from axisymmetric (high resistance to large deformation) to non-axisymmetric (low resistance to crushing) response is also less significant. Also, the period of dynamic progressive crushing is much longer which is an evidence of a strong interaction between the tube walls and reinforcements in preventing the formation of complete folds. On the other hand, the fluctuations in crushing loads is less significant for the 10T samples probably due to a low frequency of local buckling, but in general the period of progressive crushing is much limited and this could be attributed to accelerated sequence of complete fold formations. Concerning the 15T tubes, the onset of unstable buckling load response of the tube-wall probably led to non-uniform folds and plastic instability at the beginning of the crushing process based on the inherent fluctuations in the elastic region, nonetheless, the formation of inward and outward folds occurred over a much more elongated period of time as indicated by the smoother load fluctuations which is an evidence of a superior elastic-plastic deformation behavior under progressive crushing due to a stronger frictional resistance interaction between successive folds [1,3]. Also, at the densification stage where the extensional mode of deformation is dominant, there is an evidence of continuous in-plane deformation due to a rise in axial loads, hence by comparing the area under the load-displacement curves for the extensional mode of deformation, the 20T and 15T samples were found to possess the best capacities for energy absorption. Concerning the effects of reinforcement phase, optimal energy absorption performance lies between 40-70% volume concentrations for the 20T specimens, due to the fact that there were more gentle peaks with stable crest and troughs events in the 20T samples indicating a continuous inward and outward folding of the tubes with a potential mixed mode of failure delaying the transformation to global bending to indicate a better capacity for buckling resistance.

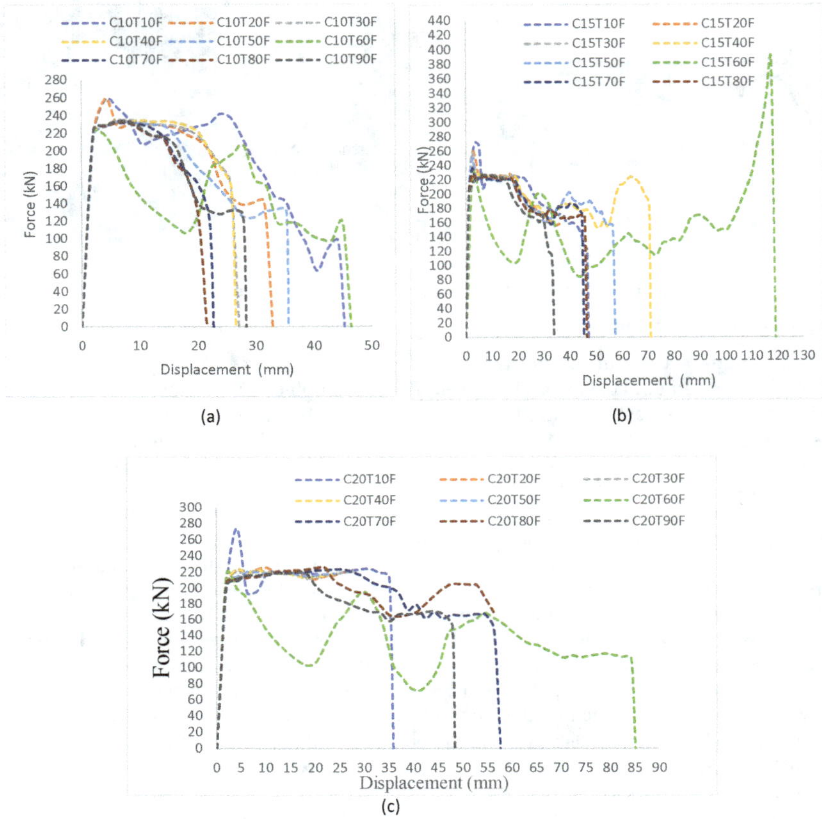

Fig.3. Graphical representations of the deformation profile of (a) 10T (b) 15T (c) 20T composite tubes

3.2 Deformation and damage modes of composite tubes

The results of the simulation of the entire structure of the composite incorporating the aluminum matrix, represented by the snapshots of the stress contour plots at varying displacements indicating the damage modes are presented for the damage history of the 10T, 15T and 20T tubes in Figs. 4-6 respectively. Generally, the mechanism of energy absorption is similar for all the configurations and it is characterized by a stable deformation profile and progressive folding effects. As the axial force is applied on the tube by the top plate, the stress wave pattern varies from top to bottom in a quasi-linear manner, and the tube elastically deform under this load until a maximum stress value is attained at the top of the tube. Beyond this point and with the continuous application of load, the tube undergoes plastic deformation and a fold at the top of the tube appears due to the effects of dynamic progressive buckling. From this point, a transformation in the deformation mode of the tube from progressive buckling to bending mode begins, this is accompanied by a decrease in the stiffness of the tube and successive folding of the tube. As the load decreases further due to plastic deformation, successive folds come together in intimate contact and start to collapse into other folds instantaneously in what could be described as a plastic buckling mode [2, 5]. It is observed that the fiber-reinforced tubes undergo a circumferential breakage due to the successive internal-external bending of the tube walls. The compressive forces introduce bending strains at the contact surface of impact causing matrix and fibre failures [1]. In correlation with Fig. 3 (a), (b) and (c). it is reasonable to suggest that the failure modes of the 20T tubes were more complex

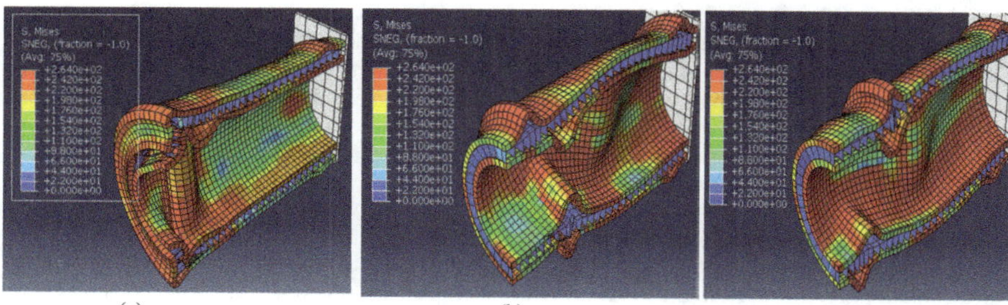

(a) (b) (c)

Fig.4. Deformation profile for the 10T composite tubes at (a) 10% (b) 60% (c) 90% volume fraction of coir-fibre reinforcements

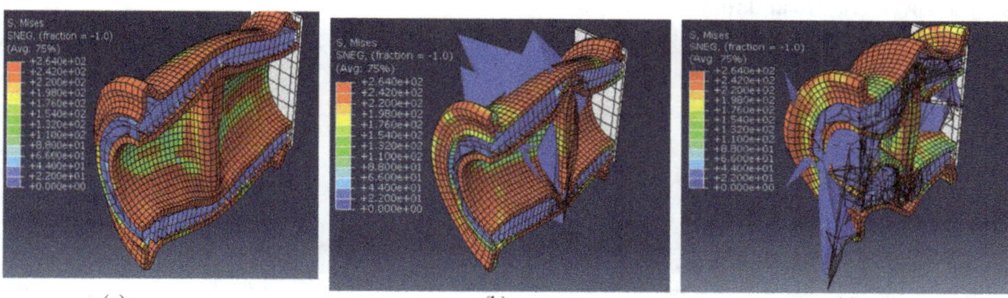

(a) (b) (c)

Fig.5. Deformation profile for the 15T composite tubes at (a) 10% (b) 60% (c) 90% volume fraction of coir-fibre reinforcements

(a) (b)

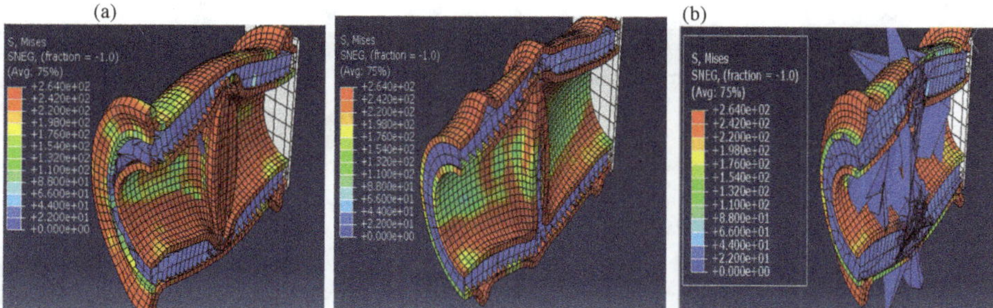

(c)

Fig.6. Deformation profile for the 20T composite tubes at (a) 10% (b) 60% (c) 90% volume fraction of coir-fibre reinforcements because the initial peaks were high due to high stiffness of the tubes, also a gradual reduction in peak forces after the plateau deformation region can be readily observed indicating a propensity for significant energy absorption through extensional collapse modes with progressive delamination damage. For the 10T samples the range of best performance lies within 30-60% samples by virtue of the progressive patterns of the associated curves, but for the 10-20% samples, the effect of stiffness degradation is responsible for their limited crushing distance to the collapse of the tube walls translating to an inferior energy absorption performance, while for the 80-90% reinforced samples, the effect of accumulated matrix damage indicated by their non-linear responses retarded their energy absorption performances at the self-contact stage. For the 15T samples, the onset of matrix damage is prominent for 10-30% as indicated by their non-linear response, but the range of best performance is between samples with 40%-70% volume fraction of reinforcement where associated family of curves indicated steady fluctuating patterns probably due to the effects of a more progressive crack propagation events occurring along the longitudinal axes of the tubes during compression that consequently introduced a transitional mode of deformation through progressive delamination damage. Lastly, concerning the response of the composite models at high volume fractions (80-90%), the onset of early densification can be attributed to the influence of the dense fibre content in accelerating matrix damage, causing a detachment of the fibres from the matrix, even though the capacity to absorb more loads was still retained [5]. Hence, for optimal design performance, it may be impracticable to incorporate high-volume fraction of coir-fibres by the hybridization concept.

3.3 Experimental validations

Quasi-static experiments were conducted on 10T, 15T and 20T tubes constituted with 60% coir-fibre content and the results as presented in Fig. 7 showed that the deformation profiles were

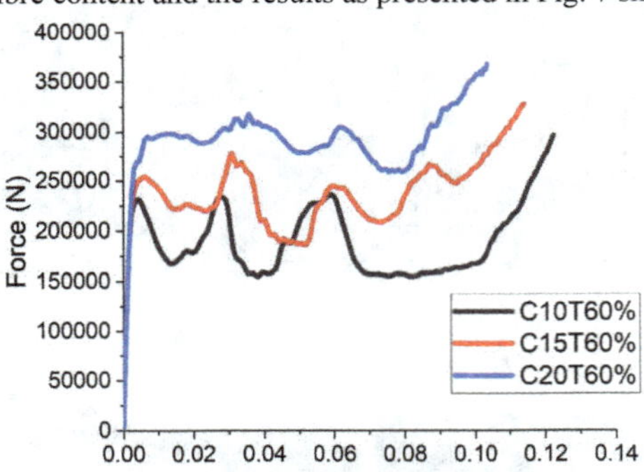

(a)

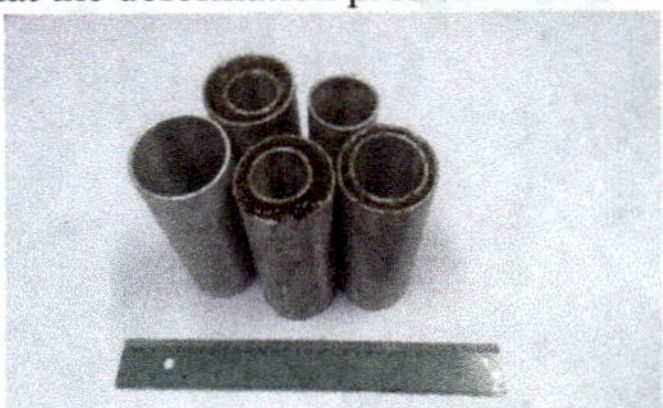

(b)

(c)

Fig.7. Experimental validation (a) stress-strain plots (b) composite tubes (c) cross-section of tube

consistent with simulated performance indicating a relatively wider range of progressive crushing of the 20T reinforced tubes to underscore its superior energy absorption capacity.

Conclusions

1. The mechanism of deformation and damage was similar for the tubes at 10-30% reinforcement concentration characterized by circumferential fiber breakage and matrix damage while at 80-90%, unsteady non-linear response by global bending degraded the tubes' resistance to buckling. The elongated plateau regions of the tubes at 40%-70% indicate the range of best performance.
2. The resistance to formation of complete folds increased significantly with the evolution of reinforcement thickness where-in the 20T tubes indicated complex profiles after initial peak stresses that constrained axisymmetric to global bending transformations.
3. The 20T tubes were found to demonstrate the best energy absorption performance due to a superior resistance to buckling occasioned by a stable dynamic progression during crushing and by controlled delamination damage under extensional collapse during densification.
4. The results on the performance of the tubes were consistent with experimental observations.

References

[1] Weirzbicki, T and Abramowicz, W (1983). On the crushing mechanism of thin-walled structures. Journal of applied mechanics, 50 (4), 727-734. https://doi.org/10.1115/1.3167137

[2] Wang, Z., Jin, X., Li, Q and Sun, G (2020). On crashworthiness design of hybrid metal-composite structures, Int. J. of Mechanical Sciences, 171, 105380, Elsevier Science Ltd. https://doi.org/10.1016/j.ijmecsci.2019.105380

[3] Cui, Z., Liu, Q., Sun,Y and Li, Q (2020). On crushing responses of filament winding CFRP/aluminum and GFRP/CFRP/aluminum hybrid structures, Composites Part B: Engineering, 200, 108341, Elsevier Science Ltd. https://doi.org/10.1016/j.compositesb.2020.108341

[4] Paz, J., Diaz, J., Romera, L and Costas, M (2014). Crushing analysis and multi-objective crashworthiness optimization of GFRP honeycomb-filled energy absorption devices, Finite Elements in Analysis and Design 91, 30-39, Elsevier Science Ltd. https://doi.org/10.1016/j.finel.2014.07.006

[5] Jung, D., Kim, H and Choi, N (2009). Aluminum-GFRP hybrid square tube beam reinforced by a thin composite skin layer, Composites Part A: Applied Science and Manufacturing, 40 (10), 1558-1565, Elsevier Science Ltd. https://doi.org/10.1016/j.compositesa.2009.06.015

[6] Farley, G (2016). Effect of specimen geometry on the energy absorption capability of composite materials, Journal of Composite Materials, 20 (4), Sage Journals. https://doi.org/10.1177/002199838602000406

[7] Islam, M; Hasbullah, N; Hasan, M; Tahb, Z; Jawaid, M and Haafiz, M (2015). Physical, mechanical and bio-degradable properties of kenaf/coin hybrid fiber reinforced polymer nanocomposites. Materials Today Communication, 4, 69-76. https://doi.org/10.1016/j.mtcomm.2015.05.001

[8] Jawaid, M; Thariq, M and Saba, N (2019). Mechanical and physical testing of bio-composites, fiber-reinforced composite and hybrid composites wood herd publishing, Elsevier, 43-57.

[9] Benveniste, Y (1987). A new approach to the application of Mori-Tanaka's theory in composite materials. Mechanics of materials 6 (2): 147-157. Elsevier Science Ltd. https://doi.org/10.1016/0167-6636(87)90005-6

Stabilization of a swelling soil using three different chemical additives

Tahar AYADAT [1,a*], Ammar NOUI[2,b], Layal JRADI[1,c], Danish AHMED[1,d], Muhammad AJMAL[3,e], Andi ASIZ[1,f]

[1] Prince Mohammad Bin Fahd University, Saudi Arabia

[2] Mohamed Al Bachir El Ibrahimi University, Algeria

[3] University of Bahrain, Bahrain

[a]tayadat@pmu.edu.sa, [b]ammar.noui@univ-bba.dz, [c]lseifeldine@pmu.edu.sa, [d]dahmed@pmu.edu.sa, [e]majmal@uob.edu.bh, [f]aasiz@pmu.edu.sa

Keywords: Expansive Soil, Stabilizing, Nano-Silica, White Cement, Waste Granite, Liquid Granite

Abstract. An expansive soil expands when water is added, and shrinks when it dries out. On construction sites, expansive soils can pose geotechnical engineering challenges. More financial loss is caused by expansive soil damage than by floods, hurricanes, tornadoes, and earthquakes combined. The aim of this study is to analyze the effect of stabilization of an expansive soil using three different chemical additives, in this case: white cement and nano-silica as admixture, waste granite dust, and a new liquid polymer soil stabilizer (i.e. liquid granite). This study investigated the effect of these stabilizing agents on the variation of soil consistency, linear shrinkage, and soil swelling behaviour. Three groups of tests were carried out as part of the experimental program. As part of the first group, tests were performed on expansive soil reinforced with 2%, 4%, 8%, and 10% of white cement combined with 2% of nano-silica. Another group of tests consisted of samples mixed with different percentages of granite powder (5%, 10%, 15%, and 20%). In the third group, liquid granite as a soil stabilizer was blended with soil specimens at various concentrations (1%, 2%, 3%, and 4%). For all admixtures, the results showed a marked improvement in soil consistency, an important decrease in linear shrinkage, and a considerable reduction on the expansion index. Compared with the original bentonite clay, the swelling behaviour (expansion index) was reduced considerably by adding these three different chemical additives. Moreover, a correlation between the expansion index and the dosage of the different stabilizers was proposed.

Introduction

Expanding soils can shrink or swell when their environment changes and moisture migrates. A clay or soil that expands or contracts under changing moisture conditions is known as expansive soil. A worldwide problem, swelling soil causes more damage to structures, particularly light buildings and pavements, than earthquakes and floods combined. Soil swelling is caused by the expansion of clay materials in its composition, a process that is induced by a variety of different phenomena, including the elastic rebound of soil grains, the attraction of clay minerals to water, the electrical repulsion between clay particles and their cations absorbed from one another, and the expansion of air trapped in soil voids. Many factors influence the amount of expansion in a soil. Mineralogical composition of soil particles, soil suction, soil structure and fabric, initial soil conditions, and thickness of expansive soil layer are the most important factors. It is more common to find expansive soils in arid and semi-arid climates. It is important to note, however, that expansive clay activities are most likely to occur in areas where evaporation or evapotranspiration rates are very high, rainfall is sufficient to wet the ground thoroughly to a depth of at least (76 cm), and the weather is often dry and then wet. Expansive soil covers vast areas of the Kingdom of

Saudi Arabia. This soil type causes many cracks and disorders in buildings, as well as large deformations and protrusions in roads and sidewalks. In some regions of Saudi Arabia, this resulted in huge losses that cost millions of riyals [1].

These types of soils require different types and amounts of treatment depending on the depth of the expansive soil layer and the structural support requirements. This problem was treated using various methods including prewetting by ponding, silt slurry injection, grouting, prewetting, electro-osmosis, heat treatment, and chemical additives. It is often practical and economical to remove the expansive soil layer and replace it with a compacted layer of suitable soil if it is superficial (less than 5 m). It is recommended that soil exchange materials be coarse, inorganic, and require moderate compacting effort. However, this method requires the availability of the borrow material near the project, as well as a large amount of water (corresponding to the optimum water content of the soil exchange) and adequate compacting efforts. Consequently, these conditions increase the project's cost. Additionally, in some cases, some difficulties are recorded due to the lack of water in arid and semi-arid areas.

It is more common today to use chemicals for soil stabilization to increase soil strength parameters and bearing capacity as well as reducing settlement [2,3,4,5]. It is convenient and low-cost improvement method, especially for geotechnical projects that require a significant amount of soil improvement [6]. A chemical improvement is a way of improving subgrade, sub-base layers, and any other unsatisfactory or poor materials in place. It is believed that the more suitable and low-cost method for treating superficial expansive layer consists on excavating the expansive soil layer and then re-mixes the soil with additives composed mainly by chemical agents. The treated expansive soil is then replaced in place in layers of 30 cm thick by moderate compaction without low to moderate water content.

The objectives of this experimental investigation are to analyze the effect of stabilization of an expansive soil using three different chemical additives, namely nano-silica and white cement as admixture, waste granite dust, and a new liquid polymer soil stabilizer (i.e. liquid granite). This study investigated the effect of these stabilizing agents on the variation of soil consistency, linear shrinkage, and soil swelling behaviour.

Materials and Test Methods

For analyzing the effect of stabilization of an expansive soil using chemical additives, an experimental program was carried out using the consolidation test, the motorized Liquid Limit testing apparatus, a mold for measuring linear shrinkage.

Materials:

The Materials used in this study are: i)- Commercial bentonite commonly used as component for compacted clay liner applications, ii)- Nano-silica also called quartz dust or silica dust, iii)- White Portland cement with a compressive strength after 28 days of 4 MPa and a fineness of 395 m^2/kg), iv)- Granite waste material produced from granite cutting in the industry of granite polishing, and v) Liquid granite which is a self-contained, pre-mixed cement, blended with specially selected aggregates. The physical and chemical characteristics of the bentonite are grouped in Table 1.

Table 1. *Chemical and Physical Characteristics of Bentonite*

Chemical composition		**Physical Characteristics**	
SiO_2 (%)	58	Specific gravity	2.6
Al_2O_3 (%)	19	Liquid limit (%)	187
MgO (%)	1	Plastic limit (%)	47
Fe_2O_3 (%)	3	Plasticity index (%)	140
CaO (%)	2	Optimum moisture content (OMC) (%)	25
Na_2O_3 (%)	4	Maximum dry unit weight (γ_{dmax}, kN/m^3)	16
Loss on ignition (%)	13	Expansive index, *EI* (%)	117.8

Test Method:
An experimental program based on standard geotechnical tests was designed to investigate the effects of stabilizing expansive bentonite clay with chemical agents. Studying the interaction between expansive soil and material amendments is the main focus of the study. A total of 26 soil stabilizer batches were prepared for the performance evaluation of different stabilizers and stabilizer combinations in treating expansive soils. Stabilizers such as white cement and granite powder were added to air-dried bentonite soil by weight, and specimens were mixed at 12.5% and 25% of moisture content (corresponding to 50% and 100% of the optimum moisture content of the bentonite). Whereas the specimens of bentonite mixed with liquid granite were prepared at low moisture content. The quantity of water added to the mixture was 0.08 ml/g of stabilizer, equivalent to 0.8 liters per 10 kg of stabilizer [7]. Inclusion rates of white cement were 2%, 4%, 8%, and 10% combined to 2% of Nano-silica. This is because the Ordinary Portland Cement requirement is higher for stabilizing expansive soil [8]. Liquid granite addition rates were kept lower (1%, 2%, 3% and 4%) because of self-cementing capabilities. Moreover, granite powder was used exclusively in stabilizing soils at 5%, 10%, 15% and 20% addition rates.

Index Properties Tests
As per ASTM standards, Bentonite clay and Bentonite mixed with different chemical additives were tested for their index properties (liquid limit, plastic limit, and linear shrinkage). Tests were conducted using different assays of additives to examine the influence of the stabilizer on the expansive bentonite clay consistency (i.e. limit of consistency). Distilled water was used to determine these limits for the different mixtures previously listed. The liquid limit (*LL*) is defined as the moisture content, in percent, required to close a distance of 12.7 mm along the bottom of a groove after 25 blows in a liquid limit device (ASTM D 4318-00). In general, high liquid limits indicate a high compressibility and shrinkage/swelling potential. A high-plasticity index *IP* generally results in a low shear strength. The plastic limit (*PL*) test was conducted according to ASTM D 4318-10 standards. Linear shrinkage test was conducted by using a fabricated mold with 139.7 mm in length, 25.4 mm in diameter and 12.7 mm in depth [9]. The linear shrinkage value was calculated using the following equation:

$$LS = \frac{L_s}{L} \times 100\% \tag{1}$$

Where:
LS = linear shrinkage in percent,
L_s = linear shrinkage length after oven dried,
L = length of the sample (i.e. the mold).

Free Swell Tests

Laboratory tests to measure the magnitude of one-dimensional wetting-induced free swell (i.e. the expansion index, *EI*) of unsaturated reconstituted soils were conducted by simple oedometer test apparatus according to ASTM D4546-90, Method A.

Unstabilized or stabilized bentonite clay was compacted into a standard mold with an internal diameter of 10.2 cm using two layers (15 blows per layer with a 2.5 kg tamper). For the white cement and waste granite stabilizers, two different moisture contents were considered 12.5% and 25% equivalent to 50 and 100% the optimum moisture content of bentonite clay (as determined by ASTM D-1557). Then a specimen was trimmed (5.0 cm diameter and 2.0 cm thickness). The test specimen was mounted in the oedometer apparatus with porous plates on the top and bottom of the specimen. Under a pressure of 7 KPa, the soil sample was submerged with distilled water, and the vertical displacement reading was recorded. After the specimen was swelled, the expansion index (*EI*) was estimated as:

$$EI = \frac{\Delta H}{H_o} \quad (2)$$

Where:

ΔH = change in the specimen height due to inundation,
H_o = initial height of the soil specimen.

The expansion index, *EI*, provides an indication of swelling potential of a soil as indicated in Table 2 (ASTM D-4829).

Table 2. *Expansion index, EI, guiding values (ASTM D-4829)*

EI	**Expansion Potential**
0 – 20	Very low
21 – 50	Low
51 – 90	Medium
91 – 130	High
> 130	Very high

Test Results

As mentioned previously, the aim of this study is to analyze the effect of stabilization of bentonite clay using three different chemical additives, including: nano-silica and white cement mixture, waste granite dust, and a new liquid polymer soil stabilizer (i.e. liquid granite).

Index Properties Test Results:

Liquid limit (*LL*) and plastic limit (*PL*) tests were conducted to determine how the chemical additives used affect the moisture content of stabilized soils. Table 3 indicates the Atterberg limits (*LL* and *PL*) of bentonite clay after adding various dosages of chemical additives. A significant decrease in liquid limit value was observed with the addition of chemical agents to bentonite clay, which had an original liquid limit value of 187%. With the addition of nano-silica and white cement to bentonite clay, the liquid limit value changed from 179.2 to 108.5%. Moreover, when granite powder was added, the liquid limit value changed from 174.1 to 89.7%. This is because waste granite is non-cohesive in nature and slightly coarse-grained compared to bentonite. But in case of liquid granite and bentonite clay mixture the change in liquid limit were from 182.7 to 136.1%. Liquid granite had good water holding capacity and chemicals present in liquid granite play a significant role, so the change in liquid limit was slightly less than the other two chemical stabilizers.

A further observation from Table 3 is that the original bentonite clay had a plastic limit of 47.5%, but with the addition of white cement and non-silica that plastic limit value decreased to 26.1%; however, the reduction was 24.6% and 30.2%, respectively, when granite powder and liquid granite were added. Using the maximum dosage of granite powder, the plasticity index (*PI*) of bentonite clay dropped to a minimum level. When compared to the untreated specimen, the *PI* was reduced by 53%. Accordingly, the increase of chemical additives in bentonite clay helped to decrease the *PI* value to a certain extent.

Linear shrinkage test was conducted to find out the percentage change in length when the soil sample shrinks linearly. The linear shrinkage percentage of bentonite is 40%, indicating a very high linear shrinkage potential. Table 3 shows that linear shrinkage decreases with increasing chemical agent dosage. In contrast to liquid granite, granite powder and white cement have a greater percentage decrease since these materials are pozzolanic in nature. In case of bentonite mixed with white cement and nano-silica, percentage change was between 35.4 and 20.2%; whereas, in bentonite granite powder mixes, it was between 30.7 and 19.3%. Likewise, bentonite mixed with liquid granite showed a percentage change between 37.2 and 24.5%. There was a reduction in linear shrinkage properties when the different chemical additives were added, which could be explained by changes in moisture content, drying process, surface phenomena, interaction between clay particles, and soil unit weight. Previous studies also reported a decrease in *LL* and *PI* values of expansive soils due to chemical treatments [9,10,11,12]. It was speculated that the stabilizers might have decreased the *LL* and *PI* values, as they might have reduced the diffuse double layer and caused clay particles to flocculate [13].

Table 3. Index Properties for Bentonite and Mixtures

<table>
<tr><th colspan="2" rowspan="2">Mixture</th><th rowspan="2">Dosage (%)</th><th colspan="3">Liquid limit and Plastic limit</th><th>Linear Shrinkage</th></tr>
<tr><th>LL (%)</th><th>PL (%)</th><th>PI (%)</th><th>LS (%)</th></tr>
<tr><td colspan="2">Bentonite</td><td>-</td><td>187</td><td>47</td><td>140</td><td>40</td></tr>
<tr><td rowspan="12">Chemical Additives</td><td rowspan="4">White Cement & 2% Nano-Silica</td><td>2</td><td>179.2</td><td>42.6</td><td>136.6</td><td>35.4</td></tr>
<tr><td>4</td><td>162.3</td><td>36.3</td><td>126.0</td><td>29.2</td></tr>
<tr><td>6</td><td>132.6</td><td>31.4</td><td>101.2</td><td>24.8</td></tr>
<tr><td>8</td><td>108.5</td><td>26.1</td><td>82.4</td><td>20.2</td></tr>
<tr><td rowspan="4">Granite Powder</td><td>5</td><td>174.1</td><td>40.4</td><td>133.7</td><td>30.7</td></tr>
<tr><td>10</td><td>152.4</td><td>34.5</td><td>117.9</td><td>26.8</td></tr>
<tr><td>15</td><td>121.2</td><td>28.3</td><td>92.9</td><td>23.5</td></tr>
<tr><td>20</td><td>89.7</td><td>24.6</td><td>65,1</td><td>19.3</td></tr>
<tr><td rowspan="4">Liquid Granite</td><td>1</td><td>182.7</td><td>44.1</td><td>138.6</td><td>37.2</td></tr>
<tr><td>2</td><td>175.8</td><td>39.7</td><td>136.1</td><td>33.6</td></tr>
<tr><td>3</td><td>166.2</td><td>34.6</td><td>131.6</td><td>29.7</td></tr>
<tr><td>4</td><td>136.1</td><td>30.2</td><td>105.9</td><td>24.5</td></tr>
</table>

Free Swell Test Results:

Bentonite clay used for this experiment program is highly expansive in nature. From Table 2, it can be seen that the expansive index for pure bentonite soil is 117.8, which indicates that it is highly expansive. As defined by ASTM D 4829 standard, soils with expansive indexes between 91 and 130 are considered highly expansive soils. For water content corresponding to 50% and 100% of optimum moisture content, bentonite mixed with 20% of granite powder reduced the expansion index (*EI*) by 60% and 48%, respectively. As a result of the pozzolanic effect of cement, *EI* was reduced by 71% and 59% for 10% of white cement and 2% nano-silica. In contrast, when 4% of liquid granite was used, the expansion index was reduced by 87%. Liquid granite dosages greater than 4% are suspected to result in a higher reduction of *EI*. Based on the curves shown in Fig. 1, Fig. 2 and Fig. 3, the relationship between expansion index and the dosage of the different stabilizers can be presented by the following equation:

$$EI = EI_0 \times e^{-b} \tag{3}$$

Where:

EI_0 = Expansion index of original expansive clay

b = constant which depend on the type of the stabilizer and the compacting water content (for granite powder $b = 0.03$ for 50% OMC and 0.04 for 100% OMC. For white cement $b = 0.08$ for 50% OMC and 0.11 for 100% OMC. Whereas for the case of liquid granite $b = 0.46$).

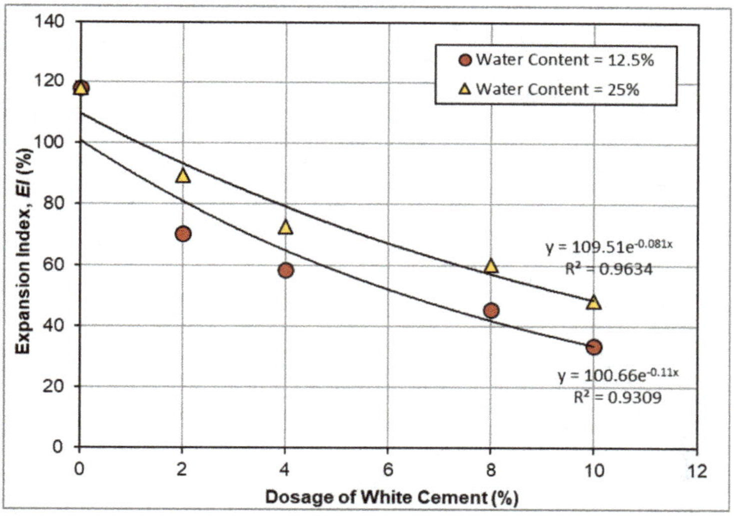

Figure 1. *Variation of Expansion Index with the dosage of white cement and nano-silica*

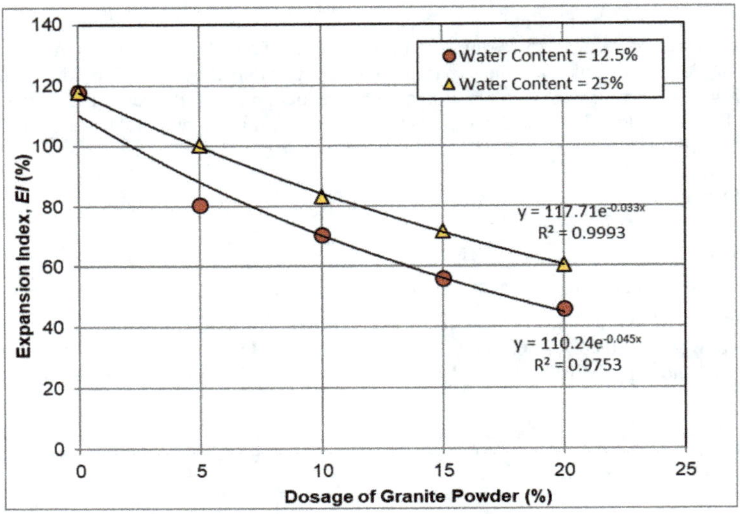

Figure 2. *Variation of Expansion Index with the dosage of granite powder*

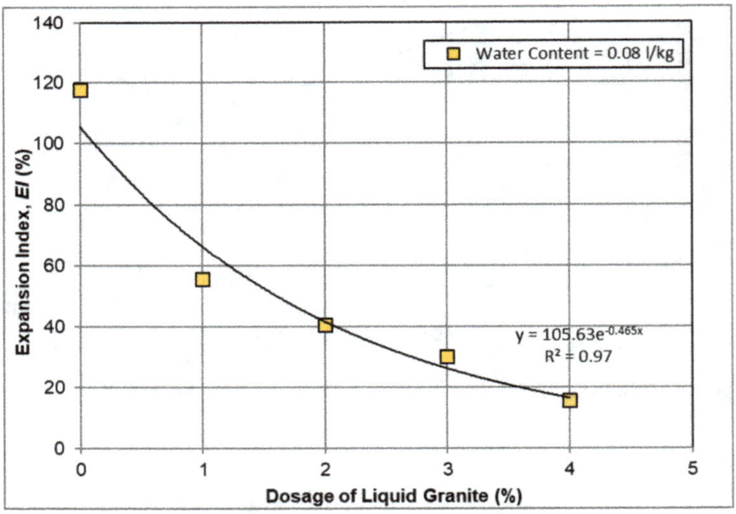

Figure 3. *Variation of Expansion Index with the dosage of liquid granite*

Conclusion

An experimental program was carried out to investigate the effect of three stabilizing additives on the variation of soil consistency, linear shrinkage, and soil swelling behaviour of bentonite clay. The chemical agents used were nano-silica and white cement as admixture, waste granite dust, and a new liquid polymer soil stabilizer (namely liquid granite). A significant decrease in the liquid limit value was observed with the addition of these chemical agents to the bentonite clay, which had an original liquid limit value of 187%. Furthermore, the increase of the dosage of the chemical additives in bentonite clay helped to decrease the PI values to a certain extent. When compared to the untreated specimen, the PI was reduced by approximately 53%. The linear shrinkage decreased with increasing chemical agents' dosage. In contrast to liquid granite, granite powder and white cement have a greater percentage decrease since these materials are pozzolanic in nature. Bentonite mixed with 20% of granite powder reduced the expansion index (*EI*) by 60%. As a result of the pozzolanic effect of cement, *EI* was reduced by about 71% for 10% of white cement and 2% nano-silica. However, when 4% of liquid granite was used, the expansion index was reduced by 87%. It is believed that for dosages greater than 4% of liquid granite, higher reduction values of *EI* will be observed. Moreover, a correlation between the expansion index and the dosage of the different stabilizers was proposed.

References

[1] Al-Mhaidib, A.I. (2002). Characteristics of Expansive Soil in the Kingdom of Saudi Arabia, King Saud University, https://www.researchgate.net/publication/273443379.

[2] Ashraf, M.A., Rahman, S.S., Faruk, M.O., Bashar, M.A. Determination of optimum cement content for stabilization of soft soil and durability analysis of soil stabilized with cement. Am. J. Civ. Eng. 6 (1), 39. (2018). https://doi.org/10.11648/j.ajce.20180601.17

[3] Begum, A.S., Raju, G.P., Prasad, D.S.V., Kumar, M.A. Influence of TerraZyme on compaction and consolidation properties of expansive soil. In: Problematic Soils and Geo-environmental Concerns. Springer, pp. 525-535. (2021). https://doi.org/10.1007/978-981-15-6237-2_44

[4] Pooni, J., Robert, D., Giustozzi, F., Setunge, S., Venkatesan, S. Stabilization of expansive soils subjected to moisture fluctuations in unsealed road pavements. Int. J. Pavement Eng. 23 (3). (2022). https://doi.org/10.1080/10298436.2020.1762083

[5] Barman, D. And Dash, S.K. Stabilization of expansive soils using chemical additives: A review, Journal of Rock Mechanics and Geotechnical Engineering, Volume 14, Issue 4, pp. 1319-1342. (2022). https://doi.org/10.1016/j.jrmge.2022.02.011

[6] Ramaji, A.E. A Review on the Soil Stabilization Using Low-Cost Methods. Journal of Applied Sciences Research 8(4):2193-2196. (2012).

[7] IMG. Liquid Granite: Indispensable, Multi-Purpose, Emergency Repair Material, Tero Tech, TS 25.09.19. (2019).

[8] PCA (Portland Cement Association). Soil-cement laboratory handbook. Skokie, IL: PCA. (1992).

[9] Kolay P.K. and Ramesh, K.C. Reduction of Expansive Index, Swelling and Compression Behavior of Kaolinite and Bentonite Clay with Sand and Class C Fly Ash, February 2016, Geotechnical and Geological Engineering, (2016). https://doi.org/10.1007/s10706-015-9930-4

[10] Ruwaih, I. A. Experiences with Expansive Soils in Saudi Arabia, Proceedings of Sixth International Conference on Expansive Soils, New Delhi, India, pp. 317-322. (1987).

[11] Bhattacharja, S. and Bhatty J. I. Comparative performance of Portland cement and lime stabilization of moderate to high plasticity clay soils. Skokie, IL: Portland Cement Association. (2003).

[12] Li, S., White, D. and Vennapusa, P. Cement stabilization of embankment materials. Ames, IA: Iowa State Univ. Digital Repository. (2015).

[13] Beetham, P., T. Dijkstra, N. Dixon, P. Fleming, Hutchison, R. and Bateman, J. Lime stabilization for earthworks: A UK perspective. Proc. Inst. Civ. Eng. Ground Improv. 168 (2): 81-95. (2015). https://doi.org/10.1680/grim.13.00030

One dimensional oedometer laboratory testing for expansive clay submerged with hydrocarbon fluids

Mousa Bani Baker[1,a,*], Adel Hanna[2,b], Maria Elektorowicz[2,c], Tahar Ayadat[3,d], Batool Al-Shorman[4,e] Raed Abendeh[1,f]

[1] Department of Civil and Infrastructure Engineering, Al-Zaytoonah University of Jordan, Amman, Jordan

[2] Department of Building, Civil and Environmental Engineering, Concordia University, Montreal, Canada

[3] Department of Civil Engineering, Prince Mohammad Bin Fahd University, Al-Khobar, Saudi Arabia.

[4] Department of Civil Engineering, Jordan University of Science and Technology, Irbid, Jordan

[a]m.banibaker@zuj.edu.jo, [b]hanna@civil.concordia.ca, [c]mariae@civil.concordia.ca, [d]tayadat@pmu.edu.sa, [e]byalsharman17@eng.just.edu.jo, [f]r.abendeh@zuj.edu.jo

Keywords: Hydrocarbon Liquids, Expansion, Fuels, Bentonite Clay, Swelling

Abstract. Landfills are currently one of the most effective ways to dispose of waste. Underground storage tanks (UGST) are also used to store hydrocarbon fluids that include different types of fuels. The bottom part of the landfills and UGST is critical. This liner material and its composition prevent heavy metals and leachate from infiltrating the groundwater table. Failure of this layer presumably causes most landfill failures. Bentonite clay is used to build such liner because of its properties including high cation exchange capacity and swelling index. The swelling of bentonite is sensitive to the type of liquid and load. It swells under low loads when submerged with water and to a lesser extent for ethanol. However, it undergoes consolidation when penetrated by biofuel. Test results indicate that bentonite undergoes swelling in water under high load (40 kPa) and consolidates for both alternative fuels (biofuel and ethanol). Under very high loads (100 kPa) bentonite consolidates for all kinds of liquids including water.

1. Introduction

Environmental engineering investigates hazardous waste contaminants, their pathways, transport, fates and disposition. It also explores ways of protecting groundwater, thereby protecting humans and the environment from hazardous wastes. Petroleum products, for instance, account for sixty-nine percent of soil contamination in Quebec [1].

Many liners are made watertight with clayey materials, such as sand-bentonite mixtures, that retain liquid and solid toxic wastes. Although many cases were reported for high leakage of bentonite liners [2], they remain in use. Over time, the principal function of liners is reduced and contaminants leak through them. As these toxic contaminants infiltrate the subsoil and the groundwater, they have serious ramifications on the stability of constructions and the safety of humans and animals.

Failure of landfill liners is a problematic issue for engineers and a costly one for governments, societies and the environment. To protect groundwater, the structure of a natural landfill liner needs to preserve its properties in harsh conditions for a sustained period. In addition, the type of disposal materials in landfills and storage tanks is changing and it is starting to include alternative fuels such as biofuel or ethanol fuel.

Containment barriers commonly use clays for their low permeability as liners in landfills and settling ponds. The properties of the fluid passing through liner material affect the liner's structure.

Thus, in the presence of alternative fuels, the dispersed orientation of clay particles alters the clay's permeability. The percolation of fluids might result in changes in the microstructure of clay fabrics. Leaking fluids destroy clay barriers, even though the liners are designed to prevent these infiltrations. In practice, when subjected to organic liquids, the clay liner leaks and destroys the clay microstructure, thereby increasing the permeability of the liner. [3] conducted a test on clayey soil permeated by liquid hydrocarbon. They reported a significant increase in hydraulic conductivity from 10-8 cm/s to 10-4 cm/s. [4] conducted research using a batch test to determine the adsorption coefficient of benzene, toluene, and 2-fluorotoluene on three soils. They found that hydrophobicity is an important factor in the relative adsorption of benzene and toluene in the materials investigated. They also supported the idea of incorporating organophilic clay to sand-bentonite mixtures to prevent compacted liner composed of these materials from desiccating when exposed to diesel fuel.

[5] investigated the correlation between permeability; microstructure and surface chemistry of geosynthetic clay liner (GCL) interacted with leachate. They concluded that when (GCL) containing sodium bentonite is in contact with fluids containing other cations, the latter exchanges with the sodium present between clay layers. This modification in clay surface chemistry changes the clay microstructure, therefore, changing the hydraulic conductivity. They also suggested that the cation selectivity of clay surface exchange sites depends also on the swelling and site occupation history of the clay. [6] conducted tests to evaluate how ammonium attenuates municipal solid waste (MSW) landfill leachate by adsorption into bentonite in a landfill liner. The bentonite adsorption capacity was measured in two types of laboratory tests. The first was a standardized test ASTM D4646 (1987) using bentonite in the dust state. The second test used compacted bentonite to reproduce actual landfill bottom conditions. Parameters were monitored such as; ammonium, heavy metals (Cd, Cr, Cu, Fe, Mn, Ni, Pb, and Zn) and COD. Based on risk simulations, they concluded that the maximum capacity of adsorption in traditional landfill liners is exhausted in a few years (2-10 years). Thereafter, chemicals pass through the liner at the concentration found in the leachate.

As the chemical concentration increase, the Diffuse Double layer (DDL) shrinks in bentonite clay, then the bentonite clay swelling decreases as a result of flocculation of the clay particles. Some of the clay tests show that there are volume changes of bentonite clay when exposed to the NaCl, CaCl or KCl solutions and then reexposed it to the water, large volume decrease was investigated as exposure the specimens to the saturated solutions [7]. It is worth mentioning that there are some numerical modeling recently used to study the swelling of clay layers such as the model of Alonso which was proposed by [8] which showed a great economic value to study the parameters of soil's swelling.

Brazilian bentonite clay was prepared to utilize quaternary ammonium salt by [9] to study its effect on the orientation of petroleum-derived fuels, the results of swelling tests indicate increasing in basal spacing and formation of absorption bands associated with the CH2 and CH3 groups. Some techniques can be used to improve the role of the bentonite clay layer, one of them is increasing the adsorption capacity of the bentonite clay, [10] studied the effect of organic modification on the adsorption behavior of bentonite and found that the adsorption capacity increases with increase in the concentration of the organic contaminant up to 500 mg/l for the modified clays. The Functionalized Bentonite clay-sPEEK was used as composite membranes for direct methanol fuel cells and found that these composite membranes exhibit lower methanol permeability in comparison to pristine membrane. [11] found that there is a long term effect of water chemistry (Distilled water and synthetic water) on the swelling pressure of a bentonite-based material experimentally, their research study results show that the swelling pressure decrease for the samples specially for the sample saturated with synthetic water. The hydraulic performance of organo clay enhanced sand bentonite as the secondary liner was investigated by [12] and the results indicate that the permeability coefficient was assessed both for loading and unloading cases. A decrease in the coefficient of permeability with diesel by an order of two was observed on increasing the organo clay content from 0 to 10% in SB mix, whereas a reverse trend was noted with water as permeant. The adsorption of ionic liquids can also change the aggregation behaviors of the bentonite particles by which the "house of cards" structures in the bentonite scuttled are improved or destroyed [13]. The compacted bentonite swelling behavior can be enhanced due to the aging effect, the axial strain and the swelling pressure decreased as the aging time increased for samples prepared at both dry and wet sides [14]. The technological voids also affect the compacted bentonite by decreasing the swelling pressure anisotropy with its increase [15].

[16] studied the volume changes of bentonite which was exposed to NaCl, KCl or CaCl solutions and re-exposed to water. Presentation of specimens to saturated solutions resulted in great volume decreases. [17] investigated the effect of water chemistry on the swelling pressure and found that there is no clear effect of it in the short term (100h) due to the high dry density of the bentonite. On the other side, considering the long term period (700 days) the swelling pressure decreases with a non-negligible ratio (9%). Other studies such as [18], which are concerned with exploring the permeability variation of the Sabkha soil during the distilled water leaching and found that there is a direct relation between the rate of ion dissolution and soil permeability. The raise of soil permeability is because of the dissolution of Cl^{-1}, $SO4^{-2}$ and Na^{+1} ions into the Sabkha soil layer.

2. Materials

2.1 Bentonite clay

Montmorillonite clay (bentonite) was used which is some commercial powdered rich in Na-montmorillonite. It was used in this study for its high swelling capacity, high cation exchange capacity (CEC) and high surface area [19]. The isomorphous substitution occurs mainly in the alumina sheet of the montmorillonite minerals, with magnesium or iron substituting for aluminium in the dioctahedral minerals. Since there are no potassium ions to bond the layers together, therefore, water can enter easily between layers causing swelling.

The used bentonite possesses the following properties: Liquid limit of 440%, Plastic limit of 65%, Shrinkage limit of 8.2%, Free swelling index is 487.5%, the optimum moisture content is 25.6%, CEC (meq/100g) equals to 95, pH in water is 9.96, size passing sieve number 325 (D=44 micrometre), the specific surface area is 600 m^2/g. The mineralogical composition of natural bentonite (X-Ray Analysis) is given in Table 1.

Table 1. *Mineralogical composition of natural bentonite (X-Ray Analysis)*

Element	Percentage (by weight)
Montmorillonite	85
Quartz	5
Feldspars	2
Cristobalite	0.35
Illite	2
Calcium and Gypsum	1
LOI	4.65

2.2 Liquids

Water was used as the permeate reference, ethanol (100%), and biofuel was used as permeates with dynamic viscosities of 0.89, 1.071 and 61 cP respectively, as measured in the laboratory at room temperature.

3. Methodology

Bentonite is the material that undergoes swelling. It is, therefore, responsible for the swelling of the liner composition when permeated with different liquids. The one-dimensional free swelling test was conducted as per the ASTM D 2435-90 using the fixed-ring oedometer test as shown in figure 1. After preparing the compacted bentonite at the optimum moisture content of 25.5% by weight using the proctor, undisturbed samples were taken from the compacted bentonite using consolidation ring cells with sharp edges from one side, a 60 mm diameter and a 20 mm height. Vertical confining pressures of 7 kPa, 40 kPa, and 100 kPa were applied on the specimen with different liquids (water, ethanol, and biofuel). Three samples were tested using the same kind of liquid and load (a total of 27 samples were prepared).

At first, the load was applied on the samples until consolidation ceased and the consecutive readings were close. Then, the liquid was added to the cell. Deformations started in the sample and displacements were obtained using a linear variable differential transformer (LVDT). It was connected to the computer and three cells to perform three simultaneous tests using the same pressure and the same liquid in each set of samples. Each time, new samples were prepared and tested using the desired liquid. Results were compiled in an Excel datasheet and an average of the three readings was calculated accordingly. The percent swell was defined as the ratio of a percent increase in the thickness to the actual thickness of the specimen.

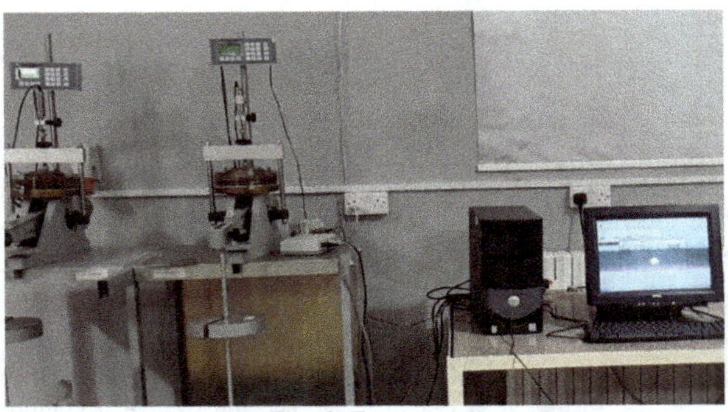

Fig. 1. Test setup.

4. Test Results

Figures 2, 3, and 4 show the swelling/compression of bentonite under low pressure of 7 kPa, medium pressure of 40 kPa, and high pressure of 100 kPa using water, biofuel, and ethanol fuel. The highest swelling of bentonite during 4 hours was achieved when water was used for both loads 7 kPa and 40 kPa. Swelling of bentonite submerged with ethanol was also observed at the lowest load (7 kPa). The highest load (100 kPa) prevented the swelling of the bentonite samples, a compression of bentonite for all liquids was observed. The bentonite submerged with ethanol under 100 kPa expressed the highest compression. Results indicate that for all types of pressures biofuel causes the bentonite clay to compress during loading.

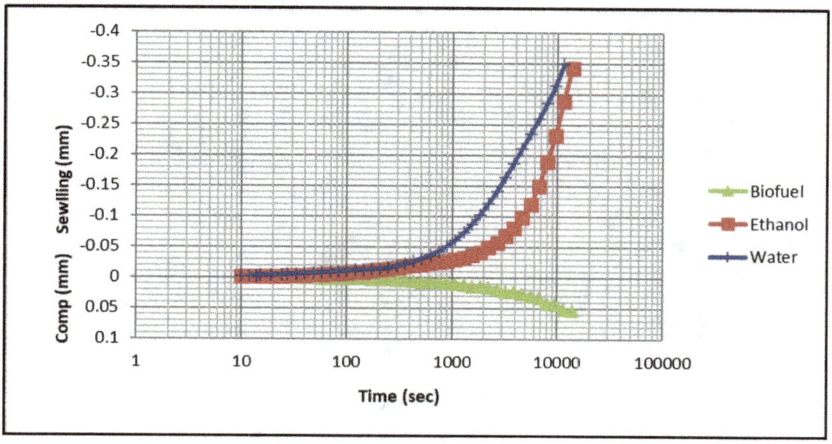

Fig.2. Swelling of bentonite under a pressure of 7 kPa using different liquids.

5. Analysis and Discussions

The results showed that bentonite swelling behaviour depends on the type of permeate as well as pressure. Many physic-chemical phenomena can be responsible for such behavior. Under low pressure (7 kPa), bentonite swelled the most in the contact with water. The swelling was also observed in the case of ethanol. However, when submerged by biofuel, the bentonite behaved differently. It underwent compression (consolidation). Under an increased load of 40 kPa, only the samples tested by water swelled. The ones tested by biofuel and ethanol underwent compression (consolidation) and biofuel had a higher rate of consolidation than ethanol. When applying a high pressure of 100 kPa, all samples are compressed. The most consolidation occurred in samples interacting with ethanol, then biofuel, leaving the samples that came in contact with water with the least compression. Very high pressure prevents the expansion of montmorillonite parts for all types of liquids.

The bentonite material submerged with ethanol (polar liquid) expressed the highest sensitivity to compression. Montmorillonite (the major component of bentonite) consists of tetrahedral and octahedral sheets. The space between these sheets can be expanded up to 10 water molecules thickness (10-9 m). In no load or low load conditions, the ethanol can provoke a similar expansion like water. However, the process of swelling, in this case, is lower.

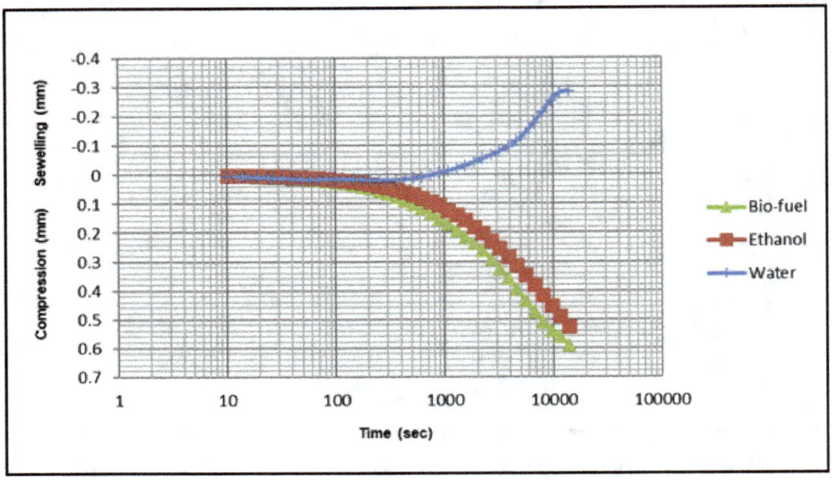

Fig.3. Swelling of bentonite under a pressure of 40 kPa using different liquids.

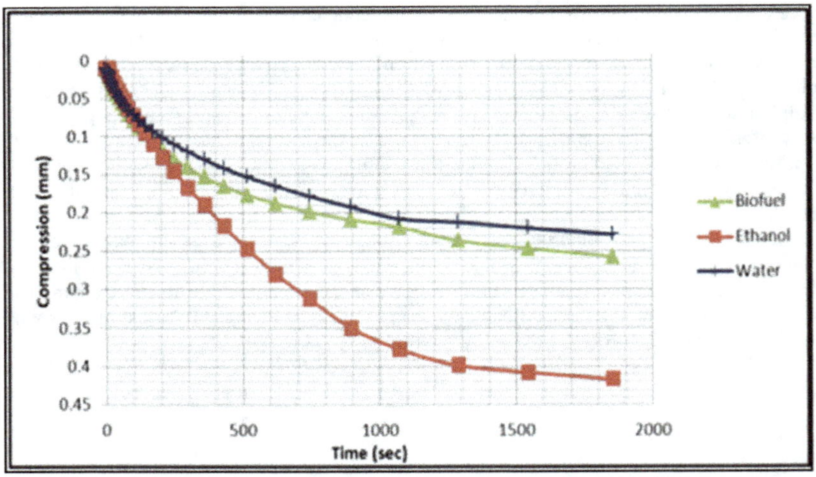

Fig.4. Swelling of bentonite under a pressure of 100 kPa using different liquids.

The bentonite swelling occurs when polar water molecules enter between the adjacent platelets and separate them so that van der Waals forces are no longer sufficient to keep the montmorillonite particles intact. This, in turn, causes the swelling of bentonite, decreasing pore voids, thus, reducing the hydraulic conductivity.

As water enters between the particles, it may displace the high-valent cations, which help the clay particles to the cluster. In turn, this may create repulsive forces between clay particles, causing the dispersion and additional swelling of bentonite and resulting in the swelling of the whole liner mixture.

The surface capillary forces, which are involved in attracting the soil particles decrease when an additional volume of water fills the pores (e.g. during saturation) and the water film surrounding the grain increases. Subsequently, clay can expand.

Biofuel operates differently, it acts oppositely, it prevents the expansion of clay, it does not permit for the tetrahedral and octahedral sheet to be expanded; montmorillonite submerged with biofuel undergoes compression with time in all tested conditions. Under the load condition, liquid films around the grains undergo deformation. This deformation is enhanced when biofuel is present. Biofuel causes the montmorillonite particles adhesion and the formation of flocs, consequently, it makes the bentonite liner coagulate and consolidate increasing pore voids, leading to fracture or tunnelling phenomena.

Biofuel has a convex meniscus (contrary to water concave meniscus) which creates higher surface tension forces (strong attractive forces between the grains) with the increase of additional biofuel liquid in pore voids. Subsequently, oil-particle flocs can be created. An extension of this phenomenon depends on temperature due to biofuel viscosity dependence.

Some studies indicated that polar hydrocarbons are required for flocculation to occur. However, experimental tests in this study indicate that both polar (ethanol fuel) and non-polar (biofuel) liquids cause the flocculation of clay fine particles. Furthermore, clay particles normally have negative charges in a stable clay solution then the repulsive forces are predominant. However, when biofuel is in contact with negatively charged bentonite in the liner's mixture, a thin layer is formed around the clay particles making them isolated, reducing the repulsive forces and allowing particles to form flocs. Therefore, two phenomena are present in this research: one for polar organic liquids (ethanol) and the other for non-polar (biofuel) forming of oil film filling the pores.

The results permitted us to conclude that the hydraulic conductivity of liners can increase under a load of waste after the percolation of alternative fuel leachates. This increase is due to many factors about the changes in the structure and composition of sand-bentonite mixtures. For instance, erosion of fine particles, as liquids infiltrate through the liner, affects the hydraulic conductivity. Also, the amount of fine particles flushed out of the mixtures increased with higher pressures, during the leaching of liquids, especially for water.

The fact that particles wash out of a liner exposed to high hydraulic pressure had already been observed. While water causes the highest rate of erosion due to its high solubility, its low density and viscosity enhance the erodibility of fine clay particles during the leaching process. The washing out of clay particles from the soil matrix increases with hydraulic pressure. This phenomenon, known as suffusion or piping, increases the final hydraulic conductivity of the liner due to the creation of larger pores (Kaoser et al., 2006).

In contact with the liner, water causes swelling of the liner's mixture through the swelling of bentonite clay in the liner composition and due to the dispersion of negatively charged clay particles. As water enters between the particles, it may displace the high-valent cations, which help the clay particles to the cluster. In turn, this may create repulsive forces between clay particles, causing the dispersion and additional swelling of bentonite and resulting in the swelling of the whole liner mixture.

Biofuel has a convex meniscus (contrary to water concave meniscus) which creates higher surface tension forces (strong attractive forces between the grains) with the increase of additional biofuel liquid in pore voids. Subsequently, oil-particle flocs can be created. An extension of this phenomenon depends on temperature due to biofuel viscosity dependence.

It was also observed that biofuel caused more cracks and higher surface fractures when percolating liners compared to ethanol fuel and water. In this study, biofuel grain size distribution and the mean size of particles increased drastically when liners were percolated by alternative fuels (ethanol or biofuel). This is due to the coagulation of fine particles in their interaction with fuel as they adhere together forming larger oil-clay clusters that do not separate in emulsion used for particle size analysis. However, many flocculated and single mineral fine particles remained, as shown in the particle size analysis.

Clay-ethanol flocs were also observed in the case of using ethanol as a permeate to percolate through the liner sand-bentonite mixtures. In the lab experiments, the clay-fuel floc aggregates were formed, which was also found by other researchers who observed clay-oil floc aggregates under an optical microscope and concluded that many flocculated single mineral fine particles were still present. A variety of crude oils ranging from light crude oils to heavy crude oils were, therefore, able to interact with micron-sized mineral fines to form —clay-oil flocs‖ consisting of solids-stabilized oil-in-water emulsions.

In addition, it is known that the adsorption behaviour appears to be inversely proportional to the solubility of the compound and directly proportional to the percentage of organic matter in the mixture. Biofuel, the least soluble of the three liquids used (water, ethanol, biofuel), appears to have a propensity to be adsorbed in the bentonite clay particles. Thus, bigger particles form through the coagulation of fine particles, causing fractures in the sand-bentonite mixture. This might explain the higher surface fracture in the liner mixtures percolated by biofuel. It ia appears that polar hydrocarbons are required for flocculation to occur. However, experimental tests in this study indicate that both polar (ethanol fuel) and non-polar (biofuel) liquids cause the flocculation of clay fine particles.

Clay particles normally have negative charges, and in a stable clay solution, repulsive forces predominate. But when biofuel is in contact with negatively charged bentonite in the liner's mixture, a thin layer forms around the clay particles making them isolated, reducing the repulsive forces and allowing particles to form flocs. Distinctive phenomena can be related to polar organic liquids (ethanol) and non-polar liquids (biofuel) forming oil films [20, 21].

During lab experiments, biofuel flow through the 100% sand liner took more time to reach the column outlet. High biofuel viscosity creates higher friction forces between the liner and the fluid.

Conclusions

The swelling of bentonite is sensitive to the type of liquid and load. It swells under low loads (e.g. 7 kPa) when submerged with water and to a lesser extent for ethanol. However, in the case of submerging the bentonite by biofuel, it undergoes consolidation. For higher load (e.g. 40 kPa) bentonite undergoes swelling in water, however, it consolidates for both alternative fuels (biofuel and ethanol). Under very high loads (e.g. 100 kPa) bentonite consolidates for all kinds of liquids including water.

References

[1] Enviroment Québec, 1994, Dix ans des restauration des terrains contamines‖, Enviro.doc EN950098.

[2] ASTM D2435-90 Standard test method for one-dimensional consolidation properties of soils. Chapuis R. P., (2002), The 2000 R.M. Hardy Lecture: Full-scale hydraulic performance of soil-bentonite and compacted clay liners. Canadian Geotech. J. 39: 417-439. https://doi.org/10.1139/t01-092

[3] Fernandez F., and Quigley R. M., (1985), Hydraulic conductivity of natural clays permeated with simple liquid hydrocarbons‖. Canadian Geotechnical Journal. 22: 205-214. https://doi.org/10.1139/t85-028

[4] Headley V., Boldt-Leppin E.J., Haug D., and Peng J., (2001), Determination of diffusion and adsorption coefficients for volatile organics in an organophilic clay-sand-bentonite liner‖ Canadian Geotechnical Journal. 38: 809-817. https://doi.org/10.1139/t01-017

[5] Guyonnet D., Gaucher E., Gaboriau H., Pons C.-H., Clinard C., Norotte V., and Didier G., (2005), Geosynthetic clay liner interaction with leachate: correlation between permeability, microstructure, and surface chemistry‖. Journal of Geotechnical and Geoenvironmental Engineering ASCE June 2005: 740-749. https://doi.org/10.1061/(ASCE)1090-0241(2005)131:6(740)

[6] Pivato A. and Raga R., (2005), Tests for the evaluation of ammonium 163 attenuation in MSW landfill leachate by adsorption into bentonite in a landfill liner‖. Waste Management 26: 123-132. https://doi.org/10.1016/j.wasman.2005.03.009

[7] Arasan S., (2010), Effect of Chemicals on Geotechnical Properties of Clay Liners: A Review‖. Research Journal of Applied Sciences, Engineering and Technology 2(8): 765-775.

[8] Aissa Mamouns, S.M., and Bekkouche, A., (2011). Modeling of the Behavior of Expansive Soils. Jordan Journal of Civil Engineering, Volume 5, No. 1, 2011.

[9] Bertagnolli C., and Carlos da Silva M. G., (2012), Characterization of Brazilian Bentonite Organoclays as Sorbents of Petroleum-derived Fuels‖. Materials Research; 15(2): 253-259. https://doi.org/10.1590/S1516-14392012005000017

[10] Sreedharan V., and. Sivapullaiah P. V., (2012), Effect of Organic Modification on Adsorption Behaviour of Bentonite‖. Indian Geotech J 42(3):161-168. https://doi.org/10.1007/s40098-012-0014-6

[11] Wang Q. et al., (2014), Long-term effect of water chemistry on the swelling pressure of a bentonite-based material. Applied Clay Science 87: 157-162. https://doi.org/10.1016/j.clay.2013.10.025

[12] Sreedharan V., and Sivapullaiah P. V., (2017), Hydraulic Performance of Organo Clay Enhanced Sand Bentonite as Secondary Liner. Indian Geotech J. https://doi.org/10.1007/s40098-017-0232-z

[13] Ren Y. et al., (2019), Adsorption of imidazolium-based ionic liquid on sodium bentonite and its effects on rheological and swelling behaviors. Applied Clay Science 182: 105248. https://doi.org/10.1016/j.clay.2019.105248

[14] Estabragh A.R., Amini M., and Soltanian M.R., (2019), Aging effects on the swelling behavior of compacted bentonite. Bulletin of Engineering Geology and the Environment. https://doi.org/10.1007/s10064-019-01708-7

[15] Zeng, Z., Cui, Y.-J., Zhang, F., Conil, N., & Talandier, J. (2020). Effect of technological voids on the swelling behaviour of compacted bentonite/claystone mixture. Canadian Geotechnical Journal. 9. https://doi.org/10.1139/cgj-2019-0339

[16] Di Maio, C., (1996). Exposure of bentonite to salt solution: Osmotic and mechanical effects. Geotechnique, 46(4): 495-707. https://doi.org/10.1680/geot.1996.46.4.695

[17] Sasikala, S., Meenakshi, S., Bhat, S.D., Sahu, A.K. (2014). Functionalized Bentonite clay-sPEEK based composite membranes for direct methanol fuel cells, Electrochimica Acta. https://doi.org/10.1016/j.electacta.2014.04.180

Materials Research Proceedings 31 (2023) 504-513 https://doi.org/10.21741/9781644902592-52

[18] Fahad A.A., (2020), Variation of Sabkha Soil Permeability Associated with Ions' Dissolution during Distilled Water Leaching. Jordan Journal of Civil Engineering, Volume 14, No. 4, 2020.

[19] Bani Baker, M., Abendeh, R., Sharo, A., Hanna, A. Stabilization of Sandy Soils by Bentonite Clay Slurry at Laboratory Bench and Pilot Scales. Coatings/Functional Building Materials and Pavements Coatings, 2022, 12 (12). https://doi.org/10.3390/coatings12121922

[20] Bani Baker, M., Elektorowicz ,M., Hanna, A., (2018), Elctrokienetic Nondestructive In-situ Technique for Rehabilitation of Liners Damaged by Fuels, Journal of Hazardous Materials. Vol. 359. Pp. 510-515. https://doi.org/10.1016/j.jhazmat.2018.07.113

[21] Sharo, A.A., Khasawneh, M.A., Bani Baker, M., Al Tarawneh, D.M. Sonicated waves procedure effect on stabilizing expansive soil by nano-clay: Treat with cause. Frontiers in Built Environment, 2022, 8, 975993 https://doi.org/10.3389/fbuil.2022.975993

Design and simulation of a large-scale telescopic paraboloidal solar concentrator: Dual receiver

Barena Bekalo Betela*[1] Venkata Ramayya Ancha[1] and Lingala Syam Sundar[2]

[1]Faculty of Mechanical Engineering, Jimma University, Jimma, Ethiopia

[2]Mechanical Engineering, College of Engineering, Prince Mohammad Bin Fahd University, Al-Khober, Saudi Arabia

* newyr29@yahoo.com

Keywords: Beam-Downsystem, Dual Receiver, Gregorian Telescope, Cassegrain Telescope, Solar Thermal Energy, Parabolic Dish, Secondary Reflector, Mounted Receiver, Ground-Fixed Receiver

Abstract. Beam-down solar concentrators with a secondary reflector are receiving a lot of attention at present. Large telescopic dual-receiver solar concentrators with Gregorian and Cassegrain alignments have been modelled and investigated in the present study with each telescopic design having a unique set of receivers that are mounted and anchored to the ground. A comparative assessment of both of the telescopic alignments have been carried out along with minimum image radii The results reveal that both telescopic designs are capable of splitting incoming sunlight and facilitating the use of two receivers. For the design and simulation of telescopic designs, Tonatiuh and Soltrace have been employed for a comparative evaluation. Both of the conventional and telescopic designs using Soltrace and Cassegrain, as well as conventional designs by Tonatiuh, produced identical results in the simulation of total power on the receiver, However, a sizable peak flux discrepancy was seen between the results from Tonatiuh and Soltrace.

Introduction

Overhead positioning of a large heat receiver of the tower, strilling engine, and furnace has difficulties during operation, maintenance, and construction. Heat is lost during the transfer procedure because the receiver location on the tower requires a lot of energy to push up. To solve this issue, research is being done on a beam-down system with a ground-fixed receiver [1].

Beam-down concentrating solar technologies supply useful heat for extremely efficient power cycles and direct solar fuels with receiver reactor technology at high temperatures. Because of the high-temperature heat production and chemical reactions occur on the ground rather than at a great height, the system is safer [2-12][7, 13, 14]. A secondary reflector is a tool for beam-down technologies. Solar collector with a secondary reflector is the fastest-growing technology [15-25]. One of the solar thermal energy technologies with a secondary reflector is a parabolic dish [26-29].

Using secondary reflectors in parabolic dish design is highly related to the design of Gregorian and Cassegrain telescopes because these telescopes are a two-mirror system [30]. Designing and aberration correction are the main works during designing telescopes [31-33]. Applying a telescopic design for a parabolic dish may be used to increase the concentration ratio and then reduce optical and thermal losses and increase collector efficiency by correcting aberration. Optical aberration prevents the conventional parabolic dish from achieving the 46,000 geometric concentration ratio that is thermodynamically feasible. Due to the aberration, only a maximum of 11,000 is obtained with a 1.0 intercept factor [26]. This shows that optical aberration is the main problem of the solar thermal technology. Therefore, an optical aberration decreases the geometric concentration ratio, increases re-radiation and other optical losses, and decreases collector efficiency. One of the important issues in telescopic design is the control of optical aberration [34-37].

There are no studies reported so far, to use telescopes for solar thermal energy technology and correcting optical aberration during design of a conventional parabolic dish. Additionally, current research does not consider sunlight blockage by secondary reflectors. This study proposes consideration of light blockage by secondaryreflectors as an opportunity for the division of incoming sunlight into two approximately equal parts. Utilization of solar energy requires multi-operation such as during daytime and nighttime. Dividing the receiver into two parts will help to reduce the challenges related to a single receiver. Moreover, there is a research gap noticed to use beam-down and beam-up solar collection simultaneously.

Therefore, paraxial ray tracing-based research was carried out in this study on the parabolic dish.The study is based on the design principle of the Gregorian and Cassegrain telescopes (two-mirror optics) and named telescopic paraboloidal solar concentrator. The main advantage of this new concept is: (i) to reduce the receiver load of the conventional parabolic dish by dividing it into two receivers;(ii) to place some challenging receiver configuration such as concentrated photovoltaics on the ground;(iii) to make telescopes to function both as solar thermal energy technology and telescope. Therefore, the tertiary reflector was added to a Gregorian telescope, whilefront and back sides simultaneous reflecting secondary reflector was added to Cassegrain telescope without disturbing the existing telescopes operations.

Methodology

Large-scale design of Gregorian and Cassegrain telescopic dual receiver system

A large-scale Gregorian and Cassegrain telescopic dual receiver paraboloidal solar concentrators were designed by paraxial ray tracing (Gaussian optics) method. Then the system was modeled and simulated by Monte Carlo ray tracing Tonatiuh, Tonatiuh-Mathematical combination and Soltrace software. The telescopes' existing design was modified to make it a solar thermal energy technology. Consequently, tertiary parabolic reflector was added to Gregory telescope, while the secondary in Cassegrain telescope was made with front and back sides simultaneously reflecting. Secondary and tertiary reflectors had the same diameter and opposite in direction. i.e., secondary views primary concavely and tertiary set up as conventional parabolic dish or were made back-to-back connected.

Axial obscuration ratio and f-Number selected were 0.5 and 1.8 respectively. Firstly, Gregory telescopc was designed and then the Cassegrain telescopic was designed depending upon Gregorian design. As indicated in Table 1, Cassegrain was designed by converting Gregorian telescopic by removing tertiary reflector from Gregory and reducing the distance between the two concave mirrors in Gregory by half.

Table 1: large-scale dimensions of Gregorian and Cassegrain telescopic dual receiver paraboloidal solar concentrator.

	Gregorian design			Cassegrain design		
	Diameter(m)	Focal length(m)	Distance between mirrors(m)	Diameter(m)	Focal length(m)	Distance between Mirrors(m)
Primary reflector	30	54		30		
Secondary reflector	21.21	20.4414	81	21.21	20.4414	40.5
Tertiary reflector	21.21	20.4414	81	-	-	-

Results and discussion

Gregorian telescopic paraboloidal dual receiver solar concentrator

This study was performed by considering light blockage by secondary reflectors as an opportunity for dual receiver solar thermal energy technology design.

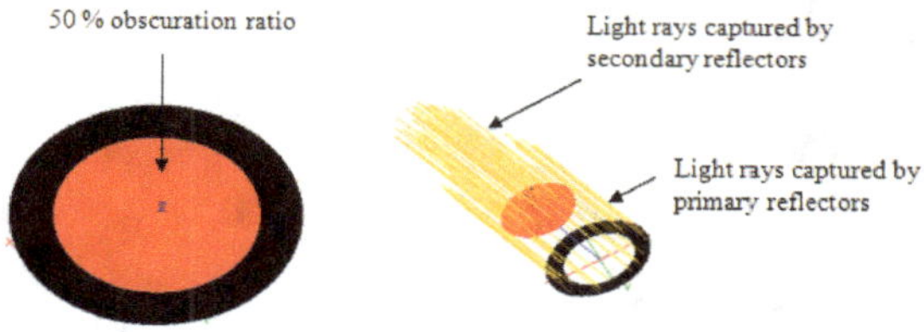

Figure 1: *Secondary reflector blocking the sun light, as drawn by Soltrace*

As Figure 1 indicates 29,994,611 light rays from 60 million released rays were captured secondary reflectorwhileprimary reflector captured 30,005,389 rays. Therefore, incoming rays were divided into two equal parts with axial obscuration ratio of 0.5.

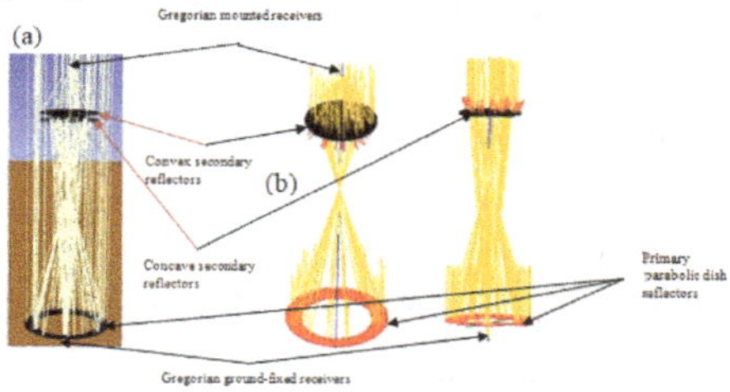

Figure 2: *Gregorian telescopic paraboloidal solar concentrator with mounted and ground-fixed receiver: (a) ray graphics in Tonatiuh software and (b) ray graphics in Soltrace.*

According to *Figure 2*above in Gregorian design, back-to-back connected secondary and tertiary parabolic dish reflectors were used to divide incoming sunlight to mounted and ground-fixed receivers in Tonatiuh. In soltrace separate modeling was made for secondary and tertiary reflectors.

Cassegrain telescopic paraboloidal dual receiver solar concentrator

As shown in Figure 3, the Cassegrain telescope design was modified by making secondary reflector to reflect at both back and front sides simultaneously.

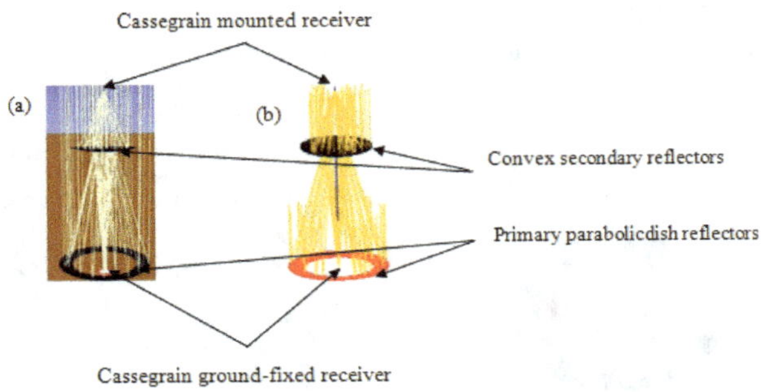

Figure 3: *Cassegrain telescopic paraboloidal solar concentrator with mounted and ground-fixed receiver: (a) ray graphics in Tonatiuh software and (b) ray graphics in Soltrace.*

As shown in *Table 3 in case of Tonatiuh, a* 30-meter diameter and 54-meter focal length parabolic dish has a total power of 706.354 kW, which is slightly greater (weak convergency or accuracy) than from the sum of the total power of the Cassegrain design and significantly different from the sum of the total power of the Gregorian design. Therefore, telescopic designs are better than conventional designs. According to Table 5 the sum of errors in the Gregorian design is 0.00104877818, which is smaller than the errors in the Cassegrain and conventional designs. Therefore, the smallest error observed in Gregorian design when comparing with the errors of Cassegrain and conventional designs indicates that Gregorian design is more accurate.

During sunshape simulation, as the distance from the primary mirror to the receiver increases, the accuracy increases. This is confirmed by the optical tool verification study of Wang et al.[38]. According to the authors, as the distance increase between the primary mirror and the target, the radiance distribution will be extremely similar to the corresponding statistical distribution of the slope error or sunshape. Therefore, Gregorian telescopic designs show better convergence than conventional and Cassegrain designs.

As indicated in Table 2, the sum of total power in case of Soltrace is equal in Gregorian and Cassegrain designs. In two cases it is 706.819kW.

Table 2: total power and peak flux Gregorian and Cassegrain telescopic paraboloidal solar concentrator

	Gregorian telescopic						Cassegrain telescopic					
	Mounted receiver			Ground-fixed receiver			Mounted receiver			Ground-fixed receiver		
	Tonatiuh	Tonatiuh-mathematica	soltrace	Tonatiuh	Tonatiuh Mathematica	soltrace	Tonatiuh	Tonatiuh Mathematica	Soltrace	Tonatiuh	Tonatiuh Mathematica	Soltrace
Total ower(kW)	335.199	331.625	353.346	339.383	339.424	353.473	352.449	352.456	353.346	350.951	353.474	353.473
Peak flux(Kw/m2)	5020.55	-	22,254,000,000	68.0677	-	3811.23	1934.11	-	22,254,000,000	740.605	-	3112410

Table 3: Tonatiuh's total power and peak flux of telescopic and conventional paraboloidal solar concentrators

	Gregorian design			Cassegrain design			Conventional design
	Mounted receiver	Ground-fixed receiver	added total power	Mounted receiver	Ground-fixed	Added total power	Receiver
Total power(kW)	335.199	339.383	674.582	352,449	350.951	703.4	706.354
Peak flux(kW/m²)	1.0382*10^7	170.514	-	1.09102*10^7	173.725	-	3.44605*10^6

As *Table 4* describes, the simulation of conventional design gave 706.861kW. Therefore, there is no total power difference between conventional and telescopic designs by soltrace simulation.

Table 4: Total power and peak flux on the receiver of 30-meter diameter conventional parabolic dish

	Tonatiuh	Tonatiuh-Mathematica	Soltrace
Total power(kW)	706.354	706.332	706.861
Peak flux(kW/m²)	3,444.95	-	1,569,190,000

Table 5: Errors of telescopic design

Gregorian design			Cassegrain design		
Tonatiuh error			Tonatiuh error		
Mounted receiver	Ground-fixed receiver	Sum of error	Mounted receiver	Ground-fixed reciever	Sum of error
0.000093384	0.00095539	0.0010487781	0.000256412	0.000112354	0.00036887

Generally, the simulation by Tonatiuh was very fast. It required 11 minutes to trace 60 million, while Soltrace used 2 hours for single run. Better convergence was obtained in case of Gregorian design by Tonatiuh. Minimum sun image radius was easily and clearly obtained by Tonatiuh-Mathematica combination.

To get total power and peak flux on receivers for comparison, modeling and simulation of a parabolic dish with a 30-meter diameter and 54-meter focal length were carried out. As previously indicated, all designs have nearly identical residual total powers, with the exception of the Gregorian design. However, a significant variation was seen in each case for the peak flux. The truth is closely akin to the study of [39] in that there was basically no difference between the Soltrace and the Tonatiuh estimates in terms of the solar total power influencing the target or the reception tube.

Conclusion

Designing Gregorian and Cassegrain telescopic dual receiver paraboloidal solar concentrators allows for the simultaneous usage of beam-down and beam-up sun beams. Both telescoping systems can produce two receivers by dividing the incoming light. The new design used the chance to be built ground-fixed and mounted receivers in both telescopic forms as a result of sunlight being blocked by secondary reflectors in beam-down concentrators. In order to avoid interfering with a Gregorian telescope's current operations, a tertiary reflector was added. While Cassegrain's existing functions are unaffected by simultaneous back and front side reflections. Reflecting telescopes can serve so as solar thermal energy technologies in addition to telescopes.

Acknowledgements
This study was supported by JiT Center of Excellence and Ministry of education in Ethiopia.

Declaration of interest statement
The authors affirm that they have no known financial or interpersonal conflicts that would have appeared to have an impact on the research presented in this study.

Reference

[1] A. S. Wardhana, M. Ashari, and H. Suryoatmojo, "Designing and modeling a novel dual parabolic concentrator with three degree of freedom (DOF) robotic arm," *Solar Energy,* vol. 194, pp. 436-449, 2019/12/01/ 2019, doi: https://doi.org/10.1016/j.solener.2019.10.057.

[2] E. Bellos, "Progress in beam-down solar concentrating systems," *Progress in Energy and Combustion Science,* vol. 97, p. 101085, 2023/07/01/ 2023, doi: https://doi.org/10.1016/j.pecs.2023.101085.

[3] F. Nie, Y. Yu, F. Bai, and Z. Wang, "Experimental and numerical investigation on thermal performance of a quartz tube solid particle solar receiver," *Solar Energy,* vol. 207, pp. 1055-1069, 2020/09/01/ 2020, doi: https://doi.org/10.1016/j.solener.2020.07.013.

[4] M. Tawfik, "A review of directly irradiated solid particle receivers: Technologies and influencing parameters," *Renewable and Sustainable Energy Reviews,* vol. 167, p. 112682, 2022/10/01/ 2022, doi: https://doi.org/10.1016/j.rser.2022.112682.

[5] C. Tregambi *et al.*, "Directly irradiated fluidized bed reactor for thermochemical energy storage and solar fuels production," *Powder Technology,* vol. 366, pp. 460-469, 2020/04/15/ 2020, doi: https://doi.org/10.1016/j.powtec.2020.02.045.

[6] F. Di Lauro, C. Tregambi, F. Montagnaro, P. Salatino, R. Chirone, and R. Solimene, "Improving the performance of calcium looping for solar thermochemical energy storage and CO2 capture," *Fuel,* vol. 298, p. 120791, 2021/08/15/ 2021, doi: https://doi.org/10.1016/j.fuel.2021.120791.

[7] G. Moumin *et al.*, "CO2 emission reduction in the cement industry by using a solar calciner," *Renewable Energy,* vol. 145, pp. 1578-1596, 2020/01/01/ 2020, doi: https://doi.org/10.1016/j.renene.2019.07.045.

[8] S. Kiwan and Q. R. Soud, "Experimental investigation of the thermal performance of a sand-basalt heat storage system for beam-down solar concentrators," *Case Studies in Thermal Engineering,* vol. 19, p. 100609, 2020/06/01/ 2020, doi: https://doi.org/10.1016/j.csite.2020.100609.

[9] D. Saldivia, J. Bilbao, and R. A. Taylor, "Optical analysis and optimization of a beam-down receiver for advanced cycle concentrating solar thermal plants," *Applied Thermal Engineering,* vol. 197, p. 117405, 2021/10/01/ 2021, doi: https://doi.org/10.1016/j.applthermaleng.2021.117405.

[10] X. Wei, Z. Lu, W. Yu, and W. Xu, "Ray tracing and simulation for the beam-down solar concentrator," *Renewable Energy,* vol. 50, pp. 161-167, 2013/02/01/ 2013, doi: https://doi.org/10.1016/j.renene.2012.06.029.

[11] S. Yang, J. Wang, P. D. Lund, C. Jiang, and X. Li, "High performance integrated receiver-storage system for concentrating solar power beam-down system," *Solar Energy,* vol. 187, pp. 85-94, 2019/07/15/ 2019, doi: https://doi.org/10.1016/j.solener.2019.05.041.

[12] S. Yang, J. Wang, P. D. Lund, C. Jiang, and X. Li, "Modelling and performance evaluation of an integrated receiver-storage for concentrating solar power beam-down system under heterogeneous radiative conditions," *Solar Energy,* vol. 188, pp. 1264-1273, 2019/08/01/ 2019, doi: https://doi.org/10.1016/j.solener.2019.07.031.

[13] M. Natarajan, Y. Rajasekhar, C. Chiranjeevi, and B. Murali, "Energy Analysis of the Solar Water Heater with Secondary Reflectors," in *Energy and Exergy for Sustainable and Clean Environment, Volume 1*, V. Edwin Geo and F. Aloui Eds. Singapore: Springer Nature Singapore, 2022, pp. 93-103.

[14] J. Juodkazytė *et al.*, "Solar water splitting: Efficiency discussion," *International Journal of Hydrogen Energy,* vol. 41, no. 28, pp. 11941-11948, 2016/07/27/ 2016, doi: https://doi.org/10.1016/j.ijhydene.2016.05.079.

[15] A. Minaeian, A. Alemrajabi, M. Chavoshi, A. Mostafaeipour, and Z. Seifi, "Effect of secondary reflector on solar flux intensity and uniformity of a Fresnel concentrator," *Journal of Renewable and Sustainable Energy,* vol. 12, no. 3, 2020, doi: 10.1063/5.0007604.

[16] X. Y. Tang, W. W. Yang, Y. Yang, Y. H. Jiao, and T. Zhang, "A design method for optimizing the secondary reflector of a parabolic trough solar concentrator to achieve uniform heat flux distribution," *Energy,* vol. 229, p. 120749, 2021/08/15/ 2021, doi: https://doi.org/10.1016/j.energy.2021.120749.

[17] B. V and S. S, "Secondary Reflector and Receiver Positions for Uniform Heat Flux Distribution in Parabolic Trough Solar Thermal Collector," *Journal of Solar Energy Engineering,* vol. 144, no. 6, 2022, doi: 10.1115/1.4054660.

[18] S. Wang, C.-A. Asselineau, Y. Wang, J. Pye, and J. Coventry, "Performance enhancement of cavity receivers with spillage skirts and secondary reflectors in concentrated solar dish and tower systems," *Solar Energy,* vol. 208, pp. 708-727, 2020/09/15/ 2020, doi: https://doi.org/10.1016/j.solener.2020.08.008.

[19] A. Vouros, E. Mathioulakis, E. Papanicolaou, and V. Belessiotis, "On the optimal shape of secondary reflectors for linear Fresnel collectors," *Renewable Energy,* vol. 143, pp. 1454-1464, 2019/12/01/ 2019, doi: https://doi.org/10.1016/j.renene.2019.05.044.

[20] J.-h. Gong *et al.*, "Optical, thermal and thermo-mechanical model for a larger-aperture parabolic trough concentrator system consisting of a novel flat secondary reflector and an improved absorber tube," *Solar Energy,* vol. 240, pp. 376-387, 2022/07/01/ 2022, doi: https://doi.org/10.1016/j.solener.2022.05.044.

[21] R. Abbas, A. Sebastián, M. J. Montes, and M. Valdés, "Optical features of linear Fresnel collectors with different secondary reflector technologies," *Applied Energy,* vol. 232, pp. 386-397, 2018/12/15/ 2018, doi: https://doi.org/10.1016/j.apenergy.2018.09.224.

[22] S. Shajan and V. Baiju, "Designing a novel small-scale parabolic trough solar thermal collector with secondary reflector for uniform heat flux distribution," *Applied Thermal Engineering,* vol. 213, p. 118660, 2022/08/01/ 2022, doi: https://doi.org/10.1016/j.applthermaleng.2022.118660.

[23] J. Ma, C.-L. Wang, Y. Zhou, and R.-D. Wang, "Optimized design of a linear Fresnel collector with a compound parabolic secondary reflector," *Renewable Energy,* vol. 171, pp. 141-148, 2021/06/01/ 2021, doi: https://doi.org/10.1016/j.renene.2021.02.100.

[24] S. Bellan, K. Matsubara, C. H. Cheok, N. Gokon, and T. Kodama, "CFD-DEM investigation of particles circulation pattern of two-tower fluidized bed reactor for beam-down solar concentrating system," *Powder Technology,* vol. 319, pp. 228-237, 2017/09/01/ 2017, doi: https://doi.org/10.1016/j.powtec.2017.06.060.

[25] S. Taramona, P. Á. González-Gómez, J. V. Briongos, and J. Gómez-Hernández, "Designing a flat beam-down linear Fresnel reflector," *Renewable Energy,* vol. 187, pp. 484-499, 2022/03/01/ 2022, doi: https://doi.org/10.1016/j.renene.2022.01.104.

[26] L.D.Jaffe and P. T. Poon, Secondary and Compound Concentrators for Parabolic-Dish Solar-Thermal Power Systems. USA, 1981.

[27] O. J. Nydal, "Ray Tracing for Optimization of a Double Reflector System for Direct Illumination of a Heat Storage," *Energy Procedia,* vol. 57, pp. 2211-2220, 2014, doi: 10.1016/j.egypro.2014.10.188.

[28] F. Dähler *et al.*, "Optical design and experimental characterization of a solar concentrating dish system for fuel production via thermochemical redox cycles," *Solar Energy,* vol. 170, pp. 568-575, 2018/08/01/ 2018, doi: https://doi.org/10.1016/j.solener.2018.05.085.

[29] H. Xu *et al.*, "A beam-down solar concentrator with a fixed focus — Design and performance analysis," *Solar Energy,* vol. 241, pp. 428-436, 2022/07/15/ 2022, doi: https://doi.org/10.1016/j.solener.2022.06.017.

[30] R. N. Wilson, Reflecting Telescope Optics II: Manufacture, Testing, Alignment, Modern Techniques. Springer Berlin Heidelberg, 2013.

[31] S. Chang and A. Prata, "Geometrical theory of aberrations near the axis in classical off-axis reflecting telescopes," *JOSA A,* vol. 22, no. 11, pp. 2454-2464, 2005.

[32] R. N. Wilson, Reflecting Telescope Optics II: Manufacture, Testing, Alignment, Modern Techniques. Springer, 2011.

[33] R. N. Wilson, Reflecting telescope optics 1 basic design theory and its historical development. Springer, 2007.

[34] S. Chen, H. Feng, D. Pan, Z. Xu, Q. Li, and Y. Chen, "Optical aberrations correction in postprocessing using imaging simulation," *ACM Transactions on Graphics (TOG),* vol. 40, no. 5, pp. 1-15, 2021.

[35] A. P. Krishnan, C. Belthangady, C. Nyby, M. Lange, B. Yang, and L. A. Royer, "Optical aberration correction via phase diversity and deep learning," *bioRxiv,* p. 2020.04. 05.026567, 2020.

[36] X. Li, J. Suo, W. Zhang, X. Yuan, and Q. Dai, "Universal and flexible optical aberration correction using deep-prior based deconvolution," in *Proceedings of the IEEE/CVF International Conference on Computer Vision*, 2021, pp. 2613-2621.

[37] A. E. Conrady, *Applied Optics and Optical Design, Part Two*. Dover Publications, 2014.

[38] Y. Wang *et al.*, "Verification of optical modelling of sunshape and surface slope error for concentrating solar power systems," *Solar Energy,* vol. 195, pp. 461-474, 2020/01/01/ 2020, doi: https://doi.org/10.1016/j.solener.2019.11.035.

[39] M. Blanco, A. Mutuberria, P. Garcia, R. Gastesi, and V. Martin, *Preliminary validation of Tonatiuh*. 2009.

Advanced Topics in Mechanics of Materials, Structures and Construction - AToMech1-2023 Materials Research Forum LLC
Materials Research Proceedings 31 (2023) 522-530 https://doi.org/10.21741/9781644902592-54

The Innovation in wire arc additive manufacturing (WAAM): A review

Saleh Fahad Alaboudi[1,a *], Muhammad Azhar Ali Khan[2,b], Muhammad Asad[2,c], Mushtaq Khan[2,d] and Faramarz Djavanroodi[2]

[1]Dammam, Saudi Arabia

[2]617 AlJawhara, Khobar, Saudi Arabia

[a]201902168@pmu.edu.sa, [b]mkhan6@pmu.edu.sa, [c]masad@pmu.edu.sa, [d]mkhan7@pmu.edu.sa, [e]fdjavanroodi@pmu.edu.sa

Keywords: Wire Arc Additive Manufacturing (WAAM), Hybrid Manufacturing (HM), Additive Manufacturing (AM), Metal Additive Manufacturing (MAM), Hybrid Additive Manufacturing (HAM), Metal Additive Manufacturing (MAM), Direct Energy Deposition (DED)

Abstract. This review or research paper is illustrated to analytically assess and address one of potential industrial revolutions, which is Wire Arc Additive Manufacturing (WAAM). WAAM is classified from Hybrid Manufacturing (HM) processes. Thus, one of the Hybrid Manufacturing ultimate goals has always been to transcend the limitation aspects associated with the tradition process. As Artificial Intelligence (AI) has evolved and expanded all over the globe, Additive Manufacturing (AM) has been gradually developed and introduced to the world to be one of a distinguished innovative impact in the history of manufacturing. Additive Manufacturing (AM) has been improved over the conventional methods in the manufacturing world due to its advanced complexity, consistency, quality of work, and various advantages and contribution that satisfy the costumers needs and requirements. Various applications in the industry have proven the AM applicability to replace the conventional processes such as casting and machining as it can deal with very coplex shapes [3]. In spite of the fact that there are numerous materials that can be manufactured in the modern technologies of AM, such as polymers, metals, ceramic, and composites, the contribution of Metal Additive Manufacturing (MAM) arguably has been a significant influence in the industries in comparison to the others [1]. In this review paper the detailed deliverable information and materials which will be established and communicated in this paper will concentrate on the history of (WAAM) including its pros and cons, latest contribution to the industries, AM classifications, materials, and primary materials and practices in industry.

Introduction

Referring to the technological developments that have been occurring to the old age of manufacturing industry, additive manufacturing (AM) has distinguished its self from and among the rest of conventional manufacturing methods by gradually eliminating human errors and interference, increase quality of work, and increase productivity. AM technology goes all the way back to the 1980s. HM has contributed to manufacture very complex and complicated materials with different geometries including internal structures to potentially allow to achieve higher productivity, quality, and mass production. In this paper, the definition of Hybrid Manufacturing (HM) has been a major controversy among researchers as it has gained its popularity back in 1990's. HM can be defined as the incorporation and merging of two manufacturing processes or technologies into one for the benefit of exceeding the limitation associated and gaining the ultimate and fundamental advantages [2].

History and classification of mam

Any produced metallic materials or wire-feed AM, it definitely has been produced by one of the three types of energy sources, which are electron beam, electric arc, or laser [31]. However, electric arc welding has defeated the other two types due to its benefits of higher energy efficiency and deposition rate. The arc welding AM deposition rate is 50-130g/min, whereas the deposition rate for both electron beam and laser is 2-10 g/min [32]. Additive manufacturing has provided various types of real life examples in the industries since its beginning of revolution, such as Selective Laser Melting (SLM). SLM has been classified as a type of metal additive manufacturing (MAM) and the way SLM works is that a 3D model is being produced via melting the particles of metallic powder through a complete laser procedure. The part manufactured via this process can have percentage of porosity less than 0.1% [9]. In addition, SLM has the ability to produce material with high efficiency as well as the ease of building complex geometries free of constrains, which the conventional manufacturing processes could not have [8]. However, the researchers have agreed that initially SLM could not contribute to the huge demand in aerospace and transportation industries due to limitation of size in the building chamber with approximate dimension of 300 mm x 300 mm for the single laser system. When it comes to the widely used materials in SLM technology, 24 CrNiMo has great tribology, structural, and mechanical properties as it one type classified from High Strength Low Alloy (HSLA). Another example of that has been evolving drastically over the past years, which is the 3D printing using natural fiber. The agricultural term of natural fiber is referring nowadays to the environmental friendly materials, such as rice shells, nut shells, reeds, wood fiber, gels, solutions, thermoplastics, polymer, polylactide, polypropolene, and crop straws, which they have been widely utilized in health care, machinery, and aerospace. Natural fiber composite materials are made from the addition of natural fibers as reinforcers and other agents as compatibilizers and flexibilizers into the polymer or gel matrix. Various methods can be applied in the 3D printing industry, but there are countable common ones, such as Fused Deposition Modelling (FDT), Stereo lithography Apparatus (SLA), Direct Ink Writing (DIW), Selective Laser Sintering (SLS), and Digital Light Processing (DLP) [10].

Discussion

Classification

Various industries have been benefiting from AM mostly in the medical, automotive, railway, and aerospace sectors and the popularity was gained during the late 20th century [26, 45]. An interesting fact about AM in aerospace, the heaviest part produced via AM weighed roughly 750 kg, its width was 1.7 m, its length was 4.7m, and the height was 0.5 m [44]. In 2016, AM aerospace sales was $ 6.7 billion and was classified as one of fastest growing among others [54]. Also, AM has been widely used in the medical field to fabricate part for their patients, such as respiratory parts [52]. AM is also coming to be part of the railway environment as the spare parts are requiring immediate alternative fabrication methods like AM since light weight, low cost, and lead time is the ultimate goal [60]. Metal additive manufacturing (MAM) has been classified as Direct Additive manufacturing with four processes, such as powder bed fusion (PBF), binder jetting (BJ), direct energy deposition (DED), and sheet lamination (SL). (fig.7) illustrates the categories for metal materials AM [29]. Also the other classification is Indirect Additive Manufacturing with three processes, such as material jetting (MJ), vat photopolymerization (VP), and material extrusion (ME) (fig.1). As per the standards of ISO/ASTM 52900, materials used, feedstock, and deposition methods are the main differences among them [57]. Also, there is a major difference between direct and indirect MAM. MAM can be classified as direct whenever the parts are fabricated directly based on provided design, specifications, or certain requirements, whereas, the indirect MAM involves certain outlines to be followed in order to eventually acquire ultimate metal parts. Some examples to indirect MAM are sand casting, injection moulding, and die casting [2,56].

DIRECT MAM

Powder bed fusion (PBF) is one of the techniques utilized in metal additive manufacturing (MAM) where metal powder particles (ranges from 20 – 100 micrometres) are positioned in a bed are being melted layer by layer through a thermal source such as electron beam or laser [42]. Additionally, few practices are associated with PBF that play a big role in producing metals and plastics, such as Selective Laser Sintering (SLS), Selective Laser Melting (SLM), and Electron Beam Melting (EBM). Having said that, PBF has several advantages in terms of materials, such as the powder is recyclable, the availability of wide variety of materials (plastics, metals, and alloys), and it is low in cost [25]. Schematic is presented in (fig 2).

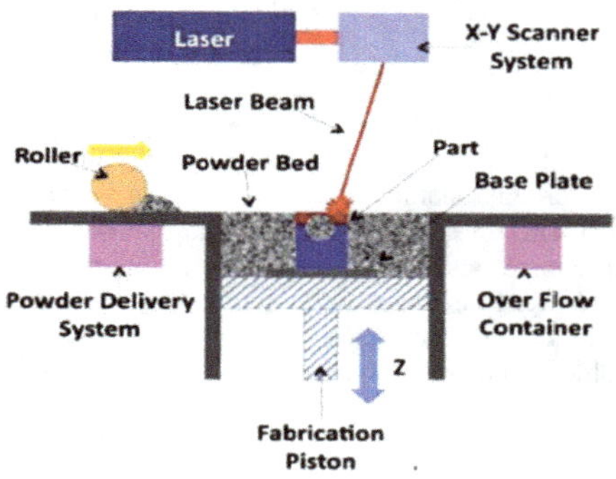

fig.2 – PBF process schematic

Binder jetting (BJ) is a liquid binder selectively deposited on a powder bed with a print head, which produces a final product of cohesive powder particles. It is a growing process that allows the production of part for the manufacturing, medical, and dental industries. This technique enables the production of metallic and ceramic parts as well as the sand moulds for castings. Furthermore, the materials used in this process is the form of powder particles that is being spread across the platform in a certain thickness and being evened out by the help of a roller. Then the liquid binder is being deposited over the build material via the inkjet printhead to form the desired layer by linking the substrate particles. Then, the operator lowers the platform to build the next layer, and the process keeps repeating until the desired object is completed. Thus, there are numerous distinct advantages for BJ, which are production of coloured objects, greater shape volume, high speed, discreet residual stresses, and astronomical material compatibility. On the other hand, BJ can have particular disadvantages linked to it such as, object shrinkage, non-sustainability for physical segments, and porosity defects [27]. Schematic is presented in (fig 3).

Materials Research Proceedings 31 (2023) 522-530 https://doi.org/10.21741/9781644902592-54

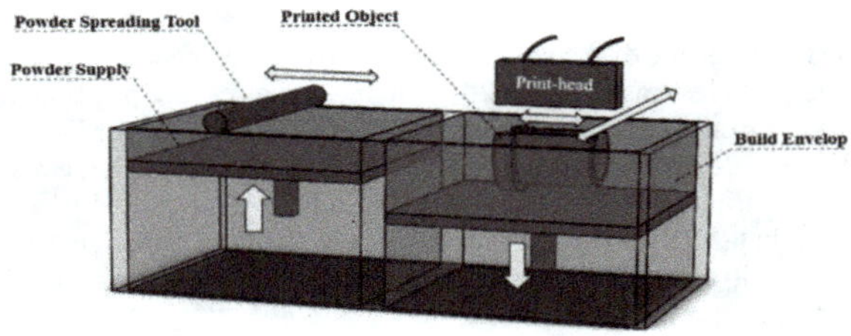

fig 3 – schematic of binder jetting

Direct energy deposition (DED) is one of MAM processes that is fully reliant on the assistance of other multi axis equipment, such as Computer Numerical Control (CNC) or robots [26]. DED is a 3D method used to fabricate metals, ceramics, and polymers objects via melting powder or wire through a thermal energy source (electron beam, laser beam, and electric arc) This kind of method has also various alternative names since the 90s, such as, Laser metal deposition (LMD), laser solid forming (LSF), and laser engineered net shaping (LENS) [55]. Although DED can manufacture large metallic materials due to its advantage of higher deposition rate, DED is only able to produce parts with lower finishing resolution than PBF [29, 48]; in the case of DED, nozzle is in use here, likewise of electron beam, laser, or plastic arc. It is important to shed the light on the fact that DED is capable of repairing existing part, or deposit additional layers on existing part through the backing of CAD software [2].

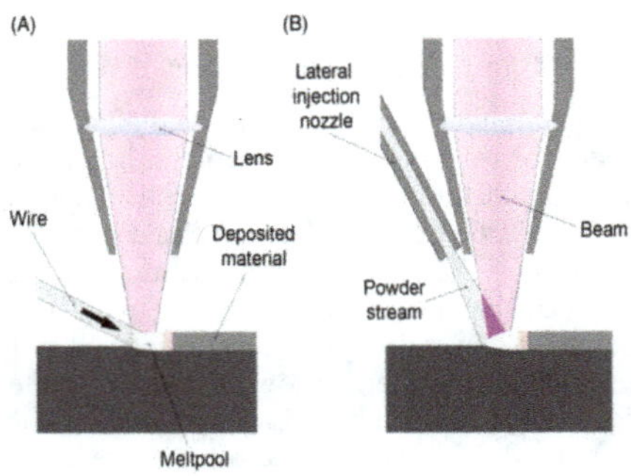

fig 4 – schematic DED process (a) wire based & (b) powder stream based DED

INDIRECT MAM

Additive manufacturing is utilized to fabricate prototypes in a direct approach, however, this process can be used and considered an indirect method whenever tools, or master patterns are manufactured, which are designated as the assistance to produce the required final parts. Favourable examples are, die casting, injection moulding tools, sand casting, casting, and vacuum casting [17]. Moreover, when it comes to indirect MAM, highly filled polymer and polymer with ceramic powders and metals incorporated are the main materials associated with MAM, which also consist of three main categories; material jetting (MJ), vat photopolymerization (VP), and material extrusion (ME) (Fig 1). material jetting (MJ), is a 3D printing method that produces very precise objects with massive variety of materials and colours. MJ jets small drops of liquid plastics and ultraviolet light exposure instantly cures the plastics solidifying it, and so, layer by layer, compound objects take shape. Nevertheless, in this process, material properties can be adjusted, such as durability or heat resistance, therefore, designers can predict future needs and serve them now. Vat photopolymerization (VP), produces a 3D models. VP uses a Vat (large tank to hold plastic liquid) to melt and solidify the resins through ultraviolet light exposure to finally form polymers (photopolymer) [2,47]. Material extrusion (ME), it is 3D process in which the material is selectively dispensed through a nozzle to form a three dimensional object, normally at elevated temperature. Also, material extrusion (ME) is also know as fused filament fabrication (FFF) with a wide range of materials to be fabricated, such as ceramics, metals, composites, and biomaterials. [2]

Wire arc additive manufacturing (WAAM)

The first invention of arc welding goes back to 1881[39]. WAAM involves only low investment compared to the rest of AM methods [51]. Wire and arc additive manufacturing (WAAM) is a three dimensional production process that works by rapidly melting metal layer by layer through a thin wall component, which has a high efficiency in deposition rate and low cost process. Usually, the standard thickness of the wire ranges between 0.8 mm to 1.2 mm [41]. Thus, the material usage can exceed 90%, 85 % - 90% of energy efficiencies, and high deposition rate that can reach 4 kg/h [40]. WAAM possesses the capability to manufacture massive part with zero reliance on moulds via the use of CAD / CAM software. However, WAAM production is highly dependent upon additional positioning systems, such as a robot, or CNC milling machine [16]. Additionally, WAAM main components are motion system, material deposition system, and heat source [28]. Studies have shown various ways of adopting and implementing heat sources during the WAAM operation, such as Cold metal transfer (CMT), Plasma Arc Welding (PAW), Gas Tungsten Arc Welding (GTAW), and Gas Metal Arc Welding (GMAW) [15].

CMT technology used in WAAM has the advantage of fabricating large scale structural parts as the the cost of materials is low and the deposition rate is high. CMT technology in WAAM and its products are widely used in the industries, especially in biomedical, aerospace, and chemical fields. Nevertheless, when it comes to the best suited materials to be produced via CMT-based WAAM, a high strength with low density titanium alloy (Two-phase α/β Ti-6Al-4V) is produced. However, the quality surface improvement after machining is lacking due to the methodology implemented (layer-by layer deposition) and it has become one of the main challenges out there, whereas the the recent studies and experimentation were concentrating mainly on tensile properties, evolution, and microstructure [15,58].

Moving on to Plasma Arc Welding (PAW), PAW is one of the most modern and advanced manufacturing process and it is well known for its high efficiency and low production cost, lower cost in operation as well as lower equipment cost compared to other advanced welding technologies. Plasma Arc Welding (PAW) is a modified type of TIG Welding and it is nothing but the ionized gases. The word Plasma is nothing but the forth state of matter after solid, liquid, and gas. PAW has proven its ability to exert an extreme high welding currents and deep penetration during the appearance of keyhole mode. Furthermore, Plasma Arc Welding (PAW) has the ability to produce gas as a heat source with an exceeding gas temperature of 20,000 K [20], which creates a narrow constricted arc pattern and the heat can easily transfer to the bottom of the work piece via the keyhole, also PAW has the tendency to concentrate on high flux and pressure [21]. Refer to below (fig 11).

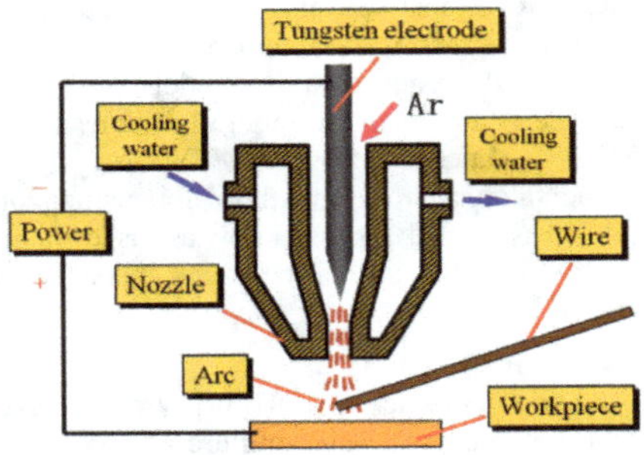

fig 11 – PAW [29]

Gas Tungsten Arc Welding (GTAW) is also same type as Tungsten Inert Gas (TIG). GTAW is an arc welding process where arc is constituted between a non-consumable electrode and conductive base metal. Additionally, the main function of inert shielding gas such as helium or argon is to eliminate any possible atmospheric pollution and oxidation in the welding area and contribute to solidify the work piece. Also, the gas plays a vital role in increasing the electric arc strike due to conductivity increase between the work piece and the electrode. Having said that, the main function of electrode is to melt the metal with a melting temperature up to 3422 Celsius. Moreover, this specific process is strictly applicable to certain materials, such as Carbon, magnesium, stainless steel, and titanium. However, when it come to the aspect of heat addition of this process, welding speed, current, and voltage are main three factors that GTAW is extensively dependant upon to achieve extraordinary excellent welds, spatter free, and low distortion. On the other hand, there are few disadvantages associated with GTAW, which are the method is sensitive for airflow, time consuming, and pricy [22]. Refer to below (fig 10).

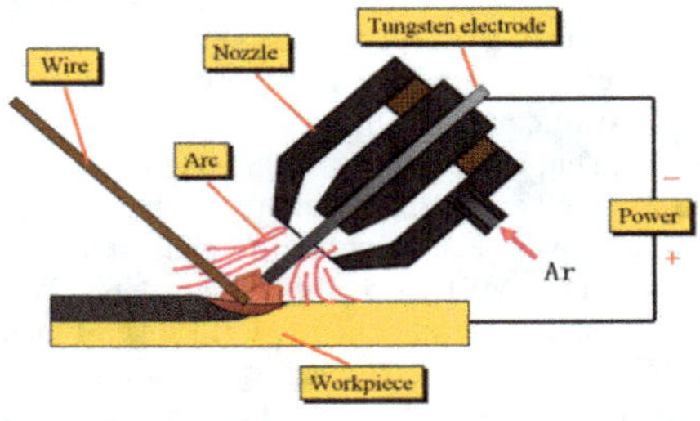

fig 10 – GTAW [29]

Gas Metal Arc Welding (GMAW) is one such arc welding process where the arc is constituted between a consumable wire electrode and the conductive base plate. This process also known as GMAW-DED due to the use of the technology of direct energy deposition, which can be deposited layer by layer in the form of beads. One of the biggest advantages of GMAW is its high deposition rate (3 – 4 kg/h) [14]. Nevertheless, this process is popularly used in the industry and could be applied in a wire form shape non-ferrous and ferrous metals, such as nickel, titanium and aluminium alloys, as well as stainless steel due to its reliability, flexibility, and high efficiency [24]. Refer to below figures (8 & 9&14).

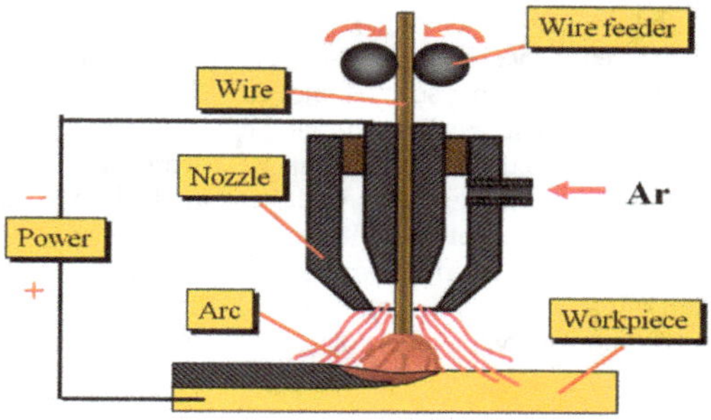

fig 8 – GMAW

Table 1
Approximate cost of different metal materials in wire and in powder [4].

Feedstock	Ti-6Al-4V	Inconel 718	Inconel 625	Stainless Steel 316L
Wire (£/kg)	120	58	49	12
Powder (£/kg)	280	80	80	40

fig 9 – Approximate cost of materials

fig 14 – Real world example of fabricated part (WAAM)

Conclusion

In conclusion, this paper's main objective was to introduce the readers to the establishment and development of AM and WAAM. Looking over the innovative idea behind this part of AM/WAAM, it could be clearly perceived that as AM/WAAM continues to expand, various tangible benefits are gradually appearing, such as, the reduction of human error and interference, reduction of exposure to polluted chemicals or dirt, and the decrease in materials usage. There is no doubt that WAAM is considered one of the tremendous transition of AM in the industrial revolution 4.0. As science and technology evolves, there is no doubt of what AM/WAAM could bring further to our future to make it a better world for us. As I went through uncountable journals and e-books, I found so much of rich information about this topic, and there are yet more to come. I am very optimistic and excited to dip deep into the development of this topic and consider it as part of my thesis. There is no suspicion that it is going to be certainly a challenging topic to deliver, but it should be fun.

References

[1] Mark Armstrong, Hamid Mehrabi, Nida Naveed, (2022). An overview of modern metal additive manufacturing technology. Journal of Manufacturing Processes. Vol: 84, pp: 1001-1029 https://doi.org/10.1016/j.jmapro.2022.10.060

[2] J.P.M. Pragana, R.F.V Sampaio, I.M.F. Praganca, C.M.A. Silva, P.A.F. Martins, (2021). Hybrid metal additive manufacturing: A state-of-the-art review. Advances in industrial and manufacturing engineering. Vol: 2 https://doi.org/10.1016/j.aime.2021.100032

[3] Jacopo Lettori, Roberto Raffaeli, Margherita Peruzzini. (2020). Additive manufacturing adoption in product design: an overview from literature and industry. Procedia Manifacturing. Vol: 51, pp: 655-662 https://doi.org/10.1016/j.promfg.2020.10.092

[4] Bert Lauwers, Nataliia Chernovol, Benjamin Peeters, (2020). Hybrid Manufacturing based on the combination of Mechanical and Electro Physical-Chemical Processes. Procedia CIRP. Vol: 95, pp: 649-661 https://doi.org/10.1016/j.procir.2020.11.003

[5] Mehmet Erdi Korkmaz, Saad Waqar, A. Garcia-Collado, (2022). A technical overview of metallic parts in hybrid additive manufacturing industry. Journal of material research and technology. Vol: 18, pp: 384-395 https://doi.org/10.1016/j.jmrt.2022.02.085

[6] Bert Lauwers, Firtz Klocke, Andreas Klink, A. Erman Tekkaya, (2014). Hybrid processes in manufacturing. CIRP Annals-manufacturing technology. Vol: 63, pp:561-583 https://doi.org/10.1016/j.cirp.2014.05.003

[7] Kathrin Pfahler, Dominik Morar, Hans-Georg Kemper, (2019). Exploring Application Fields of Additive Manufacturing Along the Product Life Cycle. Procedia CIRP. Vol: 81, pp: 151-156 https://doi.org/10.1016/j.procir.2019.03.027

[8] Miguel Seabra, Jose Azevedo, Aurelio Araujo, (2016). Selective laser melting (SLM) and topology optimization for lighter aerospace components. Procedia Structural Integrity. Vol: 1, pp: 289-296 https://doi.org/10.1016/j.prostr.2016.02.039

[9] Jue Liu, Shiyun Dong, Xin Jin, Pengyue Wu, (2022). Quality control of large-sized alloy steel parts fabricated by multi-laser selective laser melting (ML-SLM). Vol: 223 https://doi.org/10.1016/j.matdes.2022.111209

[10] Xiaoyu Bi, Runzhou Huang, (2022). 3D printing of natural fiber and composites: A state-of-the-art review. Vol:222 https://doi.org/10.1016/j.matdes.2022.111065

[11] Greymi Tan, Nicole Ioannou, Essyrose Mathew, Aristides D. Tagalakis, (2022). 3D printing in Ophthalmology: From medical implants to personalised medicine. International Journal of Pharmaceutics. Vol: 625 https://doi.org/10.1016/j.ijpharm.2022.122094

[12] Xiaoyan Xu, Alvaro Goyanes, Sarah J. Trenfield, (2021). Stereolithography (SLA) 3D printing of a bladder device for intravesical drug delivery. Materials Science ans Engineering. Vol: 120 https://doi.org/10.1016/j.msec.2020.111773

[13] Andre Mussatto, (2022). Research progress in multi-material laser-powder bed fusion additive manufacturing: A review of the state-of-the-art techniques for depositing multiple powders with spatial selectivity in a single layer. Results in Engineering. Vol: 16 https://doi.org/10.1016/j.rineng.2022.100769

[14] Jian Gou, Zhijiang Wang, Shengsun Hu, (2022). Effect of cold metal transfer mode on the microstructure and machinability of Tie6Ale4V alloy fabricated by wire and arc additive manufacturing in ultra-precision machining. Journal of material and research technology. Vol:21, pp: 1581-1594 https://doi.org/10.1016/j.jmrt.2022.10.011

[15] J. Gonzalez, I. Rodriguez, J-L. Prado-Cerqueira, (2017). Additive manufacturing with GMAW welding and CMT technology. Procedia manufacturing. Vol: 13, pp: 840-847 https://doi.org/10.1016/j.promfg.2017.09.189

[16] Joaquin Montero, Pablo Vitale, Sebastian Weber, (2020). Indirect Additive Manufacturing of resin components using polyvinyl alcohol sacrificial moulds. Procedia CIRP. Vol: 91, pp: 388-395. https://doi.org/10.1016/j.procir.2020.02.191

Technical and economic feasibility of solar PV systems supported by energy storage in hospitals in KSA

Mounir Bouzguenda[1,a*], Abdulaziz Almulhim[2,b] , Abdulrahman Shawki[2,c], Ahmed Al-Baa'dani[3,d], Ibrahim Ahmed[4,e], Mohammed Alkulaib[5,f], and Mohammed Al-Aqil[1,g]

[1]College of Engineering, King Faisal University, Eastern Province, AlAhsa, Saudi Arabia

[2]Al-Salmaniah, Hofuf, Saudi Arabia

[3]Al-Khaldiah, Hofuf, Saudi Arabia

[4]Al-Khars, Almubarraz, Saudi Arabia

[5]Al-Oyun, Oyun, Saudi Arabia

[a]mbuzganda@kfu.edu.sa, [b]Abdulazizmm99@hotmail.com, [c]Shawki291@outlook.com, [d]Smoo.1998@gmail.com, [e]Bulloain@gmail.com, [f]alkulaib2013@gmail.com, [g]malaqil@kfu.edu.sa

Keywords: Payback, Cost of Energy, KSA 2030 Vision, Community Energy Storage, Load Factor, Solar PV

Abstract. The ambitious KSA 2030 Vision has among others targeted the deployment of renewable energies to meet KSA's primary electric load in the residential, commercial, and industrial sectors. Within this framework, the authors have evaluated the economic and technical feasibility of using a combination of solar PV systems and central energy storage systems in the health sectors. The combination of the two sources means reducing gas emissions, and increasing the instantaneous power from renewable energy sources while using energy storage to counterattack the intermittency of solar PV energy that is known to cause stability issues for the grid. In addition, it reduces the dependency on standby diesel generators that emit high pollution levels. The paper addresses the available energy storage devices and the design of a hybrid solar PV system combined with battery-stored energy. Through the study, it was found that up to 15 percent of the electric consumption can be provided by the solar PV system. It was found that the cost of energy competes with conventional electricity. In addition, the cost of storage devices has a significant impact on the cost of energy. Moreover, low payback periods were obtained for some systems.

Introduction

Batteries have been around since the early 1800s, and hydropower with pumped storage has been employed in the US since the 1920s. However, there has been a noticeable increase in the construction of new energy storage projects as well as the creation of advanced and improved energy storage technologies as a result of the desire for a more dynamic and cleaner grid. One of the projects and technologies revolves around solar energy.

In terms of solar energy potential, Saudi Arabia is one of the richest nations in the world. The nation has installed more than 3 GW of photovoltaic (PV) capacity in just six years. This will supply energy to more than 600,000 dwellings. By installing an additional 7.2 GW of PV capacity by 2022, Saudi Arabia is on pace, according to a new estimate from Bloomberg New Energy Finance (BNEF). The installed capacity would then increase to 10.2 GW. [1]

There are several reasons that contribute to the Saudi Solar Market's rapid growth. For example, solar panels are now more reasonably priced for customers thanks to a recent sharp decline in their

Materials Research Proceedings 31 (2023) 531-537 https://doi.org/10.21741/9781644902592-55

cost. Second, the Saudi government has made significant investments in solar and other renewable energy sources. Also, solar will play a role in achieving the government's objective of having 9.5GW of renewable energy capacity by 2023. And finally but not least, people are becoming more and more conscious of solar energy's advantages for the environment. People are looking for solutions to lower their carbon footprint as their worry over climate change grows.

Community Solar, Storage, and Microgrids

A community solar project is a sizable solar power plant in the center of town whose electricity is used by numerous properties. A single community solar project can power hundreds or even thousands of houses because it is measured in megawatts, whereas residential solar installations are measured in kilowatts. Other names for community solar include shared solar, solar gardens, and roofless solar. Participants in community solar projects lock in monthly savings on their electricity bills, which is one of the two main advantages of community-shared solar. Community solar also enables more people to access the advantages of solar power. Community solar also benefits utility companies by allowing them to build projects in their regions.

What is the community microgrid?

The local bulk (i.e., high-voltage) transmission system, commonly referred to as the "microgrid," is a relatively small, controllable power system that is made up of one or more generation units connected to nearby users. Microgrids fall under the category of distributed generation because the energy (power and heat) is produced close to where they are needed. In the past, combined heat and power (CHP) systems powered by fossil fuels and generators with reciprocating engines were used to power microgrids. Today, projects are utilizing more environmentally friendly resources, such as solar energy and energy storage. When developing technologies like fuel cells and even tiny modular nuclear reactors become commercially viable, microgrids may be powered by renewable energy sources, combustion turbines powered by natural gas, or other sources. Microgrid demand has increased in the Saudi Arabian market due to the rising demand for reliable power supply and rapid infrastructure development. The need for microgrids is rising in the healthcare and military sectors, and there are many chances for growth in the microgrid industry. The development of renewable energy-based microgrid technology has also been boosted by several government strategies and technological developments.

System Design

The research team have planned to use solar PV to feed a local hospital equipped with energy storage banks so that this energy usage can be shifted during the day. Moreover, energy exchange with the local grid company where excess energy is sold back to the grid at an agreed upon tariff.

HOMER software is used in order to simulate the power system design. HOMER is capable of calculating the optimized cost for the system. It requires input data such as daily load demand, peak load, load factor, and renewable resources. Meanwhile, fuel costs, conventional generation units, maintenance costs, and storage devices characteristics are also considered in the optimization process.

Figure 1 shows how HOMER operation principles, the required input data, and optimal generation configuration.

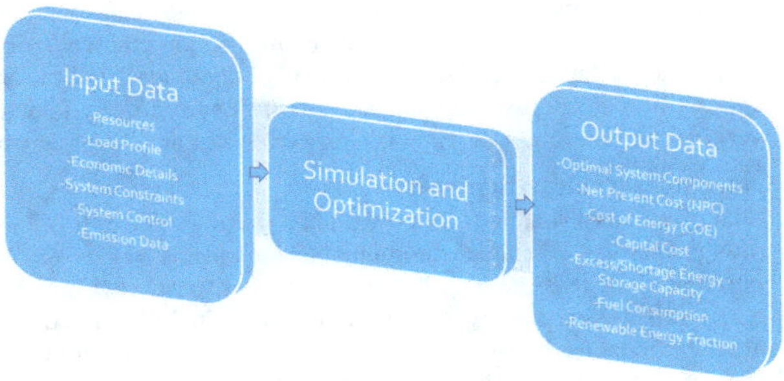

Figure 1: HOMER operation principles

System Location

The authors chose one of the most important and respected hospitals in Saudi Arabia. Figure 2 shows the layout of the hospital and potential areas for installing a rooftop solar PV system. There are 8 areas marked in red.

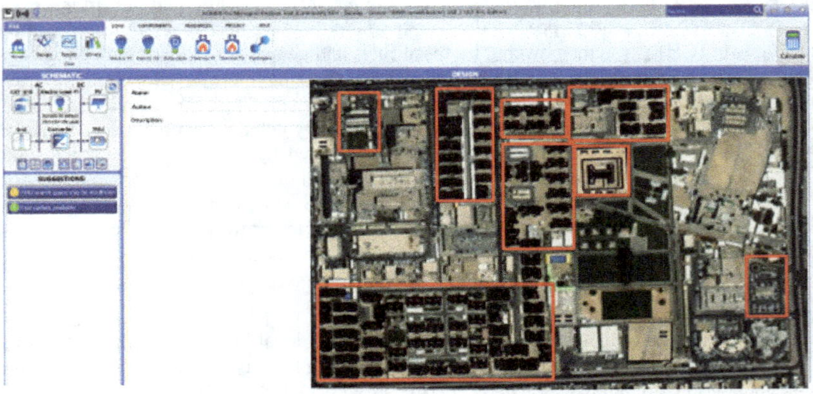

Figure 2: The layout of the public hospital under study

System Components

Figure 2 displays the various components of the proposed power system for the hospital on the top left. Table 1 displays the generation specifications and hospital electric demand.

*Table 1: **Generation and Load Constraints***

Item	Description		Notes
Load	Peak load Load factor Daily energy	30 MW 52% 362 MWh	
Grid	Cost of Energy (purchase/sale)	0.64 SAR/kWh	Capacity constraint 19 MW
Storage Unit	Lifetime Capital cost O&M cost Replacement cost	5 years 300,000 SAR/MWh 3,000 SAR 30,000 SAR	Minimum storage 1 MWh Maximum storage 6 MWh
Solar PV	Lifetime Capital cost O&M cost Replacement cost	25 years 4,000 SAR/kW 110 SAR 400 SAR	Minimum capacity 1000 kW Maximum capacity 15000 kW
Converters	Lifetime Capital cost O&M cost Replacement cost	15 years 330 SAR / 1kW 100 SAR 330 SAR	Capacity 100 kVA-5000 kVA
Diesel Generators	Lifetime Capital cost O&M cost Replacement cost	50,000 Hours 154,000 SAR / Unit 318,000 SAR 100,000 SAR	Size 500kVA/1000kVA (already installed and manually scheduled)

To find the optimal generation mix for the hospital, the Homer software was run using the above generation and load constraints found in Table 1. The search domain was extended to include additional options such as wind turbines and hydro storage and the results are included in Figure 5. However, considering that the KSA vision requires the installation of as much PV as possible, the authors selected option One which is pointed with the arrow below. The performance and the economic analysis of the selected option are discussed below. Table 2, the cost summary is found of the selected optimal system.

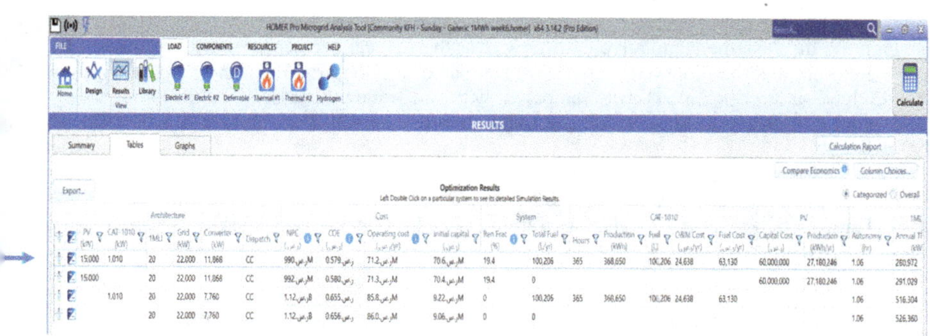

Figure 5: The Optimal Generation Mix Scenarios for the Hospital.

Figure 6 lists the project summary for the proposed solar PV and storage power system versus the existing system. In particular, the operating cost is reduced by approximately 14.88 MSAR and an annual saving of 14.8 MSAR. The designed system performance, economic analysis, and generation breakdown are shown in Table 2.

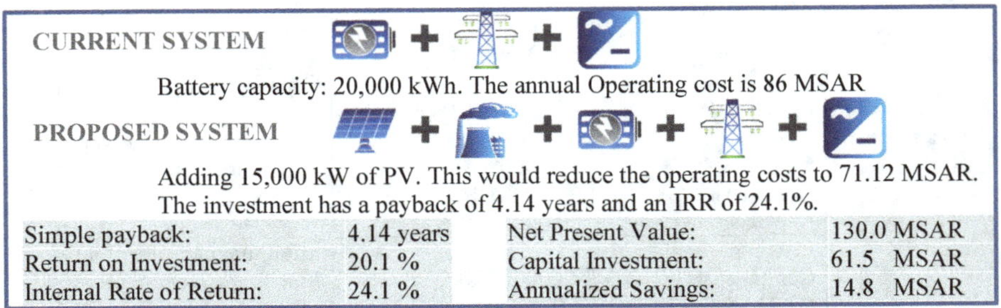

Figure 6. Project summary for the proposed solar PV and storage power system.

Table 2: Performance, economic analysis, and generation breakdown

Solar PV System		
Rated Capacity	15,000	*kW*
Capital Cost	60	*MSAR*
Specific Yield	1,812	*kWh/kW*
PV Penetration	20.6	*%*
Total Production	27,180,246	*kWh*
Maintenance Cost	1,650,000	*SAR*
LCOE	0.231	*SAR/kWh*
Storage System		
Rated Capacity	20,000 kWh	*kWh*
Annual Throughput	280,972	*kWh/yr*
Maintenance Cost	60,000	*SAR*
Autonomy	1.06	*Hour*
Expected Life	15.0	*Year*
Capital Costs	6.0	*MSAR*
Losses	29,574	*kWh/yr*

Figure 7 displays the cumulative cash flow over the 25-year project lifetime. The cash flows for the current and the proposed systems are compared. It is observed that the proposed system would have a higher initial cost compared with the reference case. The hospital energy consumption is shown in Figure 8. The hospital requires 362 MWh/day and has a peak of 29.12 MW.

Finally, the annual energy purchased from the grid is approximately 106 GWh and the annual energy sold to the grid is 116 MWh. Table 3 displays the energy interchange with the grid and total charges. The annual energy purchased from the grid and that sold to the grid are 106.20 GWh and 116.13 MWh, respectively.

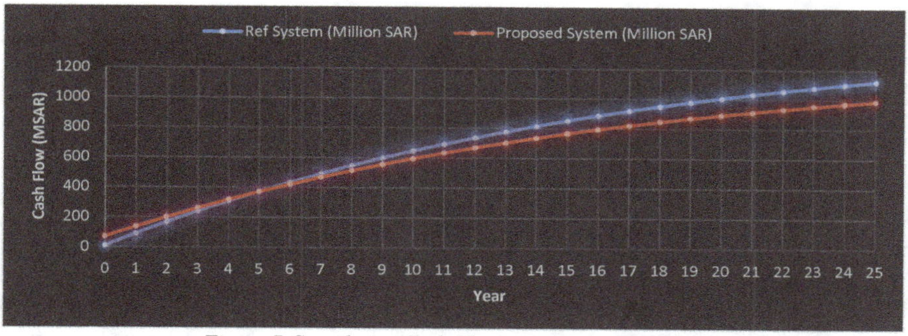

Figure 7 Cumulative Cash Flow over Project Lifetime

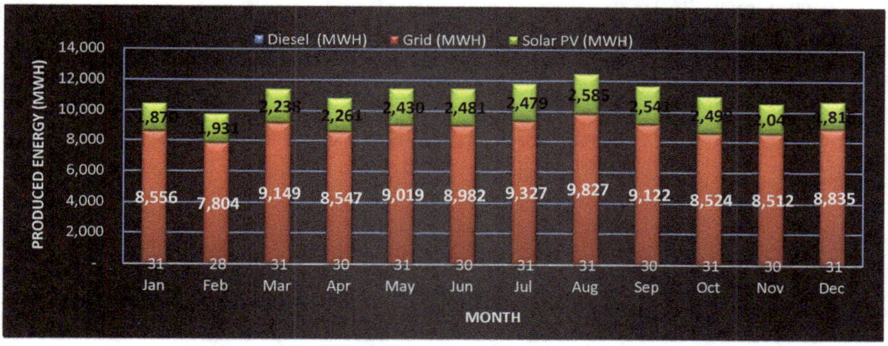

Figure 8. Monthly Grid, Solar PV and diesel Energy

Table 3 Energy Production, exchange, and charges.

Month	Solar PV Energy (GWh)	Energy Purchased (MWh)	Energy Sold (MWh)	Net Energy Purchased (MWh)	Energy Charge (MSAR)	Total Charges (MSAR)
Jan	1.870	8,556	8,524	8.524	5.460	5.465
Feb	1.931	7,804	7,796	7.796	4.990	4.995
Mar	2.238	9,149	9,138	9.138	5.850	5.855
Apr	2.261	8,547	8,536	8.536	5.460	5.465
May	2.430	9,019	9,008	9.008	5.770	5.775
Jun	2.481	8,982	8,977	8.977	5.750	5.755
Jul	2.479	9,327	9,321	9.321	5.970	5.975
Aug	2.585	9,827	9,823	9.823	6.290	6.295
Sep	2.541	9,122	9,118	9.118	5.840	5.845
Oct	2.499	8,524	8,511	8.511	5.450	5.455
Nov	2.049	8,512	8,508	8.508	5.450	5.455
Dec	1.815	8,835	8,827	8.827	5.650	5.655
Annual	27.18	106,205	106,089	106.089	67.930	67.989

Conclusion and Future Work

In this study, the optimal hybrid hospital power supply mix is obtained. It will achieve a cost of energy of 0.57 SAR/kWh and the payback is 4.14 years. Furthermore, the PV energy will reach

27.18 GWh with more than a 20% PV penetration level. Such a high penetration level is achieved in other parts of the kingdom due to the significant solar resources. On the other hand, the emission savings of CO_2 for the case without PV energy will reach 83,602 tons/yr. However, for the case that includes PV energy, it will reach 67,000 tons/yr.

Acknowledgment
The authors would like to thank King Faisal University for funding this research.

References

[1] S. Shetty, "Saudi Arabia's Solar Market Outlook for 2022," *SolarQuarter*, 27-May-2022. [Online]. Available: https://solarquarter.com/2022/05/27/saudi-arabias-solar-market-outlook-for-2022/ . [Accessed: 12-Sep-2022].

[2] S. E. I. A. "Community Solar," "Community solar," *SolSmart*, 20-Aug-2019. [Online]. Available: https://solsmart.org/solar-energy-a-toolkit-for-local-governments/community-solar/ . [Accessed: 12-Sep-2022].

[3] Amrouche, S. O., Rekioua, D., & Rekioua, T. (2022, March 8). *General idea of energy storage in renewable energy systems*. HAL Open Science. Retrieved September 11, 2022, from https://hal.archives-ouvertes.fr/hal-03601393

[4] "SMEs as a backup power supply Molokac Rybar Rybarova Grega - Indico.": https://indico.cern.ch/event/16708/sessions/126101/attachments/203820/286078/SMES_as_a_backup_power_supply_Molokac_Rybar_Rybarova_Grega.pdf . [Accessed: 19-Sep-2022].

[5] Investing, T. (2021, April 12). *Energy storage is now booming as the world moves faster to renewable energy*. SeekingAlpha. Retrieved 11-Sep-2022, from https://seekingalpha.com/article/4418625-energy-storage-is-now-booming-world-moves-faster-to-renewable-energy.

[6] "Advantages and disadvantages of pumped-storage hydropower plants ..." [Online]. Available: https://www.researchgate.net/figure/Advantages-and-Disadvantages-of-Pumped-Storage-Hydropower-Plants-developed-by-the_tbl1_341799344 . [Accessed: 19-Sep-2022].

[7] Systems, "5 benefits of compressed air energy storage," *Compressed Air Systems, Inc.*, 23-Nov-2020. [Online]. Available: https://www.compressedairsystems.com/blog/5-benefits-compressed-air-energy-storage/ . [Accessed: 20-Sep-2022].

[8] *Thermal energy storage*. [Online]. Available: http://www.eseslab.com/ESsensePages/TES-page . [Accessed: 20-Sep-2022].

[9] "Hydrogen Energy Storage," *Energy Storage Association*, 07-Apr-2021. [Online]. Available: https://energystorage.org/why-energy-storage/technologies/hydrogen-energy-storage/ . [Accessed: 20-Sep-2022].

[10] L. Richardson, "History of solar energy: Timeline & invention of solar panels: Energysage," *EnergySage Blog*, 27-Jun-2022. [Online]. Available: https://news.energysage.com/the-history-and-invention-of-solar-panel-technology/#:~:text=In%20theory%2C%20solar%20energy%20was,light%20torches%20for%20religious%20ceremonies . [Accessed: 22-Sep-2022].

[11] SEIA, "About solar energy," *SEIA*. [Online]. Available: https://www.seia.org/initiatives/about-solar-energy . [Accessed: 22-Sep-2022].

Advanced Topics in Mechanics of Materials, Structures and Construction - AToMech1-2023 Materials Research Forum LLC
Materials Research Proceedings 31 (2023) 538-549 https://doi.org/10.21741/9781644902592-56

Design of brake pads from economical materials

NOUARI Abdessamed[1], TAFRAOUI Ahmed[1,*], TAFRAOUI Salima[1], MALLEM Mohamed[1]

[1] Laboratory of Reliability mechanical engineering of TAHRI Mohamed University Bechar

* tafraoui.ahmed@univ-bechar.dz

Keywords: Design, Brake Pads, Molding, Lining, Materials

Abstract. The aim of this work is to manufacture brake pads from a local ecological eco-material. The most commonly used materials in brake linings are asbestos, metals, and ceramics. However, asbestos releases hazardous gases upon application, making it carcinogenic. For the production of this eco-material, the powder molding technique is used. The brake pads were produced using dune sand as the base material, following the standard procedure used by manufacturers, with the aim of exploiting the chemical and physical characteristics of dune sand, which is abundant in the Algerian desert. This represents 70% of the brake pad production, in addition to 30% of polyester resins in compression molding. The properties examined were microstructure analysis, compression strength, roughness, temperature testing, and water absorption. The preliminary results show that sand can be used as a material in brake manufacturing. The results obtained in this work were compared to those of other studies and commercial brake pads and showed a close correlation, indicating that sand can be used in brake pad production.

1. Introduction

Brake pads are essential components of braking systems used in modern automobiles. They are often made from materials such as asbestos, metals, or ceramics. However, the use of asbestos is now avoided due to its carcinogenic properties. Brake pads are designed to convert kinetic energy into thermal energy through friction, which allows the vehicle to slow down or stop [1, 2].

The two most common types of automotive brakes are drum brakes and disc brakes. Drum brakes are housed within a drum, while disc brakes are exposed to the environment. Brake systems have made great progress in recent decades, and car manufacturers are continuously seeking to improve the performance of their brakes.

Standards for brake pads vary by country and local regulations. In general, brake pads must meet certain requirements in terms of performance, safety, and the environment. For example, in the United States, Federal Motor Vehicle Safety Standard (FMVSS) No. 105 establishes minimum requirements for brake performance on passenger vehicles, while in Europe, brake pads must comply with the ECE R90 standard, which defines requirements for brake safety and performance [3, 4]

When it comes to brake pad performance, some important characteristics to watch for are the coefficient of friction, lifespan, heat resistance, thermal stability, and performance in wet conditions. For example, in Europe, brake pads must have a minimum coefficient of friction of 0.35 to comply with the ECE R90 standard [5, 6].

It is important to note that standards and requirements can vary depending on the type of vehicle and intended use. For example, racing vehicles may require brake pads with higher performance than standard touring vehicles.

The most commonly used materials for brake discs are steel and grey cast iron, but these materials may not always be suitable for high loads. Therefore, car manufacturers are looking to find more performance-enhancing brake disc materials, such as composites and ceramics [7].

The production of brake pads using eco-friendly materials is a process that uses environmentally friendly materials to replace traditional components that often contain harmful chemicals. The eco-friendly materials used for brake pad production may vary, but often include materials such as bamboo fibers, wood powder, coconut shell, recycled textiles, and bio-resins [8, 9, and 10].

In this context, research is being conducted to create brake pads from innovative, environmentally friendly eco-materials. For example, sand dunes have been used as a base material for brake pad production, leveraging the physical and chemical characteristics of this material that is abundant in the Algerian desert [11].

The properties of sand-based brake pads have been examined and compared to those of commercial asbestos-based brake pads. The results showed a close correlation, indicating that sand can be used in brake pad production. In addition to their low density and high wear resistance, these innovative local materials also present an optimal cost for consumers and manufacturers.

In summary, the search for alternative materials for brake pad production is ongoing to improve performance and reduce environmental impact. Local and innovative eco-materials offer sustainable solutions for the automotive industry and contribute to a more responsible use of natural resources.

2. Materials and Methods

2.1 Materials

The materials used in the formulation of a brake pad, (fig. 1), (fig.2).

- Silicones and their hardener.
- Dune sand from the Western Erg region of Bechar, with chemical and physical characteristics (see tables 1 and 2), (fig.2).
- Ground dune sand with a fineness modulus of 0.8 (S.D.B) (fig. 2);
- Ground red clay (A.B) (fig.2);
- Binders (accelerated isophthalic polyester resin) (fig. 2).

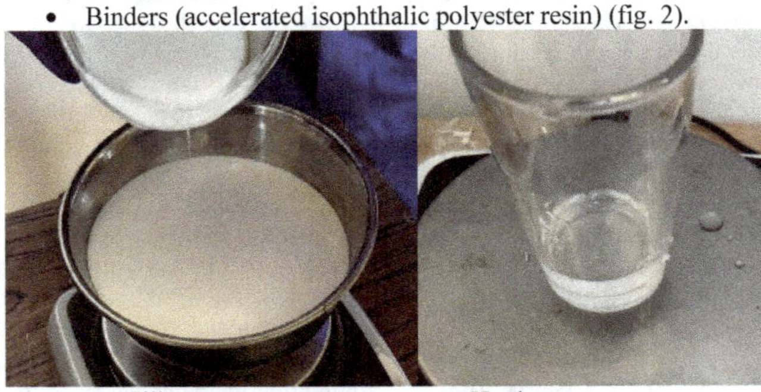

Silicone Hardener

Figure 1. Materials used in casting.

Resin Red clay Dune sand (SD) Ground dune sand (SDB)

Figure 2. Materials used in formulation.

Tableau 1. Chemical characteristics of dune sand in %.

Oxides	SiO_2	AL_2O_3	Fe_2O_3	MgO	CaO	Na_2O	K_2O	SO_3	Fire loss
(%)	97.33	0.830	0.24	0.41	0.07	0.09	0.04	0.18	0.40

Tableau 2. Physical characteristics of dune sand.

Specific surface BET(cm2/g)	Average size, D50(µm)	Real density	Hardness
115	200	3	7

Tableau 3. Chemical characteristics of crushed dune sand.

Oxides	SiO_2	AL_2O_3	Fe_2O_3	MgO	CaO	SO_3	K_2O	CI	Na_2O	Ti O_2
(%)	87.65	1.02	0.75	0.25	0.29	0.02	0.03	0.018	0.06	0.04

Tableau 4. Chemical characteristics of clay.

Oxides	SiO_2	AL_2O_3	Fe_2O_3	MgO	CaO	SO_3	K_2O	CI	Na_2O
(%)	54.64	16.82	5.64	1.73	6.88	0.05	2.47	0.018	0.05

2.2 Methods

2.2.1 Formulation

Brake pads are critical components of a vehicle's braking system, and their formulation depends on many factors such as the type of vehicle, weight, type of braking system, required performance, driving conditions, etc. However, generally, brake pads are made from composite materials, which may include the following elements:

Friction material: The outer layer of the brake pad is typically composed of a friction material, which is responsible for generating braking force. Commonly used friction materials include ceramic, sintered metal, carbon-ceramic, carbon-carbon, etc.

Backing material: The backing material is used to hold the friction material in place and to provide a solid base for the brake pad. Commonly used backing materials include fiberglass, carbon fiber, steel, etc [12, 13].

Other additives: Additives such as lubricants, bonding agents, reinforcement materials, etc., may be added to enhance the properties of the brake pad, such as durability, corrosion resistance, thermal conductivity, etc.

It should be noted that the exact formulation of brake pads varies depending on the manufacturer and car model, and it is therefore important to follow the manufacturer's specifications to ensure optimal performance and maximum safety. The formulation used is presented in Table 5.

Tableau 5. Formulation of molding sand.

Materials	Quantity (%)	Resin	Quantity (%)
dune sand	71.42		30 à 35
crushed dune sand	63.68		35 à 38
red clay	58.55		40 à 43

To prepare the mixture for molding, we slowly mix the material with resin (dune sand, crushed dune sand, clay) in order to avoid the formation of lumps and ensure that the material is completely moistened to guarantee a certain permeability during degassing. Once the mixing is complete, we retrieve the material and place it in a tray. At this point, the material is ready to be used for molding.

2.2.2 Making silicone molds

The manufacturing process may vary depending on the materials used, but it generally starts with the preparation of the material mixture. The materials are mixed until they are homogeneous and the mixture is poured into molds for the production of brake pads. Then, polymerization is carried out at a high temperature to harden the materials and make them solid.

We made wooden molds with a thickness of 25 mm and dimensions of 145 mm long by 85 mm wide. Afterwards, we placed brake pads on their outer side before proceeding with the preparation of a homogeneous mixture of silicone and hardener (see fig. 3a and 3b). Then, we poured this silicone mixture into the wooden molds (see fig. 3c) on the outside (see fig. 3) and on the inside (see fig. 4a) and left them to rest for 24 hours before removing the molds (Fig. 3d, Fig. 4b).

(a) (b) (c)

(d)

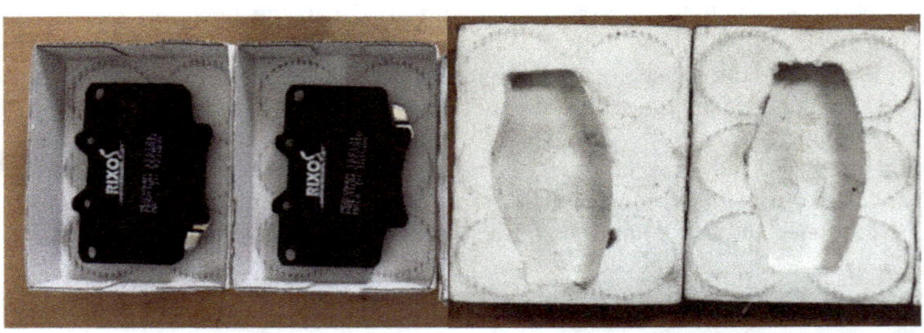

(a) Brake pad model (b) Silicone Brake Pad Mold.

Figure 3. Realizations of silicone molds outside and inside.

2.2.3 Manufacturing of a brake pad.

Depending on the chosen formulation, we mix different materials such as dune sand, crushed dune sand, or clay until the mixture is homogeneous (see fig. 4b). Then, we pour the cold mixture into the molds (see fig. 4c) and let it rest for 24 hours. After removing the pads from the silicone molds, we obtain the finished product (fig. 4e).

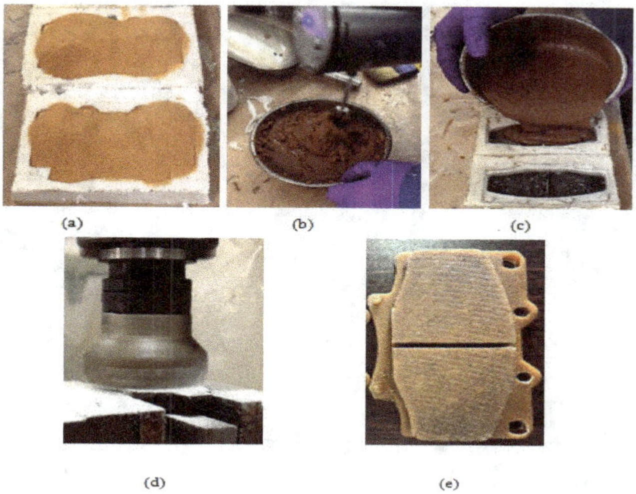

Figure 4. manufacture of a brake pad.

Alternatively, we have manufactured brake pad linings without the base or backing plate (fig. 5, 6). The preparation of the material mixture is the same as in the first experiment, but this time we use an oven at 175 degrees for the polymerization of the composite resins (fig. 7) [14, 15]. After this, we apply a special glue on the linings and backing plates (fig. 8), and then assemble the available pieces to bind the brake pads together in a cross shape, inside a metal mold (fig. 9). Next, we place the mold in an oven at 220 degrees for one hour for the bonding process. Figure 10 shows the final product.

Figure 5. Brake pad linings. Figure 6. Base or backing plate. Figure 7. Polymerization oven..

Figure 8. Application of the glue. Figure 9. clamping support. Figure 10. Finished brake pads..

Another method for producing eco-polymer brake pads involves placing the brake pad supports into silicone molds (fig. 11), then pouring in the polymer made from the studied materials and polymerizing it at 175°C (fig. 12) [16].

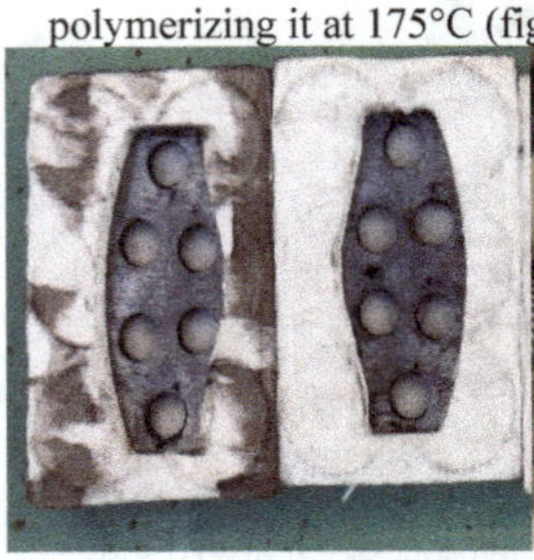
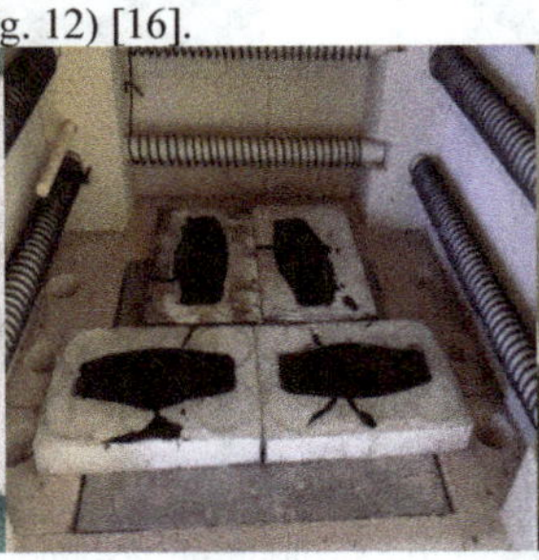

Figure 11. Brake pad backing plate. Figure 12. Polymerization Figure 13. Grinding of brake pads

Once the brake pads are manufactured, they must be tested to ensure they meet quality and safety standards [17]. Brake pads must be able to withstand heat and friction forces during braking, while offering efficient braking performance and a reasonable useful life (fig. 13).
The manufacture of eco-friendly brake pads is a more environmentally friendly alternative to traditional brake pad manufacturing methods. Eco-friendly materials are often more durable and have a reduced environmental footprint compared to traditional materials.

2.2.4 Tests

2.2.4.1 Compression strength.
The compression strength test method for brake pad samples involves using a tensile testing machine to apply an increasing load on the sample until it breaks or deforms permanently. The maximum load sustained by the sample before breaking is recorded to evaluate the compression strength of the material. However, it should be noted that compression strength is not the only important factor for overall brake pad performance, as other characteristics such as friction and thermal conductivity also play an essential role [18, 19, 20].

Table 5 summarizes the measured mechanical resistance values obtained in bending and compression for the 3 mixtures.

Tableau 5. Mechanical strengths of the mixtures..

Materials	S.D	S.D.B	A.B
Bending (MPa)	25,64	45,11	34,71
Compression (MPa)	107,28	114	129,6

2.2.4.2 Water absorption test.

The samples were weighed on a digital balance and immersed in water at room temperature for 24 hours. The samples were then removed, cleaned, and weighed again. The water absorption rate was calculated as follows [21]:

$$\textbf{Water absorption} = \frac{M_2 - M_1}{M_1} \times 100\% \quad (1)$$

où M_1= Sample mass (g),

M_2 = Sample mass after water absorption (g).

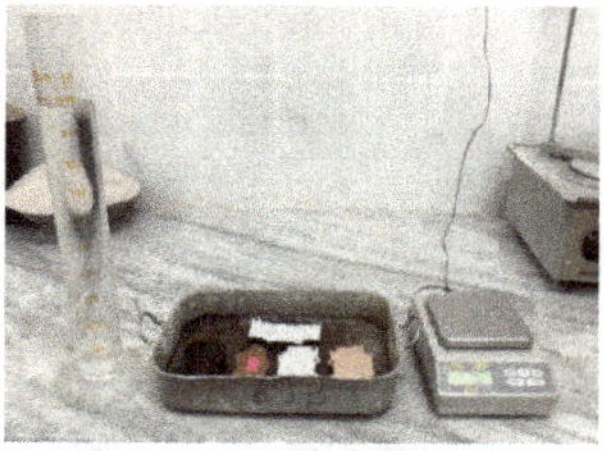

Figure 14. Capillary absorption test

Tableau 6. Results of water absorption of brake pads.

Materials	Dry weight (g)	Weight after 24H (g)	Weight after 48H (g)
dune sand	131	131.2	131.3
crushed dune sand	126.5	126.5	126.5
red clay	144.6	144.8	144.9

It appears that the results of the capillary absorption test did not show a significant difference. This may indicate that the brake pad linings are not capable of absorbing water, as they are made of resin and do not contain any absorbent materials.

However, it should be noted that capillary absorption is only one aspect of brake pad performance, and there are other important characteristics such as wear resistance and friction that can also affect their operation. Therefore, it is important to evaluate the overall performance of brake pads using a comprehensive range of tests and evaluation criteria.

2.2.4.3 Roughness testing

Medium roughness Ra :

Arithmetic mean deviation refers to the average deviation of all points on the roughness profile from a mean line over the length. In other words, it is the average difference between peaks and valleys [22, 23].

$$\mathbf{Ra} = \frac{1}{N}\sum_{j=1}^{N} | rj | \quad (2)$$

Root Mean Square roughness Rq :
Height deviations over the evaluation length measured with respect to the mean line.
Average maximum height Rz
Mean of the absolute values of the five highest peaks and the five lowest valleys over the evaluation length, Normes : ISO 4287-1997.

Tableau 7. Roughness test results.

sample	V_e[m/min]	N[Tr/min]	f [min/tr]	a[mm]	Measure	$R_{a(\mu m)}$	$R_{z(\mu m)}$	R_q (µm)
dune sand	30.14	320	0.03	1	1	1.91	8.41	2.38
					2	1.98	9.68	2.47
					3	1.01	4.61	1.22
					mean	1.63	7.56	2.02

3. Technical control
Vehicle inspection is a mandatory regular examination conducted on vehicles to ensure they are safe and comply with road safety standards. During this inspection, several elements are checked, such as the brakes. The brake pads must be in good working condition for the vehicle to come to a safe stop in case of an emergency. Therefore, the results of the braking and wear tests performed on the brake pads can play an important role in the success of the vehicle inspection. The innovative brake pads designed in this study were installed on a vehicle and successfully passed the braking and wear tests, demonstrating their ability to meet the required safety standards for vehicle inspection (Fig. 15, Fig. 16, Fig.17).

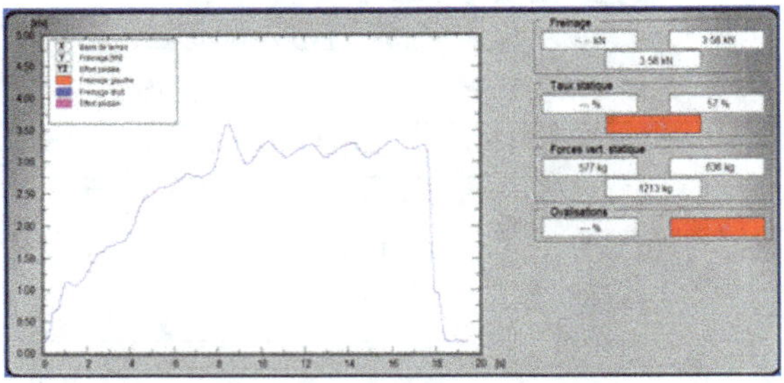

Figure 15. Technical inspection of red clay brake pads manufactured by bonding the lining to the backing plates.

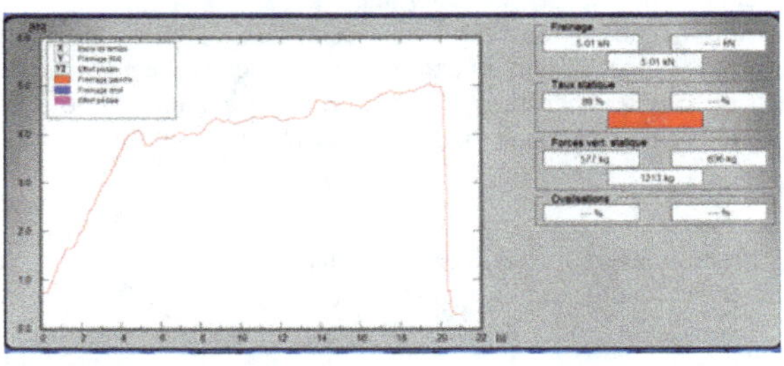

Figure 16. Technical inspection of brake pads made from crushed dune sand and manufactured by bonding the lining to the backing plates.

Materials Research Proceedings 31 (2023) 538-549 https://doi.org/10.21741/9781644902592-56

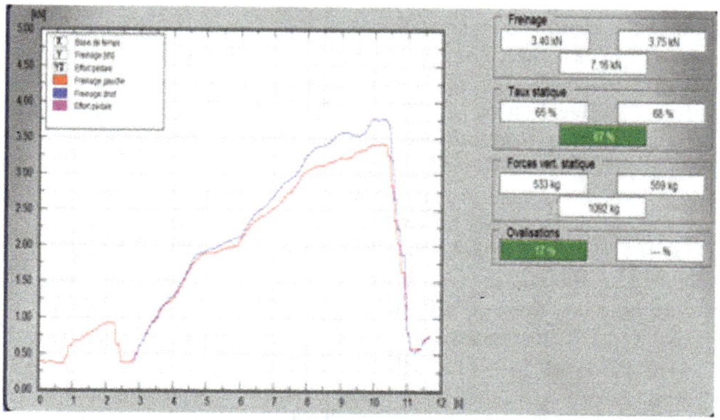

Figure 17. Technical inspection of commercial brake pads that have already been used.

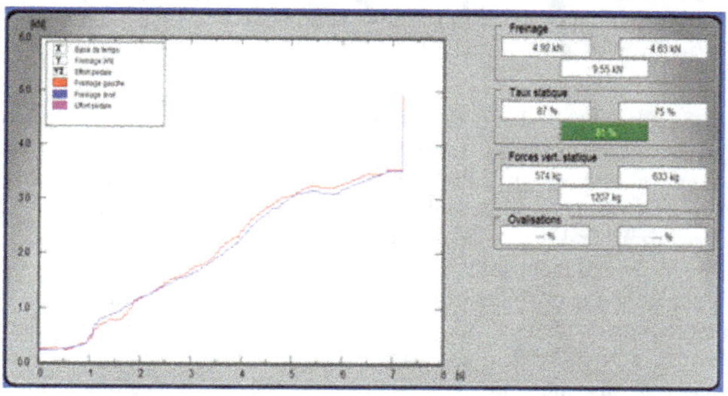

Figure 18. Technical inspection of brake pads made from crushed dune sand and manufactured by polymerization.

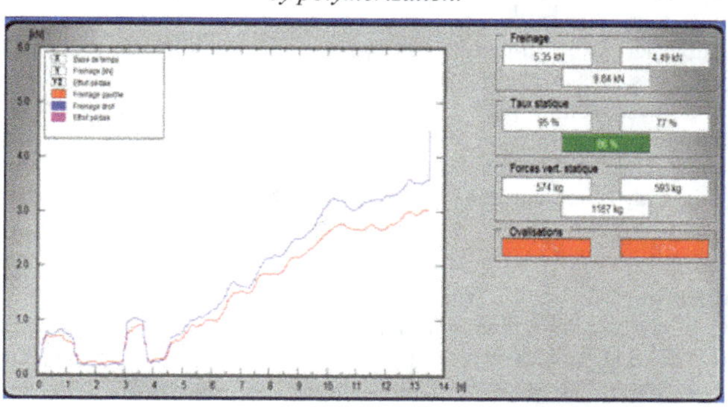

Figure19. Technical inspection of new commercial brake pads.

4. Interpretation of the results

The results presented by the technical inspection show a significant difference between brake pads made from alternative materials and commercial pads already in use. Brake pads made from red clay showed only 30% effectiveness (Fig. 15) compared to commercial pads at 67% (Fig. 17). Similarly, brake pads made from crushed dune sand glued to supports showed only 42% effectiveness (Fig. 16) compared to commercial pads at 67% (Fig. 17).

However, brake pads made from crushed dune sand that were directly cast onto the supports and subjected to high-temperature polymerization (175°C) for 45 minutes showed 81% effectiveness (Fig. 18) compared to 86% for new commercial pads (Fig. 19). These results show that the manufacturing method has a significant impact on the effectiveness of brake pads.

It is important to note that these results are only a first step in the development of alternative materials for brake pads. Researchers must continue their work to understand the reasons for differences in effectiveness and to optimize the manufacturing processes. They must also consider other important characteristics of brake pads, such as their heat resistance and durability.

It is also important to note that the use of alternative materials for brake pads can have significant environmental benefits by reducing dependence on petroleum-based materials. However, it is also important to ensure that these alternative materials do not compromise driver safety or brake effectiveness.

Conclusions

In conclusion, the results obtained in this study demonstrate the feasibility of designing innovative brake pads from local eco-materials, which exhibit encouraging mechanical properties and a perfect exterior quality without air bubbles on the surface. The tests performed confirmed that these brake pads were capable of withstanding high temperatures, were wear-resistant, and offered reliable behavior during braking. The results of various mechanical and chemical tests also confirmed the reliability and mechanical properties of these brake pads.

This study represents a significant advancement in the development of more ecological and sustainable technologies for the automotive industry. The innovative brake pads designed in this study have the potential to reduce the environmental impact of the automotive industry by using local materials and reducing the amount of waste generated during manufacturing.

It appears that the results of the technical inspection show that brake pads made from red clay and crushed dune sand are not as effective as commercially used brake pads. However, it is interesting to note that brake pads made from crushed dune sand that were cast directly onto supports and subjected to high-temperature polymerization showed higher efficiency than commercial brake pads. It is possible that this manufacturing method improved the quality of the crushed dune sand-based brake pads by allowing them to polymerize more completely and uniformly. It is important to continue research to understand the reasons for these results and to optimize the manufacturing of brake pads using alternative materials.

Ultimately, this study offers a promising alternative to brake pads made from crushed dune sand with polymerization at 175°C for 45 minutes and may contribute to making vehicles safer and more environmentally friendly.

Acknowledgments

The authors wish to thank the Directorate General for Scientific Research and Technological Development (D.G.R.S.D.T., Algeria) for financial support of this work. Thanks to the professors and students members of Laboratory of Eco-Materials: Innovations & Applications (EMIA ex. LFGM).

References

[1] W. Zhang, Y. Ren, S. Luo, S. Sun, S. Huang, A review of brake pad materials. Materials, 12 (7) (2019), 1070. https://doi.org/10.3390/ma12071070

[2] R. Banerjee, B. B. Saha, An overview on frictional behavior of automotive brake pad materials, Journal of the Mechanical Behavior of Materials, 26 (5-6), 263-276 (2017). https://doi.org/10.1515/jmbm-2016-0024

[3] B. Liu, X, Wang, Y. Zhu, X. Zhang, Friction and wear properties of brake pad materials, Tribology International, 118, 263-276 (2018). https://doi.org/10.1016/j

[4] M. Ziegler, P. Ebersbach, M. Priebe, T. Hecht, C. Schmiedel, K. Schubert, Investigation of the tribological behavior of different brake pad materials under dry and wet conditions, Wear, 398-399, 36-46, (2018). https://doi.org/10.1016/j

[5] X. Wu, Z. Wu, L. Gao, Tribological properties of friction materials under different braking conditions, A review. Friction, 7 (1), 1-20 (2019). https://doi.org/10.1007/s40544-018-0257-9

[6] Jiang, L., Wang, S., & Ma, S. (2022). Effect of pore structure on friction and wear properties of brake pads. Wear, 476-477, 203024. https://doi.org/10.1016/j. wear.2021

[7] Zou, Y., Zhu, X., Xu, C., & Yu, M. (2022). Tribological properties and wear mechanisms of non-asbestos organic brake pad materials containing SiC and CNTs. Tribology International, 164, 107035. https://doi.org/10.1016/j.triboint

[8] Zhan, Y., Li, W., Xie, Q., & Chen, L. (2022). Tribological properties of automotive brake pads based on mesocarbon microbeads (MCMBs) and Kevlar fiber. Materials Today Communications, 32, 101401. https://doi.org/10.1016/j.mtcomm

[9] Chen, Y., He, X., & Ding, L. (2022). Research on the wear behavior of automotive brake pads under complex working conditions. Wear, 478-479, 203098. https://doi.org/10.1016/j.wear.2021.203098

[10] Dong, X., Wu, H., Wu, J., & Chen, J. (2022). Effects of fiber length and fiber loading on the tribological properties of polyacrylonitrile-based brake friction materials. Polymer Composites, 43(2), 682-691. https://doi.org/10.1002/pc.26758

[11] A. Tafraoui, Contribution to the Valorization of Dune Sand from the Western Erg (Algeria) Application to new concretes; University of Toulouse, 2009.

[12] M. Collignon, Comprehension des mecanismes de degradation de disques de frein pour véhicule « POIDS LOURD » et definition de nouvelles solutions matériaux. thèse de doctorat, Ecole Doctorale SPI 072, Université Lille Nord-de-France N° ordre: 214.

[13] Z. Chang, P. Huang, Y. Hu, Study on the performance of automotive brake pads under different temperatures, Journal of Physics, Conference Series, 1944 (1), (2021), 012066, https://doi.org/10.1088/1742-6596/1944/1/012066.

[14] K. Hu, Y. Wang, X. Zhang, L. Liu, Effects of temperature and sliding speed on tribological properties of brake pad materials, Journal of Materials Engineering and Performance, 29 (10) (2020), 6575-6583. https://doi.org/10.1007/s11665-020-05277-w

[15] P. Wicker, Influence des garnitures de frein sur les sollicitations thermiques des disques TGV et conséquences sur les risques de fissuration, Ecole Centrale de Lille, (2009), <NNT:2009 ECLI 0025>.

[16] T.P. Newcomb, Transient temperatures attained in disk brakes, British Journal of Applied Physics, vol. 10, 1959, pp. 339-340.

[17] R. Limpert; Brake design and safety, Editions Elsevier, 1992.

[18] P.K, Balguri, D G Harris Samuel, C. Indira, G. Dussa, T. Rao Penki, Udayabhaskararao Thumu, Tensile, Flexural and impact properties of properties of ultra small TiO_2 nanotubes reinforced epoxy composites, Composites theory and practice, ISSN: 2084-6096, ISSN (online): 2299-128X, 2020.

[19] J. Huang, X. Zeng, Y. Chen, J. Zeng, X. Wang, Tribological performance of a brake pad system with a newly developed ceramic friction material, Tribology International, 147, (2020), 106313. https://doi.org/10.1016/j.triboint.2020.106313

[20] L. Jiang, S. Wang, S. Ma, Tribological properties of porous brake pad materials with different pore structures, Tribology International, 150, (2020), 106397. https://doi.org/10.1016/j.triboint.2020.106397

[21] X. Wu, Y. Li, B. Li, W. Li, Effects of water absorption on the tribological behaviour of automotive brake pad materials. Wear, 476-477, (2021), 203025. https://doi.org/10.1016/j.wear. 2021. 203025

[22] P. Szymański, D. Czarnecka-Komorowska, K. Gawdzińska, A. Trubas, E. Kostecka. A REVIEW OF COMPOSITE MATERIALS USED IN BRAKE DISC PADMANUFACTURING PROCESS, Composites theory and practice, ISSN: 2084-6096, ISSN (online): 2299-128X, 2020.

[23] M. Maniana, A. Azim, F. Erchiqui, A. Tajamouati, Etude du transfert thermique dans un frein à disque, Laboratoire CAMEL, UQAT, Canada.

Advanced Topics in Mechanics of Materials, Structures and Construction - AToMech1-2023 Materials Research Forum LLC
Materials Research Proceedings 31 (2023) 550-563 https://doi.org/10.21741/9781644902592-57

Recycling of machining waste fibers in the formulation of new concrete

Salima TAFRAOUI[1,a], Ahmed TAFRAOUI[1,b*]and Naima SEGHOUANI[1,c]

[1] Laboratoire de Fiabilité du Génie MécaniqueBP 417 route de Kenadsa, Bechar Algeria

[a]tafraoui.salima@univ-bechar.dz, [b,*]tafraoui.ahmed@univ-bechar.dz, [c]tafraatt@yahoo.fr

Keywords: UPFC, Resistance, Metallic Fibers, UHPC, Dune Sand, Silica Fume

Abstract. The use of waste fibers from the mechanical machining of parts in the formulation of new concretes is of great importance for the protection of the environment because, on the one hand, it makes it possible to recover the materials already used and to on the other hand, it allows nature to be protected from excessive exploitation of the reserve of artificial metallic fibers. The new concretes have high performance (high compressive strength, low porosity and permeability, durability, etc.), however these concretes are fragile and have low tensile strength, which limits their use. The objective of this study is to see the influence of the introduction of waste fibers on the behavior of new concretes. It is necessary to study the physical-mechanical characteristics of these concretes composed of these wastes in the hardened state.

1. Introduction

Concrete has always been considered the most widely used construction material in civil engineering works. UHPC are materials with a cementitious matrix, with compressive strength. These materials are supplemented with metal fibers (UPFC) in order to obtain a ductile behavior in tension.

Indeed, the mechanical performance of UHPC or UPFC encourage their use in various fields such as mechanics and civil engineering, the absence of large aggregates in UHPC makes it possible to produce very thin facing elements with a thickness of 15mm.

Fibers from machining waste are chosen for reasons of availability and economy, come from a renewable source and can be integrated in a rational way in the field of construction. The objective of this work is to contribute to the recovery of the resource, which are the fibers of machining waste thanks to their low cost and their availability. In this pretext, we thought of the use of fibers from machining waste, as concrete reinforcements.

2. Materials and Methods

2 .1 Materials used

- CPJ 42.2 cement from the GICA cement plant in Bechar (Algeria)
- Dune sand from the Taghit region (Bechar Algeria)
- Fumed silica
- Crushed Quartz
- Admixtures: sikaViscoCrete 522 RMX (super-plasticizer)
- Drinking water
- Machining metal fibers.

2.1.1 Cement

UHPC are materials with a very low W/C (water/cement) ratio. Consequently, the rheology of UHPC is closely linked to the reactivity of the cement used.

We have thus seen that, from the point of view of chemical composition and rheology, cements with a low C_3A content and a low specific surface area give the best results [8]. They have a lower water demand.

The cement used in this research is Portland cement (Mâtine, GICA companyBéchar), (CPJ - CEM II/B 42.5) gray cement for high performance concrete intended for the construction of engineering structures (bridges, viaducts, tunnels, etc.), the chemical and physical characteristics of which are presented respectively in Table 1 and Table 2.

Table 1. Chemical composition in % of materials used

	SiO_2	Al_2O_3	Fe_2O_3	CaO	MgO	Na_2O	K_2O	SO_3	Fire loss
Cement	20,70	4,75	3,75	62,92	1,90	0,09	0,19	1,98	3,00
fumed silica	>9 5,00	-	-	-	-	0,60	-	-	-
Crushed Quartz	98,50	7500 ppm	450 Ppm	300 Ppm	-	-	5500 ppm	-	0,20
Dune sand	97,33	0,83	0,24	0,07	0,41	0,09	0,04	0,18	0,40

Table 2: The physical characteristics of the materials used

	Cement	fumed silica	Crushed Quartz	Dune sand
specific surface (cm^2/g)	3390	230000	6900	115
Average size D50(µm)	-	0,1	28	250
Actual density	3,16	2,24	2,65	2,65
Water absorption coefficient (%)				0,15

2.1.2 Additions

• Silica fume

The silica fume used in this work is a gray powder (rather dark), marketed under the name of Condensil S95 DM (Sika company), and derived from the ferro-silicon manufacturing industry. Its chemical and physical characteristics are given in Table 1 and Table 2 respectively.

Figure 1: Silica fume

• Crushed quartz

The crushed quartz used for the formulations undergoing heat treatment is marketed under the name MILLISIL C400 by the Sifraco Company. The physical and chemical characteristics are given in Table 1 and Table 2.

• Dune sand

According to the bibliographical research, and the physico-chemical analysis, the quartz sand of the dune of the western erg of the region of the south-west of Algeria is promising in the formulation of UPFC. However, for practical reasons supply data t in Table 1 and Table 2 respectively.

Absolute density according to the Standard [NFP 18-301]
This test is defined as being the mass per unit volume of the material that constitutes the aggregate without taking into account the voids that may exist between the grains (Figure 3).
The absolute density is determined by the following formula:

$$\rho = \frac{M}{V_2 - V_1} \qquad (1)$$

With:
ρ: Absolute density
M: Mass of solid grains
V_1: Volume of water
V_2: Total volume (solid grains + water).

Apparent density according to the Standard [NFP 18-554]
The apparent density is defined as being the mass of the unit of apparent volume of the body, that is to say that of the volume constituted by the matter of the body and the voids it contains. The apparent density is determined by the following formula:

$$\rho_d = \frac{\text{Mass of dry aggregate}}{\text{Container volume}} = \frac{M_1 - M_0}{V} \qquad (2)$$

With:
V: Volume of the container
M_0: Mass of the clean and empty container
M_1: Mass of the filled container

Sand equivalent according to the Standard [NFP 18-598]
This test makes it possible to measure the cleanliness of the sand. It gives an overall account of the quantity and quality of the fine elements, by expressing a conventional volumetric ratio between the sandy elements which sediment and the fine elements which flocculate.
The value of the equivalent of sand (ES) is the multiple ratios by 100, of the height of the sediment sandy part on the total height of the flocculate. The formula that allows the determination of the sand equivalent is:

$$ES = \frac{h_1}{h_2}\ 100\% \qquad (3)$$

With:
h_1: clean sand + fine elements
h_2: clean sand only
The physical characteristics of the sands at the end of the tests are shown in Tables 3.

Table 3: Physical characteristics of dune sand (0/2)

Features	Results	Unit
Apparent volumetric mass	1.56	[gr/cm^3]
Absolute Density	2.50	[gr/cm^3]
Sand Equivalent (NF EN 933-8)	91.22	[%]

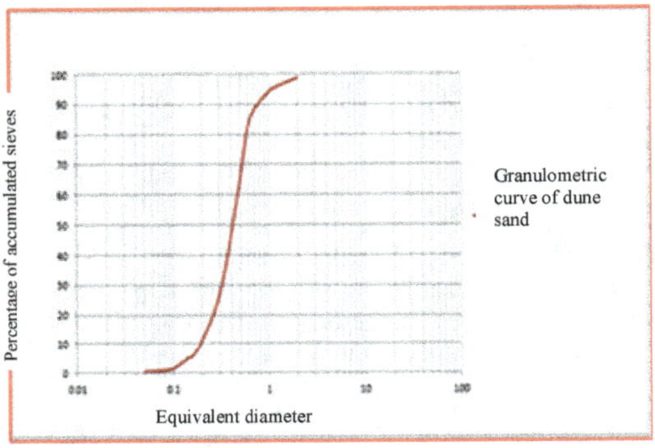

Figure 3: Granulometric curve of dune sand

- **Water**

The water used to make the mortars is drinking water from the network.

• Additive

The super high water-reducing plasticizer used during this study. This is a super plasticizer, in accordance with standard NF EN 934-2 [29], based on acrylic copolymer of new generation, non-chlorinated and intended for the prefabrication industry.

• Fibers

The fibers selected for this study are metallic steel machining fibers, 13 mm in length and 160 µm in diameter (Figure 4)

Figure 4: Machining metal fibers (length 13 mm, diameter 160 µm)

2.2 Experimental methods

2.2.1 Formulation

Table 4: Formulation of UHPC with dune sand (noted DS) (kg/m³)

	UHPC with Silica Fume	
	CDSF	CDSFCQF
Cement	828	691
silica sand	911	759
fumed silica	207	172
Crushed Quartz	0	276
Metallic fibers	0	138
Super plasticizer	26,4	22,0
water efficient	224,6	187,3
Volumic mass (kg/m³)	2153	2363
W/C	0,27	0,27
SF/C%	25	25

With:
CDSF: Concrete with Dune (SD), and Silica Fume (FS)
CDSFCQF: Concrete with Dune, Silica Fume, Crushed Quartz and Metal Fiber

2.2.2Mortier Ordinaire

- **Materials used**

✓ CPJ 42.2 cement from the GICA cement plant in Béchar (Algeria)
✓ Oued lakhdar sand from the Béchar region (Algeria)
✓ Laboratory drinking water
✓ The use of metal fibers to increase ductility.

The physical characteristics of dune sand (0/3) at the end of the tests are represented in table 5

Table 5: Physical characteristics of dune sand (0/3)

Features	Results	Unit
Apparent volumetric mass	1.53	[gr/cm³]
Absolute Density	2.7	[gr/cm³]
Sand Equivalent (NF EN 933-8)	77.65	[%]

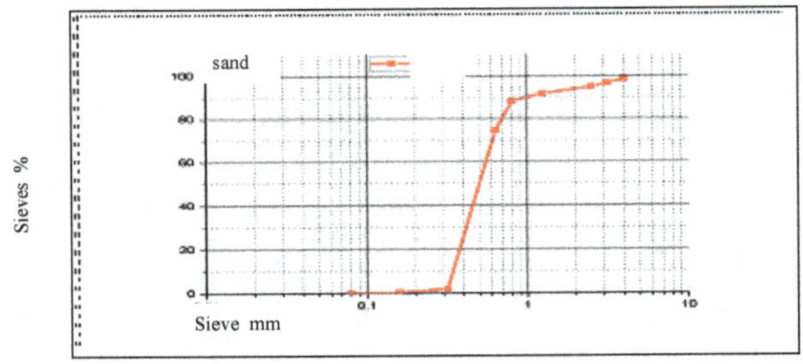

Figure 5: Particle size analysis of sand (0/3)

- **Formulation and conservation of test specimens**

The formulation of fiber-reinforced mortars (5% of metal fibers) for 3 test specimens (40x40x160) made up is shown in the following table:

Table 6: Formulation of a fiber mortar

component	Cement(g)	water(ml)	Sand(g)	metal fibers(g)
M0	450	225	1350	135

- **The mixingmethod**

Mixing plays a valuable role in the manufacture of mortars, hence the need to master and respect its mode and time, as shown in Table 7:

Table7: Formulation and mixing sequence

Materials	Mixing sequence
cement +water	Mixing Materials 30s (Slow Speed)
Sand	Added sand for 30s (slow speed) and 30s (fast speed)
Metallic fibers	Possible introduction of fibers and mixing for 1 minute
W/ C= 0.5	Stop the mixer for 30s Resume mixing for 1min (fast speed)

- **Filling and conservation of molds**

The filling and conservation of the molds is carried out as follows:

- ✓ After mixing, one proceeds to the filling at the rate of two layers, in metal molds which were coated with oil in advance.
- ✓ One proceeds to the vibration with the vibrating table for the two layers; the total duration of vibration 10s (5s for the first layer and 5s for the last)
- ✓ Level and smooth the surface of the concrete with a trowel
- ✓ The preservation of the molds containing the test specimens is done in the laboratory, and in order to avoid any Initial desiccation, the molds are protected with plastic film
- ✓ After 24 hours the specimens are removed from the mold, recorded and immersed in a basin of water until they are 60 days old (Figure 6).

Figure 6: Casting of specimens

2.2.3 Behavior when fresh

2.2.3.1 Spreading and flow test

The spreading test is carried out using the Abrams mortar cone. The spread and flow values measured for the two formulations are presented in Table 8. Figure 7 gives an illustration of the spread obtained in shape and color.

Table 8: Rheological results of UPFC

Samples	CDSQF	CDS
Spreads in cm	45	50

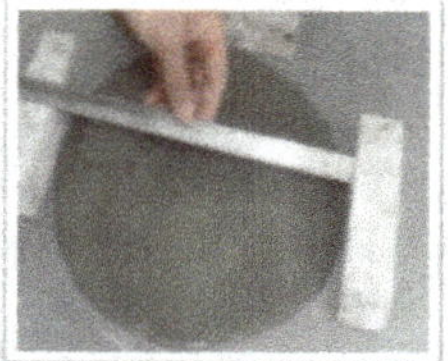

Figure 7: UHPC spread with fumed silica

CDSQF: concrete with sand dune fumed silica, quarter crushed and metallic fiber
CDS: concrete with dune sand and fumed silica

We note that the spread of CDS concrete is between 48 and 52 cm, with the same water and admixture content, the UHPC with fiber have spreads of between 42 and 46 cm.

Regarding the effect of sand type on workability, there is greater fluidity with dune sand. This is certainly to be linked to the slight differences in shape and size of the grains which have an influence on the friction with the fibers and therefore on the flow. It can be deduced from this that the use of dune sand will be beneficial with regard to the installation of UHPC.

We did not find excessive heterogeneous fiber distributions. This can be explained by their small size of the fibers (13 mm) and the limited friction generated with aggregates that are also small in size.

2.2.3.2. Flow test on workability

The flow time measured on mortar workability was about 2 seconds according to the standard (NFP 18-452) [34] for all the formulations studied.

This very short delay reflects a significant fluidity of our materials without being able to distinguish any influence of the type of fines or sand on the flow, contrary to what had been noted during the spreading test. The UHPC studied have a rheological behavior similar to that described by Bonneau [8]. They are easy to mold materials and well suited for prefabrication. They can be used to produce shapes of varying complexity and achieve excellent detail reproduction due to their fluidity and bubble-free surface quality.

2.2.3.3. Measurement of the amount of entrained air

The quantity of entrained air for all the formulations studied in the fresh state is between 1.8% and 2.0% respectively for the samples with fibers and without fibers. These values are of the same order as those of ordinary concretes. They also correspond to the orders of magnitude of the values measured on UPFC by Rougeau (3%) [35] and the BPR200 by Roux et al (1%) **[36]**.

2.2.3.4. Determination of bulk density and porosity

We note that the apparent densities of our materials are almost identical. On the other hand, a slight difference in porosity is observed between the concretes incorporating fiber (average of 6.3%) and those which do not have fibers.

(6.0%). The greater amount of void in fiber UHPC, and the higher density of this ultrafine Silica fume, at the same dosage, could explain the fact that the densities are almost equivalent (Table 9).

Table 9: Bulk density and porosity

Samples	CDSQF	CDS
Apparent Density (g/cm3)	2,382	2.361
Porosity (%)	6,3	6.0

2.2.4. Behavior in hardened state

2.2.4.1 Compressive strength

The UHPC compressive strength results correspond to the average of 2 tests carried out on the two pieces of the 3 prismatic specimens (40x40x160 mm) tested beforehand in bending. They are presented in Table 10 for the 4 maturities of 1, 7, 14 and 28 days.

Table 10: Mechanical compressive strengths of the different compositions

Samples	CDS	CDSQF
Time(days)	Compressive strength in MPa	
1	19.68	27.49
7	76.76	52.60
14	83.47	56.79
28	85.73	73.90

From the expiry of 7 days, the concrete with fiber exceeded the concrete without fibers 8 MPa of compressive strength. At 28 days, the concrete without fibers exceeded the concrete with fibers, the maximum mechanical strength that is 85.73 MPa.

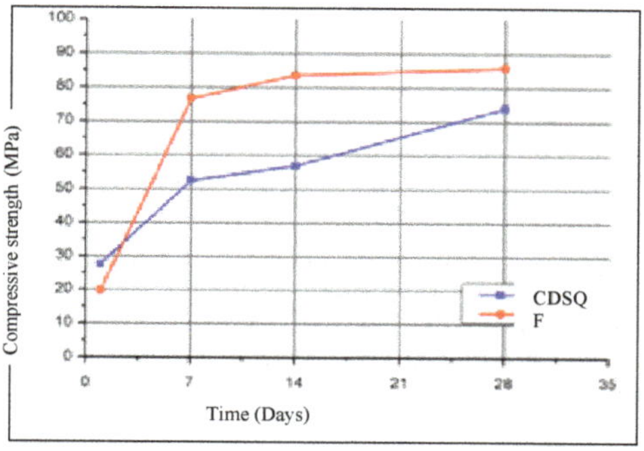

Figure 8: Hardening kinetics of the different fiber UHPC studied

Figure 8 shows the kinetics of compressive strength evolution for the different concretes (with the Eurocode 2 notations [26],fcm average strength at 28 days and fcm(t) average strength at time(t).

The analysis of the kinetics of the resistances in compression highlights the differences between the UPFC.

Concrete with dune sand (CDS and CDSQF), concrete with fibers in 24 hours is the most resistant, compared to the 3 deadlines 7,14,28 because the mixing problems and the addition of water reduce the resistors.

2.2.4.2 Bending strength

Table 11 presents the results of the mechanical bending tests as a function of the age of the material obtained for the UHPC studied according to standard NF EN 12390-5 [32] on 40×40×160 mm specimens. These bending resistances are calculated from the maximum force applied by the press.

Table 11: Mechanical resistances in bending of the various compositions at various deadlines

Samples	CDS	CDSQF
Time(days)	bending strength in MPa	
1	3.76	4.01
7	5.55	6.34
14	7.94	7.27
28	8.10	7.79

The concretes studied show similar flexural strength values, regardless of the end of the test. The most significant differences appear at 1 day (values between 3 and 4.01 MPa). At 28 days, the strengths are slightly higher than 7 MPa.

Comparisons with the results of the literature require careful reading because the formulations, type of cure and heat treatment vary according to the studies carried out on these new materials. Thus, it can be estimated that the bending strengths can evolve between 7 MPa. One of the main reasons for this wide range of values is most likely the type (geometrical and mechanical characteristics) and quantity of fibers used.

Figure 9 presents the evolutions of the flexural strengths at the normalized instant compared to their values at 28 days.

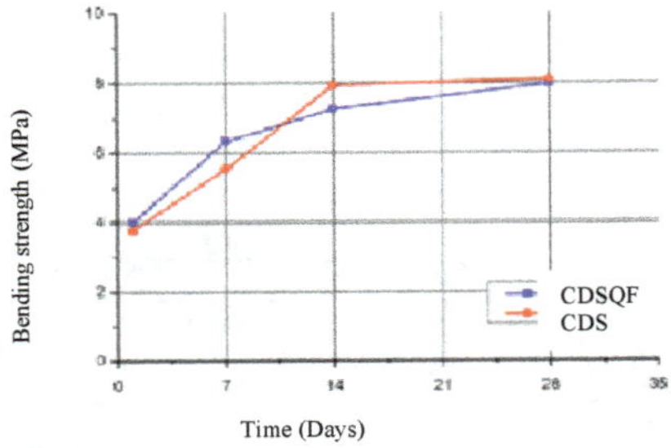

Figure 9: Kinetics of evolution of the flexural strength of the different formulations

The concretes with dune sand (CDS and CDSQF), the concrete with the fibers in 24h and 7days it was the most resisted, against for the deadlines 14.28 because the problems of mixing and the addition of water decrease resistance.

From the results of the bending tests, the tensile strength can be deduced. Indeed, the bending test is relatively simpler to implement than a direct tensile test. The AFGC [27] proposes, on the basis of the approach of the CEB-FIP calculation code, to evaluate the tensile strength from the results of the bending test on a prismatic specimen not notched by the intermediary of the following formula to take account of the scale effect.

$$R_t = R_{fl} \cdot \frac{2{,}0.\left(\frac{h}{h_0}\right)^{0{,}7}}{1 + 2{,}0.\left(\frac{h}{h_0}\right)^{0{,}7}} \qquad \textbf{(4)}$$

With
R_t: Tensile strength due to bending (MPa)
R_{fl}: Bending strength (MPa)
h: Height of the prism (mm) and h_0 = 100 mm

The bending strength R_{fl} is equal to the value of the fiber tensile stress lower than mid-span, calculated from the value of the applied force F_{fiss} corresponding to the loss of linearity of the elastic behavior. Unfortunately, the test device of the press used did not allow us to set up a deflection sensor. The exploitation of the tensile strengths by bending cannot therefore be carried out, for lack of precise evaluation of the force F_{fiss}. The breaking force indicated by the press is obviously greater than this value.

If we apply the formula taking into account the geometry of our specimens, we see that the scale effect coefficient is equal to 0.51. This therefore means that the tensile strength is less than half of the measured bending strength. One could therefore estimate with restraint that the tensile strengths are approximately in a range between 4 and 8 MPa, values significantly higher than those of common concretes.

2.2.5 The results of the fiber-reinforced mortar in compressive

Table 12: Mechanical compressive strengths of Mortar and UPFC

Samples	Fiber mortar	CDSQF
Time(days)	Compressive strength in MPa	
1	7.32	27.49
7	14.35	52.60
14	23.83	56.79
28	30.37	73.90

From the expiry of 1 day, the UHPC exceeded the fiber mortar 20 MPa of compressive strength. At 28 days, the UPFC exceeded the Fiber Mortar more than 40 MPa of compressive strength, the maximum mechanical strength of Fiber Mortar is 30 MPa.

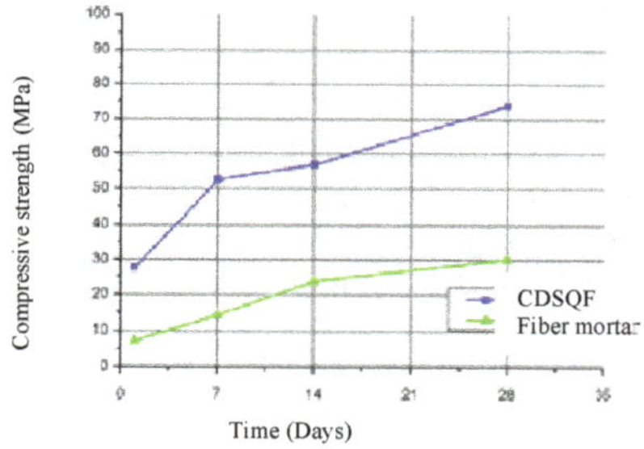

Figure 10: Mechanical compressive strengths of Mortar and UPFC

Note that concrete with dune sand (CDSQF) is more resistant than fiber mortar in the 4 deadlines 1, 7, 14, 28.

2.2.6 The results of the fiber-reinforced mortar in bending

Table 13: Mechanical resistances in bending of Mortar and UHPC

Samples	Fiber mortar	CDSQF
Time(days)	Bending resistance in MPa	
1	1.53	4.01
7	2.41	6.34
14	4.01	7.27
28	5.33	7.79

From the expiry of 1 day, the UHPC exceeds the Fiber Mortar more than 2 MPa of resistance in bending, the same superiority at 28 days for UHPC. The maximum mechanical strength of UHPFR exceeded 7 MPA.

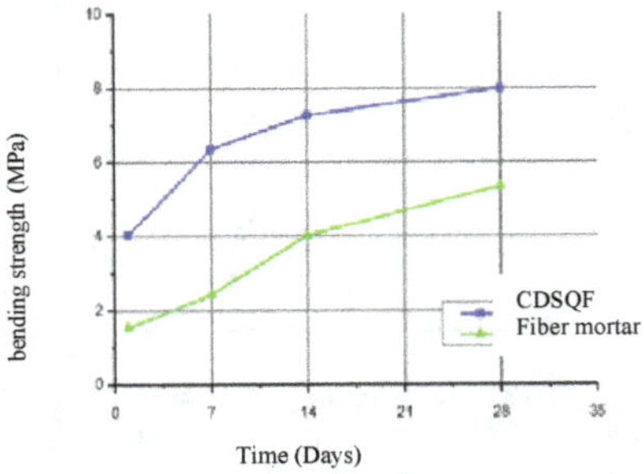

Figure 11: Mechanical resistances in bending of Mortar and UHPFRC

Deduce that the fiber mortar is less resistant for the 4 maturities than the CDSQF.

3. General conclusion

The use of artificial metal fibers as reinforcements in new concretes (UHPC) has impacts on the environment. In this case, we have chosen fibers from machining waste to reduce the problem of excessive exploitation of the reserve of metal fibers.

According to the results of the tests on the various UHPC specimens reinforced with machining waste fibers in 24 hours it was the most resistant compared to the specimens without fiber, on the other hand for the 3 deadlines 7, 14, 28 the specimens without fibers they are more resistant than UPFC because the problems of mixing and the addition of water to decrease the resistances.

The incorporation of machining waste fibers in the ordinary mortar made it possible to increase its resistance to bending relatively if looking at the results obtained in UHPC. We added the mill waste fibers into the regular mortar to see how active it is in the regular mortar and if it has any other adaptation or other UPFC results.

The perspective of using fibers from machining waste in the field of Civil Engineering has allowed us to contribute in a rather humble way to the recovery of these materials.

The addition of machining waste fibers in UHPC and ordinary mortar modifies the behavior of concrete and considerably improves the mechanical characteristics of the material. The experimental results show an efficiency of the fibers for bending stresses, an improvement in the resistance to bending and compression.

The properties of new concretes are not influenced by the length, dosage and treatment of fibers only but with other factors such as aggregates, cement and fillers.

Acknowledgments

The authors wish to thank the Directorate General for Scientific Research and Technological Development (D.G.R.S.D.T., Algeria) for financial support of this work. Thanks to the professors and students members of Laboratory of Eco-Materials: Innovations & Applications (EMIA ex. LFGM).

References

[1] A.Picon,The art of the engineer, Editions du Center Georges Pompidou, Paris ISBN 2-85850-911-5, 1997, p. 311.

[2] Sablocrete, Characteristics and practices of use, Sablocrete, National Research and Development Project, Presses of the National School of Bridges and Roads, Edition Lavoisier, 12/1994, p.236.

[3] D.M.Roy, Gouda G.R., High strength generation in cement pastes. Cement and ConcreteResearch,vol.3,no.6,1973,p.807-820. https://doi.org/10.1016/0008-8846(73)90013-6

[4] H.H.Bache, Densified cement/ultrafine particle-based materials. 2nd international conferenceonsuperplasticizersinconcreteOttawa,Canada,june10-12,1981.

[5] P.Richard, M.H.Cheyrezy, Composition of reactive powder concretes. Cement and Concrete Research,vol.25,no.7,1995,p.1501-1511. https://doi.org/10.1016/0008-8846(95)00144-2

[6] A.Tafraoui, Doctoral thesis, contribution to the recovery of dune sand from the accidental erg, University of Toulouse 13feb 2009 .

[7] M.Cheyrezy, Structural applications of BPR. AFPC-AFREM Conference on High Performance Concretes and Ultra High Performance Concretes, INSA Lyon, June 3, 1998.

[8] O.Bonneau, Study of the physico-chemical effects of superplasticizers in order to optimize the rheological behavior of Ultra High Performance concretes. Doctoral thesis, Department of Civil Engineering, University of Sherbrooke, Quebec, Canada, 1997.

[9] AFNOR,NFEN12620 ,Aggregates for concrete, AFNOR, Classification index: P18-601, August 2003.

[10] AFNOR,XPP18-545,Aggregates Elements of definition, conformity and codification AFNOR, February 2004.

[11] A.M Neville, Properties of concrete. Fourth Edition, England: Longman Edition, 1995, p844.

[12] AFNOR, NF EN 13263-1(Classification index: P18-502-1), Silica fume for concrete. Part 1: definition, requirement and compliancecriteria, September 2005.

[13] P.Rossi, S.Renwez, A.Belloc, Ultra High Performance fiber-reinforced concretes. Liaison bulletin from the central laboratory for bridges and roads, no.196, March April 1995, p. 61-66.

[14] AFGC,Ultra-high performance fiber-reinforced concretes. Interim Recommendations, January 2002.

[15] M.Behloul, Definition of a BPR behavior law. Annals of the ITBTP, no.532, March-April 1995, p. 122-127.

[16] M.Alpardi, Admixtures used to enhance placing characteristics of concrete. Cement and concrete composites,vol.20,no2/3,April-June,1998,p.103-112 https://doi.org/10.1016/S0958-9465(98)00071-7

[17] J.Dugat, N.Roux, G.Bernier, Experimental study of the deformation under stress and the fracture behavior of reactive powder concrete (RPC). Annals of the ITBTP, n°.532, March-April 1995, p. 112-121.

[18] S.Alpardi, R.Coppolla, R.Troli, M.Alpardi, Mechanical properties of modified reactive powder concrete.Proceedings,5thCANMET/ACI international conference on superplasticizer and other chemical admixtu resin concrete,SP-173,Rome,Italy,1997,p. 1-21.

[19] P.Rossi,S.Renwez,F.Guerrier,Ultra High Performance fiber-reinforced concretes.

[20] The current experience of LCPC, Bulletin des Laboratoires des Ponts et Chaussées, no.204, July August 1996, p.87-95.

[21] G.Lim, S.Hong, Slump loss control of cement paste by adding polycarboxylic types lump-releasing dispersant. Cement and Concrete Research,vol.29,1999,p.223-229. https://doi.org/10.1016/S0008-8846(98)00188-4

[22] E.Hanna, K.Luke, D.Perraton,P.C.Aitcin, Rheological métakaolin of Portlandcementinthepresenceofasuperplasticizer,Proceedings3rdInt.Conferenceonsuperplasticizers and other chemical admixtures in concrete. SP-119, Ottawa, Canada, 1989, p.171-188.

[23] G. DREUX,J. FESTA, New concrete guide, Seventh Edition Eyrolles, 1995.

[24] S. MOHELLEBI, Characterization and Modeling of the physico-hydro mechanical parameters of a lightweight concrete based on composite aggregates, Doctoral thesis, Mouloud Mammeri University of Tizi-Ouzou Algeria.

[25] Technical sheets Volume 2, Concretes: formulation, manufacture and implementation, information center on cement and its applications, January 2013.

[26] Y.Fritih, Contribution of a reinforcement of Fibers on the behavior of reinforced self-placing concrete elements, Doctoral Thesis, University of Toulouse, July 08, 2009.

[27] Eurocode2, EN1992-1-1,General rules and rules for buildings, Part 1-1. October 2005.

[28] AFNOR, NF P 18-342, Admixtures for concrete, mortar and grout - Part 2: Admixtures for concrete - definitions and requirements, Sept 2002.

[29] F.DeLarrard, T.Sedran, A new approach to concrete formulation. Annals of building and public works, no.6 p. 39-54, December 1999.

[30] AFNOR, NF EN 934-2/A2, Admixtures for concrete, mortar and grout-Part 2: admixtures for concrete-Definitions, requirements, conformity, marking and labelling, AFNOR April 2006.

[31] K.Ozawa, N.Sakata, H.Okamura, Evaluation of self compatibility of fresh concrete using the Funnel test. Concrete Journal of JSCE, no.25, June 1995.

[32] AFNOR, NF P 18-452, Concretes Measurement of the flow time of concretes and mortars with workability meters, May 1988.

[33] AFNOR, NF EN 197-1,Cement - Part 1: Composition, specifications and conformity criteria, February 2001

[34] AFPC-AFREM, Durability of concretes: Recommended methods for measuring the quantities associated with durability, Report of Technical Days LMDC, INSA-UPS, Toulouse. December 1997.

[35] AFNOR, NF P 18-452, Concretes - Measurement of the flow time of concretes and mortars with workability meters, May 1988.

[36] P. Rougeau, D.Boulet, M.Shink, Ultra High Performance concretes Synthesis of knowledge and feasibility study on laboratory prototype, CERIB document DDI 002, 2000.

[37] N.Roux,M.Barranco,Implementation of reactive powder concretes in the construction industry, Annales de L'ITBTP, no.532, 1995.

Numerical simulation of high-pressure gas-jet impingement on the adjacent equipment

Esam I Jassim[1], Muhammad Asad[1], and Bashar Jasem[2]

[1]Department of Mechanical Engineering, Prince Mohammad Bin Fahd University, Khobar, KSA

[2]Department of Computer Technology, Al-Hadba University College, Mosul, Iraq

Keyword: High-Pressure Gas, Cylinder Rupture, CFD, Jet Impingement, Thermal Stresses

Abstract. Cold jet is a result of a high-pressure leakage through a wall crack, valve stem, or any other opening caused by an accident or failure to a high-pressure device. The phenomena and its impact on the adjacent and surrounding equipment are simulated using Computational Fluid Dynamics (CFD) techniques. For the preliminary study, the working fluid used in the simulation is methane. The simulation is linked to appropriate thermal and fluid property estimation software using parameter-tuned equations of state to predict the real multi-component natural gas flow conditions. Through the simulation, a spot of condensation and/or nucleation was predicted. These spots are crucial since their effect appear on the jet simulation in which multiphase properties must be considered. The results showed that the temperature variation on the adjacent surface is divided into three regions, namely: vicinity of stagnation point, sharp variation region, and mild variation region. Such distribution results in inception of thermal stresses that could cause catastrophic incident if rapture occurs. Hence, wall thickness of the equipment that exposed to cold jet should be cautiously selected to sustain the extra thermal stress as well as localised non-uniform stress distribution as detailed in the Finite Element based coupled thermo-mechanical analysis.

Introduction

Industries with complex equipment that deal with hazardous substances mandate that a Quantitative Risk Assessment (QRA) is implemented. The need of simulation of accidental scenarios to measure the risk quantity requires efficient and accurate model to accommodate all aspects of the scenario such as compressibility effect, interaction with the surroundings, and geometrical complexity. Empirical models pertaining to jet phenomena, however, lack the presentation of jet-obstacles interaction, resulting in overestimation of the damage area. Hence, an accurate and more realistic method for the jet phenomena becomes essential. Computational Fluid Dynamics (CFD) emerged as a state-of-the-art technique aiming to simulate accidental pressurized gas releases in congested industrial environments.

In the present work, a high-pressure methane leakage from a small hole in a congested environment is considered. When high-pressure fluid leaks to the ambient due to tiny hole or crack, supersonic flow is induced and shock would be incepted. Numerous researches have addressed the shock location experimentally [1, 2] as well as numerically [3, 4]. Jassim studied the nozzle geometry impaction on the shockwave location using CFD FLUENT code. He reported that elliptical shape, among six different geometries, predicts the shock farthest from the throat. Jassim et al. studied numerically the effect of real gas behavior in supersonic nozzle on shock location [5] and vorticity potential [6]. The outcome of the simulation revealed that shock inception is significantly altered when an accurate equation of state is employed. Their results also concluded that the vorticity increases rather sharply near the shock.

The phenomenon is complex and for simplicity, it splits into two parts: release and impact. Near the rupture, the released gas is under-expanded and tends to adjust to the ambient conditions through some expansion and compression waves (shock waves).

The resulting flow is supersonic and highly compressible with strong discontinuities of the flow field variables, clearly represented by the presence of a Mach disk: this is the "release" phase. Release and dispersion of pressurized gas have been addressed in many literatures. Wilkening and Baraldi [7] modeled the jet formed from the release of pressurized hydrogen. They realized that a small-time step and dense mesh near the shock inception region are necessary. Deng et al. [8] modified the simulation by splitting the jet release and dispersion phases. Initially, the expansion process is modelled using pseudo-source. Then, it is used as an input to the modeling of dispersion process.

Dharavath et al. [9] numerically investigated two turbulence models. They figured out that k–ε model performed better in predicting the variation of fluid properties. The outcome of the simulation reported that fine meshing is mandatory for accurate predicting near shock region.

Theory

The geometry of the crack considered in the study is nozzle with 2D axisymmetric geometries. The general form of the continuity equation for steady state compressible flow, in deferential form, is given by Eq.1:

$$\frac{\partial(\rho u)}{\partial x}+\frac{\partial(\rho v)}{\partial y}=0 \tag{1}$$

The 2D Eular Equations for compressible, inviscid flow can be simplified to the following expressions:

$$\frac{\partial(\rho u)}{\partial t}+u\frac{\partial(\rho u)}{\partial x}+v\frac{\partial(\rho u)}{\partial y}=-\frac{\partial P}{\partial x} \tag{2}$$

$$\frac{\partial(\rho v)}{\partial t}+u\frac{\partial(\rho v)}{\partial x}+v\frac{\partial(\rho v)}{\partial y}=-\frac{\partial P}{\partial y} \tag{3}$$

Energy equation in deferential form is obtained from Eq 4:

$$\rho\frac{\partial(e)}{\partial t}=\frac{p}{\rho}\frac{d\rho}{dt}+k\left(\frac{\partial^2 T}{\partial x^2}+\frac{\partial^2 T}{\partial y^2}\right)+\rho\frac{dQ}{dt} \tag{4}$$

CFD modelling

To construct the nozzle geometries, unstructured cells were used. The grid density was higher near the outlet of the nozzle so that the resolution for capturing shocks can be improved. Accuracy of any simulation is significantly dependent on the mesh construction. Structure of supersonic flow is complex due to large speed gradient and sudden drop in the flow velocity across the shock. Hence, art of mesh size, number, and density construction should account for grid resolution and time consumption.

The essence of grid independency is to test different resolutions in a way that the computational grid should be able to capture the actual flow features. Since the accuracy of flow structure and properties of the fluid rely on grid resolution of the entire system, the near exit of the nozzle is chosen for measuring the resolution. Previous studies on shock capturing analysis have shown that the thickness of the shock wave is of order of 0.1 mm when the gas undergoes severe gradients in the thermo-fluid properties [1, 5, 6]. Hence, sufficient cell resolution was maintained in the region where the shock induced. From an optimization perspective, the analysis of grid resolution concludes that 0.5 mm grid spacing is firmly acceptable in the convergent portion while 0.1 mm spacing is more adequate to capture the flow feature (including shock position) in the divergent portion of the nozzle.

Results and Discussion

Jet simulation

A jet of natural gas developing outside a tank hitting a stationary wall some distant apart is simulated using FLUENT code. Figure 1 illustrates the contours of the velocity dispraise of the Jet. Near the impingement region, the flow direction deviates due to the effect of the stagnation point. The altered stream then strikes the wall of the adjacent component (in this case another tank with the same height) leaving unaffected area under less strength. Thus, the temperature of such area becomes relatively higher.

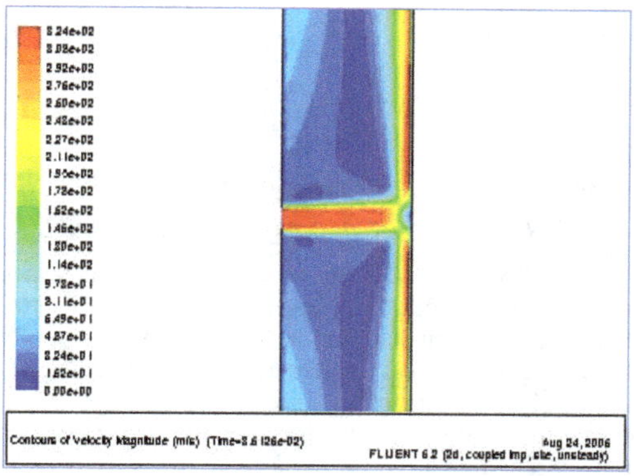

Fig. 1: Velocity contours of jet impingement

Temperature distribution along the adjacent wall is depicted in Fig. 2. The temperature variation can be divided into three sub-regions, namely: stagnation point vicinity, sharp variation region, and mild variation region.

As discussed in the jet velocity profile near the adjacent wall, the flow slightly alters then severely changes its direction to become parallel to the wall as it moves closer towards the surface. Thus there is a certain distance on the wall that stagnation or circulated flow (eddies) are in contact with the wall. Hence the temperature of this region decreases sharply from (stagnation value) to the minimal then increases gently along the rest of distance on the wall.

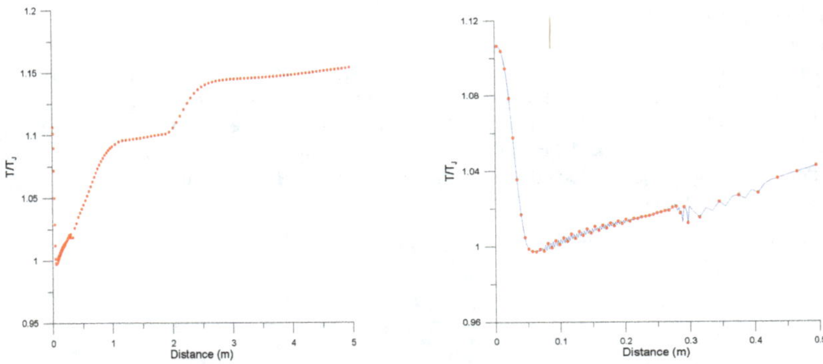

a) The Three Regions b) Enlargement near stagnant point

Fig. 2: Temperature distribution on the adjacent wall

Distance between the crack and the adjacent tank is also of interest and is addressed in the present work. Three distances are selected between the two vessels (18 cm, 30 cm, and 500 cm).

A dimensionless temperature parameter for the adjacent wall is plotted for three different distances and for the case of very high heat transfer coefficient inside the adjacent tank. As shown in Fig. 3, when the distance between the two adjacent walls is short (i.e. stronger impinging), a non-uniform temperature variation develops. Such variation decays as the distance becomes larger because the impact of the impinging jet is proportion inversely to the distance. Hence, it gradually diminishes.

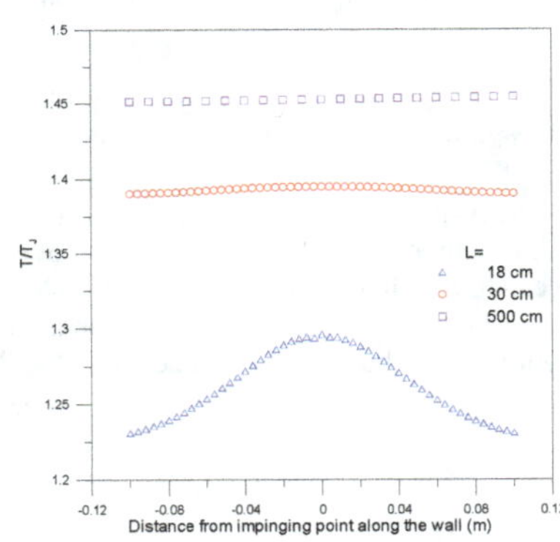

Fig. 3: Temperature Distribution along the Adjacent Wall when the inside Wall Temperature=265K

FE analysis and investigations of adjacent surface

Abaqus implicit software was exploited to perform steady-state coupled temperature displacement calculations to investigate the cold jet effect on an adjacent pressure vessel. The conceived axisymmetric model for pressure vessel (with standard dimensions [10]) was meshed with 4 nodes thermally coupled axisymmetric quadrilateral, bilinear displacement and temperature, reduced integration, hourglass control- CAX4RT elements, as shown in Fig. 4. The thermomechanical

Materials Research Proceedings 31 (2023) 564-570 https://doi.org/10.21741/9781644902592-58

properties of the simulated pressure vessel material; 34CrMo4 Chrome Molybdenum steel, are shown in Table 1.

Table 1 Thermomechanical Properties of 34CrMo4 Chrome Molybdenum steel [11]

Properties	Thermal Conductivity	Thermal Expansion (μm/m-°C)	Specific Heat (J/g-°C)	Density (g/cm³)	Modulus of elasticity (GPa)	Poisson's Ratio
	30-60	12	0.46	7.8	210	0.3

Figure 5, shows the von-Mises stresses, nodal temperatures and thermal strains profiles on the pressure vessel with and without considering cold jetting. It can be noticed that by implementing thermal loading in the form of cold jetting the stresses are increased, though still are well below yield strength of the material [12]. While, temperature variation through the cylinder thickness are also evident that has caused thermal strains. In the light of the numerical results, it can be stated that under this cold jetting, though there will not be any considerable effects on the cylinder strength and safety, however a cylinder with lower thickness may result in localized softening of the material leading to rupture and failure.

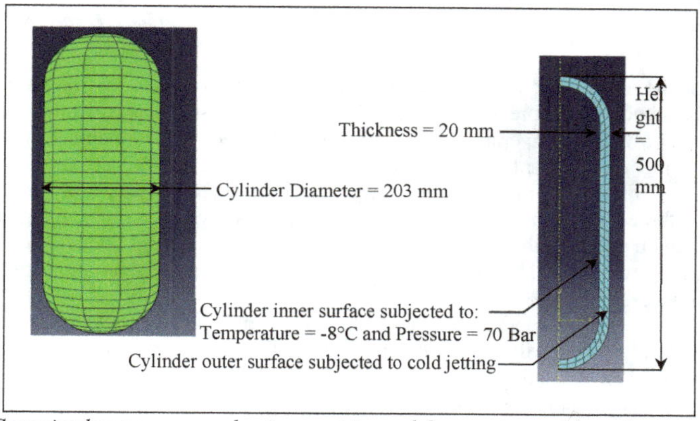

Fig 4: Conceived pressure vessel axisymmetric model geometry and boundary conditions.

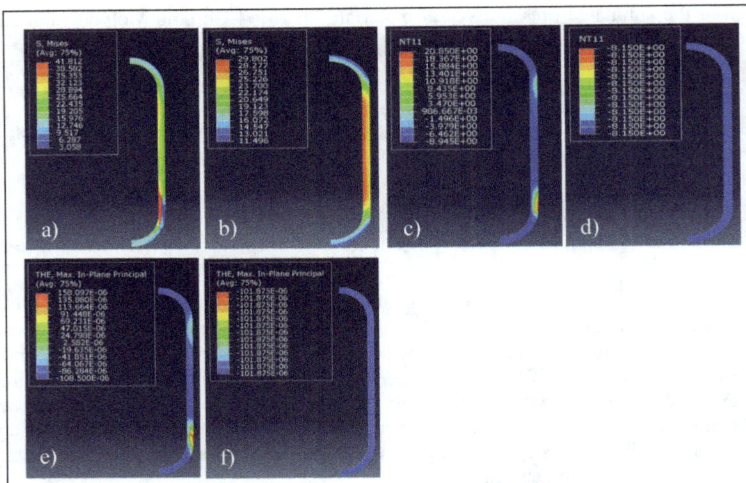

Fig 5: Von-Mises stresses (a, b), nodal temperatures (c, d) and thermal strains (e, f) profiles on pressure vessel with and without considering cold jetting effects, respectively.

Conclusion

In this paper, the jet evolution from ruptured high-pressure gas cylinder is modeled to study its impact on the adjacent high-pressure-vessel. Influence of the distance between the jet source and the adjacent part on the temperature distribution is addressed. Thermal strains of the adjacent surface due to the impingement are also reported. It is concluded that the jet starts supersonic and slows down to subsonic regime. The variation of the surface temperature of the adjacent vessel revealed that three sub-regions can be recognized with steep variation in temperature near the stagnation domain and smooth variation at the area far from the stagnant point. When the distance between the hole rupture and the adjacent surface augments, the sharp variation (the second sub region) diminishes.

Results of numerical simulation revealed that cold jet impingement could incept extra stresses that may result in localized softening of the material leading to rupture and failure. Hence caution must be taken when designing the thickness of the adjacent equipment, particularly if the equipment is a high-pressure container.

References

[1] Esam I. Jassim, 'Geometrical Impaction of Supersonic Nozzle on the Dehumidification Performance During Gas Purification Process: An Experimental Study', Arabian Journal for Science and Engineering, 44 (2019), pp.1057-1067. https://doi.org/10.1007/s13369-018-3340-x

[2] Jassim, E. 'Experimental Study on Dehumidification Performance of Supersonic Nozzle', World Academy of Science, Engineering and Technology, Open Science Index 121, International Journal of Aerospace and Mechanical Engineering (2017), 11(1), 182 - 185.

[3] Jassim, E.I. (2016) 'CFD study on particle separation performance by shock inception during natural gas flow in supersonic nozzle', Progress in Computational Fluid Dynamics, Vol. 16, No. 5, pp.300-312. https://doi.org/10.1504/PCFD.2016.078755

[4] Jassim E.I., Awad, M.M. (2013), 'Numerical investigation of nozzle shape effect on shock wave in natural gas processing', In Proceedings of World Academy of Science, Engineering and Technology (Vol. 78, p. 326). World Academy of Science, Engineering and Technology (WASET).

[5] Jassim E., Abedinzadegan Abdi M., and Muzychka Y., 'Computational Fluid Dynamics Study for Flow of Natural Gas through High Pressure Supersonic Nozzles: Part 1- Real Gas Effects and Shockwave', Journal of Petroleum Science and Technology, Vol. 26, issue 15, 1757-1772, 2008. https://doi.org/10.1080/10916460701287847

[6]. Jassim E., Abedinzadegan Abdi M., and Muzychka Y., 'Computational Fluid Dynamics Study for Flow of Natural Gas through High Pressure Supersonic Nozzles: Part 2- Nozzle Geometry and Vorticity', Journal of Petroleum Science and Technology, Vol. 26, issue (15), 1773-1785, 2008. https://doi.org/10.1080/10916460701304410

[7] H. Wilkening, D. Baraldi, 'CFD modelling of accidental hydrogen release from pipelines', International Journal of Hydrogen Energy, Volume 32, Issue 13, 2007, Pages 2206-2215. https://doi.org/10.1016/j.ijhydene.2007.04.022

[8] Yajun Deng, Hongbing Hu, Bo Yu, Dongliang Sun, Lei Hou, Yongtu Liang, 'A method for simulating the release of natural gas from the rupture of high-pressure pipelines in any terrain', Journal of Hazardous Materials, Volume 342, 2018, Pages 418-428. https://doi.org/10.1016/j.jhazmat.2017.08.053

[9] Dharavath, M., Sinha, P. and Chakraborty, D. (2010) 'Simulation of supersonic base flow: effect of computational grid and turbulence model', Proceedings of IMechE in Journal of Aerospace Engineering, pp.311-319. https://doi.org/10.1243/09544100JAERO600

[10] https://www.yongancylinder.com/13-4L-ISO11439-Standard-High-Pressure-CNG-1-Vehical-Seamless-Steel-CNG-Cylinder-pd43004559.html

[11]https://www.matweb.com/search/DataSheet.aspx?MatGUID=638f2c3469bc46f3b4ef0ebd06867ce6&ckck=1

[12] https://virgamet.com/35hm-34crmo4-1-7220-35crmo4-34cd4-aisi-4135-structural-steel

Advanced Topics in Mechanics of Materials, Structures and Construction - AToMech1-2023 Materials Research Forum LLC
Materials Research Proceedings 31 (2023) 571-580 https://doi.org/10.21741/9781644902592-59

Ultrasonic pulse velocity (UPV) and initial rate of waterabsorption (IRA) of foam concrete containing blendedcement

Y. SUNARNO[1,a*], M.W. TJARONGE[2,b], R. IRMAWATY[2,c], A.B. MUHIDDIN[2,d], M.A. CARONGE[2,e] and M. TUMPU[3,f]

[1]Civil Engineering Department, Faculty of Engineering, University of Christian Indonesia Toraja, Indonesia

[2]Civil Engineering Department, Faculty of Engineering, Hasanuddin University, Makassar, Indonesia

[3]Disaster Management Study Program, The Graduate School, Hasanuddin University, Makassar, Indonesia

[a*]yohanssunarno@gmail.com, [b]tjaronge@yahoo.co.jp, [c]rita_irmaway@yahoo.co.id, [d]bakrim@unhas.ac.id, [e]caronge_eng@yahoo.co.id, [f]tumpumiswa@gmail.com

Keywords: Blended Cement, Fly Ash, NDTs, UPV, IRA

Abstract. The adverse environmental effects caused by cement production activities could be minimized by reducing the use of cement in the concrete mix, replacement with eco-friendly materials, and making low-unit weight cement-derived products. During the cement production activities, fly ash can also be used by mixing it in form of Portland Composite Cement (PCC)/blended cement products or directly in the concrete mixing process. Furthermore, in low-unit weight concrete, foam concrete is usually obtained using an agent that forms random air bubblesin the cement paste and is applied in construction according to its structural function. Therefore, this research aims to study and analyze the correlation between unit weight and non-destructive tests (NDTs) valueon foam concrete with variations in the use of cement, namely OPC, PCC-1, PCC-2, and OPC+FA.The quality of the foam concrete was evaluated by non-destructive tests, there are Ultrasonic PulseVelocity (UPV) and Initial Rate of Water Absorption (IRA). The results show that UPV value of all specimens is directly proportional to its unit weight and varied inversely to the IRA value.

Introduction

Portland cement, which is the most important ingredient in concrete, contributes significantly to global warming. This is due to the cement production process contributing nearly 7% of the total carbon dioxide gas into the earth's atmosphere each year which causes greenhouse emissions [1-3]. Therefore, there is a need to minimize the adverse environmental effects by reducing the use of cement in concrete mixtures, replacement with eco-friendlymaterials, and making low-unit weight cement derivative products. Currently, fly ash which is waste from burning coal at the power plant has been widely used as one of the cement substitute materials. This material is usually added to the cement manufacturing process at the factory to form Portland Composite Cement (PCC)/blended cement products or directly to the fresh concrete mixing process. The manufacture of lightweight concrete is to make cement-derived products with the use of less material, as indicated by a lighter unit weight, which ranges from 400 to1850 kg/m^3. One of the developments of these products is foam concrete which uses a foaming agent to form random air in the cement paste. It has the ability as self-compacting concrete, has very low cement content, and reduces the use of aggregates [4,5]. Blended cement produced by a local cement producer has been experimentally tested in the form of foam concrete composite beams [8], and reinforced concrete-filled prefabricated foam concrete walls [7,8].

Generally, the two methods of testing concrete to ensure the quality of the concrete made as planned include the compressive strength which is destructive and the non-destructive test. A destructive test uses the Universal Testing Machine (UTM) and is carried out in the laboratory by testing several samples to obtain the compressive strength value. The samples can also be tested directly in the field using a tool that measures the compressive strength of hard concrete quickly, practically, and without damage. Meanwhile, a non-destructive method is carried out in the workplace (in situ) to obtain the approximate concrete strength data. The commonly used methods are hammer test, Ultrasonic Pulse Velocity (UPV), and Initial Rate of Water Absorption (IRA).The UPV is a reliable Non-Destructive Testing (NDT) method for monitoring and evaluating concrete quality and detecting damage to its structural components [9]. Several researchers have studied and published the characteristics of concrete using the UPV method [10][11]. In the test procedure according to ASTM C 597-2016 [12], the main principle is to quantify the travel time of the ultrasonic pulse to pass through the tested concrete structure. The ultrasonic waves are channeled from the transmitter transducer on the surface of the concrete to propagate through the material to the receiver transducer. To measure the time taken by a waveform the PUNDIT Read-Out unit (Portable Non-Destructive Indicator Unit Tester) is used in microseconds (msec). Moreover, UPV can also detect several conditions of concrete such as the integrity and uniformity of the concrete, cracks, and depth, honeycombing or voids, the unit weight, and its equivalent in compressive strength. Duringtesting, lower speeds of concrete are indicated by cracking, while better quality in terms of unit weightand homogeneity increases the speed.

The Initial Rate of Water Absorption (IRA) is the total mass of water absorbed for 60 seconds on a 30-square-inch concrete/brick surface area [13]. It is used to determine the level of water absorption from the concrete surface. The test is carried out by calculating the mass increase of the concrete sample due to water absorption at a certain time when one side of the specimen surface is submerged in water. As the hydration occurs and the mortar dries, a chemicalbond forms between the mortar and brick. The brick also needs to absorb enough water and cementfrom the mortar to form the chemical bond within its pores and allow hydration in the joint. The IRA is an important trait that affects the bond and value that can be accepted based on the standards required. If the brick is too dry, it will increase the volume of water absorbed from the mortar, thereby weakening the bond [14].

Therefore, this research aims to study and analyze the correlation between unit weight value and non-destructive test results value (UPV and IRA) of foam concrete with variations in the use of cement, namelyOrdinary Portland Cement (OPC), Portland Composite Cement brand 1 (PCC-1), Portland Composite Cement brand 2 (PCC-2), and Ordinary Portland Cement added fly ash (OPC+FA).

Materials and MethodologyMaterials

This study used OPC and PCC-type cement taken from different brands in South Sulawesi, Indonesia. The types of cement used are OPC, PCC-1, and PCC-2, respectively. Meanwhile, refer to [15] the fly ash used was class F from power plant waste in Jeneponto Regency, South Sulawesi, Indonesia. The data on the physical characteristics and chemical characteristics of all cementitious (OPC, PCC, and fly ash) are shown in Tables 1 and 2. The fine aggregate in this study used siliceous sand sourced from the Pinrang River in Pinrang District, South Sulawesi, Indonesia. The results of their physical characteristics test are shown in Table 3. All materials used in this research met the standard specifications requirement referred to American Standard Testing Method (ASTM).

Mix Design

The mix design of the foam concrete mixture was made with 4 types of variations using Ordinary Portland Cement (OPC), Portland Composite Cement brand 1 (PCC-1), Portland Composite

Cement brand 2 (PCC-2), and Ordinary Portland Cement added fly ash (OPC+30% FA), respectively. The specimen used was a cube-shaped concrete 150 mm x 150 mm x 150 mm. All specimens were treated by air curing in the laboratory with a minimum temperature of 25°C and a maximum of 32°C with a humidity of 60 RH – 74 RH and kept in constant condition until testing.

Unit Weight Tested

The unit weight test is based on [13], which describes the unit weight of structural lightweight concrete standards. The specimens were taken from the curing treatment after 6 days. Specimens were taken from the soaking tub after curing for 6 days. The specimen used is weighed in water and given the initial "C" which is the weight of the specimen in water in a fully submerged condition. Then removed from the water and left for 1 minute by placing the cylinder on a sieve measuring 9.5 mm or more. The water is dried using a damp cloth, its weight is measured, and it is given the initial "B", which is the weight of the specimen in a saturated surface dry state. Then completely dried on the surface in a chamber with humidity at 50% ± 5% and 21°C - 25°C until the weight loss of the specimen is not more than 0.5% at 28 days of age. The dry weight of the specimen is determined and registered in kg with the initial "A". While the weight of the balanced state is calculated according to Equation (1):

$$E_m = (A)/(B-C)\ (kg/m^3) \quad (1)$$

A is the weight of the specimen at drained condition (kg)

B is the weight of the specimen at saturated surface dry (SSD) condition (kg)

C is the weight of the specimen in water until it is fully submerged condition (kg).

***Table 1**. Physical characteristics of cementitious*

Properties	Result			
	OPC	PCC-1	PCC-2	Fly Ash
Fineness/Blaine meter, m^2/kg	345	448	381	-
Water Content, % volume	-	-	-	-
Autoclave expansion, %	0.10	0.06	-	-
Compressive strength				
3 days, kg/cm^2	190	155	184	-
7 days, kg/cm^2	267	228	260	-
28 days, kg/cm^2	359	285	408	-
Time of setting (Vicat test)				
Initial Set, minute	125	135	132.5	-
Final Set, minute	263	260	195	-
False set, final penetration, %	83.58	86.50	-	-
Air Content, % volume	4.53	4.97	-	-
Specific Gravity	3.10	2.94	3.11	2.05

Table 2. Chemical characteristics of cementitious

Properties	(%)			
	OPC	PCC-1	PCC-2	Fly Ash
MgO	2.58	1.79	0.99	-
SO_3	2.10	1.72	1.80	-
SiO_2	-	19.66	18.39	44.56
Al_2O_3	-	5.89	5.15	-
Al_2SO_3	-	-	-	14.55
Fe_2O_3	-	4.29	3.12	11.66
SiO_2+Al_2SO_3+Fe_2O_3	-	-	-	69.94
CaO	-	-	63.29	12.69
Loss on ignition	3.38	-	4.60	0.29
Insoluble residue	0.78			
Alkalies	0.30			

Table 3. Physical characteristics of sand

No.	Properties	Result
1.	Unit weight - Dry Unit weight - SSD	 2.60 2.62
2.	Clay Content (%)	0.96
3.	Water absorption (%)	0.88
4.	Fineness Modulus	1.30
5.	Unit weight (kg/lt) - Loose - Congested	 1.29 1.46
6.	Water Content (%)	3.80
7.	Organic Impurities	No. 1

Ultrasonic Pulse Velocity (UPV) Tested

The UPV test used was the direct method, where two transducers were placed on two parallel surfaces as shown in Figures 1 and 2. This method was based on ASTM C 597 [12] using an Electrical Pulse Generator, Transducer, and Amplifier Electronic Timing Device. According to the standard, the test procedures are (a) calibrating the instrument using a calibration cylinder, the ultrasonic pulse speed is set using a calibration cylinder of 57.8 microseconds, (b) smearing both surfaces of the concrete specimen with vaseline or similar material to level the transducer surfaces, and (c) recording the reading of the travel time figures, calculating the speed (v) using the $v=s/t$ relationship, where s is the spacing between the two transducers and t is the time required..

Initial Rate of Water Absorption (IRA) Tested

IRA testing procedures were based on ASTM C67 [17]. In this method, the specimens were dried in an oven at a temperature of 110 - 115°C for 24 hours, removed, and cooled to room temperature. Subsequently, they were weighed when dry (Wd) and their dimensions were measured (L and B) to obtain the surface area of the bricks. The specimens were placed in a partially submerged position in the water until the entire bottom surfaces were in contact with the water and maintained until the end of the test, as shown in Figure 3. When the surface of the specimen is irregular, water for immersion needs to be added. After soaking for 1 minute ± 0.1 seconds, the specimens were taken from the soaking tub and weighed (Ww).

$$X = 30\ W / BL \tag{2}$$

X = gain in weight corrected to the basis of 30 in.^{2} areasW= gain weight of specimen (g)

B = width of specimen (cm)L = length of specimen (cm)

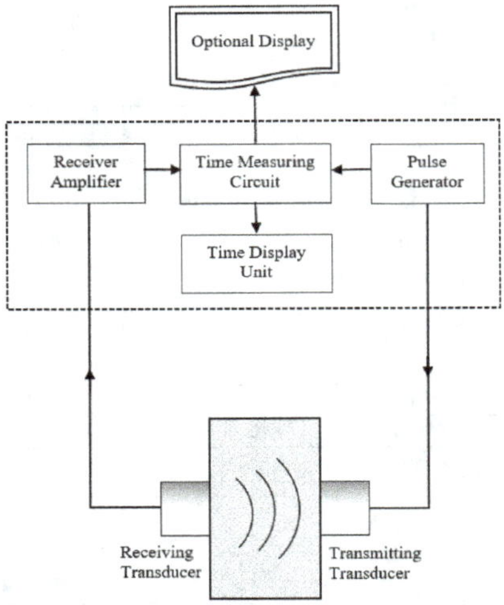

***Figure 1**. Schematic diagram of UPV test circuit*

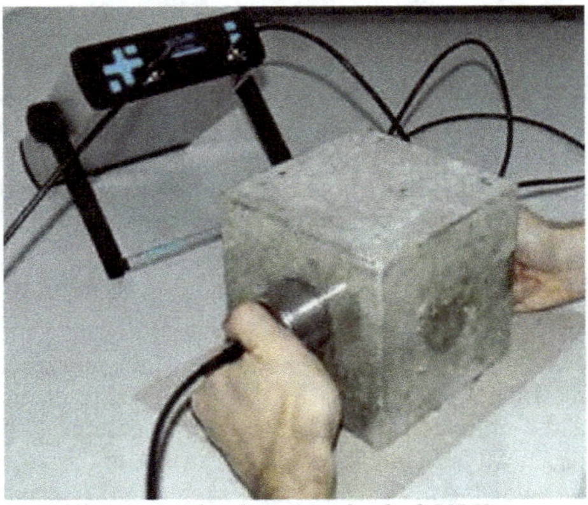

***Figure 2**. The direct method of UPV test*

***Figure 3**. Initial Rate of Water Absorption (IRA) test*

Results and DiscussionMixed Design

The foam concrete mix design for the test specimens used in this study can be seen in Table 4. All variations of the mix design use fine aggregate, water, and additives of the same type and volume. Unless cementitious uses OPC, PCC-1, PCC-2, and OPC+FA. The mix design is made in units of 1 m^3.

***Table 4**. Mixed design of foam concrete per m^3*

Material	Type			
	OPC	**PCC-1**	**PCC-2**	**OPC+FA**
Cement, kg	663.4	663.8	663.7	464.8
Sand, kg	1334.7	1325.4	1325.4	1332.5
Fly Ash, kg	-	-	-	199.2
Water, kg	233	232	232	175.5
Admixture, kg	16.7	16.6	16.6	12.5
Total, kg	2247.8	2237.8	2237.7	2237.8
LWC unit weight, gr/ltr	1325	1325	1325	1325
Mortar portion, %	43.9	43.9	43.9	42.9
Foam portion, %	56.1	56.1	56.1	57.1
Foam agent/water ratio	3 : 10			

Unit Weight

Figure 4 shows that the average unit weight of hard concrete for each specimen using OPC is 1300 kg/m^3. Furthermore, 1295 kg/m^3 was obtained with PCC-1, while PCC-2 is 1287 kg/m^3. The average unit weight of hard concrete test objects using fly ash (OPC+FA) is 1279 kg/m^3. Referring to ACI 213R3 [11], the unit weight of all test objects can be categorized as light structural concrete, where the unit weight in range between 800 kg/m^3 - 1400 kg/m^3.

Ultrasonic Pulse Velocity (UPV)

From Figure 5, we can show that the results of the UPV test of hard concrete for each specimen using OPC is 2.781 km/s. The PCC-1 is 2.625 km/s, while PCC-2 is 2.599 km/s. Furthermore, 2.552 km/s was obtained using fly ash (OPC+FA). According to the International ACI 228.2R-13 [18], all specimens are included in the category of concrete with doubtfully quality, which is below 3.000 km/s.

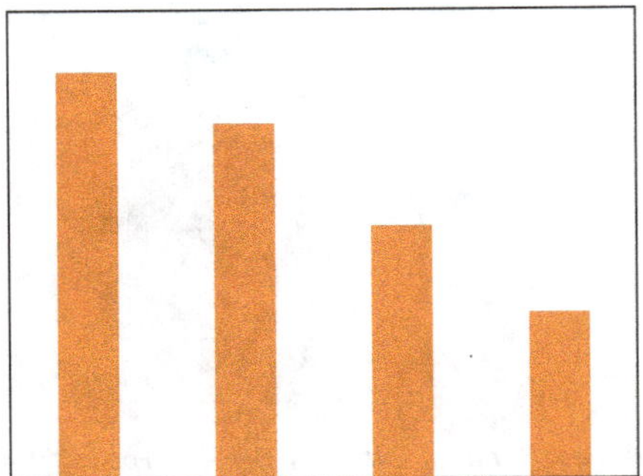

	OPC	PCC-1	PCC-2	OPC+FA
UW	1300	1295	1287	1279

***Figure 4**. Unit weight result*

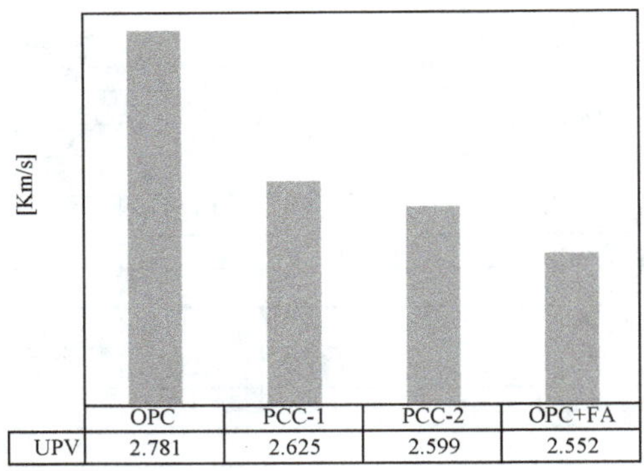

	OPC	PCC-1	PCC-2	OPC+FA
UPV	2.781	2.625	2.599	2.552

***Figure 5**. UPV tested result*

Initial Rate of Water Absorption (IRA)
Based on Figure 6, the values of the Initial Rate of Water Absorption (IRA) test on each specimen, hard concrete using OPC is 14.947 gr/cm^2. The use of PCC-1 obtained 15.178 gr/cm^2, while PCC-2 is 17.093 gr/cm^2. Moreover, fly ash (OPC+FA) is 18.627 gr/cm^2. Based on ASTM C90 [19], as lightweight concrete, the IRA values for all specimens meet the requirements, namely below 20 gr/cm^2.

Relationship Between Unit Weight, UPV, and IRA
Figure 7 indicates that a decrease in the unit weight of the specimen occurs when the UPV test value of the specimen also decreases. This showed that the smaller the value, the greater the concrete unit weight, thereby, reducing the Ultrasonic Pulse Speed (UPV value). However, with a decrease in the specimen's unit weight, the IRAvalue will be greater. This is because the unit weight increases the speed of water absorption by the concrete surface.

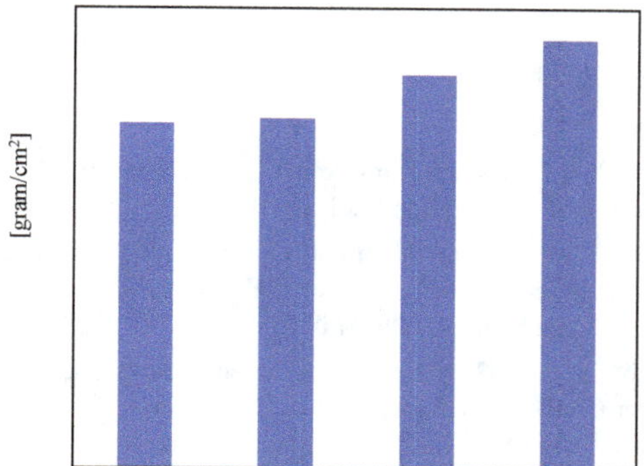

	OPC	PCC-1	PCC-2	OPC+FA
IRA	14.947	15.178	17.093	18.627

Figure 6. *IRA tested result*

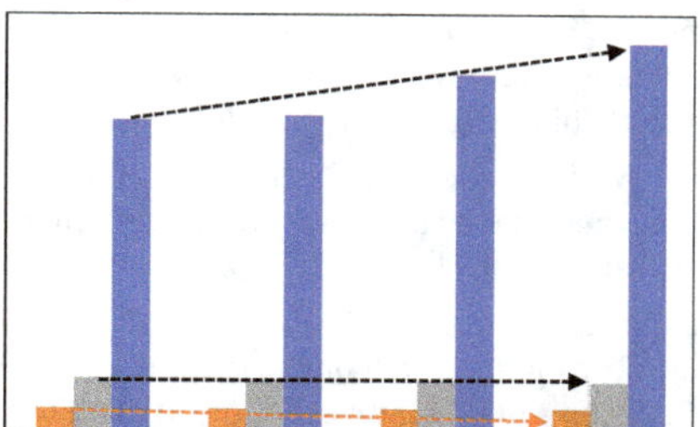

	OPC	PCC-1	PCC-2	OPC+FA
UW	1300	1295	1287	1279
UPV	2.781	2.625	2.599	2.552
IRA	14.947	15.178	17.093	18.627

Figure 7. *Relationship between unit weight, UPV, and IRA*

Conclusion

Referring to the results, the following conclusions are obtained:

1. OPC specimen has the highest unit weight of 1300 kg/m^3, while the OPC+FA has the lowest, with a value of 1279 kg/m^3. Based on ACI 213R3, all test specimens were categorized as lightweight structural concrete, which unit weight values from 800 kg/m^3 to 1400 kg/m^3.
2. The OPC specimen has the highest UPV value of 2.781 km/s, while the OPC+FA has the lowest,which is 2.552 km/s. Therefore, according to ACI 228.2R-13, all specimens are in the category of concrete with doubtfully quality, which is below 3.000 km/s.
3. The OPC specimen has the lowest IRA value of 14.947 gr/cm^2, while the OPC+FA test objecthas the highest, which is 18.627 gr/cm^2. Referring to ASTM C90, as lightweight

concrete, the IRA values for all specimens meet the requirements, namely below 20 gr/cm^2.

4. The UPV value of the specimen is directly proportional to its unit weight and varied inversely to the IRA value.

References

[1] Deja, J., et al. 2010. CO2 Emissions from Polish Cement Industry. International Journal of Greenhouse Gas Control 4: 583-588. https://doi.org/10.1016/j.ijggc.2010.02.002

[2] Ramamurthy K.E.K., Kunhanan and Nambiar, G. Indu Siva Ranjani. 2009. Classification of studies on properties of foam concrete, Cem. Concr. Compos. 31 (6): 388-396 https://doi.org/10.1016/j.cemconcomp.2009.04.006

[3] Tumpu M. and Mabui D. S. 2022. Effect of Hydrated Lime (Ca(OH)2) on Compressive Strength of Geopolymer Concrete. AIP Conference Proceedings 2391, 070011 (2022). https://doi.org/10.1063/5.0086702

[4] Sunarno Y., Tjaronge M. W. and Irmawaty R. 2020. Preliminary study on early compressive strength of foam concrete using Ordinary Portland Cement (OPC) and Portland Composite Cement (PCC). IOP Conf. Ser.: Earth Environ. Sci. 419. https://doi.org/10.1088/1755-1315/419/1/012033

[5] Sunarno Y., Tjaronge M. W., Irmawaty R. and Muhiddin A.B. 2022. (2022). Unit Weight of Foam Concrete Containing Blended Cement and Fly Ash. Jurnal Teknik Sipil & Perencanaan 24 (1) (2022) p 36 - 41. https://doi.org/10.15294/jtsp.v24i1.34664

[6] Syahrul, Tjaronge M.W., Djamaludin R., Amirudin A.A. 2021. Flexural behavior of normal and lightweight concrete composite beams Civil Engineering Journal (Iran), 7(3), pp-559. https://doi.org/10.28991/cej-2021-03091673

[7] Amirudin A.A, Parung H., Tjaronge M.W., Mansyur M., Tumpu M. 2022. Influence of prefabricated foam concrete as infilled wall on the strength due to cyclic loading. International Journal of GEOMATE. 22, pp 114-121. https://doi.org/10.21660/2022.93.j2343

[8] Mansyur M., Tjaronge M.W., Irmawaty R., Amirudin A.A. 2022. Performance of prefabricated foam concrete as an infilled wall under cyclic lateral loading. International Journal of Engineering, Transaction B: Applications, 35(2), pp 332-343. https://doi.org/10.5829/IJE.2022.35.02B.09

[9] Hannachi dan Guetteche, 2012. Application of the Combined Method for Evaluating the Compressive Strength of Concrete on Site. Open Journal Civil Engineering. Vol.2 No. 1, March 2012.

[10] Gebretsadik, B.T. 2013. Ultrasonic Pulse Velocity Investigation of Steel Fiber Reinforced Self Compacted Concrete. University of Nevada, Las Vegas. UNLV Theses, Dissertations, Professional Papers, and Capstones.

[11] Lin, Y., Kuo, S.-F., Hsiao, C.3., Lai, C.-P., 2007. Investigation of pulse velocity- strength relationship of hardened concrete. ACI Mater. J. 104 (No. 4), 344-350. https://doi.org/10.14359/18823

[12] ASTM C 597, 2016. 2016. Standard Method for Pulse Velocity Through Concrete.

[13] ASTM C642-97. 1997. Standard Test Method for Unit Weight, Absorption, and Voids in Hardened Concrete.

[14] Raimondo M., Dondi M., Gardini D., Guarini G., Mazzanti F. 2009. Predicting the initial rate of water absorption in clay bricks. Construction and Building Materials 23 (2009) 2623-2630 https://doi.org/10.1016/j.conbuildmat.2009.01.009

[15] ASTM C 618-05. 2005. Standard Specification for Coal Fly Ash and Raw or Calcined Natural Pozzolan for Use in Concrete.

[16] ACI 213R-87.1999. Guide for Structural Lightweight Aggregate Concrete, 1999.

[17] ASTM C 67. Standard Test Method for Sampling and Testing Brick and Structural Clay.

[18] ACI 228.2R-13. 2013. Report on Nondestructive Test Methods for Evaluation of Concrete in Structures

[19] ASTM C90. 2010. Standard Specification for Load-bearing Concrete Masonry Units.

Advanced Topics in Mechanics of Materials, Structures and Construction - AToMech1-2023 Materials Research Forum LLC
Materials Research Proceedings 31 (2023) 581-588 https://doi.org/10.21741/9781644902592-60

Comparison study of performance and heat leak factor of three types of heat exchangers operated with nanofluid

Esam I Jassim[1,a *] and Bashar I Jasem[2,b]

[1]College of Engineering, Prince Muhammad Bin Fahd University, 31952 Al-Khobar, Kingdom of Saudi Arabia

[2]Department of Computer Technology, Al-Hadba University College, Mosul, Iraq

[a]ejassim@pmu.edu.sa, [b]adamalzuhairee@gmail.com

Keywords: Heat Exchanger, Nanofluid, Heat Leak Factor, System/Environment Energy Interaction

Abstract. Nanofluid are categorized under class of fluids that have extreme potential to enhance the heat transfer in heat exchangers. However, the interaction between the exchanger and the ambient (heat leak) is an essential issue since it could deteriorate the exchanger performance. The quantity of the heat loss is directly related to the temperature of the hot fluid, the Reynolds number, and the type and volume fraction of the nanoparticle. The experimental study presents the heat leak when distilled water is mixed with different concentration of Al_2O_3 (range from 1% to 3%) and streamed into three different types of heat exchangers, namely concentric, shell and tube, and plate type. The results emphasize that presence of nanoparticles improves the Nusselt number (convection coefficient) and the NTU number (exchanger performance). The analysis also shows that heat leak factor augments with the increase in VoF of nanofluid in all types. However, plate type exchanger has the lowest losses to the ambient, followed by shell and tube and finally concentric type.

Introduction

Heat exchangers are type of equipment that are commonly used in numerous industries such as petrochemical sector, ore refining sector, power plants and so on. Heat exchangers can be customized based on geometric structure suitability, maintenance feasibility, and operation durability [1]. Although shell and tube heat exchangers are most commonly used in this field, other types such as double pipe and plate heat exchangers are still in demand for many applications.

Recently, numerous research studies were implemented on the impact of utilization nanometallic on the performance of energy equipment such as the heat exchanger [2, 3, 4] and engine lubricant [5].

R. Dharmalingam [6] summarized the experimental study of the forced convective heat transfer and flow characteristics of a nanofluid consisting of water as base fluid and 1% Al_2O_3 (volume concentration) nanoparticle flowing in a parallel flow, counter flow, and shell and tube heat exchanger under laminar flow conditions. He proved that shell and tube exchanger provides more enhancements in heat transfer coefficient than the other two flow arrangements. Nusselt number was also shown to be augmented for the Al_2O_3/water nanofluid, which eventually enhances the convective heat transfer coefficient.

Jaafar Albadr [7] investigated experimentally the thermal performance of propylene glycol/water with a fixed concentration of (10/90) % and Al_2O_3/water nanofluid with a various concentration (0.1, 0.4, 0.8, 1.5, and 2.5) % by volume under turbulent flow inside a horizontal shell and tube heat exchange. The researcher reported that thermal conductivity and viscosity increase due to the dispersion of the nanoparticles into the base liquid. However, friction factor augments with the increase of a particle volume concentration.

Wael I.A. Aly [8] studied the heat transfer and pressure drop characteristics of water-based Al_2O_3 nanofluid flowing inside coiled tube-in-tube heat exchangers. The design parameters of the

CFD modelling were in the range of 0.5%-2.0% VoF, coil diameters 0.18-0.3 m, inner tube and annulus sides flow rates from 2 to 5 LPM and 10 to 25 LPM, respectively. The results obtained showed a different behavior depending on the parameter selected for the comparison with the base fluid. Also for same Re or Dn, the heat transfer coefficient increases by increasing the coil diameter and nanoparticles volume concentration. The friction factor, on the other hand, increases with the increase in curvature ratio.

D. Madhesh a,b, R. Parameshwaran b,c, S. Kalaiselvam [9] study to address the heat transfer and rheological characteristics of HyNF were investigated experimentally using a tube in-tube counter flow heat exchanger. The conclusions explained that the convective heat transfer coefficient of nanofluids increased with HyNC concentration and the Reynolds number. the Nusselt number, and the overall heat transfer coefficient were enhanced by 52%, 49% and 68%, respectively. The authors also noticed that the friction factor and pressure drop of HyNF obtained for 1% up to 2.0% volume concentration were 1.7% and 14.9% respectively.

Heydar Maddah, Nahid Ghasemi [10] experimentally investigated the heat transfer efficiency of water and iron oxide nanofluid in a double pipe heat exchanger equipped with a typical twisted tape. Experiments were conducted under the laminar and turbulent flow for Reynolds numbers in the range of 1000 to 6000. Three the concentration by weight of nanofluid was selected, namely: 0.01, 0.02 and 0.03 wt%. They reported that heat transfer efficiency increases as opposed to that of water (the base fluid). Numerically speaking, at Reynolds number was equal to 3000 and 45 °C, the heat transfer efficiency value for nanofluid was enhanced approximately by 21%. At identical Reynolds number and temperature, the heat transfer efficiency was ameliorated up to 25.9%.

Jassim et al [11, 12] are separately investigated the thermal performance of two types of heat exchangers when operate at concentration of 1%, 2%, and 3% of titanium dioxide–water (TiO_2–water) nanofluids by dispersing 20 nm diameter nanoparticles in distilled water. They also experimentally compared the overall heat transfer coefficient and NTU of water vs. nanofluids in laboratory-scale plate and shell-and-tube heat exchangers. Experimental results showed both augmentation and deterioration of heat transfer coefficient for nanofluids depending on the flow rate and nanofluid concentration through the heat exchangers.

Alireza Falahat, Mohsen Shabani, Mohsen Maleki [13] examined theoretically the effects of water- Al_2O_3 nanofluid on exergy destruction, exergy efficiency and pumping power in the helically coiled tube heat exchanger under turbulent flow and subjected constant wall condition. The main parameters considered in this study are nanoparticles volume concentration, nanoparticle size, Reynolds number, curvature ratio and dimensionless inlet temperature. It is found that when the Reynolds number increases, dimensionless total exergy destruction decreases. It is observed that by increasing the nanoparticles volume concentration from 2% to 6%, the dimensionless thermal exergy destruction reduces by 3.64% to 20.21 % compared to pure water. Also, it is seen that when nanoparticles dimensions increases, the exergy efficiency increases and pumping power decreases. Finally, the exergy efficiency increases with increasing of curvature ratio and pumping power decreases with increasing of curvature ratio.

There are ample researches on employing nanofluid in heat exchanger; however, studies on the exchanger/ambient energy interaction are still scarce. This work attempts to fill the gap by emphasizing on the energy interaction of three types of heat exchangers with the environment.

Theory

Nanofluid properties are formulated as a function of the concentration of nanoparticles. Density and heat capacity of the nanofluid is computed from Eqs. 1 and 2, respectively:

$$\rho_m = (1-\phi)\rho_f + \phi\rho_p \quad (1)$$

$$(\rho Cp)_m = (1-\phi)(\rho Cp)_f + \phi(\rho Cp)_p \quad (2)$$

Maxwell correlation is employed for nanofluid thermal conductivity while Batchelor empirical model is considered to evaluate the nanofluid viscosity, as listed below:

$$k_m = \frac{k_p+2k_f+2\phi(k_p-k_f)}{k_p+2k_f-\phi(k_p-k_f)}k_f \tag{3}$$

$$(\mu)_m = (1+2.5\phi+6.2\phi^2)\mu_f \tag{4}$$

k_p, k_f are the thermal conductivity of the nanoparticle material and the base fluid, respectively. k_m is the effective thermal conductivity of the nanofluid. μ_m is the nanofluid viscosity, μ_f is the fluid viscosity, and ϕ is the volume concentration of the suspended particles.
to convert mass fraction into volume fraction of the nanofluid, the following expression stated below is adopted:

$$\% \text{ volume concentration, } \phi = \left[\frac{\left(\frac{w_{CuO}}{\rho_{CuO}}\right)}{\left(\frac{w_{CuO}}{\rho_{CuO}}+\frac{w_{bf}}{\rho_{bf}}\right)}\right]\times 100 = \left[\frac{\left(\frac{w_{CuO}}{6300}\right)}{\left(\frac{w_{CuO}}{6300}+\frac{100}{1000}\right)}\right]\times 100 \tag{5}$$

Nusselt number of nanofluid can be determined from the following expression:

$$\boldsymbol{Nu_m} = \frac{h_o D_h}{k_m} \tag{6}$$

where h_o is the convection heat transfer coefficient of the nanofluid evaluated from Eq.7:

$$\frac{1}{h_o} = \frac{1}{U_o} - \left\{\frac{D_o}{D_i}\times\frac{1}{h_i} + \frac{D_o \ln(D_o/D_i)}{2k_w}\right\} \tag{7}$$

the overall heat transfer coefficient U_o, is based on outer surface area and is calculated using the LMTD method:

$$U_o = \frac{Q_c}{A_o \times \left(\frac{\Delta T_1 - \Delta T_2}{\ln(\Delta T_1/\Delta T_2)}\right)} \; ; \; \Delta T_1 = T_{h,i} - T_{c,o} \text{ and } \Delta T_2 = T_{h,o} - T_{c,i} \tag{8}$$

The surface area A_o is calculated from different expressions, depending on the type of the heat exchangers, [11, 12, 14]

$\boldsymbol{T_{ci}}$ and $\boldsymbol{T_{co}}$ are the inlet and outlet mean temperature of the nanofluid; and D_i, D_o, D_h are the inner, outer, and hydraulic diameter, respectively.

The heat transfer coefficient of the hot fluid streamed through the tube-side is evaluated from the Dittus-Boelter empirical correlation:

$$Nu = 0.023\,\mathrm{Re}^{0.8}\,\mathrm{Pr}^{0.3} \tag{10}$$

$$\text{where } \mathrm{Pr} = \frac{\mu \times C_p}{k} \tag{11}$$

Experimental Set-up

The experimental setup, illustrates in Fig.1 consists of two units namely, the base unit and the heat exchanger unit which are computer controlled through PLC-SCADA data logging and process monitoring system, Fig.(2). The dimensions of the base unit are 1100 x 650 x 400 mm. The water in the hot fluid tank is heated by means of an electric heating element to a prescribed temperature and is then streamed to the the heat exchanger unit by means of an impeller pump. The second fluid which is the nano fluid in the cold fluid tank is pumped into the heat exchanger unit by means of another impeller pump. Two flowmeters and valves are located along the hot and cold circuits to control and measure the flow rates of hot and cold fluids. Al_2O_3 /water nanofluid is circulated in the annulus cold zone. The major components and specifications of the base unit is listed in Table 1.

Table 1: Components and Specifications of Base Unit

Component	Sub-components	Specification
Tank	1. Electric heating element 2. Level switch 3. Temperature sensor 4. Drain valve	Tank material : Stainless Steel Tank capacity: 30 liter Heater power: 3000W
Impeller Pump	Range: 0 - 2.5 liter/min	
Pressure Regulation Valve	Maximum pressure: 0.7 bar	
Flow Sensors	Quantity: 2 Range: 0 – 6.5 liter/min	
Ball Valves	Quantity: 6	
Flexible Tubes	Quantity: 4	

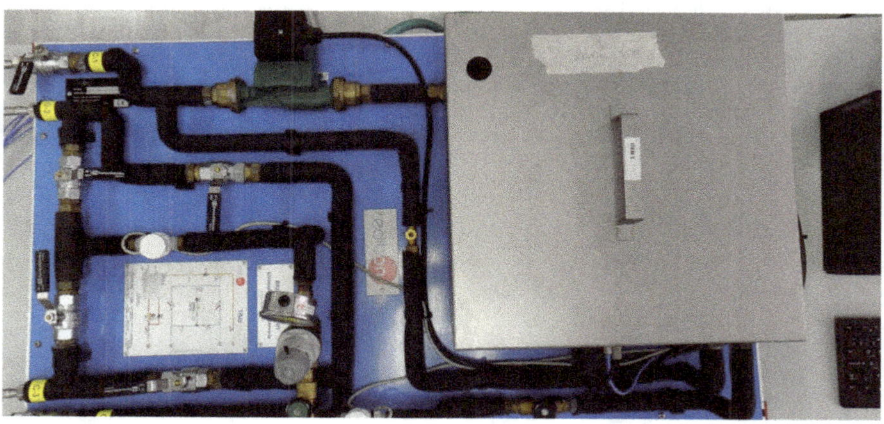

Figure 1. Base unit includes: hot fluid tank, sensors, valves, pump and pressure regulator

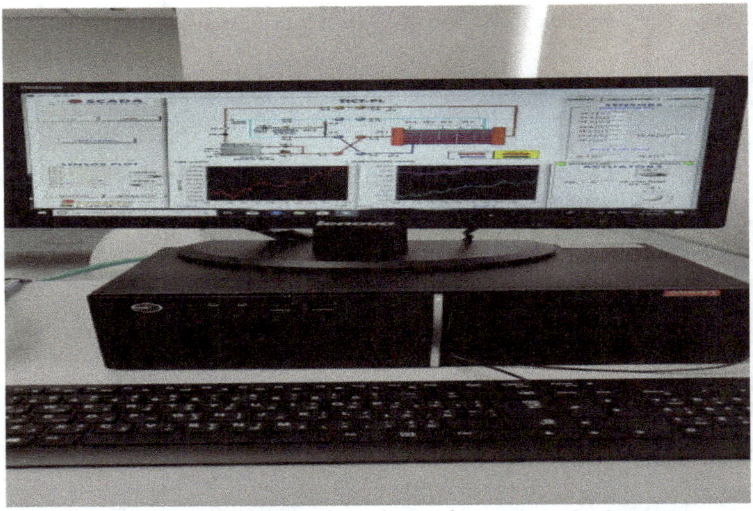

Figure 2. PLC-SCADA Data recording and process monitoring system

Results and discussion

Fig. (3) depicts the variation of Nusselt Number with the Volume Fraction of Al_2O_3 nanofluid for the three types of heat exchangers. Observation shows that Nu is augmented as the VoF of nanofluid increases. This can be attributed to the enhancement in the convection heat transfer coefficient of the nanofluid due to the presence of metallic nanoparticles. The figure also illustrates that magnitude of Nu of the plate type of heat exchanger is the largest and the concentric type in the lowest regardless of the nanofluid VoF.

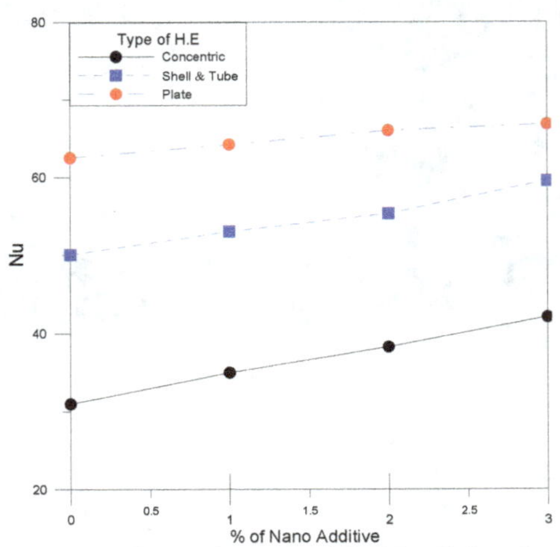

Figure 1: Variation of Nu with nanofluid VoF for Al_2O_3 volume fraction

The NTU is a prominent tool that frequently utilized to assess and evaluate the performance of any heat exchanger. Plate type is shown to operate at higher performance compared to the other types and its performance even enhances exponentially with the presence of nanofluid, figure (4). Shell and tube type enhancement is observed to be slight but always better than the concentric type.

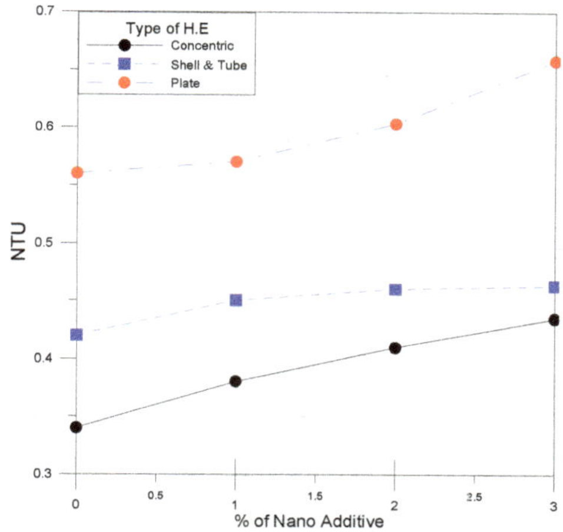

Figure 4: Variation of NTU with nanofluid VoF for Al_2O_3 volume fraction

Jassim and Ahmed [11] proposed a new parameter, so called "Leak Factor", as a benchmark to feasibly assess such the exchanger/environment intereaction. Defined as the ratio of the heat rate lost to the ambient from the entire system to the actual rate transferred from the hot fluid, the leak factor is plotted against the Reynolds number of the nanofluid for the three selected volume fractions. Fig. (5) illustrates the variation of the leak factor with the Reynolds number at various volume fractions of Al_2O_3 for shell and tube exchanger. The higher the Reynolds number, the larger the leak factor. Also, the figure concludes that leak factor is substantially elevated with the presence of nanoparticle, particularly at high Reynolds number.

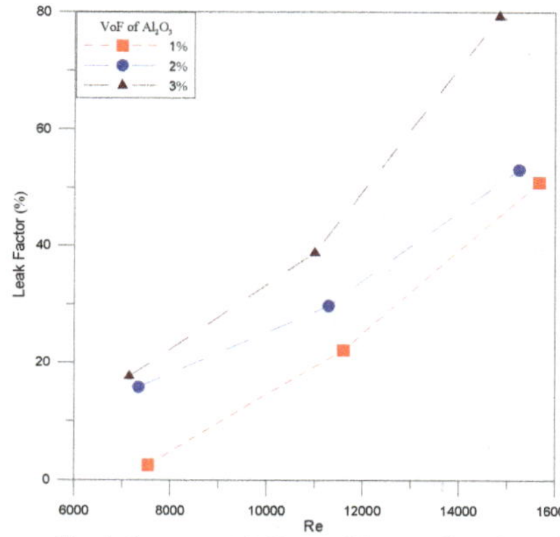

Figure 2: Variation of leak factor with Reynolds number for Al_2O_3 nanofluid, [11]

Heat leak factor is plotted against the concentration of Al_2O_3 for the three exchangers under study. Figure 6 summarizes that for all exchangers, the heat leak factor is proportion to the concentration of nanofluid. However, plate heat exchanger leak factor is observed to be the minimal. The heat loss factor predicted from the experiments increases with the increase of the concentration. The figure also concludes that using nanofluid intensifies the energy interaction of the exchanger with the surroundings.

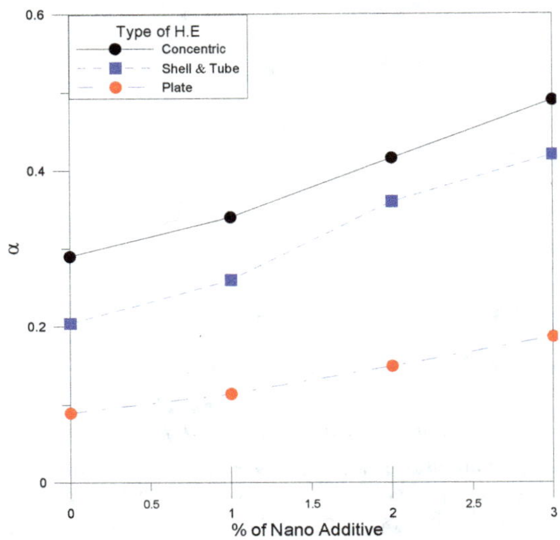

Figure 6: Heat Leak factor against nanofluid VoF for Al_2O_3 volume fraction

Conclusion

Experimental study on the behavior of Al_2O_3 nanometallic/water at various VoF for three types of heat exchangers are analyzed. Results indicated that the presence of nanofluid ameliorates the heat transfer characteristics significantly. Adding more nanoparticle to the base fluid enhances the Nusselt Number as well as the heat transfer performance.

By increasing the nanoparticle concentrations, the heat leak to the environment is shown to be augmented, resulting in reduction in the system availability.

Reference

[1] Jassim, Esam I. "Effect of coil geometry on the heat transfer effectiveness in spiral coil heat Exchanger." In ASME International Mechanical Engineering Congress and Exposition, Nov.3-9. American Society of Mechanical Engineers, 2017.

[2] Jassim, Esam I. "Experimental study on transient behavior of embedded spiral-coil heat exchanger." Mechanical Sciences 6, no. 2 (2015): 181-190. https://doi.org/10.5194/ms-6-181-2015

[3] Jassim, Esam I., Faizan Ahmed, and Bashar Jasem. "Evaluating the influence of nanofluid type and volume fraction on the performance of heat exchanger." In ASME International Mechanical Engineering Congress and Exposition, vol. 84591, p. V011T11A042. American Society of Mechanical Engineers, 2020. https://doi.org/10.1115/IMECE2020-23132

[4] Jassim, Esam. "Spiral Coil Heat Exchanger-Experimental Study." In Proceedings of the 3rd International Conference on Fluid Flow, Heat and Mass Transfer. 2016. https://doi.org/10.11159/ffhmt16.107

[5] Jassim, Esam I. "Exergy analysis of petrol engine accommodated nanoparticle in the lubricant system." International Journal of Exergy 35, no. 3 (2021): 406-420. https://doi.org/10.1504/IJEX.2021.115907

[6] R.Dharmalingam, K.K.Sivagnanaprabhu, J. Yogaraja,S. Gunasekaran, R. Mohan, "Experimental investigation of heat transfer characteristics of nanofluid using parallel flow, counter flow and shell and tube heat exchanger", Archive of mechanical engineering , VOL.LXII 2015. Pp: 509-522. https://doi.org/10.1515/meceng-2015-0028

[7] J. Albadr, "Thermal Performance of shell and tube heat exchanger using PG/water and Al2O3 nanofluid", in: L. Castro Gómez, V. Manuel Velázquez Flores (Eds.), Advances in Heat Exchangers, IntechOpen, 2019. https://doi.org/10.5772/intechopen.80082

[8] W.I. Aly, "Numerical study on turbulent heat transfer and pressure drop of nanofluid in coiled tube-in-tube heat exchangers" Energy Conversion Management, 79 (2014), pp. 304-316. https://doi.org/10.1016/j.enconman.2013.12.031

[9] Madhesh D, Parameshwaran R, Kalaiselvam S., "Experimental investigation on convective heat transfer and rheological characteristics of Cu-TiO2 hybrid nanofluids", Exp Therm Fluid Sci. 2014;52:104-15. https://doi.org/10.1365/s35147-014-0992-7

[10] Maddah H., Alizahed M., Ghasemi N.,"Experimental study of Al2O3/water nanofluid turbulent heat transfer enhancement in the horizontal double pipes fitted with modified twisted tapes. Int J Heat Mass Transf 78:1042-1054. https://doi.org/10.1016/j.ijheatmasstransfer.2014.07.059

[11] Jassim, Esam I., and Faizan Ahmed. "Experimental assessment of Al2O3 and Cu nanofluids on the performance and heat leak of double pipe heat exchanger." Heat and Mass Transfer 56, no. 6 (2020): 1845-1858. https://doi.org/10.1007/s00231-020-02826-9

[12] Jassim, Esam I., and Faizan Ahmed. "Assessment of nanofluid on the performance and energy-environment interaction of Plate-Type-Heat exchanger." Thermal Science and Engineering Progress 25 (2021): 100988. https://doi.org/10.1016/j.tsep.2021.100988

[13] A. Yousefnejad, Ali Reza, A Heyhat, Mohammad Mahdi, A Meghdadi, Amir Homayoun, "Experimental study of nanofluid effects on heat transfer in closed cycle system in shell and tube heat exchangers at Isfahan power plant", J Environmental Engineering Science V 4, P 95-109, 2016.

[14] Saad Alshahrani, Esam I. Jassim, Faizan Ahmed, Bashar I. Jasem, "Performance and environment interactivity of concentric heat exchanger practicing TiO2 nanofluid and operated near heat capacity ratio of unity", Case Studies in Thermal Engineering, Volume 28, 2021, 101702. https://doi.org/10.1016/j.csite.2021.101702

Advanced Topics in Mechanics of Materials, Structures and Construction - AToMech1-2023 Materials Research Forum LLC
Materials Research Proceedings 31 (2023) 589-595 https://doi.org/10.21741/9781644902592-61

Dynamic deformation analysis of a spherical cavity explosion

LAYENI Olawanle Patrick[1,a *] and AKINOLA Adegbola Peter[1,b]

[1]Department of Mathematics, Obafemi Awolowo University, 220005 Ile-Ife, Nigeria

[a]olayeni@oauife.edu.ng, [b]aakinola@oauife.edu.ng

Keywords: Explosion, Laurent Series, Spherical Cavity, Differential-Difference Equation, Explicit Closed-Form Solution

Abstract. The problem of rapid explosion of a spherical cavity in an infinite elastic media of Achenbach and Sun, Israilov and Hamidou is revisited in this study under central symmetric considerations. The governing partial differential equation is reformulated as a differential-difference equation, with compatibility conditions for the unknown cavity's radial evolution, with the unknown displacement being the exponential generating function of a formal radial Laurent series. It is shown that for cavity explosion at constant speed, the modified problem admits self-similar displacement profiles of the inverse hyperbolic type for conjugate decreasing cavity pressures inversely proportional to the time.

Introduction

This short paper provides an effective exact method of lines for the dynamic response of a rapidly expanding surface in an infinite elastic space consisting of a linear homogeneous and isotropic Hookean material. The mechanical problem is that of the determination of the deformation ensuing exterior to a rapidly expanding spherical cavity in an infinite elastic media. The transition into anelastic constituent properties is excluded from this consideration, and only infinitesimal strain is assumed in the media which is also assumed homogeneous.

Moreover, the mathematical problem is that of the simultaneous determination of the displacement field in the region exterior to that cavity and the dynamics of the exploding (or rapidly evolving) surface in the time. Further simplifying assumptions are that the process is isothermal and the central symmetry is invoked within a spherical polar system of coordinates. More precisely, the constitutive relation of the material is given by

$$\tilde{\sigma} = \lambda \operatorname{tr}(\tilde{\varepsilon})\tilde{\mathrm{I}} + 2\mu\tilde{\varepsilon},$$

where $\tilde{\sigma}$ is the Cauchy stress tensor; $\tilde{\varepsilon}$ is the linearized Green-St.Venant strain tensor; $\tilde{\mathrm{I}}$ is the unit second rank tensor; and λ, μ are the Lame viscosity constants. It should be emphasized also that the strain considered here is infinitesimal.

Given the more recent interesting consideration by Israilov and Hamidou [1] in which a thorough review of the earlier work by Achenbach and Sun [2] is presented, we have decided to give yet a different view of the same problem by using a method of lines and thereby obtaining exact explicit solutions for a rapidly exploding surface moving at constant speed. It should be noted that in the exposition herein, the initial displacement is taken as non-zero and radially dependent, no *a priori* assumption of the nature of the evolution of the surface, and neither is the shape of the pressure in the cavity aforehand specified, the intent being for the compatible trio of displacement, surface evolution, and cavity pressure to naturally present itself through the analytical manipulations.

This study is by no means exhaustive, and has the limitations of being able to address only a class of aforestated compatible *trio*, but it has the strength of being able to give explicit solutions which offer the utility of being a benchmark for numerical and analytical solutions for problems of this class. Specifically, the problem in context is governed by the following partial differential equation, and attendant initial and boundary conditions:

$$\frac{\partial^2 u(r,\bar{t})}{\partial r^2}+\frac{2}{r}\frac{\partial u(r,\bar{t})}{\partial r}-\frac{2}{r^2}u(r,\bar{t})=\frac{1}{c^2}\frac{\partial^2 u(r,\bar{t})}{\partial \bar{t}^2} \qquad r>s(\bar{t})>0,\bar{t}>0; \tag{2$_1$}$$

$$u(r,0)=\omega r,\ \frac{\partial u(r,\bar{t})}{\partial \bar{t}}\Big|_{\bar{t}=0}=0 \quad r>s(0)>0; \tag{2$_2$}$$

$$\left(\frac{\partial u(r,\bar{t})}{\partial r}+\frac{2\nu}{1-\nu}\frac{u(r,\bar{t})}{r}\right)\Big|_{r=s(\bar{t})}=-\frac{1}{\rho c^2}\mathsf{P}(\bar{t}), \tag{2$_3$}$$

where $u(r,\bar{t})$ is the radial displacement field in time $\bar{t}$; $c=\sqrt{\frac{\lambda+2\mu}{\rho}}$; ν the Poisson ratio of the body; and $\mathsf{P}(\bar{t})$ is time-dependent pressure in the cavity $r<s(\bar{t})$.

Despite the innocuous outlook of the problem, it is observed that there are inherent nonlinearities in the formulation of the problem, essentially being the moving surface $r=s(\bar{t})$. The current literature is replete with techniques for handling formulations of this kind, and it should be noted that in spite, problems of this class do not readily yield themselves to exact solutions.

Differential-Difference Reformulation

Motivated by the solution of the quasistatic state equation $(2)_1$

$$u(r,\bar{t})=\frac{c_1(\bar{t})}{r^2}+c_2(\bar{t})r \tag{3}$$

we assume the displacement evolution of the elastic media in the region $r>s(\bar{t})$ which is ahead of the rapidly expanding surface in the form of a spatial Laurent series

$$u(r,t)=\sum_{j=0}^{\infty}\mathsf{M}_j(t)\frac{r^{-j}}{\Gamma(j+1)}, \tag{4}$$

where $t:=c\bar{t}$. By elementary calculations, we have the representations

$$\begin{aligned}\frac{\partial^2 u(r,\bar{t})}{\partial r^2}+\frac{2}{r}\frac{\partial u(r,\bar{t})}{\partial r}-\frac{2}{r^2}u(r,\bar{t})&\equiv\sum_{j=0}^{\infty}(j-2)(j+1)\mathsf{M}_j(t)\frac{r^{-(j+2)}}{\Gamma(j+1)};\\ \frac{1}{c^2}\frac{\partial^2 u(r,\bar{t})}{\partial \bar{t}^2}&\equiv\ddot{\mathsf{M}}_0(t)+\ddot{\mathsf{M}}_1(t)r^{-1}+\sum_{j=0}^{\infty}\ddot{\mathsf{M}}_j(t)\frac{r^{-(j+2)}}{\Gamma(j+3)}.\end{aligned} \tag{5}$$

From Eq. (5) it is further observed that $\mathsf{M}_j(t)$ verifies the differential-difference equation

$$\begin{cases}\ddot{\mathsf{M}}_{j+2}(t)=(j^2-4)(j+1)^2\mathsf{M}_j(t);\\ \ddot{\mathsf{M}}_0(t)=0=\ddot{\mathsf{M}}_1(t),\end{cases} \tag{6}$$

with the boundary condition of the problem suggesting the shape of $M_j(t)$ in the form

$$\mathsf{M}_j(t) \propto \mathsf{P}(t)s^{j+1}(t), \tag{7}$$

while, defining $s_0 := s(0)$, the zero initial conditions enforce the following constraints

$$\begin{cases} 0 < \sum_{j=0}^{\infty} \frac{\mathsf{M}_j(0)}{\Gamma(j+1)} s_0^{-j}; and \\ 0 < \sum_{j=0}^{\infty} \frac{\dot{\mathsf{M}}_j(0)}{\Gamma(j+1)} s_0^{-j}. \end{cases}$$
(8)

Solution to differential-difference equation
Assuming that $\mathsf{M}_j(t) = \mathsf{P}(t)s^{j+1}(t)m_j$, m_j a yet-to-be-determined sequence, the constraints in Eq. $(6)_2$ imply that

$$\begin{cases} \left(\ddot{\mathsf{P}}(t)s(t) + 2\dot{\mathsf{P}}(t)\dot{s}(t)\right)m_0 = 0; \\ \left(\ddot{\mathsf{P}}(t)s^2(t) + 4\dot{\mathsf{P}}(t)\dot{s}(t)s(t) + 2\dot{s}^2(t)\mathsf{P}(t)\right)m_1 = 0, \end{cases}$$
(9)

from which it is observed that for non-vanishing m_0 and m_1, the admissible surface evolution pattern and the relation

$$s(t) = \tau t + s_0; \mathsf{P}(t) = \beta s^{-1}(t), \tag{10}$$

$\beta \in$ obtains.

Correspondingly, Eq. $(6)_1$ yields the following equations

$$\begin{cases} \left(\ddot{\mathsf{P}}(t)s^{j+3}(t) + 2\dot{\mathsf{P}}\,\dot{s}(t)s^{j+2}(t)(j+3) + (j+2)(j+3)\mathsf{P}(t)\dot{s}^2(t)s^{j+1}(t)\right)m_{j+2} \\ = (j^2-4)(j+1)^2\mathsf{P}(t)s^{j+1}(t)m_j; \end{cases}$$
$(11)_1$

$$\begin{cases} \left(\ddot{\mathsf{P}}(t)s^2(t) + 2\dot{\mathsf{P}}(t)\dot{s}(t)s(t)(j+3) + (j+2)(j+3)\mathsf{P}(t)\dot{s}^2(t)\right)m_{j+2} \\ = (j^2-4)(j+1)^2\mathsf{P}(t)m_j. \end{cases} \tag*{$(11)_2$}$$

Resulting from the relation $\ddot{\mathsf{P}}(t)s(t) + 2\dot{\mathsf{P}}(t)\dot{s}(t) = 0$, Eq. $(11)_2$ in turn yields

$$\left(2\dot{\mathsf{P}}(t)\dot{s}(t)s(t)(j+2) + (j+2)(j+3)\mathsf{P}(t)\dot{s}^2(t)\right) = (j^2-4)(j+1)^2\mathsf{P}(t)s^{j+1}(t)m_j. \tag{12}$$

Finally, upon a reflection of Eq. (10) on Eq. (12), the governing equation for the sequence $\{m_j\}_0^\infty$ is obtained as

$$\tau^2 m_{j+2} = (j-2)(j+1)m_j,\ j \geq 0, \tag{13}$$

which has the solution

$$m_j = \begin{cases} m_0 & j=0 \\ m_1 & j=1 \\ -\dfrac{2m_0}{\tau^2} & j=2, \\ -\dfrac{(j-1)}{2}\Gamma(j-2)\tau\left(\left(\dfrac{1}{\tau}\right)^j - \left(-\dfrac{1}{\tau}\right)^j\right)m_1 & j \geq 3 \end{cases} \tag{14}$$

with m_0, m_1 and τ being undetermined constants.

Convergence analysis
From difference equation (13), one can make some *a priori* convergence analysis of series (4). Firstly, it is observed that for $j \geq 3$,

$$\tau^2 \left|\frac{m_{j+2}}{m_{j+1}}\right|\left|\frac{m_{j+1}}{m_j}\right| = |(j-2)(j+1)| \tag{15}$$

from which it is further inferred that

$$\lim_{j\to\infty}\left|\frac{m_{j+1}}{m_j}\right| = \infty. \tag{16}$$

Moreover,

$$\lim_{j\to\infty}\frac{1}{|(j+1)|}\left|\frac{m_{j+1}}{m_j}\right| = |\tau|^{-1}. \tag{17}$$

By the ratio test, the region of convergence of series (4) is that which is determined, per the time t, by the limiting inequality

$$\left|\frac{s(t)}{r}\right| \times \lim_{j\to\infty}\left|\frac{m_{j+1}}{(j+1)m_j}\right| < 1, \tag{18}$$

which, by Eq. (17), is $r > |\tau|^{-1} s(t)$.

Explicit Solutions
In this section, we shall exploit the derived solution of the differential-difference equation, up to arbitrary constants, to explicitly determine the displacement evolution in the region exterior to the rapidly expanding surface.

The displacement of the spherical body ahead of the rapidly expanding surface, therefore, has an evolution of the form

$$u(r,t)=\beta m_0-\beta\frac{(s_0+\tau t)}{r}m_1-\beta\frac{(s_0+\tau t)^2}{\tau^2 r^2}m_0+ \\ +\beta\sum_{j=3}^{\infty}\frac{\tau m_1}{2j(j-2)}\left(\left(-\frac{1}{\tau}\right)^j-\left(\frac{1}{\tau}\right)^j\right)\left(\frac{s_0+\tau t}{r}\right)^j; \tag{19$_1$}$$

$$u(r,t)=\left(1-\left(\frac{s_0+\tau t}{\tau r}\right)^2\right)\beta m_0+ \\ +\frac{\beta}{2\tau r^2}\left(\tau r(s_0+\tau t)-(s_0+\tau(t-r))(s_0+\tau(t+r))\mathrm{ArcCoth}\left(\frac{\tau r}{s_0+\tau t}\right)\right)m_1. \tag{19$_2$}$$

Converting the temporal variable back to $\bar{t}$ gives the solution as

$$u(r,\bar{t})=\left(1-\left(\frac{s_0+\tau c\bar{t}}{\tau r}\right)^2\right)\beta m_0+ \\ +\frac{\beta}{2\tau r^2}\left(\tau r(s_0+\tau c\bar{t})-(s_0+\tau(c\bar{t}-r))(s_0+\tau(c\bar{t}+r))\mathrm{ArcCoth}\left(\frac{\tau r}{s_0+\tau c\bar{t}}\right)\right)m_1. \tag{20}$$

Initial conditions

Firstly, we take into account the zero initial speed condition $\frac{\partial u(r,0)}{\partial \bar{t}}=0$ for Eq. (20) to realize that $m_0=0$ simultaneously with the compatibility condition

$$r\tau-s_0\mathrm{ArcCoth}\left(\frac{\tau r}{s_0}\right)=0. \tag{21}$$

Set $\alpha:=\frac{r}{s_0}>1$. One observes that the relationship between the dimensionless parameters α and τ through the transcendental equation (21) is approximately

$$\alpha\tau\left(:=\frac{r}{s_0}\tau\right)\approx 1.199678. \tag{22}$$

This further constrains τ in the interval $(0,\infty)$ given the positivity of α. Noting the range of admissibility of τ, Fig.1 below gives a graphical relationship between α and τ. Secondly, an application of the initial displacement condition $u(r,0)=\omega r$ reveals that $m_1=\frac{2s\omega}{\beta\tau^2}$.

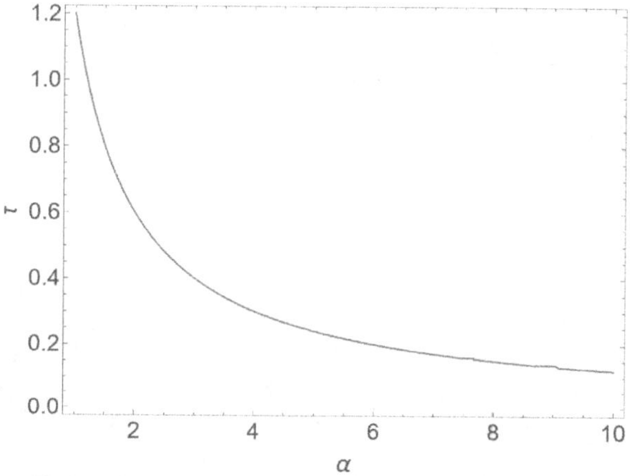

Fig.1: Relationship between dimensionless parameters τ and α .

Now the displacement reads

$$u(r,\overline{t})=\frac{s_0\omega}{r^2\tau^3}\left(r\tau(c\overline{t}\tau+s_0)-(c\overline{t}\tau-r\tau+s_0)(\tau(c\overline{t}+r)+s_0)\mathrm{ArcCoth}\left(\frac{r\tau}{c\overline{t}\tau+s_0}\right)\right). \quad (23)$$

Boundary condition

The boundary condition at the rapidly evolving surface supplies information for the determination of the parameter τ :

$$\left(1+\nu(\tau^2-2)\right)\mathrm{ArcCoth}(\tau)+\frac{(\nu-1)\beta}{2s_0c^2\rho\omega}\tau^3+(2\nu-1)\tau=0. \quad (24)$$

Due to the branch cut discontinuity of the ArcCoth function, and in order to maintain the meaningfulness of Eq. (24) , τ is only admissible in the interval $(1,1.996784)$.

Finally, the triple of equations (23) , (24) and compatibility condition (21) constitute the solution to the explosion problem in the region $r>s(\overline{t})=\tau c\overline{t}+s_0$.

Numerical Illustration

For illustration purposes, we consider an isotropic linear elastic infinite media with Poisson ratio $\nu=0.3$, $\lambda=\frac{3}{2}\mu$, $\mu=0.2$, $s_0=5$, $\beta=1$, $\rho=1500$, all in appropriate units. In this instance, it is calculated from the above relations that $\omega=-0.354452\beta$, and $\tau=1.5$ approximately. The resulting displacement field, spherical surface evolution, and pressure within the cavity are given by

$$\begin{cases} u(r,\bar{t}) = -\dfrac{1.16529}{r^2}\begin{pmatrix} 1.15r(0.0248428\bar{t}+5)-(1.15(r+0.0216025\bar{t})+5) \\ \times(-1.15r+0.0248428\bar{t}+5)\text{ArcCoth}\left(\dfrac{1.15r}{0.0248428\bar{t}+5}\right) \end{pmatrix} \\ \mathrm{P}(t) = \left(0.0248428\bar{t}+5\right)^{-1} \\ s(\bar{t}) = 0.0248428\bar{t}+5. \end{cases} \tag{25}$$

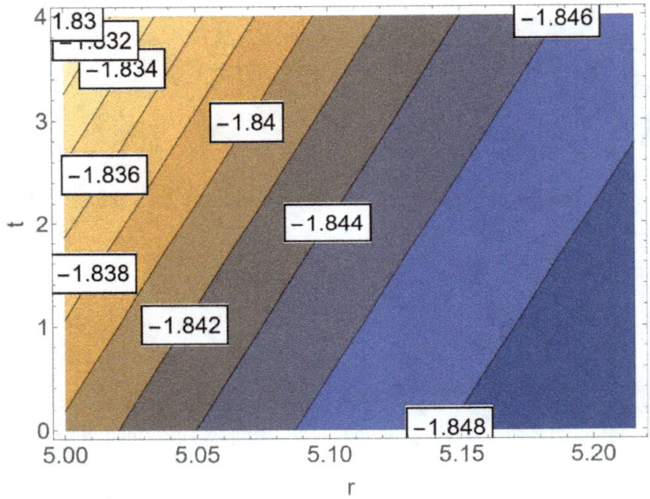

Fig. 2: Illustrative displacement field evolution in the region ahead of a rapidly expanding spherical cavity as given by Eq. (25)

The rapidly evolving surface and the ensuing compaction are demonstrated in the contour plot given in Fig.2 above. It can be deduced from Fig.2 that the spherical cavity's surface evolves linearly in the time, while lesser absolute values of the displacement $u(r,t)$ as progression is made into the far field, per time, depict decreasing compression.

References

[1] M. Sh. Israilov, H. Hamidou, Underground explosion action: Rapid expansion of a spherical cavity in an elastic medium, Mechanics of Solids 56 (2021) 376--391. https://doi.org/10.3103/S0025654421030043

[2] Jan D. Achenbach, C. T. Sun, Propagation of waves from a spherical surface of time-dependent radius, The Journal of the Acoustical Society of America 40, 4 (1966) 877--882. https://doi.org/10.1121/1.1910160

Advanced Topics in Mechanics of Materials, Structures and Construction - AToMech1-2023 Materials Research Forum LLC
Materials Research Proceedings 31 (2023) 596-613 https://doi.org/10.21741/9781644902592-62

Linear response of thin axysimmetric cross-ply structure under a static load: Numerical and analythical comparisons

Salvatore Saputo [1]*, Erasmo Carrera [1], Volodymyr V. Zozulya [2]

[1]Department of Aeronautics and Aerospace Engineering, Politecnico di Torino, Corso Duca degli Abruzzi, 24, 10129 Torino, Italy

[2]S. P. Timoshenko Institute of Mechanics, National Academy of Sciences of Ukraine, Nesterov Str. 3, 252680 Kyiv, Ukraine

*salvatore.saputo@polito.it

Keywords: Carrera Unified Formulation, Higher-Order Theories, Analytical Solution, Bending, Laminated Composite Plate

Abstract. Thin-walled mechanical components, such as beams, plates and shells, are widely used as structural components in several engineering fields, in particular mechanical, aeronautical and aerospace sectors. The purpose of this work is to analyse the cross-ply bending behaviour of cylindrical and spherical shell structures using the finite element method. Hence, numerical models, realized using commercial software, were realized using the shell and solid approaches and were compared with numerical and analytical methods to appreciate their advantages. In this research, a Navier solution in close form for high-order theories, developed using the Carrera Unified Formulation (CUF) approach, has been reported, where the high-order elastic shell model has been developed using the variational principle of virtual work for three-dimensional linear theory equations and the analytical results were obtained using the Mathematica software. The results furnished by the numerical method such as the elasticity solutions given in the literature using Navier's method are used as a benchmark for comparing the finite element method results in terms of maximum displacement and stress distribution along the principal structure direction. However, the numerical shell model cannot provide sufficient data to describe the tensional and deformational state at all points and especially along the laminate thickness. Wishing to obtain a complete description of the plate's mechanical behaviour, it is necessary to use a three-dimensional approach with the associated increase in calculation time. In contrast, the numerical solution based on the CUF approach shows a very efficient description of the composite structure behaviour and its use should be preferred to the classical lamination approach if an accurate description of the structure is necessary.

Introduction

Layered structures are being employed more and more in automotive, aerospace and naval vehicles. Throughout the latter half of the 20th century, composite materials were developed and adopted in areas where a high stiffness-to-weight ratio was required. There are modern examples of aircraft for military and civil uses as well as boats, helicopters, ultralights and gliders, whose structure is made exclusively of composite material. Layered structures choice, noticeably, complicates the design, analysis, and manufacturing processes. All these aspects adding to the difficulties already known regarding the use of conventional isotropic materials. Moreover, the high discontinuity in the mechanical properties of layered structures, due to their nature, requires the ZZ (zig zag) theory for the description of stress and displacement along thickness as well as interlaminar continuity (in-plane and out-of-plane stresses) [1][2] that greatly complicate the analysis of such structures. Analytical closed-form solutions are only available for extremely simple geometries. The resolution of more complex layered structures geometry and/or boundary conditions is left to numerical methods with different approximation values. Many authors proposed different numerical technique in order to evaluate the deformation and the stress distribution in every point of the structure. Noor and Raig [3] proposed a 3D finite difference technique for axial symmetric multilayer structure. Malik and Liew [4][5] proposed a differential quadrature technique for complex structure. A meshless collocation method, set on radial basis function, have been adopted by Ferreira et al [6][7] to analyse laminated plates and shells. For a comprehensive overview of different computational methods and their uses related to layered structure can be found in [8][9][10]. A particular focus should be posed on the finite element method (FEM) that is widely adopted to study the mechanical behaviour of the composite structure. Moreover, different finite element methods are based on axiomatic-type theories where the unknown variables are suggested along the thickness. In particular, the earliest FEM calculation are based on the classical Kirchhoff-Love theory [11][12][13][14][15]. However, it was challenging to meet the compatibility requirements in thin shell analysis because rotations were derived from transversal displacement, and to avoid this issue, plate/shell elements based on the first-order shear deformation theory (FSDT) were developed by several authors [16][17][18]. Severe stiffening limits for thin plate are showed by newest FSDT finite elements, known as shear locking. However, early FSDT-type elements showed severe stiffening in thin plate/shell limits, which resulted in a numerical mechanism known as shear or membrane locking. This locking was initially countered by implementing numerical tricks, operating on the integration schemes [19], but they introduced spurious zero energy modes. A solution to this problem is proposed by Naghdi [20], where the shell finite element could counteract the locking problem in its displacement formulation. On the other hand, if the element is not of high enough degree and the thickness is very small, the numerical solution may exhibit a loss in the rate of convergence due to locking. To avoid both the mentioned problem the mixed interpolation of tensorial component approach (MITC) can be adopted as treated by several authors [21][22][23][24]. In the past two decades, numerous plate and shell finite element approaches have been suggested, using higher-order theories (HOT). Kant et al. [25] introduced HOT- finite elements that necessitate continuity only for the unknown variables and not for their derivatives. Otherwise, Polit et al [26]. developed a C^1 six-node triangular finite element using cosine functions to represent transverse shear strains. With this approach the continuity of displacement and shear stress was ensure between the interface of the layered structure. Tessler [27] has provided a comprehensive analysis of HOT-type theories and their suitability for finite element applications. Several other studies have implemented HOTs for plates and shells, and additional information can be found in the literature [28]. There have been many proposals for finite elements based on zigzag theories, such as Rao and Meyer-Piening that suggested using the Reissner mixed variational theorem (RMVT) to develop standard finite elements.[29]. Lastly, several authors, including Noor and Burton [30], Reddy [31], and Rammerstorfer et al. [32] have proposed finite element implementations of layer-wise theories in the framework of axiomatic-type theories.

A new and improved shell finite element is introduced for the analysis of composite structures [33][34], based on Carrera's Unified Formulation (CUF), which was originally developed by Carrera for multi-layered structures [35][36]. The new shell finite element includes both equivalent single layer (ESL) and layer-wise (LW) theories found in CUF. In this work, several shell-based finite element theories were introduced and compared to analyse the mechanical response of layered structures with different geometry. Comparisons are made between classical theories such as CST and FSDT, analytical type solutions, commercial code and CUF high order approaches. Three different geometries were used, such as a plate, a cylindrical section and a cylinder. Finally, different thicknesses were evaluated for each geometry to highlight the limits of the various theories.

Unified form building up of plate theories in terms of a few "fundamental nuclei" with variational statements.

Considering theories that assume displacement and transverse shear and normal stress in the axiomatic approach, the displacement or stress fields are postulated in the plate's z-direction. Typically, two-dimensional theories are build up following four steps: assigning material behavior (e.g. Hooke's Law); suppose a geometrical relation (e.g. strain-displacement relation); postulating displacement and/or stress distributions in the thickness of the plate's z-direction by stating to a set of base functions; and finally, utilizing an appropriate variational statement (PVD or RMVT) to determine governing equations and boundary conditions which are variationally consistent with the hypotheses introduced in the first three steps. The aim of this discussion is to explore theories within the scope of the PVD and the RMVT.

Hooke's law:

The multilayered plates composed of N_l layers, shown in Figure 1, have a geometry and Cartesian coordinate system *x, y, z*. The laminae are assumed to be homogenous and to be functioning within the linear elastic range. Standard form for Hooke's law is applied for the anisotropic k-lamina stiffness coefficients.

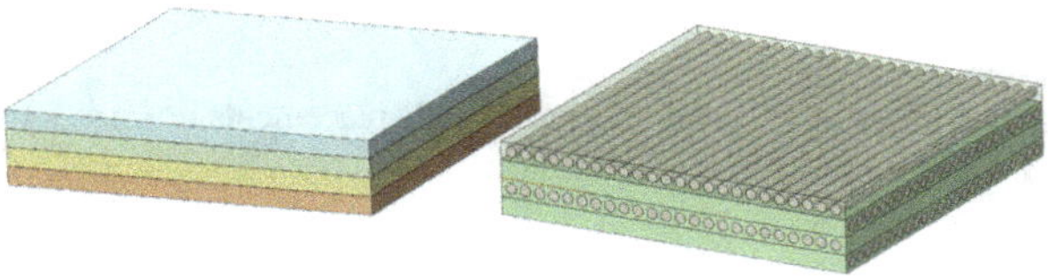

Figure 1 Multi-layered structure

The Hooke's Law reads $\sigma_i = \tilde{C}_{ij}\epsilon_j$ where sub-indices *i* and *j*, ranging from 1 to 6, stand for the index couples 11, 22, 33, 13, 23 and 12, respectively. The material is supposed to be orthotropic as specified by $\tilde{C}_{14} = \tilde{C}_{24} = \tilde{C}_{34} = \tilde{C}_{64} = \tilde{C}_{15} = \tilde{C}_{25} = \tilde{C}_{35} = \tilde{C}_{65} = 0$. This implies that ${\sigma^k}_{xz}$ and ${\sigma^k}_{yz}$ depend only on ${\epsilon^k}_{xz}$ and ${\epsilon^k}_{yz}$. In matrix form:

$$\begin{aligned} {\boldsymbol{\sigma}^k}_{p\mathrm{H}} &= \tilde{\boldsymbol{C}}^k_{pp}\boldsymbol{\epsilon}^k_{p\mathrm{G}} + \tilde{\boldsymbol{C}}^k_{pn}\boldsymbol{\epsilon}^k_{n\mathrm{G}} \\ {\boldsymbol{\sigma}^k}_{n\mathrm{H}} &= \tilde{\boldsymbol{C}}^k_{np}\boldsymbol{\epsilon}^k_{p\mathrm{G}} + \tilde{\boldsymbol{C}}^k_{nn}\boldsymbol{\epsilon}^k_{n\mathrm{G}} \end{aligned} \tag{1}$$

where,

$$\tilde{\boldsymbol{C}}^k_{pp} = \begin{bmatrix} \tilde{C}^k_{11} & \tilde{C}^k_{12} & \tilde{C}^k_{16} \\ \tilde{C}^k_{12} & \tilde{C}^k_{22} & \tilde{C}^k_{26} \\ \tilde{C}^k_{16} & \tilde{C}^k_{26} & \tilde{C}^k_{66} \end{bmatrix}, \tilde{\boldsymbol{C}}^k_{pn} = \tilde{\boldsymbol{C}}^{k\mathrm{T}}_{np} = \begin{bmatrix} 0 & 0 & \tilde{C}^k_{13} \\ 0 & 0 & \tilde{C}^k_{23} \\ 0 & 0 & \tilde{C}^k_{36} \end{bmatrix}, \tilde{\boldsymbol{C}}^k_{nn} = \begin{bmatrix} \tilde{C}^k_{44} & \tilde{C}^k_{45} & 0 \\ \tilde{C}^k_{45} & \tilde{C}^k_{55} & 0 \\ 0 & 0 & \tilde{C}^k_{66} \end{bmatrix}$$

Bold letters represent arrays. The superscript 'T' indicates array transposition. The subscripts *n* and *p* denote transverse (out-of-plane, normal) and in-plane values, respectively. Therefore

$$\boldsymbol{\sigma}^k_p = \{\sigma^k_{xx}, \sigma^k_{yy}, \sigma^k_{xy}\}, \qquad \boldsymbol{\sigma}^k_n = \{\sigma^k_{xz}, \sigma^k_{yz}, \sigma^k_{zz}\}$$

$$\boldsymbol{\epsilon}^k_p = \{\epsilon^k_{xx}, \epsilon^k_{yy}, \epsilon^k_{xy}\}, \qquad \boldsymbol{\epsilon}^k_n = \{\epsilon^k_{xz}, \epsilon^k_{yz}, \epsilon^k_{zz}\}.$$

Subscript 'H' denotes stresses assessed by Hooke's law whereas subscript 'G' denotes strain from the geometrical relation Eq. (3).

Eq. (1) is employed together with a regular displacement formulation based on the principles of PVD, and the stress-strain relationships are expressed in a blended form for the integrated solution procedure:

$$\begin{aligned} \boldsymbol{\sigma}_{p\mathrm{H}}^{k} &= \boldsymbol{C}_{pp}^{k}\boldsymbol{\epsilon}_{p\mathrm{G}}^{k} + \boldsymbol{C}_{pn}^{k}\boldsymbol{\sigma}_{nM}^{k} \\ \boldsymbol{\epsilon}_{n\mathrm{H}}^{k} &= \boldsymbol{C}_{np}^{k}\boldsymbol{\epsilon}_{p\mathrm{G}}^{k} + \boldsymbol{C}_{nn}^{k}\boldsymbol{\sigma}_{nM}^{k} \end{aligned} \tag{2}$$

using both stiffness and compliance coefficients. To link the two expressions of Hooke's law, the following can be deduced:

$$\boldsymbol{C}_{pp}^{k} = \tilde{\boldsymbol{C}}_{pp}^{k} - \tilde{\boldsymbol{C}}_{pn}^{k}\tilde{\boldsymbol{C}}_{nn}^{k^{-1}}\tilde{\boldsymbol{C}}_{np}^{k}, \qquad \boldsymbol{C}_{pn}^{k} = \tilde{\boldsymbol{C}}_{pn}^{k}\tilde{\boldsymbol{C}}_{nn}^{k^{-1}}$$
$$\boldsymbol{C}_{np}^{k} = -\tilde{\boldsymbol{C}}_{nn}^{k^{-1}}\tilde{\boldsymbol{C}}_{np}^{k}, \qquad \boldsymbol{C}_{nn}^{k} = \tilde{\boldsymbol{C}}_{nn}^{k^{-1}}$$

Superscript '-1' denotes an inversion of the array.

Geometrical relation:
The strain components $\boldsymbol{\epsilon}_p^k, \boldsymbol{\epsilon}_n^k$ are linearly related to the displacements $\boldsymbol{u}^k(\{u_x^k, u_y^k, u_z^k\})$ according to the following geometrical (subscript G) relations:

$$\boldsymbol{\epsilon}_{p\mathrm{G}}^{k} = D_p\boldsymbol{u}^k, \qquad \boldsymbol{\epsilon}_{n\mathrm{G}}^{k} = D_n\boldsymbol{u}^k \tag{3}$$

D_p and D_n denotes in-plane and out-of-plane differential operators:

$$D_p = \begin{bmatrix} \partial_x & 0 & 0 \\ 0 & \partial_y & 0 \\ \partial_y & \partial_x & 0 \end{bmatrix}; \quad D_n = \begin{bmatrix} \partial_z & 0 & \partial_x \\ 0 & \partial_z & \partial_y \\ 0 & 0 & \partial_z \end{bmatrix}.$$

Displacement and transverse assumptions:
The performance of a displacement and/or strain parameter f are assumed to be in accordance with a given expansion in the z-direction of the plate.

$$f(x,y,z) = F_i(z)\, f_i(x,y) \quad i = 0,1,\ldots,N. \tag{4}$$

The sum of the iterative indices i is computed over its domain. The polynomials $F_i(z)$ generate a group of individual functions and, this selection can be made arbitrarily. The magnitude of the projected expansion is expressed by N.

The use of displacement and transverse normal stress assumptions will result in this equation:

$$\begin{aligned} \boldsymbol{u}(x,y,z) &= F_i(z)\boldsymbol{u}_i(x,y) \\ \boldsymbol{\sigma}_{nM}(x,y,z) &= F_i(z)\boldsymbol{\sigma}_{n_i}(x,y) \end{aligned} \qquad i = 0,1,N. \tag{5}$$

M (as in model) has been introduced to separate the stresses assumed from those obtained from Hooke's law. In the numerical analysis, $N \leq 4$ will be accounted for. N can be unique for each variable as discussed in [37]. The conditions in Eq. (5) can be applied to either a single layer (LW) or multiple layers (ES).

Governing equations via PVD and RMVT:
For a multilayered plate subjected to static loadings,
PVD states

$$\sum_{k=1}^{N_l} \int_{\Omega^k}^{\Gamma^k} \int_{A_k}^{V} \left(\delta \boldsymbol{\epsilon}_{pG}^{k^T} \boldsymbol{\sigma}_{pHd}^{k} + \delta \boldsymbol{\epsilon}_{nG}^{k^T} \boldsymbol{\sigma}_{nH_d}^{k}\right) \, d\Omega^k dz = \delta L_e \tag{6}$$

δ is the variational symbol. A_k and V denote the layer thickness domain and volume; Ω^k is the layer middle surface bounded by Γ^k. The variation of the internal work has been split into in-plane and out-of-plane parts and involves stress from Hooke's Law and strain from geometrical relations. δL_e is the virtual variation of the work made by the external layer-forces $\boldsymbol{p}^k = \{p_x^k,\ p_y^k,\ p_z^k\}$. By replacing the variables in Eq.(1), Eq.(3), and the first Eq.(5) with suitable terms, a variational statement can be formulated that yields a set of equilibrium equations and boundary conditions. These equilibrium equations can be expressed concisely in the form:

$$\delta \boldsymbol{u}_\tau^k : \boldsymbol{K}_d^{k\tau s} \boldsymbol{u}_s^k = \boldsymbol{p}_\tau^k \tag{7}$$

The related boundary conditions are:

$$\boldsymbol{u}_\tau^k = \bar{\boldsymbol{u}}_\tau^k \ or\ \boldsymbol{\Pi}_d^{k\tau s} \boldsymbol{u}_s^k = \boldsymbol{\Pi}_d^{k\tau s} \bar{\boldsymbol{u}}_s^k \tag{8}$$

The number of equations derived is equal to the number of variables introduced: τ and s vary from 0 to N and k range from 1 to N_l. $\boldsymbol{K}$ and $\boldsymbol{\Pi}$ are arrays constitute by differential operators.
Reissner's mixed theorem [38][39] expresses both equilibrium and compatibility in relation to the unknowns $\boldsymbol{u}^k$ and $\boldsymbol{\sigma}_n^k$. This is achieved by means of the variational equation given below:

$$\sum_{k=1}^{N_l} \int_{\Omega^k}^{\Gamma^k} \int_{A_k}^{V} \left(\delta \boldsymbol{\epsilon}_{pG}^{k^T} \boldsymbol{\sigma}_{pH}^{k} + \delta \boldsymbol{\epsilon}_{nG}^{k^T} \boldsymbol{\sigma}_{nM}^{k} + \delta \boldsymbol{\sigma}_{nM}^{k^T} \left(\boldsymbol{\epsilon}_{nG}^{k} - \boldsymbol{\epsilon}_{nH}^{k}\right)\right) d\Omega^k dz = \delta L^e \tag{9}$$

The LHS incorporates the changes of the internal force within the plate, with the initial two terms deriving from the displacement formulation, which lead to a balance of forces. The third term is a combined term which ensures the compatibility of the transverse strain components. The whole equation, expressed in terms of displacement and stress, is summarized in a concise form as follows:

$$\begin{aligned} \delta \boldsymbol{u}_\tau^k &: \boldsymbol{K}_{uu}^{k\tau s} \boldsymbol{u}_s^k + \boldsymbol{K}_{u\sigma}^{k\tau s} \boldsymbol{\sigma}_{ns}^k = \boldsymbol{p}_\tau^k \\ \delta \boldsymbol{u}_{n\tau}^k &: \boldsymbol{K}_{\sigma u}^{k\tau s} \boldsymbol{u}_s^k + \boldsymbol{K}_{\sigma\sigma}^{k\tau s} \boldsymbol{\sigma}_{ns}^k = \boldsymbol{0} \end{aligned} \tag{10}$$

with boundary conditions

$$\boldsymbol{u}_\tau^k = \bar{\boldsymbol{u}}_\tau^k \ or\ \boldsymbol{\Pi}_u^{k\tau s} \boldsymbol{u}_s^k + \boldsymbol{\Pi}_\sigma^{k\tau s} \boldsymbol{\sigma}_{ns}^k = \boldsymbol{\Pi}_u^{k\tau s} \bar{\boldsymbol{u}}_s^k + \boldsymbol{\Pi}_\sigma^{k\tau s} \bar{\boldsymbol{\sigma}}_{ns}^k \tag{11}$$

When LW descriptions are used, the equations controlling the behavior of the layers are first derived at the individual layer level. The equations for the entire multilayer structure are then obtained by assuring the stresses and displacements remain continuous between the layers.

Navier-type closed form solution:
The aforementioned boundary value problems, in the broadest range of configurations and constraints, could be solved approximately. When the material displays orthotropic behavior, Navier-type analytical solutions can be derived by taking the harmonic forms of the applied loads and the unknown variables into account:

$$\begin{aligned}
\left(u_{x_\tau}^k, \sigma_{xz_\tau}^k, p_{x_\tau}^k\right) &= \sum_{m,n}\left(U_x^k, S_{xz_\tau}^k, P_{x_\tau}^k\right) \cos\frac{m\pi x}{a} \sin\frac{n\pi y}{b} \\
\left(u_{y_\tau}^k, \sigma_{yz_\tau}^k, p_{y_\tau}^k\right) &= \sum_{m,n}\left(U_y^k, S_{yz_\tau}^k, P_{y_\tau}^k\right) \sin\frac{m\pi x}{a} \cos\frac{n\pi y}{b} \\
\left(u_{z_\tau}^k, \sigma_{zz_\tau}^k, p_{z_\tau}^k\right) &= \sum_{m,n}\left(U_z^k, S_{zz_\tau}^k, P_{z_\tau}^k\right) \sin\frac{m\pi x}{a} \sin\frac{n\pi y}{b}
\end{aligned} \quad (12)$$

while m and n are the correspondent wave numbers. The maximum amplitudes corresponding to the RHS of Eq. (12) are represented by capital letters. This method has been implemented for different theories and the outcomes will be covered in the following sections. Any type of imposed force can be adjusted to the Navier form unless a suitable Fourier expansion is implemented. The benchmarks discussed in the following sections are related to various transverse pressure distributions p_{zT}, and the suitable Fourier expansion for these cases should be written as:

$$(p_{zT}) = \sum_{m,n}^{R,Q}(p_{zT}^{mn}) \sin\frac{m\pi x}{a} \sin\frac{n\pi y}{b} \quad (13)$$

Where R and Q are the maximum values of the considered m and n, while p_{zT}^{mn} are the Fourier series coefficients. See Figure 2

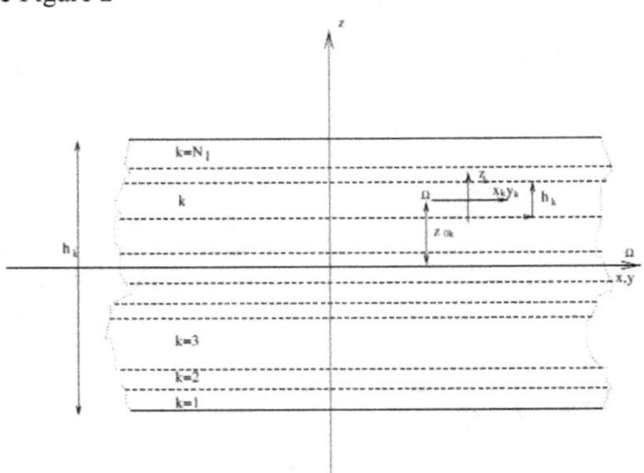

Figure 2 Plate geometry and notations

Numerical results and discussion
Three different geometrical test cases are introduced in this section: shell plate, shell cylindrical section and shell cylinder. For each test case different stacking sequence are analysed. Different thicknesses and lamination configurations will be considered for each test case.

Convergence Analysis
A preliminary study of convergence mesh is conducted for the commercial code to reduce the influence of the mesh on the results. The choice of element size made at this stage will also be adopted in the following cases. In Figure 1, the convergence analysis configuration is illustrated, where a render shell thickness is adopted to better appreciate the imposed boundary condition. The

plate is subjected to a uniform force acting along the z axes in opposite direction, a simply supported BC are applied on the four edge of the square plate and a ratio between the edge and the thickness equal 4 is considered.

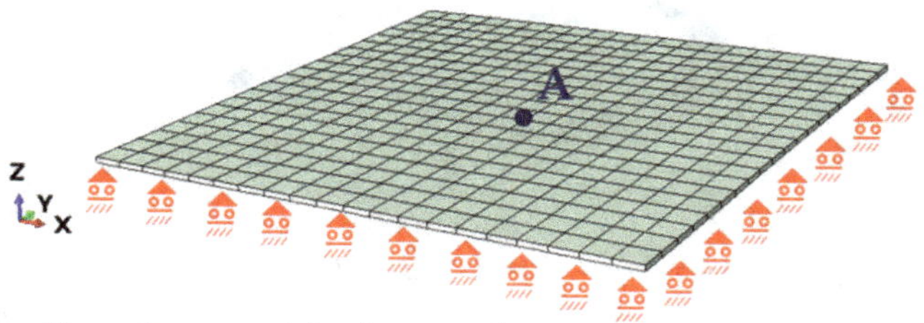

Figure 3 commercial software mesh convergence configuration

The results of the convergence analysis are shown in Table 1 where the first column reports the dimensions of the elements expressed as the ratio of the plate side size to the number of elements, the second column reports the displacement along the z-axis with respect to the maximum deflection. In accordance with Table 1, the choice of the dimension element equal to 20 was chosen to obtain a good mitigation of the effects due to the size of the elements and the computational cost.

Table 1 Mesh convergence analysis

Mesh Convergence	
Dimension Element $\left[\frac{a}{n}\right]$	Max Deflection $\left[\frac{z}{z_{max}}\right]$
3	0.812
5	0.879
10	0.976
20	0.998
30	0.999
50	1

Simply supported cross-ply plate under pressure

The cross ply square base plate is subjected to a pressure with a bi-harmonic distribution expressed as a Fourier series with m and n equal to one (Figure 2). To assess the influence of thickness on the mechanical response of the structure, three different ratios of side length to thickness are considered $\left(\frac{a}{Th} = 4; 10; 20\right)$, where Th is the plate thickness.

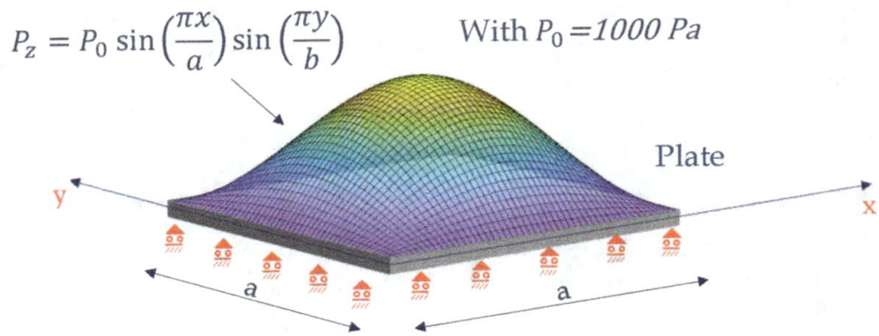

Figure 4 Square cross ply plate under biharmonic pressure load

The square cross-plate has simply supported condition on the four edges and a symmetrical stacking sequence of [0/90/0], where the mechanical properties adopted for the layered plate are expressed in accordance with Table 2.

Table 2 Mechanical Properties of the material

Material Properties		
Young's Modulus	E_L/E_T	30
Shear Modulus	G_{LT}/E_T	0.5
Shear Modulus	G_{TT}/E_T	0.35
Poisson	ν_{LT}	0.3
Poisson	ν_{TT}	0.49

The values of transverse displacement, tension in the L, T and LT directions of the central point of the plate were dimensionless in accordance with the equations 14a-d

$$\bar{U}_z = \frac{U_z 100 E_T h^3}{p_z a^4}; \quad (a) \qquad \bar{S}_{11} = \frac{S_{11}}{p_z (a/Th)^2}; \quad (b)$$

$$\bar{S}_{22} = \frac{S_{22}}{p_z (a/Th)^2}; \quad (c) \qquad \bar{S}_{12} = \frac{S_{12}}{p_z (a/Th)^2}; \quad (d) \tag{14}$$

Table 3 and Table 4 show the results obtained in accordance with Equations 14 a-d and pertain respectively the displacement values and the stress values. The values were obtained considering different theories. L refers to a Layer-Wise model, ED to an equivalent single layer model, FSDT to a first shear deformation theory model, CST to classic laminate theory. Furthermore, for the L and ED models, id 1 to 4 refer from linear to fourth order implemented expansions function in the plate/layer thickness z-direction [34]. The data on the maximum dimensionless plate displacement show that for thick structures values close to the 3D solution [37] are only equated by the Layer-wise model of order 4 and 3, while the other theories register an error ranging from 82% of the CST, to about 7.5 % of the commercial code through to about 27% of the FSDT. Moving from a thick to a very thin layered structure, the differences recorded between the various theories are greatly reduced, with a maximum variation of around 28% for the CST. The layer-wise approach allows exact displacement values to be reproduced even with small expansion orders. The commercial code, in the case of thin plate, reduces its error to (in the configuration considered) 3.5%, representing a viable alternative to the CST, FSDT and ED theories with small expansion order values.

Table 3 Maximum value of displacement in z direction

	$\bar{U}_z$		
	4	***10***	***20***
3D	2.820	0.919	0.610
L4	2.821	0.919	0.609
L3	2.821	0.919	0.609
L2	2.798	0.918	0.609
L1	2.720	0.898	0.604
ED4	2.625	0.866	0.592
ED3	2.627	0.866	0.595
ED2	2.035	0.750	0.565
ED1	2.051	0.750	0.565
FSDT	2.051	0.750	0.563
CST	0.501	0.501	0.439
ABAQUS	2.611	0.865	0.589

The tension values in the longitudinal, transverse and shear directions are reported in Table 4. The values are referred to the midpoint of the plate, and the superscript '+' denotes the tension values evaluated on the top of the plate, while the superscript '-' denotes the values obtained on the bottom face of the laminate. Similar considerations to those made for displacement can be made in the evaluation of stresses. However, the greatest discrepancy in the evaluation of tension occurs in the longitudinal direction for the ED1 approach. Again, the values obtained with the commercial code, although close to the 3D solution [37], do not reach the degree of accuracy shown by the LW (1-4) and ED (2,3) approaches. Furthermore, the commercial code reports the same absolute tension value for the top and bottom of the laminate, while the other theories manage to record a variation in tension as shown by the benchmark.

Table 4 Stress distribution in longitudinal, transversal and shear distribution on the top and bottom of the plate

	$\bar{s}_{11}^{+}$	$\bar{s}_{12}^{+}$	$\bar{s}_{22}^{+}$	$\bar{s}_{11}^{-}$	$\bar{s}_{12}^{-}$	$\bar{s}_{22}^{-}$
	4	***10***	***20***	***4***	***10***	***20***
L4	15.52	1.374	0.4982	-15.02	1.385	0.4930
L3	15.53	1.373	0.4982	-15.02	1.293	0.4955
L2	15.33	1.357	0.4982	-14.99	1.293	0.4955
L1	14.27	1.367	0.4945	-13.86	1.290	0.4954
ED4	15.39	1.366	0.4977	-15.05	1.307	0.4953
ED3	15.51	1.360	0.4977	-14.22	1.325	0.4959
ED2	11.35	1.447	0.4977	-10.80	1.375	0.4956
ED1	10.98	1.443	0.4840	-12.25	1.385	0.5000
FSDT	11.17	1.443	0.5000	-12.22	1.402	0.5000
CST	13.48	1.691	0.500	-12.22	1.402	0.5000
ABAQUS	14.08	1.298	0.4936	-14.08	1.298	0.4936

Shell cylindrical section

The second benchmark analysed and reported on in this paper is the layered cylindrical section shown in Figure 5 A-B. The upper surface of the structure is subjected to a sinusoidal pressure distributed according to Figure 5-B, while simply supported conditions are applied along the extreme sides parallel to the generators. Both the pressure distribution and the boundary conditions are described considering a cylindrical reference system whose origin coincides with the central point of the upper surface, and whose three axes coincide with the direction of the generatrix (longitudinal direction), directrix (transverse direction) and normal to the surface, respectively (as shown in Figure 5-A), respectively.

To mitigate edge effects, a section length equal to four times the radius was considered.

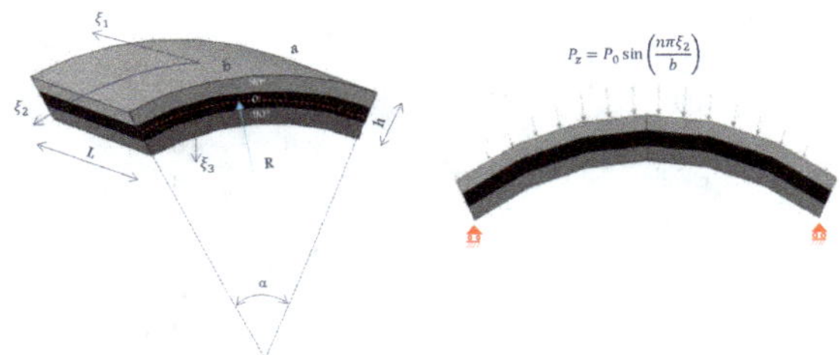

Figure 5 Cylinder shell section

The axisymmetric geometry shown in this section and studied in [33] is a cylindrical section consisting of three layers of orthotropic material with equal thickness. The cylindrical section was analysed considering four different thicknesses, where the configurations due to thickness variation will be identified through the parameter a, which expresses the ratio of the cylinder radius to the thickness of the laminate, the values being equal to $a = \frac{R}{Th} = 2; 4; 50; 500$.

According with [33] the stacking sequence for the cylinder section is equal to [0/90/0], while the mechanical properties of the material and some geometrical data are given in Table 5.

Table 5 Mechanical Properties

Material and Geometrical Properties		
Young's Modulus	E_L/E_T	25
Shear Modulus	G_{LT}/E_T	0.5
Shear Modulus	G_{TT}/E_T	0.2
Poisson's ratio	ν_{LT}/ν_{TT}	0.25
Radius	R	10
Angle span	α	$\pi/3$

The results shown in Table 4 refer to the radial displacement of the cylinder section evaluated at the origin of the cylindrical reference system. The results were obtained in accordance with Equation 14-a by evaluating the same theories discussed in the previous section. For very high thickness values, in contrast to the square plate, the variations with respect to the 3D solution are remarkably small. The ED1-4 solution underestimates the maximum value of the deflection just as it does for the CST theory, for some configurations of the LW theory and for the first shear order deformation theory. In contrast, approaching the problem with commercial software even in a small way returns a higher deflection value than the exact solution. The decrease in thickness greatly mitigates the simplifications in dowry of the various theories used for this study. For very thin structures even the CST theory, which usually significantly underestimates the deflection values, provides results in line with the 3D solution.

Table 6 Maximum value of displacement in z direction. Section cylinder

	$\bar{U}_z$			
	2	4	50	500
3D	1.436	0.457	0.0808	0.773
L4	1.435	0.4581	0.08083	0.07767
L3	1.459	0.4614	0.08084	0.07767
L2	1.411	0.4576	0.08083	0.07767
L1	1.363	0.4407	0.08067	0.07764
ED4	1.383	0.4284	0.08051	0.07767
ED3	1.369	0.4272	0.08051	0.07767
ED2	1.111	0.3310	0.07982	0.07766
ED1	1.129	0.3324	0.07982	0.07766
FSDT	1.169	0.3329	0.07976	0.07766
CST	0.09625	0.08712	0.07834	0.07766
ABAQUS	1.4852	0.4561	0.08015	0.07739

Cylinder

The last test case discussed in this article consists of a cylinder made of composite material with a constant cross-section whose geometrical characteristics are shown in Figure 7 Each layer of the cylinder made of square symmetric unidirectional orthotropic fibrous material and the adopted properties are reported in Table 5. The results, applied load and boundary conditions were imposed in accordance with a cylindrical reference system whose axes coincide with the radial, tangential and axial direction of the cylinder (axial direction coincides with the longitudinal axis of the fibre for ply oriented at 0°)

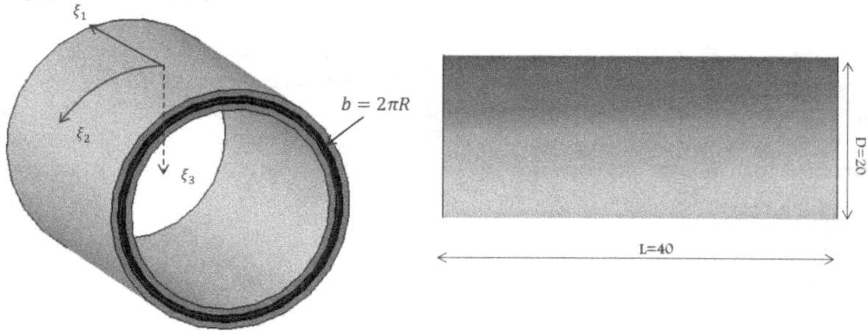

Figure 6 Geometrical description of cylinder case

The geometric variations in terms of thickness investigated are the same as reported in the previous section. In addition, two different stacking sequences are investigated, and for each of the two configurations 4 variations in thickness were analysed. For the [0/90] case, the first ply, oriented along the axis of the cylinder is positioned in the inner layer of the cylinder. A harmonic pressure is applied to the inner face of the cylinder, where the distribution is described by Equation 15, and the parameters m and n are respectively equal to1 and 8.

$$P_z = P_0 \sin\left(\frac{m\pi\xi_1}{L}\right) \sin\left(\frac{n\pi\xi_2}{b}\right) \quad (15)$$

The pressure distribution and boundary conditions are illustrated in Figure 8. In detail, the boundary conditions are of the simply supported type in accordance with the reference system previously illustrated and applied to the extreme edges of the cylinder.

Figure 7 Applied boundary conditions.

Firstly, the results for the [90/0] stacking sequence are discussed and then the three plies configuration outcomes will be introduced. The values of maximum displacement are shown in Table 7, and evaluated following the equation 13-A, where the z direction is the radial direction of the cylinder or in a simpler way the ξ_3 of Figure 7. For the [0/90] plies configuration angle, Abaqus commits an error in the evaluation of the maximum displacement value, in the case of thick structures (a=2), of approximately 22.5% by overestimating the value obtained from the 3D solution. Of the used theories, the one underlying the commercial software is the only one that overestimates the value of the deflection so markedly. Except for the LD4 configuration, all other theories underestimate the exact value with more or less marked errors.

By reducing the thickness by a factor of two, the percentage errors committed by the various considered theories become significantly lower (except for CST). In addition, for a thickness value identified by a=4 configuration, the values given in the table are always lower than the exact value except for the L4 configuration, which tends to overestimate the exact solution, albeit with an error of about 1%.By increasing the term a (decreasing the value of the thickness), the structural response in terms of maximum deflection is at the exact value for all theories, with the exception of the value obtained with the Abaqus software, which again tends to overestimate the exact value and records the value that deviates the most from that reported in [33].

Table 7 Cylinder axial displacement

	$\bar{U}_z$			
	2	**4**	**50**	**500**
3D	14.034	6.100	2.242	0.1005
L4	14.33	6.164	2.242	0.1007
L2	13.80	5.921	2.241	0.1007
ED4	14.08	6.075	2.242	0.1007
ED2	13.07	5.717	2.240	0.1007
FSDT	12.41	5.578	2.240	0.1007
CST	2.781	2.802	2.227	0.1007
ABAQUS	17.197	5.944	2.234	0.1011

The values of the stresses in the L (Longitudinal) and LT directions are shown in Table 7. For each theory, the first line shows the stress value obtained on the external surface while the second line shows the stress values evaluated on the internal surface. Comparing the results obtained from the theories considered, for both directions 11 and 12, with the 3D reference values shows a marked underestimation for both classical theories and the Abaqus model for thick structures. The decrease in thickness induces a progressive reduction in the error. In fact, for very thin structures, the stress values tend to the correct value regardless of the theory considered. This case study, unlike the previous cases, shows a less pronounced variation of results in the case of thick structures.

Table 8 Stresses value for direction L and LT

	S_{11}				S_{12}			
	2	**4**	**50**	**500**	**2**	**4**	**50**	**500**
3D	-2.660	-0.9610	1.610	0.9436	-0.5016	-0.2812	-0.3449	-0.1045
	0.2511	0.2120	0.2189	0.0449	0.2685	0.2007	-0.0784	-0.0925
L4	-2.678	-0.9557	1.606	0.9484	-0.4910	-0.2859	-0.3606	-0.1099
	0.2578	0.2210	0.2241	0.04536	0.3067	0.2216	-0.08282	-0.09736
L2	-2.610	-0.9386	1.605	0.9484	-0.4631	-0.2732	-0.3603	-0.1099
	0.19686	0.1732	0.2204	0.04534	0.2861	0.2103	-0.08276	-0.09736
ED4	-2.649	-0.9580	1.605	0.9486	-0.4812	-0.2831	-0.3605	-0.1099
	0.2302	0.2181	0.2216	0.04516	0.3032	0.2199	-0.08280	-0.09736
ED2	-2.172	-0.8725	1.606	0.9483	-0.3677	-0.2521	-0.3602	-0.1099
	0.1049	0.1156	0.2226	0.04567	0.2541	0.2025	-0.08275	-0.09736
FSDT	-1.216	-0.6911	1.603	0.9484	-0.2994	-0.2431	-0.3604	-0.1099
	0.2256	0.2018	0.2236	0.04535	0.2532	0.1946	-0.08313	-0.09736
CST	-0.5690	-0.4752	1.594	0.9484	-0.1534	-0.1761	-0.3588	-0.1099
	0.1464	0.1661	0.2230	0.04535	0.1504	0.1516	-0.08235	-0.09736
ABAQUS	-1.6875	-0.79688	1.6076	0.9432	0.3153	0.2338	-0.3442	-0.1045
	0.3120	0.23544	0.20310	0.04508	0.2945	-0.2015	0.1567	-0.0926

Furthermore, with the classical CST theories, FSDT and commercial codes such as Abaqus, it is not possible to evaluate stresses in the 33 direction (radial direction) and 23 directions. Using high order theories allows both the layer-wise approach and the equivalent single layer approach to evaluate the stress distribution in the 23 and 33 directions, thus obtaining more accurate numerical models.

Table 9 Stresses value for 23 and 33 directions on cylinder case.

	S_{23}				S_{33}			
	2	***4***	***50***	***500***	***2***	***4***	***50***	***500***
3D	-2.931	-4.440	-4.785	-0.227	-0.31	-0.7	-6.29	-3.09
L4	-3.216	-4.791	-5.024	-0.2441	-0.3408	-0.7358	-6.549	-3.082
ED4	-2.928	-4.274	-3.395	0.2764	-0.3358	-0.7126	-5.072	4.793

For the cylinder case, as previously introduced at the beginning of this section, two different stacking sequences were considered. For the case [0/90], the results have already been discussed, while the results for the stacking configuration of [0/90/0] will now be introduced. Results from previous case studies show that the best results for high order theories are provided by the layer-wise and equivalent single layer approach. Noting this aspect, to make the results clearer and to highlight the discrepancies in the values obtained from the exact solution [9] and finite element approach with the code introduced in [33] and Abaqus code, the comparisons was only made between 3D, L4, ED4 and Abaqus.

Table 10 Radial displacement for [0/90/0] cylinder configuration

	$\bar{U}_z$			
	2	**4**	**50**	**500**
3D	10.1	4.009	0.5495	0.1027
L4	10.1	4.032	0.5495	0.1027
ED4	9.1582	3.7197	0.5458	0.1027
ABAQUS	19.045	5.2111	0.5538	0.1031

Table 9 shows the values of the radial displacements calculated according to the kk formula. The results again show that for thick structures, the approach using commercial software, in this case, Abaqus, yields a value of approximately twice the reference value. Higher-order theories, albeit with different approaches, provide values very similar to the 3D solution. By moving from thicker to thinner structures, it is possible to obtain a maximum cylinder deflection that tends very closely to the 3D solution. Obviously, high order theories can perfectly reproduce the maximum radial displacement value for the cylinder.

Table 11 Stresses in direction 11 and 12 for [0/90/0] cylinder test case

	S_{11}				S_{12}			
	2	**4**	**50**	**500**	**2**	**4**	**50**	**500**
3D	-0.8428	-0.2701	-0.0225	0.0379	-0.2922	-0.1609	-0.0760	-0.0889
	0.1761	0.1270	0.0712	0.0559	0.1797	0.1081	-0.0181	-0.0766
L4	-0.8604	-0.2733	-0.0241	0.0377	-0.2918	-0.1642	-0.0795	-0.0935
	0.1841	0.1330	0.0734	0.0565	0.2015	0.1175	-0.0124	-0.0806
ED4	-0.9447	-0.3011	-0.0240	0.0381	-0.2770	-0.1568	-0.0791	-0.0935
	0.1433	0.1167	0.0730	0.0568	0.1957	0.1127	-0.0123	-0.0806
ABAQUS	-0.21450	-0.1117	-0.02136	0.03793	-0.2883	-0.1504	-0.0755	-0.0887
	0.14583	0.07656	0.07152	0.05604	0.2094	0.0968	-0.0352	-0.0769

Table 10 shows the values of the stresses evaluated in direction 11 and direction 12. Unlike the radial displacement trend, in the case of thick structures the use of high order theories with a layer-wise approach can provide values close to the reference solution. In fact, using an equivalent single-layer approach commits a percentage error of approximately 10%. Finally, using the commercial Abaqus code commits a percentage error of around 75%. The percentage errors evaluated above refer to the s11 tension case and evaluating the maximum error percentages, i.e. not taking into account whether the maximum tension error is evaluated on the outer or inner surface of the cylinder. Moreover, in Table 10, the values of the voltages in the direction 12 are given in the section of the table. A trend of the values of the voltages is noticed like that described by the direction 11 even if with different percentage variations.

Finally, the results of the stresses in Directions 23 and 33 are given in Table 12. It is important to note that in the table the corresponding values for the commercial software Abaqus are not reported, this is due to the impossibility of the commercial code to provide a stress state in all directions, with the use of shell elements (S4). Instead, high order theories allow us to obtain the stresses in the considered directions and in the radial direction (thickness direction) comparable with the 3D reference values. In detail, once again the layer-wise approach makes it possible to correctly assess the stress state both for thick structures and for structures with low thickness. While the equivalent single layer approach provides approximate values like the reference values for thick structures, it tends to underestimate the value of tension in the case of thin structures.

Table 12 Stresses in direction 11 and 12 for [0/90/0] cylinder test case

	S_{23}				S_{33}			
	2	**4**	**50**	**500**	**2**	**4**	**50**	**500**
3D	-1.379	-2.349	-3.491	-0.691	-0.34	-0.62	-4.85	-9.12
L4	-1.442	-2.464	-3.659	-0.7287	-0.343	-0.627	-5.026	-9.468
ED4	-1.280	-2.025	-2.613	-0.5195	-0.358	-0.684	-5.184	-12.26

Conclusion

In this paper, different static shell and plate finite element analyses based on the CUF formulation were presented. The analysis involved three different benchmarks with three different geometries of different complexity. Different theories were compared to evaluate the performance of the shell elements considering the classical theories and the refined ones (high order) [33-36], in addition to the comparison was introduced a model made with the commercial software Abaqus. The results were provided for different thickness values (from extremely thick structures to extremely thin structures) in terms of displacement, tension in the plane and out of the plane where it could be evaluated. In accordance with what has been shown, shell elements based on the unified formulation of Carrera both with layer-wise approach and with equivalent single layer approach are not subject to shear locking phenomena, even for extremely thin structures. Also, for any thickness value. Obviously for excessively thick structures can be used a shell modelling provided that the expansion orders of the functions of the displacement in the direction of the thickness are increased. However, the use of classical theories such as CST and FSDT can only be used with very thin layered structures. On the other hand, commercial software can provide acceptable results for relatively thin structures while for thick structures, although it can provide results affected by a minor error compared to the solutions obtained by classical theories, are however not acceptable because suffering from an error too high. In addition, layer-wise models work better than equivalent single layer models, which in turn provide more accurate results when compared to business codes. To conclude the use of LW models is required for both thick and thin shells if the distribution of transverse tresses in the thickness is to be correctly described and the interlaminar continuity requirements are to be met. Neglecting that the high order models manage to provide a situation of tensions both in the plan and out of the plan unlike the commercial codes that neglect the trend of tensions outside the plan, providing therefore an approximate resolution of the analysed problem.

References

[1] Ambartsumyan VA. In: Ashton JE, editor. Theory of anisotropic plates. Tech Pub Co; 1969. Translated from Russian by T Cheron.

[2] Reddy JN. Mechanics of laminated composite plates, theory and analysis. CRC Press; 1997.

[3] Noor AK, Rarig PL. Three-dimensional solutions of laminated cylinders. Computer Methods in Applied Mechanics and Engineering 1974; 3:319-334. https://doi.org/10.1016/0045-7825(74)90017-6

[4] Malik M. Differential quadrature method in computational mechanics: new development and applications. Ph.D Dissertation, University of Oklahoma, Oklahoma, 1994.

[5] Liew KM, Han B, Xiao M. Differential quadrature method for thick symmetric cross-ply laminates with first-order shear flexibility. International Journal of Solids and Structures 1996; 33:2647-2658. https://doi.org/10.1016/0020-7683(95)00174-3

[6] Ferreira AJM, Roque CC, Carrera E, Cinefra M. Analysis of thick isotropic and cross-ply laminated plates by radial basis functions and unified formulation. Journal of Sound and Vibration 2011; 330:771-787. https://doi.org/10.1016/j.jsv.2010.08.037

[7] Ferreira AJM, Roque CC, Carrera E, Cinefra M, Polit O. Analysis of laminated shells by a sinusoidal shear deformation theory and radial basis functions collocation, accounting for through-the-thickness deformations. Composites Part B 2011; 42:1276-1284. https://doi.org/10.1016/j.compositesb.2011.01.031

[8] Reddy JN, Robbins DH. Theories and computational models for composite laminates. Applied Mechanics Review 1994; 47:147-165. https://doi.org/10.1115/1.3111076

[9] Varadan TK, Bhaskar K. Review of different theories for the analysis of composites. Journal of Aerospace Society of India 1997; 49:202-208.

[10] Carrera E. Developments, ideas and evaluations based upon Reissner's mixed variational theorem in the modeling of multilayered plates and shells. Applied Mechanics Review 2001; 54:301-329. https://doi.org/10.1115/1.1385512

[11] Argyris JH. Matrix displacement analysis of plates and shells, Prolegomena to a general theory, part I. Ingeniur-Archiv 1966; 35:102-142. https://doi.org/10.1007/BF00536183

[12] Sabir AB, Lock AC. The application of finite elements to the large deflection geometrically non-linear behaviour of cylindrical shells. In Variational Methods in Engineering 2, Southampton University Press: 1972; 7/66-7/75.

[13] Wempner GA, Oden JT, Kross DA. Finite element analysis of thin shells. Journal of Engineering Mechanics ASCE 94 1968; 94:1273-1294. https://doi.org/10.1061/JMCEA3.0001039

[14] Abel JF, Popov EP. Static and dynamic finite element analysis of sandwich structures. Proceedings of the Second Conference of Matrix Methods in Structural Mechanics, Vol. AFFSL-TR-68-150, Wright-Patterson Air Force Base, Ohio, USA, 1968; 213-245.

[15] Monforton GR, Schmidt LA. Finite element analyses of sandwich plates and cylindrical shells with laminated faces. Proceedings of the Second Conference of Matrix Methods in Structural Mechanics, Vol. AFFSL-TR-68-150, Wright-Patterson Air Force Base, Ohio, USA, 1968; 573-308.

[16] Noor AK. Finite element analysis of anisotropic plates. American Institute of Aeronautics and Astronautics Journal 1972; 11:289-307. https://doi.org/10.1002/nme.1620110206

[17] Hughes TJR, Tezduyar T. Finite elements based upon Mindlin plate theory with particular reference to the four-node isoparametric element. Journal of Applied Mechanics 1981; 48:587-596. https://doi.org/10.1115/1.3157679

[18] Parisch H. A critical survey of the 9-node degenerated shell element with special emphasis on thin shell application and reduced integration. Computer Methods in Applied Mechanics and Engineering 1979; 20:323-350. https://doi.org/10.1016/0045-7825(79)90007-0

[19] Zienkiewicz OC, Taylor RL, Too JM. Reduced intergration technique in general analysis of plates and shells. International Journal for Numerical Methods in Engineering 1973; 3:275-290. https://doi.org/10.1002/nme.1620030211

[20] Naghdi PM. The theory of shells and plates. In Handbuch der Physic, Vol. 4. Springer: Berlin, 1972; 425-640. https://doi.org/10.1007/978-3-642-69567-4_5

[21] Bathe KJ, Dvorkin EN. A four node plate bending element based on Mindlin/Reissner plate theory and mixed interpolation. International Journal for Numerical Methods in Engineering 1985; 21:367-383. https://doi.org/10.1002/nme.1620210213

[22] Arnold D.N., Brezzi F. Locking-free finite element methods for shells. Mathematics of Computation 1997; 66(217):1-14 https://doi.org/10.1090/S0025-5718-97-00785-0

[23] Bischoff M, Ramm E. On the physical significance of higher-order kinematic and static variables in a threedimensional shell formulation. International Journal of Solids and Structures 2000; 37:6933-6960. https://doi.org/10.1016/S0020-7683(99)00321-2

[24] Simo JC, Rafai S. A class of mixed assumed strain methods and the method of incompatible modes. International Journal for Numerical Methods in Engineering 1990; 29:1595-1638. https://doi.org/10.1002/nme.1620290802

[25] Kant T, Owen DRJ, Zienkiewicz OC. Refined higher order C0 plate bending element. Computer & Structures 1982; 15:177-183 https://doi.org/10.1016/0045-7949(82)90065-7

[26] Polit O, Touratier M. A new laminated triangular finite element assuring interface continuity for displacements and stresses. Composite Structures 1997; 38(1-4):37-44. https://doi.org/10.1016/S0263-8223(97)00039-1

[27] Tessler A. A higher-order plate theory with ideal finite element suitability. Computer Methods in Applied Mechanics and Engineerings 1991; 85:183-205. https://doi.org/10.1016/0045-7825(91)90132-P

[28] Reddy JN. Mechanics of Laminated Composite Plates and Shells, Theory and Analysis. CRC Press: New York, USA, 1997

[29] Rao KM, Meyer-Piening HR. Analysis of thick laminated anisotropic composites plates by the finite element method. Composite Structures 1990; 15:185-213. https://doi.org/10.1016/0263-8223(90)90031-9

[30] Noor AK, Burton WS. Assessment of computational models for multi-layered composite shells. Applied Mechanics Review 1990; 43:67-97. https://doi.org/10.1115/1.3119162

[31] Reddy JN. An evaluation of equivalent single layer and layer-wise theories of composite laminates. Composite Structures 1993; 25:21-35. https://doi.org/10.1016/0263-8223(93)90147-I

[32] Rammerstorfer FG, Dorninger K, Starlinger A. Composite and sandwich shells. In Nonlinear Analysis of Shells by Finite Elements, Rammerstorfer FG (ed.). Springer-Verlag: Vienna, 1992; 131-194. https://doi.org/10.1007/978-3-7091-2604-2_6

[33] Cinefra M., Carrera E. Shell finite elements with different through-the-thickness kinematics for the linear analysis of cylindrical multilayered structures. International journal for numerical methods in engineering. 93:(2013), pp. 160-182. https://doi.org/10.1002/nme.4377

[34] Carrera E., Ciuffreda A., A unified formulation to assess theories of multilayered plates for various bending problems. Composite Structures 69 (2005) 271-293 https://doi.org/10.1016/j.compstruct.2004.07.003

[35] Carrera E. Theories and finite elements for multilayered anisotropic composite plates and shells. Archives of Computational Methods in Engineering 2002; 9:87-140. https://doi.org/10.1007/BF02736649

[36] Carrera E. Theories and finite elements for multilayered plates and shells: a unified compact formulation with numerical assessment and benchmarking. Archives of Computational Methods in Engineering 2003; 10(3):215-296. https://doi.org/10.1007/BF02736224

[37] Carrera E. A class of two-dimensional theories for multi-layered plates analysis. Atti Accad Sci Tor, Mem Sci Fis 1995;19-20:49-87.

[38] Murakami H. Laminated composite plate theory with improved in-plane response. J Appl Mech 1986; 53:661-6. https://doi.org/10.1115/1.3171828

[39] Meyer-Piening HR. Experiences with 'Exact' linear sandwich beam and plate analyses regarding bending, instability and frequency investigations. In: Proceedings of the Fifth International Conference On Sandwich Constructions, Zurich, Switzerland, September 5-7, vol. I.; 2000. p. 37-48.

[40] Pagano NJ. Exact solutions for composite laminates in cylindrical bending. J Compos Mater 1969;3:398-411. https://doi.org/10.1177/002199836900300304

Interaction of a rectangular jet with a slotted plate in presence of a control mechanism: Experimental study of the aeroacoustic field

Marwan Alkheir[1,a*], Hassan Assoum[2], Nour Eldin Afyouni[1,b*], Bilal El Zohbi[1], Kamel Abed Meraim[1], Anas Sakout[1], Mouhammad El Hassan[3,c*]

[1] LASIE UMR CNRS 7356, University of La Rochelle, La Rochelle, France

[2] Beirut Arab University, Mechanical Engineering Department, Tripoli, Lebanon

[3] Prince Mohammad Bin Fahd University, Mechanical Engineering Department, Al Khobar, KSA

[a] marwanalkheir@gmail.com, [b] nour.afyouni@gmail.com, [c] melhassan@pmu.edu.sa

Keywords: Aeroacoustics, Rectangular Jet, SPIV, Passive Control, Acoustic Comfort, Fluid-Structure Interaction

Abstract. The acoustic comfort inside residential buildings is of high interest. HVAC systems employ different shapes of diffusers to ensure air mixing. The interaction between the airflows and the blades of these terminals may result in intense noise radiation. In this work, an experimental study was carried out to investigate the aero-acoustic production of a rectangular jet impinging on a rectangular plate with a slot. For certain flow regimes, such configuration results in whistles with high acoustic levels, called "self-sustaining tones". These tones result from the interaction between the Aerodynamic modes of the jet and acoustic modes. The impact of the vortical structures on the rectangular plate results in pressure waves that re-excite the jet near its exit. This feedback mechanism and the aero-acoustic coupling are responsible for the high-energy tones and can lead to structural fatigue through vibrations. A control mechanism consisting of a thin rod was introduced between the jet nozzle and the impinging wall to disturb the vortex dynamics responsible for the loop of the self-sustaining tones. A total of 1085 positions of the rod were tested between the nozzle and the impinged plate to identify positions of optimal noise reduction. Simultaneous Stereoscopic Particle Image Velocimetry (SPIV) and unsteady pressure measurements were conducted to characterize both the kinematic and the acoustic fields. Two zones were distinguished in terms of control efficacy. In the first one, the sound pressure level dropped by 19 dB, while in the second zone, the sound pressure level increased by 14 dB. The velocity fields show that the presence of the rod divides the main jet into two lateral jets from both sides of the axis of the convergent. The presence of the cylinder creates an artificial expansion of the jet and divides it into two shear flows or jet-like flows. The outer part of these flows expands radially with less interaction with the plate as compared to the case without control. This behavior affects the deformation of vortices against the slot and results in a disappearance of the loop of self-sustaining tones. The main novelty of this work relates to the implementation and analysis of a control mechanism using 2D3C (SPIV) velocity measurements simultaneously with the acoustic radiation produced by the interaction of this flow with a slotted impinging wall.

Introduction

Impinging jets received great interest for several industrial applications; they are used in cooling of electronic parts, in building HVAC systems, in drilling process, and various other applications. The interaction of a jet with a solid surface may, in certain configurations listed by [1], generates whistles with a considerable acoustic level, which are called "Self-sustaining tones". The self-sustaining tones have been studied by [1]–[3]: When a turbulent jet impacts a solid surface, the surface pressure fluctuations create a periodic wave with a fundamental frequency f_0. The wave's oscillations pass upstream towards the nozzle, disturb the shear layer, and interact with the formed structures at the jet exit. These oscillations are called feedback loops. Several studies focused on the impinging jet noise in different configurations, such as the hole tone [4], an airflow around a cylinder [5], the slot tone [6]–[10], and others. The understanding of the source of noise in impinging jet is of high importance to develop a control mechanism that would lead to reducing the sound level in these jets. Two types of control methods are presented: Passive and Active control. When compared to active control, the passive control doesn't require external energy source, such as changing the jet geometry. [11] studied the noise reduction of a twin jet nozzle using flexible filaments; they demonstrated that the filaments consistently eliminated screech tones and reduced overall sound pressure level by 3 dB or more depending on the configuration. Using circular cylinders with soft porous cover in the flow was also used to reduce jet noise. [12] studied the materials of the porous cover experimentally and found that the use of low airflow resistivity materials leads to a noticeable flow noise reduction. Circular cylinders were also used to reduce noise generated by rectangular cavities that are characterized by similar Aero-Acoustic coupling found in rectangular jets [13], [14].For active flow control, [15] placed a circular array of 400 μm diameter supersonic microjets around the periphery of the main jet, which reduced the near-field noise by about 8 dB. A similar method was proposed by [16] who suggested two control strategies using microjets: the first method consisted of a steady microjet injection, while the second was based on a pulsed microjet injection, motivated by the need to further improve the noise suppression. The authors found that the pulsed microjet was able to bring about the same noise reduction as a steady injection using approximately 40% of the corresponding mass flow rate of the steady microjet case. [17] studied experimentally how plasma actuators based on surface barrier high-frequency discharge affect jet noise characteristics. They showed that jet excitation in the case of $St \approx 0.5$ using the barrier discharge plasma actuator leads to broadband amplification of jet sound radiation. The jet excitation in the case of $St > 2$ led to broadband noise reduction if the action is sufficiently intensive.

In this paper, a passive control mechanism comprising a 4 mm diameter rod installed in the flow to disturb the vortex dynamics responsible for the loop of self-sustaining tones installed along the jet is studied experimentally using acoustic measurements and SPIV technique. The configuration consists of a rectangular air jet impinging on a slotted plate, with an impact ratio $L/H = 4$, where L is the distance between the plate and the nozzle distance, and H is the height of the slot. This configuration corresponds to a flow producing self-sustaining tones. In the present study, a Reynolds number Re = 6700 is considered. This Reynolds number corresponds to a flow regime with a significantly high acoustic level. In order to find the optimal position of the rod, the distance between the jet outlet and the slotted plate was swept with a step of 1 mm along the X and Y axes, thus a total of 1085 rod positions were tested. An innovative technique that called Combined Stereoscopic PIV (C-SPIV) was used to resolve some experimental difficulties caused by the introduction of the rod in the flow. A detailed explanation of this technique is present in [18].

Experimental Setup and Metrology

The experimental device (shown in Fig. 1) consists of a compressor located outside of the experimental chamber. This compressor is controlled by a chopper frequency (Siemens MICROMASTER 420, three-phases) which allow us to vary the outlet velocity from 0 m/s up to 33 m/s ($M \approx 0.1$). The pulsed air passes through a large enclosure ($V = 1\ m^3$) equipped with grids before the airflow is ejected into the flow duct of 1250 mm length, with a rectangular section of dimensions 190 mm × 90 mm. The latter contains honeycombs to parallelize the air streamlines before it arrives at a 4th-degree polynomial convergent form, of exit height $H = 10\ mm$ and width $Lz = 190\ mm$. The jet coming out of the convergent will strike a slotted aluminum plate with 4 mm thickness and 250 mm x 250 mm section. This plate has a slot of the same size as the jet outlet ($H = 10\ mm$, $Lz = 190\ mm$). The slot is beveled at 45° downstream and aligned with the blowing mouth. The impact distance L between the jet outlet and the plate is taken equal to 4 times the nozzle height. Thus, the dimensionless distance $L/H = 4$ will be used in this study. To study the influence of the rod positions on the noise emission, specific fixing and displacement systems are needed to control the position of the rod. Following a control program developed in "LabVIEW", the rod can move in the x and y directions; in the x-direction, the rod can move from $X = -15mm$ (below the jet) to $X = 15mm$ (above the jet), however, in the y-direction, the rod starts at $Y = 3mm$ (where the rod has a radius of 2 mm with an initial safety distance of 1 mm) and moves up to the border of the split plate at position $Y = 37$ mm (we leave 2 mm radius of the rod and 1 mm of safety). Then, the rod occupies the horizontal positions $-15\ mm \leq X \leq 15\ mm$ and $3\ mm \leq Y \leq 37\ mm$ to fit a total of 1085 positions. For the acoustic measurements, a microphone B&K of type 4189 was placed at a 15 cm distance behind the plate. We carried out our acoustic measurements with a sampling frequency of 15 kHz, for 3 seconds. These acoustic signals were acquired by LabVIEW software installed on a National Instruments $NI\ PXI$ 1036 workstation equipped with a dynamic card $NI\ PXI$ 4496. A spectral analysis is then performed in order to understand the effect of the rod on the frequency activities. To correlate the acoustic behavior to the turbulent activity, the vortex structures are identified using the Combined Stereoscopic Particle Image Velocimetry technique (C-SPIV). To seed the flow, olive oil was used as tracers, using an oil particle generator that has an oil-air atomizer sprayed by a "Laskin Nozzle" aerographic system marketed by "Lavision". For the lighting system, Litron Nd: $YLF\ LDY$ 304-PIV laser having two heads of 30 $mJ/pulse$ at a frequency of 1 KHz and a wavelength $\lambda = 527$ nm is used (shown in Fig. 1). The laser light beam is carried by a laser arm to illuminate the area between the jet nozzle and the plate. This laser arm contains 7 mirrors capable to transmit 96% of the light intensity. This beam is transformed into a laser sheet by a sheet generator. Thus, the thickness of the laser sheet and its opening angle is adjusted depending on the used divergent lens with a focal length of 10 mm, which is chosen to withstand the high light power generated by the laser. Once the laser sheet has been generated between the nozzle and the slotted plate, images can be acquired. The synchronization between the laser pulses and the camera apertures is controlled by a high-speed controller (HSC) from "Lavision.

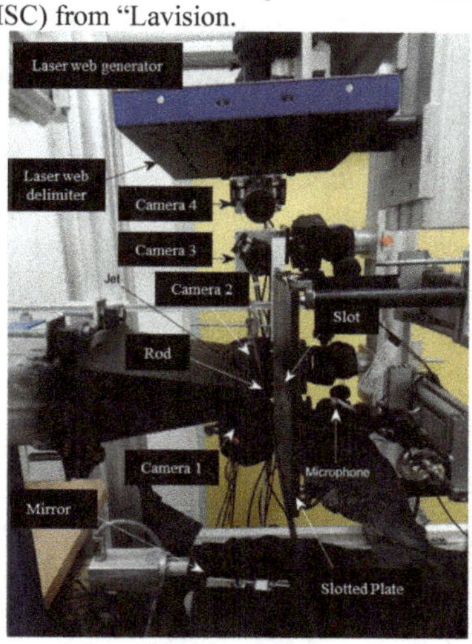

Fig. 1: Experimental setup consisting of a rectangular nozzle, slotted plate, PIV systems and Microphones

Results and Discussion

The distribution of the sound pressure levels as a function of rod positions is shown in Fig. 2. This figure shows the results of the acoustic pressure measurements for 1085 positions occupied by the rod. It shows the influence of flow disturbance introduced by the rod on the radiated sound field. For Reynolds number $Re = 6700$ with an impact ratio $L/H = 4$, in the absence of the rod, the sound pressure level produced by the self-sustaining sound is $Lp = 83\ dB$. The analysis of results indicates the presence of three different zones: The first zone (blue color) where the flow disturbance generated by the presence of the rod cause a reduction in the sound pressure level Lp, which can reach up to 19 dB. This area has the height of the nozzle and is located on either side of the axis of symmetry of the jet. This symmetry degrades from a longitudinal distance H from the outlet of the nozzle. However, the efficiency remains significant up to a distance of 26 mm from the jet exit. The second zone (orange and yellow color) corresponds to an increase in the sound field, which can reach up to +14 dB. This zone is not symmetrical with respect to the axis of the jet, it occupies longitudinally (y axis) about 28 mm from the jet's outlet and the vertical positions (x axis) are approximately between $-0.65 < X/H < -0.45$. It can be noted that this zone gradually extends in a non-symmetrical way from the longitudinal position $Y/H = 1.5$ to $Y/H = 3$. One can also notice a small zone at the level of the plate which represents a small rather significant increase in the level of the acoustic field. The third zone, apart from zones 1 and 2, where the rod does not disturb the radiated acoustic field. In this zone the acoustic level frequency characteristics are the same as those found in the absence of the rod.

To understand the effect of the rod on the flow dynamics as well as the noise production, the acoustic and PIV data are considered simultaneously. A Fast Fourier Transform is applied on the acoustic signal to find the frequency of the sound emission. Fig. 3 shows the acoustic spectrum with and without the rod. Two positions of the rod were chosen: (X/H = 0, Y/H=0.8) and (X/H= -0.5, Y/H=0.4), which correspond to arbitrary points in zone 1 and zone 2, respectively. It is found that the presence of the rod in these positions is accompanied with a reduction of 19,3 dB and an increasing of 8,6 dB respectively. The spectrum shows that, when there was no rod in the flow, the acoustic spectrum presents three energetic frequencies: 168 Hz, 212 Hz, and 380 Hz. However, when the rod is installed at the first position, one can noticed the total disappearance of the peaks corresponding to the frequencies 168 Hz and 320 H which characterize the two self-sustaining sound loops in the absence of the rod. One can also observe that the frequency activity is reduced and the broadband spectrum is very weak for this rod position. This explains the dramatic reduction in the sound pressure level of 19.3 dB. When the rod occupies the second position, the spectrum indicates the disappearance of the self-sustaining sound loops. In addition, it has a broadband appearance with high-frequency activity, which explains the 8.6 dB increase in sound pressure level. One can also notice the absence of any particular frequency peaks.

Materials Research Proceedings 31 (2023) 614-621 https://doi.org/10.21741/9781644902592-63

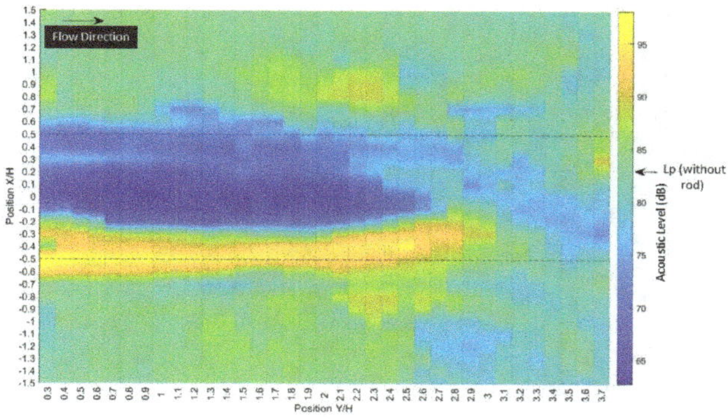

Fig. 2: Variation of the sound level as a function of the position of the rod for a Reynolds number Re = 6700, an impact ratio L/H = 4 and for a rod of 4 mm

In order to understand the origin of the variation in the acoustic spectrum, the $\boldsymbol{\lambda_2}$ criteria are calculated from the velocity field obtained by SPIV measurement. Fig. 4a shows that, when the rod occupies the position 1, the rod divides symmetrically the main jet on two jets: upper jet (in the positive y side) and lower jet (in negative y side). The two jets present an antisymmetric character dominating the vortex organization. Moreover, one note that the vortex structures coming from the upper and the lower jet strike the wall of the plate, with a strong vortex activity creating two zones of recirculation on either side of the slot, without passing through. This explains the disappearing of the frequency activity in the acoustic spectrum. However, when the rod occupies the second position (Fig. 4b), the lower jet shows no vortex activity, however the flow of the upper jet presents an antisymmetric character and the presence of the rod delays the formation of vortex structures. The coherent structures developed in the lower part of the upper jet pass completely through the slot, without interaction with the slot edge. This passage does not create feedback loop, and then no self-sustaining tones are created, which explain the disappearance of the energic frequency in the acoustic spectrum.

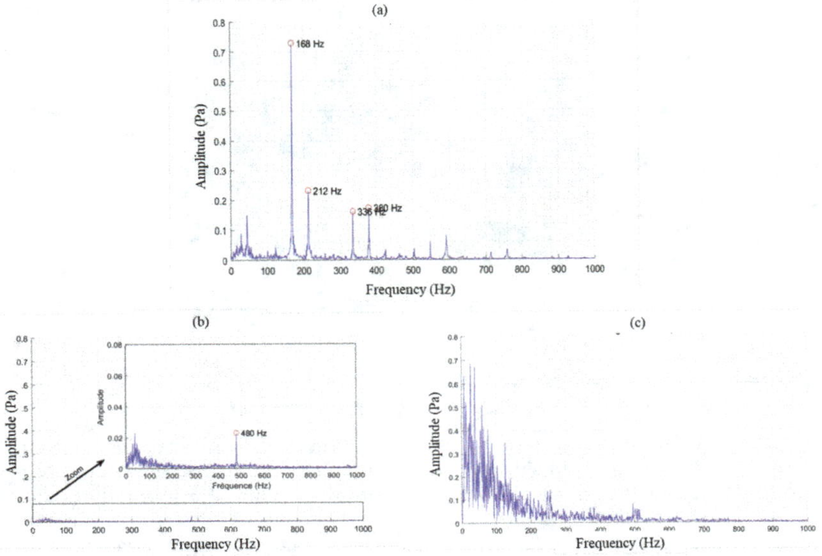

Fig. 3: Acoustic Pressure Spectra (a) in the absence of the rod, (b) in the presence of the rod at position 1: (X/H = 0, Y/H=0.8), (c) in the presence of the rod at position 2: (X/H= -0.5, Y/H=0.4)

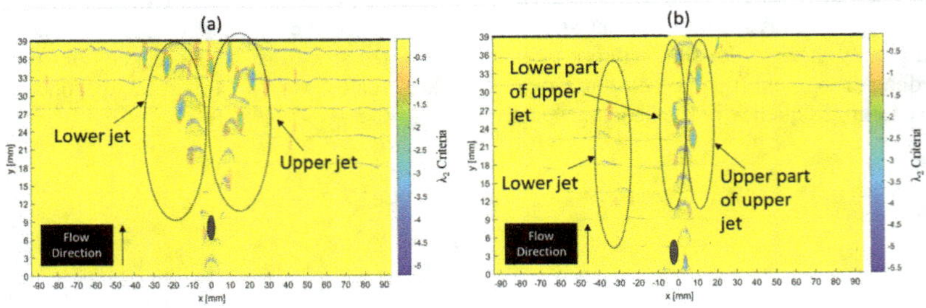

Fig. 4: The λ_2 criteria (a) in the presence of the rod at (X/H = 0, Y/H=0.8), (b) in the presence of the rod at (X/H= -0.5, Y/H=0.4)

Conclusion

In this paper, the correlations between the vortex dynamics and the radiated sound fields are investigated experimentally using Combined Stereoscopic PIV and acoustic measurements. The analyzed flow is characterized by Reynolds number of Re = 6700 with an impact ratio L/H = 4. The flow control consisted of positioning a rod of 4 mm diameter in the free jet part of the impinging jet. Two zones are distinguished: Zone 1, in which the presence of the rod leads to decreasing the sound level, and zone 2, in which the presence of the rod increases the sound level. In Zone 1, the presence of the rod causes the disappearance of self-sustaining sound loops and a low-frequency activity that characterizes the broadband spectrum appeared. It was also found that the rod splits the flow into two jets which settle in the upper and lower part of the flow, respectively. These two jets are almost symmetrical about the main axis of the jet. In addition, in this configuration, the rod totally destabilized the flow, and lead to destroying the self-sustaining sound loops that were installed when there was no control device within the flow. In zone 2, the results show that, the presence of the rod causes the disappearance of the self-sustaining sound loops, and the appearance of a broadband spectrum with high-frequency activity, despite an increase in the sound pressure level. Dynamically, the presence of the rod divided the flow into two streams. The activity of the vortex structures resulting from the upper jet is antisymmetric, while the lower jet shows no vortex activity.

References

[1] D. Rockwell and E. Naudascher, "Self-Sustained Oscillations of Impinging Free Shear Layers," *Annu. Rev. Fluid Mech.*, vol. 11, no. 1, pp. 67–94, Jan. 1979. https://doi.org/10.1146/annurev.fl.11.010179.000435

[2] A. Powell, "On the Mechanism of Choked Jet Noise," *Proc. Phys. Soc. Sect. B*, vol. 66, no. 12, pp. 1039–1056, Dec. 1953. https://doi.org/10.1088/0370-1301/66/12/306

[3] C.-M. Ho and N. S. Nosseir, "Dynamics of an impinging jet. Part 1. The feedback phenomenon," *J. Fluid Mech.*, vol. 105, no. 1, p. 119, Apr. 1981. https://doi.org/10.1017/S0022112081003133

[4] R. C. Chanaud and A. Powell, "Some Experiments concerning the Hole and Ring Tone," *J. Acoust. Soc. Am.*, vol. 37, no. 5, pp. 902–911, May 1965. https://doi.org/10.1121/1.1909476

[5] A. Henning, K. Kaepernick, K. Ehrenfried, L. Koop, and A. Dillmann, "Investigation of aeroacoustic noise generation by simultaneous particle image velocimetry and microphone measurements," *Exp. Fluids*, vol. 45, no. 6, pp. 1073–1085, Dec. 2008. https://doi.org/10.1007/s00348-008-0528-y

[6] K. Abed-Meraïm, H. Assoum, and A. Sakout, "TRANSFERTS ENERGETIQUES ENTRE LE CHAMP TURBULENT D'UN JET IMPACTANT DE VENTILATION ET LE CHAMP ACOUSTIQUE GENERE," 2016.

[7] H. H. Assoum, J. Hamdi, K. Abed-Meraïm, M. El Hassan, A. Hammoud, and A. Sakout, "Experimental investigation the turbulent kinetic energy and the acoustic field in a rectangular jet impinging a slotted plate," *Energy Procedia*, vol. 139, pp. 398–403, Dec. 2017. https://doi.org/10.1016/j.egypro.2017.11.228

[8] J. Hamdi, H. Assoum, K. Abed-Meraïm, and A. Sakout, "Analysis of the effect of the 3C kinematic field of a confined impinging jet on a slotted plate by stereoscopic PIV," *Eur. J. Mech. - BFluids*, vol. 76, pp. 243–258, Jul. 2019. https://doi.org/10.1016/j.euromechflu.2019.02.012

[9] H. H. Assoum, M. El Hassan, J. Hamdi, M. Alkheir, K. A. Meraim, and A. Sakout, "Turbulent Kinetic Energy and self-sustaining tones in an impinging jet using High Speed 3D Tomographic-PIV," *Energy Rep.*, vol. 6, pp. 802–806, Feb. 2020. https://doi.org/10.1016/j.egyr.2019.12.018

[10] H. H. Assoum *et al.*, "Turbulent kinetic energy and self-sustaining tones: Experimental study of a rectangular impinging jet using high Speed 3D tomographic Particle Image Velocimetry," *J. Mech. Eng. Sci.*, vol. 14, no. 1, pp. 6322–6333, Mar. 2020. https://doi.org/10.15282/jmes.14.1.2020.10.0495

[11] N. Lucas, M. Doty, L. Taubert, and I. Wygnanski, "Reducing the noise emanating from a twin jet nozzle using flexible filaments," *Exp. Fluids*, vol. 54, no. 4, p. 1504, Apr. 2013. https://doi.org/10.1007/s00348-013-1504-8

[12] T. F. Geyer, "Experimental evaluation of cylinder vortex shedding noise reduction using porous material," *Exp. Fluids*, vol. 61, no. 7, p. 153, Jul. 2020. https://doi.org/10.1007/s00348-020-02972-0

[13] L. Keirsbulck, M. E. Hassan, M. Lippert, and L. Labraga, "Control of cavity tones using a spanwise cylinder," *Can. J. Phys.*, vol. 86, no. 12, pp. 1355–1365, Dec. 2008. https://doi.org/10.1139/p08-086

[14] M. El Hassan and L. Keirsbulck, "Passive control of deep cavity shear layer flow at subsonic speed," *Can. J. Phys.*, vol. 95, no. 10, pp. 894–899, Oct. 2017. https://doi.org/10.1139/cjp-2016-0822

[15] F. S. Alvi, C. Shih, R. Elavarasan, G. Garg, and A. Krothapalli, "Control of Supersonic Impinging Jet Flows Using Supersonic Microjets," *AIAA J.*, vol. 41, no. 7, pp. 1347–1355, Jul. 2003. https://doi.org/10.2514/2.2080

[16] J. J. Choi, A. M. Annaswamy, H. Lou, and F. S. Alvi, "Active control of supersonic impingement tones using steady and pulsed microjets," *Exp. Fluids*, vol. 41, no. 6, pp. 841–855, Nov. 2006. https://doi.org/10.1007/s00348-006-0189-7

[17] V. F. Kopiev *et al.*, "Jet noise control using the dielectric barrier discharge plasma actuators," *Acoust. Phys.*, vol. 58, no. 4, pp. 434–441, Jul. 2012. https://doi.org/10.1134/S1063771012040100

[18] M. Alkheir, H. H. Assoum, N. E. Afyouni, K. Abed Meraim, A. Sakout, and M. El Hassan, "Combined Stereoscopic Particle Image Velocimetry Measurements in a Single Plane for an Impinging Jet around a Thin Control Rod," *Fluids*, vol. 6, no. 12, p. 430, Nov. 2021. https://doi.org/10.3390/fluids6120430

[19] P. Miron, J. Vétel, A. Garon, M. Delfour, and M. E. Hassan, "Anisotropic mesh adaptation on Lagrangian Coherent Structures," *J. Comput. Phys.*, vol. 231, no. 19, pp. 6419–6437, Aug. 2012. https://doi.org/10.1016/j.jcp.2012.06.015

Advanced Topics in Mechanics of Materials, Structures and Construction - AToMech1-2023 Materials Research Forum LLC
Materials Research Proceedings 31 (2023) 622-633 https://doi.org/10.21741/9781644902592-64

Investigation of the influence of external edge restraint on reinforced concrete walls

Muhammad Kashif Shehzad[1,a*], John P. Forth[1,b], Nikolaos Nikitas[1,c], Imogen Ridley[1,d], Robert Vollum[2,e], Abo Bakr Elwakeel[2,f], Karim El Khoury[2,g], Bassam Izzuddin[2,h]

[1]School of Civil Engineering, University of Leeds, Woodhouse Lane, LS2 9JT, United Kingdom

[2]Imperial College London, Exhibition Rd, South Kensington, London SW7 2BX, United Kingdom

[a]m.k.shehzad@leeds.ac.uk, [b]j.p.forth@leeds.ac.uk, [c]n.nikitas@leeds.ac.uk, [d]cn17ijr@leeds.ac.uk, [e]r.vollum@imperial.ac.uk, [f]a.elwakeel15@imperial.ac.uk, [g]k.khoury20@imperial.ac.uk, [h]b.izzuddin@imperial.ac.uk

Keywords: Edge Restraint, Crack Width, Aspect Ratio, Volume Change, Imposed Strain, Finite Element Analysis

Abstract. Externally restraining volume changes of concrete, i.e., thermal effects and shrinkage, may result in tensile stresses and eventually cracking. Such cracking risk is controlled / mitigated by the provision of steel reinforcement, which presumes correct understanding of the cracking patterns under different types of restraint conditions. Reinforced concrete (RC) members may be restrained at their edges or end, or in many cases a combination of the two. Existing guidance on the subject is mostly based on end restrained members, however it is applied to predict the behaviour under edge restraint too. Researchers have identified that the mechanisms of cracking associated to edge and end restraints are quite different. To this purpose, findings from an experimental investigation aiming to understand the behaviour of edge restrained RC walls were utilized to validate a finite element (FE) model. Subsequently, this FE model was used to study the edge restrained walls having different aspect ratios. Cracking patterns, widths and extent appeared to greatly depend on the wall aspect ratio. The study provides clear evidence on why similar studies related to all forms of restraint are needed to support engineers in designing against cracking due to restraints.

Introduction

The design of members subjected to restraint of imposed strains has traditionally been based on the principles of end restraint i.e., members fully restrained at both ends [1]. Whilst appropriate for elements subjected to end restraint, this is wholly inappropriate for members subjected to edge restraint as each restraint mechanism produces totally different cracking patterns and mechanisms [2]. For instance, for end restraint: 1) the restraint is uniform throughout and when a crack occurs, stress in the entire member is relieved; 2) the crack can occur anywhere along the length of the member depending mainly on the variation of the concrete tensile strength; 3) a crack, when formed, reaches its maximum width before another new crack forms. Whereas, for edge restraint: it is not uniform; it is maximum at the base and reduces over the height of the member; stress relief upon cracking is local to the area around crack, and the variation in edge restraint depends on the ratio of length to height (L/H), the location of the restraint within the member, the roughness of the concrete surface (between the restraining and the restrained panel) and the wall and base dimensions [3, 4].

Knowledge of the likely cracking patterns is essential while deciding on the steel reinforcement. Not much guidance on the cracking patterns in RC members under restraint exists. Typical cracking patterns for restrained members are proposed in Annex L to BS EN 1992-3 [5] (see Fig.1).

According to this guidance, no crack is formed within 2.4 m from the free ends of the wall (9, 10), however Al-Rawi [6] suggested that no crack would form within a distance equal to the wall height, from the free end. ACI Committee 207 [7] describes the method of estimation of degree of restraint but does not provide any guidance on the location and orientation of the cracks under different types of restraint.

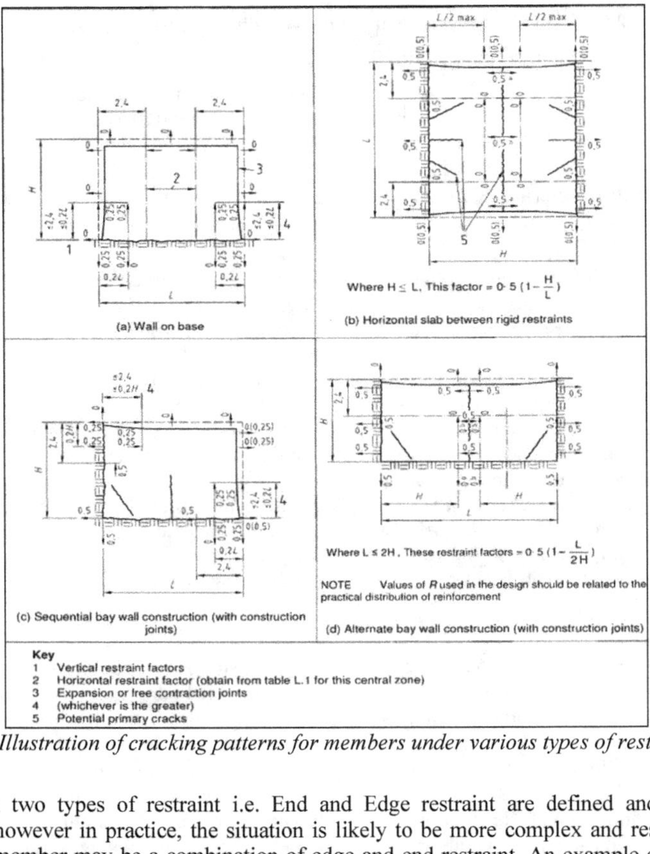

Fig. 1: Illustration of cracking patterns for members under various types of restraint [5]

Although two types of restraint i.e. End and Edge restraint are defined and considered separately, however in practice, the situation is likely to be more complex and restraint of the newly cast member may be a combination of edge and end restraint. An example of a situation where both forms of restraint may exist can be found by considering a new section of concrete cast between two pre-existing concrete wall sections and onto a pre-existing concrete base [9]. Whilst a greater understanding of edge restraint in isolation is being gained [10, 11], the optimum problem constituted by a combination of the two restraints is still little understood. Forth [12] suggested that in such a scenario, both types of restraint will dominate in different parts of the wall. Edge restraint will be dominant closer to the base, however, moving further up from the base, the effect of edge restraint will get less significant, and the influence of end restraint becomes more pronounced. At a point within the height of the wall, end restraint becomes dominant and edge restraint becomes insignificant. As part of the ongoing EPSRC UK funded research being conducted jointly at the University of Leeds and Imperial College London, the combined restraint is being investigated experimentally to confirm the above distribution of restraint in RC members.

This paper presents an experimental study which was purposefully designed [11] to understand the behaviour of RC walls when their imposed strains are subjected to edge restraint. Findings from this experimental investigation are utilized to validate a finite element (FE) model, which is used to study walls having different aspect ratios and subjected to external edge restraint.

Experimental Investigation

An experimental study of RC walls restrained along their edge has been undertaken at the University of Leeds, UK. As part of this experimental investigation, four walls were tested, and each test continued for a period of 12 weeks. In each test, a RC wall was cast onto an already cast and hardened RC base. Initially, the RC base slab was cast and cured for up to 14 days. Subsequently, once the slab was at least 28 days old, the wall was cast on it; hence, the existing slab imposed a restraint to the volume changes occurring in the newly cast wall. The walls had an aspect ratio (L/H) of 4, which is representative of the upper values of commonly used aspect ratios in practice. The ratio of vertical steel reinforcement was varied for each test, i.e., 0.07%, 0.9%, 0.1% and 1.4% in tests 1, 2, 3 and 4, respectively. This was done to evaluate the influence of vertical steel on the degree of restraint [11]. Moreover, the wall thickness was 300 mm in test 1 and 2 but was reduced to 200 mm in tests 3 and 4. The heights of the base and the wall were, however, kept constant during all the tests as 300 mm and 800 mm, respectively. Thus, in the first two tests the walls had a larger cross-sectional area (240000 mm^2) than that of the base slab (204000 mm^2) whereas in the last two tests, the cross-sectional area of the walls (160000 mm^2) was less than that of the base slab (204000 mm^2). This enabled the influence of relative cross-sectional areas on the degree of restraint to be established. The geometric and reinforcement details of the tested specimens are given in Fig. 2 and a pictorial illustration of test 4 is given in Fig. 3.

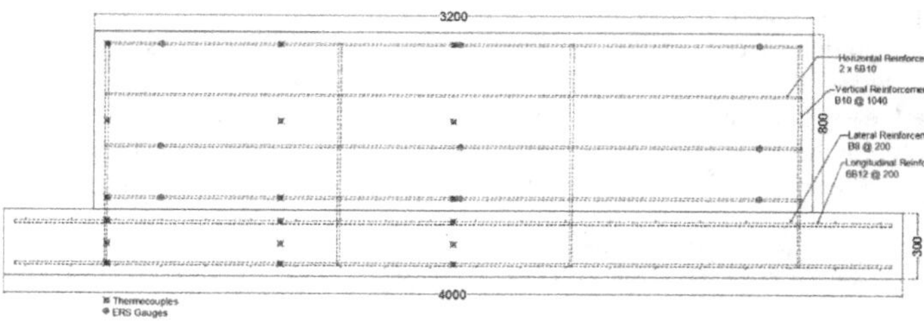

(a)

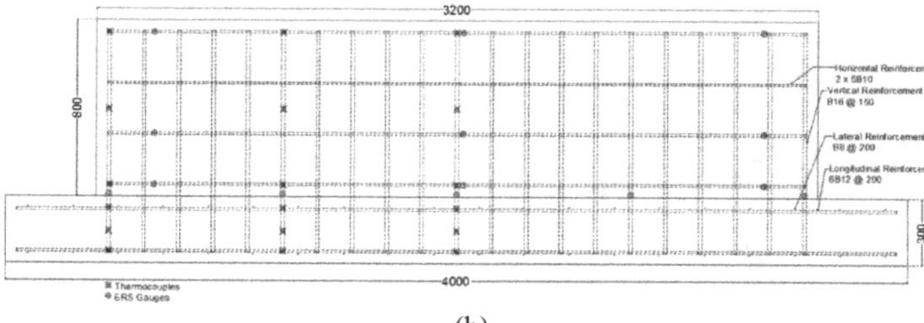

(b)

Fig. 2: Geometric and reinforcement details of tested walls: (a) Test 1 & 3; (b) Test 2 & 4 (all dimensions in mm)

Fig. 3: Illustration of the experimental work setting

Temperature development in both the wall and slab was monitored using welded tip Type K thermocouples. The temperature profiles recorded during each test are presented in Fig. 4. The vertical dotted lines in Fig. 4(b) depict the time that the formwork was removed. A sharp decline in temperature can be seen in all tests on removal of the formwork. Temperatures within the concrete dropped to ambient values after 72 to 80 hours from initial casting and the temperature drop remained gradual except at the time of formwork removal. The maximum temperatures observed in tests 1, 2, 3 and 4 were 55.4°C, 50.5°C, 48.8°C and 42.2°C, respectively; these values occurred near the top of the wall along the wall centerline. Temperatures observed in the wall close to the joint between the wall and the base were significantly lower than those at the higher locations in the wall due to flow f heat into adjoining base slab. The temperature drop (T_1) from the peak to the ambient temperature was 33.5°C in test 1, 33.3°C in test 2, 25.4°C in test 3 and 25°C in test 4 which compare reasonably well with the values predicted by CIRIA C766 [8].

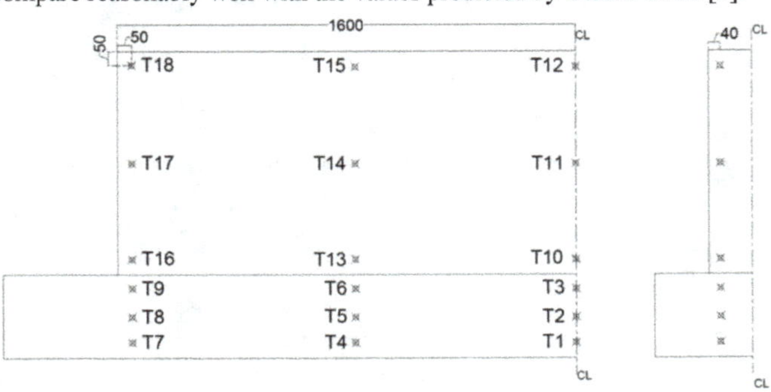

(a)

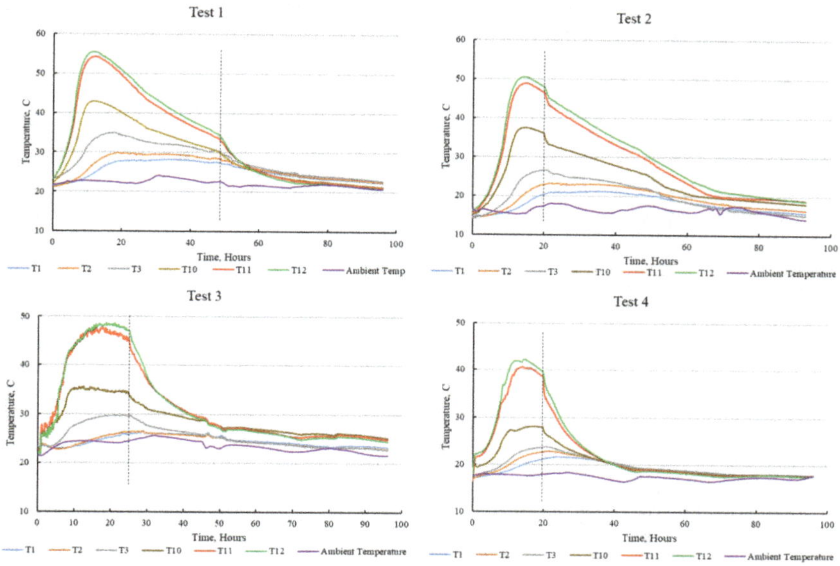

(b)

Fig. 4: Temperature profile observed in base slab and wall: (a) location and naming of thermocouples (all dimensions in mm); (b) temperature profile at mid-section, for different tests.

The strain occurring at the surface of the concrete wall and base was monitored using demountable mechanical gauges (DEMECs). In terms of edge restraint, the middle of the wall was considered to be the most critical section and as such, a 150 mm DEMEC gauge was used in the central one third of the walls; and a 400 mm gauge length was used towards the ends. Strain monitoring on the walls commenced within two hours after removal of the formwork. The free strain of the base and wall elements was equivalent to the summation of the unrestrained shrinkage and thermal strain of the concrete. The unrestrained shrinkage for each batch of concrete was obtained from small prisms, cast from the concrete used to make the base and wall elements and which were monitored throughout the duration of the tests. A comparison between the predicted and measured shrinkage revealed that the experimentally obtained shrinkage profiles from the prisms were most accurately predicted using the fib Model code 2010 [13]. Thermal strain was calculated using the temperature profiles (Fig.4) and a coefficient of thermal expansion of concrete equal to 10 microstrain per degrees C [13, 14]. By subtracting the measured strains from the free or unrestrained strain, the restrained strain could be calculated. The degree of restraint was then determined as the ratio of restrained to free strain. The restraint values obtained during each test were compared to those calculated using the methods available in both ACI and the Eurocode as well as to those worked out by Schleeh [15]. The restraint profiles obtained at the 7th and last day of each test are given in Fig. 5. It is evident that the degree of restraint calculated using the various methods is very different to the values which were obtained during this experimental investigation. The steel reinforcement continuing from restrained to restraining member has been identified as a key influencing factor for this variation. No account of this important factor is made in the current

Materials Research Proceedings 31 (2023) 622-633 https://doi.org/10.21741/9781644902592-64

guidance for determination of the degree of restraint. This suggests the need to develop a more accurate methodology for the estimation of the degree of restraint.

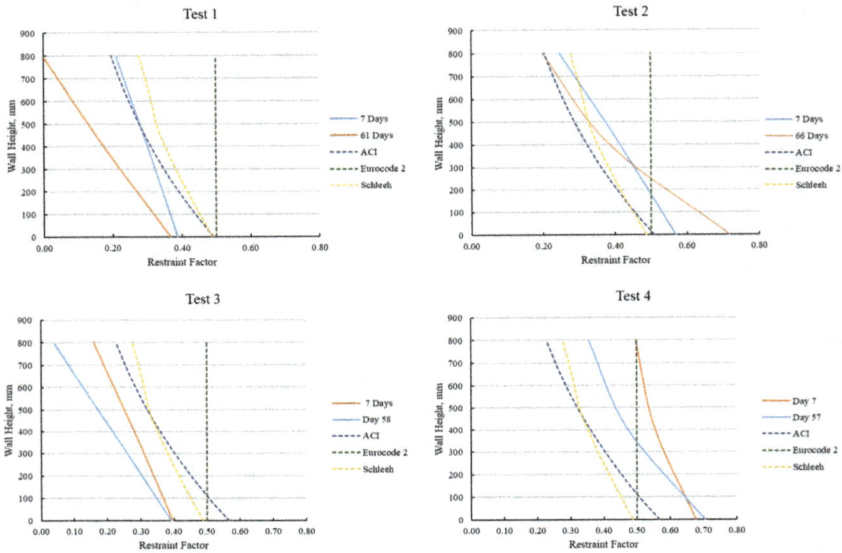

Fig. 5: Edge restraint profiles and their comparison with available estimation methods.

Finite Element (FE) Analysis

As part of this research, the FE models of the experimentally tested walls developed using MIDAS FEA (a commercially available FE based software), were prepared, and validated (7) prior to extending the parameter space. The FE study was undertaken by performing 3D nonlinear and linear analysis using MIDAS FEA. The material properties and the thermal strain and shrinkage obtained from the experimental study were used as input data for the models. The FEA study investigated walls with aspect ratios 1, 2, 4, 6 and 12 subjected to external edge restraint. The dimensions of the restraining concrete members and the restrained walls are presented collectively in Table 1. Taking advantage of the symmetry of the specimens, the reinforcement, loading and boundary conditions, a quarter of the wall and base elements were modelled.

Table 1: Geometric dimensions of restrained and restraining members.

Dimensions [m]	Restraining Slab	Restrained Wall
Length	15	12
Thickness	1	0.4
Height	0.5	1, 2, 3, 6, 12

The material properties of the concrete and steel obtained during the experimental study, given in Table 2, were used and concrete was modelled using the Total strain crack model [16]. Since the restrained contraction induces tension and ultimately results in cracking of the concrete, modelling of the tensile behaviour of concrete accurately is of great significance for the ensuing analysis. In this case, the nonlinear behaviour of concrete in tension was modelled using the Hordijk model [17]. Concrete was modelled using 3D solid elements. After performing a mesh sensitivity analysis, and carefully considering the balance between computing effort and the effect of mesh size on the results, mesh size for the restraining elements was selected as 100 mm. The reinforcement was modelled as fully embedded bar in a solid hence the stiffness of the reinforcement elements is added to the surrounding concrete elements [16]. The wall reinforcement detail was as follows: 12 mm bars spaced at 200 mm in the horizontal direction, along both faces; 16 mm bars spaced at 150 mm in the vertical direction. The area of steel provided was, therefore, 1130 mm^2/meter and 2680 mm^2/meter in the horizontal and vertical directions, respectively.

Table 2: Material properties of concrete and steel

Material Properties	Restraining Members	Restrained Wall
Compressive Strength, f_{cm} [Mpa]	43.3	31.0
Tensile Strength, f_{ctm} [Mpa]	3.4	2.4
Modulus of Elasticity, E_c [Mpa]	31000	24500
Coefficient of thermal expansion, α [$\mu\epsilon$ / ℃]	10	10
Poisson Ratio, γ	0.18	0.16
Fracture Energy, G_f [N/mm]	0.143	0.135
Thermal Conductivity, k [W/m]	1.76	2.03
Specific Heat, c [KJ/Kg C]	0.95	0.74
Steel Yield Strength, f_y [Mpa]	460	460

The end nodes of the base slab were constrained for translation along all three directions. Respective constraints were also applied along the symmetrical faces of the models for each case. Due to the limitations of the software, shrinkage could not be applied simultaneously with the thermal strain, therefore, the thermal strain and shrinkage was simulated by applying an equivalent thermal contraction, which represented the thermal strain and shrinkage, to the wall elements. It was noticed from the experimental investigation and also from the findings of other researchers [9, 18], that the thermal strain in the wall is lower adjacent to the restraining element and at the free edges of the member. Accordingly, the imposed thermal strain applied to the models was varied along the length and height of the wall. The nonlinear analysis was performed using the Newton Raphson solution technique. Force and Energy Norms for the convergence criteria, with a tolerance level of 0.001 for both norms, were specified in the iterative solution.

Results and Discussion

From the analysis, it was identified that the tensile stress in the restrained member increases with the increase in equivalent thermal contraction, until a crack occurs. On occurrence of cracking, the stress in the vicinity of the crack drops and contraction further increases. The stress in the remaining member, away from the crack, continues to increase until a second crack forms; again, this second crack relieves the stress in its vicinity. This pattern of crack formation and localized stress relief is in line with the guidance provided by Bamforth [8] in CIRIA C766. Observation of the stress development in the members also indicated that as the wall height increased, less stress developed close to the free edges, mainly because of the absence of restraint in those parts. This phenomenon prevented the propagation of the cracks to the top of the wall and to the free ends, however, this was only noticed in walls with an aspect ratio of less than 2.

The cracking sequence and patterns for all wall aspect ratios were obtained from the strain profiles. The variation of strains in the wall clearly indicates the initiation and propagation of the cracks. The sequence in which the cracks appeared was noted and the cracks were accordingly numbered (see Figs. 6 to 10). One half of the wall was modelled due to symmetry along length. The location of the maximum crack width, which was obtained from the analysis, is also marked on the figures with a red circle.

L/H - 12: First crack appeared close to the centerline, indicated by the circled 1, of the wall and the cracking pattern indicates that all of the cracks propagated vertically and reached the top of the wall (Fig. 6). The location of the maximum crack width in the wall occurred at 900mm (8% of the wall length) above the base.

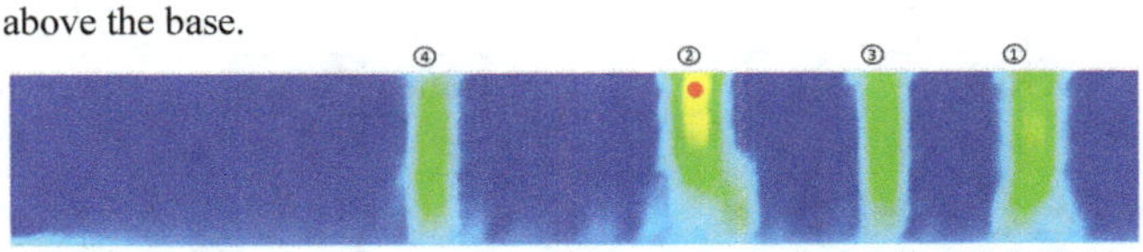

Fig. 6: Cracking patterns and marked sequences for wall with L/H 12

L/H - 6: In the case of wall with aspect ratio 6, the first crack appeared close to the center of the wall and most of the cracks reached to the top of the wall (Fig. 8). The number of cracks is the same as found in the edge restrained wall with an aspect ratio of 12. All cracks initiated at the base and propagated upwards, mostly remaining vertical. No significant cracking was seen within the zone of no cracking specified in the Annexure L of BS EN 1992-3. The location of the maximum crack width occurred close to the wall top, 1500mm (13% of the wall length) above the base.

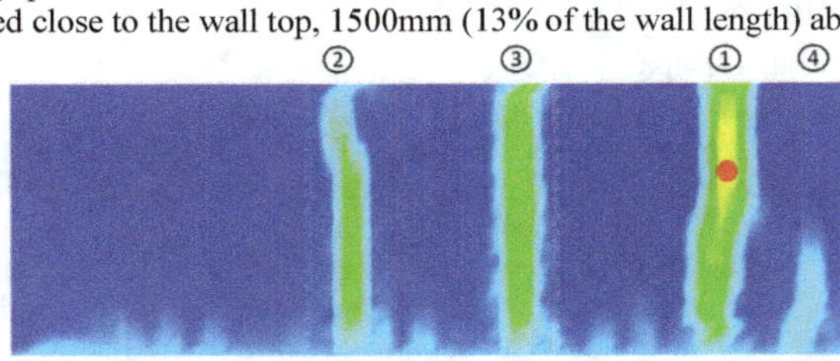

Fig. 7: Cracking patterns and marked sequences for wall with L/H 6

L/H - 4: The first crack still appeared close to the center of the wall and reached the full height of the wall (Fig. 8). None of the other cracks appearing subsequently reached the top of the wall. The cracks appearing in the outer one third of the wall length were inclined towards the free ends and prominent cracking was seen close to the end of the wall, within 2.4m of the end which is contrary to the specification of BS EN 1992-3. In this case, the maximum crack width occurred 1200mm (10% of the wall length) above the base.

Materials Research Proceedings 31 (2023) 622-633 https://doi.org/10.21741/9781644902592-64

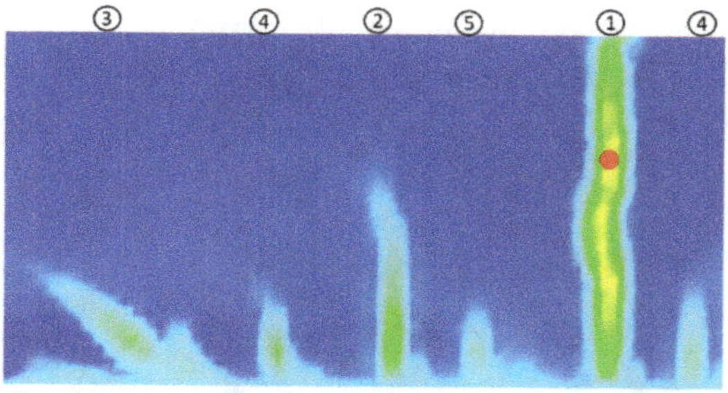

Fig. 8: Cracking patterns and marked sequences for wall with L/H 4

L/H - 2: The results suggest that when the aspect ratio of wall is reduced to 2, the cracking patterns are considerably different to their equivalents in higher aspect ratio walls. The first crack in the edge restrained wall (Fig. 9) appeared close to the center and reached the maximum height among all other cracks at this location, however, no crack reached the top of the wall. The cracks were inclined as they propagated and a noticeable amount of cracking close to the free edge was observed. The maximum crack width occurred 800 mm (7% of the wall length) above the base.

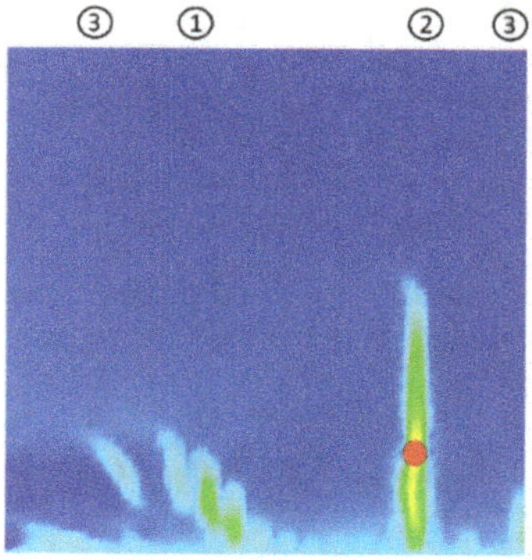

Fig. 9: Cracking patterns and marked sequences for wall with L/H 2

L/H - 1: The cracking behaviour of these walls again continued to deviate from the behaviour of the walls with aspect ratios greater than 2. The first crack, in Fig. 10, appeared at the center of the wall and propagated to approximately one third of the wall height. Subsequent cracks were less in height and did not propagate vertically but were inclined at different angles. The maximum crack width appeared 1800 mm (15% of the wall length) above the base. Although a number of cracks appeared, none of the cracks spanned the entire height of the wall.

Fig. 10: Cracking patterns and marked sequences for wall with L/H 1

As indicated above, the initiation and propagation of the cracks in the walls, as well as the location of the maximum crack width varies. It was also noticed that when a new crack was formed, the width of the existing cracks, in some instances, often decreased. The maximum crack widths for each wall examined in this study are summarized in Table 3. Crack width for the considered wall dimensions and steel reinforcement was also calculated according to BS EN 1992-3 and was found to be 0.249 mm. In all cases the maximum crack width was found to be significantly higher than the calculated value. As explained above, the location of the maximum crack width in the case of edge restraint lies at a height above the base, which in the considered cases varied from 7-15% of the wall length. This slightly differs from the findings of other researchers [6, 8] who found the location of maximum crack width to be at a height equal to 10-12% of the wall length.

Table 3: Maximum crack width for each aspect ratio

Aspect Ratio	**Maximum Crack Width (MCW) [mm]**	**Location of MCW above base [mm]**
12	0.516	900
6	0.672	1800
4	0.493	1400
2	0.480	900
1	0.689	1900

Conclusions

Results from the experimental and FE based investigation of the influence of edge restraint on the behaviour of reinforced concrete walls have been presented and discussed. The location of cracks in the region of the free ends, particularly in the case of edge restraint alone, indicates that the 'No crack zone' proposed in the Eurocode is not always correct. It is observed that in walls with aspect ratios of less than 6, significant cracking can occur close to the free end. Formation of a crack results in localized relief in stress and formation of another crack in the member at a point where the stress level remained unaffected. In higher walls, the cracks due to edge restraint do not reach the full height of the wall. With an increase in the wall height (i.e. decrease in aspect ratio), the number of cracks appearing in the wall decreases.

Acknowledgements
The authors kindly acknowledge financial support from the Engineering and Physical Sciences Research Council (EPSRC) UK, under grants EP/T004142/1 and EP/T004185/1 titled Understanding the cracking behaviour of reinforced concrete elements subjected to the restraint of imposed strains.

References

[1] E. Evans and B. Hughes, "SHRINKAGE AND THERMAL CRACKING IN A REINFORCED CONCRETE RETAINING WALL," in ICE Proceedings, 1968, vol. 39, no. 1, pp. 111-125: Thomas Telford. https://doi.org/10.1680/iicep.1968.8172

[2] J. P. Forth, A.J. Martin, Design of Liquid Retaining Concrete Structures, Third ed. UK: Whittles Publishing, 2014, p. 175.

[3] H. Stoffers, Cracking due to shrinkage and temperature variation in walls. Stevin Laboratory, Department of Civil Engineering, Delft University of Technology and IBBC Institute TNO for Building Materials and Building Structures, 1978.

[4] G. Kheder and R. Al Rawi, "Control of cracking due to volume change in base-restrained concrete members," ACI Structural Journal, vol. 87, no. 4, 1990. https://doi.org/10.14359/2747

[5] BS EN 1992-3, Eurocode 2: Design of Concrete Structures: Part 3: Liquid retaining and containment structures. British Standards Institution, 2006.

[6] G. Kheder, R. Al-Rawi, and J. Al-Dhahi, "A study of the behaviour of volume change cracking in base restrained concrete walls," Materials and Structures, vol. 27, no. 7, pp. 383-392, 1994. https://doi.org/10.1007/BF02473441

[7] ACI Committee 207, "Effect of restraint, volume change, and reinforcement on cracking of mass concrete," ACI Materials Journal, vol. 87, no. 3, 2007.

[8] P. Bamforth, Control of cracking caused by restrained deformation in concrete. CIRIA, 2018.

[9] M. Micallef, Vollum, RL, Izzuddin, BA, "Crack development in transverse loaded base-restrained reinforced concrete walls," Engineering Structures, vol. 143, pp. 522-539, 2017. https://doi.org/10.1016/j.engstruct.2017.04.035

[10] M. Micallef, R. L. Vollum, and B. A. Izzuddin, "Cracking in walls with combined base and end restraint," Magazine of Concrete Research, vol. 69, no. 22, pp. 1170-1188, 2017. https://doi.org/10.1680/jmacr.17.00026

[11] M. K. Shehzad, J. P. Forth, and A. Bradshaw, "Imposed Loading Effects on Reinforced Concrete Walls Restrained at Their Base," Proceedings of the Institution of Civil Engineers-Structures and Buildings, pp. 1-29, 2018.

[12] J. P. Forth, "Chapter 7 Edge Restraint, The Concrete Society Technical Report No 67," in "Accommodating Movement in Concrete Structures," 2008.

[13] Fib model code for concrete structures 2010, 2010.

[14] BS EN 1992-1-1, Eurocode 2: Design of Concrete Structures: Part 1-1: General Rules and Rules for Buildings. British Standards Institution, 2004.

[15] W. Schleeh, "Imposed stresses in the walls one-side restrained," Beton- Und Stahlbetonbau, vol. 57, no. 3, pp. 64-72, 1962.

[16] F. Midas, MIDAS FEA: Analysis and Algorithm (Advanced Nonlinear and Detailed Analysis System). MIDAS IT, 1989.

[17] D. A. Hordijk, Local approach to fatigue of concrete. TU Delft, Delft University of Technology, 1991.

[18] B. Klemczak and A. Knoppik-Wróbel, "Analysis of Early-Age Thermal and Shrinkage Stresses in Reinforced Concrete Walls (with Appendix)," ACI Structural Journal, vol. 111, no. 2, 2014. https://doi.org/10.14359/51686523

Passive vibration damping of a plate interacting with a flowing fluid using shunted piezoelectric element

Sergey Lekomtsev[a] *, Valerii Matveenko[b] and Alexander Senin[c]

Institute of Continuous Media Mechanics, Ural Branch of the Russian Academy of Sciences, 1, Acad. Korolev street, Perm, 614013, Russian Federation

[a]lekomtsev@icmm.ru, [b]mvp@icmm.ru, [c]senin.a@icmm.ru

Keywords: Vibrations Damping, Plate, Piezoelectric Elements, Electric Circuit, Flowing Fluid, Finite Element Method

Abstract. This paper considers a thin plate with a single piezoceramic element located on the outer surface of the structure and connected to a series electric circuit. The dependence of complex eigenvalues of the electromechanical system on the resistance and inductance of the electric circuit is analyzed to select their optimal values for suppressing the resonant vibrations of the plate interacting with a flowing fluid. The numerical studies show that in contrast to the known analytical expression, these values lead to a smaller change in the frequency spectrum of the original system and provide more effective damping of vibrations in terms of maximum rate of vibration decay. The actual decrease in the amplitude of vibrations is demonstrated by the frequency response curves.

Introduction

The technique of passive damping of structure vibrations by means of attaching a piezoelectric element to it and connecting its electroded surfaces to an external electric circuit with impedance has been known long enough in various fields of advanced technologies [1, 2]. At present, this approach has found wide practical application due to the possibility of controlling the dynamic behavior of objects located in hard-to-reach places, limited space or under water. By selecting the parameters for different elements of the electric circuit, one can achieve a considerable decrease in the amplitude of the resonance or an increase in the rate of damping of free vibrations [2–9]. A comprehensive review of publications devoted to passive damping is presented in works [10–12] and monograph [13]. Worthy of mention are articles [14–21], which consider the problems of passive suppression of noise produced by underwater objects or vibrations of structures interacting with a fluid. In these papers, with the exception of [20], the question of selecting optimal parameters of the electric circuit, providing the most effective damping of a given vibration mode of the structure, is not addressed directly. In most cases, such selection is performed using the analytical expressions proposed in [2]. Of the above publications, only [19] investigates a structure that interacts with a flowing fluid. This paper considers a lifting surface in the form of a plate, which is a simplified analogue to a submerged wing, stabilizer, rudder, etc. The experiments showed that the dependence of its natural frequencies on the Reynolds number is very weak. This fact led to the assumption about the immobility of the fluid and allowed using the derived formulas [2] to determine parameters of an electric *RL*-circuit connected to piezoelectric elements. Although the results obtained have demonstrated the effectiveness of the proposed method, its applicability in the case of dependence of natural frequencies of plate vibrations on the velocity of the fluid flow or in the presence of a hydrodynamic damping mechanism in the system is not so obvious. In this connection, it is important to develop an algorithm for selecting parameters of the passive electric circuit, which will provide the most effective damping of resonant vibrations of the plate interacting with a flowing fluid.

Governing equations and finite element formulation

We consider a thin rectangular plate with P piezoelectric elements attached to its upper surface and polarized in the direction of the z-axis. The lower surface of the structure interacts with a fluid, which flows with velocity U in the direction of the x-axis (Fig. 1a). Each k-th piezoelectric element ($k = 1, 2, \ldots, P$) is connected through the electroded surfaces to its own electrical circuit, consisting of the series-connected resistor R_k and the inductance coil L_k.

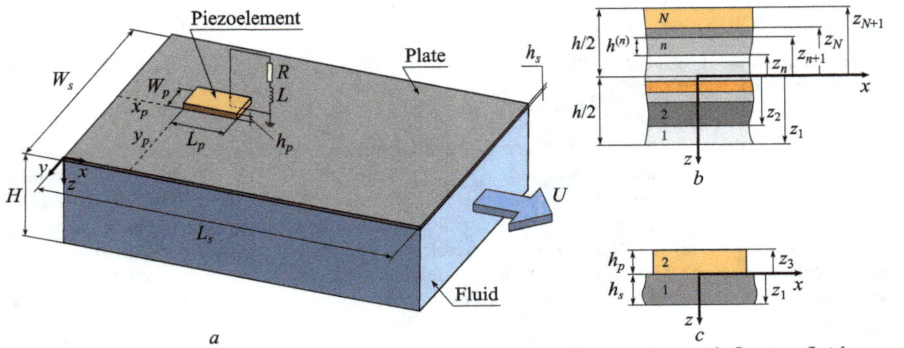

Fig. 1. Rectangular plate with a piezoelectric element and interacting with flowing fluid

An electroelastic body of small thickness is modeled using the classical laminated plate theory [22]. Its constitutive relations are based on the equations of the linear theory of piezoelasticity [23], which are written for the case of a plane stress state [22]. Based on the assumptions and hypotheses presented in [24], they are formulated for each n-th layer ($n = 1, 2, \ldots, N$) in the following form [25, 26]:

$$\begin{Bmatrix}\sigma_{xx}\\ \sigma_{yy}\\ \sigma_{xy}\end{Bmatrix}^{(n)} = \begin{bmatrix}\tilde{Q}_{11} & \tilde{Q}_{12} & 0\\ \tilde{Q}_{21} & \tilde{Q}_{22} & 0\\ 0 & 0 & \tilde{Q}_{66}\end{bmatrix}^{(n)} \begin{Bmatrix}\varepsilon_{xx}\\ \varepsilon_{yy}\\ \varepsilon_{xy}\end{Bmatrix} - \begin{bmatrix}0 & 0 & \tilde{e}_{31}\\ 0 & 0 & \tilde{e}_{32}\\ 0 & 0 & 0\end{bmatrix}^{(n)} \begin{Bmatrix}0\\ 0\\ E_z\end{Bmatrix}^{(n)} \quad \text{or } \boldsymbol{\sigma}^{(n)} = \tilde{\mathbf{c}}^{(n)}\boldsymbol{\varepsilon} - \tilde{\mathbf{e}}^{(n)}\mathbf{E}^{(n)}, \tag{1}$$

$$\begin{Bmatrix}0\\ 0\\ D_z\end{Bmatrix}^{(n)} = \begin{bmatrix}0 & 0 & 0\\ 0 & 0 & 0\\ \tilde{e}_{31} & \tilde{e}_{32} & 0\end{bmatrix}^{(n)} \begin{Bmatrix}\varepsilon_{xx}\\ \varepsilon_{yy}\\ \varepsilon_{xy}\end{Bmatrix} + \begin{bmatrix}0 & 0 & 0\\ 0 & 0 & 0\\ 0 & 0 & \check{\mathbf{o}}_{33}\end{bmatrix}^{(n)} \begin{Bmatrix}0\\ 0\\ E_z\end{Bmatrix}^{(n)} \quad \text{or } \mathbf{D}^{(n)} = (\tilde{\mathbf{e}}^{(n)})^{\mathrm{T}}\boldsymbol{\varepsilon} + \check{\mathbf{o}}^{(n)}\mathbf{E}^{(n)}. \tag{2}$$

Here: $\boldsymbol{\sigma}$ and $\boldsymbol{\varepsilon}$ are the vectors containing components of the stress and small strain tensors; $\mathbf{E}$, $\mathbf{D}$ are the vectors of electric field intensity and electric displacements; $\tilde{\mathbf{c}}$ is the reduced stiffness matrix evaluated at a constant electric field; $\tilde{\mathbf{e}}$ is the matrix of reduced piezoelectric constants; $\check{\mathbf{o}}$ is the matrix of the reduced dielectric constants evaluated at a constant mechanical strain. When the piezoelectric effect in n-th layer is absent, the part containing matrix $\check{\mathbf{o}}$ should be excluded.

The elements of the matrices $\tilde{\mathbf{c}}^{(n)}$, $\tilde{\mathbf{e}}^{(n)}$ and $\check{\mathbf{o}}^{(n)}$ are calculated as follows [27, 28]:

$$\tilde{Q}_{ij}^{(n)} = Q_{ij}^{(n)} - \frac{Q_{i3}^{(n)}Q_{j3}^{(n)}}{Q_{33}^{(n)}}, \quad \tilde{Q}_{66}^{(n)} = Q_{66}^{(n)}, \quad \tilde{e}_{3j}^{(n)} = e_{3j}^{(n)} - \frac{e_{33}Q_{j3}^{(n)}}{Q_{33}^{(n)}}, \quad \check{o}_{33}^{(n)} = \check{o}_{33}^{(n)} + \frac{e_{33}^{(n)}e_{33}^{(n)}}{Q_{33}^{(n)}}, \quad i, j = \overline{1,2}. \tag{3}$$

The coefficients $Q_{ij}^{(n)}$ can be written in terms of the engineering constants of the n-th layer:

$$Q_{11}^{(n)} = \frac{E_1^{(n)}}{1-\nu_{12}^{(n)}\nu_{21}^{(n)}},\ Q_{12}^{(n)} = \frac{\nu_{12}^{(n)}E_2^{(n)}}{1-\nu_{12}^{(n)}\nu_{21}^{(n)}} = \frac{\nu_{21}^{(n)}E_1^{(n)}}{1-\nu_{12}^{(n)}\nu_{21}^{(n)}},\ Q_{22}^{(n)} = \frac{E_2^{(n)}}{1-\nu_{12}^{(n)}\nu_{21}^{(n)}},\ Q_{66}^{(n)} = G_{12}^{(n)}, \tag{4}$$

where E_1 and E_2 are the Young moduli of the material in the x- and y- directions, ν_{ij} are the Poisson ratios, G_{12} is the shear module in the xy plane.

The electric field intensity vector is normal to the electrodes and its intensity in the n-th piezoelectric layer is uniform [24]:

$$E_z^{(n)} = -\frac{\psi^{(n)}}{h^{(n)}}, \tag{5}$$

where $\psi^{(n)}$ is the potential difference between the top and bottom electrodes of the n-th piezoelectric layer, $h^{(n)}$ is the thickness of the n-th layer.

The generalized vector $\boldsymbol{\varepsilon}$ containing middle surface strains and curvatures is written as [22]:

$$\boldsymbol{\varepsilon} = \left\{\varepsilon_{xx}^0, \varepsilon_{yy}^0, \gamma_{xy}^0, \varepsilon_{xx}^1, \varepsilon_{yy}^1, \gamma_{xy}^1,\right\}^{\mathrm{T}} = \left\{\frac{\partial u}{\partial x}, \frac{\partial v}{\partial y}, \frac{\partial u}{\partial y}+\frac{\partial v}{\partial x}, -\frac{\partial^2 w}{\partial x^2}, -\frac{\partial^2 w}{\partial y^2}, -2\frac{\partial^2 w}{\partial x\partial y}\right\}^{\mathrm{T}}, \tag{6}$$

where u, v, w are the displacements of the points on the middle surface of the plate in the direction of the corresponding axes of the Cartesian coordinates x, y, z.

Using expressions (1), (5) the vector of the forces and moments $\mathbf{t}$ is defined as:

$$\mathbf{t} = \left\{N_{xx}, N_{yy}, N_{xy}, M_{xx}, M_{yy}, M_{xy}\right\}^{\mathrm{T}} = \mathbf{S}\boldsymbol{\varepsilon} - \mathbf{G}\mathbf{E}_z, \tag{7}$$

$$\mathbf{S} = \begin{bmatrix} A_{11} & A_{12} & A_{16} & B_{11} & B_{12} & B_{16} \\ A_{12} & A_{22} & A_{26} & B_{12} & B_{22} & B_{26} \\ A_{16} & A_{26} & A_{66} & B_{16} & B_{26} & B_{66} \\ B_{11} & B_{12} & B_{16} & D_{11} & D_{12} & D_{16} \\ B_{12} & B_{22} & B_{26} & D_{12} & D_{22} & D_{26} \\ B_{16} & B_{26} & B_{66} & D_{16} & D_{26} & D_{66} \end{bmatrix},\quad \mathbf{G} = \begin{bmatrix} g_{11} & g_{12} & \dots & g_{1n} & \dots & g_{1N} \\ g_{21} & g_{22} & \dots & g_{2n} & \dots & g_{2N} \\ 0 & 0 & \dots & 0 & \dots & 0 \\ g'_{11} & g'_{12} & \dots & g'_{1n} & \dots & g'_{1N} \\ g'_{21} & g'_{22} & \dots & g'_{2n} & \dots & g'_{2N} \\ 0 & 0 & \dots & 0 & \dots & 0 \end{bmatrix},$$

where the coefficients entering into the matrices $\mathbf{S}$ and $\mathbf{G}$ are calculated as follows:

$$A_{ij} = \sum_{n=1}^{N}\tilde{Q}_{ij}^{(n)}\left(z_n - z_{n+1}\right),\ B_{ij} = \frac{1}{2}\sum_{n=1}^{N}\tilde{Q}_{ij}^{(n)}\left(z_n^2 - z_{n+1}^2\right),\ C_{ij} = \frac{1}{3}\sum_{n=1}^{N}\tilde{Q}_{ij}^{(n)}\left(z_n^3 - z_{n+1}^3\right),\ i = 1,2,6, \tag{8}$$

$$g_{in} = \left(z_n - z_{n+1}\right)\tilde{e}_{3i}^{(n)} = h^{(n)}\tilde{e}_{3i}^{(n)},\quad g'_{in} = \frac{1}{2}\left(z_n^2 - z_{n+1}^2\right)\tilde{e}_{3i}^{(n)},\quad i = 1,2.$$

According to equation (5), the components E_z of the electric field intensity vector $\mathbf{E}$ are determined in each n-th layer of the material using the following expressions:

$$\mathbf{E}_z = -\mathbf{B}_\psi \boldsymbol{\psi} = -\begin{bmatrix} 1/h^{(1)} & 0 & 0 & 0 \\ 0 & 1/h^{(2)} & 0 & 0 \\ 0 & 0 & \ddots & 0 \\ 0 & 0 & 0 & 1/h^{(N)} \end{bmatrix} \begin{Bmatrix} \psi^{(1)} \\ \psi^{(2)} \\ \vdots \\ \psi^{(N)} \end{Bmatrix}. \tag{9}$$

The equation (2) can also be rewritten taking into account (5):

$$\mathbf{D}_z = \mathbf{G}^{\mathrm{T}}\boldsymbol{\varepsilon} + \mathbf{H}\mathbf{E}_z, \quad \mathbf{H} = \mathrm{diag}\left(\breve{\epsilon}_{33}^{(1)}, \breve{\epsilon}_{33}^{(2)}, \ldots, \breve{\epsilon}_{33}^{(n)}, \ldots, \breve{\epsilon}_{33}^{(N)}\right). \tag{10}$$

The above relations allow us to describe the behavior of a monolithic composite plate, which in the general case has N orthotropic elastic and piezoelectric layers (Fig. 1*b*). When modeling a structure with piezoelectric elements on its upper surface it is assumed that the host plate consists of M elastic orthotropic layers, and because of the presence of the piezoelectric element one more piezoelectric layer is added to already existing ones (Fig. 1*c*). In this case, expression (7) goes to:

$$\begin{aligned} &\mathbf{T} = \mathbf{S}_s\boldsymbol{\varepsilon}, && n = \overline{1, M}, \\ &\mathbf{T} = \mathbf{S}_p\boldsymbol{\varepsilon} - \mathbf{G}\mathbf{E}_z, && n = N, \end{aligned} \tag{11}$$

and relation (10) is written only for $n = N$.

The piezoelectric element is a single layer of transversely isotropic material of thickness h_p, which has only two electrical unknown quantities: the potential difference ψ (voltage) and the electrical charge q. The matrices entering into expressions (9) and (10) take the following form:

$$\mathbf{G} = \{g_{11}, g_{21}, 0, g'_{11}, g'_{21}, 0\}^{\mathrm{T}}, \quad \mathbf{B}_\psi = 1/h_p, \quad \mathbf{H} = \breve{\epsilon}_{33}, \tag{12}$$

The mathematical formulation of the dynamics problem for an electroelastic plate is based on the variational principle of virtual displacements, which is written in the matrix form taking into account the work of inertia and external surface forces [24, 29]:

$$\sum_{k=1}^{P} \int_{S_{Pk}} \left(\delta\boldsymbol{\varepsilon}^{\mathrm{T}}\mathbf{S}_{p_k}\boldsymbol{\varepsilon} + \delta\boldsymbol{\varepsilon}^{\mathrm{T}}\mathbf{G}\frac{\psi_k}{h_{P_k}} + \frac{\delta\psi_k}{h_{P_k}}\mathbf{G}^{\mathrm{T}}\boldsymbol{\varepsilon} \right) \mathrm{d}S - \delta\boldsymbol{\psi}^{\mathrm{T}}\mathbf{C}_p\boldsymbol{\psi} + \sum_{k=1}^{P} \int_{S_{Pk}} \delta\mathbf{d}^{\mathrm{T}}\mathbf{J}_{p_k}\ddot{\mathbf{d}}\,\mathrm{d}S + \delta\boldsymbol{\psi}^{\mathrm{T}}\mathbf{q} + \\ + \int_{S_s} \delta\boldsymbol{\varepsilon}^{\mathrm{T}}\mathbf{S}_s\boldsymbol{\varepsilon}\,\mathrm{d}S + \int_{S_s} \delta\mathbf{d}^{\mathrm{T}}\mathbf{J}_s\ddot{\mathbf{d}}\,\mathrm{d}S - \int_{S_f} \delta w p\,\mathrm{d}S = \int_{S_\sigma} \delta\mathbf{d}^{\mathrm{T}}\mathbf{f}\,\mathrm{d}S \tag{13}$$

and is supplemented by the equation for a series *RL*-circuit connecting the voltage ψ_k at each piezoelectric element to the corresponding electric charge q_k [24]:

$$\delta\mathbf{q}^{\mathrm{T}}\boldsymbol{\psi} - \delta\mathbf{q}^{\mathrm{T}}\mathbf{R}\dot{\mathbf{q}} - \delta\mathbf{q}^{\mathrm{T}}\mathbf{L}\ddot{\mathbf{q}} = 0. \tag{14}$$

Hereinafter, the subindexes "*s*", "*p*" and "*f*" denote the belongingness of the quantity to the structure, piezoelectric element or fluid; $\mathbf{d} = \{u, v, w, \theta_x, \theta_y\}^{\mathrm{T}}$ is the generalized vector of plate displacements, including rotation angles $\theta_x = \partial w/\partial x$ and $\theta_y = \partial w/\partial y$; $\boldsymbol{\psi} = \{\psi_1, \psi_2, \ldots, \psi_P\}^{\mathrm{T}}$ and $\mathbf{q} = \{q_1, q_2, \ldots, q_P\}^{\mathrm{T}}$ are the vectors of potential differences and electrical charges; S_s is the whole plate area; S_f and S_σ are the regions of the plate, which interact with the flowing fluid and are under the static load $\mathbf{f}$, respectively; p is the hydrodynamic pressure of the fluid; $\mathbf{R} = \mathrm{diag}(R_1, R_2, \ldots, R_P)$

and $\mathbf{L} = \text{diag}(L_1, L_2, \ldots, L_P)$ are the diagonal matrices containing the values of resistance and inductance of the RL-circuits connected to each of the piezoelectric elements The matrices of the inertia $\mathbf{J}$ and capacitances $\mathbf{C}_p$ are defined using expressions:

$$\mathbf{J} = \begin{bmatrix} J_0 & 0 & 0 & -J_1 & 0 \\ 0 & J_0 & 0 & 0 & -J_1 \\ 0 & 0 & J_0 & 0 & 0 \\ -J_1 & 0 & 0 & J_2 & 0 \\ 0 & -J_1 & 0 & 0 & J_2 \end{bmatrix}, \quad \mathbf{C}_p = \begin{bmatrix} C_1 & 0 & 0 & 0 & 0 \\ 0 & C_2 & 0 & 0 & 0 \\ 0 & 0 & C_3 & 0 & 0 \\ 0 & 0 & 0 & \ddots & 0 \\ 0 & 0 & 0 & 0 & C_P \end{bmatrix}, \tag{15}$$

$$J_0 = \sum_{n=1}^{N} \rho^{(n)} \left(z_n - z_{n+1}\right), \quad J_1 = \frac{1}{2}\sum_{n=1}^{N} \rho^{(n)} \left(z_n^2 - z_{n+1}^2\right), \quad J_2 = \frac{1}{3}\sum_{n=1}^{N} \rho^{(n)} \left(z_n^3 - z_{n+1}^3\right), \quad C_k = \eth_{33}^{(k)} A_k / h_{p_k},$$

where $\rho^{(n)}$ is the density of the n-th layer of a composite material, C_k and A_k are the capacitance and the area of the k-th piezoelectric layer.

In the case of small perturbations the basic relations, describing the vortex-free dynamics of an ideal compressible fluid of volume V_f, are formulated in terms of the perturbation velocity potential φ [30–32]. The corresponding second-order differential equation written in the Cartesian coordinate system (x, y, z) associated with the elastic plate is transformed together with the impermeability condition and boundary conditions to the weak form using the Bubnov – Galerkin method [32]:

$$\int_{V_f} \nabla F_m \cdot \nabla\hat{\varphi}\, \mathrm{d}V + \int_{V_f} F_m \frac{1}{c^2}\frac{\partial^2\hat{\varphi}}{\partial t^2}\mathrm{d}V + \int_{V_f} F_m \frac{2U}{c^2}\frac{\partial^2\hat{\varphi}}{\partial t \partial x}\mathrm{d}V - \int_{V_f} F_m \frac{U^2}{c^2}\frac{\partial^2\hat{\varphi}}{\partial x^2}\mathrm{d}V - \tag{16}$$
$$-\int_{S_f} F_m \frac{\partial \hat{w}}{\partial t}\mathrm{d}S - \int_{S_f} F_m U \frac{\partial \hat{w}}{\partial x}\mathrm{d}S = 0, \quad m = \overline{1, m_f}.$$

Here: $\hat{\varphi}$ and $\hat{w}$ are the trial solutions for the velocity potential φ and normal displacements of the plate w; c is the speed of sound in a fluid; t is time; F_m and m_f are the basis functions and their number.

The hydrodynamic pressure p in equation (13) is calculated using linearized Bernoulli's formula

$$p = -\rho_f \left(\frac{\partial\varphi}{\partial t} + U\frac{\partial\varphi}{\partial x}\right), \tag{17}$$

where ρ_f is the density of fluid.

The following boundary conditions are used to solve equation (16):

$$\varphi = 0 \quad \text{for} \quad x = 0; \qquad \partial\varphi/\partial x = 0 \quad \text{for} \quad x = L_s. \tag{18}$$

The finite-element model of the system of equations (13), (14), (16) is constructed taking into account relations (6) – (12), (15) and (17). The perturbation velocity potential φ, the basis functions F_m, the electric potential difference ψ, the electric charge q, and the plate membrane displacements u and v are described using Lagrange bilinear shape functions. The bending displacements w, rotation angles θ_x and θ_y are approximated by nonconforming cubic Hermite polynomials [33]. Since the plate interacts with the fluid over its entire surface, each node of the single finite element contains six unknowns: u, v, w, θ_x, θ_y and φ. Each k-th piezoelectric element has only two unknowns: the potential difference ψ_k and the charge q_k. They are common to all its nodes. The discretization of the computational domains of the fluid and the plate is carried out using the spatial 8-node prismatic and flat quadrilateral finite elements, respectively. After implementing the known procedures of the finite element method, we obtain [33, 34]:

$$\mathbf{M}\ddot{\mathbf{u}} + \mathbf{C}\dot{\mathbf{u}} + \mathbf{K}\mathbf{u} + \mathbf{A}\mathbf{u} = \mathbf{r}, \tag{19}$$

$$\mathbf{M} = \begin{bmatrix} \mathbf{M}_f & 0 & 0 & 0 \\ 0 & \mathbf{M}_s & 0 & 0 \\ 0 & 0 & 0 & 0 \\ 0 & 0 & 0 & -\mathbf{L} \end{bmatrix}, \quad \mathbf{C} = \begin{bmatrix} \mathbf{C}_f & \mathbf{C}_{fs} & 0 & 0 \\ 0 & \mathbf{C}_{sf} & 0 & 0 \\ 0 & 0 & 0 & 0 \\ 0 & 0 & 0 & -\mathbf{R} \end{bmatrix},$$

$$\mathbf{K} = \begin{bmatrix} \mathbf{K}_f & 0 & 0 & 0 \\ 0 & \mathbf{K}_s & \mathbf{K}_{sp} & 0 \\ 0 & \mathbf{K}_{ps} & -\mathbf{K}_p & \mathbf{I} \\ 0 & 0 & \mathbf{I} & 0 \end{bmatrix}, \quad \mathbf{A} = \begin{bmatrix} \mathbf{A}_f & \mathbf{A}_{fs} & 0 & 0 \\ 0 & \mathbf{A}_{sf} & 0 & 0 \\ 0 & 0 & 0 & 0 \\ 0 & 0 & 0 & 0 \end{bmatrix},$$

where $\mathbf{u} = \{\varphi, \mathbf{d}, \psi, \mathbf{q}\}^{\mathrm{T}}$, $\mathbf{r} = \{0, \mathbf{f}, 0, 0\}^{\mathrm{T}}$ and typical finite element matrices are determined in a well-known manner:

$$\mathbf{M}_f^e = \int_{V_f} \frac{1}{c^2}\mathbf{F}^{\mathrm{T}}\mathbf{F}\mathrm{d}V, \quad \mathbf{M}_s^e = \int_{S_s} \mathbf{N}^{\mathrm{T}}\mathbf{J}_s\mathbf{N}\mathrm{d}S + \int_{S_{Pk}} \mathbf{N}^{\mathrm{T}}\mathbf{J}_{P_k}\mathbf{N}\mathrm{d}S, \quad \mathbf{C}_f^e = \int_{V_f} \frac{2U}{c^2}\frac{\partial \mathbf{F}^{\mathrm{T}}}{\partial x}\mathbf{F}\mathrm{d}V, \tag{20}$$

$$\mathbf{C}_{sf}^e = \int_{S_f} \rho_f \mathbf{N}_w^{\mathrm{T}}\mathbf{F}\mathrm{d}S, \quad \mathbf{C}_{fs}^e = -\int_{S_f} \mathbf{F}^{\mathrm{T}}\mathbf{N}_w\mathrm{d}S, \quad \mathbf{K}_f^e = \int_{V_f} (\nabla\mathbf{F})^{\mathrm{T}}\nabla\mathbf{F}\mathrm{d}V, \quad \mathbf{K}_{P_k}^e = C_k,$$

$$\mathbf{K}_{sp}^e = \int_{S_{Pk}} \mathbf{B}^{\mathrm{T}}\mathbf{G}\frac{1}{h_{P_k}}\mathrm{d}S, \quad \mathbf{K}_{ps}^e = \int_{S_{Pk}} \frac{1}{h_{P_k}}\mathbf{G}^{\mathrm{T}}\mathbf{B}\mathrm{d}S, \quad \mathbf{K}_s^e = \int_{S_s} \mathbf{B}^{\mathrm{T}}\mathbf{S}_s\mathbf{B}\mathrm{d}S + \int_{S_{Pk}} \mathbf{B}^{\mathrm{T}}\mathbf{S}_{P_k}\mathbf{B}\mathrm{d}S,$$

$$\mathbf{A}_f^e = -\int_{V_f} \frac{U^2}{c^2}\frac{\partial \mathbf{F}^{\mathrm{T}}}{\partial x}\frac{\partial \mathbf{F}}{\partial x}\mathrm{d}V, \quad \mathbf{A}_{sf}^e = \int_{S_f} \rho_f U\mathbf{N}_w^{\mathrm{T}}\frac{\partial \mathbf{F}}{\partial x}\mathrm{d}S, \quad \mathbf{A}_{fs}^e = -\int_{S_f} U\mathbf{F}^{\mathrm{T}}\frac{\partial \mathbf{N}_w}{\partial x}\mathrm{d}S.$$

Here: $\mathbf{F}$, $\mathbf{N}$, $\mathbf{N}_w$ are the matrices of the shape function for the perturbation velocity potential φ, the displacement vector of the plate $\mathbf{d}$ and its normal component w; $\mathbf{J}$ are the inertia matrices calculated according to expression (15); $\mathbf{B}$ is the gradient matrix, determining the relationship between the deformation and the nodal displacements; $\mathbf{I}$ is the unit matrix.

The formulation of the problem of natural and forced harmonic vibrations of a piecewise homogeneous electroelastic body interacting with a flowing fluid is based on the representation of the solution in the exponential form. In the first case, we use expression (21) and in the second case, we use expression (22):

$$\mathbf{u}(\mathbf{x},t) = \{\boldsymbol{\varphi}(\mathbf{x},t),\, \mathbf{d}(\mathbf{x},t),\, \boldsymbol{\psi}(\mathbf{x},t),\, \mathbf{q}(\mathbf{x},t)\}^{\mathrm{T}} = \tilde{\mathbf{u}}(\mathbf{x})e^{\mathrm{i}\lambda t}, \tag{21}$$

$$\mathbf{u}(\mathbf{x},t) = \{\boldsymbol{\varphi}(\mathbf{x},t),\, \mathbf{d}(\mathbf{x},t),\, \boldsymbol{\psi}(\mathbf{x},t),\, \mathbf{q}(\mathbf{x},t)\}^{\mathrm{T}} = \tilde{\mathbf{u}}(\mathbf{x})e^{\mathrm{i}\Omega t}, \quad \mathbf{f}(\mathbf{x},t) = \tilde{\mathbf{f}}(\mathbf{x})e^{\mathrm{i}\Omega t}, \tag{22}$$

where $\tilde{\mathbf{u}}$ is the function depending only on the coordinates $\mathbf{x}$, Ω is the angular frequency of imposed forces; i is the imaginary unit; $\lambda = \omega + i\gamma$ is the characteristic index, ω is the natural frequency of vibrations, γ is the value, characterizing damping of the system.

Substituting expressions (21) and (22) into the system of equations (19) allows us to obtain:

$$\left(-\lambda^2\mathbf{M} + i\lambda\mathbf{C} + \mathbf{K} + \mathbf{A}\right)\tilde{\mathbf{u}} = 0, \tag{23}$$

$$\left(-\Omega^2\mathbf{M} + i\Omega\mathbf{C} + \mathbf{K} + \mathbf{A}\right)\tilde{\mathbf{u}} = \tilde{\mathbf{r}}. \tag{24}$$

Equation (23) is reduced to the generalized eigenvalue problem (25) for the asymmetric matrices of doubled size [35], which is solved using the implicitly restarted Arnoldi method [36]:

$$\left(\begin{bmatrix} \mathbf{C} & \mathbf{K}+\mathbf{A} \\ -\mathbf{I} & 0 \end{bmatrix} + i\lambda \begin{bmatrix} \mathbf{M} & 0 \\ 0 & \mathbf{I} \end{bmatrix}\right)\begin{Bmatrix} i\lambda\tilde{\mathbf{u}} \\ \tilde{\mathbf{u}} \end{Bmatrix} = 0. \tag{25}$$

The location of the piezoelectric element on the surface of the structure, which will provide its most effective working, is determined numerically from the condition of maximum electromechanical coupling coefficient K_i [2]. It is usually calculated based on i-th natural frequencies of vibrations for system with open circuit (hereinafter referred to as "o/c") and closed electrodes (short circuit, hereinafter referred to as "s/c").

Analysis of changes in the complex eigenvalues of the electromechanical system depending on the resistance R and the inductance L of the electric circuit allows us to find their optimal values, providing the maximum rate of natural vibration damping. The condition [37] is used as a criterion:

$$\left|\Delta\lambda_i\right| = \left|\lambda_i - \lambda_{ic}\right| \to \min, \tag{26}$$

where λ_i and λ_{ic} are the complex eigenvalues corresponding to the damped (i-th) mode and mode of electrical circuit connected to the piezoelectric element.

Verification of the numerical algorithm

In this section, we demonstrate the performance of the developed finite-element algorithm and the reliability of the solutions to various dynamic problems. Unless otherwise indicated, the calculations were performed using the following parameters: $L_s = 150$ mm, $W_s = 115$ mm, $h_s = 0.5$ mm, $E_s = E_1 = E_2 = 68.5$ GPa, $\nu_s = \nu_{12} = \nu_{21} = 0.3$, $\rho_s = 2714$ kg/m^3 (isotropic plate); $L_p = 50$ mm, $W_p = 20$ mm, $h_p = 0.3$ mm, $x_p = 50$ mm, $y_p = 47.5$ mm, $Q_{11} = Q_{22} = 109$ GPa, $Q_{33} = 93$ GPa, $Q_{44} = Q_{55} = Q_{66} = 24$ GPa, $Q_{12} = 61$ GPa, $Q_{13} = Q_{23} = 54$ GPa, $e_{33} = 14.9$ C/m^2, $e_{31} = e_{32} = -4.9$ C/m^2, $e_{15} = e_{24} = 10.6$ C/m^2, $\epsilon_{11} = \epsilon_{22} = 820\epsilon_0$, $\epsilon_{33} = 840\epsilon_0$, $\epsilon_0 = 8.85\times10^{-12}$ F/m; $\rho_p = 7500$ kg/m^3 (piezoelectric element); $H = 10$ mm, $\rho_f = 997$ kg/m^3, $c = 1500$ m/s (ideal fluid).

Table 1. Natural vibration frequencies of a plate with piezoelectric element connected to an external RL-circuit

i	Plate in vacuum, $R = 2800\ \Omega$, $L = 10.2$ H				Plate interacting with quiescent fluid, $R = 6000\ \Omega$, $L = 61$ H			
	3D formulation [20]		Present solution		3D formulation [20]		Present solution	
	ω_i	γ_i	ω_i	γ_i	ω_i	γ_i	ω_i	γ_i
1	**236.033**	**10.907**	**235.494**	**10.951**	96.970	3.870	96.769	3.895
2	257.511	10.883	257.870	10.837	**104.708**	**3.939**	**104.837**	**3.913**

3	442.502	0.000	442.589	0.000	168.688	0.000	168.763	0.000
4	616.049	0.000	617.012	0.000	228.540	0.000	229.013	0.000
5	806.896	0.036	808.484	0.037	310.618	0.012	311.553	0.012
6	840.763	0.000	841.315	0.000	311.945	0.000	312.373	0.000

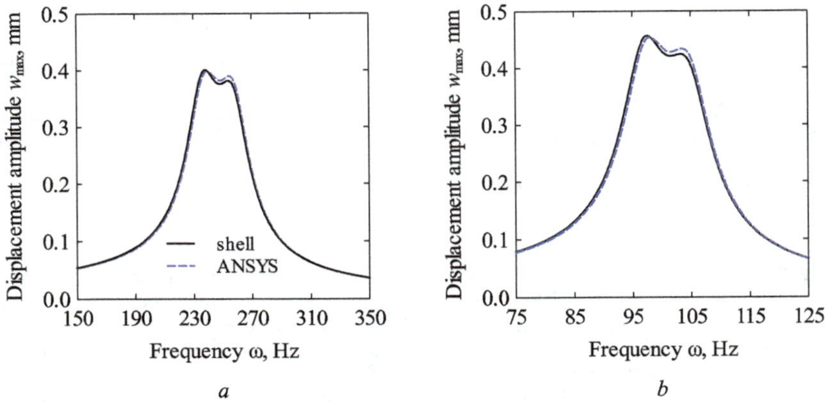

Fig. 2. Comparison of frequency response curves of a plate with a piezoelectric element connected to an external RL-circuit: a — plate in vacuum (R = 2800 Ω, L = 10.2 H), b — plate interacting with a quiescent fluid (R = 6000 Ω, L = 61 H)

We consider the natural vibrations of a rectangular plate rigidly clamped at all edges in vacuum (fluid is absent). A piezoceramic element, whose electroded surfaces are connected to a series *RL*-circuit, is attached to the upper surface of the plate at the point with coordinates (x_p, y_p). Table 1 shows the values of the real and imaginary parts of the complex eigenvalues λ obtained using the above approach *versus* the results of calculations done with the use of three-dimensional equations of the linear theory of piezoelasticity [20]. A similar comparison was made for a plate interacting with a layer of a quiescent fluid of height *H*. The following boundary conditions for the velocity potential φ were used: $\partial\varphi/\partial n = 0$ at the edges $x = 0$, $x = L_s$, $y = 0$, $y = W_s$ and $\varphi = 0$ at $z = H$. In both cases considered, the relative error does not exceed 1% for the vibration frequencies ω and 4% for the damping ratio γ. In this case, the formulation used provides a qualitatively correct description of the appearance of a new natural frequency in the spectrum. It occurs as a result of the formation of a resonant *RLC*-circuit when the piezoelectric element is connected to an external shunt *RL*-circuit. The corresponding mode of vibrations is highlighted in bold type.

The results of solving the problem of forced harmonic vibrations are shown in Fig. 2, which presents the frequency response curves in the vicinity of the first resonance obtained when a uniform pressure of 100 Pa is applied to the plate. All calculation parameters are similar to those considered in the above examples. The dashed line corresponds to the values calculated with the commercial software ANSYS using a three-dimensional formulation. The depicted curves are qualitatively similar, and their maximum quantitative difference does not exceed the value of the relative error of 4%.

Numerical results

We consider the plates with two kinds of boundary conditions: rigidly clamped along the perimeter (CCCC) and free only at the edge $x = L_s$ (CCFC). Their distinguishing feature is a significant difference in the imaginary parts γ of complex eigenvalues obtained at the same fluid flow velocities. This allows us to analyze the performance of the shunt *RL*-circuit in the system with high dissipative characteristics. The calculations were performed for the parameters given at the beginning of the previous section.

The first example demonstrates the possibility of vibration damping for a rectangular plate rigidly clamped at all edges and interacting with the fluid flowing with a velocity of 7 m/s. The parameters of the *RL*-circuit connected to the piezoelectric element ($x_p = 54$ mm, $y_p = 47.5$ mm) were determined numerically using condition (26) and are given in Table 2. It also contains the values calculated with the formulas proposed in [2]. Hereinafter, abbreviation "pp" corresponds to the values obtained using the pole placement techniques and abbreviation "tf" – to expressions using the transfer function.

In papers [20, 37], it has been shown that the analytical expressions formulated in [2] do not allow us to obtain the values of inductance and resistance, which would be truly optimal in terms of the rate of vibration damping. Of all the variants considered, only the condition (26) provides the smallest difference between the natural frequencies ω_1 and ω_{1c} (less than 0.2%). In this case, an increase in the imaginary part γ_1, relative to the open circuit mode indicates an increase in the damping rate of free vibrations.

Table 2. Optimal parameters of the external RL-circuit calculated in various ways and complex eigenvalues (Hz) (plate, CCCC, 7 m/s)

No.	Solution case	Circuit parameters		Eigenvalues λ	
		R, Ω	L, H	λ_1	λ_{1c}
1	Condition (26)	324252	18781	5.5523 + 0.6818	5.5605 + 0.6922
2	Hagood, pp [2]	342682	19855	5.7071 + 0.4438	5.2439 + 0.9298
3	Hagood, tf [2]	249497	21050	5.9372 + 0.3590	4.9434 + 0.5843
4	Open circuit	1×10^{15}	1×10^{15}	5.5986 + 0.0018	—

In the second example, a plate with CCFC boundary conditions is considered. With such kind of boundary condition, the imaginary part of the complex eigenvalues γ, which is responsible for the dissipative characteristics of the system, is several orders of magnitude larger than that in the variant considered above (see rows "Open circuit" in Table 2 and Table 3). In spite of this fact, the connection of a series *RL*-circuit with optimal parameters to the piezoelectric element makes it possible to suppress vibrations even at high modes. The results that confirm this conclusion are given in Table 3. The highest rate of vibration damping is achieved using condition (26).

Table 3. Optimal parameters of the external RL-circuit calculated in various ways and complex eigenvalues (Hz) (plate, CCFC, 1 m/s)

No.	Solution case	Circuit parameters		Eigenvalues λ	
		R, Ω	L, H	λ_7	λ_{7c}
1	Condition (26)	4296.32	13.3203	217.209 + 13.253	217.283 + 13.313
2	Hagood, pp [2]	3889.40	13.1780	212.758 + 10.654	223.185 + 13.722
3	Hagood, tf [2]	2766.20	13.3310	208.476 + 8.095	226.924 + 9.187
4	Open circuit	1×10^{15}	1×10^{15}	217.704 + 1.627	—

The frequency response curves, demonstrating the suppression of forced harmonic vibrations, are shown in Fig. 3. They were obtained for the plate under the action of a uniform pressure of 100 Pa. Although in all variants considered, the resonance amplitude decreases by several times.

The best result is attained with the parameter, which are determined in terms of the transfer functions according to the formulas [2] (see curve 3).

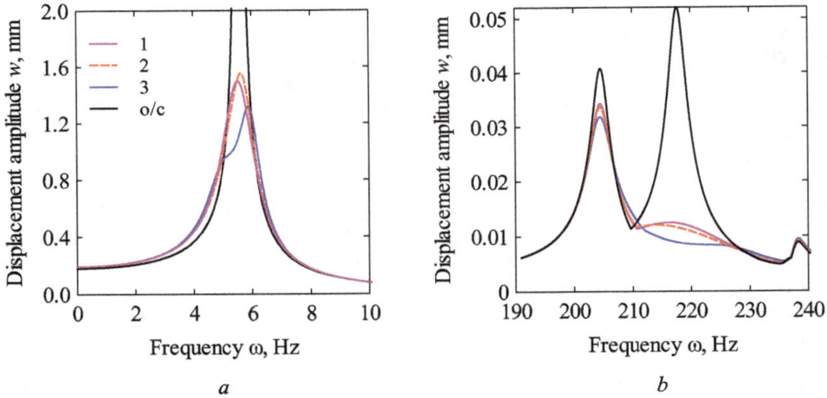

Fig. 3. Frequency response curves of a plate with a piezoelectric element connected to an external RL-circuit: a — CCCC (7 m/s), b — CCFC (1 m/s)

Conclusion

A mathematical formulation of the problem and a finite-element algorithm for its numerical implementation have been developed to analyze natural and forced harmonic vibrations of a plate interacting with the flowing fluid and containing the piezoelectric elements connected to external electric circuits.

The dependence of changes in the complex eigenvalues of electromechanical system on the parameters of the electric circuit consisting of a series-connected resistor and inductance has been analyzed. The optimal values of these parameters, providing the most effective suppression of free vibrations in terms of the rate of their decay have been selected. It is shown that compared to the analytical expressions traditionally used for this purpose, the proposed approach makes it possible to obtain higher damping ratios γ.

The frequency response curves, demonstrating a decrease in the amplitude of forced harmonic vibrations at different velocities of the fluid flow, have been obtained. They have been used to compare different ways of calculating the optimal parameters of the *RL*-circuit.

The study was supported by the grant of the Russian Scientific Foundation (project No. 18-71-10054).

References

[1] R.L. Forward, Electronic damping of vibrations in optical structures, J. Appl. Optics, 18 (1979) 690–697. https://doi.org/10.1364/AO.18.000690

[2] N.W. Hagood, A.H. von Flotow, Damping of structural vibrations with piezoelectric materials and passive electrical networks, J. Sound Vib., 146 (1991) 243–268. https://doi.org/10.1016/0022-460X(91)90762-9

[3] C.H. Park, D.J. Inman, Enhanced piezoelectric shunt design, Shock Vib., 10 (2003) 127–133. https://doi.org/10.1155/2003/863252

[4] A.J. Fleming, S.O.R. Moheimani, Control orientated synthesis of high-performance piezoelectric shunt impedances for structural vibration control, IEEE Trans. Contr. Syst. Technol., 13 (2005) 98–112. https://doi.org/10.1109/TCST.2004.838547

[5] M. Porfiri, C. Maurini, J.-P. Pouget, Identification of electromechanical modal parameters of linear piezoelectric structures, Smart Mater. Struct., 16 (2007) 323–331. https://doi.org/10.1088/0964-1726/16/2/010

[6] O. Thomas, J. Ducarne, J.-F. Deü, Performance of piezoelectric shunts for vibration reduction, Smart Mater. Struct., 21 (2012) 015008. https://doi.org/10.1088/0964-1726/21/1/015008

[7] P. Soltani, G. Kerschen, G. Tondreau, A. Deraemaeker, Piezoelectric vibration damping using resonant shunt circuits: an exact solution, Smart Mater. Struct., 23 (2014) 125014. https://doi.org/10.1088/0964-1726/23/12/125014

[8] O. Heuss, R. Salloum, D. Mayer, T. Melz, Tuning of a vibration absorber with shunted piezoelectric transducers, Arch. Appl. Mech., 86 (2016) 1715–1732. https://doi.org/10.1007/s00419-014-0972-5

[9] J.F. Toftekær, A. Benjeddou, J. Høgsberg, General numerical implementation of a new piezoelectric shunt tuning method based on the effective electromechanical coupling coefficient, Mech. Adv. Mater. Struct., 27 (2020) 1908–1922. https://doi.org/10.1080/15376494.2018.1549297

[10] J.A.B. Gripp, D.A. Rade, Vibration and noise control using shunted piezoelectric transducers: A review, Mech. Syst. Signal Process., 112 (2018) 359–383. https://doi.org/10.1016/j.ymssp.2018.04.041

[11] A. Presas, Y. Luo, Z. Wang, D. Valentin, M. Egusquiza, A review of PZT patches applications in submerged systems, Sensors, 18 (2018) 2251. https://doi.org/10.3390/s18072251

[12] K. Marakakis, G.K. Tairidis, P. Koutsianitis, G.E. Stavroulakis, Shunt piezoelectric systems for noise and vibration control: A review, Front. Built Environ. 5 (2019) 64. https://doi.org/10.3389/fbuil.2019.00064

[13] A.J. Fleming, S.O.R. Moheimani, Piezoelectric transducers for vibration control and damping, 1st ed., Springer, London, 2006.

[14] J.M. Zhang, W. Chang, V.K. Varadan, V.V. Varadan, Passive underwater acoustic damping using shunted piezoelectric coatings, Smart Mater. Struct., 10 (2001) 414–420. https://doi.org/10.1088/0964-1726/10/2/404

[15] J. Kim, J.-H. Kim, Multimode shunt damping of piezoelectric smart panel for noise reduction, J. Acoust. Soc. Am., 116 (2004) 942–948. https://doi.org/10.1121/1.1768947

[16] J. Kim, Y.-C. Jung, Broadband noise reduction of piezoelectric smart panel featuring negative-capacitive converter shunt circuit, J. Acoust. Soc. Am., 120 (2006) 2017–2025. https://doi.org/10.1121/1.2259791

[17] W. Larbi, J.-F. Deü, R. Ohayon, Finite element formulation of smart piezoelectric composite plates coupled with acoustic fluid. Compos. Struct., 94 (2012) 501–509. https://doi.org/10.1016/j.compstruct.2011.08.010

[18] Y. Sun, Z. Li, A. Huang, Q. Li, Semi-active control of piezoelectric coating's underwater sound absorption by combining design of the shunt impedances, J. Sound Vib., 355 (2015) 19–38. https://doi.org/10.1016/j.jsv.2015.06.036

[19] L. Pernod, B. Lossouarn, J.-A. Astolfi, J.-F. Deü, Vibration damping of marine lifting surfaces with resonant piezoelectric shunts, J. Sound Vib., 496 (2021) 115921. https://doi.org/10.1016/j.jsv.2020.115921

[20] S.V. Lekomtsev, D.A. Oshmarin, N.V. Sevodina, An approach to the analysis of hydroelastic vibrations of electromechanical systems, based on the solution of the non-classical eigenvalue problem, Mech. Adv. Mater. Struct., 28 (2021) 1957–1964. https://doi.org/10.1080/15376494.2020.1716120
[21] M.A. Iurlov, A.O. Kamenskikh, S.V. Lekomtsev, V.P. Matveenko, Passive suppression of resonance vibrations of a plate and parallel plates assembly, interacting with a fluid, Int. J. Struct. Stabil. Dynam., 22 (2022) 2250101. https://doi.org/10.1142/S0219455422501012
[22] J.N. Reddy, Mechanics of laminated composite plates and shells: theory and analysis, 2nd ed., CRC Press, Boca Raton, 2004.
[23] IEEE Standard on Piezoelectricity, ANSI/IEEE Std176-1987, IEEE, New York, 1988. https://doi.org/10.1109/IEEESTD.1988.79638
[24] O. Thomas, J.-F. Deü, J. Ducarne, Vibrations of an elastic structure with shunted piezoelectric patches: efficient finite element formulation and electromechanical coupling coefficients, Int. J. Numer. Meth. Eng. 80 (2009) 235–268. https://doi.org/10.1002/nme.2632
[25] S.H. Moon, S.J. Kim, Active and passive suppressions of nonlinear panel flutter using finite element method, Appl. Math. Model. 39 (2001) 2042–2050. https://doi.org/10.2514/2.1217
[26] G. Yao, F.-M. Li. The stability analysis and active control of a composite laminated open cylindrical shell in subsonic airflow, J. Intel. Mat. Sys. Struct. 25 (2014) 259–270. https://doi.org/10.1177/1045389X13491020
[27] A. Benjeddou, J.-F. Deü, S. Letombe, Free vibrations of simply-supported piezoelectric adaptive plates: an exact sandwich formulation, Thin-Walled Struct. 40 (2002) 573–593. https://doi.org/10.1016/S0263-8231(02)00013-7
[28] G.G. Sheng, X. Wang, Thermoelastic vibration and buckling analysis of functionally graded piezoelectric cylindrical shells, Appl. Math. Model. 34 (2010) 2630–2643. https://doi.org/10.1016/j.apm.2009.11.024
[29] H. Allik, J.R. Hughes, Finite element method for piezoelectric vibration, Int. J. Numer. Meth. Eng. 2 (1970) 151–157. https://doi.org/10.1002/nme.1620020202
[30] M.A. Il'gamov, Vibrations of elastic shells containing liquid and gas, Nauka, Moscow, 1969.
[31] M.P. Païdoussis, Fluid–structure interactions: slender structures and axial flow, Vol. 2, 2nd ed., Elsevier Academic Press, London, 2014.
[32] S.A. Bochkarev, S.V. Lekomtsev, V.P. Matveenko, Hydroelastic stability of a rectangular plate interacting with a layer of ideal flowing fluid, Fluid Dyn. 51 (2016) 821–833. https://doi.org/10.1134/S0015462816060132
[33] O.C. Zienkiewicz, R.L. Taylor, The finite element method. The basis, Vol. 1, 5th ed., Butterworth-Heinemann, Oxford, 2001.
[34] J.N. Reddy, An introduction to nonlinear finite element analysis, 2nd ed., Oxford University Press, London, 2015.
[35] F. Tisseur, K. Meerbergen, The quadratic eigenvalue problem, SIAM Rev., 43 (1988) 235–286. https://doi.org/10.1137/S0036144500381988
[36] R.B. Lehoucq, D.C. Sorensen, Deflation techniques for an implicitly restarted Arnoldi iteration, SIAM J. Matrix Anal. Appl., 17 (1996) 789–821. https://doi.org/10.1137/S0895479895281484
[37] V.P. Matveenko, N.A. Iurlova, D.A. Oshmarin, N.V. Sevodina, M.A. Iurlov, An approach to determination of shunt circuits parameters for damping vibrations, Int. J. Smart Nano Mater., 9 (2018) 135–149. https://doi.org/10.1080/19475411.2018.1461144

Advanced Topics in Mechanics of Materials, Structures and Construction - AToMech1-2023 Materials Research Forum LLC
Materials Research Proceedings 31 (2023) 646-655 https://doi.org/10.21741/9781644902592-66

Predicting the strength of recycled glass powder-based geopolymers for improving mechanical behavior of clay soils using artificial intelligence

Abolfazl Baghbani[1,a *], Firas Daghistani[2,b], Hasan Baghbani[3,c], and Katayoon Kiany[4,d]

[1]School of engineering, Deakin University, Melbourne, VIC, Australia

[2]Department of Civil Engineering, La Trobe University, Melbourne, VIC, Australia; Also with Civil Engineering Department, University of Business and Technology, Jeddah, Saudi Arabia

[3]School of Engineering, Ferdowsi University, Mashhad, Iran

[4]School of Design, University of Melbourne, Melbourne, VIC, Australia

[a]abaghbani@deakin.edu.au, [b]f.daghistani@latrobe.edu.au, [c]hasanbaghbani1998@gmail.com, [d]kianyk@student.unimelb.edu.au

Keywords: Recycled Glass Powder, Ordinary Portland Cement, Clayey Soil, Artificial intelligence, Soil Stabilizer, CRRF, SVM

Abstract. The paper investigates the use of artificial intelligence (AI) methods to predict the strength of recycled glass powder (RGP) and soil mixtures based on different input parameters. The study utilized a database of 57 sets with 5 inputs, including RGP percentage, ordinary Portland cement (OPC) percentage, molar concentration, curing temperature and time, and one output, mixed UCS. There were two artificial intelligence models used in this study, a support vector machines (SVM) and classification and regression random forest (CRRF). The results demonstrate the potential of RGP-based geopolymers to improve the mechanical behavior of clay soils, and the use of AI methods to predict the strength of RGP and soil mixtures with high accuracy. Using SVM model, the testing dataset had a mean absolute error (MAE) and R^2 of 0.072 and 0.978, respectively. Also, CRRF had an accurate performance with a MAE of 0.075 and the R^2 of 0.979. These results suggest that the AI models fits well with the data. Also, by analyzing the results of the SVM and CRRF models, it is found that curing time is the most important input parameter, while RGP and OPC are the least significant.

Introduction

Increasing environmental concerns have led to an interest in reusing waste materials as a partial replacement for traditional construction materials [1-6]. Recycled glass powder (RGP) has been investigated for its potential use in geopolymer applications [7]. As an inorganic polymer, geopolymers can be produced from a variety of waste materials, including RGP [8]. A geopolymer based on RGP may improve the mechanical properties of clay soils, as it has been demonstrated that it enhances the strength and durability of other construction materials [5].

A geopolymer is an inorganic material that is formed by the reaction of an aluminosilicate precursor with an alkali activator [9]. A three-dimensional network of Si-O-Al bonds results in a material with unique mechanical and chemical properties [10]. In addition to concrete, ceramics, and composites, geopolymers have been used in a variety of applications. Geopolymers have been shown to improve the mechanical properties of construction materials, such as their strength, durability, and resistance to corrosion [11].

During the recycling of glass, a waste material called recycled glass powder is generated. In the construction industry, it has shown potential as a partial replacement for cement or as an additive to concrete. As well as reducing the amount of waste sent to landfills, the use of RGP has been shown to reduce the carbon footprint of construction materials. Clay soils have poor mechanical properties, such as low strength and high compressibility [12]. Geopolymer has been investigated as a means of improving the mechanical performance of clay soils. The addition of geopolymer to clay soils has been shown to improve their compressive strength, stiffness, and durability.

Recent studies have investigated the use of RGP-based geopolymers to improve the mechanical behavior of clay soils. Bilondi et al. [5] investigated the effects of RGP-based geopolymer on the mechanical properties of expansive soils. According to the results, the addition of RGP-based geopolymer enhanced the unconfined compressive strength and reduced the compressibility of the soil. In a study by Ashiq et al. [13], RGP-based geopolymer was examined for its effect on the strength and deformation behavior of soft clay. Based on the results, RGP-based geopolymer increased the soil's unconfined compressive strength and reduced its deformation.

It has been shown that various factors, such as the glass content, curing time and temperature, can affect the effect of adding glass. It has not yet been possible to develop a comprehensive model for determining the strength of a mixture of glass and soil. One of the reasons for this problem is the multiplicity of effective factors and the non-linearity of their effects. Artificial intelligence is one method that can be used to solve this problem. Based on AI methods, it is possible to predict the output with high accuracy without knowing the relationship between the parameters in advance [14]. In the last two decades, AI methods were used in geotechnical engineering applications include slope stability [15-17], tunnelling [18-19], road construction [20-21], and soil cracking [22-23], soil dynamics [24-26] and recycled material [27-31]. There has not yet been an article published on artificial intelligence methods for determining the strength of RGP and soil mixtures based on different input parameters.

In this study, for the first time, two AI methods, namely, the support vector machines (SVM) and classification and regression random forest (CRRF) method, are used to predict RGP and soil mixtures mixture strength using differing parameters. Input parameters include RGP percentage, ordinary Portland cement (OPC) percentage, molar concentration, curing temperature and time. There are 57 data sets included in the database. After AI modelling, selection of the best artificial intelligence model, sensitivity analysis, and parameter importance have been conducted.

Database Collection and Processing

Experiment and data collection

This study utilized a database of 57 datasets with 5 inputs, including RGP percentage, ordinary Portland cement (OPC) percentage, molar concentration, curing temperature and time, and one output, the strength of RGP and soil mixtures. Database was collected from the study conducted by Bilondi et al. [5]. Table 1 displays descriptive statistics for five variables collected from 57 observations. The variables include:

- UCS: This variable refers to the unconfined compressive strength of a material in mega-pascals (MPa). The observations range from 0.15 MPa to 2.2 MPa, with a mean of 1.003 MPa and a standard deviation of 0.613 MPa.
- Molar Concentration: This variable measures the concentration of a certain compound in solution in moles per liter. The observations range from 0 to 7 moles per liter, with a mean of 2.719 moles per liter and a standard deviation of 1.800 moles per liter.
- RGP: This variable represents the percentage of recycled glass powder used in the material. The observations range from 0% to 25%, with a mean of 8.211% and a standard deviation of 6.681%.
- OPC: This variable refers to the percentage of ordinary Portland cement in the material. The observations range from 0% to 5%, with a mean of 0.263% and a standard deviation of 1.126%.
- Curing Temperature (°C) and Curing Time (Days): These variables represent the curing conditions for the material, with the temperature measured in degrees Celsius and the time measured in days. The observations for curing temperature range from 25 °C to 70 °C, with a mean

of 28.158 °C and a standard deviation of 11.597 °C. The observations for curing time range from 7 days to 91 days, with a mean of 32.053 days and a standard deviation of 33.543 days.

Table 1. Statistical information of database

Variable	Observations	Minimum	Maximum	Mean	Std. deviation
UCS (MPa)	57	0.150	2.200	1.003	0.613
Molar Concentration	57	0.000	7.000	2.719	1.800
RGP (%)	57	0.000	25.000	8.211	6.681
OPC (%)	57	0.000	5.000	0.263	1.126
Curing Temperature (°C)	57	25.000	70.000	28.158	11.597
Curing Time (Days)	57	7.000	91.000	32.053	33.543

Preparation of the data for AI modelling

In the database, the parameters have different units. The accuracy and performance of artificial intelligence models can be adversely affected by this issue. Therefore, linear normalization has been used to normalize the database. Eq. 1 shows the linear normalization equation.

$$X_{norm} = \frac{X - X_{min}}{X_{max} - X_{min}} \tag{1}$$

where X_{max}, X_{min}, X and X_{norm} are maximum, minimum, actual, and normalized values, respectively.

A linear normalization technique is a commonly used technique in data pre-processing that aims to scale the values of parameters in a database to a common range (in this study from 0 to 1). Data is transformed linearly so that it falls within a specified range using this technique. The units of the parameters are thus standardized, which makes it easier for the AI models to process the data and make accurate predictions.

Additionally, dividing the database into training and testing parts is an important part of the preparation process. As part of this study, randomly 20% (12 datasets) of the total database was used for testing, while 80% (45 datasets) was used for training. Tables 2 and 3 provide statistical information about these two databases. As shown in Tables 2 and 3, the statistical information of the two databases is quite similar, which can lead to more accurate performance of artificial intelligence models. By using similar statistical information in the training and testing data sets, the model is less likely to overfit to the training data and can make better predictions on new, unseen data.

Table 2. Statistical information of training database

Variable	Observations	Minimum	Maximum	Mean	Std. deviation
UCS (MPa)	45	0.150	2.200	1.026	0.617
Molar Concentration	45	0.000	7.000	2.733	1.851
RGP (%)	45	0.000	25.000	8.800	7.086
OPC (%)	45	0.000	5.000	0.222	1.042
Curing Temperature (°C)	45	25.000	70.000	28.000	11.352
Curing Time (Days)	45	7.000	91.000	32.667	32.867

Table 3. Statistical information of testing database

Variable	Observations	Minimum	Maximum	Mean	Std. deviation
UCS (MPa)	12	0.250	2.200	0.918	0.615
Molar Concentration	12	0.000	6.000	2.667	1.670
RGP (%)	12	0.000	12.000	6.000	4.431
OPC (%)	12	0.000	5.000	0.417	1.443
Curing Temperature (°C)	12	25.000	70.000	28.750	12.990
Curing Time (Days)	12	7.000	91.000	29.750	37.414

Data-driven modeling

Support vector machine (SVM)

Support vector machine (SVM) is a powerful and popular machine learning algorithm used in classification and regression analyses. A group of scientists led by Vladimir Vapnik developed the technique in the 1990s. A SVM is particularly useful when dealing with complex, high-dimensional, or nonlinear data [32]. This technique is widely used in many fields, including image and text classification, bioinformatics, and many others. In SVM, the basic idea is to find a hyperplane that divides the data into two classes with the greatest margin. Margin refers to the distance between the hyperplane and the nearest data points in each class. It is the hyperplane that maximizes the margin that is most robust to new data points and has the best generalization capability.

In SVM, the data is transformed into a high-dimensional feature space, where it is easier to find separate hyperplanes. During the transformation, a kernel function measures the similarity between pairs of data points in the original space. There are several types of kernel functions, including linear, polynomial, and radial basis functions. In the case of nonlinearly separable data, SVM uses a technique known as the kernel trick. By using this trick, the algorithm is able to find a hyperplane in a high-dimensional feature space without having to calculate the coordinates of the data. As a result, SVM is computationally efficient and scalable.

In addition to binary classifications, SVMs can also be used to analyze regression data and perform multi-class classifications. SVM uses several binary classifiers to separate each pair of classes in the case of multi-class classification. These classifiers determine the final decision. SVMs have several advantages over other machine learning algorithms, such as decision trees and artificial neural networks. As a result, it is less prone to overfitting, requires fewer data preprocessing, and can be used with both numerical and categorical data. There are, however, some limitations associated with SVM, including the selection of the kernel function and the difficulty of interpreting the results.

Classification and regression random forest (CRRF)

Classification and Regression Random Forest (CRRF) is a powerful machine learning algorithm that combines decision trees and random forests to perform both classification and regression tasks. To create a robust and accurate model, random forests are ensembles of decision trees trained on different subsets of data and feature sets. The CRRF algorithm can be applied to both classification and regression problems, making it a versatile algorithm with a wide range of applications. Based on different features, the CRRF algorithm divides the data into smaller and smaller subsets using decision trees. Each decision tree divides the data into subsets based on different rules, and each subset corresponds to a particular decision. To predict the output of the classification or regression problem, the algorithm uses an ensemble of decision trees.

In the case of classification, CRRF uses a combination of decision trees to classify data points into different categories. A random subset of the data and a random subset of the features are used to train each decision tree. In this way, the model is able to reduce overfitting and improve accuracy. As a result of the ensemble of decision trees, CRRF predicts the class label of a new data point based on the results from the ensemble of decision trees. The class with the highest number of votes is considered to be the final prediction. As for regression, CRRF uses a similar approach to predict continuous values rather than class labels. A combination of decision trees is used to predict the output value for a given set of input features. Based on a random subset of the data and a random subset of the features, each decision tree is trained, and the final prediction is derived from the average of all the predictions.

In comparison with other machine learning algorithms, CRRF has several advantages, including its ability to handle both classification and regression tasks, its accuracy and robustness, and its ability to handle missing or noisy data. In addition, CRRF is less likely to overfit than other models, making it a good choice for complex datasets.

Results

Support vector machine (SVM)

By trial and error, different models with opposite values of effective parameters were tested in order to determine the most optimal SVM model. In Fig. 1, the predicted UCS values are compared with the actual UCS values. Using the obtained results, SVM model has successfully determined the UCS values for the RGP and soil mixture.

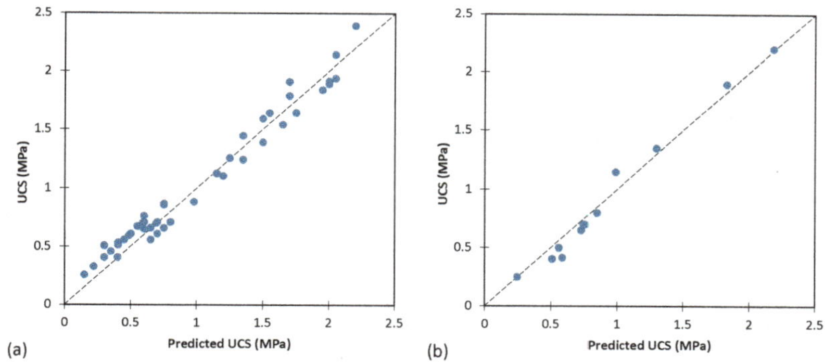

Fig. 1. Results of SVM

The performance metrics for SVM model are discussed in Table 4, including the mean absolute error (MAE) and R-squared (R^2) values for both the training and testing datasets. The MAE for the training dataset is 0.098, whereas the MAE for the testing dataset is 0.072. Having a lower MAE value indicates that the model performs better on the testing dataset. For the training dataset, the R^2 value is 0.969, while for the testing dataset, the R^2 value is 0.978. This indicates that the model fits well with the data, particularly with the testing data.

Table 4. The performance of SVM model

Performance metrics	Training	Testing
MAE	0.098	0.072
R^2	0.969	0.978

Classification and regression random forest (CRRF)

To find the most optimal CRRF model, various CRRF models were constructed by changing the effective parameters. This process involved testing different combinations of parameter values to find the combination that results in the best CRRF model performance. The results of these experiments are presented in Table 5, which shows the specifications of the best CRRF model. The parameters that were optimized for the best model include the minimum node size, minimum son size, maximum depth, Mtry, CP, sampling method, sample size, and number of trees. These parameters were selected based on their impact on the model performance, with the goal of maximizing accuracy while minimizing computational resources.

Table 5. The specifications of the best CRRF.

Trees parameters					Forest parameters		
Min. node size	Min. son size	Max depth	Mtry	CP	Sampling	Sample size	Number of trees
2	1	7	2	0.00001	Random with replacement	44	200

Fig. 2 shows the predicted values of UCS compared to their actual values for both the training and test databases. According to the figure, the CRRF model is performing well, since the predicted values are close to the actual values.

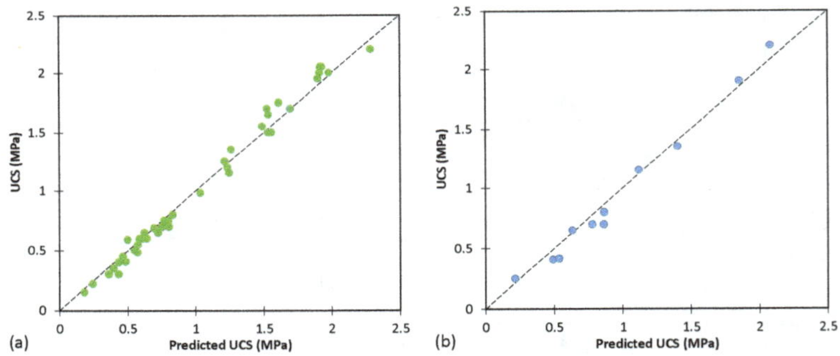

Fig. 2. The results of CRRF for predicting UCS

Table 6 provides additional information regarding the accuracy and error values of the CRRF model for both the training and test databases. According to the table, the R^2 value for predicting the training database is 0.986, indicating a good fit between the model and the data. MAE for the training database is 0.060, indicating that the predicted values are generally accurate.

Also, Table 6 shows the results for the test database. R^2 for predicting the UCS in the test database is 0.979, which indicates that the model is also well suited to the test data. Nevertheless, the MAE value for the test database is slightly higher than that of the training database, indicating that the magnitude of errors in the predicted values is greater for the test dataset.

Table 6. The performance of CRRF model

Performance metrics	Training	Testing
MAE	0.060	0.075
R^2	0.986	0.979

The variable importance of input parameters

It is important to investigate the sensitivity of artificial intelligence models to their input parameters in order to evaluate their importance. The error was calculated by varying one input parameter at a time from -100% to +100% while keeping the other parameters constant. The purpose of this analysis is to identify the input parameters that are most important for the accuracy of the AI model. Fig. 3 shows the results of this analysis for two different models. According to the results of the SVM model, curing time is the most important input parameter, while RGP is the least significant. As a result, changes in curing time have a greater impact on the accuracy of the SVM model than changes in RGB. Similarly, curing time is identified as the most significant input parameter in the CRRF model, while OPC is identified as the least significant.

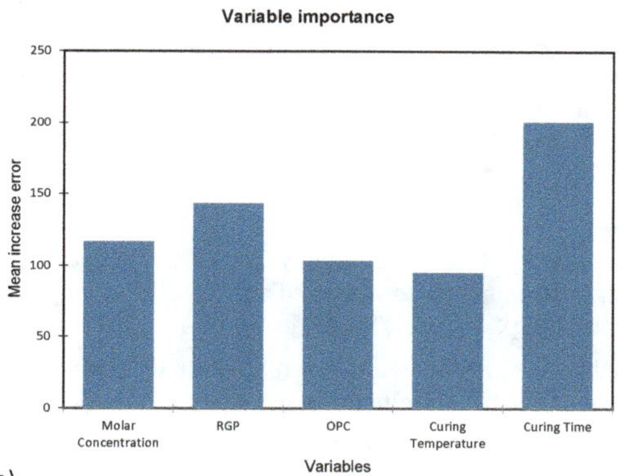

(a)

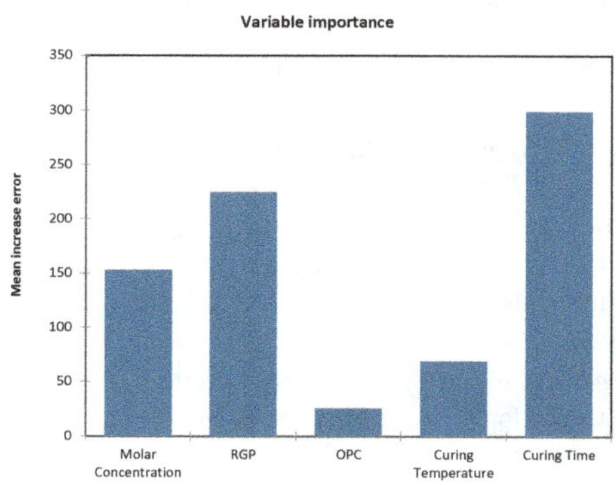

(b)

Fig. 3. The importance of parameters to predict UCS, based on (a) SVM and (b) CRRF

Conclusion

The use of recycled glass powder in geotechnical applications has become increasingly important due to the significant benefits it provides. As a byproduct of the glass manufacturing process, glass powder can be recycled to reduce landfill waste, conserve natural resources, and reduce greenhouse gas emissions. The study aimed to predict soil and glass mixture strength using two

AI methods, namely the artificial intelligence method and the vector machine method, using different input parameters. The database used in the study contained 57 items with 5 inputs, including density, and one output, mixed resistance. The input parameters used in the study were RGP percentage, ordinary Portland cement (OPC) percentage, molar concentration, curing temperature, and time.

The SVM model was successful in determining the UCS values for the soil and glass mixture. According to the results, the MAE for the training dataset was 0.098, while the MAE for the testing dataset was 0.072. In the training dataset, the R2 value was 0.969, while in the testing dataset, the R2 value was 0.978, indicating that the model is well suited to the data.

The results showed that the model, called Classification and Regression Random Forest (CRRF), performed better than SVm model in predicting the UCS, with a high R2 value and low mean absolute error (MAE) for both the training and test databases. The study also conducted a sensitivity analysis to identify the most important input parameters for the accuracy of the CRRF model. The analysis showed that curing time was the most significant input parameter for both CRRF and Support Vector Machine (SVM) models, while RGP and OPG were the least significant.

References

[1] AlBiajawi, M.I., Embong, R. and Muthusamy, K., 2022. An overview of the utilization and method for improving pozzolanic performance of agricultural and industrial wastes in concrete. Materials Today: Proceedings, 48, pp.778-783. https://doi.org/10.1016/j.matpr.2021.02.260

[2] Nguyen MD, Baghbani A, Alnedawi A, Ullah S, Kafle B, Thomas M, Moon EM, Milne NA. Investigation on the suitability of aluminium-based water treatment sludge as a sustainable soil replacement for road construction. Transportation Engineering. 2023 Mar 28:100175. https://doi.org/10.1016/j.treng.2023.100175

[3] Sahebzadeh S, Heidari A, Kamelnia H, Baghbani A. Sustainability features of Iran's vernacular architecture: A comparative study between the architecture of hot-arid and hot-arid-windy regions. Sustainability. 2017 May 4;9(5):749. https://doi.org/10.3390/su9050749

[4] Baghbani A, Costa S, O'Kelly BC, Soltani A, Barzegar M. Experimental study on cyclic simple shear behaviour of predominantly dilative silica sand. International Journal of Geotechnical Engineering. 2022 Oct 23:1-5. https://doi.org/10.1080/19386362.2022.2135226

[5] Bilondi, M.P., Toufigh, M.M. and Toufigh, V., 2018. Experimental investigation of using a recycled glass powder-based geopolymer to improve the mechanical behavior of clay soils. Construction and building Materials, 170, pp.302-313. https://doi.org/10.1016/j.conbuildmat.2018.03.049

[6] Baghbani A, Baumgartl T, Filipovic V. Effects of Wetting and Drying Cycles on Strength of Latrobe Valley Brown Coal. Copernicus Meetings; 2023 Feb 22. https://doi.org/10.5194/egusphere-egu23-4804

[7] Nguyen MD, Baghbani A, Alnedawi A, Ullah S, Kafle B, Thomas M, Moon EM, Milne NA. Experimental Study on the Suitability of Aluminium-Based Water Treatment Sludge as a Next Generation Sustainable Soil Replacement for Road Construction. Available at SSRN 4331275.

[8] Komnitsas, K., Zaharaki, D. and Perdikatsis, V., 2009. Effect of synthesis parameters on the compressive strength of low-calcium ferronickel slag inorganic polymers. Journal of Hazardous Materials, 161(2-3), pp.760-768. https://doi.org/10.1016/j.jhazmat.2008.04.055

[9] Abdullah, M.M.A., Hussin, K., Bnhussain, M., Ismail, K.N. and Ibrahim, W.M.W., 2011. Mechanism and chemical reaction of fly ash geopolymer cement-a review. Int. J. Pure Appl. Sci. Technol, 6(1), pp.35-44.

[10] Wu, Y., Lu, B., Bai, T., Wang, H., Du, F., Zhang, Y., Cai, L., Jiang, C. and Wang, W., 2019. Geopolymer, green alkali activated cementitious material: Synthesis, applications and challenges. Construction and Building Materials, 224, pp.930-949. https://doi.org/10.1016/j.conbuildmat.2019.07.112

[11] Koushkbaghi, M., Alipour, P., Tahmouresi, B., Mohseni, E., Saradar, A. and Sarker, P.K., 2019. Influence of different monomer ratios and recycled concrete aggregate on mechanical properties and durability of geopolymer concretes. Construction and Building Materials, 205, pp.519-528. https://doi.org/10.1016/j.conbuildmat.2019.01.174

[12] Tatlisoz, N., Benson, C.H. and Edil, T.B., 1997. Effect of fines on mechanical properties of soil-tire chip mixtures. ASTM Special Technical Publication, 1275, pp.93-124. https://doi.org/10.1520/STP15645S

[13] Ashiq, S.Z., Akbar, A., Farooq, K. and Mujtaba, H., 2022. Sustainable improvement in engineering behavior of Siwalik Clay using industrial waste glass powder as additive. Case Studies in Construction Materials, 16, p.e00883. https://doi.org/10.1016/j.cscm.2022.e00883

[14] Baghbani A, Choudhury T, Costa S, Reiner J. Application of artificial intelligence in geotechnical engineering: A state-of-the-art review. Earth-Science Reviews. 2022 May 1;228:103991. https://doi.org/10.1016/j.earscirev.2022.103991

[15] Bardhan A, Samui P. Application of artificial intelligence techniques in slope stability analysis: a short review and future prospects. International Journal of Geotechnical Earthquake Engineering (IJGEE). 2022 Jan 1;13(1):1-22. https://doi.org/10.4018/IJGEE.298988

[16] Baghbani A, Daghistani F, Naga HA, Costa S. Development of a Support Vector Machine (SVM) and a Classification and Regression Tree (CART) to Predict the Shear Strength of Sand Rubber Mixtures. InProceedings of the 8th International Symposium on Geotechnical Safety and Risk (ISGSR), Newcastle, Australia 2022. https://doi.org/10.3850/978-981-18-5182-7_00-08-004.xml

[17] Baghbani A, Costa S, Choudhury T. Developing Mathematical Models for Predicting Cracks and Shrinkage Intensity Factor During Clay Soil Desiccation. Available at SSRN 4408164.

[18] Zhang X, Wu X, Park Y, Zhang T, Huang X, Xiao F, Usmani A. Perspectives of big experimental database and artificial intelligence in tunnel fire research. Tunnelling and Underground Space Technology. 2021 Feb 1;108:103691. https://doi.org/10.1016/j.tust.2020.103691

[19] Baghbani A, Baghbani H, Shalchiyan MM, Kiany K. Utilizing artificial intelligence and finite element method to simulate the effects of new tunnels on existing tunnel deformation. Journal of Computational and Cognitive Engineering. 2022 Aug 15.

[20] Sharma S, Ahmed S, Naseem M, Alnumay WS, Singh S, Cho GH. A survey on applications of artificial intelligence for pre-parametric project cost and soil shear-strength estimation in construction and geotechnical engineering. Sensors. 2021 Jan 11;21(2):463. https://doi.org/10.3390/s21020463

[21] Baghbani A, Costa S, Lu Y, Soltani A, Abuel-Naga H and Samui P. Effects of Particle Shape on Shear Modulus of Sand Using Dynamic Simple Shear Testing. Arabian journal of geosciences. 2023.

[22] Xu JJ, Zhang H, Tang CS, Cheng Q, Tian BG, Liu B, Shi B. Automatic soil crack recognition under uneven illumination condition with the application of artificial intelligence. Engineering Geology. 2022 Jan 1;296:106495. https://doi.org/10.1016/j.enggeo.2021.106495

[23] Baghbani A, Costa S, Choundhury T, Faradonbeh RS. Prediction of Parallel Desiccation Cracks of Clays Using a Classification and Regression Tree (CART) Technique. InProceedings of the 8th International Symposium on Geotechnical Safety and Risk (ISGSR), Newcastle, Australia 2022. https://doi.org/10.3850/978-981-18-5182-7_00-08-003.xml

[24] Njock PG, Shen SL, Zhou A, Lyu HM. Evaluation of soil liquefaction using AI technology incorporating a coupled ENN/t-SNE model. Soil Dynamics and Earthquake Engineering. 2020 Mar 1;130:105988. https://doi.org/10.1016/j.soildyn.2019.105988

[25] Baghbani A, Choudhury T, Samui P, Costa S. Prediction of secant shear modulus and damping ratio for an extremely dilative silica sand based on machine learning techniques. Soil Dynamics and Earthquake Engineering. 2023 Feb 1;165:107708. https://doi.org/10.1016/j.soildyn.2022.107708

[26] Baghbani A, Costa S, Faradonbeh RS, Soltani A, Baghbani H. Experimental-AI Investigation of the Effect of Particle Shape on the Damping Ratio of Dry Sand under Simple Shear Test Loading, 2023, preprint. https://doi.org/10.20944/preprints202303.0021.v1

[27] Baghbani A, Daghistani F, Baghbani H, Kiany K, Bazaz JB. Artificial Intelligence-Based Prediction of Geotechnical Impacts of Polyethylene Bottles and Polypropylene on Clayey Soil. EasyChair; 2023 Feb 19.

[28] Baghbani A, Costa S, Faradonbeh RS, Soltani A, Baghbani H. Modeling the Effects of Particle Shape on Damping Ratio of Dry Sand by Simple Shear Testing and Artificial Intelligence. Applied Sciences. 2023 Mar 29;13(7):4363 https://doi.org/10.3390/app13074363

[29] Baghbani A, Daghistani F, Kiany K, Shalchiyan MM. AI-Based Prediction of Strength and Tensile Properties of Expansive Soil Stabilized with Recycled Ash and Natural Fibers. EasyChair; 2023 Feb 19.

[30] Baghbani A.; Abuel-Naga H.; Shirani Faradonbeh R.; Costa S.; Almasoudi R. Ultrasonic Characterization of Compacted Salty Kaolin-Sand Mixtures Under Nearly Zero Vertical Stress Using Experimental Study and Machine Learning, Geotechnical and geological engineering, 2023. https://doi.org/10.1007/s10706-023-02441-5

[31] Baghbani, A.; Nguyen, M.D.; Alnedawi, A.; Milne, N.; Baumgartl, T.; Abuel-Naga, H. Improving Soil Stability with Alum Sludge: An AI-Enabled Approach for Accurate Prediction of California Bearing Ratio. Preprints 2023, 2023030197. https://doi.org/10.20944/preprints202303.0197.v1

[32] Vapnik V. The nature of statistical learning theory. Springer science & business media; 1999 Nov 19. https://doi.org/10.1007/978-1-4757-3264-1

Advanced Topics in Mechanics of Materials, Structures and Construction - AToMech1-2023 Materials Research Forum LLC
Materials Research Proceedings 31 (2023) 656-665 https://doi.org/10.21741/9781644902592-67

Utilizing machine learning techniques for predictive modelling of absorptivity in I-shaped metamaterials

Md Adil[1,a*], Pratik Punj[2,b]

[1]Department of Mechanical Engineering, Jadavpur University, Kolkata, India

[2]Department of Mining Machinery Engineering, IIT (ISM) Dhanbad, Dhanbad, India

[a]adil.alam45@gmail.com, [b]pratik10singh@gmail.com

Keywords: Metamaterials, Machine Learning, Absorptivity, Simulation, Predictive Modelling

Abstract. Metamaterials are artificially engineered materials that have properties not found in naturally occurring materials. They are designed to have specific electromagnetic or other physical properties, such as negative refraction, superconductivity or high absorptivity. They are often composed of structures on a scale much smaller than the wavelength of the phenomena they are intended to manipulate. Metamaterials have a wide range of potential applications, including in antennas, cloaking devices, and super resolution imaging. In this paper we have simulated and validated an L shaped meta material to make a data set of its absorptivity by varying different input parameters and then used these data to predict the absorptivity of any L shaped metamaterial using machine learning and it gave satisfactory results.

Introduction

Metamaterials, artificially engineered substances that possess uncommon electrical and magnetic characteristics, have become the focus of much interest due to their potential uses in fields such as negative refraction [1-2], superlensing [3,4], and optical cloaking [5,6].

The idea behind metamaterials, where properties are obtained by controlling the boundary conditions or the phase and shape of its components, can be traced back to early explorations in the fields of acoustics and vibrations, such as Newton's research on sound transmission through air and Riley's examination of alternate structures. In recent times, Acoustic Metamaterials (AMs) have become a subject of growing interest among the scientific community, due to advancements in the measurement of sound waves and the control of feedback vibrations [7].

Metamaterials have found significant industrial applications, particularly in the field of antenna engineering, where their ability to control electromagnetic waves has led to the creation of smaller and more efficient antennas for various applications such as mobile communication, satellite communication, and radar. The use of metamaterials in antenna design has also led to the development of antennas with improved bandwidth, gain and radiation patterns, resulting in more efficient communication systems. Additionally, metamaterials have been utilized in the field of imaging and sensing, leading to the development of sensors with enhanced resolution and sensitivity. These sensors are used in applications such as medical imaging, remote sensing and security imaging, allowing for early detection and diagnosis of diseases, and remote monitoring of environmental conditions.

Metamaterial other's industrial applications, includes waveguiding, energy harvesting, as well as in public safety, sensor identification, high-frequency battlefield communications, improved ultrasonic sensors, solar energy management for high-gain antennas and remote aerospace applications [8-10]. Moreover, researchers from the army and air force employ metamaterials for detecting explosives, biological materials, and contamination [11-12]. The capability of metamaterials to manipulate sound wavelengths, which are much larger than light, has led navy researchers to investigate their potential use in hiding submarines in military operations. In addition, metamaterials have found applications in civilian domains, such as the creation of sound chambers. Army engineers also utilize metamaterials to manufacture small and high-speed photonic equipment, as light is becoming increasingly important in the development of future circuit boards [13].

Metamaterials have the ability to create a "perfect" absorber with close to full absorption, a concept first introduced by Landy et al. [14]. By incorporating intrinsic loss and carefully designing the structure of the metamaterial, various types of perfect metamaterial absorbers can be created for specific wavelengths, including the microwave range , optical wavelengths [14], and infrared [15]. These absorbers enhance sensitivity in chemical and biological sensing applications [16].

There are two main approaches to creating metamaterials: forward design and inverse design [17-18]. The forward design method involves obtaining material responses through theoretical analysis, simulations, and experiments . However, this method can become expensive as the design complexity increases, as it requires repeatedly modifying material parameters and recalculating responses. In contrast, inverse design [19] has gained popularity as it involves constructing appropriate structures through optimization algorithms in a large parameter space.

The progress in Artificial Intelligence (AI) has made the concept of inverse design a practical reality. AI includes optimization algorithms like simulated annealing, genetic algorithms, particle swarming optimization, and topology optimization, which are well- established and widely employed for generating inverse designs as per the specific needs. These algorithms typically depend on intermediate results obtained through an iterative forward design process. Despite this, performing inverse design under multiple constraints remains a challenge due to the restrictions of a purely random search.

The field of artificial intelligence (AI) has seen significant advancements in recent years, with machine learning (ML) becoming a prevalent approach. The origins of ML can be traced back to the 1940s and 1950s, with the introduction of the artificial neural network and the concept of variable connection strength between neurons. However, it wasn't until the 1980s that ML truly began to gain traction, with the development of the back-propagation algorithm for training neural networks. Deep learning, a branch of machine learning, has been widely used to predict the properties of materials and design (nano)photonic devices. Researchers have used deep neural networks (DNN) to approximate the electromagnetic response of a given structure, a process known as forward prediction. For example, Peurifoy et al. trained a neural network to predict the scattering of light by multilayer nanoparticles [20]. However, when it comes to inverse design of photonic devices, which involves finding the structure that corresponds to a specific electromagnetic response, DNNs often face challenges due to the one-to-many mapping problem.

The application of machine learning (ML) in the fields of electromagnetic and light waves has led researchers to explore the use of ML in the design of other materials that manipulate, providing a powerful tool for characterizing artificial material properties and structuring materials.

Therefore, it's not surprising that Machine Learning (ML) methods have gained increasing attention for analyzing the transmission and dispersion characteristics of periodic acoustic metamaterials, which are characterized by the presence of local resonators. Reviews on the application of ML in acoustics can be found in studies such as those by Bianco et al. [21] and Michalopoulou et al. [22]. Using ML techniques, novel functional applications have been proposed, such as the optimal design of tunable mechanical filters and directional waveguides, as seen in the works of Bacigalupo et al. [23] and Gurbuz et al. [24]. In this research paper, we will explore the possibility of predicting the absorptivity of metamaterial on any particular wavelength based on its parameter for which a simulation model has been created and validated to create the dataset. The results are promising absorptivity estimation performance.

NARX Neural Network

The NARX, or Nonlinear Autoregressive model with Exogenous inputs, is a type of artificial neural network that mirrors the structure of the biological neural system. It comprises interconnected nodes, each of which receives one or multiple inputs and processes them by summing them up and then passing the result through a nonlinear activation function. These nodes perform like artificial neurons. NARX networks are distinct from other types of artificial neural networks because they facilitate information flow in both directions. This allows for connections between neurons within the same layer and between current and previous layers, resulting in faster optimization of connection weights and needing fewer layers to be calibrated. As a result, NARX is a highly effective model.

Governing equation for basic NARX NN model:

$$z(m+1) = f[z(m), \dots, z(m-d_z+1);\ v(m), v(m-1), v(m-d_v+1)]$$

Or can be written as,

$$z(m+1) = f[Z(m); V(m)]$$

Here, z(m) and v(m) represent the output and input of a model, respectively. The terms dz and dv indicate the output and input memory orders, respectively, with $dz > 1$ and dv > 1. The vectors T(m) and V(m) represent the input and output regressors, respectively.

NARX (Nonlinear Autoregressive with exogenous inputs) is a highly capable type of dynamic model that has been shown to have computational capabilities equivalent to those of a Turing machine.

Training Algorithm

Levenberg–Marquardt (LM):

The Levenberg-Marquardt (LM) algorithm is a well-known method utilized for forecasting time series through artificial neural networks (ANNs). This method incorporates the benefits of both the Gauss-Newton and steepest descent methods to optimize a non-linear function that represents the sum of squares of the error. The objective of the LM algorithm is to identify the lowest possible value of this function, thus enhancing the prediction's accuracy.

$$F(t) = \frac{1}{2}\sum_{i}^{m} [f_i(t)]^2$$

Scaled Conjugate Gradient (SCG):

The Scaled Conjugate Gradient (SCG) algorithm is frequently utilized in the training of feedforward neural networks. It is a general optimization method that has been modified to enhance the determination of step size and direction of search. It is based on a second-order approximation as represented by an equation.

$$E(a+b) \approx E(a) + E'(a)^T + \frac{1}{2} v^T E''(a) b$$

The objective of the algorithm is to determine the optimal distance in each iteration through line search, with the goal of finding the most suitable distance to move in the search direction using equation.

$$a_{k+1} = a + l_k * p_k$$

After that, it performs the next search direction which is conjugate to the previous search instructions.

Bayesian regularization(BR):
Bayesian regularization is a method employed in artificial neural networks (ANNs) for adjusting the values of weights and biases through the use of Levenberg-Marquardt optimization. The goal is to find the best combination of these values by first reducing the square error and weight, and then incorporating the weights into the objective function of the training procedure, as represented by an equation.

$$F(\omega) = \alpha\varepsilon\omega + \beta\varepsilon D$$

The Bayesian method is then employed to optimize the values of the objective parameters α and β.

$$P(\alpha, \beta|E, N) = \frac{P(E|\alpha, \beta, N)P(\alpha, \beta|N)}{P(E|N)}$$

Dataset
The illustration of the unit cell for the proposed single L-shaped metamaterial absorber, which was simulated and validated to gather data for training the neural network, is displayed in Fig. 1. Unit cell of the proposd single L shaped metamaterial absorber which we have used to gather data for neural network can be seen in Fig. 1. Proposed metamaterial has been simulated on COMSOL Mutiphysics and has been validated wih the experimental data[25]. In the suggested metamaterial, gold has been used as a material for the L shaped patch and also for the bottom metal layer. Drude model has been used to model the dielectric behaviour of the gold with theplasma and collision frequency of 16 x 10^1.2 rad/s and 13 x 10^10.5 rad/s respectively[44]. SiC has been used as the dielectric spacer with dielectric constant and loss tangent of 10.8 and 0.003 respectively. Thickness of gold and SiC has been taken as 0.27 µm(td) and0.1µm(tm)m(td) and 0.1 µm(td) and 0.1µm(tm) m(tm) respectively.

Absorption capability has been calculated using the relation A = 1 - |S11|^2 , where |S11|^2 represent the square of the magnitude of reflection coefficient.

The absorption capability of the proposed single L-shaped metamaterial absorber is determined by the amount of reflection it experiences, which is calculated by taking the difference of 1 and the square of the magnitude of the reflection coefficient, represented as |S11|^2. The absorption can be represented mathematically as A = 1 - |S11|^2.

The absorption spectrum of the single L-shaped metamaterial absorber, as measured experimentally, is presented in Fig. 2. The graph also displays a comparison between the theoretical calculations and experimental results for the case of transverse electric incidence. The theoretical calculations were made by using the dispersive model of silicon carbide's dielectric constant. As seen in Fig. 2, the simulation was able to accurately capture the presence of two strong peaks and their corresponding wavelengths.

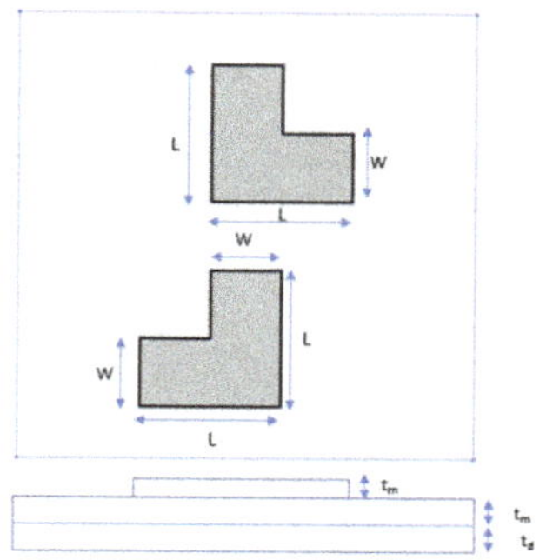

Fig. 1 Schematics of L-Shaped metamaterial

Now two genarated data set, dielectrci constatnt, loss tangent and wavelength has been choosen as an input parameter to predict the absorptivity of the designed metamaterial. In total we have generated 1000 data points using Comsole Multiphysics.

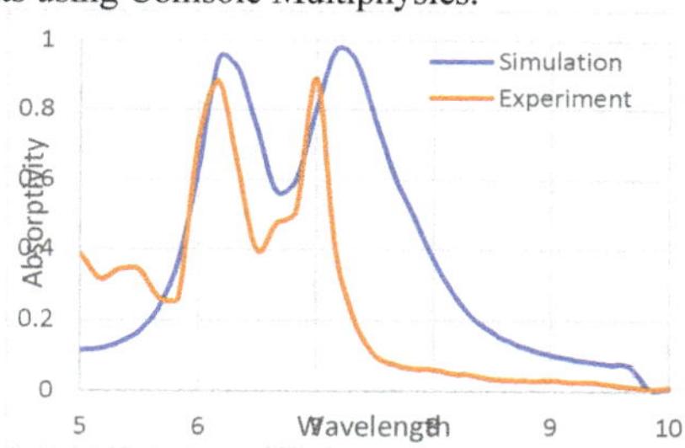

Fig . 2 Comparision of Simulated and Experimental results

Performance Parameter

In this paper, following matrices are employed to evaluate the precision of each model and to compare the performance of various training algorithm.

Mean Square Error (MSE):

$$MSE = \frac{1}{n}\sum_{t=1}^{n}(error)^2$$

Root Mean Square Error (RMSE):

$$RMSE = \sqrt{\frac{1}{n}\sum_{t=1}^{n}(error)^2}$$

Where,

$$error = actual\ value - estimated\ value$$

Results and Discussion

FEA analysis has been done with the following parameters using Comsole Multiphysics simulation software as shown in TABLE 1.

As can be seen in Fig. 3, with increase in dielectric constant value, wavelength for maximum absorptivivty and that for all other local maxima start to shift toward a longer wavelength.

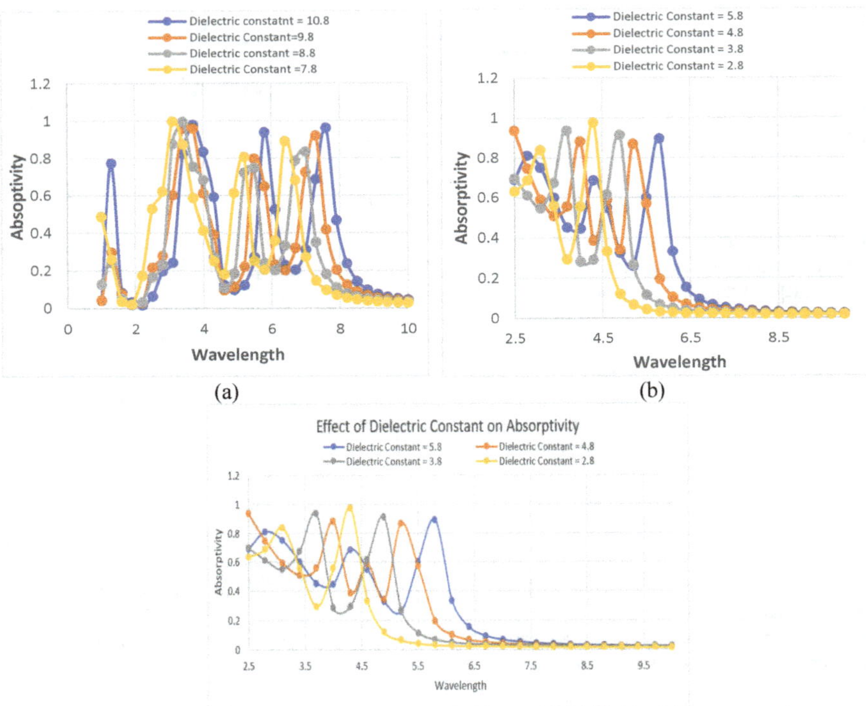

(a) (b) (c)

Fig. 3 : Effect of dielectric constant on absorptivity at loss tamgent (a) 0.001 (b) 0.002 (c) 0.003

When changing the loss tangent value for a given dielectric cpnstant we haven't found any change in either of the maximum absorptivity or the corresponding wavelength as can be seen in Fig. 4.

The proposed NARX NN used data collected from multiple simulations for training, testing, and validation. Whole data set has been divided in two sets (1:3) the larger set (Set 1)has further been divided in 70:15:15 ratio in random manner to train, test and validate the network while the input parameter from smaller set (Set 2) has been used to predict the value and then the result of both has been compared. The network was first trained using the Levenberg-Marquardt, Scaled Conjugate Gradient, and Bayesian Regularization algorithms, then tested, and finally used to predict the absorptivity for any arbitary input of dielectric constant, loss tangent and wavelength.

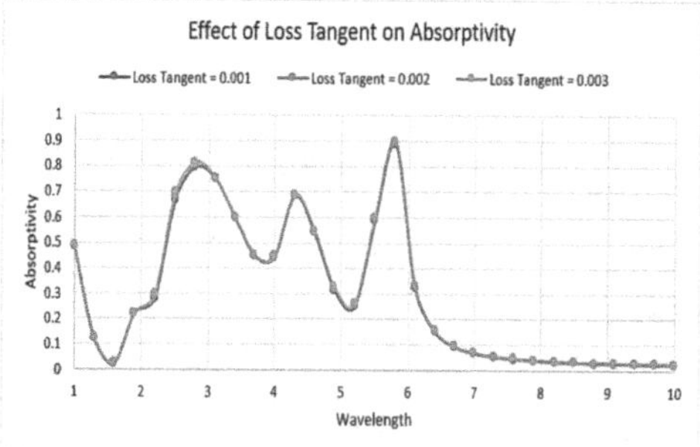

Fig. 4 : Effect of loss tamgent on absorptivity at dielectric constant value of 5.8

The proposed NARX NN used data collected from multiple simulations for training, testing, and validation. Whole data set has been divided in two sets (1:3) the larger set (Set 1)has further been divided in 70:15:15 ratio in random manner to train, test and validate the network while the input parameter from smaller set (Set 2) has been used to predict the value and then the result of both has been compared. The network was first trained using the Levenberg-Marquardt, Scaled Conjugate Gradient, and Bayesian Regularization algorithms, then tested, and finally used to predict the absorptivity for any arbitary input of dielectric constant, loss tangent and wavelength.

The network has been trained using the mean squared error (MSE) as the criteria, in which the network calculated the gradient and updated the weights to reach a point of minimum error.To validate the network performance, autocorrelation was used as a metric and was observed to be influenced by the value of delays. Thus, the delays were chosen in a way to ensure that the values remained within a confidence level. the autocorrelation remained within the 95% confidence level all the three algorithm. Correlation between input and errors, has also been found to be within the confidence limit at all lags, indicating that the model had captured all the features of the system and that the input and output were modeled accurately.

TABLE.1 Performance parameter for different training algorithm on set 1

ALGORITHM	TRAINING	VALIDATION	TESTING
Bayesian regularization	0.92644	0.86774	0.91776
Levenberg-Marquardt	0.89644	0.86196	0.81638
Scaled Conjugate Gradient	0.86635	0.85201	0.81736

As can be seen in the Table. 1 that *Bayesian regularization* was mos accurate while testing the network on Set 1 while other two i.e *Levenberg-Marquardt* and *Scaled Conjugate Gradient* performed almost similarly. On Set 2 *Scaled Conjugate Gradient* performed much better than the other two algorithms as can be seen in Table. 2.

TABLE.2 Performance parameter for different training algorithm on set 2

ALGORITHM	**RMSE**	**MSE**
Bayesian regularization	0.46673	0.21784
Levenberg-Marquardt	0.40905	0.16732
Scaled Conjugate Gradient	0.27871	0.07768

NARX NN has been able to capture the peaks and valleys pretty accurately but the corresponding wavelength is little offset. SCG training algorithm has fared comparitively better as compared to other two when predicting the absorptivity for wavelength above 8 μm. BR algorithm resulted in most noisy response in predicting absorptivity above 8 μm wavelength while at the same time it has also predicted the absorptivity most accurately for wavelength value of upto 6μm. SCG algorithm has also been able to capture the upper and lower limit of the absortivity value as compared to other two training algorithms.

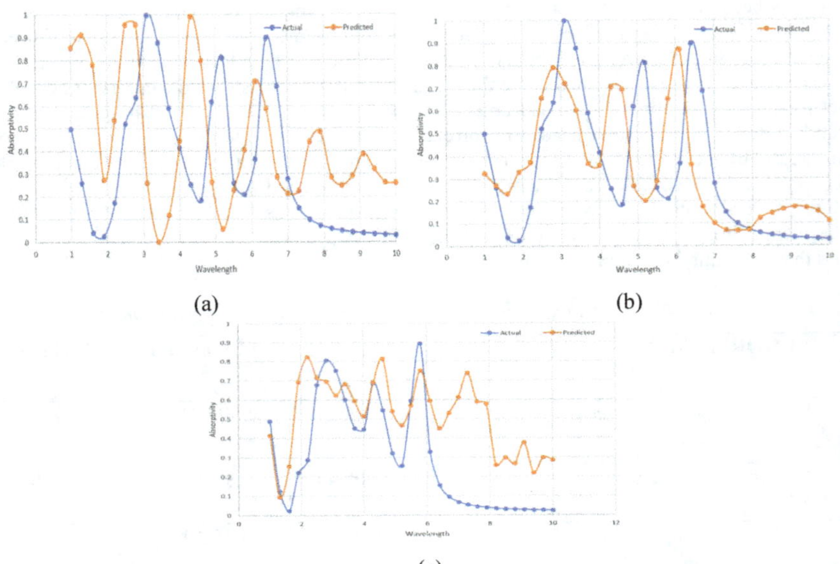

(a) (b)

(c)

Fig.5 Comparison of the forecasted and actual absorptivity value for (a)LM at DC = 7.8 and LT =0.002, (b) SCG at DC= 7.8 and LT = 0.002 (c) BR at DC=6.8 and LT=0.002

Conclusions

Designing metamaterial and to have desired process is a iterative process which requires time and money. Simulation of these type of materials also requires huge computing power and the validation of simulated result is another hurdle. The rise of ML has given another approach for designing these type of materials. We have simulated and validated a model metamaterial which is capable of absorptivity up to 1 using COMSOL Multiphysics to create a data set for our NARX NN training.

We have used three different training algorithms and found that BR algorithm has predicted the absorptivity most accurately for wavelength value of upto 6 μm after which it has given somewhat noisy results. SCG has been able to capture the absorptivity level more accurately in the later half of wavelength i.e after 6 μm.

The results demonstrate that the NARX NN is effective in predicting the absorptivity behaviour of the designed metamaterial. However, like any ANN, the accuracy of the predictions is heavily influenced by the amount of training data available. To improve prediction accuracy, increasing the size of the training data set would be beneficial.

References

[1] R. A. Shelby, D. R. Smith, and S. Schultz, "Experimental verification of a negative index of refraction," Science 292(5514), 77-79 (2001). https://doi.org/10.1126/science.1058847

[2] D. R. Smith, D. C. Vier, T. Koschny, and C. M. Soukoulis, "Electromagnetic parameter retrieval from inhomogeneous metamaterials," Phys. Rev. E Stat. Nonlin. Soft Matter Phys. 71(3), 036617 (2005). https://doi.org/10.1103/PhysRevE.71.036617

[3] J. B. Pendry, "Negative refraction makes a perfect lens," Phys. Rev. Lett. 85(18), 3966-3969 (2000). https://doi.org/10.1103/PhysRevLett.85.3966

[4] D. R. Smith, J. B. Pendry, and M. C. K. Wiltshire, "Metamaterials and negative refractive index," Science 305(5685), 788-792 (2004). https://doi.org/10.1126/science.1096796

[5] J. B. Pendry, D. Schurig, and D. R. Smith, "Controlling electromagnetic fields," Science 312(5781), 1780-1782 (2006). https://doi.org/10.1126/science.1125907

[6] W. Cai, K. U. Chettiar, A. V. Kildishev, and V. M. Shalaev, "Optical cloaking with metamaterials," Nat. Photonics 1(4), 224-227 (2007). https://doi.org/10.1038/nphoton.2007.28

[7] Hussein MI, Leamy MJ and Ruzzene M. Dynamics of phononic materials and structures: historical origins, recent progress, and future outlook. Appl Mech Rev 2014; 66: 040802. https://doi.org/10.1115/1.4026911

[8] Apaydin N, Sertel K and Volakis JL. Nonreciprocal leaky-wave antenna based on coupled microstrip lines 38 Proc IMechE Part L: J Materials: Design and Applications 0(0) on a non-uniformly biased ferrite substrate. IEEE Trans Antennas Propag 2013; 61: 3458-3465. https://doi.org/10.1109/TAP.2013.2257646

[9] Hu G. Metastructure with piezoelectric element for simultaneous vibration suppression and energy harvest_ing. J Vib Acoust 2016; 139: 011012. https://doi.org/10.1115/1.4034770

[10] Erb RM, Libanori R, Rothfuchs N, et al. Composites reinforced in three dimensions by using low magnetic fields. Science 335; 2012: 199. https://doi.org/10.1126/science.1210822

[11] Courjal N, Bernal MP, Ulliac G, et al. LiNbO3 acousto-optical and electro-optical micromodulators. J Eur Opt Soc 2009; 4: 09018. https://doi.org/10.2971/jeos.2009.09018

[12] Shen SH, Fang W and Young ST. Design considera_tions for an acoustic MEMS filter. Microsyst Technol 2004; 10: 585-591. https://doi.org/10.1007/s00542-003-0335-6

[13] Hofmann T, Helbig T, Schindler F, et al. Reciprocal skin and its realization in a topolectrical circuit. arXiv Prepr 2019; 2: 023265. https://doi.org/10.1103/PhysRevResearch.2.023265

[14] C. Hu, Z. Zhao, X. Chen, and X. Luo, "Realizing near-perfect absorption at visible frequencies," Opt. Express 17(13), 11039-11044 (2009). https://doi.org/10.1364/OE.17.011039

[15] N. Liu, M. Mesch, T. Weiss, M. Hentschel, and H. Giessen, "Infrared Perfect Absorber and Its Application As Plasmonic Sensor," Nano Lett. 10(7), 2342-2348 (2010). https://doi.org/10.1021/nl9041033

[16] W. L. Barnes, A. Dereux, and T. W. Ebbesen, "Surface plasmon subwavelength optics," Nature 424(6950), 824- 830 (2003). https://doi.org/10.1038/nature01937

[17] Y. Wang, F. Li, Y. Wang, K. Kishimoto, and W. Huang, "Tuning of band gaps for a two-dimensional piezoelectric phononic crystal with a rectangular lattice," Acta Mech. Sin., vol. 25, pp. 65-71, 2008. https://doi.org/10.1007/s10409-008-0191-9

[18] J. Zhao, Y. Li, and W. K. Liu, "Predicting band structure of 3D mechanical metamaterials with complex geometry via XFEM," Comput. Mech., vol. 55, pp. 659- 672, 2015. https://doi.org/10.1007/s00466-015-1129-2

[19] S. Molesky, Z. Lin, A. Y. Piggott, W. Jin, J. Vucković, and A. W. Rodriguez, "Inverse design in nanophotonics," Nat. Photonics, vol. 12, pp. 659-670, 2018. https://doi.org/10.1038/s41566-018-0246-9

[20] J. Peurifoy, Y. Shen, L. Jing, et al., "Nanophotonic particle simulation and inverse design using artificial neural networks," Sci. Adv., vol. 4, p. eaar4206, 2018 Bianco, M.; Gerstoft, P.; Traer, J.; Ozanich, E.; Roch, M.; Gannot, S.; Deledalle, C. https://doi.org/10.1126/sciadv.aar4206

[21] Machine learning in acoustics: Theory and applications. J. Acoust. Soc. Am. 2019, 146, 3590-3628. https://doi.org/10.1121/1.5133944

[22] Michalopoulou, Z.; Gerstoft, P.; Kostek, B.; Roch, M. Introduction to the special issue on machine learning in acoustics. J. Acoust. Soc. Am. 2021, 150, 3204-3210. [CrossRef] https://doi.org/10.1121/10.0006783

[23] Bacigalupo, A.; Gnecco, G.; Lepidi, M.; Gambarotta, L. Machine-Learning Techniques for the Optimal Design of Acoustic Metamaterials. J. Optim. Theory Appl. 2020, 187, 630-653. https://doi.org/10.1007/s10957-019-01614-8

[24] Gurbuz, C.; Kronowetter, F.; Dietz, C.; Eser, M.; Schmid, J.; Marburg, S. Generative adversarial networks for the design of acoustic metamaterials. J. Acoust. Soc. Am. 2021, 149, 1162-1174. https://doi.org/10.1121/10.0003501

[25] Yang Bai, Li Zhao, Dongquan Ju, Yongyuan Jiang, and Linhua Liu: "Wide-angle, polarization-independent and dual-band infrared perfect absorber based on L-shaped metamaterial," Optics Express. Vol. 23, Issue 7, pp. 8670-8680 (2015). https://doi.org/10.1364/OE.23.008670

Advanced Topics in Mechanics of Materials, Structures and Construction - AToMech1-2023 Materials Research Forum LLC
Materials Research Proceedings 31 (2023) 666-673 https://doi.org/10.21741/9781644902592-68

Modeling and simulation of modular agricultural robot flexible production systems

Mohammed Aldossary[1,a*], Yousuf Alsuliman[1,b], Abdul Aziz Afzal[1,c], Stephen Limbos[1,d] and Megdi Eltayeb[1,e]

[1]Department of Mechanical Engineering, Prince Mohammad Bin Fahd University

[*a]mdossary2920@hotmail.com, [*b]Yousuf.alsuliman@gmail.com, [*c]aafzal@pmu.edu.sa, [d]slimbos@pmu.edu.sa, [e]meltayeb@pmu.edu.sa

Keywords: Modeling, Simulation, Agricultural, Robot, Finite Element Analysis

Abstract. This paper outlines the stages in the development of an integrated modular system for agricultural robots through distinct concepts, designs, and analyses of the outcomes. The integrated modular system must be developed and produced to the highest standards for all probable situations and obstacles in diverse agricultural activities in order to improve an efficient autonomous agricultural robotic modular system. The agricultural robot is designed in accordance with design principles to produce a final product. Design concepts were made, and then a virtual prototype was created in SolidWorks with the necessary dimensions to simulate the dynamic simulation and working space, and the prototype. The finite element analysis of the integrated modular system for agricultural robots test based on the stress, strain, deformation, and mechanical analysis simulations carried out in SolidWorks, uses the virtual model as an input to verify the impact and mechanical properties. The primary objective of this study is to create a high-quality integrated modular system useful for developing supplementary agriculture tools and operations. In this study, modeling and analysis of an integrated modular system for agricultural robots are reported. The modular system is modeled in 3D using the software SolidWorks, and its static structural analysis is performed using SolidWorks. To determine the deformation, stresses, strains, and forces that the loads have on the structures, a static-structural analysis is carried out. Under static structural analysis, the following results for a modular system made were obtained: equivalent von-mises stress, total deformation, and equivalent von-mises strain.

Introduction

Modular agricultural robots are becoming increasingly popular in the agricultural industry. These robots are designed to help farmers increase their efficiency and productivity while reducing costs. They can be used for a variety of tasks such as planting, harvesting, and weeding. With the help of modeling, design, and simulation, these robots can be customized to meet the needs of specific crops or farms. The use of these agricultural robots can enable farmers to maximize their yields while minimizing their labor costs. Moreover, a robot is considered a mechatronic system consisting of several electronically controlled mechanisms. Based on the application of the robot, specific design parameters have to be taken place. The primary objective is to determine the effectiveness of the proposed project. It provides an estimate of the resources obtained to achieve the project requirement and its constraints for completion. Designing a farming robot includes the ability to perform agriculture tasks in a farming field, where the robotics concept aims to improve the processes involved in the agriculture cycle. The chassis of an agricultural robot is a crucial-component in its design and mechanics. It needs to be able to withstand the load or stress encountered during operation while performing its tasks. This requires careful consideration of the structural design, geometry and material used in the construction of the chassis.

Methodology and simulation

The robot's chassis is the structure that supports and contains the primary mechanisms and all of the instrumentation and electronics that power it. Also, the chassis must accommodate all the stresses caused by the applied loads by the robot operations. The structure manages the different forces experienced by the robot, either internally or by conveying them to the ground. The frame must handle the driving and steering motors within the wheels. Furthermore, the robot chassis is a foundation that connects the upper mechanism to the motors and conveying forces applied. These parameters can be controlled by several aspects while designing a chassis for a farming robot, such as structure, size, material, modeling, and analysis that simulate the performance.

Structure. Chassis structure designs were obtained based on comparing commercial and prototype agricultural robots, aiming to reduce total mass while increasing the volume for the required components. However, one of the most critical factors that must be considered when designing the chassis is the overall structure capable of distributing the load applied statically and dynamically while operating. Also, other essential factors that are generally needed to be considered while optimizing the chassis are the overall weight, economic aspects, performance and sustainable chassis design, in which extra features may be installed on the chassis without affecting stability or load distribution. Moreover, from comparing different chassis structures such as square, rectangular, circular, Y-shape and truss-shaped, an implemented H-shaped structure has been observed that is less consuming material, lighter weight, sufficient stress distribution and sustainable modular design. In addition, the mechanical construction of the agricultural robot was established by utilizing the virtual prototyping approach, which is a phase in the product development process. Furthermore, it includes utilizing CAD and CAE tools to evaluate a design before committing to producing actual artifacts.

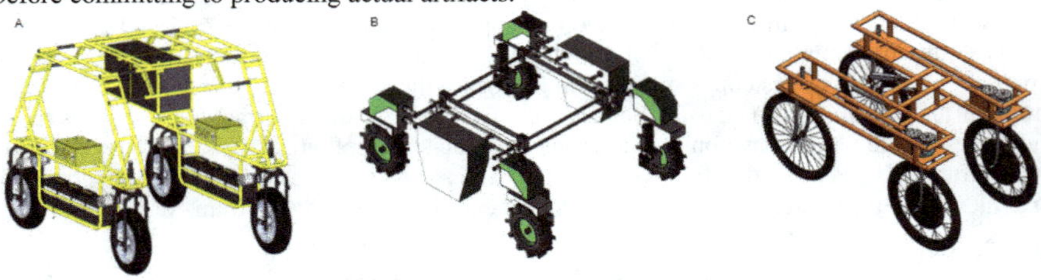

Figure 1: (A) Truss-shaped chassis, (B) Square shaped chassis, (C) Optimized H-shaped chassis

The first design in Fig. 1(A), acquire strong structure and a large space where modular equipment design is possible. In the other hand, it has a lot of tubes, heavy and too highly spaced form soil which requires excess of material to carry the agricultural implements. The second design shown in Fig. 1(B), is the most common mobile robot structure, which provide critical spacing for tools, but less in term of strength and rigidity since the design has to many joints such as binary, ternary, and quaternary joints. The chassis optimized for the project as illustrated in Fig. (C), achieves the objectives of the design, such as handling the load applied, less material requirement, and being sustainable in terms of integrating futuristic features to the design. Also, the robot's operating speed is modest, so there is no need to construct a suspension system. Overall, the design process is done by creating computer-generated geometrical shapes, combining them into an assembly, and testing different mechanical motions, fit and function. Additionally, the assembly or individual parts could be optimized in CAD software to simulate the different stresses the product may encounter in the real world. However, the mechanical structure was designed symmetrically between the right and left sides, allowing the homogenous weight distribution and simplifying the project's development. Moreover, designing a chassis requires consideration of the beam structure. It is important to ensure that the beams are strong enough to support the weight of the vehicle and its contents, while at the same time not being overly bulky and heavy. The profile of the beam also needs to be taken into account, as this will dictate the amount of space available for other components within the chassis. Furthermore, the profile needs to be well-suited for any welding and fastening that may encounter during manufacturing processes.

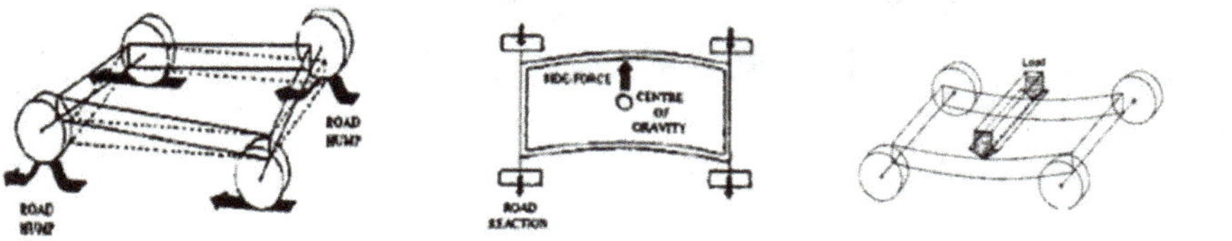

Longitudinal torsion Lateral bending Vertical bending

Figure 2: Chassis operation conditions

Structural beams are an essential element of construction, providing strength and support for the various structures that is built. This is because a beam profile determines its strength and rigidity, which can be useful when designing a structure with complex requirements. A wide range of beam profiles can be used depending on the project's needs, such as square beams, rectangular beams, and circular beams. Furthermore, when choosing the structural beam profile, the overall dimensions must be defined to make sure that the constructed chassis is rigid and doesn't deform within unaccepted range during operation, such as those illustrated in Fig. 2.

Figure 3: Structural beam profiles

The structural profile shown in Fig. 3, demonstrates the chosen cylindrical beam profile. Cylindrical beams are becoming increasingly popular for the construction of vehicle chassis due to their strength and durability. However, these beams offer superior torsional stiffness, which allows them to withstand twisting forces without distortion. Additionally, they provide superior lateral stiffness which keeps them from buckling under pressure. Cylindrical beams are also lightweight, making them ideal for reducing overall vehicle mass.

Design. The dimensions are chosen after considering the numerous issues that might arise from sizes. The constructed chassis was implemented based on field operation requirements, weight distribution, and stress concentration. The chassis size is critical to the stress distribution process and determines the robot's kinematics. This robot's chassis has an overall dimension of 1.5x0.9x0.85 m, as illustrated in Fig. 4.

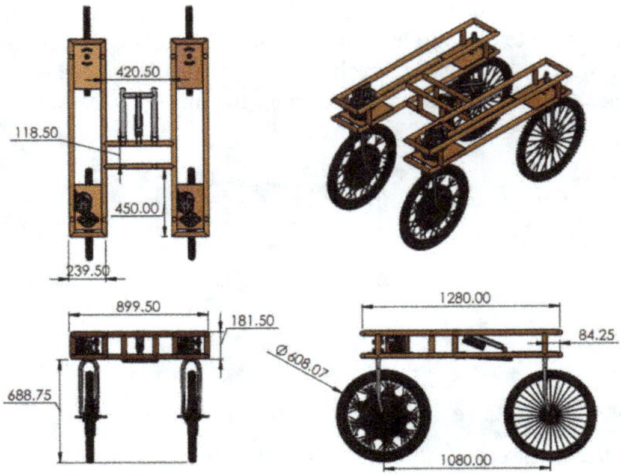

Figure 4: Modular agricultural robot dimension

Likewise, the structure must function in crops with varying spacing of up to 90 cm in height. The chassis structure implemented as a result of design optimization was H-shaped, which is light and flexible and decreases weight while enhancing efficiency when compared to commercial agricultural vehicles. Furthermore, because of the height of the robot, the distance between axles must allow the vehicle to overcome slopes, which are common in crop fields. The short wheelbase helps the vehicle's agility but makes it more difficult to handle. However, because our goal is to create a flexible vehicle, the wheelbase is selected to provide only sufficient space for tools while still allowing for maximum mobility.

Materials. When deciding the material of constructing the chassis and its component, it required to sort all the parameters that need to addressed, mechanical properties, stress concentration, manufacturability, weight and cost. Those parameters can be controlled by choosing the suitable material, by utilizing three different materials that have been observed within the study such as those typically used in vehicle chassis. The selection criteria were mainly subjected to mechanical properties of the material behavior within application characteristics that meets the design requirements, Table 1 illustrate the three materials studied.

Table 1: Material properties comparison

Parameters	ASTM A36 Steel	AISI 1020 Steel	AISI 4310 Steel
Yield Strength [MPa]	250	351	460
Ultimate Strength [MPa]	400.0	420.5	731.0
Modulus of Elasticity [GPa]	200	200	205
Modulus of Rigidity, [GPa]	79.3	72.0	80.0
Poisson's Ratio	0.260	0.290	0.285
Density [kg/m^3]	7850	7900	7850

From the table above, an observation based on material characteristics has been specified to compare between those material, such as yield strength, modulus of rigidity, Poisson's ratio and material density. ASTM A36 is a common structural steel grade used in construction and fabrication applications. It is an economical, general-purpose steel with good strength, formability, and weldability characteristics. However, it is particularly well-suited for applications where strength and toughness are needed in combination. It can be used in many areas such as buildings, bridges, highways, and other structures, as well as in industrial maintenance, agricultural applications, transportation equipment, and general fabrication. In addition, it achieves the requirement of being ductile and difficult to deform under various stress that might encounter during operation. These conditions are controlled by specific mechanical properties targeted such as yield strength, modulus of elasticity and modulus of rigidity. Ductile materials are essential for robot chassis building, they provide a highly durable and lightweight structure that robots need to operate effectively. Also, its easily manipulated to fit into any shape and size desired, allowing for customization of the chassis design to suit different needs. Additionally, ductile materials are extremely resilient and can withstand harsh exterior environments, protecting the sensitive internal components from damage. Moreover, the robot chassis is best constructed from a material that is both strong and ductile, such as those that can be stretched and hammered out into desired shapes without breaking. These materials like aluminum, steel and titanium, are perfect for constructing a robot's chassis as they can be shaped to meet any desired form. Furthermore, they can withstand greater forces and pressures than other materials, making them more durable and reliable in the long run. Additionally, ductile materials boast a high modulus of elasticity, which gives them the ability to flex and bend when forces are applied, without becoming permanently deformed. This property makes them an ideal fit for constructing a chassis, as it can absorb impacts and shocks from the environment while safely protecting its internal components. In conjunction with its high modulus of elasticity, the ductile material used to construct a robot's chassis also has a significant modulus of rigidity. This means that, while the material is still pliable, it will not deform irreversibly when subjected to pressure. High modulus of rigidity allows the chassis to remain rigid and stable in its structure, further protecting its internal components from damage. This makes the material selected perfect to be utilized, as it can remain rigid enough to protect its components, while still being resilient enough to withstand impacts and shocks from operational conditions.

Results and discussions. The material properties which were utilized in finite elements analyses were obtained from SolidWorks as shown in Table 1 for the ASTM-A36 steel. In order to simplify the model, elements that have no structural function as electronic controls, batteries, motors and steering system have been removed from analysis. The removed items have been replaced by resulting forces.

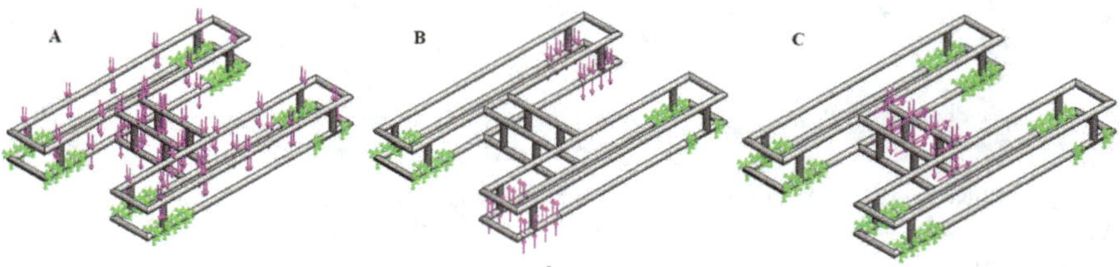

Figure 5: Loads applied on the structure: (A) Vertical bending, (B) Longitudinal torsion, (C) Combined bending

The details of the structure with the forces applied are presented in Fig. 5. Simulation has been performed considering the structure under condition of variable strain in three cases: (A) Vertical bending due to maximum weight of 80 kg; (B) Longitudinal torsion due to crossing obstacles with 20 kg applied at each corner; and (C) Combined bending due to farming implements with maximum mass of 30 kg.

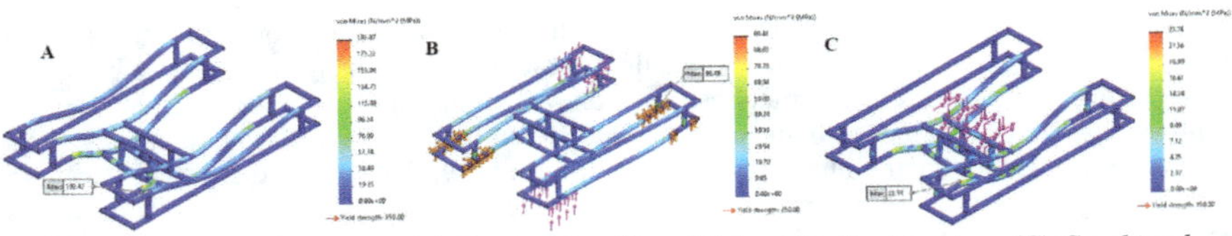

Figure 6: Maximum Stress: (A) Vertical bending, (B) Longitudinal torsion, (C) Combined bending

Fig. 6, shows the results of the stress analysis. It is clear to see that the majority of the deformation happened in the lower section of the chassis, but without affecting the set's performance. In terms of tension, there was some stress deposition at the joint of the bottom and inner side frames, but not enough to cause permanent deformation.

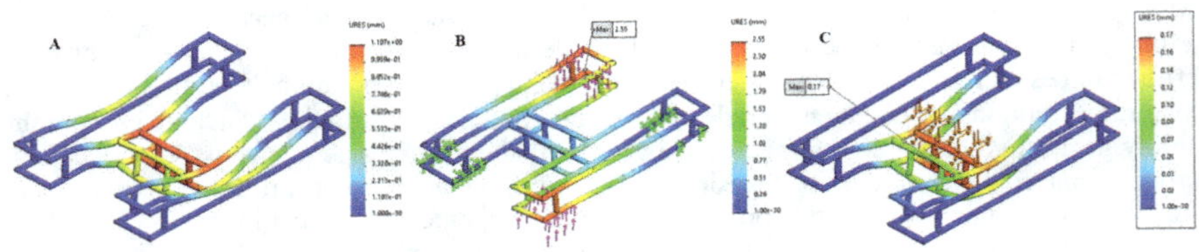

Figure 7: Maximum Deformation: (A) Vertical bending, (B) Longitudinal torsion, (C) Combined bending

Fig. 7, shows the deformations obtained from the analysis for each of the cases investigated. The maximum deformation is found at the rear corner of the chassis, and it decreases as it shifts toward the middle cross members of the chassis. This characteristic is expected to occur, which is due to longitudinal torsion and the distance between the wheel modules and the center connection between the side frames.

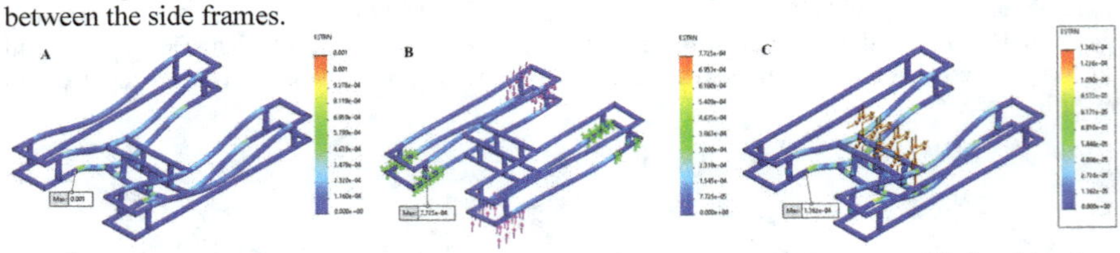

Figure 8: Maximum Strain: (A) Vertical bending, (B) Longitudinal torsion, (C) Combined bending

Fig. 8, illustrate the results of maximum strain that would occur during operational conditions. As a result, most of strain in the structure would initiate from the fixed supports, due to the ductility of the material selected.

Table 2: Simulation results

Design Parameters	Case [a]	Case [b]	Case [c]
Maximum stress [MPa]	192.47	98.48	23.74
Maximum deformation [mm]	1.1	2.55	0.17
Maximum strain [mm/mm]	0.001	0.0008	0.00014
Factor of safety	1.30	2.54	10.53
Average stress [MPa]	104.9		
Average factor of safety	2.38		

Table 2, shows parameters obtained from the simulation results. In maximum stress detected, the analysis shows the design would obtain 1.3 factor of safety, which is almost critical for a ductile material. Maximum deformation characteristics in a robot vehicle chassis play an important role in its overall performance. Through the simulation, maximum deformation observed is 2.55 mm, acceptable based on the prototype geometrical scale. Such characteristics are evaluated through dynamic tests and are important for evaluating the overall robustness of a design.

Conclusion

This paper presents the modelling and simulation of an integrated modular system for agricultural robot operations in the agricultural industry. This study presents modeling and analysis of the integrated modular system for agricultural robots. The modular system is modeled in the finite element analysis and the computer-aided design using the software SolidWorks, and its static structural analysis is performed using SolidWorks under different loading conditions. The integrated modular system is not expected to break under various loading situations, according to the analysis's findings, which were assessed using the deformation, stresses, and strain failure analysis criterion. The integrated modular system will be able to carry out the tasks during the agricultural processes with accuracy and precision.

References

[1] Diary Ali, R. (2020). Design and Development of a Chassis Concept for an Autonomous Airport Shuttle (Dissertation). Retrieved from http://urn.kb.se/resolve?urn=urn:nbn:se:kth:diva-280558

[2] Tabile, R. A., Godoy, E. P., Pereira, R. R. D., Tangerino, G. T., Porto, A. J. V., & Inamasu, R. Y. (2010). Design of the mechatronic architecture of an agricultural mobile robot. IFAC Proceedings Volumes, 43(18), 717–724. https://doi.org/10.3182/20100913-3-us-2015.00102

[3] Vijayaragavan, E., Bhat, S., Patel, A., & Rana, D. (2018). Design and analysis of a mobile robot for storage and retrieval system. IOP Conference Series: Materials Science and Engineering, 402, 012205. https://doi.org/10.1088/1757-899x/402/1/012205

[4] P. (2021). Structural & dynamic analysis and simulation of mobile transportation robot. International Journal of 3D Printing Technologies and Digital Industry. https://doi.org/10.46519/ij3dptdi.949803

[5] Agarwal, A., & Mthembu, L. (2022). Structural Analysis and Optimization of Heavy Vehicle Chassis Using Aluminium P100/6061 Al and Al GA 7-230 MMC. Processes, 10(2), 320. https://doi.org/10.3390/pr10020320

[6] Rothmund, M., & Institut Für Landtechnik, B. U. U. (2006). Automation Technology for Off-road Equipment 2006: Proceedings of the 1 - 2 September International Conference, Bonn, Germany. Landtechnik Weihenstephan.

[7] Michaud, S., Richter, L., Thüer, T., Gibbesch, A., Huelsing, T., Schmitz, N., Weiss, S. T., Krebs, A., Patel, N. D., Joudrier, L., Siegwart, R., Schäfer, B., & Ellery, A. (2006). Rover Chassis Evaluation and Design Optimisation using the RCET. International Conference on Robotics and Automation. https://doi.org/10.3929/ethz-a-010043426

Numerical study on the effect of real gas model on the flow structure and shock location of Laval nozzle

Bashar I Jasem[1,a *] and Esam I Jassim[2,b]

[1]Department of Computer Technology, Al-Hadba University College, Mosul, Iraq.

[2]College of Engineering, Prince Muhammad Bin Fahd University, 31952 Al-Khobar, Kingdom of Saudi Arabia.

[a]adamalzuhairee@gmail.com, [b]ejassim@pmu.edu.sa

Keywords: CFD, Real Gas, Supersonic Flow, Laval Nozzle, Shock Wave

Abstract. Design of supersonic nozzle requires accurate and robust procedure since the flow becomes subtle during shifting from subsonic to supersonic. This article presents some aspects of the fluid features at supersonic region when it behaves as real gas using 3D-numerical simulation. The objectives include the variation in the shock position, the fluid properties, and the real gas model at different Nozzle-Pressure-Ratio (NPR). Results of CFD simulation showed that ideal gas model predicts higher Mach number than any real gas model. Also, the prediction is different between SRK and BWR models. The erroneous in predicating Mach number approaches 21% for SRK and 43% for BWR. For the range of NPR 2-3, shock position is found to be proportion to NPR; however, significant discrepancy in the shock location is observed when ideal gas verses real gas is assumed.

Introduction

Natural Gas is one of the prominent energy sources in industries and domestic consumption. The transportation from its sources of production to processing areas is one of the greatest challenges that transportation technology encountered [1]. Hence, research on enhancement the transportation equipment and process treatment to minimize the energy loss is still in demand. One of these devices used for purification natural gas during transportation is the convergent-divergent type nozzle, usually known as a De-Laval nozzle.

De-Laval nozzle was named after Gustaf de Laval, a Swedish inventor in 1888, [2]. The convergent divergent nozzle consists of three parts: the first part is known as the convergent or the subsonic part, where the gas velocity is accelerating, but at subsonic zone (M<1). The fluid in this part is accompanied by a decrease in pressure and temperature. The second part is the throat area, where the gas velocity approaches the sound velocity (M=1). The last part is the divergent part, where the velocity of the gas exceeds the sound speed and becomes supersonic (M>1), [3].

Recent industrial on gas purification processes are in demand in compact design of equipment to overcome the perpetual problems of traditional means [4]. Supersonic nozzle was introduced to be an excellent alternative equipment due to its flexibility and feasibility to prevent any fouling or deposition of undesired particles, [5].

The features of shockwave region have been reported by numerous authors; however, most of these methods were developed mainly for perfect gas assumption, [4]. The assumption that the thermodynamic properties of the fluid do not significantly affect the performance of the system could lead to mislead in predicting design parameters, particularly when the supersonic fluid is made of multi-component gas mixture [6].

A good numerical model to account for real gas effects should satisfy some basic requirements, including accuracy in predicting thermodynamic properties of the fluid, capability of dealing with a wide range of state parameters, and effective and time-efficient implementation for computational fluid dynamics (CFD) analyses.

Kouremenos (1986) [7] study on normal shock reported that significant discrepancy in fluid properties was obtained when ideal gas is considered. Arina (2004) [8] numerically tested three models of real gas using CFD code. Van der Waals (VW) equation was developed in 1873 and was used by Ginzburg (1939) and later by Bai-Shi-I (1961) [9] during investigation of gas stream in nozzles. After development of Redlich-Kwong (RK) equation in 1949, a third parameter have been added to increase the accuracy of the compressibility factor. The new equation developed by Soave, Redlich, and Kwong (Soave, 1972) and Peng and Robinson (1976) is nowadays called "the cubic EOS". To encompass wide range of pressure and temperature, Benedict-Webb-Rubin (BWR) equation and its modification (MBWR) are introduced and have proven to be quite accurate [2].

Mathematical modeling

Analysis of shock location in a Laval nozzle has been reported in numerous literatures. Jassim [10] derived a theoretical 1-D expressions that relates the Nozzle Pressure Ratio (NPR), which refers to the ratio between the local static pressure and the total pressure, to the shock location w.r.t. the nozzle throat. Here we lists the final equation (Eq.1) and one can return to ref.[10] for more details.

$$M_2 = \left[\frac{1+[\gamma-1/2]M_1^2}{\gamma M_1^2-(\gamma-1)/2}\right]^{1/2} \qquad (1)$$

Numerical modeling

When selecting a numerical model to represent the state of gas flow through the nozzle, must be cover the basic requirements of the fluid including accuracy in predicting thermodynamic properties of the fluid, capability of dealing with a wide range of state parameters, and effective and time-efficient implementation for computational fluid dynamics (CFD) analyses. To analyze the shock wave that occurs, six properties of the gas flowing through the nozzle must be calculated, pressure, temperature, enthalpy, entropy, density, and velocity.

ANSYS 18 is employed to simulate the convergent-divergent nozzle, illustrated in Fig. (1), and its dimensions is listed in table (1). The 3D compressible steady state flow and energy equations are discretized to evaluate the air thermal properties while streaming along the convergent and the divergent parts.

The continuity equation in 3D form can be written as:

$$\frac{\partial}{\partial x_i}(\rho u_i) = 0 \qquad (2)$$

Conservation of momentum for steady compressible flow is defined as:

$$\frac{\partial}{\partial x_i}(\rho u_i u_j) = -\frac{\partial p}{\partial x_i} + \left[\mu\left(\frac{\partial u_i}{\partial x_j} + \frac{\partial u_j}{\partial x_i} - \frac{2}{3}\delta_{ij}\frac{\partial u_l}{\partial x_l}\right)\right] + \frac{\partial}{\partial x_j}(\rho \acute{u}_i \acute{u}_j) \qquad (3)$$

The 3D energy equation takes the form:

$$\frac{\partial}{\partial x_i}[u_i(\rho E + p)] = \frac{\partial}{\partial x_j}\left[\mu\left(k + \frac{C_p\mu_t}{0.85}\frac{\partial T}{\partial x_j} + u_i(-\rho \acute{u}_i \acute{u}_j)\right)\right] \qquad (4)$$

The abovementioned 3 equation are solved by ANSYS-FLUENT for each grid together with the following *k-ε* equations:

$$\frac{Dk}{Dt}=\frac{\partial}{\partial x_j}\left[\left(\nu+\frac{\nu_t}{\sigma_k}\right)\frac{\partial k}{\partial x_j}\right]-\langle u_i'u_j'\rangle\frac{\partial u_i}{\partial x_J}-\varepsilon$$

$$\frac{D\varepsilon}{Dt}=\frac{\partial}{\partial x_j}\left[\left(\nu+\frac{\nu_t}{\sigma_\varepsilon}\right)\frac{\partial \varepsilon}{\partial x_j}\right]-C_{\varepsilon 1}\frac{\varepsilon}{k}\langle u_i'u_j'\rangle\frac{\partial u_i}{\partial x_J}-C_{\varepsilon 2}\frac{\varepsilon^2}{k} \quad (5)$$

$$\nu_t=C_\mu\frac{k^2}{\varepsilon}$$

System Design and fluid properties:
The Laval nozzle shown in figure (1) is constructed using DesignModeler software, a modeler embedded in ANSYS. The convergent part is 8 cm inlet diameter and 4 cm long while the divergent part is 4 cm diameter at outlet and 11 cm long. The throat diameter is 2 cm. Air is used in the simulation as working fluid and streamed axially from the convergent inlet. The properties of the working fluid is calculated based on the gas model selected, namely: Ideal gas, SRK, and BWR models.

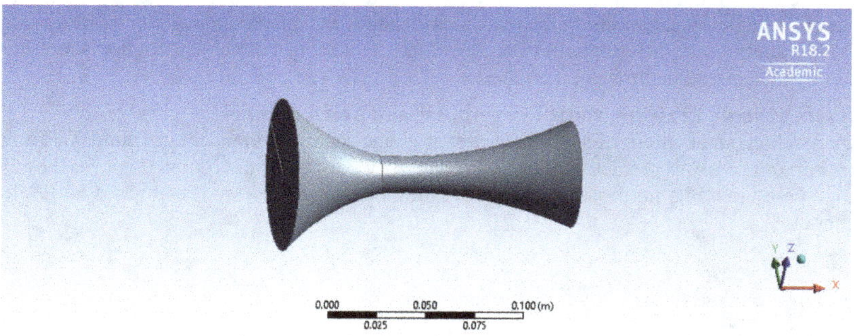

Figure 1. Geometry of the Laval Nozzle used in the simulation

Mesh architecture and dependency
Unstructured tetrahedral cell is used to construct the nozzle, considering higher grid density in the divergent part (Fig. 2). Such dense guarantees improvement of capturing the shock at the right location. The number of elements for optimal resolution was 135054, which captured the fluid structure near shock region at reasonable resolution. Mesh dependency near shock was performed and it was found that the recommended grid size should be within 0.1 mm while up to size of 1 mm was acceptable for the other regions. Since the flow is turbulent, a k-ε two turbulent equation was activated and solved simultaneously with the continuity, momentum, and energy equation for each element at steady state condition.

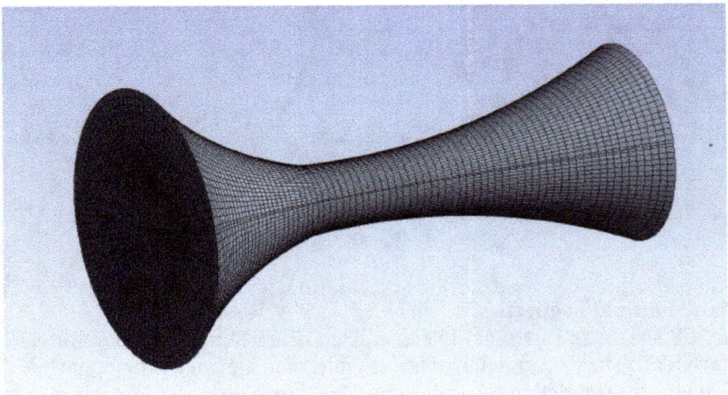

Figure 2. Unstructured mesh for better resolution

Results and discussion
Simulation is performed for Nozzle Pressure Ratio NPR of 3/1 at three cases the ideal-gas model and two real-gas models namely SRK, and The Benedict-Webb-Rubin (BWR) equation models. The target parameters of the study are the shock location as well as the quantity of Mach number and static temperature at various NPRs. However, it is more convenient to firstly present the contours of the abovementioned parameters.

Contours of static Pressure, static temperature and Mach number
Figure 3 depicts the contour of static pressure, in gauge scale, for the ideal gas and BWR models. The minimum pressure predicted when the model of ideal gas is incorporated approaches 6% of the total pressure while not less than 30% the minimum static pressure is estimated by BWR real gas model.

Materials Research Proceedings 31 (2023) 674-682 https://doi.org/10.21741/9781644902592-69

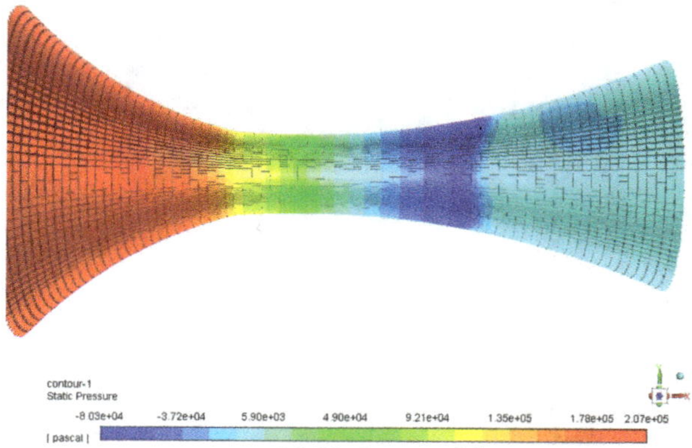

(a) Ideal gas model

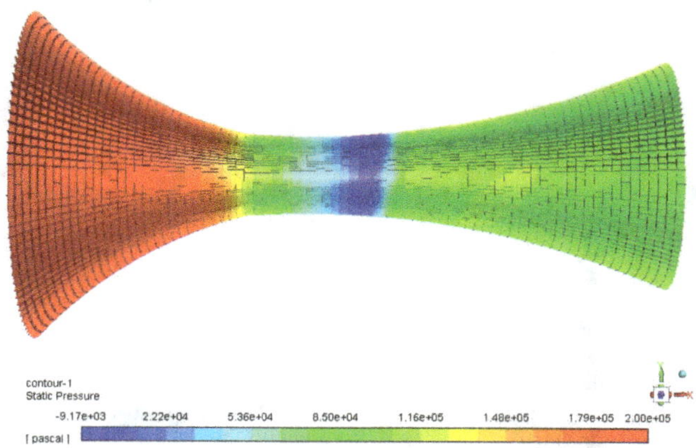

(b) BWR model

Figure 3. Contours of static pressure along the nozzle

Since the shock location predicted by employing the ideal gas model is the farthest from the throat in comparison with the two real gas models, it is rational to anticipate that the maximum value of supersonic Mach number exists in the ideal gas case. The error in the magnitude of Mach number is about 21% extra for SRK model and 43% for BWR, as shown in figure 4.

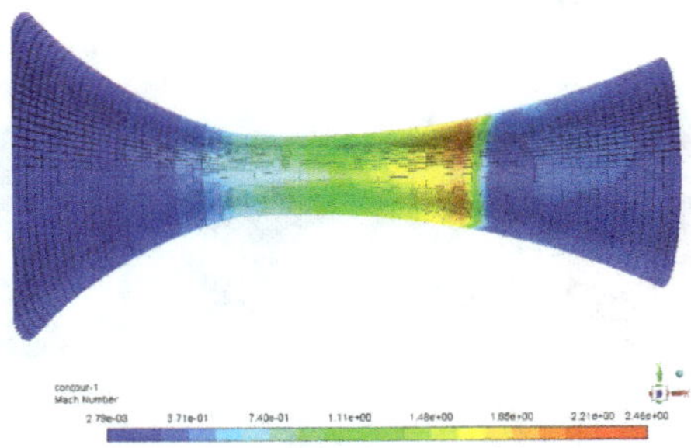

(a) Ideal gas model

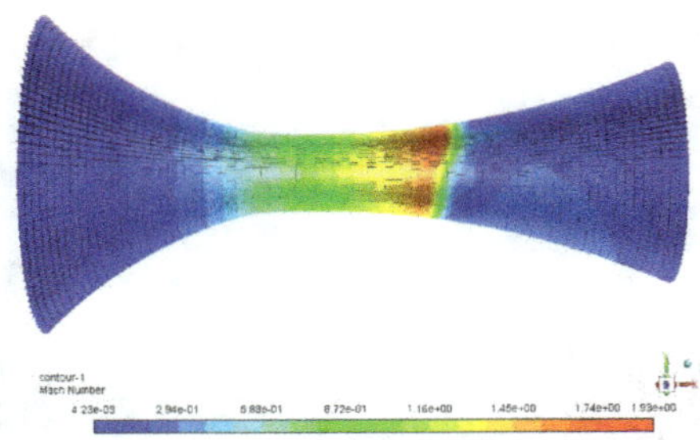

(b) SRK model

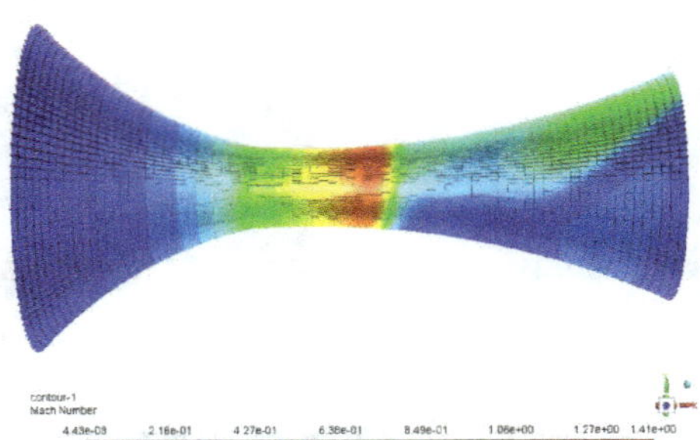

(c) BWR model

Figure 4. Contours of Mach number

Variation of static temperature for the three cases is illustrated in figure 5. Ideal gas model predicts the shock location farther than the real gas model. However, incorporating real gas model in the simulation predicts different location depending on the model adopted. The contour shows that shock inception occurs earlier for BWR model compared to SRK model. However, the minimum temperature for ideal-gas model is shown to be the lowest in comparison with the real gas models (0. 453T_o for the ideal gas, 0. 573T_o for SRK, and 0.71T_o for BWR model).

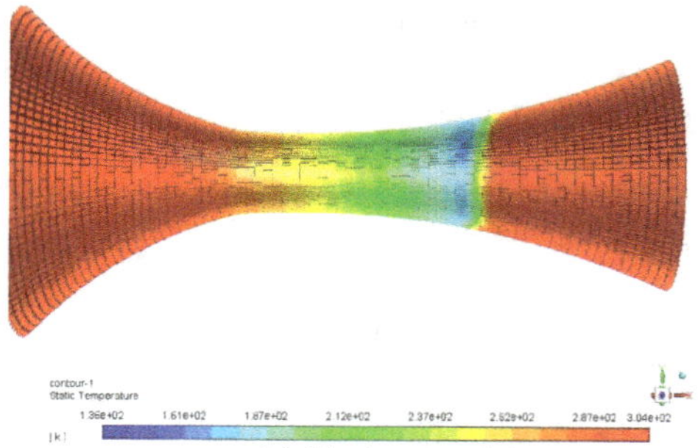

(a) Ideal gas model

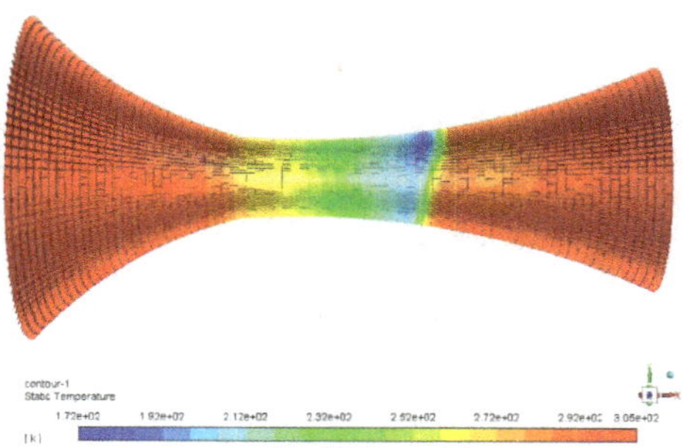

(b) SRK model

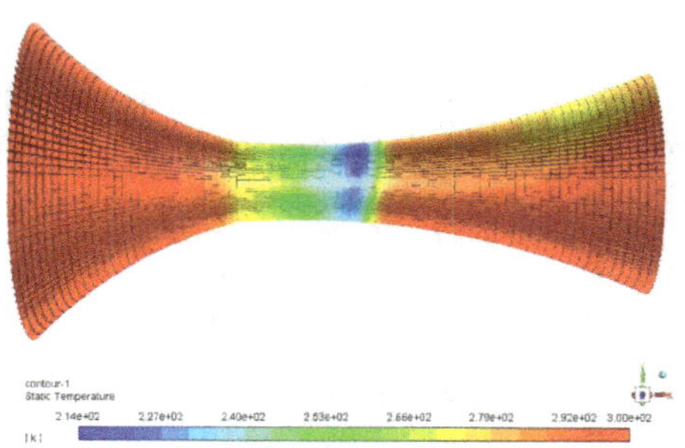

(c) BWR model

Figure 5. Contours of static temperature

Shock location

Normalized shock location is presented in Figure 6 against dimensionless NPR for the three models. The shock location is normalized w.r.t. the length of divergent part of the nozzle (L). The figure concluded that the shock position occurs near the throat when NPR is small for all models. However, Ideal gas model predicts the location farthest at prescribed NPR. BWR model, on the other hand, predicts the position of the shock closer to the throat.

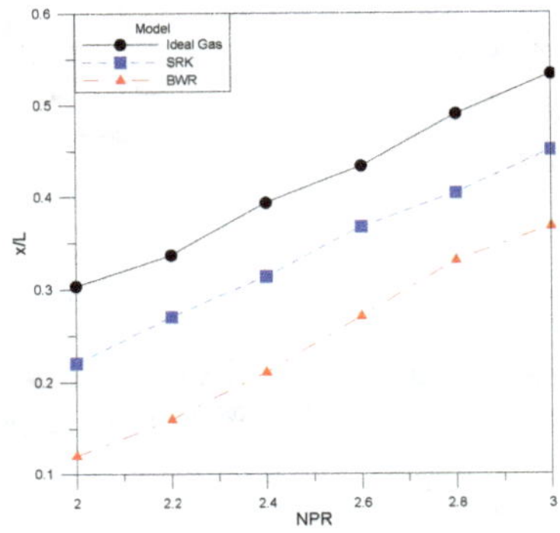

Conclusion

Effect of two real gas models on the fluid properties and shock position is reported in this article by using numerical approach. The two models are compared with the ideal gas model to assess the erroneous in the prediction of shock location once assuming the gas behaves as perfect gas. The results of the simulation found that it is crucial to beware of the real gas fluid behavior when analyzing gas flow systems, particularly high-pressure flow. Comparison between perfect and real gas models showed that the ideal gas assumption would lead to serious misrepresentation of the flow field, including the position of shockwave and severe miscalculations in the design of supersonic devices.

References

[1] Jassim, E., Abdi, M. and Muzychka, Y. (2008b) 'Computational fluid dynamics study for flow of natural gas through high pressure supersonic nozzles: part 2 - nozzle geometry and vorticity', Journal of Petroleum Science and Technology, Vol. 26, No. 15, pp.1773-1785. https://doi.org/10.1080/10916460701304410

[2] Robert, M. N. (1903). The Steam Turbine. Longmans, Green, and Company, London.

[3] Garrett, S. (2000) From Galaxies to Turbines: Science, Technology, and the Parsons Family. Taylor & Francis Group, England.

[4] Jassim, E., Abdi, M. and Muzychka, Y. (2008a) 'Computational fluid dynamics study for flow of natural gas through high pressure supersonic nozzles: part 1 - real gas effects and shockwave', Journal of Petroleum Science and Technology, Vol. 26, No. 15, pp.1757-1772. https://doi.org/10.1080/10916460701287847

[5] Jassim, E.I. (2018), 'Geometrical Impaction ofsupersonic nozzle on the dehumidification performance during Gas purification process -experimental study', Arabian Journal for Science and Engineering. https://doi.org/10.1007/s13369-018-3340-x

[6] Jassim, E. and Awad, M. (2013) 'Numerical investigation of nozzle shape effect on shock wave in natural gas processing', World Academy of Science, Engineering and Technology, Vol. 78, No. 6, pp.347-352.

[7] Kouremenos, D. A. (1986). Normal shockwaves of real gases and the generalized isentropic exponent. Forschung im Ingenieurwesen 52:23-31. https://doi.org/10.1007/BF02558430

[8] Arina, R. (2004). Numerical simulation of near-critical fluids. Appl. Numl. Math. 51:409-426. https://doi.org/10.1016/j.apnum.2004.06.002

[9] Bai-Shi-I, An Introduction into the Theory of Compressible Liquid Flow, Moscow: Izd. Inostrannoi Literatury, 1961.

[10] Jassim, E.I. (2016) 'CFD study on particle separation performance by shock inception during natural gas flow in supersonic nozzle', Progress in Computational Fluid Dynamics, Vol. 16, No. 5, pp.300-312. https://doi.org/10.1504/PCFD.2016.078755

Investigating the influence of alternative fuels on the properties of sand-bentonite liners

Mousa Bani Baker[1,a,*], Adel Hanna[2,b], Maria Elektorowicz[2,c], Tahar Ayadat[3,d]
Batool Al-Shorman[4,e]

[1] Department of Civil and Infrastructure Engineering, Al-Zaytoonah University of Jordan, Amman, Jordan

[2] Department of Building, Civil and Environmental Engineering, Concordia University, Montreal, Canada

[3] Department of Civil Engineering, Prince Mohammad Bin Fahd University, Al-Khobar, Saudi Arabia

[4] Department of Civil Engineering, Jordan University of Science and Technology, Irbid, Jordan.

[a]m.banibaker@zuj.edu.jo, [b]hanna@civil.concordia.ca, [c]mariae@civil.concordia.ca, [d]tayadat@pmu.edu.sa, [e]byalsharman17@eng.just.edu.jo

Keywords: Landfill, Sand-Bentonite Liner, Biofuel, Ethanol-Fuel, Hydraulic Conductivity, Performance

Abstract. Landfills are one of the most effective ways utilized to dispose the wastes and usable up today, most of their failures were suspected due to the failure of their protecting natural liners. Landfill liners failure is one of the most problematic issues that face engineers and expose governments, societies, and the environment to high costs. Protection of groundwater requires that natural landfill liner structure be able to reserve its properties in harsh conditions and over a long period. However, the composition of disposed of residuals might change in the following contemporary trends and it could contain alternative fuels and their impact on natural liners requires further investigations. This paper focuses on the changes in sand-bentonite liner due to the interaction and infiltration of hydrocarbon liquids (alternative fuels); biofuel and ethanol-fuel. In this investigation, an experimental program was carried out to examine the influence of chemical properties of alternative fuels on the hydraulic conductivity, erodibility, swelling potential and shrinkage behaviour of natural liners. Series of laboratory tests were conducted in 20 cm-high PVC columns to investigate the hydraulic conductivity of the liners, swelling behaviour of bentonite when subjected to water, biofuel or ethanol-fuel was assessed by oedometer tests, and shrinking behaviour and cracking patterns of samples taken from the PVC columns were examined employing digital photo analysis. The results can be useful in designing liners, barriers as well in assessing the behaviour of clayey soil in case of accidental spills or intentional discharges.

1. Introduction

Many liners are made watertight with clayey materials, such as sand-bentonite mixtures, that retain liquid and solid toxic wastes. Although many cases were reported for high leakage of bentonite liners [1], they remain in use. In Quebec, for example, 450 projects built since 1980 involved more than 1000 liners. Over time, the principal function of liners is reduced and contaminants leak through them. As these toxic contaminants infiltrate the subsoil and the groundwater, they have serious ramifications on the stability of constructions and the safety of humans and animals. This undoubtedly leads to the elimination of the principal function of the liners over a while, leading to the migration of contaminants through these liners. Tunnelling and the migration of fine particles throughout the liner mixture might be one of the reasons for increasing the hydraulic conductivity. Despite the existing municipal solid waste (MSW) landfill disposal practices, leaching still occurs, leading to heavy metal migration towards the groundwater, whether the liners are built of synthetic materials or sand and bentonite mixtures [2].

Failure of landfill liners is a challenging issue for engineers and a costly one for governments, societies and the environment. The increase in hydraulic conductivity may lead to the failure of the liner. The infiltration of liquid hydrocarbon has a significant increase in hydraulic conductivity from 10-8 cm/s to 10-4 cm/s [3]. To protect groundwater, the structure of a natural landfill liner needs to preserve its properties in harsh conditions for a sustained time. Also, the type of disposing of materials in landfills and storage tanks is changing and it is starting to include alternative fuels such as biofuel or ethanol fuel. [4], studied the function of liquid chemistry on the compressibility and swell characteristics. Two types of bentonites are mixed with pure silica sand and the results indicate that swell potential, swell pressure, swell time, and volume compressibility decreases with the increase of chemical concentrations of the sand-bentonite mixture. As research shows that the coefficient of hydraulic conductivity changes when soil barriers are infiltrated with organic fluids [5]. Various problems appeared in several case histories such as extreme leakage of a stable soil–bentonite mix, internal erosion of a natural clay liner through an insufficient sand filter, washing of bentonite in a soil–bentonite mix, and effects of freezing on the total leakage of a clay liner, [1] and these leading to misuse of the landfill due to decreasing of its linear function. [6] results showed that hydraulic conductivity of bentonite saturated with gasoline was two orders of magnitude 10-5 cm/s higher than the requirement for earthen liner10^{-7} cm/s. Soil/clay liner was suspected to deteriorate when exposed to organic fluids, resulting in a major increase in hydraulic conductivity. [7] observed that compaction conditions of the sand-bentonite liners affect values of the coefficient of permeability of the liner (k). The compaction conditions are resultant in moisture content slightly wet of optimum lead to the smallest values of (k). The particle size of the sand may affect the behaviour of the sand-bentonite liners, [8] found that unheeding of the sand particle size, the sand which mixed with bentonite content less than 20 % showed a general lack of considerable swelling, for the same bentonite content, mixtures with fine sand displayed relatively higher swelling pressure and lower hydraulic conductivity. [9] studied the strength and hydraulic conductivity of the sand-bentonite linear and found that the safest liner carries an 8% of bentonite with sand mixture had a hydraulic conductivity below 1×10^{-7} cm/s. At times researchers typically left out volumetric shrinkage in their researches, but field studies have shown that desiccation can induce harsh cracking of unprotected soil barriers, [10].

Therefore, in this research, the coefficient of hydraulic conductivity is used as an indicator of liner performance (including failure, disintegration, shrinking, fracture, etc). Hence, this paper focuses on the changes within the sand-bentonite liner structure due to the infiltration of biofuel and ethanol fuel.

2. Methodology

The research examines the infiltration of leachate through sand-bentonite barriers. Three liquids were used as leachate: first, water, as a control; then, ethanol fuel, and biofuel. The sand-bentonite barriers (i.e. liners) were composed of sand and bentonite in proportions varying by weight and mixed using the optimum moisture content. Then, each liquid was leached through each liner under specific pressure. The research consisted of laboratory work to observe the infiltration of the three different liquids through the sand-bentonite liners using column leaching test, under different pressures, to test hydraulic conductivity, erodibility, particle size distribution, surface fracture, and changes of some geotechnical properties of the clay materials when exposed to water and alternative fuels. Three different mixtures (100% sand, 95% sand: 5% bentonite, and 90% sand: 10% bentonite) were prepared and tested. First of all, preparing a set of samples (18 liners) tested under 40 kPa and 100 kPa to study the hydraulic conductivity and particle size distribution changes before and after the infiltration of the liquid through the liners. 6 liners were studied using biofuel as permeate and the other 6 ones were studied using ethanol fluid and the last 6 infiltrations by water.

Then set of tests were done as the water was introduced from the bottom part until saturation is achieved, then water was introduced from the top part under the pressure of 20 kPa. The same test will be done for unsaturated samples. Four samples will be taken from different depths 2, 8, 12, and 18 cm for analysis. Digital images will be taken from the top part, middle parts (after cutting the liner into two parts using a guitar wire) and the bottom part; hydraulic conductivities will be tested along the liner depths after the percolation using tubes.

2.1. Preparation of Materials

Two types of minerals were used: silica sand and bentonite clay (Dried sand from Quebec was sieved and sand passed mesh # 40 was used, as per the ASTM standards, ASTM E 11-70 -1995). Bentonite commercial powdered rich in Na-montmorillonite from Houston TX was used in this study for its high swelling capacity, high cation exchange capacity (CEC) and high surface area. The properties of sand were: CEC (meq/100g) =3 as measured in the lab, pH in water (measured in lab) =8.52, size (mesh# =40, D= 0.42 mm), specific surface area (m^2/g, [11]) = 0.1, and percentage of organic matter =0.07 (as measured in the laboratory using ASTM D 2974). The properties of bentonite were: 95, 9.96, 325 (D= 0.044mm), 600, and 4.1, respectively. The mineralogical composition of natural bentonite (X-Ray Analysis: [11]) were: 85% Montmorillonite, 5% Quartz, 2% Feldspars, 0.35% Cristobalite, 2% Illite, and 1% Calcium and Gypsum. Liquids: Water (tap water-Montreal) was used as the permeate reference for all tests, 100% ethanol (source: Quebec) and biofuel (originated from corn oil- Quebec) were used as permeates as well with dynamic viscosities of 0.89, 1.071 and 61 cP respectively, as measured in the laboratory at room temperature. The properties of ethanol were:density (g/cm^3) =0.789,Solubility in water= miscible, and surface tension (mN/m) at 20° C = 22.39. The properties of biofuel were: 0.992, immiscible, and 20-25, respectively.

Materials were blended in different proportions by weight. Then water was added to soil mixtures to reach the optimum moisture content. The optimum moisture contents are 20%, 19%, and 10% for (85%:5%), (90%:10%), and (95%:5%); (sand:bentonite), respectively. Stopping at a maximum of 10% bentonite was deemed sufficient as it satisfies the hydraulic conductivity required to construct liners. A rammer similar to the standard proctor test was used to compact the liner samples. The soil was compacted in the PVC column leaching mold in three equal layers. It delivered 19 blows to each layer as per the ASTM D 698-78 to obtain the equivalent compaction energy due to the change of column diameter (the diameter of the column is less than that in the standard proctor used to find the optimum moisture content). Constructing liners of 100% sand is impractical: it was only used here for comparison purposes. Since sand alone is hard to compact, it was fully saturated to be placed later in the column with compaction by a metallic rod. Liners were tested for hydraulic conductivity for leachate infiltrating downward, representing most cases in the field. Samples were collected from different depths along the liner samples to test for grain size distribution.

2.2. Experimental Method

Column leaching test simulates a clay liner in a landfill or an underground storage tank. Liner materials were prepared and placed in the column in two parts. Each layer was compacted for optimum moisture content using the standard proctor test procedure. Then, the column was leached using different mixtures, liquids and pressure values.

To perform the test, a PVC wall permeameter was used. A PVC column consists of two equal parts; each has a height of 98.00 mm and an inside diameter of 96.00 mm used to contain the liner samples during the experiments. In addition, two PVC heavy-duty flanges were used and held to continuously threaded stainless steel rods and the PVC cell-wall cylinder, as explained in Fig 1. Both ends of the column were equipped with perforated stainless steel plates to support the liner sample and facilitate the flow. Another stainless steel mesh was placed between the stainless steel plate and the liner to confine the liner and prohibit it from swelling. A filter paper (opening diameter of 110 μm) was placed between the mesh and the liner, facing the liner sample to prevent migration of fine particles. The stainless steel plates rest on another permeameter composed of two parts: 1. A bottom part with the same characteristics and a height of 40 mm includes cylinder support open from both sides with openings on the wall bottom sides to facilitate the flow of fluids, support the bottom plate and consequently support the liner. 2. A top part consists of a PVC permeameter with the same diameter and a height of 40 mm. These two parts can be fixed to the rest of the column using O-N Buna rubber rings. The top plate consists of thick PVC heavy-duty flanges with a valve on the top to control the flowing fluid. Another opening, which can be blocked, allows air to be released when starting the flow (bleeding valve). The bottom plate is equipped with an elbow to be used as an outlet opening.

The upper and bottom PVC plates can be fixed to the rest of the column using the rubber rings and threaded stainless steel rods by assembling all parts of the column and pressuring with a hydraulic pressure machine. This prevents leakage from the column.

Three typical cells were manufactured and connected with a pressure vessel, designed using two PVC heavy-duty flanges and a PVC pipe (20.3 cm inside diameter and 50 cm height). The flanges and the pipe were connected by threaded stainless steel rods and rubber rings to be placed between the plates, the top and the bottom parts of the pipe to prevent the fluid from leaking. The top part of the vessel has a pressure regulator and is connected to a pressure cylinder and a safety valve. An inlet opening-ball valve, which can be blocked, is used to fill the liquid in the vessel. The bottom part is connected to the cell via the outlet (valve). The pressure vessel is connected to a transparent tube, from bottom to top, that measures the elevation of liquid in the vessel and displays it during experiments. Fig. 2 shows the experimental setup of the column leaching test.

3. Test Results and Analysis

3.1. Coefficient of Hydraulic Conductivity

The liners of pure sand and sand-bentonite composition with different proportions which infiltration by water, biofuel, and ethanol-fuel; 18 tests were done and the following results were found (Table 1).

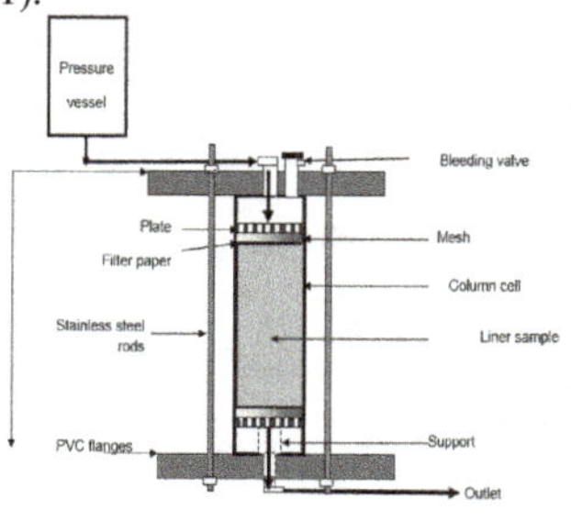

Fig.1. *Setup for column leaching test*

Fig.2. *Column leaching test cells with pressure vessel setup*

The coefficient of permeability or hydraulic conductivity is defined as a constant proportional to the ease with which a fluid passes through a porous medium. The constant head method was used to test for hydraulic conductivity according to ASTM D 2434-68.

Table 1 shows the coefficient of hydraulic conductivity for different liners saturated downward (with the same kind of permeate used as leachate) then leached by water, ethanol and biofuel under pressures of 40 kPa and 100 kPa.

Table 1. *Coefficient of hydraulic conductivity (k, cm/s) for the liners infiltrated by water, ethanol and biofuel under pressures of 40 kPa and 100 kPa*

Liquid / %Bentonie	Water	Ethanol	Biofuel	Pressure (kPa)
0	6.35×10^{-05}	5.72×10^{-05}	1.69×10^{-05}	40
5	1.57×10^{-07}	2.42×10^{-06}	5.10×10^{-06}	
10	1.37×10^{-11}	1.53×10^{-11}	3.12×10^{-08}	
0	7.20×10^{-05}	1.17×10^{-04}	2.11×10^{-05}	100
5	2.30×10^{-07}	8.13×10^{-06}	4.52×10^{-06}	
10	1.90×10^{-11}	1.39×10^{-11}	2.73×10^{-08}	

3.2. Grain Size Distribution

After dismantling the column, disturbed samples were taken from different depths to be tested for particle size distribution using the HORIBA LA-950V2 laser scattering particle size distribution analyzer. Four samples from each liner were tested to evaluate the changes in particle size distribution in the liner under different pressures infiltrated by water, ethanol and biofuel as illustrated in the following figure (Fig. 3).

Fig. 3 showed the variation of the particle size distribution of a liner composed of 90% sand and 10% bentonite.

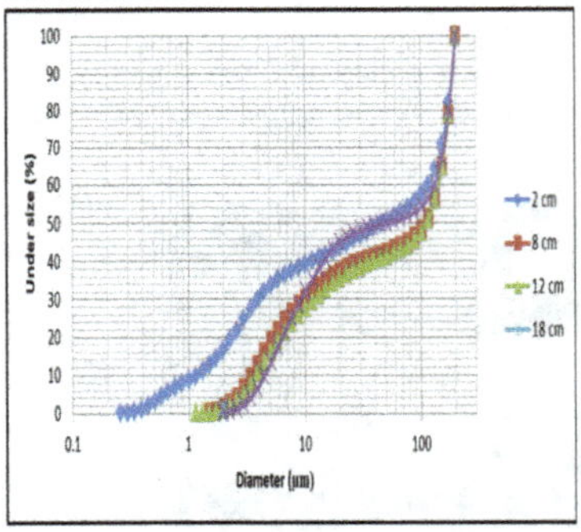

a.

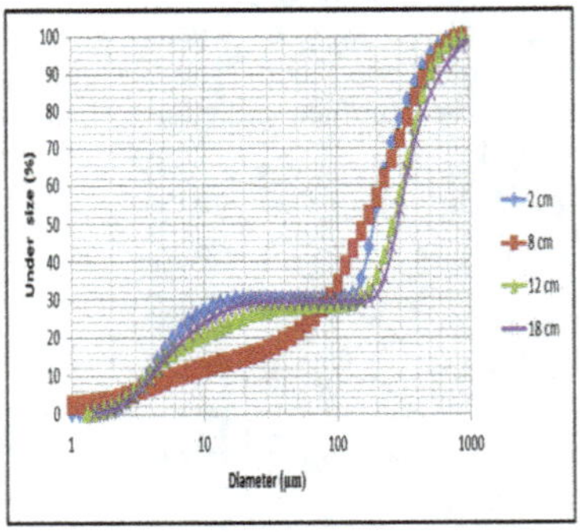

b.

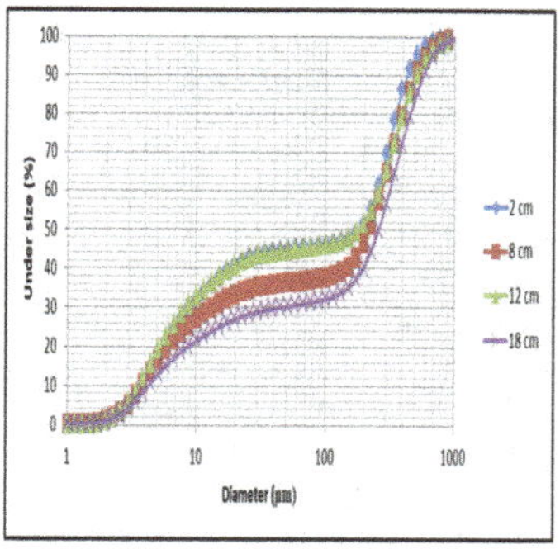

c.

Fig. 3. *Particle size distribution for the liner of 90% sand and 10% bentonite permeated by water, biofuel and ethanol under 40 kPa (a.water, b. biofuel, and c. ethanol)*

Under 40 kPa, the particle size distribution for a liner infiltrated by water ranged from 0.1 to 200 µm at different depths (2, 8, 12 and 18 cm). For a liner infiltrated by biofuel or ethanol, the particle size distribution ranged from 1 to 1000 µm. Particle size distribution curves along the liner's depths behaved in the same manner under 100 kPa. Black spots were detected at 10 to 14 cm down from the liner surface. This could be attributed to the presence of anoxic microorganisms which used ethanol as a carbon source causing ethanol biodegradation, and therefore, no free droplets of an organic liquid (ethanol) were detected by the laser particle size analyzer. The maximum particle size detected was 250 µm. It could be concluded that fine particles in the sand-bentonite mixture were highly susceptible to agglomeration when interacting with fuels. As a result, the sand grains and bentonite clay particles formed bigger flocs which lead to an increase in the liner hydraulic conductivity, an increase in the rate of infiltration, and thus a remarkable fracture along with the liner's depth as shown in Fig 4. The lab test demonstrated changes in liner behaviour depending on bentonite content and the type of permeate. Several Physico-chemical phenomena can be responsible for such behaviour. The results permitted us to conclude that the hydraulic conductivity of liners can increase under a load of waste after the infiltration of alternative fuel leachates.

Fig 4. *Digital image for the upper surface of a 95% sand and 5% bentonite liner leached by biofuel under a pressure of 100 kPa (cracks are showed in circles)*

This increase is due to many factors pertaining to the changes in the structure and composition of sand-bentonite mixtures. For instance, erosion of fine particles, as liquids infiltrate through the liner, affects the hydraulic conductivity. Also, the amount of fine particles flushed out of the mixtures increased with higher pressures, during the leaching of liquids, especially for water.

The fact that particles wash out of a liner exposed to high hydraulic pressure had already been observed [12]. While water causes the highest rate of erosion of soil due to its high solubility, besides, water has low density and viscosity which enhance the erodibility of fine clay particles during the leaching process. The washing out of clay particles from the soil matrix increases with hydraulic pressure. This phenomenon, known as suffusion or piping, increases the hydraulic conductivity of the liner due to the creation of larger pores [13]. In contact with the liner, water causes swelling of the liner's mixture through the swelling of bentonite clay [14] in the liner composition and due to the dispersion of negatively charged clay particles.

As water enters between the particles, it may displace the high-valent cations, which help the clay particles to cluster. In turn, this may create repulsive forces between clay particles, causing the dispersion and additional swelling of bentonite and resulting in the swelling of the whole liner mixture. The surface capillary forces (Fig. 5), which are involved in attracting the soil particles decrease when the additional volume of water fills the pores (e.g. during saturation) and the water film surrounds the grain increases leading to the point of rupture of the liquid bridge due to the increase of its volume. Subsequently, clay can expand. The concave meniscus of water creates repulsive forces between clay particles and leads to the expansion of the clay soils. Biofuel has a convex meniscus (contrary to water concave meniscus) which creates higher surface tension forces (strong attractive forces between the grains and clay particles) with the increase of additional biofuel liquid in pore voids. Subsequently, oil-particle flocs can be created. An extension of this phenomenon depends on temperature due to biofuel viscosity dependence on it.

It was also observed that biofuel caused higher cracks and surface fractures when infiltrating liners compared to ethanol fuel and water. In this study, the grain size distribution of particles increased drastically when liners were infiltrated by alternative fuels (ethanol or biofuel). This is due to the coagulation of fine particles in their interaction with fuel as they adhere together forming larger oil-clay clusters that do not separate in emulsion used for particle size analysis. However, many flocculated and single mineral fine particles remained, as shown in the particle size analysis.

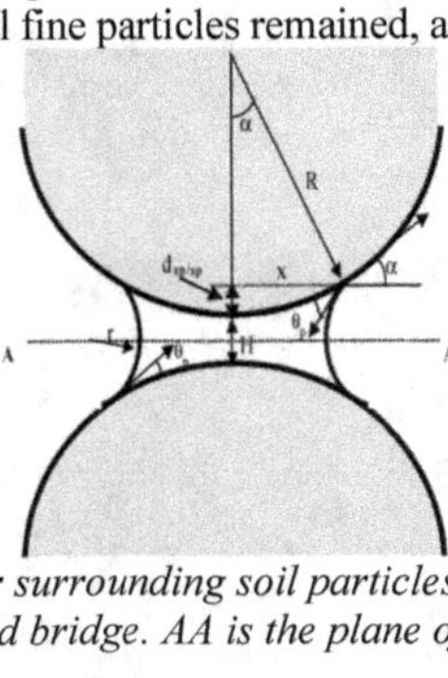

Fig 5. *Capillary forces of water surrounding soil particles, Geometry of the sphere/sphere interaction with a liquid bridge. AA is the plane of symmetry. (after [15])*

Where: γ is the liquid surface tension, R is the particle radius, X is the radius of curvature, θ_p is the contact angle between grain and liquid surface, α is the "embracing angle", and H is the shortest distance between the two spheres.

Clay-ethanol flocs were also observed in the case of using ethanol as a permeate to infiltrate through the liner composed of sand-bentonite mixtures. In the lab experiments, the clay-fuel floc aggregates were formed, which was also found by [16] who observed clay-oil floc aggregates under an optical microscope and concluded that many flocculated single mineral fine particles were still present. A variety of crude oils ranging from light crude oils to heavy crude oils were, therefore, able to interact with micron-sized mineral fines to form "clay-oil flocs" consisting of solids-stabilized oil-in-water emulsions. Also, it is known that the adsorption behaviour appears to be inversely proportional to the solubility of the compound [17] and directly proportional to the percentage of organic matter in the mixture [18]. Biofuel, the least soluble of the three liquids used (water, ethanol, biofuel), appears to have a propensity to be adsorbed in the bentonite clay particles. Thus, bigger particles form through the coagulation of fine particles, causing fractures in the sand-bentonite mixture. This might explains the higher surface fracture in the liner mixtures leached by biofuel. [19] indicated that polar hydrocarbons are required for flocculation to occur. However, experimental tests in this study indicate that both polar (ethanol fuel) and non-polar (biofuel) liquids cause the flocculation of clay fine particles. Clay particles normally have negative charges, and in a stable clay solution, repulsive forces predominate. But when biofuel is in contact with negatively charged bentonite in the liner's mixture, a thin layer forms around the clay particles resulting in their isolation, reducing the repulsive forces and allowing particles to form flocs. Distinctive phenomena can be related to polar organic liquids (ethanol) and non-polar liquids (biofuel) forming oil films. During lab experiments, biofuel flow through the 100% sand liner took more time to reach the column outlet [20, 21]. High biofuel's viscosity creates higher friction forces between the liner and the fluid. Formation of microorganism colonies in the case of using ethanol fuel as permeate was observed in an anoxic part of the leaching column. These microorganisms used organic carbon as a source of energy to multiply. It might influence the performance of the liner.

Conclusion

Alternative fuels have adverse impacts on sand-bentonite liners through which pores are filled with miscible and immiscible or polar and non-polar liquids/residuals. Thus, the properties of the soil matrix were subject to several physical changes such as surface tension and capillary forces. These changes would lead to the formation of oil-clay flocs, provoking fractures, erodibility, and consequently increased the hydraulic conductivity and finally liners failure. Based on the experimental results the following conclusions can be drawn:

1- The voids in sand-bentonite mixtures increase due to the infiltration of alternative fuels, creating channels that enhance the flow of liquids, resulting in a higher coefficient of permeability of the liner.
2- The biological growth of anoxic microorganisms was observed according to the reported substrate (ethanol) consumption and hence influenced the liner's performance.
3- The swelling of bentonite is sensitive to the type of leachate.
4- The migration of fine particles depends on the flow direction and has a direct influence on the hydraulic conductivity of the liner.Fine particles migrate the most in the case of water used as the permeate, less for ethanol and then biofuel.
5- Fine particles flushed out the liner is reversely proportional to the amount of bentonite. In the case of biofuel, there are almost no fines flushed out of the liners, when bentonite is higher than 5% of the liner composition.
6- Liner's surface has more fracture when infiltrated by biofuel compared with ethanol and water. The hydraulic conductivity increases drastically when the liner leached by biofuel.

References

[1] Chapuis R. P., (2002), "The 2000 R.M. Hardy Lecture: Full-scale hydraulic performance of soil-bentonite and compacted clay liners". Canadian Geotech. J. 39: 417-439. https://doi.org/10.1139/t01-092

[2] Lee J.-M., Shackelford D., Benson H., Jo H.-Y., and Edil B., (2005), "Correlating index properties and hydraulic conductivity of geosynthetic clay liners". Journal of Geotechnical and Geoenvironmental Engineering, November 131: 1319-1329. https://doi.org/10.1061/(ASCE)1090-0241(2005)131:11(1319)

[3] Fernandez F., and Quigley R. M., (1985)," Hydraulic conductivity of natural clays permeated with simple liquid hydrocarbons". Canadian Geotechnical Journal. 22: 205-214. https://doi.org/10.1139/t85-028

[4] Alawaji, H. (1999). Swell and compressibility characteristics of sand-bentonite mixtures inundated with liquids. Applied Clay Science, 15(3-4), 411-430. doi:10.1016/s0169-1317(99)00033-2. https://doi.org/10.1016/S0169-1317(99)00033-2

[5] Katsumi T., Benson C. H., Foose G. J., and Kamon M., (2001), "Performance-based design of landfill liners".Engineering Geology 60: 139-148. https://doi.org/10.1016/S0013-7952(00)00096-X

[6] Yang X. and Lo M. C., (2004), "Flow of Gasoline through Composite Liners". Journal of Environmental Engineering ASCE, pp: 886-890. https://doi.org/10.1061/(ASCE)0733-9372(2004)130:8(886)

[7] Amadi, A. A., & Eberemu, A. O. (2012). Delineation of compaction criteria for acceptable hydraulic conductivity of lateritic soil-bentonite mixtures designed as landfill liners. Environmental Earth Sciences, 67(4), 999-1006. doi:10.1007/s12665-012-1544-z https://doi.org/10.1007/s12665-012-1544-z

[8] Srikanth, V., & Mishra, A. K. (2016). A Laboratory Study on the Geotechnical Characteristics of Sand-Bentonite Mixtures and the Role of Particle Size of Sand. International Journal of Geosynthetics and Ground Engineering, 2(1). doi:10.1007/s40891-015-0043-1 https://doi.org/10.1007/s40891-015-0043-1

[9] Ojuri, O.O. and Oluwatuyi, Opeyemi E. (2017) Strength and Hydraulic Conductivity Characteristics of Sand-Bentonite Mixtures Designed as a Landfill Liner. Jordan Journal of Civil Engineering, 11 (4). pp. 614-622.

[10] Yamusa, B. Y., Kamarudin, A. ,& Norhan, A. (2017). Hydraulic Conductivity And Volumetric Shrinkage Properties Review On Gradation Effect Of Compacted Laterite Soil Liner. Malaysian Journal of Civil Engineering 29 Special Issue (1):153-164

[11] Baroid Industrial Drilling Products Company (bentonite data sheet). (2008). Product Service Line, Halliburton 3000 N. Sam Houston Pkwy E. Houston, TX 77032.

[12] Marcotte D., Marron J.C., and Fafard M., (1994), "Washing of bentonite in laboratory hydraulic conductivity tests". Journal of Environmental Engineering 120 (3): 691-698. https://doi.org/10.1061/(ASCE)0733-9372(1994)120:3(691)

[13] Kaoser S., Barrington S., Elektorowicz M., and Ayadat T., (2006), "The influence of hydraulic gradient and rate of erosion on hydraulic conductivity of sand-bentonite mixtures". Soil & Sediment Contamination, 15: 481-496. https://doi.org/10.1080/15320380600847815

[14] Bani Baker, M., Abendeh, R., Sharo, A., Hanna, A. Stabilization of Sandy Soils by Bentonite Clay Slurry at Laboratory Bench and Pilot Scales. Coatings/Functional Building Materials and Pavements Coatings, 2022, 12 (12). https://doi.org/10.3390/coatings12121922

[15] Rabinovich Y. I., Esayanur M. S., and. Moudgil B. M., (2005), "Capillary Forces between Two Spheres with a Fixed Volume Liquid Bridge: Theory and Experiment". Langmuir 2005, 21, pp: 10992-10997. https://doi.org/10.1021/la0517639

[16] Weise M., (1997), "The significance of clay-oil flocculation to oil biodegradation".Department of Botany at the University of Toronto.Toronto-Ontario-Canada.

[17] Meyers A. and Oas G., (1978). "Comparison of associations of different hydrocarbons with clay particles in simulated seawater".Environmental Science Technology. 12: 934-937. https://doi.org/10.1021/es60144a005

[18] Kirchhoff W., Brown S., and Scott A., (1978), "Sorption of hydrophobic pollutants on natural sediments". Water Resources, 13: 241 -248. https://doi.org/10.1016/0043-1354(79)90201-X

[19] Bragg, J.R. and S.H. Yang, (1995), "Clay-oil flocculation and its role in natural cleansing in Prince William Sound following the Exxon Valdez oil spill". "Exxon Vaidez oil spill: Fate and Effects in Alaskan waters, ASTM STP-1219." Peter G. Wells, James. J. Butler, and Jane S. Hugues, Eds., American Society for Testing and . Materials. Philadelphia. (pp 178-214). https://doi.org/10.1520/STP19864S

[20] Bani Baker, M., Elektorowicz ,M., Hanna, A., (2018), Elctrokienetic Nondestructive In-situ Technique for Rehabilitation of Liners Damaged by Fuels, Journal of Hazardous Materials. Vol. 359. Pp. 510-515. https://doi.org/10.1016/j.jhazmat.2018.07.113

[21] Sharo, A.A., Khasawneh, M.A., Bani Baker, M., Al Tarawneh, D.M.Sonicated waves procedure effect on stabilizing expansive soil by nano-clay: Treat with cause. Frontiers in Built Environment, 2022, 8, 975993 https://doi.org/10.3389/fbuil.2022.975993

On the deformation of layered composite arches using exponential shear and normal deformation theory

Valmik M. Mahajan[1,a*] and Amit Sharma[2,b]

[1] Research Scholar, Department of Civil Engineering, Oriental University Indore-453555, Madhya Pradesh, India

[2] Professor, Department of Civil Engineering, Oriental University Indore-453555, Madhya Pradesh, India

[a]valmikmahajan001@gmail.com, [b]sgsitsamit25@gmail.com

Keywords: Deformation, Stresses, Displacements, Layered Composite Arches, ESNDT

Abstract. In the present study, the stresses and displacements are analyzed for layered composite arches of various lamination schemes subjected to uniformly loading. The present work is majorly highlighted the effects of transverse normal stress and transverse normal strain using exponential shear and normal deformation theory (ESNDT). Governing equations are derived using Hamilton's principle with application of Navier's method subjected to simply supported end conditions. Present theory is free from use of any shear correction factor and it satisfies the zero traction free end boundary condition at the top and bottom surfaces of the layered composite arches. In the present work symmetric and antisymmetric lamination scheme have been studied to obtain the numerical results for four layered composite arches and is validated through results available in prior literature.

Introduction

Composite arches are widely used across the world due to their exceptional qualities such as excellent strength, an admirable stiffness-to-weight ratio, and outstanding fatigue-resistance. Now a day, there is enormous demand in architectural view of the structures. Composite arches are most suitable for such innovative constructions. Hence, design of layered composite arches is very useful because of their superior properties compared to available materials. Subsequently, these composite materials are mostly used in mechanical and civil engineering, marine, automobile private sector, aerospace engineering, aircraft, ships, bridges and spacecraft. Curved and arched surfaces are commonly used in advanced architectural structures and military engineering for fighter jet, rocket launchers, antiballistic missiles, aircraft carriers, antitank-mines etc. The vast range of literature is available on straight beams/surfaces subjected to several loadings. Bernoulli-Euler [1, 2] established beam theory universally named as classical beam theory (CBT). But authors ignored the effect of thickness stretching. Further study has modified and improved by Timoshenko [3] with accounted the effect of shear deformation but it requires shear correction factor. These theories well-known as Timoshenko beam theory (TBT) or first order shear deformation (FSDT) theory in 1921. Classical and Timoshenko beam theory doesn't capture the transverse normal strain and shear deformation effects. In few decades, it is found that a very limited research is completed on arches or curved beams.

Reddy [4] has developed higher-order theory for composite-laminated plates and obtained results are validated through first-order-shear-deformation theory and 3-D elasticity solutions. Carrera [5] studied thermal-stress analysis for layered and isotropic (homogeneous) plates by adopting thickness stretching effect. Carrera *et al.* [6] addressed a model for static responses of FGM plates known as variable kinematic model subjected to mechanical loads. Zenkour [7] established the 3-D elasticity solution for sandwich and cross-ply laminates exposed to sinusoidally distributed (SDL) and uniformly distributed (UDL) loads. Kant and Shiyekar [8] presented model on cylindrical bending for piezoelectric-laminates plates using higher order theory. Tornabene [9] and his co-authors investigates the responses of static behavior for curved panels using DQM, differential geometry and Carrera-unified-formulation approach and also investigates recovery of repossession of stresses, shear-strains and transverse-normal through-the-thickness variations for functionally graded sandwich panels [10]. Sayyad and Ghugal [11] studied cylindrical bending for multilayered composite laminate plates using theory of higher-order. Present exponential shear and normal deformation theory captures excellent structural behavior of layered composite arches due to consideration of effect of transverse normal strain and transverse shear deformation. Present theory will be valuable asset in the research field of aerospace, civil and mechanical structures.

Mathematical-formulation for layered arches

Considered layered composite arches with one end is roller supported and another is hinged support with radius of curvature (*R*), for a length (*L*), total thickness (*h*) and unit width of the arch (*b*) $\{0 \le x \le L;\ -b/2 \le y \le b/2;\ -h/2 \le z \le h/2\}$ as shown in Fig. 1.

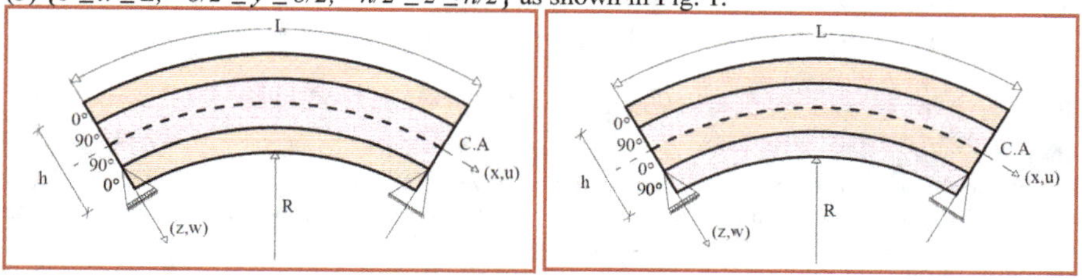

(a) Symmetric layered ($0^0/90^0/90^0/0^0$) arch (b) Antisymmetric layered ($0^0/90^0/0^0/90^0$) arch

Fig. 1 Geometry and coordinates system for layered composite arches

Displacement field

Displacement field for layered composite arches with consideration of transverse normal strain and shear deformations are given below,

$$u_{(x,z)} = \left(1+\frac{z}{R}\right)u_0 - z\frac{\partial w_0}{\partial x} + f(z)\phi \quad \text{and} \quad w_{(x,z)} = w_0 + f'(z)\psi . \tag{1}$$

Strain displacement relationship

$$\varepsilon_x^k = \left[\frac{\partial u}{\partial x}+\frac{w}{R}\right],\ \varepsilon_z^k = \left[\frac{\partial w}{\partial z}\right],\ \gamma_{xz}^k = \left[\frac{\partial u}{\partial z}+\frac{\partial w}{\partial x}-\frac{u_0}{R}\right]. \tag{2}$$

$$\varepsilon_x^k = \left[\frac{\partial u_0}{\partial x} - z\frac{\partial^2 w_0}{\partial x^2} + f(z)\frac{\partial \phi}{\partial x} + \frac{w_0}{R} + \frac{f'(z)}{R}\psi\right],\ \varepsilon_z^k = f''(z)\psi,\ \gamma_{xz}^k = f'(z)\left(\phi+\frac{\partial \psi}{\partial x}\right). \tag{3}$$

$$\text{where, } f(z) = z e^{-2\left(\frac{z}{h}\right)^2},\ f'(z) = e^{-2\left(\frac{z}{h}\right)^2}\left(1-\frac{4z^2}{h^2}\right),\ f''(z) = e^{-2\left(\frac{z}{h}\right)^2}\left(\frac{16z^3-12zh^2}{h^4}\right). \tag{4}$$

where, u_0, w_0, ϕ and ψ be the four unknown functions at mid-plane for composite arches. *f(z)* and *f'(z)* is shear and normal deformations. In present theory transverse normal strain is not equal to zero i.e. $\varepsilon_z \neq 0$. Shear deformation considered at any point on the arch as stated in Eq. (3).

Hooke's law

Two dimensional Hooke's law is applied layerwise to obtained equations for axial bending stresses and shear stresses with reference axes (x,z) from Eq. (5).

$$\begin{Bmatrix}\sigma_x\\ \sigma_z\\ \tau_{xz}\end{Bmatrix}^k=\begin{bmatrix}Q_{11} & Q_{13} & 0\\ Q_{13} & Q_{33} & 0\\ 0 & 0 & Q_{55}\end{bmatrix}^k\begin{Bmatrix}\varepsilon_x\\ \varepsilon_z\\ \gamma_{xz}\end{Bmatrix}^k \text{ or } \{\sigma\}^k=\left[Q_{ij}\right]^k\{\varepsilon\}^k. \tag{5}$$

where, $\{\sigma\}^k$ be normal stresses, $\{\varepsilon\}^k$ be transverse-shear-strain, and $\left[Q_{ij}\right]^k$ be transformed-rigidity-matrix w.r.t. (x,z) axes. Reduced stiffness coefficients are given below,

$$Q_{11}^k=\left[\frac{E_1^k}{1-(\mu_{13}*\mu_{31})}\right],\ Q_{13}^k=\left[\frac{E_3^k*\mu_{13}}{1-(\mu_{13}*\mu_{31})}\right],\ Q_{33}^k=\left[\frac{E_3^k}{1-(\mu_{13}*\mu_{31})}\right],\ Q_{44}^k=G_{23}^k,\ Q_{55}^k=G_{13}^k. \tag{6}$$

where, μ_{13} and μ_{31} be the Poisson's ratio, E_1^k and E_3^k are elasticity-modulus about (x,z) axes and G_{13}^k and G_{23}^k are the shear-modulus. Axial-bending stresses is stated as per Hooke's law,

$$\sigma_x^k=Q_{11}^k\,\varepsilon_x^k+Q_{13}^k\,\varepsilon_z^k,\ \sigma_z^k=Q_{13}^k\,\varepsilon_x^k+Q_{33}^k\,\varepsilon_z^k,\ \tau_{xz}^k=Q_{55}^k\gamma_{xz}^k,\ \varepsilon_y^k=\gamma_{xy}^k=\gamma_{yz}^k=0. \tag{7}$$

Hamilton's principle

Using virtual work principle with application of integration by-parts for traction free simply supported boundary condition are given below, where, (δ) called as variational-operator. The complete expression for Eq. (8) generated from Eq. (2). we have,

$$\therefore \int_0^L\int_{-h/2}^{+h/2}\left(\sigma_x^k\,\delta\varepsilon_x+\sigma_z^k\delta\varepsilon_z+\tau_{xz}^k\,\delta\gamma_{xz}\right)dx\,dz-\int_0^L\left(q\,\delta w\,dx\right)=0. \tag{8}$$

$$\begin{aligned}\int_{-h/2}^{+h/2}\int_0^L\left(\sigma_x^k\,\delta\varepsilon_x\right)dx\,dz=\int_{-h/2}^{+h/2}\int_0^L&\left[Q_{11}\left(\frac{\partial u_0}{\partial x}-z\frac{\partial^2 w_0}{\partial x}+f(z)\frac{\partial\phi}{\partial x}+\frac{w_0}{R}+\frac{f'(z)}{R}\psi\right)+Q_{13}\left(f''(z)\right)\psi\right].\\ &*\left[\frac{\partial\delta u_0}{\partial x}-z\frac{\partial^2\delta w_0}{\partial x}+f(z)\frac{\partial\delta\phi}{\partial x}+\frac{\delta w_0}{R}+\frac{f'(z)}{R}\delta\psi\right]dx\,dz\end{aligned} \tag{9}$$

$$\int_{-h/2}^{+h/2}\int_0^L\left(\sigma_z^k\,\delta\varepsilon_z\right)dx\,dz=\int_{-h/2}^{+h/2}\int_0^L\left\{\begin{bmatrix}Q_{13}\left(\frac{\partial u_0}{\partial x}-z\frac{\partial^2 w_0}{\partial x}+f(z)\frac{\partial\phi}{\partial x}+\frac{w_0}{R}+\frac{f'(z)}{R}\psi\right)\\ +Q_{33}\left(f''(z)\right)\psi\end{bmatrix}*\left[f''(z)\,\delta\psi\right]\right\}dx\,dz. \tag{10}$$

$$\int_{-h/2}^{+h/2}\int_0^L\left(\tau_{xz}^k\,\delta\gamma_{xz}\right)dx\,dz=\int_{-h/2}^{+h/2}\int_0^L\left\{\left[Q_{55}\,f'(z)\left(\phi+\frac{\partial\psi}{\partial x}\right)\right]*\left[f'(z)\left(\delta\phi+\frac{\partial\delta\psi}{\partial x}\right)\right]\right\}dx\,dz. \tag{11}$$

From Eq. (9) - (11) integrate individuals one by one and solve the expression by-parts rule and finally gathering the terms of δu_0, δw_0, $\delta\phi$ and $\delta\psi$ to build governing equations.

$$\begin{bmatrix}A_{11},B_{11},C_{11},D_{11},E_{11},F_{11}\\ G_{11},H_{11},I_{11},J_{11}\end{bmatrix}=Q_{11}^k\int_{-h/2}^{+h/2}\begin{bmatrix}1,z,f(z),f'(z),z^2,z\,f(z)\\ z\,f'(z),f(z)^2,f(z)\,f'(z),f'(z)^2\end{bmatrix}dz. \tag{12}$$

$$\left[L_{13},M_{13},N_{13},O_{13}\right]=Q_{13}^k\int_{-h/2}^{+h/2}\left[f''(z),z\,f''(z),f(z)\,f''(z),f'(z)\,f''(z)\right]dz. \tag{13}$$

$$[L_{33}]=Q_{33}^{k}\int_{-h/2}^{+h/2}\left[f''(z)^{2}\right]dz,[J_{55}]=Q_{55}^{k}\int_{-h/2}^{+h/2}\left[f'(z)^{2}\right]dz. \tag{14}$$

Governing equation achieved in the normalized form with integration constants are listed below from Eq. (15) to Eq. (18).

$$\delta u_0:-A_{11}\left(\frac{\partial^2 u_0}{\partial x^2}\right)+B_{11}\left(\frac{\partial^3 w_0}{\partial x^3}\right)-\frac{A_{11}}{R}\left(\frac{\partial w_0}{\partial x}\right)-C_{11}\left(\frac{\partial^2\phi}{\partial x^2}\right)-\left(\frac{D_{11}}{R}+L_{13}\right)\left(\frac{\partial\psi}{\partial x}\right)=0. \tag{15}$$

$$\begin{aligned}\delta w_0:&-B_{11}\left(\frac{\partial^3 u_0}{\partial x^3}\right)+\frac{A_{11}}{R}\left(\frac{\partial u_0}{\partial x}\right)+E_{11}\left(\frac{\partial^4 w_0}{\partial x^4}\right)-2\frac{B_{11}}{R}\left(\frac{\partial^2 w_0}{\partial x^2}\right)+\frac{A_{11}}{R^2}(w_0)-F_{11}\left(\frac{\partial^3\phi}{\partial x^3}\right)+\frac{C_{11}}{R}\left(\frac{\partial\phi}{\partial x}\right)\\&+\left(\frac{D_{11}}{R^2}+\frac{L_{13}}{R}\right)(\psi)-\left(\frac{G_{11}}{R}+M_{13}\right)\left(\frac{\partial^2\psi}{\partial x^2}\right)=q_0\end{aligned}. \tag{16}$$

$$\begin{aligned}\delta\phi:&-C_{11}\left(\frac{\partial^2 u_0}{\partial x^2}\right)+F_{11}\left(\frac{\partial^3 w_0}{\partial x^3}\right)-\frac{C_{11}}{R}\left(\frac{\partial w_0}{\partial x}\right)+J_{55}\phi-H_{11}\left(\frac{\partial^2\phi}{\partial x^2}\right)-\frac{I_{11}}{R}\left(\frac{\partial\psi}{\partial x}\right)-N_{13}\left(\frac{\partial\psi}{\partial x}\right)\\&+J_{55}\left(\frac{\partial\psi}{\partial x}\right)=0\end{aligned}. \tag{17}$$

$$\begin{aligned}\delta\psi:&\left(\frac{D_{11}}{R}+L_{13}\right)\left(\frac{\partial u_0}{\partial x}\right)+\left(\frac{D_{11}}{R^2}+\frac{L_{13}}{R}\right)w_0-\left(\frac{G_{11}}{R}+M_{13}\right)\left(\frac{\partial^2 w_0}{\partial x^2}\right)+\frac{I_{11}}{R}\left(\frac{\partial\phi}{\partial x}\right)+N_{13}\left(\frac{\partial\phi}{\partial x}\right)\\&-J_{55}\left(\frac{\partial\phi}{\partial x}\right)+\frac{J_{11}}{R}\psi+2\frac{O_{13}}{R}\psi+L_{33}\psi-J_{55}\left(\frac{\partial^2\psi}{\partial x^2}\right)=0\end{aligned}. \tag{18}$$

Navier's method

This technique is applied for simply supported (SS) boundary condition for layered composite arch under the action of transverse uniformly distributed loading (UDL).

$$u_0=w_0=\phi=\psi=0,\ \text{at}\ x=L, x=0. \tag{19}$$

In the form of trigonometric, unknown variables are listed below,

$$u_0=\sum_{m=1}^{\infty}u_m\cos(\alpha x),\ w_0=\sum_{m=1}^{\infty}w_m\sin(\alpha x),\ \phi=\sum_{m=1}^{\infty}\phi_m\cos(\alpha x),\ \psi=\sum_{m=1}^{\infty}\psi_m\sin(\alpha x) \tag{20}$$

$$q_0=q_m\sin(\alpha x),\qquad \left(\because\alpha=\frac{m\pi}{L}\right). \tag{21}$$

where, u_m, w_m, ϕ_m and ψ_m be the unknown-factors. Transverse UDL as stated below,

$$q_{(x)}=\sum_{m=1}^{\infty}\left(\frac{4q_0}{m\pi}\sin(\alpha x)\right). \tag{22}$$

where, m=Positive integer-variables from odd numbers to the infinity (∞). Substituting the values of unit loading from Eq. (22) and, (u,w,ϕ,ψ) unknown variables of Eq. (20) and Eq. (21) by putting in the governing equations. Bending stresses for layered composite arches are presented in matrix form of Eq. (23) is given below.

$$[K]\{\Delta\}=\{f\} \text{ or } \begin{bmatrix} K_{11} & K_{12} & K_{13} & K_{14} \\ K_{21} & K_{22} & K_{23} & K_{24} \\ K_{31} & K_{32} & K_{33} & K_{34} \\ K_{41} & K_{42} & K_{43} & K_{44} \end{bmatrix} \begin{Bmatrix} u_m \\ w_m \\ \phi_m \\ \psi_m \end{Bmatrix} = \begin{Bmatrix} 0 \\ q_m \\ 0 \\ 0 \end{Bmatrix}. \tag{23}$$

where, $[K]$= known as stiffness-matrix; $\{\Delta\}$= called as unknown-variables and $\{f\}$= known as force-vector. The each elements of stiffness matrix are stated below,

$$K_{11}=A_{11}\alpha^2 \quad K_{12}=-\left(\frac{A_{11}}{R}\alpha+B_{11}\alpha^3\right) \quad K_{13}=C_{11}\alpha^2 \quad K_{14}=-\left(\frac{D_{11}}{R}\alpha+L_{13}\alpha\right)$$

$$K_{21}=K_{12} \quad K_{22}=\left(\frac{A_{11}}{R^2}+2\frac{B_{11}}{R}\alpha^2+E_{11}\alpha^4\right) \quad K_{23}=-\left(\frac{C_{11}}{R}\alpha+F_{11}\alpha^3\right) \quad K_{24}=\left(\frac{D_{11}}{R^2}+\frac{L_{13}}{R}\right)+\left(\frac{G_{11}}{R}\alpha^2+M_{13}\alpha^2\right)$$

$$K_{31}=K_{13} \quad K_{32}=K_{23} \quad K_{33}=\left(H_{11}\alpha^2+J_{55}\right) \quad K_{34}=-\left(\frac{I_{11}}{R}\alpha+N_{13}\alpha-J_{55}\alpha\right)$$

$$K_{41}=K_{42} \quad K_{42}=K_{24} \quad K_{43}=K_{34} \quad K_{44}=\left(\frac{J_{11}}{R^2}+2\frac{O_{13}}{R}+L_{33}+J_{55}\alpha^2\right)$$

Present theory considered the normalized displacements and stresses relation are given below,

$$\bar{w}=\frac{100E_3h^3}{q_0L^4}w\left(\frac{L}{2},0\right),\ \bar{u}=\frac{E_3}{q_0h}u\left(0,-\frac{h}{2}\right),\ \bar{\sigma}_x=\frac{h}{q_0}\sigma_x\left(\frac{L}{2},-\frac{h}{2}\right),\ \bar{\tau}_{xz}=\frac{\tau_{xz}}{q_0}(0,0). \tag{24}$$

where $\bar{w}$, $\bar{u}$, $\bar{\sigma}_x$ and $\bar{\tau}_{xz}$ be dimensionless parameters.

Numerical results with discussions

Numerical results are presented in tabular form consists of Table 2 to Table 5 and variations of displacements and stresses for layered composite arches through the thickness are plotted using Grapher as shown in Figure 2 to Figure 5. Present theory analyzed the different lamination schemes for arches viz, symmetric and antisymmetric layered composite arches. The material properties for various arches are given below in Table 1.

Table 1 Materials property for layered composite arches.

Theory	Source	Lamination scheme	Properties
Present	Sayyad and Ghugal [11]	$(0^0/90^0/90^0/0^0)$ symmetric arch	E_1 =181 GPa; E_3 =10.3 GPa; G_{13} =7.17 GPa; G_{23} =2.87 GPa; μ_{13} =0.25; μ_{31} =0.01.
		$(0^0/90^0/0^0/90^0)$ antisymmetric arch	E_1=172.5 GPa; E_3=6.9 GPa; G_{13}=3.45 GPa; G_{23}=1.38 GPa; μ_{13} =0.25; μ_{31} =0.01.

Table 2 presented the non-dimensional results for four layered symmetric straight beams when subjected to transverse uniformly distributed loadings. Present results are in good-agreement with earlier published results by Sayyad and Ghugal [11] of normalized stresses and displacements for aspect ratio L/h = 4, 10 and 100. Present numerical results are compared and closely matches with well-known theory of Reddy [4] for transverse and axial deformation at aspect ratio L/h = 100.

Normalized axial displacements and transverse deflections are plotted through the thickness of symmetric layered composite arch as shown in Figure 2. With the application of constitutive relation, it is observed that at the interlaminar surfaces of arches shows two values for shear stress and axial bending through the thickness as shown in Figure 3 for symmetric layered and Figure 5 for antisymmetric layered composite arch.

Table 2 Normalized displacements and stresses for four layered ($0^0/90^0/90^0/0^0$) straight beams.

Theory	L/h	Model	$\bar{w}$ (L/2,0)	$\bar{u}$ (0,-h/2)	$\bar{\sigma}_x$ (L/2,-h/2)	$\bar{\tau}_{xz}$ (0,0)
Present ($\varepsilon_z \neq 0$)		ESNDT	3.4548	1.4850	18.1369	2.1863
Sayyad and Ghugal [11]		SSNPT	3.4354	1.4648	17.9237	2.3734
Reddy [4]	4	HSDT	3.4033	1.4401	17.6119	2.3250
Timoshenko [3]		FSDT	2.6074	1.0285	13.6041	2.5586
Bernoulli-Euler [1,2]		CBT	1.0044	1.0285	13.6041	2.5586
Present ($\varepsilon_z \neq 0$)		ESNDT	1.4073	17.3079	89.4987	5.9314
Sayyad and Ghugal [11]		SSNPT	1.3986	17.1903	89.3921	6.0174
Reddy [4]	10	HSDT	1.3939	17.1607	89.0567	6.0575
Timoshenko [3]		FSDT	1.2609	16.0703	85.0255	6.3965
Bernoulli-Euler [1,2]		CBT	1.0044	16.0703	85.0255	6.3965
Present ($\varepsilon_z \neq 0$)		ESNDT	1.0091	16094.00	8499.900	62.0397
Sayyad and Ghugal [11]		SSNPT	1.0057	16039.93	8509.589	63.8913
Reddy [4]	100	HSDT	1.0084	16082.14	8507.056	63.9154
Timoshenko [3]		FSDT	1.0069	16070.26	8502.562	63.9646
Bernoulli-Euler [1,2]		CBT	1.0044	16070.26	8502.562	63.9646

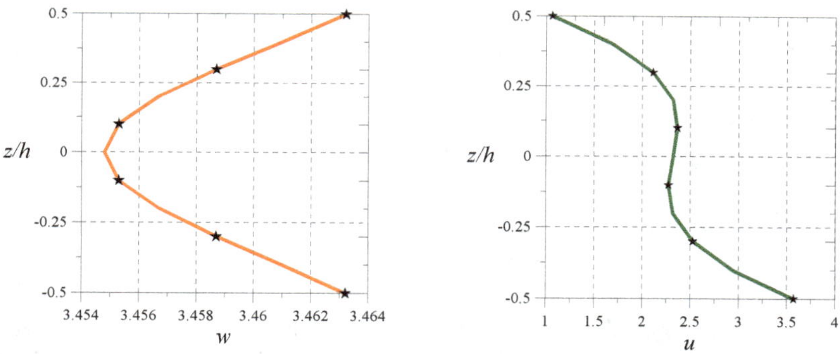

Fig. 2 Normalized transverse and axial deformations through-the-thickness for four layered symmetric ($0^0/90^0/90^0/0^0$) composite arch due to uniformly distributed loading [R/h =5, L/h =4].

Table 3 presented the normalized transverse ($\bar{w}$) deformation, axial ($\bar{u}$) deformation, axial bending stress ($\bar{\sigma}_x$) and shear stress ($\bar{\tau}_{xz}$) deformation for four-layered symmetric composite arch subjected to uniformly loading for aspect ratio (L/h = 4, 10, 100). It is observed that, maximum non-dimensional value of axial deformation and axial bending stress have noted at the top fibre of layered arch i.e. *(z/h = -h/2)* due to placing of fibers in 0^0 horizontal direction along the length of arches. While minimum non-dimensional bending stress and axial displacement have been reported at bottom surface of the arch i.e. *(z/h = +h/2)*, it means that fibers are laid in 90^0 direction or perpendicular to zero degree layer of the arch. From Table 3 it is observed that transverse deflection and shear stress deformation remains constants with varying radius of curvature. Table 4 presented numerical results of stresses and displacements for four-layered antisymmetric composite arch. It is found that normalized shear stress and transverse deflection are remains constant with varying radius of curvature for aspect ratio L/h = 4, 10, 100 (R/h = 1.0 to ∞).

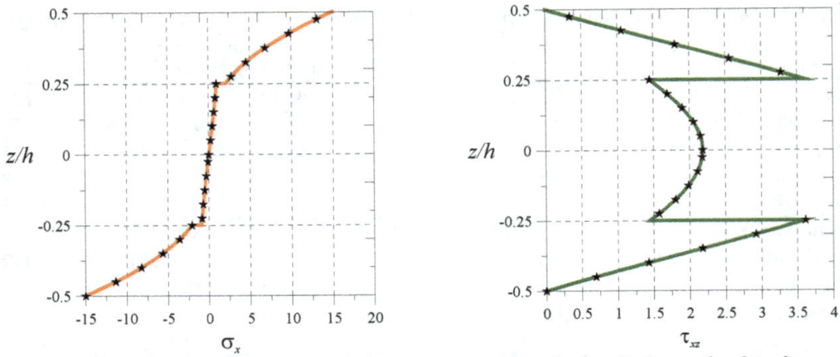

Fig. 3 Normalized axial-bending and shear stresses through-the-thickness for four layered symmetric ($0^0/90^0/90^0/0^0$) composite arch due to uniformly distributed loading [R/h =5, L/h =4].

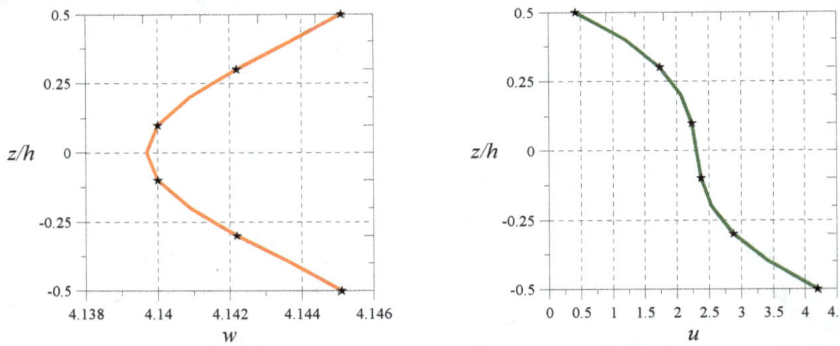

Fig. 4 Normalized transverse and axial deformations through-the-thickness for four layered antisymmetric ($0^0/90^0/0^0/90^0$) composite arch due to uniformly distributed loading [R/h=5,L/h=4]

It is observed that the variations of bending stress through the thickness is nearly equal to zero when the fibers are placed in 90^0 direction of 2nd and 4th layer of antisymmetric composite arch under the action of transverse uniformly distributed loading as shown in Figure 5. It is also observed that bending stress is increasing parabolically in 1st layer from maximum non-dimensional to zero and 3rd layer varying from minimum to maximum non-dimensional of the layered arch. But in case of shear stress is linearly increasing in the 1st layer, parabolic nature in 2nd and 3rd layer and nearly equal to zero variations in the 4th layer of four layered antisymmetric composite arch subjected to uniformly loading through the thickness variations.

Table 5 shows that, present theory have been great-agreement to Sayyad and Ghugal [11] for stresses and displacements at aspect ratio (*L/h* = 10). Present numerical results of transverse deformation are closely-matches with well-known theory of Reddy [4] at aspect ratio (*L/h* = 4, 10 and 100). In the present investigation it is found that transverse shear stress is slightly improved for aspect ratio (*L/h* = 4, 10 and 100). Present theory well captures the effect of normal deformation which is not considered in some prior available literature and numerical results are exceptional matches with Sayyad and Ghugal [11] and Reddy [4].

Table 3 Normalized displacements and stresses for four layered symmetric composite arch.

Theory	L/h	R/h	$\bar{w}$ (L/2,0)	$\bar{u}$ (0,-h/2)	$\bar{\sigma}_x$ (L/2,-h/2)	$\bar{\tau}_{xz}$ (0,0)
Present ($\varepsilon_z \neq 0$)	4	1	3.4549	7.2830	60.0227	2.1865
Present ($\varepsilon_z \neq 0$)		2	3.4548	5.8335	1.4377	2.1864
Present ($\varepsilon_z \neq 0$)		3	3.4548	4.7061	9.4218	2.1864
Present ($\varepsilon_z \neq 0$)		4	3.4548	4.0216	13.2260	2.1864
Present ($\varepsilon_z \neq 0$)		5	3.4548	3.5723	14.9884	2.1864
Present ($\varepsilon_z \neq 0$)		10	3.4548	2.5866	17.3429	2.1863
Present ($\varepsilon_z \neq 0$)		25	3.4548	1.9395	18.0065	2.1863
Present ($\varepsilon_z \neq 0$)		50	3.4548	1.7146	18.1029	2.1863
Present ($\varepsilon_z \neq 0$)		100	3.4548	1.6003	18.1277	2.1863
Present ($\varepsilon_z \neq 0$)		∞	3.4548	1.4850	18.1369	2.1863
Present ($\varepsilon_z \neq 0$)	10	1	1.4073	244.9949	1149.700	5.9316
Present ($\varepsilon_z \neq 0$)		2	1.4073	188.0736	220.2640	5.9315
Present ($\varepsilon_z \neq 0$)		3	1.4073	143.8010	48.1529	5.9314
Present ($\varepsilon_z \neq 0$)		4	1.4073	116.9212	12.0813	5.9314
Present ($\varepsilon_z \neq 0$)		5	1.4073	99.2755	39.9590	5.9314
Present ($\varepsilon_z \neq 0$)		10	1.4073	60.5685	77.1231	5.9314
Present ($\varepsilon_z \neq 0$)		25	1.4073	35.1586	87.5231	5.9314
Present ($\varepsilon_z \neq 0$)		50	1.4073	26.3243	89.0066	5.9314
Present ($\varepsilon_z \neq 0$)		100	1.4073	21.8388	89.3766	5.9314
Present ($\varepsilon_z \neq 0$)		∞	1.4073	17.3079	89.4987	5.9314
Present ($\varepsilon_z \neq 0$)	100	1	1.0091	1.6164E+7	8.8801E+6	62.0398
Present ($\varepsilon_z \neq 0$)		2	1.0091	1.2127E+7	2.2137E+6	62.0397
Present ($\varepsilon_z \neq 0$)		3	1.0091	8.9875E+6	9.7912E+5	62.0397
Present ($\varepsilon_z \neq 0$)		4	1.0091	7.0811E+6	5.4704E+5	62.0397
Present ($\varepsilon_z \neq 0$)		5	1.0091	5.8296E+6	3.4704E+5	62.0397
Present ($\varepsilon_z \neq 0$)		10	1.0091	3.0843E+6	8.0384E+4	62.0397
Present ($\varepsilon_z \neq 0$)		25	1.0091	1.2821E+6	5720.6000	62.0397
Present ($\varepsilon_z \neq 0$)		50	1.0091	6.5558E+5	4945.2000	62.0397
Present ($\varepsilon_z \neq 0$)		100	1.0091	3.3745E+5	7611.4000	62.0397
Present ($\varepsilon_z \neq 0$)		∞	1.0091	16094.000	8499.9000	62.0397

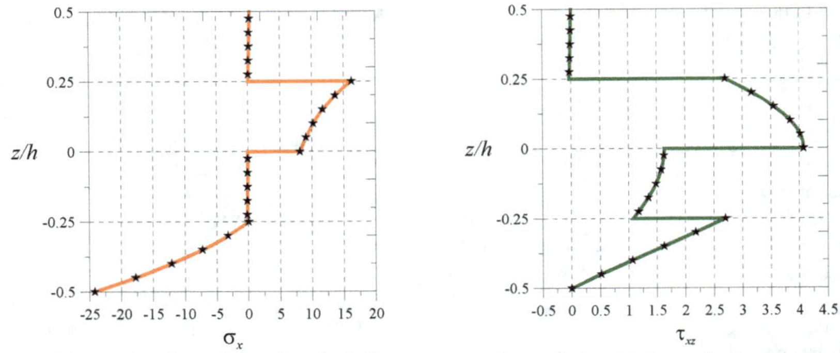

Fig. 5 Normalized axial-bending and shear stresses through-the-thickness for four layered antisymmetric ($0^0/90^0/0^0/90^0$) composite arch due to uniformly distributed loading [R/h=5,L/h=4]

Table 4 Normalized displacements and stresses for four layered antisymmetric composite arch.

Theory	L/h	R/h	$\bar{w}$ (L/2,0)	$\bar{u}$ (0,-h/2)	$\bar{\sigma}_x$ (L/2,-h/2)	$\bar{\tau}_{xz}$ (0,0)
Present ($\varepsilon_z \neq 0$)	4	1	4.1602	8.8736	101.0799	4.0791
Present ($\varepsilon_z \neq 0$)		2	4.1441	6.9762	2.5958	4.0781
Present ($\varepsilon_z \neq 0$)		3	4.1412	5.5826	15.2512	4.0780
Present ($\varepsilon_z \neq 0$)		4	4.1402	4.7432	21.3900	4.0780
Present ($\varepsilon_z \neq 0$)		5	4.1397	4.1939	24.1841	4.0780
Present ($\varepsilon_z \neq 0$)		10	4.1391	2.9924	27.7744	4.0781
Present ($\varepsilon_z \neq 0$)		25	4.1390	2.2056	28.6508	4.0781
Present ($\varepsilon_z \neq 0$)		50	4.1390	1.9323	28.7301	4.0781
Present ($\varepsilon_z \neq 0$)		100	4.1390	1.7936	28.7298	4.0782
Present ($\varepsilon_z \neq 0$)		∞	4.1390	1.6535	28.7030	4.0782
Present ($\varepsilon_z \neq 0$)	10	1	1.8864	326.7604	2205.500	11.1041
Present ($\varepsilon_z \neq 0$)		2	1.8745	247.5209	437.7593	11.1005
Present ($\varepsilon_z \neq 0$)		3	1.8727	187.9800	114.4785	11.1000
Present ($\varepsilon_z \neq 0$)		4	1.8723	151.9904	2.0108	11.0999
Present ($\varepsilon_z \neq 0$)		5	1.8721	128.4004	49.7849	11.0998
Present ($\varepsilon_z \neq 0$)		10	1.8721	76.71360	118.1390	11.0999
Present ($\varepsilon_z \neq 0$)		25	1.8723	42.80650	136.6160	11.0999
Present ($\varepsilon_z \neq 0$)		50	1.8724	31.01990	139.0198	11.1000
Present ($\varepsilon_z \neq 0$)		100	1.8724	25.03570	139.5175	11.1000
Present ($\varepsilon_z \neq 0$)		∞	1.8725	18.99070	139.5457	11.1000
Present ($\varepsilon_z \neq 0$)	100	1	1.4431	2.3113E+7	1.8069E+7	116.3519
Present ($\varepsilon_z \neq 0$)		2	1.4321	1.7206E+7	4.4723E+6	116.3114
Present ($\varepsilon_z \neq 0$)		3	1.4305	1.2736E+7	1.9779E+6	116.3044
Present ($\varepsilon_z \neq 0$)		4	1.4301	1.0030E+7	1.1064E+6	116.3022
Present ($\varepsilon_z \neq 0$)		5	1.4300	8.2561E+6	7.0321E+5	116.3013
Present ($\varepsilon_z \neq 0$)		10	1.4301	4.3660E+6	1.6581E+5	116.3003
Present ($\varepsilon_z \neq 0$)		25	1.4304	1.8121E+6	1.5397E+4	116.3003
Present ($\varepsilon_z \neq 0$)		50	1.4305	9.2408E+5	6071.9000	116.3004

Present ($\varepsilon_z \neq 0$)	100	1.4305	4.7313E+5	11430.000	116.3005
Present ($\varepsilon_z \neq 0$)	∞	1.4306	17567.000	13203.000	116.3005

Table 5 Normalized displacements and stresses for four layered ($0^0/90^0/0^0/90^0$) straight beams.

Theory	*L/h*	Model	$\bar{w}$ *(L/2,0)*	$\bar{u}$ *(0,-h/2)*	$\bar{\sigma}_x$ *(L/2,-h/2)*	$\bar{\tau}_{xz}$ *(0,0)*
Present ($\varepsilon_z \neq 0$)		ESNDT	4.1390	1.6535	28.7030	4.0782
Sayyad and Ghugal [11]		SSNPT	4.1737	1.6454	28.4633	3.7133
Reddy [4]	4	HSDT	4.1744	1.6222	28.0294	3.6618
Timoshenko [3]		FSDT	4.1055	1.1239	21.1274	3.5289
Bernoulli-Euler [1,2]		CBT	1.4269	1.1267	21.1274	3.5289
Present ($\varepsilon_z \neq 0$)		ESNDT	1.8725	18.9907	139.5457	11.1000
Sayyad and Ghugal [11]		SSNPT	1.8722	18.9205	139.4595	8.48720
Reddy [4]	10	HSDT	1.8731	18.8824	138.9880	8.50550
Timoshenko [3]		FSDT	1.8555	17.5615	132.0461	8.82230
Bernoulli-Euler [1,2]		CBT	1.4269	17.6055	132.0461	8.82230

Table 5 continued.....

Theory	*L/h*	Model	$\bar{w}$ *(L/2,0)*	$\bar{u}$ *(0,-h/2)*	$\bar{\sigma}_x$ *(L/2,-h/2)*	$\bar{\tau}_{xz}$ *(0,0)*
Present ($\varepsilon_z \neq 0$)		ESNDT	1.4306	17567.00	13203.00	116.3005
Sayyad and Ghugal [11]		SSNPT	1.4286	17540.51	13216.65	88.18290
Reddy [4]	100	HSDT	1.4314	17575.30	13211.91	88.17670
Timoshenko [3]		FSDT	1.4312	17561.83	13204.90	88.22490
Bernoulli-Euler [1,2]		CBT	1.4269	17605.47	13204.61	88.22490

Conclusions

Present scientific study mainly contributes the precise exponential shear and normal deformation theory for symmetric and antisymmetric four-layered composite arches when subjected to uniform load. The effects of transverse normal-stress and transverse normal-strain have been taken into account by present theory. The present theory meets the zero traction free end boundary condition and does not require any shear correction factors. Present theory is very accurate estimation for displacements and stresses which are rarely found in the literature. The obtained numerical results can be use for accurate design of such complex engineering structures. These innovative results will obviously set the benchmark for upcoming researchers in the area of composite arches.

References

[1] J. Bernoulli, Curvatura laminae elasticae, Acta Eruditorum Lipsiae. 34 (1694) 262-276.

[2] L. Euler, Methodus inveniendi lineas curvas maximi minimive proprietate gaudentes, Apud Marcum-Michaelem Bousquet and Socios. 24 (1744) 1-322. https://doi.org/10.5479/sil.318525.39088000877480.

[3] S.P. Timoshenko, On the correction for shear of the differential equation for transverse vibrations of prismatic bars, Philos. Mag. Series 1. 41 (1921) 744-746. https://doi.org/10.1080/14786442108636264.

[4] J.N. Reddy, A simple higher-order theory for laminated composite plates, J. Appl. Mech. 51(4) (1984) 745-752. https://doi.org/10.1115/1.3167719.

[5] E. Carrera, Transverse normal strain effects on thermal stress analysis of homogeneous and layered plates, AIAA J. 43(10) (2005) 2232-2242. https://doi.org/10.2514/1.11230.

[6] E. Carrera, S. Brischetto and A. Robaldo, Variable kinematic model for the analysis of functionally graded material plates, AIAA J. 46(1) (2008) 194-203.

https://doi.org/10.2514/1.32490.

[7] A.M. Zenkour, Three-dimensional elasticity solution for uniformly loaded cross-ply laminates and sandwich plates, J. Sandw. Struct. Mater. 9(3) (2007) 213-238. https://doi.org/10.1177/1099636207065675.

[8] T. Kant and S.M. Shiyekar, Cylindrical bending of piezoelectric laminates with a higher order shear and normal deformation theory, Comput. Struct. 86 (2008) 1594-1603. https://doi.org/10.1016/j.compstruc.2008.01.002.

[9] F. Tornabene, N. Fantuzzi, E. Viola and E. Carrera, Static analysis of doubly-curved anisotropic shells and panels using CUF approach, differential geometry and differential quadrature method, Compos. Struct. 107 (2014) 675-697. http://dx.doi.org/10.1016/j.compstruct.2013.08.038.

[10] F. Tornabene, N. Fantuzzi, E. Viola and R.C. Batra, Stress and strain recovery for functionally graded free-form and doubly-curved sandwich shells using higher-order equivalent single layer theory, Compos. Struct. 119 (2015) 67-89. http://dx.doi.org/10.1016/j.compstruct.2014.08.005.

[11] A.S. Sayyad and Y.M. Ghugal, Cylindrical bending of multilayered composite laminates and sandwiches, Adv. Aircr. Spacecr. Sci. 3(2) (2016) 113-148. http://dx.doi.org/10.12989/aas.2016.3.2.113.

Advanced Topics in Mechanics of Materials, Structures and Construction - AToMech1-2023 Materials Research Forum LLC
Materials Research Proceedings 31 (2023) 704-714 https://doi.org/10.21741/9781644902592-72

Mechanics and failure analysis of stitched sandwich structures damaged by the impact of various energies

Amina ASSOULI[1,a*], Bounoua TAB[1,b], Fodil HAMMADI[2,c] and Izzeddine BEKKAR[1,d]

[1] Mechanical Engineering Reliability Laboratory, TAHRI Mohammed University, Bechar, Algeria

[2] Mechanics laboratory: Modeling and Experimentation L2ME, TAHRI Mohammed University, Bechar, Algeria

[a] assouli.amina@univ-bechar.dz , [b] bounoua.tab@univ-bechar.dz ,
[c] fodil.hammadi@univ-bechar.dz, [d] izzeddine.bekkar@univ-bechar.dz

Keywords: Mechanical, Failure Analysis, First Impact, Stitched Sandwich, Damage, Shock Accumulation, Shock without Rebound

Abstract. In the fields of structural mechanics and failure analysis, the presence of damage to structural materials requires several studies. Expertise, damage, is crucial to understand the behavior of structures under various loading conditions. Generally, this damage affects the most stressed area of the structure and can lead to the accumulation and propagation of damage, leading to the partial or total collapse of the structure. On the other hand, stitched sandwich plates are an innovative structural material with extremely high performance in the fields of building mechanics and others. However, when these plates are damaged, it is essential to study their behavior under subsequent loading conditions to assess their ability to withstand a second impact. After having been damaged by a first impact at low speed, a diagnosis was carried out which showed the capacity of these plates to undergo a new shock. The present work focuses on the study of the behavior of these said plates under the effect of a second impact. First, the plates subject to damage in previous works were subjected to a new constant energy impact in order to observe the cumulative effect. In the second part, repeated impact tests were applied to healthy plates according to two different modes. The diversity of impact modes is based on the cycle (time) factor. An immediate repeated shock whereas for the other mode the specimen only suffers a second shock after a rest of 24 hours. An analysis of the behavior of the sandwich plates under the effect of different shocks (without rebound) was carried out. The results obtained showed once again the effectiveness of these structural materials in the face of the effect of impact compared to traditional sandwiches. The damage seen is characterized by positioning and pivoting around the point of impact and is limited to the edges of the collider's diameter and depth. Fracture analysis of these plates indicated that the damage resulted in a decrease in their bearing capacity.

Introduction

Sandwich composites are defined as two thin layers with high resistance to tensile and compressive loads, called skins, comprising a thick layer of low density, bonded together by an adhesive to resist transverse shear loads called the core [1], and all this with a necessary and important characteristic in various fields of modern engineering (aeronautical engineering, maritime engineering, civil engineering...) where the reduction of structural weight is of utmost importance and priority. It has a high strength-to-weight ratio [2], high bending stiffness, and a high energy absorption capacity. This design is what has given the diversity and the possibility of having several classifications of sandwich composites, where we find in the specialized literature that there are two divisions. The first concerns the nature of the two skins that make up the sandwich.

"Composite skin sandwich _non-composite skin sandwich," while the second core classification is divided into two parts: a solid core (ex: balsa wood...) and a structural core (honeycomb...).

Since their creation, sandwich composites have undergone several tests in order to study their mechanical properties under different loads, and their effectiveness has been proven on several occasions. However, the biggest hurdle was the separation of the "soul" core from the two skins due to adhesive peeling, which was solved by adding a suturing technique to strengthen the bond between the skins and the core [3] , known as a stitched sandwich. The latter gave a new and elegant concept for the use of sandwich composites.

The stitched sandwich composites were initially the subject of several experiments to compare their function with ordinary sandwich panels, and it was found that the majority of the experiments conducted on them were focused on studying the effect of adding welds on the mechanical behavior of the sandwich [4][5][6][7][8][9][10] , and despite its effectiveness compared to the regular sandwich, it has not escaped various tests, perhaps the most important of which is the effect test. On the other hand, this characteristic is carried out in two forms of low-speed impact and is carried out in two ways: Izod and Charpy impact tests, the content of which is aimed at testing the impact resistance of materials and comparing composite materials with different layers, including woven and unidirectional laminates. The second form is the high velocity impact test, which is often used to study aviation-related structures.

Impact testing is of great importance in the characterization of sandwich compounds and is influenced by several factors: "fiber properties, sample thickness, projectile shape and mass and projectile velocity" [11].

In recent years, several studies have been conducted to understand the impact behavior of stitched sandwich structures, including those with carbon fiber-reinforced polymer skins, aluminum foam cores, 3D woven composite sandwich panels with stitched Z-pins, and sandwich structures with stitched composite skins [12][13][14]. These studies have significantly contributed to our understanding of the behavior of sandwich composites under different loads and their potential for use in various engineering applications. In this context, tests were carried out for the second time on stitched sandwich panels made of glass fiber-reinforced polymer, which had already been damaged by impacts of varying intensities. Therefore, in this work, tests were carried out on damaged samples to study their response to repeated shock loads, paying attention to the damage mechanisms that occur and their impact on the dynamic behavior of the stitched sandwich.

The paper consists of four sections that explore the structural behavior of stitched sandwich plates under the effect of a second impact after being damaged by a first impact at low speed. Section I provides an introduction to the problem being studied and a brief literature review. Section II focuses on the experimental study of the impact on samples. In Section III, the effect of the impact on samples is analysed and discussed. Finally, Section IV concludes the paper by summarizing the findings and presenting the conclusion.

Experimental research

Methodology

In this study, a test rig based on a falling weight impact test, was performed involving impact energy provided by an impact head collider that falls and collides with the target "in our cases are stitched sandwich panels' thus realizing the principle of soft shock. First, the stitched sandwich panel samples were subjected to a low velocity impact test with progressively increasing energies. This study not only considered and resolved the damage observed at the upper skin level resulting from the first was interested in the vibrational behavior of the lower skin due to the progression of the damage (ie; diameter and depth of the indentation) of the impacted skin. In order to evaluate the effect of the cumulative shock phenomenon over time, the samples were subjected to a second impact test with a lower energy than that of the first impact. Finally, in order to study the effect of time on the behavior of these panels, two samples were subjected to a shock test repeated without and after relaxation for 24 hours. The details of the experimental methodology are as follows:

Materials and specimen preparation

The specimens shown in Figure 1 of this study consist of stitched sandwich panels to woven composite skins. These panels measure 200 mm by 150 mm and are 23 mm thick. The skins themselves are composed of two identical layers of woven glass fibers 1.5 mm thick each, with a density of 400 kg/ m3. They envelop a polyurethane foam core 20 mm thick and with a density of 35 kg/ m3. The integrity of the "skins-foam core" suit is reinforced by a seam composed of 2400TEX glass rovings.

Fig. 1: Specimen

Test device and impact method

The operating mode shown in Figure 2 consists of a test bench ensuring the principle of soft impact by falling weight (steel disc + impact head) totaling a mass of 11 kg. The collider thus formed is equipped with a hemispherical impact head with a diameter = 25 mm which allows the impact to be focused. The support system is made to ensure a perpendicular application of the load on the embedded plate in order to avoid any parasitic movement of the specimen Figure 3.

Fig.2: Experimental impact setup

Fig.3: Collision between the collider and the embedded sample

•The energy is given using a graduated disk which makes it possible to lift the mass to the height corresponding to the desired energy Figure 4.

Fig.4: Graduated disk

•The bench is equipped with a braking system which makes it possible to avoid the rebound of the collider.

•The acquisition mode is ensured by Pulse software which is linked to the non-impacted face of the specimen by a sensor Figure 5 bellow.

•For the survey of the deformations of the impacted face (diameter of impact and depth of indentation) they are carried out by an electronic caliper.

•The tests, as can be seen in the previous figures, were carried out according to the above methodology.

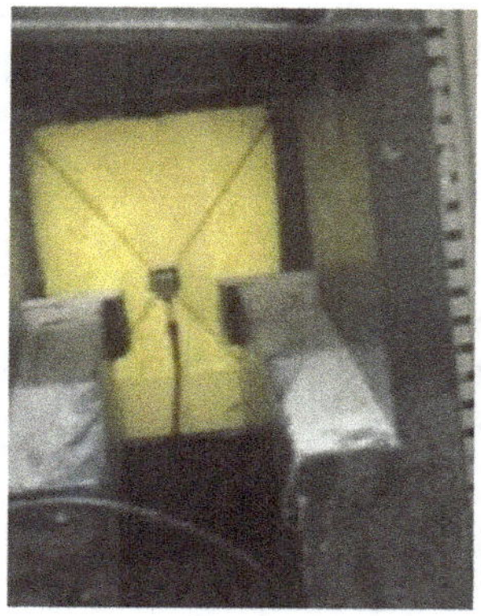

Fig.5: A sensor linked to the non-impacted face of the specimen

Analysis of experimental results

Constant impact at 10 [J] on already damaged plates

After evaluating the behavior of the stitched sandwich under the effect of the first impact, a second impact was applied with a constant energy of 10 J for all the specimens already impacted (13.75-24,75- 35,75 and 55 J). In this experiment the effect of the accumulation of shocks was observed. The following figures 6, 7, 8 and 9 explain the results where the graphs showing the vibration behavior of the non-impacted skin are superimposed, while the photo shows the cumulative effects of the two shocks on the impacted face.

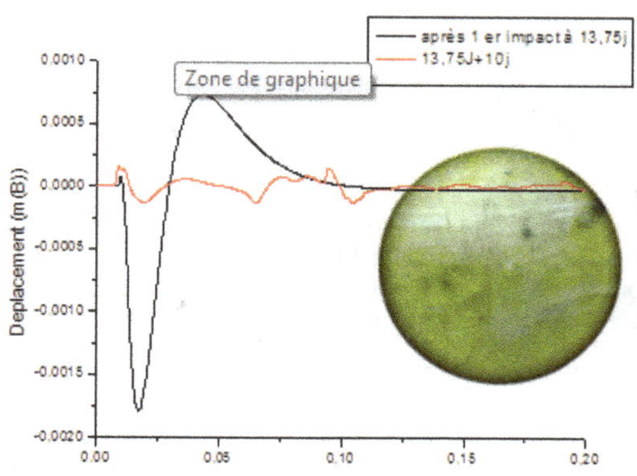

Fig.6: The vibrational response (displacement history) of the non-impacted skin resulting after two impacts: 13,75 J and 10 J

The photo in figure 6, show the vibrational response (displacement history) of the non-impacted skin resulting from the first impact at 13.75 J and the second impact with an energy of 10 J where we notes that the deformation effect of the second impact is negligible. In fact, this is well expressed on the graph of the vibrational responses of the non-impacted skin, where lower amplitudes are observed for the second impact. On the other hand, the impacted skin did not show an increase in the diameter or depth of the impact table 1.

Tab.1: the variation of the parameters of the indentation after two impacts: 13,75 J and 10 J

First Impact energy : 13,75 J		Second Impact energy : 10 J	
Diameter [mm]	Depth [mm]	Diameter [mm]	Depth [mm]
0	0	0	0

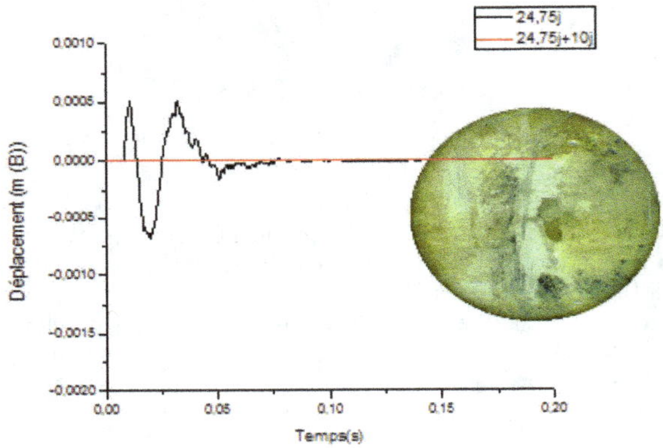

Fig.7: The vibrational response (displacement history) of the non-impacted skin resulting after two impacts: 24,75 J and 10 J

For the specimen in figure 7, it is noted that the vibrational response of the non-impacted skin to the second impact is weaker compared to its response to the first. The phenomenon is certainly due to the absorption of the shock by the fiber debris and resignation resulting from the first damage. On the other hand, the impacted skin shows a 28% increase in the impact diameter (table 2). For the depth, the exaggerated increase (5 times) is indeed justified by the low resistance of the foam.

Tab.2: the variation of the parameters of the indentation after two impacts: 24,75 J and 10 J

First Impact energy : 24,75 J		Second Impact energy : 10 J	
Diameter [mm]	Depth [mm]	Diameter [mm]	Depth [mm]
14,63	1,84	18,8	11,16

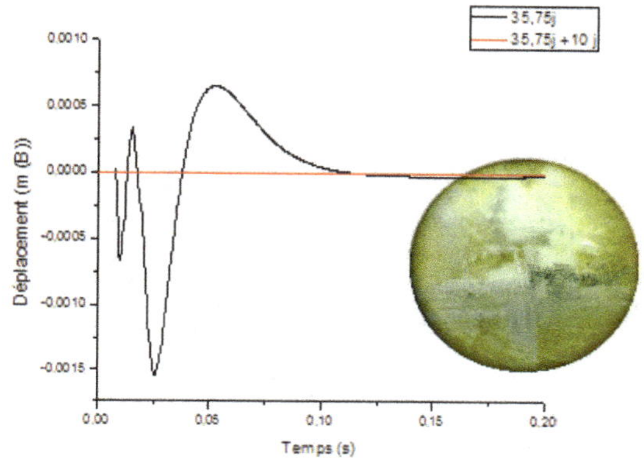

Fig.8: The vibrational response (displacement history) of the non-impacted skin resulting after two impacts: 35,75 J and 10 J

For the test specimen in figure 8, we find ourselves in the particular case where the plate undergoes two impacts on a sewing knot. The vibrational response of the non-impacted skin to the second impact is weaker compared to its response to the first. The phenomenon is due to the absorption of shock by the presence of the seam. On the other hand, the impacted skin shows an increase of 4% in the diameter of the impact (table 3) and 35% in the depth of indentation.

Tab. 3: the variation of the parameters of the indentation after two impacts: 35,75 J and 10 J

First Impact energy : 35,75 J		Second Impact energy : 10 J	
Diameter [mm]	Depth [mm]	Diameter [mm]	Depth [mm]
15,96	3,12	16,62	4,24

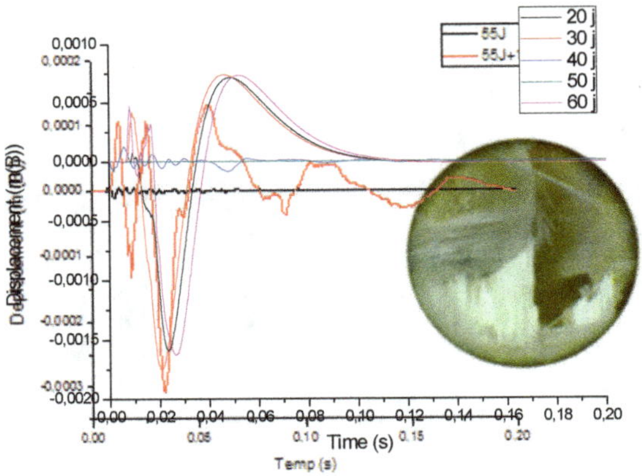

Fig.9: The vibrational response (displacement history) of the non-impacted skin resulting after two impacts: 55 J and 10 J

Figure 9 shows the vibration response of the non-impacted skin resulting from the first and second impact at 55 and 55+10 d. It is noted that unlike the previous plates, the response of the non-impacted face in this case is greater. In fact, at the first impact, the impacted face has been completely perforated, which causes the non-impacted skin to undergo the second shock directly. The impacted skin on its side showed a 9% increase in impact diameter (Table 4) and 15% in indentation depth. It can be seen that the compression of the debris prevents the sinking of the collider head.

Tab. 4: the variation of the parameters of the indentation after two impacts: 55 J and 10 J

First Impact energy : 55 J		Second Impact energy : 10 J	
Diameter [mm]	Depth [mm]	Diameter [mm]	Depth [mm]
17,29	14,94	18,92	17,25

Repeated shock

In this experiment, two healthy plates were impacted according to two impact modes with an energy step of 10 Joules. The first mode in Figure 10 consisted in repeating a series of successive impacts without rest of the plate. For the second mode in Figure 11, the succession of impacts is spaced 24 hours apart in order to allow the specimen to rest. The results were recorded in the same way in order to obtain comparable results.

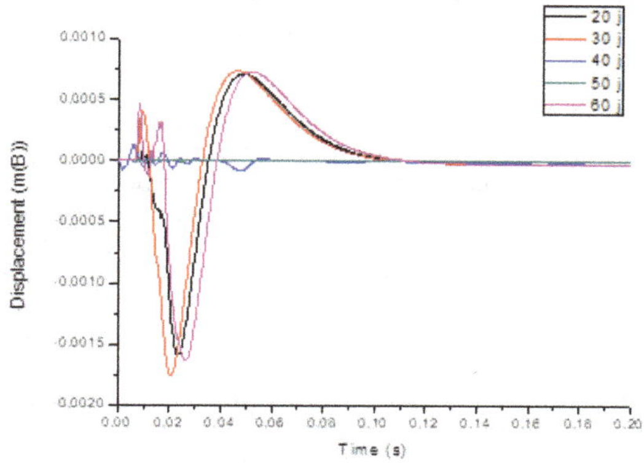

Fig.10: Non-impacted skin vibrational response (displacement history) resulting from successive impacts without rest

As can be seen in Figure 10, the resulting curves of the different energies form a spindle which gives the same behavioral appearance. The shock without rest gives very similar response curves in amplitude with very small differences. The phenomenon is certainly due to the fact that the successive shocks without rest do not allow the material to regain its strength to receive the next impact.

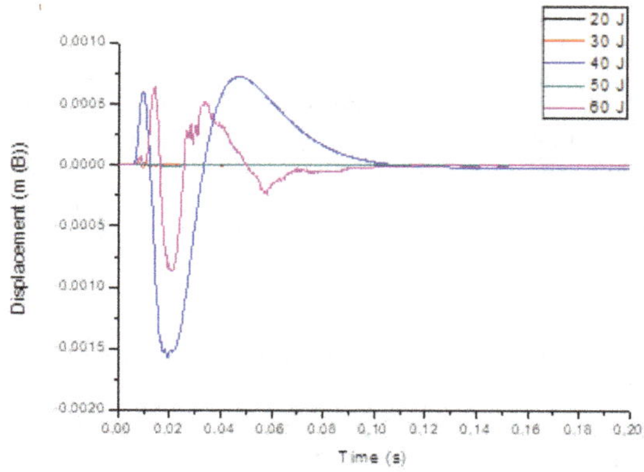

Fig.11: Non-impacted skin vibrational response (displacement history) resulting from successive impacts spaced 24 hours of rest

In figure 11 one can notice the influence of the rest of 24 hours given to the specimen. Although the resulting curves of the different energies have the same behavioral appearance, they differ in the amplitude values. Indeed the phenomenon, noticed in the first experience of the plates impacted at different energies, manifests itself once again and we note the role of the debris which dampens the shock and gives a lesser amplitude in spite of the increase in energy.

Conclusion

In this work, we studied the performance of shock-loaded stitched sandwich plates when subjected to a second shock with an energy level lower or higher than the first. The study was conducted using a test bench under real operating conditions. The performance of these shock-loaded panels was measured after receiving a shock with lower energy. It was observed that the panel charged with higher energy did not respond to the shock direction when a higher-energy shock occurred.

The results of the study revealed several important findings. Firstly, despite the perforation of the impacted skin, the second skin displayed elastic behavior. This indicates that the shock-loaded stitched sandwich plates have the ability to maintain their structural integrity even after being impacted.

Secondly, the debris from the previous shocks formed a layer that acted as a damping mechanism during the second shock. This layer effectively absorbed some of the energy from the subsequent shock, thereby reducing its impact on the plate.

To complete the study, we compared the response to shocks with the same intensity for two stitched sandwich panels. One panel was subjected to repeated shocks without interruption, while the second panel received the same number of shocks with the same intensity but with a 24-hour break between each shock. The results showed that the inclusion of a rest period between shocks proved to be beneficial. The rest allowed the plate to regain its strength and better resist the next shock. This finding suggests that the stitched insulation panels can recover and adapt to repeated shocks when given sufficient time for recovery.

Overall, the results highlight the high performance of the stitched insulation panels, emphasizing their elastic behavior, the dampening effect of debris, and the positive impact of rest periods. These findings also underscore the need for further studies in this area, as they simulate real-world conditions more accurately than studies that solely investigate the behavior after the first impact.

Acknowledgements

- The authors would like to thank the Directorate General for Scientific Research and Technological Development (DGRSDT, Algeria) for the financial support of this work.
- Thanks to the professors and student members of the LFGM

Funding

The authors received no financial support for the research, authorship and/or publication of this article.

References

[1]Noor, A.K., Burton, W.S., Bert, C.W. (1996). Computational models for sandwich panels and shells, Appl. Mech. Rev., 49(3), pp. 155–99. https://doi.org/10.1115/1.3101923

[2]Ogorkiewicz, R.M. (1970). Analysis and design of structural sandwich panels, Composites, 1(6), pp. 378. https://doi.org/10.1016/0010-4361(70)90244-2

[3]Hosur, M. V., Karim, M.R., Jeelani, S. (2003). Experimental investigations on the response of stitched/unstitched woven S2-glass/SC15 epoxy composites under single and repeated low velocity impact loading, Compos. Struct., 61(1–2), pp. 89–102. https://doi.org/10.1016/S0263-8223(03)00032-1

[4]Drake, D.A., Sullivan, R.W., Lovejoy, A.E., Clay, S.B., Jegley, D.C. (2021). Influence of stitching on the out-of-plane behavior of composite materials – A mechanistic review, J. Compos. Mater., 55(23), pp. 3307–21. https://doi.org/10.1177/00219983211009290

[5]Mouritz, A.P., Leong, K.H., Herszberg, I. (1997). A review of the effect of stitching on the in-plane mechanical properties of fibre-reinforced polymer composites, Compos. Part A Appl. Sci. Manuf., 28(12), pp. 979–91. https://doi.org/10.1016/S1359-835X(97)00057-2

[6]Lascoup, B., Aboura, Z., Khellil, K., Benzeggagh, M. (2006). On the mechanical effect of stitch addition in sandwich panel, Compos. Sci. Technol., 66(10), pp. 1385–98. https://doi.org/10.1016/j.compscitech.2005.09.005

[7]Mouddene, B., Labbaci, B., Missoum, L., Abdeldjebar, R. (2017). Mechanical properties identification of stitched sandwich by compression, Int. J. Mech. Eng. Technol., 8(11), pp. 376–84.

[8]Shigang, A., Yiqi, M., Yongmao, P., Daining, F., Liqun, T. (2013). Effect of stitching angle on mechanical properties of stitched sandwich panels, Mater. Des., 50, pp. 817–24. https://doi.org/10.1016/j.matdes.2013.03.058

[9]Santhanakrishnan, R., Samlal, S., Joseph Stanley, A., Jayalatha, J. (2018). Impact study on sandwich panels with and without stitching, Adv. Compos. Mater., 27(2), pp. 163–82. https://doi.org/10.1080/09243046.2017.1410376

[10]Che, L., Xu, G. dong., Zeng, T., Cheng, S., Zhou, X. wei., Yang, S. cai. (2014). Compressive and shear characteristics of an octahedral stitched sandwich composite, Compos. Struct., 112(1), pp. 179–87. https://doi.org/10.1016/j.compstruct.2014.02.012

[11]Safri, S.N.A., Sultan, M.T.H., Yidris, N., Mustapha, F. (2014). Low velocity and high velocity impact test on composite materials – A review, Int. J. Eng. Sci., 3(9), pp. 50–60. https://doi.org/10.1177/1464420711409985

[12]Ma, Q., Rejab, M.R.M., Siregar, J.P., Guan, Z. (2021). A review of the recent trends on core structures and impact response of sandwich panels, J. Compos. Mater., 55(18), pp. 2513–55. https://doi.org/10.1177/0021998321990734

[13]van Hoorn, N., Kassapoglou, C., Turteltaub, S., van den Brink, W. (2022). Experimental damage tolerance evaluation of thick fabric carbon/epoxy laminates under low-velocity and high-velocity impact and compression-after-impact, J. Compos. Mater., 56(5), pp. 761–78. https://doi.org/10.1177/00219983211060501

[14]Papa, I., El Hassanin, A., Langella, T., Lopresto, V. (2021). Experimental and analytical model for the penetration and indentation prediction on BFRP laminates under low velocity impacts, J. Compos. Mater., 55(10), pp. 1331–8. https://doi.org/10.1177/0021998320969790

Magnetoelastic deformation of conductive semilinear hyperelastic solids

Odunayo Olawuyi Fadodun

Department of Mathematics, Obafemi Awolowo University, Ile-Ife 220005, Nigeria

ofadodun@oauife.edu.ng

Keywords: Magnetoelasticity, Radial Deformation, Hyperelastic Materials, Cylinder, Conductive Elastomer

Abstract. This work investigates the radial deformation of conductive magneto-hyperelastic solid cylinder subject to azimuthal magnetic field. It shows effect of the current density on the radial deformation of the solid. A simple magnetoelastic energy function is proposed for the cylinder under consideration such that its purely elastic part corresponds to the strain energy of the well-known semilienar hyperelastic materials. The consequent magnetoelasticity field equations, in conjunction with the accompanying boundary conditions, are specialized for application to the problem of radial deformation of solid cylinder. The obtained magnetoelastic constitutive model shows that the stress distribution in the solids is sensitive to the magnetic induction while the associated magnetic field at point within the cylinder is deformation-dependent. Furthermore, it is revealed that the azimuthal magnetic induction produced by steady current within the solid cylinder increases along its radius. Finally, and among other things, the graphical illustration shows that the effect of steady axial current density on the magnitude of the displacement function at points within the cylinder is significantly pronounced.

Introduction

Conductive magnetoactive elastomers (CMEs) are magneto-sensitive materials that conduct electricity. They are manufactured by mixing micron/nano -size magnetic and conductive particles into nonmagnetic rubber-like matrices. CMEs exhibit change in mechanical response when subject to applied magnetic field and/ or current of electricity. The widespread applications of these materials have continued to instigate the needs for the development of new magnetoelasticity theories. In the fundamental formulation, magnetoelasticity field equations govern magneto-mechanical interaction of solids. These equations consist of magnetostatic and elasticity fields equations, and are used to construct solutions to problems involving magnetoelastic deformation.

Finite magnetoelastic interaction of solids has long been a subject of interest since the classic studies of Maugin [1], Eringen and Maugin [2] and Pao [3]. Recently, Pei et al. [4] investigated nonlinear magnetoelastic deformation of porous solids; Reddy and Saxena [5] studied instabilities in axisymmetric magnetoelastic deformation of a cylindrical membrane; Garcia-Gonzalez and Hossain [6] proposed a microstructural-based approach to model magneto-viscoelasticity materials at finite strain; Ren et al. [7] studied multi-functional soft-bodied jellyfish-like swimming; Bostola and Hossain [8] gave a review on magneto-mechanical charactirizations of magnetorheological elastomers; Dorfman and Ogden [9] studied nonlinear theory of electroelastic and magnetoelastic interactions; Nedjar [10] proposed a modelling framework for finite strain magnetoviscoelasticity; and Saxena et al. [11] developed a finite deformation theory for magneto viscoelasticity.

In view of Fadodun et al. [12], this work proposes a simple magnetoelastic energy function for conductive semilinear magnetohyperelastic solids. Using the laws of thermodynamic, Coleman-Noll procedure and tensor calculus, the study develops a magnetoelastic constitutive model for the solids under consideration. The consequent magnetoelasticity field equations together with the accompanying boundary conditions are specialized for applications to the problem of radial deformation of a conductive magnetoelastic cylinder subject to steady current of electricity. The rest of the paper is as follow: the first sections present magnetoelasticity field equations and constitutive relations while the remaining sections detail the solution to the radial deformation problem of magneto-sensitive solid cylinder subject to steady axial current density.

Kinematics

Consider a stress-free conductive magnetohyperelastic solid occupying the reference configuration $\Omega_0 \subset \mathbb{R}^3$ with smooth boundary $\partial\Omega_0$ and surface outward unit normal vector $\vec{N}$. When subject to magnetic field and /or mechanical surface load the body deforms onto deformed configuration Ω with boundary $\partial\Omega$ and surface outward unit normal vector $\vec{n}$. The deformation of the body is defined by vector function $\vec{\varphi}$

$$\vec{\varphi}: \bar{\Omega}_0 \to \bar{\Omega}, \tag{1}$$

such that $\vec{x} = \vec{\varphi}(\vec{X})$ where $\vec{X}$ denotes position vector of material points in Ω_0 and $\vec{x}$ represents position vector of the corresponding material points in Ω. The closures Ω_0 and Ω in Eq. (1) are defined by

$$\bar{\Omega}_0 = \Omega_0 \cup \partial\Omega_0 \quad and \quad \bar{\Omega} = \Omega \cup \partial\Omega.$$

The deformation gradient $\boldsymbol{F}$ is defined by

$$\boldsymbol{F} = Grad\, \vec{x} = Grad\, \vec{\varphi}(\vec{X}), \tag{2}$$

where $Grad$ is the gradient operator with respect to Ω_0. At an arbitrary point $\vec{X}$, the determinant $\det(\boldsymbol{F}) > 0$ measures the local volume change.
Applying the polar decomposition theorem, the deformation gradient $\boldsymbol{F}$ is decomposed into product of second-rank tensors $\boldsymbol{O}^D$ and U

$$\boldsymbol{F} = \boldsymbol{O}^D\boldsymbol{U}, \tag{3}$$

where $\boldsymbol{O}^D$ is the orthogonal rotation tensor and $\boldsymbol{U}$ is the right stretch symmetric tensor.
The tensors $\boldsymbol{U}$ and $\boldsymbol{O}^D$ are obtained by the relations

$$\boldsymbol{U} = \sqrt{\boldsymbol{F}^T\boldsymbol{F}} = \sqrt{\boldsymbol{C}} \quad and \quad \boldsymbol{O}^D = \boldsymbol{F}\boldsymbol{U}^{-1}, \tag{4}$$

where $\boldsymbol{F}^T$ is the transpose of $\boldsymbol{F}$, $\boldsymbol{U}^{-1}$ is the inverse of $\boldsymbol{U}$ and $\boldsymbol{C} = \boldsymbol{F}^T\boldsymbol{F}$ is the right Cauchy-Green deformation tensor [12].

Eulerian Form/Description: Magneto-Mechanical Field Equations

Let $\vec{H}, \vec{B}$ and $\vec{M}$ denote the Eulerian forms of the magnetic field, magnetic induction and effective magnetization vectors respectively. For a purely magnetostatic field produced by steady current, and in the absence of electric interaction and surface current the Maxwell's field equations read

$$curl\, \vec{H} = 4\pi\vec{J}, \quad div\, \vec{B} = 0, \tag{5}$$

where the operators div and $curl$ are defined in the deformed configuration Ω and $\vec{J}$ is the current density in Eulerian form. The current density $\vec{J}$ satisfies the equation

$$div\, \vec{J} = 0. \tag{6}$$

In magnetic materials, the vectors $\vec{H}, \vec{B}$ and $\vec{M}$ are related by the constitutive law

$$\vec{B} = \mu\vec{H} = \mu_0\overrightarrow{(H + M)}, \tag{7}$$

where μ is the magnetic permeability of the material and μ_0 is the magnetic permeability of free space.
In free space exterior to the body, the corresponding magnetostatic fields are denoted by vectors $\vec{H}^*$ and $\vec{B}^*$, which are governed by the equations

$$curl\, \vec{H}^* = \vec{0},\ \ div\, \vec{B}^* = 0,\quad \vec{B}^* = \mu_0\vec{H}^*. \tag{8}$$

At the bounding surface of the considered material in the deformed configuration Ω, the standard boundary conditions associated with Eq. (5) are

$$\vec{n} \times \left(\vec{H}^* - \vec{H}\right) = \vec{0},\ \ \vec{n} \cdot \left(\vec{B}^* - \vec{B}\right) = 0 \qquad on \quad \partial\Omega, \tag{9}$$

where $\vec{n}$ is the unit outward normal vector on $\partial\Omega$.
Let $\boldsymbol{T}$ denote the total stress tensor which incorporates magnetostatic body forces. In the absence of mechanical body forces the mechanical equilibrium equation reads

$$div\, \boldsymbol{T} = \vec{0}. \tag{10}$$

The standard boundary condition accompanying the equilibrium equation is

$$\boldsymbol{T}\vec{n} = \vec{t}_a + \vec{t}_m \qquad on \quad \partial\Omega, \tag{11}$$

where $\vec{t}_a$ is the mechanical traction on $\partial\Omega$ per unit area, $\vec{t}_m = \boldsymbol{T}^*\vec{n}$ is the load due to the Maxwell stress

$$\boldsymbol{T}^* = \mu_0\left(\vec{H}^* \otimes \vec{H}^*\right) - \tfrac{1}{2}\mu_0(\vec{H}^* \cdot \vec{H}^*)\boldsymbol{I}, \tag{12}$$

and $\boldsymbol{I}$ is the second-order unit tensor in Ω [9, 12].

Lagrangian Form/Description: Magneto-Mechanical Field Equations

Let $\vec{H}_L,\ \vec{B}_L$ and $\vec{M}_L$ denote the Lagrangian magnetic field, magnetic induction and effective magnetization vectors respectively. The Lagrangian variables $\vec{H}_L,\ \vec{B}_L$ and $\vec{M}_L$ are related to the Eulerian quantities $\vec{H}, \vec{B}$ and $\vec{M}$ by

$$\vec{H}_L = \boldsymbol{F}^T\vec{H},\ \vec{B}_L = \det(\boldsymbol{F})\, \boldsymbol{F}^{-1}\vec{B},\ \vec{M}_L = \boldsymbol{F}^T\vec{M}, \tag{13}$$

where the tensor $\boldsymbol{F}^{-1}$ is the inverse of $\boldsymbol{F}$ and $\det(\boldsymbol{F})$ is the determinant of $\boldsymbol{F}$.
The magnetostatic field equations in Lagrangian forms read

$$Curl\, \vec{H}_L = 4\pi\vec{J}_L,\qquad Div\, \vec{B}_L = 0, \tag{14}$$

where the operators Div and $Curl$ are defined in the reference configuration Ω_0, and $\vec{J}_L = \det(\boldsymbol{F})\,\boldsymbol{F}^{-1}\vec{J}$ is the Lagrangian current density satisfying the equation

$$Div\,\vec{J}_L = 0. \tag{15}$$

Similarly, the vectors Let $\vec{H}_L,\ \vec{B}_L$ and $\vec{M}_L$ are related by

$$\vec{B}_L = \mu_0 \det(\boldsymbol{F})\boldsymbol{C}^{-1}(\vec{H}_L + \vec{M}_L). \tag{16}$$

In addition, the vectors $\vec{H}_L$ and $\vec{B}_L$ satisfy the standard boundary conditions

$$\vec{N} \times \left(\overline{\boldsymbol{F}^T\vec{H}^*} - \vec{H}_L\right) = \vec{0},\quad \vec{N}\cdot\left(\det(\boldsymbol{F})\,\boldsymbol{F}^{-1}\vec{B}^* - \vec{B}_L\right) = 0 \quad on \quad \partial\Omega_0. \tag{17}$$

Let $\boldsymbol{P}$ denote the total first Piola-Kirchhoff's stress tensor. The total stress tensor $\boldsymbol{T}$ and first Piola-Kirchhoff's stress tensor $\boldsymbol{P}$ are related by

$$\boldsymbol{P} = \det(\boldsymbol{F})\,\boldsymbol{T}\boldsymbol{F}^{-T}, \tag{18}$$

where $\boldsymbol{F}^{-T}$ is the inverse of $\boldsymbol{F}^T$.
In term of $\boldsymbol{P}$ the equilibrium equation assumes the equivalent form

$$Div\,\boldsymbol{P} = \vec{0}. \tag{19}$$

The corresponding boundary condition reads

$$\boldsymbol{P}\vec{N} = \vec{t}_F + \vec{t}_{mF} \quad on \quad \partial\Omega_0, \tag{20}$$

where $\vec{t}_F$ is the mechanical traction on $\partial\Omega_0$ per unit area, $\vec{t}_{mF} = \boldsymbol{P}^*\vec{N}$ and $\boldsymbol{P}^* = \boldsymbol{P} = \det(\boldsymbol{F})\,\boldsymbol{T}^*\boldsymbol{F}^{-T}$ is the pull back version of the Maxwell stress $\boldsymbol{T}^*$ [12].

Magnetoelastic Energy Function and Constitutive Model

In order to complete the mathematical equations formulation for the study, we choose the deformation gradient $\boldsymbol{F}$ and magnetic induction vector $\vec{B}_L$ as independent variables; and model magnetoelastic constitutive laws that give first Piola-Kirchhoff stress tensor $\boldsymbol{P}$ and magnetic field vector $\vec{H}_L$ in terms of $\boldsymbol{F}$ and $\vec{B}_L$. Consequently, we take the magnetoelastic Helmholtz free energy function $\Phi = \Phi(\boldsymbol{F}, \vec{B}_L)$ to depend on $\boldsymbol{F}$ and $\vec{B}_L$, and ensure the objectivity condition

$$\Phi(\mathbf{Q}\boldsymbol{F}, \vec{B}_L) = \Phi(\boldsymbol{F}, \vec{B}_L),$$

is satisfied for all proper orthogonal second-rank tensor $\mathbf{Q}$.
Using the laws of thermodynamics and Coleman-Noll procedure, the first Piola-Kirchhoff stress tensor $\boldsymbol{P}$ and the Lagrangian magnetic field vector $\vec{H}_L$ are obtained through the relations

$$\boldsymbol{P} = \frac{\partial\Phi(\boldsymbol{F},\vec{B}_L)}{\partial \boldsymbol{F}}, \qquad \vec{H}_L = \frac{\partial\Phi(\boldsymbol{F},\vec{B}_L)}{\partial\vec{B}_L}. \tag{21}$$

Now, we recall and state the elastic strain energy function Φ^* per unit volume

$$\Phi^*(\boldsymbol{F}) = \frac{1}{2}\lambda_e \mathbb{I}_1^2(\boldsymbol{U} - \boldsymbol{I}_0) + \mu_e \mathbb{I}_2(\boldsymbol{U} - \boldsymbol{I}_0), \tag{22}$$

for an isotropic semilinear hyperelastic solid, where $\mathbb{I}_1(\boldsymbol{U} - \boldsymbol{I}_0)$ is the first invariant of the second-rank tensor $(\boldsymbol{U} - \boldsymbol{I}_0)$, $\mathbb{I}_2(\boldsymbol{U} - \boldsymbol{I}_0) = \mathbb{I}_1(\boldsymbol{U} - \boldsymbol{I}_0)^2$, λ_e, μ_e are the Lame's constants and $\boldsymbol{I}_0$ is the second-rank unit tensor in the reference configuration [12]

Following Fadodun el al. [12], Melnikov and Ogden [13] and Dorfmann and Ogden [14], we generalize and consider a simple energy function of the form

$$\Phi(\boldsymbol{F}, \vec{B}_L) = \frac{1}{2}\lambda_e \mathbb{I}_1^2(\boldsymbol{U} - \boldsymbol{I}_0) + \mu_e \mathbb{I}_2(\boldsymbol{U} - \boldsymbol{I}_0) + \frac{1}{2\mu}\vec{B}_L \cdot \boldsymbol{U} \cdot \vec{B}_L, \tag{23}$$

for the magnetoelastomeric solid under consideration such that its purely elastic part corresponds to the semilinear hyperelastic energy function in Eq. (22), where the scalar μ is the permeability of the solid.

The Frechet derivatives of invariants $\mathbb{I}_1(\boldsymbol{U} -)^2$ and $\mathbb{I}_1^2(\boldsymbol{U} - \boldsymbol{I}_0)$ with respect to $\boldsymbol{F}$ are [12]

$$\frac{\partial \mathbb{I}_1(\boldsymbol{U} - \boldsymbol{I}_0)^2}{\partial \boldsymbol{F}} = 2(\boldsymbol{U} - \boldsymbol{I}_0)\frac{\partial \boldsymbol{U}}{\partial \boldsymbol{F}} = 2(\boldsymbol{U} - \boldsymbol{I}_0)\boldsymbol{O}^{DT} = 2(\boldsymbol{F}^T - \boldsymbol{O}^{DT}), \tag{24}$$

and

$$\frac{\partial \mathbb{I}_1^2(\boldsymbol{U} - \boldsymbol{I}_0)}{\partial \boldsymbol{F}} = 2\mathbb{I}_1(\boldsymbol{U} - \boldsymbol{I}_0)\boldsymbol{I}_0\frac{\partial \boldsymbol{U}}{\partial \boldsymbol{F}} = 2\mathbb{I}_1(\boldsymbol{U} - \boldsymbol{I}_0)\boldsymbol{I}_0\boldsymbol{O}^{DT} = 2\mathbb{I}_1(\boldsymbol{U} - \boldsymbol{I}_0)\boldsymbol{O}^{DT}, \tag{25}$$

respectively. The tensors $\boldsymbol{O}^{DT}$ and $\boldsymbol{F}^T$ are transposes of $\boldsymbol{O}^D$ and $\boldsymbol{F}$.
Next, the Fréchet derivative of invariant $\vec{B}_L \cdot \boldsymbol{U} \cdot \vec{B}_L$ with respect to $\boldsymbol{F}$ is

$$\frac{\partial \vec{B}_L \cdot \boldsymbol{U} \cdot \vec{B}_L}{\partial \boldsymbol{F}} = (\vec{B}_L \otimes \vec{B}_L)\frac{\partial \boldsymbol{U}}{\partial \boldsymbol{F}} = (\vec{B}_L \otimes \vec{B}_L)\boldsymbol{O}^{DT}. \tag{26}$$

In view of Eqs. (24)-(26), and substituting Eq. (23) into Eq. (21) gives the total first Piola-Kirchhoff's stress tensor $\boldsymbol{P}$

$$\boldsymbol{P} = \frac{\partial \Phi(\boldsymbol{F}, \vec{B}_L)}{\partial \boldsymbol{F}} = 2\mu_e \boldsymbol{F}^T + \left((\lambda_e \mathbb{I}_1(\boldsymbol{U} - \boldsymbol{I}_0) - 2\mu_e)\boldsymbol{I}_0 + \frac{1}{2\mu}(\vec{B}_L \otimes \vec{B}_L)\right)\boldsymbol{O}^{DT}, \tag{27}$$

and deformation-dependent magnetic field vector $\vec{H}_L$

$$\vec{H}_L = \frac{\partial \Phi(\boldsymbol{F}, \vec{B}_L)}{\partial \vec{B}_L} = \frac{1}{\mu}\boldsymbol{U}\vec{B}_L, \tag{28}$$

as the magnetoelastic constitutive model for the magnetoelastic solids under consideration, where ⊗ denotes the tensor product.
In view of Eq. (18), the corresponding Eulerian total stress tensor $\boldsymbol{T}$ is

$$\boldsymbol{T} = (\det(\boldsymbol{F}))^{-1}\left(2\mu_e \boldsymbol{F}^{2T} + \left((\lambda_e \mathbb{I}_1(\boldsymbol{U} - \boldsymbol{I}_0) - 2\mu_e)\boldsymbol{I}_0 + \frac{1}{2\mu}(\vec{B}_L \otimes \vec{B}_L)\right)(\boldsymbol{F}\boldsymbol{O}^D)^T\right).$$

Remark 1: The obtained magnetoelastic constitutive model in Eqs. (27) and (28) shows that the stress distribution is sensitive to the magnetic induction generated while the magnetic field at point within the body is deformation-dependent.

Remark 2: In the absence of externally applied magnetic field, the magnetic induction vector varnishes at points within the solids, and the derived constitutive equations degenerate to $\boldsymbol{P} = 2\mu_e \boldsymbol{F}^T + (\lambda_e \mathbb{I}_1(\boldsymbol{U} - \boldsymbol{I}_0) - 2\mu_e)\boldsymbol{O}^{DT}$ and $\vec{H}_L = \vec{0}$ which yield purely mechanical stress in the solids [12]. In addition, if the stretch symmetric tensor $\boldsymbol{U} = \boldsymbol{I}_0$ the derived model degenerates to $\boldsymbol{P} = \frac{1}{2\mu_L}(\vec{B}_L \otimes \vec{B}_L)\boldsymbol{O}^{DT}$ and $\vec{H}_L = \frac{1}{\mu}\vec{B}_L$ which implies that the body exhibits purely magnetic behaviour.

Application: Magnetoelastic Deformation of Conductive Hyperelastic Cylinder

In view of the constitutive Eqs. (27) and (28), it is convenient to solve the problem of the radial deformation of conductive semilinear magneto-hyperelastic solid cylinder in the Lagrangian frame of reference. The theory of magnetoelastcity presented in the previous sections is now specialized for application to the problem of magnetoelastic deformation of a solid cylinder. The cylinder under consideration has radius A and is subject to uniform axial current density. The geometry of the cylinder is assumed to be sufficiently long/thin such that the edge effect is neglected.

Let the cylindrical coordinates (R, Θ, Z) with associated unit basis vectors $\vec{E}_R, \vec{E}_\Theta, \vec{E}_Z$ describe the position vector $\vec{R} = R\vec{E}_R + Z\vec{E}_Z$ of material point of the cylinder in the reference configuration defined by

$$0 \leq R \leq A, \quad 0 \leq \Theta \leq 2\pi, \quad 0 \leq Z \leq L, \tag{29}$$

where A and L are the radius and length of the cylinder respectively.

Invoking the constraint of circular symmetry, and let the cylindrical coordinates (r, θ, z) with unit basis vectors $\vec{e}_r, \vec{e}_\theta, \vec{e}_z$ give the position vector $\vec{r} = r\vec{e}_r + z\vec{e}_z$ of the corresponding material point in the deformed configuration, the deformation of the cylinder is defined by

$$r = r(R), \quad \theta = \Theta, \quad z = \lambda_z Z, \tag{30}$$

where $r(R)$ is a function of R only and λ_z is the uniform axial stretch.

Using Eqs. (2) and (30), the deformation gradient $\boldsymbol{F}$ is

$$\boldsymbol{F} = Grad\,\vec{r} = \frac{\partial r}{\partial R}\vec{e}_r \otimes \vec{E}_R + \frac{r}{R}\vec{e}_\theta \otimes \vec{E}_\Theta + \frac{\partial z}{\partial Z}\vec{e}_z \otimes \vec{E}_Z,$$

$$\boldsymbol{F} = Grad\,\vec{r} = \frac{\partial r}{\partial R}\vec{e}_r \otimes \vec{E}_R + \frac{r}{R}\vec{e}_\theta \otimes \vec{E}_\Theta + \lambda_z \vec{e}_z \otimes \vec{E}_Z, \tag{31}$$

where $\vec{e}_r, \vec{e}_\theta, \vec{e}_z$ and $\vec{E}_R, \vec{E}_\Theta, \vec{E}_Z$ are the orthonormal basis vectors in Ω and Ω_0 respectively.

Using Eqs. (3), (4) and (31) gives

$$(\boldsymbol{U})_{ij} = (\boldsymbol{F})_{ij} \;\; and \;\; (\boldsymbol{O}^D)_{ij} = (\boldsymbol{I}_0)_{ij}, \tag{32}$$

where for any second-rank tensor $\boldsymbol{A}$, $(\boldsymbol{A})_{ij}$ denotes the components of $\boldsymbol{A}$.

Solution of magnetostatic field equations

Let the axis of the solid cylinder of radius A and constant conductivity σ be taken along the Z axis and let $\vec{J}_L = J_L\vec{E}_Z$be the uniform axial current density along the axis of the cylinder, where J_L is the magnitude of $\vec{J}_L$ and $\vec{E}_Z$ is the unit vector along the axis of the cylinder.

For this problem, Eq. (15) is satisfied for uniform $\vec{J}_L$. The solution of Eq. 14(b) is obtained by introducing a uniquely defined vector (magnetic vector potential) $\vec{G}_L$ such that

$$\vec{B}_L = Curl\,\vec{G}_L,\ Div\,\vec{G}_L = 0. \tag{33}$$

Using Eq. 33(b) and $\vec{B}_L = \mu\vec{H}_L$ in Eq. 14(a) gives

$$Curl\left(Curl\,\vec{G}_L\right) = \begin{cases} 4\pi\mu J_L\vec{E}_Z, & R \le A \\ \vec{0}, & R > A \end{cases}. \tag{34}$$

The form of Eq. (34) suggests that $\vec{G}_L = G(R,\Theta,Z)\vec{E}_Z$ where $G(R,\Theta,Z)$ is a function of cylindrical coordinates $R,\ \Theta\ and\ Z$. Meanwhile Eq. 33(b) shows that G is independent of Z and by symmetry, G is independent of Θ, thus, $\vec{G}_L = G(R)\vec{E}_Z$ is a function of R only [15].

The resolute of $\left(Curl\,\vec{G}_L\right)$ and $Curl\left(Curl\,\vec{G}_L\right)$ are

$$\begin{cases} Curl\,\vec{G}_L = \left(0, -\frac{dG(R)}{dR}, 0\right) \\ Curl\left(Curl\,\vec{G}_L\right) = \left(0,0,-\frac{1}{R}\frac{d}{dR}\left(R\frac{dG(R)}{dR}\right)\right) \end{cases}. \tag{35}$$

Substituting Eq. 35(b) into Eq. (34) gives

$$\begin{cases} \frac{1}{R}\frac{d}{dR}\left(R\frac{dG(R)}{dR}\right) + 4\pi\mu J_L, & R \le A \\ \frac{1}{R}\frac{d}{dR}\left(R\frac{dG(R)}{dR}\right) = 0, & R > A \end{cases}. \tag{36}$$

The solution of Eq. (36) yields

$$G(R) = \begin{cases} C_1 \ln R + C_2 - \pi\mu J_L R^2, & R \le A \\ C_3 \ln R + C_4, & R > A \end{cases}, \tag{37}$$

where $C_i, i = 1,2,3,4$ are constants to be determined.

Since $G(R)$ must be finite along the axis of the tube ($R = 0$), $C_1 = 0$. Thus,

$$G(R) = \begin{cases} C_2 - \pi\mu J_L R^2, & R \le A \\ C_3 \ln R + C_4, & R > A \end{cases}. \tag{38}$$

The constant C_3 is obtained by using the Maxwell's first circuital relation

$$\int \vec{H}\cdot d\vec{s} = 4\pi I \Longrightarrow \frac{1}{\mu_L}\int_0^{2\pi} -\frac{dG(R)}{dR} R\,d\Theta = 4\pi I, \tag{39}$$

where I is the current flowing in the tube. Hence,

$$C_3 = -2\pi\mu J_L A^2. \tag{40}$$

Substituting Eq. (40) into Eq. (38) gives

$$G(R) = \begin{cases} C_2 - \pi\mu J_L R^2, & R \leq A \\ -2\pi\mu J_L A^2 \ln R + C_4, & R > A \end{cases}. \tag{41}$$

Recall that the magnetic vector potential is continuous at the surface of seperation, thus,

$$C_2 - \pi\mu J_L A^2 = -2\pi\mu J_L A^2 \ln A + C_4. \tag{42}$$

and setting $C_4 = 0$ (without loss of generality) gives

$$C_2 = \pi\mu J_L A^2 - 2\pi\mu J_L A^2 \ln A. \tag{43}$$

Substituting Eq. (43) into Eq. (41) gives the solution

$$G(R) = \begin{cases} \pi\mu J_L (A^2 - R^2) - 2\pi\mu J_L A^2 \ln A, & R \leq A \\ -2\pi\mu J_L A^2 \ln R, & R > A \end{cases}. \tag{44}$$

In view of Eqs. 33(a) and 35(a), the magnetic induction $\vec{B}_L = \left(0, -\frac{dG(R)}{dR}, 0\right) = (0, B_\Theta, 0)$ has non-varnishing azimuthal resolute [15]

$$B_\Theta = -\frac{dG(R)}{dR} = \begin{cases} 2\pi\mu J_L R, & R \leq A \\ \frac{2\pi\mu J_L A^2}{R}, & R > A \end{cases}. \tag{45}$$

Using Eqs. (28), (31), (32) and (45), the corresponding magnetic field strength $\vec{H}_L = (0, H_\Theta, 0)$ within the tube has non-varnishing azimuthal component H_Θ

$$H_\Theta = \frac{1}{\mu}\frac{r(R)}{R} B_\Theta = 2\pi J_L r(R), \tag{46}$$

where $r(R)$ is a function of R only.

Solution of equilibrium equation

Using Eqs. (31), (32) and 45(a) in Eq. (27) gives the non-zero components $P_{RR}, P_{\Theta\Theta}, P_{ZZ}$ of the first Piola-Kirchhoff's stress tensor $\boldsymbol{P}$:

$$P_{RR} = 2\mu_e\left(\frac{dr}{dR} - 1\right) + \lambda_e\left(\frac{dr}{dR} + \frac{r}{R} + \lambda_z - 3\right), \tag{47}$$

$$P_{\Theta\Theta} = 2\mu_e\left(\frac{r}{R} - 1\right) + \lambda_e\left(\frac{dr}{dR} + \frac{r}{R} + \lambda_z - 3\right) + 2\mu(\pi J_L R)^2, \tag{48}$$

$$P_{ZZ} = 2\mu_e(\lambda_z - 1) + \lambda_e\left(\frac{dr}{dR} + \frac{r}{R} + \lambda_z - 3\right). \tag{49}$$

Materials Research Proceedings 31 (2023) 715-725 https://doi.org/10.21741/9781644902592-73

Using Eqs.(47)-(49), the equilibrium equation in Eq. (19) reduces to

$$\frac{dP_{RR}}{dR}+\frac{1}{R}(P_{RR}-P_{\Theta\Theta})=0. \tag{50}$$

Substituting Eqs. (47)-(48) into Eq. (50) gives

$$\left(\frac{d^2r}{dR^2}+\frac{1}{R}\frac{dr}{dR}-\frac{r}{R^2}\right)+\left(\frac{2\mu\pi^2J_L^2}{2\mu_e+\lambda_e}\right)R=0. \tag{51}$$

The solution of Eq. (51) gives the deformation function

$$r(R)=C_5R+\frac{C_6}{R}+\varpi_J R^3, \tag{52}$$

where C_5, C_6 are constants and the scalar $\varpi_J=\frac{1}{4}\left(\frac{\mu\pi^2J_L^2}{2\mu_e+\lambda_e}\right)$ depends on the axial current density. The displacement field u(R) at points within the tube is defined by

$$u(R)=r(R)-R=C_5^*R+\frac{C_6}{R}+\varpi_J R^3, \tag{53}$$

where the constant $C_5^*=C_5-1$.
Since the displacement u(R) must be finite at $R=0$, then $C_6=0$. Furthermore, using the condition $P_{RR}=0$ at the tube surface $R=A$ yields
$C_5^*=\left(\frac{3\mu_e+2\lambda_e}{\mu_e+\lambda_e}\right)\varpi_J A^2-\left(\frac{\lambda_e}{2(\mu_e+\lambda_e)}\right)(\lambda_z-1)$,
where λ_z is the uniform axial stretch.
Thus, the displacement $u(R)$ at point within the tube

$$u(R)=\varpi_J R^3+\left(\frac{3\mu_e+2\lambda_e}{\mu_e+\lambda_e}\right)\varpi_J A^2R-\left(\frac{\lambda_e}{2(\mu_e+\lambda_e)}\right)(\lambda_z-1)R. \tag{54}$$

Using Eqs. (46), (54) and knowing that $r(R)=u(R)+R$, the magnetic field strength H_Θ at points within the tube is

$$H_\Theta=2\pi J_L\left(\varpi_J R^3+\left(\frac{3\mu_e+2\lambda_e}{\mu_e+\lambda_e}\right)\varpi_J A^2R-\left(\frac{\lambda_e}{2(\mu_e+\lambda_e)}\right)(\lambda_z-1)R+R\right). \tag{55}$$

In the absence of current density J_L (i.e. $J_L=0$), the displacement function u(R) and magnetic field function H_Θ in Eqs. (54) and (55) reduce/degenerate to

$$u(R)=\left(\frac{\lambda_e}{2(\mu_e+\lambda_e)}\right)(1-\lambda_z)R,\ H_\Theta=0, \tag{56}$$

respectively.

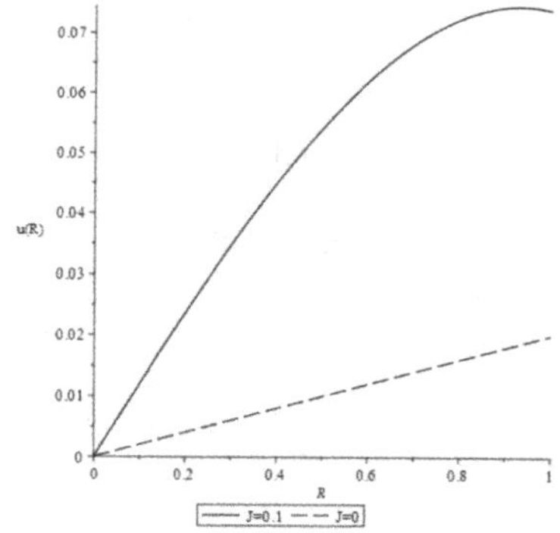

Fig. 1: This plot shows that the effect of steady current density on the magnitude of displacement function at point within solid cylinder is significantly pronounced.

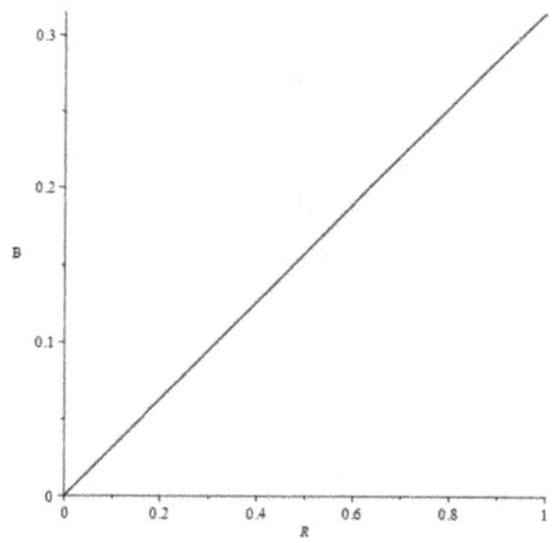

Fig. 2: This plot shows that the azimuthal magnetic induction produced by the steady current within the solid cylinder increases linearly along the radius of the cylinder.

Conclusion

The study develops a new magnetoelastic constitutive theory for modelling magneto-mechanical interaction of solids. The theory is specialized for application to the problem of radial deformation of a solid circular cylindrical made of conductive semilinear magnetohyperelastic materials. It is obtained that the stress propagation in the solid cylinder is sensitive to the magnetic induction produced by uniform axial current density while the associated magnetic field is deformation-dependent. Furthermore, it is shown that the effect of uniform axial current density on the deformation of the tube is significantly pronounced. Finally, the results in this study find applications in design of soft actuators, sensors and energy harvesters to mention a few.

References

[1] G. A. Maugin, Continuum mechanics of electromagnetic solids, Asterdam: North-Holland, 1988.

[2] A. C. Eringen and G. A. Maugin, Electrodynamics of continuum I, New York: Springer, 1990. https://doi.org/10.1007/978-1-4612-3236-0

[3] Y. H. Pao, Electromagnetic forces in deformable continua, In: Nemat-Nasser, S (ed) Mechanics today, (1978) vol 4, 209-306. https://doi.org/10.1016/B978-0-08-021792-5.50012-4

[4] Z. Pei, T. Xiong and D. Ding, Nonlinear magnetoelastic deformation of porous solids, Math. Mech. Solids, (2021) 1-21. doi: 10.1177/108:2865219865

[5] N. H. Reddy and P. Saxena, Instabilities in axisymmetric magnetoelastic deformation of a cylindrical membrane, Int. J. Solids and Struc. (2018) 136-137: 203-219. https://doi.org/10.1016/j.ijsolstr.2017.12.015

[6] D. Garcia-Gonzalez and M. A. Hossain, A microstructural-based approach to model magneto-viscoelastic materials at finite strain, Int. J. Solids and Struc. (2021) 208-209, 119-132. https://doi.org/10.1016/j.ijsolstr.2020.10.028

[7] Z. Ren, W. Hu, X. Dong et al., Multi-functional soft-bodied jellyfish-like swimming, Nat. Commun. (2019) 10: 2703. https://doi.org/10.1038/s41467-019-10549-7

[8] A. K. Bostola and M. A. Hossain, A review on magneto-mechanical charactirazation of magnetorheological elastomers, Compos, Part B, Eng. (2020) 200: 108348 https://doi.org/10.1016/j.compositesb.2020.108348

[9] Ll. Dorfmann AND R. W. Ogden, Nonlinear model of electroelastic and magnetoelastic interaction, New-York: Springer, 2014. https://doi.org/10.1007/978-1-4614-9596-3

[10] B. A. Nadjar, A modelling framework for finite strain magnetoviscoelasticity, Math. Mech. Solids, (2020) 25 (2) 288-304. https://doi.org/10.1177/1081286519873963

[11] P. Saxena, M. Hossain and P. Steinmann, A theory of finite deformation magneto-viscoelasticity, Int. J. Solids Struc. (2013) 50: 3886-3897. https://doi.org/10.1016/j.ijsolstr.2013.07.024

[12] O. O. Fadodun, O. G. Fadodun, A. S. Borokinni, B. A. Olokuntoye, O. P. Layeni and A. P. Akinola, Electroelastic constitutive model for dielectric semilinear hyperelastic solids with application to radial deformation of a rotating tube, Meccanica, (2022) vol. 57, 2355-2363. https://doi.org/10.1007/s11012-022-01580-y

[13] A. Melnikov and R. W. Ogden, Finite deformations of an electroelastic circular cylindrical tube, Z. Angew. Math. Phys. (2016) 69, 60. https://doi.org/10.1007/s00033-016-0733-0

[14] L. Dorfmann and R. W. Ogden, The effect of deformation dependent permittivity on the elastic response of a finitely deformed dielectric tube. Mech Res Commun. (2018) 93, 47-57. https://doi.org/10.1016/j.mechrescom.2017.09.002

[15] V. Ferraro, Electromagnetic Theory. The Athlone Press, University of London, 1965.

Tensile behavior of functionally graded sandwich PLA-ABS produced via fused filament fabrication process

Umut Caliskan[1,a], Caglar Sevim[2,b] and Munise Didem Demirbas[3,c*]

[1]Department of Mechanical Engineering, Erciyes University, Kayseri 38039, Turkey

[2]Department of Mechanical Engineering, Nigde Omer Halisdemir University, Nigde 51240, Turkey

[3]Department of Mechanical Engineering, Erciyes University, Kayseri 38039, Turkey

[a]ucaliskan@erciyes.edu.tr, [b]caglar.sevim@ohu.edu.tr, [c]mddemirbas@erciyes.edu.tr

Keywords: Additive Manufacturing, Functionally Graded Material, Sandwich Structure, Tensile Behavior, Acrylonitrile Benzidine Styrene, Polylactic Acid

Abstract. The study investigated the tensile behavior of Sandwich Functionally Graded Material (SFGM) fabricated using Additive Manufacturing (AM) technology. SFGMs are characterized by a gradual variation in composition and structure with respect to the forming volume from the lower and upper surfaces of the structure towards the center, resulting in a corresponding change in material properties. Fused Filament Fabrication (FFF), a widely used AM process, was used in the present work to fabricate the thermoplastic polymer-based SFGM specimens. SFGM were produced by the FFF method using ABS and PLA materials and subjected to tensile tests according to ASTM D638.

Introduction

Since the history of humanity, science and technology have been making progress in an ever-increasing manner. The tools and equipment that humanity wants to acquire in line with the needs of humanity with the instinct of self-realization have been the main factors in the development of science and technology. In the development of mechanical properties in industry; Factors such as lightness, strength, production method, raw material supply, temperature effect and cost make the need for new materials and designs permanent.

In the last two decades, many studies have been conducted on the determination of the mechanical behavior of functionally graded materials (FGMs). However, the use of FGMs in the industry has been limited due to the difficulties encountered in production [1,2]. The production, which is generally made by adding materials layer by layer with a 3D model, is called the AM technique or the AM in technical literature [3-5]. With the developing production methods, the necessity of obtaining better material properties has arisen and multi-material Additive Manufacturing (MMAM) technologies have gained importance for materials that need to be printed by combining more than one material in a single process, such as composites and FGMs[2,6,7]. Among these multi-material Additive Manufacturing technologies, techniques such as Fused Deposition Modelling (FDM), Fused Filament Fabrication (FFF), and Vat photopolymerization (VP) have gained importance. With the spread of these technology, the difficulties experienced in the production of this FGM have disappeared and these materials with certain variations in the volume/surface of the components have been started to be produced. Thus, studies investigating the mechanical behavior of FG materials have entered the literature and continue to do so.

First of all, studies on polymers planned to be produced using the AM technique in FG structure will be reviewed. Afterward, the studies that have been made and what will be done in the case of producing these polymers as FG will be emphasized. Kumar and Narayan [8] produced PLA, a biodegradable material, by the AM technique according to ASTM D638 Type I standards and investigated its behavior under tensile load. They stated that the properties of PLA material produced with the AM technique show similar properties to those produced traditionally, and they confirmed the data they obtained with FEM.

In this study, a tensile test was applied to FG-Sandwich structure samples produced by the FFF technique, and the gradation effect on mechanical strength was investigated. The basic parameters affecting the layer structures determined to create the sandwich structure; layer thickness, volume ratio and total thickness were investigated.

Experimental Study

PLA filament, made entirely from PLA granules by Filameon company [9] and ABS filament, containing 100% ABS granules, used in this study. The filaments were in the form of a continuous wire and fed from reels into a 3D printer that had a 300mm x 300mm heated printing area and linear sliding ball bearings in each movement axis. The printer had two input and one output printing nozzles that allowed for printing with different materials and adjusting mixing ratios using G-CODEs. Figure 1 shows the printing principle of the 3D printer with two inputs and one output. The printer had linear plain bearings and ball carriages in the X, Y, and Z axes, providing stable and accurate movements. The filament feeding was done directly with the feeder motor unit on the movable print head on the X-axis. Each test sample was printed individually and under identical conditions. Table 1 provides details on the printing parameters used for each test sample.

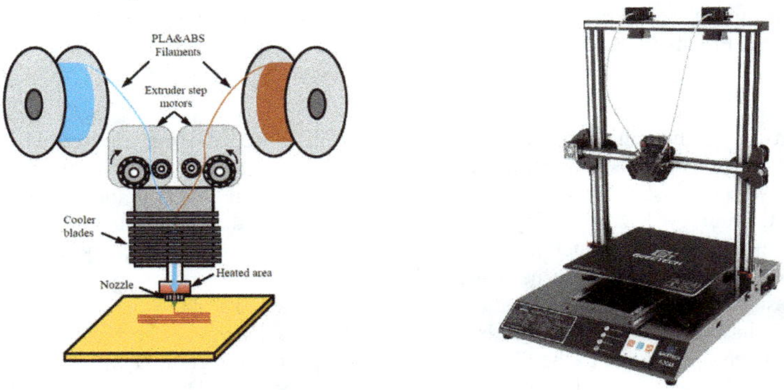

Figure 1. The design of the 3D printer used in this study.

Since the printing conditions of PLA and ABS are partially different from each other, the printing conditions must be adjusted according to the material structure when printing the composite structure. This situation is neglected by many researchers in the literature. For the best printing properties, as the ABS ratio in the composite structure increased, the printing nozzle temperature and the table temperature were increased gradually, and the cooling fan was turned off when the ABS ratio exceeded 20%. The designed test samples were converted to .stl format via the CAD program and their GCODEs were created with the Ultimaker CURA V4.10.0 slicing program [10]. Mixing ratios and temperature adjustments are adjusted by editing GCODEs by us.

Table 1. Printing parameters

Layer thickness	0.2 mm
Filler fiber thickness	0.4 mm
Wall thickness	0.4 mm
Print speed	30 mm/sn
Nozzle temperature	200-230 ^{0}C
Table temperature	60-100 ^{0}C

The properties of PLA and ABS materials produced under the conditions described above are detailed in Table 2. In this study, the tensile strengths of the samples produced by determining the functionally specific volume ratios were investigated.

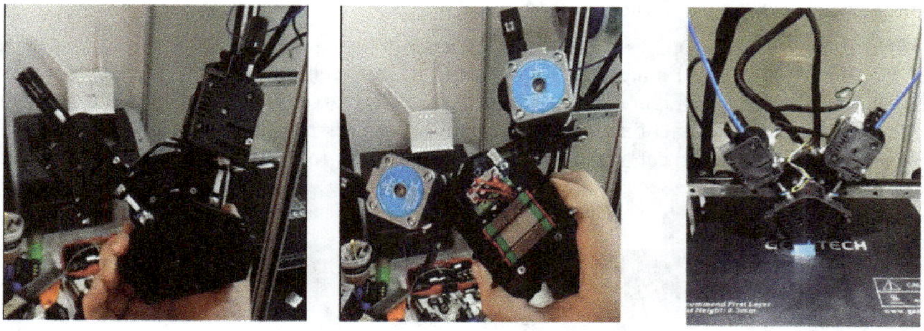

Figure 2. Designed double nozzle system

Table 2. Thermal and mechanical properties of ABS and PLA [11]

Properties	PLA	ABS
Filament diameter (mm)	1.75	1.75
Density (gr/cm^3)	1.24	1.04
Bed temperature (°C)	60	80-100
Nozzle temperature (°C)	190-230	230-260
Melt Flow Index (210°C/2.16kg)	6	80-120
Tensile strength (MPa)	53	45
Elongation (%)	6	10
Bending strength (MPa)	83	73
Rackwell hardness	108	108
Max service temperature (°C)	55	85

The variation of the compositional gradient exponent was considered linear in the grading of ABS and PLA materials. Functionally Graded samples for the 11 layers given in Figure 3 were produced in all ratios. However, delamination occurred in compositions with less than 70% PLA (ie more than 30% ABS) (Figure 4). Therefore, in the sample production made according to the ASTM D638 standard, the volume ratios in the Sandwich FG production were made as given in Figure 5.

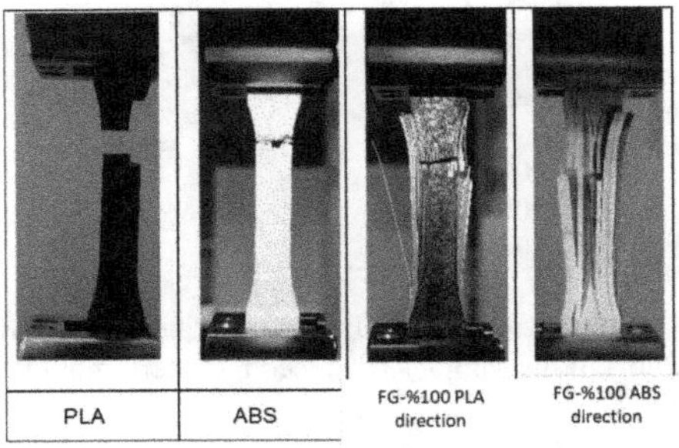

Mixing Ratio		
Layer	PLA	ABS
1	%100	%0
2	%90	%10
3	%80	%20
4	%70	%30
5	%60	%40
6	%50	%50
7	%40	%60
8	%30	%70
9	%20	%80
10	%10	%90
11	%0	%100

Figure 3. Functionally Graded samples and mixing ratios

Considering the ASTM D638 standard, all dimensions change with changing thicknesses. Due to the fact that the produced materials are not isotropic, different parameters were planned and productions were made. In Figure 3, layer thicknesses were given as 0.8, 1.3, 1.6 and 2.6 mm according to the planned functional change. These thicknesses represent two different types of ASTM standards. In this respect, samples were produced both for samples of the same type and for different types of thicknesses and subjected to tensile testing.

Layer	PLA	ABS
1. Layer	100%	0%
2. Layer	90%	10%
3. Layer	80%	20%
4. Layer	70%	30%
5. Layer	70%	30%
6. Layer	80%	20%
7. Layer	90%	10%
8. Layer	100%	0%

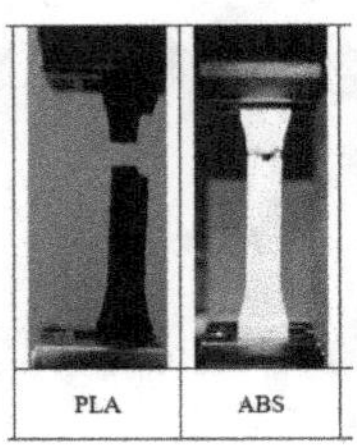

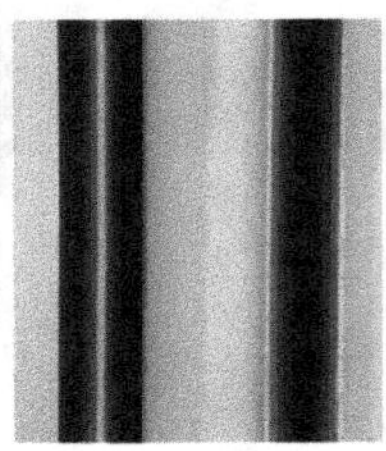

Layer and Sample Thickness of 8-Layer Sandwich Structures Produced by 3D Printer	
Layer	**Thickness**
0,8 mm	6,4 mm
1,3 mm	10,4 mm
1,6 mm	12,8 mm
2,6 mm	20,8 mm

Figure 4. FG-Sandwich structures samples, which were subjected to tensile tests according to ASTM D638 standard.

In order to determine the sizing in Figure 4, the 11-layer FG structure was fabricated and tensile tested as shown in Figure 3. According to this test, delamination was observed in the FG structure. Therefore, this situation is prevented by reducing the number of stratified layers. As seen in the after-tensile test shape, Reducing the ABS layers prevented the splitting considerably.

Samples of four different layer thicknesses were produced for the FG structure for sandwich type. These layer thicknesses are 0.8, 1.3, 1.6, and 2.6 mm. The number of layers without delamination is 4 and for the sandwich structure is 8 accordingly, total sample thicknesses were produced as 0.8x8=6.4 mm (Type I), 1.6x8=12.8 mm (Type III), 1.3x8= 10.4 mm (Type III), 2.6x8=20.8 mm (Type III). For these thicknesses, production was made on the basis of Type 1 and III dimensions according to the ASTM D638 standard given in Figure 5. While choosing the layer thicknesses, both the limitations in production and the standards were taken into account in order to compare the samples. Figure 5 shows the FG sandwich samples which are manufactured FFF method in different dimensions. The color change refers to the functionally changing material distribution. Specimen descriptions indicate filament thickness, layer thickness, total thickness, ASTM standard, and overall length.

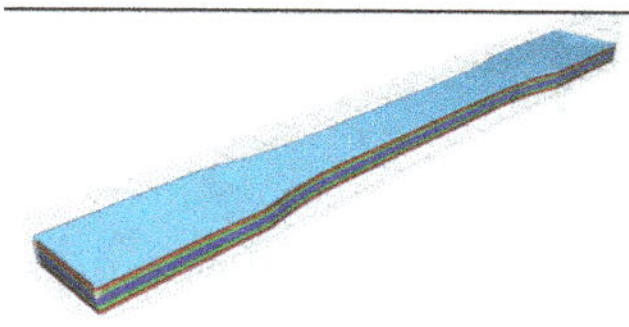	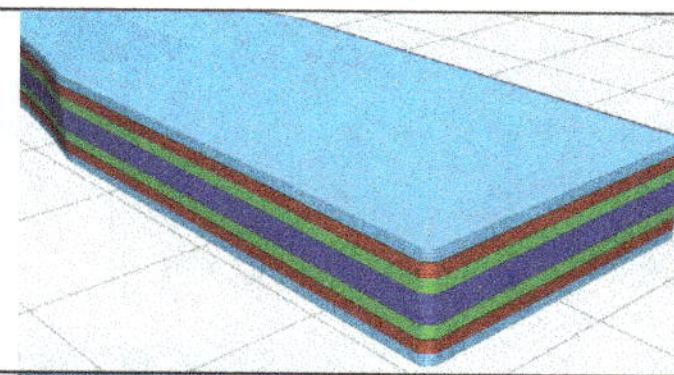	**Filament Thickness: 0.2 mm** **Layer Thickness: 0.8 mm** **Total Thickness: 6.4 mm** **ASTMD638-Type –I** **Total Length: 165 mm**
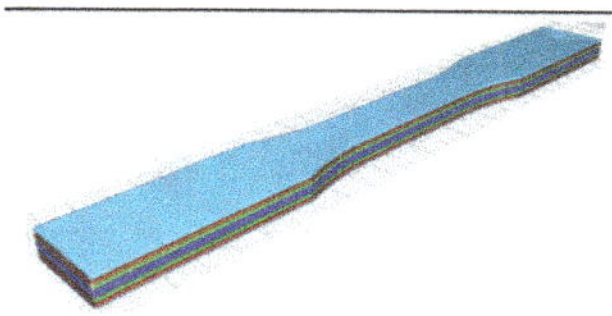	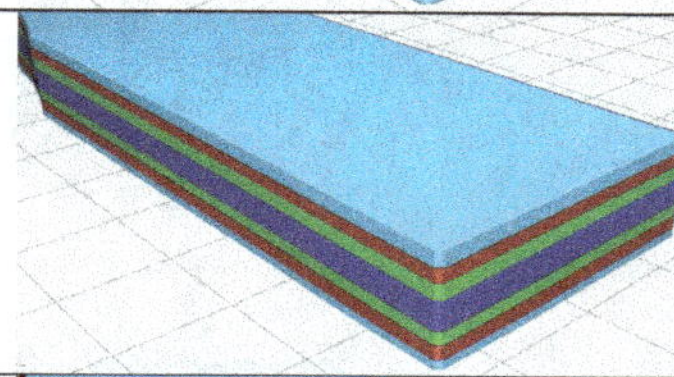	**Filament Thickness: 0.2 mm** **Layer Thickness: 1.3 mm** **Total Thickness: 10.4 mm** **ASTMD638-Type – II** **Total Length: 246 mm**
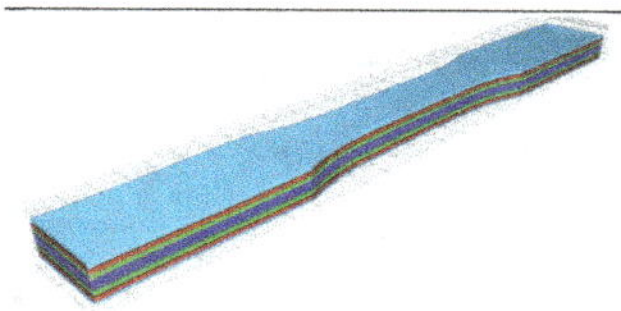	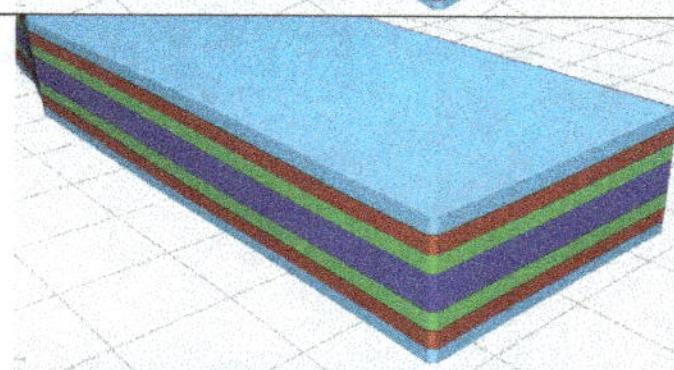	**Filament Thickness: 0.2 mm** **Layer Thickness: 1.6 mm** **Total Thickness: 12.8 mm** **ASTMD638-Type – II** **Total Length: 246 mm**
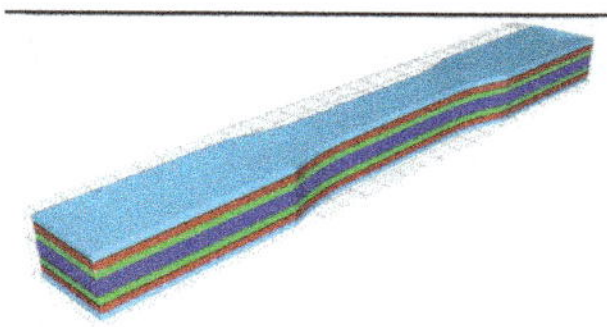	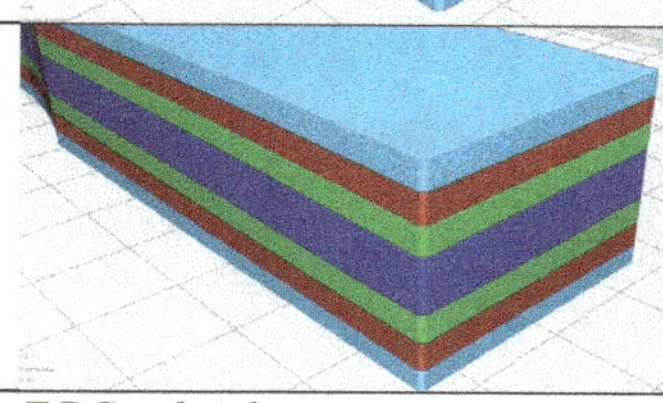	**Filament Thickness: 0.2 mm** **Layer Thickness: 2.6 mm** **Total Thickness: 20.8 mm** **ASTMD638-Type – II** **Total Length: 246 mm**

Figure 5. FG Sandwich structure types

Results & Conclusion

The present study investigate the tensile test behavior of FG-Sandwich structure samples produced by the FFF technique, and the gradation effect on mechanical strength. The basic parameters affecting the tensile test results such as layer thickness, volume ratio and total thickness were also investigated. As it started with planned production with additive manufacturing, there were many unknown parameters. One of them was the printing speed in the FFF method. First, pure PLA material and FG sandwich structure were produced at different printing speeds and subjected to tensile testing. Figure 6 shows the effect of the printing speed for PLA and FG sandwich structure for 30 and 40 mm/s. In pure PLA material, the slower-produced material resulted in slightly greater elongation, and the strength levels remained similar. In the material produced as sandwich, while the fast produced material revealed higher force, it created more elongation.

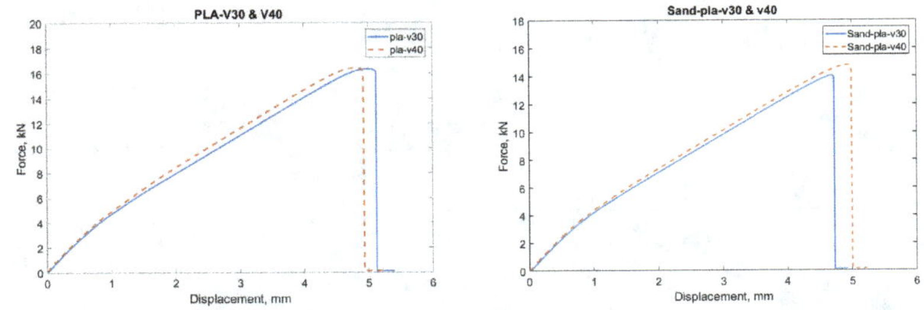

Figure 6. The effect of the printing speed for PLA and FG Sandwich materials.

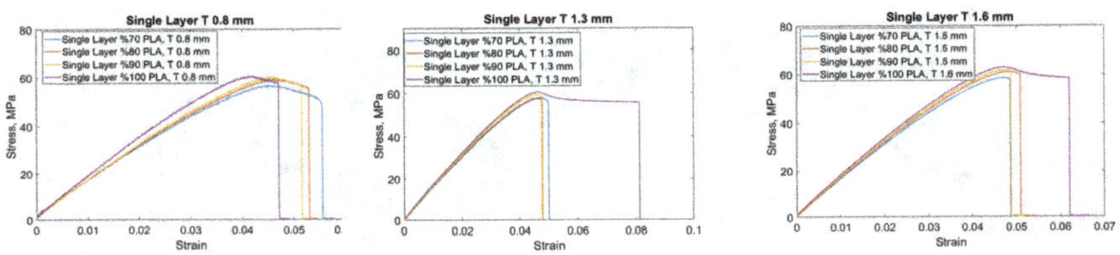

ASTM-D638-Type I

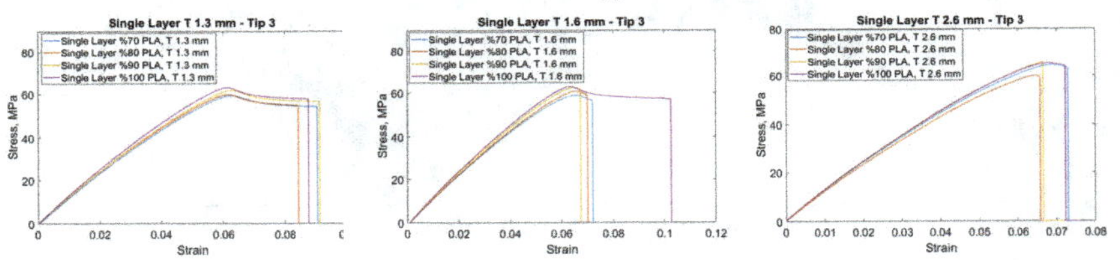

ASTM-D638-Type III

Figure 7. The effect of the changing functional gradient for single-layer specimens

In Figure 7, tensile test results of single-layer samples with different functional gradients produced according to ASTM D638 Types I and III are given. Single-layer samples have thicknesses of 0.8, 1.3, 1.6 and 2.6 mm. The functional gradient ratios are 70% PLA-30% ABS, 80% PLA-20% ABS, 90% PLA-10% ABS and 100% PLA-0% ABS. While the thicknesses of the sample produced for Type I are 0.8, 1.3 and 1.6 mm, the lengths are 165 mm. In the sample produced for Type III, the thickness is 1.3, 1.6 and 2.6 mm, while the length is 246 mm. Compared to Type I, strain increased at similar stress levels with ABS reinforcement in the 0.8 mm thick sample. In the 1.3 mm thick sample, the strain decreased with ABS additive. Similar results were obtained in the 1.6 mm thick sample. There was no significant change in the results with ABS reinforcement in the 1.3 mm thick sample compared to Type III. However, for the 1.6 mm thick sample, the strain value decreased from 0.1 to 0.07 with ABS reinforcement. Figure 8 shows the stress-strain behavior of FG sandwich structures as a result of the tensile test according to two different ASTM standards. The first comparison shows the thickness variation for Type I and Type III. The first sample Type I- had a layer thickness of 0.8 mm and an overall thickness of 6.4 mm.

The other sample, Type-III, has a layer thickness of 1.6 mm and a total thickness of 12.8 mm. When the layer and total thickness were doubled, there was no significant decrease in the stress level, while the strain value almost doubled. This result shows that the material behavior is not isotropic. And it shows that the standard gives different results in this respect. In the other comparison, it gives the tensile test result of samples of the same size (Type III). While the layer thickness of the first sample is 1.3 mm and the total thickness is 10.4 mm, the layer thickness of the other sample is 2.6 mm and the total thickness is 20.8 mm. When the layer thickness was doubled, the strain value also doubled and increased from 0.06 to 0.17. Stress levels are around 60 MPa. This result showed that the tensile behavior standard of sample ASTM D638 rigid plastics differ in terms of the results of plastics produced by additive manufacturing.

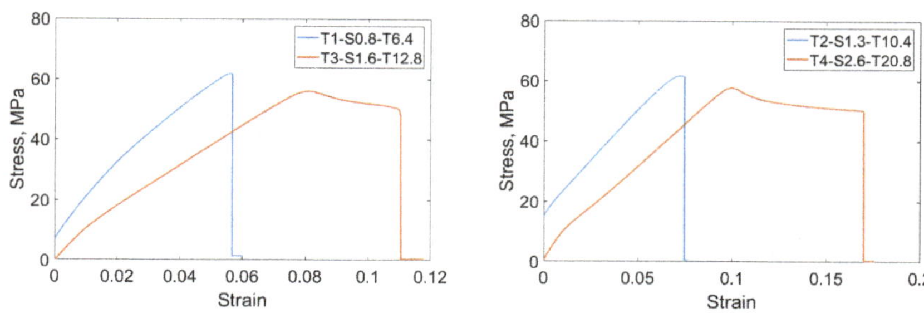

ASTM-D638-Type I & III ASTM-D638-Type III

Figure 8. The effect of ASTM standard types for FG Sandwich structure.

References

[1] R. M. Mahamood, E.T. Akinlabi, M. Shukla, S. Pityana, Functionally Graded Material: An overview, Proceedings of the World Congress on Engineering, III, (2012).

[2] S. Nohut, M. Schwentenwein, Vat Photopolymerization Additive Manufacturing of Functionally Graded Materials: A Review. Journal of Manufacturing and Materials Processing. 6 (2022)17. https://doi.org/10.3390/jmmp6010017

[3] T.D. Ngo, A. Kashani, G. Imbalzano, K.T.Q. Nguyen, D. Hui, Additive manufacturing (3D printing): A review of materials, methods, applications and challenges, Composites Part B: Engineering, 143 (2018), 172-196 https://doi.org/10.1016/j.compositesb.2018.02.012

[4] K.W.A. Van der Elst , Industry 4.0: The new production paradigm and its implications for EU policy (2017)

[5] A. Sartal, R. Bellas, A.M. Mejías, A. García-Collado, The sustainable manufacturing concept, evolution and opportunities within Industry 4.0: a literature review

[6] B. Saleh, J. Jiang, R. Fathi, T. Al-hababi, Q. Xu, L. Wang, D. Song, A. Ma, 30 Years of functionally graded materials: An overview of manufacturing methods, applications, and future challenges. Compos. Part B Eng. 201 (2020), 108376. https://doi.org/10.1016/j.compositesb.2020.108376

[7] A. Al Rashid, W. Ahmed, M.Y. Khalid, M. Koç, Vat photopolymerization of polymers and polymer composites: Processes and applications, Additive Manufacturing, 47(2021), 102279, https://doi.org/10.1016/j.addma.2021.102279

Materials Research Proceedings 31 (2023) 726-733 https://doi.org/10.21741/9781644902592-74

[8] S. Anand Kumar, Y. Shivraj Narayan, Tensile Testing and Evaluation of 3D-Printed PLA Specimens as per ASTM D638 Type IV Standard, Innovative Design, Analysis and Development Practices in Aerospace and Automotive Engineering (2019). https://doi.org/10.1007/978-981-13-2718-6_9

[9] Riiz Machinery ltd. co. FILAMEON, www.filameon.com.

[10] Ultimaker CURA (https://ultimaker.com/software/ultimaker-cura) (accessed Dec 10, 2021).

Artificial intelligence-based prediction of geotechnical impacts of polyethylene bottles and polypropylene on clayey soil

Abolfazl Baghbani[1,a *], Firas Daghistani[2,b], Hasan Baghbani[3,c]
Katayoon Kiany[4,d] and Jafar Bolouri Bazaz[5,e]

[1]School of engineering, Deakin University, Melbourne, VIC, Australia

[2]Department of Civil Engineering, La Trobe University, Melbourne, VIC, Australia; Also with Civil Engineering Department, University of Business and Technology, Jeddah, Saudi Arabia

[3]School of Engineering, Ferdowsi University, Mashhad, Iran

[4]School of Design, University of Melbourne, Melbourne, VIC, Australia

[5]School of Engineering, Ferdowsi University, Mashhad, Iran

[a]abaghbani@deakin.edu.au, [b]f.daghistani@latrobe.edu.au, [c]hasanbaghbani1998@gmail.com, [d]kianyk@student.unimelb.edu.au, [e]bolouri@um.ac.ir

Keywords: Plastic Waste, Clayey Soil, Artificial Intelligence, Soil Stabilizer, CRRF, ANN

Abstract. This study aims to investigate the application of artificial intelligence (AI) methods in predicting the resilient modulus of soil mixtures with polyethylene (PE) bottles and polypropylene (PP). The AI methods used in the study are artificial neural network (ANN) and classification and regression random forest (CRRF), and the modeling was conducted using a database of 160 datasets. The study also evaluated the importance of different input parameters on the accuracy of the models. The results show that the CRRF model is more accurate than the ANN model in predicting the effects of materials PE and PP on soil resilient modulus. Additionally, the study found that the number of hidden layers and neurons in the ANN model should be optimized for the best performance and increasing their number does not always lead to increased accuracy. Finally, the study identified the most and least important input parameters for predicting the effect of PE and PP on the resilient modulus of the mixture using both AI models.

Introduction

Globally, plastic waste has become a significant environmental concern [1]. The disposal of plastic waste has become a significant challenge, resulting in negative environmental impacts, including soil pollution [1-3]. Recently, there has been a growing interest in recycling plastic waste and using it as a partial replacement for traditional construction materials. Plastic waste is one of these materials that can be used in geotechnical applications.

A clayey soil is characterized by high plasticity, high compressibility, and low shear strength. As a result of these characteristics, the soil is prone to instability and erosion. A clayey soil's mechanical behavior is affected by several factors, including its mineral composition, water content, and confining pressure [4-6]. In addition to being non-biodegradable, plastic waste materials can persist in the environment for decades or even centuries. As a result of the accumulation of plastic waste in the soil, soil pollution can occur, and the soil's geotechnical properties can be adversely affected. In order to reduce plastic waste and improve soil mechanical properties, the use of plastic waste materials in geotechnical applications has been investigated [7-9].

Recently, studies have investigated the effects of plastic waste materials on clayey soil's geotechnical properties. Niyomukiza et al. [10] investigated the effects of polyethylene terephthalate (PET) waste on clayey soil's shear strength and compressibility. According to the results, the addition of PET waste reduced the shear strength of the soil and increased its compressibility. According to Bandyopadhyay and Sharma and Sharma [11], plastic waste affects clayey soil's deformation characteristics and shear strength. As a result of the addition of plastic waste, the soil's shear strength was reduced and its deformation was increased. However, it was observed that plastic waste improved the bearing capacity of the soil and reduced its settlement.

There are several parameters that influence the strength of soil and waste bottles mixtures. Due to the multiplicity and nonlinearity of these parameters, a comprehensive equation for predicting the strength of soil and waste bottles mixtures has not yet been developed. Artificial intelligence has been used as a method of solving this problem. A number of fields, including soil mechanics [12-14], soil dynamics [15-16], soil cracking [17-18], road construction [19-20], recycled material [21-24] and slope stability [25] have successfully used artificial intelligence methods [26-27]. In spite of this, no study has yet been conducted to investigate the use of artificial intelligence methods in predicting the resilient modulus of two soil- of polyethylene (PE) bottles and polypropylene (PP) mixtures. To predict the parameters of resilient modulus, two artificial intelligence methods, artificial neural network (ANN) and classification and regression random forest (CRRF), were used in this study for the first time. Modeling was conducted using a database consisting of 160 datasets. The inputs included confining pressure, cyclic stress, constant stress, load cycle number, length of waste materials, fiber content, UCS and CBR of mixtures. The importance of input parameters was evaluated after evaluating the AI models.

Database Collection and Processing

Experiment and data collection

In this study, a database consisting of 160 sets with eight inputs and one output was utilized. Collection data for two types of recycled waste bottles, PE and PP, was collected from Hassan et al. [28]. The statistical parameters of this database are shown in Tables 1 and 2 for PE and PP, respectively. According to Tables 1 and 2, the database has a proper distribution.

Table 1. Statistical information of database for PE

Variable	Observations	Minimum	Maximum	Mean	Std. deviation
Resilient modulus (Mpa)	160	128.000	188.000	144.833	10.507
Confining pressure (kPa)	160	13.790	41.370	28.442	11.437
Cyclic stress (kPa)	160	12.410	62.050	36.454	17.311
Constant stress (kPa)	160	1.280	6.890	4.046	1.931
Load cycle no.	160	100.000	1000.000	156.250	218.539
length (CM)	160	1.000	2.000	1.500	0.502
Fiber content (%)	160	0.000	4.000	2.000	1.419
UCS (kPa)	160	148.000	291.000	239.500	48.139
CBR (%)	160	4.000	7.200	5.540	1.015

Table 2. Statistical information of database for PP

Variable	Observations	Minimum	Maximum	Mean	Std. deviation
Resilient modulus (Mpa)	160	130.000	150.100	137.073	4.091
Confining pressure (kPa)	160	13.790	41.370	28.442	11.437
Cyclic stress (kPa)	160	12.410	62.050	36.454	17.311
Constant stress (kPa)	160	1.280	6.890	4.046	1.931
Load cycle no.	160	100.000	1000.000	156.250	218.539
Length (CM)	160	1.000	2.000	1.500	0.502
Fiber content (%)	160	0.000	4.000	2.000	1.419
UCS (kPa)	160	148.000	256.000	217.400	36.457
CBR (%)	160	4.000	6.000	4.960	0.682

Preparation of the data for AI modelling

In order to prepare the database for AI modeling, the values of different parameters were linearly normalized using the following equation. Normalizing the database will increase the accuracy of the model since each parameter in the database has a specific unit.

$$X_{norm} = \frac{X - X_{min}}{X_{max} - X_{min}} \tag{1}$$

where X_{max}, X_{min}, X and X_{norm} are maximum, minimum, actual, and normalized values, respectively.

As part of the preparation process, the database was divided into two training and testing databases. A total of 20% (26 sets) of the database was used for testing and 80% (134 sets) for training purposes. Tables 3 and 4 illustrate the statistical information for these two databases. To increase the accuracy of the model, it is better to have the statistical information of these two databases close to each other, as is the case in this study.

Table 3. Statistical information of training database

Variable	Observations	Minimum	Maximum	Mean	Std. deviation
Resilient modulus (MPa)-PE	134	128.000	188.000	144.619	10.463
Resilient modulus (MPa)-PP	134	130.000	150.100	137.213	4.365
Confining pressure (kPa)	134	13.790	41.370	28.712	11.722
Cyclic stress (kPa)	134	12.410	62.050	37.323	17.048
Constant stress (kPa)	134	1.280	6.890	4.142	1.903
Load cycle no.	134	100.000	1000.000	160.448	226.121
length (CM)	134	1.000	2.000	1.485	0.502
Fiber (%)	134	0.000	4.000	1.993	1.438
UCS-Kpa	134	148.000	291.000	238.507	48.964
CBR (%)	134	4.000	7.200	5.525	1.029

Table 4. Statistical information of testing database

Variable	Observations	Minimum	Maximum	Mean	Std. deviation
Resilient modulus (MPa)-PE	26	132.100	170.000	145.941	10.877
Resilient modulus (MPa)-PP	26	132.100	142.000	136.356	2.107
Confining pressure (kPa)	26	13.790	41.370	27.050	9.929
Cyclic stress (kPa)	26	12.410	62.050	31.980	18.297
Constant stress (kPa)	26	1.280	6.890	3.551	2.036
Load cycle no.	26	100.000	1000.000	134.615	176.505
length (CM)	26	1.000	2.000	1.577	0.504
Fiber (%)	26	0.000	4.000	2.038	1.341
UCS-Kpa	26	148.000	291.000	244.615	44.172
CBR (%)	26	4.000	7.200	5.615	0.959

Data-driven modeling

Artificial neural network (ANN)

Artificial Neural Networks, or ANNs, are a type of machine learning model that has been inspired by the structure and function of the human brain. They have proven to be highly effective at processing complex data sets and making predictions based on them. ANNs consist of multiple interconnected nodes, which process information and adjust the connections between them to learn from the data. They have been used in a wide variety of applications, ranging from image and speech recognition to natural language processing and predictive modeling. The power of ANNs lies in their ability to learn and generalize from large amounts of data, making them a key technology in the field of artificial intelligence.

Classification and regression random forest (CRRF)

Random Forest is a machine learning algorithm that can be used for both classification and regression tasks. Random Forest works by constructing multiple decision trees during training and then aggregating the results of each tree to make a final prediction.

For classification tasks, Random Forest constructs a set of decision trees, where each tree predicts the class of a given input based on a set of features. During training, each decision tree is built using a random subset of the training data and a random subset of the features. When making a prediction, the input is passed through each tree, and the predicted class is determined by taking a majority vote of the predictions made by each tree.

For regression tasks, Random Forest works similarly, but instead of predicting a class, it predicts a continuous numerical value. Each decision tree is constructed to predict the output value based on a set of features, and the final prediction is made by averaging the predictions made by each tree.

Random Forest is a popular machine learning algorithm due to its high accuracy and ability to handle large datasets with a large number of features. It also has the advantage of being less prone to overfitting compared to single decision trees, as the aggregation of multiple trees helps to reduce the effects of individual trees that may be overfitting to the data.

Results

Artificial neural network (ANN)

The artificial neural network model was used and investigated with a variety of architectures. Several factors influence the architecture of a ANN, including the number of hidden layers and the number of neurons within each layer and type of algorithm. This study examined the number of hidden layers ranging from one layer to five layers. Additionally, two algorithms Bayesian Regularization (BR) and Levenberg-Marquardt (LM) were examined in order to determine their performance. At the end of the process, the number of neurons in each layer was changed in order to achieve the best and most optimal architecture. In Tables 5 and 6, the results of ANN modeling were presented for PE and PP, respectively. For testing database, to predict the effects of adding PE to soil, the average accuracy (R^2) of the ANN models for algorithm BR is equal to 0.977, and for algorithm LM is equal to 0.943. Therefore, algorithm BR has performed better than algorithm LM. Based on Table 6, algorithm BR performed better than algorithm LM for predicting the effects of PP.

Table 5. Results of ANN for PE

	The number of hidden layers	R^2-Test	R^2-Train	MAE-Test	MAE-Train
Bayesian Regularization	1H	0.954	0.972	1.964	2.431
	2H	0.981	0.974	1.031	1.453
	3H	0.993	0.988	0.743	0.959
	4H	0.983	0.986	0.983	1.432
	5H	0.973	0.987	1.731	1.923
	Average	0.977	0.981	1.290	1.640
Levenberg-Marquardt	1H	0.932	0.943	2.893	2.564
	2H	0.945	0.949	2.230	2.021
	3H	0.949	0.953	1.455	1.421
	4H	0.953	0.955	1.321	1.342
	5H	0.934	0.940	2.102	2.054
	Average	0.943	0.948	2.000	1.880

Table 6. Results of ANN for PP

	The number of hidden layers	R^2-Test	R^2-Train	MAE-Test	MAE-Train
Bayesian Regularization	1H	0.864	0.923	1.495	1.453
	2H	0.873	0.938	1.239	1.254
	3H	0.880	0.951	0.742	0.831
	4H	0.881	0.956	0.629	0.735
	5H	0.872	0.938	0.902	0.972
	Average	0.874	0.943	1.001	1.049
Levenberg-Marquardt	1H	0.842	0.905	2.593	2.019
	2H	0.852	0.920	2.442	1.875
	3H	0.874	0.934	2.002	1.549
	4H	0.878	0.949	1.730	1.349
	5H	0.856	0.921	2.021	1.892
	Average	0.860	0.926	2.158	1.737

It has also been determined that the number of 25 neurons for each hidden layer is the optimal number after many trials and errors. It was found that the optimum number of hidden layers for material PE was three, and the optimum number of hidden layers for material PP was four. It appears that increasing the number of hidden layers and neurons will not always increase the accuracy of the ANN model, and there is a point at which there is an optimal number of hidden layers and neurons.

Figs 1 and 2 show diagrams of ANN predicted values versus actual values for PE and PP, respectively. According to the results, the ANN model predicts well the performance of both materials.

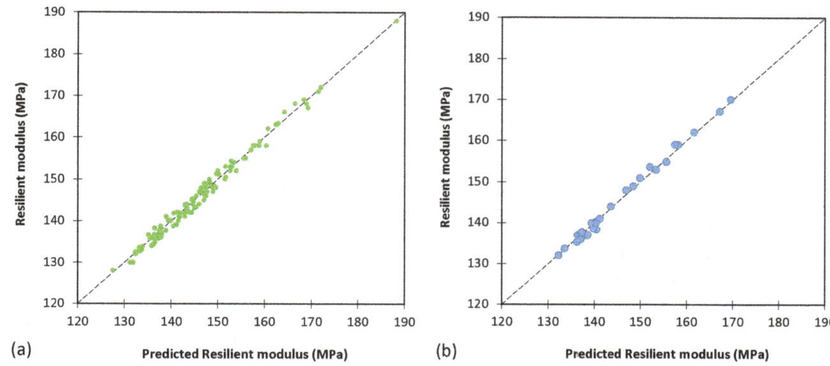

Fig. 1. The results of ANN for predicting resilient modulus for PE

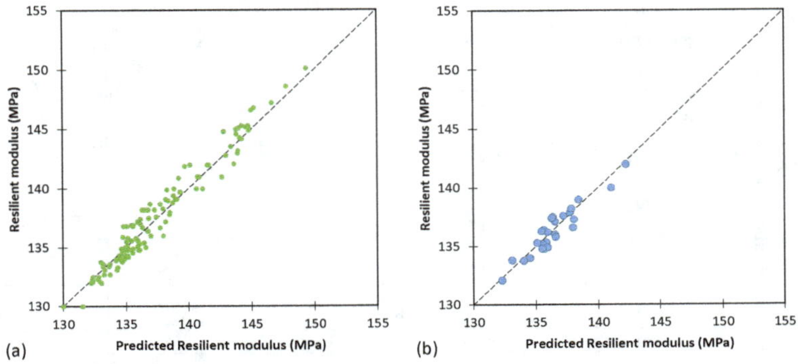

Fig. 2. The results of ANN for predicting resilient modulus for PP

Classification and regression random forest (CRRF)

Among the important parameters of the CRRF method are parameters max depth of trees and the number of trees. The effective parameters of the CRRF model were repeatedly changed and a number of trials and errors were conducted in order to obtain the best CRRF model. Table 7 shows the values of the effective parameters in the CRRF model. On the basis of the obtained results, the CRRF method is most effective when the number of trees is equal to 300 and the max depth of trees is 8.

Table 7. The specifications of the best CRRF.

Trees parameters					Forest parameters		
Min. node size	Min. son size	Max depth	Mtry	CP	Sampling	Sample size	Number of trees
2	1	8	2	0.00001	Random with replacement	78	300

Figs 3 and 4 illustrate the values predicted by the CRRF model in comparison with the actual resistance values for materials PE and PP, respectively. The CRRF model is able to predict the effect of both materials on soil resilient modulus based on the results obtained.

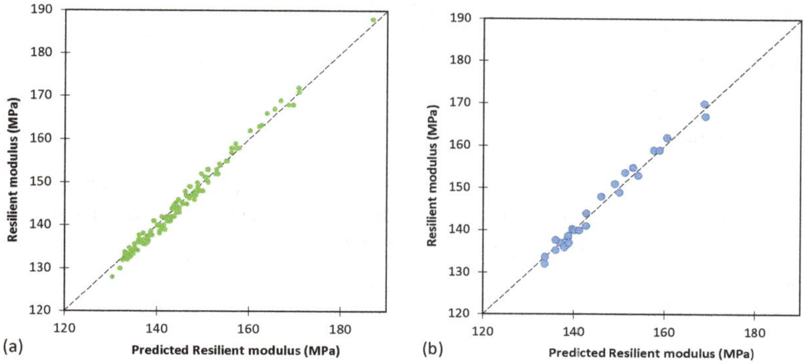

Fig. 3. The results of CRRF for predicting resilient modulus for PE

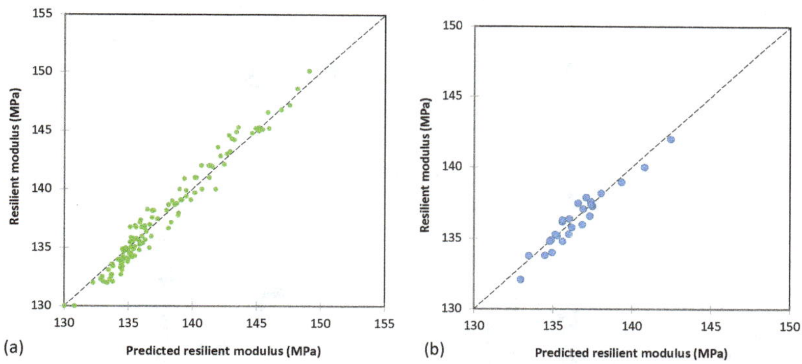

Fig. 4. The results of CRRF for predicting resilient modulus for PP

Table 8 illustrates the performance of the CRRF model for predicting soil resilient modulus after adding materials PE and PP. For predicting the training database for PE, the accuracy (R^2) and error (MAE) of the CRRF model are equal to 0.986 and 1.047, respectively, and for predicting the effect of PP, they are equal to 0.961 and 0.705. According to the testing database, the accuracy (R^2) of the CRRF model for predicting the effect of PE and PP on soil resilient modulus is 0.961 and 0.926, and its error (MAE) is 0.705 and 0.470, respectively. Based on these results, the CRRF model is capable of predicting the effects of materials PE and PP on soil resilient modulus, and CRRF performed better than ANN method.

Table 8. The performance of CRRF model

Performance metrics	PE		PP	
	Training	Testing	Training	Testing
MAE	1.047	1.237	0.705	0.470
R^2	0.986	0.982	0.961	0.926

The variable importance of input parameters

Figs 5 and 6 show the importance of input parameters on soil resilient modulus prediction for both artificial intelligence models and for PE and PP, respectively. In the ANN model, parameter confining pressure was determined to be the most important parameter to predict the effect of PE on soil resilient modulus, while parameter load cycle number was determined to be the least important parameter. Although, according to Fig. 5b, in the CRRF model, parameters CBR and load cycle number have the highest and lowest importance, respectively in predicting rffect of PE on the resilient modulus of the mixture. According to Fig. 6, in the ANN and CRRF model, the most and least important parameters to predict effect of PP on the resilient modulus of mixture are parameters confining pressure and length, respectively.

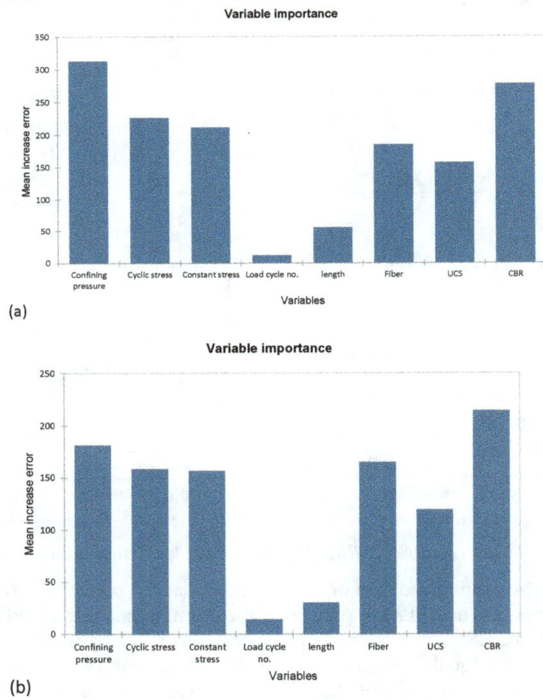

Fig. 5. The importance of parameters to predict resilient modulus for PE, based on (a) ANN and (b) CRRF

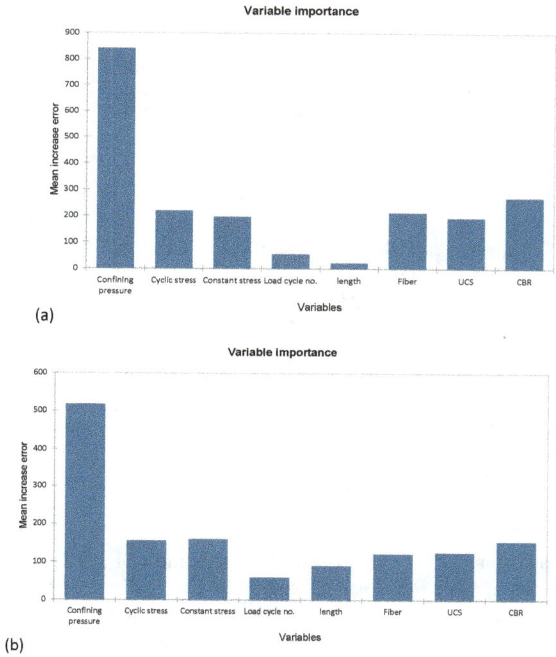

Fig. 6. The importance of parameters to predict resilient modulus for PP, based on (a) ANN and (b) CRRF

Conclusion

In this study, artificial intelligence methods were employed to predict the resilient modulus of two soil mixtures containing polyethylene (PE) and polypropylene (PP). For the first time, artificial neural networks (ANNs) and classification and regression random forests (CRRFs) were used to predict resilient modulus parameters. Results indicated that the CRRF model outperformed the ANN model in predicting the effects of PE and PP on soil resilient modulus. In addition, the study found that the number of hidden layers and neurons for the ANN model is optimal and that increasing them beyond this point will not necessarily increase accuracy. It was also found that the ANN model was able to accurately predict the performance of both materials, with the optimal number of hidden layers and neurons being three and 25 for PE, and four and 25 for PP. In summary, the study demonstrated the potential of using artificial intelligence methods to predict soil properties and to assist with the design and construction of soil structures..

References

[1] Huang S, Wang H, Ahmad W, Ahmad A, Ivanovich Vatin N, Mohamed AM, Deifalla AF, Mehmood I. Plastic waste management strategies and their environmental aspects: A scientometric analysis and comprehensive review. International Journal of Environmental Research and Public Health. 2022 https://doi.org/10.3390/ijerph19084556

[2] Sahebzadeh S, Heidari A, Kamelnia H, Baghbani A. Sustainability features of Iran's vernacular architecture: A comparative study between the architecture of hot-arid and hot-arid-windy regions. Sustainability. 2017 May 4;9(5):749. https://doi.org/10.3390/su9050749
[3] Baghbani A, Baumgartl T, Filipovic V. Effects of Wetting and Drying Cycles on Strength of Latrobe Valley Brown Coal. Copernicus Meetings; 2023 Feb 22. https://doi.org/10.5194/egusphere-egu23-4804
[4] Baghbani A, Costa S, Lu Y, Soltani A, Abuel-Naga H and Samui P. Effects of Particle Shape on Shear Modulus of Sand Using Dynamic Simple Shear Testing. Arabian journal of geosciences. 2023.
[5] Leng Y, Peng J, Wang S, Lu F. Development of water sensitivity index of loess from its mechanical properties. Engineering Geology. 2021 Jan 1;280:105918. https://doi.org/10.1016/j.enggeo.2020.105918
[6] Baghbani A, Costa S, O'Kelly BC, Soltani A, Barzegar M. Experimental study on cyclic simple shear behaviour of predominantly dilative silica sand. International Journal of Geotechnical Engineering. 2022 Oct 23:1-5. https://doi.org/10.1080/19386362.2022.2135226
[7] Abukhettala M, Fall M. Geotechnical characterization of plastic waste materials in pavement subgrade applications. Transportation Geotechnics. 2021 Mar 1;27:100472. https://doi.org/10.1016/j.trgeo.2020.100472
[8] Hassan HJ, Rasul J, Samin M. Effects of plastic waste materials on geotechnical properties of clayey soil. Transportation Infrastructure Geotechnology. 2021 Sep;8(3):390-413. https://doi.org/10.1007/s40515-020-00145-4
[9] Baghbani A, Daghistani F, Naga HA, Costa S. Development of a Support Vector Machine (SVM) and a Classification and Regression Tree (CART) to Predict the Shear Strength of Sand Rubber Mixtures. InProceedings of the 8th International Symposium on Geotechnical Safety and Risk (ISGSR), Newcastle, Australia 2022. https://doi.org/10.3850/978-981-18-5182-7_00-08-004.xml
[10] Niyomukiza JB, Bitekateko A, Nsemerirwe J, Kawiso B, Kiwanuka M. Investigating the effect of PET plastic bottle strips on the strength and compressibility properties of clayey soil. InIOP Conference Series: Earth and Environmental Science 2021 Nov 1 (Vol. 894, No. 1, p. 012021). IOP Publishing. https://doi.org/10.1088/1755-1315/894/1/012021
[11] Sharma A, Sharma RK. Effect of addition of construction-demolition waste on strength characteristics of high plastic clays. Innovative Infrastructure Solutions. 2019 Dec;4(1):27. https://doi.org/10.1007/s41062-019-0216-1
[12] Nguyen MD, Baghbani A, Alnedawi A, Ullah S, Kafle B, Thomas M, Moon EM, Milne NA. Investigation on the suitability of aluminium-based water treatment sludge as a sustainable soil replacement for road construction. Transportation Engineering. 2023 Mar 28:100175. https://doi.org/10.1016/j.treng.2023.100175
[13] Inazumi S, Intui S, Jotisankasa A, Chaiprakaikeow S, Kojima K. Artificial intelligence system for supporting soil classification. Results in Engineering. 2020 Dec 1;8:100188. https://doi.org/10.1016/j.rineng.2020.100188
[14] Nguyen MD, Baghbani A, Alnedawi A, Ullah S, Kafle B, Thomas M, Moon EM, Milne NA. Experimental Study on the Suitability of Aluminium-Based Water Treatment Sludge as a Next Generation Sustainable Soil Replacement for Road Construction. Available at SSRN 4331275.
[15] Baghbani A, Costa S, Faradonbeh RS, Soltani A, Baghbani H. Experimental-AI Investigation of the Effect of Particle Shape on the Damping Ratio of Dry Sand under Simple Shear Test Loading, 2023. https://doi.org/10.20944/preprints202303.0021.v1

[16] Baghbani A, Choudhury T, Samui P, Costa S. Prediction of secant shear modulus and damping ratio for an extremely dilative silica sand based on machine learning techniques. Soil Dynamics and Earthquake Engineering. 2023 Feb 1;165:107708. https://doi.org/10.1016/j.soildyn.2022.107708
[17] Xu JJ, Zhang H, Tang CS, Cheng Q, Tian BG, Liu B, Shi B. Automatic soil crack recognition under uneven illumination condition with the application of artificial intelligence. Engineering Geology. 2022 Jan 1;296:106495. https://doi.org/10.1016/j.enggeo.2021.106495
[18] Baghbani A, Costa S, Choundhury T, Faradonbeh RS. Prediction of Parallel Desiccation Cracks of Clays Using a Classification and Regression Tree (CART) Technique. InProceedings of the 8th International Symposium on Geotechnical Safety and Risk (ISGSR), Newcastle, Australia 2022. https://doi.org/10.3850/978-981-18-5182-7_00-08-003.xml
[19] Yaseen ZM, Ali ZH, Salih SQ, Al-Ansari N. Prediction of risk delay in construction projects using a hybrid artificial intelligence model. Sustainability. 2020 Feb 18;12(4):1514. https://doi.org/10.3390/su12041514
[20] Baghbani A, Baghbani H, Shalchiyan MM, Kiany K. Utilizing artificial intelligence and finite element method to simulate the effects of new tunnels on existing tunnel deformation. Journal of Computational and Cognitive Engineering. 2022 Aug 15.
[21] Baghbani A, Daghistani F, Kiany K, Shalchiyan MM. AI-Based Prediction of Strength and Tensile Properties of Expansive Soil Stabilized with Recycled Ash and Natural Fibers. EasyChair; 2023 Feb 19.
[22] Baghbani, A.; Nguyen, M.D.; Alnedawi, A.; Milne, N.; Baumgartl, T.; Abuel-Naga, H. Improving Soil Stability with Alum Sludge: An AI-Enabled Approach for Accurate Prediction of California Bearing Ratio. Preprints 2023, 2023030197. https://doi.org/10.20944/preprints202303.0197.v1
[23] Baghbani A, Daghistani F, Baghbani H, Kiany K. Predicting the Strength of Recycled Glass Powder-Based Geopolymers for Improving Mechanical Behavior of Clay Soils Using Artificial Intelligence. EasyChair; 2023 Feb 19.
[24] Baghbani A, Costa S, Faradonbeh RS, Soltani A, Baghbani H. Modeling the Effects of Particle Shape on Damping Ratio of Dry Sand by Simple Shear Testing and Artificial Intelligence. Applied Sciences. 2023 Mar 29;13(7):4363. https://doi.org/10.3390/app13074363
[25] Baghbani A, Costa S, Choudhury T. Developing Mathematical Models for Predicting Cracks and Shrinkage Intensity Factor During Clay Soil Desiccation. Available at SSRN 4408164.
[26] Baghbani A, Choudhury T, Costa S, Reiner J. Application of artificial intelligence in geotechnical engineering: A state-of-the-art review. Earth-Science Reviews. 2022 May 1;228:103991. https://doi.org/10.1016/j.earscirev.2022.103991
[27] Baghbani A.; Abuel-Naga H.; Shirani Faradonbeh R.; Costa S.; Almasoudi R. Ultrasonic Characterization of Compacted Salty Kaolin-Sand Mixtures Under Nearly Zero Vertical Stress Using Experimental Study and Machine Learning, Geotechnical and geological engineering, 2023. https://doi.org/10.1007/s10706-023-02441-5
[28] Hassan HJ, Rasul J, Samin M. Effects of plastic waste materials on geotechnical properties of clayey soil. Transportation Infrastructure Geotechnology. 2021 Sep;8(3):390-413. https://doi.org/10.1007/s40515-020-00145-4

Static analysis of cross-ply laminated spherical shells using a new hyperbolic shear deformation theory

Anuja S. Jape[1,2,a *], Atteshamuddin S. Sayyad[2,b]

[1]Research Scholar, Department of Civil Engineering, Sanjivani College of Engineering, Savitribai Phule Pune University, Kopargaon-423603, Maharashtra, India

[2]Department of Structural Engineering, Sanjivani College of Engineering, Savitribai Phule Pune University, Kopargaon-423603, Maharashtra, India

[a]anujajape@gmail.com, [b]attu_sayyad@yahoo.co.in

Keywords: Spherical Shell, Laminated Composites, Hyperbolic Shear Deformation Theory, Static Analysis

Abstract. Using a new hyperbolic shear deformation theory, higher-order closed-form solutions to the static bending analysis of laminated composite spherical shells are derived in this study. The current theory accurately predicts the distributions of transverse shear stresses across the thickness of the shell. The governing equations and related boundary conditions are obtained using the principle of virtual work. The Navier type semi-analytical closed-form solutions are obtained for the simply supported boundary conditions. The results obtained using the present theory are compared with previously published results to verify the accuracy and efficiency of the present hyperbolic shear deformation theory.

Introduction

Due to their improved strength-to-weight and stiffness-to-weight ratios, laminated composite materials have seen a significant rise in a variety of engineering applications over the past several decades. They have found their way into a wide range of products, including innovative spacecraft and aircraft components, boat and scull hulls, swimming pool panels, racing car bodies, sports goods, sensor or actuator, catalysts etc. Transverse shear deformation is substantially more significant in the kinematics of thick laminated composite shells than it is in homogeneous metallic ones because advanced composite materials have low transverse shear modulus. Since 3-D elasticity solutions for the laminated shells are involve complex mathematics, laminated shell theories have been developed by researchers to make these problems mathematically more simple. Laminated shell theories are approximate in nature as they are based on assumptions and hypotheses that reduce a three-dimensional problem to a two-dimensional one. Sayyad and Ghugal [1,2] provided a thorough study of displacement-based shear deformation theories for laminated composite beams, plates, and shells. It is well-known that the classical shell theory is suitable for the analysis of thin shell only due to neglect of transverse shear deformation. Mindlin [3] considered the impact of shear deformation in the first order shear deformation theory, in which transverse shear stress is constant across the thickness of shell and does not satisfy the traction free boundary conditions at the top and bottom surfaces of the shell. Revisions to shell theories are therefore necessary in order to take transverse shear and normal deformations into account. Reddy [4] developed well-known parabolic shear deformation theory for the static and dynamic analysis of laminated composite beams, plates and shells satisfying traction free boundary conditions. Liew and Lim [5] presented the higher-order shell theory for the vibration analysis of doubly-curved shallow shells. Tornabene and Ceruti [6], have conducted different studies on the static and dynamic analysis of doubly curved shells and panels using refined shear deformation theories. Using an extended higher-order shell theory, Sayyad and Ghugal [7] found higher-order closed-form solutions for static bending and free vibration analysis of laminated composite and sandwich spherical shells. Shinde and Sayyad [8] have presented a new higher-order shear and normal deformation theory for the free vibration analysis of laminated shells. In the present work, a new hyperbolic shear deformation theory is developed for the static analysis of cross-ply laminated composite spherical shells.

Laminated Shell under Consideration

As illustrated in Fig. 1, consider a differential element of a spherical shell in the *(x, y, z)* coordinate systems where, *x* and *y* curves represents principal curvatures on the mid-plane of laminate. R_x and R_y denote the principal radii of curvature of the mid-plane along *x* and *y* axes respectively. A cross-ply laminated shell element is made up of fibrous composite materials and composed of a *N* number of layers which are perfectly bonded together. A laminate is subjected to transverse load *q (x, y)* on the top surface i.e. $z=-h/2$ because the downward *z*-direction is assumed as positive.

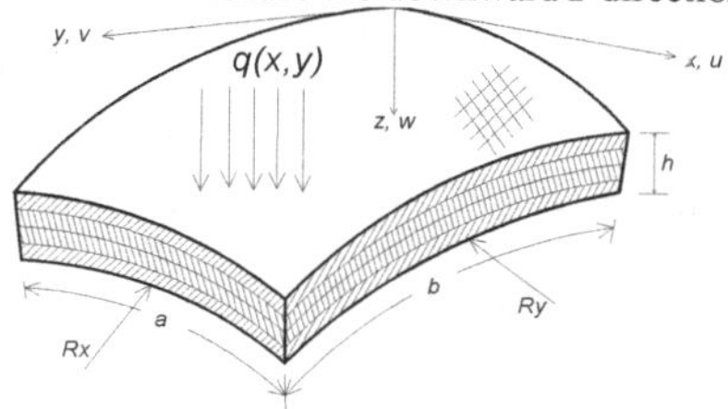

Fig. 1. Laminated shell geometry and coordinate system

Kinematics of the Present Theory

The present theory is built upon the classical shell theory and considers the effects of transverse shear and normal deformations in the in-plane and transverse displacements. Following is the displacement field assumed for the present theory.

$$\begin{aligned} u(x,y,z) &= (1+z/R_x)u_0(x,y) - z\,\partial w_0/\partial x + f(z)\theta_x(x,y),\\ v(x,y,z) &= (1+z/R_y)v_0(x,y) - z\,\partial w_0/\partial y + f(z)\theta_y(x,y),\\ w(x,y,z) &= w_0(x,y) + C_1\, g(z)\theta_z(x,y). \end{aligned} \quad (1)$$

where, *u, v, w* are the displacements of any point of the shell in *x, y, z* directions respectively. $\theta_x, \theta_y, \theta_z$ are the shear slopes in *x, y* and *z* direction respectively. u_0, v_0, w_0 are the mid-plane displacements of any point of the shell in *x, y, z* direction.

Strain Displacement Relationship

Using the linear theory of elasticity, the normal and shear strains associated with the present theory can be obtained as,

$$\begin{aligned} \varepsilon_x &= (\partial u_0/\partial x + w_0/R_x) - z\,\partial^2 w_0/\partial x^2 + f(z)\,\partial\theta_x/\partial x + C_1\left(f'(z)/R_x\right)\theta_z,\\ \varepsilon_y &= (\partial v_0/\partial y + w_0/R_y) - z\,\partial^2 w_0/\partial y^2 + f(z)\,\partial\theta_y/\partial y + C_1\left(f'(z)/R_y\right)\theta_z,\\ \varepsilon_z &= C_1 f''(z)\theta_z,\\ \gamma_{xy} &= \partial u_0/\partial y + \partial v_0/\partial x - 2z\,\partial^2 w_0/\partial x\partial y + f(z)\left(\partial\theta_x/\partial y + \partial\theta_y/\partial x\right),\\ \gamma_{xz} &= f'(z)\theta_x + C_1 f'(z)\,\partial\theta_z/\partial x;\ \gamma_{yz} = f'(z)\theta_y + C_1 f'(z)\,\partial\theta_z/\partial y,\\ f(z) &= z\cosh(\xi/2) - h/\xi\,\sinh(\xi z/h), \text{ where } \xi = 2.634. \end{aligned} \quad (2)$$

Stress- Strain Relationship

Using the Hooke's law, stresses for the kth layer of laminated shell can be obtained.

Materials Research Proceedings 31 (2023) 745-754 https://doi.org/10.21741/9781644902592-76

$$\begin{Bmatrix} \sigma_x \\ \sigma_y \\ \sigma_z \\ \tau_{xy} \\ \tau_{yz} \\ \tau_{xz} \end{Bmatrix} = \begin{bmatrix} Q_{11} & Q_{12} & Q_{13} & 0 & 0 & 0 \\ Q_{12} & Q_{22} & Q_{23} & 0 & 0 & 0 \\ Q_{13} & Q_{23} & Q_{33} & 0 & 0 & 0 \\ 0 & 0 & 0 & Q_{66} & 0 & 0 \\ 0 & 0 & 0 & 0 & Q_{44} & 0 \\ 0 & 0 & 0 & 0 & 0 & Q_{55} \end{bmatrix} \begin{Bmatrix} \varepsilon_x \\ \varepsilon_y \\ \varepsilon_z \\ \gamma_{xy} \\ \gamma_{yz} \\ \gamma_{xz} \end{Bmatrix}. \quad (3)$$

where, $\sigma_x, \sigma_y, \sigma_z, \tau_{xy}, \tau_{xz}, \tau_{yz}$ are the stress components, $\varepsilon_x, \varepsilon_y, \varepsilon_z, \gamma_{xy}, \gamma_{yz}, \gamma_{xz}$ are the strain components. $(Q_{11}, Q_{12}, Q_{13}, Q_{22}, Q_{23}, Q_{33}, Q_{44}, Q_{55}, Q_{66})$ are the reduced stiffness coefficients as follows.

$$\begin{aligned} &Q_{11} = E_1(1-\mu_{23}\mu_{32})/\Delta; Q_{12} = E_1(\mu_{21}+\mu_{31}\mu_{23})/\Delta; Q_{13} = E_1(\mu_{31}+\mu_{21}\mu_{32})/\Delta; Q_{44} = G_{23}; \\ &Q_{22} = E_2(1-\mu_{13}\mu_{31})/\Delta; Q_{23} = E_2(\mu_{32}+\mu_{12}\mu_{31})/\Delta; Q_{33} = E_3(1-\mu_{12}\mu_{21})/\Delta; Q_{55} = G_{13}; Q_{66} = G_{12}. \end{aligned} \quad (4)$$

Principle of Virtual Work

$$\int_0^a\int_0^b\int_{-h/2}^{h/2} \left(\sigma_x\delta\sigma_x + \sigma_y\delta\sigma_y + \sigma_z\delta\sigma_z + \tau_{xy}\delta\tau_{xy} + \tau_{xz}\delta\tau_{xz} + \tau_{yz}\delta\tau_{yz}\right)dzdydx = \int_0^a\int_0^b q\delta w dydx. \quad (5)$$

Substituting the expressions of stresses and strains from Eqs. (2)-(4), into the Eq. (5), integrating by parts, collecting the coefficients of unknowns and setting them equal to zero, one can derive the following governing equation in terms of stress resultants as shown in Eq. (6).

$$\begin{aligned} &\boldsymbol{\delta u_0}: \partial N_x/\partial x + \partial N_{xy}/\partial y = 0; \quad \boldsymbol{\delta v_0}: \partial N_y/\partial y + \partial N_{xy}/\partial x = 0, \\ &\boldsymbol{\delta w_0}: \partial^2 M_x^b/\partial x^2 + \partial^2 M_y^b/\partial y^2 + 2\partial^2 M_{xy}^b/\partial x\partial y - N_x/R_x - N_y/R_y + q = 0, \\ &\boldsymbol{\delta\theta_x}: \partial M_x^S/\partial x + \partial M_{xy}^S/\partial y - Q_{xz} = 0, \\ &\boldsymbol{\delta\theta_y}: \partial M_y^S/\partial y + \partial M_{xy}^S/\partial x - Q_{yz} = 0, \\ &\boldsymbol{\delta\theta_z}: \partial S_{xz}/\partial x + \partial S_{yz}/\partial y - S_x/R_x - S_y/R_y - S_z' = 0. \end{aligned} \quad (6)$$

where,

$$\begin{aligned} &\left(N_x, N_y, N_{xy}, M_x^b, M_y^b, M_{xy}^b\right) = \int_{-h/2}^{h/2}\left[\sigma_x, \sigma_y, \tau_{xy}, z\sigma_x, z\sigma_y, z\tau_{xy}\right]dz; \\ &\left(M_x^S, M_y^S, M_{xy}^S\right) = \int_{-h/2}^{h/2}\left\{\left[f(z)(\sigma_x, \sigma_y, \tau_{xy})\right]\right\}dz; \quad \left(Q_{xz}, Q_{yz}\right) = \int_{-h/2}^{h/2}\left\{\left[f'(z)(\tau_{xz}, \tau_{yz})\right]\right\}dz; \\ &\left(S_x, S_y, S_{xz}, S_{yz}\right) = \int_{-h/2}^{h/2}\left[f'(z)(\sigma_x, \sigma_y, \tau_{xz}, \tau_{yz})\right]dz; \quad \left(S_z'\right) = \int_{-h/2}^{h/2}\left\{\sigma_z\left[f''(z)\right]\right\}dz. \end{aligned} \quad (7)$$

Further substituting the expression of stress resultants from Eq. (7) into Eq. (6), the governing equations can be written in the following forms as stated in Eqs. (8)-(13).

$$\begin{aligned} &\boldsymbol{\delta u_0}: \left[A_{11}\left(\partial^2 u_0/\partial x^2 + \partial w_0/R_x\partial x\right) - B_{11}\partial^3 w_0/\partial x^3 + C_{11}\partial^2\theta_x/\partial x^2 + (F_{11}C_1/R_x)\partial\theta_z/\partial x\right] \\ &+\left[A_{12}\left(\partial^2 v_0/\partial x\partial y + \partial w_0/R_y\partial x\right) - B_{12}\partial^3 w_0/\partial x\partial y^2 + C_{12}\partial^2\theta_y/\partial x\partial y + (F_{12}C_1/R_y)\partial\theta_z/\partial x\right] \\ &+J_{13}C_1\partial\theta_z/\partial x + \left[A_{66}\left(\partial^2 u_0/\partial y^2 + \partial^2 v_0/\partial x\partial y\right) - 2B_{66}\partial^3 w_0/\partial x\partial y^2 + C_{66}\left(\partial^2\theta_x/\partial y^2 + \partial^2\theta_y/\partial x\partial y\right)\right] = 0. \end{aligned} \quad (8)$$

$$\begin{aligned} &\boldsymbol{\delta v_0}: \left[A_{21}\left(\partial^2 u_0/\partial x\partial y + \partial w_0/R_x\partial y\right) - B_{21}\partial^3 w_0/\partial x^2\partial y + C_{21}\partial^2\theta_x/\partial x\partial y + (F_{21}C_1/R_x)\partial\theta_z/\partial y\right] \\ &+\left[A_{22}\left(\partial^2 v_0/\partial y^2 + \partial w_0/R_y\partial y\right) - B_{22}\partial^3 w_0/\partial y^3 + C_{22}\partial^2\theta_y/\partial y^2 + (F_{22}C_1/R_y)\partial\theta_z/\partial y\right] + J_{23}C_1\partial\theta_z/\partial y \\ &+\left[A_{66}\left(\partial^2 u_0/\partial x\partial y + \partial^2 v_0/\partial x^2\right) - 2B_{66}\partial^3 w_0/\partial x^2\partial y + C_{66}\left(\partial^2\theta_x/\partial x\partial y + \partial^2\theta_y/\partial x^2\right)\right] = 0. \end{aligned} \quad (9)$$

$$
\begin{aligned}
&\boldsymbol{\delta w_0} : \left[B_{11}\left(\partial^3 u_0/\partial x^3 + \partial^2 w_0/R_x\partial x^2\right) - I_{11}\,\partial^4 w_0/\partial x^4 + O_{11}\,\partial^3\theta_x/\partial x^3 + \left(L_{11}C_1/R_x\right)\partial^2\theta_z/\partial x^2 \right]\\
&+\left[B_{12}\left(\partial^2 v_0/\partial x^2\partial y + \partial^2 w_0/R_y\partial x^2\right) - I_{12}\,\partial^4 w_0/\partial x^2\partial y^2 + O_{12}\,\partial^3\theta_y/\partial x^2\partial y + \left(L_{12}C_1/R_y\right)\partial^2\theta_z/\partial x^2 \right]\\
&+\left[B_{12}\left(\partial^3 u_0/\partial x\partial y^2 + \partial^2 w_0/R_x\partial y^2\right) - I_{12}\,\partial^4 w_0/\partial x^2\partial y^2 + O_{12}\,\partial^3\theta_y/\partial x\partial y^2 + \left(L_{12}C_1/R_x\right)\partial^2\theta_z/\partial y^2 \right]\\
&+\left[B_{22}\left(\partial^3 v_0/\partial y^3 + \partial^2 w_0/R_y\partial y^2\right) - I_{22}\,\partial^4 w_0/\partial y^4 + O_{22}\,\partial^3\theta_y/\partial y^3 + \left(L_{22}C_1/R_y\right)\partial^2\theta_z/\partial y^2 \right]\\
&+\left[2B_{66}\left(\partial^3 u_0/\partial x\partial y^2 + \partial^3 v_0/\partial x^2\partial y\right) - 4I_{66}\,\partial^4 w_0/\partial x^2\partial y^2 + 2O_{66}\left(\partial^3\theta_x/\partial x\partial y^2 + \partial^3\theta_y/\partial x^2\partial y\right) \right]\\
&-\left[A_{11}/R_x\left(\partial u_0/\partial x + w_0/R_x\right) - \left(B_{11}/R_x\right)\partial^2 w_0/\partial x^2 + \left(C_{11}/R_x\right)\partial\theta_x/\partial x + \left(F_{11}C_1/R_x^2\right)\theta_z \right]\\
&-\left[A_{12}/R_x\left(\partial v_0/\partial y + w_0/R_y\right) - \left(B_{12}/R_x\right)\partial^2 w_0/\partial y^2 + \left(C_{12}/R_x\right)\partial\theta_y/\partial y + \left(F_{12}C_1/R_xR_y\right)\theta_z \right]\\
&-\left[A_{12}/R_y\left(\partial u_0/\partial x + w_0/R_x\right) - \left(B_{12}/R_y\right)\partial^2 w_0/\partial x^2 + \left(C_{12}/R_y\right)\partial\theta_x/\partial x + \left(F_{12}C_1/R_xR_y\right)\theta_z \right]\\
&-\left[A_{22}/R_y\left(\partial v_0/\partial y + w_0/R_y\right) - \left(B_{22}/R_y\right)\partial^2 w_0/\partial y^2 + \left(C_{22}/R_y\right)\partial\theta_y/\partial y + \left(F_{22}C_1/R_y^2\right)\theta_z \right]\\
&+\left(M_{13}C_1\,\partial^2\theta_z/\partial x^2\right) + \left(M_{23}C_1\,\partial^2\theta_z/\partial y^2\right) - \left(J_{13}C_1/R_x\right)\theta_z - \left(J_{23}C_1/R_y\right)\theta_z + q = 0.
\end{aligned}
\tag{10}
$$

$$
\begin{aligned}
&\boldsymbol{\delta\theta_x} : \left[C_{11}\left(\partial^2 u_0/\partial x^2 + \partial w_0/R_x\partial x\right) - O_{11}\,\partial^3 w_0/\partial x^3 + P_{11}\,\partial^2\theta_x/\partial x^2 + \left(R_{11}C_1/R_x\right)\partial\theta_z/\partial x \right]\\
&+\left[C_{12}\left(\partial^2 v_0/\partial x\partial y + \partial w_0/R_y\partial x\right) - O_{12}\,\partial^3 w_0/\partial x\partial y^2 + P_{12}\,\partial^2\theta_y/\partial x\partial y + \left(R_{12}C_1/R_y\right)\partial\theta_z/\partial x \right]\\
&+\left[C_{66}\left(\partial^2 u_0/\partial y^2 + \partial^2 v_0/\partial x\partial y\right) - 2O_{66}\,\partial^3 w_0/\partial x\partial y^2 + P_{66}\left(\partial^2\theta_x/\partial y^2 + \partial^2\theta_y/\partial x\partial y\right) \right]\\
&+S_{13}C_1\,\partial\theta_z/\partial x - U_{55}\theta_x - C_1U_{55}\,\partial\theta_z/\partial x = 0.
\end{aligned}
\tag{11}
$$

$$
\begin{aligned}
&\boldsymbol{\delta\theta_y} : \left[C_{21}\left(\partial^2 u_0/\partial x\partial y + \partial w_0/R_x\partial y\right) - O_{21}\,\partial^3 w_0/\partial x^2\partial y + P_{21}\,\partial^2\theta_x/\partial x\partial y + \left(R_{21}C_1/R_x\right)\partial\theta_z/\partial y \right]\\
&+\left[C_{22}\left(\partial^2 v_0/\partial y^2 + \partial w_0/R_y\partial y\right) - O_{22}\,\partial^3 w_0/\partial y^3 + P_{22}\,\partial^2\theta_y/\partial y^2 + \left(R_{22}C_1/R_y\right)\partial\theta_z/\partial y \right] +\\
&+\left[C66\left(\partial^2 u_0/\partial x\partial y + \partial^2 v_0/\partial x^2\right) - 2O66\,\partial^3 w_0/\partial x^2\partial y + P66\left(\partial^2\theta_x/\partial x\partial y + \partial^2\theta_y/\partial x^2\right) \right]\\
&S_{23}C_1\,\partial\theta_z/\partial y - U_{44}\left(\theta_y + C_1\,\partial\theta_z/\partial y\right) = 0.
\end{aligned}
\tag{12}
$$

$$
\begin{aligned}
&\boldsymbol{\delta\theta_z} : -\left(F_{11}/R_x\right)\left(\partial u_0/\partial x + w_0/R_x\right) + L_{11}\partial^2 w_0/R_x\partial x^2 - R_{11}\partial\theta_x/R_x\,\partial x - U_{11}C_1/R_x^2\,\theta_z\\
&-F_{12}/R_x\left(\partial v_0/\partial y + w_0/R_y\right) + L_{12}\partial^2 w_0/R_x\partial y^2 - R_{12}\partial\theta_y/R_x\,\partial y - U_{12}C_1\theta_z/R_xR_y\\
&-2Y_{13}C_1\theta_z/R_x - F_{12}/R_y\left(\partial u_0/\partial x + w_0/R_x\right) + L_{12}\partial^2 w_0/R_y\partial x^2 - R_{12}\partial\theta_x/R_y\,\partial x - U_{12}C_1\theta_z/R_xR_y\\
&-F_{22}/R_y\left(\partial v_0/\partial y + w_0/R_y\right) + L_{22}\partial^2 w_0/R_y\,\partial y^2 - R_{22}\partial\theta_y/R_y\,\partial y - U_{22}C_1\theta_z/R_y^2 - 2Y_{23}C_1\theta_z/R_y\\
&-J_{13}\left(\partial u_0/\partial x + w_0/R_x\right) + M_{13}\,\partial^2 w_0/\partial x^2 - S_{13}\,\partial\theta_x/\partial x - J_{23}\left(\partial v_0/\partial y + w_0/R_y\right) + M_{23}\,\partial^2 w_0/\partial y^2\\
&U_{55}\left(\partial\theta_x/\partial x + C_1\,\partial^2\theta_z/\partial x^2\right) + U_{44}\left(\partial\theta_y/\partial y + C_1\,\partial^2\theta_z/\partial y^2\right) - S_{23}\,\partial\theta_y/\partial y - Z_{33}C_1\theta_z = 0.
\end{aligned}
\tag{13}
$$

where,

$$\left(A_{ij},B_{ij},I_{ij},C_{ij},O_{ij},F_{ij},J_{ij}\right)=Q_{ij}\int_{-h/2}^{h/2}\left[1.0,z,z^2,f(z),zf(z),f'(z),f''(z)\right]dz;$$

$$\left(P_{ij}\right)=Q_{ij}\int_{-h/2}^{h/2}\left[f(z)\right]^2dz;\ \left(U_{ij}\right)=Q_{ij}\int_{-h/2}^{h/2}\left[f'(z)\right]^2dz;$$

$$\left(L_{ij},R_{ij},Y_{ij}\right)=Q_{ij}\int_{-h/2}^{h/2}f'(z)\left[z,f(z),f''(z)\right]dz;$$

$$\left(M_{ij},S_{ij}\right)=Q_{ij}\int_{-h/2}^{h/2}f''(z)\left[z,f(z)\right]dz;\left(Z_{ij}\right)=Q_{ij}\int_{-h/2}^{h/2}\left[f''(z)\right]^2dz.$$

Eq. (14) expresses the boundary condition associated with the present theory.

$$\begin{aligned}&\partial u_0: N_x=0, N_{xy}=0;\ \partial v_0: N_y=0, N_{xy}=0;\\&\partial w_0: M_x^b=0, M_y^b=0, M_{xy}^b=0, \partial w_0/\partial y=0;\\&\partial M_x^b/\partial x=0, \partial M_y^b/\partial y=0, \partial M_{xy}^b/\partial x=0;\\&\partial\theta_x: M_x^s=0, M_{xy}^s=0;\ \partial\theta_y: M_y^s=0, M_{xy}^s=0;\\&\partial\theta_z: S_{xz}=0, S_{yz}=0.\end{aligned} \tag{14}$$

The Navier Solution Method

Laminated composite spherical shells subjected to transverse load is considered for the static analysis. The top surface of the shell is subjected to a transverse load that is represented in terms of double trigonometric series.

$$q(x,y)=q_0\sin\alpha x\sin\beta y. \tag{15}$$

where, $\alpha=\pi/a, \beta=\pi/b$ and q_0 represents the intensity of the load. The unknown variables $u_0, v_0, w_0, \theta_x, \theta_y, \theta_z$ are assumed in the following double trigonometric forms, which satisfy the simply supported boundary conditions exactly:

$$\begin{Bmatrix}u_0\\v_0\\w_0\\\theta_x\\\theta_y\\\theta_z\end{Bmatrix}=\begin{Bmatrix}u_{mn}\cos\alpha x\sin\beta y\\v_{mn}\sin\alpha x\cos\beta y\\w_{mn}\sin\alpha x\sin\beta y\\\theta_{xmn}\cos\alpha x\sin\beta y\\\theta_{ymn}\sin\alpha x\cos\beta y\\\theta_{zmn}\sin\alpha x\sin\beta y\end{Bmatrix}. \tag{16}$$

where, $u_{mn}, v_{mn}, w_{mn}, \theta_{xmn}, \theta_{ymn}, \theta_{zmn}$ are the unknown parameters to be determined. Now, substituting Eq. (16) into Eqs. (8)-(13), can be written in following compact form for the static analysis of laminated composite spherical shell as given in Eq. (17).

$$[K]_{6\times6}\{\Delta\}_{6\times1}=\{f\}_{6\times1}. \tag{17}$$

where, $[K]$ is the stiffness matrix, $\{\Delta\}$ and $\{f\}$ are the matrix of unknown displacements and matrix of force vectors respectively. Elements of stiffness matrix are written in Eq. (18).

$$
\begin{aligned}
&K_{11} = -A_{11}\alpha^2 - A_{66}\beta^2,\quad K_{12} = -A_{12}\alpha\beta - A_{66}\alpha\beta,\\
&K_{13} = A_{11}/R_x\,\alpha + A_{12}/R_y\,\beta + B_{11}\alpha^3 + B_{12}\alpha\beta^2 + 2B_{66}\alpha\beta^2,\\
&K_{14} = -C_{11}\alpha^2 - C_{66}\beta^2,\ K_{15} = -C_{12}\alpha\beta - C_{66}\alpha\beta,\\
&K_{16} = \left(F_{11}/R_x + F_{12}/R_y + J_{13}\right)C_1\alpha,\ K_{22} = -A_{22}\beta^2 - A_{66}\alpha^2,\\
&K_{23} = \left(A_{12}/R_x + A_{22}/R_y\right)\beta + B_{12}\alpha^2\beta + B_{22}\beta^3 + 2B_{66}\alpha^2\beta,\\
&K_{24} = -C_{12}\alpha\beta - C_{66}\alpha\beta, K_{25} = -C_{66}\alpha^2 - C_{22}\beta^2,\\
&K_{26} = \left(F_{12}/R_x + F_{22}/R_y + J_{23}\right)C_1\beta,\\
&K_{331} = -I_{11}\alpha^4 - I_{22}\beta^4 - 2\alpha^2\beta^2\left(I_{12} + 2I_{66}\right)\\
&K_{332} = -2\alpha^2\left(B_{11}/R_x + B_{12}/R_y\right) - 2\beta^2\left(B_{12}/R_x + B_{22}/R_y\right) - A_{11}/R_x^2 - A_{22}/R_y^2 - 2A_{12}/R_xR_y,\\
&K_{33} = K_{331} + K_{332}\\
&K_{34} = O_{11}\alpha^3 + O_{21}\alpha\beta^2 + 2O_{66}\alpha\beta^2 + \alpha C_{11}/R_x + \alpha C_{12}/R_y,\\
&K_{35} = O_{22}\beta^3 + O_{12}\alpha^2\beta + 2O_{66}\alpha^2\beta + \beta C_{12}/R_x + \beta C_{22}/R_y,\\
&K_{361} = C_1\left(-\alpha^2\left(L_{11}/R_x + L_{12}/R_y\right) - \beta^2\left(L_{12}/R_x + L_{22}/R_y\right) - M_{13}\alpha^2\right)\\
&K_{362} = C_1\left(-M_{23}\beta^2 - F_{11}/R_x^2 - F_{22}/R_y^2 - 2F_{12}/R_xR_y - J_{13}/R_x - J_{23}/R_y\right),\\
&K_{36} = K_{361} + K_{362},\\
&K_{44} = -P_{11}\alpha^2 - P_{66}\beta^2 - U_{55},\ K_{45} = -P_{12}\alpha\beta - P_{66}\alpha\beta,\\
&K_{46} = C_1\alpha\left(R_{11}/R_x + R_{12}/R_y + S_{13} - U_{55}\right), K_{55} = -P_{66}\alpha^2 - P_{22}\beta^2 - U_{44},\\
&K_{56} = C_1\beta\left(R_{12}/R_x + R_{22}/R_y + S_{23} - U_{44}\right),\\
&K_{66} = C_1^2\left(-U_{55}\alpha^2 - U_{44}\beta^2 - U_{11}/R_x^2 - 2U_{12}/R_xR_y - U_{22}/R_y^2 - 2W_{13}/R_x - 2W_{23}/R_y - Z_{33}\right).
\end{aligned}
\tag{18}
$$

Numerical Result and Discussion

The following material properties stated in Eqs. (19) and (20) are considered to obtain the static deformation quantities of isotropic and orthotropic laminated spherical shells using the current hyperbolic shell theory.

$$E_1 = E_2 = E_3 = 210\ \text{GPa}, G_{13} = G_{23} = G_{12} = G = E/2(1+\mu),\ \mu_{12} = \mu_{32} = \mu_{31} = \mu = 0.3. \tag{19}$$

$$E_1/E_2 = 25,\ E_3/E_2 = 1,\ E_3/E_2 = 1, G_{12}/E_2 = G_{13}/E_2 = 0.5,\ G_{23}/E_2 = 0.2,\ \mu_{12} = \mu_{13} = \mu_{23} = 0.25. \tag{20}$$

The non-dimensional form stated in Eq. (21) is used to present the numerical results in tabular as well as graphical form.

$$
\begin{aligned}
&\bar{u}(0,b/2,h/2) = h^2E_3u/q_0a^3,\quad \bar{w}(a/2,b/2,0) = 100h^3E_3w/q_0a^4,\\
&\left(\bar{\sigma}_x,\bar{\sigma}_y\right)(a/2,b/2,h/2) = \left(h^2/q_0a^2\right)\left(\sigma_x,\sigma_y\right),\ \bar{\tau}_{xy}(0,0,h/2) = h^2\tau_{xy}/q_0a^2,\\
&\bar{\tau}_{xz}(0,0,h/2) = h\tau_{xz}/q_0a,\ \bar{\tau}_{yz}(0,0,h/2) = h\tau_{yz}/q_0a.
\end{aligned}
\tag{21}
$$

Table 1 *Non-dimensional displacement and stresses in isotropic spherical shell under the sinusoidal mechanical load (a/h=10, $R_x = R_y = R$).*

R/a	Theory	$\bar{u}$	$\bar{w}$	$\bar{\sigma}_x$	$\bar{\sigma}_y$	$\bar{\tau}_{xy}$	$\bar{\tau}_{xz}$	$\bar{\tau}_{yz}$
5	Present	0.0434	2.6020	0.1437	0.1437	0.1225	0.2076	0.2076
	Shinde and Sayyad [8]	0.0508	2.6099	0.1490	0.1490	0.1228	0.2272	0.2272
	Reddy [4]	0.0506	2.6472	0.1519	0.1519	0.1225	0.2130	0.2130
	Mindlin [3]	0.0502	2.6262	0.1506	0.1506	0.1230	0.2137	0.2137
50	Present	0.0383	2.9357	0.1901	0.1901	0.1093	0.2340	0.2340
	Shinde and Sayyad [8]	0.0454	2.9407	0.1967	0.1967	0.1096	0.2559	0.2559
	Reddy [4]	0.0456	2.9572	0.1962	0.1962	0.1102	0.2379	0.2379
	Mindlin [3]	0.0452	2.9310	0.1944	0.1944	0.1109	0.2385	0.2385
100	Present	0.0377	2.9386	0.1918	0.1918	0.1078	0.2342	0.2342
	Shinde and Sayyad [8]	0.0447	2.9435	0.1985	0.1985	0.1081	0.2562	0.2562
	Reddy [4]	0.0450	2.9598	0.1979	0.1979	0.1088	0.2381	0.2381
	Mindlin [3]	0.0446	2.9336	0.1961	0.1961	0.1095	0.2387	0.2387
∞	Present	0.0371	2.9395	0.1935	0.1935	0.1062	0.2343	0.2343
	Shinde and Sayyad [8]	0.0441	2.9445	0.2001	0.2001	0.1065	0.2563	0.2563
	Reddy [4]	0.0444	2.9607	0.1994	0.1994	0.1074	0.2382	0.2382
	Mindlin [3]	0.0440	2.9345	0.1976	0.1976	0.1080	0.2387	0.2387
	Pagano [9]	0.0443	2.9425	0.1988	0.1988	---	0.2383	0.2383

Table 2 *Non-dimensional displacement and stresses in ($0^0/90^0$) spherical shell under the sinusoidal mechanical load (a/h=10, $R_x = R_y = R$).*

R/a	Theory	$\bar{u}$	$\bar{w}$	$\bar{\sigma}_x$	$\bar{\sigma}_y$	$\bar{\tau}_{xy}$	$\bar{\tau}_{xz}$	$\bar{\tau}_{yz}$
5	Present	0.0097	1.1141	0.6515	0.0760	0.0691	0.1044	0.1266
	Shinde and Sayyad [8]	0.0151	1.1200	0.6450	0.0759	0.0692	0.0930	0.1230
	Reddy [4]	0.0151	1.1164	0.6530	0.0754	0.0694	0.0823	0.1382
	Mindlin [3]	0.0148	1.1096	0.6262	0.0747	0.0686	0.0839	0.1402
50	Present	0.0160	1.2125	0.7388	0.0854	0.0552	0.1245	0.1269
	Shinde and Sayyad [8]	0.0098	1.2186	0.7354	0.0856	0.0552	0.1159	0.1192
	Reddy [4]	0.0100	1.2148	0.7424	0.0847	0.0555	0.1170	0.1230
	Mindlin [3]	0.0096	1.2070	0.7116	0.0840	0.0546	0.1189	0.1250
100	Present	0.0163	1.2133	0.7410	0.0857	0.0541	0.1251	0.1264
	Shinde and Sayyad [8]	0.0095	1.2194	0.7378	0.0858	0.0542	0.1168	0.1184
	Reddy [4]	0.0096	1.2156	0.7447	0.0850	0.0545	0.1186	0.1216
	Mindlin [3]	0.0092	1.2078	0.7138	0.0842	0.0536	0.1205	0.1235
∞	Present	0.0165	1.2136	0.7428	0.0858	0.0530	0.1258	0.1258
	Shinde and Sayyad [8]	0.0091	1.2197	0.7398	0.0860	0.0531	0.1176	0.1176
	Reddy [4]	0.0092	1.2158	0.7466	0.0851	0.0534	0.1201	0.1201
	Mindlin [3]	0.0088	1.2081	0.7156	0.0843	0.0525	0.1220	0.1220
	Pagano [9]	--	1.2250	0.7302	0.0886	0.0535	0.1210	0.1250

The non-dimensional displacements and stresses for an isotropic, ($0^0/90^0$) and ($0^0/90^0/0^0$) laminated spherical shells for aspect ratio 10 and *R/a*=5, 50, 100, ∞ subjected to sinusoidal loading are shown in Tables 1 through 3 using the present theory. The numerical findings are compared with higher-order shear deformation theories that have been previously published. Additionally, first-order shear deformation theory [3] and Pagano's exact elasticity solution [9] are used to compare the displacements and stresses for plate obtained using the present theory. As can be seen from Table 1, the displacement and stresses obtained by the present theory are in good agreement with those other higher-order theories. For plates, there is a less percentage of inaccuracy in the precise and present theory results. Due to the influence of normal and shear deformations, the findings obtained using the present theory are also superior to those obtained using other higher order theories in the case of ($0^0/90^0$) and ($0^0/90^0/0^0$) laminated spherical shells. For ($0^0/90^0$) and ($0^0/90^0/0^0$) laminated composite spherical shells exposed to sinusoidal load, through-the-thickness distributions of in-plane and transverse shear stresses are shown in Figs. 2 and 3.

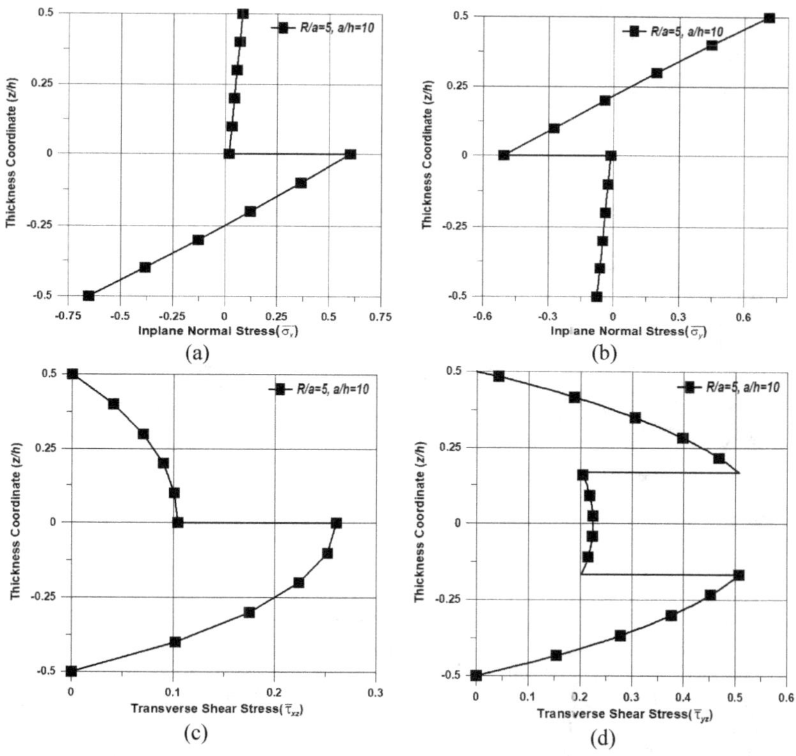

Fig. 2 Stress distributions in ($0^0/90^0$) laminated composite spherical shells

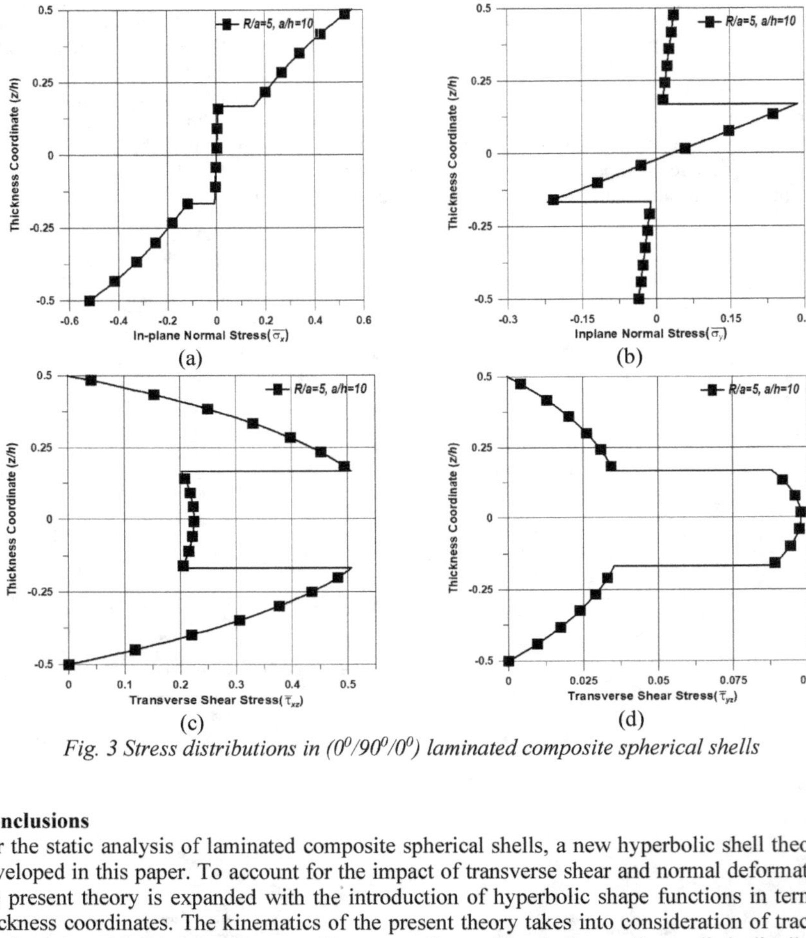

Fig. 3 Stress distributions in ($0^0/90^0/0^0$) laminated composite spherical shells

Conclusions

For the static analysis of laminated composite spherical shells, a new hyperbolic shell theory is developed in this paper. To account for the impact of transverse shear and normal deformations, the present theory is expanded with the introduction of hyperbolic shape functions in terms of thickness coordinates. The kinematics of the present theory takes into consideration of traction-free boundary conditions on the top and bottom surfaces of the shell as well as realistic distribution of the transverse shear stresses throughout the thickness of the shell. Using the principle of virtual work, the governing equations and accompanying boundary conditions are obtained. Using Navier's solution method, higher-order closed form solutions for static analysis of simply supported spherical shells are presented. By taking both curvature radii as infinite, these solutions also apply to plates. The numerical findings for the plate using the present hyperbolic shell theory are in good agreement with the exact elasticity solutions, demonstrating the validity and accuracy of the present theory.

Table 3 *Non-dimensional displacement and stresses in ($0^0/90^0/0^0$) spherical shell under the sinusoidal mechanical load ($a/h=10$, $R_x = R_y = R$).*

R/a	Theory	$\bar{u}$	$\bar{w}$	$\bar{\sigma}_x$	$\bar{\sigma}_y$	$\bar{\tau}_{xy}$	$\bar{\tau}_{xz}$	$\bar{\tau}_{yz}$
5	Present	0.0124	0.6713	0.5538	0.0394	0.0384	0.2246	0.0391
	Shinde and Sayyad [8]	0.0112	0.6972	0.5455	0.0360	0.0399	0.3005	0.0395
	Reddy [4]	0.0108	0.6769	0.5218	0.0352	0.0388	0.3508	0.1109
	Mindlin [3]	0.0098	0.6025	0.4780	0.0311	0.0346	0.3658	0.1018
50	Present	0.0088	0.7062	0.5659	0.0399	.0287	0.2362	0.0411
	Shinde and Sayyad [8]	0.0079	0.7345	0.5923	0.0396	0.0299	0.3165	0.0416
	Reddy [4]	0.0075	0.7121	0.5662	0.0385	0.0290	0.3691	0.1167
	Mindlin [3]	0.0069	0.6303	0.5153	0.0338	0.0257	0.3826	0.1064
100	Present	0.0086	0.7064	0.5652	0.0398	0.0280	0.2363	0.0411
	Shinde and Sayyad [8]	0.0077	0.7348	0.5935	0.0397	0.0293	0.3167	0.0416
	Reddy [4]	0.0073	0.7124	0.5674	0.0386	0.0283	0.3692	0.1167
	Mindlin [3]	0.0067	0.6305	0.5163	0.0339	0.0252	0.3828	0.1064
∞	Present	0.0084	0.7065	0.5644	0.0397	0.0274	0.2364	0.0411
	Shinde and Sayyad [8]	0.0074	0.7349	0.5946	0.0398	0.0286	0.3167	0.0416
	Reddy [4]	0.0071	0.7125	0.5684	0.0387	0.0277	0.3693	0.1168
	Mindlin [3]	0.0065	0.6306	0.5172	0.0340	0.0246	0.3828	0.1065
	Pagano [9]	--	0.7528	0.5898	0.0418	0.0289	0.3570	0.1200

References

[1] A.S. Sayyad, Y.M. Ghugal, On the free vibration analysis of laminated composite and sandwich plates: a review of recent literature with some numerical results, Compos. Struct. 129(2015)177-201. https://doi.org/10.1016/j.compstruct.2015.04.007

[2] A.S. Sayyad, Y.M. Ghugal, Bending, buckling and free vibration of laminated composite and sandwich beams: a critical review of literature, Compos. Struct. 171(2017)486-504. https://doi.org/10.1016/j.compstruct.2017.03.053

[3] R.D. Mindlin, Influence of rotatory inertia and shear on flexural motions of isotropic elastic plates, ASME J. Appl. Mech. 18(1951)31-38. https://doi.org/10.1115/1.4010217

[4] J.N. Reddy, Exact solutions of moderately thick laminated shells, ASCE J. Eng. Mech. 110(1984)794-809. https://doi.org/10.1061/(ASCE)0733-9399(1984)110:5(794)

[5] K.M. Liew, C.W. Lim, A higher-order theory for vibration of doubly curved shallow shells. ASME J. Appl. Mech. 63(1996)587-593. https://doi.org/10.1115/1.2823338

[6] F. Tornabene and A. Ceruti, Mixed static and dynamic optimization of four-parameter functionally graded completely doubly curved and degenerate shells and panels using GDQ method, Mathematical Problems in Engineering, (2013). https://doi.org/10.1155/2013/867079

[7] A.S. Sayyad and Y.M. Ghugal, Static and free vibration analysis of laminated composite and sandwich spherical shells using a generalized higher-order shell theory, Compos. Struct., 219(2019)129-146. https://doi.org/10.1016/j.compstruct.2019.03.054

[8] B. M. Shinde and A. S. Sayyad, A new higher-order shear and normal deformation theory for the free vibration analysis of laminated shells, Mechanics of Advanced Composite Structures, 9(2022)89-104, doi:10.22075/macs.2022.22741.1327

[9] N.J. Pagano, "Exact solutions for bidirectional composites and sandwich plates", Journal of Composite Materials, 4(1970)20-34. https://doi.org/10.1177/002199837000400102

Advanced Topics in Mechanics of Materials, Structures and Construction - AToMech1-2023 Materials Research Forum LLC
Materials Research Proceedings 31 (2023) 755-764 https://doi.org/10.21741/9781644902592-77

A comprehensive analysis of the erosion in a carbon steel boiler tube elbow through the use of 3D mapping of the corroded surface and CFD modelling

Nikolay Bukharin[1,2*], M.A. Farrokhzad[1], Mouhammad El Hassan[3]

[1] The Southern Alberta Institute of Technology, Calgary, Alberta, Canada

[2] Thermofluid Solutions, Calgary, Alberta, Canada

[3] Prince Mohammad Bin Fahd University, Mechanical Engineering Department

* nikolay.bukharin@sait.ca

Keywords: Erosion, Corrosion, Erosion Pattern, CFD, Particulate Matter

Abstract. Erosion corrosion is a common problem that affects boiler tubes, particularly those in power plants and industrial settings where hard water and abrasive particles are present in the flow. These particles can cause physical erosion to the surface of the tubes, which can then lead to further corrosion. This type of corrosion is often accelerated by the high temperatures and pressures present in a boiler system, as well as the presence of oxygen. The combination of physical erosion and chemical corrosion attack can cause significant damage to the tubes, reducing their ability to efficiently transfer heat and potentially leading to system failure. Therefore, it is important to predict rates of erosion to prevent costly and potentially dangerous failures. The focus of this paper is an investigation into the effects of erosion caused by hard water particles on a carbon steel boiler tube elbow (ANSI 16.9). A semi-empirical procedure, which considers properties of the material and flow parameters, is developed for predicting erosion rates. The study revealed that the primary erosion damage occurred on the extrados of the bend. The findings indicated that particles within the flow began to separate from the front wall surface, resulting in significant erosion along the lateral sides. The disappearance of erosion from the front surface of the bend was also consistent with the erosion patters observed on the eroded pipe sample, which was extracted from the line. Moreover, it was demonstrated the presence of two different erosion patterns in the separation region, which matched qualitatively the erosion pattern observed on the sample wall.

Introduction

The phenomenon of erosion, or sometimes, erosion-corrosion in metals is defined as the synergic effect of surface degradation of metals caused by moving corrosive fluids containing erosive undissolved solid particle. Erosion in pipes can be also related to the effect cavitation [1 – 3]. Typically, the high velocity of the fluid containing hard particles induces mechanical wear and abrasion effects, leading to a rapid decay of the metallic surfaces. Additionally, this form of erosion-corrosion effect can be caused by turbulent flow regime at localized site, often due to disruption in the steady flow pattern, such as a flow-directional change at a bend or a mechanical obstacle in a pipe. The complex phenomenon of erosion-corrosion in steels impacts multiple industrial sectors, including but not limited to water injection systems, boilers, oil and gas pipelines, the nuclear power generation sector, slurry transportation (hydrotransport), etc. [4 – 7] Understanding the underlying causes and mechanisms of erosion corrosion is essential for industries that rely on the proper functioning of equipment and infrastructure, as it can help prevent costly and potentially catastrophic accidents.

Various methods are employed to predict the rates, locations, and patterns of erosion-corrosion. In [8] Computational Fluid Dynamics (CFD) simulations of a laminar flow with solid particles were performed for erosion prediction in 90° elbow. The main goal of this research was to study how different dimensions of sand affect tube corrosion depending on the different aperture of the valve during water flow with solid particles. It was shown that the erosion rate of a pipe elbow when water flows through a valve opening that is conditioned by specific angles (15°, 30°, 45°, and 90°) and injected with particles of varying sizes can exhibit various different scenarios. It has been concluded that a decrease in moving undissolved solid particle size leads to a decrease in erosion rate. Medium-sized particles reduce erosion rate by approximately twice as much as larger particles and ten times as much as smaller particles. In [9], the effect of different parameters such as particles size, stream velocities and elbows diameter on the erosion rate for an elbow in light crude oil-solid flow was studied. Three distinct models were used in this study: continuous flow modeling, Lagrangian particle tracking, and an empirical erosion equation. The parameter ranges evaluated were particle size of 100-500 μm, stream velocities of 3-7 m/s, and elbow diameters of 0.0762-0.1778 m. The numerical results showed that the highest erosion rate increases with bigger particle size and stream velocities, but decreased with increasing elbow bending radius. However, the position of the peak erosion rate in the elbow area was unaffected by these factors. In [10] a new erosion model, which is referred to as SIEM (shear impact energy model), is used to investigate elbow erosion under different working conditions using numerical simulations. The fluid motions are predicted by CFD, and the particle movements are calculated using DEM (discrete element method) in the simulations. Both a one-way coupling method and a two-way coupling method in CFD-DEM are adopted to calculate the gas-solid interaction. Experiments were carried out in [11] with flows comprising of water, air, and sand particles having an average diameter of 300 μm, as well as liquid-sand flows. Wall thickness losses were measured at 63 locations on the outer radius of elbows, across various flow conditions. To explore different flow regimes, Computational Fluid Dynamics (CFD) simulations were conducted using the Euler-Lagrange approach to model turbulence, and the Eulerian Multiphase model to handle multiphase flow scenarios. The Reynolds Stress Model (RSM) was also implemented in the simulations as a turbulence model. The results showed that the maximum erosion increased with increasing mixture velocity. The CFD simulations using the Euler-Lagrange and Euler-Euler-Lagrange approaches accurately predicted the observed maximum erosion rates, erosion locations, and erosion patterns for liquid-sand and dispersed-bubble-sand flows. A numerical study was conducted in [12], to estimate the erosion of solid particles in angle-cutting elbows (ACEs) where there is a flow of gas and solid particles. The simulation models used a combination of the two-way coupled Eulerian-Lagrangian approach and the E/CRC model, along with particle-wall rebound models and a user-defined function in the dispersed phase model. The accuracy of the simulation results was tested by comparing them with previously reported experimental results. The study's outcomes revealed that significant erosion occurs in the downstream section of the concave wall of the ACE, while only minor erosion occurs on the convex wall of the ACE. In [13], a new approach for modeling erosion and corrosion in hydrocarbon pipes using Computational Fluid Dynamics (CFD) has been introduced. This approach was based on high-order Discontinuous Galerkin Spectral Element Method (DGSEM) to approximate the incompressible Navier-Stokes/Cahn-Hilliard model. This technique enables a highly detailed three-dimensional representation of the flow regime, phases distribution, and contact surfaces within the pipe, resulting in precise calculations of erosion and corrosion rates and their distribution over the pipeline surface. The effectiveness of this methodology has been verified with experiments relevant to the oil and gas industry, specifically, by simulating erosion in a one-phase ascending pipe with two elbows and corrosion in a two-phase pipe under various flow regimes. In [14], a process has been developed to forecast erosion rates in pipe elbows and tees, for use in estimating safe operating conditions and velocities during oil and gas production in the presence of sand. The procedure introduces a new concept that enables the calculation of erosion rates for different pipe geometries. It establishes a connection between the erosion rate and the impact velocity of sand particles on a pipe fitting wall based on empirical observations.

In this paper we are studying the nature of erosion pattern in the sample of carbon steel boiler tube elbow (ASME/ANSI 16.9) using CFD numerical modeling. In the given sample description, it was stated, that the damaged occurred from the solid particles in the hard water flow at the average velocity of 5 m/s. Using numerical calculation in combination with material properties measurements and 3D scanning of the given sample we recreate erosion damage mechanisms. To the far of our knowledge such combination of numeric models and material and geometric measurements in the given geometry is not described in the literature and might be used in development of accurate semi-empirical procedures for erosion prediction and prevention.

CFD Model

Geometry and Mesh

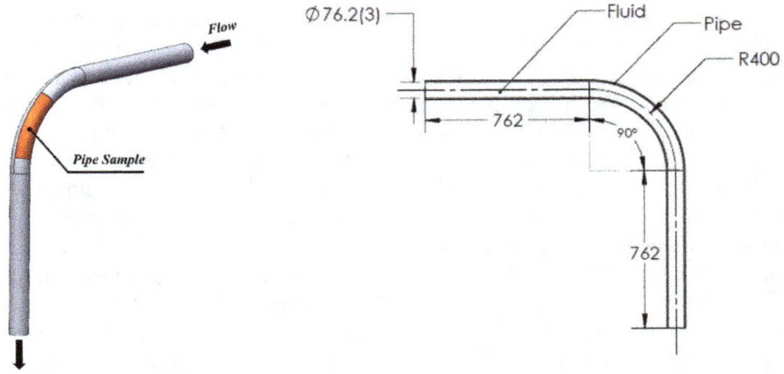

Figure 1. Geometry and location of extracted pipe sample

Figure 1 displays the geometry employed for CFD modeling. Additionally, it shows the flow direction and illustrates the position of the extracted sample from the eroded pipe, which was analyzed in this study.

Figure 2 shows the mesh structure. In the mesh independence analysis, mesh adaptation was used, which utilizes the simulation results to iteratively enhance the mesh structure. This process generates a mesh that is tailored to the specific simulation, with finer mesh elements in regions with high gradients and coarser elements elsewhere (Figure 2d). The total number of elements in the final 3D mesh was 987,201 (when the change in average velocity and pressure between the cycles was less than 1.5%)

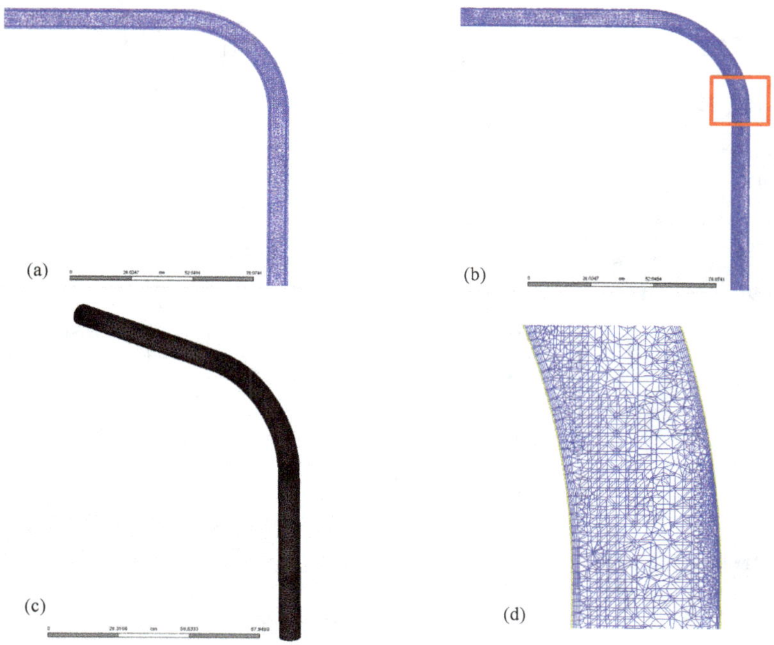

Figure 2. Pipe section. Mesh structure. (a) Initial mesh, (b) Final refined mesh based on mesh adaptation (red square shows zoomed region in (d)), (c) 3D mesh structure, (d) Zoomed mesh region

Turbulence model and boundary conditions.

For the CFD analysis a RANS (Reynolds-averaged Navier–Stokes) RNG k–ε turbulence model was employed. This model was developed using re-normalization group (RNG) methods to re-normalize the Navier–Stokes equations, thereby accounting for the impact of smaller scales in the flow [15],[16]. The RNG k-ε turbulence model is analogous in structure to the standard k-ε model and yields an analytical derivation of the model constants that differ from those in the standard model. Additionally, it introduces new terms in the transport equations for k and ε, allowing for more accurate predictions. The reduced dependence on empirical data and the incorporation of these additional terms make the RNG k-ε model more accurate for a wider range of flows than the standard k-ε model.

The eddy viscosity and eddy conductivity are calculated as follows:

$$\mu_t = C_\mu \rho \frac{K^2}{\varepsilon}\,;\; k_t = \frac{\mu_t C_p}{\sigma_t}$$

where σ_t is a turbulent Prandt number, usually taken to be 1.0 and C_μ is an empirical constant. The momentum equations are transformed to wave-number space and re-normalization group theory is used to derive the equations for calculating eddy viscosity. The equation used for K (Turbulent Kinetic Energy) is:

$$\rho\frac{\partial K}{\partial t}+\rho U\frac{\partial K}{\partial x}+\rho V\frac{\partial K}{\partial y}+\rho W\frac{\partial K}{\partial z}=\frac{\partial}{\partial x}\left[\left(\frac{\mu_t}{\sigma_K}+\mu\right)\frac{\partial K}{\partial x}\right]+\frac{\partial}{\partial y}\left[\left(\frac{\mu_t}{\sigma_K}+\mu\right)\frac{\partial K}{\partial y}\right]+\frac{\partial}{\partial z}\left[\left(\frac{\mu_t}{\sigma_K}+\mu\right)\frac{\partial K}{\partial z}\right]-\rho\varepsilon+\mu_t\left[2\left(\frac{\partial U}{\partial x}\right)^2+2\left(\frac{\partial V}{\partial y}\right)^2+2\left(\frac{\partial W}{\partial z}\right)^2+\left(\frac{\partial U}{\partial y}+\frac{\partial V}{\partial x}\right)^2+\left(\frac{\partial U}{\partial z}+\frac{\partial W}{\partial x}\right)^2+\left(\frac{\partial V}{\partial z}+\frac{\partial W}{\partial y}\right)^2\right]$$

The equation for ε (Turbulent Energy Dissipation)

$$\rho\frac{\partial \varepsilon}{\partial t}+\rho U\frac{\partial \varepsilon}{\partial x}+\rho V\frac{\partial \varepsilon}{\partial y}+\rho W\frac{\partial \varepsilon}{\partial z}=\frac{\partial}{\partial x}\left[\left(\frac{\mu_t}{\sigma_\varepsilon}+\mu\right)\frac{\partial \varepsilon}{\partial x}\right]+\frac{\partial}{\partial y}\left[\left(\frac{\mu_t}{\sigma_\varepsilon}+\mu\right)\frac{\partial \varepsilon}{\partial y}\right]+\frac{\partial}{\partial z}\left[\left(\frac{\mu_t}{\sigma_\varepsilon}+\mu\right)\frac{\partial \varepsilon}{\partial z}\right]-C_2\rho\frac{\varepsilon^2}{K}+C_1\mu_t\frac{\varepsilon}{K}\left[2\left(\frac{\partial U}{\partial x}\right)^2+2\left(\frac{\partial V}{\partial y}\right)^2+2\left(\frac{\partial W}{\partial z}\right)^2+\left(\frac{\partial U}{\partial y}+\frac{\partial V}{\partial x}\right)^2+\left(\frac{\partial U}{\partial z}+\frac{\partial W}{\partial x}\right)^2+\left(\frac{\partial V}{\partial z}+\frac{\partial W}{\partial y}\right)^2\right]$$

C_1 is calculated using the following expression:

$$C_1=C_0-\frac{\eta\left(1-\frac{\eta}{\eta_0}\right)}{1+\beta\eta^3}$$

Where η is defined as:

$$\eta=\frac{\sqrt{G}K}{\varepsilon}$$

$$G=2\left[\left(\frac{\partial U}{\partial x}\right)^2+\left(\frac{\partial V}{\partial y}\right)^2+\left(\frac{\partial W}{\partial z}\right)^2\right]+\left(\frac{\partial U}{\partial y}+\frac{\partial V}{\partial x}\right)^2+\left(\frac{\partial U}{\partial z}+\frac{\partial W}{\partial x}\right)^2+\left(\frac{\partial V}{\partial z}+\frac{\partial W}{\partial y}\right)^2$$

Other constants were adjusted directly in the solver.

We used velocity boundary condition on the inlet of the pipe and “Unknown” boundary condition on the outlet of the pipe. The Unknown boundary condition is a mixed Neumann-Dirichlet-type (specified value) boundary condition applied to the pressure variable. It is implemented into the solution in a two-part process. During the matrix solution of the pressure equation, nodes assigned an Unknown boundary condition are treated as fixed or specified (Dirichlet) and after the matrix solution, the values on these nodes are re-calculated as the average of the neighboring values, effectively enforcing a zero gradient (Neumann) condition on the pressure equation (Figure 3).

Autodesk CFD software package was used for all numerical calculations in the present study.

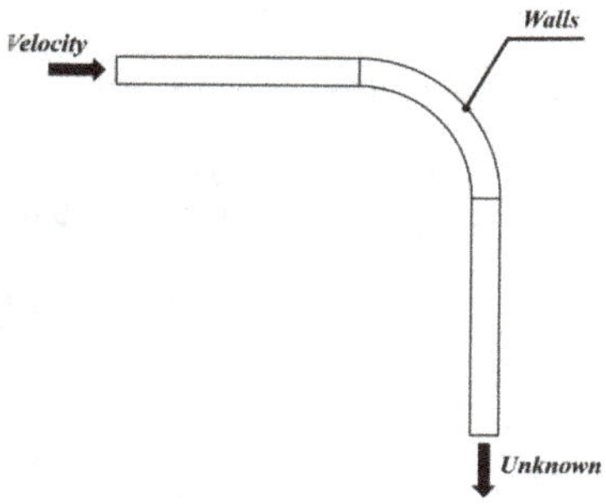

Figure 3. Boundary conditions

Temperature affects were not studied in this paper and temperature was considered 20°C within computation domain.

Erosion model

Lagrangian particle tracking with the Edwards Model [17] is used to compute erosion. A low particle concentration assumption (not a slurry erosion model) is employed, and results are presented as a scalar result quantity. This facilitates design comparison, and removes the guesswork from interpreting erosion predictions. The erosion model uses angle of attack bounce data and the Brinell material hardness to compute the material volume removal rate. This approach qualitatively identifies areas subject to erosion. It illustrates the relationship between the flow and erosion trends, which can lead to erosion reduction through design improvements. Spherical particles were considered in this study with the radius of 0.36mm and density of 2750 kg/m^3. In order to accurately measure all geometrical parameters of the given sample and to identify its exact location in the elbow it was 3D scanned using the high-end HandySCAN 3D (BLACK Series) 3D scanner with the accuracy of 0.025mm and 0.1mm of mesh resolution. A pipe sample and a scanned 3D image with hardness measurements locations are shown in Figure 4.

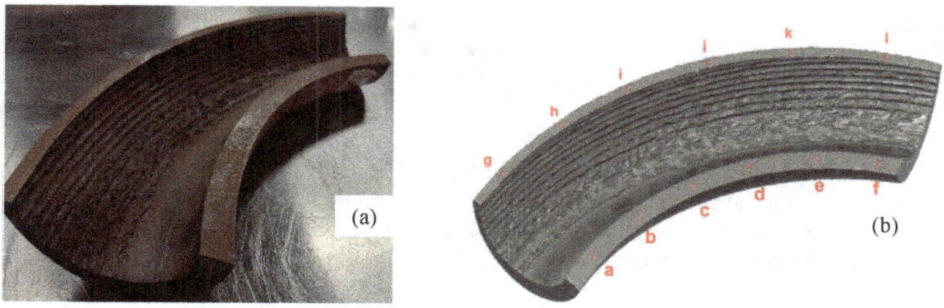

Figure 4. (a) Eroded pipe, (b) 3D Scan of eroded pipe with shown hardness measurements locations.

In order to include the surface hardness of the bend interior in the erosion simulation models, hardness measurements (indentations) were conducted using Rockwell Hardness Testing method

(HRB). 36 indentations were placed alongside the bend to represent its hardness map. 18 of these hardness indentations were placed in group of three at 15° angle increments at the bend's intrados cross-section and their average values are shown in Table 1. The other 18 indentation were placed at the extrados cross-section of the bend (in group of three and at 15° angle increments as well) and their average per location is also shown in Table 1.

For the Rockwell test, the test was carried out with a Mitutoyo, Model ARK-600 (Japan) Hardness Tester using 1/16" diameter tungsten carbide (WC). The applied load was set at 100 kg for duration of 8 seconds.

The higher hardness values at the extrados were expected and are believed to be the result of work-hardening occurred during the bend process from pipe to elbow.

It shall be noted that the CFD model used in this study uses Brinell hardness values, however due to the size limitations of elbow thickness versus the size of Brinell hardness indenter (10 mm), applying the Brinell hardness measurements was not possible. Therefore, the hardness tests were conducted using Rockwell hardness tests (smaller indenter) and for the purpose of CFD modeling, the Rockwell hardness values (HRB) were converted to Brinell hardness (HB) values. The average value (226) was used as one of the inputs in CFD model.

Table 1: Rockwell-Ball (HRB) Hardness Measurements, the values shown in the brackets are the Brinell Hardness (HB)values calculated from hardness conversion tables.

		Intrados		Extrados
15°	a	92.3 (eq. 194HB)	g	97.1 (eq. 221 HB)
30°	b	93.9 (eq. 203 HB)	h	98.0 (eq. 227 HB)
45°	c	93.6 (eq. 200 HB)	i	98.4 (eq. 230 HB)
60°	d	94.8 (eq. 208 HB)	j	98.4 (eq. 230 HB)
75°	e	91.3 (eq. 189 HB)	k	98.2 (eq. 229 HB)
90°	f	88.7 (eq. 177 HB)	l	96.6 (eq. 218 HB)

CFD Results

In order to investigate the relationship between the flow dynamics and the erosion pattern inside the bend, the velocity distribution and the erosion rate are investigated using the CFD results. Figure 5 shows the velocity field inside the pipe in the midplane section and erosion on the front wall of a bend. As the flow approaches the bend, it becomes fully developed and the solid particles gain enough momentum to deviate from the intrados due to the sudden change in flow direction. This results in the solid particles collide directly with the extrados. Particles on extrados are being dragged along the wall with the flow, producing erosion. When they reach the mark shown in Figure 5, particles start to separate from pipe surface and continues to travel along the wall on the lateral sides (Figure 6). It can be also seen (Figure 5) that the velocity distribution is not uniform across the bend and that the velocity magnitudes are higher on the external wall which correlates well with the erosion pattern. The flow begins to gradually recover downstream of the bend to almost fully re-establish at the pipe exit.

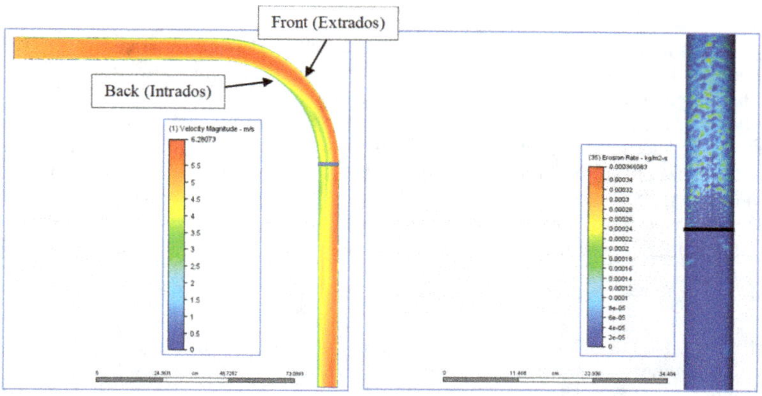

Figure 5 (a) Velocity magnitude in the midsection view, (b) Front erosion pattern.

Figure 6b shows two types of erosion pattern that are formed in the particle separation region. The major mass of particles moves to the lateral sides producing same pattern as prior to separation on extrados and separating particles on front produce less erosion and generate more regular pattern. A similar behavior can be observed on the pipe sample (Figure 4), as well as absence of erosion on intrados (Figure 6c).

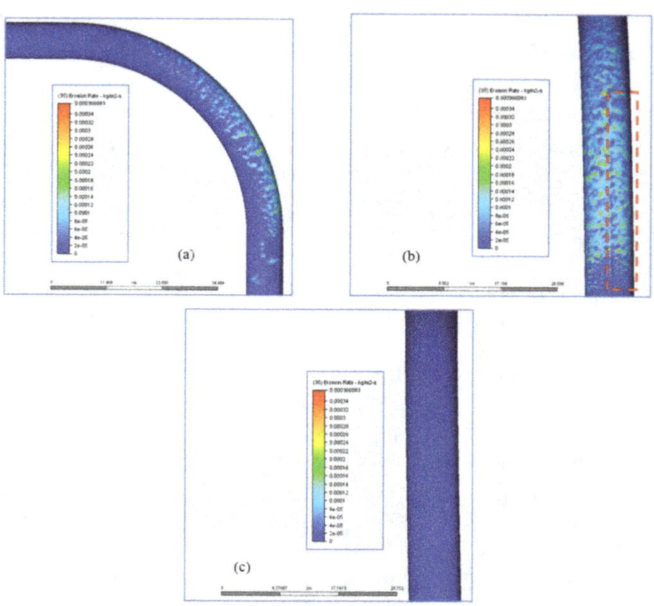

Figure 6. Elbow erosion. (a) Side view, (b) Magnified front view, sample location is marked with red, (c) Intrados

Conclusion

In this study, a numerical analysis of 90 degrees carbon steel boiler tube elbow (ASME/ANSI 16.9), combined with geometrical and material hardness measurements, was conducted. It was shown that the main erosion damage was produced along the extrados of the bend. It was shown that particles in the flow at some point start to separate from the front wall surface and start to travel along the lateral sides producing significant erosion in those regions. It was also found that erosion disappears from the front surface of the bend. These results corroborated well with the experimental results. It was also demonstrated that in the separation region two different erosion pattern existed which also correlates with erosion pattern observed on the sample wall. Finally, it should be noted that the straight lines erosion pattern is not completely clear and needs additional investigation.

References

[1] Farrokhzad, M. A., Saha, G.C., Khan, T.I. (2013) Wear performance of co-electrodeposited cermet coatings, Surface and Coatings Technology 235(1-2):75–85 https://doi.org/10.1016/j.surfcoat.2013.07.015

[2] Farrokhzad, M. A., Saha, G.C., Khan, T.I. (2014) Three-body wear performance of co-electrodeposited cermet coatings, Wear Volume 313(Issues 1-2):34–42, https://doi.org/10.1016/j.wear.2014.02.013

[3] El Hassan, M.; Bukharin, N.; Al-Kouz, W.; Zhang, J.-W.; Li, W.-F. (2021). A Review on the Erosion Mechanism in Cavitating Jets and Their Industrial Applications. Appl. Sci., 11, 3166. https://doi.org/10.3390/app11073166

[4] M.C. Roco, G.R. Addie, (1987). Erosion wear in slurry pumps and pipes, Powder Technology, 50, 1, 35-46, ISSN 0032-5910, https://doi.org/10.1016/0032-5910(87)80081-5

[5] H., Wael, (2012). Flow Accelerated Corrosion in Nuclear Power Plants, Nuclear Power - Practical Aspects, InTech, https://doi.org/10.5772/51346

[6] P. C. Okonkwo, A. M. Mohamed, (2014). Erosion-corrosion in oil and gas industry: a review, International Journal of Metallurgical & Materials Science and Engineering, 4, 3, 7-28

[7] A.V. Levy, (1993). The erosion-corrosion of tubing steels in combustion boiler environments, Corrosion Science, 35, 5–8,1035-1043, ISSN 0010-938X, https://doi.org/10.1016/0010-938X(93)90322-8

[8] M. Amara, B.G.N. Muthanna, M. Tahar Abbes, M. Hadj Meliani, (2018). Effect of sand particles on the Erosion-corrosion for a different locations of carbon steel pipe elbow, Procedia Structural Integrity, Volume 13, Pages 2137-2142, ISSN 2452-3216, https://doi.org/10.1016/j.prostr.2018.12.151

[9] M. A. H. Yusof, Z. Zakaria, A. Supee, M. Z. M. Yusop, N. B. Haladin, (2019). Prediction of Erosion Rate in Elbows for Liquid-Solid Flow via Computational Fluid Dynamics (CFD), Applications of Modelling and Simulation, 3(1), 28-38

[10] L. Xu, Q. Zhang, J. Zheng, Y. Zhao, (2016). Numerical prediction of erosion in elbow based on CFD-DEM simulation, Powder Technology, Volume 302, Pages 236-246, ISSN 0032-5910, https://doi.org/10.1016/j.powtec.2016.08.050

[11] T. A. Sedrez, S. A. Shirazi, Y. R. Rajkumar, K. Sambath, H. J. Subramani, (2019). Experiments and CFD simulations of erosion of a 90° elbow in liquid-dominated liquid-solid and dispersed-bubble-solid flows, Wear, Volumes 426–427, Part A, Pages 570-580, ISSN 0043-1648, https://doi.org/10.1016/j.wear.2019.01.015

[12] R. Liu, W. Yu, M. Liu, (2023). Numerical prediction of solid particle erosion in angle-cutting elbows with gas–solid flow, Proceedings of the Institution of Mechanical Engineers, Part C: Journal of Mechanical Engineering Science, 06-320, 237, 2, https://doi.org/10.1177/09544062221121986

[13] Redondo, C., Chávez–Modena, M., Manzanero, J., Rubio, G., Valero, E., Gómez–Álvarez, S., & Rivero–Jiménez, A. (2021). CFD–based erosion and corrosion modeling in pipelines using a high–order discontinuous Galerkin multiphase solver. Wear, 478-479, 203882, https://doi.org/10.1016/j.wear.2021.203882

[14] Shirazi, S. A., Shadley, J. R., McLaury, B. S., & Rybicki, E. F. (1995). A Procedure to Predict Solid Particle Erosion in Elbows and Tees, Journal of Pressure Vessel Technology, 117(1), 45. https://doi.org/10.1115/1.2842089

[15] Yakhot, V., Orszag, S.A., Thangam, S., Gatski, T.B., Speziale, C.G., (1992), Development of turbulence models for shear flows by a double expansion technique., Phys. Fluids A, 4, 1510–1520, https://doi.org/10.1063/1.858424

[16] Omelyanyuk M, Pakhlyan I, Bukharin N, El Hassan M. (2021). Reduction of Energy Consumption for Water Wells Rehabilitation. Technology Optimization. Fluids., 6(12):444. https://doi.org/10.3390/fluids6120444

[17] Edwards, J. K., McLaury B. S., Shirazi S. A. (2000). Evaluation of Alternative Pipe Bend Fittings in Erosive Service. Proceedings of FEDSM 2000. ASME Fluid Engineering Division Summer Meeting: 959-966

Structural and environmental safety study of the holy mosque area using CFD

Mohamed Farouk [1,2]*

[1]College of Engineering, Imam Mohammad Ibn Saud Islamic University, IMSIU, Riyadh 11432, Saudi Arabia

[2]Faculty of Engineering, Ain Shams University, Cairo 11517, Egypt

*miradi@imamu.edu.sa

Keywords: Holy Mosque, The Comfort of Pedestrians, Air Quality, Gust Wind, Calm Wind & Wind Envelope

Abstract: A CFD model was developed, covering a square area of 3.64 km^2 and comprising the Holy Mosque near its center, the actual terrain, and the main buildings surrounding the Mosque. The gust wind effects on the existing cranes, the comfort of the pedestrians, and the air quality were studied for the first time in this Holy area. The air quality in this study was related to calm speed and creating a fertile environment for spreading infectious diseases. The study revealed that wind comfort levels are achieved in nearly all selected locations. Also, the wind speeds are generally low in the area. However, gusting wind currents appeared from limited directions, causing more wind loads on cranes. It is recommended to finalize some cranes' work shortly in specific locations or change their places, lower the crane boom at rest, and stop working on windy days. The air quality in some locations may be considered relatively poor. New tall buildings surrounding the Holy Mosque from the North and the East are not recommended except after studying their impacts on the air quality. Pruning North and/or East Mounts can remarkably improve natural ventilation. Large-scale fans are a solution after detailed study and simulation.

Introduction

The studied area is the holiest place for all Muslims worldwide, with at least a million daily visits for praying. While in seasons such as Hajj and Ramadan, millions daily visit the Mosque, which may be considered the most crowded area in the world, especially during the seasons. Millions of Muslims worldwide visit the Holy Mosque with different cultures, languages, and health. Saudi authorities are keen to increase the capacity to achieve 15 million external visitors to the holy areas annually. Because of the regularly expanding number of Muslims who want to perform Hajj, Omera, and pray in the Holy Mosque. Some new tall structures surrounding the Holy Mosque have been constructed to serve visitors in the last few years. That affects the wind envelope, which has structural and environmental impacts on the buildings and people.

Three primary wind effects must be considered [1]: wind loads on structural, environmental studies, and wind loads for the façade. This research discussed the structural and environmental impact and excluded the wind loads for the façade.

The structural effects of the wind include:

- The wind forces and moments on the building [2-5]
- The vibration of the structures is one of the significant effects of wind, especially on bridges and tall buildings [6]. It could lead to substantial displacements, accelerations, and resulting forces [7-9].

In this study, the first point was only tackled. Some environmental wind issues in cities can be tackled, such as

1. The pedestrian-level wind, where it is in the current case study, will be the comfort of the prayers, visitors, and pilgrims.
2. Calm wind speed affects the air quality.
3. Accumulating vehicular emissions at the street level [10] affects air quality.
4. Degradation of outdoor thermal comfort [11-14].

In the current study, two environmental points were studied: the comfort of the pedestrians and the area of low wind speed for the first time regarding the Holy Mosque area. In comparison, vehicular emissions at the street level nearby the Mosque were excluded since the Saudi authorities kept the vehicles from the Holy Mosque. Also, the degradation of outdoor thermal effects was excluded from the study because of insufficient data.

The environmental wind design criteria:
This study considers two environmental wind issues: the comfort of pedestrians and the calm wind. Each has different wind criteria. A summary of each subject was briefed, and the design criteria used in this study were stated.

Comfort of pedestrians
Some unpleasant incidents occurred for pedestrians nearby tall buildings in cities since the beginning of their construction of tall buildings in cities. Many such incidents were recorded in countries including Japan [15-17], the United Kingdom [18], the United States [19], and Canada [20]. The strong wind flows near tall buildings result from intense downwash, bringing high-speed winds from higher altitudes down to the ground level. Strong separation layers form at the sharp corners of tall buildings [21]. A taller building catches the more upper-level wind and directs it to the pedestrian level. Hence it poses high wind speed conditions but improves near-field air ventilation [22-24]. At the same time, turbulence intensity is not significantly influenced by the height variation of the building [24]. As the buildings' width increases, the incoming wind's sheltering effect enlarges the extent of the low wind speed zone on the downstream side of the building [25]. Therefore, some researchers invented techniques to assess the wind environment near tall buildings [21, 26-28] and developed evaluation criteria to estimate the wind comfort of pedestrians in built-up areas [29-32]. City authorities had also attempted to regulate adverse wind flow in built-up areas by stipulating wind ordinances and building design guidelines [33-36].

Table 1 represents the various pedestrian-level wind criteria converted to the common frequency of return (20%) and the mean wind velocity (km/h) [37]. The average group criteria were selected in this research. The wind speeds for the comfort level of sitting, standing, and walking pedestrians were 9.5, 14,5, and 19.00 km/h, respectively.

Table (1): Various pedestrian-level wind criteria converted to the typical frequency of return (20%) and mean wind velocity (km/h) [37]

	Sitting	Standing	Walking
Lawson and Penwarden [38]	8.5	13.9	20.3
Penwarden and Wise [18]	----	----	18
Hunt et al. [31]	10.4	15.6	15.5
Melbourne [30]	9.7	12.6	----
Davenport and Isyumov [39]	8	14.5	19.5
Lawson [40]	10.6	15.8	21.1
Group Avarge	9.44	14.48	18.88

Calm Wind speed:
The quick spreading of airborne pathogens [41-42] arises from low wind speed; subsequently, it causes deterioration of the urban quality of life. The study [43] reveals that in China, the viral agent of SARS-CoV-2 may be suspended in the air for various minutes. This fact can explain the transmission dynamics and high number of infected people and deaths from COVID-19 in many regions.

Based on low wind speed, atmospheric stability reduces the dispersion of gaseous and particulate matter (air pollution), which can act as a carrier of the SARS-CoV-2 in the air to sustain the diffusion of COVID-19 in the environment, generating problems for public health in society. [44-45]. The average low wind speed of 2-4km/h may cause a massive increase in infected people by SARS and Corona. Still, infectious diseases are expected to increase the number of infected people in calm atmospheres at different rates.

There are no specified worldwide criteria for ventilation requirements in outdoor suburban except national standards in a few countries. Hong Kong has Air Ventilation Assessments, which are mandatory for new developments. The recommended minimum wind speed at the pedestrian level is 1 to 2 m/s to ensure sufficient outdoor ventilation in Hong Kong. In the current study, the minimum mean speed is assumed to be 1.5m/s.

The studied area of the Holy Mosque:
To the author's knowledge, no research studied the comfort of pedestrians or the gusting wind in the Holy Mosque area. Also, no studies dealt with the stagnant area in the Holy Mosque. Therefore, the air quality articles due to emissions were only mentioned below. Then, the daily speed and the gusting wind in Holy Makkah are capsulated. Finally, the features of the studied area are described.

The air quality study of Holy Makkah:
Some researchers focused on the air quality assessment inside Holy Makkah during the Hajj season, such as Hajj (2005) [46-47], Hajj 2012 [48-49], and Hajj 2019-2020 [50]. Some others have studied one or more pollutants for a relatively long time, such as [51], where the fine particulate matter (PM2.5) sampling was performed from February 26, 2014 – January 27, 2015, in four cycles/seasons. Also, PM10, NO_x, SO_2, and CO have been analyzed over the study periods (1997- 2012) in Makkah [52]. NO_2, SO_2, CO, O_3, CH_4, and THC were studied from November 2002 to October 2003 [53]. The stagnant areas in the Holy Mosque area have been mentioned briefly by research [54].

The studies may require validation nowadays to be used around the Holy Mosque since the main reason for the air pollution was the emission from Hajj daily activities of pilgrims, accompanied by the increased demands of transportation means, which are currently prohibited nearby the Holy Mosque in Hajj season. Also, modern vehicles have better emission quality besides the increase in the number of pilgrims throughout the last decades [55]. Finally, many new structures have developed, especially around the Holy Mosque, which affects the wind envelope at the Holy Mosque.

The daily wind speed in Holy Makkah:
The wind rose for the Hajj months is presented in research [56]. In comparison, the windrose in Holy Makkah for the entire year [57] shows that the wind speed in Holy Makkah is less than 10Km/h for nearly 22.9% of the year and between 10-20Km/h for almost 32% of the year and more than 40Km/h for 3.5% of the year. Also, the mean wind speed is nearly 32 km/h throughout the year. The prevailing wind directions in Holy Makkah are northwest and southwest [57].

The gust wind in Holy Makkah:
According to the Saudi Building Code [58], the maximum wind gust speed in Jeddah and Taif cities is 152km/h. No gust wind speed is specified in Holy Makkah. Therefore it can be considered that it is the same in Jeddah and Taif. The terrain exposure constants are relevant to category B.

The studied area:
The model area is square, with a side length of 1.8km. The actual terrain of the studied area was considered by using the digital elevation model (DEM) for the Holy Mosque area to create the existing terrain surface in the model. Then the main buildings were added to the model, such as the old Holy Mosque, the third expansion of the Holy Mosque, the terraces, the piazza of the old Holy Mosque, the piazza between the terraces and the third expansion, Jabal Omar structures, the royal palace on Abu Kubais mount, Dar El-Tauheed, Abraj Makkah, a part of the Abu Kubais mount, a part of the Qiqaan Mount & other hotel buildings as shown in Figure 1. a.

1. Old Holy Mosque
2. The third (new) expansion
3. Terraces
4. Abraj Al-Bait
5. Jabal Omar
6. Abu Kubais Mount
7. Qiqaan Mount
8. Masaa
9. Piazza of the old Holy Mosque
10. Piazza between terraces and the third expansion
11. Piazza of the third expansion
12. Piazza near Masaa

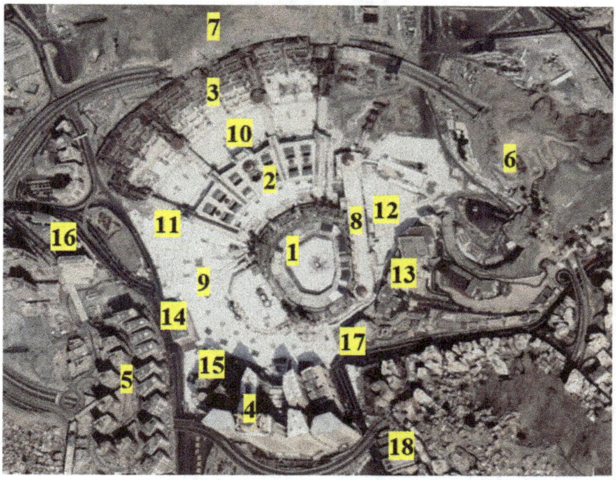

a. The central regions of the studied area as per Google Map October 2022

13. Royal Palace on Abu Kubais Mount
14. Dar El-Tauheed Hotel
15. Abraj Makkah
16. Some Hotels
17. Structure nearby the royal palace
18. Structures nearby Abraj Al-Bait

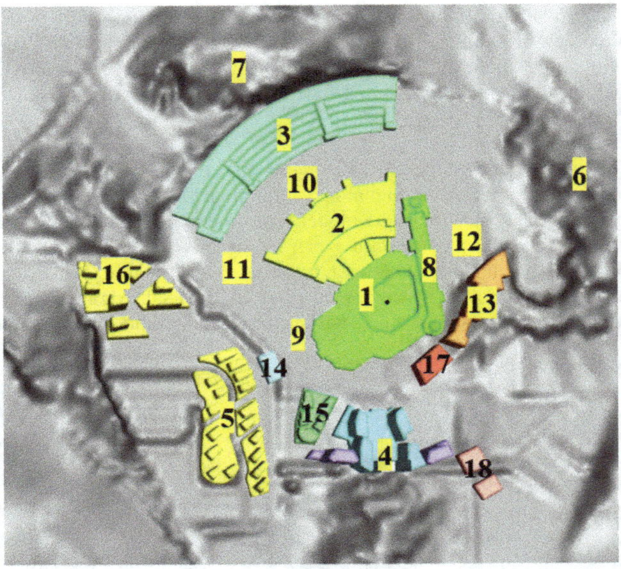

b. The central regions of the studied area as per the CFD Model

Figure 1: The geometry of the studied area

Goals, objectives, and scope:

The goals, objectives, and scope of the current study are listed below:

The goals:

In this study, four goals were tackled, two structural and two environmental goals as follows:

1. The wind gust speeds were calculated for almost currently working cranes to check their safety against gust wind.
2. Study the case of the tower crane accident in 2015.
3. The comfort of the pedestrians (prayers and visitors) was checked.
4. The calm wind areas (stagnation areas) were defined.

The objectives:

The objectives of this study are:

1. Study the gusting wind speed at each currently existing crane location.
2. Comparing the gusting wind speeds at each location with those in open space.
3. Determining the crane locations that are most critical regarding the gusting wind.
4. Recommend suitable precautions to increase cranes' safety.
5. Study the case of the collapsed tower crane to check the gusting wind loads before falling.
6. Check the comfort of the prayers/visitors in the Holy Mosque area.
7. Study the alternative to protect the prayers/visitors if they felt discomfort in some locations.
8. Check the stagnant wind locations in the exterior areas of the Holy Mosque.
9. Give recommendations to improve the air quality if there is a needy.

The scope of the study:

Twenty-four locations of current working cranes inside the Holy Mosque area were determined. The locations included the old Holy Mosque, the Massa, the third expansion of the Mosque, and the terraces. Also, the place of the collapsed tower crane in 2015 was selected. On the other side 18 locations on the top surface of the terraces, the roof of the Holy Mosque, the piazza, and the lower level of the terraces. The locations for the structural and environmental studies are discussed below.

Structural study of the crane's locations:

Points 1-24 represent the current locations of almost all working cranes inside the Holy Mosque area. While point number 25 represents the site of the collapsed tower crane in 2015. Nearly all operating cranes are found on different levels. Figure 2.a represents the locations on the third expansion and the terraces. In contrast, Figure 2.b illustrates the places on the old Mosque, the Massa, and the collapsed tower crane location. Cranes 1-13 are placed on the terraces on different levels. Cranes 14. 15 and 16 are used to complete the ceremonial dome construction. Cranes 17, 19, 20, and 21 are on the top of the Massa (Safa-Marwa). Working cranes 18, 22, and 23 are fixed to establish the new minarets. Crane 24 is set on the top of the Holy Mosque. The gust wind speeds at all the studied locations are compared to those in an open area.

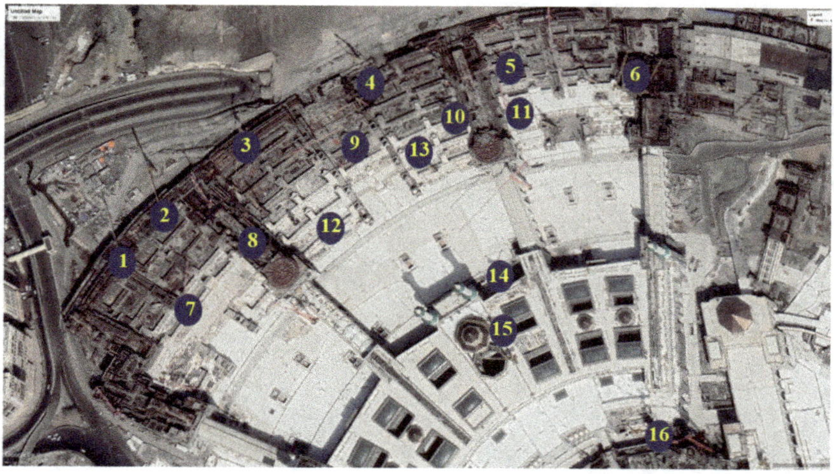

(a) Terraces and the third expansion of the Holy Mosque

(b) The old Holy Mosque

Figure 2: The selected locations of the wind gust (GoogleMap on October – 2022)

Environmental study of the crane's locations:

For the environmental studies, 18 locations are selected, as shown in Figure 3. Six points are chosen on the roofs (R1 to R6), three points on the first level of the terraces (T1 to T3), and nine are selected on the piazza level (P1 to P9). Points R1 to R3 are found on the top of the terrace step, while R4 to R6 4 exist on the roof of the old Holy Mosque. On the other hand, locations P1 to P3 are located on the piazza between terraces and the third extension of the Holy Mosque. P4 and P5 are located on the piazza outside the third expansion. Point 9 is located on the piazza adjacent to the Massa. Finally, points P6 to P8 are located on the piazza of the old Holy Mosque.

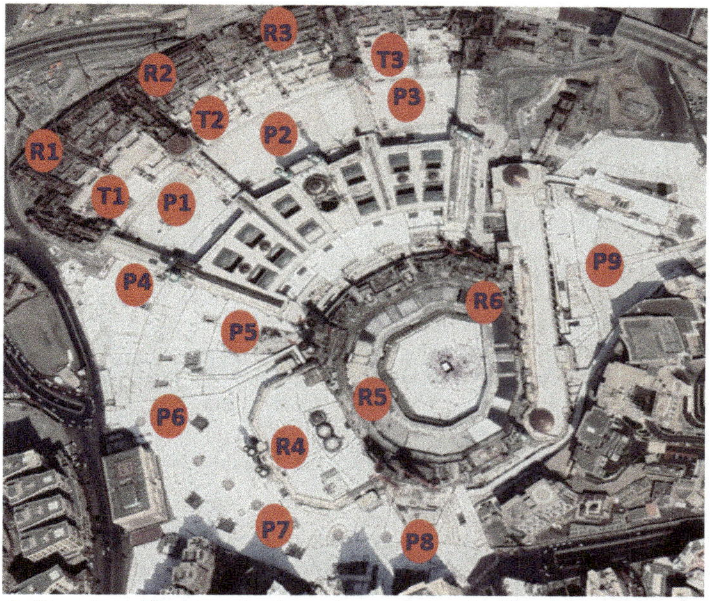

Figure 3: The selected locations for the environmental study (GoogleMap in October 2022)

The CFD model

ANSYS Fluent has been used to develop the CFD simulation of the Holy Mosque area. The following items are briefed below: the model's geometry, the grid generation and sensitivity analysis, the CFD model validation, the wind directions, assumptions, simplifications, boundary parameters, and solver parameters. The gust wind speed and the speed for the environmental studies are discussed under the studied area of the Holy Mosque and the environmental wind design criteria, respectively.

The geometry of the model:

1. b shows the main features of the studied area in the CFD simulation compared to the actual case study in Figure 1. a. Any new nearby tall buildings or features in the studied area can be added to the current CFD model to keep it valid for any future studies.

The grid generation and sensitivity analysis:

The accuracy of the results achieved from the CFD modeling significantly depends on the mesh quality, which also has implications for model convergence. This research applied a non-uniform mesh to the volumes of all CFD models. The grid was smoothened around the Holy Mosque structures and its piazza, which were the main areas of the current study. Different mesh sizes were generated to investigate the solution of independence from the grid. Grid sensitivity analysis was used to validate the CFD model. Four mech sizes were assumed to select the suitable size. These meshes were 977,408, 1,572,575, 2,403,713, and 3,625,124 tetrahedral cells. The mesh of 2.4 million cells was chosen with a maximum expected difference in the velocity results of less than 1.5%.

The CFD model validation:
The CFD model has been validated by comparing its results with wind tunnels by Irwin Inc.(RWDI) and Wacker-Ingenieure- Wind Engineering [confidional studies]. The wind velocities and the wind loads were validated with maximum discrepancies of less than 10%.

The wind directions:
Sixteen wind directions have been studied with an equal interval of 22.5° to cover all the possible directions. These sixteen wind directions are N, NNE, NE, ENE, E, ESE, SE, SSE, S, SSW, SW, WSW, W, NWN, NW, and NNW.

Assumptions and Simplifications:
The assumptions and the simplifications, which were assumed in this study, are listed below:

- There is no thermal effect considered.
- All short buildings around the Holy Mosque were ignored.
- Any interior locations inside the old Holy Mosque and their expansions were excluded.
- All areas nearby the existing internal mechanical ventilation were excluded from the present study
- Details of the Holy Mosque, such as the minarets and the domes, were discarded.

The boundary conditions and the solver parameters:
The main solver parameters and the boundary condition of the current CFD model are listed in Table 2.

Table 2. Summary of the boundary conditions and the solver parameters of the CFD models.

Parameters	Settings
Solver	Pressure-based, steady
Velocity formulation	Absolute
Turbulence model	Standard k-ε
Gravity	Not applied
The inlet boundary	Velocity inlet
The outlet boundary	outflow
The top of the model	symmetry
The bottom of the model	wall
The other boundary	symmetry
Cell Shape	Tetrahedral

The results and discussion
Different models were developed in this research study, where sixteen wind directions were conducted for each model. The first model was performed to study the wind velocities on the entire Holy Mosque and its piazza as its current status. Another model was used after adding the collapsed tower crane in 2015 to the Holy Mosque Model. The last model was the collapsed tower crane in an open area. The results of all the models were briefed hereunder.

The wind gust results:
All sixteen wind directions have been studied for the twenty-four locations of the current working cranes and the location of the collapsed tower crane, as shown in Figure 2. The wind velocities in the Holy Mosque area were generally reduced because of the development of new buildings around the Holy Mosque. But the wind gust risk increases in some wind directions due to the downwashing over the tall buildings and channeling between buildings. The wind velocities were determined at all studied places from all wind directions. Then, the maximum wind velocity at each location was compared to the wind velocity in the case of an open area. The discussion below can be divided into the existing cranes on the terraces, the third expansion, and the old Holy.

The cranes on the terraces
Thirteen cranes exist on the terraces, as shown in Figure 2. a. Three exist on the top of the terraces, which are 1, 2, and 3. All the other cranes are found on different step levels. The wind speeds at 5.0m higher than the base level of each crane were calculated for each wind direction. Then they compared to those in the case of open areas. In general, the wind speeds for all the cranes inside the Holy Mosque area are remarkably less than the speeds in the case in open space except for some limited wind directions, where the wind speeds in the Holy Mosque are higher than those in open space. Crane 11 is the only exception, where the maximum wind speed inside the Holy Mosque is nearly equal to that in an open space. The worst-case scenarios for the wind directions for most of the thirteen cranes are the south wind, as shown in Figure 4, and the south-southwest. The west-southwest wind is critical for cranes at locations 7, 10, and 13. Cranes at locations 4, 6, and 9 have the most fundamental forces of wind increase by nearly 30% compared with the wind speeds in an open area. All Cranes except those at 7, 10, 11, and 13 have an increment in the actual wind speeds at Holy Mosque than those in an open area by 10-20%

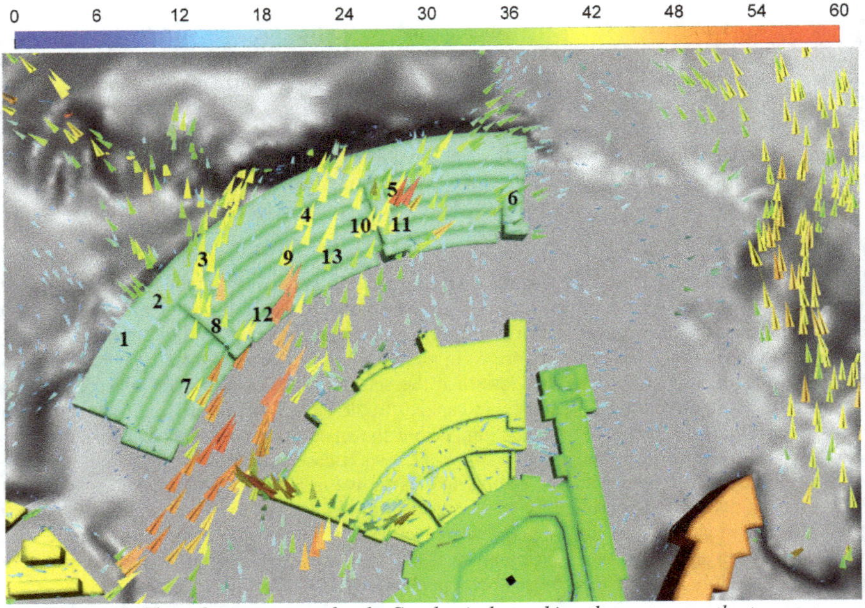

Figure 4: The velocity vectors for the South wind attacking the cranes on the terraces

The cranes on the third expansion

Three crane locations were considered in the current study for the third expansion of the Holy Mosque at locations 14, 15, and 16. A strong wind current is expected to attack the roof surface of the third expansion of the Holy Mosque in case of blowing wind from the south-southwest direction for the three locations. The channeling between the Jabal Omer buildings and the Abraj Al-Bait increases the wind velocity, and downwashing over the Abraj Makkah tower directs the strong wind current to attack the crane at location 15 and consequently, the ceremonial dome, as shown in Figure 5. The wind velocity is increased by about 25%, then the wind speed in an open area. Cranes at locations 14 and 16 have lower wind impact since their bases rested on the piazza, not on the roof of the expansion as crane 15.

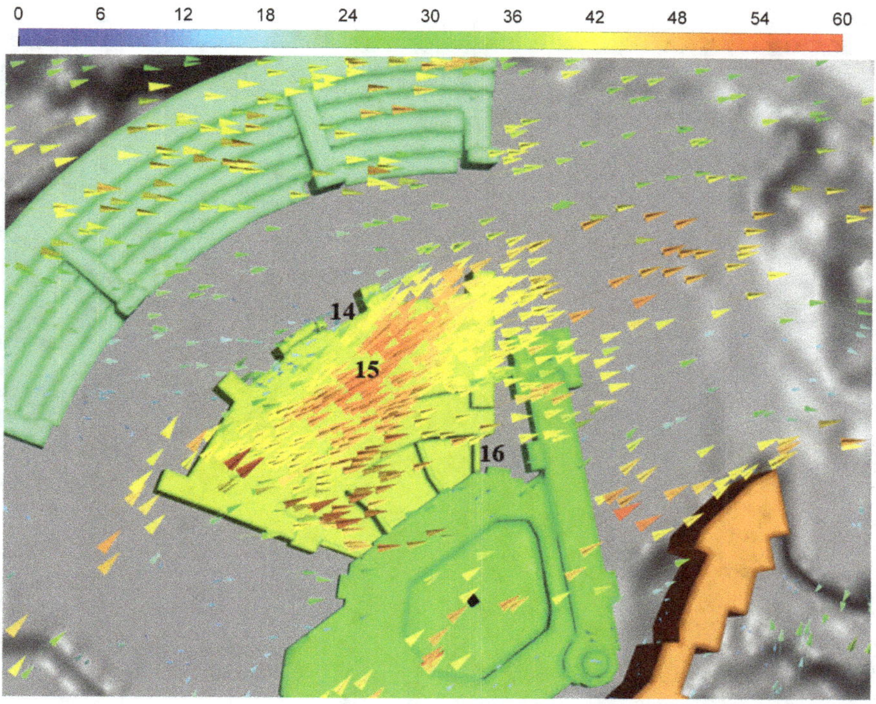

Figure 5: The velocity Vectors map for the third expansion of the Holy Mosque in case of the south-southwest wind 5m above the top surface of the expansion

The cranes on the old Holy Mosque and Massa

Cranes at locations 18, 22, 23, and 24 are considered on the old Holy Mosque. On the other hand, the cranes at locations 17, 19, 20, and 21 are placed on Massa. The worst-case scenarios for almost all cranes are the south-southwest and west-southwest wind directions when the wind current develops from channeling between Abraj Makkah and Abraj Al-Bait and down-washing over both structures. Figure 6 illustrates the velocity vectors for the south-southwest wind direction attacking the old Holy Mosque and Massa. The wind velocities at the cranes on the Massa increased by 25-30% compared to the wind in an open area. The wind speeds at cranes 18, 23, and 24 increase by up to 30% compared to the speeds in open space. The crane at location 22 has an increase in wind velocity of nearly 20% compared to that in open space.

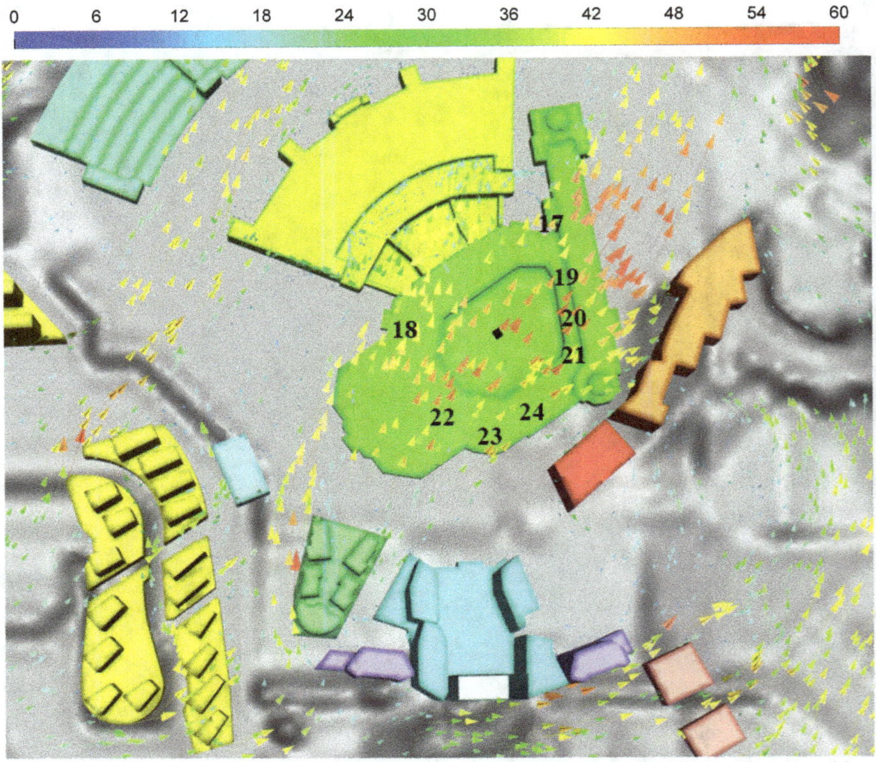

Figure 6: The velocity vectors for the South-southwest wind attacking the cranes on the old Holy Mosque and Massa

The crane location 25 for the case of the tower crane accident in 2015:
On September 11, 2015, 27th Thul-Qida 1436, 107 people died, and at least 238 were injured after the collapse of a tower crane at the Holy Mosque piazza during rest due to a strong wind. The tower crane was on the east side of the Holy Mosque beside the Royal Palace on Abu Kubais Mount. The long boom turned in the opposite direction, and the tower crane collapsed on the Massa.

This case is considered the worst disaster ever for the tower crane collapse. However, it happened during the off-peak hours. There were only two pieces of the collapsed tower crane model worldwide; one was at the Holy Mosque area. Its height is more than 200 meters. The tower crane overturned in the opposite direction and fell over the Massa.

In this research, the case of the collapsed tower crane was studied. First, the geometry of the tower crane was simulated. The effect of the gust on the tower crane was studied in an open area from all wind directions in a CFD model for the tower crane only. The bending (overturning) moments were calculated at the crane's base for each wind direction. Then, a new CFD model inserted the tower crane in the Holy Mosque area. The location and the direction of the crane while collapsing were precisely simulated. Next, the gust wind effects on the tower crane inside the Holy Mosque was studied for all wind directions. The wind forces acting on the tower crane were considerably limited to the wind blowing from 11 out of the 16 known directions, as shown in Figure 7. However, the bending moment (overturning moment) values in the actual case inside the Holy Mosque were more than those in an open area in four wind directions. These four critical wind directions were north-northeast, east-northeast, northwest, and west-southwest. Although, the ratio between the bending moment in an open area to that in an open space for the actual case is maximum in the west-southwest direction. The ultimate overturning moment acting on the tower crane occurred in the north-northeast wind direction. The overturning moment on the tower crane was 30% higher than in an open area. Figure 8 illustrates the contour of the wind velocity at the top of the Massa level.

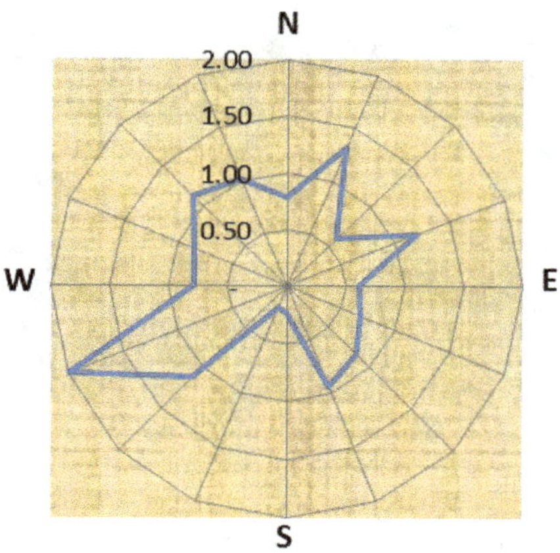

Figure 7: The ratio between the overturning moment on the crane in Holy Mosque to that in an open area

Advanced Topics in Mechanics of Materials, Structures and Construction - AToMech1-2023 Materials Research Forum LLC
Materials Research Proceedings 31 (2023) 765-786 https://doi.org/10.21741/9781644902592-78

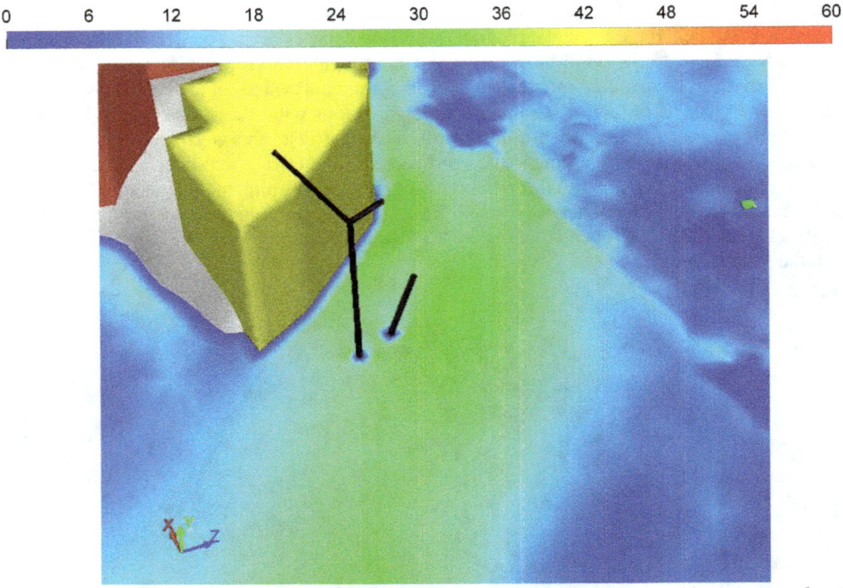

Figure 8: the velocity contour map for the North-northeast wind direction (the contour plan is at the top of the Massa level)

The comfort of the prayer:
The pedestrian comfort level has been checked in all wind directions for all mentioned locations in Figure 3. The study shows that all the places have different comfort levels (sitting, standing, and walking) to an expected return frequency of 20%. As shown in Figure 9, green dots represent the locations that fulfill the sitting comfort level, yellow for the standing comfort level, and orange for areas of the walking comfort level. The results of this study can be capsulated in the following points:

- The wind speeds are at the walking comfort level in all locations on the piazza except P5, P6, and P9.
- The locations P6 and P9 are at a standing comfort level.
- The location P5 is at sitting comfort level.
- P7 and P8 have the maximum wind speed out of the eighteen points.
- The wind speed at point R1 on the terraces is at the walking comfort level, while points R2 and R3 are at the standing comfort level.
- The wind speeds on the top of the old Holy Mosque at locations R4 and R5 are at the standing comfort level. On the other hand, location R6 is at the sitting comfort level.
- The wind speeds on the first level of the terraces T1, T2, and T3 are at the sitting comfort level.

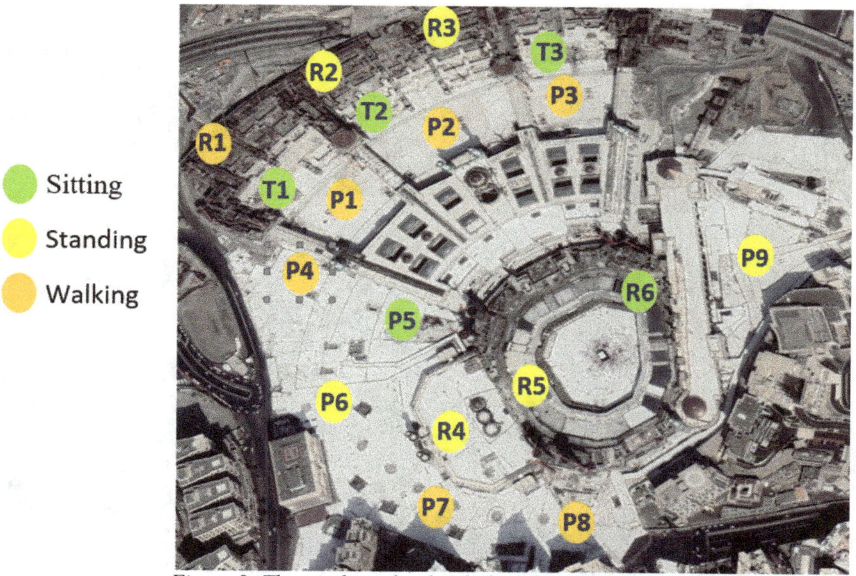

Figure 9: The wind comfort levels for all studied points.

The calm wind areas (stagnation areas) for the air quality study:
All areas nearby the existing internal mechanical ventilation are excluded from the present study using the Holy Makkah wind data, the average wind speed throughout the year in Holy Makkah is classified into four categories. As shown in Figure 10, red dots represent locations where the criterion mean wind speed of 1.5 m/s is exceeded less than 20% of the time; Orange for 20% to 30%, yellow for 30% to 45%, and green for areas where the mean wind speed criterion of 1.5 m/s is exceeded more than 45% of the time. Since there is no Saudi Standard for such a case, the last category (>45%) is recommended in this research to maintain the best air quality in the Holy Mosque area. The results show that all the selected points on the piazza have good ventilation except location P5 has relatively poor ventilation of 20-30% because it is on one of the confined corners. Such a site may need a source of mechanical ventilation. Also, locations P6 and P9 have relatively good ventilation, 30-45%. Location R4 on the Holy Mosque roof has good ventilation, while Point R5 has reasonably good ventilation, 30-45%, and Point R6 has poor ventilation. Location R1 on the top of the terraces also has good ventilation (>45%), while R2 and R3 have relatively poor ventilation, 20-30%. The ventilation is lacking on the lower terraces levels T1, T2, and T3.

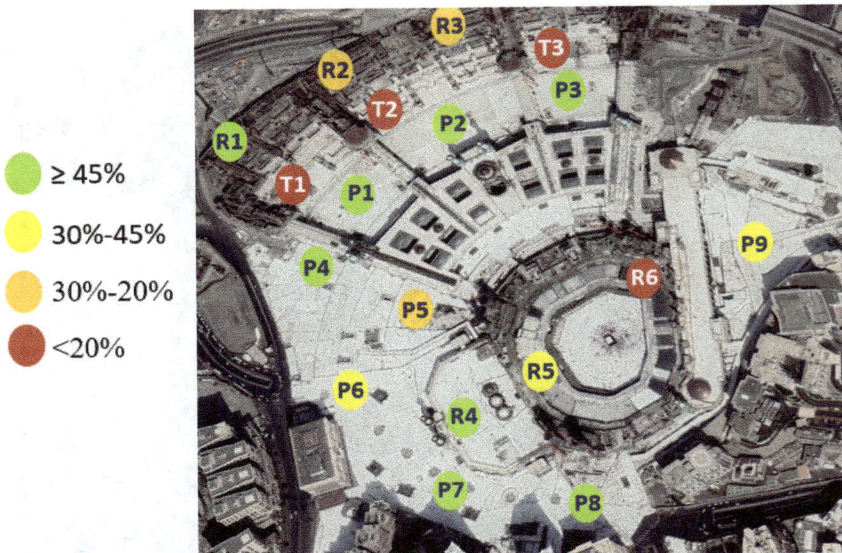

Figure 10: The Air Ventilation Assessment for the selected eighteen locations inside the Holy Mosque and its piazza.

Conclusions:

In this research, four goals were tackled, the safety of the working cranes (24 locations), the case of the collapsed tower crane at the Holy Mosque piazza in 2015, the wind comfort degree of the prayers/visitors in eighteen places inside the area and locations having calm wind, which expose to lower air quality.

Twenty-four locations are selected for the gusting effects for almost all currently working cranes. Thirteen cranes were placed on the terraces on different levels. Four crane sites are on the top of the Massa (Safa-Marwa). Another three cranes place to serve the third expansion of the Mosque, and the last four cranes are on the roof of the old Holy Mosque. Generally, the gust wind speeds for all the crane locations inside the Holy Mosque area are remarkably lower than those in open space except for some limited wind directions. However, almost all the cranes are exposed to higher wind speeds than the open area values for a few wind directions. The increment ratios vary depending on the crane location.

The worst-case scenarios for the wind directions for most of the thirteen cranes on the terraces are the south-southeast and the south wind. The wind speeds increase at the crane locations between 10-30% compared to those in open areas. The south-southwest wind direction is the critical direction for the cranes on the Mosque expansion, where the wind velocity is increased up to 25% compared to those in open space. The south-southwest and west-southwest are the critical wind directions for most of the locations of the eight cranes on the top of the old Holy Mosque and Massa, where the wind speeds increase at the crane locations between 20-30% compared to those in open areas. It is recommended to finalize the work of the cranes on Massa, the minarets, and the ceremonial dome before others or to change their locations. Also, it is suggested to lower the crane boom if it is at rest and stop working on windy days.

The case of the collapsed tower crane in 2015 was studied. The study revealed that the gusting wind was only critical from 5 directions out of 16. The gusting wind causing the collapse was blowing from one of them when the overturning moment increased by 30% compared to the case in open space.

The study shows that all the places have different comfort levels (sitting, standing, and walking) to an expected return frequency of 20%. All the areas on the top of the piazza have walking/standing comfort levels, except areas in partially closed areas have sitting comfort levels. The northwest side on the top surface of the terraces has a walking comfort level, while the rest has a standing comfort level. The top of the Holy Mosque has a standing/sitting comfort level. The wind speeds on the first level of the terraces T1, T2, and T3 are at the sitting comfort level.

Regarding the air quality, the results show that all areas on the piazza have good ventilation, except areas in confined corners with poor ventilation. Also, the places at the back of Massa and nearby Dar El-Tauheed Hotel have relatively good ventilation. The west side of the Holy Mosque roof has good ventilation, while its northeast has relatively poor ventilation. The northwest of the upper level of the terraces has good ventilation. In contrast, the north and east sides have relatively poor ventilation. The ventilation is lacking on the lower terrace levels.

Recommendations

The recommendations can be capsulated as follows:

1. Finalize the work of the cranes on Massa, the minarets, and the ceremonial dome before others.
2. Change the locations of cranes exposed to strong wind and/or move their bases to lower levels, if possible.
3. Lower the crane boom if it is at rest.
4. Stop working on windy days.
5. Channeling and downwashing may cause a collapse of the cranes, as happened to the collapsing tower crane in 2015.
6. The comfort of pedestrians satisfies international standards.
7. The air quality in different places is lower than the target level; more ventilation is required.
8. Using natural ventilation is much better than mechanical.
9. Study the possibility of improving the natural ventilation of the Holy Mosque by pruning the Abu Kubais Mount in the East and/or Qiqaan Mounts in the north.
10. Study the effect of using mechanical ventilation by using large-scale fans. A wind tunnel or CFD should simulate the case to ensure sufficient mechanical ventilation.
11. Using large-scale fans without prior modeling may help spread infectious diseases.
12. It is not recommended to have new tall buildings surrounding the Holy Mosque, especially from the North and the East, except after studying the impact on the air quality using either wind tunnel or CFD simulations.

References

[1] Mendis, Priyan, Tuan Ngo, N. Haritos, Anil Hira, Bijan Samali, and John Cheung. "Wind loading on tall buildings." *Electronic Journal of Structural Engineering* (2007). https://doi.org/10.56748/ejse.641

[2] Thordal, M.S., Bennetsen, J.C. and Koss, H.H.H., 2019. Review for practical application of CFD for the determination of wind load on high-rise buildings. *Journal of Wind Engineering and Industrial Aerodynamics*, *186*, pp.155-168. https://doi.org/10.1016/j.jweia.2018.12.019

[3] Farouk, M. I., S. A. Mourad, and A. S. Salaheldin. "Numerical simulation of wind effects on an airport air traffic control tower." WIT Transactions on The Built Environment 92 (2007). https://doi.org/10.2495/FSI070171

[4] Dagnew, Agerneh, and Girma T. Bitsuamlak. "Computational evaluation of wind loads on buildings: a review." *Wind Struct* 16, no. 6 (2013): 629-660. https://doi.org/10.12989/was.2013.16.6.629

[5] Tamura, Tetsuro, Kojiro Nozawa, and Koji Kondo. "AIJ guide for numerical prediction of wind loads on buildings." Journal of Wind Engineering and Industrial Aerodynamics 96, no. 10-11 (2008): 1974-1984. https://doi.org/10.1016/j.jweia.2008.02.020

[6] Farouk, Mohamed I. "Check the comfort of occupants in high rise building using CFD." Ain Shams Engineering Journal 7.3 (2016): 953-958. https://doi.org/10.1016/j.asej.2015.06.011

[7] Hou, Fangwei, and Mohammad Jafari. "Investigation approaches to quantify wind-induced load and response of tall buildings: A review." *Sustainable Cities and Society* 62 (2020): 102376. https://doi.org/10.1016/j.scs.2020.102376

[8] Kwok, Kenny CS, Peter A. Hitchcock, and Melissa D. Burton. "Perception of vibration and occupant comfort in wind-excited tall buildings." *Journal of Wind Engineering and Industrial Aerodynamics* 97, no. 7-8 (2009): 368-380. https://doi.org/10.1016/j.jweia.2009.05.006

[9] Elias, Said, and Vasant Matsagar. "Wind response control of tall buildings with a tuned mass damper." Journal of Building Engineering 15 (2018): 51-60. https://doi.org/10.1016/j.jobe.2017.11.005

[10] Fu, Lixin, Jiming Hao, Dongquan He, Kebin He, and Pei Li. "Assessment of vehicular pollution in China." *Journal of the Air & Waste Management Association* 51, no. 5 (2001): 658-668. https://doi.org/10.1080/10473289.2001.10464300

[11] Rupp, Ricardo Forgiarini, Natalia Giraldo Vásquez, and Roberto Lamberts. "A review of human thermal comfort in the built environment." *Energy and Buildings* 105 (2015): 178-205. https://doi.org/10.1016/j.enbuild.2015.07.047

[12] Cheng, Vicky, Edward Ng, Cecilia Chan, and Baruch Givoni. "Outdoor thermal comfort study in a sub-tropical climate: a longitudinal study based in Hong Kong." *International Journal of Biometeorology* 56 (2012): 43-56. https://doi.org/10.1007/s00484-010-0396-z

[13] Lai, Dayi, Zhiwei Lian, Weiwei Liu, Chaoran Guo, Wei Liu, Kuixing Liu, and Qingyan Chen. "A comprehensive review of thermal comfort studies in urban open spaces." *Science of the Total Environment* 742 (2020): 140092. https://doi.org/10.1016/j.scitotenv.2020.140092

[14] Potchter, Oded, Pninit Cohen, Tzu-Ping Lin, and Andreas Matzarakis. "Outdoor human thermal perception in various climates: A comprehensive review of approaches, methods and quantification." *Science of the Total Environment* 631 (2018): 390-406. https://doi.org/10.1016/j.scitotenv.2018.02.276

[15] Kamei, I., and E. Maruta. "Study on wind environmental problems caused around buildings in Japan." *Journal of Wind Engineering and Industrial Aerodynamics* 4, no. 3-4 (1979): 307-331. https://doi.org/10.1016/0167-6105(79)90010-2

[16] Murakami, S., K. Uehara, and H. Komine. "Amplification of wind speed at ground level due to construction of high-rise building in urban area." *Journal of Wind Engineering and Industrial Aerodynamics* 4, no. 3-4 (1979): 343-370. https://doi.org/10.1016/0167-6105(79)90012-6

[17] Murakami, Shuzo, Yoshiteru Iwasa, and Yasushige Morikawa. "Study on acceptable criteria for assessing wind environment at ground level based on residents' diaries." *Journal of Wind Engineering and Industrial Aerodynamics* 24, no. 1 (1986): 1-18. https://doi.org/10.1016/0167-6105(86)90069-3

[18] Penwarden, Alan David, and Alan Frederick Edward Wise. *Wind environment around buildings*. London, UK: HM Stationery Office, 1975.

[19] Durgin, Frank H. "Pedestrian level wind studies at the Wright brothers facility." Journal of Wind Engineering and Industrial Aerodynamics 44, no. 1-3 (1992): 2253-2264. https://doi.org/10.1016/0167-6105(92)90016-4

[20] Isyumov, N. "Studies of the pedestrian level wind environment at the boundary layer wind tunnel laboratory of the University of Western Ontario." Journal of Wind Engineering and Industrial Aerodynamics 3, no. 2-3 (1978): 187-200. https://doi.org/10.1016/0167-6105(78)90009-0

[21] Beranek, W. J. *Wind Environment Around Single Buildings of Rectangular Shape; And, Wind Environment Around Building Configurations*. Stevin-Laboratory of the Department of Civil Engineering, Delft University of Technology, 1984.

[22] Stathopoulos, Ted. "Wind environmental conditions around tall buildings with chamfered corners." *Journal of Wind Engineering and Industrial Aerodynamics* 21, no. 1 (1985): 71-87. https://doi.org/10.1016/0167-6105(85)90034-0

[23] Tsang, C. W., Kenny CS Kwok, and Peter A. Hitchcock. "Wind tunnel study of pedestrian level wind environment around tall buildings: Effects of building dimensions, separation and podium." *Building and Environment* 49 (2012): 167-181. https://doi.org/10.1016/j.buildenv.2011.08.014

[24] Xu, Xiaoda, Qingshan Yang, Akihito Yoshida, and Yukio Tamura. "Characteristics of pedestrian-level wind around super-tall buildings with various configurations." *Journal of Wind Engineering and Industrial Aerodynamics* 166 (2017): 61-73. https://doi.org/110.1016/j.jweia.2017.03.013

[25] Mittal, Hemant, Ashutosh Sharma, and Ajay Gairola. "A review on the study of urban wind at the pedestrian level around buildings." *Journal of Building Engineering* 18 (2018): 154-163. DOI: 10.1016/j.jobe.2018.03.006

[26] Irwin, Hamlyn Peter Anthony Hugh. "A simple omnidirectional sensor for wind-tunnel studies of pedestrian-level winds." *Journal of wind engineering and industrial aerodynamics* 7, no. 3 (1981): 219-239. https://doi.org/10.1016/0167-6105(81)90051-9

[27] Uematsu, Toshihiko, Ken-ichi Yamada, Hiroyuki Matsuno, and Mitsuyoshi Nakashima. "The measurement of haloperidol and reduced haloperidol in neonatal hair as an index of placental transfer of maternal haloperidol." *Therapeutic drug monitoring* 13, no. 2 (1991): 183-187. https://doi.org/10.1097/00007691-199103000-00016

[28] Wu, Hanqing, and Theodore Stathopoulos. "Wind-tunnel techniques for assessment of pedestrian-level winds." *Journal of engineering mechanics* 119, no. 10 (1993): 1920-1936. https://doi.org/10.1061/(ASCE)0733-9399(1993)119:10(1920)

[29] Jackson, Peter S. "The evaluation of windy environments." *Building and Environment* 13, no. 4 (1978): 251-260. https://doi.org/10.1016/0360-1323(78)90016-1

[30] Melbourne, W. H. (1978). Criteria for environmental wind conditions. *Journal of Wind Engineering and Industrial Aerodynamics*, *3*(2-3), 241-249. https://doi.org/10.1016/0167-6105(78)90013-2

[31] Hunt, J. C. R., E. C. Poulton, and J. C. Mumford. "The effects of wind on people; new criteria based on wind tunnel experiments." *Building and Environment* 11, no. 1 (1976): 15-28. https://doi.org/10.1016/0360-1323(76)90015-9

[32] Durgin, Frank H. "Pedestrian level wind criteria using the equivalent average." *Journal of Wind Engineering and Industrial Aerodynamics* 66, no. 3 (1997): 215-226. https://doi.org/10.1016/S0167-6105(97)00130-X

[33] Arens, Edward, D. Ballanti, C. Bennett, S. Guldman, and Bryon White. "Developing the San Francisco wind ordinance and its guidelines for compliance." *Building and Environment* 24, no. 4 (1989): 297-303. DOI: 10.1016/0360-1323(89)90023-1

[34] Durgin, F.H., 1989. Proposed guidelines for pedestrian level wind studies for Boston—Comparison of results from 12 studies. *Building and Environment, 24*(4), pp.305-314. https://doi.org/10.1016/0360-1323(89)90024-3

[35] Willemsen, Eddy, and Jacob A. Wisse. "Design for wind comfort in The Netherlands: Procedures, criteria and open research issues." *Journal of Wind Engineering and Industrial Aerodynamics* 95, no. 9-11 (2007): 1541-1550. https://doi.org/10.1016/j.jweia.2007.02.006

[36] Ng, E., 2009. Policies and technical guidelines for urban planning of high-density cities–air ventilation assessment (AVA) of Hong Kong. *Building and Environment, 44*(7), pp.1478-1488. https://doi.org/10.1016/j.buildenv.2008.06.013

[37] Soligo, Michael J., Peter A. Irwin, Colin J. Williams, and Glenn D. Schuyler. "A comprehensive assessment of pedestrian comfort including thermal effects." *Journal of Wind Engineering and Industrial Aerodynamics* 77 (1998): 753-766. https://doi.org/10.1016/S0167-6105(98)00189-5

[38] Lawson, T. V., and PENWARDEN AD. "The effects of wind on people in the vicinity of buildings." (1977).

[39] Isyumov, N., and Davenport, A.G. "The ground level wind environment in built-up areas." In *proc. of 4^< th> Int. Conf. on Wind Effects on Buildings and Structures, London, 1975.* 1975.

[40] Lawson, T. V. "The determination of the wind environment of a building complex before construction." *Department of Aerospace Engineering, University of Bristol, Report Number TVL* 9025 (1990). https://doi.org/10.1016/0360-1323(76)90015-9

[41] Yip, Cleo, Wen L. Chang, K. H. Yeung, and Ignatius TS Yu. "Possible meteorological influence on the severe acute respiratory syndrome (SARS) community outbreak at Amoy Gardens, Hong Kong." *Journal of environmental health* 70, no. 3 (2007): 39-47.

[42] Zhang, Yu, Kenny CS Kwok, X-P. Liu, and J-L. Niu. "Characteristics of air pollutant dispersion around a high-rise building." *Environmental Pollution* 204 (2015): 280-288. https://doi.org/10.1016/j.envpol.2015.05.004

[43] Van Doremalen, Neeltje, Trenton Bushmaker, Dylan H. Morris, Myndi G. Holbrook, Amandine Gamble, Brandi N. Williamson, Azaibi Tamin et al. "Aerosol and surface stability of SARS-CoV-2 as compared with SARS-CoV-1." *New England Journal of Medicine* 382, no. 16 (2020): 1564-1567. https://doi.org/10.1056/NEJMc2004973

[44] Coccia, Mario. "How do low wind speeds and high levels of air pollution support the spread of COVID-19?." Atmospheric pollution research 12, no. 1 (2021): 437-445. DOI: 10.1016/j.apr.2020.10.002

[45] Coccia, Mario. "The effects of atmospheric stability with low wind speed and of air pollution on the accelerated transmission dynamics of COVID-19." International Journal of Environmental Studies 78, no. 1 (2021): 1-27. https://doi.org/10.1080/00207233.2020.1802937

[46] Al-Jeelani, H. A. "Evaluation of air quality in the Holy Makkah during Hajj season 1425 H." *Journal of Applied Sciences Research* 5, no. 1 (2009): 115-121.

[47] Seroji, A. R. "Particulates in the atmosphere of Makkah and Mina Valley during the Ramadan and Hajj seasons of 2004 and 2005." *Air Pollution XIX (Eds: Brebbia CA, Longhurst JWS, Popov V). Wessex Institute of Technology, UK* (2011): 319-327.

[48] Habeebullah, Turki M. "An Analysis of Air Pollution in Makkah-a View Point of Source Identification." *EnvironmentAsia* 6, no. 2 (2013).

[49] Simpson, Isobel J., Omar S. Aburizaiza, Azhar Siddique, Barbara Barletta, Nicola J. Blake, Aaron Gartner, Haider Khwaja, Simone Meinardi, Jahan Zeb, and Donald R. Blake. "Air quality in Mecca and surrounding holy places in Saudi Arabia during Hajj: initial survey" *Environmental science & technology* 48, no. 15 (2014): 8529-8537. https://doi.org/10.1021/es5017476

[50] Farahat, Ashraf, Akshansha Chauhan, Mohammed Al Otaibi, and Ramesh P. Singh. "Air quality over major cities of Saudi Arabia during hajj periods of 2019 and 2020." *Earth Systems and Environment* 5 (2021): 101-114. https://doi.org/10.1007/s41748-021-00202-z

[51] Nayebare, Shedrack R., Omar S. Aburizaiza, Azhar Siddique, David O. Carpenter, Mirza M. Hussain, Jahan Zeb, Abdullah J. Aburiziza, and Haider A. Khwaja. "Ambient air quality in the holy city of Makkah: A source apportionment with elemental enrichment factors (EFs) and factor analysis (PMF)." *Environmental pollution* 243 (2018): 1791-1801. https://doi.org/10.1016/j.envpol.2018.09.086

[52] Munir, Said, Turki M. Habeebullah, Abdulaziz R. Seroji, Safwat S. Gabr, Atif MF Mohammed, and Essam A. Morsy. "Quantifying temporal trends of atmospheric pollutants in Makkah (1997–2012)." *Atmospheric Environment* 77 (2013): 647-655. https://doi.org/10.1016/j.atmosenv.2013.05.075

[53] Al-Jeelani, Hesham A. "Air quality assessment at Al-Taneem area in the holy Makkah City, Saudi Arabia." *Environmental Monitoring and Assessment* 156 (2009): 211-222. https://doi.org/10.1007/s10661-008-0475-3

[54] Farouk, Mohamed. Three-Dimensional Dynamic Numerical Simulation of the Holy Mosque Area. No. 9865. EasyChair, 2023.

[55] Ascoura, Ibrahim Elsayed. "Impact of pilgrimage (Hajj) on the urban growth of the Mecca." *Journal of Educational and Social Research* 3, no. 2 (2013): 255. https://doi.org/10.5901/jesr.2013.v3n2p255

[56] Farahat, A., Chauhan, A., Al Otaibi, M. and Singh, R.P., 2021. Air quality over major cities of Saudi Arabia during hajj periods of 2019 and 2020. *Earth Systems and Environment*, *5*, pp.101-114. https://doi.org/10.1007/s41748-021-00202-z

[57] Shehata, Ahmed, and Sahl Waheeb. "Thermal Performance of Contemporary Residential Buildings in Hot-Arid Climates." *Journal of Advanced Research in Dynamical and Control Systems. Special* (2018): 1295-1303.

[58] Saudi Building Code Requirements (SBC301)

Physical-chemical characterization and exploitation of an ultra fine local in mortars

ZAOUI Miloud[1,a *], TAFRAOUI Ahmed[1,b] and BEKKAR Izzeddine[1,c]

[1]Laboratoire de Fiabilité du Génie Mécanique BP 417route de kenadsa 08000 Bechar Algérie

[a] miloud_zaoui@yahoo.fr, [b] tafraoui.ahmed@univ-bechar.dz,
[c] bekkar.izzeddine@univ-bechar.dz

Keywords: Physico-Chemical Characterization, Formulation, Ultrafine, Clay, Dune Sand, Mortar

Abstract. The research aims to find a suitable replacement for the current ultrafine material used in concrete and mortar formulation. The study will focus on the physical, chemical, and mechanical properties of the new ultrafine material in the cement matrix. The goal is to improve the performance of mortars through the use of local ultrafine. The study will include physico-chemical characterization of the ultrafine material, which may include clay or dune sand. The research will provide insights into the impact of using local ultrafine on cement-based mortars. The findings will inform the development of better formulations for concrete and mortar. The research will contribute to the use of locally sourced materials in the construction industry.

1. Introduction

Currently, research on new construction materials has shifted towards the use of local materials to produce cost-effective and high-performing mortars and concrete. The utilization of industrial by-products is justified due to their abundance and the lack of recycling and recovery programs.

Our work is part of an experimental study on the physical-chemical characterization and utilization of a local ultra-fine material in mortars. Mortars are widely used and provide benefits such as strength, workability, and porosity.

However, some of these properties are still subject to research and improvement. For example, ultra-fines significantly improve the strength of mortar. [1], [2]

The incorporation of additives (ultra-fines) is now an important technique in improving the properties of mortar such as strength and durability. These additives significantly affect the rheology of fresh cementitious materials, which is directly related to the development of strength and durability of hardened materials.

However, to fully benefit from these advantages and choose the best solution for optimization, it is necessary to understand the characteristics of these additives and their effects on mortar properties.

Can we achieve acceptable performance by using locally obtained ultra-fine additives from dune sand (Taghit erg).

The main objective of this study is to:

- Physico-chemical characterization of local ultra-fines (clays from Tabalbala and El-Outa).
- Utilization of a local ultra-fine material in dune sand-based mortar.
- Development of a new, economical mortar.

2. Materials and Methods

2.1 Materials

The characterization of the materials used in the composition of a mortar plays a very important role on its properties and subsequent performance, the essential properties of the mortar are largely influenced by the characteristics of its constituents. As a result, the standardization of testing methods and identification of mortar components is necessary. The different materials to be used in the preparation of the mortar to be studied, as well as the tests to be performed according to French standards and current operating procedures, are presented. [3]

2.2.1 sand dune:

Fig. 1, sand dune

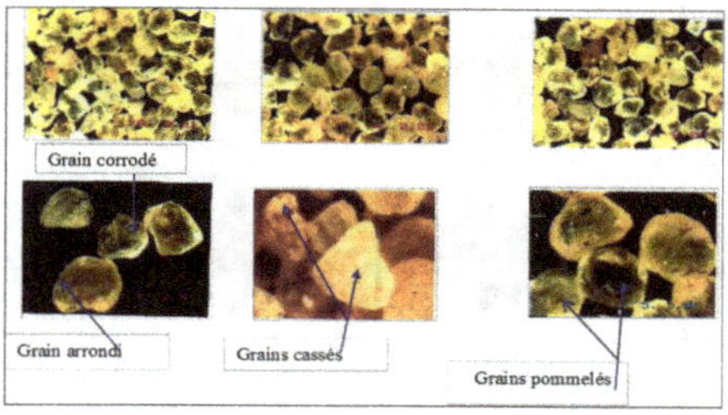

Fig. 2, Observations of sand grains under the optical microscope [Tafraoui, 2009].

The optical microscope (Fig. 2) identifies and qualifies the following minerals:
- quartz: it appears in a white color or under a yellow tint if the slide is a little thick (> 0.03 mm). The grains are mostly rounded but can also be dull or broken.
- iron oxides : they are either in the form of independent grains or they coat the quartz grains. or they are in the form of inclusions in the quartz. They are opaque and appear under a black to reddish color.

It was observed other minerals such as calcite, the anhydrite or the dolomite (but in evidence elsewhere), but this is due to their very low proportions in this sand.

The results of elemental analysis by XRD on the Taghit sand show (Fig. 3) a peak of about 100% silica which reflects the dominance of SiO2 in the analyzed sand. The other elements revealed are $CaCO_3$ and Fe_2O_3 present at low percentages. To detect the fraction of the sands richest in silica.

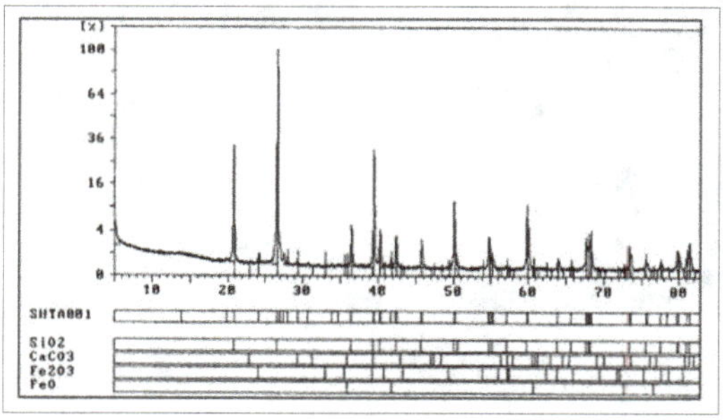

Fig. 3: XRD analysis of a dune sand (Savd of Taghit) [9]

The results are presented as oxides in the following Table 1,

Table 1, Results of the elemental chemical analyses DRX (in %) (Sand of Taghit)

Fraction (mm)	**0-0,04**	**0,04-0,10**	**0,10-0,12**	**0,12-0,16**	**0,16-0,20**	**0,20-0,25**
%SiO_2	81,61	92,42	95,18	96,33	97,33	**97,15**
%Al_2O_3	3,78	2,05	1,41	1,00	0,83	**0,79**
%Fe_2O_3	2,24	0,99	0,59	0,30	0,24	**0,21**
%CaO	3,92	0,87	0,27	0,33	0,07	**0,11**
%MgO	0,63	0,17	0,02	0,47	0,41	**0,05**
%SO_3	0,18	0,19	0,16	0,18	0,18	**0,14**
%K_2O	1,08	0,59	0,33	0,10	0,04	**0,02**
%Na_2O	0,48	0,20	0,09	0,09	0,09	**0,18**
%P_2O_5	0,10	0,02	0,01	0,01	0,01	**0,00**
%TiO_2	0,96	0,56	0,25	0,07	0,05	**0,05**
%MnO	0,04	0,03	0,02	0,01	0,02	**0,01**
%Cr_2O_3	0,01	0,01	0,01	0,01	0,01	**0,01**
% Loss on Fire	4,58	1,61	1,09	0,75	0,65	**0,58**
%TOT	**99,6**	**99,7**	**99,4**	**99,6**	**99,9**	**99,3**

2.1.1 Particle size analysis by sieving [NFP 18-560].

a- Objective

The grain size analysis determines the size and weight percentages of the different grain families in the sample. It applies to all aggregates with a nominal dimension of 63mm or less, excluding fillers. It's important to avoid confusion between grain size, which is concerned with determining the grain dimension, and granularity, which deals with the dimensional distribution of the grains in an aggregate.

b- Principle

The test involves sorting the grains in the sample using a series of sieves stacked on top of each other with decreasing opening sizes. The material is placed on the upper sieve and the sorting is achieved by vibrating the sieve column.

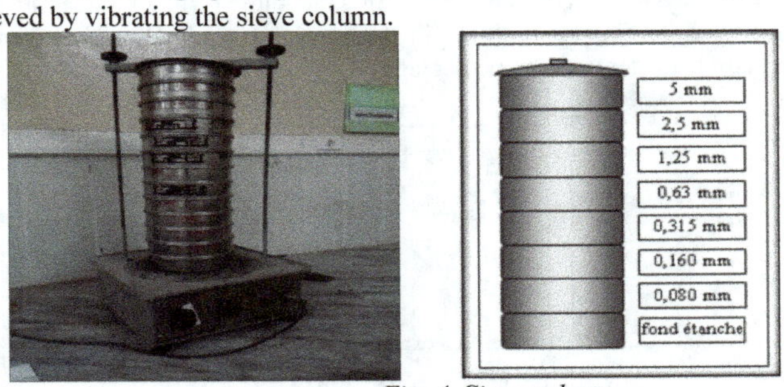

Fig. 4, Sieve column

The results of the granulometric analysis are assembled in the following Table II.2:

Table. 2, Results of particle size analysis of dune sand

Sample mass(g)		Refusal weights	Refusal weights	Cumulative refusal weights	Sieve(%)
1000	0,5	0	0	0	100
	0,4	0	0	0	100
	0,315	36,80	36,80	3,68	96,32
	0,25	614,10	650,90	65,09	34,91
	0,2	276,10	927,00	92,70	7,30
	0,16	47,30	974,30	97,43	2,57
	0,125	11,50	985,80	98,58	1,42
	0,1	6,70	992,50	99,25	0,75
	0,08	3,00	995,50	99,55	0,45
	Fond	4,5	1000	100	0

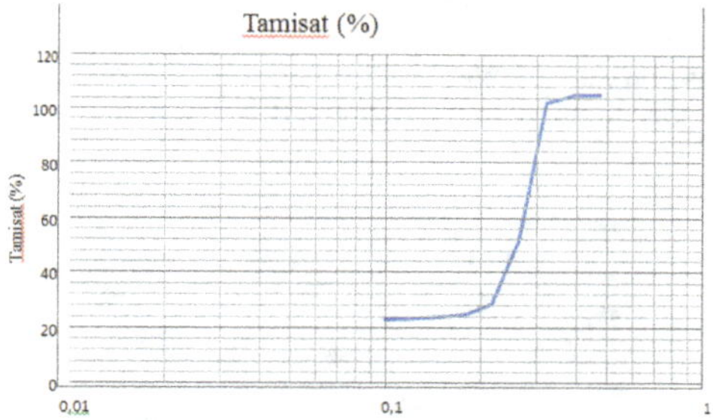

Fig. 5, Grain size curve of a dune sand

2.1.2. Fineness module

The sands must have a granulometry such as the fine elements are neither in excess, nor in too weak proportion, the character more or less fine of a sand can be quantified by the calculation of the modulus of fineness (Mf).
The modulus of fineness is all the smaller that the aggregate is rich in fine elements.
Eq. 2.1, is

$$Mf = \frac{\sum Rc\%}{100} \quad (2.1)$$

RC: Refuse cumulus in (%) under the sieves of module 0,16 to 5
When Mf is between :
1.8 And 2.2: the sand is mostly fine
2.2 And 2.8 : we are in presence of a preferential sand
2.8 And 3.3: the sand is a little coarse, it will give concrete resistant but less workable.
The results of modulus of fineness are given in the following Table. 3,

Table. 3, The results of modulus of fineness

Material	Module of finesse
dune sand	1,01

2.1.3 Equivalent of sand [NF P 18-598]

a- Objective

The sand cleanliness test is commonly used to assess the quality of sand in concrete. It separates fine particles from coarse sand to quantify the cleanliness using a standardized sand equivalent coefficient.

b- Principle

The test is performed on sand. The sample is washed according to a standardized process and allowed to rest. After 20 minutes, the following elements are measured:

Height h1: clean sand + fine elements.

Height h2: clean sand only.

The equivalent sand content is then deduced, which, by convention in the Eq. 2.2, is

$$ES = \frac{h2}{h1} * 100\% \qquad (2.2)$$

c - Method of operation

Fill each of the three test tubes with water to a level 100 ml lower than the marking. Then weigh a sand sample of 120g and pour one into each cylinder and let it settle for 10 minutes. After this time, fill the test tubes with a little water and shake them, repeating this until the upper marking is reached. Finally, let these mixtures settle for 20 minutes. Fig. 6 A. The test tube is then placed horizontally in an automatic vibration machine. Fig. 6 B. The test tube is vibrated 90 times and left for 20 minutes.

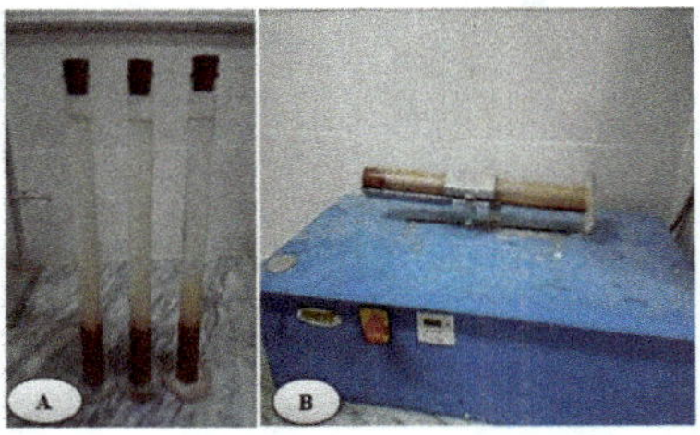

Fig. 6, Sand equivalent test

The sand equivalent values in Table .4, indicate the nature and quality of the sand according to the measuring means and allow to appreciate its quality for the composition of a concrete.

Table. 4, the values of sand equivalent indicate the nature and quality of the sand

N°	S.E. to visual	S.E. at the piston	Nature of the sand quality
01	ES< 65%	ES<60%	Clayey sand : risk of shrinkage or swelling. Sand to be rejected for quality concretes or more precise verification of the nature of the fines by a test with
02	65%≤ES<75%	60%≤ES<70%	Slightly clayey sand of acceptable cleanliness for standard quality concrete when the shrinkage has no significant effect on the quality of the concrete.
03	75%≤ES<85%	70%≤ES<80%	Clean sand with a low proportion of clay fines perfectly suited for high quality concrete.
04	ES≥85%	70≤ES<80%	Very clean sand. The almost total absence of fine clay clayey fines may lead to a plasticity defect of the concrete that will have to be compensated by an

From the interpretation table Table. 4, we concluded that the dune sand is very clean. The results obtained are established in the following Table. 5,

	Sand equivalent		Value limit	Observation
Dune sand	E.S.V%	100	E.S.V≥85%	Very clean sand

According to the NF P18 598 standard, this sand is very clean and acceptable for the composition of mortar and concrete.

2.1.4 Density (NF P 18-301)

a- Objective

The aim of this essay is to determine the mass of a granular fraction when developing a concrete composition. This parameter is used to determine the mass or volume of different granular classes mixed to obtain concrete with specified characteristics.

b- Principe

The principle involves measuring the weight of a sample in a given volume. It is an indirect measurement used in construction sites as a quantity control argument. Two types of bulk density are generally distinguished.

2.1.4.1 Absolute density

a- Definition

The absolute density ρs is the mass per unit volume of the material that constitutes the aggregate, without taking into account the voids that may exist in or between grains.

Fill a graduated cylinder with a volume (V1=300) of water.

b- Method of operation

Weigh a dry sample M of aggregates (about 300 g) and introduce it into the test tube, taking care to eliminate all air bubbles.

The liquid rises in the test tube. Read the new volume V2, Fig. 7,

The absolute density in the Eq. 2.3, is :

$$\rho_s = \frac{M_S}{V_2 - V_1} \quad (2.3)$$

ρ_s : absolute density.
M_s : mass of solid grain.
V_1 : volume of water.
V_2 : total volume (solid grain + water).

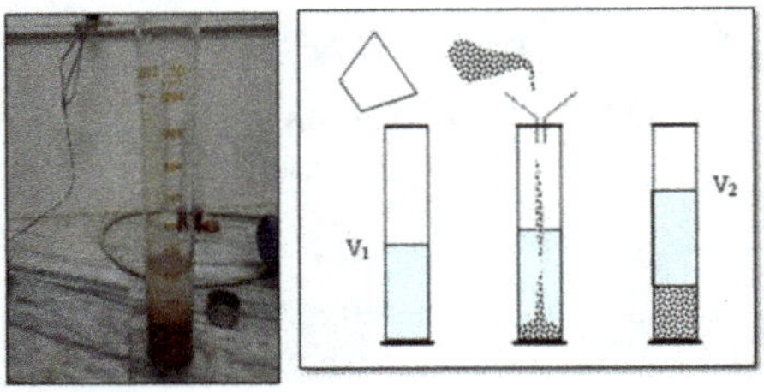

Fig. 7, Absolute density test

2.1.4.2 The apparent density

a- Definition
This is also indicated in the name (bulk density) and is a mass of dry constituents per unit volume constituents per unit volume where the voids between the grains are included.

b- Procedure
Take the sand (for example) in the 2 hands forming a funnel. Place these 2 hands at about 10 cm above a one liter measure and let fall this stable, neither too fast, nor too slow.
Pour the body in this way, always in the center of the measure, until it overflows all the center of the measure.
Pass to the rule. Weigh the contents Fig. 8,
The apparent density in the Eq. 2.4, is :

$$\rho_a = \frac{M_t - M_0}{v} \quad (2.4)$$

ρ_a: apparent density
M_0: the mass of the empty measuring vessel
M_t: the mass of the vessel with the sample
V: the volume of the vessel
The results of the tests of the apparent density ρa and the absolute density ρs are recorded in Table. 4, follows:

	Apparent density	Absolute density
Sand dune	1,48	2,53

2.2 The cement

Two different types of cement have been chosen in this study (CPA CEM I and CPJ CEMII/B), [6].

2.2.1 Chemical analysis of cements

The results of chemical analysis of cement are interpreted in Table. 5 following :

Table. 5, Chemical analysis of cements (CEM I) [LAFARGE ALGERIE].

Characteristics	Results (%)
Loss of iron	0,5 à 3
Sulphate content (SO3)	1,8 à 3
Magnesium oxide content (MgO)	1,2 à 3
Chloride content	0,01 à 0,05
Iron loss	10,0 ± 2
Sulphate content (SO3)	2,5 ± 0,5
Magnesium oxide content (MgO)	1,7 ± 0,5
Chloride content	0,02 – 0,05

2.3 Ultra-fines

The ultrafines used are local materials found in the Tabelbala and El-ouata regions of southwestern Algeria. They are found in a raw state and must be ground and screened with a 80 μm sieve before being added to the other components of the mortar. (Fig. 8).

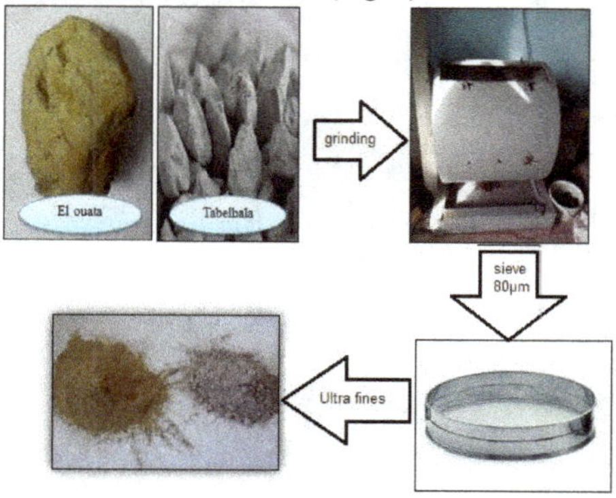

Fig. 8, Preparation steps of the ultras fines

2.3.1 Density

After obtaining the final state of the two materials, the role of passing through physical and chemical analyses will come. The Table. 6, below summarizes the results of the density found

Table. 6, Density results of the ultras fines

The sample	Density (g/cm³)	
	Apparent	Absolute
TAB	0,45	3,38
ELO	0,99	2,82

2.3.2 Chemical analysis of ultra fines

- **Methylene blue test (spot test) [NF P18-592].**

a- Objective

This test measures the capacity of fine elements to adsorb methylene blue.

b- The principle

It consists in introducing in an aqueous bath containing the test sample, increasing quantities of methylene blue by successive doses, and to control after each addition the adsorption of the blue by means of the test "of the spot", until the clayey particles of the material are saturated with blue.

- Method of operation
- Take a sample of 30 g
- Put it into a beaker filled with distilled water to 200 ml.
- Stir the sample using a magnetic stirrer.
- Add 5 ml of blue dye to the suspension and stir for 5 minutes.
- Use a glass stirring rod to take drops of the suspension and place them on a filter paper.
- Gradually add increasing amounts of methylene blue, in 5 ml increments.
- Stop when a light blue halo is observed around the suspension stain. If this occurs, the test is positive (Fig. 9).

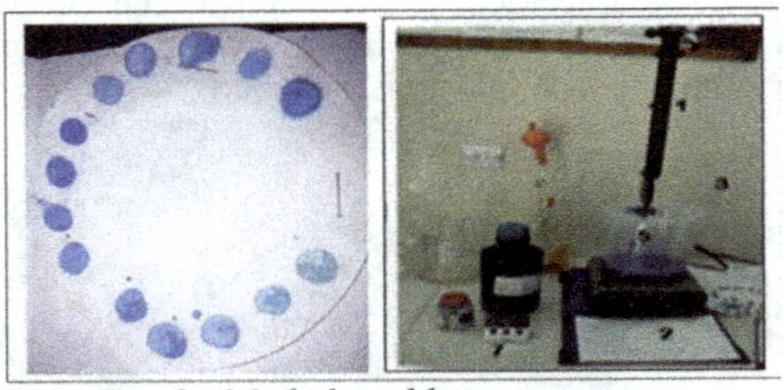

Fig. 9 , Methylene blue apparatus

Expression of the results is given by the following Eq.2.5,

$$V.B.S = \frac{V}{m} \tag{2.5}$$

V.B.S: Value of blue.
V: Volume of methylene blue.
M: The mass of test sample =30g.
Positive. The test is repeated identically, five times at one minute intervals to confirm it.
Table. 7, following:

Table. 7, methylene blue test results

Methylene blue value (VBS)	Soil category
VBS ≤ 0,2	Sandy soils
0,2 ≤ VBS < 2,5	Silty soil
2,5 ≤ VBS < 6	Clayey-silt soil
6 ≤ VBS < 8	clayey soil
VBS > 8	Very clayey soil

classification of the ultra fines are given in the table. 8,

Table. 8, test results

The sample	m (g)	VB (ml)	VBS	Observation
TAB	30	80	2,66	Clayey-silt soil
ELO	30	35	1,16	Silty soil

3. Results and discussion

We present the various results of tests conducted on mortars made according to the two types of ultras fines Tebelbala and El-ouata, such as the density, porosity, tensile strength by bending and compressive strength (7, 14 and 28 days).

3.1 Density of the mortar

The histograms represent the variation of density as a function of percentage of ultras fines.

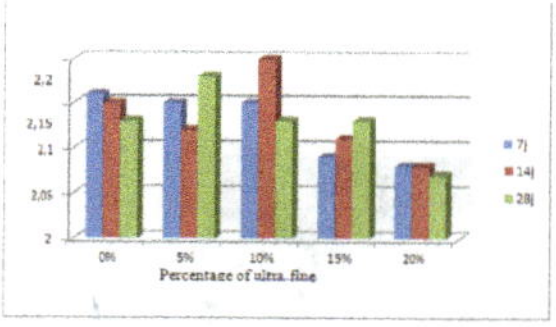

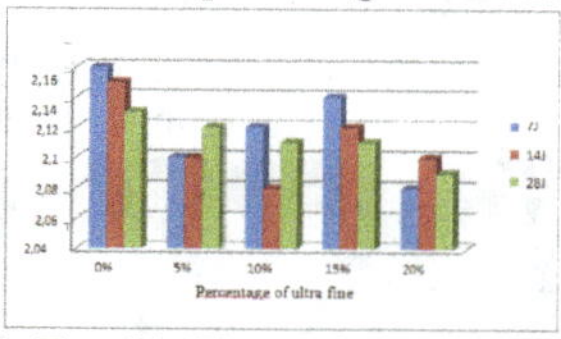

Fig. 10, The evolution of the density in (g/cm3) TAB, Fig. 11, The evolution of the density in (g/cm3) ELO

According to the results presented, despite the different percentages of ultra fines, the density has a small variation compared to the control mortar, even if the type of cement is changed.

3.2 porosity

We present the two histograms which give the variation of the porosity according to the percentage of the ultras fines used in the mortars.

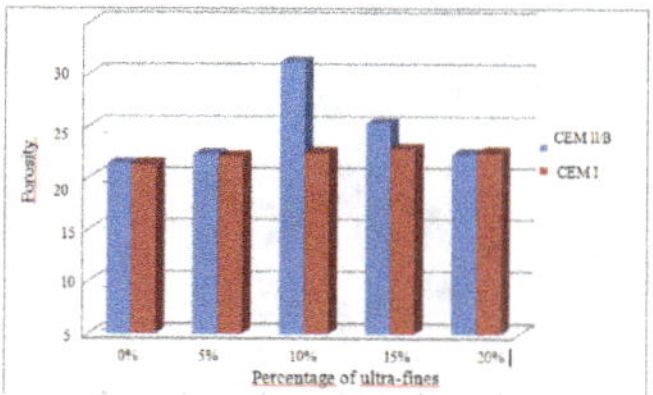

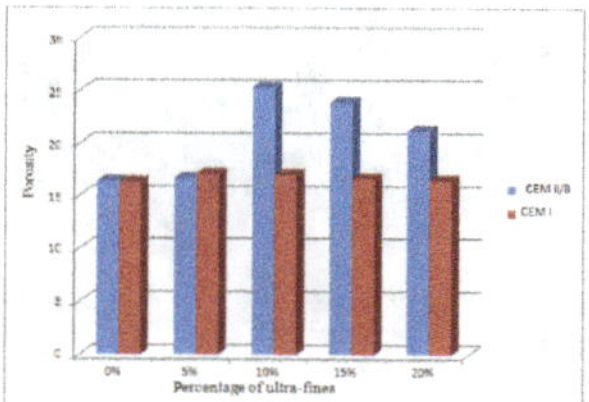

Fig. 12, The evolution of the porosity (TAB) , Fig. 13, The evolution of the porosity (ELO)

According to Fig. 12and Fig. 13 the porosity of the specimens for both cements increases with the increase of the percentage of ultra-fines (TAB; ELO)

3.3 Tensile strength by flexion

The flexural tensile test is performed on 4x4x16 prismatic specimens using a 3-point bending device. Three specimens are tested for each sample of age (7, 14 and 28) days.

The use of the different types of ultras fines as a partial replacement of cement (CEM II/B) induced an increase in flexural tensile strength. The results presented in Fig. 17 show the increase in tensile strength of a mortar at the average age (14 days). [8]

3.4 Compressive strength

The influence of the addition of the ultras fines on the compressive strength of the mortars at different ages (7; 14 and 28) days is represented on which.

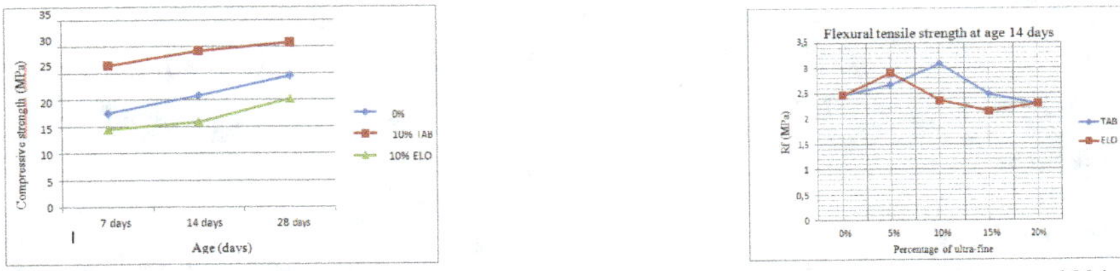

Fig. 14 Variation of compressive strength as a function of age of mortars containing 10% of ultras fines (TAB and ELO)

According to the results obtained, the compressive strength increases with time.These results confirm that the formulation of the mortar which contains 10% of the ultras fines of TABALBALA is better than that of EL-OUATA.

3.5 Scanning electron microscope (SEM) observation

magnifications x150 and x800.

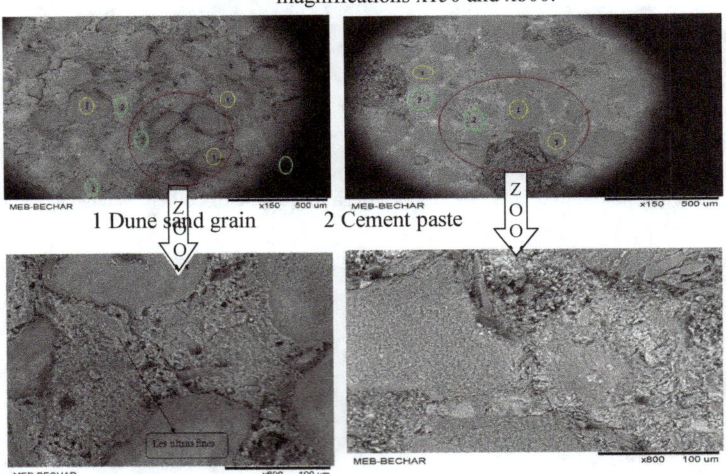

Fig. 15, SEM observation of the polished plain mortar interface at different

Fig. 15, present the interfaces of the polished mortars which shows the dune sand grains in zone 1 and the cementitious paste and the ultra fines in zone 2.

It can be seen from the above figures that there is an adhesion between the dune sand grains and the cementitious paste due to the presence of the ultra fines.

Conclusion

The present research falls under the framework of exploiting local materials, including additive elements, for the preparation of economical mortars. The new mortars are based on several high-quantity additives, which are known to be expensive and rare as they are industrial products. This drawback prompted us to find an equivalent product that plays a role in the mortar and is available in our region, and is accessible to all.

The second part of the research was carried out in the laboratory, allowing us to identify the sand dune, the cements used, and the ultra-fines from various sources, as well as chemical and physical analysis. Available testing devices enabled us to highlight the mechanical properties of the mortars made with ultra-fines.

The exploitation of local ultra-fines in sand dune-based mortars is comprised of an experimental study on the fresh mortar, workability and bulk density, as well as a study on the hardened mortar, mechanical testing (flexural and compression strength), porosity, and bulk density.

Through this study, we have achieved the production of mortars using local ultra-fines and sand dune, leading to a very interesting valorization for the sand of the western erg both technically and economically.

References

[1] A. Ouled mohamed, S. Ethmaneouled, Study of the mechanical behaviour Mortars in hot climate, Kasdi merbah ourgla University. (2017).

[2] A.Kada, Study of the deferred deformations of self-placing concretes (Shrinkage and creep).diss. Oran, (2012).

[3] A. belhocine, N. nagoudi, Experimental study of a mortar with mineral additions; (2014).

[4] B. Abdelkrim, Optimization of the quantity of silica fume by the use of crushed dune sand in dune sand in high performance concretes, Tahri Mohamed University Bechar, (2012).

[5] A. Chabane, Deferred behavior of concrete varieties, PhD thesis, University Mentouri Constantine. (2008).

[6] Cimbéton Technical COLLECTION, Prescription Guide for Cements for Sustainable Construction Case of cast-in-place concrete, T47

[7] D. Lozach, The constituents of concretes and mortars, TOME.1, (2005).

[8] J. Baron, O. Jean-Pierre , Concrete - Fundamentals and databases for their formulation; Edition Eyrolles (Third printing, (1999).

[9] A. Tafraoui, Contribution to the Valorization of Dune Sand from the Western Erg (Algeria) Application to new concretes; University of Toulouse, (2009).

[10] S. Zaoiai, Improvement of the mechanical characteristics of self-compacting concrete based on rubber aggregates through the addition of siliceous fines. PhD thesis LMD UTMB, (2016).

Advanced Topics in Mechanics of Materials, Structures and Construction - AToMech1-2023 Materials Research Forum LLC
Materials Research Proceedings 31 (2023) 799-805 https://doi.org/10.21741/9781644902592-80

The state of 3D printing technologies: A review

Omar Yahya[1,a *], Gharam Yahya[2,b], Faramarz Djavanroodi[1,c], Mushtaq Khan[1,d]

[1]Mechanical Engineering Department, College of Engineering, Prince Muhammad Bin Fahd University, 31952 Al-Khobar, Kingdom of Saudi Arabia

[2]College of Science, Imam Abdulrahman Bin Faisal University, 31952 Dammam, Kingdom of Saudi Arabia

[a]201502345@pmu.edu.sa, [b]miss-gharamy@mail.ru, [c]fdjavanroodi@pmu.edu.sa, [d]mkhan7@pmu.edu.sa

Keywords: 3D Printing, Digital Light Projection, laser beam, Stereolithography, Jet Beaming, 3D Printing Techniques

Abstract. This review paper is written to present the 3D printing technologies from all aspects which are commercially available. Since 3D printing is evolved rapidly over the past decade, there are a number of considerable applications that are well-known in almost all areas and disciplines. This paper will focus on the existing 3D printers, its advantages, and disadvantages and applications. Furthermore, this will provide a sufficient general review in the development of 3D printing technologies.

Introduction

A developed technology in the late twentieth century is termed as additive manufacturing (AM) and known as implementation of three-dimensional (3D) fabrication is growing widely in almost all areas and starting to be a key process in considerable manufactural areas. In other words, this innovation which known as the process of building layer-by-layer using solid CAD model, is a major role in the fourth industrial revolution (IR4.0) and advancing engineering future, consequently our globe. In addition, in single session it is able to conclude fabrication cycle of whole part [1]. Moreover, this advancement cannot only leave impacts on efficiency of industry, energy, economic, and etc, but it can also positively contribute in the environment, sustainability, and our interaction with our world.

The base idea of printing can be described as ink is deposited in the page in one layer, while in additive manufacturing material is deposited in manner which the layer over another. This advanced technology makes creating simple or even complex shape 3D objects with total flexibility and minimum amount of material wastage and cost. Additionally, all these are compelling reasons for replacing subtractive manufacturing with additive manufacturing.

This review will critically be evaluating 3D printing technologies in general bases going through types and scales. Moreover, by highlighting evolvement, advantages, disadvantages of each three-dimensional printer.

Discussion

This technology considered to a major key technological revolution, as it plays a huge economical role [2]. All this due to the fast-developing prototype creating and producing customizable parts, as these points considered to be the main reason for the growth and demand in almost all sectors. Further, in three-dimensional printing there are seven 3D printing techniques, and they are applicable with a huge number of materials in broad classifications such metals, ceramics, polymers, composites and hybrids in areas such automobile, aerospace, health-care and medical, construction, electronics, food and etc.

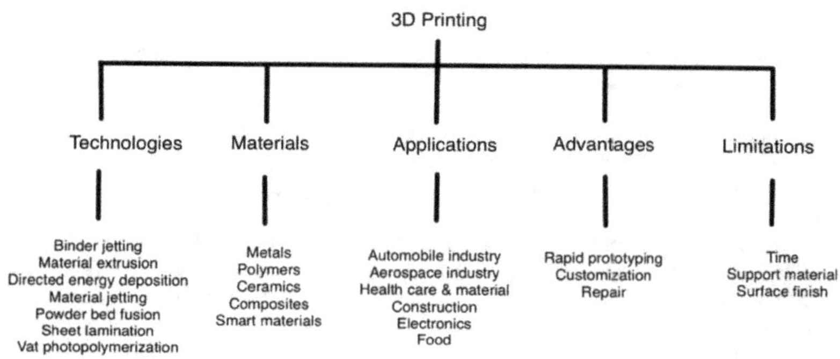

Fig. 1. *Flow chart of additive manufacturing*

One of the advantages of additive manufacturing can be summarized in rapid prototyping, due to the ability to evaluate 2D drawing of a design by fabricating 3D printed prototype using a wide range of material. Also, 3D printing is heavily used for repairing purposes. However, there some limitation in additive manufacturing such consuming size and time, as most printers fabricate small build volume of about 400x400x400 mm3 only binding jetting and directed energy deposition have a larger build volume than the rest, of about 2000x2000x2000mm3 [3].

Vat-Polymerization

Vat-polymerization method is the process that build an object by hardening a bulk liquid, as 3D photopolymerization explained as the using monomers or oligomers sated as liquid. Additionally, in the process the liquid is photopolymerized when exposed to light (UV) with a specific wavelength. Yet, this method of photopolymerization usually shrinks the material [2].

Vat-polymerization method is highly used in medical and dental devices [4]. For example, this method of additive manufacturing is being used in single layered or dual layered acrylic resin materials, however dual layered devices provide more comfort to patient, improved fit, and can be adjusted for fit compared to single layered devices. In addition, these devices can be occlusal made biocompatible photosensitive acrylic esters resins to be photopolymerized by this method [4]. Vat-polymerization technologies is divided into different categories as they differ in the light source used by the printer: stereolithography (SLA), direct light processing (DLP), and liquid crystal display based (LCD), also called daylight polymer printing (DPP).

Stereolithography (SLA)

Stereolithography (SLA) 3D printing technology processes photo-curable resins by a focused light beam with a wavelength range 300 nm-700 nm. In addition, this light is produced by the Digital Light Processing (DLP) projector and goes through the lens imposed on a layer of resin. Yet, the photo-polymerization reaction occurs and the resins transform from the liquid state to the solid state. To perform the motion needed of the Z-stage platform, an Arduino microcontroller controlling a stepper motor and ball screw is used with software to control other slicing parameters. SLA additive manufacturing technique and additionally used for rapid prototyping (RP), it uses the one at a time layering technique of 2D slices in order to create the 3D object, all this based on photo-polymerization reaction of resin [5]. The printing of an object starts from the top and going downward vertically. SLA 3DP offers the best resolution compared to any other 3DP techniques [6].

Surface roughness known to be major issue for most 3D printing techniques, because of adding discretized printed layers. Projection micro- Stereolithography (PμSLA), is one of the AM processes that known to be capable to 3D print required objects with high level of accuracy by capturing minor features of the objects compared to other methods. However, surface roughness is variable when process parameters change.

The LED transmitting a 405 nm beam through a spreading optic in order to be able to cover the micromirror array device (DMD) entire surface. On the PNG, there is a black pixel required to turn the corresponding micromirror. This micromirror is turned to a position reflect the light to the heat sink, while a white pixel is used to reflects the light on the photosensitive resin in the vat for the specified exposure time. As a result, this transmission of light photopolymerize the photosensitive resin, and the resin is squeezed to print layers. Keep in mind before starting UV exposure, the distance of squeezing between transparent PDMS window and cured layer, should be on the corresponding required thickness. The facilitating is for the polymerized layers from PDMS window. After that, the vertical translation platform goes upward by a the corresponding distance of the thickness [7, 8].

Digital Light Processing (DLP)
If compared stereolithography printing and digital light processing printing we will find that DLP has a faster print times, due to that each layer revealing at once compared to tracing intersectional field by a laser. The projection of light on the photosensitive resin is with a digital micromirror device (DMD) on the LED screen or UV light, where DMD can be explained as the Miro speakers controlling light projection on the structure surface generating light pattern [9].

Liquid Crystal Display (LCD)
LCD display as system of imaging. To explain this 3DP technique, a light is lighting onto photosensitive resin through flat LCD panel. Moreover, this technique of fabrication force light not to expand, which result in less distorted pixels and eliminates this issue, as this issue occurs in DLP 3DP. Comparing LCD with SLA, LCD does not require scanning point-by-point as you are able to process a full layer at once, which result in faster printing. LCD printers known to use UV or LED arrays as the light source in the LCD flat panel [10].

Continuous Liquid Interface Production (CLIP)
Continuous liquid interface production (CLIP) is an additive manufacturing process different than SLA as it processes a preamble widow of oxygen to the inhabitation of photopolymerization at its surface preventing adhering. In CLIP, resin able to flow to "dead zone" liquid at the surface of window leading to a layer-by-layer printing of an object.

Fused Deposition Modeling (FDM)
Material extrusion (MEX) printing, also called fused deposition modelling (FDM) or fused filament fabrication (FFF), is a popular AM process [11]. MEX is known to hold the advantage of low material and energy requirements of thermoplastics. As a result, MEX technology by far is the most economical AM technique that provides customizing across many fields; against all others [12]. FDM as the name implies, is 3D fabricating technique uses fused material to be deposited in a layer-by-layer manner in order to 3D print parts.

Materials used for FFF technology are thermoplastics polymers, Poly lactic acid (PLA), Polycarbonate (PC), and Acrylonitrile Butadiene Styrene (ABS) [13].

Material Jetting (MJ)
Material jetting (MJ) 3DP technology uses a process similar to traditional inkjet printer, as the prating head prints various photopolymer droplets to be cured by a UV lamp. In addition, this is a layer-by-layer printing process that result in a 3D printed object [9].

Drop-on-Demand Droplet Jetting (DOD)
Drop-on-demand droplet jetting is a new method in the additive manufacturing world. This new technology holds a research progress for metal, colloid, and liquid resin materials. However, since this is a new fabrication technique in 3DP, there is a limitation in researches and sources in some specific areas. With the highly increase in demand of quality, droplet jetting 3DP technique is based on the FDM process. Droplet jetting technology produces and stacks precisely the formed micro droplets.

Poly-Jet
Poly-jet 3D printing (PJ-3DP) is a method used in 3D printing that uses droplets of jetted photopolymers to be cured by a UV light source, in order to 3D print layer-by-layer a product. In PJ-3DP via inkjet, selective photopolymers resin layers are jetted in order to fabricate a build-tray. With the composition of multiple micro-jetting heads, the head of printing prints a 16 μm thick resin layer to the build-tray, bearing the cross-sectional profile of bult [14].

Nano-Particles Jetting (NPJ) by XJet
Nano-particles jetting (NPJ) solution-based deposition is a technology created by XJet under the category MJ, and the process known to additively manufacture by small solid-state substance to be hold/contained in a carrier solution liquid. Advantages of nano-particles jetting (NPJ) by XJet, able to print dense, detailed featured objects with a thickness of layer of 10 μm with 20 μm jetting resolution. However, two major limitations, which are the inability control the deposition pattern in a precise manner, and nozzle clogging formed by precipitates [15].

D. Binder Jetting (BJ)
Binder jetting (BJ) is an additive manufacturing process that joins powder feedstock by organic binder. BJ method of 3DP prints layer-by-layer in order to fabricate the desired complex shape [16]. Keep in mind, this fast-printing technique is applicable for ceramics, metals, or etc [17].

The BJ 3DP process uses selectively deposited liquid binder to selected raw material powder bed. In addition, this powder material is produced by an inkjet print head that prints layer-by-layer [5].

Powder-Based Fusion (PBF)
Powder-based fusion (PBF) technology known to be widely spread worldwide over last couple of years in industrial applications, and specially in medical sector. To explain PBF generally, the process of starts by melting fed wire in inert or vacuum by laser or electron beam, while product moves in three-dimensions. As a result, to build layer in x-y plans, scan by powder feed and melting system selectively build a layer. After that, the built layer moves down in z axis in order be able to build the next layer. All this three-dimensional printing process is controlled by computer aided design (CAD) program to fabricate desired 3D object.

In PBF technology, there four categories of Powder-based fusion additive manufacturing technology, and each one differs in printing process. These categories are selective laser sintering (SLS), selective laser melting (SLM), electron beam melting (EBM), and multi jet fusion (MJF) [19].

Sheet Lamination (SL)
Helisys company innovated sheet lamination technology in 1991, where the process uses a laser to cut fused sheets together while it is guided by a digital system to fabricate the parts [20].

Moreover, there are many types under the category of sheet lamination which varies with its materials and processes, as there are LOM for papers, CAM-LEM for alumina and other ceramics, SDL for standard type paper, UAM for a variety of material combinations. Another is composite based additive manufacturing (CBAM) process which is a first-generation assembly methodology innovated by Ropert Swartz. Technique begins by a load of fiber sheets, carbon, glass, or Kevlar to be put in machine, then mats considered layers of the object is moved to printing stage [21]. Also, selective lamination composte object manufacturing (SLCOM) process, which is an additive manufacturing process that fabricates fibre reinforced polymer (FRP) composites. This technology selectively cut and bond woven composite sheets [22].

Direct Energy Deposition (DED) / Laser Cladding

Direct energy deposition (DED) (also called laser cladding) is a metal additive manufacturing (MAM) process that uses high power laser as well as coaxal powder delivery system to 3D print 3D metal objects layer-by-layer. This promising technology able to fabricate large metal products with complex shapes [19]. DED technology is considerably very fast freeform manufacturing process. Laser cladding is efficient to manufacture near net shapes for engineering applications and medical applications [23]. Laser cladding known to offer many advantages for modern application, as when parameters are properly set the quality reach a very good level. Howeover, DED 3D printed objects have high porosity and poor mechanical properties [24, 25]. DED categories are laser engineered net shaping (LENS), aerosol jet technology, electron beam additive manufacturing (EBAM), laser deposition welding (LDW), and wire and Arc additive manufacturing (WAAM).

Conclusion

3D printing is the manufacturing technique of the present and the future. A broad spectrum of materials can be processed for pro- ducing a component and new material are being developed regu- larly. This technique has found application in industries like automobile, aerospace, health and medicine, electronics, construc- tion, food and many more for the purposes like rapid prototyping, modifications and repair. The range of opportunities in 3D Printing is extraordinary, specifically in aerospace and medicine. With regards to future perspective, high strength lightweight materials for their application in the aerospace industry could be very bene-ficial. Similarly, the growth and development of biomaterial for producing artificial organs for humans would be favorable as the number of organ donors is very less as compared to the number of recipients. A personalized, biocompatible and implantable 3D artificial liver can be printed using the patient's hepatic cells, who are suffering from the severe hepatic disorder which increase the therapeutic efficiency of liver transplantation. Moreover, 3D Printing can also be used for facial reconstruction and bone regen- eration and repair. The most useful area for research in 3D printing is reducing the time it takes for printing a component as it being one of the major drawbacks of this technique. In the present sce- nario considering the advantages and limitations this technique cannot be used as a standalone process but it should be incorpo- rated in the multistage process of manufacturing of a product. Furthermore, according to ASTM Standard F2792, 3D printing can be classi- fied into seven major categories including binder jetting, material extrusion, directed energy deposition, material jetting, powder bed fusion, sheet lamination and vat photo-polymerization.

References

[1] Krishnanand, Soni, S., & Taufik, M. (2020). Design and assembly of fused filament fabrication (FFF) 3D printers. *Materials Today: Proceedings*, *xxxx*. https://doi.org/10.1016/j.matpr.2020.08.627

[2] Att, W. (2021). *l P re of.* https://doi.org/10.1016/j.jdent.2021.103630

[3] Singh, T., Kumar, S., & Sehgal, S. (2020). Materials Today : Proceedings 3D printing of engineering materials : A state of the art review. *Materials Today: Proceedings*, *28*, 1927–1931. https://doi.org/10.1016/j.matpr.2020.05.334

[4] Piedra-Cascón, W., Sadeghpour, M., Att, W., & Revilla-León, M. (2020). A vat-polymerized 3-dimensionally printed dual-material occlusal device: A dental technique. *Journal of Prosthetic Dentistry*, 1–5. https://doi.org/10.1016/j.prosdent.2020.07.011

[5] Chavez, L. A., Ibave, P., Wilburn, B., Alexander, D., Stewart, C., Wicker, R., & Lin, Y. (2020). The Influence of Printing Parameters, Post-Processing, and Testing Conditions on the Properties of Binder Jetting Additive Manufactured Functional Ceramics. *Ceramics*, *3*(1), 65–77. https://doi.org/10.3390/ceramics3010008

[6] Suryatal, B. K., Sarawade, S. S., & Deshmukh, S. P. (2021). Journal of King Saud University – Engineering Sciences Fabrication of medium scale 3D components using stereolithography system for rapid prototyping. *Journal of King Saud University - Engineering Sciences*, *xxxx*. https://doi.org/10.1016/j.jksues.2021.02.012

[7] Mostafa, K. G., Nobes, D. S., & Qureshi, A. J. (2020). Investigation of Light-Induced Surface Roughness in Projection Micro-Stereolithography Additive Manufacturing (PμSLA). *Procedia CIRP*, *92*(ii), 187–193.

[8] Song, H., Rodriguez, N. A., Seepersad, C. C., Crawford, R. H., Chen, M., & Duoss, E. B. (2021). Development of a variable tensioning system to reduce separation force in large scale stereolithography. *Additive Manufacturing*, *38*(December 2020), 101816. https://doi.org/10.1016/j.addma.2020.101816

[9] Manoj Prabhakar, M., Saravanan, A. K., Haiter Lenin, A., Jerin leno, I., Mayandi, K., & Sethu Ramalingam, P. (2020). A short review on 3D printing methods, process parameters and materials. *Materials Today: Proceedings*, *xxxx*. https://doi.org/10.1016/j.matpr.2020.10.225

[10] Tosto, C., Pergolizzi, E., Blanco, I., Patti, A., Holt, P., Karmel, S., & Cicala, G. (2020). *Epoxy Based Blends for Additive Manufacturing by Liquid Crystal Display (LCD) Printing : The Effect of Blending and Dual Curing on Daylight Curable Resins*. *Lcd*, 1–12. https://doi.org/10.3390/polym12071594

[11] Tino, R., Yeo, A., Brandt, M., Leary, M., & Kron, T. (2021). The interlace deposition method of bone equivalent material extrusion 3D printing for imaging in radiotherapy. *Materials and Design*, *199*(February), 109439. https://doi.org/10.1016/j.matdes.2020.109439

[12] Costanzo, A., Spotorno, R., Candal, M. V., Fernández, M. M., Müller, A. J., Graham, R. S., Cavallo, D., & McIlroy, C. (2020). Residual alignment and its effect on weld strength in material-extrusion 3D-printing of polylactic acid. *Additive Manufacturing*, *36*(June), 101415. https://doi.org/10.1016/j.addma.2020.101415

[13] Wankhede, V., Jagetiya, D., Joshi, A., & Chaudhari, R. (2019). Experimental investigation of FDM process parameters using Taguchi analysis. *Materials Today: Proceedings*, *27*(xxxx), 2117–2120. https://doi.org/10.1016/j.matpr.2019.09.078

[14] Vidakis, N., Petousis, M., Vaxevanidis, N., & Kechagias, J. (2020). Surface roughness investigation of poly-jet 3D printing. *Mathematics*, *8*(10), 1–14. https://doi.org/10.3390/math8101758

[15] Oh, Y., Bharambe, V., Mummareddy, B., Martin, J., McKnight, J., Abraham, M. A., Walker, J. M., Rogers, K., Conner, B., Cortes, P., MacDonald, E., & Adams, J. J. (2019). Microwave dielectric properties of zirconia fabricated using NanoParticle Jetting™. *Additive Manufacturing*, *27*(July 2020), 586–594. https://doi.org/10.1016/j.addma.2019.04.005

[16] Cordero, Z. C., Siddel, D. H., Peter, W. H., & Elliott, A. M. (2017). Strengthening of ferrous binder jet 3D printed components through bronze infiltration. *Additive Manufacturing*, *15*(April), 87–92. https://doi.org/10.1016/j.addma.2017.03.011

[17] S, S., MP, B., MP, B., M, L. G., & H, Z. (2018). Mechanism of Bonding in Seashell Powder Based Ceramic Composites Used for Binder-Jet 3D Printing. *Bioceramics Development and Applications*, *08*(01). https://doi.org/10.4172/2090-5025.1000108

[19] Kim, D. H., Lee, J., Bae, J., Park, S., Choi, J., Lee, J. H., & Kim, E. (2018). Mechanical analysis of ceramic/polymer composite with mesh-type lightweight design using binder-jet 3D printing. *Materials*, *11*(10). https://doi.org/10.3390/ma11101941

[19] Murr, L. E. (2020). Metallurgy principles applied to powder bed fusion 3D printing/additive manufacturing of personalized and optimized metal and alloy biomedical implants: An overview. *Journal of Materials Research and Technology*, *9*(1), 1087–1103. https://doi.org/10.1016/j.jmrt.2019.12.015

[20] Molitch-Hou, M. (2018). Overview of additive manufacturing process. In *Additive Manufacturing: Materials, Processes, Quantifications and Applications*. Elsevier Inc. https://doi.org/10.1016/B978-0-12-812155-9.00001-3

[21] Nuttall, D., Chen, X., Kunc, V., Love, L., Office, A. M., Division, E., & Science, T. (2016). Feasibility and Scaling of Composite Based Additive Manufacturing. *ORNL Report*, *ORNL/TM-20*, 17.

[22] Shaikh, S., Jadhav, H., Gajinkar, A., Khare, I., & Singh, M. (2020). *SELECTION OF ADDITIVE MANUFACTURING TECHNOLOGY FOR OPTIMIZED INTAKE MANIFOLD : A REVIEW*. *17*(9), 4270–4300.

[23] Azhagarsamy, P., Sekar, K., & Murali, K. P. (2021). *An Overview of Metallic Materials Fabrication by Direct Energy Deposition An Overview of Metallic Materials Fabrication by Direct Energy Deposition. January.*

[24] Lin, P. Y., Shen, F. C., Wu, K. T., Hwang, S. J., & Lee, H. H. (2020). Process optimization for directed energy deposition of SS316L components. *International Journal of Advanced Manufacturing Technology*, *111*(5–6), 1387–1400. https://doi.org/10.1007/s00170-020-06113-z

[25] Kim, K. H., Jung, C. H., Jeong, D. Y., & Hyun, S. K. (2020). Causes and measures of fume in directed energy deposition: A review. *Journal of Korean Institute of Metals and Materials*, *58*(6), 383–396. https://doi.org/10.3365/KJMM.2020.58.6.383

Analyzing the electromagnetic forming process of SS304 sheet using AA6061-T6 driver through a fully coupled numerical model and experimentally validation

Zarak Khan[1,a*], Mushtaq Khan[2,b], Faramarz Djavanroodi [2,c], A Samad Danish [3,d]

[1]Department of Mechanical Engineering, HITEC University, Taxila, 47080, Pakistan

[2]Mechanical Engineering Department, Prince Mohammad Bin Fahd University, Al Khobar, Saudi Arabia

[3]Department of Computer Science, HITEC University, Taxila, 47080, Pakistan

[a]zarak.khan@hitecuni.edu.pk, [b]mkhan7@pmu.edu.sa, [c]fdjavanroodi@pmu.edu.sa
[d]abdul.samad@hitecuni.edu.pk

Keywords: Electromagnetic Forming, AA6061-T6 Driver, SS304 Workpiece

Abstract. Driver-based electromagnetic forming (EMF) is a process that uses electromagnetic forces to form metal components into a desired shape. The process involves the use of a conducting driver metal, which is driven by an electromagnetic coil that generates a high-frequency, high-amplitude current. The driver sheet pushes the nonmagnetic sheet metal workpiece against a die causing it to be deformed into the desired shapes. EMF is particularly useful for forming complex shapes and for creating precise forms with minimal material loss. Applications of EMF include the formation of aerospace and automotive components, medical devices, and other products that require precise shaping and minimal material waste. This research focuses on the driver-based electromagnetic forming of Austenitic Stainless Steel (SS304) sheets using Aluminum alloy (AA6061-T6) as a driver. The analysis includes the numerical results of the magnetic field, Lorentz force, sheet deformation and velocity at four points on the sheet profile. The study also compares the numerical results to experimental measurements of the sheet profile. The study found that the numerical results were in good agreement with the experimental results.

Introduction

With development of sheet metal forming industries interest in the electromagnetic forming process is catching the interest of researchers due its unique qualities like reduced wrinkling, increased formability and reduction in tooling costs [1]. Electromagnetic forming process is high-speed sheet metal forming method which uses impulsive magnetic forces to deform the workpiece [2]. Due to high strain rate (10^{-3}-10^{-5}) the formability of metals increases in electromagnetic forming as compared to conventional forming process [3]. Based on coil and workpiece configuration the electromagnetic forming can broadly be divided into tube forming with a helical coil and sheet metal forming with a spiral pan cake coil. In the later configuration the electromagnetic force distribution is not even which is the biggest challenge in this kind of electromagnetic forming. [4, 5]. Several studies have been carried out in the past to accurately estimate the magnetic force and deformation in electromagnetic sheet metal forming such as Daehn et al [6], they focused on uniform force distribution of the coil on the workpiece by changing the coil geometry. A loosely coupled numerical method was developed by Oliveira et al [7] to validated free forming experimental data however due to loose coupling the excessive bounce back phenomena was observed in elliptical closed die analysis [8]. Localized uniform magnetic force technique was used by [9] to achieve uniform deformation in a workpiece. Parametric analysis was carried out by [10] to analysis the electromagnetic forming process and identify important process parameters of the process. [11, 12] introduced driver based electromagnetic forming for sheet materials with low electrical conductivity which can not directly be deformed using Lorentz force. Good results have been achieved so far using driver based sheet metal forming but the methods, shape of the driver and coil are not unique which is a major issue.[12–14]. A fully coupled numerical model for closed die electromagnetic forming was developed by [15]. The numerical model gave better results in comparison to uncoupled model at cost of slight increase in computational time.

This research study aims to develop a fully coupled numerical model using COMSOL Multiphysics to analyze the driver-based electromagnetic forming process. Stainless steel SS304 workpiece is being deformed using AA6061-T6 alloy driver. The numerical results are validated with experimental results.

Numerical Model

A 3D numerical model was developed in COMSOL Multiphysics to analysis driver based electromagnetic forming process. To calculate the transient current, the following input variables were used: input voltage (V), capacitance of the capacitor bank (C), the sum of system inductance and coil inductance (L), and resistance of the system (R). The values of these parameters are provided in Table 1. Equations 1-5 were used to calculate the transient current (I(t)), damping coefficient (β), frequency of the current (ω), magnetic flux, and current density. The Lorentz body force was determined using Eq. 6, which was then applied as a body load on AA6061-T6.The numerical model was simplified by modeling only 1/4 of the geometry to reduce mesh elements and computational time. The model consists of air domain, die domain, driver sheet domain, driven sheet domain and coil domain. Meshed model is presented in Fig. 1.

$$I(t) = \frac{U0}{\omega L} e^{-\beta t} sin\,(\omega t) \qquad (Eq.\ 1)$$

$$\beta = \frac{R}{2L} \qquad (Eq.\ 2)$$

$$\omega = \sqrt{\frac{1}{LC} - \beta^2} \qquad (Eq.\ 3)$$

$$\nabla \mathrm{x} \vec{E} = \frac{-d\vec{B}}{dt} \qquad (Eq.\ 4)$$

$$\vec{J} = \frac{\vec{I_c}}{s} = \sigma_e \vec{E} \qquad (Eq.\ 5)$$

$$\rho \frac{d^2\overline{u}}{dt^2} - \Delta.\sigma s = \overline{fm} \qquad (Eq.\ 6)$$

Cowper-Symond model was used for high-speed nonlinear deformation of both SS304 and AA6061-T6 sheets. The constants were taken from the work of [16]. The material properties are given in table 2.

$$\bar{\sigma} = \sigma_y \,[1 + \left(\frac{\dot{\varepsilon}}{p}\right)^m] \qquad (\mathrm{Eq.\ 7})$$

Table 1 Input parameters of numerical model

Serial	Input parameters	Value (Unit)
1	Input Voltage (V)	2800 V
2	Capacitance (C)	0.006 F
3	Inductance (L)	0.5e-6 H
4	Resistance (R)	0.02 Ω

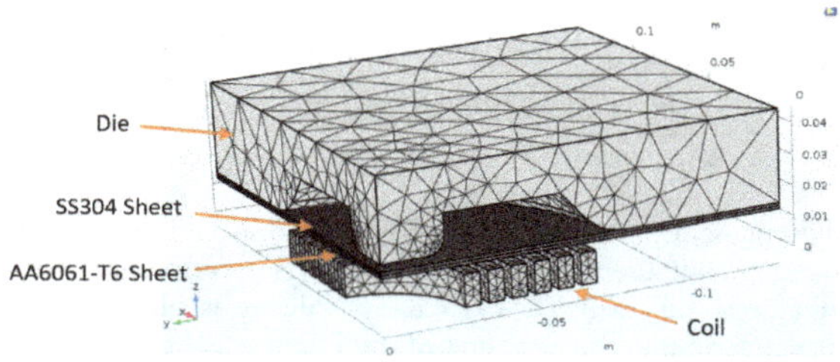

Figure 1. 3D Mesh of driver based closed die electromagnetic forming.

Table 2 Material properties of Coil, SS304 sheet and AA6061-T6 sheet

Sr	Item/material	Properties	Parameter	Values
1	Forming Coil / Copper	Resistivity	ρ	1.72e-8 m
2	Sheet / AA6061-T6	Resistivity	ρ	2.65e-8 m
		Poison's ratio	ν	0.35
		Density	D	2980 kg/m^3
		Elastic Modulus	E	69.0 GPa
		Yield strength	σ	276MPa
3	Sheet/ SS304	Resistivity	ρ	0.72e-6 Ω.m
		Poison's ratio	ν	0.29
		Density	D	8000 kg/m^3
		Elastic Modulus	E	200 GPa
		Yield strength	σ	300 MPa

Experimental Details

The basic experimental setup is shown in Fig. 2. The experimental setup consists of a capacitor bank the specifications are discussed in table 1. Furthermore, the setup includes power cables, spiral pan cake type copper coil with cross section of 10mm height and 5mm thickness having total diameter of 140mm and 6 number of turns, stainless steel die (SS304), 200mm x 200mm SS304 sheet blank of 0.71mm thickness and 200mm x 200mm AA6061-T6 driver sheet of 1.4mm thickness. The experiment was performed at 2800V and total energy of 23.530KJ.

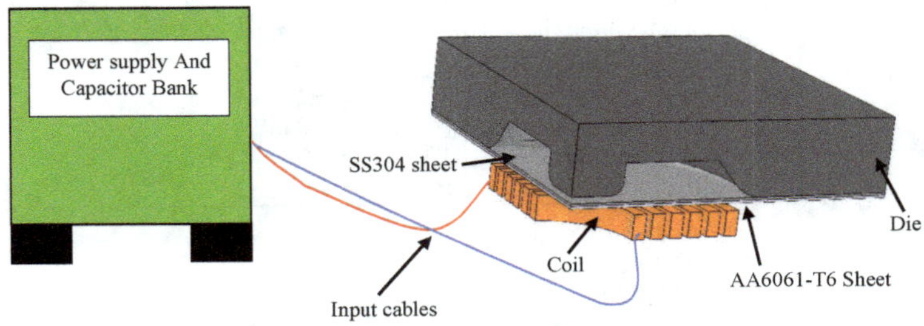

Figure 2. Electromagnetic forming setup

Results

Magnetic field :

Fig. 3 illustrates the magnetic flux density on AA6061-T6 at various time steps throughout the electroforming process. While the magnetic flux density is present until 1000 microseconds, the highest value of magnetic flux density was 4.8 T, which was observed between 50 and 100 microseconds. This suggests that the rise time of the impulse is much shorter than the fall time, which is advantageous for the electromagnetic forming process. At 200 μs the magnetic field lines and the deformation of both driver (AA6061-T6) and driven sheet can be seen in Fig. 4. It was observed that the magnetic flux due to induced current is very prominent on driver sheet which is due to its very high conductivity, however the magnetic flux in driven sheet (SS304) is almost negligible, that is because of its lower conductivity and austenitic non-magnetic characteristic. At 200 μs as evident from Fig. 4 the driver and driven sheets are still in contact while the driver sheet is forcing the driven sheet against the die.

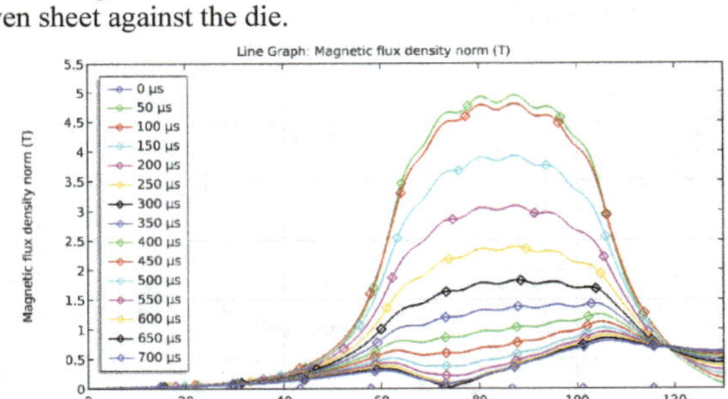

Figure 3. Magnetic flux density in AA6061-T6 sheet at various time steps during electromagnetic forming

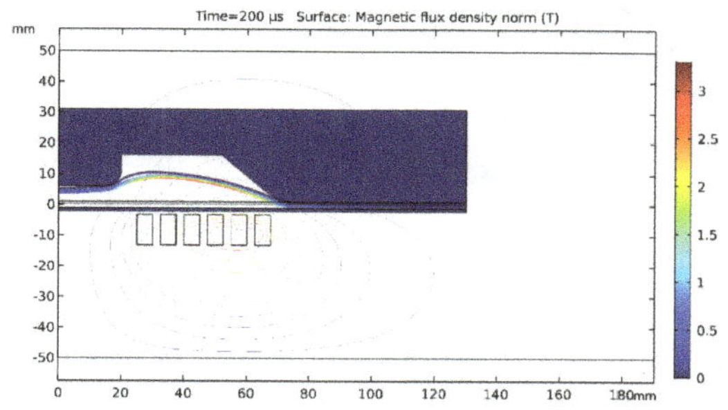

Figure 4. Magnetic flux density at 200 µs time step

Lorentz force:
Lorentz force is the repulsive magnetic force between two conducting maetrials, in this case the force was generated between the coil and driver (AA6061-T6) sheet. The Lorentz force reaches its maximum value at time step 75 µs as can be seen in Fig. 5. The Lorentz force than diminishes after 300 µs but the deformation continues till 1000 µs. The deformation after the first 300 µs is purely due to inertial forces and the duration of inertial motion is greater than the duration of motion due to impulsive magnetic force. Therefore, inertia has major contribution in electromagnetic forming. The Lorentz force originates from the area on the driver sheet which is directly above the coil. Fig. 6 presents the location where Lorentz force originates from and also gives the maximum Lorentz force value of $1.6x10^{10}$ N/m^3 at time step 75 µs.

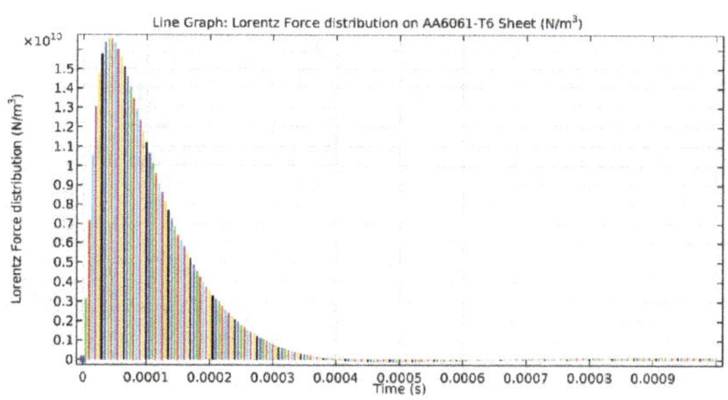

Figure 5. Lorentz force applied on AA6061-T6 on all time steps

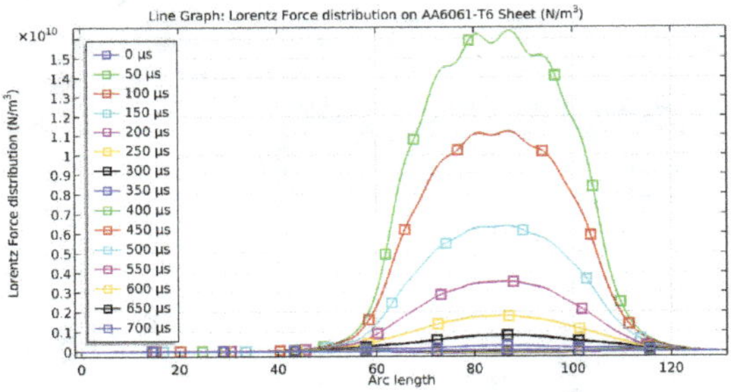

Figure 6. Lorentz force applied on length of AA6061-T6 on various time steps

Deformation:
The final deformation of the driven sheet (SS304) can be observed in Fig. 7. The highest deformation that was achieved by this sheet was 9.3 mm.The driver sheet being pushed by Lorentz force drives the driven sheet impacting it against the die. The experimental sample is shown in Fig. 8. The maximum numerical deformation and experimental deformation are compared and plotted in Fig. 9. The maximum deformation calculated numerically at peaks was 9.3 mm while the experimental results was 8.9mm which is in close approximation. The maximum error in deformation values is 4.49% which is acceptable.

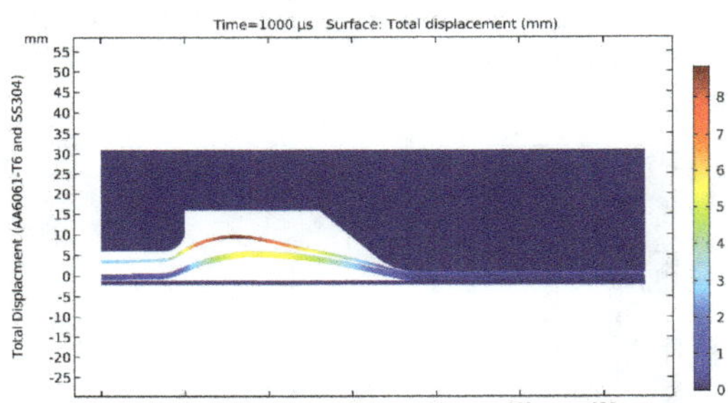

Figure 7. Numerically simulated final deformation of SS304 sheet and AA6061-T6 driver sheet

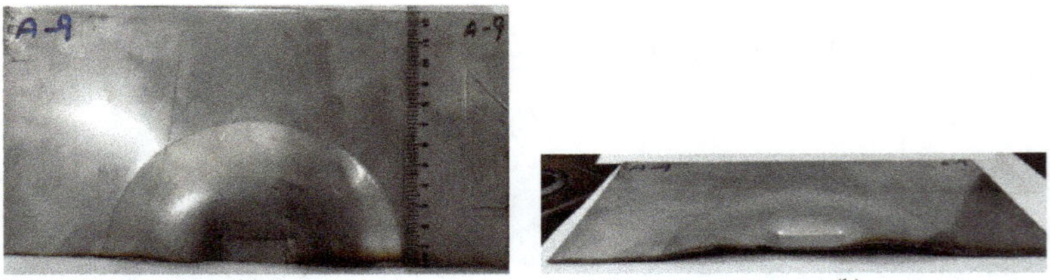

(a) (b)

Figure 8. (a) Experimentally deformed driver sheet (SS304) (b) Cross section of driver sheet

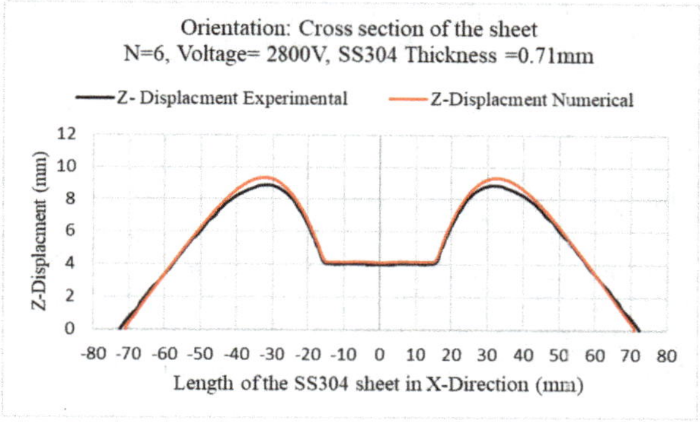

Figure 9. Comparison of numerically estimated deformation and experimental deformation of SS304 in Z-direction

Conclusion

- Driver based electromagnetic forming of material with low conductivity can be approximated and best initial conditions can be achieved using the numerical model instead of iterative experimentation saving time and resources.
- Magnetic flux density and magnetic field lines estimated by the numerical model clearly illustrates the current being induced in the driver sheet.
- The research found that the Lorentz force only plays a role in one-third of the electromagnetic forming process, while the remaining two-thirds of the deformation is caused by the inertial forces generated by the impulsive force.
- The numerical simulations were found to closely match the experimental results, with an error of only 4.49%. The error can be reduced by careful measurement and reducing approximations in the model.
- The driven sheet deformation can be enhanced if the driver sheet remains free between the driven sheet and the coil instead of fixing it at ends.

References

[1] V. Psyk, D. Risch, B. L. Kinsey, A. E. Tekkaya, and M. Kleiner, Electromagnetic forming - A review, J. Mater. Process. Technol., May 2011, vol. 211, no. 5, pp. 787–829. https://doi.org/10.1016/j.jmatprotec.2010.12.012

[2] Q. Cao et al., Investigation of the Lorentz-force-driven sheet metal stamping process for cylindrical cup forming, J. Mater. Process. Technol., 2019, vol. 271, pp. 532–541, https://doi.org/10.1016/j.jmatprotec.2019.03.002

[3] L. Qiu, Y. J. Yu, X. P. Nie, Y. Q. Yang, and P. Su, Study on material deformation performance and electromagnetic force loading in electromagnetic tube expansion process, Trans. China Electrotech. Soc., 2019, vol. 34, no. 2, pp. 212–218.

[4] Q. Cao et al., Electromagnetic attractive forming of sheet metals by means of a dual-frequency discharge current: design and implementation, Int. J. Adv. Manuf. Technol., 2017, vol. 90, no. 1, pp. 309–316.

[5] Q. Xiong et al., Design and implementation of tube bulging by an attractive electromagnetic force, J. Mater. Process. Technol., 2019, vol. 273, p. 116240.

[6] M. Kamal and G. S. Daehn, A uniform pressure electromagnetic actuator for forming flat sheets, J. Manuf. Sci. Eng. Trans. ASME, Apr. 2007, vol. 129, no. 2, pp. 369–379. https://doi.org/10.1115/1.2515481

[7] D. A. Oliveira, M. J. Worswick, M. Finn, and D. Newman, Electromagnetic forming of aluminum alloy sheet: Free-form and cavity fill experiments and model, J. Mater. Process. Technol., Dec. 2005, vol. 170, no. 1–2, pp. 350–362. https://doi.org/10.1016/j.jmatprotec.2005.04.118

[8] Y. U. Haiping, L. I. Chunfeng, and D. E. N. G. Jianghua, Sequential coupling simulation for electromagnetic-mechanical tube compression by finite element analysis, J. Mater. Process. Technol., Jan. 2009, vol. 209, no. 2, pp. 707–713. https://doi.org/10.1016/j.jmatprotec.2008.02.061

[9] L. Qiu et al., Analysis of electromagnetic force and experiments in electromagnetic forming with local loading, Int. J. Appl. Electromagn. Mech., 2018, vol. 57, no. 1, pp. 29–37.

[10] Z. Khan, M. Khan, S. H. I. Jaffery, M. Younas, K. S. Afaq, and M. A. Khan, Numerical and experimental investigation of the effect of process parameters on sheet deformation during the electromagnetic forming of AA6061-T6 alloy, Mech. Sci., Oct. 2020, vol. 11, no. 2, pp. 329–347. https://doi.org/10.5194/ms-11-329-2020

[11] M. Kamal, J. Shang, V. Cheng, S. Hatkevich, and G. S. Daehn, Agile manufacturing of a micro-embossed case by a two-step electromagnetic forming process, J. Mater. Process. Technol., 2007, vol. 190, no. 1–3, pp. 41–50.

[12] S. Golowin et al., Application of a uniform pressure actuator for electromagnetic processing of sheet metal, J. Mater. Eng. Perform., 2007, vol. 16, no. 4, pp. 455–460.

[13] S. Ouyang et al., Investigation of the electromagnetic attractive forming utilizing a dual-coil system for tube bulging. J Manuf Process 49: 102–115. 2020.

[14] L. Chen, Y. Wang, T. Ben, X. Yang, X. Zhang, and Z. Wang, Waveform conditioning problems of nanocrystalline alloys under one/two-dimensional high-frequency magnetization, IEEE Trans. Appl. Supercond., 2018, vol. 28, no. 3, pp. 1–5.

[15] Z. Khan, M. Khan, S. H. I. Jaffery, M. Younas, and A. Khan, Numerical and experimental investigation of fully-coupled and uncoupled finite element model for electromagnetic forming of Aluminium Alloy Al 3014, in IOP Conference Series: Materials Science and Engineering, 2020, vol. 999, no. 1, p. 12008.

[16] H. G. Noh, W. J. Song, B. S. Kang, and J. Kim, 3-D numerical analysis and design of electro-magnetic forming process with middle block die, Int. J. Precis. Eng. Manuf., 2014, vol. 15, no. 5, pp. 855–865. https://doi.org/10.1007/s12541-014-0409-7

Keyword Index

3D Printing Techniques 799
3D Printing 799

AA6061-T6 Driver 806
Absorptivity 656
Acoustic Black Hole 456
Acoustic Comfort 614
Acrylonitrile Benzidine Styrene 726
Additive Manufacturing (AM) 522, 726,475
Additives 109
Admixture 306
Adsorbent 194
Aeroacoustics 614
Agricultural 666
Air Quality 765
Aluminum 6063 485
Analytical Solution 596
ANN 734
Arabic Gum 306
Arduino-based Controller 408
Artificial Intelligence 146,646,734
Asaoka's Method 86
Aspect Ratio 622
Asphalt Binder 16
Asphalt Mix 248
Asymptotic Solution 99
Automated Cleaning System 408
Auxetic Materials Design 172
AWJ 212
Axial Loading 269

Bailey Gradation 248
Basement Excavation 313
Bayesian Regularization Approach 437
Beam-Downsystem 514
Bending 596
Bentonite Clay 504
Best Theory Diagram 135
Bi-Axial Behaviour 466
Biaxial Stress 385
Bioethanol 366
Biofuel 683
Biomass 366
Biomechanical 187
Blast Load 375
Bleeding 306
Blended Cement 571
Bond Energy Method 324
Bonded Joints 324
Brake Pads 538
Burgers Tensor 281

Calm Wind & Wind Envelope 765
Canard Configuration 227
Carbon Nanotube 55
Carbon-Basalt-Aramid Hybrid 356
Carrera Unified Formulation (CUF) 135,596
Carrying Capacity 269
Cassegrain Telescope 514
Cellular Glass Aggregates 66
CFD 240,564,674,755
Chemical Additives 109
Christofell Symbols 422
Clay 109,787
Clayey Soil 646,734
CLC 262
CO2 Laser 212
Coefficient of Consolidation 86
Coir-Fibre 485
Community Energy Storage 531
Composite Plates 135

Composite Structures 324
Compression 1
Compressive Strength 262,306,415
Conductive Elastomer 715
Construction 194
Controlled Release 165
Correlations 437
Corrosion 755
Cost of Energy 531
Crack Width 622
Crack 324
CRRF 146,646,734
Cruciform Specimen 385
Crumb Rubber Modifier (CRM) 16
Cylinder Rupture 564
Cylinder 715
Cylindrical Tube 336

Damage 704
Deformation 485,693
Delamination 324
Design 538
DfAM 475
DIC 1
Dielectric Material 99
Dielectric 336
Differential-Difference Equation 589
Digital Light Projection 799
Dimensional Accuracy 212
Direct Energy Deposition (DED) 522
Displacements 693
Distortion Gradient Plasticity 281
Dry Density 66
Dual Receiver 514
Dune Sand 550,787
Dynamic Strain Aging 128

Edge Restraint 622
Elastic Couplings 29
Electric Circuit 634
Electric Discharge Machining 448
Electroelasticity 336
Electrohydrodynamic 165
Electromagnetic Forming 806
Electromechanical Coupling 456
Electronic Properties 204
Electrorheological Fluids 7
Elevated Temperature 1
Encapsulation 165
Energy Absorption 485
Energy Conversion 456
Energy Harvesting 456
Environmental Impact Analysis (EIA) 16
Erosion Pattern 755
Erosion 755
ESNDT 693
Ethanol Production 366
Ethanol-Fuel 683
Expansion 504
Expansive Soil 495
Experimental 16
Explicit Closed-Form Solution 589
Explosion 589
Exterior Differential Forms 422

Failure Analysis 704
Failure Load 324
Far Field Approximation 227
Fatigue Life 248
FDM 475
FGM Shell 117
Fibre Reinforced Concrete (FRC) 155,313
Fine Aggregate 262, 415
Finite Deformation 336
Finite Element Analysis 622,666

Finite Element Method 634
Finite Element Modelling (FEM) 313
First Impact 704
Flexoelectricity 99
Flexural Strength 415
Flow 306
Flowing Fluid 634
Fluid-Structure Interaction 614
Fly Ash 398,571
Foldcore 375
Formulation 787
Fragility Curves 38
Fresh Properties 155
FSI 240
Fuels 504
Functionally Graded Material 726
Functionally Graded Shells 76

Gaussian Heat Flux 448
Gears 299
Generalized Gradient Approximation Given by Perdew-Burke-Ernzerhof 204
Generalized Gradient Approximations 204
Genetic Algorithm 135
Geopolymer Concrete 398
GFRP 1,466
Glass-Basalt-Aramid Hybrid 356
Gradient Theory 99
Gregorian Telescope 514
Grid-Connected Solar PV System 347
Ground Granulated Blast Furnace Slag 398
Ground-Fixed Receiver 514
Guo 86
Gust Wind 765
HDMR 38
HDPE 475
Heat Exchanger 581
Heat Leak Factor 581
Higher-Order Shear 117
Higher-Order Theories 596
High-Pressure Gas 564
Holy Mosque 765
Homogenization Theory 172
Hybrid Additive Manufacturing (HAM) 522
Hybrid Manufacturing (HM) 522
Hydraulic Conductivity 683
Hydrocarbon Liquids 504
Hyperbolic Method 86
Hyperbolic Shear Deformation Theory 745
Hyperbolic Shell Theory 46
Hyperelastic Materials 715
Hyperelasticity 336
Hyperelasticity 422

Imposed Strain 622
Industrial Wastes 415
Invariants 422
IRA 571
Iron Ore Tailings 415
Isogeometric Topology Optimization 172

Jet Beaming 799
Jet Impingement 564

Kirigami 375
KSA 2030 Vision 347,408,531

Laminated Composite Plate 596
Laminated Composites 745
Landfill 683
Laser Beam 799
Laurent Series 589
Laval Nozzle 674
Layered Composite Arches 693

Life Cycle Cost Analysis 347
Lining 538
Liquid Granite 495
Load Factor 531
Local Density 204
L-Shaped Bent 240
Machine Learning 135,240,656
Magnetoelasticity 715
Materials 538
Maximum Power Point 408
MCS 38
Mechanical Properties 66
Mechanical Resistance 155
Mechanical Strengths 398
Mechanical 704
Metal Additive Manufacturing (MAM) 522
Metallic Fibers 550
Metamaterials 656
Meta-Model 38
Microscopy 306
Microstructures Design 172
Mixed FEM 99
Mixed Formulation 29
Mode of Operation 7
Modeling 187,666
Modified VA Model 128
Molding 538
Monostatic RCS 227
Monotonic Shear 66
Mortar 787
Mounted Receiver 514
Moving Morphable Method 172
Multi Helices 269
Multi-Scale Model 55
Musculoskeletal 187

Nanocomposite 55
Nanofluid 581
Nanomaterials 109
Nanoparticles 165
Nano-Silica 495
Natural Sand 248
NDTs 571
Niobium 128
Non-Traditional Additives 109
Normal Deformation Theory 117
Open End Burst Test 466
OpenSim 187
Ordinary Portland Cement 646

Parabolic Dish 514
Particulate Matter 755
Passive Control 614
Payback 531
Performance 683
Physical Optics 227
Physico-Chemical Characterization 787
Piezoelectric Elements 634
Plastic Flow Rule 281
Plastic Waste 734
Plate 634
PLS 146
PM Gears 299
Polyethylene Terephthalate Fiber 155
Polylactic Acid 726
Polymers 165,475
Porosity 76
Portland Cement 306
Powder Metal (PM) Technology 299
Prediction 86
Predictive Modelling 656
Progressive Failure Analysis 29
Quasi-Static Compression 375

Radar Cross Section 227
Radial Defomation 715

Randomly Distribution 55
Real Gas 674
Rectangular Jet 614
Recycled Ash 146
Recycled Concrete 155
Recycled Fiber 146
Recycled Glass Powder 646
Recycling, Nickel (Ni) 194
Resistance 550
Response Surface Method 38
Ritz Method 29
Robot 666

Sacrificial Cladding 375
Sand-Bentonite Liner 683
Sandwich Shells 46
Sandwich Structure 726
Screw Pile 269
Secondary Reflector 514
Self-Compacting Concrete 155,415
Semi-Analytical Methods 29
Sensors 408
Shock Accumulation 704
Shock Wave 674
Shock without Rebound 704
Silica Fume 550
Simulation 656,666
Sintered PM 299
Size-Effect 281
Slag 194
Soil Settlement 86
Soil Stabilizer 146,646,734
Soil 109
Solar PV 531
Solar Thermal Energy 514
Spherical Cavity 589
Spherical Shell 745
SPIV 614
SS304 Workpiece 806
Stabilization 109
Stabilizing 495
Static Analysis 46,745
Static Response 76,117
Statistical Analysis 16
Steel Fiber 398
Stereolithography 799
Stitched Sandwich 704
Stone Dust 262
Strain Gradient Plasticity 422
Strength 212
Stress and Strain 466
Stresses 693
Stress-Strain 385
Structural Pounding 38
Structural Simulation 240
Supersonic Flow 674
Sustainable Mortar 306
SVM 646
Swelling 504
SWOT Analysis 306
System/Environment Energy Interaction 581

Taguchi 385
Tensile Behavior 726
Tg 356
The Comfort of Pedestrians 765
Theoretical Model 269
Thermal Activation Energy 128
Thermal Expansion Coefficient 356
Thermal Model 448
Thermal Stresses 564
Thermophysical Properties 437
Thermoplasticity 281
Tool Material 448
Tower Garden 287

Traditional Additives 109
Transverse Normal Strain 46
Triaxial Testing 66
Trigonometric Shear Deformation Theory 76
Tube Thickness 485
Tunnel Lining 313
Tunnel 313
Two-Way Meters 347

UHPC 550
Ultrafine 787
UPFC 550
UPV 571

van der Waals Interactions 55
Vertical Farming 287
Vertical Garden 287
VF Garden 287
Vibrations Damping 634
Volume Change 622
Volume Weight 262

Waste Granite 495
Water and Ethylene Glycol 437
White Cement 495
Wire Arc Additive Manufacturing (WAAM) 522

www.ingramcontent.com/pod-product-compliance
Lightning Source LLC
LaVergne TN
LVHW052203200926
848481LV00001B/17
* 9 7 8 1 6 4 4 9 0 2 5 8 5 *